D0930520

Patty's Industrial Hygiene and Toxicology

THIRD REVISED EDITION
In Three Volumes

Volume 1
GENERAL PRINCIPLES

Volumes 2A, 2B, and 2C
TOXICOLOGY

Volume 3
THEORY AND RATIONALE
OF INDUSTRIAL HYGIENE
PRACTICE

Patty's Industrial Hygiene and Toxicology

THIRD REVISED EDITION

Volume 2B
TOXICOLOGY

GEORGE D. CLAYTON
FLORENCE E. CLAYTON
Editors

Contributors

D. M. Aviado	C. Frank	E. E. Sandmeyer
R. R. Beard	M. S. Micozzi	H. E. Stokinger
C. M. Berry	J. T. Noe	T. R. Torkelson
W. B. Deichmann	V. K. Rowe	R. C. Wands

A WILEY-INTERSCIENCE PUBLICATION

JOHN WILEY & SONS, New York · Chichester · Brisbane · Toronto · Singapore

Copyright © 1981 by John Wiley & Sons, Inc.

All rights reserved. Published simultaneously in Canada.

Reproduction or translation of any part of this work
beyond that permitted by Sections 107 or 108 of the
1976 United States Copyright Act without the permission
of the copyright owner is unlawful. Requests for
permission or further information should be addressed to
the Permissions Department, John Wiley & Sons, Inc.

Library of Congress Cataloging in Publication Data:

Patty, Frank Arthur, 1897–
  Patty's Industrial hygiene and toxicology.

  "A Wiley-Interscience publication."
  Includes index.
  Contents: v. 1. General principles—[etc.]
  1. Industrial hygiene. 2. Industrial toxicology.
I. Clayton, George D. II. Clayton, Florence E.
III. Battigelli, Mario C. IV. Title. V. Title:
Industrial hygiene and toxicology.

RC967.P37   1978   613.6'2   77-17515
ISBN 0-471-07943-X   AACR1

Printed in the United States of America

10 9

Contributors

DOMINGO M. AVIADO, M.D., Professor of Pharmacology, New Jersey College of Medicine, Newark; and Atmospheric Health Sciences, Inc., Short Hills, New Jersey

RODNEY R. BEARD, M.D., M.P.H., Professor of Preventive Medicine, Emeritus, Stanford University School of Medicine, Stanford, California

CLYDE M. BERRY, Ph.D., Professor, College of Medicine, Institute of Agricultural Medicine, Department of Preventive Medicine and Environmental Health, University of Iowa, Oakdale, Iowa

WILLIAM B. DEICHMANN, Ph.D., M.D. (hon.), Professor of Pharmacology, Emeritus, University of Miami School of Medicine; Consultant, Miami, Florida

CLYDE FRANK, Ph.D., Chief, Environmental Chemistry, College of Medicine, Institute of Agricultural Medicine, Department of Preventive Medicine and Environmental Health, University of Iowa, Oakdale, Iowa

MARC S. MICOZZI, M.D., Department of Research Medicine, University of Pennsylvania; Department of Pathology, Pennsylvania Hospital, Philadelphia, Pennsylvania

JOSEPH T. NOE, M.D., Consultant, Occupational Health, San Juan Capistrano, California, formerly General Medical Director, Inland Steel Company

V. K. ROWE, Sc.D. (hon.), Consultant, formerly Director, Toxicological Affairs, Dow Chemical Co., Midland, Michigan

ESTHER E. SANDMEYER, Ph.D., Consultant, Transcontec, Inc., Pittsburgh, Pennsylvania

HERBERT E. STOKINGER, Ph.D., retired, formerly Chief Toxicologist, U.S. Public Health Service, Cincinnati, Ohio

T. R. TORKELSON, Sc.D., Health and Environmental Sciences, Dow Chemical Co., Midland, Michigan

RALPH C. WANDS, The Mitre Corporation, McLean, Virginia

Preface

This book (Volume 2B) is a continuation of the Patty series on industrial toxicology, the first book of this volume having been published in January 1981. The final book of this volume on toxicology is scheduled for release early in 1982.

The quest for more and "absolute" information is the story of mankind and parallels the development of industrial toxicology. The utopian state would exist if complete and absolute data were accessible for every industrial chemical in existence, and similarly for every new chemical contemplated for use. The cost of acquiring such absolute data on *all* product toxicology is prohibitive. Restraints forbidding product use prior to "absolute" data would minimize the number of new products reaching the general public for their benefit. Manufacturers and governmental agencies involved therefore must use judgment in releasing new chemicals in order to afford protection to society, yet temper legal procedures to avoid unnecessary restraints which impede technological development.

Toxicology is defined as the scientific study of poisons, their actions, their detection, and the treatment of conditions produced by them. According to Paracelsus (1493–1521) *all* substances are poisons, there is none that is not a poison, and it is the right dose that differentiates a poison and a remedy. The primary role of the toxicologist is to specify the safe use of substances.

If we had been guided by these definitions in preparing these volumes on toxicology, the substances included would have been so copious as to prohibit containment within manageable binding. Therefore the elements and compounds selected for inclusion are those of primary interest to industry at present.

Since the publication of Volume 2 of the second edition in 1963, the number of chemicals utilized in industry has expanded rapidly, as technology has flooded the market with new products and new types of industry, increasing the potential sources of exposure. Prior to 1970 the amount of research done by industry and government was not prolific, but was considered adequate by those directing such programs, and published data reflected this condition. Public awareness of the toxic effects of air- and waterborne contaminants culminated in federal legislation in 1969–1971, from which emanated the establishment of the Environmental Protection Agency (EPA) and the Occupational Safety and Health Administration (OSHA). Federal laws passed in that period establishing the mechanism for limiting air and water contaminants were based on meager technical data, and their implementation, or notice of such, often created a scientific furor. Because of the economic consequences of the established standards and the inadequate data on which many were based, both industry and government accelerated funding in an effort to resolve differences of scientific opinion.

In the 1960s most of the toxicological research was directed toward protection of physically fit, employed persons, mostly male, within a specific age bracket. Since the creation of EPA and OSHA, community air pollution standards must be concerned also with infants, the aged, and the infirm, as well as the group in the working category. This working category has also drastically changed since the 1960s, including a greatly increased number of women, with the rights of women to work during an extended period of pregnancy. Thus industrial toxicology must now be concerned with a new aspect, the effect of chemicals upon the fetus and the expectant mother, as well as male fertility.

With the burgeoning data available from expanded research, the authors of the chapters of Volume 2 were faced with the task of selecting and condensing pertinent and important information into a consolidated format. They were confronted with a monumental task. Demands for the expertise of toxicologists had increased with public awareness, and this became a factor in the delay of the revision of this volume. Although the authors made valiant attempts to condense their material, as the manuscripts were received by the editors it became apparent that if the editors were to delete any data, it would be detrimental to the readers. A wealth of references with each chapter supports professional judgments and conclusions, providing direction to those who wish to thoroughly scrutinize the subject on their own.

When approximately one-half the assigned chapters had been received, it was evident that the significant amount of data could not be contained in one binding. The publisher agreed to separate Volume 2 into three books, designated A, B, and C. Those manuscripts received by us initially and edited by September, 1979 compose Volume 2A. The manuscripts received by September 1980 constitute 2B, and the remaining chapters make up 2C. Because of our desire to expedite publication of material, there is no classification of chapters into scientific groupings such as organic and inorganic. Several of the authors have contributed to all three books of Volume 2. All have years of experience in the fields of toxicology and/or medicine, and are recognized for their competence and expertise. They have devoted an extensive period of time to the preparation of these manuscripts, often at great sacrifice, and did so as part of their continuing effort in the profession of industrial toxicology. We believe you will find their contributions to be cogent, germane, and an aid to the professional who is interested in the health of the worker.

In addition to books A and C of this volume, the reader is also directed to Volume 1 of this series, published in 1978, on *General Principles*, which provides guidance to solving the wide range of problems facing industrial hygienists and managers. A third volume, entitled *Theory and Rationale of Industrial Hygiene Practice*, edited by Lester and Lewis Cralley and published in 1979, presents the motivation and philosophy behind many of the subjects covered in Volume 1.

GEORGE D. CLAYTON
FLORENCE E. CLAYTON

San Luis Rey, California
June 1981

Contents

Contents
Volume 2A

USEFUL EQUIVALENTS AND CONVERSION FACTORS

1 kilometer = 0.6214 mile

1 meter = 3.281 feet

1 centimeter = 0.3937 inch

1 micrometer = 1/25,4000 inch = 40 microinches = 100,000 Angstrom units

1 foot = 30.48 centimeters

1 inch = 25.40 millimeters

1 square kilometer = 0.3861 square mile (U.S.)

1 square foot = 0.0929 square meter

1 square inch = 6.452 square centimeters

1 square mile (U.S.) – 2,589,998 square meters = 640 acres

1 acre = 43,560 square feet = 4047 square meters

1 cubic meter = 35.315 cubic feet

1 cubic centimeter = 0.0610 cubic inch

1 cubic foot = 28.32 liters = 0.0283 cubic meter = 7.481 gallons (U.S.)

1 cubic inch = 16.39 cubic centimeters

1 U.S. gallon = 3.7853 liters = 231 cubic inches = 0.13368 cubic foot

1 liter = 0.9081 quart (dry), 1.057 quarts (U.S. liquid)

1 cubic foot of water = 62.43 pounds (4°C)

1 U.S. gallon of water = 8.345 pounds (4°C)

1 kilogram = 2.205 pounds

1 gram = 15.43 grains

1 pound = 453.59 grams

1 ounce (avoir.) = 28.35 grams

1 gram mole of a perfect gas ≈ 24.45 liters (at 25°C and 760 mm Hg barometric pressure)

1 atmosphere = 14.7 pounds per square inch

1 foot of water pressure = 0.4335 pound per square inch

1 inch of mercury pressure = 0.4912 pound per square inch

1 dyne per square centimeter = 0.0021 pound per square foot

1 gram-caloric = 0.00397 Btu

1 Btu = 778 foot-pounds

1 Btu per minute = 12.96 foot-pounds per second

1 hp = 0.707 Btu per second = 550 foot-pounds per second

1 centimeter per second = 1.97 feet per minute = 0.0224 mile per hour

1 footcandle = 1 lumen incident per square foot = 10.764 lumens incident per square meter

1 grain per cubic foot = 2.29 grams per cubic meter

1 milligram per cubic meter = 0.000437 grain per cubic foot

To convert degrees Celsius to degrees Fahrenheit: °C (9/5) + 32 = °F
To convert degrees Fahrenheit to degrees Celsius: (5/9) (°F – 32) = °C
For solutes in water: 1 mg/liter ≈ 1 ppm (by weight)
Atmospheric contamination: 1 mg/liter ≈ 1 oz/1000 cu ft (approx)
For gases or vapors in air at 25°C and 760 mm Hg pressure:
    To convert mg/liter to ppm (by volume): mg/liter (24,450/mol. wt.) = ppm
    To convert ppm to mg/liter: ppm (mol. wt./24,450) = mg/liter

CONVERSION TABLE FOR GASES AND VAPORS[a]

(Milligrams per liter to parts per million, and vice versa; 25°C and 760 mm Hg barometric pressure)

Molecular Weight	1 mg/liter ppm	1 ppm mg/liter	Molecular Weight	1 mg/liter ppm	1 ppm mg/liter	Molecular Weight	1 mg/liter ppm	1 ppm mg/liter
1	24,450	0.0000409	39	627	0.001595	77	318	0.00315
2	12,230	0.0000818	40	611	0.001636	78	313	0.00319
3	8,150	0.0001227	41	596	0.001677	79	309	0.00323
4	6,113	0.0001636	42	582	0.001718	80	306	0.00327
5	4,890	0.0002045	43	569	0.001759	81	302	0.00331
6	4,075	0.0002454	44	556	0.001800	82	298	0.00335
7	3,493	0.0002863	45	543	0.001840	83	295	0.00339
8	3,056	0.000327	46	532	0.001881	84	291	0.00344
9	2,717	0.000368	47	520	0.001922	85	288	0.00348
10	2,445	0.000409	48	509	0.001963	86	284	0.00352
11	2,223	0.000450	49	499	0.002004	87	281	0.00356
12	2,038	0.000491	50	489	0.002045	88	278	0.00360
13	1,881	0.000532	51	479	0.002086	89	275	0.00364
14	1,746	0.000573	52	470	0.002127	90	272	0.00368
15	1,630	0.000614	53	461	0.002168	91	269	0.00372
16	1,528	0.000654	54	453	0.002209	92	266	0.00376
17	1,438	0.000695	55	445	0.002250	93	263	0.00380
18	1,358	0.000736	56	437	0.002290	94	260	0.00384
19	1,287	0.000777	57	429	0.002331	95	257	0.00389
20	1,223	0.000818	58	422	0.002372	96	255	0.00393
21	1,164	0.000859	59	414	0.002413	97	252	0.00397
22	1,111	0.000900	60	408	0.002554	98	249.5	0.00401
23	1,063	0.000941	61	401	0.002495	99	247.0	0.00405
24	1,019	0.000982	62	394	0.00254	100	244.5	0.00409
25	987	0.001022	63	388	0.00258	101	242.1	0.00413
26	940	0.001063	64	382	0.00262	102	239.7	0.00417
27	906	0.001104	65	376	0.00266	103	237.4	0.00421
28	873	0.001145	66	370	0.00270	104	235.1	0.00425
29	843	0.001186	67	365	0.00274	105	232.9	0.00429
30	815	0.001227	68	360	0.00278	106	230.7	0.00434
31	789	0.001268	69	354	0.00282	107	228.5	0.00438
32	764	0.001309	70	349	0.00286	108	226.4	0.00442
33	741	0.001350	71	344	0.00290	109	224.3	0.00446
34	719	0.001391	72	340	0.00294	110	222.3	0.00450
35	699	0.001432	73	335	0.00299	111	220.3	0.00454
36	679	0.001472	74	330	0.00303	112	218.3	0.00458
37	661	0.001513	75	326	0.00307	113	216.4	0.00462
38	643	0.001554	76	322	0.00311	114	214.5	0.00466

CONVERSION TABLE FOR GASES AND VAPORS[a] (Continued)

(Milligrams per liter to parts per million, and vice versa; 25°C and 760 mm Hg barometric pressure)

Molecular Weight	1 mg/liter ppm	1 ppm mg/liter	Molecular Weight	1 mg/liter ppm	1 ppm mg/liter	Molecular Weight	1 mg/liter ppm	1 ppm mg/liter
115	212.6	0.00470	153	159.8	0.00626	191	128.0	0.00781
116	210.8	0.00474	154	158.8	0.00630	192	127.3	0.00785
117	209.0	0.00479	155	157.7	0.00634	193	126.7	0.00789
118	207.2	0.00483	156	156.7	0.00638	194	126.0	0.00793
119	205.5	0.00487	157	155.7	0.00642	195	125.4	0.00798
120	203.8	0.00491	158	154.7	0.00646	196	124.7	0.00802
121	202.1	0.00495	159	153.7	0.00650	197	124.1	0.00806
122	200.4	0.00499	160	152.8	0.00654	198	123.5	0.00810
123	198.8	0.00503	161	151.9	0.00658	199	122.9	0.00814
124	197.2	0.00507	162	150.9	0.00663	200	122.3	0.00818
125	195.6	0.00511	163	150.0	0.00667	201	121.6	0.00822
126	194.0	0.00515	164	149.1	0.00671	202	121.0	0.00826
127	192.5	0.00519	165	148.2	0.00675	203	120.4	0.00830
128	191.0	0.00524	166	147.3	0.00679	204	119.9	0.00834
129	189.5	0.00528	167	146.4	0.00683	205	119.3	0.00838
130	188.1	0.00532	168	145.5	0.00687	206	118.7	0.00843
131	186.6	0.00536	169	144.7	0.00691	207	118.1	0.00847
132	185.2	0.00540	170	143.8	0.00695	208	117.5	0.00851
133	183.8	0.00544	171	143.0	0.00699	209	117.0	0.00855
134	182.5	0.00548	172	142.2	0.00703	210	116.4	0.00859
135	181.1	0.00552	173	141.3	0.00708	211	115.9	0.00863
136	179.8	0.00556	174	140.5	0.00712	212	115.3	0.00867
137	178.5	0.00560	175	139.7	0.00716	213	114.8	0.00871
138	177.2	0.00564	176	138.9	0.00720	214	114.3	0.00875
139	175.9	0.00569	177	138.1	0.00724	215	113.7	0.00879
140	174.6	0.00573	178	137.4	0.00728	216	113.2	0.00883
141	173.4	0.00577	179	136.6	0.00732	217	112.7	0.00888
142	172.2	0.00581	180	135.8	0.00736	218	112.2	0.00892
143	171.0	0.00585	181	135.1	0.00740	219	111.6	0.00896
144	169.8	0.00589	182	134.3	0.00744	220	111.1	0.00900
145	168.6	0.00593	183	133.6	0.00748	221	110.6	0.00904
146	167.5	0.00597	184	132.9	0.00753	222	110.1	0.00908
147	166.3	0.00601	185	132.2	0.00757	223	109.6	0.00912
148	165.2	0.00605	186	131.5	0.00761	224	109.2	0.00916
149	164.1	0.00609	187	130.7	0.00765	225	108.7	0.00920
150	163.0	0.00613	188	130.1	0.00769	226	108.2	0.00924
151	161.9	0.00618	189	129.4	0.00773	227	107.7	0.00928
152	160.9	0.00622	190	128.7	0.00777	228	107.2	0.00933

CONVERSION TABLE FOR GASES AND VAPORS[a] (Continued)

(Milligrams per liter to parts per million, and vice versa; 25°C and 760 mm Hg barometric pressure)

Molec-ular Weight	1 mg/liter ppm	1 ppm mg/liter	Molec-ular Weight	1 mg/liter ppm	1 ppm mg/liter	Molec-ular Weight	1 mg/liter ppm	1 ppm mg/liter
229	106.8	0.00937	253	96.6	0.01035	277	88.3	0.01133
230	106.3	0.00941	254	96.3	0.01039	278	87.9	0.01137
231	105.8	0.00945	255	95.9	0.01043	279	87.6	0.01141
232	105.4	0.00949	256	95.5	0.01047	280	87.3	0.01145
233	104.9	0.00953	257	95.1	0.01051	281	87.0	0.01149
234	104.5	0.00957	258	94.8	0.01055	282	86.7	0.01153
235	104.0	0.00961	259	94.4	0.01059	283	86.4	0.01157
236	103.6	0.00965	260	94.0	0.01063	284	86.1	0.01162
237	103.2	0.00969	261	93.7	0.01067	285	85.8	0.01166
238	102.7	0.00973	262	93.3	0.01072	286	85.5	0.01170
239	102.3	0.00978	263	93.0	0.01076	287	85.2	0.01174
240	101.9	0.00982	264	92.6	0.01080	288	84.9	0.01178
241	101.5	0.00986	265	92.3	0.01084	289	84.6	0.01182
242	101.0	0.00990	266	91.9	0.01088	290	84.3	0.01186
243	100.6	0.00994	267	91.6	0.01092	291	84.0	0.01190
244	100.2	0.00998	268	91.2	0.01096	292	83.7	0.01194
245	99.8	0.01002	269	90.9	0.01100	293	83.4	0.01198
246	99.4	0.01006	270	90.6	0.01104	294	83.2	0.01202
247	99.0	0.01010	271	90.2	0.01108	295	82.9	0.01207
248	98.6	0.01014	272	89.9	0.01112	296	82.6	0.01211
249	98.2	0.01018	273	89.6	0.01117	297	82.3	0.01215
250	97.8	0.01022	274	89.2	0.01121	298	82.0	0.01219
251	97.4	0.01027	275	88.9	0.01125	299	81.8	0.01223
252	97.0	0.01031	276	88.6	0.01129	300	81.5	0.01227

[a] A. C. Fieldner, S. H. Katz, and S. P. Kinney, "Gas Masks for Gases Met in Fighting Fires," U.S. Bureau of Mines, Technical Paper No. 248, 1921.

Patty's Industrial Hygiene and Toxicology

THIRD REVISED EDITION
In Three Volumes

Volume 1
GENERAL PRINCIPLES

Volumes 2A, 2B, and 2C
TOXICOLOGY

Volume 3
THEORY AND RATIONALE
OF INDUSTRIAL HYGIENE
PRACTICE

CHAPTER THIRTY-NINE

Occupational Carcinogenesis

H. E. STOKINGER, Ph.D.

1 PREFACE

This chapter is written from the viewpoint of one with long years of experience in determining thresholds of response to environmental agents, particularly those in the industrial atmosphere, and one who has observed with grave concern, along with numerous colleagues, the ever growing list during the 1970s of substances labeled environmental carcinogens in food, air, and water, whose control embraced the concept of no threshold, leading in several instances to outright banning of food components, pesticides, and drugs. Whereas this concept may be acceptable for regulating exposure of the general public to these environmental agents, in my opinion it is neither acceptable nor appropriate for controlling worker exposure to industrial carcinogens for the following contrasting reasons:

1. The substances designated as carcinogens by the Food and Drug Administration (FDA) and the Environmental Protection Agency (EPA) are "suspect" carcinogens only, as determined from animal experiments, and as such, pose great unknowns for their human counterparts, thus requiring ultraconservative control measures, whereas substances considered *occupational* carcinogens have

The author is past chairman of the Threshold Limits Committee of the American Conference of Governmental Industrial Hygienists, of the Subcommittee on Toxicology of the PHS Drinking Water Standards, and of the National Research Council on Toxicology, Chief of the Experimental and Toxicology Branch of NIOSH, Cincinnati; in addition, has served over the years on several working groups and task forces of WHO and ILO.

been determined from epidemiologic studies or from industrial surveys of workers under conditions of actual exposure.

2.   There is obviously a great range in the susceptibility of the general public because of age and varying states of health and disease. Moreover, whereas ranges and duration of such exposures not only to suspect carcinogens, but to possible cocarcinogens and promoters, are unknown, they are relatively well known and identifiable in industrial exposures and, further, are in compliance with the criteria of risk estimates set forth in the Interagency report on the identification of potential carcinogens and estimation of their risks (1). As briefly summarized, they are "(a) definition and quantification of exposures; (b) characterization of the exposed populations in quantitative terms; (c) chemical and physical properties of the substance and its chemical reactivity in relation to exposure," criteria that rarely, if ever, can be met in general, human population exposures. In short, the susceptibility variation of the industrial worker group is thus far more circumscribed than that of the general public (if women of childbearing age are persuaded to accept work in areas not involving carcinogen exposure) because of normal working ages (18 to 65 years), preplacement and periodic medical examinations, and environmental control, among other factors.

It is for these reasons, and for others to be detailed later, that the threshold concept can be applied for the regulation of occupational carcinogens. The sharp differences demarcating the two populations have been overlooked too long and the issue of "threshold—no threshold" has been debated and contested without realizing that the circumstances of the one are not those of the other. I hope this point of view will go far to quell the endless debate between the "no-thresholders" and the "thresholders."

2   INTRODUCTION

It is clear from the preceding statements that the conditions for estimating risk and determining thresholds of response to carcinogens are far more favorable in occupational environments than in nonoccupational environments. It is clear also from the Interagency report on estimation of risk that to do this some improvement in the current design of experimental animal studies is in order. Thus one of the present purposes is to suggest ways to establish end points of carcinogenic response. If such is done, the need for statistical extrapolation, the current bugaboo of cancer research, is eliminated. Included also is a proposed classification system for occupational carcinogens, and, since mechanisms of carcinogenesis presently concern only those of an organic nature, a theory for inorganic (metal) carcinogens. The overall purpose of the effort is to summarize within the limitations of space the present state of knowledge of occupational carcinogenesis and to point to directions that could improve detection, evaluation, and control of carcinogenic substances in the workplace.

Abbreviations Used. AAF, 2-acetylaminofluorene; AHH, aryl hydrocarbon hydroxylase; BaP, benzo(a)pyrene; bisCME, bis(chloromethyl) ether; CML, chronic myelocytic leukemia; DDP, diamminodichloroplatinum; DMBA, dimethylbenzanthracene; DMS, dimethyl sulfate; DNA, deoxyribonucleic acid; EDB, ethylene dibromide; EGF, epidermal growth factor; EPA, Environmental Protection Agency (U.S.); FDA, Food and Drug Administration (U.S.); MAC, maximal allowable concentration; MCA, 3-methylcholanthrene; MOCA, methylene bis-o-chloraniline; NCI, National Cancer Institute (U.S.); NIOSH, National Institute for Occupational Safety and Health (U.S.); OSHA, Occupational Safety and Health Administration (U.S.); PAH, polycyclic aromatic hydrocarbon(s); PD, protein deficiency; PG, prostaglandin(s); RLV, Rauscher leukemia virus; RNA, ribonucleic acid; TBX, thromboxanes; TPA, tetradecanoyl phorbol acetate; WHO, World Health Organization; XP, xeroderma pigmentosum.

3 IDENTIFICATION OF OCCUPATIONAL CARCINOGENS

Two main types of evidence can be used to identify substances, or occupational processes, that may pose a carcinogenic threat to workers: (1) epidemiologic studies or surveys of industrial operations where both environmental and medical data are available, and (2) experimental animal studies, encompassing long-term or life-term bioassays. Important leads may also be obtained from short-term tests of mutagenicity (2), on the basis, now generally agreed, that carcinogenicity parallels mutagenicity; or suggestive evidence can simply be obtained from comparative chemical structures (3). This cognitive procedure, however, is merely an alerting mechanism for pointing to structures with possible carcinogenic potential, and by no means assures their carcinogenic potency.

3.1 Epidemiologic Studies

On the basis that the best study of man is man himself, a well-designed and controlled epidemiologic study provides unequivocal evidence of the existence of a carcinogenic exposure, which no animal study, however well designed, can duplicate. Epidemiologic studies are of two types, *retrospective*, in which cancer incidence is related to past exposure of a fixed group of workers, and *prospective*, in which environmental data and medical records are maintained on a varying number of workers.

Unfortunately for positive retrospective studies, cancer findings are obtained *ex post facto,* come too late to help those involved, and too often, in the past, have not included adequate work on medical records, medical histories, and data on smoking and drinking habits. Prospective studies serve the dual purposes of (1) monitoring cancer-control procedures at sites where known or suspected cancer cases have occurred, or may occur, and (2) revealing sites of unsuspected sources, although far in the future. Prospective studies, however, have the defect of diluting and obscuring the true cancer incidence if worker turnover is large, and particularly if replacements are made with young workers with no past history of exposure and of an age when cancer development is not to be expected for 20 or more years.

Two main methods are used in epidemiologic studies to establish evidence of a carcinogenic hazard, cohort studies and case-control studies. Epidemiologic cohort studies involve the comparison of groups differently exposed to a substance. The comparison may involve (*a*) totally unexposed versus exposed groups; (*b*) groups having distinctly different levels of exposure; or (*c*) rates in exposed groups versus the rates prevailing in the general population. The groups must be comparable with respect to demographic factors such as age, sex, economic status, and race, and controlled for exposure to known carcinogens. Epidemiologic case-control studies involve the comparison of people with a given cancer type versus people without the disease, who are otherwise comparable with respect to appropriate demographic variables, to ascertain if they differ for exposure to the cancer hazard under investigation.

Epidemiologic studies are a "must" in situations in which the cancer-causing agent(s) is (are) unknown, associated with a process or work site comprising numbers of potential agents that cannot, because of their complexity or special characteristics of plant operation, be duplicated in laboratory animal studies. An example is the long-term mortality study of malignant neoplasms in coke-oven workers (4). This study covered 10 steel plants in diverse parts of the United States and Canada, employing thousands of workers in differing types of jobs and modes of operation. Relative risks of lung cancer varied according to work site; the highest, 6.87, was for men employed "full topside" for 5 or more years; 3.22 for those with mixed topside and side-oven experience for 5 or more years; 2.10 for men with side-oven experience for 5 or more years; and 1.70 for all men with less than 5 years experience. Apart from the impossibility of duplicating in animal studies the varying conditions of exposure at the different sites near the ovens, the exact nature and amount of particulate and gaseous substances can only be approximated even with the best of process manufacturing advice. (Animal inhalation studies, by NIOSH, of fractions of coal tar have elicited tumors at relatively low benzopyrene chamber concentrations, thus, in a manner, giving supportive evidence for the epidemiologic findings.)

3.2 Animal Studies

Evidence of the carcinogenic activity of an agent can be obtained also from bioassays in experimental animals showing that the test substance causes either (a) an increase in the incidence of neoplasms or (b) a decrease in the latency period. Although generally performed for the considerable period of the 2 to 3 years of the life term of rodents, and at times for longer periods in dogs and subhuman primates, animal studies still permit a more rapid means of screening large numbers of industrial substances and at considerably less cost than the epidemiologic method. It must be recognized that the indicated agent is only a "suspect" carcinogen, by animal test, and no amount of statistical extrapolation of data can provide a reliable estimate of carcinogenic risk for the industrial worker (see Section 3.2.2).

Part of the evidence for animal carcinogenicity consists in a critical review of the experimental design, the quality of the experimental effects, and accuracy of the results by a statistical evaluation of their significance, to assure that the only major experimental variable between control and experimental groups is the presence of the test substance.

Criteria for the conduct of animal tumor studies have been the subject of continually changing recommendations over the years. A report by four governmental agencies, CPSC (Consumer Product Safety Commission), EPA, FDA, and OSHA, issued in 1979 (1), would appear to represent a consensus of sanctioned methods of procedure. The report discusses experimental design, choice of species and numbers of animals, routes of administration, identity of test substance, dose levels, age at treatment, the conduct and duration of the bioassay, including pathological evaluation, and statistical analysis of results.

Because these criteria were developed for identifying general environmental carcinogens for the protection of the general public, comments are made only on those procedures that could be better oriented toward the industrial worker group, or to point out additional facts not mentioned in the report.

3.2.1 Identity of the Test Agent

In addition to the Interagency report requirement that ideally, for a full assessment of carcinogenicity, each component of a mixture be tested individually as well as the mixture itself, trace element analyses of supposedly pure substances should be made for determination of the amounts of those substances that are known to influence the carcinogenic response. A few disturbing examples follow from my personal observations.

1. A commercial sample of benzopyrene (BaP) used in NIOSH animal experiments, from a source typical of that used by other experimenters in the

late 1960s, contained 125 ppm of selenium (presumably the residual from its use as a ring-condensing agent). Because trace amounts of selenium are known to be antitumorigenic, such impurity could influence tumor yield from BaP in animal studies.

    2.   Various forms of asbestos contain relatively large and variable amounts of trace metals depending upon source. Cralley et al. (6) found Canadian chrysotile to contain from 0.01 to 0.2 percent nickel, average 0.10 percent; <0.01 to 0.12 percent chromium, average 0.04 percent; 0.03 to 0.08 percent manganese, average 0.04 percent; and <0.01 percent cobalt. Because certain forms of nickel and chromium are known human carcinogens, and manganese and cobalt are known to be toxic, noting that 0.04 percent is 400 ppm, the authors remarked that "the presence of metals in the etiology of this disease (cancer) cannot be ignored." Later studies by Dixon et al. (7) showed that these metals had varied but marked effects in inhibiting BaPase, the enzyme believed to be the initiating cause of the marked increase (90 percent) in asbestos cancers in those who smoke.

    3.   Platinum, a recognized tumorigen in low concentrations but an antitumorigen at high concentrations in certain complex forms, was found to the extent of 25 ppm in commercial beryllium hydroxide (8), a potent (microgram amounts) inducer of pulmonary adenomas in rats. Could the presence of platinum be responsible, in whole or in part, for the tumorigenic action of beryllium?

    The above remarks should suffice to caution experimenters in the field of chemical carcinogenesis to examine closely the chemical composition of substances undergoing test. All may not be what it seems!

3.2.2   Dose Levels

The IRLG report (1) subscribes to the recommendations of the FDA that "testing should be at doses and under experimental conditions likely to yield maximal tumor incidence." The reason for this recommendation is that the widely diverse susceptibility of the general population can best be equated in experimental animals by providing maximal detectability. The means by which this can be accomplished is to estimate the highest dose level that will be tolerated by the test animals during lifetime administration (i.e., the estimated maximum tolerated dose, or EMTD). The EMTD is defined as the highest dose that can be administered to the test animals for their lifetime, and which should not produce (*a*) *clinical signs of toxicity or pathological lesions other than those related to a neoplastic response*, which may interfere with the neoplastic response; (*b*) alteration of the normal longevity of the animals from toxic effects other than carcinogenesis; or (*c*) more than a relatively small percent inhibition of normal weight gain (not to exceed 10 percent). These well-considered specifications were apparently generated from past failure of NCI-sponsored researchers to adhere to these common-sense directives.

Unfortunately, maximal doses for maximal detection requires extrapolation from high animal doses to low doses for estimation of cancer risk for the general population, leaving it to statistical manipulation of the data to achieve this end, for low doses are not prescribed in the Interagency report. Such data manipulation up to the present has been based on highly questioned and continually changing theories, and has led to overestimation of cancer risk generally.

Amended Dose Levels.   A means to avoid the pitfall of statistical manipulation would appear to be to use a series of dosages that include lower levels of exposure, and let the data speak for themselves. This procedure, although not wholly acceptable for the population at large, is acceptable for the industrial worker group, who, as pointed out in Section 1, have a considerably narrower range of susceptibility owing to their circumscribed conditions of exposure and control, that sets them apart from the general population.

For estimating levels for control of worker exposure, ideally five dose levels, separated logarithmically (9), should be used, the selection of lower doses to be such that at least one, preferably two, should result in no tumors at the end of the life term of two rodent species, mouse and rat. Fifty animals of each sex of each species are used at each level, and 50 of each species and sex serve as untreated controls. If this requirement seems excessive in cost and effort, compare the direct, interpretable results with the not inconsiderable costs of animal care and statistical effort of the other, which yields untrustworthy and misleading conclusions! (See Section 5.3.)

This experimental design confines statistical treatment of data to analysis of the significance of the results, and avoids statistical attempts to extrapolate from responses in animals to estimates of risk for the industrial worker.

In addition to the reason just given for omitting statistical extrapolation procedures for the industrial worker group at least, there are now several lines of evidence that cancer risk estimates cannot be solved by statistical gimmicks. The mere fact that several modifications in statistical procedures have been made over the years since Mantel and Bryan (10a) in 1961 made their first proposals, to the latest refinement by Cornfield (10b) in 1977, is beginning to convince the statisticians themselves of the folly of their statistical approach (11a, b). But the scientific findings that should sound the death knell for statistical theory, that are lifting the matter out of the realm of theory into practical reality, are the pharmacokinetic studies by Gehring and associates (12). The report shows, among other things, that for chemicals requiring metabolic activation for tumor response, the amount of vinyl chloride, for example, biotransformed is linear rather than the exposure concentration on which statistical theory has been based, and shows why pharmacokinetics must be considered in design and interpretation of toxicologic data. From these data also, "There is some evidence for at least a practical threshold in both rats and man." (See also Section 8.) This is certainly a clear-cut instance for letting the data speak for themselves and an argument against the idea that by statistical theory one can protect everybody from all cancers, all the time (13); I reached this inescapable conclusion some time ago.

Some Oddities of Biometric Strategies.   An appropriate subtitle might be, "As viewed by an occupational toxicologist." The following are a few vignettes of observations made over the years of the way in which biometricians view toxicologic data, sometimes to the consternation, and sometimes to the frustration, of the toxicologist.

Item 1.   When the value of vitamin C for protection against ozone toxicity was tested, 20 treated mice and 20 controls were used. The results were clear-cut and obvious to a toxicologist: there was significant protection by the vitamin. The statistician saw it differently; 20 animals were too few to draw any conclusion with an appropriate degree of confidence, and besides, what editor would accept such data when statistical confidence is essential for publication? So, we increased the numbers tenfold, and repeated the test. Two years and several thousand dollars later, the identical result was obtained (but the article was now publishable) (14).

Item 2.   It has been common toxicologic practice now for many years to express LD_{50} values with a confidence limit. To omit it is statistical treason. But how confident can one be of the confidence limit as the years roll on? Comparison of LD_{50} values determined 25 or 30 or more years ago with those of recent date provides the answer. LD_{50} values, complete with 95 percent confidence limits, made in the 1970s by an established consulting laboratory, compared with similar determinations made by a similarly well-known toxicity laboratory 20 to 25 years earlier, bear little resemblance to each other; not only do the values differ by severalfold (1.8 to 3 times) in one direction, but they differ equally in the other.

Two examples must suffice: pyridine was found in 1951 to have an oral rat LD_{50} of 1.58 g/kg, with 95 percent confidence limits of 1.42 to 1.77; in 1970, an LD_{50} of 0.891, with confidence limits 0.683 to 1.16, or 1.8 times more toxic, was determined. For *n*-propyl alcohol, the LD_{50} of 1.87 g/kg, with confidence limits 1.34 to 2.60, determined in 1954, compares with 3.83 g/kg, with 2.32 to 5.43, determined in 1970, the latter about half as toxic!

Obviously, confidence limits have value only for the moment; changes over the years in dosing procedures, compound purity, and animal susceptibility and resistance all are subject to change, and with them the confidence limits. Notorious in regard to changing toxicities and their confidence limits are the chemical dyes; 40 to 50 years ago, dye toxicity was unreliable by today's standards because of unknown impurities and amounts. This is strikingly reflected in the false naming of α-naphthylamine as a carcinogen until quite recently, when it was shown that by reducing the β derivative to levels below 0.5 percent, α-naphthylamine was no longer demonstrably carcinogenic (Section 8.2). In like measure, it is still not known whether it is the impurities in the aromatic amine dye magenta (basic fuchsin) that cause bladder cancer in man in spite of its known toxicity confidence limits.

Item 3 draws attention to a kind of common statistical ineptness when confronted with unfamiliar areas in carcinogenicity. As part of a continuing program of the Inhalation Section of what is now NIOSH, tests for possible carcinogenic action were made on each substance undergoing toxicologic examination. This was done by exposing CAF_1/JAX mice throughout their life terms and observing whether the test substance accelerated the spontaneous rate of development of pulmonary adenomas. In the early 1950s, this lung tumor-susceptible strain had a spontaneous rate of 38 percent at 15 months of age. Following 15 months of daily exposure, 5 days/week, of 200 mice to ozone at about 1 ppm, tumor incidence had increased to 85 percent, and the average number of tumors per mouse was 1.9 versus 1.5 for unexposed controls, a significant increase. Not so, said our statistician, because the data show that at 22 months, tumor incidence in controls equalled that of the test mice! Because of statistical objection the experiment was not published in a technical journal, but details may be found in Reference 15. It was because of failure to recognize that acceleration of tumor development is an important measure of tumorigenesis that the evidence was rejected. Incidentally, the test program was abandoned 20 years later when it was found that the incidence of spontaneous tumor development dropped below 10 percent, and the mice showed increased toxicity resistance, evidence of genetic change.

Item 4. The most important question, "How far can animal data be relied upon to predict human cancer rates after they have been given the statistical treatment of extrapolating from high to low doses?" has been preliminarily answered in the case of ethylene dibromide (EDB) (16). On the basis that EDB resulted in an increased incidence of gastric tumors in rats after intubation at 40 mg/kg per day, the Carcinogen Assessment Group of EPA, using a one-hit model, estimated an almost 100 percent lifetime incidence of cancer in workers exposed to EDB at 0.4 ppm (equivalent to about 0.04 mg/kg per day per worker) at citrus fumigation centers. The one-hit model also predicted a total of 85 cancers above the normal background incidence in a group of 156 EDB production workers. A total of eight cancers has been observed thus far, in two separate plants, and in one, no increase in incidence has been observed. Although comparison must be viewed as preliminary, because only 36 deaths have yet occurred, and the remaining population ranges in age from 31 to 76 (average 55), the estimates are sufficiently divergent from the facts that we may be nearly certain that the estimates are unreliable. Because it can be shown (1) on theoretic grounds, (2) from pharmacokinetic findings, and (3) now by direct comparison that predictions from animal to man using the one-hit model bear no relation to reality, the one-hit model must go! (Fashioners of other biometric models, take note.)

The theory of the one-hit model, generated from the self-replicating nature of neoplastic growth, has the singular flaw of overlooking the fact that the probability of one hit resulting in a tumor is so astronomically small as to be nonexistent. Cases against the one-hit theory have already been made by Dinman (17a) and Stokinger (17b, c) on the basis that a finite number of molecules are required for measurable functional activity (homeostasis), estimates of required numbers of molecules varying between 10^{14} to 10^4 to 10^6 depending on the biologic system involved. Further, the one-hit theory overlooks the fact recognized for more than 100 years that the body has endogenous (built-in) antagonists that counteract the actions of substances foreign to it. (See Section 6.3.1.)

Recent pharmacokinetic studies have also shown why the predictive model has failed; EDB is metabolized in the rat by conjugation with hepatic glutathione, an endogenous substance that is anticarcinogenic. At the excessively high dose levels from which the EPA parameters were derived, glutathione levels become depleted, and hence its anticarcinogenic action overcome, a process not occurring at industrially controlled levels. This, coupled with differences in the route of exposure (intubation versus inhalation) from which the extrapolation was made, probably contributed to the failure of the predictive value of the model.

4 OCCUPATIONAL CARCINOGENS

By definition, occupational carcinogens are those substances or mixtures of substances to which workers may be exposed in or around their places of occupation. They may be discrete, readily characterized substances or, as is rather common, associated with some chemical operation or manufacturing process and as yet uncharacterized.

As noted in Section 3.1, epidemiologic studies constitute the most definitive way to identify human carcinogenesis; experimental animal studies can only suggest substances with possible carcinogenic potential for man, and are accordingly labeled suspect occupational carcinogens. The occupational cancers recognized in 1979 are listed in Table 39.1 with references to their epidemiologic reports. Because of space limitations, only those substances listed in Table 39.1 that bear references not in the open literature, or on which recent information has been developed, are further treated in brief notes and commentary.

**Table 39.1   Occupational Carcinogens—
Substances or Industrial Processes with
Carcinogenic Potential**

Carcinogen or Process	Refs.[a]
Acrylonitrile	18
4-Aminodiphenyl	19
Antimony trioxide production	20
Arsenic trioxide production	21
Asbestos (and smoking)	22
Benzene (and leukemia virus)	23a, b
Benzidine production	24
Cadmium oxide production	25
Bis(chloromethyl) ether	26
Chloromethyl methyl ether	27
Chromite ore processing	28a–c
Coal tar pitch volatiles and parti- culates (coke-oven emissions)	29
β-Naphthylamine	30a, b
Nickel sulfide roasting	31a, b
4-Nitrodiphenyl	32
Uranium mining	33
Vinyl chloride	34
Wood dusts	35

[a] Supplementary data and references on these listings may be found in *Documentation of the Threshold Limit Values for Substances in Workroom Air*, 4th ed., American Conference of Governmental Industrial Hygienists, Cincinnati, Ohio, 1980. Supplementary data and references on metals may be found also in the chapter on metals in Vol. IIA of *Industrial Hygiene and Toxicology*.

4.1   Acrylonitrile

On May 23, 1977, NIOSH was informed by E. I. du Pont de Nemours and Company of the preliminary results of an epidemiologic study demonstrating an excess of cancer among workers exposed to *acrylonitrile* at the Du Pont textile fibers plant in Camden, South Carolina. Sixteen cancer cases (living and dead) occurred among 470 male employees first exposed to acrylonitrile in a polymerization operation at a textile mill between 1950 and 1955. These cases were compared to an expected 5.8 based on company rates and 6.9 based on national rates. There were six lung cancers (1.5 expected), three colon cancers

(0.5 expected), and one cancer in each of seven other primary sites. No cases of cancer were identified in employees with initial exposures between 1953 and 1955.

4.2   Arsenic Trioxide Production

In an epidemiologic study prior to the Enterline study of copper-smelter workers (21), Lee and Fraument in 1969 (21b) noted what seemed to be "an influence of sulfur dioxide or unidentified chemicals" varying concomitantly with arsenic exposure (see section on *arsenic* in "Metals," Vol. 2A). Working from this clue, Groth (8) exposed rats by intermittent inhalation of SO_2 after intratracheal injection of arsenic-containing, smelter-flue dust. The treatment was unproductive of any tumors! . . . , indicating that SO_2 is probably not the cocarcinogenic agent. Because SO_2 in the smelting process arises from the copper–arsenic sulfide matte, the possibility exists that the sulfides of arsenic are the etiologic agents. This is not unreasonable, for the carcinogen in nickel production has been narrowed to the sulfides of nickel in the roasting process (see statement on nickel sulfide roasting below). Another influencing factor could arise not only in As_2O_3 production, but in that of Sb_2O_3 as well (see below) because of the invariable mutual presence of the one in the other, both production processes being carcinogenic; hence one could be a cocarcinogen for the other.

4.3   Antimony Trioxide Production

In March 1974, representatives of the Antimony Works of the Associated Lead Mfgrs. Ltd., Newcastle, England reported (to me and to NIOSH representatives) 15 deaths from respiratory cancers in an epidemiologic study of the Works made in 1971 on 1081 smelter workers who had had exposures to antimony dust and fume for 7 years or longer (range 7 to 43 years, average 22 years). Exposures during these years greatly exceeded the recommended TLV of 0.5 mg/m^3, and even exceeded the average range of 3 to 5 mg/m^3 found in 1963, after controls had been added (20). Three cases were considered "very heavy" smokers, three "heavy" smokers, five "moderate" smokers, and one a "very light" smoker. Two were classed as nonsmokers, and one was unaccounted for. According to estimates made by EMAS (Employment Medical Advisory Service) in 1973 on the mortality statistics then available on lung cancer deaths, there was a twofold excess over the number that would be expected and "this excess is not necessarily of consequence" because of the relatively small numbers involved. In the Newcastle factory and offices as a whole, however, the number of deaths from all causes was less than expected (56 observed, 65.8 expected), thus adding credence to the relatively small cancer death excess.

4.4 Benzene

The isolation of a human leukemia virus by Gallagher and Gallo (23b) puts a new complexion on the presumptive etiologic agent in so-called benzene leukemia. The 60 cases tabulated between 1931 and 1960 by Browning (23c), the bulk of which were one or two cases per plant, and the relatively few cases of leukemia found in McMichael's studies of tens of thousands of rubber workers (23a) create a pattern consistent with an hypothesis that benzene, rather than being a carcinogen per se, triggers increased activity of the leukemia virus, present in only an occasional worker.

4.5 Benzidine Production

The inclusion of benzidine in lists of carcinogens needs correction; Zavon et al. (24) have, after more than 15 years of surveying two plants, one engaged in production, the other in handling, clearly observed that only those workers engaged in *production* of benzidine were cancer victims, not those in its handling.

4.6 Chromite Ore Processing

The 28.9-fold incidence of lung cancer among workers processing chromite ore (28b) firmly establish. 'the association of a considerable cancer risk in the industry as it was con. olled in the 1930s and 1940s, when environmental samples were well above the recommended TLV of 0.1 mg/m^3. A water-insoluble, acid-soluble chromite–chromate mixture was believed to be the etiologic agent, for repeated attempts to elicit tumors in animals from the soluble chromates were unsuccessful (28a). Epidemiologic investigations in three chrome pigment plants in the United States (28c), which produced the insoluble lead and zinc chromates, were equivocal as to the carcinogenicity of these chrome pigments, possibly owing to the small number of workers, 584 with 53 deaths and SMRs of 236 and 313 for lung cancer. If lead and zinc chromates are ultimately found to be human carcinogens, they certainly are not potent ones, for almost one-half of the air samples exceeded the TLV manyfold.

4.7 Nickel Sulfide Roasting

The repeated reference in several lists of carcinogens to nickel carbonyl needs correction; carcinoma of the bronchi and nasal sinuses formerly attributed to exposure to Ni(CO)$_4$ has been laid to rest by three separate investigations. The cancer risk was the same in nickel refineries where Ni(CO)$_4$ was present and in plants where no Ni(CO)$_4$ was used, but which had one operation in common, the roasting of nickel sulfide ore. This, incidentally, represents a case in which positive animal results have been misleading.

5 SUSPECT OCCUPATIONAL CARCINOGENS

The NIOSH volume of *Suspected Carcinogens* for 1978 (36) contains around 2400 substances, with references to the literature, suspected of having carcinogenic potential by virtue of their tumorigenic responses in animals. In addition, some 300 or more chemical substances are undergoing animal tests for tumorigenicity in an NCI-sponsored program. Selection of substances in this program is made on the basis of human exposure, production levels, and chemical structure. Table 39.2 lists some 45 chemical substances, or substances associated with industrial processes, which are suspected of inducing cancer, based on either (1) limited epidemiologic evidence of (2) demonstration of tumorigenesis in one or more animal species *by appropriate test methods*. Some 18 have been listed by Weisberger (37); the remainder have been culled either from the 1978 TLV

Table 39.2 Suspect Occupational Carcinogens (1979)[a]

3-Amino 1,2,4-triazole[b]	Hydrazine[b]
Auramine production	Lead smelting[c]
Benzo(a)pyrene	Leather and leather products manufacture[c]
*Benzoyl chloride production[c]	Magenta (basic fuchsin) production
Beryllium and compounds	Mesidine[b]
Carbon tetrachloride[b]	4,4'-Methylenebis(2-chloraniline) (MOCA)[b]
?Chloroform[b]	Monomethylhydrazine
Chromates of lead and zinc	Nitrogen dioxide (15)
Croton oil	2-Nitronaphthalene
1,2-Dibromethane(ethylene dibromide)	2-Nitropropane
3,3'-Dichlorbenzidine	Nitrosamines[b]
Dimethylcarbamyl chloride[b]	Ozone(15)
1,1-Dimethylhydrazine[b]	Phenyl β-naphthylamine
Dimethyl sulfate[b]	Propane sultone[b]
N,4-Dinitro-N-methylaniline[b]	β-Propiolactone
1,4-Dioxane[b]	Propylene dichloride
Epichlorhydrin	Propylenimine[b]
Ethylenimine[b]	Tetramethyl thiourea
Ethylene thiourea[b]	o-Tolidine
Glycidaldehyde[b]	2,4-Toluenediamine[b]
Hexachlorobutadiene	?Trichloroethylene
Hexamethyl phosphoramide	Vinyl bromide
	Vinylcyclohexene dioxide

[a] References may be found in Reference (36), with the exception of those otherwise noted and those treated separately in the text, Section 5.2.

[b] Denotes those items with brief summaries of evidence of suspected carcinogenesis, reported by Weisberger (37).

[c] Denotes suggestive, but not firm, evidence of carcinogenic potential for man based on epidemiologic evidence (see Section 5.2 for summary of evidence).

list of industrial substances suspect of carcinogenic potential for man (36) or from recent literature reports.

5.1 The Suspect Occupational Cancer List

In developing Table 39.2 for the 45 entries that are to be viewed as having carcinogenic potential for the industrial worker, questions arose as to the suitability of some to appear there, some for very different reasons: the dyes auramine and magenta, for example, for reasons of little or no present usage; others, for example, chloroform and trichloroethylene, for lack of epidemiologic substantiation.

The implication of auramine as a carcinogen derives from the older literature, when dye purity and manufacturing processes were unlike those of today. Also these dyes are of little commercial interest; only one commercial house lists auramine O, according to *Chemical Buyers Directory 1978–1979*, and magenta is not listed as commercially available, a situation resembling that of 2-acetylaminofluorene and dimethylaminoazobenzene, which have been used only in laboratories over the years as models of tumorigenic activity in animals, but which have been included needlessly in the OSHA list of 14 industrial carcinogens, and given the same degree of concern as duly recognized carcinogens.

A matter of greater concern in selecting candidates for the suspect list is the obvious unevenness of the supporting data. Tumor yields differ widely from one substance to another, and more importantly, instances of abrogation of the acceptable experimental design have led to serious question whether inclusion in Table 39.2 is appropriate at all. Three cases will suffice, chloroform ($CHCl_3$), trichloroethylene (TCA), and carbon tetrachloride (CCl_4). The case of $CHCl_3$ is given in some detail, because it typifies the unacceptable results that are obtained when poor experimental design of past NCI-sponsored studies has been followed.

The results of a carcinogenic bioassay of $CHCl_3$ were presented at the annual meeting of the Society of Toxicology, March 1976, as part of the verbal presentation of the NCI report of March 1, 1976 of its cancer-testing program. This report showed that $CHCl_3$ elicited hepatocellular carcinomas in both sexes of mice, with 96 and 95 percent incidences for males and females, respectively, at an average daily oral dose of 277 mg/kg five times per week for 78 weeks for the males, and 477 mg/kg for the females. At the lower dose (about one-half the higher dose), the incidences were 36 percent and 80 percent for males and females, compared with 6 percent in matched and colony control males, and essentially no incidence in female controls. Renal epithelial tumors were found also in male rats only, with incidences of 24 percent and 8 percent at the higher and lower doses (180 and 90 mg/kg). *Decreased survival rate* occurred in all treated rat groups, as was true for the higher-dosed female mice. At the conclusion of the presentation, permission was granted to F. J. C. Roe, advisor to Huntingdon Research Center, Tunstall Laboratory, Kent, England, to present a preliminary report of long-term tests of $CHCl_3$ in rats, mice, and dogs. This report differed so remarkably from the above NCI report as to demolish the evidence that $CHCl_3$ is a carcinogen in rodents. With one strain of rats, four strains of mice, and the beagle dog, and toothpaste base as a vehicle for $CHCl_3$, an excess of benign tumors of the renal cortex was seen only in males in only one strain of mice studied; in one strain of mice, survival of the treated mice exceeded that of the controls, and two of the controls had liver tumors, compared with none of the chloroform-treated mice. In the other two mouse strains, no excess of lung tumors, malignant lymphomas, or any other kind of neoplasms was seen in any of the strains. In the Sprague–Dawley rats, *survival in the treated group exceeded that of the controls* with no effect on tumor incidence in either sex. At the time of presentation, after 7.5 years of daily exposure of dogs to 30 mg/kg $CHCl_3$, the only effects seen have been some transient elevations in the mean SGPT. No dog has yet developed a neoplasm of the liver.

In his verbal presentation, Roe attributed the remarkable differences in findings to the lower dosage levels used (60 mg/kg per day and 30 mg/kg per day), as compared with 477 mg/kg for the mice and 180 mg/kg for the rats at the higher levels used by NCI. In questioning the NCI report, Roe elicited the fact that heptocellular degeneration and necrosis occurred in animals at the higher dosage, along with abdominal distention. This, Roe and Stokinger pointed out, was the probable basis for the carcinogenic findings in the NCI studies as opposed to those of the Tunstall Laboratory. The abdominal distention apart from the liver damage can readily be appreciated if one calculates the volume of vaporized $CHCl_3$ in the rodents at 38°C from either of the dosage levels; it is found to be far in excess of the volume of the gastrointestinal tracts of these animals. In short, the experimental design of the NCI-sponsored studies yields results from which it is impossible to make any judgment regarding the carcinogenic potential of $CHCl_3$; hence there is a question mark before chloroform in Table 39.2. Further, no support for these animal findings appears to be forthcoming from industrial experience; to our knowledge, no reports of hepatocellular cancers or renal epithelial tumors in industrial workers (or in chlorinated-water drinkers) have been published.

The same experimental design was used in the carcinogen bioassay of TCE, with the same unacceptable results. In this case also, no human confirmation of the animal results was found; following the NCI report, NIOSH dispatched an epidemiologist to find cancer cases arising from TCE exposure. At this writing (1979), four years later, none have been found either among degreasers or dry-cleaning workers, those with the greatest exposure. Here is a clear call to abandon past practices of high-dose experimental design, that yields results in debilitated animals, and to moderate it to include a series of lower levels as suggested in Section 3.2.2.

A like case can be made also for CCl_4, where animal findings have not been substantiated in industrial experience, and where CCl_4 has been banned for many years from the workplace as a hazardous solvent, except where its use as a chlorinating agent for inorganic reactions (UCl_4) is used in hermatized processes.

5.2 Further Information on Suspect List

In addition to the questionable items listed in Table 39.2 noted above, evidence of carcinogenicity based on suggestive epidemiologic findings on five industrial manufacturing processes that have not thus far appeared in such lists is summarized here.

5.2.1 Benzoyl Chloride Production

Three cases of cancer of the lung have been reported (38a) among Japanese workers engaged in the production of benzoyl chloride (BC) in a factory that had made BC from 1953 to 1973, and in which about 20 workers had been exposed. Two were smokers; one was not. In addition, a case of malignant, maxillary lymphoma occurred at another plant manufacturing BC between 1960 and 1970, from among another group of about 20 exposed workers. Exposure conditions about the plants were described as very poor (no measurements were reported); workers suffered from chronic pharyngitis, chronic sinusitis, hyposmia or anosmia, and skin troubles classed as parachroma and warts. Because BC goes through several intermediates before its final product, exactly what substance(s) contributes to the cancers has yet to be determined.

5.2.2 β-Chloroprene (2-Chlorobutadiene-1,3)

Early studies in animals by von Oettingen in 1936, by Nystrom in 1948, followed by those of Khachatryan in 1972, in which β-chloroprene was stated to produce excess skin and lung cancer in plant workers where exposure levels were reported to be low, gave sufficient concern to generate further investigation of the question. These investigations showed that both acute and chronic daily inhalation of 10 and 50 ppm β-chloroprene by animals for as long as 24 months, and epidemiologic studies at two plants in the United States involving 270 and 1576 workers for 17 and 30 years, respectively, revealed no evidence that β-chloroprene was carcinogenic for animals or man. A Soviet study of about 7000 workers, of whom 600 had up to 40 years of β-chloroprene exposure, has thus far agreed with those in the United States (39).

5.2.3 Lead Smelting

An epidemiologic study of 7032 male workers in six lead-production facilities (one primary smelter, two refineries, three recycling plants, with 2352 workers) and 10 battery plants (4680 workers) who had been employed between 1946 and 1970 has been reported by Cooper (38b). Lead absorption was greatly in excess of accepted standards. The standard mortality ratios for malignant neoplasms was somewhat elevated in both groups, 133 for smelters and 111 for battery workers, the excesses arising largely from tumors of the digestive organs and the respiratory system. Of interest, in view of the experimental production of renal tumors in rats, were three deaths from malignant renal tumors and seven from tumors of the central nervous system, comparable to gliomas found experimentally in rats. Again, whether it is lead or other substances associated with its production has yet to be determined.

5.2.4 Leather and Leather Products Manufacture

Risks of bladder cancer of $6.3\times$ for males and $4.36\times$ for females, and risks of 3.22 and 3.58, respectively, for cancer of the buccal cavity and pharynx, and 3.31 and 5.48 for cancer of the larynx have been reported by DeCoufle (40). When the contributions from smoking were evaluated, risks for urinary bladder cancer were reduced from 6 to 4, but were still considered significantly elevated for the two groups of leather handlers, those manufacturing leather from animal skins and hides and leather products, shoes, belts, handbags, and luggage. A review of the processes and agents in leather manufacture showed several areas with exposure to potential carcinogens, including azo and other synthetic dyes that have induced cancer in animals. Again, further studies are needed to characterize the nature of the etiologic agents and their dose–response relationships.

5.2.5 Hexamethylphosphoramide

This phosphoramide (HMPA), a polymer solvent used in the manufacture of Kevlar® aramid fiber, has been found to induce squamous cell carcinomas of the nasal cavity in rats exposed daily for 3 months by inhalation at 0.05 ppm after a 13-month latent period (41a). Daily exposures at 0.01 ppm for 24 months, however, elicited no tumors on histopathological examination (41b). Because dogs also showed squamous metaplasia of the nasal cavity following inhalation exposure to HMPA for 5 months at 0.4 and 4 ppm, HMPA can thus be labeled a suspect carcinogen; however, no tumors have yet appeared in workers after more than 10 years of Kevlar fiber production (41a).

5.2.6   2-Nitronaphthalene

This unwanted by-product, to the extent of 5 percent in the production of crude nitronaphthalene, caused bladder tumors in dogs fed 100 mg daily for 8 months and then held for 6.5 to 10.5 years (41c). β-Naphthylamine, a bladder carcinogen in man, was detected in the urine.

5.2.7   2-Nitropropane

2-Nitropropane, an oily liquid with appreciable vapor pressure (10 mm Hg, 16°C) and a wide variety of industrial applications, from polymer solvent to rocket propellant, produced hepatocellular carcinomas in all 10 rats exposed daily for 6 months at 207 ppm, but no neoplasias at 27 ppm (41d).

5.2.8   Vinyl Bromide

Vinyl bromide, a low boiling liquid (b.p. 15 to 18°C), useful as a fire retardant in plastics, has induced tumors of several types in rats exposed 6 hr daily for periods up to 18 months at all levels above 10 ppm (50, 250, 500, 2500 ppm). The first angiosarcoma appeared in a rat before the end of 1 year, with hepatocellular sarcomas and mammary and brain tumors at a later date in a dose–response fashion as found in sacrifices at 12 and 18 months (41d), making vinyl bromide of about the same potency as vinyl chloride dosewise, but with relatively great capacity for tumors of different types (41e).

If we take note of the animal findings of suspect carcinogens (Table 39.2) one thing at least is certain—the list of suspect carcinogens will continue to increase. Far less certain is an improvement in predicting worker cancer risk from these sources as long as present experimental protocol is used.

5.3   Determination of Mutagens

Because of the high and constantly rising cost and 3 years of effort required to assay a substance for suspicion of its carcinogenicity for man in tests in rodents, a simple test for identifying such substances has been developed by Ames (42a). The test has been shown to detect about 90 percent (158 of 176) of the organic chemical carcinogens as determined by animal experiments from about 300 carcinogens and noncarcinogens examined (42c). A review of accomplishments of 15 years of experience with this test, and others similar, has been made by Ames (42b), who identifies by their mutagenic activity those substances known or suspected of being human carcinogens. The review details the considerable success attained in detecting mutagenic activity in certain pesticides, drugs, and particularly for our concern here, heavy chemicals such as ethylene dibromide, ethylene dichloride, vinyl chloride, and many others. Only two outstanding exceptions were found: stilbesterol and benzene. (For possible failure in the latter case, see Section 5.1.) Now, what does the track record show?

But as with all shortcut tests, there are obvious limitations, even in this case with the improved procedure of adding liver homogenates to the mutant bacterial strains to catch those substances that require metabolic activation for display of their mutagenicity. Regretfully we must note that for the one substance, ethylene dibromide, for which carcinogenic risk assessment of the industrial exposed worker has been made (16), the finding of mutagenicity bears no relation to preliminary findings of the lack of its carcinogenicity for man. (For this and other instances of lack of correspondence between the test and worker experience, see Section 5.1.)

If one reflects on the comparative biologic systems involved, it is remarkable that the test detects a mutagen for man at all. That it does is presumably because the mutagen acts directly and without antagonizing influences to damage DNA, thought to be the major cause of cancer. In so doing, the mutation test, as presently used, exhibits the maximal potential expression of a mutagen, going straight and unopposed to DNA, unopposed by any immunologic "surveillance" by the thymus and bone marrow lymphocytes, hormonal influences, and those of interferons, which in the human system counteract effects of viruses. Also lacking in the bacterial system are the prostaglandins, and a circulatory system that can renew cancer antagonists, and so on! Evidence for the counteracting influences of these systems on oncogenesis is commonly noted by cancer pathologists who see neoplastic growths in individuals who never die of cancer (43). And again, we regretfully note that one-quarter of a century ago, when concern was expressed for the seemingly endless number of air pollutants that were becoming recognized over U.S. cities in 1955, great hope was held out in some quarters (Earl and associates) for shortcut determinations of acute toxicity by unicellular organisms. Hopes were dashed when fewer than 50 percent of the tests gave results corresponding to those obtained in animals. This was the knell for shortcut toxicity tests.

Thus with the systemic deficiencies characteristic of unicellular organisms, the actual success rate in pinpointing human cancers is very low. This is unfortunate in two respects. (1) From the increasing number of substances found to be mutagenic and hence suspect carcinogens, the cry of "Wolf, Wolf" goes out, and cancer phobia is spread throughout the land. (2) Because the main virtue of the mutagen test is to put NCI dogs on the scent of the wolf trail, the return on the pathfinding mutagen effort is very low indeed. Expressed slightly differently, it was the concluding sentiment of Maltoni, Parmegiani, and Izmerov and others at the Occupational and Environmental Health Section of the Carlo Erba Foundation, meeting in Milan, December 1977, that "these tests (mutagenic) do not provide an adequate basis for quantitative evaluation of the carcinogenic risk for humans," and they might have added "and in many cases, not even a qualitative basis." So, for the estimation of human cancer risk or for pinpointing suspect carcinogens by mutagenic testing, we are left for the present with a nondisposable, almost indispensable item, which at best can make important findings, and at worst, lead animal experimenters on wild goose chases.

6 METABOLISM AND MECHANISMS

This section is restricted to the way in which chemical carcinogens are metabolically activated to initiate the process of neoplastic growth, thus excluding from consideration viral carcinogenesis per se, discussions of which may be found in References 44a and b, but their interaction with chemical carcinogens is touched upon. The biochemistry of chemical carcinogenesis was first reviewed by Miller and Miller in 1959 (45a), and 15 years later by Heidelberger (45b). Both these reviews, it should be noted, deal only with *organic* chemicals; an hypothesis of a probable mode of action of inorganic carcinogens is included here.

It has now become generally recognized by oncologists that chemical carcinogenesis consists of at least two separate stages, initiation and promotion. From the standpoint of the industrial worker, one of the most important parts of the initiation stage is the degree of exposure, absorption, and distribution of the carcinogen, all of which can be measured in the industrial setting as can activation in some cases, and elimination of the chemical and its metabolic products. The second stage consists of interaction with critical receptor sites ultimately resulting in transmittable molecular species. The next step of the promotion stage is survival and proliferation of the cells containing the transformed molecular species to clinically recognized cancer.

Interference in any of these steps can alter the outcome by either enhancing or inhibiting the response to the carcinogen; promoters, by definition, unable of themselves to initiate the neoplastic process, can greatly enhance the incidence of cancer, or conversely, damaged receptor sites may undergo spontaneous repair. Still other factors may increase or decrease the response: age, with its concomitant changes in immunosuppressive mechanisms, the lymphocytes of the thymus and bone marrow, interferons, in the case of virus-promoted cancer, and trace metal concentrations, concomitant disease, and hormonal and nutritional statuses. Thus a precancerous state may be induced without cancer ever developing, or contrariwise, cancer may develop when the precancerous process can no longer be suppressed by the body's protective mechanisms.

6.1 Metabolic Activation

It has been so repeatedly noted in reviews of chemical carcinogenesis since 1959, that it is now common knowledge that many classes of organic chemical carcinogens become covalently bound to DNA, RNA, and proteins of the cells in target tissues, and that the induction of cancer results from such covalent binding to one or more of these cellular macromolecules. Before binding occurs, however, most classes of chemical carcinogens must be metabolically activated. Because of the rather vast literature on the subject, only three classes of carcinogens of industrial interest, polycyclic aromatic hydrocarbons, aromatic amines, and halogenated hydrocarbon are discussed.

6.1.1 Polycyclic Aromatic Hydrocarbons (PAH)

Metabolic activation of PAH consists of an oxidation of the rings of unsubstituted PAH, for example, benzo(*a*)pyrene, chrysene, and dibenzanthracene, and oxidation of the alkyl group of alkyl-substituted PAH, for example, methylcholanthrene. These oxidations are carried out by "mixed function" oxidases of the liver, which contain cytochromes P-450 and P-448, and require reduced nicotine adenine dinucleotide (NADPH) and oxygen. In this oxidation, an epoxide intermediate is formed which has been shown (46a, b) to have the requisite chemical reactivity to form covalent complexes with DNA and histones and to serve as the ultimate carcinogenic form of PAH.

Potent carcinogenic methylated PAH, such as dimethylbenzanthracene, appears to be metabolically activated by oxidation to an hydroxymethyl compound by microsomal metabolism (47a, b). Another type of metabolic activation of alkylated PAH is the formation of carbonium ions, which appear to bind with DNA, and which have moderate carcinogenic activity (47b).

PAH may also be converted to radical cations capable of DNA complexing, for free radical signals have been detected when benzopyrene (BaP) was incubated with skin homogenates (48a). This signal was attributed to the formation of the 6-phenoxy radical (48b). Treatment of BaP with free radical-generating substances (I_2 or H_2O_2/Fe^{3+}) also led to covalent binding to DNA (48c), a correlation between free radical production and carcinogenic activity (48d). Thus at present it appears that metabolic activation of PAH can involve the formation of epoxides, free radicals, and for alkylated PAH, carbonium ions.

A review of the several types of binding of organic chemical carcinogens to DNA has been made by Irving (49) and the structures of DNA-bound hydrocarbon derivatives have been elucidated by Daudel and co-workers (50a, b).

6.1.2 Benzene

Though not a PAH, benzene, a single aromatic ring structure, similarly is metabolically activated to an epoxide and thus is capable of forming a covalent complex with DNA (51). Its association constant, however, must be considerably lower than that of most PAH epoxides, for it is a far less potent carcinogen, if indeed it is a carcinogen (see "Benzene," Section 4.4).

6.1.3 Aromatic Amines

Knowledge of the metabolic activation, binding to macromolecules, and mechanism of action of aromatic amines has been developed largely by Miller and Miller (45a), much of whose work, though carried out on 2-acetylaminofluorene (AAF), a carcinogen of very secondary interest to occupational health, can nevertheless serve as a model for those aromatic amines that are. AAF is activated by microsomal hydroxylation of the nitrogen atom to a proximal carcinogen, which, in (rat) liver, is converted by the enzyme sulfotransferase to a highly reactive, electrophilic sulfate, which in turn is considered to be the ultimate carcinogenic form. This ester reacts covalently with DNA, RNA, and proteins, thereby initiating the carcinogenic process. Activation by N-hydroxylation of the industrially recognized carcinogens β-naphthylamine and 4-aminobiphenyl has been shown to occur in laboratory animals (monkey and dog) (52a), and is probably the mode of activation of the other aromatic amines of industrial interest (Tables 39.1, 39.2).

Like activation mechanisms of PAH, other mechanisms for aromatic amines have also been found; one involves formation of a nitroxide free radical by a peroxidase, which then undergoes dismutation to yield N,O-acetyl AAF (52b); the other involves the participation of an acetyltransferase that acts directly on the N-hydroxylated AAF to give the same acetyl derivative as in the former case (52c).

6.1.4 Vinyl Chloride

The metabolic activation of vinyl chloride to a proximal carcinogen has been demonstrated in a dose-dependent fashion by Gehring and associates (53a) in rats exposed by inhalation for a single 6-hr period using ^{14}C-labeled vinyl chloride. It was found that it was the amount of metabolically activated vinyl chloride and not the concentration of vinyl chloride to which the animals were exposed that correlated with the incidence of hepatic angiosarcomas for any administered level of exposure (1.4 to 4600 ppm). Correlative to this finding, covalent binding to hepatic macromolecules was relative to the amount of vinyl chloride metabolized (53b), presumably as vinyl free radical.

6.1.5 Ethylene Dibromide (EDB)

Although EDB is ultimately metabolized to a detoxified form by conjugation with hepatic glutathione (54a), it is speculated that the more highly toxic bromacetaldehyde metabolized from EDB may be the ultimate carcinogen to combine with DNA, because disulfiram, a known enzyme inhibitor, increases the mortality and morbidity (hemangiosarcomas and adenocarcinomas) in rats, presumably by blocking the glutathione detoxication process at the bromacetaldehyde stage (54b).

6.2   Cocarcinogenesis: Promoting Factors and Mechanisms

A discussion of the factors affecting carcinogenesis has appeared in a recent volume, *Persons at High Risk of Cancer* (44b). The multiple factors have been grouped under those of the environment and of the host as applied to the general population. In what follows here, evaluation is made of those factors, cocarcinogens and promoters, that affect the cancer risk to industrial workers. Because these factors can be both promoters and inhibitors of the carcinogenic process, anticarcinogenic factors are included also, as well as brief summaries of the present understanding of their mechanisms.

6.2.1   Environmental Cocarcinogens

The cocarcinogens affecting the incidence of occupational cancer include smoking, alcohol, chemicals and drugs, diet, viruses, and microbes. Boutwell reviewed in 1974 the chemical nature, function, and mechanism of promoters of chemical carcinogenesis (55).

Smoking.   The extraordinary increase in the risk of lung cancer in asbestos workers who smoke is paralleled by no other cocarcinogenic factor; asbestos workers who smoke have about 92 times the risk of dying of bronchogenic cancer as men who neither are exposed to asbestos nor smoke cigarettes (22). The risk of developing mesotheliomas, both pleural and peritoneal, appears also to be enhanced, but the number of deaths from these malignancies was too small to draw firm conclusions.

Because of the large contributory role of smoking to lung cancer in asbestos workers, obviously the magnitude of this contribution should be ascertained for all occupational exposures presenting a potential for cancer development in the respiratory tract, for example, arsenic and antimony trioxide production, bischloromethyl ether exposure, and nickel sulfide roasting.

Apart from its cocarcinogenic role in occupational exposures, cigarette smoking per se increases cancer mortality in the age group 65 to 79 over that of nonsmokers by 11.59 times for lung, 8.99 times for larynx, 2.96 for bladder, and 2.93 for buccal cavity and pharynx cancers (44b).

Mechanism proposals for this cocarcinogenic action, however, have been few, ostensibly because in the present state of knowledge, it is not possible to evaluate the relative contributions of the numerous potential carcinogens and cocarcinogens and their interactions in tobacco smoke. A few suspects can be named: benzo(a)pyrene (BaP) and related PAH, radioactive elements, arsenic, irritants in tar, oxides of nitrogen, and ammonia. One serious attempt (56a) attributed at least a part of the carcinogenic action of cigarette smoke to the radioactivity of polonium^{-210}. This thesis was adopted and enlarged in great detail later to show that ^{210}Po and ^{210}Pb persist in lung tissue in insoluble smoke particles, highly localized in clusters with activities ranging from less than 10^{-6} pCi to more than 10^{-4} pCi at each focal point (56b).

Alcohol. Although there is a priori little cause to complicate alcohol as a carcinogen, its many metabolic and pharmacological effects (irritation of the mucous membranes and skin) render it a reasonable candidate as a cocarcinogen, a role suggested by animal studies. It has been known for decades, for example, that alcohol consumption increases the risk factor from smoking. Heavy drinkers experience a risk two to six times higher than nondrinkers, the effect depending on and increasing with the amount of tobacco smoked (55). The combined effect of heavy drinking and smoking results in a risk more than 15 times as great as that for those who niether smoke nor drink, whereas heavy smoking alone results in no more than a two- or threefold increase in risk.

For the effect of alcohol consumption on cancer at specific sites, heavy whiskey drinking raised the risk for laryngeal cancer more than 10 times compared to abstinence or moderate drinking (57). These data support the idea that although smoking may initiate laryngeal cancer, alcohol acts as a promoter rather than a primary carcinogen.

Although these data pertain to cancer of the oral cavity, it is not unreasonable to assume that cancer of the respiratory tract among industrial workers would be also increased from high consumption of strong alcoholic drinks, inasmuch as alcohol and its metabolites in high concentrations act as respiratory tract irritants.

Esophageal cancer, though rare in the United States, has a risk factor of 17 for alcoholics above that for nonalcoholics, and 25 for heavy drinkers who have been moderate smokers. Risk factors at other sites, stomach, liver, and rectum, are more controversial (55).

Chemicals and Drugs. Because of the large number of chemicals and drugs that can act as cocarcinogens or promoters, only prominent examples are cited here.

Alkylated Hydrocarbons. Probably the first demonstration of strong cocarcinogenic effect of chemicals was made in 1957 by Horton et al. (58a), who showed that aliphatic and related hydrocarbons accelerated the development of skin tumors in mice. Later work by Bingham and Falk (58b) showed that the incidence of skin tumors in mice from BaP was increased from 10 to 20 percent to about 50 percent when the applied solvent was changed from mineral oil to the cocarcinogen decalin (decahydronaphthalene). Similarly, the carcinogenic potency of benz(*a*)anthracene was enhanced from 5 to 21 tumors per group and the latent period reduced from 72 to 46 weeks when the solvent was changed from toluene to *n*-dodecane. Further, a thousandfold increase in the potency of low concentrations of BaP occurred when *n*-dodecane was the applied solvent. Because relatively low concentrations of these carcinogenic hydrocarbons are the prevailing exposure levels in industrial workers, a brush with cocarcinogens such as these is something to be avoided.

   Certain alkylated naphthalenes appear to enhance the carcinogenicity of BaP. The most active in terms of skin tumorigenicity were 1,4,6,7-tetramethylnaphthalene and 2,3,6-trimethylnaphthalene. When applied together with BaP, 1,5-dimethyl-, 1,4,6,7-tetramethyl-, and 2-ethylnaphthalene were most active in the development of mouse lymphomas. Thus there appears to be some correlation between the degree of alkylation of naphthalenes and cocarcinogenic activity.

Phenols.   Enhancement of tumor production by cocarcinogens is not restricted to tumors of the skin or lymphoid tissue but can develop also along the respiratory tract; Tye and Stemmer (59) found the acidic fractions of two coal tars containing phenols, 4.5 and 1.4 percent, respectively, and BaP, 0.7 and 1.1 percent, to elicit 13 percent adenocarcinomas, 60 percent intrabronchial adenomas, and 30 percent squamous metaplasias, in mice (C_3H/HeJ) after 46 weeks of intermittent, daily inhalation exposures. These incidences are to be compared with no incidence of adenocarcinoma, 55 percent intrabronchial adenoma, and 10 percent squamous metaplasia in the mice not exposed to the phenolic fraction.

Trace Metals.   Trace metals occupy the rather unusual position of being both promoters and inhibitors of chemical carcinogenesis, depending on their ambient concentration (60a). On the hypothesis that in the induction of asbestos cancers, BaP and related PAH are major determinants, and asbestos plays a passive role as carrier of trace metals, experiments with tissue homogenates of both animal and man showed that Cu, Ni, Co, Fe(II), Mg, and Zn stimulated BaP hydroxylase (AHH) significantly at low concentrations. The inhibition of AHH by these trace metals indicates that they have the potential to interfere with the metabolism (detoxication) of BaP and thus contribute to its carcinogenicity. Evidence in support of trace metal activity is (1) the relatively large amounts of Ni, Cr, Mn, and Fe that have been found associated with certain forms of asbestos, particularly chrysotile (60b), and (2) the development in rats of lung cancer after having inhaled chrysotile dust fortified with Ni, Co, and Cr (60c), not demonstrable, in rats at least, from natural chrysotile.

   In a related study, nickel carbonyl was shown to inhibit the induction of

AAH, an action more pronounced in the lung than in the liver, indicating again that nickel can act as an indirect carcinogen for the lung (61). Trace metal chelation has long been thought to be a dominant feature in chemical carcinogenesis (62). (See also hypothesis of inorganic chemical carcinogenesis, Section 7.)

Phorbol Esters. The complex, substituted, polycyclic alcohol phorbol and its ester congeners, the active principles in croton oil, long recognized as a potent promoter, have been found to be promoters *par excellence*, and as a result have been much studied to learn more about how promoters in general function. Most potent of the esters is the acetate, 12-*O*-tetradecanoyl phorbol-13-acetate, designated TPA. TPA is at least 10 times as effective in inducing tumors as the carcinogen alone.

Studies with TPA on mouse skin and cell cultures showed induction of several phenotypic changes resembling those seen in cells transformed by viruses or chemical carcinogens (63a). These changes included altered cell morphology, lipid metabolism, and cell surface glycoproteins, increased membrane transport of 2-deoxyglucose, induction of the enzymes ornithine decarboxylase and plasminogen activator, and induction of prostaglandin synthesis (63b).

In another category of effects, TPA enhances the transformation of cultured cells previously exposed to a chemical carcinogen (63c) and inhibits terminal differentiation in mouse epidermal cultures (63d).

These findings led Lee and Weinstein (64) to speculate that TPA acts by competing with an endogenous epidermal growth factor (EGF) for cell receptor site(s) that normally mediate its action. EGF has been reported to promote tumor induction on mouse skin (65), to induce ornithine decarboxylase and prostaglandin synthesis (66), and to be a potent inducer of plasminogen activator (67), all functions of TPA. Some TPA effects, however, are not shared by EGF, for some cell types lack EGF receptors, yet respond to TPA (64).

Obviously, the mechanism of phorbol esters' promoting action is even more complex than has been summarized above; Rovera et al. (68) subsequently observed that these compounds can have more than one effect on differentiation, depending on the target cell. Phorbol diester tumor promoters can not only inhibit differentiation and stimulate differentiation along the normal pathway, but can also induce differentiation along an alternate pathway, that is, in human promyelocytic leukemia cells (68a), and can induce blastogenesis (activation of lymphocytes) of human lymphocytes (68b). To add to the complexity, studies with other tumor promoters, for example, dimethyl sulfoxide, indicated separate mechanisms of action.

Help in unraveling the stages in tumor promotion can come from studies of inhibitors of the process: protease inhibitors, certain anti-inflammatory steroids, retinoids, and inhibitors of prostaglandin synthesis, the last suggesting that prostaglandins may be required for promotion. Evidence for separate mechanisms is strong; retinoids inhibit ornithine decarboxylase activity; the steroids do not, but do inhibit production of plasminogen activator. Obviously, there is much more to be learned.

Drugs. The common practice among all too many industrial workers to resort to drugs, apart from alcohol (Section 6.2.1), prompts mention of the very suspicious evidence of the interaction of these agents on carcinogenicity. Although a wide range of drugs has been implicated as cancer promoters in man, including immunosuppressive drugs for kidney transplants, steroid hormones in long-term treatment of aplastic anemia, cytotoxic and alkylating agents used in cancer chemotherapy, dilantin for the control of epileptic seizures, and amphetamines for weight reduction, human data on drug-induced cancer is sparse, sometimes questionable, and often in need of further investigation (44b).

In one study (69a) amphetamine was linked to a sixfold excess risk of Hodgkin's disease. A subsequent negative study (69b) renders this relationship questionable, however, Dilantin (diphenylhydantoin) is also under suspicion as a cancer promoter (lymphoma) (70), but again, epidemiologic testing is inadequate. Phenobarbital, a commonly used and abused sedative drug, appears from several studies to act as a promoter of liver tumors (71). The drug disulfiram, used in alcohol-control programs, has been shown not only to increase morbidity and mortality of ethylene dibromide–disulfiram treated rats, but to induce substantial numbers of hemangiomas of the liver, spleen, omentum, and kidney of males and females, adenocarcinoma of the mammary gland, and atrophy of the testis (72a). These histopathological changes were noted with greater frequency in the groups exposed to both chemical agents. The mechanism has been shown (72b) to be that of disulfiram inhibiting the action of liver dehydrogenases on the alcohol–acetaldehyde–acetic acid metabolic pathway, in this case the ethylene dibromide–bromoacetaldehyde pathway, giving rise to increased amounts of the more toxic bromacetaldehyde.

Diet. Because there is no shortage of carcinogens in the human diet, and exposure to foodstuffs is as much a part of worker existence as exposure to industrial chemicals, an attempt should be made to evaluate the potential for cocarcinogenesis from this source. We say *attempt*, for although many cancers in the United States are linked with dietary patterns, knowledge of specific carcinogens is lacking. The tenuous association is made further difficult by low host susceptibility, low biologic activity, and generally very low dosage. To compound the difficulties, in our present state of knowledge the associations are complex, and depend greatly on combinations with industrial environmental agents, other cofactors, and natural or acquired metabolic peculiarities of the worker.

Of the numerous factors in worker diets that can be considered suspect on the basis of a strong correlation with a particular cancer, only a few that appear to have the strongest correlation are considered here.

The strongest correlation between *specific agents* and human cancers are the polycyclic hydrocarbons, PAH, associated with smoked fish (73); all populations that consume large quantities of smoked fish are at high risk of cancer of the stomach. Because PAH represent a common exposure not only from the general environment, but in several occupations as well, smoked fish would seem to be a likely cocarcinogenic candidate. Barbecued beef, on the other hand, does not appear to present this risk, presumably owing to the presence of naturally occuring antagonists.

Caffeine is included not only because it is a most common beverage, but more importantly, because it has a wide range of host reactions. An association exists between heavy coffee drinking and bladder cancer in two separate epidemiologic investigations (74a, b), and caffeine per se is a potentiator for several carcinogens in animal systems (74c). The association, however, does not rule out the possibility of another variable to consider (74d).

Much attention has been given to finding the specific dietary factors causing high rates of cancers of the gastrointestinal tract in various countries, Ireland, Japan, Scotland, and China. In Ireland, although numerous candidates have been examined, no causative agent has unequivocally emerged. In Japan, talc-polished rice has been found to be the cause of gastrointestinal tumors of rice eaters (75); talc was found in the tumors as evidence of the association. In Scotland, the high bowel cancer incidence was thought to be related to the bracken fern eaten in certain areas (44b, p. 206), and "pickled vegetables," commonly eaten in a northwest province of China, have definitely been considered the cause of high rates of esophageal cancer in that area (76). People living in that area have an estimated 10 percent chance of acquiring cancer of the esophagus, a frequency tenfold that in other provinces where this dietary practice does not exist. Further evidence on the etiology is that chickens fed the moldy vegetables acquired cancer of the gullet at rates that paralleled those in humans, whereas no such condition occurred among chickens not fed the vegetables.

In addition to the cocarcinogens naturally occurring in foods or formed during their processing, certain *food additives* have now come under scrutiny as cancer promoters, particularly saccharin and cyclamate. Both compounds produce bladder cancer in mice, although in extremely high doses and under unacceptable experimental procedures. Cyclamate has been banned by the FDA. Saccharin has been shown to have cocarcinogenic properties only, and not to be a complete carcinogen when fed to animals (77), and because its active metabolites have not been identified, human cancer risks are difficult to estimate.

With respect to *food contaminants*, chemical fertilizers and pesticides are general additives to foods derived from plants, whereas antibiotics and growth stimulants get into meats and other animal products, and thence into the human diet through the food chain. Cocarcinogenic risks presumably come primarily from those substances that accumulate in the body. The chlorinated hydrocarbon pesticides are in this category. DDT is one of several of these pesticides that can produce tumors in animals, but no human cancer risks for these substances have yet been estimated; it is true similarly for aldrin and dieldrin, and all have been banned!

The best example of a food contaminant problem is diethylstilbesterol (DES) in beef. DES is a known carcinogen for women when high doses of about 2 g are given in the first trimester of pregnancy. Accordingly, DES pellets, embedded under the skin of chickens, have been banned, but because a heavy eater of beef would consume no more than a few micrograms DES per year, creating a hormone level below physiological levels of natural hormones, and well below the amount of "natural" estrogens of those adding wheat germ to their diets, there is little reason to believe that DES in beef poses a major or even a minor cancer risk (78). No estimates of cancer risk of DES have been made, however, for workers in any occupation.

Undernutrition presents some interesting facets to the problem of cocarcinogenesis, particularly with respect to vitamins, trace elements, and proteins. Undernutrition is still claimed to occur in some significant portion of the U.S. work force, and thus bears some relation to the cancer risk in this worker group.

Among the vitamins, vitamin A has been most studied in this respect because of its importance to the functioning of normal epithelium. Vitamin A deficiency, in a study designed to investigate the high frequency of salivary gland cancer in Eskimos, was found to enhance this cancer in rats (79a) and to increase bowel cancer (79b), and has been considered a probable contributor to the high level of nasopharyngeal cancer in Kenya (79c) and as a possible cofactor in gastric carcinogenesis (79d).

Other vitamins affect carcinogenesis also. Riboflavin (vitamin B_2) deficiency both inhibited and enhanced epithetial neoplasia (80a, b), and vitamin B_{12} deficiency slowed hepatic carcinogenesis (81a), but pernicious anemia, a B_{12} deficiency, is a precancerous condition with patients having increased risk of leukemia and cancer of the stomach (81b). Low vitamin C intake is associated with a region of high rate of stomach cancer (82).

In like measure, low selenium intake appears to predispose to higher cancer rates in areas of the United States than in those areas (Midwest) where higher selenium levels occur in plants, milk, and blood (83); this effect has been attributed to its antioxidant properties, which are those of vitamins C and E and ubiquinone also. Manganese deficiency has been associated with the high cancer rates in Finland (84). Other more tenuous relationships between trace elements and cancer have been reported, but because of their complex interactions with one another, none of the reports has been developed to the point of removing contradictions or presenting a clear picture regarding altered human risk.

Protein deficiency (PD) can promote the activation of some carcinogens and inactivate others (85a, b). New facts have been brought to light on these separate effects that PD can produce on different tumor types from the same carcinogen (85c). Methylene-bis-o-chloroaniline (MOCA) fed to rats on a PD diet (8 percent casein), while enhancing hepatocellular carcinoma incidence to 18 percent at the 500 ppm dietary level, compared to 4 percent on the protein adequate diet, decreased the incidence of lung adenocarcinomas, resulting in a linear dose–response for lung adenocarcinomas but not for hepatocellular carcinomas. This suggested to the authors that the mechanisms for induction of these two tumor types were different, and it is likely that the conditions responsible for promoting one tumor type will suppress the development of the other.

MOCA metabolite levels in rat urine correlated well with differences in incidence of hepatocellular carcinomas, but did not correlate with differences in lung or mammary adenocarcinomas when MOCA dietary levels increased; they correlated instead with dietary concentrations of MOCA, suggesting that there may be a threshold for the induction of hepatic carcinomas, but not for pulmonary or mammary adenocarcinomas.

Although urinary concentration of MOCA metabolites in rats at the lowest dose (125 ppm) was comparable to that of workers exposed to MOCA, how far these results are translatable to man is uncertain. The gastrointestinal route, as opposed to the respiratory, subjects MOCA first to an acid treatment in the stomach where its metabolites undergo modification, and then to an alkaline wash from pancreatic secretions with still further metabolic modification. MOCA undergoes no such treatment by the respiratory route and thus is subjected to far less metabolic activity, all of which could materially modify the findings by the gastrointestinal route.

Viruses and Other Infectious Agents.   Review of the interactions of chemical carcinogens and oncogenic viruses in transforming cells to the carcinogenic state have revealed a two-way street; chemical carcinogenesis is enhanced by the presence of viruses, just as virus transformation is enhanced by chemical carcinogens. A few examples follow.

Weisburger and associates, using high-passage rat embryo cells infected with Rauscher leukemia virus (RLV) (86a) or AKR leukemia virus (86b), induced cell transformation by treatment with a variety of chemical carcinogens, many of industrial concern. Cell transformation did not occur spontaneously, by either RLV or by chemical carcinogens alone, but occurred by the combined action of the virus and chemical as cocarcinogen (86a). Transformations were similarly induced by these chemicals in AKR leukemia virus-infected mouse cells (86b). Transformation was induced by each of eight PAHs, each of three azo dyes, six of seven aromatic amines, and each of four miscellaneous compounds known to be carcinogenic (86a, b).

In similar fashion, virus transformation was enhanced by treatment with chemical carcinogens (87a, b). Treatment of hamster embryo cells infected with simian adenovirus, SA7, with several PAHs, including BaP, resulted in enhanced transformation, an effect directly related to chemical concentration. Concentrations greater than those required for maximal enhancement resulted in decreased or complete inhibition of viral transformation (87a). Chemical carcinogen enhancement by a PAH, methylcholanthrene, of wild type C RNA viruses (oncornaviruses) a natural cause of various cancers in mice, hamsters, fowl, and cats, was found, however, to depend on the strain of mouse (Swiss or BALB/d), sex, age at treatment, carcinogen dose, and virus strain (87b). Subsequent work by Heidelberger and associates showed that the genotype of the mouse determines the expression of the oncornavirus and that the transformed phenotype does not (87c).

In this complex field of of interaction between chemical carcinogen and virus, many theories on the *cellular* mechanisms of chemical oncogenesis have been proposed. Some have been in part substantiated; others await confirmation. In our limited space, a few that have considerable supportive evidence are noted; other mechanistic aspects, membrane alterations, effects of cyclic AMP on cell properties, immunologic effects on tumor induction, the role of DNA repair, or mechanisms of mutagenesis, are either neglected or mentioned only briefly. And throughout, it should be noted that currently there are no convincing reports on the transformation of *human* cells by industrial chemical carcinogens. Such a finding would give the much needed credence to the animal, viral, and bacterial work quoted above.

One theory of cellular mechanisms, the "oncogene" hypothesis (88a, b), for which considerable evidence has been developed, is that chemical carcinogens may exert their effects solely by "switching on" oncogenic viruses to cause the malignant transformation. Supernatants from MCA-induced sarcomas, containing infectious oncornavirus, injected into mice greatly increased the incidence of lymphoid neoplasms (89).

It has been my contention, following the isolation of a human leukemia virus, a type C RNA tumor virus, by Gallagher and Gallo (23b) in 1975, that the so-called benzene leukemias arising from occupational exposure to benzene vapor conformed to the switch-on hypothesis. The consistent pattern of relative rarity of occurrence (less than 0.1 percent) among those occupationally exposed (23a) conforms to the induction of a disease only in an individual infected with a virus whose activity is triggered by a significant exposure to benzene (above 25 ppm on an intermittent daily basis), noted in Section 4.4 on benzene.

As noted above, the induction process is a two-way street; a chemical carcinogen can directly, and without aid of a virus, transform cells to malignancy.

Cells have been transformed to malignancy with several chemicals without expression of oncornavirus group-specific antigens or of infectious virus (87c). But whether virus intervention is essential or not, Heidelberger concluded that it was the genotype of the animal host that determines the expression of oncornaviruses (45b, p. 111).

Another mechanistic aspect that appears to have reached a satisfactory resolution is for those carcinogens that transform cells by themselves; it can be a perpetuated process that does *not* involve a mutation, but rather an alteration of a gene expression (45b, p. 112). However, Ames (42b) believes, based on correlations between mutagenic and carcinogenic activities, that in about 90/ percent of instances, carcinogenesis results from mutagenesis (Section 5.3). However, correlations are no proof of mechanisms, and it is still a fact that potent mutagens are not carcinogens and the mutagenesis theory of carcinogenesis remains to be proved.

The mechanistic theories of the relation of the much investigated and debated chromosomal alterations to promotion of malignancy have both proponents and disclaimers (45b, pp. 112, 113). On the one hand, it is postulated that there are chromosomes that cause and those that suppress malignancy, and whether malignancy develops depends on the balance between them (90a, b). On the other hand, chromosomal alterations are considered by others to be random and not directly concerned with the process of transformation. In view of the many reports of chromosomal aberrations found among industrial workers, it is possible that, if epidemiologic studies of the future concern themselves with these changes, resolution of these discrepancies on an ultimate human basis may occur!

6.2.2 Host Factors

Although separation of host from environmental factors is a common inclination, it should be noted that host factors often share common effects with environmental factors. The above-quoted exogenous, viral-induced, chromosome damage appears to be associated with the production of neoplasms, at least in the views of cytogeneticists (91a). (Less sanguine have been the views of cancer biochemists, immunologists, and industrial toxicologists.) A similar overlap is noted in environmentally induced immunosuppression in the form of drugs given for renal transplantation, which produces lymphoreticular neoplasia, and thus shares an effect with inborn immunodeficiency diseases.

Among the more prominent host factors are genetic diseases, familial susceptibility, immune deficiency diseases, and acquired diseases. Entire volumes (91a–e) have been recently devoted to the genetics of human cancers; only those aspects relevant to the industrial worker, and which are more positively associated, are treated here.

Genetic Diseases.   Genetic factors may be classified into three groups: (1) chromosomal, in which genetic imbalance occurs because of the absence of entire lengths of genetic material, or presence in excess; (2) single-gene locus, with disease arising from mutation either in one allele, as in a dominant trait, or in paired alleles, as in a recessive trait; (3) polygenic, indicating many genes interacting, possibly with environmental factors to cause disease, with no one gene having a major role.

1.   The finding that most patients with chronic myelocytic leukemia have a chromosomal defect consisting of the loss of several gene group systems (44a, pp. 4–6) adds another possible predisposing factor in the so-called benzene leukemias of industrial workers, in addition to the virus activation theory proposed in Section 6.2.1.

2.   Among single-gene disorders that relate to industrial cancers are gluta-thione reductase deficiency and the occurrence of aryl hydrocarbon hydroxylase (AHH), the former relative to leukemia, the latter to bronchogenic carcinoma. Patients with lung cancer tend to have more active levels of AHH, this genetically determined enzyme, which speeds the conversion of PAH to potent carcinogens. The role of GSH reductase deficiency in promoting leukemia is not so clear.

Compared to these two instances of some little concern to industry, there are 1142 proved, single-gene traits presently known, and 1194 others with suggestive but inconclusive evidence of Mendelian behavior, of which about 200 have been associated with neoplastic or preneoplastic tendencies, or manifestations or complications (91a). The magnitude of their role in the industrial setting has yet to be determined.

3.   Familial susceptibility, or polygenic predisposition, on evidence from family histories of a neoplasm, is a strong factor (two- to fourfold) in a person's risk for developing that neoplasm. Still today, it is uncertain whether the increased risks in all familial occurrences represent purely inherited suscepti-bility or common exposure to environmental influences, or the interaction of both. However, many site-specific neoplasms are known to have a genetic component, and thus the question is no longer whether cancer susceptibility can be inherited, but how the susceptibility genes act.

Immune deficiency diseases, whether genetically or drug induced, often develop into cancer more frequently and with a shorter latency period than age-matched control groups. Primary immune deficiency diseases of genetic origin are of no interest in the occupational setting since few individuals with these diseases survive to the working age, for they generally die very early of infection or malignancy. Drug-induced immune deficiency, however, is of some interest, because individuals who have received renal transplants which require long-term treatment with immunosuppressive drugs to prevent graft rejection develop cancer in significant proportion in the months and years following the transplantation.

Immunosuppressive drugs are of further interest because of their direct relation to the immune mechanisms that regulate in a major way the onset of carcinogenesis through immune "surveillance," its decline with age, and thus the determination of the latency period.

One of the immune mechanisms regulating carcinogenesis include the thymus-derived (T) lymphocytes, and the bone marrow-derived (B) lymphocytes. These lymphocytes produce gamma globulin antibodies of several types which act to suppress tumor development; this is called immunologic competence. Primary immunologic deficiency syndromes involve components of both T- and B-cell systems. The B-cell system is involved with a deficiency in the immune globulin, IgM, which is associated with increased susceptibility to bacterial infection, and thus promotion of carcinogenesis. IgA deficiency promotes infections and autoimmune disorders.

Age. Barring genetic, viral, or other factors, cancer is a disease of advancing age. Despite this knowledge, and that immunologic competence declines with age, no experimental effort appears to have been made to determine how large a factor advancing age plays in altering the onset of carcinogenesis. Such studies have now been made (92a, b), using rats of advanced age, administered two industrial chemicals, vinyl chloride, a recently recognized industrial carcinogen, and beryllium hydroxide, an experimental animal carcinogen. Because of its novelty, the study is given in some detail.

A total of 945 male and 955 female Sprague Dawley rats, which included 949 unexposed controls, were exposed by inhalation to vinyl chloride at 1000 ppm. Exposures were 7 hr/day, 5 days/week for 22 to 29 weeks. The ages of the rats at the start of exposure were 6, 17–18, 32–33, and 51–53 weeks. Ten months after the start of exposure, all surviving exposed and control animals were sacrificed and autopsied and their tissues examined microscopically.

It is clear from Table 39.3 that the older adult rats were more susceptible to the induction of angiosarcomas than were the young adult rats, and that the latent period for eliciting angiosarcoma is definitely shorter than in the young adult rats. Comparing tumor incidences in the 52- and 6-week groups, 17.6 and 15.5 percent versus 1.2 and 2.2 occurred in male and female, respectively, a 14.6- and sevenfold increase; for the 32-week old groups, eight- and 12.7-fold, respectively. Just detectable increased incidence occurred in those rats that were 1 year older than the young adults. Latency at these two oldest age groups was reduced by more than 50 percent compared with that of the 6-week group.

A similar though less pronounced decrease in latency was found in the development of pulmonary adenomas even in rats only 9 months older than young adults (3 months old) on minute doses (0.09 mg divided into two doses)

Table 39.3   Increased Frequency and Decreased Latency of Angiosarcomas in Rats after Inhaling Vinyl Chloride (92a)

Group Age (weeks)	Week of First Angiosarcoma	Exposed Angiosarcoma		Control Angiosarcoma
		Number	Percent	
51–53	21(M)	18/102	17.6	0/68
	16(F)	16/103	15.5	0/80
32–33	26(M)	9/94	9.6	1/65
	20(F)	27/98	28.0	0/64
17–18	39(M)	2/91	2.2	0/70
	28(F)	7/96	7.3	0/73
6	39(M)	1/83	1.2	0/62
	30/(F)	2/90	2.2	0/79
Total		30/370(M) 52/387(F)		1/561(M)

of beryllium hydroxide given intratracheally. Tumor incidence was 41 percent, somewhat greater than in a small number of young adult controls (92b).

From the standpoint of occupational exposures, the indications from these findings are obvious, that older workers, should *not* be preferentially selected for work in atmospheres containing potentially hazardous concentrations of carcinogens.

Acquired Disease.   Templeton lists 30 diseases that have been implicated as predisposing to cancer, in addition to viral, endocrine, and nutritional disorders that have already been discussed (44b). These come under the headings of *bacterial* (syphilis and tuberculosis), which predispose to cancer of the tongue and lung, respectively; *plasmodial* (malaria, associated with lymphoma); *parasitic* (schistosomiasis and intestinal parasites, leading to lymphoma and cancer of the bladder; nine *noninfectious, inflammatory states*, the most common of which occur in industry, gastritis, ulcerative colitis, cirrhosis, and pancreatitis, leading to cancer of the stomach, colon, liver, and pancreas; *trauma* (burns and calculi) and *postoperative states*; and *benign proliferations*, particularly adenomatous polyps (cancer of the colon and hydatidiform moles).

Among the bacterial diseases, syphilitic glossitis may predispose to lingual cancer; 40 percent of individuals with cancer of the tongue had positive serology for the spirochete. Tuberculosis and other scar-producing pulmonary infections have been implicated in the pathogenesis of adenocarcinoma of the lung, but epidemiologic evidence is slim. Malaria, leading to Burkitt's lymphoma, and schistosomiasis, to bladder cancer in Africa, are a major concern relative to cancer risk in many foreign countries where these organisms are endogenous.

The noninfectious inflammatory states, however, are among the most common conditions predisposing to cancer among workers in the temperate zones. Gastritis and intestinal metaplasia are regularly found in association with cancer of the stomach. Individuals with atrophic gastritis and pernicious anemia have a risk of developing stomach cancer four or five times normal, and 10 percent of those with pernicious anemia develop gastric cancer before they die. An excess of biliary cancer has been reported in cases of ulcerative colitis. Fifty to 90 percent of hepatocellular carcinoma occurs in association with cirrhosis; in the United States, about 15 percent of cirrhotics have this cancer at autopsy; 44 percent has been reported in South Africa. Individuals with sarcoidosis have a threefold increased risk of lung cancer, and an elevenfold excess of lymphoma, thought to be related to the reduced immunologic competence associated with this disease.

Trauma, in the form of chronic irritation, has been linked rather positively with carcinogenesis in some situations; long-standing urinary stones may predispose to cancer of the renal pelvis, and gallstones are an important risk factor in carcinoma of the gall bladder. Thermal trauma can account for the increased risk of cancer of the lip in pipe smokers, and scars following burns, bites, infections, corrosive damage, and even vaccinations have all been implicated in carcinogenesis, particularly of the skin, lung, and esophagus. Trauma tends to accelerate malignancy of an otherwise benign tumor, such as nevi on the sole of the foot.

Evidence that *postoperative states* increase the risk of cancer is sparse to nonexistent. Neither tonsillectomy, splenectomy, appendectomy, or even thymectomy, which removes the anticarcinogenic T cells, has been ultimately associated with increased cancer risk, although they have been under suspicion for many years.

Papillomatous conditions, such as villous papilloma of the colon, is complicated by cancer in 48 percent of biopsies, and the reduction in cancer risk in those undergoing removal of adenomatous polyps indicates a close association with colorectal cancers. Women with hydatidiform moles have a risk of developing choriocarcinoma at least 1000 times greater than women with normal pregnancy.

Many benign skeletal lesions undergo malignant change also. Chondrosarcoma develops in about 25 percent of patients with multiple enchondromas, and in about 3 percent of those that have been tested, but osteochondroma only rarely (<1 percent) develops into malignant tumors. In extensive Paget's disease of the bone, the risk of osteosarcoma is increased thirtyfold, which accounts for the majority of osteosarcomas seen in middle life.

Genetic and Environmental Interactions. Dominantly inherited cancers, for example, cancer of the colon, provide evidence that at least one step in carcinogenesis is mutational. There is one line of evidence also that environmental agents, such as the experimental animal carcinogen *N*-acetoxy2-acetylaminofluorene (AAF), and UV radiation cause somatic mutations; the second line of evidence comes from the Ames mutagenicity test (Section 5.3).

Cleaver has shown (91a) that UV radiation induces DNA damage, a mutational step that cells from individuals with xeroderma pigmentosum (XP), a hereditary human disease, are unable to repair, thus clearly demonstrating the direct action of an environmental agent (sunlight). Similarly, Setlow and Regan (93) have demonstrated that 2-acetylaminofluorene produces DNA damage that is not repaired by defective XP cells.

It is well known that individuals with XP disease have a genetic susceptibility to environmental carcinogens. This is highly variable in man owing to his genetic polymorphism. But there are instances of genetic susceptibility other than XP disease; individuals who smoke cigarettes are more susceptible to cancer of the lung than others (Section 6.2.1), owing to a genetically enhanced activation of PAH carcinogens by the enzyme aryl hydrocarbon hydroxylase (AHH) (94). There is also, among other genetic conditions that predispose to cancer, α_1-antitrypsin deficiency, which is mediated by industrial exposure to industrial irritants (and smoking).

Prostaglandins.   Prostaglandins have been implicated in the regulation of the tumoricidal action of macrophages which have a large role in differentiating transformed from normal cells and in selectively killing transformed cells. Schulz et al. (95) have found that prostaglandins of the E series can reversibly inhibit interferon-activated macrophages from killing lymphoblastic leukemia cells (Section 6.3.1). The capacity of prostaglandin E to control the tumor-killing, activated macrophages, and thus serve as a biologic "resistor" certainly gives considerable insight into why certain tumors continue to progress in the face of a high macrophage content. In this connection, it is to be noted that cells transformed by chemical carcinogens or viruses produce greatly increased quantities of prostaglandins compared with their normal, untransformed counterparts (95). At critical periods, as in the initiation of tumor growth, prostaglandins have been implicated in preventing the normal macrophage killing function, as shown by surgical removal of tumors when rapid *increase* of chemotaxis occurs, or conversely, by *reduced* migratory activity of macrophages in animals implanted with syngeneic tumors with their augmented prostaglandin content.

Thromboxanes (TBX), closely related structurally to prostaglandin, PGF_{2a}, and so-called because they contain an oxane ring and were first isolated from thrombocytes (platelets), have been called "the power behind PG" (95b) because their function cannot be understood apart from TBX or their related PG endoperoxides from which they are derived. PG endoperoxides' effects are believed to be due to their conversion to TBXa, but at present, these effects have been studied relative to their action on blood clotting (platelet aggregation) and smooth muscle contraction, and as regulators of fat cell response to hormones; as of 1975, no investigations of their role in cancer-promoting effects of PG by macrophages appeared to have been made (95b).

6.3 Anticarcinogenesis

Agents and systems that counteract carcinogenesis, can, like cancer promoters, be conveniently grouped into endogenous (host), and exogenous (environmental) factors. The endogenous factors are "built-in," innate body systems which have evolved in man to combat carcinogens from whatever source; they are most efficient during young and later adult life, but unfortunately gradually decline with advancing age, leading to the familiar phrase of cancer being a disease of old age. From our present knowledge, there is reason to believe that under normal, disease-free states, there is a balance between decreasing populations of target cells and increasing cell populations susceptible to tumor formation, with the latter gradually "winning out" in the declining years of most individuals.

Thus endogenous factors are primary in determining the carcinogenic process, and the exogenous factors, though increasing in numbers with industrial activity, are "at the disposal" of the endogenous; for example, increased levels of vitamins and hormones in oldsters are required to produce the same effects as in younger adults. Thus the anticarcinogenic efficacy of exogenous factors is governed by their physiological levels in the body, which can be altered in the case of dietary items, by intake; for others, such as enzyme inducers, PAH induction increases as the exposure level increases.

6.3.1 Endogenous Factors

Present information is insufficient to rank the importance of the several and varied-function endogenous factors that act to thwart carcinogenesis, but DNA repair must be placed among the most effective, at least up to the declining years.

DNA Repair. Because alterations in the structure of DNA by the constant barrage of chemicals may cause harm and even be lethal, it is not surprising that an elaborate array of mechanisms for DNA repair has evolved. The equally elaborate and tedious number of investigations that have been made to elucidate these mechanisms have been reviewed by Grossman et al. (96).

DNA repair takes many forms to reverse the changes made in the DNA molecule by the chemical carcinogens. The enzymes that reverse these changes are both exo- and endonucleases, both phosphodiesterases, the first of which requires a terminus on the DNA molecule, the other, not.

The most important of these changes is the formation of new chemical linkages between two adjacent pyrimidine bases on the DNA nucleic acid chain to produce dimers. These dimers distort the DNA molecule and, if not repaired before replication begins, interfere with the normal production of daughter DNA molecules.

In the first step in the repair mechanism, an enzyme, endonuclease, "recognizes" the distorted region and makes a single hydrolytic split (*incision*) in the dimer of the affected chain. In step two, *excision repair*, the correctional endonucleases correndonuclease I and II, for monoadduct and diadduct, respectively, attack at the site of the initial split, and excise the damaged part, leaving a gap in one of the two strands of the double-stranded DNA molecule. Following excision of the damaged region, reinsertion of nucleotides is catalyzed by DNA polymerases using the complementary strand as a template (*reinsertion* step). The excision repair step is considered by most to be an errorfree process, otherwise a mutation with carcinogenic potential.

For the sake of clarity, the above description is considerably oversimplified. In addition to its not noting a pre-excision step, which has two incision mechanisms, excision mechanisms have been found to be even more complex, as are reinsertion mechanisms with different genic control of "short-patch" and "long-patch" pathways of repair. The complex role of gene regulation throughout all these repair steps reviewed by Grossman (95) has not been mentioned. And obviously, many important issues remain unsettled. Because most of the information has been obtained from unicellular organisms (bacteria, phage, and some mammalian cells), two large questions loom. (1) How far can the analogy from cells to man be drawn? (2) How is the DNA repair process affected in human aging?

T and B Cells and Immunosurveillance. During the decade of the 70s, an explosion of information has resulted in the emergence of entirely new concepts of cancer immunology, so extensive that only the barest essentials can be mentioned here, stripped of all immunologic details including the fundamental regulation of the immune process by gene determinants.

T and B Lymphocytes and Immunologic Surveillance. Suppression of growth, or outright rejection, of tumors is primarily mediated by lymphocytes, aided by macrophages which ingest foreign tumor cells and present them to the lymphocytes, which are directly responsible for all specific immune responses (97a, b).

Lymphocytes are of two different classes; both arise from primitive stem cells in the bone marrow and migrate to organs where they differentiate into cells capable of interacting specifically with tumor antigens. One route of development is through the thymus gland, where they differentiate into so-called T cells of different classes. The other route of differentiation results in bone marrow lymphocytes, or B cells.

T cells are not homogeneous populations, but make up subclasses with differing properties. One set of T cells has the capacity to damage or destroy cells that they recognize as antigenically foreign, such as from chemically induced tumors (cytotoxic, or "killer" T cells). There are also two distinct sets of "helper" T cells (98); one is tumor-antigen specific, and is required for the activation of B cells to produce antibody; the other "recognizes" immunoglobulin on B cells. Still another property of T cells, the so-called "suppressor" function, renders the host immunologically tolerant and unresponsive to a tumor antigen. It is believed that this suppressor function is a necessary homeostatic control mechanism that provides immunologic surveillance and prevents autoimmune reactions (97b).

Great variations in effectiveness in the capacity of antitumor systems to prevent tumor growth have been found, however. Whereas natural, cell-mediated, cytotoxic T lymphocytes have been observed to kill EL4 lymphoma cells in vitro (99a), the effector cells for natural cytotoxicity were effective against cell lines from chemically induced methylcholanthrene fibrosarcomas in mice, they were not in another MCA tumor (99b), thus indicating that a group of functionally related cells that may be operative in some form of antitumor surveillance for one tumor cell line may not so act in another. Other forms of tumor cell cytotoxicity have been observed (99c); macrophage-mediated tumor-cell cytotoxicty against Lewis lung carcinoma cells has been demonstrated to occur by reductive cellular division, whereby tumor cell count doubles in the absence of DNA synthesis, leading to tumor cell lethality. Many other instances of cytostatic or cytotoxic effects against target tumor cells by activated peritoneal macrophages have been noted.

It must be clear, however, that inasmuch as occupational cancers do occur, tumor cytotoxicity and repression of tumor growth either must be incomplete, by reason of natural in vivo blocking mechanisms, or, if effective, wane with advancing worker age. Evidence for the former is ample; evidence for the latter, declining immunocompetence, is surmised, but mechanisms have not yet been adduced.

Serum-mediated blocking of lymphocyte-mediated cytotoxicity (LMC) has been demonstrated in patients with primary intracranial tumors (gliomas) (100a). The blocking activity of serum antibody binding sites and those of LMC are believed directed at separate antigens on the tumor cell for which several mechanisms have been proposed. In steric hindrance by the blocking antibody for LMC, the blocking antibody may so alter the tumor cell surface as not to be "recognized" by LMC.

Another piece of evidence for the blocking of LMC was demonstrated in MCA-induced fibrosarcomas (MCA_2), in which cytotoxic T lymphocytes were present in situ in the tumor, yet the tumor progressed (100b); the cytotoxicity of the effector cells from the MCA_2 tumor was abrogated by treatment with anti-θ serum and complement. Cells from SAD_2 fibrosarcomas were not so affected, indicating that not all MCA tumors display the same degree of host cell infiltration, or the same degree of immunogenicity.

A preliminary overview of the mechanisms regulating generation, differentiation, and activity of tumor-specific cytolytic T lymphocytes (CTL) for a murine sarcoma virus (MSV) has finally been reported (100c). MSV-specific CTLs generated in vivo undergo a transition in size from large to small cells as the tumor regresses, but most of the CTL *precursors* occur among small to medium-sized MSV-immune cells. Two suppressor cell populations occur in MSV-immune spleens that are capable of inhibiting the generation of CTLs. One population appeared to consist of macrophages, the other, of T cells which inhibited the differentiation of tumor-immune CTL precursors selectively.

Relating these complex and often confusing newer findings to the real world of occupational carcinogenesis, a reasonable interpretation would appear to be that at the relatively low level of worker exposure to carcinogens, at a time of life when in vivo levels of cytotoxicity and DNA repair are actively functioning, the ratio of cytotoxic lymphocytes to tumor cells favors the former and thus prevents carcinogenesis, or at least, once started, inhibits further progression. (Cancer pathologists commonly see neoplasms in individuals who do not succumb to cancer.) This cytotoxic immunosurveillance could account for the long latent period of most cancers, and places the time of their clinical appearance at a worker's age when his immunocompetence is waning.

Interferon.   There is now considerable evidence (101a, b) that interferon, a glycoprotein of low molecular weight (15,000 to 20,000) long considered to be the body's first line of defense against viral infections, has anticarcinogenic action at least against certain types of cancer. Among the human cancers that have been reported to show some response to interferon are osteogenic sarcoma, multiple myeloma, melanoma, breast cancer, and certain types of leukemia and lymphoma. Leukemias, such as are thought to arise from benzene exposures in industry, may respond to interferon treatment if sufficient dosage is given. Antitumor action by interferon against several viral types has also been demonstrated experimentally in certain mouse strains (102).

Interferon is produced by all cells in the body when properly stimulated, such as by viral infection. Human cells produce at least three types (103). The principal interferon is made by leukocytes and differs from the major form of interferon made by fibroblasts. The third form produced by T cells of the immune system is called T or immune interferon.

Interferon has been suggested to be the ultimate inducer of macrophage activation because all of the macrophages that are tumoricidal either are inducers of interferon or contain interferon (95a). Because many diverse substances in addition to viruses can induce cells to make interferon, such as the chemical antibiotics, it is reasonable to think that chemical substances encountered in industry can do likewise, and thus offer some degree of protection against carcinogenesis.

Regarding the mechanism of action of interferon, as far as it is presently known, interferon molecules, after they have been produced, move out of their synthesizing cells, and diffuse to neighboring cells where they bind to receptors on the cell surface. In an unknown manner the binding triggers the synthesis of at least three cellular proteins which, in turn, act to prevent viral reproduction.

Glutathione and Other SH-containing Substances. Prominent among the body's natural substances for defense against industrial carcinogens is the hepatic, nonprotein, SH-containing substance glutathione (GSH). Components of GSH have been found to be major urinary metabolites of the human carcinogen vinyl chloride (VC), following exposure in animals (104a). The urinary metabolites of N-acetyl-S-(2-hydroxyethyl)cysteine and thiodiglycolic acid methyl ester attest to the involvement of GSH in the metabolism of VC, for GSH is composed of glutamyl cysteinyl glycine. The capacity of GSH to inactivate VC is shown by the finding that VC tumor production ceased at the point where the amounts of hepatic GSH continued to be sufficient to combine with the VC dosage administered (up to levels of 50 ppm) (104a).

A compensatory synthesis of hepatic, nonprotein, SH groups following exposure to VC was found, thus showing added anticarcinogenic capacity of the body (104b, c).

6.3.2   Exogenous (Environmental) Factors

As with endogenous factors, it is not possible for several reasons to rate the numerous exogenous factors in order of their relative importance as anticarcinogens; in several instances the evidence is purely experimental (enzyme induction, certain trace metals), or, in the case of dietary anticarcinogens, depends on variable and often unknown intake levels and on the magnitude of the effect resulting from as yet undetermined interactions. Accordingly, these factors will be mentioned merely as having recognized anticarcinogenic potential which varies in effect from individual to individual.

Enzyme Induction. Much has been written on the stimulating effect of PAH on the induction of its oxidizing enzyme, aryl hydrocarbon hydroxylase (AHH), to hydroxylate such PAH compounds as BaP, AAF, 1,2-dimethylbenzanthracene, and 3-methyl-4-(dimethylaminoazo)benzene to substances with reduced or no carcinogenic activity (105). Administration of BaP or other PAH to rats induces severalfold increases in BaP hydroxylase activity in the liver, lung, gastrointestinal tract, skin, and other tissues. Similar increases in AHH (fourteen- to sixteenfold) were found in placentas of women who smoked cigarettes, but not in those who did not smoke. The degree of induction, however, varied among individual women, ascribed to genetic influence. These genetic differences in the inducibility of enzymes to detoxify PAH in the various body tissues, including the placenta, could have an important role in the susceptibilities of different individuals and their fetuses to the carcinogenic effect of environmental and occupational PAH.

The induction effect, however, cannot be an overriding factor in most cases, for cigarette smoking is the main determinant in eliciting cancer, at least in asbestos-exposed workers (22).

Trace Elements. Prominent among the trace elements that have shown anticarcinogenic activity are certain of the platinum-group metals, gallium, and selenium, often acting through special compound forms and because of special affinity for specific tumor-type tissue (platinum, gallium).

Of the *platinum-group metals*, platinum, rhodium, ruthenium, and osmium have been found to act as anticarcinogens best in the form of their amine complexes $(M(NH_3)_6Cl_x)$ through their capacity to bind RNA and DNA (106a), to bind with their constituent purines (106b), or to inhibit their synthesis (106c). The tumor-inhibiting capacity of the red dye of ruthenium, $Ru_2(OH_2Cl_4 \cdot 7NH_3 \cdot 3H_2O)$, appears to involve interference with the mitochondrial transport of calcium (106d, e), as does that of the platinum diammine complex (106f). In general, the platinum-group diammines are immunosuppressive in one system or another (106g).

An early report from a phase I study of the anticarcinogenic effects of diaminodichloroplatinum (DDP) showed responses in one patient with anaplastic thyroid carcinoma, one with transitional cell carcinoma of the bladder, and one with breast cancer. In patients with testicular tumors, responses were seen in nine of 11 patients. There were three complete regressions seen, one in seminoma, one in embryonal cell carcinoma, and one in male choriocarcinoma. Three partial regressions and three cases of objective improvement were also seen in this series. These were also distributed over different tumor types. The possibility that diaminodichloroplatinum has a specific effect on tissues arising from the genitourinary anlage is suggested (106h).

Gallium nitrate and other group IIIa metal salts, stimulated by the effectiveness of *cis*-diammine dichloroplatinum, were also found effective against several experimental tumors (107a). Gallium nitrate in particular prolonged the lifespan of rats with ascites Walker carcinosarcoma, and induced more than 90 percent tumor growth inhibition in six different solid tumors. Gallium nitrate was later found (107b) especially active against subcutaneously transplanted solid tumors, suppressing the growth of six of eight tumors more than 90 percent. In phase I clinical trial of eight patients with disseminated neoplasms refractory to DDP chemotherapy, gallium was found localized in the tumor tissue and was relatively well tolerated at intravenous doses of from 300 to 600 mg/m^2, provided renal function was intact (107c).

Platinum and gallium are not alone in exhibiting anticarcinogenic activity; generally ≤5 µg metal/mg lung protein homogenate (Cu, Fe, Mg, Ni, Co) was sufficient to stimulate BaP hydroxylase activity, and thus remove the carcinogenic BaP from the lung at an increased rate, thus acting as anticarcinogens at these concentrations (60a).

Selenium, in the form of sodium selenite, has been found an effective antitumor agent in animals with various organ-specific carcinogens. Selenium has been effective against 7,12-dimethylbenz(*a*)anthracene, DMBA–croton oil-induced skin tumors, and fluorenylacetamide-induced liver and mammary tumors (108a). There is also an inverse relationship between selenium occurrence and human cancer mortality in the United States, Canada, and New Zealand (108b), in areas where higher selenium content occurs in plants, milk, and ultimately human blood.

Antioxidants. In addition to selenium, Shamberger et al. (108a) have given evidence with references to the scientific literature, for eight other antioxidants with anticarcinogenic activity. Vitamin E tocopherols, present in all plants, especially wheat germ, either alone or fortified with selenium, reduced the number of skin tumors from DMBA–croton oil, and the number of fibrosarcomas induced by MCA, as did ascorbic acid, vitamin C.

Ascorbic acid has been shown to be related to host resistance to neoplastic growth in several significant ways in a detailed review by Cameron et al. (108c). In addition to a secondary response of increased utilization of ascorbate by cancer subjects causing serious deficiencies in blood and leukocytes, ascorbic acid is involved in (1) maintaining the integrity of the intercellular matrix that inhibits tumor growth, (2) increasing collagen in the intercellular matrix to aid tumor encapsulation, and (3) maintaining immunocompetence and (4) adrenal and pituitary hormone balance, and (5) as an antiviral agent. The review (108c) also notes specific actions against nitrosamines, PAH, and tryptophan metabolites in bladder cancer.

The artificial food additives and antioxidants butylated hydroxytoluene (BHT) and hydroxyanisole (BHA) inhibited the carcinogenicity of BaP and that of 7,12-DMBA on the mouse forestomach, along with inhibition of experimental pulmonary adenomas (BHA), hepatomas, and mammary tumors in rats (BHT). The antialcoholism drug Antabuse, tetraethylthiuram disulfide, likewise reduced the number of forestomach tumors from BaP. The combination of these dietary antioxidants, which are eaten in biologically significant amounts, is believed responsible in part for the drop in cancer of the stomach in the United States.

Since the 1974 report of Shamberger et al. on the anticarcinogenic action of antioxidants (108a), two other substances have been added, *putrescine* (109) and the retinoid *cis-retinoic acid* (110).

The carcinogenic action of BaP in mice can be postponed or completely inhibited by putrescine; of 132 of 136 mice (97 percent) with BaP tumors, only three of 38 mice (8 percent) treated with putrescine developed tumors at a single injected dose of 10 mg putrescine and 2.52 mg BaP (109). Putrescine, a diamine of the structure $NH_2(CH_2)_4NH_2$, although a naturally occurring substance formed by bacterial action on foodstuffs in the gastrointestinal tract (ornithine), is highly toxic, and thus is readily detoxified by histaminase. Accordingly, putresine, to be effective, must be added as an exogenous agent.

13-cis-Retinoic acid has been found an effective antitumor agent against methylnitrosourea-induced bladder cancer in Wistar-Lewis rats (110a), a squamous cell type. *cis*-Retinoic acid also was effective against the transitional cell carcinomas of the bladder, typical of the human cancer. *cis*-Retinoic acid, a synthetic analogue of vitamin A, became an anticarcinogen candidate because of the known dependence on retinoids of epithelial cells for normal differentiation (110b).

7 A THEORY FOR THE MECHANISM OF METAL CARCINOGENESIS

D. H. Groth, M.D., Chief, Pathologist Section, Division of Biomedical and Behavioral Science, NIOSH, in Cincinnati, has developed an hypothesis to help predict the stabilities of coordination complexes. The mechanisms of carcinogenesis that have thus far been recounted here have all dealt with carcinogens of organic nature. Now, for the first time, a molecular theory is presented on how bivalent cations, at least, may react to initiate the carcinogenic process.

Inorganic compounds of several different metal cations, for example, Be^{2+}, Ni^{2+}, and Cd^{2+}, have been shown to cause cancer in experimental animals. However, little research and still fewer hypotheses have been developed to help explain the mechanism of causation by these agents. This hypothesis is presented here as an aid to help direct scientists in determining the role of metals in carcinogenesis.

One of the prevailing theories in the field of carcinogenesis is that nucleic acids and their derivatives in DNA and/or RNA are altered by direct bonding of the carcinogens to them. Since divalent inorganic cations have been shown to bind to nucleic acids in vitro, it is possible that this also occurs in vivo. However, under in vivo conditions, there are thousands of different kinds of molecules that are competitively binding the inorganic cations. Since cancer can be caused by relatively small amounts of some inorganic compounds [e.g., $Be(OH)_2$] in a milieu of thousands of competing ligands, it is logical to assume that only specific types of inorganic compounds or complexes form sufficiently stable complexes with nucleic acids to cause cancer.

Although there are many factors that seem to influence the stabilities of inorganic complexes, there are widely varying opinions and theories on the significance of these factors and how generally they can be applied (Basolo and

Johnson) (111). In general, however, the most stable compounds formed with metal cations are chelation complexes in which at least two of the ligands are part of one organic molecule, and the ring formed by this chelation contains five atoms, including the metal cation. Although this factor is important, it is not sufficiently restrictive (as well as other theories) to appreciably increase the predictability of reactions in complex biologic systems.

7.1 The Hypothesis

Consequently, an hypothesis was developed in 1968 and published in 1975 to help predict the stabilities of coordination complexes (Groth et al.) (112). In that hypothesis the stability of coordination complexes of divalent metals was determined by comparing the sum of the first two ionization potentials of the inorganic atom with the sum of the ionization potentials of the ligands. In most instances ligands donate one or more electrons and the electrons are shared between the ligands and the inorganic cation. If the central cation attracts the electrons with the same force as the ligands, then the bond can be considered to be maximally stable. The amount of energy a central cation uses in attracting electrons is the same amount of energy that was required to remove them from the atom, and the measured values are called ionization potentials (IP). Since this hypothesis addresses only divalent cations, the energies involved in the central cation are the sums of the first and second IPs for each element. The amount of energy any ligand uses in attracting electrons is not so readily measured, and can only be approximated. In this hypothesis the first ionization potential of each ligand was used, and the ligands considered included N, O, S, Cl, Br, I, F, NH, and NH_2. Since two organic molecules are utilized in each complex, one electron is considered to be donated by each of the organic molecules. The electron affinity of an organic molecule for the shared electron is then estimated to be the average of the sum of the first IPs of the two ligands on that molecule. The total energy of the four ligands for each complex is therefore the sum of the first IPs of the four ligands divided by two, and can be called the total ligand energy (TLE). When this energy is equal to the sum of the first two IPs of the coordinated cation, the bonds are considered to be maximally stable.

All possible combinations of the energies of several ligands were determined utilizing a computer program, and were matched with the sums of the first two IPS of all the elements in the periodic table. Considering the simple treatment of complex phenomena, it was rather surprising to find that the results by this hypothesis compared favorably with existing experimental data on stabilities of coordination complexes. As an example, the TLE for four N ligands is 671.02 kcal/mol, and this matches with the cation energies for Ag^{2+} and Hg^{2+} (673.17 and 670). This means that organic molecules with only nitrogen (N) ligands will bind Hg^{2+} and Ag^{2+} better than any other cations. Likewise, the TLE for four oxygen (O) ligands is 628.08 kcal/mol, and this matches with the cation energies for Be^{2+} and Zn^{2+} (634.82 and 630.73). The TLE for two O and two N ligands is 649.55 kcal/mol, and this matches with Cu^{2+} (646.08).

Before this hypothesis can be applied to metal complex binding to DNA, however, it is necessary to know the best binding sites on DNA. On nucleic acids there are only a few sites that can function as ligands to form five-membered chelate rings. These are the N_7 and 6 NH_2 sites of adenine and the 6 O and N_7 sites on guanine. There is experimental evidence that a simple inorganic salt, for example, $ZnCl_2$ (Happe and Morales) (113) and a Pt complex, *cis*-dichlorodiammine Pt(II), (Mansy et al.) (114) bind at the N_7 and 6 NH_2 sites on adenine. Utilizing this information and applying the IP hypothesis, it is possible to construct an hypothetical coordination complex to explain the carcinogenicity of some metal cations. The one selected for this purpose is Be^{2+}, and the target nucleic acid is adenine. Since the ligands on adenine are N and NH_2 with IPs of 335.51 and 262.89 kcal/mol, respectively, the ligand energy for adenine is 299.2 kcal/mol. The Be^{2+} energy is 634.82 kcal/mol. Thus, in a highly stable complex of Be^{2+} with adenine, the other two organic ligands must have a ligand energy of 634.82 to 299.2, or 335.62 kcal/mol. The only possible combination of ligands that will provide that energy are two N ligands (335.51 kcal/mol). Therefore, if Be^{2+} is a carcinogen by virtue of its bonding to adenine, then it will do so only in the presence of an organic compound with two N ligands. The fact that simple inorganic compounds of Be^{2+} induce cancer in animals does not negate this hypothesis, since in animals there are organic compounds with N ligands that could probably participate in the complex.

Similarly, Pt compounds have recently been found to be carcinogenic in animals by Leopold et al. (115). Since the cation energy of Pt^{2+} (635.56 kcal/mol) is similar to that ofBe^{2+} (634.82), the same arguments apply regarding the type of complex that must form to cause cancer. However, since the difference between the TLE and the cation energy is smaller for Be^{2+} (0.11 kcal/mol) than it is for Pt^{2+} (0.85 kcal/mol), it is probable that Be^{2+} is a more potent carcinogen than Pt^{2+}.

7.2 Confirmation of Hypothesis

7.2.1 Synthesized Complex Positive in Ames test

In order to provide more supportive evidence for the hypothesis, quantitative dose–response studies must be performed utilizing appropriate metal ion complexes. Since whole-animal carcinogenesis experiments require years to complete, short-term mutagenesis tests have been developed to provide information which can usually be directly related to carcinogenesis. Mutagenesis testing has recently been done to test this hypothesis with platinum complexes

(116). Three compounds, K_2PtCl_4, bipyridyl $PtCl_2$, and bipyridyl Pt adenine Cl_2 were tested in the TA98 Ames test nonactivated system. All three compounds were mutagenic; however, bipyridyl $PtCl_2$ was 24 times more mutagenic than K_2PtCl_4 and 72 times more mutagenic than bipyridyl Pt adenine Cl_2, when the comparisons were made on the basis of the maximum number of revertants per plate/μg Pt. The reason that bipyridyl $PtCl_2$ was more mutagenic than K_2PtCl_4 is probably because the preformed complex with N ligands was added to the test system, whereas with K_2PtCl_4, the Pt is needed to react with existing N ligands in the test system before it could form a stable complex with adenine. The fact that bipyridyl Pt adenine Cl_2 was much less potent a mutagen than bipyridyl $PtCl_2$ can be explained on the basis of the adenine in the complex competing with the adenine in the DNA and/or RNA in the bacteria for the bipyridyl Pt part of the complex.

7.2.2 Positive Shifts in NMR Spectra

Other supporting evidence to prove that bipyridyl $PtCl_2$ binds to the 6 NH_2 and N_7 sites on adenine is derived from NMR spectra. There are shifts in the C_2H, C_8H, and NH_2 signals in adenine from 490, 490, and 427 to 498, 509, and 249 Hz, respectively, when bipyridyl $PtCl_2$ is reacted with adenine to form bipyridyl Pt adenine Cl_2 (117).

In summary, metal cations probably initiate the carcinogenic process by direct binding to nucleic acids in DNA and/or RNA. However, it is unlikely that the metal cations bind as simple inorganic salts, but more likely in coordination complexes with organic compounds. The hypothesis has some experimental support and should be of some value in predicting the nature of the ultimate carcinogens, that is, the molecules that bind directly to the nucleic acids.

8 THE CASE FOR THRESHOLDS IN OCCUPATIONAL CARCINOGENESIS

In Section 1, some of the reasons were pointed out why thresholds can be reasonably established for the control of carcinogenic substances in the air of workplaces, whereas thresholds for carcinogens for populations at large are difficult, if not impossible, to set. The question has been reviewed also by a WHO Scientific Committee in 1977 (118) which envisaged the existence of a threshold for occupational carcinogens, and pointed to efforts of Soviet scientists to establish MACs for carcinogens including those in the workplace.

8.1 Theoretical Basis

P. Kotin (119), approaching the question of thresholds theoretically, pointed out that, because there are thresholds for each of the sequences of the six steps

(Farber) of the carcinogenic process, it follows that an overall threshold exists. It may be further pointed out that cofactors (synergists and promoters) are absent in the industrial situation, or if they do occur, their nature is not unknown as it is among the general population. Thus one factor leading to uncertainties in setting thresholds has been eliminated. Another environmental factor, the carcinogenic agent, unknown quantitatively for the population at large, is under constant control through the application of industrial air standards. Uncertainties relating to host factors (Section 6.2.2) and disease, both genetic and acquired, common factors in the general population, are essentially nonexistent in the industrial situation, where preplacement job and periodic medical examinations lead to hiring and maintaining the physically fit.

8.2 The Data Basis

As shown in striking instances in Section 3.2.2, determinations of safe levels for carcinogen exposure by statistical treatment of animal data have proved sadly wanting, despite periodic attempts for many years to improve the theoretic models. Thus Kotin concluded in his discussion of threshold concepts that it is becoming clearer with the years, even to biometricians, that "the determination of 'safe' doses are problems that transcend our field of statistics" (119).

Since this is so, attention should be focused on what the data derived from recent (since 1970) biochemical and pharmacokinetic studies are saying in regard to carcinogenic and subcarcinogenic doses.

In 1977, we tabulated eight industrial substances of potential carcinogenicity as determined in three animal species including man, by three routes of entry, mostly respiratory, with those doses eliciting and not eliciting tumors (120a, b) (Table 39.4) presented as evidence for thresholds of response to occupational carcinogens.

First to be noted in Table 39.4 is a considerable variety of chemical structures, ranging from complex carcinogenic mixtures of PAH in coal tars of relatively low carcinogenic potency as judged from the latency period, to the highly potent structures of bisCME and HEMPA, with short latent periods even at ppb levels. Response thresholds are not confined to any one route of entry, but are common to all three, irrespective of whether the carcinogen was of very high or low potency. Probably the strongest case for thresholds is their occurrence in three instances of human carcinogenesis: β-naphthylamine, dimethyl sulfate, and vinyl chloride.

We start with a confirmed, high potency, human carcinogen, bisCME; in a dose–response study by inhalation, nasal tumors were elicited in rats after a few months of daily exposures at 0.1 ppm, but not at either 0.01 or 0.001 ppm. 1,4-Dioxane, however, a demonstrated animal tumorigen of low potency, with a large dose of 0.1 percent in drinking water, amounting to 94 and 148 mg/kg for male and female rats, produced variable degrees of kidney and liver

Table 39.4 Evidence for Thresholds in Carcinogenesis (120a, b)

Test Substance	Route	Species	Dose Levels Eliciting Tumors	Dose Levels Not Eliciting Tumors	Duration	Source
Bis(chloromethyl) ether (bisCME)	Inhln.	Rat	100 ppb	10 and 1 ppb	6 months daily	Leong et al., 1975
,4-Dioxane	Oral Inhln.	Rat Rat	1% H_2O >1000 ppm	0.1 and 0.01% 111 ppm	2 years 2 years daily	Torkelson et al., 1974 *Ibid.*
Coal tar	Topical	Mouse	6400; 640; 64 mg	<0.64 mg	2 weeks, 64 weeks	Bingham, 1974
3-Naphthylamine	Inhln. and skin	Man	>5% β, in α form	<0.5% β, in α form	22 years	Zapp, 1975
Hexamethyl phosphoramide (HEMPA)	Inhln.	Rat	4000; 400; 50 ppb	0.01 ppm	24 months	Zapp, 1978
Vinyl chloride	Inhln.	Rat	2500; 200; 50 ppm	<50; >10 ppm 1950–1959, 160 ppm average; 30–170 ppm	7 months	Keplinger et al., 1975 Kramer and Mutchler, 1971
(+ vinylidene chloride)	Inhln.	Man	>200 ppm	Range 1960 <50 ppm, decreasing to 10 ppm	25 years	Ott et al., 1975
Dimethyl sulfate	Inhln. Inhln.	Rat Man	10: 3 ppm (est'd) Unknown	Unknown <2–5 ppm	>10 months 15 years	Druckrey et al., 1970 Pell and DuPont, 1975
Asbestos	Inhln.	Man	>125 mppcf	<125 ppm	up to 25 or more years	Enterline, 1973

degenerative changes, but no tumors after almost 2 years of treatment. Similarly, 111 ppm average daily exposure by inhalation for 2 years resulted in no tumors, indicating a threshold somewhere between this level and below 1000 ppm. Again, application of coal tar pitch volatiles to mouse skin resulted in tumors at total doses of 6400, 640, and 64 mg, but not at three doses below 64 mg.

HEMPA, an experimental carcinogen of potency about equal to that of bisCME, induces squamous cell carcinomas of the nasal cavity of rats exposed daily for 3 months by inhalation at 0.05 ppm, and after a 13-month latent period. Daily exposures at 0.01 ppm for 24 months, however, elicited no tumors, nor have neoplasias appeared in workers after more than 10 years of Kelvar fiber production (41a), indicating a probable threshold for man as well as for animals.

As noted above, of greatest significance and interest are the three and probably four instances of apparent thresholds for man exhibited by β-naphthylamine, vinyl chloride, dimethyl sulfate, and certain insoluble inorganic chromates, lead and zinc.

From the information available to 1979, 26 years of exposure have elapsed without the appearance of bladder tumors in workers exposed to α-naphthylamine containing less than 0.5 percent of the β derivative, whereas tumors occurred in prior exposures to α-naphthylamine containing more than 5 percent β derivative.

Similarly, more than 25 years have passed since workers were exposed to vinyl chloride containing small amounts (5 ppm) of vinylidene chloride, without the appearance of tumors. As indicated in Table 39.3, vinyl chloride levels during the 1950s, although averaging 160 ppm, rose occasionally to 1000 ppm at some work operations, but later were around 50 ppm or below. This gives epidemiologic evidence of a threshold somewhere below 50 and above 10 ppm limits, interestingly enough, confirmed in animals studies (Maltoni, Keplinger) in which a few tumors were still appearing in rats at 50 ppm, but none at 10 ppm, indicating a threshold somewhere between 50 and 10 ppm.

Further evidence for these limiting thresholds has come from the biochemical and pharmacokinetic studies of Gehring and associates (104a–c), who produced indisputable evidence of a threshold for vinyl chloride, although in rats. With available hepatic nonprotein sulfhydryl used as an indicator of the body's capacity to neutralize vinyl free radicals (Section 6.3.1), a distinct dose–response relation was found for exposure levels of 2000, 1000, 250, 150, 50, and 10 ppm. Exposure at the four highest levels caused a progressive depression of the hepatic nonprotein sulfhydryl content, whereas no depression was observed at 10 ppm. Exposure at 50 ppm resulted in inconsistent depression, indicating a threshold for vinyl chloride hemangiomas (tumors) of the rat liver somewhere more than 10 ppm and less than 50 ppm.

A threshold for dimethyl sulfate (DMS) for man has emerged from an epidemiologic study made in 1972 (Pell) of three manufacturing plants in the

United States, which showed no excess incidence of cancer of the respiratory tract in workers exposed up to 26 years at levels frequently well above 1 ppm. Similarly, no overt cases of cancer of the lung attributable to DMS occurred in a German plant where workers receive annual physical examinations.

Asbestos is a special case. First, asbestos exhibits several types of cancer, in addition to the respiratory disease asbestosis, from bronchogenic cancer, which is complicated in a major way by smoking and appears to depend on heavy dosages for response, to mesotheliomas of the pleura and peritoneum, which are induced by seemingly minute dosages. And for none of these cancers has an overall mechanism been defined, although some promoting factors have been noted (60a–c).

A few studies on dose–response relationships between asbestos dust levels and respiratory cancer had already suggested a definite relation, in which below certain measurable levels there appeared to be little or no carcinogenic response (121a–c), before Enterline et al. (122) made their more definitive report. This report concluded that important increments in respiratory cancer mortality apparently occur somewhere between 100 and 200 mppcf-years exposure, and that no direct relationship between asbestos dust exposure and respiratory cancer occurred below 125 mppcf-year.

This conclusion was based on an epidemiologic study of the relation of respiratory cancer to occupational exposures among 1348 retired asbestos workers, who had an average length of employment of 25 years, and often at very high dust exposures.

These nine examples of carcinogens with thresholds should not be taken as all that can be mustered, but rather as indications of what measures must be taken to discover thresholds for carcinogens.

REFERENCES

1. "Scientific Bases for Identifying Potential Carcinogens and Estimating Their Risks," Interagency Regulatory Liaison Group, February 6, 1979, Washington, D.C., 113 pp.

2. J. McCann and B. N. Ames, *Ann. N.Y. Acad. Sci.*, **271**, (1976).

3. K. Bridbord, J. K. Wagoner, and H. P. Blejer, "Chemical Carcinogens," in *Occupational Diseases, A Guide to their Recognition*, rev. ed., U.S. Government Printing Office, Washington, D.C., June 1977.

4. C. K. Redmond et al., *J. Occup. Med.*, **14**, 621 (1972).

5. P. Kotin, *Ann. N.Y. Acad. Sci.*, **271** (1976).

6. L. J. Cralley et al., *Am. Ind. Hyg. Assoc. J.*, **28**, 452 (1967).

7. J. R. Dixon et al., *Cancer Res.*, **30**, 1068 (1970).

8. D. H. Groth, Chief, Pathology Services, Division of Biomedical and Behavioral Sciences, NIOSH, Cincinnati, unpublished results.

9. The choice of logarithmic separation of dose levels rests on the long-established fact that the

diffusion rate of substances into cells doubles with each tenfold increase in substrate concentration.

10. (a) N. Mantel and W. R. Bryan, *J. Natl. Cancer Inst.*, **27,** 455 (1961); (b) J. Cornfield, *Science*, **198,** 693 (1977), **199** (1978).

11. (a) M. A. Schneiderman, *J. Wash. Acad. Sci.*, **64,** 68 (1974); (b) N. Mantel and M. A. Schneiderman, *Cancer Res.*, **35,** 1379 (1975).

12. P. J. Gehring, P. G. Watanabe, and C. N. Park, *Tox. Appl. Pharm.*, **44,** 581 (1978), and previous reports cited herein.

13. D. B. Clayson, *Science*, **203,** 1068 (1979).

14. R. N. Matzen, *J. Appl. Physiol.*, **11,** 105 (1957).

15. H. E. Stokinger and D. L. Coffin, "Biologic Effects," in *Air Pollution*, A. C. Stern, Ed., Vol. I, 2nd ed., Academic Press, New York, 1968.

16. J. C. Ramsey, C. N. Park, M. G. Ott, and P. J. Gehring, *Toxicol. Appl. Pharm.*, **47,** 411 (1979).

17. (a) B. D. Dinman, *Science*, **175,** 495 (1972); (b) H. E. Stokinger, *Arch. Environ. Health*, **25,** 153 (1972); (c) H. E. Stokinger, *J. Am. Water Works Assoc.*, **69,** 399 (1977).

18. "Acrylonitrile," Current Intelligence Bulletin NIOSH, July 1, 1977.

19. W. F. Melick et al., *J. Urol.*, **74,** 760 (1955).

20. Discussions with J. Cunningham and I. L. McCallum, Antimony Works, Newcastle-on-Tyne, June, 1974–March 1977.

21. (a) S. S. Pinto, V. Henderson, and P. E. Enterline, *Arch. Environ. Health*, **33,** 325 (1978); (b) A. M. Lee and J. F. Fraumeni, *J. Natl. Cancer Inst.*, **42,** 1045 (1969).

22. I. V. Selikoff, E. C. Hammond, and V. Churg, *J. Am. Med. Assoc.*, **204,** 104 (1968).

23. (a) A. J. McMichael, et al., *Ann. N.Y. Acad. Sci.*, **271,** 125 (1976); (b) R. E. Gallagher and R. C. Gallo, *Science*, **187,** 350 (1975); (c) E. Browning, *Toxicity and Metabolism of Industrial Solvents* Elsevier, New York, 1965, p. 48.

24. M. R. Zavon, U. Hoegg, and E. Bingham, *Arch. Environ. Health*, **27,** 1 (1973).

25. R. A. Lemen et al., *Ann. N.Y. Acad. Sci.*, **271,** 273 (1976).

26. R. A. Lemen et al., *Ann. N.Y. Acad. Sci.*, **271,** 71 (1976).

27. N. Nelson, *Ann. N.Y. Acad. Sci.*, **271,** 81 (1976).

28. (a) A. M. Baetjer, *Arch. Ind. Hyg. Occup. Med.*, **2,** 487 (1950); (b) W. M. Gafafer, Ed., *U.S. Public Health Rep. Publ.* **192** (1953); (c) "An Epidemiologic Study of Lead Chromate Plants," Final Report, Equitable Environmental Health Inc., Berkeley, Calif., July 1976.

29. J. W. Lloyd et al., *J. Occup. Med.*, **13,** 53 (1971).

30. (a) T. S. Scott, *Carcinogenic and Chronic Toxic Hazards of Aromatic Amines*, Elsevier, New York, 1962, pp. 44, 164; (b) T. F. Mancuso and A. A. El-Attar, in *Bladder Cancer: A Symposium*, K. F. Lampe, Ed., Aesculapius Press, Birmingham, Ala., pp. 144–162.

31. (a) R. B. Sutherland, "Respiratory Cancer Mortality in Workers Employed in an Ontario Nickel Refinery, 1930–1957," Rept. Div. Ind. Hyg., Ontario, Can. Nov. 1959; (b) Discussions with Dr. Lindsay Morgan, Medical Director, INCO, and Dr. Louis Renzoni, V. P., INCO, Sudbury, Ontario, Canada, July, 1975.

32. D. W. Wallace, *Tumors of the Bladder*, E. and S. Livingstone Ltd., Edinburgh, 1959, p. 35.

33. V. E. Archer et al., *Ann. N.Y. Acad. Sci.*, **271,** 280 (1976).

34. R. J. Waxweiler et al., *Ann. N.Y. Acad. Sci.*, **271,** 40 (1976).

35. E. D. Acheson et al., *Brit. Med. J.*, **2,** 587 (1968).

36. *Suspected Carcinogens*, A Subfile of the Registry of Toxic Effects of Chemical Substances, R. J. Lewis, Project Coordinator, NIOSH, Cincinnati, Ohio, 1978.

37. E. K. Weisberger, "Industrial Cancer Risks," in *Dangerous Properties of Industrial Chemicals*, I. Sax, Ed., Van Nostrand, New York, 1975, pp. 274–288.

38. (a) H. Sakabe et al., *Ann. N.Y. Acad. Sci.*, **271,** 67 (1976); (b) W. Clark Cooper, *Ann. N.Y. Acad. Sci.*, **271,** 250 (1976).

39. Personal communication from Dr. H. J. Trochimowicz, Staff Toxicologist, E. I. du Pont de Nemours & Co., Wilmington, Del., May 25, 1979.

40. P. DeCoufle, *Arch. Environ. Health*, **34,** 33 (1979).

41. (a) DuPont memorandum, HMPA, July 26, 1978; (b) Letter to Chairman of TLV Committee from James Morgan, April 21, 1979; (c) R. M. Moore, NIOSH Memorandum, Aug. 25, 1976; (d) T. R. Lewis et al., *J. Environ. Pathol. Toxicol.*, **2,** 233 (1979); (e) W. M. Busey et al., Presentation at Meeting of Society of Toxicologists, March 1980.

42. (a) B. N. Ames et al., *Mutat. Res.*, **31,** 347 (1975); (b) B. N. Ames, *Science*, **204,** 587 (1979); (c) F. J. DeSerres, *Mutat. Res.*, **38,** 165 (1976).

43. J. Rosai and L. V. Ackerman, "Pathology of Tumors, Pt. I., Precancerous and Pseudomalignant Lesions, in Ca," A Cancer Journal for Clinicians, **28**(6) 331 (1978).

44. (a) W. Eckhart, *Ann. Rev. Biochem.*, **41,** 503 (1972); (b) J. F. Fraumeni, Jr., Persons at High Risk of Cancer, Academic Press, New York, 1975.

45. (a) E. C. Miller and J. A. Miller, *Ann. Rev. Biochem.*, **28,** 291 (1959); (b) C. Heidelberger, *Ann. Rev. Biochem.*, **44,** 79 (1975).

46. (a) P. L. Grover and P. Sims, *Biochem. J.*, **110,** 159 (1969); (b) H. V. Gelboin, *Cancer Res.*, **29,** 1272 (1969).

47. (a) A. Dipple et al., *Eur. J. Cancer*, **4,** 493 (1968); (b) A. Dipple and T. A. Slade, *Eur. J. Cancer*, **6,** 417 (1970).

48. (a) C. Nagata et al., *Gann,* **58,** 493 (1967); (b) C. Nagata ct al., *Gann*, **59,** 289 (1968); (c) S. A. Lesko et al., *Prog. Mol. Subcell. Biol.*, **2,** 347 (1971); (d) W. Caspary et al., *Biochem.*, **12,** 2649 (1973).

49. C. Irving, *Methods Cancer Res.*, **7,** 190 (1973).

50. (a) P. Daudel et al., *C. R. Acad. Sci. Paris*, **277,** 2437 (1973); (b) P. Daudēl et al., *C. R. Acad. Sci. Paris*, **278,** 2249 (1974).

51. D. M. Jerina et al., *Arch. Biochem. Biophys.*, **128,** 176 (1968).

52. (a) J. L. Radomski et al., *J. Natl. Cancer Inst.*, **50,** 989 (1973); (b) J. D. Scribner and N. K. Naimy, *Cancer Res.*, **33,** 1159 (1973); (c) H. Bartsch et al., *Biochim. Biophys. Acta*, **286,** 272 (1972).

53. (a) P. J. Gehring et al., *Toxicol. Appl. Pharm.*, **44,** 581 (1978); (b) P. G. Watanabe et al., *Toxicol. Appl. Pharm.*, **571,** 581 (1978).

54. (a) E. Nachtomi, *Biochem. Pharmacol.*, **19,** 2853 (1970); (b) H. B. Plotnick, *J. Am. Med. Assoc.*, **239** (April 21, 1978).

55. R. K. Boutwell, "Function and Mechanism of Promoters of Carcinogenesis," *Crit. Rev. Toxicol.*, **2,** 419 (1974).

56. (a) E. P. Radford, Jr. and V. R. Hunt, *Science*, **143,** 247 (1964); (b) E. A. Martell, *Am. Sci.*, **63,** 404 (1975).

57. E. L. Wynder, I. J. Bross, and E. Day, *J. Am. Med. Assoc.*, **160,** 1384 (1956).

58. (a) A. W. Horton et al., *Cancer Res.*, **17,** 758 (1957); (b) E. Bingham and H. L. Falk, *Arch. Environ Health*, **19,** 779 (1969).

59. R. Tye and K. L. Stemmer. *J. Natl. Cancer Inst.*, **39,** 175 (1967).

60. (a) J. R. Dixon et al., *Cancer Res.*, **30,** 1068 (1970); (b) L. J. Cralley, R. G. Keenan, and J. R.

Lynch, *Am. Ind. Hyg. Assoc. J.,* **28,** 452 (1967); (c) P. Gross et al., *Arch. Environ. Health,* **15,** 343 (1967).

61. F. W. Sunderman, Jr., *Cancer Res.,* **27,** 950 (1967).

62. A. Furst, *Metal Binding in Medicine,* Lippincott, Philadelphia, 1960, pp. 336–344.

63. (a) I. B. Weinstein et al., in *Origins of Human Cancer,* Hiatt, Watson, Winston, Eds., Cold Spring Harbor, New York, 1977, Vol. 4, p. 751; (b) M. Wigler and I. B. Weinstein, *Nature,* **259,** 232 (1976); (c) S. Mondal et al., *Cancer Res.,* **36,** 2254 (1976); (d) H. Yamanski et al., *Proc. Natl. Acad. Sci.,* **74,** 3451 (1977).

64. L-S. Lee and I. B. Weinstein, *Science,* **202,** 313 (1978).

65. S. P. Rose et al., *Experientia,* **32,** 913 (1976).

66. L. Levine and A. Hassid, *Biochem. Biophys. Res. Commun.,* **76,** 1181 (1977).

67. L-S. Lee and I. B. Weinstein, *Nature,* **274,** 696 (1978).

68. (a) G. Rovera et al., *Science,* **204,** 868 (1979); (b) J. Abb et al., *J. Immunol.,* **122,** 1639 (1979).

69. (a) G. R. Newell et al., *J. Natl. Cancer Inst.,* **51,** 1437 (1973); (b) "Boston Collaborative Drug Surveillance Program," *J. Am. Med. Assoc.,* **229,** 1462 (1974).

70. R. A. Gams et al., *Ann. Intern. Med.,* **69,** 557 (1968).

71. C. Peraino et al., *Food Cosmet. Tox.,* **15,** 93 (1977).

72. (a) H. B. Plotnick, *J. Am. Med. Assoc.,* **239,** 1609 (1978); (b) H. B. Plotnick, *J. Am. Med. Assoc.,,* **239,** 2783 (1978).

73. J. Sigurjonsson, *Brit. J. Cancer,* **21,** 651 (1967).

74. (a) P. Cole, *Lancet,* **1,** 1335 (1971); (b) R. Schmauz and P. Cole, *J. Natl. Cancer Inst.,* **52,** 1431 (1974); (c) P. J. Donovan and J. A. DiPaolo, *Cancer Res.,* **34,** 2720 (1974); (d) J. F. Fraumeni, Jr., *Lancet,* **2,** 1204 (1971).

75. H. Matsudo et al., *Arch. Pathol.,* **97,** 366 (1974).

76. L. V. Ackerman et al., "Cancer of the Esophagus," in *Cancer in China,* Kaplan and Tsuchitani, Eds., Liss Press, New York, 1978.

77. R. M. Hicks et al., *Nature,* **243,** 347 (1973).

78. T. H. Jukes, *J. Am. Med. Assoc.,* **229,** 1920 (1974).

79. (a) N. H. Rowe et al., *Cancer,* **26,** 436 (1970); (b) A. E. Rogers et al., *Cancer Res.,* **33,** 1003 (1973); (c) P. Clifford, *Proc. Roy. Soc. Med.,* **65,** 682 (1972); (d) E. L. Wynder et al., *Cancer,* **16,** 1461 (1963).

80. (a) R. S. Rivlin, *Cancer Res.,* **33,** 1977 (1973); (b) E. L. Wynder and P. C. Chan, *Cancer,* **26,** 1221 (1970).

81. (a) L. A. Poirier et al., *Proc. Am. Assoc. Cancer Res.,* **15,** 51 (abstr.) (1974); (b) E. K. Blackburn et al., *Int. J. Cancer.* **3,** 163 (1968).

82. N. Dungal and J. Sigurjonsson, *Brit. J. Cancer,* **21,** 270 (1967).

83. R. J. Shamberger and C. E. Willis, *CRC Crit. Rev. Clin. Lab., Sci.,* **2,** 211 (1971).

84. H. Marjanen, *Ann. Agric. Fenn,* **8,** 326 (1969).

85. (a) N. Venkatesan et al., *Cancer Res.,* **30,** 2563 (1970); (b) P. Czygan et al., *Cancer Res.,* **34,** 119 (1974); (c) C. Kommineni et al., *J. Environ. Pathol. Toxicol.,* **2,** 149 (1979).

86. (a) A. E. Freeman et al., *J. Natl. Cancer Inst.,* **51,** 799 (1973); (b) J. S. Rhim, D. K. Park, E. K. Weisburger, and J. H. Weisburger, *J. Natl. Cancer Inst.,* **52,** 1167 (1974).

87. (a) B. C. Casto et al., *Cancer Res.,* **33,** 819 (1973); (b) B. C. Casto, *Cancer Res.,* **33,** 402 (1973); (c) U. R. Rapp et al., *Virology,* **65,** 392 (1975).

88. (a) R. J. Huebner and G. J. Todaro, *Proc. Natl. Acad. Sci.,* **64,** 1087 (1969); (b) G. J. Todaro and R. J. Huebner, *Proc. Natl. Acad. Sci.,* **69,** 1009 (1972).

89. M. A. Basombrio, *J. Natl. Cancer Inst.*, **51**, 1157 (1973).

90. (a) S. Hitosumachi et al., *Nature*, **231**, 511 (1971); (b) S. Hitosumachi et al., *Int. J. Cancer*, **9**, 305 (1972).

91. (a) "Progress in Cancer Research and Therapy," Vol. 3, *Genetics of Human Cancer*, Mulvihill, Miller, Fraumeni, Eds., Raven Press, New York, 1977; (b) Symposium on Fundamental Cancer Research, 1969. *Genetic Concepts and Neoplasia*, Williams and Wilkins, Baltimore, Md., 1970; (c) L. G. Jackson, "Genetics in Neoplastic Diseases," in *Genetic Disorders of Man*, R. M. Goodman, Ed., Little, Brown, Boston, 1970; (d) J. F. Fraumeni, Jr., *Genetic Factors in Cancer Medicine*, Holland, Frei, Eds., Lea and Febiger, Phila, Pa., 1973; (e) S. D. Lawler, *Cancer in Clinical Genetics*, Soraby, Ed., 2nd ed., Butterworth, 1973.

92. (a) D. H. Groth, "Interim Report on Effect of Vinyl Chloride Inhalation on Aged Rats," January 1979, NIOSH, Cincinnati, Ohio; (b) D. H. Groth, L. D. Scheel, H. E. Stokinger, reported at Invitational Beryllium Conference, Bureau of Occupational Safety and Health, Cincinnati, Ohio, July 9–10, 1969.

93. R. B. Setlow and J. D. Regan, *Biochem. Biophys. Res. Commun.*, **46**, 1019 (1972).

94. G. Kellerman et al., *N. Engl. J. Med.*, **289**, 934 (1973).

95. (a) R. M. Schulz et al., *Science*, **202**, 320 (1978); (b) Research News, *Science*, **190**, 770 (1975).

96. L. Grossman et al., *Ann. Rev. Biochem.*, **44**, 19 (1975).

97. (a) L. J. Old, *Sci. Am.*, **236**, 62 (1976); (b) H. Cantor, *Am. Rev. Med.*, **30**, 269 (1979).

98. C. A. Janeway, Jr., *Fed. Proc.*, **38**(7), 2071 (1979).

99. (a) T. L. Rothstein et al., *J. Immunol.*, **121**, 1652 (1978); (b) O. Stutman et al., *J. Immunol.*, **121**, 1819 (1978); (c) A. M. Kaplan et al., *J. Immunol.*, **121**, 1781 (1978).

100. (a) N. L. Levy, *J. Immunol.*, **121**, 916 (1978); (b) F. DeLustro and J. S. Haskill, *J. Immunol.*, **121**, 1007 (1978); (c) F. Plata, H. R. MacDonald, and B. Shain, *J. Immunol.*, **123**, 852 (1979).

101. (a) Research News, *Science*, **204**, 1183 (1979); (b) Research News, *Science*, **204**, 1293 (1979).

102. I. Gresser et al., *J. Natl. Cancer Inst.*, **45**, 365 (1970).

103. J. Vilcek and E. A. Havell, *In Vitro* (Monogr.), **0**(3), 47 (1974).

104. (a) P. G. Watanabe, R. E. Hefner, Jr., and P. J. Gehring, *Toxicol.*, **36**, 1 (1976); (b) R. E. Hefner, P. G. Watanabe, and P. J. Gehring, *Ann. N.Y. Acad. Sci.*, **246**, 135 (1975); (c) R. E. Hefner, P. G. Watanabe, and P. J. Gehring, *Toxicol. Appl. Pharm.*, **37**, 49 (1976).

105. R. M. Welch et al., *Science*, **160**, 541 (1968).

106. (a) I. A. Ross et al., *Chem. Biol. Interact.*, **16**, 39 (1977); (b) R. F. Fisher et al., *Environ. Health Perspect.*, **12**, 57 (1975); (c) B. Rosenberg, *Naturwiss.*, **60**, 399 (1973); (d) K. C. Reed et al., *Biochem. J.*, **140**, 143 (1973); (e) L. J. Anghileri et al., *Z. Krebsforsch.*, **88**, 213 (1975); (f) A. F. LeRoy, *Environ. Health Perspect.*, **10**, 73 (1975); (g) M. C. Berenbaum et al., *Brit. J. Cancer*, **25**, 208 (1971); (h) D. J. Higby et al., *Cancer*, **33**, 1219 (1974); (i) M. E. Madias and J. T. Harrington, *Am. J. Med.*, **65**, 307 (1978).

107. (a) M. M. Hall et al., *J. Natl. Cancer Inst.*, **47**, 1121 (1971); (b) R. H. Adamson et al., *Cancer Chemother. Rep.*, Pt. I, **59**, 599 (1975); (c) S. W. Hall et al., *Clin. Pharmacol. Ther.*, **25**, 82 (1979).

108. (a) R. J. Shamberger et al., *J. Natl. Cancer Inst.*, **53**, 1771 (1974); (b) R. J. Shamberger et al., in *Trace Substances in Environmental Health, VII*, Hemphill, Ed., University of Missouri Press, Columbia, 1973; (c) E. Cameron et al., *Cancer Res.*, **39**, 663 (1979).

109. G. Kallistratos and E. Fasske, *Z. Krebsforsch. Klin. Onkol.*, **87**, 81 (1976).

110. (a) C. J. Grubbs et al., *Science*, **198**, 743 (1977); (b) M. B. Sporn et al., *Science* **195**, 487 (1977).

111. F. Basolo and R. C. Johnson, *Coordination Chemistry*, W. A. Benjamin, New York, 1964.

112. D. H. Groth, L. Stettler, and G. Mackay, "Interactions of Mercury, Cadmium, Selenium,

Tellurium, Arsenic, and Beryllium," in *Effects and Dose Response Relationships of Toxic Metals*, Nordberg, Ed., Elsevier, Amsterdam, 1976.

113. J. A. Happe and M. Morales, *J. Am. Chem. Soc.*, **88**, 2077 (1966).

114. S. Mansy et al., *J. Am. Chem. Soc.*, **95**, 1633 (1973).

115. W. R. Leopold et al., *Cancer Res.*, **39**, 913 (1979).

116. D. H. Groth et al., unpublished results, 1978.

117. G. Brubaker, "A Study of Formation and Stabilities of Chemical Complexes Containing Beryllium and Platinum," Contract No. CDC-99-OSH-101(6) 1977.

118. "Methods for Carcinogenesis Tests at the Cellular Level and Their Evaluation for the Assessment of Occupational Cancer Hazards," *Proc. Meet. Sci. Comm., Milan, Dec. 4–6, 1977*, pp. 144–146.

119. P. Kotin, "Dose-Response Relationships and Threshold Concepts," *Ann. N.Y. Acad. Sci.*, **271**, 22 (1976).

120. (a) H. E. Stokinger, *J. Am. Water Works Assoc.*, **69**, July 1977; (b) H. E. Stokinger, *Occup. Health Safety*, March/April 1977.

121. (a) J. C. McDonald et al., *Arch. Environ. Health*, **22**, 677 (1971); (b) J. F. Knox et al., *Brit. J. Ind. Med.*, **25**, 293 (1968); (c) M. L. Newhouse, *Brit. J. Ind. Med.*, **26**, 294 (1969).

122. P. Enterline, P. DeCoufle, and V. Henderson, *Brit. J. Ind. Med.*, **30**, 162 (1973).

CHAPTER FORTY

The Halogens and The Nonmetals Boron and Silicon

H. E. STOKINGER, Ph.D.

1 HALOGENS

The halogens of industrial interest consist of the elements fluorine, chlorine, bromine, and iodine. Astatine, the heaviest member of the family, atomic weight, 211, is a radioactive element and of no industrial importance at present. The halogens exhibit the best gradation of physical properties of all the families of elements. Table 40.1 shows an almost perfect doubling of atomic weights progressing from fluorine to iodine, paralleled by increase in specific gravity, melting and boiling points, and orderly decreases in water solubility according to their decreasing chemical reactivity. Color progressively deepens in proceeding from the pale yellow of fluorine, to the yellow-green of chlorine, the dark red of bromine, and the deep violet of iodine.

Because each halogen has seven electrons in its outer shell, the normal valence is minus 1 in their binary compounds, for example, HX^{1-}; but, with the exception of fluorine, ternary halogen compounds with oxygen have valences of $1+$, $3+$, $5+$, and $7+$, and are of industrial interest, for example, KXO to KXO_4. Interhalogen compounds exist that are of some industrial interest, and are of the type XX′, where n is also 1, 3, 5, or 7, for example, ClF_3, but ternary compounds of the type XO_4X^n have not yet attained industrial importance.

Chemically, fluorine is the most powerful oxidizing agent known. The heavier halogens have progressively less oxidizing ability. Each forms an acid with hydrogen and forms salts with metals. The properties of these acids and salts

Table 40.1. Physical Properties of the Halogens

Halogen	Chemical Symbol	Atomic Weight	Sp. Gr.	M.P. (°C)	B.P. (°C)	Solubility
Fluorine	F	18.9984	1.69^{15}	−219.62	−188.14	Cold H_2O, HF + O_2; dec hot H_2O
Chlorine	Cl	35.453	3.214^{0}	−100.98	−34.6	310 kg/l H_2O, 10°C, 177 kg/l H_2O, 30°C, sol. alkali
Bromine	Br	79.904	3.119^{20}	−7.2	58.78	41.7 g/l H_2O, 0°C, 35.2 g/l H_2O, 50°C; very sol. alcohol, ether, chloroform, CS_2
Iodine	I	176.905	4.93	113.5	184.35 atm	0.29 g/l H_2O, 20°C, 0.78 g/l H_2O, 50°C, 205 g/l alcohol, 15°C, 206 g/l ether, 17°C

show as consistent a relationship as the elements themselves. Organic halogen compounds generally show progressively increased stability in the order iodine, bromine, chlorine, and fluorine.

The relative abundances of the halogens are as follows:

Element	Seawater (ppm)	Lithosphere (ppm)
Fluorine	1.4	770
Chlorine	18,980	550
Bromine	65	1.6
Iodine	0.05	0.3

Although there are more than 70 minerals containing a halogen as the sole or principal anionic constituent, only a few are common. These can be grouped according to the following formation processes.

1. *Saline deposition by evaporation of seawater or salt lakes.* Rock salt (halik), NaCl, is the most important of this type, and of the other associated minerals, sylvite, KCl, and carnallite, $KMgCl_3 \cdot 6H_2O$, are the most important.

2. *Hydrothermal deposition.* Chief representatives are fluorite and cryolite.

3. *Secondary alteration.* Chlorides, bromides, or iodides of Ag, Cu, Pb, or Hg may form as surface alterations of ore bodies of these metals; most common are cerargyrite, AgCl, and atacamite, $Cu_2(OH)_3Cl$.

4. *Deposition by sublimation.* Halides are formed as sublimation products from volcanic eruptions, such as NH_4Cl, $FeCl_3$, $PbCl_2$ (Mt. Vesuvius).

1.1 Fluorine, F_2

Because fluorine is chemically the most energetic of all the nonmetals, it combines in binary form with all other chemical elements except argon, helium, and neon. Consequently, fluorine is not found free in nature.

1.1.1 Source and Production

Many minerals contain small amounts of fluorine. The chief ore is fluorspar (CaF_2), which is found in many parts of the world in low-grade or small deposits; cryolite (Na_3AlF_3), formerly used in the metallurgy of aluminum and in ceramics to increase fluidity of the melt, is in such short supply from natural deposits that it is now mostly made synthetically from fluorspar and fluosilicic acid (H_3SiF_6). Fluorapetite $[Ca_{10}(PO_4)_6F_2]$ or rock phosphate, is the most plentiful mineral.

Free, elemental fluorine gas is produced by the electrolysis of HF made conducting by the addition of KF in a ratio of KF:2HF at a temperature of 100°C. Because of its extreme reactivity, special handling precautions are necessary. The industrial handling of fluorine has been described by Landau and Rosen (1a), including the disposal of unused gas, and by Stokinger (1b). How extreme this reactivity is can be better envisaged by noting that the jet stream from a tank of fluorine can burn asbestos as if it were paper, can burn a line of copper tubing, if slight moisture is present, and the gas under pressure can produce a skin burn equivalent to that of an oxyacetylene flame (1b).

1.1.2 Physiologic Response

Inhalation. The first studies of the effects of exposure of animals to metered dilutions of fluorine in nitrogen were made at the University of Rochester's Manhattan Project in 1944 by a team led by Stokinger (1b).

The gas was uniformly fatal to rabbits, guinea pigs, rats, and mice in exposures ranging from 5 min at 10,000 ppm to 3 hr at 200 ppm. Guinea pigs survived an exposure of 7 hr at 100 ppm, but the overall mortality among the various species was 60 percent. Respiratory damage, with pulmonary edema, was the cause of death. More prolonged exposure, up to 35 days, was made at lesser concentrations. Irritation of the eyes and nasal and buccal mucosa was noted at concentrations of 5 to 10 ppm, and dogs exhibited irrational seizures, many of which were fatal. Moderate to severe pulmonary irritation occurred at all levels down to 3 mg/m^3, and rats showed also a high degree of testicular degeneration at the 25-mg level. The tolerated exposure was taken as 1 ppm (1.7 mg/m^3), and hydrogen fluoride and fluorine were regarded as independently toxic.

Cutaneous Burns. The Rochester group also exposed the skin of the back of anesthetized rabbits for periods of 0.2 to 0.6 sec at a distance of 1 in. to fluorine under 40 lb pressure. The briefest exposure led to the appearance of a small ischemic area $\frac{1}{4}$ in. in diameter, surrounded by an erythematous area. This became a superficial eschar that sloughed off by the fourth day, disclosing normal epidermis. The longer exposures were accompanied by a flash of flame, burning the hair and causing coagulation necrosis of the burned area and charring of the epidermis. The thermal flash burns resembled those induced by an oxyacetylene flame.

Twenty-five years later Keplinger and Suissa (2) determined the LC$_{50}$ values of fluorine for several brief exposure periods in the same species of laboratory animals, as shown in Table 40.2. From these exposures, as with those of Stokinger, there was little species difference in the acute lethal effects of F$_2$. However, although strict comparison is not possible because of differing protocols, their results indicate a toxicity three to four times that found by Stokinger's group, and are now believed to represent truer values. It was recognized at the time that air-sampling procedures were often well below 50 percent, leading to gross underestimates of the exposure levels. (The

Table 40.2. LC$_{50}$ Values For Animals Exposed To Fluorine[a]

Species	Exposure Time (min)	LC$_{50}$ mg/m^3	LC$_{50}$ ppm	19/20 Confidence Limits (ppm)
Rat	5	1088	700	636–770
Mouse	5	932	600	517–696
Rabbit	5	1274	820	730–920
Rat	15	606	390	361–422
Mouse	15	583	375	344–410
Guinea pig	15	614	395	352–443
Rat	30	420	270	232–313
Mouse	30	350	225	199–254
Rabbit	30	420	270	240–310
Rat	60	287	185	142–240
Mouse	60	233	150	139–162
Guinea pig	60	264	170	152–190

[a] Ten rats or mice per group. Five rabbits or guinea pigs per group.

Willard–Winter thorium–alizarin dye method gave satisfactory results, but under the pressures of wartime, the necessary effort for improvement of sampling could not be made.) The sampling of the exposure chamber air by Keplinger and Suissa, on the other hand, had a recovery of 99.4 percent, SD 2.4 percent, and the air was analyzed by a modification of the colorimetric method of Bellack and Schoubee (2b).

According to Keplinger and Suissa (2a), signs and symptoms and gross pathology resulting from the acute and subacute exposures were similar to those previously reported (1b):

Dyspnea, lethargy, red nose and swollen eyes were observed at concentrations equivalent to 50 percent of the LC$_{50}$s. At concentrations which were 25 percent of the LC$_{50}$s there were only mild signs of intoxication, manifested by slight dyspnea and closed eyes. At lower concentrations there were no gross signs of intoxication. Complete blood counts on these animals did not show significant changes due to F$_2$.

Gross pathology following exposures near the LC$_{50}$s was congestion, hemorrhage and atelectasis in the lungs and some congestion and/or mottling in the liver. Survivors, sacrificed up to 45 days after such exposures, had congestion in the liver. There was some discoloration of the kidneys in animals sacrificed seven to 14 days after exposure.

Following sublethal exposures there was pathology in the lungs, liver and kidneys. Effects in the lung were observed immediately after exposure. Effects in the kidney and liver were observed first on the seventh to 14th day following exposure. Pathology in the lung or kidney occurred from exposure at almost the same concentration. Exposure at higher concentrations was necessary before pathology was observed in the liver.

Exposure at concentrations at or below 100 ppm for five minutes, 70 ppm for 15 minutes, 55 ppm for 30 minutes or 45 ppm for 60 minutes caused no apparent effects in the animals.

Volunteer human subjects experienced slight irritation of the eyes when exposed at 25 ppm for five minutes. Slight irritation of the nose was experienced at 50 ppm. At 100 ppm there was marked irritation of the eyes and nose after one minute. The skin of the face was very slightly irritated and felt "sticky" after exposure at 100 ppm.

Fluorine concentrations causing irritating responses from brief exposures in human volunteers were reported by Rickey (3a) to be intolerable at 40 mg/m^3, and breathing was impossible at 75 mg/m^3; Belles et al. (3b) reported that among the majority of their subjects, 25 to 40 mg/m^3 caused some nasal and eye irritation after two or three breaths, but all tolerated, without discomfort, repeated short-term exposures up to 15 mg/m^3. Skin exposures resulted in dermal irritation at levels between 150 and 300 mg/m^3, and odor thresholds for this group were between 0.15 and 0.30 mg/m^3. Lyon (3c), concerned with the health of workers exposed to fluorine, reported that during a 9-year period, intermittent exposures up to 20 mg/m^3 (ca. 10 ppm) for periods of 5 to 30 min caused no ill effects, a conclusion in general agreement with those noted above. Presumably, workers become "hardened" to nasal and eye irritation.

The TLV for fluorine gas, revised from 0.1 to 1 ppm in 1971, was adopted by the American Conference of Governmental Industrial Hygienists in 1973.

1.1.3 Nonmetallic, Inorganic Fluorine Compounds

The compounds of industrial interest are mainly those of the fluoro acids and their salts. Chief among this group is fluosilicic acid, commercially available as sodium silicofluoride (Na_2SiF_6), and fluoboric acid (HBF_4) and its sodium, potassium, magnesium, and lead salts [M^+BF_4, $M^{2+}(BF_4)_2$] (4). Although the fluorosulfonates and difluorophosphates were of some interest in the late 1950s, they appear not to have reached great commercial importance. A similar fate has befallen fluorine oxide (F_2O), considered in the early 1960s as an oxidizer for propellants of spacecraft, presumably because of its extremely difficult handling problems.

The fluoro acid group has one use in common, that of the electropolishing of metals, particularly aluminum. Fluosilicic acid, as its sodium salt, finds its largest use in the hardening of cement and ceramics, as a wood preservative, in technical paints, in the production of synthetic cryolite in aluminum production, and in fluoridation of drinking water supplies.

Fluosilicic Acid and Its Salts.   Despite long and widespread use, or perhaps because of it, few reports of the toxicity of the acid or its salts can be found, many of them in the early 1930s or earlier. Weber and Engelhardt (5a) exposed guinea pigs to air bearing sodium silicofluoride as a dust in concentrations ranging from 13 to 55 mg/m^3 and found the dust capable of causing pulmonary irritation. They concluded that the least concentration that caused death when inhaled for a period of 6 hr was 33 mg/m^3.

The sodium salt is highly toxic when ingested; numerous deaths have been recorded. The signs of poisoning resemble those seen in intoxication by fluorides, according to Sollmann (5b). When in contact with the skin, the acid and its salts cause redness and a burning sensation, sometimes followed by the formation of ulcers. A pustular rash has been observed among men who worked with the sodium salt (5c).

When it became evident in the 1960s that sodium silicofluoride might serve as a water fluoridating agent, additional toxicity data were developed (6). The oral LC_{50} and LD_{LO} values for Na_2SiF_6 ranged from 125 mg/kg for the rat, to 500 mg/kg for the potassium salt in the guinea pig, with intermediate values for the lead salt of 250 mg/kg and for the free acid of 200 mg/kg.

Water converts Na_2SiF_6 slowly to NaF and SiF_4, the conversion being favored by alkalinity, thus raising its oral toxicity more than might be expected. Its effects are identical with those of fluoride according to Sollmann (5b), whether by oral or parenteral administration, although the solubility of the fluorosilicate is but 1/150 that of NaF.

In comparative tests of chronic toxicity of Na_2SiF_6 and NaF in rats for 90 to 103 days at 5, 10, 15, 25, and 50 ppm F in the drinking water, McClure (7) found no differences in the quantity of fluorine deposited in the incisor and molar teeth, mandibles, and femurs, or in the percent of the ingested fluorine which was retained in the rat's body. There was no difference in the ash, calcium, and phosphorus content of the incisor teeth, molar teeth, mandibles, and femurs that could be related to the kind of fluoride ingested.

There were no differences in the appearance of the characteristic striations on the rats' incisor teeth that could be attributed to the Na_2SiF_6 versus NaF.

As a result of this study and because of cost benefits, the silicofluoride is used in fluoridating large public water supplies, and because fluosilicic acid solution has greater ease in handling, can be metered directly, and eliminates potential hazards from dust exposure to the Na salt, H_2SiF_6 is the fluoridating agent of choice.

Fluoboric Acid and Salts. Fluoboric acid (HBF_4), derived from boric and sulfuric acid on fluorspar (CaF_2), is a strongly acid liquid that serves as a source of fluoborates, chief of which in commercial use are the salts of Na and K, Mg^{2+}, Pb^{2+}, and Sn^{2+} [$M^{2+}(BF_4)_2$]. Along with HBF_4 as a solute, the divalent salts are used in the electroplating of metals, and the monovalent metal salts in the sand casting of aluminum and magnesium, fluxes for nonferrous metals, and for soldering and brazing.

Fluoboric acid has a strong caustic action on the skin and mucous membranes, and is irritating to the eyes and respiratory tract. Although no quantitative animal toxicity data on the free acid were found, neutralization of the acid with metal base reduces the salts to a relatively low toxicity. KBF_4 has an intraperitoneal LD_{50} ranging from 240 to 600 mg/kg in three small laboratory animal species, indicating that once absorbed into the bloodstream, it has a moderately low toxicity. No effects were noted in the eyes of rabbits exposed at 1 to 3 mg/m^3, and injury from skin penetration was reported to be negligible (8).

From the metabolic standpoint, the fluoborates stand apart from other inorganic fluorides (9). According to Largent, all of a total intake of 6.4 mg/day of $NaBF_4$ that he ingested was recovered in the urine in a 14-day period, and less than 10 percent of the amount absorbed into the bloodstream was retained when ingestion was extended to 7 to 38 weeks in other subjects. Contrary to the F^- ion, the complex BF_4^- ion is not stored in the bone, and because of the slow rate of hydrolysis, excretion outruns hydrolysis, leaving little opportunity for F^- ion storage.

The BF_4^- ion does accumulate specifically in the thyroid gland, as do the fluosulfonate and difluorophosphate, preventing the uptake of iodine (10a, b). It should be noted, however, that the trapping mechanism of the thyroid involves a factor (a protein) that scarcely distinguishes among different anions with the same charge and comparable volume, such as perchlorate (ClO_4^-) and thiocyanate ($CNS-$) (10b). The BF_4^- has found use diagnostically to detect intracranial, space-occupying lesions (10c).

Fluorine Oxide.   Fluorine oxide (F_2O), oxygen difluoride, came into prominence in the United States in the early 1960s, when it was considered a candidate oxidizer for missile propellant systems. Its greater density at liquid nitrogen storage temperatures, and its higher boiling point (b.p. $-144.8°C$) commended it to easier handling. On the other hand, F_2O is about fiftyfold more acutely toxic than fluorine in 5- to 15-min exposures of rats according to studies of Lester and Adams (11a) and LaBelle and Stokinger (11b).

For the above reason, F_2O never gained acceptance as a space-propellant booster; however, sufficient toxicologic information has been developed on this uniquely potent oxidizer to warrant summation here. Industrially, F_2O readily forms during scrubbing of F_2 with NaOH.

Fluorine oxide proved lethal to mice, rats, and rabbits in brief, 5- to 15-min exposures at all concentrations from 10 ppm to 0.5 ppm (11b). An approximate LC_{50} for Sprague–Dawley rats for these brief exposures was 100 ppm-min (11a). Below 0.5 ppm, a sharp cutoff point in lethality occurs; no deaths occurred among 50 mice, 70 rats, 20 guinea pigs, four rabbits, and two dogs following 5,6-hr, and 1,3-hr periods/week for 30 days at 0.1 ppm F_2O, as estimated from metered dilutions of F_2O in pressurized cylinders; higher levels were determined by sampling in KI, measuring the liberated I_2 colorimetrically

(11b). The gas had a purity range of 70 to 90 percent, the impurities being chiefly O_2 and N_2, whereas that of Lester and Adams approximated 97 percent, with O_2 and CF_4 being the main impurities (2.0 and 0.54 percent, respectively).

Signs and symptoms associated with lethal exposures were widespread damage to the lungs, which did not appear during exposure. Nor was there evidence at this time of irritation to the external mucosa or respiratory embarrassment. Only immediately prior to death, which followed a latent period of from 9 to 66 hr, was overt respiratory distress seen. Damage ranged from slight to moderate hemorrhage to swelling, edema, and consolidation of entire lobes (11a). No detectable effects were noted in any species exposed for 30 days at 0.1 ppm, however, in body weight, in the formed elements of the blood, in blood Ca and NPN, or in urinary sugar and protein; nor was there evidence of histological changes in lung, liver, kidneys, thyroid, or gonads (11b). At higher levels, F_2O proved extremely irritating to the respiratory tract, the corticomedullary zone of the kidney, and internal genitalia.

A limited tolerance to F_2O develops in rats exposed at sublethal concentrations, and given a few days for recovery before challenge (11b), a tolerance similar to that to the triatomic molecule O_3.

The TLV Committee of the American Conference of Governmental Industrial Hygienists adopted in 1966 a limit of 0.05 ppm. The National Research Council Committee on Toxicology has recommended short-term exposure limits for 10, 30, and 60 min of 0.5, 0.2, and 0.1 ppm.

Fluorine oxide is just detectable by smell at 0.1 ppm and is obvious at 0.5 ppm, but tolerance to its odor develops rapidly, so that odor does not serve as a good warning device for exposed workers. Technicians operating the exposure chamber experienced drowsiness and intractable headache at fractions of a part per million.

Hydrogen Fluoride and Hydrofluoric Acid. Hydrogen fluoride (HF) is a corrosive, fuming, nearly colorless liquid at ordinary pressures below 19°C; above 19°C, it is gaseous. HF has a monomolecular weight of 20.01, but at 1 atm, and at temperatures below 100°C, it exists as an associated molecule up to H_6F_6, with an average molecular weight of 50 to 55. Although toxicity may be related to its degree of polymerization, because all determinations are made at about the same temperature and pressure, the results are comparable if expressed as mg HF/m^3. The liquid has a specific gravity of 0.987; the gas, a density of 0.921 g/liter at 0°C. Hydrogen fluoride is infinitely soluble in cold water. For this reason, it fumes strongly in moist air. Its aqueous solutions dissolve glass, reacting to form gaseous SiF_4.

Sources and Occupational Exposure. Active volcanoes are the only known natural source of HF, but many industrial processes contribute HF to the atmosphere. The gas is an effluent of processes such as aluminum reduction, phosphate fertilizer manufacturing, petroleum refining, manufacture of fluorocarbon compounds, the making of brick, pottery, glass, and ceramics, ferroenamel production, metal-fluxing agents used in foundries and metal-fabricating plants, welding processes, and the burning of coal. Coal may contain 40 to 295 ppm of fluoride depending on its source, some of which is released as HF on burning.

Increased use of HF for this variety of applications similarly increases the need to transport the material from place of manufacture to place of use. Temporary storage facilities at both locations and transfer from storage containers to other containers are operational requirements. In each instance, there is danger of accidental spills that could result in exposure of occupational personnel.

Toxicity. Early determinations of acute toxicity of HF gas in laboratory animals provided ball-park figures for later studies (12a). Guinea pigs and rabbits died within 5 min when they inhaled air containing 1500 mg/m^3. The inhalation of air containing 1000 mg/m^3 for 30 min killed no animals, but did cause damage to the tissues. All animals exposed at the concentration of 500 mg/m^3 for 15 min or more showed signs of weakness and ill health; concentrations below 100 mg/m^3 could be tolerated for 5 hr without causing death; and a concentration of 24 mg/m^3 was tolerated for a total of 41 hr without fatality, although the animals subsequently lost weight. The concentration of 15 mg/m^3 was found tolerable. When inhaled at a concentration of not more than 50 mg/m^3, HF induced signs of mild irritation, such as coughing and sneezing, which appeared to lessen after 5 to 15 min. Inhaled in greater concentrations, it acted as a severe irritant: the eyes were kept closed, paroxysms of coughing and sneezing were frequent, respiration was slowed, and there was a copious discharge from the nose and eyes.

Rosenholz and co-workers (12b) exposed rats, rabbits, guinea pigs, and dogs to provide a better estimate of the LC$_{50}$. The values shown below indicate a rather low acute, lethal toxicity for laboratory rodents compared with those for fluorine.

Species	Exposure Time (min)	LC (mg/ m^3)[a]
Rat	5	4060
Rat	15	2200
Guinea pig	15	3540
Rat	30	1670
Rat	60	1070

[a] 1 mg/m^3, approximately equivalent to 1.223 ppm as HF.

Signs of toxicity in the animals included irritation of the conjunctiva, nasal tissues, and respiratory system. The survivors ceased to show these signs about 1 week after exposure. Pathological lesions were observed in the kidney and liver, the severity of which was directly related to the dosage received. The external nares and nasal vestibules were black and, at dosages causing considerable mortality, those areas showed zones of mucosal and submucosal necrosis. The skin of animals exposed at high (lethal) concentrations showed superficial subcutaneous and deep dermal zones of acute inflammation. The hair of these animals could be pulled out with ease and the skin ruptured under minimal tension. The data show that the rat was the most susceptible species of those tested.

Subacute, 30-day exposures of five laboratory animal species at levels that bracketed the maximal and minimal effects were performed at 7 and 25 mg/m^3 in 6-hr daily exposures by Stokinger and co-workers (12c).

Exposure at the higher concentration was lethal to all the rats and mice, but not to guinea pigs, rabbits, and dogs. Among the surviving animals, the rabbits showed a slight loss in weight, the dogs were apparently unaffected, and the guinea pigs began to lose weight after the third week of exposure. Exposure at the low concentration did not interfere with normal weight gains in any of the animals except the rabbits.

This order of susceptibility to HF differs from that to F$_2$, in which dogs and guinea pigs were equally susceptible as rats and mice, and thus provides evidence for the independent toxic action of these two gases. That is, owing to the extremely rapid reaction of F$_2$ with HOH to yield HF, it might be assumed that the toxicity of F$_2$ would resemble that of HF. This evidence contradicts this assumption for acute, toxic exposures.

Histological examination of the dog, rabbit, and rat subacutely exposed revealed moderate hemorrhage and edema of the lungs in all three species at the 25 mg/m^3 level. Renal cortical degeneration and necrosis were found in the rat, and ulceration of the scrotum in the dogs. At the 7 mg/m^3 level, localized hemorrhages were found in the lung of one of five dogs examined, and no changes were found in the rabbit or rat (12c).

Limited exposures of two human volunteers (12a) caused smarting of the skin, conjunctival and respiratory irritation, and recognition of the flat, sour taste of HF at 100 mg/m^3 for 1 min. Exposure at 50 mg/m^3 was exceedingly uncomfortable, as shown by irritation of mucous membranes, but there was no smarting of the skin. At 26 mg/m^3 irritation decreased and the ability to taste HF was delayed, but the atmosphere was uncomfortable throughout the 3-min period of exposure. No apparent tolerance results from repeated brief exposures.

Contact of the skin with the anhydrous liquid produces severe burns that are felt immediately. Concentrated aqueous solutions also cause an early sensation of pain, but more dilute solutions may give no warning of injury. If the solution is not promptly removed, the skin may be penetrated by the fluoride ion, leading to the later development of painful ulcers, which heal slowly.

After any contact with a solution of the acid, even though there may be no immediate pain, the area should be flushed with copious amounts of water, after which it may be swabbed with cotton moistened with a 10 percent solution of 28 percent aqueous ammonia and then allowed to remain immersed in a bath of water for a prolonged period. Prolonged washing, when carried out without any delay following contact, may succeed in restoring a pink color to any initially blanched area, thereby saving the tissue. Finally an ointment containing 20 percent magnesium oxide in glycerin should be applied (13a). In the case of more serious burns, other procedures have been suggested (13b, c), such as setting up a barrier to the spread of the fluoride ion by injecting a 10 percent solution of calcium gluconate subcutaneously around and underneath the affected area.

Precautions necessary for the safe handling of liquid hydrogen fluoride have been described by the Chemical Manufacturers Association (13d).

Industrial Experience. Even though there are numerous industrial processes that can cause HF injury to workers, there are few records of such injuries. However, there are several well-documented reports of injuries resulting from accidental exposures to ruptured containers of anhydrous HF (14). The major effects from such accidents are associated with liquid HF, although gaseous HF plays some part. In three separate events, a total of eight men were splashed with liquid HF. In one event, involving four men, one died of pulmonary edema approximately 2 hr after the accident. The other three survived, but suffered severe chemical burns. In the other two events, all four men died, two of them in 2 hr, one in 4 hr, and one 10 hr after the accident. Consciousness remained until death, which was usually sudden, from respiratory distress and cardiac arrest. Characteristically, the respiratory tree is inflamed and the lungs moderately to severely congested.

There are other reported cases of severe poisonings and deaths among superphosphate and fertilizer factory workers, welders, garage workers, water-works employees, and enamel-factory workers (15a, b). In most of these cases the clinical picture in part resembles HF poisoning, but other fluoride-containing particulates and gases such as SiF_4, phosgene (from the high temperature decomposition of halogenated hydrocarbons), and a variety of fluoride dusts also were present.

Threshold Limits. The TLV Committee of the American Conference of Governmental Industrial Hygienists adopted the values of 3 ppm and 2 mg/m^3 for gaseous HF among its first activities around the end of the 1940s, and in 1976 adopted the same values for its STELs.

Ammonium and Sodium Bifluorides. Ammonium bifluoride, or ammonium acid fluoride (NH_4HF_2), and sodium bifluoride, or sodium acid fluoride

($NaHF_2$), are segregated from the simple fluorides of the alkalies and alkaline earths because of their different physical and chemical properties and hence their different industrial uses. In their toxicological properties and industrial hazards, they resemble the alkalies and alkaline earths (Section 1.1.2).

The number of commercial outlets in the late 1970s indicate that NH_4HF_2 has a greater number of industrial applications than $NaHF_2$, 28 versus 18. Among the many uses of NH_4HF_2 are in the electrolytic processing of beryllium, electroplating, ceramics, etching of glass, and sterilizing brewery, dairy, and other equipment. Sodium bifluoride, in addition to sharing the uses of the ammonium salt as an etcher of glass, antiseptic, and preservative, is used in tinplate production. No toxicity data on either bifluoride appears to have reached the open literature.

Fluorides of the Alkali and Alkaline Earths. The chief fluorine salts of the alkali metals that find major industrial use are those of sodium and potassium (NaF, KF), with the uses of NaF far outstripping those of KF. Calcium fluoride (CaF_2) and magnesium fluoride (MgF_2) represent those of the alkaline earth series with major uses. The fluoride salts of the remaining alkaline earths, barium and strontium, have relatively minor uses (16), and beryllium fluoride is an intermediate in the production of beryllium metal.

Sources and Uses. Sources of these fluorine salts have been noted in Section 1.1. The greatest use of CaF_2 along with NaF is in steelmaking, where it prolongs the fluidity of the ingot, thus facilitating the release of gaseous products. Fluorides are also used as a flux in the smelting of nickel, copper, gold, and silver. Other applications are in the opacification of glass and enamels and in coatings for welding rods. Considerable quantities of CaF_2 are treated with sulfuric acid for the preparations of HF. Sodium and potassium fluorides are used as rodenticides and fungicides, in electroplating, in the manufacture of glass and vitreous enamels, as a wood preservative (NaF), and in many other processes and operations (16). The complex NaAl fluoride, cryolite, is the electrolyte used in reducing alumina (Al_2O_3) to aluminum. It is also used as a binder for abrasives, in electric insulation, and in polishes. Potassium, but not sodium, forms complexes with the fluorides of zirconium (K_2ZrF_6) and tantalum (K_2TaF_7) which have minor industrial uses (16).

Industrial Exposures. The ubiquity of fluorine in nature is no less paralleled by its ubiquity in industrial processes and operations, and because this makes for an almost endless array of industrial exposures, only the more prominent can be mentioned here. Table 40.3 shows some of these industrial operations with their fluoride source and type of exposure.

NIOSH, in 1976, estimated that approximately 350,000 workers in the United States alone are exposed to some form of fluoride. Some concept of how

Table 40.3. Industrial Processes or Operations Presenting Worker Exposure to Fluorides

Operation and Fluoride Type	Exposure
Steelmaking, fluorspar, CaF_2	Fluorides, dust and fume
Fertilizer industry, F in phosphate rock	HF acid, H_2SiF_6, SiF_4, F fume
Chemical industry, fluorspar	HF acid, fluorides, dust and fume
Aluminum industry, cryolite, Na_3AlF_6	Fluorides, dust and fume
Foundry operations, F fluxes	Fluorides, dust and fume
Beryllium production, NH_4HF_2	NH_4F, F dust and fume
Welding, F-coated rods, CaF_2	Fluorides, dust and fume
Charging water-fluoridating hoppers	Na_2SiF_6, H_2SiF_6

widespread exposures are can be gained by noting that one of the major sources, fluospar, CaF_2, in the United States in 1975 amounted to 1,245,000 short tons, of which the iron and steel industry used 44 percent, the chemical industry 40 percent, and the aluminum industry most of the remainder (17).

For magnitude of the fluorine exposure, Shilen found fluorine concentrations in the air of a beryllium production plant to range as high as 10.6 mgF/m^3 in 1943 (18); conditions since have considerably improved. At about the same time, Williams found fluorine concentrations in magnesium foundries, where it is sprayed on cores, mixed with core sand, or added to melting pots as fluxes, to average 0.7 mg/m^3 at core-spraying, 1.26 at melting, 1.88 at molding, and 8.77 mg/m^3 at shake-out (19). Using urinary fluorine excretion as a measure of exposure, Largent and Ferneau found the greatest exposure in the core-spraying and pouring area, indicating that factors other than air concentration determine degree of exposure. Very similar results have been obtained in British magnesium foundries (20). Air concentrations of fluorine during shipyard welding averaged just above the TLV of 2.5 mg F/m^3 (3.7 mg/m^3), and in four large industrial plants, from 0.4 to 1.1 mg/m^3 (21). Experimentally determined fluorine levels from three different welding rod types, E70S-3, E70T-1, E11018, showed values ranging from 0.0424 to 7.13 mg F/m^3 outside the helmet, to 0.0129 to 0.472 mg F/m^3 inside the helmet (21).

Physiologic Response. The physiological effects and toxicity of this group of salts, as far as the industrial worker is concerned, are marked for their chronic rather than their acute responses. A limited amount of acute toxicity data has been developed on NaF and CaF_2 for the purpose of establishing the acute toxic dose of NaF as a rodenticide, and for accidental poisonings in man. No acute toxicity data could be found for KF.

The Acute Toxic Response. The lowest oral lethal dose (LD_{LO}) of NaF reported for man is 75 mg/kg (6), the same dose as reported for the dog. That for the guinea pig was 250 mg/kg, with the oral LD_{50} for rat and mouse 180 and 97 mg/kg, respectively. Somewhat greater toxicity resulted from intraperitoneal administration; the intraperitoneal LD_{50} for the mouse of NaF was 49 mg/kg. All determinations were reported in the 1960s with the exception of the oral LD_{LO} for the dog. Calcium fluoride, because of its lesser solubility, is far less toxic than NaF; rats survive the largest dose that can be orally administered (22).

Subacute exposure of rats and rabbits for 3 hr daily, for periods up to 110 days, to welding fumes from electrodes containing about 17 percent CaF_2 resulted in mild increases (twice normal) in blood fluorine levels of rabbits, and "significant" increases in fluorine content of muscles, liver, kidney, and brain after 1 month, which remained at this level, then decreased to near control levels during the 30-day recovery period (23). Urinary fluorine excretion increased after a month to 6.2 mg F/liter, dropping to a steady state of 4.2 mg/ liter for the duration. The fume content of fluorine to which the animals were exposed was 8.9 percent, of which 3.4 mg/m^3 was particulate fluorine, and 1.45 mg/m^3, volatile fluorine.

The symptoms that follow the ingestion of soluble fluorides by man are listed in the order of diminishing frequency of occurrence: vomiting, abdominal pain, diarrhea, convulsions, generalized and muscular weakness, collapse, dyspnea, paresis, difficulty in articulation, thirst, weakness of the pulse, disturbed color vision, loss of consciousness, and motor unrest (24). Albuminuria is frequently present. Acute toxic nephritis, hemorrhagic gastroenteritis, and more or less definite pathological damage to other organs are found on examination. The calcium content of the blood is reduced following the ingestion of large amounts of fluorides (25). Fluoride acts as an inhibitor of certain intracellular enzymes concerned in the anaerobic glycolysis of many types of cells, plant as well as mammalian (26). It interferes with enzymes concerned with the conversion of phosphoglyceric to phosphopyruvic acids, an essential line in the chain of reactions (27). A number of other enzymes, particularly those concerned with processes of phosphorylation, are also affected by the fluoride ion.

The therapy of acute fluoride poisoning has been reviewed by Peters (28).

Immediate intoxication by the ingestion of fluorides is rare in industry, but various degrees of respiratory irritation may result from the inhalation of fluorides in the form of dusts. Some magnesium founders complain of a severe biting sensation in the nose when the concentration of fluorides in the air exceeds 10 mg/m^3. This is accompanied after a few minutes by a discharge from the nose or by nosebleed. No such effects are noted when the concentration does not exceed 2.5 mg/m^3 (29).

Fluorosis—Chronic Fluoride Intoxication. Chronic fluoride intoxication is not restricted to long-term exposure to salts of the alkali or alkaline earth metals, but can result from long-term, even low grade exposures to fluorides in whatever chemical combination, and whether particulate or gaseous.

Roholm (Denmark), in what is considered now a classic account of fluorosis among heavily exposed *cryolite* workers, described digestive disturbances of nausea, vomiting, loss of appetite, diarrhea, or constipation in more than 80 percent of a worker population exposed at an estimated concentration of 15 to 20 mg F/m^3 (30). The characteristic increased bone density was found among slightly more of the workers than those showing digestive disturbances, presumably resulting from levels greater than 20 mg F/m^3. Minimal changes occurred in 38.2 percent, moderate changes in 34.2, and marked changes in 10.3 percent. Digestive tract disturbances were found also in a group of *Al furnace-room workers* in Great Britain exposed at F levels far lower, 0.14 to 3.43 mg/m^3 compared to a control group (14.7 versus 5.2 percent) (31). Nearly all of a group of aluminum potroom workers in the United States had increased bone density following time-weighted, average 8-hr exposures of 2.4 to 6.0 mg F/m^3; minimal osteosclerotic changes occurred in 33 percent (32).

From the standpoint of trying to determine the TLV that will prevent the onset of fluorosis, all the foregoing epidemiologic studies have the common fault of reporting effects from past cumulative exposures which were undoubtedly greater than those measured at the time of the studies. The study of Derryberry et al. in the United States (33) is free of this fault. At fluorine levels between 1.78 and 7.73 mg/m^3 (TWA, 3.38 mg/m^3), minimal or questionable skeletal fluorosis was found in 23 percent of a group of 74 phosphate rock *fertilizer manufacturing workers*. Of those found to have minimal or questionable increases in bone density, 71 percent were exposed at a TWA F level greater than 2.5 mg/m^3. The average urinary F excretion of this group was 5.18 mg F/liter (range, 2.2 to 8.9) compared to 4.53 mg F/liter (range, 2.1 to 14.7) in 57 workers showing no increase in bone density. When increases in bone density were compared to urinary fluorine levels in 1 mg F/liter increments, increases in urinary fluorine levels paralleled the percentage of suspected cases of osteofluorosis.

This study provides a firm basis for the TLV of 2.5 mg F/m^3 that was recommended by the TLV Committee of the American Conference of Governmental Industrial Hygienists in its earliest listings many years before.

For other characteristic changes in fluorosis, the metabolism of the fluorine ion, its absorption, distribution, storage, and excretion, see Vol. II of *Patty's Industrial Hygiene and Toxicology*, 2nd ed. These features have been brought up to date in NIOSH's *Criteria Document on Inorganic Fluorides*, 1975, which also contains details of biologic monitoring for urinary fluoride in industry.

FLUORINE REFERENCES

1. (a) R. Landau and R. Rosen, *Ind. Eng. Chem.*, **39,** 281 (1947); (b) H. E. Stokinger, in *Pharmacology and Toxicology of Uranium Compounds*, Vol 2., Natl. Nucl. Energy Ser. McGraw-Hill, New York, 1949.

2. (a) M. L. Keplinger and L. W. Suissa, *Am. Ind. Hyg. Assoc. J.*, **29,** 10 (1968); (b) E. Bellack and P. J. Schoubee, *Anal. Chem.*, **30,** 2032 (1958).

3. (a) R. P. Rickey, "Decontamination of Large, Liquid Fluorine Spills," USAF Flight Training Command Tech. Rept. 59-21 (1959); (b) F. Belles et al., Eds., *Fluorine Handbook*, NASA, Lewis Research Center, Cleveland, 1965; (c) J. S. Lyon, *J. Occup. Med.*, **4,** 199 (1962).

4. *Chemical Buyers Directory for 1978–1979*, 66th Am. ed., Schneel, New York.

5. (a) H. H. Weber and W. E. Engelhardt, *Zentr. Gewerbehyg. Unfallverhut.*, **10,** 41 (1933); (b) T. Sollmann, *Manual of Pharmacology*, 6th ed., W. B. Saunders, Philadelphia, 1944; (c) R. P. White, *The Dermatergoses*, 4th ed., Lewis, London, 1934.

6. *NIOSH Registry of Toxic Effects of Chemical Substances for 1976*, Rockville Md.

7. F. J. McClure, *Public Health Rep.*, **65,** 1175 (1950).

8. C. T. Blaisdell, Chem. Corps Med. Lab. Army Chem. Center Res. Rept. MLRR 351, Mar. 1955.

9. E. J. Largent, "Fluoridation as a Public Health Measure" American Association for the Advancement of Science, Washington, D.C., 1954.

10. (a) M. Anbar et al., *Endocrinol.*, **66,** 888 (1960); (b) M. Anbar et al., *Nature*, **183,** 1517 (1959); (c) Askenasy et al., *Am. J. Roentgenol.*, **88,** 350 (1962).

11. (a) D. Lester and W. R. Adams, *Am. Ind. Hyg. Assoc. J.*, **26,** 562 (1965); (b) C. W. LaBelle and H. E. Stokinger, "Studies on the Toxicity of Oxygen Difluoride at Levels from 10 to 0.1 ppm," University of Rochester, January 1945.

12. (a) W. Machle et al., *J. Ind. Hyg.*, **16,** 129 (1934); (b) M. J. Rosenholz et al., *Am. Ind. Hyg. Assoc. J.*, **24,** 253 (1963); (c) H. E. Stokinger, in *Pharmacology and Toxicology of Uranium Compounds*, C. Voegtlin and H. C. Hodge, Eds., McGraw-Hill, New York, 1949, pp. 1021–1057.

13. (a) Booklet issued by Universal Oil Products Co.; (b) F. Flury, *J. Ind. Hyg. Toxicol.*, **24,** 92A (1942); (c) A. Paley and J. Seifter, *Proc. Soc. Exp. Biol. Med.*, **46,** 190 (1941); (d) Chemical Manufacturers' Association, Chemical Safety Data Sheet S.D.-25, 1948; Manual Sheet, H-10 (1948).

14. R. Y. Eagers, *Toxic Properties of Inorganic Fluorine Compounds*, Elsevier, New York, 1969.

15. (a) K. K. Huffstutler, "Fluoride Concentrations in Various Receptors Near Phosphate Industries," presented at *63rd Ann. Meet. A.P.C.A., St. Louis, Mo., June 14–18, 1970*; (b). J. Krechniak, "Fluoride Hazards Among Welders," *Fluoride Q. Rep. (Int. Soc. Fluoride Res.)*, **2,** 13 (1969).

16. G. G. Hawley, Ed., *The Condensed Chemical Dictionary*, 9th ed., Van Nostrand Reinhold, New York, 1978.

17. *The Minerals Yearbook*, Vol. 1, Bureau of Mines, U.S. Dept. Interior, Washington, D.C., 1975.

18. J. Shilen et al., *Ind. Med.*, **13,** 464 (1944).

19. C. R. Williams, *J. Ind. Hyg. Toxicol.*, **26,** 113 (1944).

20. R. G. Bowler et al., *Br. J. Ind. Med.*, **4,** 216, 231 (1947).

21. *The Welding Environment*, American Welding Society, Miami, Fla., 1973.

22. E. J. Largent, *J. Ind. Hyg. Toxicol.*, **30,** 92 (1948).

23. J. Krechniak, *Fluoride Q. Rep.*, **2,** 13 (1969).

24. K. Roholm, In Heffter-Huebner, *Handbuch Exp. Pharmakol. Erg.*, **7,** 1 (1938); *Z. Ges. Gerichtl. Med.*, **27,** 174 (1936); F. McClure, *Physiol. Revs.*, **13,** 277 (1933); D. A. Greenwood, *Physiol. Revs.*, **20,** 582 (1940).

25. H. Wieland and G. Kurtzahn, *Arch. Exp. Pathol. Pharmakol.*, **97,** 488 (1923); A. Jodlbauer, *Arch. Exp. Pathol. Pharmakol.*, **164,** 464 (1931).

26. F. Lipmann, *Biochem. J.*, **206,** 171 (1929); K. Lohmann, *Biochem. J.*, **222,** 324 (1930); F. Lipmann and K. Lohmann, *Biochem. J.*, **222,** 389 (1930); F. Dickens and F. Simer, *Biochem. J.*, **23,** 936 (1929).

27. C. W. Bishop and E. Roberts, *Natl. Nucl. Energy Ser.,* Div. VI, **1,** Book 4, 1867 (1953).
28. J. H. Peters, *Am. J. Med. Sci.,* **216,** 278 (1948).
29. C. R. Williams, *J. Ind. Hyg. Toxicol.,* **24,** 277 (1942).
30. K. Roholm, *Fluorine Intoxication,* H. K. Lewis, London, 1937.
31. J. N. Agate et al., "Industrial Fluorosis," *Med. Res. Council Memo. 22,* London, HMS Office, 1949.
32. N. L. Kaltreider et al., *J. Occup. Med.,* **14,** 531 (1972).
33. O. M. Derryberry et al., *Arch. Environ. Health,* **6,** 503 (1963).

1.2 Chlorine, Cl$_2$

Chlorine displays the largest array of industrially useful compounds of all the halogens. (Table 40.4). Chemically, chlorine occupies a place between fluorine and bromine and iodine in its reactivity; it replaces bromine and iodine from their salts, and enters into substitution and addition reactions in both inorganic and organic substances. Moist, but not dry, chlorine unites directly with most elements. The physical properties of chlorine are given in Table 40.1.

1.2.1 Source and Production

The reactivity of chlorine is so great that none is found free in nature. Instead, it is found in landlocked lakes as NaCl (Great Salt Lake, Utah) and in underground deposits mainly as rock salt (NaCl) and brines in southern Michigan, Central New York, Louisiana, Mississippi, along the Gulf Coast of Texas, Stassfurt, Germany and elsewhere, in almost inexhaustible supplies.

Chlorine is produced by electrolyzing NaCl brines in diaphragm or mercury cells (34). In 1973 (35a) chlorine was produced in 65 plants in the United States operated by 33 companies, including six pulp mill plants which produced their own chlorine. Of the 65 plants, 29 used diaphragm cells, 23 used mercury cells, five used both, four had the fused-salt process, one had combined fused-salt and diaphragm process, one had an HCl electrolysis process, one used a diaphragm and magnesium cell, and one was a nonelectrolytic plant. The bulk of the chlorine produced in the United States and in Europe is by the diaphragm and mercury cell processes, with the latter supplanting the diaphragm in an increasing number of plants. 1976 estimates placed U.S. chlorine production at 11 million short tons (35b), and U.S. demand is forecast for the year 2000 at 39.5 million tons (36).

1.2.2 Uses and Industrial Exposures

The largest end use for chlorine, according to the Bureau of Mines for 1975 (36) approximately 33 percent, is the manufacture of "chemicals" not included in the classifications plastics, agriculture, or sanitary services. The chemicals

Table 40.4. Properties of Industrially Useful Inorganic Chlorine Compounds

Compound	Formula	Mol. Wt.	Physical Properties	Sp. G.	M.P. (°C)	B.P. (°C)	Solubility
Hydrogen chloride	HCl	36.46	Colorless gas	$1.187^{-84.9}$	-114.8	-84.9	823 g/l H_2O 0°C, 561 g/l H_2O 60C; very sol. alcohol; sol. ether
Sodium chloride (halite)	NaCl	58.44	Colorless cubic	2.165^{25}	801	1,413	357 g/l H_2O 0°C, 391.2 g/l H_2O 100°C; sl. sol. alcohol, liq. NH_3; insol. HCl
Sodium hypochlorite	NaClO	74.44	In soln. only	—	—	—	
Sodium chlorate	$NaClO_3$	106.44	Colorless cubic	2.490^{15}	248–261	Dec.	790 g/l H_2O 0°C, 2.3 kg/l H_2O 100°C; sol. alcohol, liq. NH_3, glycerin
Sodium perchlorate	$NaClO_4$	122.44	White and brown, deliq.	—	482 (dec.)	Dec.	Sol. cold H_2O, very sol. hot H_2O, sol. alcohol
Calcium chloride	$CaCl_2$	110.99	Colorless cubic deliq.	2.15^{25}	782	>1600	745 g/l H_2O 20°C, 1.59 kg/l H_2O 100°C; sol. alcohol, acetone
Chlorine dioxide	ClO_2	67.45	Yellow-red gas; expl.	3.09^{11} g/l	-59.5	$9.9^{7.31}$ (expl.)	20 l H_2O 4°C; dec. $HClO_3$, Cl_2, O_2; sol. alkali. H_2SO_4
Chlorine trifluoride	ClF_3	92.45	Colorless gas	1.77^{13}	-83	11.3	Dec. cold, hot H_2O

classification includes chlorine-containing inorganic chemicals, such as HCl, chlorine dioxide (ClO_2) for pulp bleaching, hydrazine intermediate from Cl_2, NaOH, and NH_3 as a rocket fuel, chlorinated organic chemicals (halocarbons), chlorinated solvents, and ethylene dichloride, a gasoline antiknock additive.

The second most important use, about 22 percent of the total, is in the production of plastics, such as vinyl chloride, which is polymerized into polyvinyl chloride. Similarly vinylidene chloride, $CH_2 = CCl_2$, is polymerized into a textile as a flexible food wrap. Plastics containing chlorine are used as protective films, floor coverings, electric insulation, pipe and tubing, foam, upholstering, draperies, synthetic rubber (chlorobutadiene), and numerous other uses.

The third most important use, about 18 percent, is in the pulp and paper industry. Most of the chlorine is in elemental form or as ClO_2, both of which, are oxidizing bleaching agents to make white paper.

The chlorination of drinking water supplies and sewage uses only 6 percent of the total. Another small use goes for pesticides, germicides, rodenticides, and herbicides (economic poisons) amounting to about 5 percent.

Other miscellaneous uses include household bleaches, pharmaceuticals, cosmetics, metal extractants, lubricants, flameproofers, adhesives, food additives, and hydraulic fluids. Exposure to chlorine is entailed in any of the uses. NIOSH estimated in 1973 that 15,000 workers had potential exposure to chlorine (37).

1.2.3 Analytic Determination

For a substance requiring strict environmental control for worker safety, it is no surprise that there are a multiplicity of analytic methods for chlorine. Colorimetric methods of analysis involving air sampling have used the following reagents for the most part: arsenic trioxide, neutral and acidic iodide, *o*-tolidine, and methyl orange. Instrumental methods are primarily various types of gas chromatography, UV spectrophotometry, colorimetry, amperometry, mass spectrometry, and catalytic combustion. Chlorine-specific tubes for semiquantitative analysis have been certified by NIOSH.

Colorimetric methods in general are not specific for chlorine, but are affected by other oxidizing agents, and often by metals and reducing agents. Instrumental methods, however, often require specialized equipment and expertise, or prove difficult with mixtures of halides. Both types of methods have been briefly described or referenced in a NIOSH criteria document on chlorine (37).

The method of choice selected by NIOSH (37) for the sampling and analysis of chlorine gas in the workroom atmosphere is a methyl orange procedure (38). In this procedure, 10 ml methyl orange sampling solution is placed in a fritted bubbler, and a 30-liter air sample is drawn through at a rate of 1 to 2 liters/min, corresponding to about 0.05 to 1.0 ppm Cl_2 in air. The method claims an accuracy of ± 5 percent. Reagent stability is good and samples remain stable for 24 hr.

1.2.4 Physiologic Response

In spite of long years of widespread exposures to chlorine, its use as a war gas in World War I, and innumerable accidental exposures, much of the information relating dose to response still must be derived from experimental exposures of animals and man.

Acute Effects. Chlorine is a recognized primary irritant to the mucous membranes of the eyes, nose, and throat, and to the linings of the entire respiratory tract. Experimental exposures of human volunteers in five separate studies from three countries including the United States, in the 1960s resulted in general agreement that the threshold of odor perception was between 0.2 and 0.4 ppm, with considerable variation among subjects, and 0.06 ppm when the odor of chlorine is first detected (37).

The effect of chlorine on olfactory centers was marked; among 13 workers previously exposed for 2 to 5 years, 11 had severe olfactory deficiency, one, moderate, and one, slight; and for four, exposed for 1 year or less, one deficiency was severe, one, moderate, and two slight. This would indicate that odor detection does not serve as a good warning device for "chlorine-hardened" workers.

Discomfort in these subjects resulting from irritation of the nose, eye, and throat, first in the form of itching of the nose, at 0.2 ppm for 4 to 20 min; and at 1 ppm, burning in the conjunctivas, scratchiness and dryness of the throat, coughing, and a sense of difficulty in inhaling. At 1.3 ppm, some had severe shortness of breath and violent headaches after 30 min.

Severe, acute effects from accidental exposures have been documented since World War I, when 1843 of 70,742 total casualties resulted from gassing by chlorine. Of 838 studied, 28 died, four of which were attributed to "late" effects of gassing: bronchopneumonia, lobar pneumonia, purulent pleurisy, and tubercular meningitis. Nine were discharged because of disabilities due to gassing, which included pulmonary tuberculosis, a flare-up from former disease, bronchitis, pleurisy, tachycardia, dyspnea, and nephritis. Thirty-nine others were disabled at the time of discharge with similar conditions due to gassing (39).

Other reported, severe, accidental exposures have resulted from rupture of chlorine tank cars and cylinders, and from industrial operations, causing deaths and hospitalization, with symptoms of pulmonary congestion and edema developing only after a latent period of several hours. Moist rales developed later in all severely exposed, but respiratory distress generally subsided within 72 hr, with some exceptions among those with preexisting cardiac conditions (37).

The lowest toxic concentration (TC_{LO}) producing respiratory distress in man has been reported to be 15 ppm, and the lowest lethal concentration (LC_{LO}), 430 ppm for 30 min exposure (6). By comparison, the 1-hr LC_{LO} for the rat and mouse was of a similar magnitude, 138 ppm. Animals surviving sublethal exposures for 15 to 193 days after gassing showed marked emphysema, which was associated with an organizing exudate in the bronchioles leading to bronchiolitis obliterans and organizing pneumonia with desquamated cells found in the more chronic forms of pneumonia in man (40).

An accepted theory of the action of chlorine is based on its strong oxidizing capacity, in which it splits hydrogen from the water in moist tissue, causing release of nascent O_2 and HCl. The O_2 produces the major tissue damage, which is enhanced by the HCl. The result is the tissue irritation observed and/ or damage upon excessive exposure.

Chronic Effects. Several epidemiologic studies have been made of chlorine exposures in chemical plants, pulp and paper mills, and in chlor-alkali plants. All but one of these studies were designed to determine the health of the workers under the various exposure conditions associated with the operations, and hence resulted in descriptions of effects without providing any information on the chlorine dosage relating to these effects.

An in-depth study of 25 chlor-alkali plants by Patil et al. was reported in 1970 (41a). This was a study involving 332 male workers on diaphragm cells on whom information about chlorine dose–response relations was available as a result of bimonthly air-sampling program at representative plant locations, and from which TWA exposure data were calculated for each worker on an 8-hr basis. A group of 382 workers not exposed to chlorine made up the controls, well matched for age, ranging from 19 to 69 years, with 60 percent workers between 30 and 49 years, average 31.2 years. About 60 percent of the workers smoked at the time of the study.

The TWA exposures to chlorine ranged from 0.006 to 1.42 ppm with a mean of 0.15 ppm. All but six of the 332 had TWA exposures below 1 ppm; 21 had TWAs above 0.52 ppm, and the average number of chlorine-exposure years for all workers was 10.9.

From these dosages (exposure concentration × years of employment) medical histories showed no dose–response correlation between prevalence of colds, dyspnea, palpitation, or chest pain.

Of chest X-rays taken on 544 chlorine-exposed workers, fewer (21.3 percent) had abnormalities compared with those of controls (26.8 percent), which were mainly (75 percent) hilar or parenchymal calcifications. No significant dose–response correlation was found when chlorine exposure was related to ventilatory capacity, maximal ventilatory volume, or forced expiratory volume at 3 sec; thus there was no evidence of permanent lung damage from chlorine at the reported exposures.

However, of 329 electrocardiograms from 332 workers, 9.4 percent were abnormal compared to 8.5 percent in controls, a possibly significant difference. The incidence of fatigue was greater in those exposed above 0.5 ppm, but not below, and anxiety and dizziness showed a modest correlation ($p = .02$) with exposure level. Leukocytosis ($p < .05$) and lower hematocrit ($p < .017$) showed some relation to chlorine exposure.

No neoplasia or serious, acute pulmonary disease was reported. This is in accord with other studies which found no evidence of chlorine as a carcinogen or as a mutagen (37).

Another, more limited study of 52 Italian electrolytic cell workers with about the same average employment (10 years) but at twice the average chlorine exposure (0.298 ppm, SD 0.181) and reported in the same year (1970), found no clinically significant incapacity, even temporarily (41b). Using five tests of respiratory function, VC, FEV, RV, DL_{CO}, and He washout, only DL_{CO} showed a significant difference between exposed and control workers. This value was significantly lower in chlorine-exposed smokers than in nonexposed smokers ($p < .02$), lower in exposed smokers than in exposed nonsmokers, ($p < .04$) and lower in exposed smokers than in nonexposed nonsmokers ($p < .003$), indicating a slightly significant effect on respiratory function of chlorine over and above that from smoking after 10 years exposure at an average level around 0.3 ppm.

1.2.5 Inorganic Compounds

The inorganic compounds of chlorine are of three types, those formed as binary salts with metals, such as NaCl, $CaCl_2$, and $AlCl_3$, as ternary compounds with the alkali metals with oxygen, such as NaClO and $NaClO_4$, and as binary compounds with nonmetals, such as ClO_2 and ClF_3 (Table 40.4). Binary and ternary salts are represented in this table as Na salts, but chlorides of NH_4, K, Ba, and Al, for example, are common articles of commerce.

Hydrogen Chloride, HCl. *Sources, Production, Uses, and Industrial Exposure.* Hydrogen chloride gas is made by passing hot "burner" gases, (containing H_2) over anhydrous $CaCl_2$. The great solubility of the gas in water (Table 40.4) is the basis of hydrochloric acid, commercially called muriatic acid, an impure, aqueous solution of 38 percent HCl. It has a wide variety of uses for acidizing of petroleum wells, to "pickling" of metals and metal cleaning, to uses in food processing. Because of the strongly irritant nature of the gas, when in moist, humid air, and the acid, serious injuries to skin, eye, and lung are rare, except in accidental exposures from gas leaks or acid splashes.

Analytic Determination. HCl can be collected by passing contaminated air through either water or alkaline solution. Although several methods are available for analysis of HCl or the chloride ion, no single method is free from interference by other contaminants that might be present in the air.

In the absence of other acids in the air, HCl levels can be determined by collecting it in a known quantity of alkali solution and titrating the excess alkali.

However, serious errors can occur at low levels of HCl if there are also relatively high concentrations of CO_2 in the atmosphere. By collecting the HCl in a $1:1$ mixture of glycerol and water, those errors can be avoided. A small amount of silicone antifoamer will prevent foaming of the solution during sampling (47a).

Elkins (43b) describes two methods for the analysis of HCl in air: (1) The HCl is collected in 0.005 N NaOH, acidified with acetic acid, and titrated with 0.01 N AgNO$_3$, using chromate indicator (10 percent K_2CrO_4); (2) the HCl is collected in 0.005 N NaOH and acidified with 2 N HNO$_3$ (without shaking), 1 ml of 0.1 N AgNO$_3$ is added, and after 30 min the transmission of light through the solution is measured on a photometer; sensitivity to 1 ppm (1.5 mg/m^3) is reported for this method. Jacobs (47b) describes several methods of analysis for HCl and chlorides, one of which (Volhard method) is based on the reaction of thiocyanate with excess ferric ions to form a pink or red color that can be measured spectrophotometrically or organoleptically. A modification of this method involves the reaction of mercury thiocyanate. In this case the chloride ion releases thiocyanate, which reacts with ferric ions, forming the red complex, hexacyanatoferrate.

A direct method for measuring chloride ions in solution utilizes a chloride ion electrode. As in the case of glass pH electrodes, the chloride ion electrode measures electric potential across a layer of water-immiscible ion exchanger held in place by an inert porous membrane. The chloride ion electrode can be used for either individual determinations or continuous monitoring. A sensitivity down to 0.35 ppm is reported (47c).

The potential interference from other contaminants in any air sample may make it necessary to employ more than one method of analysis for the determination of HCl. The determination of HCl in the presence of chlorine is described in detail in Reference 47d.

Physiologic Response.   Human, male volunteers exposed to HCl gas found concentrations of 50 to 100 ppm barely tolerable for 1 hr, 35 ppm caused irritation of the throat on brief exposures, and 10 ppm was considered the maximal concentration acceptable for prolonged exposures (43a). Although immediately irritating to nose and throat at 5 ppm or above, concentrations below 5 ppm resulted in no lasting effects (43b).

When inhaled in high concentrations, the gas causes necrosis of the tracheal and bronchial epithelium as well as pulmonary edema, atelectasis, and emphysema, and damage to the pulmonary blood vessels. Detailed descriptions of the lesions in the liver and other organs have been given by Machle and co-workers (44). The insufflation of weak aqueous solutions of hydrochloric acid into the bronchi of rabbits sets up inflammatory processes resembling those seen in influenza or after exposure to certain chemical warfare agents (45).

Prolonged exposure at low concentrations causes erosion of the teeth. Mists of heated metal-pickling solutions may cause bleeding of the nose and gums, as well as ulceration of the nasal and oral mucosa, and render the skin of the face so tender that shaving becomes painful (46). Exposure of the skin to gaseous hydrogen chloride, escaping from leaks in apparatus or piping, has caused severe burns. Contact with concentrated solutions of hydrochloric acid in cleaning metal gives rise to small burns and ulcerations of the hands (5c).

Acute, lethal exposures of small laboratory animals for 30 min to HCl vapor range from 2640 to 4700 ppm, and for HCl mist, from 2142 to 5666 ppm (6).

No reports of prolonged exposure of workers appear to have been made, but exposures of 6 hr daily at 100 ppm repeated for 50 days caused only slight unrest and irritation of the eyes and nose of rabbits, guinea pigs, and pigeons (48). The hemoglobin content of the blood was only slightly diminished. Twenty exposures, each of 6 hr, at the concentration of 33 ppm caused no harm to a monkey or to smaller animals (44). Repeated exposure at higher concentrations resulted in a loss of weight that paralleled the severity of the exposure.

Threshold Limit. The TLV Committee of the American Conference of Governmental Industrial Hygienists assigned a ceiling value of 5 ppm to HCl gas (approximately 7 mg/m^3) in 1963, after it had been a TWA value since the early listings in the late 1940s. The TWA value permitted an undesirably large excursion for working comfort.

Binary Salts of the Alkali and Alkaline Earth Metals. The chief items of industrial concern are rock and common table salt, NaCl, and calcium chloride, CaCl$_2$. NaCl has multifold industrial uses, chief of which is the production of chlorine already mentioned, as well as the production of NaOH, HCl, and metallic Na, ceramic glazes, in hide curing, in salting out dyestuffs, and numerous others. The main uses of CaCl$_2$ are for deicing and dust control of roads, in mud drilling, dust- and freezeproofing, thawing coal, coke, stone, and sand, concrete conditioning, as a drying and desiccating agent, and many more.

The toxicity of NaCl is so low that only mild nasal irritation is experienced by drillers in salt mines, even when dust levels exceed the TLV for "nuisance" dusts of 10 mg/m^3 (author's note). The main systemic effect of excess NaCl is on blood pressure elevation; intake of NaCl from all sources, air, water, and food, must be carefully regulated in those with high blood pressure. The lowest toxic dose (TD$_{LO}$) for an adult man with normal blood pressure is reported to be 8.2 g/kg (6); in the rat, the oral LD$_{50}$ is 3.0 g/kg. The corresponding LD$_{50}$ for CaCl$_2$ is 1.0 g/kg (6).

Ternary Salts of Alkali Metals Containing Oxygen. Sodium and calcium hypochlorites, sodium chlorate, and perchlorate are the chief items of industrial manufacture in this group, with the chlorate and chlorite having the largest number of commercial outlets, 23 and 17, respectively, compared with three for the perchlorate (4). Chlorinated lime, approximately Ca(ClO)$_2$, and NaClO are the bleaching agents used on an industrial scale. The main use of $NaClO_3$ is as an oxidizing agent and bleach, especially to make chlorine dioxide (see below) for paper pulp; other uses include ore processing, in matches, in explosives, flares, and pyrotechnics, and in recovery of bromine from natural brines (Section 3). $NaClO_4$, a substance with still greater oxidizing capacity, is used in explosives and in jet fuel.

No specific toxicity data on NaClO have been developed because of its unstable nature in aqueous solutions, but pulmonary edema can occur from chlorine vapors developing after mixing clorox, a 5.25 percent solution of NaClO, and Saniflush (80 percent $NaHSO_4$) (48a, b). The toxicity of $NaClO_3$ and $NaClO_4$ stems from their strong oxidizing reaction on body tissues, particularly their destructive action on red blood cells. The lowest oral toxic dose (TD_{LO}) reported for this action in a woman is 800 mg/kg; the rat LD_{50} is 1.2 g/kg (6). The only toxicity information on $NaClO_4$ is an intraperitoneal LD_{50} for the mouse of 551 mg/kg (6).

Nonmetallic, Inorganic, Chlorine Compounds.   Of this group, two are industrially useful compounds, chlorine dioxide, ClO_2, chlorine trifluoride, ClF_3, and possibly, chlorine pentafluoride, ClF_5. Chlorine dioxide, a strong oxidizer, is used as a bleaching agent for wood pulp, fats, oils, and flour, and in other situations where a strong bleaching agent is needed. Chlorine trifluoride, also an extremely reactive substance comparable to fluorine, explodes in contact with organic materials, and is used as a fluorinating agent and oxidizer in propellants. Chlorine pentafluoride, whose properties are not shown in Table 40.4, is very unstable at ordinary pressures and temperatures, has a lower melting point and a higher vapor pressure than ClF_3 (49a), and has been investigated as a candidate jet propellant aid (49b).

Administration of microliter amounts of ClF_5 to four species of laboratory animals caused traumatic explosions and death in most animals and, in those that survived, inhibition or absence of enzyme (SGOT) activity, and alterations of protein (macroglobulins) structure immediately after topical, parenteral, or dermal application at or near the site of administration. Dermal application of 10 µl cause severe irritation and massive destruction of the skin. Inhalation of ClF_5 for brief periods showed graded responses; no rats survived longer than 10 min at 400 ppm; 30 percent, exposed at 200 ppm, survived for 24 hr; and almost all survived exposure at 100 ppm for 15 min daily for 5 days, but lost weight by the end of that period. No pharmacological action was noted, however, at tolerated levels from single or multiple doses; where effects were noted, they were attributed to rapid evolution of energy from reaction of ClF_5 with body tissues and fluids (49b).

Chlorine dioxide has the toxic effects of an acute severe respiratory irritant. Elkins found that a concentration of 5 ppm is definitely irritating and that 19 ppm of the gas was more than sufficient to cause the death of a worker inside a bleach tank (43b). Bronchitis and emphysema also developed in a chemist repeatedly exposed for several years to ClO_2; symptoms of increasing dyspnea and asthmatic bronchitis were evident even without further exposure (50).

Clinical investigations have been reported (51a) on a small group of men who showed respiratory symptoms after having worked in a sulfite-cellulose plant in Sweden in the mid 1950s, where spot samples indicated that exposures to ClO_2 and Cl_2 were normally each below 0.1 ppm, with low exposures on rare occasions to SO_2. However, higher concentrations did occur from technical faults such as leakage, and these were the cases investigated for possible sequelae.

Bronchoscopy with biopsy revealed slight chronic bronchitis in seven of 12 workers and physical signs were absent in all but two who had experienced acute exposure shortly before examination. In one case, an earlier-observed bronchitis disappeared, indicating regression can occur. Complaints of irritation of the eyes and respiratory organs, and in certain cases of the gastrointestinal tract, indicated that exposure to ClO_2 is worse than to other respiratory irritant gases, but symptoms from the central nervous system noted by previous clinicians were not observed in this investigation. It was my impression that the observed irritant responses resulted from the unusual exposures above 0.1 ppm (51a).

No significant differences were found by Ferris et al. (51b) in respiratory symptoms or in ventilatory functions in a sample of 147 pulp mill workers exposed at times to ClO_2, Cl_2, and SO_2 compared with 124 paper mill workers where the average exposure years were but one-sixth of those in the pulp mill, but those exposed to chlorine had poorer respiratory function and shortness of breath.

A TLV of 0.1 ppm for ClO_2 was adopted by the American Conference of Governmental Industrial Hygienists in 1960, based chiefly on the five-year study of Gloemme and Lundgren of workers in a sulfite-cellulose plant (51a).

Chlorine trifluoride is a highly reactive gas and acts as a fluorinating and oxidizing agent. Despite its instability in moist air to form varying proportions of hydrolysis products, ClF, Cl_2, $ClOF$, ClO_2F, ClO_3F, ClO_2, and HF, the respiratory damage is considered to result mainly from the gas itself with Cl_2, ClO_2, and HF contributing to the damage to the degree associated with the rate and depth of breathing (52a).

Inhalation exposure of rats at 800 ppm for 15 min was uniformly lethal, but 13 min resulted in survival. Exposure at 400 ppm for 35 min was usually lethal. Chlorine trifluoride, given intraperitoneally at doses higher than by inhalation, was not lethal, indicating that the action of ClF_3 on the respiratory tract is the main damaging event. The toxic effect of ClO_3F was ruled out on the basis that no methemoglobin formed during ClF_3 exposures, and its lethal effects exceeded those by ClF_3. Chlorine trifluoride exposures caused severe mucosal inflammation and burning of the skin, followed by corneal ulceration. Pulmonary release of $^{14}CO_2$ from injected ^{14}C bicarbonate decreased after ClF_3 inhalation, as did blood pH, indicating pulmonary impairment. Chlorine trifluoride lethality was considered by the authors to be comparable to that of HF on an F-equivalent basis. Unfortunately, the exposure resulting in no demonstrable acute responses was not determined (52a).

Chronic exposures of ClF_3 performed on two dogs and 20 rats at an average concentration of 1.17 ppm for 6 hr daily, 5 days/week for a period of 6 months resulted in coughing, sneezing, rhinorrhea, and lacrimation within a few hours after exposure, and became progressively worse upon each reexposure. Six rats and one dog died during the course of the study (52b). Unpublished data on the effects of ClF_3 on the skin suggest that the concentration tolerable by inhalation would not produce burns or ulceration.

The TLV Committee recommended in 1956, and the American Conference of Governmental Industrial Hygienists adopted, a TLV for ClF_3 of 0.1 ppm (approximately 0.4 mg/m^3), the same value as for ClO_2, subsequently found to have the same lethality as ClF_3 on a Cl-equivalent basis (52a).

CHLORINE REFERENCES

34. R. T. MacMillan, "Chlorine," in *Bur. Mines Bull. 650*, U.S. Dept. Interior, Washington, D.C., 1970.

35. (a) *North American Chlor-alkali Industry Plants & Production Data Book*, CI pamphlet 10. The Chlorine Institute, New York, January 1976; (b) D. M. Kiefer, *Chem. Eng. News*, **55**, 25 (1975).

36. "Mineral Facts and Problems," *Bur. Mines, Bull. 667*, U.S. Dept. Interior, Washington, D.C., 1975.

37. "Criteria for a Recommended Standard for Occupational Exposure to Chlorine," *HEW Pub. No. (NIOSH) 76-170*, May 1976.

38. "Tentative Method of Analysis for Free Chlorine Content of the Atmosphere (Methyl Orange Method)," *Health Lab. Sci.*, **8**, 53 (1971).

39. H. L. Gilchrist and P. B. Matz, *Med. Bull. Vet. Admin.*, **9**, 229 (1933).

40. M. C. Winternitz et al., *Collected Studies on the Pathology of War Gas Poisoning*, Yale University Press, New Haven, Conn., 1920.

41. (a) L. R. S. Patil, R. G. Smith, A. J. Vorwald, and T. F. Mooney, *Am. Ind. Hyg. Assoc. J.*, **31**, 678 (1970); (b) E. Capodaglio et al., *Med. Lav.*, **60**, 192 (1970).

42. E. C. White, *J. Am. Chem. Soc.*, **50**, 2148 (1928).

43. (a) Y. Henderson and H. W. Haggard, *Noxious Gases* Reinhold, New York 1943, p. 126; (b) H. B. Elkins, *The Chemistry of Industrial Toxicology*, Wiley, New York, 1959, p. 79.

44. W. Machle et al., *J. Ind. Hyg. Toxicol.*, **24**, 222 (1942).

45. M. C. Winternitz et al., *J. Exp. Med.*, **32**, 199, 205 (1920).

46. (a) W. Ludewig, *Arch. Gewerbepath, Gewerbehyg.*, **11**, 296 (1942); (b) L. Carozzi, *Occupation and Health*, International Labor Office, Geneva, 1930.

47. (a) F. Miller et al., *Am. Ind. Hyg. Assoc. Q.*, **17**, 221 (1956); (b) M. B. Jacobs, *The Analytic Chemistry of Industrial Poisons*, 2nd ed. Interscience, New York, 1949, pp. 378–382; (c) Orion Research Laboratories, Cambridge, Mass. Data Sheet: Model 92-17 (Liquid Ion Exchange). Chloride Ion Activity Electrode. 1967; (d) I. Iwasaki et al., *Bull. Chem. Soc. Jap.*, **29**, 860 (1956).

48. (a) F. L. Jones, Jr., *J. Am. Med. Assoc.*, **222**, 1312 (1972); (b) H. C. Faigel, *N. Engl. J. Med.*, **271**, 618 (1964).

49. (a) D. F. Smith, *Science*, **141**, 1039 (1963); (b) M. S. Weinberg and R. E. Goldhamer, *Pharmacology, Toxicology and Metabolism of ClF$_5$*, AMRL-TR-66-238, Wright-Patterson AFB, Ohio, 1966.

50. H. Petry, *Arch. Gewerbepath. Gewerbehyg.*, **13**, 363 (1954).

51. (a) J. Gloemme and K. D. Lundgren, *Arch. Ind. Health*, **16**, 169 (1957); (b) B. G. Ferris et al., *Br. J. Ind. Med.*, **24**, 26 (1964).

52. (a) F. N. Dost et al., *Toxicol. Appl. Pharm.*, **27**, 527 (1974); (b) H. J. Horn and R. J. Weir, *Arch. Ind. Health*, **12**, 515 (1955).

1.3 Bromine Br$_2$

Bromine, a dark red, low-boiling liquid, is the only nonmetallic element that is liquid at normal temperatures and pressures (Table 40.1). The most stable valence states of bromine in its salts are 1 − and 5 +, although 1 +, 3 +, and 7 + are known. Within wide limits of temperature and pressure, molecules of the liquid and vapor are diatomic. The atomic number is 35, and there are two stable isotopes (^{79}Br and ^{81}Br) that occur naturally in nearly equal proportion. A number of radioisotopes are also known.

1.3.1 Source and Production

Although it is estimated that from 10^{15} to 10^{16} tons of bromine is contained in the earth's crust, the element is widely distributed and found only in low concentrations in the form of its salts. The bulk of the recoverable bromine, however, is found in the hydrosphere. Seawater contains an average of 65 ppm, equal to 308,000 tons Br$_2$ per cubic mile of ocean. At this concentration, 15,000 tons of seawater must be processed to obtain 1 ton Br$_2$. With only a few known exceptions (terrestrial molds), the sea is the main source of natural organic bromine compounds, of which Tyrian purple (6,6'-dibromindigo) is a classic example.

Recovery of bromine from brines and from seawater involves the oxidation of the bromide ions in solution to free elemental bromine, which is then varporized from the solution either by air or by stream. Although oxidation was accomplished at one time by electrolysis, all current processes utilize chlorine.

With seawater, bromine must first be concentrated before the steaming-out process becomes economically practical. Consequently, air is used to vaporize the bromine from chlorinated seawater, and then sulfur dioxide is introduced into the dilute bromine-laden air. Subsequent absorption of the hydrogen bromide in a controlled amount of water produces a much more concentrated bromide solution, according to the equation:

$$SO_2 + Br_2 + 2\,H_2O \rightarrow 2\,HBr + H_2SO_4$$

Bromine must then be released again by chlorine and steamed from the solution. The by-product hydrochloric and sulfuric acids are recycled to acidify the incoming seawater to the pH necessary for efficient chlorination.

The air-blowing process can also be used when an alkali bromide is the desired end product. Absorption of the bromine from the air stream by sodium carbonate, for instance, produces sodium bromide and sodium bromate, as shown:

$$3 \, Br_2 + 3 \, Na_2CO_3 \rightarrow 5 \, NaBr + NaBrO_3 + 3 \, CO_2$$

The sodium bromate may then be crystallized from solution or reduced with iron, if NaBr is the only desired product. One process utilizes NaBr solution as the absorbing medium for bromine from bromine-laden air.

Typical, freshly prepared bromine from a modern plant is likely to be at least 99.9 percent pure. Probable impurities are chlorine moisture and organic material at levels of less than 50 ppm each. Specifications of the U.S. Pharmacopoeia and of the American Chemical Society for reagent-grade bromine allow up to 0.3 percent Cl_2, 0.05 percent I_2, 0.002 percent S, and nonvolatile matter 0.015 percent maximum.

World bromine production in 1973 amounted to 617.2 million 1b, with an estimated production in 1980 of 776 million 1b, the United States producing an estimated 65 percent, the United Kingdom, 11.6 percent, Japan, 6.5 percent, and the Soviet Union, Israel, and France each about 4 to 4.5 percent (36).

1.3.2   Uses and Industrial Exposures (36, 53)

The largest single industrial use of Br_2 is for ethylene dibromide, a gasoline additive that functions as a scavenger for the lead (tetraethyl- or tetramethyllead) added to gasoline to reduce "knocking." In the United States the portion of bromine production going to the manufacture of ethylene dibromide has been as high as 71 percent in 1970, but was reduced to 66 percent in 1974, and is continually being reduced further as leadfree gasoline is replacing leaded gas.

Use in fire retardants has become the second largest outlet for bromine. With greater emphasis on safety, there is an increasing demand for flameproof or self-extinguishing fabrics and plastics. Growth has been rapid, and no letup in demand is evident. In 1974 possibly 15 percent of the bromine went to flameproofing and fire retardants.

Sanitation preparations in 1974 accounted for an estimated 10 percent of bromine demand. A large part of this usage was elemental bromine added directly to water in swimming pools, cooling towers, and the like for control of bacteria, algae, and odors.

Agricultural chemicals utilize considerable but undisclosed amounts. Miscellaneous uses for bromine together account for about 8 percent of overall demand. Among the products containing bromine or requiring it as an intermediate in their manufacture are photographic film and paper, dyes, inks, sedatives, anesthetics, antispasmodic agents, hydraulic fluids, refrigerating and dehumidifying agents, hair-waving preparations, and laboratory reagents.

In some countries in which the automobile industry is not well developed, the major uses for bromine are for disinfecting agents and fire extinguishants. Production facilities where bromine is manufactured or used should be designed to rapidly dispose of liquid bromine spills. Solutions or slurries of 10 to 50 percent potassium carbonate, 10 to 13 percent sodium carbonate, and 5 to 10 percent sodium bicarbonate or saturated "hypo" solution (prepared by dissolving 4 kg of technical-grade sodium thiosulfate in 9.5 liters of water and adding 113 g of soda ash) are preferred neutralizing agents for liquid bromine spills (4). A 5 percent lime slurry or a 5 percent sodium hydroxide solution may be used, but heats of reaction are higher for these reagents. Ammonia solutions should not be applied to liquid spills because of the high heat of reaction and nitrogen evolution. Anhydrous ammonia gas is useful for neutralization of bromine fumes.

Full body protection constructed of resistant materials should be worn when handling bromine in significant quantities. Major manufacturers of bromine furnish information on the safe handling procedures for bromine.

Because liquid bromine and its vapors are highly corrosive and irritating to skin and mucous membranes of eyes, nose, throat, and respiratory tract, spills of liquid bromine or leaks of its vapors constitute the major industrial exposures.

1.3.3 Analytic Determination

The determination of bromine and inorganic bromides in air may be determined by passing a measured volume through a solution of potassium iodide and titrating the liberated iodine with a standard solution of sodium thiosulfate (54). It also may be absorbed in alkali, oxidized to bromate by sodium hypochlorite, and estimated iodometrically (55). After absorption in alkali, bromine may also be liberated by chlorine water and determined colorimetrically, either in aqueous solution or after extraction with carbon tetrachloride (56).

1.3.4 Physiologic Response

Information on the acute toxic response of man to bromine vapor is just sufficient to gain an impression of the limiting exposures which are fatal, produce dangerous consequences, or are acceptable for the workplace; corresponding chronic exposure data are nonexistent. Henderson and Haggard (43a) state that 1000 ppm (approximately 7000 mg/m^3) is rapidly fatal; 40 to 60 ppm is dangerous for brief exposures; 4 ppm is the maximal allowable concentration for $\frac{1}{2}$– to 1 hr; and 0.1 to 0.15 is the maximal allowable for prolonged exposure.

The symptoms arising in man following the inhalation of bromine in small amounts include coughing, nosebleed, a feeling of oppression, dizziness, and headache, followed after some hours by abdominal pain and diarrhea, and sometimes by a measleslike eruption on the trunk and extremities.

Oppenheim (57) mentioned the frequency with which discharging pustules and furuncles appear in exposed areas of the skin of those who handle bromine. Brief contact of the liquid with the skin leads to the formation of vesicles and pustules. If not removed at once, it induces deep, painful ulcers.

Necropsy of guinea pigs and rabbits following a 3-hr exposure to bromine at 300 ppm revealed the presence of edema of the lungs, a pseudomembranous deposit on the trachea and bronchi, and hemorrhages of the gastric mucosa. Foci of bronchopneumonia were found in animals that died several days after exposure, and there was evidence of functional disturbances in the central nervous system (58).

Treatment. For the treatment of bromine burns to the skin, the bromine should be removed from the skin as soon as possible with aqueous sodium bicarbonate. For relief of pulmonary edema from overexposure to bromine vapor, the primary treatment is bed rest and oxygen (59).

Limiting Values. Following a review of available human and animal data, the TLV Committee of the American Conference of Governmental Industrial Hygienists lowered the TLV for bromine vapor in 1959 to 0.1 ppm from 1 ppm, where it had been for years among the early listings.

The lowest concentration that has a detectable odor is about 3.5 ppm, according to Fieldner et al. (60); however, in view of the fact that lacrimation may result from prolonged exposure at concentrations below 3.5 ppm, the odor-warning properties may be of little value.

1.3.5 Inorganic Compounds

The inorganic compounds of industrial interest with their physical properties are listed in Table 40.5. By comparison with those of chlorine, their number is considerably smaller, and in tonnage production, somewhat less than one-fiftieth that of chlorine.

Hydrogen Bromide, HBr (Hydrobromic Acid). Gaseous HBr and aqueous hydrobromic acid, usually as the constant-boiling solution (48 percent HBr), are the major intermediates in introducing bromine into a molecule. The principal means of producing HBr is by reacting H_2 and Br_2 at high temperatures in the absence of air, although HBr is formed also as a by-product of an

Table 40.5. Properties of Industrially Useful Inorganic Bromine Compounds

Compound	Formula	Mol. Wt.	Physical Properties	Sp. Gr.	M.P. (°C)	B.P. (°C)	Solubility
Hydrogen bromide (hydrobromic acid)	HBr	80.92	Colorless gas, pale yellow liq.	Gas 3.5°; Liq. 2.77^{-67}	-88.5	-67.0	2.21 kg/l H_2O 0°C, 1.30 kg/l H_2O 100°C
Sodium bromide	NaBr	102.90	Colorless cubic, hygr.	$3.2(3^{25}$	747	1390	1.16 kg/l H_2O 50°C, 1.21 kg/l H_2O 100°C; sl. sol. alcohol
Ammonium bromide	NH_4Br	97.95	Cubic, hygr.	2.429	Subl. 452	235 vac.	970 g/l H_2O 25°C, 1.456 kg/l H_2O 100°C, 100 g/l alcohol; 78°C sol. acetone, ether
Potassium bromate	$KBrO_3$	167.01	Colorless trig.	$3.27^{17.4}$	434d370	—	133 g/l H_2O 40°C, 4.975 kg/l H_2O 100°C; sl. sol. alcohol; insol. acetone
Bromine chloride	BrCl	115.36	Red liq., gas	—	~66 Dec. 10	~5	Sol. cold H_2O (dec.); sol. ether CS_2
Bromine trifluoride	BrF_3	136.90	Gray-yellow liq.	2.49^{135}	$(-2)8.8$	135	Dec. viol. to O_2, cold H_2O; dec. alkali
Bromine pentafluoride	BrF_5	174.90	Colorless liq.	2.466^{25}	-61.3	40.5	Dec. cold H_2O

organic substitution reaction. Liquefied hydrogen bromide is available commercially in cylinders, and various strengths of aqueous hydrobromic acid, usually 48 percent, are available in carboys, drums, or tank cars.

Human response to HBr vapors appears to be limited to one report (61). Subjective responses of six volunteers exposed at unit HBr concentrations from 2 to 6 ppm for several minutes indicated that 5 and 6 ppm cause nasal irritation in all individuals and throat irritation in only one of six, with eye irritation negative at all concentrations. The breakpoint in nose and throat irritation occurred at 3 and 4 ppm, where one of six at 3, and three of six at 4 ppm experienced positive effects. Detectable odor was registered at all levels.

One-hour rat and mouse LC_{50} values for HBr vapor are 2858 and 814 ppm, respectively (6).

Based on the Connecticut Health Department study, a TLV of 3 ppm for HBr vapor was recommended by the TLV Committee in 1956.

Metal Salts, Bromides and Bromates.   Physiological effects of the soluble *bromides*, Na, K, and NH_4, are similar, and are all attributable to the bromide ion. Most of the available data comes from their medical use as oral sedatives, diuretics, and antiepileptics. An oral dose of 3 g/kg (30 to 60 mg/kg adult), causing the blood bromide to rise to 50 mg percent, was considered a "no-ill effect" level (62), and typical signs of bromide intoxication may be absent when blood levels attain 200 mg percent or more (59). Acute overdoses may cause vomiting or profound stupor, and chronic, prolonged use often leads to depression, ataxia, and psychoses, especially with low sodium intake. Bromo Seltzer contains about 320 mg NaBr and 160 mg acetanilide.

The oral lethal dose of NaBr for man is high, if judged from the rat LD_{50} of 3500 mg/kg (6).

Metabolically, bromide has a biologic half-life of about 12 days, is not incorporated into fat or blood proteins, and none is extractable from plasma or hemolyzed blood cells by ether (63a, b). Nor does the bromide ion interfere with thyroid activity even at large daily doses for extended periods of time (64a, b).

The *bromates* of chief industrial interest are those of sodium and potassium, although those of the alkaline earth metals, calcium and barium, have limited use as oxidizers, maturing agents in flour, and analytic reagents, uses similar to those of sodium and potassium.

The scant toxicologic information that exists has been developed for in-house use by the Dow Chemical Company for the sodium and potassium salts only (65a). The oral rat LD_{50} of $KBrO_3$ is about 400 mg/kg, indicating a moderate acute toxicity by this route. When introduced into the rabbit eye, $KBrO_3$ produces mild pain, moderate conjunctival irritation, and a very slight, transitory corneal injury. Healing occurs in 24 to 48 hr, and is complete. When confined to intact rabbit skin for short periods, it produces mild irritation, and for repeated, prolonged periods, superficial burns.

Skin penetration of $NaBrO_3$ is slight as judged by blood bromide levels of somewhat less than 1 mg percent from a 7-hr moistened application of the salt to rabbit's skin.

Potassium bromate dust is irritating to the mucous membranes of the upper respiratory tract, but there was no evidence of any cumulative effects on the lung of workers after 18 years of exposure at past "nuisance" dust levels up to 15 mg/m^3 (65a).

Analytic determination of these soluble bromates is made by collection in water and applying the Vandermullen procedure for bromide (65a) or the bromide ion electrode (65b).

Bromine Halides.   The commercially available bromine halides are $BrCl$, BrF_3, and BrF_5. Their physical properties, given in Table 40.5, indicate high chemical reactivity. Bromine chloride finds use in both addition and substitution reactions in organic synthesis to form bromochloro compounds by addition and organic bromides by substitution; it is superior to chlorine as a disinfectant in wastewater treatment, for the bromamines formed are less stable and less toxic to fish than are the chloramines.

The tri- and pentafluorides are strong fluorinating agents, useful in organic synthesis and in forming uranium fluorides both for isotopic enrichment and for full element reprocessing. With alkali fluorides they form double salts; $MBrF_4$, $MBrF_6$, and various quaternary ammonium compounds are used pharmaceutically as bromides because of their suitable solubility, crystalline character, and ready preparation (53).

No toxicity data on any of the bromine halides appear to have reached the open literature, but it is obvious from their high reactivity that they are highly toxic and corrosive to biologic tissues.

REFERENCES

53. Kirk-Othmer *Encyclopedia of Chemical Technology*, 3rd ed., Wiley-Interscience, New York, 1978.
54. F. H. Goldman and J. M. DallaValle, *Public Health Rep. U.S.*, **54**, 1728 (1930).
55. I. M. Kolthoff and H. Yutzy, *Ind. Eng. Chem. Anal. Ed.*, **9**, 75 (1937).
56. M. Lane, *Ind. Eng. Chem. Anal. Ed.*, **14**, 149 (1942).
57. M. Oppenheim, *Wien Klin. Wochenschr.*, **28**, 1273 (1915).
58. K. B. Lehmann and R. Hess, *Arch. Hyg.*, **7**, 335 (1887).
59. Deichmann and Gerarde, *Toxicity of Drugs and Chemicals*, Academic Press, New York, 1969.
60. A. C. Fieldner et al., *U.S. Bur. Mines Tech. Paper*, 248 (1921).
61. Connecticut State Department of Health, Hartford, unpublished data, 1955.

62. F. B. Flinn, *J. Lab. Clin. Med.*, **26**, 1325 (1941).

63. (a) R. Söremark, *Acta Physiol. Scand.*, **50**, 119 (1960); (b) R. Söremark, *Acta Physiol. Scand.*, **50**, 306 (1960).

64. (a) R. R. Grayson, *Am. J. Med.*, **28**, 397 (1960); (b) I. Rosenblum et al., *Arch. Pathol.*, **75**, 591 (1963).

65. (a) Personal communication from V. K. Rowe, Dow Chemical Co. Midland, Mich., February 9, 1970; (b) F. Poser, *Z. Klin. Chem. Biochem*, **12**, 350 (1974).

1.4 Iodine, I_2

Iodine is the heaviest of the halogens that are of industrial interest. Under ordinary conditions, it exists as a gray-black solid with a metallic crystalline luster. Heating yields a sublimed, violet vapor. Its physical properties are given in Table 40.1.

Although iodine resembles other members of the halogens, it is the least electron negative; hence it is the least chemically reactive and forms the weakest bonds with the more electropositive elements. Like other halogens, iodine unites with all elements except S, Se, and the noble gases, reacts directly with most, except C, N, O, and some of the more unreactive metals, forms compounds with the other halogens that are of some industrial interest (see Table 40.6 and below), and reacts with numerous organic compounds that are of pharmaceutic interest. The most stable of the positive iodine compounds are the iodates, in which the valence of iodine is $5+$, and the periodates, in which it is $7+$.

1.4.1 Source and Production (36)

The major U.S. sources of iodine are natural and oil field brines, available in brines near Shreveport, Louisiana (oil field), in the Los Angeles basin of California, and in the natural brines of Midland, Michigan, and Woodward, Oklahoma.

Large resources of iodine exist in foreign countries. Japan's natural gas-well brines are credited with as much as four-fifths of the world's iodine reserve. An estimate of the indicated reserve of Chilean nitrate minerals, from which iodine is obtained as a by-product, is about 1 billion short tons, analyzing approximately 0.04 percent I_2. Unmeasured quantities of iodine are contained in brines in Indonesia, West Germany, France, Italy, the United Kingdom, Norway, Ireland, and the Soviet Union. Reserves in the Soviet Union might be rather extensive. The vast and self-replenishing iodine content of seaweed is excluded from this assessment (36).

In the United States the principal method of recovery from oil brines involves oxidation of *iodide* to free, elemental iodine by chlorine and removal of the volatile iodine from solution with an air stream. The iodine is reabsorbed in solution and reduced to HI with SO_2. The solution is then chlorinated to precipitate free iodine, which is further purified by treatment with concentrated sulfuric acid. The same process is used to recover I_2 from natural brines, which also contain Br_2, Ca, Mg, and K.

In the recovery of iodine from Chilean nitrate deposits, solutions containing the *iodates* are reduced with sodium bisulfite, precipitating the iodine, which is purified by sublimation.

The U.S. output in 1974 approximated 600,000 lb of a total world output of about 26 million lb, of which Japan contributed the major part, 16.5 million lb, or about three times the output of Chile, and almost six times that of the Soviet Union.

1.4.2 Uses and Industrial Exposures

Iodine consumed in the United States for manufacturing immediate downstream products, such as resublimed iodine, potassium iodide, sodium iodide, and organic iodine-containing compounds, has not shown radical changes in recent years. In 1973, about 2,500,000 lb iodine went into organic compounds, 1,500,000 lb into potassium iodide, 700,000 lb into resublimed iodine, 1,000,000 lb into various other inorganic compounds, and possibly 1,500,000 lb into intermediates for manufacture of catalysts.

Downstream uses for iodine, catalysts and inks (and colorants), have shown substantial increases during the last five years; pharmaceuticals and stabilizers have held their own, and food products (mainly for cattle feed) have declined noticeably. A million pounds or more of iodine is generally consumed annually in each of these five categories.

No reports have appeared showing that the production and handling of iodine have incurred any unusual or serious injuries to workers.

1.4.3 Analytic Determination

Fritted-glass scrubbers have been recommended for the absorption of iodine from air samples. Standard volumetric procedures are available for its determination by titration with sodium thiosulfate, using starch–iodide indicator.

1.4.4 Physiologic Response

Iodine is an essential element in nutrition, being required by the thyroid for the elaboration of its hormone, thyroxin. Because of its use in medicine in the treatment of hypothyroidism and other diseases, and because of the addition of iodides to salt for the prevention of endemic goiter, the pharmacology and metabolism of iodine have been the subjects of many investigations, which are described in texts on pharmacology and in reviews (66). Industrial poisoning by iodine is uncommon.

Iodine vapor is more irritant and corrosive than either bromine or chlorine according to Matt, cited by Flury and Zernik (67), and pulmonary edema has been observed in dogs exposed to the vapor. Systemic poisoning may result from its absorption from the respiratory tract. The lethal oral dose for man is between 2000 and 3000 mg/kg (68).

Lacrimation and a burning sensation in the eyes, blepharitis, rhinitis, catarrhal stomatitis, and chronic pharyngitis have been noted following industrial exposures, and excessive exposures result in pulmonary conditions resembling those from chlorine gassing (69).

Human response studies found that a concentration of 0.57 ppm was tolerated for 5 min without eye irritation, which was experienced in 2 min at 1.63 ppm (70a). Determinations made in a Massachusetts plant (70b) showed concentrations ranging from 0.07 ppm at the nearest work area, to 1 ppm directly over the tank containing the iodine solution; exposure at 0.07 ppm caused no complaints, but concentrations of 1 ppm were highly irritating.

Average oral doses may cause skin lesions, and prolonged oral therapy may produce iodism. Application of strong iodine tincture produces an acute dermatitis. The antidote is sodium thiosulfate or cornstarch (68).

The average daily urinary excretion of iodine by five normal German volunteers was 173 µg (71). Administered iodine is rapidly excreted in the urine and in smaller quantities in saliva, milk, sweat, bile, and other secretions. The storage of iodine in the thyroid depends upon the functional state of the gland.

Threshold Limit. The American Conference of Governmental Industrial Hygienists in 1961 adopted 0.1 ppm (approximately 1 mg/m^3) as the threshold concentration for iodine. On information later submitted to the TLV Committee, 0.1 ppm was recommended as a ceiling value.

1.4.5 Inorganic Compounds

The inorganic compounds available commercially with their physical properties are listed in Table 40.6. By comparison with bromine their number is considerably larger and somewhat more diversified, but in world production, iodine is but one-thirtieth that of bromine (1974). Iodine forms industrially useful and important compounds with hydrogen, metals, the other halogens, and oxygen. The ones presented below are typical.

Hydrogen Iodide, HI (Hydriodic Acid). Hydrogen iodide gas dissolves in water at 10°C and 1 atm pressure to the extent of 70 percent by weight to form hydriodic acid. The technical grade, 47 percent HI, is a highly corrosive liquid which fumes in moist air. Its solutions, like others containing the iodide ion, can dissolve large quantities of I_2 (e.g., tincture of I_2, KI·I_2). The HI acid is more stable than the gas and is one of the strongest acids; it dissolves metals, oxides, carbonates, and salts of other weak, nonoxidizing acids with the formation of iodides. Thus among its uses is the preparation of iodine salts, organic iodides and pharmaceuticals, disinfectants, and analytic reagent.

No toxicity data appear to have been reported on either the HI gas or acid, and no TLV has been recommended.

Iodides. Of the commercially available iodides, KI is the most important as judged by its 26 outlets, compared with 14 for NaI and four for NH_4I (4). The uses of KI range from photography, where it precipitates silver, to dietary supplement in table salt, to feed additive and analytic reagent. All are highly water soluble, stable, and high melting solids (Table 40.6), and have very low oral toxicities. The oral, mouse LD_{LO} for KI is 1862 mg/kg and 4340 mg/kg for NaI for the rat oral LD_{50} (6). The tetraiodides of titanium and zirconium, however, decompose to their elements at elevated temperatures, yielding very high purity zirconium metal by this procedure.

Iodates and Periodates. These are the most stable and well known of the iodine compounds in which iodine is positive. Except for the salts of the alkali metals, most iodates are sparingly soluble in water. Both iodates and periodates are powerful oxidizers in acid solution and are thus used as disinfectants. Other uses are as feed additives and in medicine.

From what meager information is available (6), the acute toxicity of the iodates is of the order of from three to four times that of the corresponding iodides.

Iodine Halides. The four iodine halides of interest are listed in Table 40.6 with their physical properties. It is noted that the monohalides IBr and ICl are colored crystalline compounds, whereas the polyhalides IBr_3 and IF_5 are colored liquids at ordinary temperatures. All are high density substances.

The chief use of these halides is in organic synthesis and as halogenation catalysts. Iodine pentafluoride is a fluorinating and incendiary agent. As a fluorinating agent, reactions with organic substances must be carefully controlled because these substances carbonize on contact, sometimes explosively! When hot, IF_5 attacks glass.

Toxicity data are almost nonexistent on these halides; only ICl has been tested for its acute effects in the rat (6). The oral, rat LD_{LO} is 50 mg/kg; the corresponding skin LD_{LO} is ten times this value. No information could be found on the toxicity of any of the halides by inhalation. In general, however, the toxicities appear to resemble that of their individual constituents, a statement based on their chemical reactivity with organic compounds.

Organic Compounds. There are two important classes: the organic iodides, as represented by methyl iodide (CH_3I) and iodobenzene (C_6H_5I), and those

TABLE 40.6. Properties of Industrially Useful Inorganic Iodine Compounds

Compound	Formula	Mol. Wt.	Physical Properties	Sp. Gr.	M.P. (°C)	B.P. (°C)	Solubility
Hydrogen iodide (Hydriodic acid)	HI	127.91	Colorless gas	Gas 5.66^0 g/l	-50.8	-35.88^4	425 cm^3/l H$_2$O 0°C, sol. alcohol
Ammonium iodide	NH$_4$I	144.94	Colorless, cubic, hygr.	2.514^{25}	Subl. 551	220vac.	1.542 kg/l H$_2$O 0°C, 2.503 kg/l H$_2$O 100°C, very sol. alcohol acetone, NH$_3$
Potassium iodide	KI	166.01	Colorless or white, cubic	3.13	681	1330	1.275 kg/l H$_2$O 0°C, 2.08 kg/l H$_2$O 100°C, 18.8 g/l alocohol 25°C; sl. sol. ether
Sodium iodide	NaI	149.89	Colorless cubic	3.667^{25}	661	1304	1.84 kg/l H$_2$O 25°C, 3.02 kg/l H$_2$O 100°C, 425 g/l 25°C alcohol
Iodic acid	HIO$_3$	175.91	Colorless or pale yellow, cryst.	4.629^0	Dec. 110	—	2.8 kg/l H$_2$O 0°C, 4.73 kg/l H$_2$O 80°C, very sol. acohol 87%

Name	Formula	Mol. wt.	Color, form	Density	M.p. °C	B.p. °C	Solubility
Potassium iodate	KIO_3	214.0	Colorless, monoclinic	3.9_3^{32}	560	Dec. >100	47.4 g/l H_2O 0°C, 323 g/l H_2O 100°C, sol. KI; insol. alcohol acid
Sodium iodate	$NaIO_3$	197.89	White, rhombic	$4.27_7^{17.5}$	Dec.	—	90 g/l H_2O 20°C, 340 g/l H_2O 100°C; insol. alcohol; sol. acetic acid
Periodic acid	HIO_4	191.91	Colorless	—	Subl. 110	Dec. 138	Very sol. (dec.) cold H_2O
Iodine bromide	IBr	206.81	Dark gray cryst.	4.4157^0	(42) Subl. 50	Dec. 116	Sol. (dec.) cold H_2O; sol. alcohol, ether, chloroform, CS_2
Iodine chloride	ICl	162.36	Dark red needles	3.1822^0	27.2	97.4	Dec. to HIO_3, cold H_2O; sol. alcohol, ether, CS_2, HCl
Iodine tribromide	IBr_3	366.63	Brown liq.	—	—	—	Sol. cold H_2O; sol. alcohol
Iodine pentafluoride	IF_5	221.90	Colorless liq.	3.75	9.6	98	Dec. cold H_2O; Dec. acid, alkali

in which iodine is in a positive oxidation state. The former resemble the other halides in their properties. The carbon–iodine bond is the weakest of the halogen–carbon bonds, and the organic iodides are the least stable and the most reactive. Because of the heavy iodine atom, the iodides are also the densest and the least volatile. The latter are characteristic of iodine alone among the halogens, and occur only if an aromatic or olefinic group is attached to the iodine, as in iodoso compounds (RIO) and iodoxy compounds (RIO_2). These oxygenated iodine compounds have presently little or no commercial interest.

IODINE REFERENCES

66. R. E. Remington, *J. Chem. Educ.*, **7**, 2590 (1930); T. von Fellenberg, *Ergeb. Physiol.*, **25**, 176 (1926).
67. L. Matt, Dissertation, Würzburg, 1889, cited in F. Flury and F. Zernik, *Schädliche Gase*, Springer, Berlin, 1931, p. 124.
68. *The Merck Index*, 7th ed. Merck & Co., Rahway, N.J., 1960.
69. A. B. Luckhardt et al., *J. Pharm. Exp. Therap.*, **15**, 1 (1920).
70. (a) American Industrial Hygiene Association, Hygienic Guide: Iodine (1965); (b) J. P. Fahey, Communication to TLV Committee member, (1960).
71. T. von Fellenberg, *Biochem. Z.*, **184**, 85 (1927).

2   BORON, B

Boron, with three valence electrons, is classified as a metalloid, and has the distinction of being the only nonmetallic element with fewer than four electrons in its outer shell. In addition to the well-known inorganic compounds of boric acid and its salts, the oxide, carbide, metallic borides, halides, and hydrides, boron forms numerous organic derivatives, boric acid esters, boron alkyls, and organic B–N compounds, all of which are industrially useful.

2.1   Source and Production (36, 53)

Essentially all U.S. boron production, and about three-fifths of the world production, comes from bedded deposits and lake brines in California. Turkey is the only other boron-producing country of significance besides the United States.

The principal boron minerals are tincal, $Na_2B_4O_7 \cdot 10H_2O$; kernite, $Na_2B_4O_7 \cdot 4H_2O$; colemanite (borocalcite), $Ca_2B_6O_{11} \cdot 5H_2O$; ulexite (boronatrocalcite), $CaNaB_5O_9 \cdot 8H_2O$; priceite (pandermite), $5CaO \cdot 6B_2O_3 \cdot 9H_2O$; boracite (stassurite), $Mg_7Cl_2B_{16}O_{30}$; and sassolite (natural boric acid), H_3BO_3.

U.S. production in 1973 was 207,000 short tons as boron, with a projected capacity in 1980 of 250,000 tons. Comparable figures for Turkey were 80,000 and 130,000 tons. Total world production figures for the periods were 342,000 and 480,000 tons, with the Soviet Union contributing about 12.5 percent of the world total. Chile is expected to contribute about 1 percent of the total by 1980.

Boron is prepared from its compounds by chemical reduction with reactive metals, by nonaqueous electrolytic reduction, or by thermal decomposition.

Reduction of boron compounds, including borates, boron oxides and halides, fluoborates, and borohydrides, is most commonly performed by the exothermic (energy-saving) reduction of boron trioxide with magnesium. Reduction with hydrogen at high temperatures of boron halides is the conventional method of obtaining high purity boron of 99 percent or better. When boron is prepared by electrolysis, fused melts of boron trioxide in potassium halides or KBF_4–KCl melts with boron carbide anode are used, yielding a product purity of 87 to 98.8 percent boron.

Direct thermal decomposition of boron compounds to high purity boron is limited to halides and hydrides. Boron tribromide or triiodide and boron hydrides (from diborane to decaborane) have been decomposed on a wide variety of substrates ranging from glass to tungsten at temperatures from 800 to 1500°C. Ultrapure boron is purified by zone refining or other thermal techniques; that is, impurities are removed by progressive recrystallization or volatilization at high temperatures.

2.2 Uses and Industrial Exposures

Possibly no other element has so many diversified uses as boron. Of the U.S. consumption of 330,000 tons in 1974, two-fifths was used in the manufacture of various kinds of glasses, accounting for 5 to 10 percent of many special glasses. About 15 percent of all boron consumed went into insulating fiberglass, 10 percent into textile fiberglass, and 15 to 20 percent into all other glasses. Manufacture of enamels, frits, and glazes for protective and decorative coatings on sinks, stoves, refrigerators, and many other household and industrial appliances accounted for another 10 percent of the boron consumption. Approximately one-sixth of the B compounds consumed in the United States went into soaps and cleansers.

Possibly 5 percent of boron used went into agriculture and another 2 to 3 percent into herbicides. Minor amounts of boron compounds were consumed as fluxing materials in welding, soldering, and metal refining. Some elemental boron was used as a deoxidizer in nonferrous metallurgy, as a grain refiner in aluminum, as a thermal neutron absorber in atomic reactors, in delayed-action fuses, as an ignitor in radio tubes, and as a coating material in solar batteries. Use of boron compounds in abrasives is increasing, particularly cubic boron nitride produced by synthetic diamond producers. Use of boric acid as a catalyst in the air oxidation of hydrocarbons accounts for possibly 1 to 3 percent of boron consumption in chemicals, conditioning agents or precursors to chemicals, plasticizers, adhesive additives for latex paints, fire retardants, antifreeze, textile and paper products, biocides in jet fuels, photography, and composite materials.

Industrial Exposures. The numerous and varied boron compounds of industrial interest present, as would be expected, a wide spectrum of hazardous exposures. High on the list are the borohydrides, deca-, penta-, and diborane, followed closely by dimethylamine borane and the boron halides. Sodium borohydride on moist skin presents a hazard. Boric acid and its sodium tetraborate salts offer a low exposure hazard as do the boron alkyl esters, but not the aryl esters. In the "nuisance" category are the boron carbide, nitride, and oxide, and boron dust itself.

2.3 Physical and Chemical Properties

The physical properties of boron and representative industrially important compounds are listed in Table 40.7. As can be seen, boron has a wide variety of compounds with widely varying properties, from the highly insoluble oxide and carbide, to the soluble borates, to the liquid alkyls, halides, and boric acid esters, and the gaseous hydrides.

2.4 Physiologic Response

2.4.1 Boron

The early studies of Stokinger and Spiegel (72, pp. 2307–2308), incomplete as they were, constitute about the only reported toxicologic information on amorphous boron. When these studies were performed in the early 1940s, only limited amounts were available for study; hence only 15 mice could be exposed on an inhalation schedule of 7 hr/day, 5 days/week, for 30 days. Exposure levels averaged 72.8 mg/m^3 with a wide SD of 42.4. The median particle size was 0.67 μm, SD 2.0.

Under these exposure conditions, there was scant evidence of toxicologic effects, even from these exposure levels, which were at the time at least five times that of the recommended "nuisance" dust level. A mean weight loss of 7 percent during the first week was later recovered in steady increases, and could be attributed to confinement in the exposure chamber. No mice died during exposure, and no gross pathological abnormalities were found on sacrifice of the animals. Oddly, boron was found widely distributed in kidney, liver, gastrointestinal tract, and lung despite a solubility of only 0.72 μg/ml water and 1.3 μg/ml plasma. On a post-exposure weekly sacrifice schedule of three mice each, lung boron concentration reached 792 μg/g tissue, kidney, 252, liver, 106, and gastrointestinal tract, 73 μg/g, decreasing gradually, but still measurable at the fifth sacrifice by the spectrographic procedure.

The oral mouse LD_{50} for presumably amorphous boron is reported to be 2000 mg/kg. Hence, unless contrary information is developed, amorphous boron dust can be considered for control purposes, a "nuisance" dust with a TLV of 10 mg/m^3, or 30 mppcf.

Analytic Determination. *Normal values* of boron in blood and urine have been determined in almost 300 working men in various industries and locales, throughout the continental United States, by the spectrographic method after ashing with Li_2CO_3 (74a). The median boron concentration in blood of 298 individuals was 9.76 µg/100 g, range 3.90 to 36.50; in urine, 720 µg B/liter, range 40 to 6,600, with insignificant variation in the medians from one part of the country to the other. Similar values for blood boron were obtained by Parr bomb destruction followed by carmine analysis (74b–d). As noted above, the spectrographic method has been used for the determination of boron in solid biologic tissues. Using the boron line of 2497.7 Å and the tin line at 2495.7 Å as internal standard, the sensitivity of the spectrographic method was reported to be 0.02 µg B/ml (74a), thus permitting detection of as little as 5.0 µg B in 100 g blood and 40 µg/liter urine.

The boron content of 116 ashed samples of autopsied bone from 33 California individuals, aged from 5 months to 75 years, ranged from 16 to 138 ppm bone ash, with a mean of 61 ppm, SE ± 1.7, as determined by a spectrographic procedure using a cathode layer technique to increase sensitivity (92).

Based on analysis of parietal, vertebral, femoral, and costal samples, no significant differences were found in any of the specimen types or between the ages of 5 months to 17 years, or from 17 to 75 years. This seems odd in view of the known cumulation of boron in the skeleton with age. Whether the ashing of the samples at 600°C for 12 hr reduced all samples to a common base line is unknown. No discussion of the matter was given by the authors.

2.4.2   Boron Oxide

Similar control measures have been recommended for this oxide, B_2O_3, as for boron. These recommendations are based on the absence of deaths or signs of intoxication other than mild nasal irritation in rats exposed at 470 mg/m^3 daily for 10 weeks. In dogs exposed at 57 mg/m^3, the urine volume and acidity increased, as well as the creatinine coefficient (73).

Intragastric intubation of a 10 percent slurry of B_2O_3 in water to rats daily for 3 weeks caused no retardation in growth rate or other observable effects; however, rats refused to drink water containing 1 to 1.5 percent B_2O_3 and consequently lost weight. Topical application of B_2O_3 dust to the clipped backs of rabbits resulted in erythema that persisted for several days. Likewise, conjunctivitis immediately followed ocular instillation, effects attributed to the exothermic reaction of B_2O_3 with water (73).

TABLE 40.7. Physical Properties of Boron and Some of its Industrially Important Compounds

Compound or Element	Formula	Mol. wt.	Physical Properties	Sp. Gr.	M.P. (°C)	B.P. (°C)	Solubility
Boron	B	10.811	Yellow monoclinic or, brown amorph.	2.34, 2.37 amorph.	2300	2550 subl.	Insol. cold and hot H_2O; very sl. sol. HNO_3
Boron oxide	B_2O_3	69.62	Rhombic cryst.	2.46 ± 0.01	460	~1860	Sl. sol. cold H_2O; sol. hot H_2O
Hydrogen ortho borate (boric acid)	H_3BO_3	59.81	white, waxy pl.	1.5172^{14}	170.9 $-H_2O$, HBO_2	—	25.2 g/l H_2O 0°C (53), 275.3 g/l H_2O 100°C, 173.9 g/l MeOH 25°C, 94.4 g/l EtOH 25°C
Sodium tetraborate (borax decahydrate)	$Na_2B_4O_7$ $\cdot 10H_2O$	381.37	Colorless monoclinic, effl.	1.73	75, $-8H_2O$ 60	320 $-$ 10 H_2O	20.1 g/l H_2O 0°C, 1.70 kg/l H_2O 100°C; very sl. sol. alcohol
Diborane	B_2H_6	27.67	Colorless gas	Liq: 0.447^{-11}	−165.5	−92.5	Sl. sol. cold H_2O; dec. to H_3BO_3
Decaborane	$B_{10}H_{14}$	122.22	White cryst.	0.94^{25}	99.5	213	Sl. sol. cold H_2O; dec. hot H_2O; very sol. CS_2; sol. alcohol ether, benzene

Name	Formula						Solubility
Sodium borohydride	$NaBH_4$	37.83	White cubic	1.074	Dec. 400	—	550 g/l 25°C H_2O; 40 g/l alcohol; 164 g/l MeOH
Boron carbide	B_4C	55.26	Black rhombohedral	2.52	2350	>3500	Insol. cold or hot H_2O; Insol. acohol; sol. fused alk
Boron tribromide	BBr_3	250.54	Colorless fuming liq.	$2.6431^{18.4}$	−46	91.3 ± 0.25	Dec. cold H_2O; sol. alcohol, CCl_4, S_2Cl_2, $SiCl_4$, SO_2
Boron trifluoride	BF_3	67.81	Colorless gas	2.99 g/l	−126.7	−99.9	1.06 kg/l cold H_2O; dec. hot H_2O; dec. alcohol sol.conc. H_2SO_4
Potassium fluoborate (native avogadrite)	KBF_4	125.91	Colorless rhombic or cubic	2.498^{20}	Dec. 350	Dec.	4.4 g/l H_2O 20°C, 62.7 g/l H_2O 100°C; sl. sol. alcohol; ether
Trimethyl borate (a boric acid ester)	$B(OCH_3)_3$	103.92	Colorless liq.	0.915	−29	68.7 (65)	Dec. cold H_2O; sol. alcohol, ether
Triethyl boron (a boron alkyl)	$B(C_2H_5)_3$	98	Colorless liq.	0.6961^{23}	−92.9	$0^{12.5}$	Insol. cold and hot H_2O; sol. alcohol, ether

The TLV for B_2O_3 of 15 mg/m^3 or 50 mppcf was adopted by the American Conference of Governmental Industrial Hygienists in 1960 as that of a "nuisance" dust and remained there until 1971, when it was reduced to 10 mg/m^3 or 30 mppcf, in keeping with the general trend to improve industrial hygiene conditions.

2.4.3   Boric Acid and Inorganic Borates

Boric Acid, H$_3$BO$_3$.   Three grades of granular and powdered H_3BO_3 are produced in the United States, technical, NF grade, and special quality grade. The purity associated with each of the grades is not recorded in the open literature, but is probably available from the manufacturer.

Boric acid has a large variety of applications in both industrial and consumer products. It serves as a source of B_2O_3 in many fused products, including textile fiberglass, optical and sealing glasses, heat-resistant borosilicate glass, ceramic glazes, and porcelain enamels. It also serves as a component of fluxes for welding and brazing.

A number of boron chemicals are prepared directly from boric acid. These include synthetic, inorganic borate salts, the phosphate, fluoborates, borate esters, and metal alloys such as ferroboron. Boric acid catalyzes the air oxidation of hydrocarbons and increases the yield of alcohols by forming esters that prevent further oxidation of hydroxyl groups to ketones and carboxylic acids.

The bacteriostatic and fungicidal properties of boric acid have led to its use as a preservative in natural products such as lumber, rubber latex emulsions, leather, and starch products. It is also used in washing citrus fruits to inhibit mold, and in mildew-resistant latex paints.

NF grade boric acid serves as a mild, nonirritating antiseptic in mouthwashes, hair rinses, talcum powder, eyewashes, and protective ointments. Although relatively nontoxic to mammals, boric acid is quite poisonous to insects, and has been used to control cockroaches and to protect wood against insect damage.

Analytic Determination.   A rapid, sensitive, and simple method has been devised for the determination of the borate ion in biologic and other materials by Hill et al. (75a). It is based on the migration of the borate ion BO_3^{3-} to the anode of an electric field and its visualization as the red boron-turmeric complex (75b), which is developed on a paper strip after ionophoresis. The sensitivity is claimed to be of the order of 0.01 μg B.

Toxicity.   There is no dearth of acute toxicity information on H_3BO_3. This stems from its systemic and lethal effects in infants because of its ease of penetration through the skin. (It is not apocryphal that within minutes, boron can be detected in increased amounts in the blood from soaking the feet in a tub of H_3BO_3 solution.) Unfortunately for industrial hygiene control, no toxicity information has been developed from its inhalation.

The lowest oral lethal dose LD_{LO} for man is reported to be 640 mg/kg (6); the LD_{LO} by skin, 8600 mg/kg, and the lowest toxic dose affecting the eye, 143 mg/kg. Intravenously, H_3BO_3 is highly toxic, the LD_{LO} being 29 mg/kg. Death has occurred from 5 to 20 g in adults, and from less than 5 g in infants.

Signs and symptoms from ingestion or absorption of H_3BO_3 are nausea, vomiting, diarrhea, abdominal cramps, erythematous lesions on skin and mucous membranes, and from lethal or near-lethal doses, circulatory collapse, tachycardia, cyanosis, delirium, convulsions, and coma. Chronic use may lead to dry skin, eruptions, and gastric disturbances (79).

Chronic, 2-year dietary feeding in dogs and rats showed that H_3BO_3 was tolerated at 2000 ppm (350 ppm B equivalent) (76). Rats fed 1170 ppm levels (B equivalent) showed growth depression, decreased food utilization efficiency, degeneration of gonads, and skin desquamation of paws and tails. Testicular degeneration occurred in both dogs and rats at this level, and rats became sterile as observed in reproduction studies, but at 350 ppm B equivalent, there was no adverse effect on fertility, lactation, litter size, weight, and appearance.

In a previous metabolism study in rats (77) the percent of dietary boron retained in soft tissues and skeleton was inversely proportional to its concentration in the diet; at normal dietary levels, 19 ppm B, total boron retained was 0.49 percent as an average of 4 and 8 weeks on the diet; when H_3BO_3 was added to the diet at 73, 104, and 198 ppm, retention decreased regularly to 0.17 percent of the amount administered at the 198 ppm level. The percent of total boron retained by the skeleton in growing rats, however, increased on the addition of H_3BO_3, from 19 percent in 8 weeks on the normal 19 ppm in the diet to 30 percent from other dietary additions; in adult rats, the percent of skeleton retention decreased from 26 percent at normal dietary boron levels to 16 percent at the 198 ppm H_3BO_3 level.

No TLV has been recommended for H_3BO_3, presumably because of little significant exposure to workers during its production and handling.

Inorganic Borates. There are four inorganic borates of major industrial interest, sodium borate, $NaBO_2$; sodium tetraborate anhydrous, $Na_2B_4O_7$; sodium tetraborate decahydrate, $Na_2B_4O_7 \cdot 10H_2O$ (borax); and sodium tetraborate pentahydrate, $Na_2B_4O_7 \cdot 5H_2O$. Of lesser commercial use are calcium and potassium borates.

The natural sodium tetraborate decahydrate, tincal (borax), by far outweighs all other borates in number and kinds of uses. First of all, it is the starting substance from which the other inorganic borates are derived. Besides this, borax has many other uses which include it as a constituent of heat-resistant glass and porcelain enamel, in starch and adhesives, as detergent, as herbicide, in fertilizers, rust inhibitors, pharmaceuticals, leather, photography, bleaches, and paint, as a flux for smelting, as a component in insulation materials and antifreeze, and as a laboratory reagent.

Analytic Determination. The alkali borates can be conveniently analyzed for their B_3O_3 content by complexing with mannitol followed by titration with dilute NaOH (78a). A review of the methods for analysis of industrial borates is found in Reference 78b.

Toxicity. Systemic toxicity of borax when introduced into the lung would appear to be low as judged from the lack of reactions in the lung, the gastrointestinal tract, or any other organ of the guinea pig when given in three intratracheal injections of 50 mg each of borax dust of particle size below 5 μm (79). During the 12 weeks following injection, there was no impairment of appetite, diarrhea, or retention of urine. The reaction to stimuli, shown by all the animals, was alert and normal, and there was no evidence of any central nervous system involvement. Microscopic examination of sections of the lungs, liver, spleen, kidneys, adrenal glands, and cerebral cortex showed no abnormalities in any of the animals.

A 2-year feeding study in dogs and rats, however, was not without adverse effects, although at high intake levels (76); rats fed borax at 1170 ppm B equivalent (1 percent borax in the diet) showed growth suppression, decreased food utilization efficiency, degeneration of gonads, and skin desquamation on paws and tails. At this level also, both dogs and rats showed testicular atrophy. Reproduction studies at this level resulted in rat sterility, but at 350 ppm B equivalent, there was no adverse effect on fertility, lactation, litter size, weight, and appearance in the rat. Based on a 90-day feeding study of dogs, at which they displayed no adverse effects from borax at 175 ppm B equivalent, whereas 525 ppm boric acid was the "no-effect" level, borax would appear to be somewhat more toxic than boric acid subchronically (76). In either case, 350 ppm B equivalent is to be considered as the "no-effect" level.

Alteration of borate toxicity has been demonstrated in young, 4- to 6-week old mice by simultaneous administration of D-glucose with the borax (80a). Least toxic were the molar ratios of 1:1.5 and 1:2 boraxid–glucose, which reduced mortality from 100 percent to 45 and 37.5 percent, respectively. On the other hand, ratios of 1:3 and 1:4 glucose showed steadily *increasing* toxicity reaching almost 100 percent. It is well known that borax and boric acid form complexes with polyhydroxy compounds in aqueous solution, resulting in altered toxicities. Such complexing occurs also with other inorganic and organic borates (80b).

Threshold Limits. Threshold limits were recommended for three sodium tetraborates in 1975. The TLV for the decahydrate is 5 mg/m³ for the salt, and

1 mg/m^3 for both the pentahydrate and the anhydrous form, razorite. Each bears the "SKIN" notation, indicating that, owing to appreciable skin absorption, the TLVs should be reduced when appreciable skin contact occurs.

The TLVs are based on their occupationally important toxic property of acute, irritant effects when in contact with skin and mucous membranes of the eyes, nose, and other sites in the respiratory tract. The hazard of this irritant property increases with decreasing water of hydration owing to the exothermic effect of hydration. Further background information on the basis of the threshold values may be found in Reference 81.

2.4.4 Boron Hydrides

The commercially significant compounds in this group are decaborane, diborane, and sodium borohydride. Of lesser interest are dimethylamineborane and pentaborane-9, the latter having been formerly used along with deca- and diborane as a high energy fuel propellant. It appears now to be no longer available in commercial quantities (4). Because the early industrial hygiene literature on the boranes concerned impure mixtures, only later literature on the discrete compounds is summarized here.

Decaborane, B$_{10}$H$_{14}$ A white crystalline substance (Table 40.7) with a characteristic odor of unusual persistence, decaborane is a by-product of the hydrolysis of diborane, B$_2$H$_6$. Because of its unusual chemical structure of four divalent hydrogen atoms, it is highly reactive and can explode on contact with heat or flame, or with oxygenated or halogenated solvents. Among its many uses as a solid propellant, in polymer synthesis, corrosion inhibitor, fuel additive, mothproofing agent, dye-stripping agent, reducing agent, fluxing agent, and oxygen scavenger, its use as a solid propellant was the contributing element that first brought to the attention of the military the highly toxic nature of the high molecular weight boranes. Unfortunately, case histories, though registering early signs of CNS effects of dizziness, drowsiness, and headache (82a), with immediate decrease in cardiac output with profound and persistent bradycardia (dogs) (82b), failed to associate the CNS symptoms with the degree of worker exposure. Hence dose–response information derives from animal experimentation.

Animal Toxicity.   A series of studies on the toxicity of decaborane has been made by Svirbely (83a–d) at the U.S. Army Chemical Center. The 4-hr LD$_{50}$ value for mice by inhalation was 25.7 ppm at 48 hr after exposure (83a). Rats were far more resistant; 4-hr exposures at concentrations up to 95 ppm were not lethal to any of the exposed rats, which showed normal activity during exposure. The mice, however, showed signs of restlessness, depressed breathing, generalized weakness, and jerkiness of the head, all signs resembling those seen in exposed workmen. In addition, corneal opacities developed, and in the later stages transient spasms, but partial recovery occurred by the next day.

Irrespective of whether administration was oral or intraperitoneal, decaborane was highly toxic (83b); the mouse oral (stomach tube) LD_{50} was 40.9 mg/kg, and intraperitoneal, 33.2 mg/kg; for rats, the values were 64.3 mg/kg and 23.4 mg/kg, respectively. By percutaneous application, the rabbit LD_{50} was 113 mg/kg. Histopathological examination of various organs of dogs given decaborane by intraperitoneal injection gave suggestive evidence of liver and kidney damage, again signs of intoxication seen in some overexposed industrial workers (82a).

Cumulative toxic effects occurred in rats, rabbits, and dogs receiving repeated small doses of decaborane by oral, intraperitoneal, or cutaneous routes (83c) or by inhalation of its vapors in repeated subacute exposures for 4 weeks (83d). Evidence of CNS effects noted in the acute experiments was intensified in rats and rabbits after cutaneous applications, but was not found in dogs or rabbits after oral or intraperitoneal administration, although dogs showed evidence of possible liver and kidney damage. The rate of recovery was markedly delayed in those animals surviving repeated doses in comparison with those that had received large, single doses.

This observation on rate of animal recovery brings to light the striking and fortunate recovery of workers exposed not only to decaborane, but to other boranes as well; of 13 of 83 patients requiring hospitalization of 3 to 7 days, some of whom were so severely affected that they remained comatose for days, not one died, and all eventually recovered, those from chronic exposures requiring longer recovery periods (82a).

Chronic, 5- to 6-hr daily inhalation exposures for as long as 6 months at 4.5 ppm (rats) were fatal to rabbits after a few exposures, to the dog and monkey after 4 to 15 exposures, to mice after 10 to 100, and to rats after 135 exposures, showing wide differences in susceptibility among species. Pathological changes in the liver were prominent in the areas of active cellular metabolism, the central zones of the liver, and the tubules of the kidney.

Mechanism.   The finding by Levinskas et al. (84) of the marked difference in metabolism of decaborane from that of dimethylaminoborane (DMAB), whereby decaborane disappeared from the blood more rapidly and was not found in the urine as was DMAB, indicated an unusual metabolism. A series of subsequent findings on the mechanism of action of decaborane eventually led to the conclusion that the mode of action of decaborane was indeed unusual (85). Decaborane had been found to lower levels of norepinephrine (86a), dopamine (86b), and serotonin (86c) in rat brain and histamine in several organs of the rat (86d). Explanation of these effects came from the finding that decaborane is an inhibitor of three decarboxylases and one transaminase, all of which require pyridoxal phosphate for their activity. Evidence was finally adduced that the toxicity of decaborane attendant on the inhibition of pyridoxal phosphate-requiring enzymes was not due to decaborane per se, but to reduction of the phosphate to presumably the $B_{10}H_{13}^-$ anion, formed nonenzymatically from decaborane in the presence of water (85).

Behavioral Effects.   In a study to alert industrial hygiene personnel to performance changes in workers exposed to decaborane, Reynolds et al. (87) examined behavioral decrements in adult macaque monkeys injected with small doses (2 or 4 mg/kg) of decaborane. Of the eight negatively reinforced (low-level shock) tasks and two positively reinforced (food reward) tasks, all animals showed a performance decrement on at least one task, with 75 percent of the animals exhibiting a decrement on the remaining tasks. In more than one-half the cases, performance changes preceded clinical symptoms. Although there was no significant difference between the 2- and 4-mg doses when a subject showed a decrement on negatively reinforced tasks, those subjects that received 1 mg followed by 2 mg the next day (positively reinforced animals) did not noticeably improve or return to base line for a much longer period than the 2 and 4 mg negatively reinforced groups.

Accordingly, it was pointed out that workers exposed at such levels of decaborane may be expected to show a performance decrement or clinical symptoms during the first 50 hr. Usually, the first indication is a performance decrement on a task requiring continuous motor behavior, or a series of discriminations within the first 30 hr. Tasks of a discrete nature may, at lower levels, reflect no decrement or clinical symptoms. When, however, performance decrements do occur, return to base line performance levels may be expected within 3 to 10 days.

Therapy.   Because therapy for overexposure of workers to boranes had been nonspecific in the past, consisting of oxygen, chlorpromazine, meperidine HCl, and the like, treatment of borane poisoning was directed toward their common property of reducing potential (88). Accordingly, methylene blue (MB), a stable, readily available oxidizing agent, was infused into rabbits continuously for 2 days after having been injected intraperitoneally with decaborane at lethal doses (30 mg/kg). All animals treated with MB lived longer than 24 hr and 50 percent lived longer than 48 hr, whereas all the untreated rabbits died in less than 24 hr.

An impressive feature of MB is the prevention of brain and heart norepinephrine depletion following decaborane exposure (88).

Analytic Determination.   When it became apparent in the early 1950s that the boron hydrides were useful as high energy fuels, a rash of colorimetric, microanalytic methods for the determination of these hydrides in workroom air appeared, many of which were developed by W. H. Hill and associates. One of these, the direct determination of decaborane, which obviated the conversion of borane to boric acid in previous methods, reported by Hill and Johnston (89), used a quinoline–decaborane complex that could be determined colorimetrically. This method was used by Svirbely to monitor the exposure concentrations in his series of animal toxicity studies mentioned previously.

Decaborane is sampled in all-glass scrubbers containing aqueous 1:1 triethanolamine. Analysis is performed by combining decaborane with quinoline in xylene to form a red-colored complex that is read at its maximal absorption of 490 mμ. The red complex is specific for decaborane; neither di- nor pentaborane interfere, and the color has great stability. The method is useful in the range of about 1 to 20 μg B as debaborane per milliliter of solution.

A method analogous to the quinoline procedure, but which uses dipyridyl ethylene as the color-developing agent, has been reported by Pfitzer and Seals (90). Its claimed virtues are increased sensitivity, more rapid color development, and lower reagent blank readings.

A continuous air-monitoring device has been developed based on Hill's reduction of triphenyltetrazolium chloride to form the red-colored formazan (91).

For the determination of boron in body tissues and fluids, spectrographic methods are available (74a, 92).

Odor Detection. The median detectable "nominal" concentration of decaborane has been reported to be 0.35 mg/m^3 (approximately 0.06 ppm) (93).

Threshold Limit. The TLV for decaborane adopted in 1957 by the American Conference of Governmental Industrial Hygienists on the recommendation of the TLV Committee is 0.05 ppm, approximately 0.3 mg/m^3. When recommended, the limit was derived by analogy with di- and pentaborane, occupying a position intermediate between the more toxic pentaborane and the less toxic diborane (83).

Diborane, B_2H_6. In physical, chemical, and toxicologic properties diborane stands in distinct contrast to decaborane. Diborane is a gas at ordinary temperatures and pressures (Table 40.7). It is spontaneously flammable in air at ordinary temperatures, whereas decaborane is unreactive in air. In its reaction with water, diborane instantly hydrolyzes, in contrast to the slow hydrolysis of decaborane. Even more marked are differences in its toxicologic properties; the toxic effects of diborane are chiefly on the lungs, whereas decaborane affects the central nervous system.

Diborane is of special interest because it is the starting material for the preparation of various other borohydrides and because of its synthetic uses. It can be prepared in essentially quantitative yields by reacting metal hydrides (sodium borohydride) with boron trifluoride.

Diborane is an extremely versatile reagent for the preparation of organoboranes, which, in turn, are useful intermediates in organic synthesis. Diborane also has other uses apart from being a high energy fuel for air-breathing machines; it is a polymerization catalyst for ethylene, a strong reducing agent, and a doping agent in semiconductors.

Records of industrial exposure to diborane have, like those of decaborane, been descriptive of effects only, without the environmental data that caused them (82a). Hence dose–response data derive from animal experimentation.

Animal Toxicity.   The acute data, such as exist, indicate diborane to be a highly toxic compound (6). The 4-hr rat inhalation LC_{50} is about 50 ppm (approximately 40 mg/m^3); the corresponding LC_{50} for the mouse is 30 ppm; and the 10-hr LC_{LO} for the guinea pig is 53 ppm, indicating a lower susceptibility of this species for the gas.

Chronically, diborane showed the same order of toxicity among the animal species (94). Daily, 6-hr exposures of dogs at about 7 ppm resulted in their death after 10 to 25 exposures, but exposures at about 1 to 2 ppm resulted in some dogs surviving 6 months of exposure. On the other hand, guinea pigs survived 95 exposures at these levels without showing any changes attributable to diborane. Rats showed intermediate effects; 17 of 18 rats died during the course of a 6-month exposure at 7 ppm, with deaths being distributed between the seventh and 113th exposure.

Pharmacologic Effects.   In a study of the acute pharmacological effects of diborane, Kunkel et al. (95) observed respiratory embarrassment during exposure in dogs and rabbits. This was followed by a slight fall in blood pressure, increased activity of the intestinal smooth muscle, and an increase and subsequent diminution of cortical activity, the last effects concluded to be associated with anoxia. Bradycardia occurred late in the poisoning and was followed terminally by ventricular fibrillation or complete disappearance of ventricular activity. The primary effect of diborane poisoning was concluded to be the production of pulmonary edema, resulting from the local irritation of the compound set off by the exothermic reaction of hydrolysis.

No therapeutic drug was found among seven tested, which included respiratory stimulants (atropine), depressants (pentabarbital), and adrenal stimulatory agents (cortisone). The apparently effective therapeutic agent methylene blue was not to be reported until 10 years later (88).

Human Toxicity.   Observations made over a 5-year period (1956 to 1960), involving 26 cases of acute diborane toxicity and 33 with subacute exposures, showed that acute diborane intoxication is associated predominantly with bronchopulmonary involvement, a finding paralleled by the animal studies. Pneumonitis was encountered in two cases of acute poisoning. Chronic respiratory distress was present in two patients from recurrent diborane exposure.

Because this appears clinically as an asthmatic bronchitis, the authors concluded that the chronic disability was on a hypersensitivity basis (96).

Analytic Determination. Diborane may be determined in air spectrophotometrically at 270 mμ after collection in an all-glass bubbler containing bromine in glacial acetic acid saturated with potassium bromide (97a). The method is specific for diborane.

When there are no other boron hydrides present, diborane may be determined by the nonspecific method of collecting it in water and the resulting boric acid determined by the colorimetric carmine method of Hatcher and Wilcox (74c) and read at 625 mμ in a spectrophotometer (97b). The solution follows Beer's law over the range of 0 to 20 μg B/ml.

Odor Detection. The median detectable concentration by odor for man as determined analytically was in the range of 2 to 4 mg/m^3 (98). The nauseating odor is not a reliable warning of toxic exposure.

Threshold Limit. The TLV for diborane adopted in 1956 by the American Conference of Governmental Industrial Hygienists on the recommendation of the TLV Committee is 0.1 ppm (approximately 0.08 mg/m^3). The TLV is based on the animal studies of Comstock and co-workers (94) in 1954 and is designed to prevent acute injury to the lung and cumulative effects on the nervous system. No new information has come to the attention of the Committee since that time that would suggest altering this value.

Pentaborane-9, B_5H_9. Pentaborane-9 is so designated in order to differentiate it from the other pentaborane structured with 11 H atoms, B_5H_{11}. Although B_5H_9 is presently not a commercial item (4), mention is made here of its toxicologic effects chiefly because a considerable amount of information has been developed which may prove useful in the future.

As may be judged from its intermediate structure between the gas B_2H_6 and the solid $B_{10}H_{14}$, B_5H_9 is a liquid. It is spontaneously flammable in air and highly explosive, and though B_5H_9 is insensitive to shock, it forms shock-sensitive mixtures with chlorinated hydrocarbons and carbonyl-containing compounds. Minor spills can result in acute toxic effects in nearby workers. Presumably for these reasons, and the lack of a specific, sensitive analytic method, industrial interest in B_5H_9 has lagged.

As with other borohydrides, no environmental data were collected during periods of acute worker exposure; accordingly, all dose–response data must be derived from animal studies.

Animal Toxicity. The highly acute toxic nature of B_5H_9 prompted exposure of animals for unusually brief periods of 0.5, 2, 5, and 15 min (99). The calculated LD_{50} values for mice, monkeys, and dogs after 2-min inhalation exposures were respectively 342, 640, and 734 mg/m^3 (approximately 170, 320, and 365 ppm), indicating the mouse to be the most sensitive species on an acute basis.

Toxic signs in order of onset were tremors, ataxia, convulsions, and death, all deaths occurring within 24 hr after exposure. In experiments in which dogs were exposed to progressively lower concentrations, severe signs of toxicity were seen only at doses approximating one-half of the LC_{50} minute exposures, and minimal or no toxic signs at one-fourth of each LC_{50}. In monkeys in which convulsions and tremors followed 2-min exposures at 368 mg/m^3, no abnormal signs were observed on the next day, and no lesions were found grossly or microscopically that could be attributed to B_5H_9 exposure, nor were significant changes in blood components or in BSP retention seen. These results emphasize that B_5H_9 appears to act solely on the CNS *in acute exposures*.

In chronic exposures, a 6-month, daily inhalation of B_5H_9 vapor at 0.2 ppm (approximately 0.4 mg/m^3) by monkeys, dogs, rabbits, rats, and hamsters showed conspicuous differences in toxic signs and symptoms from those seen in acute studies (100).

All species showed (1) weight loss or no weight gain due to loss of appetite, and in the monkeys, nausea and vomiting, (2) apathy and insensitivity to pain bordering on stupor, or (3) loss of limb mobility, muscle tremors, and impaired coordination. One monkey died on day 4; the other was sacrificed on day 15.

Similar toxic signs were seen in dogs, rabbits, and rats, but not in hamsters, though delayed and less severe symptoms occurred. Eye and nasal discharges were prominent, as were violent reactions to handling as exposure progressed. Death rates at the end of the 6-month exposure (124 exposure days) were remarkably similar, averaging about 25 percent for the three species. Hamsters were the least affected of all species, with behavior similar to that of controls. Typical of the boranes, partial recovery from these toxic signs occurred over weekends when there was no exposure.

Another striking feature of the animal responses versus those of industrial workers is the noticeable death rate, whereas among all overexposed and even hospitalized workers, none died or were left permanently disabled, with one possible exception. This could well be due to metabolic differences in man. In any case, when the toxic effects of the three borohydrides are compared, pentaborane-9 emerges as the most toxic of the three.

In an attempt to determine the way in which pentaborane-9 exerts its extreme toxicity, Reed et al. (101) found that it rapidly formed a nonvolatile hydrolysis intermediate product in the bloodstream (rats) as molecular hydrogen was evolved. This product slowly disappeared, and the hydrogen from the product slowly exchanged with that of body water. Histidine appeared to stabilize the intermediate, and its disappearance seemed to be enzymatic.

Although no attempt was reported to have been made to locate the site of toxic action (the central nervous system), a clue to its action was found in the inhibition of glycolysis, upon which nervous tissue is highly dependent. In this hyperglycemic response, pentaborane-9 resembled decaborane (102a, b), whose chief site of action is also the central nervous system.

Using conditioned avoidance response as a behavioral measure of toxic response, Weeks et al. (99) showed that significant increases in mean response times occurred only at pentaborane-9 doses of at least one-half the LC_{50} values, indicating that the ratio between lethal and incapacitating doses is not greater than 4 in dogs.

Analytic Determination. No methods that are specific and highly sensitive that have been used in the reports summarized herein exist. Where specificity or short sampling times are not required, collection in a hydrolytic solution and analysis for boric acid by the carmine method of Hatcher and Wilcox (74c) may be used.

The usefulness of analyzing pentaborane-9 in biologic tissues and fluids has not been established, although Miller et al. (103) reported a correlation between serum levels of a similar borane and severity of toxicity in animals. Urinary boric acid can be determined with reasonable sensitivity, but is of very questionable usefulness in the case of pentaborane-9, where levels of intoxication may not exceed normal boron values in the population in which there is considerable variation.

Odor Detection. The median detectable concentration by odor for man is reported to be in the range of 2.5 mg/m^3 (approximately 1.3 ppm) (93). This limit is too high to provide a satisfactory warning (see Threshold Limits).

Threshold Limits. The TLV for pentaborane-9 was placed on the list of Tentative Values in 1956 without a limit, suggesting that the air concentration be kept as near zero as possible throughout the work shift. On the Tentative List in 1961, with the recommended value of 0.005 ppm, it was adopted by the American Conference of Governmental Industrial Hygienists 2 years later without change and has remained at this value to date. Its basis as the lowest TLV of the three boranes is given in the Documentation of the TLVs in 1980 (81).

Emergency exposure limits proposed by the Committee on Toxicology of the National Research Council in 1965 for exposures of 5, 15, 30, and 60 min are 25, 8, 4, and 2 ppm, respectively.

Sodium Borohydride, $NaBH_4$, and Potassium Fluoborate KBF_4. These two inorganic boron compounds are classed together because of structural similarity, but there resemblance ends. Apart from differing conventional nomenclature (potassium fluoborate, not borofluoride) their physical and chemical properties (Table 40.7) are widely different, as are their toxicities. Sodium borohydride is flammable and a dangerous fire risk, whereas KBF_4 is stable in air at ordinary temperatures, and does not hydrolyze readily in water as does $NaBH_4$, in which it is 125 times more soluble.

Sodium borohydride is a source of hydrogen and other borohydrides, a reducer of aldehydes, ketones, and acid chlorides, a bleacher of wood pulp, and a blowing agent for plastics. Potassium fluoborate is used in sand casting of aluminum and magnesium, as a grinding aid, and as a flux for soldering and brasing.

From the scant toxicity information that exists (6, 104), $NaBH_4$ is from 30 to four to six times more acutely toxic than KBF_4 as judged by intraperitoneal administration in the mouse, rat, and rabbit, respectively (104). By dermal application to the moist back of rabbits six of 10 died when 400 mg was applied in two doses in the same area; however, when the same application was made in different areas, no deaths resulted. In comparison, injury to the skin or penetration of KBF_4 was negligible. Sodium borohydride on dry skin, however, produced no signs of irritation (104).

When instilled into the eye of rabbits, 1 mg $NaBH_4$ produced irreversible damage, whereas up to 3 mg KBF_4 was not irritating to the eye (104).

The obvious industrial hygiene recommendations from this limited information are that (1) $NaBH_4$ is flammable and exothermic in the presence of moisture and should be handled with special care to prevent serious burns, and (2) measures should be taken to prevent workers form ocular, cutaneous, or inhalation exposure to $NaBH_4$.

2.4.5   Boron Carbide, B_4C, and Boron Nitride, BN

These two boron compounds are classed together because of their common property of great hardness comparable to that of the diamond, their great insolubility, high melting point, and general chemical inactivity (Table 40.7) and consequent lack of toxicity.

The major uses of B_4C relate to its hardness or its high neutron absorptivity (^{10}B isotope) and thus it is used in the shielding and control of nuclear reactors. Hot-pressed B_4C finds use as wear parts, sandblast nozzles, seals, ceramic armor plates, and as a dressing for grinding wheels.

Because of its general inertness, it could be classed as a "nuisance" dust, although no specific studies of its toxicity have been made.

The major uses of BN are as a refractory and furnace insulator, in crucibles, rectifying tubes, self-lubricating brushes, and nose-cone windows, as a metal-working abrasive, and in heat-resistant fibers. Like B_4C, its general inertness under ordinary conditions would indicate the classification of BN as a "nuisance" dust, although the hexagonal form is attacked slowly by acids as well as some organic solvents (53). In any case, no reports of its toxicity have reached the open literature.

2.4.6 Boron Halides

Of the four boron halides, only three boron tribromide, BBr_3, boron trichloride, BCl_3, and boron trifluoride, BF_3 are presently in commercial use (4). The boron trihalides are colorless, volatile compounds, and sensitive to moisture, hydrolyzing to boric acid and the corresponding halogen acid at ordinary temperatures.* Much of their physical behavior resembles that of a covalent compound, hence their solubility in organic solvents (Table 40.7). Their chemical behavior is also governed by their Lewis acidity, being electron acceptors, and are thus called Lewis acids (e.g., BF_3). The order of decreasing acidity is BBr_3, BCl_3, and BF_3 (53).

Uses of boron halides range from acid catalysts in organic synthesis to soldering fluxes and extinguishing magnesium fires in sand molds and furnaces (BCl_3, BF_3).

Industrial exposures can occur from the release of these halides into the worker atmosphere as well as from the production and handling of these substances.

Boron Trifluoride, BF_3. This boron halide is selected as the prototype of the three commercial halides for the reason that BF_3 has the greatest amount of toxicologic and analytic information; little or no such information is in the open literature on either BCl_3 or BBr_3.

As seen in Table 40.7, BF_3 is a heavy, colorless gas, which on contact with air rapidly reacts with its water vapor to form a visible mist, if in sufficiently high concentration.

Analytic Determination. A large number of analytic and sampling methods have been tried in attempts to develop a monitoring procedure that will accurately assess the concentration of BF_3 in workroom air (105), without success (106). Small wonder that under the conditions of animal and worker exposure alike, no pure BF_3 species exists and measurement must be made on those that do, whatever their nature.

Horton and Weil (107) developed a semiquantitative spot test to measure the "BF_3" exposure levels in the early animal exposure studies for the Manhattan Project in the mid-1940s. The method involved drawing the "BF_3"-laden air through a filter which consisted of a paper saturated with alcoholic turmeric

---

* This statement is based on the observation of a coating of white, waxy, crystal-line plates on the entrance duct to the exposure chamber, and red, inflamed, swollen paws of animals at the end of the exposure day.

solution and the resulting color compared visually with prepared standards. The method has limited application, because color development is subject to interferences by other compounds. The method was used as a check on the metered chamber concentrations.

Torkelson et al. (108), in connection with long-term animal exposure studies, used the colorimetric procedure of Hatcher and Wilcox (74c) of carminic acid in H_2SO_4 for determining the boric acid hydrolysis product, which was then analyzed spectrophotometrically. Boron concentrations as low as 0.48 ppm could be measured, which was more than adequate for the exposure levels used of about 6, 3.5, and 1.5 ppm. This was one of four analytic methods that were evaluated for NIOSH for the improved determination of BF_3 (106). In this evaluation, accuracy of the measurement was claimed between 0.4 and 5.0 µg B/ml. Unfortunately, no reproducible sampling procedure could be found.

Similar difficulties were found with the other three methods that were evaluated. In the tetrafluoborate specific ion electrode method, the rate of conversion to tetraborate ion was so slow as to require unacceptably long sampling times. With the infrared technique, inconsistent absorbance due to BF_3 reaction with cell windows ruled this method out. In the third method tested, an atomic absorption procedure, the limit of detection (20 ppm) again required too long collection times, leaving no entirely satisfactory monitoring procedure available for animal exposure studies or for the workplace. These deficiencies in analytic monitoring led to the unfortunate decision of NIOSH not to recommend a specific value for worker exposure to BF_3 (105).

A possible, though not entirely satisfactory, way out of this predicament would be to use the method of Torkelson et al. (108) but under rigidly defined sampling and humidity conditions. In this way, irrespective of the precise knowledge of the boron structures from which the dose–response data were obtained in animal studies, these data could be used in the workplace. For it is clear from past investigations that there is no exposure to pure BF_3 either in animal chamber studies or in industrial operations. Hence measurement must be made of the boron structures such as they exist, under the same specified conditions of sampling rate and humidity.

Toxicity.   The first determination of the toxicity of BF_3 was made in the early 1940s under the direction of Stokinger and Spiegel (109) and was based on 30-day inhalation exposures of six animal species at metered concentrations approximating 100 and 15 ppm. Boron trifluoride was metered into the exposure chamber (in dry nitrogen) because under wartime pressures, sufficient time was not allowed for the development of an analytic method (107).

Under the exposure conditions, BF_3 and its atmospheric products were found to be pulmonary irritants predisposing the animals to pneumonia and dental fluorosis. The 100 ppm level proved fatal to all animals (100 mice and rats, 40 guinea pigs, 12 rabbits, 6 cats, and 4 dogs). At 15 ppm, significant mortality occurred only in the mouse (19 percent, 18 of 93), with questionable deaths in guinea pigs or rats (1 in 30, and 2 in 100, respectively), and no deaths in the larger species, rabbits, cats, and dogs.

At the higher level, weight response was subnormal as would be expected; unexpected was the weight loss (12 percent) in dogs and cats, while normal growth was observed in mice, rats, guinea pigs, and rabbits. Weight loss possibly could be attributed to the more adult, mature status of the dogs and cats.

Although definite histopathological pulmonary changes (edema, hemorrhage, and congestion) were noted in the two species examined, dogs and rats, no pathology characteristic of BF_3 poisoning was found among dogs, rats, and guinea pigs at the 15 ppm level. At this level, no changes in serum calcium or phosphorus were seen in dogs and rabbits, although definite decreases in blood phosphorus were found in dogs at the 100 ppm level.

Progressive dental fluorosis was observed in the rats following the first week of exposure at 100 ppm, which became moderately severe at the third week, with the degree of dental hypoplasia paralleling the amount of tooth fluoride; no marked dental hypoplasia was seen at the 15 ppm level.

The Dow toxicology group headed by Torkelson some 20 years later extended and improved upon the earlier studies with exposure periods up to 6 months and with analytic monitoring of the chamber concentrations (108). Using the carminic acid method of Hatcher and Wilcox for chamber air concentrations, they found that, as before, repeated, daily inhalation exposure of rats, rabbits, and guinea pigs resulted in respiratory irritation to such a degree as to cause the death of guinea pigs from respiratory failure, which occurred after the nineteenth exposure day at the nominal concentration of 12.8 ppm as BF_3 (calculated concentration ca. 6.5 ppm). Deaths still occurred in guinea pigs, but not in rats, exposed at an analyzed concentration of 3 to 4 ppm, but all three species exposed at an analyzed concentration of 1.5 ppm were only minimally affected, with average body weights of the guinea pigs only 85 percent of that of controls, and showing only occasional pneumonitis. Rabbits did not differ histologically from controls. Like the previous findings, fluorosis of rat teeth was evident at the highest concentration, but was of doubtful significance at the next lowest level. The average fluorine content of rat teeth and bone was elevated at all exposures levels but not in the soft tissues analyzed, lung, liver, and blood.

On the basis of the results of this longer-term study, the authors recommended a reduction in the TLV to 0.3 ppm from the 1 ppm previously set.

A concentration of 1.5 ppm was detectable by smell, but was felt to be insufficient as a warning of overexposure. The odor was characterized "as a rather pleasant acidic odor" (108).

Human Experience.   Exposure of workers to BF_3 vapors is limited for the most part to a description of physical signs and symptoms with few or no environmental data from which insight into the relation of the response of workers to that in animals may be gained, or from which a validation of the TLV based on animal responses may be made.

In one BF_3 manufacturing plant in the United States, physical examinations initiated in 1974 for a small group of workers included seven with present and six with past exposures to BF_3 or other fluorides, ranging from 1 to 27 years (110). Five of the seven currently exposed, and three of the six with previous exposures showed lower pulmonary function (forced VC and FEV) than predicted for a normal population. Lowered pulmonary function was severe in one, moderate in three, and minimal in one of the seven currently exposed who had an average exposure of 13 years. X-rays were negative, and no urinary fluorine concentrations were above the acceptable preshift concentration of 4 mg/liter.

An air sampling survey of five points throughout the plant showed boron trifluoride concentrations ranging from 0.27 to 0.69 ppm (0.75 to 1.9 mg/m^3) during one 24-hr period in May 1974, and from 0.1 to 1.8 ppm in another 24-hr period in August 1974 (110). Here for the first time an attempt was made to correlate dose with response for any boron compound.

A description of effects on 78 workers exposed from 10 to 15 years to BF_3, as reported in a USSR abstract (111), consisted of complaints of dryness and bleeding of the nasal mucosa, bleeding gums, dry and scaly skin, and pain in the joints. Typical of Soviet industrial hygiene reporting, no specific BF_3 concentrations were reported, merely that "BF_3 concentrations were quite high," and inasmuch as there was concurrent exposure to ethylene and isobutylene with no reported concentrations also, there is no way of knowing what contribution to the reported effects these additional exposures made.

A very limited test of the application of the turmeric filter paper method of Horton and Weil (107) in a U.S. industrial plant showed levels of 12 ppm and below in the general work area without mention of any adverse effects on the workers.

In an attempt to characterize the irritating effect on the skin of BF_3, cotton-soaked aqueous BF_3 was placed on the skin for a day or two. The resulting acid-like burn was less severe than that of the typical HF burn (112).

Threshold Limit.   A TLV for BF_3 of 1 ppm (approximately 3 mg/m^3 as a ceiling value), recommended by the TLV Committee on the basis of dose–response studies in animals, was adopted by the American Conference of Industrial Hygienists in 1961.

Boron Trichloride, BCl_3.   Boron trichloride is a colorless, corrosive liquid that fumes in moist air, liberating HCl. In 1976, U.S. consumption was about 225 tons, of which about 96 percent was used in the manufacture of boron fibers, the remainder as an acid catalyst in a variety of organic reactions. Boron fibers have extremely high tensile strength (elastic modules, 55 million psi); they can be woven into fabrics and are used in spacecraft and sporting goods. Boron trichloride is used as a catalyst for the polymerization of cyclepentadiene and styrene, for the production of alkyl and aryl halosilanes, and for the preparation of boron compounds.

Toxicity.   Only one rather preliminary determination of the toxicity of BCl_3 for laboratory animals appears to exist in the open literature (109). Performed for the Manhattan Project along with BF_3, BCl_3 was found to be more highly irritant and corrosive owing to its hydrolysis to nascent HCl, consistent with its Lewis acid order for boron halides. All the rats and mice, but not guinea pigs, died after a 7-hr exposure at concentrations down to and including 20 ppm. When clean cages were replaced every 2 hr, all rats, mice, and guinea pigs survived two 7-hr exposures at 50 ppm, but most mice and guinea pigs succumbed at 100 ppm, while rats survived. It was evident that the sticky, oily liquid adhering to the cages contributed to the greater mortality.

Gross pathological findings in rats and guinea pigs indicated that BCl_3 acted as a typical chemical irritant; guinea pigs and, to a lesser degree, rats showed extensive hemorrhages of the lungs and some consolidation. An occasional mouse showed mottled kidneys and punctate hemorrhages of the lung. Blood was present in the gastrointestinal tract in a few individuals of all three species. Striking evidence of the irritant nature of the decomposition products of BCl_3 was the grossly swollen feet and mouths of rats and mice exposed by cage contact with the oily liquid and from licking the oil-coated fur (109).

No TLV has been established for BCl_3, but judging from its greater corrosiveness than BF_3, the limit should be lower than 1 ppm.

Boron Tribromide, BBr$_3$.   Boron tribromide is also a colorless, dense, corrosive liquid that fumes in moist air, hydrolyzing rapidly in contact with water and decomposing violently witn the liberation of heat, HBr, and boric acid. Boron tribromide is corrosive to rubber and Bakelite, but not to Teflon or stainless steel. Compared to BCl_3, only about 11 tons of BBr_3 was consumed in the United States in 1976, 95 percent of which was used for catalysis, the remainder for semiconductors.

No specific toxicity determinations appear to have been made, but judging from the resemblance of BBr_3 to BCl_3, its toxicity should be of a similar order.

In the absence of specific toxicity data, the following precautionary and emergency measures should be taken when handling BBr_3 (113). These measures should apply equally to BCl_3 and BF_3.

Precautions.   Personnel handling or exposed to BBr_3 should wear eye, face, hand, and body protection. Whenever possible BBr_3 should be used under a hood with adequate ventilation. Respiratory equipment such as air-supplied or oxygen-supplied masks and drenching emergency showers should be readily available.

Emergency Measures. Personnel exposed to the fumes of BBr_3 should shower immediately and change into fresh clothing, since clothing absorbs the fumes rapidly.

Personnel brought into accidental contact with liquid BBr_3 should immediately attempt to blot up as much of the liquid as possible with an absorbent material before drenching with water. If no suitable absorbent is readily available, exposed personnel should without delay enter a drenching shower and remove contaminated clothing under the shower. All areas of the body in contact with the liquid should be washed with copious quantities of water. No attempt at neutralization should be made.

If even minute quantities of BBr_3 come into contact with the eyes, irrigation with copious quantities of water should be initiated immediately and continued for at least 15 min. The eyelids should be held apart forcibly to ensure effective irrigation. A physician, preferably an eye specialist, should be summoned immediately.

The TLV for BBr_3, adopted in 1969, of 1 ppm (approximately 10 mg/m^3) is based on its decomposition to 3 mol HBr and assumes no additional independent toxicity from BBr_3, an assumption that may not be wholly justified in all circumstances. Accordingly, the limit, equivalent to that of HBr, or 1 ppm, may not be sufficiently conservative.

2.4.7 Organic Compounds

Organic compounds of boron are numerous and varied. They may be broadly subdivided into two classes: (1) alkyl and aryl derivatives of boron, in which boron is directly bonded to carbon; and (2) the boron–oxygen–carbon compounds, such as *borate esters* and *boroxines*. These latter substances may be regarded as derivatives of boric acid or boric oxide.

Boron alkyls, such as trimethylboron (a gas, b.p. $-21.8°C$) and triethylboron (liquid, b.p. 95°C), are extremely reactive materials. Most of the boron alkyls are spontaneously inflammable in air. Controlled oxidation yields alkyl boron oxides, $R \cdot BO$, which dissolve in water to form alkyl boric acids, $R \cdot B(OH)_2$. Boron alkyls are usually prepared from boron halides and metal alkyls, such as zinc, or Grignard reagents, RMgX, or by the reaction of diborane with unsaturated hydrocarbons.

Boron aryls, such as triphenylboron (m.p. 142°C), are usually solids. The aryl compounds are also prepared by the Grignard method from boron halides. The boron aryls are somewhat less sensitive to oxygen than their alkyl analogues.

The aryl and the alkyl compounds of boron are potentially useful intermediates for the synthesis of other boron compounds. Sodium tetraphenyl borohydride, an analytical reagent, is used for the estimation of potassium, rubidium, and cesium.

Borate esters are derivatives of boric acid (or oxide) and alcohols or phenols. They may be prepared either directly from parent materials or indirectly from the reaction of boron halides with alcohols or phenols, or by ester interchange reactions. In general borate esters are extremely sensitive to hydrolysis, but if the alcohol is sufficiently hindered sterically, this hydrolytic tendency can be appreciably reduced. The simplest borate ester, trimethyl borate, is a valuable intermediate for the preparation of metal borohydrides. In addition to synthetic possibilities for borate esters, a number of miscellaneous uses have been described. Some of these include the use of borate esters in cosmetic preparations, as antioxidants for rubber and alcohols, as curing agents for epoxy resins, as petroleum additives, in pharmaceutic preparations, in plasticizers, and as surface-active agents.

Borate esters readily dissolve boric oxide to form boroxines,

$$
\begin{array}{c}
\text{O} \\
\text{RO—B} \quad \text{B—OR} \\
\text{O} \quad \text{O} \\
\text{B} \\
\text{O} \\
\text{R}
\end{array}
$$

where R is an alkyl group, most of which will burn in air, with their combustion product being a glassy boric oxide. For this reason they are being employed as extinguishing agents for metal fires. Boroxines can also be used as polymerization catalysts.

Trialkyl borate esters (in the newer IUPAC nomenclature, trialkoxyboranes), in their anhydride, cyclic form are produced in relatively large quantities and marketed as biocides, epoxy curing agents, and gasoline additives ("Boron" gas, Standard Oil of Ohio).

Because there is a complete lack of toxicity information from usual sources on the borate esters and the organic boron–oxygen compounds from which precautionary procedures may be formulated, the following procedures are given for handling and shipping these compounds (53), and from which measures to be taken for worker protection may be deduced.

Procedures for shipping of boric acid esters vary according to the particular compound being handled. Aryl borates produce phenols when contacted with water, and they must be labeled as corrosive chemicals and are subject to the shipping regulations governing such materials. Lower alkyl borates are flammable (flash points of methyl, ethyl, and butyl borates, are 0, 32, and 94°C, respectively) and must be stored in approved areas. All the low boiling esters from methyl to butyl must also be labeled as flammable. Other compounds, such as hexylene glycol biborate, offer no hazard and may be shipped or stored in any convenient manner.

Organic boron–oxygen compounds are difficult to handle because of their propensity to hydrolyze. The more sensitive compounds should be stored and transferred in an inert atmosphere.

Of the several other organic boron compounds of commercial interest, only three trimethyl borate (in the trade, methyl borate), triphenyl borate, and dimethylamino borate (DMAB, $(CH_3)_2NHBH_3$) have sufficient toxicity information on which to judge their hazards to workers and develop precautionary procedures.

Methyl borate (Table 40.7), the liquid methyl ester of boric acid, is of low toxicity by all routes tested; the oral rat LD_{50} has been reported to be as low as 6.14 ml/kg in a range-finding test (114), and the percutaneous LD_{50} for the rabbit, 1.98 ml/kg. Earlier work (104) placed the oral, rat LD_{50} at 2.82 ml/kg, and the intraperitoneal LD_{50} at 3.2 ml/kg. The reported oral LD_{50} for the mouse is 1.41 ml/kg (115), and the corresponding mouse LD_{50} for the aryl derivative, triphenyl borate, is 0.220 ml/kg, indicating the greater susceptibility of the mouse for these boron compounds.

Tests of methyl borate for ocular toxicity in rabbits showed it to be moderately irritating (104, 114), but only mildly so for the skin (114).

The acute toxicity of DMAB, a white, crystalline solid of limited commercial interest, is very high in comparison with that of methyl borate, or by any toxicity standard (100), the rat intraperitoneal LD_{50} for males and females being 50.5 and 39 mg/kg, respectively. If a density of 0.82 is assumed, the acute toxicity of DMAB by this route is about 50 times that of methyl borate. The oral rat LD_{50} of 59.2 mg/kg DMAB would place it again about 100 times that of methyl borate by this route.

For obvious reasons, no TLVs have been established for any organic boron compound.

BORON REFERENCES

72. H. E. Stokinger and C. J. Spiegel, in *Pharmacology and Toxicology of Uranium Compounds*, Voegtlin and Hodge, Eds., Vol. IV, McGraw-Hill, New York, 1953.

73. J. L. Wilding et al., *Am. Ind. Hyg. Assoc. J.*, **20**, 284 (1959).

74. (a) H. R. Imbus et al., *Arch. Environ. Health*, **6**, 286 (1963); (b) W. H. Hill and R. C. Smith, *Am. Ind. Hyg. Assoc. J.*, **20**, 131 (1959); (c) J. T. Hatcher and L. V. Wilcox, *Anal. Chem.*, **22**, 567 (1950); (d) D. L. Callicoat and J. D. Wolson, *Anal. Chem.*, **31**, 1434 (1959).

75. (a) W. H. Hill et al., *Arch. Ind. Health*, **15**, 152 (1957); (b) J. A. Naftel, *Ind. Eng. Chem. Anal. Ed.*, **11**, 407 (1939).

76. R. J. Weir, Jr. and R. S. Fisher, *Toxicol. Appl. Pharmacol.*, **23**, 351 (1972).

77. R. M. Forbes and H. H. Mitchell, *Arch. Ind. Health*, **16**, 489 (1957).

78. (a) I. M. Kolthoff and E. B. Sandell, *Textbook of Quantitative Inorganic Analysis*, Macmillan, New

York, 1952, p. 534; (b) F. D. Snell and C. L. Hilton, Eds., *Encyclopedia of Industrial Chemical Analysis*, Wiley-Interscience, New York, 1968, p. 368.

79. Letter communication to U.S. Borax and Chemical Corp. from Bio-Research Labs., Berkeley, Calif, July 18, 1957.

80. (a) O. D. Easterday and L. E. Farr, *J. Pharm. Exp. Therap.*, **132**, 392 (1961); (b) O. D. Easterday and L. E. Farr, unpublished results.

81. *Documentation of the Threshold Limit Values*, 4th ed., American Conference of Governmental Industrial Hygienists, Cincinnati, Ohio, 1980.

82. (a) H. J. Lowe and G. Freeman, *Arch. Ind. Health*, **16**, 523 (1957); (b) A. S. Tadepalli and J. P. Buckley, *Toxicol. Appl. Pharm.*, **29**, 210 (1974).

83. (a) J. L. Svirbely, *Arch. Ind. Hyg. Occup. Med.*, **10**, 298 (1954); (b) J. L. Svirbely, *Arch. Ind. Health*, **11**, 132 (1955); (c) *ibid.*, p. 138; (d) J. L. Svirbely, *Arch. Ind. Health. Occup. Med.*, **10**, 305 (1954); (e) J. L. Svirbely and J. C. Roberts, *Arch. Ind. Health*, **14**, 163 (1956).

84. G. J. Levinskas et al., *Arch. Ind. Health*, **14**, 346 (1956).

85. L. L. Naeger and K. C. Liebman, *Toxicol. Appl. Pharm.*, **22**, 517 (1972).

86. (a) J. H. Merritt and T. S. Sulkowski, *Biochem. Pharmacol.*, **16**, 369 (1967); (b) R. W. Schayer and M. A. Reilly, *J. Pharm. Exp. Therap.*, **177**, 177 (1971); (c) J. H. Merritt and E. J. Schultz, *Life Sci.*, **5**, 27 (1966); (d) W. N. Scott et al., *Proc. Soc. Exp. Biol. Med.*, **127**, 697 (1968); (e) W. N. Scott et al., *Proc. Soc. Exp. Biol. Med.*, **134**, 348 (1970).

87. H. H. Reynolds, H. W. Brunson, K. C. Back, and A. A. Thomas, Rep. AMRL-TDR-64-74, Wright-Patterson AFB, Ohio, August 1964.

88. J. A. Merritt, Jr., *Arch. Environ. Health*, **10**, 452 (1965).

89. W. H. Hill and M. S. Johnston, *Anal. Chem.*, **27**, 1300 (1955).

90. E. A. Pfitzer and J. M. Seals, *Am. Ind. Hyg. Assoc. J.*, **20**, 392 (1959).

91. L. J. Kuhns et al., *Anal. Chem.*, **28**, 1750 (1956).

92. G. V. Alexander et al., *J. Biol. Chem.*, **192**, 489 (1951).

93. C. C. Comstock and F. W. Oberst, *U.S. Army Chem. Corps Med. Lab. Res. Rep.*, **206**, August 1953.

94. C. C. Comstock et al., *U.S. Army Chem. Corps Med. Lab. Rep.*, **258**, *March 1954*.

95. A. M. Kunkel et al., *Arch. Ind. Health*, **13**, 346 (1956).

96. E. M. Cordasco et al., *Dis. Chest*, **41**, 68 (1962).

97. (a) L. Feinsilver, *U.S. Army Chem. Corps Med. Lab. Res. Rep.*, **170**, 1953; (b) J. E. Long et al., *Arch. Ind. Health*, **16**, 393 (1957).

98. E. H. Krackow, *Arch. Ind. Hyg. Occup. Med.*, **8**, 335 (1953).

99. M. H. Weeks et al., *J. Pharm. Exp. Therap.*, **145**, 382 (1964).

100. G. J. Levinskas et al., *Am. Ind. Hyg. Assoc. J.*, **19**, 46 (1958).

101. D. J. Reed, F. N. Dost, and C. H. Wang, "Fate of Pentaborane-9-H^3 in small Animals and Effects on Glucose Catabolism in Rats," AMRL-TR-64-112, Wright-Patterson AFB, Ohio, December, 1964.

102. (a) A. A. Tamas, "State of the Art Report—Health Hazards of Borane Fuels and Their Control," Aero Med. Lab. Wright-Patterson AFB, Ohio, October 1958; (b) D. L. Hill, "Some Aspects of the Chemistry and Biochemistry of Decaborane," CWL Spec. Pub. 2-15, U.S. Army Chem. Warfare Labs. Army Chemical Center, Md., 1958.

103. D. F. Miller et al., *Toxicol Appl. Pharm.*, **2**, 430 (1960).

104. C. T. Blaisdell, "Toxicity of Sodium Borohydride and Potassium Fluoborate and Trimethyl Borate," *Chem. Corps. Med. Lab. Res. Rep.*, **351**, Army Chemical Center, Md., March 1955.

105. "NIOSH Criteria Document for a Recommended Standard for Occupational Exposure to Boron Trifluoride," December 1976.

106. "Evaluation of Four Analytic Methods for Determining Boron Trifluoride Concentrations," Report to NIOSH, Industrial Bio-Test Labs, 663-05569, November 24, 1975.

107. C. A. Horton and C. S. Weil, in *Pharmacology and Toxicology of Uranium Compounds*, Vol. IV, McGraw-Hill, New York, 1953, pp. 2328–2334.

108. T. L. Torkelson et al., *Am. Ind. Hyg. Assoc. J.*, **22**, 263 (1961).

109. H. E. Stokinger and C. J. Spiegl, in *Pharmacology and Toxicology of Uranium Compounds*, Vol. IV, McGraw-Hill, New York, 1953.

110. "Boron Trifluoride—Information Concerning Industrial Hygiene Practices," Allied Chem. Corp. Specialty Chemicals Division, Buffalo, N.Y., 1975.

111. V. G. Kirii, *Biol. Abstr.*, **49**, 4068 (1966).

112. H. S. Halbedel, "Acid Fluorides and Safety," Tech. Bull. AF and S, 673, Harshaw Chem. Co., Cleveland, Ohio.

113. Technical Data Sheet, "Boron Tribromide," *Bull. No. DB-26*, American Potash and Chemical Corp., Los Angeles, Calif.

114. H. F. Smyth, Jr. et al., *Am. Ind. Hyg. Assoc. J.*, **23**, 95 (1962).

115. Roy Adams, *Boron Compounds*, Wiley, New York, 1964.

3   SILICON, Si

Silicon, a stable, relatively light metalloid, is the most abundant electropositive element in the earth's surface, second only to the electronegative element oxygen.

3.1   Source and Production (36)

Mining of the raw materials quartz (crystalline SiO_2), quartzite, and sandstone (quartz cemented by Fe oxide or $CaCO_3$) as sources of silicon is essentially a domestic industry. However, silicon-bearing raw materials are imported when production costs, high chemical quality, or physical specifications indicate an economic advantage for a specific use. A few Canadian deposits of quartzite and quartz sand, owned or operated by U.S. interests, are mined and imported for domestic production of silicon products.

U.S. domestic production of silicon in 1973 was estimated at 138,000 short tons compared with 400,000 tons of the world total. By 1980, a rise in the U.S. production capacity is estimated to attain 178,000 tons, and the world capacity, 600,000 tons.

Methods of mining the various siliceous materials differ widely, depending on the nature and location of the deposit and the use to which the product will be put (4). Commercial sources of silica for use in the production of ferrosilicon and silicon usually are quartzite, vein quartz, or gravel deposits. Most deposits are worked by open-pit or quarrying methods, often in a very crude way. Gravel deposits may be worked by hydraulicking, by dredging, or by the use of heavy equipment such as power shovels, pumps, jackhammers, and compressors.

In mining quartzite or well-cemented sandstone, drilling and blasting are normal requirements. Sometimes secondary hand drilling is required if fragmentation is not complete. Because quartzite is brittle, it is relatively easy to blast and crush.

In the processing of silicon, silica raw material to be used for silicon-alloy smelting requires only simple mill beneficiation such as washing, crushing, and screening. High grade quartzite requires only crushing and sizing.

The smelting processes for silicon and ferrosilicon are the same except that shredded iron or steel scrap is added to the charge when producing ferrosilicon. Beneficiated quartzite, coke, or charcoal as reductant and scrap iron are measured and blended in proportions to give the various grades of silicon and ferrosilicon desired. This charge is fed into electric arc furnaces, the furnace is periodically tapped, and the molten silicon (or ferrosilicon) is drawn and cast into ingots. After cooling, the ingots are crushed and sized for shipping.

The production of high purity silicon, of semiconductor grade, involves hydrogen reduction of silicon tetrachloride or trichlorosilane, $SiHCl_3$. The use of $SiHCl_3$ is preferred because of faster decomposition rates, and greater ease in removing phosphorus and boron contaminants. Apparent consumption of high purity Si by the electronics industry during 1974 was estimated as 280 tons in the United States and 230 tons each in Japan and Europe.

3.2 Uses and Industrial Exposures

Silicon is used for deoxidizing and as a strengthening alloy in the production of iron, steel, and nonferrous metals. Silicon supply and production are related primarily to the requirements of the steel and aluminum industries in an alloying agent. Silicon's low cost, almost limitless resources, and inertness make it a prime subject for research to develop new uses and as a substitute for more expensive materials.

As a result, silicon has been employed in the production of silicones (see Section 6.7.9). Because of its high latent heat of fusion, silicon is the element of choice for heat energy storage in solar-energy-conversion systems. Silicon is also widely used in infrared optical instruments because of its high infrared transmissivity, hardness, and chemical inertness.

Silicon is used to produce intermediate products such as silanes from which several hundred silicone resins, lubricants, plastomers, antifoaming agents, and water-repellent compounds are formulated.

Silicon of high purity is used extensively in the electronics industry, more specifically as a semiconductor in computers, calculators, and communications equipment to control and amplify electrical signals. Silicon rectifiers are more reliable than most other types because they can operate efficiently at high temperatures. A more recent use is in photovoltaic cells in which pure silicon converts sunlight directly into electricity.

Specifications and standards for the various grades of silicon have been thoroughly detailed by the American Society for Testing Materials (ASTM) (116a) and recommends that the producer furnish with each silicon shipment an analysis showing its Si content, and on request, to include content of C, S, P, Al, and Mn. At present, there are 10 different grades based primarily on Si content and/or additive elements (36).

3.3 Physical and Chemical Properties

The physical and chemical properties of silicon are given in Table 40.8. Being a tetravalent element like carbon it crystallizes in the diamond lattice, which again is the basis for forming chains of polymers with alternating atoms of silicon and oxygen with organic substituents attached to the silicon atoms (silicones).

Naturally occurring silicon contains 92.2 percent of the isotope of mass number 28, 4.7 percent of ^{29}Si, and 31 percent of ^{30}Si. In addition to these stable, natural isotopes, artificially radioactive isotopes of masses 27 and 31 are known.

In very pure form silicon is an intrinsic semiconductor, although the extent of its semiconduction is greatly increased by the introduction of minute amounts of impurities. Elements of the third group, such as boron, introduce atoms with a deficiency of electrons into the crystal structure, which conduct electric current by migration of electron vacancies or "holes."

Silicon resembles metals in its chemical behavior; it is about as electropositive as tin and more positive than germanium or lead. Silicon oxidizes rapidly at room temperature to form a protective layer of silica about 10 Å thick. More complete oxidation begins at 650°C, but is not rapid up to about 1200°C. The oxide layer is amorphous to about 1200°C, crystalline (tridymite or cristobalite) about 1200°C, and somewhat volatile about 1600°C. Silicon semiconductor devices are generally protected with a silica layer by oxidizing at 1100 to 1300°C.

Silicon forms several series of hydrides, a variety of halides (some of which contain silicon-to-silicon bonds), and also many series of oxygen-containing compounds which may be either ionic or covalent-acting in their properties.

3.4 Analytic Determination

Silicon dust may be determined by standard X-ray powder diffraction methods using α-Al_2O_3 in a microcrystalline state (0.3 mμ average particle size) and a reference intensity ratio as described by Hubbard (116b), or by destructive methods, emission spectrographic or neutron activation analysis (NAA). Silicon

TABLE 40.8. Physical and Chemical Properties of Silicon and Some of Its Industrial Compounds

Element or Compound	Formula	At. or Mol. Wt.	Physical Properties	Sp. Gr.	M.P. (°C)	B.P. (°C)	Solubility
Silicon	Si	28.086	Steel gray cubic cryst., large to microsc.	2.32–2.34	1410	2355	Insol. cold, hot H_2O sol. HF + HNO_3; insol. HF, molten metals
Silicon dioxide (Native quartz)	SiO_2	60.08	Colorless hex.	2.635–2.660	1610	2230	Insol cold, hot H_2O; sol. HF; very sl. sol. alkali
Silicon dioxide (Native crystobalite)	SiO_2	60.08	Colorless cubic or tetr.	2.32	1713±5	2230	Insol cold, hot H_2O; sol. HF; very sl. sol. alkali
Silicon dioxide (Native tridymite)	SiO_2	60.08	Colorless rhombic	2.26	1703	2230	Insol cold, hot H_2O; sol. HF; very sl. sol. alkali
Sodium meta-silicate	Na_2SiO_3	122.06	Colorless monoclinic	2.4	1088	—	Sol. cold H_2O; dec. hot H_2O; insol. alcohol, K and Na salts

Name	Formula	Mol. wt.	Color, form	Density	M.P.	B.P.	Solubility
Sodium metasilicate, hydrate	$Na_2SiO_3 \cdot 9H_2O$	284.20	Colorless rhomb. bipyramid	—	40–48	$-6H_2O$ 100°C	Very sol. cold, hot H_2O: sol. (dec.) 11 NaOH; insol. alcohol, acid
Sodium orthosilicate	Na_4SiO_4	184.04	Colorless hex.	—	1018	—	Sol. cold, hot H_2O
Silicon carbide	SiC	40.10	Colorless to black, hex. or cubic	3.217	~2700	—	Insol. cold, hot H_2O; insol. acid; sol. fused KOH
Silicon tetrachloride	$SiCl_4$	169.90	Colorless fuming liq.	Liq: 1.483^{20} Gas: 7.59	−70	57.57	Dec. cold, hot H_2O; dec. alcohol
Silane	SiH_4	32.12	Colorless gas	Liq: 0.68^{-85} Gas: 1.44 g/l	−132.5	−14.5	Sl dec. cold H_2O: sol. alcohol, benzene, CS_2
Trichlorosilane (silicochloroform)	$SiHCl_3$	135.45	Colorless liq.	1.34	−126.5	33 (758 mm)	Dec. cold, hot H_2O; sol. CS_2, CCl_4, Chl, benzene
Ethyl silicate (ortho)	$Si(OC_2H_5)_4$	124.2	Colorless liq.	0.9356	110	168.1	Dec. cold, hot H_2O; sl. sol. benzene

has very strong lines at 2516.123 and 2881.578 Å in the 1 to 10 ppm range, and NAA uses the n–p reaction to ^{28}Al, which has a half-life of 2.3 min.

Silicon in biologic tissues and fluids can be determined by the method of proton-induced X-ray emission (PIXE) as developed by Grant and Buckle (116c). After tissue digestion with HNO_3, a $HCl–HNO_3$ mixture, or H_2O_2, a thin film of the digested sample is bombarded with protons from a Van de Graaf accelerator. Analysis time is 15 min, using only a few milligrams of sample. Sensitivity is of the order of <1 ppm, silicon being at the upper limit because of its lower atomic weight. Data are analyzed from a computer printout. Checks with atomic absorption spectrometry show reasonably good agreement on samples containing 40 to 60 μg Si/g sample. PIXE also simultaneously determines 39 elements from Al to U.

PIXE is equally applicable for determining silicon, or any inorganic silicon compound in air samples collected on millipore or nuclepore filters. Because the method depends on the availability of a Van de Graaf accelerator, samples must be submitted to an institution that has such equipment (116c).

3.5 Physiologic Response

No determination of the toxicity of pure, elemental silicon dust has been found, but judging from its general inertness, it could be classed as a "nuisance" dust. Some forms treated at around 1200°C, however, have an amorphous layer that becomes crystalline which could alter its toxicologic inertness.

3.6 Hygienic Standard of Permissible Exposure

The TLV for silicon appears in Appendix E of the TLVs for Chemical Substances as a "nuisance" particulate with the limits of 30 mppcf or 10 mg/m^3 of total dust containing less than 1 percent quartz, or 5 mg/m^3 respirable dust. The STEL is 20 mg/m^3.

3.7 Silicon Alloys and Compounds

There are three great classes of silicon-containing substances: (1) alloys and silicides; (2) inorganic compounds and substances, both naturally occurring as the oxide, SiO_2, and silicates, as found in asbestos, cement, mica, and soapstone, for example, and manufactured, such as silicon halides, hydrides (silanes, Si_nH_{2n+2}), a carbide and nitride, and metal silicates; and (3) organic compounds, silicon ethers and esters, silylating, agents and a multitude of various silicones.

3.7.1 Alloys

Alloys containing 6 to 95 percent silicon are used extensively by the iron and steel industry, normally in the form of various grades of ferrosilicon. In the

TABLE 40.9. Some Commercial Silicon Alloys

Name	Composition
Standard 50% ferrosilicon	47–51% Si, remainder mainly Fe
Standard 65% ferrosilicon	65–70% Si, remainder mainly Fe
Standard 75% ferrosilicon	73–78% Si, remainder mainly Fe
Standard 85% ferrosilicon	83–88% Si, remainder mainly Fe
Silicon metal	97.75% Si min, 0.07% Ca max, 0.51–1.00% Fe max
Calcium–silicon	30–33% Ca, 60–65% Si, 1.50–3.00% Fe
Calcium–manganese–silicon	16–20% Ca, 14–18% Mn, 54–59% Si
Ferrochrome silicon	38–42% Cr, 38–42% Si, 0.05% C max
Magnesium ferrosilicon	44–48% Si, 8–10% Mg, 1.00–1.50% Ca, 0.50% Ce
Simanal	approx. 20% each of Si, Mn, Al, balance mainly Fe
Silicomanganese	65–68% Mn, 12.5–18.5% Si, 1.50–3.00% C, balance mainly Fe
SMZ alloy	60–65% Si, 5–7% Zr, 5–7% Zr, 5–7% Mn, 3–4% Ca, balance mainly Fe

iron and steel industry, silicon alloys, often referred to as silicides, are used for alloying, deoxidizing, and reducing other alloying agents such as Mn, Cr, W, and Mo. In the nonferrous metal industry, silicon is used primarily as an alloying agent for Cu, Al, Mg, and Ni. In the form of 75 percent ferrosilicon, it is used as a reducing agent in the production of magnesium by the Pidgeon process. The reaction between ferrosilicon and caustic soda can be used to make hydrogen.

Table 40.9 lists some of the more important commercial silicon alloys.

U.S. Domestic production of ferrosilicon in 1973 was 447,000 short tons of a total world production of 1.543 million tons, European countries contributing 55 percent of the total (36).

For further details on the types, production and uses of these alloys, see the chapter on silicon in Reference 53.

3.7.2 Silicon Dioxide, SiO_2

Silicon dioxide exists in two varieties, amorphous and crystalline. A number of natural noncrystalline varieties of SiO_2 exist, such as the hydrated form, opal, and an unhydrated form, flint. In recent years, synthetic amorphous silicas have come on the market, in three forms according to their method of preparation, SiO_2 gel (silica G), precipitated SiO_2 (silica P), and fumed SiO_2 (silica F).

The chief representative of crystalline SiO_2 is quartz. Three other forms of crystalline SiO_2 are cristobalite, tridymite, and tripoli, which, though identical chemically, differ from quartz in crystalline form. They occur naturally and synthetically.

Amorphous Silica. By definition (117) synthetic amorphous SiO_2 is silica produced either by vapor-phase hydrolysis, precipitation, or other processes which assure the absence of crystalline particles. The fumed variety is derived from the vapor-phase hydrolysis of a silicon-bearing halide, silicon chloride; silica gels by reacting sodium silicate and an acid in solution; and precipitated silica by destabilizing sodium silicate in solution, causing precipitation of very fine silica particles.

Because of the many and varied special use requirements, these three varieties have proliferated into numerous types, presently 64, each with company-designated trade names such as Hi-Sils, Aerosils, Cab-O-Sils, Silcrons, and Syloids. These types have been further grouped into five classes according to their physiological effects when inhaled by human subjects, based on a few standard test procedures that identify dusts having similar origins or methods of manufacture.

The most striking property of these silica powders is their enormous surface areas, which in turn determine their applications; precipitated SiO_2 has a surface area (B.E.T.) of 125 to 150 m^2/g; fumed SiO_2, 175 to 225; silica gels, from 275 to 325 m^2/g, for a mean particle diameter of 0.014 to 0.022 to 0.010 to 0.018 μm, respectively. Their density varies around 2 g/ml. Oil absorption is high, from 170 to 190 g/100 g for precipitated SiO_2 to 275 to 325 g/100 g for the gels. Their pH at 5 g/15 g H_2O is between 6.5 and 7.5 for precipitated gels, but considerably lower for the fumed, 3.5 to 4.2 (117).

These amorphous silicas have as many as 20 applications in rubber and rubber goods, 25 in paper and paper products, 31 in anticaking, conditioning, and carriers in food, feeds, and chemicals, and 14 in paints, varnishes, and protective coatings. Their diversity of uses results in the rubber industry from their reinforcement qualities, giving increased tensile strength and hardness, abrasion resistance, increased stiffness, and better color. In the paper industry, the qualities of increased opacity and brightness allow better printability and improved smoothness and bulk. In their anticaking and conditioning qualities, improved flowability and dispersion of mixes commend them for such use, and their faster wetting of chemicals, foods, and feeds gives an outlet for these silicas in this area. They also provide controlled polishing and cleaning as well as acting as carrier for encapsulating catalysts and antifoams and defoamers. In their applications in paints, varnishes, and protective coatings, they provide higher hiding power, efficient flatting and extension of prime colors, improved thixotropy, and protection from the elements and corrosion.

Pathologic Effects. The newer methods of producing ultrafine amorphous SiO_2 particles with their enormous surface areas (SA) gave cause for reinvestigating their toxic potential. (The comparison of the SA of 55 m^2/g of the fluorescent grade of BeO dust, which is highly injurious when inhaled, with the SA of the newer silicas of 2.5- to sixfold greater, is sufficient reason to prompt a reinvestigation.) Such an investigation was made by a team of researchers in inhalation toxicology, particle size and electron probe analysis, animal pulmonary function studies, and animal pathology in the Division of Biomedical Science, NIOSH, Cincinnati, during 1978 and 1979 (118a). Because of the great health significance of the results, and their present unpublished form, they are given in some detail.

An animal inhalation toxicity study was performed on 80 to 90 rats, 20 guinea pigs, and 10 monkeys for daily periods of 5.5 to 6 hr for up to 18 months at 7 to 10 mg/m^3 respirable dust (15 mg/m^3, total SiO$_2$ dust). Exposures were to the three types of synthetic amorphous silicas, precipitated (P), gelled (G), and fumed (F) (pyrolyzed). All three dusts were analyzed by proton-induced, X-ray fluorescence. Some of the rat lungs were analyzed for silicon by plasma emission spectroscopy. Anderson impactor studies revealed that 46 percent of silica P, 62 percent silica G, and 65 percent silica F were less than 4.7 μm in diameter, and therefore to these extents were respirable. Autopsies were performed on rats at 3, 6, and 12 months and on guinea pigs and monkeys at 10 to 18 months after the start of exposures.

The most significant morphological alterations from all three silicas were confined to the lungs of the monkeys, which contained large numbers of macrophage and mononuclear cell aggregates, and which, in the respiratory bronchioles, appeared to reduce the size of their lumina significantly.

In the cellular aggregates, there was a difference in type and quantity of extracellular components. Reticulin fibers were uniformly present in the aggregates from all three silica types, but collagen was present in significant amounts only in those monkeys exposed to silica F; none were seen in the aggregates of lungs of monkeys exposed to silica G, and very few in those exposed to silica P. Histopathological examination of the lungs of rats and guinea pigs showed far fewer and smaller aggregates than those of the monkeys. A probable reason was the estimate made from light microscopy that less than 10 percent silica was in the lungs of these animals than in the monkeys. Microprobe analysis revealed the presence of silicon in all the aggregates examined (116c).

The great significance of the presence of collagen fibers in the cellular aggregates in the monkey lungs exposed to amorphous silica F is that it is *fibrogenic*. The much smaller amounts of collagen in the monkey lungs exposed to silica P indicate that it is much less fibrogenic than silica F. The reticulin fibers found in all aggregates is not without toxicologic significance also, for its potential for producing collagen in silica-containing aggregates is greater than in those areas without it. The full potential effect of the presence of reticulin and collagen fibers from exposure to these silicas is probably not expressed in these studies of 10 to 18 months duration, which is relatively brief for monkeys.

The silicon content of the freeze-dried lungs of the exposed animals, as analyzed, was, as expected, considerably greater than those of controls, which in the case of the rats were six, three, and two times greater for silica P, G, and F, respectively. In view of the fact that silica F was the most fibrogenic, the finding of the lowest amounts of the three provides further evidence for the far greater fibrogenic potential of silica F.

Pulmonary Function.   Eleven tests of pulmonary function were made to determine the degree of respiratory impairment in the monkeys exposed separately for 1 year to the three amorphous silicas in an apparatus uniquely designed for the purpose (118b). Of the functions that were statistically significantly different from controls at the end of 1 year were the lung volume measurements of forced vital capacity (FVC), inspiratory capacity (IC), and total lung capacity (TLC) in those exposed to silica F and P. In the measurements of lung ventilatory mechanisms, resistance, compliance, forced expiratory flow at the last 10 percent ($FEF_{10\%}$), and closing volume were statistically different from controls in the F and G groups; the greatest differences from controls occurred in compliance, $FEF_{10\%}$, and in residual volume/total lung capacity, the magnitude of the differences again being greatest in the silica F group. These significant decrements in lung function after an exposure of only 1 year further demonstrate the injurious effects of these amorphous silicas. Heretofore, amorphous silicas previously encountered were considered to produce a benign pneumoconiosis. Whether effects of greater severity would result from long-term exposure must be left for future study. Another intriguing question from these studies is, why is silica F restrictive of lung size, yet obstructive in getting respiratory flow expired?

Although no proven explanation is forthcoming on the greater, early fibrogenic action of silica F, the explanation may be in its far greater surface area than its counterparts, silicas P and G, its greater solubility resulting in a far lower pH value (see above), thus providing greater degree of tissue irritation, and the greater content of aluminum (67- and elevenfold) in silicas P and G. Aluminum is recognized as an abettor of silicosis and thus conceivably reduces the fibrogenic action of these two silicas.

A pathological study of rats inhaling Degussa dust, a submicron, high surface area, amorphous SiO_2 by Schepers (118c) 20 years before, showed that at about 50 mg/m^3 (1.5 mg/ft^3) the majority of the rats died from pulmonary, vascular obstruction coupled with pulmonary insufficiency due to emphysema at three to five months. Progressive increases of SiO_2 in the lungs accompanied the pulmonary lesions. Surprisingly, the majority of the surviving rats rapidly recovered on removal from the dust exposure, the silica was largely eliminated, and the cellular nodules, perivascular infiltrations, and emphysema were almost completely resolved.

It is clear from the above findings that the TLV for these amorphous silicas must be reduced in accordance with their relative fibrogenicity from the TLV of 6 mg/m^3 for total, and 3 mg/m^3 for respirable dust.

Crystalline Silica, SiO_2. *Source and Production.* The chief source of crystalline SiO_2 is quartz, a mineral found ubiquitously in most classes of rock, and an important constituent of those igneous rocks such as granite and pegmatite which contain an excess of SiO_2. Quartz also occurs in large amounts as sand in stream beds, seashores, and deserts, and as a constituent of soils. In rocks, quartz is associated chiefly with feldspar, a potassium aluminum silicate, $KAlSi_3O_8$, and its hydrous form, $KAl_2(AlSi_3O_{10}) (OH)_2$ muscovite.

Two other minerals chemically identical to quartz are cristobalite and tridymite, differing from quartz in crystalline form and recognized by microscopic examinations, X-ray diffraction, and infrared spectrophotometry. Cristobalite and tridymite usually occur together as high temperature silicate minerals in the volcanic rocks of California, Colorado, and Mexico. Cristobalite also forms in the calcining of diatomaceous earth, the amount depending on the time and temperature of calcining. Under usual operating conditions, cristobalite may form to the extent of 60 percent of the original diatomaceous earth.

Apart from the usual mining operations of dredging and mechanical crushing to obtain impure quartz, pure crystals of quartz are prepared by mass production methods under carefully controlled conditions of temperature and concentration.

Uses and Industrial Exposures. Impure quartz, as silica sand, is used in the manufacture of glass and silica brick, in mortar, and as an abrasive. In powdered form, as silica flour, it is used in paints, porcelain, scouring soaps, and as a wood filler. The clear rock crystals of the colored varieties are valued as gems and used ornamentally (amethyst, onyx, rose quartz, agate). Pure, synthetic quartz is used in electronic components, for piezoelectric control in filters, oscillators, frequency standards, wave filters, and for radio and TV components.

The sources of industrial exposures are many. Probably the most common source of SiO_2 dust exposures leading to silicosis is the drilling of free silica-bearing rock. Presenting even more extreme hazards is abrasive blasting. Pottery workers come under significant SiO_2 exposure if dust-dirty work clothes are not properly cleaned. Likewise, dust generated from handling and transporting SiO_2-containing dust from crushed rock, sand, gravel, clay, or other minerals can pose a hazard. Similarly the handling of carloads and bags of SiO_2 flour has an associated exposure hazard. Finally, flux-calcined diatomaceous earth containing substantial amounts of cristobalite and/or tridymite can generate very hazardous dust, as does the dust associated with maintenance and repair of equipment.

Physical and Chemical Properties. The physical and chemical properties of three forms of crystalline SiO_2 are given in Table 40.8, but the physical properties of quartz are far more complex than can be presented in such a table. Twenty-two different phases of silica have been identified. At 573°C the ordinary α-quartz changes over reversibly to β-quartz, which has a lower density, and at 86°C, β-quartz changes to a different crystal modification, β-tridymite. At a still higher temperature, 1470°C, β-tridymite becomes a third modification called β-cristobalite. Both β-tridymite and β-cristobalite have lower temperature α modifications which have lower optical symmetry, and all the forms can be maintained at room temperature if chilled rapidly from the stable equilibria. Each appears to have its own melting point, but the usual melting point for silica is that of β-cristobalite, about 1723°C. Rapid chilling of the liquid produces silica glass, often incorrectly called quartz glass, a vitreous modification of SiO_2 which has a very low coefficient of expansion.

Analytic Determination—Sampling and Analysis. The National Institute for Occupational Safety and Health (NIOSH) endorses the sampling and analysis procedures for evaluating worker exposure to crystalline SiO_2 (quartz, tridymite, cristobalite) recommended by the American Conference of Governmental Industrial Hygienists (ACGIH) (119a, b).

The dust is collected with a size-selective personal sampler positioned in the breathing zone of the worker. Dust penetrating the precollector is collected on a low-ashing polyvinyl chloride (PVC) filter and the free SiO_2 content is determined by X-ray diffraction, after the dust is redeposited on a silver membrane filter.

The procedures are also given in detail with regard to principle of the method, range and sensitivity, interferences, precision and accuracy, advantages and disadvantages, and apparatus (120a, b).

Physiologic Response. The pulmonary effects on workers exposed to crystalline SiO_2 are to be found in Vol. I of this edition (121). Accordingly, only those aspects of exposure relating to these effects are summarized here in an attempt to demonstrate the basis on which the control of SiO_2 exposure was made for the prevention of the disease in the many industries and occupations, where varying types and percentages of SiO_2 dust and associated nonsiliceous dusts and personal habits (smoking) interact.

Silicosis, a dust disease (pneumoconiosis) of the lungs resulting from overexposure to free SiO_2 dust, usually begins insidiously, with symptoms of coughing, dyspnea, wheezing, and repeated, nonspecific chest illnesses. Impairment of pulmonary function may be progressive. In individual cases, there may be little or no decrement when simple, discrete, nodular silicosis is present, but when nodulations become larger, or when conglomeration occurs, cardiopulmonary impairment tends to develop.

The various stages of progression of the silicotic lesions are related to the degree of exposure of free SiO_2, the duration of exposure, and the time during which the retained dust is permitted to react with the lung tissue. Epidemiologic studies show that the higher the SiO_2 dust exposure, the more rapid the development of silicosis and its prevalance. As dust is controlled, however, the frequency of occurrence of silicosis decreases, the severity of the disease lessens, and the length of time for the disease to become manifest increases. As a consequence of this extended time, it becomes increasingly difficult to establish a relation between dose and response.

To illustrate the complexities facing early investigators in attempts to find a "safe" exposure limit for free SiO_2-containing dusts, one has only to note that there were six major classes of industries which presented different degrees of exposure and differing percentages and sizes of SiO_2 dusts. To cope with this situation, early (1930) investigators assigned different "maximal, permissible, safe, dust limits" which were reported at a National Conference on Silicosis, held in Washington, D.C. in 1936 (Table 40.10) (122). Not considered at the conference were the exposures presented in the iron and steel industries, the nonferrous foundries, and the amorphous SiO_2-producing industry (cf. Section 6.7.2), which also has its individual and varied types of exposure.

Hygienic Standards of Exposure.   Hygienic standards of exposure to free SiO_2-containing dusts are based on the concept that the degree of toxicity is proportional to the concentration of free SiO_2 in the dust. When represented as a threshold limit, it is expressed as

$$\text{Threshold limit} = \frac{K}{\%SiO_2}\text{ mppcf}$$

where K is a constant related to the incidence of silicosis as determined from epidemiologic studies.

To make the threshold limit consistent with the nuisance dust TLV, adopted

TABLE 40.10   Industries with Potential Exposure to Free Silica and Permissible Limits[a]

Industry	Location	Percent SiO_2 in Dust	Permissible Safe Dust Concn. (mppcf)
Metal mining	South African gold mines, Ontario gold mines	80 ~35(in rock)	4.5 8.5
Coal mining	Pennsylvania anthracite	35	5–10
Nonmetal minerals (except fuels)	Broken Hill, Australia	10–17	14
Stone, clay, and glass products	Barre, Vt. granite Australia sandstone	31–38 90 (in rock)	10–20 6

[a] From National (U.S.) Silicosis Conference, 1937 (122).

The onset of silicosis can be hastened by the inhalation of other substances along with silica. The alkalinity from sodium carbonate added to the sand in a scouring preparation was reported (122) to have caused silicosis in from 20 to 26 months ("galloping" silicosis) in a small group of exposed individuals; the silica dust concentration was tremendously high as far as can be gathered from the original report. Rapidly progressing silicosis had been previously reported in the U.S. (123) from the same cause. Tolcosis of unusually rapid development (16 to 24 months) was reported (124) also under circumstances of excessively high talc concentrations. Fluoride, in the form of fluorspan, in admixture with quartz has been reported (125) experimentally to induce more intense fibrosis than quartz alone, although the ester may be present in relatively small amounts (1 percent) in the mixture.

in 1970, K was placed at 300 and a constant of 10 added in the denominator:

$$\text{TLV} = \frac{300}{\%\text{SiO}_2 + 10} \text{ mppcf} \tag{a}$$

Hosey et al. (123) reported in 1957 that no cases of silicosis occurred in Vermont granite workers whose exposure to dust had been subsequent to dust control in 1937 and where dust exposures had averaged less than 5 mppcf. The airborne dust averaged 25 percent quartz, so the revised TLV formula, 300/ (% quartz + 10), would have given a TLV of 9 mppcf. This was identical with the upper limit of exposure of the group of granite workers found without silicosis by Russell in 1929 (124). The count formula is also reasonably consistent with results from studies in the anthracite region of Pennsylvania (125a), an early nonferrous metal mine study (125b), and studies of pegmatite workers (125c). Studies of foundry workers (125d), pottery workers (125e), and metal miners (125f) showed chest roentgenograms consistent with silicosis in workers who were at the time of the surveys exposed below the TLV, but it was believed that previous exposures may have been considerably higher. Early limits for count concentration of dust from South Africa, Australia, and Ontario (Table 40.10) were reasonably consistent with the count formula, even though sampling methods and counting procedures were somewhat different from U.S. practice.

With the development of devices for size sampling of dusts, it became possible to express the TLV in terms of mass for respirable and total dust, which has considerable, obvious advantages over the counting procedure. Comparisons of impinger-count concentration and respirable-mass concentrations show that 9 to 10 mppcf of granite dust, suggested as a limit by Russell, contains 0.1 mg/m³ of respirable quartz (126a). The formula TLV = (10/% respirable quartz) mg/ m³ generalizes this relationship to all percentages of quartz in respirable dust. If the TLV were used only for dust containing at least 5 percent quartz, the above TLV formula would be satisfactory, but to prevent excessively high permissible respirable dust concentrations when the fraction of quartz in the dust is less than 5 percent, a constant has been added in the denominator, as with the count TLV (a), giving the formula

$$\text{TLV} = \frac{10}{\% \text{ respirable quartz} + 2} \text{ mg/m}^3 \tag{b}$$

The additive constant "2" limits the concentration of respirable dust with 1 percent quartz to 5 mg/m³. The above TLV has been demonstrated to give hazard evaluations comparable to the impinger method in foundry dust exposures (128b). Where agglomerates are a factor, the results by the respirable mass method are more closely related to the hazard.

Percent quartz in respirable dust is often quite different from the percentages in settled dust or total airborne dust, and the percent quartz for use in the respirable-mass TLV formula must be determined in a sample of respirable dust.

The concentration of respirable dust must be determined by an instrument which gives a size separation equivalent to that specified in the ACGIH criteria for size selection (126c). The ACGIH criteria are as follows:

Aerodynamic Diameter (μm) (Unit Density Sphere)	% Passing Selector
<2	90
2.5	75
3.5	50
5.0	25
10	0

For low concentrations of dust, especially where quartz percentages are high, respirable quartz may be determined directly from air samples and concentrations of quartz compared with the basic value of 0.1 mg/m³ quartz. In the use of 0.1 mg/m³ respirable quartz as a limit, care must be taken that other limits ("nuisance" dust, coal mine dust, etc.) are not exceeded.

A formula for total dust concentration expressed as mass has also been recommended by the TLV Committee:

$$\text{TLV (total dust)} = \frac{30}{\% \text{ quartz} + 3} \text{ mg/m}^3 \tag{c}$$

It is based on the limitation of "nuisance" dust to 10 mg/m³ and the fact that respirable quartz usually averages less than one-third of the total quartz. Limited data from foundries and coal mines suggest that the formula for the TLV for total dust will usually give a result on the safe side, when compared with the respirable dust TLV formula. However, respirable quartz must not average more than 0.1 mg/m³, and in those instances where respirable dust is more than one-third the total dust, the respirable dust TLV formula must be used. Thus three formulas, (a), (b), and (c), for controlling free crystalline SiO_2 in workroom air are recommended by ACGIH.

NIOSH, however, in recommending a standard exposure to crystalline SiO_2 to OSHA, took a different tack. Instead of adopting formulas that allow differences in dust concentration according to its SiO_2 content, NIOSH made a blanket recommendation of 0.05 mg/m³ as the standard of permissible exposure for all situations, the standard to be applied as a TWA concentration, "as determined by a full-shift sample for up to a 10-hour workday, 40-hour workweek" as a breathing zone sample (120b).

The reason for this choice is not made clear in the criteria document (120b). As stated, however, the basis was the 50-year study of the Vermont granite sheds by the USPHS, in which effective control was gained at 5 mppcf, which in mass is equivalent to 0.05 mg/m^3. Accordingly, "it seemed appropriate to apply this limit, in terms of respirable mass, to other operations producing dusts containing free silica."

The present federal standard for free SiO$_2$ is an 8-hr, TWA based on the 1968 ACGIH TLV formulas of 250/(%SiO$_2$ + 5) mppcf or 10/(%SiO$_2$ + 2) mg/m^3 for respirable quartz, one-half this limit for cristobalite and tridymite.

The allowable level of airborne quartz in coal mines is 0.10 mg/m^3, twice the limit recommended by the criteria document.

Silicosis Therapy.   For a discussion of aluminum in the prevention and treatment of silicosis, see Section 5.9 in the chapter "Metals," Vol. 2A of this edition.

Cristobalite, Tridymite, Tripoli, and Fused Silica.   Of the four crystalline forms, three, cristobalite, tridymite, and tripoli, exist naturally. Cristobalite, tridymite, and fused silica (quartz) can be produced by heating quartz or amorphous SiO$_2$ to elevated temperatures. The industrial source of cristobalite and tridymite, however, is from flux-calcining diatomaceous earth at regulated temperatures and times. These two minerals are used extensively as filtering and insulating media, and as siliceous refractory materials for furnace linings and silica bricks.

Cristobalite and Tridymite.   These two minerals are treated together because of their common natural (volcanic rock) and synthetic association, their common industrial applications, and toxicologic response.

Physiologic Response.   The pneunoconiosis-producing effects of these two minerals have been much studied both in industry (127a–c) and in the laboratory (128a–d) since the early (1941) report of Fulton and associates in the burned silica brick industry (129). This study drew attention to the greater capacity of cristobalite and tridymite to induce silicosis than quartz, a finding later confirmed by animal studies (128b, c) and suggested by epidemiologic studies of the diatomite industry (127b, c) and which led to a *threshold limit value* of one-half that calculated from the count or mass formulas for both silica forms (130).

Tripoli, as encountered in industry, refers to a *group* of highly porous, microcrystalline minerals with free silica contents ranging from 90 to 98 percent, found in Missouri, Oklahoma, Arkansas, and Georgia, in the United States. It is not to be confused with tripolite, a diatomaceous earth, found in Tripoli, North Africa. Rottenstone and microcrystalline silica are similar to tripoli. These minerals are used in buffing compounds, on buffing wheels, in scouring soaps and powders, and in polishes; a very finely sized white grade, "white rouge," is used for polishing optical lenses. About 35,000 tons of tripoli, including rottenstone and microcrystalline silica, was used for abrasives in the United States in 1960.

Early (1943) physiological response studies by McCord et al. showed that direct implantation of tripoli in animals led to tissue proliferation similar to that from quartz (131a), but later (1962) investigation of tripoli workers by McCord (131b) failed to find any clinical evidence of silicosis among the workers. He attributed this lack of response to a relative lack of exposure brought about by the dielectric properties of tripoli particles, which, because of rapid agglomeration and settling of the airborne particles, reduced exposure to a minimum. Gardner had similarly suggested particle flocculation as a reason for the absence of serious lung pathology in diatomaceous earth workers, whereas animal studies showed that this substance produced progressive nodular fibrosis (131c).

Because of particle agglomeration, conventional counting procedures by light-field microscopy are of questionable value, and the *threshold limit* is based on *respirable mass* and is the same as that for quartz.

Fused silica is quartz that has been melted to a glass-like substance on cooling. According to Gross (132), although fused silica may appear amorphous, when submitted to X-ray diffraction it is actually shown to consist of microcrystals of quartz too small to be detected by direct X-ray. Although King et al. (128b) found fused silica considerably less active than quartz when injected intratracheally in rats, his particle size of 87 percent <2.6 μm may account in some part for his results. Be that as it may, the lack of sufficient industrial experience with fused silica handling and use moved the TLV Committee to recommend the same *threshold limit* as that for quartz.

Fused silica finds use in apparatus and equipment, such as vacuum tubes, where its high melting point, ability to withstand large and rapid temperature changes, chemical inertness, and transparency including to UV light are requirements. It is produced as fibers and fabrics where heat resistance, low expansion coefficient, and high dielectric strength are needed.

3.7.3 Silicates

Silicates can be grouped into two great classes, natural and synthetic. Of major industrial importance in the natural group are asbestos, mica, mineral wool, perlite, portland cement, soapstone, talcs, and tremolite, all of which have recommended threshold limits. The synthetic, inorganic group comprises for the most part the soluble silicates formed with the alkali metal and quaternary ammonium base such as the sodium silicates (Table 40.8). Insoluble silicates such as lead silicates are formed from the alkali metal silicates. Insoluble silicates also exist in nature, for example, beryllium aluminum silicate, beryl.

Insoluble, Inorganic Silicates. *Asbestos.* Limitations of space confine the treatment of these silicates, particularly asbestos, to brief summaries and references to the literature for further details.

Asbestos is a generic term that applies to a group of naturally occurring, hydrated mineral silicates that are separable into fibers. Chrysotile, with the theoretical formula 3 $MgO \cdot 2SiO_2 \cdot H_2O$, is the variety wanted by more than 95 percent of the world's consumers. Other types include amosite, $(FeMg)SiO_3$, crocidolite, $NaFe(SiO_3)_2 \cdot FeSiO_3 \cdot H_2O$, anthophyllite, $(MgFe)_7Si_8O_{22}(OH)_2$, and tremolite, $Ca_2Mg_5Si_8O_{22}(OH)_2$.

Total world production in 1974 was 4.5 million tons of all grades and varieties (36). Of this, Canada's output was 40 percent, the Soviet Union, an estimated 33 percent, South Africa 8 percent, Republic of China 5 percent, Italy 4 percent, and the United States 2 percent.

Asbestos is adaptable to more than 2000 uses, all as processed fiber. Chrysotile is graded and grouped according to fiber length, the longest fibers being in groups 1 to 3, whose major uses include textiles, different types of packings, woven brake linings to clutch facings, and electric insulation. The major use of fibers in group 4 is in asbestos cement pipe for transporting water, and in group 5, for asbestos cement sheets, low pressure cement pipes, and molded products; the main consumption of fibers in groups 6 and 7 is in asbestos cement mixes such as gaskets, vinyl sheet backing, joint and insulation cements, roof coatings, plastics, and caulking compounds.

Amosite is used mainly for felted insulation for high temperature service up to 900°F and as covering for marine turbines and jet engines; long fiber crocidolite (blue asbestos) is woven into fabrics for locomotive boiler lagging and is used for acid-resistant packings and gaskets, shorter fibers being used for asbestos cement pipe; anthophyllite and tremolite are used for chemical-resistant filters, as welding-rod castings, and as fillers in various products.

It can be seen from these many and varied uses that potential industrial (and environmental) exposures are almost ubiquitous.

Analytic Determination. Because fibers, not the motes, are generally, but not completely, agreed to be the injurious agent, fiber counting is the approved way of estimating exposure to airborne asbestos (133). The preferred index of asbestos exposure is the concentration of fibers longer than 5 μm counted on a membrane filter at 430× with phase contrast illumination.

The recommended method for taking airborne samples and counting fibers is collection on a 37-mm Millipore type AA filter mounted in an open-face filter holder fastened to the worker's lapel, with air drawn through at a flow rate of from 1 to about 2.5 liter/min.

Details of counting procedure may be found in References 133 or 134.

Physiologic Response. The pulmonary aspects of fibrogenesis (asbestosis) and bronchogenesis, as well as mesothelioma, have been discussed by Wright in Vol. I (121). The aspects treated are under the following headings: (1) fiber aerodynamic behavior and pulmonary penetration, (2) fate of fibers in respiratory tract, (3) development of fibrogenesis and its organization relative to fibers and motes, (4) role of immunologic processes in pulmonary tissue reaction, (5) diagnosis radiographically and by physical examination, (6) risk of bronchogenic carcinoma in relation to smoking, and diffuse pulmonary fibrosis (asbestosis), and (7) risk of developing peritoneal mesotheliomas according to types of asbestos and relation to dose.

Control of Exposure.   This aspect has been discussed by Burgess in Vol. I of this edition.

Hygienic Standards of Exposure.   The history of the development of air standards for the prevention of asbestos disease is a history of change as reviews and closer examination of the several facets of the problem unfolded. In 1938 Dreessen et al. (135a) recommended a threshold limit of 5 mppcf after studying employees in four asbestos textile plants where massive exposures to *chrysotile* occurred. It was recognized at the time that dust counts by impinger collection give only an indirect measure of the risk of asbestosis because of collection of particulates other than asbestos.

Later evaluation and review, paticularly by Balzer and Cooper (135b), who reported asbestosis in insulation workers at levels that were felt highly unlikely to have exceeded 5 mppcf, indicated that the 5 mppcf limit was not sufficiently low to protect workers exposed for 30 years. At the same time, Selikoff et al. (135c) added a further consideration for lowering the threshold after a retrospective study found lung cancer associated with asbestos workers who smoked. Following these disclosures an interim TLV of 5 fibers/cc longer than 5 μm was adopted by ACGIH.

In 1979, after considerably more discrete information had accumulated, the TLV Committee listed the following TLVs for each of the industrial types of asbestos in the Notice of Intended Changes: chrysotile, 2 fibers/ml; crocidolite, 0.2; amosite, 0.5; tremolite, 0.5; and "other forms", 2 fibers/ml, with the notation directing the reader to the section on human carcinogens.

The development of standards by OSHA for asbestos is given in Vol. III of this edition. In September 1975 OSHA issued a proposed revised standard that revised the existing standard by reducing the permissible exposure level to 0.5 fibers/ml for all employments covered by the act except the construction industry, for which a separate proposal will be issued later.

Tremolite.   This mineral is a variety of asbestos with a theoretical formula corresponding to $Ca_2Mg_5Si_8O_{22}(OH)_2$. Its color is from white to light green owing to trace metal impurities, and has a vitreous to silky luster. Its specific gravity is somewhat greater than that of chrysotile asbestos, being 3.0 to 3.3; its hardness is from 5 to 6. Like other forms of asbestos, it is resistant to acids and is noncombustible. Some tremolite is sold as "fibrous talc."

Because of the striking similarity in the fibrotic reaction produced by tremolite and asbestos, the recommended TLV is 0.5 fibers/ml for those exceeding 5 μm in length.

Mica. Mica is a nonfibrous, naturally occurring silicate, found in plate form in nine different species. Muscovite, a hydrous aluminum silicate with the theoretical formula $KAl_2(AlSi_3O_{10})\cdot(OH)_2$, and phlogopite, a magnesium silicate (amber mica), $KMg_3AlSi_3O_{10}(OH)_2$, are the chief micas of commerce. Other forms include biotite, lepidolite, roscoetite, and zinnwaldite, all hydrous silicates, but differing in their content of metal type, Fe, Li, V, and LiFe, respectively.

Muscovite, the principal form of commercial sheet mica, is mined by hand from pegmatities (mixed rare metal silicates). Phlogopite is mined from pyroxenites, chiefly calcium and magnesium silicates, and undergoes the same hand processing as muscovite. In the processing and preparation of sheet mica for fabricators located mainly in the eastern United States, the sheets are stamped or punched to required sizes and shapes. Scrap and flake mica is recovered from pegmatites as a sole product, or as scrap, as a by-product from processing sheet mica (36).

U.S. production of sheet mica has fallen to essentially zero compared to a world production capacity of 45,000,000 lb (1954), 55 percent of which came from India and the remainder chiefly from South America and Africa. U.S. production of scrap and flake mica, however, constituted 60 percent of the world production, which continues to expand owing to low cost, mass-mining operations and large reserves (U.S.).

The foreign sheet mica product is fabricated in the United States which employs about 1400 workers. The majority of the production of scrap and flake comes from the beneficiation of open quarry ore. Flake is also a coproduct of many feldspar or kaolin operations, both aluminum silicates. About 400 workers are employed in the United States in these operations (36).

The various uses of sheet mica relate to its unique electric and thermal insulating properties, and its capacity to be cut, punched, or stamped to very fine tolerances. Primary users of sheet mica are the electronic and electric industries, which use block and film mica as vacuum tube spacers, and backing mica is used in the manufacture of capacitors. Very high quality mica has many important uses including linings, diaphragms, washers, disks, and plates in various gauges and dials, and communication devices, the latter of which take advantage of its dielectric and insulating properties. A new use of high quality sheet is in helium–neon lasers. Other uses include flexible plate for electric motor and generator armatures and field coil, magnet, and communication core insulation.

Uses of scrap and flake, generally processed as ground mica, is primarily in gypsum plasterboard as a filler, in the production of rolled roofing and asphalt shingles, and as a surface coater to prevent sticking.

Physiologic Response. At one time mica was considered to be in the "nuisance" dust category, on the basis of its inert reaction when injected intraperitoneally in guinea pigs according to the technique of Miller and Sayers (136). Dreessen and co-workers (137), however, found evidence of pneumoconiosis in 8 of 57 workers exposed to dust associated with mica-scrap grinding. Of five workers exposed at less than 10 mppcf, none had pneumoconiosis; cases were found in three men exposed at 18 mppcf for 18, 20, and 26 years, respectively; three cases resulted from exposure at 40 mppcf for 10, 17, and 23 years; and two cases were found in workers exposed at 50 ppm for 24 and 46 years. Only one of six workers exposed more than 10 years at concentrations in excess of 25 mppcf failed to show evidence of pneumoconiosis.

Although the signs and symptoms in these cases resembled those of silicosis, the X-ray pattern of the lung field markings differed somewhat, and tuberculosis was not a complication in any of the mica cases. No cases were found among workers exposed to mica dust in concentrations averaging 3 mppcf.

Vestal and associates (138) confirmed the presence of pneumoconiosis in workers exposed only to mica dust. They also found that a group of mica miners showed a higher incidence of pneumoconiosis, and also of tuberculosis, than miners of other minerals. Quartz was present in the dust to which both mica and other miners were exposed, however.

Heimann et al. (139) reported that mica workers in India, with exposures corresponding to 18 years at 20 mppcf, showed mild pneumoconiosis as evidenced by readings of chest X-rays. It was concluded from this study that the TLV of 20 mppcf is reasonable. Workers in the mica mines of India showed a very high incidence of silicosis (140a–c).

Hygienic Standards of Exposure.   The above information led the TLV Committee to propose a TLV for mica containing less than 5 percent free silica of 20 mppcf, with the caution that it should prevent disabling pneumoconiosis, but may not be sufficiently low to eliminate positive chest X-ray findings in workers with many years of exposure.

The present TLV restricts the silica content to less than 1 percent. This is the standard for mica adopted in 1968 by OSHA.

Portland Cement.   Portland cement refers to the most common type of cement used the world over. It derives its name from its resemblance to a well-known English building stone on the Isle of Portland. It is made by blending lime, alumina, silica, and iron oxide as tetracalcium aluminoferrate (with the theoretical formula $4CaO \cdot Al_2O_3 \cdot Fe_2O_3$), tricalcium aluminate ($3CaO \cdot Al_2O_3$), tricalcium silicate ($3CaO \cdot SiO_2$), and dicalcium silicate ($2CaO \cdot SiO_2$). Small amounts of magnesia (MgO), Na, K, and S are also present. Sand is added as a diluent (concrete). Cement may also be modified with various plastics to improve adhesion, strength, flexibility, and curing properties, and water evaporation can be retarded by adding resins such as methyl cellulose.

Portland cement is produced in the United States by grinding the raw materials, usually dry (to conserve fuel later), and calcining the mixture at about 1400°C in rotary kilns equipped with suspension heaters, and fueled mostly by coal. The cooled clinker thus formed is ground and mixed with certain additives, including gypsum ($CaSO_4 \cdot 2H_2O$) to form the cement, which is combined with sand or gravel to form concrete. The quartz content of most finished cements is low, usually below 1 percent.

U.S. Domestic production in 1973 was about 87.5 million short tons of a world total of 780 million (11 percent) (36).

The hazards of operating cement plants and their evaluation are presented by Burgess in Vol. I of this edition.

Physiologic Response. Gardner et al. (141) found no pneumoconiosis due to exposure to finished Portland cement in 17 cement plants with 2278 workers, despite heavy and prolonged exposures. This has been confirmed by other studies (142a–c). Conflicting reports from Italy (143a, b), however, appear related to exposures which occurred in mining, quarrying, or crushing silica-containing raw materials.

Threshold Limit. The TLV for portland cement is that for a "nuisance" dust, 30 mppcf, and is based on the lack of findings in the U.S. cement industry. A limit is needed, however, to prevent excessive exposures that result in caking in the nose and ears owing to its absorptive reaction with moist surfaces. The OSHA standard adopted from the TLV list of 1968 is 50 mppcf for cement containing <1 percent SiO_2.

Perlite. Perlite is a natural glass formed from volcanic action. It is essentially a metastable, amorphous mineral composed of sodium potassium aluminum silicate. It has the unusual characteristic of expanding to about 20 times its volume when heated to temperatures within its softening range (between 1400 and 2000°F, depending on water content and rate of heating). It is this expanded form that represents nearly 70 percent of consumption. A typical, average chemical analysis shows a range of 71 to 75 percent SiO_2, 12.5 to 18.0 percent Al_2O_3, 4 to 5 percent K_2O, 1 to 4 percent Na and Ca oxides, and trace amounts of other metal oxides (36).

Perlite is chemically inert with a pH of about 7. Its specific gravity is from 2.2 to 2.4 (139 to 150 lb/ft^3) and its hardness between 5.5 and 7 (Mohs scale). Crude perlite varies in color from transparent, light gray to glassy black, but when expanded is snowy white to grayish white. Expanded perlite is either a fluffy, highly porous substance, or glassy-white particulate with low porosity, depending on heating conditions. The bulk weight of the expanded perlite ranges from 3 to 20 lb/ft^3. Specifications of size and bulk density for specific uses have been established by ASTM.

U.S. sources of perlite are from deposits in New Mexico, which supplied 88 percent of the crude perlite mined in 1974. Thirty plants produce crude, and about 75 produce expanded perlite, distributed over 30 states.

Crude perlite mining is by open-pit methods. Most commercial deposits are in the form of massive lava that extend over a wide area close to the surface. Most deposits can be mined by bulldozers or tractors equipped with rippers. To ensure uniformity, initial mining is done in lateral sections across the open face rather than at depth.

Before perlite can be expanded, the crude must be crushed to produce particles approximating a cubic shape and the required particle size gradation (secondary crushing) and then sized to specified particle gradation. Dry processing is done in preference to wet to conserve energy.

The first processing step is to reduce the crude ore to approximately $\frac{5}{8}$-in. size in a primary jaw crusher and, if necessary, a secondary roll crusher. To facilitate further processing and expansion, the crude is usually passed through an oil-fired rotary dryer to reduce the moisture content. Secondary grinding is normally accomplished in a closed circuit system using screens, air classifiers, hammer mills, and rod mills. The various sized materials are stored for later blending and shipment. Any oversized material produced from the secondary grinding circuit is returned. Large quantities of fines—up to 25 percent of the mill feed—are produced throughout the processing stages but are removed by air classification at designated stages. This fine material is either collected and bagged, if salable, or dumped as waste.

In a study of the composition of 16 samples of perlite ores collected from 19 deposits in 16 western states, the free silica content was found to range from less than 1.0 to 2.0 percent in 15 of the samples to 3.0 percent in one sample. The free silica in these samples was reported to be crystalline (142a). The free silica content of typical expanded perlite was shown to vary from 0 to 2 percent (142b). Quantitative analysis by X-ray diffraction of the crystalline silica content of a series of 24 samples of perlite ore processed crude and expanded perlite showed a maximum of 0.4 percent quartz and 0.2 percent cristobalite (142c). In a plaster aggregate manufacturing plant samples of perlite ore and aggregate were found to contain from 1.6 to 2.5 percent free silica (142d).

U.S. domestic production in 1974 of crude processed perlite was 555,000 tons, or 32 percent of the estimated worldwide production.

Industrial uses for perlite (expanded) are many and varied. The more important applications include the following: abrasives, acoustical plaster and tile, charcoal barbecue base, cleanser base, concrete construction aggregates, filter aid, fertilizer extender, foundry ladle covering and sand additive, inert carrier, insulation board filler, loose-fill insulation, molding filler medium, packaging medium paint texturizer, pipe insulator, plaster aggregate and texturizer, propagating cuttings of plants, refractory products, soil conditioner, tile mortar aggregate and lightweight insulating concrete for roof-decks, and wallboard core filler.

In the construction industry, perlite's incombustibility and low water absorption make it a superior insulating material. Perlite plaster aggregate is used extensively to fireproof structural steel construction and to reduce the weight of interior walls and ceilings. Perlite concrete aggregate roof-decks also insulate and save weight. Expanded perlite is an important component of roof insulation (gypsum) board, masonry (cavity fill), and floor and wall tiles. Expanded perlite used as a filler also finds applications in adding bulk to paper and paint, and as a carrier for insecticides, pesticides, and chemical fertilizers.

For agricultural applications such as soil conditioning, perlite is chemically inert and long lasting, it does not alter the pH value of the soil, and its cellular structure enables it to retain several times its own weight in water, thereby providing roots with a ready supply of moisture.

Some other important applications of perlite, to name just a few, include its use as an insulator (in cryogenic technology) to hold solidified gases such as liquid oxygen at extremely low temperatures, to absorb oil spills on water and wet surfaces, to clean up effluents containing oily wastes, and as an additive in molding sands.

No data were found in the literature to indicate that exposure to perlite dust, either to the ore or to the expanded form, has resulted in adverse physiological effects. A threshold limit at the level set for inert dust is recommended for perlite containing less than 1 percent crystalline silica.

Talc, Soapstone, and Pyrophyllite.   The mineral talc is a soft, hydrous magnesium silicate, $3MgO \cdot 4SiO_2 \cdot H_2O$. Commercial talcs range from something approaching the theoretical mineral composition to mineral products that have properties in common with pure talc but contain very little of the actual minerals. Soapstone is a term used for a massive form of rock containing the mineral talc in quantities ranging from near-theoretical to as little as 50 percent. Ordinary usage usually restricts the term "soapstone" to impure massive talcose rock; the high purity massive talc is called steatite. Pyrophyllite is similar to talc in most of its physical characteristics and has the formula $Al_2O_3 \cdot 4SiO_2 \cdot H_2O$.

Talc or soapstone was produced domestically in 1974 from 46 mines in 14 states from Vermont to California, and pyrophyllite from five mines in North Carolina and one in Pennsylvania. The largest bodies of domestic talc ore known are those in New York State. Pyrophyllite reserves in North Carolina alone have an estimated total of at least 12 million tons, and deposits of commercial value are in California and Pennsylvania, with discoveries in other states regarded as excellent.

A large share of the output comes from open-pit operations, with the talc for processing being shuttled from one facility to another as convenient or advantageous. Talc deposits often contain impurities of iron minerals such as magnetite (Fe_3O_4), pyrite (FeS_2), and limonite ($FeO(OH) \cdot \eta H_2O$), which for most uses are objectionable, and are removed to the extent feasible.

There are several different grades of talc with specifications according to end use. *Steatite*, a mixture of talc, clay, and alkaline earth oxides, is used chiefly as a ceramic insulator in electronic devices. *Lava*, a frequently used trade name for block talc, has uses similar to those of steatite. *French chalk* is a soft, massive variety of talc used for marking cloth. *Soapstone* refers to all massive gray to bluish or green talcose rocks which have a slippery feeling and can be hand carved. *Pyrophyllite*, a hydrous aluminum silicate, has properties and uses similar to talc. *Wonderstone*, a massive block pyrophyllite, is essential for manufacturing synthetic diamonds.

From the standpoint of industrial health concern, the division of talcs into fibrous and nonfibrous types is of paramount importance (see the following section on hygienic standards of permissible exposure).

The largest use of the talc group minerals is in the manufacture of ceramics. Next in order is as a filler and/or pigment for paints. Third is for coating and/ or loading of high quality papers. The fourth largest use is in agricultural chemicals, where it acts as a carrier and/or diluent for insecticides. In fifth and sixth place are roofing materials and cosmetic and pharmaceutical preparations. It is in these areas that much concern has been given to the type and amount of the fibrous variety in these products for their possible health effects (baby powders). Refractories and rubber products are seventh and eighth, respectively, among the major outlets for talc. The remaining uses in the United States are distributed over such an extensive and diverse range of applications from incorporation in chemical warfare agents, floor waxes, and shoe polishes to such items as peanut polishing and salami dusting!

Microgrinding in fluid-energy mills enables talc to compete with alternative materials in the manufacture of paint, paper, plastics, and rubber. An attrition-grinding process, originally developed for the beneficiation of kaolin, has been applied advantageously to talc to yield an ultramicronized product especially suited for cosmetic and pharmaceutical preparations.

Hygienic Standards of Permissible Exposure. The TLV of 20 mppcf for *nonasbestiform talc* is based largely on the work of Dreessen and Dalla Valle (143), who studied 66 individuals who had been exposed to dust in two mills and mines handling Georgia steatite talc (soapstone) with a 10 percent tremolite content; they found no pneumoconiosis in those who worked at an average dust concentration of 17 mppcf, but severe and disabling cases were detected in groups working at average dust concentrations of 135 and 300 mppcf. Porro et al. (144a), on the basis of 15 cases with five postmortem examinations, reported that asbestotic bodies were almost invariably present in the fibrotic areas in cases of talcosis and noted a degree of similarity between asbestosis and pneumoconiosis due to talc. Siegal and associates (144b) found an advanced fibrosis incidence rate of 14.5 percent in a study covering 221 tremolite talc miners and millers. The dust to which these workers were exposed was largely *fibrous talc*, and this physical characteristic was considered to be responsible for the pathology of the lung lesion, particularly because of its resemblance to asbestosis. Findings by McLaughlin and co-workers (144c) indicated that talc pneumoconiosis could be caused only by the fibrous varieties of talc. Asbestos was considered to be more actively fibrogenic, possibly owing to the higher proportion of fibers in asbestos. In postmortem examinations on eight talc industry workers, Schepers and Durkin (144d) found histological features which suggested that tremolite was the main pathogenic agent in provoking the characteristic talc lung lesion. There was reason to suspect that the fibrous amphiboles might cause a reaction of the asbestotic type. In studying the effects of talc dust on animal tissue, the same authors (144e) found that the degree to which the fibrosis supervenes depends on the length of the fibers rather than on their chemical composition. Kleinfeld et al. (144f) reported postmortem findings on six talc industry workers in which the bodies found in respiratory bronchioles or embedded in fibrous tissues were indistinguishable from asbestos bodies as seen in asbestosis.

Because of the striking similarity in the fibrotic reaction produced by talc and asbestos fibers, the suggested threshold limit of 0.5 fibers exceeding 5 μm in length, per milliliter of air, is recommended for fibrous talc.

Miller and Sayers (136) tested two samples of *soapstone*, one 65 percent talc, 30 percent tremolite, and 5 percent dolomite and the other 55 percent talc, 30 percent dolomite, 15 percent tremolite (no quartz). Inert reactions were observed after intraperitoneal injection in guinea pigs.

The threshold limit for soapstone prior to 1949 was set at the level of a nuisance or inert dust, 50 mppcf, but in 1949 the limit was reduced to the present 20 mppcf. This was presumably based on the work of Dreessen and Dalla Valle (143), who studied workers in two Georgia mills in which dust exposures were to massive or steatite talc, also called soapstone, with a 10 percent tremolite content.

In 1968 OSHA adopted the TLV for nonasbestiform talc of 20 mppcf; for fibrous talc, the asbestos limit is to be used. OSHA's soapstone standard is the same as that for nonfibrous talc.

Neither agency has established an air standard for pyrophyllite, but from its resemblance to nonfibrous talc the TLV for this variety should be the same.

Synthetic, Soluble Silicates. The synthetic inorganic silicates of industrial interest, the so-called "soluble silicates" are formed with the alkali metals sodium, potassium, and lithium. The soluble silicates are viscous, aqueous, alkaline solutions with a high proportion of silica in an ionized form. In the anhydrous state, they are vitreous and crystalline. The best way to view these silicates is as varying ratios of alkali metal oxides to silica (M_2O/SiO_2) forming glasses of a broad range of these ratios. These soluble silicates are also referred to as salts of the hypothetical metasilicic and orthosilicic acids, H_2SiO_3 and H_4SiO_4 (Table 40.8).

Typically, sodium and potassium silicates are manufactured in ordinary glass furnaces by melting sand with soda ash (crude sodium carbonate) at about

1450°C. The ratio of the glass, or the aqueous liquid, is determined by the proportion of sand and alkali added to the furnace. The reaction follows the simple equation

$$Na_2CO_3 + n\ SiO_2 \rightarrow Na_2O{\cdot}nSiO_2 + CO_2$$

The usual sources of No. 1 glass sand are the bank sands of New Jersey, the sandstones of the Allegheny mountains, and the Mississippi Valley deposited by ancient seas. This sand, crushed and screened to about 20 to 100 mesh and washed with water (and acid if necessary), has less than 0.03 percent Fe_2O_3. The silicates thus formed are marketed as hydrated and anhydrous products (Table 40.8). Solutions of the hydrous varieties form colloidal micelles which determine many properties not found in the strictly ionic silicates, especially when the ratio rises above about $2SiO_2/Na_2O$. The aqueous liquids can be stored in tightly closed steel drums or in other nonreactive containers, but will react slowly with glass and absorb CO_2, or lose water even though corked tightly. Crystallization may occur if solutions are kept hot or are seeded. The more alkaline liquids when concentrated form a sticky or tacky gum, but at the more siliceous end of the range are sufficiently plastic to form into balls of great elasticity.

There are more than 50 well-recognized applications of the soluble silicates requiring quantities of industrial significance. These economic, incombustible, nonflammable, colorless, odorless, and unputrescible products are used in industry by the millions of pounds, but in most cases, they require special care and effort in application.

Detergents for commercial and household use consume a large proportion of the production of sodium silicate. Large quantities are also used alone or in mixture with other alkalies for metal and heavy-duty cleaning, adding up to about 15 percent of U.S. production. About 13 percent of U.S. production of soluble silicates is used as adhesives in corrugated board manufacture. For desiccant silica gels and catalysts for oil cracking, another 35 percent of U.S. production is consumed. Their property as highly insoluble binders for weather-resistant shingles and roofing granules accounts for about 9 percent. Potassium and lithium silicates are used as binders in welding rods and for bonding phosphors for oscilloscopes and television screens for molds for casting metal. The textile industry requires large tonnages for stabilizing bleach baths and for separating and cleaning fibers. The silicates are added to the pulp in the manufacture of paper to improve printability and handling, and grease-resistant films are formed by coating paper with soluble silicates. Bleaching in the paper mill also requires soluble silicate stabilizers. Such chemical reactions and treatments take about 6 percent of U.S. production. The treatment of boiler water, water-based paints, concrete hardening, clay refining, zeolites, fireworks, soil solidification, and the fabrication of electrical components where the dielectric properties can be controlled by the composition, as well as the impregnation of wood, and use in pottery slips and glazes, and enamels, make up the final 10 percent of U.S. consumption.

Physiologic Response. Although very few experimental toxicity determinations in animals have been made, long industrial experience has shown that the ordinary liquid, soluble silicates are of very low toxicity. There is a record of a man who allegedly drank 200 ml of a *neutral* liquid silicate with resulting severe gastrointestinal upset, but with no fatal or serious consequences (53). Commercial liquids at about $3SiO_2/Na_2O$ have an oral LD_{50} of about 2500 mg/kg. About 0.002 percent powdered sodium silicate has been added to dairy feed since the early 1900s with no reported injurious effects. Neither potassium nor lithium silicate solutions have proved more toxic, and the organic ammonium silicates have been found to be essentially nontoxic according to experience and standard test methods (53).

The *hazards* involved in handling liquid solutions, however, should be strictly noted.

None of these compounds should be allowed in contact with the eyes, and the liquids more alkaline than the 2.4 SiO_2/Na_2O should be treated with caution. When a film of liquid silicate is allowed to dry it becomes brittle, and the conchoidal fracture produces sharp edges which may readily cut the skin and permit infection from outside sources or adhering dirt.

The orthosilicates and anhydrous sesquisilicates are strong alkalies and release about 50 percent of the heat of solution of caustic soda. The ordinary precautions for handling caustic soda should be followed. The Manufacturing Chemists Association has suggested proper labels for such containers, as well as for the other commercial silicates. Goggles or a face shield should be used in handling these very alkaline products, and the granules should be added slowly to the solution when dissolving. They should be flushed off the skin immediately, and eyes should be flushed for at least 15 min in case of contact with a particle of the alkali. Sodium metasilicate pentahydrate has a negative heat of solution, and sodium sesquisilicate pentahydrate releases only about 2 percent of the heat which is released by flaked caustic, so that the rate of corrosion of the skin is greatly moderated. These products are considered corrosive in contact with the skin but if handled with caution and not allowed to remain in contact with the wet skin, they are not dangerous. The removal of natural oils from the skin by alkaline solutions can produce dermatitis and should be avoided by the use of protective creams or an acid detergent following exposure.

Because it is the hazard, not the toxicity of these soluble silicates, that is the problem, no TLVs have been established.

3.7.4 Silicon Carbide, SiC

Silicon carbide is a crystalline substance with a color that varies from almost clear to pale yellow, or green to black (Table 40.8), depending on the amount of impurities. It occurs naturally only in moissanite, a meteoric iron mineral, found in the Diablo Canyon in Arizona. The commercial product is usually obtained as an aggregate of iridescent crystals, caused by a thin layer of silica by superficial oxidation of the carbide (145). Silicon carbide is sold under a number of trade names that include Carborundum, Crystalon, Carbofrax, Carbonite, and Electrolon.

Silicon carbide is produced by the electrochemical reaction of high grade silica (quartz) and carbon in an electric resistance furnace. The overall reaction is

$$SiO_2 + 3\,C \rightarrow SiC + 2\,CO$$

with the carbon either petroleum coke or anthracite. Sawdust may be added to increase porosity and prevent furnace blowout. A temperature of 2040 to 2600°C is required for the melt which yields a cylinder of high grade SiC and a crust of 30 to 35 percent SiC.

Physical properties tend to be the dominant characteristics of SiC as far as usage is concerned, which vary according to crystal structure (dense, direct-bonded, or sintered alpha), whether single crystal or polycrystalline, and amount of impurities (Al). Hardness, thermal conductivity and expansion, electric conductivity, resistivity, and semiconducting properties all depend on the purity, density, and even thermal history of material. From electron bonding studies, SiC is considered to be predominantly covalent, but with some ionic character (146).

Chemical reactions with SiC do take place, however, but at greatly elevated temperatures, for SiC is highly stable and inert at ordinary temperatures. Sodium silicate attacks it about 1300°C, and it reacts with calcium and magnesium oxides above 1000°C and copper oxide at 800°C to form the metal silicide. Silicon carbide is resistant to chlorine below 700°C but forms carbon and silicon tetrachloride at high temperature. It dissociates in molten iron and the silicon reacts with oxides present in the melt, a reaction of use in the metallurgy of iron and steel. A silicon nitride bonded type shows improved resistance to cryolite.

The property of extreme hardness provides use for SiC as an abrasive (wheels and knives) and for wear surfaces (brake linings). Its low coefficient of expansion, high thermal conductivity, and general stability make it a valuable substance for refractory use; its heat stability, for heating elements in electric furnaces; and its semiconducting property, for thermistors and lightning arresters. Its extensive use in ferrous metallurgy has been mentioned above. All these special properties give rise to a host of other applications including catalyst carriers, tower packings, and fluidized-bed heaters. Its relatively low neutron cross-section and resistance to radiation damage make it useful in nuclear reactors, and its oxidation and erosion resistance, for coatings for support of electrodes and metal rods.

Analysis of SiC includes identification, chemical analysis, and physical testing.

For identification, X-ray diffraction (147) and petrographic microscopy are used (148a, b); for chemical analysis of abrasive grain and crude SiC, a standard, wet chemical analysis, and determination of metals by conventional procedures of spectrographic analysis, mass spectrometry, or activation analysis are the methods of choice; total Si is determined by atomic absorption, and free Si by spectrophotometry after selective acid dissolution.

Physical tests performed on SiC in the United States are the standard methods approved by the Abrasive Grain Association of particle size (sieve) analysis bulk density, wettability, friability, and sedimentation. In Europe, it is the Federation of European Producers of Abrasive Products (FEPA) that issues standards.

Physiologic Response. Gardner in 1923 (149a) showed that SiC produced no fibrosis of the lungs in normal experimental animals, but that it profoundly altered the course of inhalation tuberculosis, leading to extensive fibrosis and progressive disease. Miller and Sayers (149b) found that an inert reaction resulted when SiC was injected intraperitoneally in guinea pigs.

The only published evidence of pulmonary disease associated with inhalation of SiC dust is the reports of Smith and Perina (150a) and Brunsgaard (150b). The former's patients had both SiC and alumina exposures; the latter observer described slight radiographic changes in 10 of 32 workers exposed exclusively to SiC. Most of the affected individuals had worked for 15 years or more in dusty atmospheres with an average dust count of 1200 particles/cm^3 (34 mppcf) and an average particle diameter of 1 μm. All cases were tuberculin positive. Those presenting pulmonary changes had only slight respiratory symptoms.

Threshold Limit. Because of its chemical and toxicologic inertness, a TLV of 30 mppcf, the same as for "nuisance" particulates, has been recommended for SiC.

3.7.5   Silicon Nitride, Si$_3$N$_4$

Si$_3$N$_4$ can be prepared as a grayish white powder by direct combination of its elements at elevated temperatures in an electric furnace, or as crystals which sublime at 1900°C. The crystals have a specific gravity of 3.44, and a bulk density of 70 to 75 lb/ft^3. Its hardness of 9+ Mohs is only slightly greater than that of SiC. The nitride is unreactive chemically at ordinary temperatures, but reacts with carbon at high temperatures to form SiC. It is highly refractory and resistant to oxidation and various corrosive media, but is soluble in hydrofluoric acid. Like SiC, it finds uses in refractory coatings, as an abrasive, and in rocket nozzles, as high strength fibers, and for many other purposes.

No toxicologic information appears to have been developed on Si$_3$N$_4$, and hence no TLV has been established, but its general inertness would place it along with SiC with a "nuisance" dust limit of 30 mppcf.

3.7.6   Silicon Halides

Of the four silicon tetrahalides (also known as tetrahalosilanes) only three, SiF_4, $SiCl_4$, and $SiBr_4$, are of commercial significance; of these, the tetrachloride has the greatest commercial interest. All are covalent and, as such, are readily hydrolyzed and react with alcohols and Grignard reagents. They have an orderly progression of melting and boiling points, beginning with the tetrafluoride as a colorless gas, through the tetrachloride (Table 40.8) and bromide as fuming liquids, to the tetraiodide, a white solid. They can be prepared by direct halogenation of quartz or silicon carbide.

The main industrial uses of $SiCl_4$ are the production of high purity silicon, amorphous fumed silica, and ethyl silicate and for smoke screens. The main uses of SiF_4 are the manufacture of fluosilic acid (H_2SiF_6) for water fluoridation and to seal out water from oil wells during drilling.

All silicon tetrahalides are highly toxic by inhalation and ingestion, and extremely irritating to skin and mucous membranes owing to their corrosive nature resulting from hydrolysis in moist tissue to liberate halogen acid. No TLVs have been set for any of the three industrial halides owing to lack of toxicologic or industrial use information, but from their general instability when in contact with moist tissue, the limits should be at least one-quarter of those of the corresponding halogen acid.

3.7.7   Silanes

The hydrides of silicon are named silanes and have the general formula Si_nH_{2n+2} analogous to the carbon alkane series, and are thus capable of forming an almost endless variety of derivatives simply by replacing one or more of the hydrogen atoms with any one of a number of chemical elements (halogens) or organic groupings. The first member of the series is silane, SiH_4 (Table 40.8); the second, disilane, Si_2H_6; the third, trisilane, Si_3H_8; etc. All other covalent compounds of silicon may be considered to be derived from these silanes and are named according to their substituent groups and their placement along the silicon chain or ring. Thus we have trichlorosilane (Table 40.8), hexachlorodisilane, Si_2Cl_6, octamethylcyclotetrasiloxane, $[(CH_3)_2SiO]_4$, triethylsilanol, $(C_2H_5)_3SiOH$, etc.

The unsubstituted silanes, prepared from the corresponding chloride and $LiAlH_4$, must be handled with extreme care, and thus are prepared in a vacuum system because all the silanes are readily oxidized by air and form spontaneously flammable or explosive mixtures with air. All are also attacked by water in the presence of even minute traces of hydroxyl ion to evolve hydrogen and form silicic acid or hydrated SiO_2. The reaction with water is further accelerated by larger amounts of inorganic or organic bases.

The silanes decompose at elevated temperatures to liberate H_2 and deposit a high purity silicon, which leads to some of the principal uses of silanes. The most stable, SiH_4, decomposes rapidly at 500°C, and slowly at 250°C, and all undergo reactions such as addition to the double bond of olefins, although limited by the competing decomposition reaction, forming compounds such as ethyl silane, $C_2H_5SiH_3$. Similar compounds form with more unsaturated carbon compounds, yielding an endless variety of organosilanes. The *NIOSH Registry of Toxic Substances* for 1976 lists 64 of these organosilanes, *The Handbook of Chemistry and Physics*, 82.

Of the substituted silanes, trichlorosilane (Table 40.8) trichlorophenylsilane ($C_6H_5SiCl_3$), and hexachlorodisilane are of some industrial interest. Trichlorosilane is representative of a simple, halogenated silanes; in addition to its property of ready decomposition in water to yield the corresponding halogen acid, it is flammable and a dangerous fire risk with a flash point of 7°F. Its acute toxicity is relatively low, however, the rat inhalation LC_{50} being 1000 ppm and the oral rat LD_{50}, 1030 mg/kg.

Silane itself has the disagreeable property of being a gas with a repulsive odor and of igniting spontaneously in air, and is thus a dangerous fire hazard, but for a volatile inorganic hydride, has a low acute toxicity by inhalation in laboratory animals, the 4-hr mouse LC_{LO} being 9600 ppm and that for the rat, 4000 ppm; on the rabbit skin, the LD_{50} is 3540 mg/kg (6).

The TLV for silane (silicon tetrahydride, SiH_4) of 0.5 ppm (approximately 0.7 mg/m^3) was adopted by ACGIH in 1974 on the basis of its acute toxicity in laboratory animals relative to other hydrides of metalloids.

3.7.8 Silicon Esters

These organic compounds of silicon have the general formula $Si(OR)_4$, where R may be the same or different alcohol radical groups. The esters are produced by the same general reaction of $SiCl_4$, or other silicon halide, on virtually any alkyl or aryl alcohol. Although referred to in the organic chemistry literature as organic ethers of silicon, for example, tetraethoxysilane, the term used in industry for these substances is alkyl silicates, such as ethyl silicate $(C_2H_5O)_4Si$.

The best known, and industrially most useful, of these esters is ethyl silicate, some of whose properties are given in Table 40.8. It is marketed in two grades, a 29 percent and 40 percent SiO_2, and in a "condensed" form consisting of 85 percent ethyl silicate and 15 percent polyethoxysiloxanes,

$$\begin{array}{c|c|c} & C_2H_5 & C_2H_5 \\ & | & | \\ O-\!\!\!\!&Si-O&\!\!\!\!-Si-O- \\ & | & | \\ & C_2H_5 & nC_2H_5 \end{array}$$

Both have faint but not unpleasant odors and are flammable, but present only a moderate fire hazard. They are generally thermostable but unstable in water.

Ethyl silicate is widely used in industry for weather- and acid-proofing mortar and cement, for refractory bricks and other molded objects, in heat- and chemical-resistant paints, protective coatings for buildings and castings, in lacquers, and as a bonding agent. The condensed form is used for precision casting of high-melting alloys, pigment binder for paints, and surface hardener for sandstones.

Ethyl silicate is only moderately to mildly toxic acutely for laboratory rodents, the 4-hr rat inhalation LC_{LO} being 1000 ppm, the corresponding 6-hr LC_{LO} for the guinea pig, 700 ppm, and the oral rat LD_{50}, 1000 mg/kg (6). In man, 1200 ppm is lacrimatory, 250 ppm produces slight eye irritation, and 85 ppm is detectable by odor (151). Ethyl silicate is readily absorbed with resultant effect on the red blood cells giving etherlike hemolysis.

In chronic studies Pozzani and Carpenter (152) showed that exposure of rats at 400 ppm ethyl silicate for 7 hr a day for 30 days resulted in significant mortality and kidney, liver, and lung damage in the survivors. Exposure of rats, guinea pigs, and mice, however, at 88, 50, and 23 ppm for 7 hr/day, 5 days/week, for 90 days resulted only in a decrease in the kidney weights of mice exposed at the 88 ppm level, a change of questionable health significance.

Kasper et al. (153) showed that animals exposed at 164 ppm ethyl silicate, 8 hr/day for 17 days, did not show weight increases equal to those of the controls.

Rowe et al. (154) found that rats died or suffered severe kidney injury from three to five 7-hr exposures at 500 ppm of ethyl silicate. Rats exposed at 250 ppm for up to 10 7-hr periods lost weight and showed kidney and lung changes. Five to 10 exposures at 125 ppm caused slight to moderate kidney damage and an increase in kidney weight; these changes apparently did not progress after an additional 20 exposures.

Compared to ethyl silicate, methyl silicate is far more toxic; the 4-hr rat inhalation LC_{LO} is 250 ppm. Its toxicity is further distinguished by its severe effects on the eye, including early ulceration of the cornea following necrosis of the epithelium; a single exposure precipitated the event (53). It is perhaps by reason of its high toxicity that methyl silicate is no longer an article of commerce (4), although a TLV of 5 ppm, recommended by the Committee, was adopted by ACGIH in 1969 when the silicate was in production.

The *threshold limit* for ethyl silicate of 100 ppm was in the first of the ACGIH published lists, based on the not inconsiderable information noted above. Patty had cautioned, however, that this limit had not been confirmed by human experience. After review of the information, which indicated that significant kidney damage could occur in rats at concentrations below 100 ppm (154), the TLV for ethyl silicate was reduced in 1979 to 10 ppm (approximately 85 mg/m^3) with an STEL of 35 ppm (approximately 255 mg/m^3).

The only other silicon ester presently of industrial interest is 2-ethyl butyl silicate $[CH_3CH_2CH(C_2H_5)CH_2O]_4Si$ (53). It is, like ethyl silicate, a colorless liquid, but far less volatile (b.p. 164°C at 1 mm Hg). It is formed in the conventional manner from $SiCl_4$ and 2-ethylbutanol. Its main use is as an hydraulic fluid and heat transfer liquid, owing to its general stability and high boiling point.

The only toxicity information that could be found was an oral rat LD_{50} of 20 mg/kg (6), which places this silicate in the highly acute toxicity range. Because no industrial experience with this compound has come to the attention of the TLV Committee, no TLV has been set for this compound. Its hazard should be slight, however, because of its high boiling point and the enclosed conditions surrounding its preparation and use.

3.7.9   Silicones and Siloxanes

By definition, silicones are any of a large group of organic siloxane polymers (see formula) where the R groups represent various organic groupings which replace the oxygens of the side chains of the siloxanes thus:

$$
\begin{array}{ccc}
(R)O & (R)O & (R)O \\
| & | & | \\
-O-Si- & -O-Si- & -O-Si-O- \\
| & | & | \\
(R)O & (R)O & n(R)O
\end{array}
$$

the polymers may be either linear or cyclic, and depending on the magnitude of n, are liquids, semisolids, or solids.

The industrial production of silicones has as its starting materials organochlorosilanes, RCl_3Si, made from an alkyl halide and silicon catalyzed by copper at about 300°C. These silanes are purified by distillation, and by carefully controlled hydrolysis, the chain length of the polymer (n) is determined. By substituting, say, one $Ch_3(R)$ group per monomer unit by a trifluoropropyl group, a polymer insoluble in common solvents results. Boron may be incorporated in the monomer as boric acid, and other inorganic elements may be introduced in the chain.

Silicone Fluids.   Dimethyl silicone fluids are made by reacting dimethyl silicone stock with hexamethyldisiloxane, $(CH_3)_6Si_2$. For relatively low viscosity fluids, the process is run for several hours at 180°C with acid clay catalysts. Alkaline catalysts are used to produce high viscosity fluids or gums. Polymerization is continued for gums of polymers of more than 500,000 average molecular weight and 10_6 cSt viscosity.

Fluids with chlorinated phenyl groups attached to silicone have improved lubricating properties, as does methyl trifluoropropyl silicone.

Silicone Rubber.   Heat-curing types, made up of gums, fillers, additives, and catalysts, are commercially available as general purpose, (methyl and vinyl), low shrink (devolatilized), and solvent resistant (fluoro silicone).

Silicone Resins are highly cross-linked siloxane systems, the cross-linking components being introduced as trifunctional or tetrafunctional chloro silanes in the first stage of manufacture. The second stage consists of hydrolyzing the mixture of silanes, separating the HCl, washing the resin solution, and heating with a mild condensation catalyst to bring the resin to the proper viscosity and cure time. It is finally adjusted to specifications by distilling off or adding solvents. For molding resins, fillers such as glass fibers may be added and solvent removed to yield a solid resin or resin–glass mixture. This is also the procedure for preparing silicone-resin *intermediates* used for making silicone–organic copolymers. Paints are made from silicone resin solutions by mixing with stable pigments such as aluminum flakes and TiO_2.

The great diversity of silicone applications is shown by the 14 end group uses in Table 40.11. Several types of greases are made from mixtures of silicone fluids and fillers. For insulating and water-repellent greases, silica filler is used, and the fluid may be dimethyl, dimethyl copolymerized with methyl phenyl, or methyl trifluoropropyl silicone. For lubricating grease the fillers are generally lithium soaps, and the preferred fluids are methyl phenyl, chlorinated phenyl methyl, or methyl trifluoropropyl silicones. Other fillers, such as carbon black, indanthrone blue, or aryl ureas have also been used in such greases. Silicone films deposited on glass from solutions of carbon-functional alkoxysilanes improve adhesion of the glass to organic resins.

Release resins are baked on metal surfaces to prevent materials from sticking to these surfaces. Water-repellent resins applied to masonry surfaces coat the pores and act by repelling water which otherwise would wet the walls of these pores.

Emulsions of silicone fluids in water are made for convenience in applying small amounts of silicone to textiles, paper, or other surfaces. Concentrations are usually in the range of 20 to 40 percent silicone, except for foam-control agents, which may contain as little as 10 percent silicone. Various emulsifying agents are used, in general nonionic. The emulsified silicone fluids may be of any type, but are usually dimethyl silicone. Those based on methyl hydrogen fluid (MeHSiO) can react with water to generate hydrogen and must be stabilized by buffers when stored.

TABLE 40.11. End Uses of Industrial Silicones and Siloxanes

Cosmetics	Monomers
Curing agents	Oils
Emulsions	Polishing additives
Fluids (adhesives, dielectrics)	Release agents
Intermediates	Rubber compounds (Silicone rubber)
Leveling agents	Surfactants, wetting agents
Lubricants and grease	Water repellants

Antifoams are made from fluids of 350 to 12,000 cSt viscosities, which may be compounded with fillers and dispersing agents. Fluorosilicone and dimethyl silicone fluids are used in antifoams. Emulsified fluids are especially convenient as antifoams, because they disperse rapidly.

For other applications, the manufacturers' specification brochures should be consulted.

Analytic and Test Methods.   Qualitative identification of silicones can be made through their infrared or UV spectra. Quantitative determination, for example, on paper, on textiles, or in formulations, can be done by ashing and determination as silica, or by solvent extraction and measurement by infrared absorption.

Special groups are determined by specific reactions. For example, chlorosilanes can be hydrolyzed and the halogen determined by titration with alkali or silver nitrate. Other types of halogen substitution may require more drastic methods of decomposition. SiH is determined by infrared absorption at 2100 to 2250 cm^{-1} or by evolution of hydrogen on hydrolysis or alcoholysis catalyzed by bases.

Gel permeation chromatography has been used to determine the distribution of molecular weights in silicone gums (155a). Gas chromatography is satisfactory for separation and determination of small linear and cyclic polymers. Mass spectrometry and gas chromatography are useful in identifying and measuring volatile silicones and intermediates.

The general analytic chemistry of silicones has been reviewed by McHard (155b).

Physiologic Response.   The NIOSH *Registry of Toxic Effects of Chemical Substances* for 1978 lists 30 commercially available silicone products that have been tested either for their acute oral or cutaneous toxicity or, when indicated, their ocular toxicity for laboratory animals. They are the results of range-finding tests performed for Union Carbide Corporation by Smyth and his associates between the years 1963 to 1973.

Review of these data show oral LD$_{50}$ values for the rat of from slightly less than 2 to 49 g/kg, indicating their essentially nontoxic nature by this route. (Comparatively, a 70-kg worker would have to ingest from about 100 g to 3 kg to be seriously affected.) Dermal toxicity in the rabbit was also very slight, with one exception (product Y-1806); LD$_{50}$ values expressing percutaneous absorption were again expressed in grams per kilogram (from 1.5 to 16 g/kg), the exception being less than 1 g/kg. The rabbit dermal irritation dose of 500 mg produced a very mild response, again with the one exception of a moderate response at 10 mg. Rabbit open-eye irritation tests, made on five products, ranged from a mild response at 100 mg to moderate at 1 mg in two rabbits, to severe in three, in one at 1 mg and in two at 50 mg, indicating that with these products at least eye splashes would require immediate attention.

Two lots of dimethyl siloxane (Dow Corning), tested for subcutaneous toxicity in the mouse, showed identical values of 120 mg/kg as the lowest toxic dose (TD_{LO}).

A review (156a) of the toxicology of the silicones in 1961 pointed out that, in addition to their low or negligible toxicity by all routes except the eye in some cases, the hazard by inhalation is also negligible because the vapor pressure in most silicone fluid vapors is quite low. Hexamethyldisiloxane, $(CH_3)_6Si_2O$, a silicone fluid with the highest vapor pressure, was found by Rowe et al. (156b) to have a relatively low vapor toxicity. On the other hand, when $(CH_3)_6Si_2O$ was administered by other parenteral routes (intraperitoneal, subcutaneous, and intradermal), it proved quite irritating to rabbits.

A Dow Corning 555 fluid administered as a spray to rabbits as a 2 percent solution, 30 sec twice daily for 90 days, showed no sign of irritation or foreign body reaction in the lung (156a). Similarly, a fog of silicone antifoam presented to animals for 2 hr resulted in no pulmonary edema or other harmful effects (156c), and Dow Corning 200 Fluid produced transitory conjunctivitis in rabbits, which persisted a few hours but left no permanent residual (156a).

Therapeutic Use. Silicones of various types have found applications in several medical fields, in the treatment of burns of the hand (Dow-Corning 360 Medical Fluid, 100 CS) (157a); aerosols have proved of protective value in cement dermatitis (157b), and certain rubber stocks, RTVs, can be readily vulcanized to desired shapes as prostheses in head and neck surgery (157c).

Si REFERENCES

116. (a) American Society for Testing Materials, Ann. Bk. Pt. 2, 1973 pp. 79–81; (b) C. R. Hubbard, in *Advances in X-Ray Analysis*, Vol. 20, 1977, pp. 27–39, and personal communication Nov. 27, 1979; (c) G. C. Grant and D. C. Buckle, unpublished results, Virginia Associated Research Campus, Newport News, Va., December 1979.

117. Report on Synthetic Amorphous Silica, Silica Manufacturers' ad hoc Committee, R. O. Treat., J. M. Huber Corp., Havre de Grace, Md., April 30, 1975.

118. (a) D. H. Groth, C. Kommineni, L. E. Stettler, W. D. Wagner, W. J. Moorman, and R. W. Hornung, "Pathologic Effects of Inhaled Amorphous Silicas in Animals," presented at Symposium on Health Effects of Synthetic Silica Particulates, sponsored by ASTM Committee E-34TF16, Marbella, Spain, November 5–7, 1979; (b) W. J. Moorman, T. R. Lewis, and W. D. Wagner, *J. Appl. Physiol.*, **39,** 444 (1975); (c) G. W. H. Schepers et al., *Arch. Ind. Health*, **16,** 125 (1957).

119. (a) M. Lippman, "Respirable Dust Sampling," *Air Sampling Instruments for Evaluation of Atmospheric Contaminants*, 4th ed., Am. Conference of Governmental Industrial Hygienists, Cincinnati, Ohio, 1972, pp. G1–G17; (b) Aerosol Technology Committee, H. J. Ettinger, Chairman, "Guide for Respirable Mass Sampling," *Am. Ind. Hyg. Assoc. J.*, **31,** 133 (1970).

120. (a) *NIOSH Manual of Analytic Methods*, Cincinnati, Ohio, 1974, pp. 109–1 to 109–7; (b) "NIOSH Criteria Document for a Recommended Standard for Occupational Exposure to Crystalline Silica," 1974, Appendix II pp. 107–114.

121. G. W. Wright, in *Patty's Industrial Hygiene and Toxicology*, G. D. and F. E. Clayton, eds., Vol. I, 3rd ed., Wiley, New York, 1978, pp. 185–190.

122. National Silicosis Conference, *U.S. Dept. Labor, Div. Labor Standards, Bull. No. 13*, February 3, 1937.

123. (a) A. D. Hosey, H. B. Ashe, and V. M. Trasko, *U.S. Public Health Serv. Bull. No. 557*, Washington, D.C. 1957. (b) H. Desoille, et. al., Arch. Maladies Prof. Med. Travail et Securite Sociale, **12**, 279 (1953).

124. A. E. Russell et al., *U.S. Public Health Serv. Bull. No. 187*, Washington, D.C. 1929. (b) R. J. Ritterhoff, Am. Rev. Tuherc., **43**, 117 (1941).

125. (a) R. R. Sayers et al., *Penn. Dept. Labor Ind. Spec. Bull. No. 41*, 1934; (b) W. C. Dreessen at al., *U.S. Public Health Bull. No. 277*, Washington, D.C., 1942; (c) W. C. Dreessen et al., *U.S. Public Health Bull. No. 244*, Washington, D.C., 1940; (d) L. E. Renes, H. J. Paulus, A. D. Hosey, et al., *U.S. Public Health Serv. Publ. No. 31*, Washington, D.C., 1939; (e) R. H. Flinn et al., *U.S. Public Health Serv. Bull. No. 244*, Washington, D.C. 1939; (f) R. H. Flinn, L. J. Cralley, R. L. Harris et al., *U.S. Public Health Serv. Publ. 1076*, Washington, D.C., 1963. (g), G. P. Alivisatos, et. al., Brit. J. Ind. Med., **12**, 43 (1955).

126. (a) H. E. Ayer, *Am. Ind. Hyg. Assoc. J.*, **30**, 117 (1969); (b) H. E. Ayer et al., *Am. Ind. Hyg. Assoc. J.*, **29**, 336 (1968); (c) Aerosol Technology Committee, *Am. Ind. Hyg. Assoc. J.*, **31**, 133 (1970). (d) A. Policard and A. Collet, Arch. Maladies Prof. Med. Travail et Securite Sociale, **14**, 117 (1953).

127. (a) E. C. Vigliani and G. Mottura, *Brit. J. Ind. Med.*, **5**, 149 (1948); (b) R. H. Smart and W. M. Anderson, *Ind. Med. Surg.*, **21**, 509 (1952); (c) W. C. Cooper and L. J. Cralley, "Pneumoconiosis in Diatomite Mining and Processing", *U.S. Public Health Serv. Publ. No. 601*, 1958.

128. (a) L. U. Gardner, *Am. Inst. Min. Metall. Eng. Tech. Publ. No. 929*, 1938; (b) E. J. King et al., *Brit. J. Ind. Med.*, **10**, 9 (1953); (c) W. D. Wagner, D. A. Fraser, P. G. Wright, O. J. Dobrogorski, and H. E. Stokinger, *Am. Ind. Hyg. Assoc. J.*, **29**, 211 (1968).

129. W. B. Fulton et al., *A Study of Silicosis in the Silica Brick Industry*, Commonwealth of Pennsylvania Dept. Health, Bureau of Industrial Hygiene, Harrisburg, 1941, 60 pp.

130. *Documentation of the Threshold Limit Values for Substances in Workroom Air*, 4th printing, American Conference of Governmental Industrial Hygienists, Cincinnati, Ohio, 1980.

131. (a) C. P. McCord, S. F. Meek, and G. C. Harrold, *Ind. Med.*, **12**, 373 (1943); (b) C. P. McCord, *Ind. Med.*, **31**, 104 (1962); (c) L. U. Gardner, Annual Report, Saranac Laboratory for Study of Tuberculosis, Saranac Lake, N.Y., 1942, p. 12.

132. P. Gross, communication to TLV Committee, 1970.

133. *Criteria for a Recommended Standard for Occupational Exposure to Asbestos*, NIOSH, 1972.

134. G. H. Edwards and J. R. Lynch, *Ann. Occup. Hyg.*, **11**, 1 (1968).

135. (a) W. C. Dreessen et al., *U.S. Public Health Bull. No. 24*, Washington, D.C., 1938; (b) J. L. Balzer and W. C. Cooper, *Ind. Hyg. News Rep.*, December 1967; (c) I. V. Selikoff et al., *J. Am. Med. Assoc.*, **204**, 104 (1968).

136. J. W. Miller and R. R. Sayers, *U.S. Public Health Rep.*, **56**, 264 (1941).

137. W. C. Dreessen et al., *U.S. Public Health Bull. No. 250*, 1940.

138. T. F. Vestal et al., *Ind. Med.*, **12**, 11 (1943).

139. H. Heimann et al., *Arch. Ind. Hyg. Occup. Med.*, **8**, 531 (1953).

140. (a) Government of India, Ministry of Labor, "Health Hazards of Mica Processing," Report No. 4, New Delhi, Office of the Chief Adviser Factories, Ministry of Labor and Employment, 1954, 11 mimeo. pp. *Abstr. Bull. Hyg.*, **33**, 1149 (1958); (b) Government of India, Ministry of Labor, "Silicosis amongst Hand Drillers in Mica Mining in Bihar" (M. N. Gupta) Report No. 12, New Delhi, Office to the Chief Adviser Factories, Ministry of Labor and Employment, 1956, 12 mimeo. pp. *Abstr. Bull. Hyg.*, **32**, 1168 (1957); (c). Government of India, Ministry of Labor, "Silicosis amongst Supervisory Staff in Mica Mining in Bihar" (M. N. Gupta), Report No. 8, New Delhi, Office of the Chief Adviser Factories, Ministry of Labor and Employment, 1955, 6 mimeo. pp. *Abstr. Bull. Hyg.*, **32**, 1168 (1957).

141. L. U. Gardner et al., *J. Ind. Hyg. Tox.*, **21**, 297 (1939).

142. (a) F. G. Anderson et al., *U.S. Bur. Mines, Rep. Invest. No. 5199*, 1956; (b) Technical Data Sheet, No. 1-1, 1962, Perlite Institute, New York, (c) Johns-Manville Corp. (C. L. Sheckler), personal communication to Threshold Limit Committee member, April 7, 1971; (d) New York State Department of Labor, Division of Industrial Hygiene, unpublished reports.

143. W. C. Dreessen and S. M. Dalla Valle, *U.S. Public Health Rep.*, **50**, 131 (1935).

144. (a) F. W. Porro et al., *Am. J. Roentgenol.*, **47**, 507 (1942); (b) W. Siegal et al., *Am. J. Roentgenol.*, **49**, 11 (1943); (c) A. I. G. McLaughlin et al., *Brit. J. Ind. Med.*, **6**, 184 (1949); (d) G. W. H. Schepers and T. M. Durkin, *Arch. Ind. Health*, **12**, 182 (1955); (e) G. W. H. Schepers and T. M. Durkin, *Arch. Ind. Health*, **12**, 317 (1955); (f) M. Kleinfeld at al., *Arch. Envir. Health*, **7**, 101 (1963).

145. S. G. Clark and P. F. Holt, *J. Chem. Soc.* 5007 (1957).

146. J. Dowart and G. DeMaria, in *Silicon Carbide*, O'Connor and Smiltens, Eds. Pergamon Press, New York, 1960, pp. 366–370.

147. Powder Diffraction File, Inorganic Section, Joint Committee Powder Diffraction Standards, Swarthmore, Pa.

148. (a) A. N. Winchell and H. Winchell, *Microscopic Characters of Artificial Inorganic Solid Substances*, 3rd ed., Academic Press, New York, 1974; (b) W. C. McCrone and J. G. Delly, *Particle Analysis*, 2nd ed., Ann Arbor Science Publications, 1973, pp. 407–408.

149. (a) L. U. Gardner, *Am. Rev. Tuberc.*, **7**, 344 (1923); (b) J. W. Miller and R. R. Sayers, *U.S. Public Health Rep.*, **56**, 264 (1941).

150. (a) A. R. Smith and A. E. Perina, *Occup. Med.*, **5**, 396 (1948); (b) A. Brunsgaard, *Proc. 9th Int. Cong. Ind. Med.*, London, p. 676 (1949).

151. H. F. Smyth, Jr. and J. J. Seaton, *J. Ind. Hyg. Toxicol.*, **23**, 288 (1940).

152. U. C. Pozzani and C. P. Carpenter, *Arch. Ind. Hyg. Occup. Med.*, **4**, 465 (1951).

153. J. A. Kasper et al., *J. Ind. Med.*, **6**, 660 (1937).

154. V. K. Rowe et al., *J. Ind. Hyg. Toxicol.*, **30**, 332 (1948).

155. (a) F. Rodriquez et al., *Ind. Eng. Chem. Prod. Res. Dev.*, **5**, 121 (1966); (b) J. A. McHard, "Silicones," in *Analytical Chemistry of Polymers*, Kline, Ed., Interscience, New York, 1959.

156. (a) R. R. McGregor, *The Bulletin*, Dow Corning, Midland, Mich., 1961; (b) V. K. Rowe et al., *J. Ind. Hyg. Toxicol.*, **30**, 332 (1948); (c) M. Nickerson and C. F. Curry, *J. Pharm. Expt. Therap.*, **114**, 138 (1955).

157. (a) J. Miller et al., *J. Occup. Med.*, **9**, 183 (1967); (b) G. Desmichelle et al., *Arch. Mal. Prof.*, **24**, 330 (1963); (c) C. G. Mullison, *Arch. Otolaryngol.*, **84**, 91 (1966).

CHAPTER FORTY-ONE

Alkaline Materials

RALPH C. WANDS

The occupational hazards from exposure to the alkaline materials discussed in this chapter are primarily those of irritation and corrosion of tissues coming in direct contact with the chemical. The tissues most susceptible to rapid, severe, and often irreversible damage are the eyes. Accidental entry into the eyes by way of splashes of solid or liquid materials, or strong solutions of them, should be prevented by the use of eye protection covering all angles of entry. Facilities should be available for immediate and prolonged washing of the eyes with water wherever there is an opportunity for such accidents to occur.

Contact of alkaline materials with the skin or respiratory tract may result in irritation, corrosion, or erosion. These materials react with tissue proteins to form albuminates and gelatinized tissues, resulting in deep injuries. Acclimatization or "hardening" of the upper respiratory tract to prolonged or repeated inhalation of these materials does occur. This may be the result of an increased protective mechanism or a decrease in the sensory protective system, in which case the exposure may lead to chronic injury. There is insufficient information available for resolving the dilemma.

As with all toxic effects there is a dose–response relationship, but it should be remembered that prolonged contact of the skin with dilute solutions of the strong alkalies can cause irritation, warranting the use of protective clothing.

1 AMMONIA, NH₃, AMMONIUM HYDROXIDE, NH₄OH, and AMMONIUM SALTS

1.1 Sources, Uses, Occupational Exposures

Ammonia is a naturally occurring substance that plays a vital role in protein metabolism in almost all species including man. In man it also is an important component of the balanced acid–base, electrolyte system (1). The mean blood level in normal man is 0.08 mg percent measured as N (2). Excess ammonia in man is detoxified in the liver by conversion to urea (3). In the event an individual's liver function is greatly reduced, any source of ammonia, such as protein catabolism, ingestion of ammonium salts, or inhalation of ammonia, can lead to hepatic coma with increased circulating ammonia.

Ammonia is manufactured primarily by a modified Haber reduction process using atmospheric nitrogen and a hydrogen source, for example, methane, ethylene, or naphtha, at high temperatures (400 to 6500°C) and pressures (100 to 900 atm) in the presence of an iron catalyst (4, 5). It is a by-product of coal carbonization. Calcium cyanamide, formed by reacting nitrogen with calcium carbide, can be reacted with water to produce ammonia. In 1975 U.S. production of ammonia was between 13 and 15 million tons. Approximately 4 and 10 million tons are produced annually in Japan and Russia, respectively (5, 6). Worldwide production is estimated at 69 million tons currently and projected to 84 million tons in 1985 (5).

Most of the ammonia produced worldwide is used for fertilizer. Much of it is applied either by injection of the anhydrous gas directly into the soil or as an aqueous solution. Some is applied as various salts as the nitrate, sulfate, or diphosphate. The state of Iowa alone used about 700,000 tons of anhydrous ammonia in 1970. Ammonia is used extensively as a refrigerant gas in commercial installations (5, 7). It is also used in the manufacture of chemicals such as plastics, explosives, nitric acid, urea, hydrazine, pesticides, and detergents (4, 5).

In 1974 it was estimated that 500,000 workers were potentially exposed to ammonia (8). These exposures may occur at points of manufacture, transfer, use, or disposal. For example, 1.6 kg NH_3 is released at ammonia plants per ton produced, and another 0.2 kg is lost during loading for shipment. When ammonia is used as a developer in photocopying processes, for example, blueprint and diazo, it may be released into the workplace (8). Ammonia is transported by way of pipelines, barges, trucks, and cylinders, and may be stored under high pressure, refrigerated at low pressure, or as aqueous ammonia in low pressure tanks (4).

In addition to being handled as a compressed gas, ammonia is commonly encountered as aqueous solutions of 28 percent (aquammonia), called ammonium hydroxide, and 10 percent, called household ammonia.

1.2 Physical and Chemical Properties

Ammonia is a colorless gas at ambient temperatures and pressures with a strong, irritating odor. It has the following characteristics:

Molecular weight	17.03
Melting point	− 77.7°C

Boiling point	$-33°C$
Specific gravity (liquid)	0.682 ($-33.35°C/4°C$)
Specific gravity (28 percent aqueous)	0.90 ($25°C/25°C$)
Vapor density	0.59 (air = 1) at $25°C$, 760 mm Hg
Solubility	90 g in 100 ml water at $0°C$
	13.2 g in 100 ml ethanol at $20°C$
Alkalinity	pH of 1 percent aqueous solution is about 11.7
Odor threshold	3.5 to 37 mg/m^3 (5 to 53 ppm)
Autoignition temperature	$651°C$ ($1204°F$)
Explosive limits	16 percent to 25 percent by volume in air
Critical temperature	$132.9°C$
Pressure at critical temperature	111.5 atm
Ionization constants	K_b 1.774×10^{-5}, K_a 5.637×10^{-10} at $25°C$

In addition to the foregoing, extensive thermodynamic data have been tabulated (4, 9, 10).

1 mg/liter = 1438 ppm and 1 ppm = 0.7 mg/m^3 at $25°C$ and 760 mm Hg

Ammonia is classified as a flammable gas by the National Fire Protection Association. The fire hazard of high concentrations of ammonia is increased when other combustibles such as oil are also present. The critical temperature of $133°C$ is easily exceeded in fires so that containers of liquefied ammonia may explode unless their rupture strength is safely in excess of 112 atm (11). Concentrations of 28,800 mg/m^3 or greater are considered by the Occupational Safety and Health Administration (OSHA) to be a potential fire or explosion hazard. Dry chemical or carbon dioxide are recommended extinguishing media (8).

The strong, pungent, penetrating odor of low levels of ammonia (± 35 mg/m^3) becomes increasingly irritating as concentrations exceed 70 mg/m^3 (12, 13).

1.3 Atmospheric Analysis

Analytical procedures for ammonia in a variety of media have been reviewed and extensively discussed (4). Absorption of the ammonia in a standard acid by means of an impinger followed by back-titration with standard base is generally applicable but is subject to interference from acids and bases (14). The traditional method using Nessler's reagent is sensitive to 3 ppm in a 50-liter air sample (14). Alternatives to the Nessler procedure with fewer interferences are the indophenol (15) and the pyridine–pyrazolone (16) techniques. Infrared spectrophotometry may be used for direct air analysis at 10.34 nm (17). Specific ion electrodes are now available commercially for the ammonium ion in aqueous systems. Gas detector tubes for use in conjunction with handheld pumps are commercially available. These are certified by the National Institute for Occupational Safety and Health (NIOSH) to have an accuracy of ± 35 percent at one-half the exposure limit, and ± 25 percent at one to five times the limit (18). They are useful and convenient for exploratory surveys but more precise techniques are essential for assuring compliance with standards. Numerous techniques have been critically reviewed (19).

1.4 Biologic Effects

1.4.1 Animal Exposures

Static exposures of cats and rabbits for 1 hr to ammonia at 7000 mg/m³ resulted in the death of approximately 50 percent of the animals. Postmortem examination showed severe effects on the upper respiratory tract, indicating high absorption by these tissues. Less severe effects in the lower respiratory tract included damage to the bronchioles and alveolar congestion, edema, atelectasis, hemmorhage, emphysema, and fluid (20). Alpatov (21) and Mikhailov (22) found the LC_{50} for 2-hr exposures of rats and mice to ammonia to be 7.6 mg/liter and 3.31 mg/liter respectively. Coon et al. (17) exposed 15 guinea pigs, 15 rats, three rabbits, two squirrel monkeys, and two beagle dogs in 30 repeated exposures to ammonia of 8 hr/day, 5 days/week at concentrations of 155 and 770 mg/m³. There were no signs of toxicity, no hematologic changes, and no gross or histopathological changes at necropsy from the lower concentration. The higher concentration caused moderate lacrimation in the dogs and rabbits initially, but this was not observed after the first five exposures. Continuous exposure of rats for 114 days at 4 mg/m³ ammonia resulted in no signs of toxicity, and at necropsy the only finding was clinically insignificant lipid-filled macrophages in the lungs of both dogs, one monkey, and one rat. Rats were also exposed continuously for 90 days at 127 and 262 mg/m³ and for 65 days at 455 mg/m³. The 127 mg/m³ ammonia exposure induced no changes for the 48 rats in gross or microscopic pathology, hematology, or liver histochemistry for NADH, NADPH, or dehydrogenases for succinate, isocitrate, lactate, and β-hydroxybutyrate. The exposure at 262 mg/m³ also was without specific effect other than mild nasal discharge in about 25 percent of the 49 rats. All the 51 rats exposed at 455 mg/m³ showed mild dyspnea and nasal irritation. There were 32 deaths by day 25 and 50 by day 65. These authors also exposed rats, dogs, guinea pigs, rabbits, and monkeys to 470 mg/m³ ammonia continuously for 90 days. Mortalities were 13/15 rats, 4/15 guinea pigs, 0/3 rabbits, 0/2 beagles, and 0/3 squirrel monkeys. The dogs had heavy lacrimation and nasal discharge. There were erythema, discharge, and corneal opacity in the rabbits.

There were no hematologic effects. The gross necropsies revealed moderate lung congestion in two of three rabbits and one of two dogs. Histopathology examinations found focal or diffuse interstitial pneumonitis in all animals with epithelial calcification in the renal tubules and bronchi, epithelial proliferation of the renal tubules, myocardial fibrosis, and fatty liver plate cell changes in several of the exposed animals of each species. Control animals showed less severe similar changes (17). Stolpe and Sedlag (23) exposed rats continuously for 50 days at concentrations of 20, 35, and 63 mg/m^3 of ammonia. There were no effects on weight gained or hematologic parameters at 35 mg/m^3 at normal temperature, 22°C. There was slight weight reduction compared to controls at 10°C.

Swine exposed for 2 to 6 weeks at 100 ppm developed conjunctival irritation and a thickening of the nasal and tracheal epithelium without injury to the bronchi or alveoli (24).

Mayan and Merilan (25) reviewed the literature on the effect of ammonia on respiratory rates in connection with their studies of the biologic effect in rabbits. Exposures of 2.5 to 3.0 hr at 35 and 70 mg/m^3 decreased the respiratory rate by 33 percent. Blood pH was not affected but blood urea was elevated by about 25 percent, and blood CO_2 increased by 32 percent. These latter determinations are consistent with the known rapid conversion of absorbed ammonia to ammonium carbonate and urea in most mammals. Postmortem examination revealed no adverse effects of the exposure on the lungs, liver, spleen, or kidneys.

In its review of the toxicity of ammonia to aquatic organisms, the U.S. Environmental Protection Agency (EPA) emphasized that the adverse effects are due to the un-ionized ammonia (26).

Intraperitoneal or intravenous injections of ammonium salts produce neurotoxicity, evidenced by increased respiration, tremors, convulsions, and coma in proportion to the ammonia content in the blood and brain. Death is apparently due to cardiotoxicity (27).

1.4.2 Human Exposures

Accidental exposures of humans to ammonia may arise from failure of equipment containing either liquid or gaseous ammonia. The chemical injuries are the same, however; liquid ammonia exposures may be complicated by freezing of tissues and by injection of a liquid stream under high pressure. The biologic effects of ammonia in humans clearly depend on concentration. Six volunteers inhaled ammonia at 21 and 35 mg/m^3 for 10 min. Five reported faint to moderate irritation and one reported no irritation at 35 mg/m^3. This exposure was described as penetrating but not discomforting or painful (28). Another group of volunteers was exposed for 5 min to 22, 35, 50, and 94 mg/m^3. The 35 mg/m^3 was not irritating to the eyes, nose, throat, or chest, whereas the 94 mg/m^3 exposure caused eye irritation with lacrimation, nose and throat irritation, and in one volunteer, chest irritation (29). Six subjects were exposed to ammonia at 17, 35, and 70 mg/m^3 for 2, 4, and 6 hr. They reported no discomfort during any of the exposures. Medical examination revealed mild nasal irritation in all at 35 and 70 mg/m^3. Two had nasal irritation at 17 mg/m^3. The examining physician could not detect any difference in the degree of irritation at any of the levels. Three individuals showed no eye irritation from any exposure and two had no throat irritation (30).

All of seven volunteers reported upper respiratory irritation; two described it as "severe," from 30 min exposure to 350 mg/m^3 by means of a nose-only mask. There was lacrimation in two persons. There were no changes in blood urea nitrogen and nonprotein nitrogen, nor any changes in urinary urea or ammonia (31).

At the other end of the exposure spectrum severe injuries and death have resulted from accidental exposures to anhydrous ammonia or solutions of ammonia at high concentrations. A bank teller received an ammonia solution directly into the eyes, nose, and throat during the course of a robbery. Severe damage to one eye consisted of gross chemosis, corneal staining, loss of pupillary reaction, lens pigmentation, and uveitis, resulting in greatly decreased vision. The nasopharynx and glottis were extremely swollen so as to prevent swallowing. The trachea and lungs had chemical pneumonitis. There were extensive burns of the face and mouth. All lesions except the eye healed slowly following treatment (32). Zygladowski (33) has described 44 cases of exposure to ammonia vapors and Walton (34) has reported on seven patients. When first seen such patients are near collapse in great pain or unconscious. The exposed surfaces, including the eyes, nose, mouth, and throat show chemical burns, blisters which rupture and bleed, severe local edema, coughing, dyspnea, and progressive cyanosis. Examination reveals characteristic ecchymoses of the soft palate and swelling of the larynx and glottis to the point of respiratory obstruction. Shock and chemical pneumonitis may ensue. If death occurs it is usually due to suffocation or pulmonary edema. Depending on the severity of exposure and the promptness of treatment, there may be full or partial recovery. Residual effects may include visual impairment, decreased respiratory function, or hoarseness. In contradistinction to the animal studies by Ballantyne et al. (35), White reports a case of near fatal poisoning and ocular damage with no increase in ocular tension (36). Sugiyama et al. report a patient developing acquired bronchiectasis 1 year after ammonia inhalation (37). Verberk (38) exposed 16 volunteers to ammonia vapor for 2 hr at concentrations of 50, 80, 110, and 140 ppm. Subjective responses and pulmonary function parameters were reported. There were no effects on ventilatory capacity or 1-sec forced expiratory and inspiratory volumes. Eight of the volunteers found the irritation from 140 ppm so severe that they terminated the exposure prematurely. Other subjective responses such as smell and irritation of the eyes and nose increased with exposure concentration. Frosch and Kligman (39) describe an experimental dermatology technique wherein a 1:1 aqueous solution of ammonia is applied to human skin under a plastic closure. In 13 min an intra-epidermal blister results, which is virtually painless and which heals rapidly without scarring. The technique provides a means of comparing the irritancy of chemicals contacting the skin.

1.4.3 Absorption, Metabolism, Excretion

Ammonia is absorbed by inhalation, ingestion, and probably percutaneously at concentrations high enough to cause skin injury. Data are not available on absorption of low concentrations through the skin. Once absorbed, ammonia is converted to the ammonium ion as the hydroxide and as salts, especially as carbonates. The ammonium salts are rapidly converted to urea, thus maintaining an isotonic system. Ammonia is also formed and consumed endogenously by the metabolism and synthesis of amino acids. Excretion is primarily by way of the kidneys, but a not insignificant amount is passed through the sweat glands.

1.5 Hygienic Standards

The more recent data cited above generally support the effects in the classic tabulation by Henderson and Haggard (40) (Table 41.1), taken from earlier editions of these volumes.

The NIOSH recommended occupational health standards for ammonia are 35 mg/m^3 (50 ppm) as a ceiling for a 5-min exposure (41). The 1974 U.S. federal standard is 35 mg/m^3 (50 ppm) as a ceiling limit (42). The American Conference of Governmental Industrial Hygienists (AGGIH) recommends a threshold limit value of 18 mg/m^3 and a 15-min limit of 27 mg/m^3 (43). The Japanese (1971) standard is the same as OSHA's (44). Winell (45) reports the following standards: Federal Republic of Germany (1974), 35 mg/m^3; German

Table 41.1. Physiological Response to Ammonia (40)

Response	Concentration (mg/m^3)
Maximum concentration for prolonged exposure	70
Maximum amount for 1 hr	210–350
Least amount causing immediate irritation of eyes, nose, and throat	280–490
Dangerous for as little as 30 min	1750–4550
Rapidly fatal for short exposures	3500–7000

Democratic Republic (1973), 25 mg/m^3; Sweden (1975), 18 mg/m^3; Czechoslovakia (1969), 40 mg/m^3; and the Soviet Union (1972), 20 mg/m^3.

2   CALCIUM HYDROXIDE, Ca(OH)$_2$ (Slaked Lime, Hydrated Lime)

2.1   Sources, Uses, Occupational Exposures

Calcium hydroxide is formed by adding calcium oxide to water. This is an exothermic reaction. Calcium hydroxide is used to neutralize acids, for example, as a medicinal antacid, and in the manufacture of foodstuffs such as acid-hydrolyzed polysaccharides. Its alkalinity makes it useful in dehairing hides. Its reactivity with carbon dioxide in air leads to its use in the manufacture of mortar, plaster, whitewash, and the like. Commercial material is usually 95 percent or more calcium hydroxide.

2.2   Physical and Chemical Properties

It is commonly encountered as a white, microcrystalline powder.

Molecular weight	74.09
Specific gravity	2.34
Solubility	1 g in 590 ml water at 25°C
	1 g in 1300 ml water at 100°C
Alkalinity	pH of saturated solution at 25°C is 12.8

2.3   Biologic Effects

2.3.1   Animal Exposures

The single oral intubation LD$_{50}$ for rats is between 4.83 and 11.14 g/kg (46). Male rats were given tap water containing 50 and 350 mg/liter (47). At 2 months they were restless and aggressive and had a reduced food intake. At 3 months there were a loss in body weight, decreased counts for erythrocytes and phagocytes, and decreased hemoglobin. At sacrifice the gross necropsy showed inflammation of the small intestine and dystrophic changes in the stomach, kidneys, and liver. Rabbits were exposed for 1 min to a paste of Ca(OH)$_2$ in the eyes followed by cleaning and rinsing with a physiological salt solution. This resulted in a gradual decrease in mucopolysaccharides of the cornea, reaching a maximum at 24 hr, which did not return to normal levels in three months (48).

2.3.2   Human Effects

Direct contact with calcium hydroxide can result in skin and eye irritation. Calcium is an essential dietary element involved in formation of bony structures and in maintaining the normal ionic balance.

2.4   Hygienic Standards

The ACGIH recommended time-weighted threshold limit value for calcium hydroxide is 5 mg/m^3 (43). The EPA has removed calcium oxide and calcium hydroxide from their list of hazardous substances discharged into surface waters (49).

3   CALCIUM OXIDE, CaO (Lime, Burnt Lime, Quicklime)

3.1   Sources, Uses, and Occupational Exposures

Calcium oxide is produced by kiln-roasting of limestone, CaCO$_3$, to drive off carbon dioxide. It is used in the production of calcium hydroxide, mortar, plaster, and chlorinated lime for bleaching. It is also used in the refining of ores to remove silica, in the manufacture of glass, for dehairing hides for leather, and in food manufacturing. Commercial grades are 95 percent or more CaO.

3.2   Physical and Chemical Properties

Calcium oxide comes from the kiln as white or grayish white, porous lumps which may be crushed or powdered for distribution.

Molecular weight	56.08
Specific gravity	3.35
Solubility	1 g in 835 ml water at 25°C
	1 g in 1670 ml water at 100°C
Alkalinity	Same as Ca(OH)$_2$ (q.v.)

Calcium oxide reacts exothermically with water to form the hydroxide (q.v.). It also reacts readily with carbon dioxide from the air to form calcium carbonate as the matrix in mortar.

3.3   Biologic Effects

Calcium oxide in direct contact with tissues can result in burns and severe irritation owing to the combined effect of its reactivity with water and its high alkalinity. The Pennsylvania Department of Health has reported that strong nasal irritation was observed from a mixture of dusts containing calcium oxide in the range of 25 mg/m^3 but levels of 9 to 10 mg/m^3 produced no observable irritation (50). Dust exposures ordinarily are self-limiting by irritation of the upper respiratory tract.

3.4 Hygienic Standards

The ACGIH recommends a time-weighted average TLV of 2 mg/m^3 for calcium oxide (43). The Italian standard is 5 mg/m^3 (51).

4 NaK ("NACK")

NaK is an alloy in any proportions of sodium and potassium metals (q.v.) which may be encountered in the molten state as a heat exchanger. NaK is strongly caustic and highly reactive, especially with water. It is readily combustible.

5 POTASSIUM, K

5.1 Source, Uses, Occupational Exposures

Potassium metal is made by thermal reduction of the chloride with sodium or by electrolysis of molten salts. It is used in the production of NaK (see above) and of KO_2, the superoxide, used as an oxygen source in self-contained breathing apparatus.

5.2 Physical and Chemical Properties

When freshly cut, this ductile metal is soft and waxy with a metallic sheen which is soon lost by reaction with atmospheric oxygen, carbon dioxide, or moisture.

Molecular weight	39.10
Specific gravity	0.86 (20°C)
Melting point	62.3°C
Boiling point	758°C
Vapor pressure	8 mm Hg at 432°C

Potassium reacts readily with most gases and liquids. It does not react with the noble gases such as helium or argon or with hydrocarbons such as hexane or higher alkanes. It reacts violently with water to release hydrogen, which may ignite from the heat of the reaction. The resulting solution of potassium hydroxide (q.v.) may have a pH of 13 or greater.

5.3 Biologic Effects

Potassium's high reactivity makes it strongly caustic and corrosive in contact with tissues as described at the beginning of this chapter. The products of potassium combustion include oxides (see sodium peroxide). It is an essential element, commonly found in most foods as a salt, which regulates osmotic pressure within cells, maintains the acid–base balance, and is necessary for many enzymatic reactions, especially those involving energy transfer.

6 POTASSIUM HYDROXIDE, KOH (Caustic Potash)

6.1 Source, Uses, Occupational Exposures

Potassium hydroxide is produced by electrolysis of potassium chloride solution. Its principal use is in the manufacture of soft and liquid soaps. It is also used to make high purity potassium carbonate, K_2CO_3, for use in the manufacture of glass.

6.2 Physical and Chemical Properties

Potassium hydroxide is a white, crystalline, deliquescent solid which may be encountered in flakes, sticks, pellets, or cake form.

Molecular weight	56.10
Specific gravity	2.044 (20°C)
Melting point	360°C
Boiling point	1320°C
Vapor pressure	1 mm Hg at 719°C
Solubility	100 g in 90 ml of water at 25°C
	100 g in 375 ml of ethanol at 25°C
Alkalinity	A 1 percent aqueous solution has a pH around 13

6.3 Atmospheric Analysis

Dusts, mists, or vapors of potassium hydroxide or similar alkaline materials in the atmosphere may be determined by passing a measured volume of air through an impinger containing a measured volume of a standard solution of sulfuric acid. The excess acid is titrated with a standard alkaline solution using methyl red or other means to indicate the end point.

6.4 Biologic Effects

Potassium hydroxide, when inhaled in any form, is strongly irritating to the upper respiratory tract. Severe injury is usually avoided by the self-limiting sneezing, coughing, and discomfort. Contact with eyes or other tissues can produce serious injury as described at the beginning of this chapter. Rubber gloves should be worn when handling potassium hydroxide in order to prevent irritation, burns, or contact dermatitis. When exposed to air, potassium hydroxide forms the bicarbonate and carbonate. Very little is known of their biologic effects. Since they are less alkaline in aqueous solutions they may be expected to be less irritant or corrosive to skin and eyes. Bailey and Morgareidge (52), in a study for the Food and Drug Administration, found potassium carbonate to be nonteratogenic in mice when they were given daily oral intubations of up to 290 mg/kg on days 6 through 15 of gestation. Accidental ingestion of a solution of potassium hydroxide may be expected to produce rapid corrosion and perforation of the esophagus and stomach (53). Frequent applications of aqueous solutions (3 to 6 percent) of potassium hydroxide to the skin of mice for 46 weeks produced tumors identical to those from coal tar (54).

6.5 Hygienic Standards

The ACGIH recommends a ceiling threshold limit value of 2 mg/m^3 (43). No TLVs have been recommended for the carbonates.

7 SODIUM, Na

7.1 Source, Uses, Occupational Exposures

Sodium is manufactured by electrolysis of a molten mixture of sodium and calcium chlorides. It is a soft, waxy material having a silvery sheen on freshly cut surfaces. These surfaces quickly change to a coating of the white peroxide, Na_2O_2, by reaction with oxygen in the air. One-pound bricks and smaller amounts may be encountered in chemistry laboratories, where it is usually protected from the air by immersion in an aliphatic oil. Larger amounts are shipped in drums or tank cars. The manufacture of organometallic compounds, such as tetraethyllead, consumes the bulk of the production. Significant amounts are also used as a heat-exchange medium, frequently as the alloy with potassium known as NaK (q.v.). Metal descaling baths also use a sodium–sodium hydride mixture.

7.2 Physical and Chemical Properties

Metallic sodium is a light, ductile material having a high electrical conductivity.

Molecular weight	23.00
Specific gravity	0.9684 (20°C)
Melting point	97.83°C
Boiling point	886°C
Vapor pressure	1 mm Hg at 432°C

Sodium reacts violently with carbon dioxide, water, and most oxygenated and halogenated organic compounds. It may ignite spontaneously in air at temperatures above 115°C producing a sodium peroxide fume (q.v.) which is strongly alkaline and is thus a serious hazard for inhalation or skin and eye contact. Procedures for handling sodium and similar materials are described in the Atomic Energy Commission *Liquid Metals Handbook* (55). Firefighting and waste disposal methods are included. The behavior of the aerosol from a sodium fire had been described by Clough and Garland (56).

7.3 Biologic Effects

Vapors and fumes arising from sodium are strongly alkaline and are highly irritating and corrosive to the respiratory tract, eyes and skin. Physiologically, sodium is an essential element encountered as a salt in most foodstuffs. Its ion is the principal electrolyte in extracellular fluids which is excreted in the urine. Prolonged dietary excess may lead to renal hypertension.

7.4 Atmospheric Analysis

The analysis is similar to that of potassium hydroxide (q.v.).

7.5 Hygienic Standards

The threshold limit value set by OSHA is 2 mg/m^3 in moist air as sodium hydroxide.

8 SODIUM TETRABORATE (BORAX)

8.1 Sources, Uses, and Occupational Exposures

Sodium tetraborate, $Na_2B_4O_7$, usually occurs as borax, which is the decahydrate mineral deposited by evaporation of salt lakes in the Tertiary Period. It is also mined as the minerals colemanite, the pentahydrate, and kernite, the tetrahydrate, both similarly formed. Its primary uses are in soaps or detergents. Lesser amounts are used in fertilizers, especially apple orchards, since boron is an essential element for plants. A minor amount, but a commonly encountered source, goes into pharmaceuticals as such, or as one of the boric acids. Borax is also used as a tumescent component of fire-retardant paints.

8.2 Physical and Chemical Properties

Borax is readily soluble in water to form moderately alkaline solutions.

Molecular	381.37
Melting point	75°C (-8 H_2O at 60°C)
Boiling point	320°C (-10 H_2O)
Solubility	2.01 g/100 ml at 20°C
	170 g/100 ml at 100°C

Borax and other borates readily form complexes with polyhydroxy organic compounds such as glycerol, glucose, and mannitol.

8.3 Atmospheric Analyses

Airborne borax and related borates may be collected as aqueous solutions in impingers. The metallic ions may be determined by titration with dilute hydrochloric acid. The acidic B_2O_3 moiety may then be measured by complexing it with mannitol and titrating with dilute sodium hydroxide (57). Elemental boron may be determined by flame spectrophotometric absorption at 546 nm or in the presence of other alkali metals at 518 or 492 nm (58).

8.4 Biologic Effects

Borax is readily absorbed from the gastrointestinal tract and excreted in the urine with a half-life of about 24 hr. Borax and organoboron compounds concentrate shortly after adsorption in the brain, liver, adipose tissue, and cerebrospinal fluid (59). Brain tumors have been treated by introducing boron compounds followed by neutron irradiation to produce an in situ source of alpha radiation. Symptoms from high doses producing acute toxicity include shock, kidney damage, and atrophy of cells in the central nervous system (60). Chronic exposures to borax lead to accumulation in bone.

8.4.1 Animal Exposures

In rats, ingestion of borax and of boric acid give essentially the same LD_{50} values based on boron content. The acute toxicities are low, ranging from 3.16

to 6.08 g/kg, depending on the strain and sex of the test animals (61). In 90-day feeding studies, Weir and Fisher found that 525 ppm boron as borax or boric acid was tolerated by rats, but levels of 1750 and 5250 ppm were toxic. In 2-year feeding studies with dogs and rats the tolerated dose of borax or boric acid was 350 ppm as boron. This same dose had no effect on fertility, lactation, litter size, weight, or appearance in a three-generation feeding study of rats (61).

8.4.2 Human Exposures

Excessive doses to man during medical procedures have caused severe gastrointestinal upset, headache, erythematous rash, and peeling of the skin. Fatal doses for humans are variously estimated to be 5 to 6 g for children and 10 to 25 g for adults. Gosselin et al. estimate the LD_{50} for adult humans to be in excess of 30 g for borax or boric acid (62). Autopsies have revealed gastroenteritis, nephrosis, hepatitis, and cerebral edema. Occupational poisoning has not been reported but workers exposed to boric acid dust at levels greater than 31 mg/ m^3 showed atrophic and subatrophic changes in the respiratory mucous membranes (60).

Borax, or its related compounds, are not significantly absorbed through intact skin or mucosa. Penetration occurs rapidly through open wounds, burned skin, areas of active dermatitis when exposed to the dry materials, or high concentrations of aqueous solutions (63). Kátó and Gözsy found that borax was nonirritating to the skin as measured by the lack of phagocytic activity in the endothelial cells of capillaries serving the underlying skin (64).

9 SODIUM CARBONATE, Na₂CO₃ (SODA ASH)

This compound is usually encountered as the decahydrate, $Na_2CO_3 \cdot 10H_2O$, commonly called sal soda or washing soda.

9.1 Sources, Uses, Occupational Exposures

Sodium carbonate occurs naturally in large deposits in Africa and the United States either as the carbonate or trona, a mixed ore of equal molar amounts of the carbonate and bicarbonate. Soda ash is manufactured primarily by the Solvay process whereby ammonia is added to a solution of sodium chloride and carbon dioxide is then bubbled through to precipitate the bicarbonate, $NaHCO_3$. Calcination of the bicarbonate produces sodium carbonate. It may also be produced by injecting carbon dioxide into the cathode compartment, containing sodium hydroxide, of the diaphragm-electrolysis of sodium chloride.

The glass industry consumes about one-third of the total production of sodium carbonate. About one-fourth is used to make sodium hydroxide by the double decomposition reaction with slaked lime, Ca(OH)$_2$. Large amounts are also used in soaps and strong cleansing agents, water softeners, pulp and paper manufacture, textile treatments, and various chemical processes.

9.2 Physical and Chemical Properties

This hygroscopic, white powder is strongly caustic.

Molecular weight	106.0
Specific gravity	2.53 (20°C)
Melting point	851°C
Boiling point	Decomposes
Solubility	7.1 g/100 ml water at 0°C
	45.5 g/100 ml water at 100°C
Alkalinity	pH of 11.5 for a 1 percent aqueous solution

9.3 Atmospheric Analysis

Sodium carbonate and similar strongly alkaline materials when airborne as mists or dusts may be sampled with an impinger containing standard sulfuric acid. Quantitative determination is by forward titration of the excess acid using methyl red as an indicator.

9.4 Biologic Effects

Male rats were exposed to an aerosol of a 2 percent aqueous solution of sodium carbonate, 4hr/day, 5 days/week, for 3½ months. The particle size of the aerosol was less than 5 μm diameter. A concentration of 10 to 20 mg/m^3 did not cause any pronounced effect. In observations from exposure at 70 ± 2.9 mg/m^3, the weight gain of the exposed group was 24 percent less than that of controls. There were no differences in hematologic parameters. Histological examination of the lungs showed thickening of the intra-alveolar walls, hyperemia, lymphoid infiltration, and desquamation (65).

An aqueous solution, 50 percent w/v, of sodium carbonate was applied to the intact and abraded skins of rabbits, guinea pigs, and human volunteers. The sites were examined at 4, 24, and 48 hr and scored for erythema, edema, and corrosion. The solution produced no erythema, edema, or corrosion of the intact skins. The abraded skins of the guinea pigs were neglibly affected, but the rabbit and human skins showed moderate erythema and edema. In humans, one-third of the volunteers showed tissue destruction at the abraded sites (66).

Pregnant mice were dosed daily by oral intubation with aqueous solutions of sodium carbonate at levels of 3.4 to 340 mg/kg on days 6 through 15 of gestation. There were no effects on nidation or survival of the dams or fetuses. The number of abnormalities in soft and skeletal tissues in the experimental group did not differ from sham-treated controls. Positive controls gave the expected results (67). Similar studies in rats at doses up to 245 mg/kg and in rabbits at doses up to 179 mg/kg produced similar negative results.

Sodium bicarbonate, NaHCO$_3$, was evaluated for teratologic effects by the same procedures as for sodium carbonate. Maximum dose levels were as follows: mice, 580 mg/kg; rats, 340 mg/kg; and rabbits, 330 mg/kg. No effects were found in any of these species (68).

Twenty-seven Army inductees assigned to dishwashing immersed their bare hands for 4 to 8 hr in hot water containing a strong detergent blend of sodium carbonate, sodium metasilicate, and sodium tripolyphosphate. All developed irritation of the exposed surfaces. Six developed vesicles and giant bullae within 10 to 12 hr after exposure. Three also had subungual purpura. Secondary infections were noted in several individuals (69).

9.5 Hygienic Standards

No standard has been established in the United States. The level of 5 mg/m^3 has been tentatively recommended in the USSR for sodium carbonate (66).

10 SODIUM HYDROXIDE, NaOH (CAUSTIC SODA, CAUSTIC-FLAKE, LYE, LIQUID CAUSTIC)

10.1 Sources, Uses, Occupational Exposures

The primary source of sodium hydroxide is the electrolysis of sodium chloride solutions, which also yields chlorine. In this process the anode may be surrounded by an asbestos diaphragm to isolate the chlorine. The caustic soda so produced may contain a significant amount of asbestos fibers. As noted above, sodium hydroxide may also be produced from sodium carbonate.

The millions of tons of sodium hydroxide produced annually in the United States are used in the manufacture of chemicals, rayon, soap and other cleansers, pulp and paper, petroleum products, textiles, and explosives. Caustic soda is also used in metal descaling and processing and in batteries.

As indicated in the synonyms, sodium hydroxide may be encountered as solids in various forms (pellets, flakes, sticks, cakes) and as solutions, usually 45 to 75 percent in water. Mists are frequently formed when dissolving sodium hydroxide in water, which is an exothermic process.

10.2   Physical and Chemical Properties

Molecular weight	40.01
Specific gravity	2.13 (20°C)
Melting point	318.4°C
Boiling point	1390°C
Vapor pressure	1 mm Hg at 739°C
Solubility	42 g in 100 ml H_2O at 0°C
	347 g in 100 ml H_2O at 100°C
	Soluble in aliphatic alcohols
Refractive index	1.3576
Alkalinity	The pH of a 1 percent aqueous solution is about 13

10.3   Atmospheric Analysis

See the procedure for potassium hydroxide.

10.4   Biologic Effects

This strong alkali is irritating to all tissues and requires extensive washing to remove it. Eye splashes are especially serious hazards. See the opening discussion in this chapter and Section 6 on potassium hydroxide. Protective equipment is essential and treatment must be prompt. A 5 percent aqueous solution of sodium hydroxide produced severe necrosis when applied to the skin of rabbits for 4 hr (70). Rats were exposed to an aerosol of 40 percent aqueous sodium hydroxide whose particles were less than 1 μm in diameter. Exposures were for 30 min, twice a week. The experiment was terminated after 3 weeks when two of the 10 rats died. Histopathological examination showed mostly normal lung tissue with foci of enlarged alveolar septae, emphysema, bronchial ulceration, and enlarged lymph adenoidal tissues (71). Nagao and co-workers (72) examined skin biopsies from volunteers having 1 N sodium hydroxide applied to their arms for 15 to 180 min. There were progressive changes beginning with dissolution of the cells in the horny layer and progressing through edema to total destruction of the epidermis in 60 min.

10.5   Hygienic Standards

The ACGIH recommends a ceiling threshold limit value of 2 mg/m^3 (43).

11   SODIUM PEROXIDE, Na_2O_2 (SODIUM DIOXIDE, SODIUM SUPEROXIDE)

11.1   Source, Uses, Occupational Exposures

Metallic sodium reacts in dry air to form sodium monoxide and sodium peroxide. It may be encountered as an oxidant in chemical processes or as a bleaching agent, for example, of textiles. Its reactivity with carbon dioxide finds utility in self-contained breathing apparatus. It may also be encountered in the aerosol from sodium fires.

11.2 Physical and Chemical Properties

This white powder is a very strong oxidizing agent. A vigorous exothermic reaction takes place with water to form sodium hydroxide and oxygen.

Molecular weight	77.99
Specific gravity	2.805 (20°C)
Melting point	460°C with decomposition

11.3 Atmospheric Analysis

See the procedure for potassium hydroxide and observe personal protection precautions when collecting samples.

11.4 Biologic Effects

In order to simulate the products of an alkali metal fire, sodium vapor was used to form a fresh aerosol composed mainly of sodium peroxide with some sodium monoxide. It was passed into an aging chamber to reach equilibrium with atmospheric carbon dioxide and water and to stabilize its particle size. The resulting aerosol was thought to represent the product of an accidental sodium metal fire in a nuclear reactor. Juvenile and adult rats were exposed for up to 2 hr to various dilutions of the aerosol. The final particle size was 2.5 μm or less and was predominantly sodium carbonate with some sodium hydroxide. At necropsy following sacrifice, the only lesion observed was necrosis of the surface of the larynx; the area of affected epithelium and the depth of penetration were related to increased concentrations of aerosol. The ED_{50}, based on the number of animals affected, was about 510 μg/liter for adults and 489 μg/liter for juveniles. The severity of the injury was significantly greater in the juvenile rats. Animals sacrificed 4 to 7 days post-exposure had no exposure-related lesions, suggesting a healing process (73). The characteristics of the experimental aerosol are in accord with those projected by Clough and Garland (56). Hughes and Anderson (74) measured the decreased visibility encountered in a sodium fire. They collected data on volunteers exposed for a short time to the fumes. They concluded that, "Short term exposure of unprotected workers up to 40 mg/m^3 NaOH in air is unlikely to result in any serious discomfort. A concentration of 100 mg/m^3 produced serious discomfort promptly and precluded continuing work."

11.5 Hygienic Standards

No recommendation has been made but an acceptable level probably should not exceed the ceiling value of 2 mg/m^3 of sodium hydroxide.

12 TRISODIUM PHOSPHATE, Na$_3$PO$_4$·12H$_2$O (TSP, SODIUM ORTHOPHOSPHATE)

12.1 Sources, Uses, Occupational Exposures

Trisodium phosphate is produced by neutralization of disodium phosphate with sodium hydroxide. The disodium salt is produced from phosphoric acid and sodium carbonate. Well over 100 million lb of trisodium phosphate is consumed annually in the United States. Trisodium phosphate is an important ingredient in soap powders, detergents, and cleaning agents. It is also used as a water softener to remove polyvalent metals and in the manufacture of paper and leather. Products for removing or preventing boiler scale often contain trisodium phosphate as do those for removing insecticide residues from fruit and inhibiting mold.

12.2 Physical and Chemical Properties

This strongly basic, tertiary salt is usually seen as colorless crystals.

Molecular weight	380.21
Specific gravity	1.62 (20°C)
Melting point	75°C
Solubility	25.8 g/100 g water at 20°C
	157 g/100g water at 70°C
Alkalinity	The pH of a 1 percent aqueous solution is 11.6

12.3 Atmospheric Analysis

It is suggested that dusts of trisodium phosphate be collected in an impinger containing distilled water which may then be titrated with a standard solution of sulfuric acid. They may also be analyzed for P$_2$O$_3$ content by the precipitation of ammonium phosphomolybdate followed by redissolving in standard alkali and back-titrating with acid. Gravimetric procedures are also available, as are nuclear magnetic resonance spectroscopy and ion exchange chromatography (75).

12.4 Biologic Properties

The toxicity of trisodium phosphate has not been investigated but it may be expected to be related only to its alkalinity since its ions are normal constituents of all living matter. Its alkalinity is close to that of sodium carbonate (q.v.).

12.5 Hygienic Standards

There has been no recommendation of an acceptable amount of trisodium phosphate in air. However, its alkalinity, which is the source of its toxicity, is close to that of sodium carbonate (q.v.).

13 SODIUM METASILICATE, $Na_2SiO_3 \cdot nH_2O$

13.1 Sources, Uses, Occupational Exposures

The various hydrates of sodium metasilicate range from the anhydrous to the nonahydrate, with the anhydrous, penta- and nonahydrates being the most common. The sodium metasilicates are differentiated from other sodium salts of silicic acid by the molar ratio of the Na_2O and SiO_2 components. In the metasilicates this ratio is 1:1. A continuum of sodium silicates of other ratios may also be commonly encountered. Their chemical and biologic properties are essentially similar to those of the metasilicates. Fusing silica (sand) with sodium carbonate at 1400°C produces sodium metasilicate. A major use is as a builder in soaps and detergents. It is also used extensively as an anti-corrosion agent in boiler-feed water. The metasilicates should not be confused with other, less alkaline silicates used as adhesives for corrugated paper and as an additive to alfalfa cattle feed. Annual U.S. production of the metasilicates exceeds 400 million lb.

13.2 Physical and Chemical Properties

The metasilicates are highly water soluble. The anhydrous and the pentahydrate are produced as amorphous beads, whereas the nonhydrate appears as efflorescent sticky crystals (76).

Molecular weight	122.07 anhydrous
	212.15 pentahydrate
	284.21 nonahydrate
Specific gravity	2.614 anhydrous
	1.749 pentahydrate
	1.646 nonahydrate

Melting point	1089°C anhydrous
	72.2°C pentahydrate
	47.8°C nonahydrate
Refractive index	1.49 anhydrous
	1.447 pentahydrate
	1.451 nonahydrate
Alkalinity	pH of a 1 percent aqueous solution is about 13

Solutions of sodium metasilicate, when heated or acidified, are hydrolyzed to free sodium ions and silicic acid. The latter polymerizes through oxygen bridges to form amorphous silica. At high initial concentrations a gel is formed, whereas silica sols arise from dilute solutions. When dried, these colloidal forms of silica have great absorbing power.

13.3 Atmospheric Analysis

Neat material collected in an impinger may be titrated for total alkalinity to a methyl orange end point. Silicon dioxide in the titrate is determined by evaporating to dryness by ignition, then treating the weighed residue with $HF-H_2SO_4$ and calculating the weight loss as SiF_4 (77). The theoretical ratio of percent Na_2O: percent SiO_2 is 1.032.

Sodium metasilicate in soaps and detergents is determined by ashing to remove organic matter, dissolving the residue in HCl, igniting and redissolving twice, and measuring SiO_2 as above (78).

For microgram quantities in air, as from escaping steam of treated boiler water, a colorimetric method using an ammonium molybdate–sulfuric acid reagent has been described (79).

13.4 Biologic Effects

13.4.1 Animal Studies

The acute oral toxicity (LD_{50}) of sodium metasilicate to rats is 1280 mg/kg as a 10 percent aqueous solution, and for mice the LD_{50} is 2400 mg/kg (76). The intraperitoneal injection in rats of the nonahydrate as a neutral solution in amounts of 300 mg on day 1 and 200 mg on days 2 and 3 produced lesions in the spleen and lymph nodes and caused mitotic changes in the nuclei of cells resembling those from ionizing radiation or hypoxia (80). Weaver and Griffith (81), in their studies of detergentemesis in 11 dogs by gastric intubation, found that 8 mg/kg as a 10.5 percent aqueous solution of a sodium silicate (SiO_2:Na_2O ∷ 3.2:1) produced emesis in 6 min which continued for up to 33 min. Dogs were fed sodium silicate in their diet at a dose of 2.4 g/kg per day for 4 weeks (82). Polydipsia and polyuria were observed in some animals. Damage to renal tubules was observed in 15/16 dogs.

Radiolabeled (^{31}Si) sodium metasilicate, partially neutralized, was given orally to male guinea pigs for metabolism and excretion studies. Most of the silica was rapidly absorbed and excreted in the urine but a significant amount was retained in the tissues (82). These findings are consistent with the recognition that silicon is an essential trace element for bone formation in animals (83).

As might be expected, detergents containing sodium metasilicate and other alkaline materials are strong irritants to the skin, eyes, and respiratory tract (84–86). Seabaugh (84) has also shown that, of 134 detergents containing alkaline silicates, 81 percent were irritant or corrosive. Automatic dishwashing detergents were the most frequently corrosive.

"Soluble silica" in the drinking water of rats at 1200 ppm as silicon dioxide from weaning through reproduction reduced the numbers of offspring by 80 percent and decreased the number of pups surviving to weaning by 24 percent (87).

13.4.2 Human Effects

Sodium metasilicates have not been evaluated in man. Experience has shown that skin contact with solutions of strong detergents containing this builder produces severe skin irritation (69). However, other components of these detergents undoubtedly contribute to the irritancy.

Inhalation of dusts from soluble silicate powders is irritating to the upper respiratory tract (88). Exposure to such dusts is not related to the development of silicosis since their solubility permits them to be readily eliminated. Confirmation of this is found in the work of Svinkina (89), who was unable to produce immunologic sensitization in rabbits with sodium silicate bound to protein.

13.5 Atmospheric Standards

No recommendations have been made for occupational exposures to airborne sodium metasilicates. In view of their high alkalinity, approaching that of sodium hydroxide, it is suggested that exposures not exceed the molar equivalent of the sodium hydroxide standard of 2 mg/m^3.

REFERENCES

1. J. M. Lowenstein, *Physiol. Rev.*, **52,** 382 (1972).
2. K. Diem, Ed., *Documenta Geigy. Scientific Tables*, 6th ed., Geigy Pharmaceuticals, Ardsley, N.Y., 1962.

3. W. J. Visek, *Fed. Proc.*, **31,** 1178 (1972).

4. *Ammonia*, National Academy of Sciences, Washington, D.C., 1977.

5. U.S. Dept. Labor, *Fed. Reg.*, **40,** 54864 (1975).

6. Anonymous, *Jap. J. Ind. Health*, **14,** 45 (1972).

7. E. S. White, *J. Occup. Med.*, **13,** 549 (1971).

8. *Criteria Document for Occupational Exposure to Ammonia*, Pub. No. HEW (NIOSH) 74-136, U.S. Dept. Health, Education and Welfare (NIOSH), Cincinnati, Ohio, 1974.

9. N. A. Lange, Ed., *Handbook of Chemistry*, 8th ed., Handbook Publishers, Sandusky, Ohio, 1952.

10. R. C. Weast, Ed., *CRC Handbook of Chemistry and Physics*, 59th ed., CRC Press, West Palm Beach, Fla., 1978.

11. *Fire Protection Guide on Hazardous Materials*, 6th ed., National Fire Protection Association, Boston, Mass., 1975.

12. *Guides for Short-Term Exposures of the Public to Air Pollutants. IV. Guide for Ammonia*, Committee on Toxicology, National Academy of Sciences, Washington, D.C., 1972.

13. *Hygienic Guide Series*, "Anhydrous Ammonia," *Am. Ind. Hyg. Assoc. J.*, **32,** 139 (1971).

14. H. B. Elkins, *The Chemistry of Industrial Toxicology*, Wiley, New York, 1959.

15. M. C. Rand, A. E. Greenberg, and M. J. Taras, Eds., *Standard Methods for the Examination of Water and Wastewater*, 14th ed., American Public Health Association, Washington, D.C., 1976.

16. T. Okita and S. Kanamori, *Atm. Environ.*, **5,** 621 (1971).

17. R. A. Coon, R. A. Jones, L. J. Jenkins, Jr., and J. Siegel, *Toxicol. Appl. Pharmacol.*, **16,** 646 (1970).

18. CFR 42 Chap. 1, Subchapter G, Part 84, *Certification of Gas Detector Tube Units*, 1973.

19. D. N. Kramer and J. M. Sech, *Anal. Chem.*, **44,** 395, (1972).

20. E. M. Boyd, M. L. MacLachlan, and W. F. Perry, *J. Ind. Hyg. Toxicol.*, **26,** 29 (1944).

21. I. M. Alpatov, *Gig. Tru. Prof. Zabol.*, **8**(2), 14 (1964).

22. V. I. Mikhailov, *Probl. Kosm. Biol. Akad. Nauk. SSR*, **4,** 531 (1965).

23. J. Stolpe and R. Sedleg, *Arch. Exp. Vet. Med. Leipzig*, **30**(4), 533 (1976).

24. P. A. Doig and R. A. Willoughby, *J. Am. Vet. Med. Assoc.*, **159,** 1353 (1971).

25. M. H. Mayan and C. P. Merilan, *J. Animal Sci.*, **34,** 448 (1972).

26. W. T. Willingham, EPA-908/3-76-001 National Technical Information Service, Springfield, Va., Doc. No. PB-256447, 1976.

27. National Research Council, Committee on Medical and Biological Effects of Environmental Pollutants, *Ammonia*, 1977.

28. J. D. MacEwen, J. Theodore, and E. H. Vernot, Eds., *Proceedings First Annual Conference on Environmental Toxicology*, AMRL-TR-70-102, Wright Patterson Air Force Base, Ohio, 1970.

29. Industrial Bio-Test Laboratories, Inc., IBT 663-0 3161, *Report to the International Institute of Ammonia Refrigeration of Irritational Threshold Evaluation Study*, 1973.

30. Allied Chemical Corp., *Preliminary Report to OSHA of Ammonia Test Program*, 1975.

31. L. Silverman, J. L. Whittenberger, and J. Muller, *J. Ind. Hyg. Toxicol.*, **31,** 74 (1949).

32. A. H. Osmund and C. J. Tallents, *Brit. Med. J.*, **3,** 740 (1968).

33. J. Zygladowski, *Otolaryngol. Polska*, **22,** 773 (1968).

34. M. Walton, *Brit. J. Ind. Med.*, **30,** 78 (1973).

35. B. Ballantyne, J. F. Gazzard, and P. W. Swanston, *J. Physiol.*, **226,** 12P (1972).

36. E. S. White, *J. Occup. Med.*, **13,** 549 (1971).

37. K. Sugiyama, M. Yoshida, and M. Koyamada, *Jap. J. Chest. Dis.*, **27,** 797 (1968).

38. M. M. Verberk, *Int. Arch. Occup. Environ. Health*, **39,** 73 (1977).

39. P. J. Frosch and A. M. Kligman, *Brit. J. Dermatol.*, **96,** 461 (1977).

40. Y. Henderson and H. W. Haggard, *Noxious Gases*, Reinhold, New York, 1943.

41. National Institute for Occupational Safety and Health, *NIOSH Recommended Standard for Occupational Exposure to Ammonia*, U.S. Department of Health, Education, and Welfare, Rockville, Md., 1974.

42. Occupational Safety and Health Administration, U.S. Department of Labor, *Fed. Reg.*, **40**(228), 54692 (1975).

43. American Conference of Governmental Industrial Hygienists, *Threshold Limit Values for Chemical Substances in Workroom Air*. Adopted by ACGIH, Cincinnati, Ohio, 1979.

44. Japanese Association of Industrial Health, Subcommittee on Permissible Concentrations, *Sangyo Igaku*, **14,** 45 (1972).

45. M. Winell, *Ambio*, **4**(1), 34 (1975).

46. H. F. Smyth, Jr., C. P. Carpenter, C. S. Weil, U. C. Pozzani, J. A. Striegel, and J. S. Nycum, *Am. Ind. Hyg. Assoc. J.*, **30,** 470 (1969).

47. L. I. El'piner and A. M. Voitenko, *Tr. Nauch. Issled. Inst. Gig. Vod. Transp.*, **1,** 304 (1968).

48. J. Teterwak, *Klin. Oczna*, **39**(4), 543 (1969).

49. *Fed. Reg.* (Sect. 311 Clean Water Act), **44,** 65400 (November 13, 1979).

50. Pennsylvania Department of Health, personal communication to the Division of Occupational Health, P.O. Box 90, Harrisburg, Pa., by N. E. Whitman.

51. Italian Association of Industrial Hygienists, *Med. Lavoro*, **66,** 361 (1975).

52. D. E. Bailey and K. Morgareidge, *Teratologic Examination of FED 73-76 (K_2CO_3) in Mice*, National Technical Information Service PB-245522 (1975).

53. National Information Service PB-265507, 1976.

54. J. K. Narat, *J. Cancer Res.*, **9,** 135 (1925).

55. C. B. Jackson, *Liquid Metals Handbook*, Sodium, NaK Supplement, Atomic Energy Commission, July, 1955.

56. W. S. Clough and J. A. Garland, *J. Nucl. Energy*, **25,** 425 (1971).

57. Kirk-Othmer *Encyclopedia of Chemical Technology*, 3rd. ed., Vol. 4, Wiley, New York, 1978, p. 86.

58. B. W. Bailley and F. C. Lo, *J. Environ. Anal. Chem.*, **1,** 267 (1972).

59. W. J. Underwood, *Trace Elements in Human and Animal Nutrition*, Academic Press, New York, 1971.

60. A. A. Kasparov, *Encyclopedia of Occupational Health and Safety*, Vol. 1, McGraw-Hill, New York, 1971, p. 203.

61. R. J. Weir, Jr. and R. S. Fisher, *Toxicol. Appl. Pharmacol.*, **23,** 351 (1972).

62. R. E. Gosselin, H. C. Hodge, R. P. Smith, and M. N. Gleason, *Clinical Toxicology of Commercial Products*, 4th ed., Williams and Wilkins, Baltimore, 1976.

63. D. J. Birmingham, in *Cutaneous Toxicity*, V. A. Drill and P. Lazar, Eds., Academic Press, New York, 1977, p. 58.

64. L. Kátó and B. Gözsy, *Can. Med. Assoc. J.*, **73,** 31 (1955).

65. A. L. Reshetyuk and L. S. Shevchenko, *Hyg. Sanit.* **33**(1–3), 129 (1968).

66. G. A. Nixon, C. A. Tyson, and W. C. Wertz, *Toxicol. Appl. Pharmacol.*, **31,** 481 (1975).

67. K. Morgareidge, *Teratologic Evaluation of Sodium Carbonate in Mice, Rats, and Rabbits*, PB-234 868, National Technical Information Service, Springfield, Va., 1974.

68. K. Morgareidge, *Teratologic Evaluation of Sodium Bicarbonate in Mice, Rats and Rabbits*, PB-234 871, National Technical Information Service, Springfield, Va., 1974.

69. N. Goldstein, *J. Occup. Med.*, **10**, 423 (1968).

70. E. Horton, Jr. and R. R. Rawl, *Toxicological and Skin Corrosion Testing of Selected Hazardous Materials*, PB-264 975, National Technical Information Service, Springfield, Va., 1976.

71. M. Dluhos, B. Sklensky, and J. Vyskocil, *Vnitr. Lek.*, **15**(1), 38 (1969).

72. S. Nagao, J. D. Stroud, T. Hamada, H. Pinkus, and D. J. Birmingham, *Acta Dermatovener* (Stockholm), **52**, 11 (1972).

73. G. M. Zwicker, M. D. Allen, and D. L. Stevens, *J. Environ. Pathol. Toxicol.*, **2**, 1139 (1979).

74. G. W. Hughes and N. R. Anderson, "Visibility in Sodium Fume," presented at International Atomic Energy Agency International Working Group on Fast Reactor Meeting, March 17–19, 1971, IAE-NPR-12.

75. Kirk-Othmer *Encyclopedia of Chemical Technology*, 2nd ed., Vol. 15, Wiley, New York, 1968, p. 267.

76. A. Weissler, "Monograph on Sodium Metasilicate," PB287766, National Technical Information Service, Springfield, Va., 1978.

77. Philadelphia Quartz Company, *Technical Bulletin on Sodium Metasilicate and Industrial Alkalies and Detergents*, Philadelphia Quartz Company, Valley Forge, Pa., 1978.

78. *1976 Annual Book of ASTM Standards*, "Part 30, Soaps and Detergents," American Society for Testing and Materials, Philadelphia, Pa., 1976, pp 14, 27–28, 70–72.

79. *Standard Methods for the Examination of Water and Wastewater*, 14th ed., American Public Health Association, Washington, D.C., 1975, pp. 484–492.

80. L. Nanetti, *Zacchia*, **9**(1), 96 (1973).

81. J. E. Weaver and J. F. Griffith, *Toxicol. Appl. Pharmacol.*, **14**, 214 (1969).

82. F. Sauer, D. H. Laughland, and W. M. Davidson, *J. Biochem. Physiol.*, **37**, 1173 (1959).

83. E. M. Carlisle, *Science*, **178**, 619 (1972).

84. V. M. Seabaugh, *Detergent Survey Toxicity Testing*, PB264698/AS, National Technical Information Service, Springfield, Va., 1977.

85. G. E. Morris, *Arch. Ind. Hyg. Occup. Med.*, **7**, 411 (1953).

86. L. G. Scharpf, Jr., I. D. Hill, and R. E. Kelly, *Food Cosmet. Toxicol.*, **10**, 829 (1972).

87. G. S. Smith, A. L. Newmann, A. B. Nelson, and E. E. Ray, *J. Animal Sci.*, **36**, 271 (1973).

88. Philadelphia Quartz Company, *Soluble Silicates Bulletin T-17-65*, Philadelphia Quartz Company, Valley Forge, Pa., 1965.

89. N. V. Svinkina, *Labor Hyg. Occup. Diseases (USSR)*, **10**, 20 (1966).

CHAPTER FORTY-TWO

Fluorine-Containing Organic Compounds

DOMINGO M. AVIADO, M.D., and
MARC S. MICOZZI, M.D.

1 INTRODUCTION

The authors of this chapter have reservations about the timing of its preparation. They appreciate that 17 years have elapsed since Don D. Irish reviewed the fluorine-containing compounds for the second edition of this book, and that a considerable amount of toxicological information has since been accumulated. The several fluorocarbons (or fluorinated hydrocarbons) that make up a large majority of the chemicals covered in this chapter have been the subject of global concern over depletion of the ozone layer. The ozone problem is specifically related to the fully halogenated, or *nonhydrogenated* fluorocarbons, which cause free radical reactions with ozone by photodissociation in the upper atmosphere. The ozone depletion allows increased levels of ultraviolet radiation to reach the surface of the earth. This phenomenon may lead to a higher incidence of malignant melanoma and other solar radiation-related diseases among the world population. This supposition has led to prohibition of fluorocarbon aerosol products in the United States. However, other essential applications of fluorocarbons are in the process of being examined by the U.S. Environmental Protection Agency. In the event that such uses are similarly prohibited, then the most commonly used fluorocarbons that are nonhydrogenated may become of little concern in the United States.

The authors are confident that such a drastic prohibition of the manufacture of *nonhydrogenated* fluorocarbons will not occur abruptly. The fluorocarbon manufacturers are preparing for gradual prohibition of the manufacture of fully halogenated or *nonhydrogenated* fluorocarbons, and a probable continuation of manufacture of partially halogenated or *hydrogenated* fluorocarbons. New and *developmental* fluorocarbons, also *nonhydrogenated,* are being considered to substitute for the hydrogenated ones when these ultimately become obsolete.

The uncertain regulatory statutes for the fluorocarbons make it difficult to summarize their toxicity and occupational health factors. Our presentation highlights the unique features of the fluorine-containing compounds. This group of compounds is used in therapeutic agents, to propel anti-asthmatic drugs, and as general anesthetics. There is considerably more toxicologic information derived from human observations on fluorocarbons than for any other group of chemicals covered in this book. The medical uses of fluorine-containing compounds require amounts that are negligible compared to those for the industrial uses (refrigerants, plastic foaming agents, solvents), so that only nonmedical applications are being questioned environmentally. The toxicity of fluorocarbons on the cardiovascular and bronchopulmonary systems is a topic that is covered in detail in this chapter. Information on comparative toxicity in man and animal species is more extensive than that of any other class of chemicals.

More than 30 fluorine-containing compounds are reviewed here. It is not possible to discuss each chemical separately for its toxicity, occupational hazard, and industrial safety. Most fluorocarbons have similar patterns of toxicity and have interchangeable uses. This chapter is separated into two sections: toxicology (Sections 2 to 5), and occupational health (Sections 6 to 8), and although the chapter is almost entirely devoted to fluorocarbons, other fluorine-containing compounds are also covered, such as fluorinated ethers used as general anesthetics and blood substitutes (Section 8) and bromine and fluorine compounds used in fire extinguishers (Section 7).

The fluorocarbons are separated into those that are commercially available (Sections 2 to 4) and those that are developmental (Section 5). The latter are all *hydrogenated* fluorocarbons which have less of an effect on depleting the ozone layer than do the *nonhydrogenated* fluorocarbons. The developmental fluorocarbons, although extensively studied, are only briefly discussed, for the detailed results of toxicity studies have not yet been published. However, nonscientific publications have emphasized desirable features of selected fluorocarbons, and undesirable toxicities of other fluorocarbons, which are impossible to verify. The authors have ignored such claims and recognized that toxicity information has been used to promote the continuous popularity of fluorocarbons in general, in favor of alternatives such as chlorinated solvents, hydrocarbon propellants, and nonhalogenated gases.

In view of the above uncertainties, we have decided to discuss one selected prototype in detail, trichlorofluoromethane (fluorocarbon 11, or FC 11). This selection is based on its widespread use, as well as the considerable information available on its toxicological features (1). By detailing the methodology for the study of FC 11, the reader would be prepared to evaluate new and alternative fluorocarbons. Likewise, if any one of the commercially available hydrogenated fluorocarbons were to be used continually or even replace FC 11, then the necessary toxicologic data will become apparent to justify widespread use.

2 PROTOTYPE FLUOROCARBON 11: TRICHLOROFLUOROMETHANE

When the fluorocarbons were introduced in the 1940s, they were regarded as "inert" refrigerants compared to the sulfur dioxide, ammonia, carbon tetra-chloride, and chloroform that were then in use. The first aerosol application of the fluorocarbons was as an insecticide bomb used in World War II to protect troops in tropical areas against malaria and other vector-borne diseases (2). As other forms of aerosols were introduced, it became apparent that two fluoro-carbons were needed to propel the active ingredient, one at high pressure (such as dichlorodifluoromethane, or fluorocarbon 12) and one at low pressure. Of the latter, trichlorofluoromethane (fluorocarbon 11, or FC 11) proved to be the most popular, because of its vapor depressant effect and also its solvent and flame retardant actions. Following the reports of fatalities from the misuse and abuse of aerosol products, toxicologic studies were conducted; it became apparent that fluorocarbons are not "inert" and that the most widely used fluorocarbon (FC 11) was also the most toxic.

2.1 Aerosol Use

In the mid-1970s, the annual worldwide production of fluorocarbons exceeded 2 billion lb. The allocation for their uses was as follows: 50 percent aerosol propellants; 28 percent refrigerants; 10 percent for manufacture of fluoropo-lymers; 7 percent plastic foaming agents; and 5 percent solvents and other uses. On October 15, 1978, the U.S. Environmental Protection Agency (EPA) prohibited the manufacture of fluorocarbons for aerosol propellant uses. Medicated aerosols are exempted from the ban, and inkless fingerprinting aerosols have been permitted a 2-year extension in order to allow development of a substitute nonfluorocarbon propellant. Another exemption is the World War II insecticide aerosol because the manufacturers were unable to use the large supply of aerosol cans purchased prior to the date of banning, resulting from a shortage in Africa of the insecticide pyrethrin. As a historical note, the first fluorocarbon-containing aerosol product continues to be the only consumer product that is still manufactured at the time of writing this chapter.

2.1.1 Medicated Aerosols

The major use of FC 11 in therapeutics is in the administration of drugs by inhalation into the respiratory tract. The fluorocarbons (both FC 11 and FC 12) are used to propel bronchodilator sympathomimetic drugs or corticosteriods for the treatment of bronchial asthma. Of the two classes of antiasthmatic drugs delivered by aerosol, the sympathomimetic drugs are more potentially hazardous to the heart. The asthmatic patient inhales the aerosol sprayed directly into the oral cavity, holds his breath to permit the sympathomimetic drug to be absorbed, and then exhales the unabsorbed drug and aerosol. During the course of this maneuver, a certain amount of FC 11 is absorbed, which is suspected of sensitizing the heart to the proarrhythmic activity of the sympathomimetic drug. The propellant, although useful in delivering the desired bronchodilator drug, also stimulates receptors in the respiratory passages, which initiate cardiac arrhythmia as well. The details of the cardiotoxic action of FC 11 are discussed in Sections 2.3 to 2.5.

2.1.2   Cosmetic Products

Although cosmetic products no longer contain FC 11, it is important to retrospectively review the toxicologic information, in the event that another fluorocarbon (hydrogenated rather than nonhydrogenated) is introduced in the future. The health hazards of unintentionally inhaling these products have not been as completely elucidated as those described above for the medicinal aerosols. The amounts of fluorocarbons contained in some cosmetic products are known, but their long-term toxicity has not been investigated. There is negligible information on the fate of cosmetic particles inhaled in the lungs.

The cosmetic aerosol products include deodorants and antiperspirants, breath fresheners, women's personal hygiene products, skin lotions, foot hygiene preparations, hair shampoos, and hair sprays. In the past, the major problem was in identification of the role of FC 11 in the overall toxicity of the hair spray aerosol products. The observation of Zuskin and Bouhuys (3), that placebo aerosol-containing propellants cause a reduction in mean expiratory flow rates when inhaled, signifies a bronchoconstrictor action. However, the effect is less than that elicited by the hair spray aerosol. These authors suggested that the hair spray particles and aerosol propellants release histamine since pretreatment with atropine (vagal blockade) or chlorpheniramine (antihistaminic) protected the subjects. Valic et al. (4) observed a reduction in forced expiratory volume following exposure of human subjects to hair sprays. They did not test the effect of propellants alone but postulated that the interaction between FC 11 and alcohol would form aldehydes and hydrochloric acid, which are irritating to the airways.

The only published long-term study of toxicity of the propellant relative to hair sprays was completed by Giovacchini et al. (5), who exposed beagle dogs for a 10-sec period twice daily with the dogs remaining in the chamber for 15 min after each spray period. There were no pathological lesions after 1 and 2 years of exposure. In an editorial in 1966 that commented on the study (6), it was pointed out that since beagle dogs do not develop pulmonary granulomatous lesions, this animal model may not be useful in elaboration of the pulmonary lesions seen among beauticians exposed to hair sprays. It was also pointed out that guinea pigs would have been more suitable for the inhalation study, a suggestion that has not been answered by experimentation.

2.2 Poisoning

Deaths associated with exposure to fluorocarbons have been encountered in the occupational setting, in consumer use of household aerosols, and in patient inhalation of bronchodilator drugs. As stated in the introduction, there is a considerable amount of human observations showing the dangers of exposure to high concentrations of fluorocarbons in general, and of FC 11 in particular.

2.2.1 Occupational Poisoning

The uses of FC 11 in industry include refrigeration, food processing, solvent applications, plastic foam blowing, and fire extinguishers (Section 7). The literature on poisoning from refrigerants is reviewed below because it may apply to other forms of occupational exposure.

Fatalities have been reported following acute exposure to refrigerants such as methyl chloride alone (7–9), fluorocarbon alone (10, 11), and a combination of fluorocarbon thermal decomposition products and sulfur dioxide (12). Dalhamn (13) reported two cases of phosgene poisoning from disintegration of FC 11 propellant at an open flame in an enclosure. Mendeloff (14) reported the case of a refrigeration equipment repair mechanic who claimed chronic exposure to refrigerant gases including fluorocarbons, methyl chloride, ethanol and sulfur dioxide. The illness began insidiously with malaise, chills, fever, and nausea; during the following days the patient had abdominal pain with further nausea and vomiting, backache, headache, swollen abdomen, and mild episodes of epistaxis. He was admitted to the hospital on the sixth day of the illness with initial temperature of 98.6°F that rose to 100°F. The X-ray findings and postmortem lesions in the lungs resembled those of viral pneumonia, yet Mendeloff concluded that this case represented the first reported death from chronic exposure to fluorocarbon refrigerants. Leinoff (15) reported workers exposed to refrigerants who developed coronary thrombosis. However, there is no experimental work to prove or disprove that chronic exposure to fluorocarbon refrigerants leads to coronary heart disease.

2.2.2 Aerosol Addiction and Abuse

Although there have been hundreds of deaths from inhaling aerosols containing FC 11 in a plastic bag, it is difficult to obtain an accurate estimation. The first documented report was by Baselt and Cravey (16) in 1968, who described a 15-year-old boy found dead with a plastic bag and a 9-oz aerosol can of a spray-on coating for frying pans lying adjacent to him. Both FC 11 and FC 12 were detected in the tissues removed at the autopsy:

	FC 11 (μl/100 g)	FC 12 (μl/100 g)
Blood	0.86	0.05
Kidney	1.65	0.05
Brain	1.33	0.67
Liver	0.83	0.67
Stomach contents	5.78	6.92

The two fluorocarbons were the propellants used in the aerosol product. At that time, the authors concluded that death was due to asphyxia since the prevalent opinion was that the fluorocarbons were "inert" gases.

By 1975, the explanation for the death of a teenager due to inhalation of fluorocarbon aerosols was cardiotoxicity instead of simple asphyxia. Poklis (17) reported the distribution of fluorocarbons in postmortem tissues as follows:

	FC 11 (mg/100 g)	FC 12 (mg/100 g)
Blood	3.2	0.32
Brain	6.1	0.45
Liver	4.5	0.39
Lung	3.2	0.32
Kidney	2.5	0.18
Trachea	2.1	0.16
Bile	0.6	—

The above results show concentrations in tissues higher than those reported for the 1968 case (16). There is a significant accumulation of fluorocarbons in the brain, liver, and lungs compared to blood levels, signifying a tissue distribution of fluorocarbons similar to that of chloroform.

The change in explanation for causation of death from aerosol propellant abuse between 1968 and 1975 was triggered by Bass (18), who in 1970 reviewed the case histories of 110 American youths who died from the "sudden sniffing death syndrome." His eyewitness reports included one case of a 17-year-old boy who died from inhaling a plastic bag filled with the contents of a spray can of hair shampoo; another case was that of a 15-year-old boy who inhaled an antitussive aerosol from a plastic bag; and the third case was a 15-year-old girl who sprayed frying-pan aerosol into a plastic bag. The total number of deaths from aerosols was 59 out of 110 deaths from the syndrome. This review by Bass is of special significance because he postulated the cause of death as severe cardiac arrhythmia, a phenomenon proved by subsequent human investigation and animal experimentation.

Subsequent reports of deaths from the abuse of aerosol products continued to appear in the literature. In 1970, Chapel and Thomas (19) reported six deaths in Missouri from the abuse of aerosol glass-chill or from aerosol disinfectant spray. The aerosol was either sprayed directly into the oral cavity or inhaled after collection in a plastic bag.

In 1971, Kramer and Pierpaoli (20) reported a 15-year-old girl from Connecticut, who used deodorant aerosol to produce a "flashback," in which the individual experiences the symptoms of a hallucinogenic "trip" without actually having taken the active hallucinogenic agent. Crooke (21) described a 17-year-old boy from California who inhaled a can of hair shampoo from a plastic bag. The immediate effect was a compulsion to run, and after 150 yards, he collapsed and was pronounced dead. Kamm (22) reported a 16-year-old male from Louisiana who inhaled aerosol deodorant and subsequently died of ventricular fibrillation. The only findings at autopsy were cerebral edema, pulmonary edema, and generalized visceral congestion.

The most recent reported fatality from fluorocarbon propellant poisoning is a Canadian youth who continually inhaled a lipid aerosol containing soybean extract, used to prevent food adherence to cooking pans (23). The addict developed acute adult-type respiratory distress syndrome, rather than sudden cardiac death. After performing laboratory tests, Fagan and Forrest concluded that FC 11 and FC 12 had damaged lung surfactant function, the possibility of which is discussed below (Section 2.6).

2.3 Death from Use and Misuse of Bronchodilator Aerosols

The treatment of bronchial asthma was much improved by the introduction of bronchodilator aerosols, first marketed commercially in the United States in 1956. A form whereby the sympathomimetic bronchodilators epinephrine and isoproterenol could, aided by a propellant, be dispersed from a pressurized container small enough to be conveniently carried in a purse or pocket seemed to have much to recommend it. The pressurized unit had a valve specifically designed to release a measured and reproducible dose of the drug for inhalation by the patient. The bronchodilation produced was reported to be prompt and less likely to be accompanied by tachycardia than when epinephrine or isoproterenol was administered by subcutaneous injection.

The novel bronchodilator aerosols were a commercial success both in the United States and abroad. The assumptions that underlay this modality—that the propellants were "inert" and that the dose of the drug inhaled was unlikely to do any harm—remained uncontested, and the pressure units were made available for over-the-counter sale in most countries including the United States. It became immediately apparent to certain physicians that the aerosols were being misused by some patients. The first such warnings were made in 1958 by Harris (24) in the United States and in 1961 by Pavlik (25) in Europe, and some clinicians went so far as to question the effectiveness of the aerosols even when properly employed. More specifically, Saunders (26) reported that in 46 patients tested, only 18 gained full benefit. More important were his findings that 15 of them, although experiencing partial relief, were incorrectly using the aerosols, and that 13 showed no response at all. Yet these articles were overshadowed by the large number of publications and advertisements attesting the efficacy of the aerosol bronchodilators. It was against this background that the asthmatic deaths in Great Britain, Australia, and continental Europe, and the appearance of the locked-lung syndrome in the United States, were reported.

2.3.1 Deaths from Bronchial Asthma in Great Britain

In 1966, Smith (27) reported an increase in the annual number of deaths from bronchial asthma in England and Wales. Between 1959 and 1964, the overall death rate increased 42 percent, from 2.7 per 100,000 persons per year. The increase over the same period for the age group 5 to 34 years was 165 percent, and for those aged 5 to 14 years it was 330 percent. However, the rate of death at young ages from other respiratory conditions was stable. Together with the fact that there was no evidence to indicate that greater numbers of persons were suffering from bronchial asthma, this finding suggested that the increase might be due not to vicissitudes in the criteria for diagnosis of asthma, but rather to changes in the assignment of this disease entity as cause of death. The possibility that the increase was drug-induced (iatrogenic) found partial confirmation in the coincidental introduction of bronchodilator aerosols at a time when the mortality rate among asthmatics started to increase.

Pickvance (28) and Speizer et al. (29) examined the annual number of deaths attributed to bronchial asthma in England and Wales from 1960 to 1965. The increase was more pronounced at ages 5 to 34 years than at older ages, and was greatest among those aged 10 to 14, in which group fatalities increased nearly eightfold in 7 years. Corticosteroids had been used increasingly since 1952, and pressurized aerosols containing sympathomimetics had enjoyed great popularity beginning in 1960.

Speizer et al. (30) and Fraser et al. (31) examined copies of the death certificates of 171 persons aged 5 to 34 years who, according to those records, had succumbed to bronchial asthma during a 6-month period from 1966 to 1967. Signs of severe asthma were found in 91 percent of the necropsies for which data were available. Death was sudden and unexpected in 81 percent of the patients. While two-thirds had received corticosteroids before the terminal episode, such information about their use that was available provided no suggestion that excessive use was responsible for a large proportion of the deaths. However, pressurized aerosol bronchodilators had been used by 84 percent of the patients, and several instances of aerosol abuse were described. A similar increase in mortality rates occurred in other parts of Great Britain as well. In Eire, Linehan (32) reported 130 percent increase in deaths of those aged 5 to 34 years during the years 1960 to 1968.

In December 1968, the pressurized aerosol bronchodilators ceased to be available to the public over-the-counter, and became obtainable only by prescription in Great Britain. Although the high mortality rate of asthma was reduced, it became more difficult to further evaluate what factors were involved, or to establish a direct correlation between the death rate from asthma and changes in the distribution and frequency of use of bronchodilator aerosols. Statistics collected by Inman and Adelstein (33) yielded a mortality curve for the period 1961 to 1967 that closely resembled the curves for sales of pressurized aerosols during these same years, that is, a period of steady increase followed by a period of leveling, with a terminal sharp decline.

The first deaths attributable to the use of pressurized bronchodilator aerosols were reported in 1967 by Greenberg and Pines (34). Four asthmatic patients were found unexpectedly dead, with empty pressurized aerosols either by their sides or in their hands. Eight other patients who had been hospitalized for treatment of varying degrees of severe bronchial asthma had also died suddenly. These deaths were also without obvious cause and remained unexplained after postmortem examination. The report by Greenberg and Pines cited ventricular fibrillation resulting from excessive use of aerosols as the cause of death, and was followed by reports of other deaths under similar circumstances: two adolescent asthmatics (35), a 32-year-old asthmatic (36), a 46-year-old asthmatic (37), a 50-year-old female asthmatic, a 68-year-old coal miner with chronic bronchitis (38), and three adolescent asthmatics (39).

2.3.2   Locked-Lung Syndrome in the United States

Until the 1960s, a consistent finding reported among deaths from bronchial asthma was mucus plugs in the airways. Shapiro and Tate (40) ascribed other cases of sudden death to cardiac arrhythmia, and Van Metre (41) reported 17 deaths associated with acute asthmatic attack. In none of these deaths had the individual used bronchodilator aerosol excessively. In a subsequent report, however, Herxheimer (42) reported that nine patients whose deaths were attributable primarily to asthma had used bronchodilator aerosols well beyond the point of diminishing returns, and 29 patients who became resistant to the drug actually improved when they stopped inhaling the medicated aerosol.

In a review of 20 deaths over a 9-year period, Ghannam et al. (43) concluded that no one factor could reliably predict fatal outcome in patients known to use medicated aerosols extensively. In their review of hospital admissions at a pediatric hospital from 1935 to 1968, Palm et al. (44) noted that there was an increase in the number of patients hospitalized for bronchial asthma during the preceding 10 years and there was a concomitant decline in mortality rate. In elderly patients, death was usually associated with carbon dioxide retention (45). Most of the subsequent reports in the United States related to the ineffectiveness of aerosol bronchodilators in asthmatic patients and, by illustrating both the limitations and hazards of these agents, contributed to limiting their distribution and the number of fatalities attributable to their use.

2.3.3 Explanations for Asthmatic Deaths and Locked-Lung Syndrome

One outcome of the publicity surrounding the reports of deaths and locked-lung syndrome associated with the use of bronchodilators was the prohibition of their over-the-counter sale in some countries such as Great Britain. Physicians, once informed of the situation by national publications, tended to prescribe bronchodilator aerosols more cautiously and also helped protect the public from the possible dangers inherent in their indiscriminate use. Although the causes for the deaths and cases of locked-lung syndrome observed have not been identified, such explanations as have been proposed tend to emphasize one or the other of two groups of hypotheses. The first hypothesis related to the toxicity of the bronchodilators when inhaled. Two opposite suggestions were made as to the manner by which the sympathomimetics contribute to death among asthmatics. Herxheimer (46) emphasized the ineffectiveness of the drug, whereas Stolley (47) attributed the phenomenon to the high concentrations of the drug available in certain countries, including Great Britain. The second hypothesis emphasized the toxicity of FC 11 and other fluorocarbon propellants. Although fatalities associated with the use of aerosols had been reported, the tendency was to underestimate the hazards of the propellants. The two reasons for this misconception were that (a) the propellants were claimed to be "inert" and nontoxic; and (b) the amount contained in the inspired aerosol was small, and little was thought to be absorbed into the blood (Section 2.4.3). Phillips (48) questioned the nontoxicity of the propellants, since the only available information was based on topical testing on the skin of animals. The first experimental evidence offering unequivocal proof of the cardiotoxicity of the propellants themselves was supplied by Taylor and Harris (49) and Taylor et al. (50). The spokesman for the manufacturers of aerosol bronchodilators has continually criticized the work as being relevant only for sniffing deaths and only remotely applicable to asthmatic deaths. Other laboratories have been encouraged to investigate the problem.

2.4 Human Studies

Based on results of animal inhalational experiments (see Section 2.5), a threshold limit value (TLV) of 1000 ppm (approximately 5600 mg/m^3) was recommended by the American Conference of Governmental Industrial Hygienists (ACGIH) (51). Subsequently, human studies were performed which justify the TLV on the basis of cardiopulmonary effects, uptake, and excretion of FC 11.

2.4.1 Cardiopulmonary Effects

As early as 1967, Plaiut (52) reported one subject who developed bronchospasm after inhalation of a fluorocarbon (unspecified), with no bronchodilator drug, from an aerosol unit. Subsequently, Fabel et al. (53) tested patients with chronic obstructive pulmonary disease, given 10 to 16 actuations of the propellant, each dose comprising 12.6 ml of a mixture of 4 parts FC 11 and 6 parts dichloro-fluoromethane (FC 21). A third study by Brooks et al. (54) showed that in five of 13 asthmatic patients there was a measurable increase in airway resistance following five inhalations of aerosol propellant. The publication specifies dichlorodifluoroethane, which is not commercially available, although a typo-graphical error could have meant dichlorodifluoromethane (FC 12), which is commonly used in bronchodilator aerosol units.

That bronchospasm consistently occurs in human subjects inhaling fluoro-carbons was unequivocally demonstrated in 1977 by Valic et al. (55). The propellant gases were generated from commercial aerosol units and applied to the subject from a distance of 50 cm for periods of 15 to 60 sec. At a measured concentration of 95,000 mg/m^3 (1700 ppm), there was a biphasic change in ventilatory capacity, the first reduction occurring within a few minutes after exposure, and the second delayed until 13 to 30 min after exposure. Most subjects developed bradycardia, and inversion of the T-wave. A 10 to 90 percent mixture of FC 11 and FC 12, respectively, caused more severe respiratory effects than either fluorocarbon inhaled singly.

The threshold for the cardiopulmonary effects of FC 11 was not determined by Valic et al. (55). However, from the studies of Stewart et al. (56), it is certain that human exposure to 1000 ppm, 8 hr/day, 5 days/week for a total of 18 exposures had no untoward subjective effects, and there were no changes in the electrocardiogram or pulmonary function tests. The venous blood levels of FC 11 after 8 hr were as high as 4.69 μg/ml. The gradual attainment of this level represents a low uptake of the gas, similar to that of halothane, a fluorine-containing compound widely used to induce a general anesthesia (Section 8).

2.4.2 Respiratory Uptake and Elimination

The human exposure studies by Stewart et al. (56) described in the preceding paragraph include post-exposure breath data that reveal a predictable excretion pattern. The rate of excretion of FC 11 in the expired air is a function of the duration of exposure, although there was no significant accumulation of FC 11 in the body following 8-hr exposures to 1000 ppm, repeated every 24 hr. The post-exposure breath decay curves serve to refine those reported earlier by Paulet et al. (57, 58).

Radioactive tracer techniques were used by Morgan et al. (59, 60) to measure the partition coefficient of fluorocarbons, including FC 11. As a group, the fluorocarbons have low lipid solubility compared to aliphatic chlorinated hydrocarbons. Chlorine-38-labeled fluorocarbons were poorly absorbed in the lung, with much of the inhaled vapor exhaled. After 30 min, the amount retained in the lungs was about 23 percent of the total unexpired FC 11. Presumably, the fluorocarbon remained in the lung tissue; after 5 min only a small fraction of the retained material was present in the blood.

Using fluorine-18-labeled FC 11, Williams et al. (61) derived the same partition coefficient (olive oil/air partition = 27) as that derived by Morgan et al. (59). The former group of investigators characterized the subsequent fate of FC 11 following its inhalation. The fall in pulmonary concentration was consistent with rapid uptake into the tissues followed by slow elimination into expired air (61).

2.4.3 Blood Levels of Exposed Individuals

Marier et al. (62) failed to detect fluorocarbons by random sampling of blood from users of household aerosols. Their conclusion that the use of aerosols poses no health hazard has been criticized (63). On the other hand, there are patients inhaling bronchodilator aerosols who show a significant concentration of fluorocarbons in their blood. Paterson et al. (64) observed peak concentrations ranging from 0.13 to 2.60 μg/ml venous blood in a group of nine subjects. In a group of asthmatics, Dollery et al. (65, 66) detected peak arterial levels as follows:

	Level (μg/ml) in Arterial Blood	
	FC 11	FC 12
Two actuations 30 sec apart	0.53–3.1	0.2–3.13
One actuation	0.26–2.0	0–2.03

Dollery and co-workers (65–67) have concluded that the blood levels are not sufficiently high to exert cardiotoxicity. However, they have not considered the possibility that fluorocarbons may induce reflexes from the respiratory tract which in turn influence the heart. In other words, the effects on the heart are partially triggered by the inhaled fluorocarbon, prior to its absorption. The details of cardiotoxicity of FC 11 derived from animal experiments are covered in the next section.

2.5 Cardiotoxicity in Animal Studies

In recent years, Aviado has completed a comprehensive comparison of the inhalational effects of fluorocarbons in several animal species (1, 68–71). These studies were initiated following reported fatalities from the use, misuse, and abuse of aerosol products in general and of bronchodilator aerosols in particular. The experimental procedures developed for the toxicologic evaluation of fluorocarbons are discussed below because they may prove to be useful in the risk assessment of chemical vapors and gases. It has become apparent that selected animal species are useful in anticipating the potential toxicity of inhalants in humans.

2.5.1 Acute Inhalation Toxicity

There are no published reports on estimation of the LC_{50} for FC 11 administered by inhalation. All available observations relate to the concentration that would be lethal to exposed animals. The most resistant species requiring the highest lethal concentration is the guinea pig. The rat is the most sensitive, and the mouse and cat are between the two extremes.

Guinea Pig. Nuckolls (72) reported the first investigations of the acute toxicity of FC 11. Twelve guinea pigs divided into four groups of three each were exposed for 5 min, 30 min, 1 hr, and 2 hr, respectively. Exposure to 2.5 percent for 30 min caused occasional tremors and bruxus, and the rate of respiration became irregular. Exposure to 10 percent for 1 hr resulted in coma. The guinea pigs exposed to this concentration for 2 hr were sacrificed 8 days later. Whereas their lungs were found to contain mottled areas of congestion, other organs showed no pathological changes. Scholz (73) reported that exposure to a concentration of 20 percent for 1 hr was lethal. According to Caujolle (74), inhalation of a concentration of 25 percent for 30 min was lethal in half the guinea pigs tested. A concentration of 3 percent inhaled for 2 hr, although not fatal, caused unconsciousness.

Rat. Lester and Greenberg (75) exposed rats to FC 11 in concentrations ranging from 5 to 50 percent for 30 min. Whereas a concentration of 5 percent produced no symptoms of intoxication, concentrations of 6 and 7 percent caused a loss of postural reflex, 8 percent a loss of righting reflex, and 9 percent complete unconsciousness. The following concentrations and times were lethal: 10 percent inhaled for 20 to 30 min; 15 percent for 8 min; 20 to 30 percent for 4 min; and 50 percent for 1 min. Scholz (73) reported a lethal concentration of 10 percent after 90 min. Friedman et al. (76) exposed anesthetized rats to increasing concentrations of FC 11 and observed that apnea occurred 5 min after the concentration reached 20 percent.

Cat and Mouse. Scholz (73) reported that inhalation of 10 percent of FC 11 for 1 hr was lethal to the cat. Caujolle (74) determined the lethal concentration in mice. In an atmosphere containing 15 percent FC 11, mice succumbed in a few minutes.

2.5.2 Cardiac Arrhythmia

The dog and the monkey require a lower concentration of FC 11 to provoke cardiac arrhythmia than do the rat or the mouse. In each species, the sensitivity to the proarrhythmic action of FC 11 can be altered by special procedures that simulate disease.

Mouse. Recent interest in the cardiotoxicity of propellants was initiated with the observations of Taylor and Harris (49), using aerosols released from a bronchodilator pressure unit containing three propellants, including FC 11. Experiments relating to the inhalation of FC 11 in gaseous form have been reported by Aviado and Belej (77). Mice under pentobarbital anesthesia did not show any cardiac arrhythmia following inhalation of 2 or 5 percent FC 11. However, inhalation of 10 percent FC 11 produced second-degree atrioventricular block, and inhalation of 5 percent FC 11 also caused the appearance of atrioventricular block, following a concurrent intravenous injection of epinephrine. Mice that had experimental bronchitis developed arrhythmia during inhalation of FC 11 even without injection of epinephrine.

Dog. Reinhardt et al. (78) reported sensitization of the heart to epinephrine in the unanesthetized dog. The inhalation of 0.35 to 0.61 percent FC 11 for 5 min caused ventricular fibrillation and cardiac arrest following the injection of epinephrine. Although inhalation of lower concentrations (0.09 to 0.13 percent) did not sensitize the heart, administration of higher concentrations (0.96 to 1.21 percent) resulted in a greater frequency of cardiac arrhythmia. Exercising on a treadmill, known to effect release of endogenous epinephrine, did not induce arrhythmia in dogs inhaling 0.5 to 1.0 percent FC 11 (79). Clark and Tinston (80) investigated the interaction between FC 11 and two sympathomimetic drugs in unanesthetized dogs. Inhalation of 1.25 percent FC 11 sensitized the heart to epinephrine but not to isoproterenol.

Monkey. The minimal concentration that elicited cardiac arrhythmia in the anesthetized monkey was 5 percent FC 11 inhaled for less than 5 min (81). In a group of seven monkeys, two of them developed ventricular premature beats and atrioventricular block. The sensitivity of the heart to arrhythmia was increased by infusion of epinephrine or by coronary arterial occlusion which reduced the threshold proarrhythmic doses to 2.5 and 1.25 percent, respectively.

The combination of the two procedures further reduced the threshold concentration to 0.5 percent FC 11, which marks a tenfold increase in sensitivity of the heart as a result of experimental infarction.

Rat. In the unanesthetized rat, the minimal concentration that produced arrhythmias consisting of atrial fibrillation, ventricular extrasystoles, and widening of the T-wave was 2.5 or 5 percent (82). The induction of pentobarbital anesthesia reduced the incidence of arrhythmia and increased the threshold concentration to 10 percent of the fluorocarbon (83). Rats that developed cardiac necrosis elicited by injections of large doses of isoproterenol showed a reduction in threshold concentration to 5 percent. Likewise, those that developed pulmonary arterial thrombosis showed a similar increase in the proarrhythmic activity of FC 11. The induction of pulmonary emphysema did not increase the sensitivity of the heart (82). Adrenalectomy or the injection of drugs that block cardiac adrenergic receptors protected the heart from FC 11-induced arrhythmia (83).

2.5.3 Cardiac Rate

Three animal species show tachycardia in response to the inhalation of FC 11. The threshold levels were as follows: 1.0 percent in the anesthetized dog (84); 2.5 percent in the anesthetized monkey with open chest (81) or closed chest (85); and 2.5 percent in the unanesthetized rat (82). The induction of anesthesia in rats causes a conversion of the tachycardiac response to bradycardia, and a similar influence probably occurs in a fourth species, mice. Only anesthetized mice have been used which respond with bradycardia during inhalation of 10 percent of this propellant (86). It is useful to recall that bradycardia is the usual response in human subjects inhaling low concentrations of FC 11 (Section 2.4.1). A similar bradycardia is encountered in dogs when the administration of FC 11 is limited to the upper respiratory tract, that is, oropharyngeal and nasal areas (87). It is reasonable to suggest that bradycardia in man originates from irritation of the upper respiratory tract, and that cardiac effects can be initiated prior to absorption of FC 11 in the lungs.

2.5.4 Depression of Myocardial Contractility

Two techniques in different species have been used to investigate the effects of FC 11 on myocardial contractility. In the canine heart–lung preparation, the inhalation of 2.5 percent FC 11 depressed the ventricular function curve (88). In the anesthetized monkey with myocardial ischemia, the inhalation of 0.5 percent caused a depression of the force of contraction recorded by a strain gauge sutured to the ventricular surface (80).

2.5.5   Subacute Cardiac Effects

Taylor and Drew (89) have reported that hamsters with inherited cardiomy-opathy are predisposed to FC 11 cardiotoxicity, as compared to normal hamsters. After exposure to 2 percent FC 11, the cardiomyopathic animals died of frank congestive heart failure. Balazs et al. (90) observed focal myocardial necrosis in dogs exposed for two consecutive days to aerosols containing FC 11 as propellant, and isoproterenol as the bronchodilator drug. There are no reported cardiac function studies on animals exposed to low levels of FC 11 daily for more than a few days.

2.5.6   Summary of Cardiotoxicity of FC 11

The above threshold concentrations of FC 11 that influence each animal species permit the following generalization. There is a striking similarity in threshold concentrations between the mouse and the rat on one hand, and the dog and the monkey on the other. The mouse and rat require a FC 11 concentration of 2.5 to 5.0 percent in order to affect the circulatory system. When anesthetized, these two species respond with bradycardia. The importance of the parasym-pathetic nervous system has been demonstrated by the use of atropine, which blocks the response. In the unanesthetized rat, only tachycardia results from inhalation of the propellant.

The circulatory system of the monkey and the dog can be influenced by a concentration of 0.5 percent FC 11. This level causes cardiac arrhythmia in the unanesthetized dog, and in the anesthetized monkey with coronary arterial occlusion and epinephrine infusion.

The most serious sign of toxicity to acute inhalation of FC 11 is cardiac arrhythmia which can be elicited in all four animal species. This may account for the sudden deaths associated with the use, misuse, and abuse of aerosols. There are three procedures that increase the sensitivity of the heart to arrhythmia: (*a*) the injection of epinephrine; (*b*) coronary ischemia or cardiac necrosis; and (*c*) experimental bronchitis or pulmonary thrombosis. A common feature of all three procedures is an increase in cardiac irritability caused by epinephrine directly by initiation of ectopic foci in areas of the heart, or indirectly through hypoxia and hypercapnea resulting from lesions in the lung. On the other hand, certain procedures depress the sensitivity of the heart to the proarrhythmic action of FC 11, including adrenalectomy, adrenergic blockade, and general anesthesia. There is one preparation that has not been completed for testing of FC 11 and related propellants, namely, the unanesth-etized monkey with myocardial ischemia. The preparation would resemble more closely the patient with heart disease who is liable to be exposed to bronchodilator aerosols. The threshold level of 0.5 percent in monkey may be further reduced by omission of the anesthesia. This possibility is likely because the dog and the rat have demonstrated a decrease in threshold in the unanesthetized state.

Finally, the mechanism of FC 11 cardiotoxicity requires additional comments. The adverse effects originate from irritation of the respiratory tract which in turn reflexly influences the heart rate even prior to absorption of the fluorocarbon, followed by direct depression of the heart after absorption. Since the heart is sensitized to sympathomimetic amines, the combination of fluorocarbon with a sympathomimetic bronchodilator is potentially dangerous for the treatment of bronchial asthma. For the same reason, sympathomimetic drugs are contraindicated in cardiac resuscitation of patients suffering from fluorocarbon poisoning. A cardiotonic drug that is free of proarrhythmic activity is available in Europe and is undergoing clinical study (71).

2.6 Bronchopulmonary Toxicity in Animal Studies

Although there has been a nearly comprehensive comparison between human studies and animal experiment on the toxicity of FC 11 on the cardiovascular system, the information on the respiratory or bronchopulmonary area is less extensive. Animal studies on FC 11 are useful in interpreting human observations in the following respects: (a) sublethal concentration causes reduction in respiratory movements in animals that correspond to the general anesthetic properties of fluorothane, a widely used fluorine-containing compound in medicine; (b) bronchospasm has been encountered in man exposed to FC 11, a response that can be reproduced in animal experiments; (c) animal studies show a reduction in pulmonary compliance which may relate to the reduction in pulmonary surfactant postulated to occur in man (Section 2.2.2).

2.6.1 Decreased Respiratory Minute Volume

Unlike most other fluorocarbon propellants, FC 11 causes only depression of respiratory minute volume that is not preceded by stimulation of breathing. There is ultimate cessation of respiration, which is a manifestation of generalized depression of the central nervous system by FC 11.

Monkey. The dose that causes a significant reduction in respiratory minute volume is 5 percent. In the same group of monkeys, circulation is influenced by 2.5 percent of the propellant (81). The respiratory effect is brought about by a combination of reduced respiratory rate and tidal volume.

Dog. The respective threshold doses for the dog are lower than those for the monkey. Inhalation of 1 percent FC 11 influenced heart rate and blood pressure but not respiration. The minimal dose that depresses respiratory minute volume is 2.5 percent of the fluorocarbon (84).

Rat. With administration of increasing concentrations of FC 11, a 40 percent depression of respiratory minute volume occurred at 10 percent concentration (75). In the emphysematous rat, the reduction in response to inhalation of 2.5 percent FC 11 is less than the control, indicating that lesions in the lung cause a decreased response (82).

Mouse. The only dose tested in mice is 2.5 percent FC 11. This concentration caused a 65 percent depression of respiratory minute volume, which was supported by a reduction in rate and tidal volume (86).

2.6.2 Increased or Decreased Airway Resistance

The effects of FC 11 on the airways are variable. Both bronchodilation and bronchoconstriction have been observed in those animal species so far examined.

Monkey and Dog. The minimal concentration that reduced resistance is 5.0 percent FC 11 in the anesthetized monkey (85) and 2.5 percent in the anesthetized dog (84). In the latter species the reduction in resistance is blocked by pretreatment with a sympathetic blocking drug, suggesting that the effect is mediated through adrenergic receptors.

Rat and Mouse. The anesthetized rat shows an increase in airway resistance while exposed to increasing concentrations of the fluorocarbon (75). The minimal dose is about 2.5 percent concentration in the anesthetized rat (82). However, in the emphysematous rat with an elevated airway resistance, the inhalation of 2.5 percent FC 11 had no influence, suggesting that the response appeared only in the nonemphysematous state. The minimal concentration that increases resistance in anesthetized mouse is 1 percent, which is blocked by pretreatment with atropine (86). These observations indicate that the broncho-constriction observed in this animal species is precipitated by vagal innervation to the lungs.

2.6.3 Decreased Pulmonary Compliance

There are no changes in pulmonary compliance in the monkey (85) and in the dog (75) following inhalation of 1.0 to 5.0 percent FC 11. In the rat a reduction in compliance occurs following inhalation of 2.5 percent FC 11, which is more intense in the emphysematous rat (82). The mouse is more sensitive, since 1 percent FC 11 causes a significant reduction of compliance which is not mediated through the vagus (86). The cause of changes in elasticity of the lung has not been examined. One possibility is the formation of acute pulmonary edema. In another species, the dog, inhalation of 2.5 percent FC 11 does not cause an elevation of pulmonary arterial pressure (91). Since it is technically impossible to measure pulmonary arterial pressure in the mouse, the cause of the reduction in compliance must be identified by another means. A reduction in tracheobronchial clearance as a cause of decreased compliance has been excluded in experimental observations in donkeys (92).

2.7 Suspected Toxicity Based on Animal Studies

Less conservative toxicologists have raised questions on the human hazards of exposure to FC 11 with regard to hepatoxicity, reproductive abnormalities, mutagenicity, and carcinogenecity. Since FC 11 (trichlorofluoromethane) is structurally similar to tetrachloromethane it was thought that the health hazards may be alike. This argument was silenced in 1966 and 1967 by Clayton, who contrasted the difference between chlorine and fluorine substitution to hydro-carbons. In his reviews of fluorocarbons, Clayton concluded that fluorine stabilizes the adjacent C–Cl bonds and reduces biologic activity (93–95). During the past decade, human exposures to lethal and sublethal concentrations of FC 11 show involvement only of the cardiopulmonary and central nervous systems, without involvement of other viscera (Section 2.2). This section discusses the animal and *in vitro* studies that support the general proposition that the fluorocarbons are less toxic than chlorinated hydrocarbons.

2.7.1 Hepatotoxicity

Repeated inhalational exposure to FC 11 does not produce pathological lesions of the liver and other visceral organs. The studies were conducted as follows: 1.25 or 2.5 percent FC 11 for 3.5 hr each on 20 consecutive days using guinea pigs, rats, and cats (72); 0.1 percent FC 11 continuously for 90 days, or 1.025 percent FC 11 for 8 hr for 30 days in guinea pigs, rats, dogs, and monkeys (96). Slater (97) compared the oral toxicity of FC 11 with tetrachloromethane. He concluded that the latter compound is hepatotoxic, whereas FC 11 is not. The most likely explanation is that tetrachloromethane undergoes biochemical metabolism to a potent hepatotoxin but such a transformation appears insig-nificant in the case of FC 11. In a later study, Slater suggested that the most likely route of metabolism for tetrachloromethane, but not for FC 11, is the formation of free radical products resulting in lipid peroxidation, which is requisite for liver necrosis (98). Cox et al. (99) have challenged Slater's conclusion by performing in vitro studies that show binding of FC 11 with liver microsomal enzymes. Coincidentally, Blake and Mergner (100) showed that carbon-14-labeled FC 11 is refractory to biotransformation and is rapidly exhaled in its unaltered form in beagle dogs, similar to the results of human studies (Section 2.4.2).

2.7.2 Teratogenicity

There is no published study of the effects of FC 11 inhalation on reproduction. Paulet et al. (101) used a propellant mixture of 10 percent FC 11 and 90 percent FC 12; 20 percent of the mixture in air was administered for 2 hr daily, to rats from the fourth to the sixteenth day of gestation, and to rabbits from the fifth to the twentieth day. There were no adverse effects on the offsprings of the exposed pregnant animals.

2.7.3 Mutagenicity

There is no published study on mutagenicity of FC 11. Negative results on the Ames bacterial test are cited in a review (102) without details on concentration of fluorocarbon in the media.

2.7.4 Carcinogenicity

A National Cancer Institute-sponsored bioassay on carcinogenicity has been completed. The oral administration of 488 and 977 mg/kg per day to male rats, 538 and 1077 mg/kg per day to female rats, and 1962 and 3925 mg/kg per day to mice of both sexes did not show carcinogenicity (103). This study is significant because it differentiates the noncarcinogenicity of FC 11 from the carcinogenicity of carbon tetrachloride in a similar oral bioassay study. The negative results temporarily have refuted the potential carcinogenicity of fluorocarbons in general, and of FC 11 in particular, suggested by injection studies performed by Epstein et al. (104). However, the question of carcinogenicity has been raised owing to a suspicious leukemia seen in exposed animals (Section 5) and the higher incidence of cancer among medical personnel exposed to fluorine-containing general anesthetics (Section 8).

3 COMPARATIVE TOXICITY OF NONHYDROGENATED FLUOROCARBONS

In 1974, Molina and Rowland (105) drew attention to a potential biologic hazard resulting from depletion of the ozone layer owing to the release of fluorocarbons into the atmosphere. The resulting "ozone war" among scientists and nonscientists, recently recounted by Dotto and Schiff (106), appears to have been settled in favor of the general premise that photodissociation of fluorocarbons in the stratosphere produces significant amounts of chlorine atoms and leads to the destruction of atmospheric ozone (107–109). A reduction in the ozone allows more ultraviolet light to reach the earth's surface and is anticipated to increase the incidence of malignant melanoma, a serious form of skin cancer frequently causing death, and increase incidence of basal and squamous-cell carcinomas of the skin that are less serious but much more prevalent. The effects of increased ultraviolet radiation on plants and animals due to ozone depletion are of unknown magnitude (109).

It is generally agreed by most experts that FC 11 and other *nonhydrogenated* fluorocarbons are more stable in the stratosphere, and are more likely to deplete the ozone layer, than the *hydrogenated* fluorocarbons, which are less stable and less likely to reach the ozone layer. The latter group is discussed separately (Section 4) from the following nonhydrogenated fluorocarbons covered in this section:

Dichlorodifluoromethane (FC 12)	CCl_2F_2
Trichlorotrifluoroethane (FC 113)	$CClF_2-CCL_2F$
Dichlorotetrafluoroethane (FC 114)	$CClF_2-CClF_2$
Chloropentafluoroethane (FC 115)	$CClF_2-CF_3$
Octafluorocyclobutane (FC C-318)	C_4F_8

The available information on the above nonhydrogenated fluorocarbons is disappointingly less than that on FC 11 (i.e., trichlorofluoromethane, CCl_3F).

3.1 Dichlorodifluoromethane or Fluorocarbon 12

Prior to the banning of fluorocarbons in aerosol products, FC 12 was the most widely used high pressure propellant, together with the low vapor pressure propellant FC 11. FC 12 continues to be used as a refrigerant and for preparing frozen tissue sections (110). Studies on human volunteers showed that inhalation of 10,000 ppm of FC 12 for 2.5 hr cause a 7 percent reduction in standardized psychomotor scores (111). At concentrations of 1000 ppm for 8 hr daily, 5 days/week for a total of 17 repetitive exposures, there were no untoward subjective responses and no abnormal physiological responses of the lungs or heart (112). Concentrations as high as 27,000 ppm of FC 12 for 15 or 60 sec caused an increase in airway resistance and electrocardiographic changes (55). The rate of transfer of FC 12 from blood to tissue compartments was more rapid for FC 12 than for FC 11 in human volunteers (59).

In experimental animals, FC 12 is less toxic than FC 11. The presence of a chlorine ion (in FC 11) with a fluoride (to form FC 12) is accompanied by a reduction in toxicity. The overall potency ratios of FC 12 to FC 11 are as follows:

4 to 7	Acute inhalational lethality (102)
6 to 8	CNS depression (102)
5	Sensitization of heart of unanesthetized dogs to epinephrine
4	(78)
4	Induction of cardiac arrhythmia in anesthetized monkeys (81)

Depression of myocardial contractility in dogs and monkeys (81, 84)

There is also a reported depression of the heart in rabbits (113), but comparative response to FC 11 is not known. The only reported differences between the two fluorocarbons are increased airway resistance and decreased pulmonary compliance in dogs and monkeys for FC 12 that are absent for FC 11 (70). FC 12, like FC 11, is not a teratogen in rats and rabbits (102).

3.2 Trichlorotrifluoroethane or Fluorocarbon 113

The major current use of FC 113 is as a solvent for cleaning electronic equipment and degreasing of machinery (Section 7.3). Like other fluorocarbons, the threshold limit value permissible in the work environment is 1000 ppm (7600 mg/m^3). Imbus and Adkin (114) examined 50 workers at the Kennedy Space Center exposed to levels ranging from 46 to 4700 ppm for an overall average duration of 2.77 years. There were no signs or symptoms of adverse effects.

In animal studies, FC 113 is more toxic than FC 12, but less toxic than FC 11. Although all three fluorocarbons are cardiotoxic to dogs and monkeys there is a difference in their effects on the respiratory system of monkeys. FC 11 causes early respiratory depression and FC 12 causes bronchoconstriction, whereas FC 113 does not influence either respiratory activity or airway resistance (70).

3.3 Dichlorotetrafluoroethane or Fluorocarbon 114

Human studies indicate that inhaled FC 114 is rapidly excreted as other fluorocarbons (58, 59, 102). Animal studies indicate that the lethality and cardiotoxicity of FC 114 is less than FC 11 and about equal to that of FC 12. The respiratory effects of FC 114 vary, depending on the animal species. This fluorocarbon causes respiratory depression in the monkey and stimulation in the rat, with no significant effect in the dog. Pulmonary compliance is reduced in the dog and rat but not in the monkey. For purposes of generalization, the complete respiratory profile for FC 114 is that it reduces pulmonary compliance and increases airway resistance (70).

3.4 Chloropentafluoroethane or Fluorocarbon 115

FC 115 was originally introduced as a propellant for use in food aerosol products, but has been banned. This fluorocarbon has one of the lowest level of cardiotoxicity, sensitizing the dog heart to epinephrine-induced arrhythmias in concentrations 25 to 50 times greater than that of FC 11 (70). In sublethal concentrations of FC 115, the following effects are elicited, depending on the animal species. The monkey does not show any effect on respiration or circulation when exposed to 20 percent FC 115 (81, 85). The dog shows no respiratory depression when exposed to 20 percent but shows bronchoconstriction, decreased compliance, sensitization of the heart to epinephrine, tachycardia, myocardial depression, and hypotension when inhaling 10 to 25 percent FC 115 (84, 88). Nevertheless, the potency is considerably less than that of FC 11, which produces most of these effects when inhaled in concentrations of 0.3 to 2.5 percent. The rat responds with bronchospasm, decreased compliance, and respiratory stimulation to 10 percent FC 115, whereas inhalation of 2.5 percent FC 11 causes bronchoconstriction, decreased compliance, and respiratory depression (76).

3.5 Octafluorocyclobutane or Fluorocarbon C-318

This last example of commercially available nonhydrogenated fluorocarbons, FC C-318, is the least toxic. In acute inhalation experiments, rats survive a mixture of 80 percent in oxygen for 4 hr (115). However, physiological measurements indicate that inhalation of 10 percent FC C-318 causes the following: bronchoconstriction and reduced compliance in the dog (84) and rat (76), myocardial depression in the monkey (81), and sensitization to epinephrine-induced arrhythmia in the dog and the mouse (77). Compared to FC 11, the concentrations of FC C-318 to produce the following adverse effects are higher: 250-fold in the anesthetized rat for respiratory depression and bronchoconstriction. The initial use of FC C-318 as a propellant and aerating agent for foods has been banned owing to its suspected depletion of ozone in the stratosphere.

4 COMPARATIVE TOXICITY OF HYDROGENATED FLUOROCARBONS

Present evidence indicates that hydrogenated fluorocarbons are less stable in the atmosphere than nonhydrogenated, as discussed in the preceding sections. The hydrogenated compounds react with hydroxyl radicals, resulting in the formation of water and the degradation of the fluorocarbon before it reaches the stratosphere. It has been estimated that the hydrogen-containing fluorocarbons have lifetimes about 5 to 100 times shorter than those for FC 11, FC 12, FC 113, FC 114, FC 115, and FC C-318 (all nonhydrogenated). In the event that the nonhydrogenated are all banned for industrial use, the following commercially available hydrogenated fluorocarbons can be substituted:

Chlorodifluoromethane (FC 22)	$CHClF_2$
Chlorofluoromethane (FC 31)	CH_2ClF
Chlorodifluoroethane (FC 142b)	CH_3-CClF_2
Difluoroethane (FC 152a)	CH_3-CHF_2

Of the above compounds, FC 152a is unique in that the chlorine atom is absent. Therefore it is the least likely to deplete the ozone layer since the release of chlorine in the ionosphere is involved n the conversion of ozone to oxygen.

4.1　Chlorodifluoromethane or Fluorocarbon 22

On the basis of early inhalation experiments in guinea pigs and rats, FC 22 was shown to be two to three times less toxic than FC 11 (70). More recent comparative studies indicate an eight- to ten-fold difference in cardiotoxicity in mice and dogs (102). Since FC 22 is a high pressure fluorocarbon, it is more appropriate to make a comparison with another high pressure compound, for example, FC 12. Both fluorocarbons cause early respiratory depression, bronchoconstriction, tachycardia, myocardial depression, and hypotension in approximately equivalent concentrations (5 to 10 percent) in dogs and monkeys. The difference between the two high pressure fluorocarbons is that FC 22 does not induce cardiac arrhythmia in the monkey (81), although it sensitizes the heart to epinephrine in the mouse (77), and that FC 22 does not decrease pulmonary compliance in the monkey (85).

There are no reported studies on chronic exposure to FC 22 in animals. Speizer et al. (116) have undertaken an epidemiologic study of hospital personnel exposed to FC 22 for tissue freezing. They reported a 3.5-fold excess incidence of palpitation in the exposed individuals compared to nonexposed hospital personnel. Although there are animal models to determine the influence of FC 22 on the pathogenesis of coronary heart disease, such a study has not been executed. It is more reasonable to explain the palpitation as a form of cardiac arrhythmia from noncoronary vascular etiology, provided that the cardiotoxicity of FC 22 is similar to that of FC 11.

The mutagenicity of FC 22 has not been completely evaluated. The initial studies of a positive response in the Salmonella reverse mutation assay (102) have not been pursued in other forms of mutagen testing. The results of an industry-sponsored rat teratology study have been summarized as follows: "extremely weak, atypical response" . . . "a recent evaluation concludes that the TLV of 1000 ppm satisfactorily protects the fetuses of pregnant women in the workplace" (102). It is not possible to comment further because the studies are not published.

4.2　Chlorofluoromethane or Fluorocarbon 31

There are no published toxicity studies on FC 31. Unpublished studies conducted by one company are reported as indicating that "FC 31 is too toxic to be acceptable as an aerosol propellant" (102). The summary statements on positive mutagenicity and "lingering kidney injury" cannot be evaluated here.

4.3   Chlorodifluoroethane or Fluorocarbon 142b

As a low pressure fluorocarbon, FC 142b has a level of toxicity lower than FC 11 and FC 114, but higher than FC C-318 (70). The known range of effect of FC 142b is less than that of FC 114. The following characteristics of FC 114 are not observed when FC 142b is administered: cardiac arrhythmia and tachycardia in the monkey (81), epinephrine-induced arrhythmia in the mouse (77), decreased pulmonary compliance in the dog (84), and bronchoconstriction and early respiratory depression in the monkey (85). On the contrary, FC 142b is a respiratory stimulant in the monkey (85) and the dog (84), and this fluorocarbon is the only one known to exert a nondepressant central nervous system effect in two animal species.

4.4   Difluoroethane or Fluorocarbon 152a

We regard FC 152a as the least toxic of 20 fluorocarbons that we have examined in our laboratories (1, 70). Although FC 152a has no detectable effect in the monkey, it causes sensitization to epinephrine in the dog. The mouse exposed to FC 152a shows bronchoconstriction, respiratory depression, and decreased compliance but no cardiac arrhythmia (77, 86). In the mouse that has developed bronchitis (86) and in the rat with pulmonary emphysema (82), the administration of FC 152a provoked abnormalities in the electrocardiogram. These observations are noteworthy, indicating that bronchopulmonary disease increases the cardiotoxicity to FC 152a in particular and possibly to all fluorocarbons in general.

   Foltz and Fuerst (117) reported positive mutagenicity for FC 152a in *Drosophila melanogaster*. A review of published and unpublished work conducted by a fluorocarbon manufacturer cites negative mutagenicity (Salmonella) from its own laboratory (102). However, the reviewers (102) overlooked the results of Foltz and Fuerst published in 1974 so that additional studies are required to settle the conflicting results on mutagenicity.

5   DEVELOPMENTAL HYDROGENATED FLUOROCARBONS

So far only commercially available fluorocarbons have been discussed. Manufacturers of fluorocarbons are developing substitutes for the completely halogenated compounds (FC 11, FC 12, FC 113, FC 114, FC 115, and FC C-318). The 11 developmental fluorocarbons are all hydrogenated and, like the commercially available hydrogenated fluorcarbons, are less likely to deplete the ozone layer in the stratosphere. The developmental fluorocarbons can further be grouped into the *chlorine-containing* and the *nonchlorine-containing* fluorocarbons as follows:

Chlorine-containing and hydrogenated fluorocarbons

Dichlorofluoromethane (FC 21)	$CHCl_2F$
Dichlorotrifluoroethane (FC 123)	$CHCl_2-CH_3$
Chlorotetrafluoroethane (FC 124)	$CHClF-CF_3$
Dichlorodifluoroethane (FC 132b)	$CH_2Cl-CClF_2$
Chlorotrifluoroethane (FC 133a)	CH_2Cl-CF_3
Dichlorofluoroethane (FC 141b)	CH_3-CCl_2F
Chlorodifluoroethane (FC 142b)	CH_3-CClF_2

Nonchlorine-containing and hydrogenated fluorocarbons

Difluoromethane (FC 32)	CH_2F_2
Pentafluoroethane (FC 125)	CHF_2-CF_3
Tetrafluoroethane (FC 134a)	CH_2F-CF_3
Trifluoroethane (FC 143a)	CH_3-CF_3

The above grouping is useful to emphasize the significance of the presence of the chlorine atom, which contributes to the depletion of the ozone layer and also to the degree of cardiac toxicity. The last group of four developmental fluorocarbons is more desirable than the others, in light of environmental and toxicological considerations. A commercially available nonchlorine-containing hydrogenated fluorocarbon (namely, FC 152a; Section 4.4), has been ignored by those who have been studying the toxicity of developmental fluorocarbons.

There are scattered and difficult-to-verify reports that one of the developmental fluorocarbons is a suspected leukemogen in rats, that another causes sterility in male rats, still another is a teratogen, and several are mutagenic. All these claims are based on unpublished experiments and the studies appear to have been performed without following the general principles that toxicologic evaluation of fluorocarbons in animals must include comparison of two or more substances. The interested reader may examine the summary of studies completed by one manufacturer (102) and be advised that most other manufacturers have similar studies with conflicting results.

The authors have examined the cardiotoxicity of one developmental fluorocarbon, namely, dichlorofluoromethane (FC 21), which also appears as a contaminant of commercially available FC 22. As a low pressure fluorocarbon, FC 21 is less toxic than FC 11, indicating that the removal of one chlorine atom reduces the toxicity of fluorocarbons, The minimal concentration of FC 21 which induces cardiac arrthythmia, tachycardia, myocardial depression, and hypotension in the monkey is about half that of FC 11. In the dog (84), the hypotensive dose of FC 21 is 10 times that of FC 11, whereas the tachycardiac dose is 2.5 times that of FC 11. The most significant differences between both propellants relate to respiration. In the dog, FC 21 causes bronchoconstriction and decreased compliance, whereas FC 11 does not. It should be noted that the effects of FC 21 are opposite in these two species: bronchodilation, no decreased compliance, and early respiratory depression in the monkey; and bronchoconstriction, reduced compliance, and no respiratory depression in the dog. The reponse of the rat is more like that of the monkey, so that it is more reasonable to accept the pattern of action for the monkey as characteristic of FC 21.

There is no information on the cardiotoxicity of developmental fluorocarbons, other than that of FC 21. Most manufacturers are awaiting regulatory decisions as to whether or not the nonhydrogenated fluorocarbons will be permitted for nonaerosol uses. In the event they are banned from production, then the hydrogenated fluorocarbons commercially available (FC 22, FC 31 and FC 152a), as well as the 11 developmental ones listed above, will receive considerable attention.

Major future uses of developmental fluorocarbon are as working fluids in central station or onsite power generation plants and/or in the Rankine cycle turbine engine. These fluorocarbon systems can be used alone or in conjunction with solar electric or geothermal electric generating plants, with steam turbines in dual cycle power plants, or with fossil fuel-fired boilers for apartments, motels, hospitals, and shopping centers. When we attended the UN Conference on the Human Environment held in Stockholm in June 1972, the development of energy applications for the organochlorine compounds was suggested as the second priority for both developed and developing countries.

5.1 Summary of Comparative Toxicity of Commercial and Developmental Fluorocarbons

The above discussion on the toxicity of FC 21 marks the last of the developmental fluorocarbons considered in this chapter. Compared to the commercially available fluorocarbons, FC 21 belongs to the first of three categories of decreasing order of toxicity:

High level of toxicity

Trichlorofluoromethane (FC 11)	CCl_3F
Trichlorotrifluoroethane (FC 113)	$CClF_2-CCl_2F$
Dichlorofluoromethane (FC 21)	$CHCLli2F$

Intermediate level of toxicity

Dichlorotetrafluoroethane (FC 114)	$CClF_2-CClF_2$
Dichlorodifluoromethane (FC 12)	CCl_2F_2
Chlorodifluoromethane (FC 22)	$CHClFl_2$
Chlorofluoromethane (FC 31)	CH_2ClF

Chlorodifluoroethane (FC 142b) CH_3-CClF_2
Chloropentafluoroethane (FC 115) $CClF_2-CF_3$

Low level or toxicity

Octafluorocyclobutane (FC C-318) C_4F_8
Difluoroethane (FC 152a) CH_3-CHF_2

Coincidentally, the above grouping shows a direct relationship between the number of chlorine atoms and the level of toxicity. The two examples of least toxic fluorocarbons do not contain chlorine atoms.

5.2 Animal Models for Evaluation of Developmental Fluorocarbons

Although the nonchlorine-containing hydrogenated fluorocarbons are preferred from the environmental standpoint (namely, FC 32, FC 125, FC 134a, and FC 143a), they will have to be examined to prove that they have a low level of toxicity similar to the commercially available FC 152a. After application of various techniques to measure responses to fluorocarbons of four animal species, it has become apparent that certain generalizations can be made regarding the sensitivity of each animal model. The following discussion points out the specific application of the mouse, rat, dog, and monkey in determination of the potential hazard of an inhalant to the circulatory and respiratory systems.

5.2.1 Mouse

The proarrhythmic activity of fluorocarbons has been examined by administration in various concentrations, while recording the electrocardiograms (77). The inhalant was also administered while injecting epinephrine to determine sensitization of the heart to developing arrhythmia. As a rule, most fluorocarbons that provoke spontaneous arrhythmias also sensitize the heart to epinephrine. There are also fluorocarbons that do not induce spontaneous arrhythmia but cause sensitization. Lastly, there are fluorocarbons that do not exert spontaneous or epinephrine-associated arrhythmia provided that a lack of activity is confirmed in one other species.

There is a wide range in concentrations (5 to 40 percent) for the various fluorocarbons to cause arrhythmia. The sensitivity of the mouse can be increased by experimental induction of bronchitis (86), indicating that the minimal concentration can be reduced by disease. Compared to the other species, the mouse is least sensitive for determination of threshold concentration that would provoke cardiac arrhythmia.

In the course of testing fluorocarbons, a technique for measuring airway resistance and pulmonary compliance was developed for application to the mouse (86). This was not hitherto possible so that for the first time the sensitivities of the respiratory and circulatory systems have been compared in this species. The minimal concentration of FC 11 is 1 percent to produce bronchoconstriction and 2.5 percent to depress respiration. For the same fluorocarbon, 10 percent is needed to produce arrhythmia, indicating that the airways are more sensitive than the heart in the mouse.

5.2.2 Rat

A comparison of the rat with and without general anesthesia indicated that the unanesthetized state is more suitable for the investigation of cardiotoxicity (82). Anesthesia blocks the cardio-accelerator response, or even converts it to bradycardia. Although the unanesthetized rat is more sensitive than the mouse for demonstration of cardiotoxicity, the dog and monkey show effects when exposed to even lower concentration of the fluorocarbon.

In the rat, the cardiotoxicity of fluorocarbon is reduced by adrenalectomy or prior treatment with adrenergic blocking drugs (83). This observation is significant because it demonstrates that the sympathetic nervous system particulates in the cardiac effects of inhalants.

The rat is less sensitive than the mouse in manifesting respiratory toxicity (76). However, there are forms of experimentally induced diseases applicable to the rat. In addition to pulmonary emphysema, thrombosis of the pulmonary artery and myocardial necrosis have been induced in the rat. In emphysematous animals, it has been demonstrated that the lungs become more sensitive to a reduction in pulmonary compliance (82). Rats with cardiac necrosis or pulmonary arterial thrombosis show an increase in sensitivity to cardiotoxicity of the propellants (83).

5.2.3 Dog

The dog is the most sensitive animal for eliciting hypotension, depression of myocardial contraction, tachycardia, and cardiac arrhythmia (84). For the last-mentioned effect, the unanesthetized dog with epinephrine injection is five to 10 times more sensitive to fluorocarbons than the anesthetized dog without injection of epinephrine. This preparation is sufficiently sensitive that the proarrhythmic activity of one fluorocarbon (FC 152a) can be demonstrated even though the three other animal species fail to do so. The heart–lung preparation can be used to demonstrate direct depression of contractility of the ventricles, without participation of adrenal glands or autonomic innervation to the heart (81). As a rule, the concentrations of fluorocarbons that influence respiration are higher than those for circulation. The dog appears to be the least sensitive animal for demonstrating respiratory toxicity. However, in the investigation of mode of action, there are several techniques applicable to the dog. The bronchodilation induced by some fluorocarbons is mediated by adrenergic receptors (84). In addition, the fluorocarbons stimulate the receptors in the upper and lower respiratory tract, which in turn influence respiration, bronchomotor tone, and heart rate.

5.2.4 Monkey

The advantages and disadvantages of the monkey have not been fully appreciated owing to its limited application in our laboratory. As a rule, the circulatory system is more responsive than the respiratory system (81, 85). However, there are instances in which the monkey's response is opposite to that of the other species, and the relevance to humans is unsettled. There are fluorocarbons that produce cardiotoxicity in all three species but not in the monkey, some that cause myocardial depression and hypotension in the dog but not in the monkey, and some that cause early respiratory depression, bronchoconstriction, or decreased compliance in the dog, rat, and mouse, but not in the monkey. Opposite results also have been encountered, for example, fluorocarbons that are toxic to the monkey but not to the other species. It has been generally assumed that the monkey's response closely resembles the human's, but there has been no definitive comparison. When additional human studies on fluorocarbons become available it will be possible to decide the significance of the various animal models specifically for the prediction of toxicity in man.

6 OCCUPATIONAL EXPOSURE IN THE MANUFACTURE OF FLUOROCARBONS

Manufacturing processes use hydrofluoric acid from fluorospar in the production of all fluorocarbons. Some processes use carbon tetrachloride from carbon disulfide, or as a co-product of perchloroethylene and chlorination of propylene, or chloroform from chlorination of methanol. At the outset it is important to emphasize that all fluorocarbons are less toxic than any of the process materials used in their manufacture. The major hazards relate primarily to the inadvertent release of hydrofluoric acid or carbon tetrachloride, rather than to the manufactured fluorocarbons.

More than 6000 workers in the United States today are employed directly or indirectly in the production of fluorocarbons (118). Occupational hazards result from the properties that fluorocarbons can be toxic, are heavier than air, and decompose at high temperatures.

6.1 Inhalation Hazards

The recommended TLV for five particular fluorocarbons (FC 11, FC 12, FC 22, FC 113, and FC 114) is 1000 ppm (51). This level of occupational standard may be applicable to other fluorocarbons, since there is no evidence that the others are more toxic.

Fluorocarbon vapors are four to five times heavier than air. Thus high concentrations tend to accumulate in low-lying areas, resulting in the hazard of inhalation of concentrated vapors, which may be fatal. Under certain conditions, fluorocarbon vapors may decompose on contact with flames or hot surfaces, creating the potential hazard of inhalation of toxic decomposition products (13).

Fluorocarbons are manufactured for use in charging refrigeration systems, degreasing parts, and packaging aerosol formulations. Worker exposure occurs in manufacturing plants and where fluorocarbons are used in products. Worker exposure levels (TWA) tend to be less than the current permissible levels for various jobs in the plant, although seasonal and operational changes are possible (118). There are several sources of exposure to fluorocarbons in manufacturing. Exposure in chemical plant operations and production is generally low, but highly variable, and may be high in areas without adequate ventilation. Cylinder packers and shippers have occasional high exposure. Exposure during tank farm operations, and tank and drum filling, may exceed the TLV. Tank truck and tank car fillers have potentially high exposure, which may be intermittent with the occurrence of accidents. Maintenance operators, laboratory analysts, and supervisory personnel have low exposure. The highest exposure in the plant occurs with venting of gases from returnable cylinders. The formation of high temperature thermal decomposition products may occur during storage.

Administrative controls may be applied to limit occupational exposure to fluorocarbons during manufacture, packaging, and use. Enclosure of process materials and isolation of reaction vessels and proper design and operation of filling heads for packaging and shipping are two such measures. Inhalation of fluorocarbon vapors should be avoided. Forced air ventilation at the level of vapor concentration together with the use of individual breathing devices with independent air supply will minimize the risk of inhalation. Lifelines should be worn when entering tanks or other confined spaces. Filling areas should be monitored to ensure that the ambient concentration of fluorocarbons does not exceed 1000 ppm (0.1 percent by volume). If inhalation occurs, epinephrine or other sympathomimetic amines and adrenergic activators should not be administered since they will further sensitize the heart to the development of arrhythmias. The proposed system should be tested and appropriate safety precaution taken during storage. The appearance of toxic decomposition products serves as warning of the occurrence of thermal decomposition and detection of a sharp acrid odor warns of the presence of these products. Halide lamps or electronic leak detectors may also be used. Adequate ventilation also avoids the problem of toxic decomposition products.

6.2 Dermal Contact Hazards

The degreasing effect of fluorocarbons may cause dermatologic problems, although there is negligible dermal absorption. Some fluorocarbon liquids remove natural oils from the dermis, causing irritation and development of dry, sensitive skin. These lower-boiling liquids may also be splashed onto the skin or into the eyes, causing freezing, temporary irritation, or serious damage. The freezing effect is produced upon evaporation of the fluorocarbons from the skin surface, and is a manifestation of cooling by evaporation, whereby molecules with high kinetic energy escape from the system. Thus heat energy is taken with the escaping molecules, leaving behind the low kinetic energy (low temperature) molecules. Since the temperature of a fluid is based on the concentration and average kinetic energy of its molecules, a rapid cooling effect is thus produced. Frostbite may be a complication of this freezing effect. If frostbite occurs, the exposed area should be soaked in lukewarm water within 20 to 30 min after exposure. Ice cold or hot water should not be used; body temperature is ideal. Soaking should be eliminated if treatment begins more than 30 min after exposure. A light coating of bland ointment, such as petroleum, should be applied together with a light bandage. If the frostbitten area is large or severely affected, administration of an anticoagulant or vasodilator drug may be necessary in order to avoid the onset of gangrene. The eyes may also be damaged by freezing. If such contact occurs, they should be flushed with water for several minutes. Neoprene gloves, protective clothing, and eye protection minimize the risk of topical contact. The degreasing effect on the skin can be treated with lanolin ointment.

7 OCCUPATIONAL EXPOSURE TO MANUFACTURED FLUOROCARBONS

Occupational exposure to manufactured fluorocarbons occurs in the manufacture, use, servicing, and disposal of refrigeration units, food processing, solvent applications, plastic foam blowing, and fire extinguishing. The only documented reports of occupational poisoning are from exposure to fluorocarbon refrigerants (Section 2.2). Exposure to fluorocarbons by health personnel is discussed separately (Section 8).

7.1 Refrigeration

Several fluorocarbons once used as propellants for nonmedical aerosols now find their major application as refrigerants. The widely used refrigerants include the nonhydrogenated fluorocarbons (FC 11, FC 12, FC 113, FC 114, FC 115; Sections 2 and 3), and the hydrogenated fluorocarbons (FC 22, FC 31, FC 142b, FC 152a; Section 4). Mechanical vapor compression systems use fluorocarbons for refrigeration and air conditioning, and account for the vast majority of refrigeration capability in the United States. The occupational hazard is almost entirely in the exposure to fluorocarbons during manufacturing of refrigeration equipment, while product safety hazard is negligible since fluorocarbon is held in sealed units. Flurocarbons are used as refrigerants in home appliances, mobile air conditioning units, retail food refrigeration systems, and centrifugal and reciprocating chillers.

7.1.1   Home Appliances

Fluorocarbon refrigerant is used in the manufacture of home appliances, such as refrigerators and freezers, some of which have automobile and commercial applications as well. Fluorocarbon 12 is the only refrigerant used in home appliances (114). Manufacturing involves intermediate fluorocarbon exposure in leak testing of components, during systems charging, and in reworking of defective systems. Exposure is minimal during normal product use. Servicing involves the highest occupational exposure. There is also some exposure risk in the disposal of home appliances.

7.1.2   Mobile Air Conditioning

Mobile air conditioning units for vehicles are available as original equipment or as add-on units. Most of the occupational exposure during manufacturing and installation occurs with leak testing of the products. Exposure may also occur with bulk storage, system leaks, in-plant repair purging, and holding charge purging of fluorocarbon refrigerants.

Some leakage occurs during the regular use of air conditioners, which leads to service and recharging by the owner or a qualified mechanic. The mechanic may employ routine venting and replacement, or "topping-off" of the intact system. Venting to enter the system occurs with replacement of the compressor or compressor seal, replacement of the receiver–drier bottle, and changing of the hoses, condenser, or evaporator. Leak testing and purging of the system is required for this work.

The greatest occupational exposure by volume use of refrigerants is in servicing (not including recharging), initial charging, and manufacturing and installation. Exposure during recharging is negligible. The largest volumes of work exposure occur in independent repair shops, service stations, automobile dealership, and fleet shops (121). Some exposure occurs with release of refrigerants in automobile accidents.

7.1.3   Retail Food Refrigeration

Retail food refrigeration systems are used in supermarkets, superettes, convenience, bantam, and drive-in stores, "mom-and-pop" stores, and specialty stores such as fish markets and butcher shops. Manufacturing and installation results in limited exposure to refrigerants. Leakage during use and servicing of these systems is the greatest source of occupational exposure to fluorocarbon refrigerants (122). Vibrations, poor service techniques, and abuse from loading, unloading, and cleaning of cases are the causes of this exposure. Exposure potential from leakage is greatest in the smallest stores owing to the equipment type, and limited and enclosed spaces.

7.1.4 Centrifugal and Reciprocating Chillers

Centrifugal and reciprocating chillers use a secondary refrigerant to perform the air conditioning function, such as cooled water or air. These fluids are, in turn, cooled at a central location using fluorocarbon refrigerants. As elsewhere, there is exposure during manufacturing, shipping and installation, and leakage of these units. Service-related exposure to fluorocarbons is the greatest occupational hazard (123).

7.2 Food Processing

The major use of fluorocarbon for food processing is in quick-freezing applications and production of aerosol food products, such as whipped cream (also mixed with nitrous oxide).

7.2.1 Liquid Food-Freezing Applications

Fluorocarbons are used in direct contact freezing of raw and cooked fruits and vegetables, seafood, meats, and specialty items. The quick frozen technique is preferable for such foods as berries, cob corn, raw shrimp, and clams. The food is graded, washed, blanched, and placed on a belt for conveyance through the freezing apparatus. Close operator attention is required during these steps.

Loss of fluorocarbons to evaporation occurs from each end and from the top of the apparatus by movement of the conveyor belt and by "surges" of fluorocarbons when the condensing capability of the cooling coils is exceeded. There is also adherence of fluorocarbons to the product, with evaporation outside the apparatus (analogous to solvent "drag-out"). There is also significant fluorocarbon exposure during shutdown of the apparatus for cleaning or during production runs (124).

7.2.2 Aerosol Food Products

Fluorocarbons are used in food aerosols ranging from whipped topping to dry vermouth. Whips or foams include canapes, blue cheese and cheese spreads, cake frostings, honey, mustard, maple syrup, peanut butter and jelly combinations, and whipped margarine or butter with or without garlic. Sprays include various flavorings, spices, oil and vinegar salad sprays, and popcorn spray. Liquid applications are made for flavor concentrates, vegetable oils, and special flavor mixes. Pastes and syrup applications include cake decorations, cheese spreads, and sugar candies. There are also sprays for application to crockery pots and pans to prevent sticking of cooked foods. These aerosol food products enjoy widespread use in commercial and home food preparation. It should be noted that when the aerosol is ingested, the amount of fluorocarbon absorbed does not pose a health hazard.

7.3 Solvent Applications

The use of fluorocarbons as solvents is more hazardous to human health than is their use as refrigerants, because solvents are generally employed in open containers. Fluorocarbons find wide application as cleaning and drying solvents for defluxing and electronics cleaning, degreasing, displacement drying, and in the dry-cleaning industry and miscellaneous specialy applications. As nonpolar solvents, fluorocarbons will clean metal, glass, plastic, and electroplated surfaces of manufactured goods. The occupational safety and product safety of commerical solvents are of equal concern owing to the volatility of fluorocarbons.

Fluorocarbons are important to the electrical and electronic industry for defluxing of printed circuit boards, and cleaning of electronic parts, electric motors, and delicate scientific and electronic equipment and instruments. Degreasing applications include cold cleaning, vapor degreasing, and flushing, used alone or together with ultrasonic equipment. More than 2000 industrial plants in the United States use fluorocarbons for vapor degreasing and/or cold cleaning (125). The solvent may leave cleaning tanks by evaporation at room temperature, adherence to cleaned parts when removed, and accidental loss with removal of contaminants introduced during cleaning. It has been estimated that the amounts of solvent lost to the work environment during the cleaning applications are as follows:

Cold cleaning of metal	5%
Vapor degreasing	5–10%
Large cold cleaning operations	15%
Cold cleaning of printed circuit boards	35–40%

Displacement drying is also a source of fluorocarbon exposure where a fluorocarbon drying fluid is used for removal of water from metal, glass, plastic, and plated materials in order to leave a residue-free surface. Fluorocarbons are also used in machine cutting fluids where precise tolerances are required, as in aircraft construction.

The dry-cleaning industry usually makes use of petroleum distillates for washer/extractor transfer type and coin-operated units. Fluorocarbons are used in dry-to-dry type "Valclene" units in a commercialized dry cleaning process compatible for use with plastic buttons, trims, furs, and leather. In specialty cleaning, it is used concurrently with, or has replaced, the Stoddard solvent and/or perchloroethylene. It should be noted that the most widely used fluorocarbon as solvent is nonhydrogenated, and may be banned owing to depletion of the ozone layer (FC 113, Section 3.2).

7.4 Foam Blowing Applications

Nearly 10% of the use of fluorocarbons in the United States is in the fabrication of urethane and nonurethane closed-cell plastic foams, and flexible urethane foams. Suppliers of basic raw materials and foam systems, fabricators and suppliers of flexible foams, and producers and fabricators of rigid foams make use of fluorocarbons. The plastic foams are manufacturd for application in thermal insulation and packaging. Rigid urethane foams require gas for foaming, provided by carbon dioxide released during polymerization, or from volatile fluorocarbon liquids, especially FC 11. Volatilization of the blowing agent is caused by the heat of reaction between the isocyanate and hydroxyl components. A substantial portion of fluorocarbons is trapped within the closed plastic cells and release is nil. In the production of flexible urethane foams, fluorocarbon augments blowing from carbon dioxide released by the reaction of free isocyanate groups with water. Virtually all of these fluorocarbons are released soon after manufacturing (126–128).

Raw materials for flexible and rigid urethane foams include isocyanates, polysols, blowing agents, catalysts, and additives. These raw materials are used in liquid urethane systems for the formation of bun and slab stock, which in turn is used in furniture and bedding, apparel, carpeting, packaging, and transportation. Rigid foams are manufactured for thermal insulation in building construction, refrigerators, freezers, transportation carriers (trucks, trailers, and freezer cars), and industrial insulation of tanks and pipelines. The insulating properties of fluorocarbons are based on their low thermal conductivity. The low density rigid foams are 97% fluorocarbon vapors trapped within polymer cells. The thermal conductivity of the vapors in these cells determines the overall insulation value of the foams. Other applications are in packaging and luggage, boats and flotations, and molded structural parts to replace wood in furniture, television sets, and interior and industrial fixtures. The fabrication of soft, flexible urethane foam depends on fluorocarbons. These super-soft foams are used in textile lamination and self-bonded rug backings. Polyethylenes are blown into foamed sheet and film from crystal. Polystyrene, olefin, and miscellaneous foams have minor applications from bead and board stock. Fluorocarbons are also used with polyurethanes, polystyrenes, and polyolefins. The fabrication of vinyl urea, phenolic, and epoxy polymers does not involve fluorocarbon use.

Fluorocarbon emissions occur after vapors are trapped and retained in rigid urethane and epoxy foam. Fluorocarbons diffuse rapidly from flexible foams, and slowly from nonpolar polymers such as polystyrene and polyethylene. In addition, foam production processes entail varying loss of blowing agen ᵗuring fabrication. The equipment used, as whether foam is poured or frothed, determine loss during manufacture, when most occupational exposure occurs. Exposures due to subsequent losses and ultimate disposal are variable. Urethane foam plants must work with relatively toxic isocyanates (low maximum TLV at ceiling of 0.02 ppm). Therefore, all plants are well-ventilated and equipment hooded to protect workers. Monitored fluorocarbon levels are well below allowable industrial exposures (TLV = 1000 ppm).

7.5 Fire Extinguishing Applications

Although most fluorocarbons are non-flammable, they have not been used to extinguish fires because of the availability of bromine-containing fluorocarbons, specifically bromotrifluoromethane and bromochlorodifluoromethane (Halon). The bromine-containing fluorocarbons operate by chemical interruption of the combustion chain, and are used in total flooding systems for computer rooms and telephone facilities, as well as aircraft and portable fire extinguishers, including use on the Air Force P-13 rapid intervention crash trucks. Halon emerges from the fire extinguisher nozzle as a mixture of 85 percent liquid and 15 percent vapor and is discharged over long distances. It completely vaporizes upon contact with fire. There is some exposure in charging and refilling, testing, and leakage of the fire extinguishing systems. There can be accidental in-service discharges. During use in fires, thermal decomposition products present toxicity hazards (129). Exposure to bromofluorocarbons in fire extinguishing equipment has led to parasthesia, tinnitus, anxiety reactions, electroencephalographic changes, slurred speech, and decreased performance on psychological tests.

8 OCCUPATIONAL HAZARDS IN THE HEALTH PROFESSIONS

Fluorocarbons find wide application in the health profession. Medical applications for fluorocarbons are found in hospitals, clinics, and physicians' offices, and with a wide range of reusable medical and surgical devices which can not withstand steam sterilization. They are employed in sterilization procedures throughout the hospital, and for tissue preparation and histotechnology in the clinical laboratory. Although some fluorocarbons have general anesthetic properties, none is in current use because of the availability of fluorine-containing anesthetics such as the following:

> Halothane (2-bromo-2-chloro-1,1,1-trifluoroethane)
> Methoxyflurane (2,2-dichloro-1,1-difluoroetheyl methyl ether)
> Halopropane (1,1,2,2-tetrafluoro-3-bromopropane)
> Enflurane (2-chloro-1,2-trifluoroethyl difluoromethyl ether)
> Isoflurane (1-chloro-2,2,2-trifluoroethyl difluoromethyl ether)

The use of fluorine-containing anesthetics involves significant exposure in the handling and processing of waste anesthetic gases and vapors, as well as in their initial use. Guidelines have been established for regulating exposure to these anesthetic wastes. The NIOSH criteria document on waste anesthetic gases and vapors, released in March 1977, has insisted upon a level of 5 ppm as the occupational standard for exposure to anesthetic vapors including halothane (130). Some scientists are raising the question that 5 ppm should logically be applied to fluorocarbons in general, since some are mutagens (Sections 2.7 and 4.1).

Epidemiologic studies in female hospital personnel and nurse anesthetists have shown associations between exposure to anesthetic vapors and the occurence of cancers, spontaneous abortions, and congenital anomalies (130). It is possible that the 5 ppm standard will be extended to lowering the limits for exposure to all fluorocarbons from the current standards of 1000 ppm.

8.1 Waste Anesthetic Vapors

Chloroform is a prototype for an anesthetic gas that is now obsolete and has been shown to be carcinogenic together with trichloroethylene and isoflurane (131). The unsaturated ethylenes (trichloroethylene, vinyl chloride, perchloroethylene) and ethers (isoflurane, dichloromethyl ether, chloromethyl methyl ether) have also been used in these applications. Vinyl chloride causes a rare hemangiosarcoma of the liver. Halothane behaves similarly to FC 11 and FC 12, and is currently the most widely used inhalational anesthetic agent. Halothane may manifest significant dose-related human toxicity, in addition to idiosyncratic hepatotoxicity which may be produced in the liver. Hospital workers, students, and volunteers, regardless of status, may be uniformly exposed to inhalation anesthetic agents that escape into locations associated with the administration of, or recovery from, anesthetic agents. The gaseous or volatile liquid agents are released into work areas such as operating, recovery and labor/delivery rooms, and other job-related areas. The number of hospital personnel exposed to anesthetic vapors and gases as estimated from the memberships of professional organizations are as follows (130):

Anesthesiologists	13,700
Nurse-anesthetists	17,546
Operating room nurses	21,600
Operating room technicians	12,000
Dentists and assistants	100,000
Veterinarians and employees	50,000

More than 50,000 hospital personnel are exposed every day. Surgeons are somewhat less exposed, since they usually do not operate on a daily basis. Furthermore, 20 million patients in 25,000 hospital operating rooms are exposed every year. There are an additional 4.5 million exposed in dental offices per year. Sources of exposure in the operating room include intentional outflow or discharge from the breathing circuit in administration of anesthetic gases. The main factors influencing the amount of exposure are flow rates, concentrations, the nature of the breathing circuit used, whether there is access to atmospheric pressure or rebreathing, and whether or not carbon dioxide is absorbed. Other sources are from leaks and careless work practices. Poorly fitting valves, connections, tubing, face masks, and endotracheal tube cuffs are causes of leaks. The use of mechanical ventilators and waste anesthetic gas scavenging systems also influences the amount of exposure. Without gas scavenging, the factors influencing anesthetic gas concentrations are the type and concentration of gas used, breathing system, method of administration (face mask vs. endotracheal tube), room air movement, and operating room configuration.

The types of breathing circuits without carbon dioxide absorption are (*a*) open drop; (*b*) insufflation (with the greatest risk of exposure); (*c*) Mapleson-type semiclosed Magill, and T-tube or Ayre's T-piece (for pediatric use); and (*d*) nonreturn (non-rebreathing). The circuits with carbon dioxide reabsorption are the to-and-fro system, and the closed and partial rebreathing (semiclosed), circle systems.

Analytic methods have been used for the detection of anesthetic gases to determine their concentrations in the ambient atmosphere and anesthetist breathing zones of operating and recovery rooms. These methods include manometric, combustion, gas chromatography flame ionization detection, quadrupole mass spectrometer, and infrared spectrometer techniques. The peak ambient concentrations of halothane detected are 30 ppm, with an average of 10 ppm, in a rebreathing system. The average concentration may rise to 85 ppm with a non-rebreathing system. In one operating room where breathing systems were compared, the non-rebreathing system produced 15 ppm concentrations, whereas the rebreathing system produced 10 ppm, and the semiclosed circle system had 5 ppm. Biotransformation of inhalation anesthetics may also play a role in toxicity.

The non-rebreathing technique with face mask produced a halothane level of 28 ppm at 25 cm from the discharge point, which is representative of the exposure of the anesthetist. The concentration at the discharge point was 290 ppm, and an ambient concentration of 10 ppm was constant throughout the operating room. The concentration of methoxyflurane in the operating room are 2 to 10 ppm in the breathing zone of the anesthetist, and 1 to 2 ppm in that of the surgeon. The concentration of halothane from expired air in the recovery room is 3 ppm.

8.2 Hospital Sterilization Techniques

Fluorocarbons are employed as diluents for ethylene oxide in hospital and industrial gas sterilization applications (132). The health industry makes extensive use of fluorocarbons in the preparation of prepackaged disposable medical and surgical supplies. Ethylene oxide is highly toxic, flammable, and explosive, and the fluorocarbon is added as a safety feature. It is also possible to use propylene oxides, which have a higher TLV of 1000 ppm. Sterilization media include radiation, direct heat, steam, chemical baths, and various pure or diluted gases, but gas sterilization has had increasing use in the hospital for equipment that cannot stand exposure to steam. Gas sterilization units employing fluorocarbons range from small desk-top units to 40-ft^3 autoclaves. Disposal of the gas mixture after sterilization can be accomplished by direct venting to the outside air or by adding water and flushing down the drain. The relative use of methods of disposal are 59 percent venting, 27 percent water flushing, 2 percent wet sponge, 2 percent vacuum, 17 percent landfill, and 17 percent unknown (132). All these systems are designed for the elimination of ethylene oxide and result in almost immediate release of fluorocarbons into the ambient atmosphere.

8.3   Clinical Laboratory: Histotechnology and Blood Substitutes

Clinical pathologists exposed to fluorocarbons in the preparation of frozen tissue sections have been seen to develop coronary heart disease (Section 4.1). Synthesis and biologic screening of hybrid fluorocarbons have been carried out for application as artificial synthetic blood substitutes (133). Perfluorochemical emulsions are being put into use in Japan, but American investigators are searching for more appropriate chemicals.

9   CONCLUSIONS

We began this chapter with the statement that its preparation is inappropriately timed owing to uncertainty in federal regulations on the manufacture of fluorocarbons. The primary concern continues to be the depletion of the ozone layer by fluorocarbons released to the atmosphere, leading to an increase in the incidence of malignant melanoma and other conditions in the population. This eventuality is impossible to prove or disprove at present, and the debate between regulators and manufacturers will undoubtedly continue during the 1980s.

Preoccupation with the future status of fluorocarbons as based on projected events in the stratosphere has led to several unexpected developments. The emphasis in new substitute fluorocarbons is on their environmental fate, favoring the *hydrogenated* over the *nonhydrogenated* fluorocarbons, owing to significant differences in their atmospheric lifetimes. Attention to human toxicity and animal testing has been reduced to the point where there has been no significant publication on the health effects of fluorocarbons during the late 1970s. Continued interest in the stratospheric effects of fluorocarbons has delayed the publication of animal toxicity studies conducted by manufacturers that are competing for new and environmentally acceptable fluorocarbons. There have been brief statements in nonscientific publications on certain fluorocarbons that may cause leukemia, cancer, sterility, teratogenicity, and mutagenicity. The basis for such statements cannot be verified in the open literature, so that there is an incomplete discussion of these topics in this chapter.

It is certain that most fluorocarbons are potentially toxic to the cardiovascular and bronchopulmonary systems. Deaths from the use, misuse, and abuse of aerosol products have been extensively documented. There are isolated reports of poisoning from exposure to fluorocarbon propellants and some studies showing a higher incidence of coronary heart disease among hospital personnel and refrigerant mechanics exposed to fluorocarbons. Additional investigation is required to establish causal relationship between fluorocarbons and cardiovascular and bronchopulmonary diseases among exposed workers. The high incidence of cancer among hospital personnel repeatedly exposed to fluorine-containing general anesthetics raises a fundamental need to examine other fluorocarbon-exposed workers for similar effects.

Finally, it should be recognized that there are alternatives to the fluorocarbons, in the event that they are completely banned owing to environmental and long-term health effects. The alternatives are not entirely fluorine-containing compounds but also include nonfluorocarbons, such as hydrocarbons, chlorine-containing solvents, and chemical generation of gases such as carbon dioxide. In 1976, the EPA prepared a list of technical alternatives to fluorocarbon uses (134) which undoubtedly will be revised and expanded in the 1980s. Whether nonfluorocarbons or alternative fluorocarbons will be accepted in the future, the general principles of toxicological testing, industrial hygiene, and occupational health discussed in this chapter will be applicable to either group of chemicals.

REFERENCES

1. D. M. Aviado, *Prog. Drug Res.*, **18**, 365 (1974).

2. W. N. Sullivan, *Mil. Med.*, **136**, 157 (1971).

3. E. Zuskin and A. Bouhuys, *N. Engl. J. Med.*, **290**, 660 (1974).

4. F. Valic, E. Zuskin, Z. Skuric, and M. Denich, *Acta Med. Iugosl.*, **28**, 231 (1974).

5. R. P. Giovacchini, G. H. Becker, M. J. Brunner, and F. E. Dunlap, *J. Am. Med. Assoc.*, **193**, 298 (1965).

6. Anonymous, *Food Cosmet. Toxicol.*, **4**, 73 (1966).

7. A. H. Kegel, W. D. McNally, and A. S. Pope, *J. Am. Med. Assoc.*, **93**, 353 (1929).

8. H. M. Baker, *Am. J. Public Health*, **20**, 291 (1930).

9. W. D. McNally, *J. Ind. Hyg. Toxicol.*, **28**, 94 (1946).

10. J. Cheymol, *Presse Med.*, **44,** 1123 (1936).

11. E. H. Schiotz and O. Arbeidsnemnd, *Nord. Hyg. Tidskr.*, **27,** 230 (1946).

12. T. Marti, *Annu. Med. Legal Criminol.*, **28,** 147 (1948).

13. T. Dalhamn, *Nord. Hyg. Tidskr.*, **39,** 165 (1958).

14. J. Mendeloff, *Arch. Ind. Hyg. Occup. Med.*, **6,** 518 (1952).

15. H. D. Leinoff, *Am. Heart J.*, **24,** 187 (1942).

16. R. C. Baselt and R. H. Cravey, *J. Forensic Sci.*, **13,** 407 (1968).

17. A. Poklis, *Forensic Sci.*, **5,** 53 (1975).

18. M. Bass, *J. Am. Med. Assoc.*, **212,** 2075 (1970).

19. J. L. Chapel and G. Thomas, *Mo. Med.*, **67,** 378 (1970).

20. R. A. Kramer and P. Pierpaoli, *Pediatrics*, **48,** 322 (1971).

21. S. T. Crooke, *Tex. Med.*, **68,** 67 (1972).

22. R. C. Kamm, *Forensic Sci.*, **5,** 91 (1975).

23. D. G. Fagan and J. B. Forrest, *Lancet*, (Aug. 13, 1977).

24. H. C. Harris, *Postgrad. Med.*, **23,** 170 (1958).

25. I. Pavlik, *Cas. Lek. Cesk.*, **100,** 275 (1961).

26. K. B. Saunders, *Br. Med. J.*, **1,** 1037 (1965).

27. J. M. Smith, *Lancet*, **1,** 1042 (1966).

28. W. Pickvance, *Br. Med. J.*, **1,** 756 (1967).

29. F. E. Speizer, R. Doll, and P. Heaf, *Br. Med. J.*, **1,** 335 (1968).

30. F. E. Speizer, R. Doll, P. Heaf, and L. B. Strang, *Br. Med. J.*, **2,** 339 (1968).

31. P. M. Fraser, F. E. Speizer, S. D. M. Waters, R. Doll, and N. M. Mann, *Br. J. Dis. Chest*, **65,** 71 (1971).

32. W. D. Linehan, *Br. Med. J.*, **4,** 172 (1969).

33. W. H. W. Inman and A. M. Adelstein, *Lancet*, **2,** 279 (1969).

34. M. J. Greenberg and A. Pines, *Br. Med. J.*, **1,** 563 (1967).

35. W. Pickvance, *Br. Med. J.*, **1,** 756 (1967).

36. E. M. Douglas, T. Hillier, and I. C. Johnson, *Br. Med. J.*, **2,** 53 (1967).

37. P. D. Exon, *Br. Med. J.*, **2,** 178 (1967).

38. G. S. Graham, *Scott Med. J.*, **13,** 282 (1968).

39. A. P. Norman and S. Sanders, *Practitioner*, **201,** 909 (1968).

40. J. B. Shapiro and C. F. Tate, *Dis. Chest*, **48,** 484 (1965).

41. T. E. Van Metre, Jr., *Trans Am. Clin. Climatol. Assoc.*, **78,** 58 (1966).

42. H. Herxheimer, *Lancet*, **2,** 642 (1969).

43. R. D. Ghannam, L. Schreirer, and N. A. Vanselow, *Ann. Allergy*, **26,** 194 (1968).

44. C. R. Palm, M. A. Murcek, T. R. Roberts, H. C. Mansmann, Jr., and P. Fireman, *J. Allergy*, **46,** 257 (1970).

45. W. C. Fabb and J. L. Guerrant, *J. Allergy*, **42,** 249 (1968).

46. H. Herxheimer, *Br. Med. J.*, **4,** 795 (1972).

47. P. D. Stolley, *Am. Rev. Respir. Dis.*, **105,** 883 (1972).

48. M. A. Phillips, *Lancet*, **2,** 677 (1967).

49. G. J. Taylor, IV, and W. S. Harris, *J. Am. Med. Assoc.*, **214,** 81 (1970).

50. G. J. Taylor, IV, W. S. Harris, and M. D. Bodgonoff, *J. Clin. Invest.*, **50,** 1546 (1971).

51. American Conference of Governmental Industrial Hygienists, *Documentation of the Threshold Limit Values for Substances in Workroom Air (with supplements for those substances added or changed since 1971)*, 3rd ed., 1971, Cincinnati, Ohio, approx. 458 pp.

52. S. Plaiut, *Lancet*, **2,** 721 —1967).

53. H. Fabel, R. Wettengel, and W. Hartmann, *Dtsch. Med. Wochenschr.*, **97,** 428 (1972).

54. S. M. Brooks, S. Mintz, and E. Weiss, *Am. Rev. Respir. Dis.*, **105,** 640 (1973).

55. F. Valic, Z. Skuric, Z. Bantic, M. Rudar, and M. Hecej, *Br. J. Ind. Med.*, **34,** 130 (1977).

56. R. D. Stewart et al., The Medical College of Wisconsin, Department of Environmental Medicine, *Acute and Repetitive Human Exposure to Fluorotrichloromethane*, approx. 100 pp., prepared for Cosmetic, Toiletry and Fragrance Association, Inc., Washington, D.C., U.S. Department of Commerce, National Technical Information Service, PB-279-209, Springfield, Va., (December 1, 1975).

57. G. Paulet and R. Chevrier, *Arch. Mal. Prof. Med. Trav. Secur.*, **30,** 251 (1969).

58. G. Paulet, R. Chevrier, J. Paulet, M. Duchene, and J. Chappet, *Arch. Mal. Prof. Med. Trav. Secur.*, **30,** 101 (1969).

59. A. Morgan, A. Black, M. Walsh, and D. R. Belcher, *Int. J. Appl. Radiat. Isot.*, **23,** 285 (1972).

60. A. Morgan, A. Black, and D. R. Belcher, *Ann. Occup. Hyg.*, **15,** 273 (1972).

61. F. M. Williams, G. H. Draffan, C. T. Dollery, J. C. Clark, and A. J. Palmer, *Thorax*, **29,** 99 (1974).

62. G. Marier, H. MacFarland, G. S. Wiberg, H. Buchwald, and P. Dassault, *Can. Med. Assoc. J.*, **111,** 39 (1974).

63. D. B. Rix and T. King, *Can. Med. Assoc. J.*, **111,** 645 (1974).

64. J. W. Paterson, M. F. Sudlow, and S. R. Walker, *Lancet*, **2,** 565 (1971).

65. C. T. Dollery, G. H. Draffan, D. S. Davies, F. M. Williams, and M. E. Conolly, *Lancet*, **2,** 1164 (1970).

66. C. T. Dollery, F. M. Williams, G. H. Draffan, G. Wise, H. Sahyoun, J. W. Paterson, and S. R. Walker, *Clin. Pharmacol. Ther.*, **15,** 59 (1974).

67. G. H. Draffan, C. T. Dollery, Faith M. Williams, and R. A. Clare, *Thorax*, **29,** 95 (1974).

68. D. M. Aviado, *J. Clin. Pharmacol.*, **15,** 86 (1975).

69. D. M. Aviado, *Toxicology*, **3,** 311 (1975).

70. D. M. Aviado, *Toxicology*, **3,** 321 (1975).

71. D. M. Aviado, *Environ. Health Persp.*, **26,** 207 (1978).

72. A. H. Nuckolls, *The Comparative Life, Fire and Explosion Hazards of Refrigerants*, Underwriters Lab., Chicago, 1959, approx. 113 pp.

73. J. Scholz, *Fortschr. Biol. Aerosol-Forsch.*, **4,** 420 (1962).

74. F. Caujolle, *Bull. Inst. Int. Froid*, **1,** 21 (1964).

75. D. Lester and L. A. Greenberg, *Arch. Ind. Hyg.*, **2,** 335 (1950).

76. S. A. Friedman, M. Cammarato, and D. M. Aviado, *Toxicology*, **1,** 345 (1973).

77. D. M. Aviado and M. A. Belej, *Toxicology*, **2,** 31 (1974).

78. C. F. Reinhardt, A. Azar, M. E. Maxfield, P. E. Smith, Jr., and L. S. Mullin, *Arch. Environ. Health*, **22,** 265 (1971).

79. L. S. Mullin, A. Azar, C. F. Reinhardt, P. E. Smith, Jr., and E. F. Fabryka, *Am. Ind. Hyg. Assoc. J.*, **33,** 389 (1972).

80. D. G. Clark and D. J. Tinston, *Ann. Allergy*, **30,** 536 (1972).

81. M. A. Belej, D. G. Smith, and D. M. Aviado, *Toxicology*, **2,** 381 (1974).

82. T. Watanabe and D. M. Aviado, *Toxicology*, **3,** 225 (1975).

83. R. E. Doherty and D. M. Aviado, *Toxicology*, **3,** 213 (1975).

84. M. A. Belej and D. M. Aviado, *J. Clin. Pharmacol.*, **15,** 102 (1975).

85. D. M. Aviado and D. G. Smith, *Toxicology*, **3,** 241 (1975).

86. R. S. Brody, T. Watanabe, and D. M. Aviado, *Toxicology*, **2,** 173 (1974).

87. D. M. Aviado and J. Drimal, *J. Clin. Pharmacol.*, **15,** 116 (1975).

88. D. M. Aviado and M. A. Belej, *Toxicology*, **3,** 79 (1975).

89. G. J. Taylor and R. T. Drew, *Toxicol. Appl. Pharmacol.*, **32,** 177 (1975).

90. T. Balazs, F. L. Earl, G. W. Bierbower, and M. A. Weinberger, *Toxicol. Appl. Pharmacol.* **26,** 407 (1973).

91. J. A. Simaan and D. M. Aviado, *Toxicology*, **5,** 139 (1975).

92. D. E. Bohning, R. E. Albert, M. Lippmann, and V. R. Cohen, *Am. Ind. Hyg. Assoc. J.*, **36,** 902, (1975).

93. J. W. Clayton, Jr., *Handb. Exper. Pharmacol.*, **20,** 459 (1966).

94. J. W. Clayton, Jr., *Fluorine Chem. Rev.*, **1,** 197 (1967).

95. J. W. Clayton, Jr., *J. Soc. Cosmet. Chem.*, **18,** 333 (1967).

96. L. J. Jenkins, Jr., R. A. Jones, R. A. Coon, and J. Siegel, *Toxicol. Appl. Pharmacol.*, **16,** 133 (1970).

97. T. S. Slater, *Biochem. Pharmacol.*, **14,** 178 (1965).

98. T. S. Slater and B. C. Sawyer, *Biochem. J.*, **123,** 805 (1971).

99. P. J. Cox, L. J. King, and D. V. Parke, *Xenobiotica*, **6,** 363 (1976).

100. D. A. Blake and G. W. Mergner, *Toxicol. Appl. Pharmacol.*, **30,** 396 (1974).

101. G. Paulet, S. Desbrousses, and E. Vidal, *Arch. Mal. Prof. Med. Trav. Secur. Soc.*, **35,** 658 (1974).

102. R. R. Montgomery and C. F. Reinhardt, in *Handbook of Aerosol Technology*, P. A. Saunders, Ed., 2nd ed., Van Nostrand Reinhold, New York, 1979, p. 466.

103. National Cancer Institute Carcinogenesis Technical Report Series No. 106, *Bioassay of Trichlorofluoromethane for Possible Carcinogenicity*, 1978, 46 pp.

104. S. S. Epstein, S. Joshi, J. Andrea, P. Clapp, H. Falk, and N. Mantel, *Nature*, **214,** 526 (1967).

105. M. J. Molina and F. S. Rowland, *Nature*, **249,** 810 (1974).

106. L. Dotto and H. Schiff, *The Ozone War*, Doubleday, New York, 1978, 380 pp.

107. National Science Foundation, *The Possible Impact of Fluorocarbon and Halocarbons on Ozone*, 1975, 75 pp.

108. Council on Environmental Quality, *Fluorocarbons and the Environment*, 1975, 109 pp.

109. National Research Council, *Halocarbons, Environmental Effects of Chlorofluoromethane Release*, 1976, 125 pp.

110. M. F. Casling and C. R. Tribe, *J. Clin. Pathol.*, **22,** 244 (1969).

111. A. Azar, C. F. Reinhardt, M. E. Maxfield, P. E. Smith, Jr., and L. S. Mullin, *Am. Ind. Hyg. Assoc. J.*, **33,** 207 (1972).

112. R. D. Stewart, A. A. Herrmann, E. D. Baretta, H. V. Forster, J. H. Crespo, P. E. Newton, and R. J. Soto, *Acute and Repetitive Human Exposure to Difluorodichloromethane*, NTIS-PB-279-2041, 1976, 73pp.

113. G. J. Taylor, IV, and R. T. Drew, *J. Pharmacol. Exp. Ther.* **192,** 129 (1975).

114. H. R. Imbus and C. Adkin, *Arch. Environ. Health*, **24,** 257 (1972).

115. J. W. Clayton, Jr., M. A. Delaplane, and D. B. Hood, *Am. Ind. Hyg. Assoc. J.*, **21,** 382 (1960).

116. F. E. Speizer, D. H. Wegman, and A. Ramirez, *N. Engl. J. Med.*, **292,** 624 (1975).

117. V. C. Foltz and R. Fuerst, *Environ. Res.*, **7**, 275 (1974).

118. National Institute for Occupational Safety and Health, *Fluorocarbons—Worker Exposure in Four Facilities*, 1978, 49 pp.

119. Evironmental Protection Agency, *Domestic Use and Emissions of Chlorofluorocarbons in Home Appliances*, Rand Corporation Contract 68-01-3882, 1979, 38 pp.

120. Environmental Protection Agency, *Domestic Use and Emissions of Chlorofluorocarbons in Mobile Air Conditioning*, Rand Corporation Contract 68-01-3882, 1979, 91 pp.

121. Environmental Protection Agency, *Nonaerosol Chlorofluorocarbon Emissions, Identification of Regulatory Options for Detailed Analysis*. Rand Corporation Contract 68-01-3882, 1978, 112 pp.

122. Environmental Protection Agency, *Domestic Use and Emissions of Chlorofluorocarbons in Retail Food Store Refrigeration System*, Rand Corporation Contract 68-01-3882, 1979, 44 pp.

123. Environmental Protection Agency, *Domestic Use and Emissions of Chlorofluorocarbons in Centrifugal and Reciprocating Chillers*, 1979, 66 pp.

124. Environmental Protection Agency, *Interim Report: The Use and Emissions of Chlorofluorocarbons in Liquid Fast Freezing Applications*, Rand Corporation Contract WN-10277-EPA, 1978, 23 pp.

125. Environmental Protection Agency, *Interim Report: Projecting Atmospheric Emissions from Chlorofluorocarbons Used in Solvent Applications*, Rand Corporation Contract WN-10272-EPA, 1978, 123 pp.

126. Environmental Protection Agency, *Interim Report: Emission Projections for Flexible Urethan Foams*, Rand Corporation Contract WN-10274-EPA, 1978, 45pp.

127. Environmental Protection Agency, *Interim Report: The Use and Emissions of Chlorofluorocarbons in Closed-cell Plastic Foams*, Rand Corporation Contract 68-01-3882, 1978, 97 pp.

128. Environmental Protection Agency, *Interim Report: The Use and Emission of Chlorofluorocarbons in Nonurethane Closed-cell Foams*, Rand Corporation Contract 68 01 3882, 1978, 83 pp

129. Environmental Protection Agency, *Interim Report: The Use and Emissions of Chlorofluorocarbons in Fire Extinguishing Applications*, Rand Corporation Contract 68-01-3882, 1978, 41 pp.

130. National Institute for Occupational Safety and Health, *Criteria for a Recommended Standard— Occupational Exposure to Waste Anesthetic Gases and Vapors*, 1977, approx. 255 pp.

131. Food and Drug Administration, *Proposed Guidelines for Carcinogenesis and Bioassay of Inhalation Anesthetic Agents*, 1977, 47 pp.

132. Environmental Protection Agency, *The Use and Emissions of Chlorofluorocarbons in Sterilization Applications*, Rand Corporation Contract, 1978, 24 pp.

133. National Institutes of Health, *Synthesis and Biological Screening of Novel Hybrid Fluorocarbon Hydrocarbon Compounds for Use as Artificial Blood Substitutes*, Jet Propulsion Laboratory Publication 77-80, 1978, 280 pp.

134. Environmental Protection Agency, *Chemical Technology and Economics in Environmental Perspectives, Task-1 Technical Alternatives to Selected Chlorofluorocarbon Uses*, 1976, 218 pp.

CHAPTER FORTY-THREE

N-Nitrosamines

C. W. FRANK, Ph.D., and C. M. BERRY, Ph.D.

1 GENERAL CONSIDERATIONS

1.1 Nomenclature, Sources, and Occurrence

N-Nitroso compounds are characterized by a nitrosyl group (—N=O) bonded to a nitrogen atom. The class of compounds may be divided into two distinct types, N-nitrosamides and N-nitrosamines. The nitrosamides originate from the nitrosation alkyl and aryl amides such as ureas, biurets, urethanes, and carbamates. Nitrosamines, generally a more stable species, are produced from the nitrosation of dialkyl, diaryl, alkylaryl, and cyclic amines. This chapter is directed only toward N-nitrosamines; however, the general discussion can be applied to N-nitrosamides.

N-nitroso compounds are produced when an acidic nitrite solution (HNO_2) comes in contact with a secondary amine (1, 2). Reaction of oxides of nitrogen with secondary amines have been shown to result in the formation of N-nitroso derivatives (3, 4). Furthermore, reaction of the parent amine with nitrosyl chloride (NOCl) (5–7) and nitrosyl tetrafluoroborate (8, 9) produces the nitrosamine.

Two requirements exist for the formation of nitroso compounds, the presence of a nitrite and a secondary amine. Exposure of humans to nitrite can take numerous pathways. It is used as a food additive to meat products for the inhibition of *Clostridium botulinum* growth (10–12) and to enhance food flavor (13). Nitrite is an intermediate within the nitrogen cycle; thus the anion is ubiquitous within the environmental situation. In addition, nitrite is found at elevated levels in selected leafy vegetables and roots. Production of nitrite from nitrate can be accelerated by the presence of microorganisms in soil, intestinal bacteria, and oral microflora; the latter has been estimated to result in excess of 30% of the nitrite body burden.

In the industrial setting nitrite is found in those areas where nitrate is used. Also, nitrite can be formed at high levels from decomposition processes. Secondary amines can be considered to be ubiquitous both environmentally and industrially. In particular, as metabolic or bacterial degradation products, this class of compounds is produced at high levels. In addition, the use of certain drugs or pesticides results in elevated levels of secondary amine exposure. Industrially, secondary amines are used for many purposes including the vulcanization of rubber, soap manufacture, tanning, as flotation agents, in resins, dyes, pharmaceuticals, emulsifying agents, and in the production of lubricants in textiles. Chapter 44 of this book (Vol. 2, 2nd ed.) should be consulted for further information.

The lower molecular weight amines are used in highest abundance in industry. The N-nitroso derivatives of dimethyl- and triethanolamines are of prime importance to this discussion. It should be stated however, that numerous other compounds of this type may present a possible hazard to the human. Three major sources of exposure to nitroso compounds are known:

1.   Direct exposure to industrial N-nitrosamines.
2.   Direct in vitro formation of the product (N-nitrosamine through reaction of a secondary amine and nitrite).
3.   Indirect exposure through in vivo formation of the N-nitroso compound by absorption of a secondary amine and/or nitrite.

Few examples exist where an N-nitrosamine is used in large quantities in industrial processes. Dimethylnitrosamine is used, however, as an intermediate in the production of 1,1-dimethylhydrazine, a component in spacecraft fuel and rocket fuel mixtures. In addition, N-nitrosodiphenylamine and dinitroso-pentamethylenetetramine have been used in the rubber industry as a vulcanizing retarder and blowing agent, respectively. N-Nitrosomethylurea and N-methyl-N'-nitro-N-nitrosoquanidine, both N-nitrosamides, have been used on a laboratory scale for the production of diazomethane and as a genetic mutagenic agent, respectively. N-Nitrosodimethylamine was used as a solvent in the automotive industry but has been replaced.

Direct exposure to a nitrosamine produced as a by-product through reaction of the two essential components appears to have a larger industrial population impact. Industrial areas within which a secondary amine and nitrite (or the nitrogen oxides) are present can be presumed to have elevated levels of the associated N-nitrosamine. Industries such as automotive, dye, rubber, and leather have been observed to have elevated levels of certain nitrosamine derivatives. Most notable in this case is the presence of N-nitrosodiethanolamine in cutting fluids. On aging and use, diethanolamine levels increase, ultimately resulting in increased levels of *N*-nitrosodiethanolamine by reaction with nitrite. Although the effect of such exposure can only be presumed, the quest for substitutes for these compounds has been initiated.

Exposure to amines followed by in vivo formation of the respective *N*-nitroso derivatives has been shown to be a viable contribution to the human body burden of this class of compounds. Industries that produce or use amines and/or nitrites (nitrates) feasibly may have elevated *N*-nitrosamine exposure. Table 43.1 presents a partial list of media in which nitroso compounds have been detected at appreciable levels.

1.2 Toxicology of *N*-Nitrosamines

Exposure to *N*-nitroso compounds is known to produce both acute and chronic effects in animals. A partial list of experimental animal observations is presented in Table 43.2. In general, many of the *N*-nitrosamines have been observed to be carcinogens. Nitrosamines require a metabolic enzymatic transformation to active intermediates. Accidental acute exposure of humans to the lower molecular weight amine derivatives has resulted in severe liver damage. Chronic exposure of animals has been found to produce cancer in a wide variety of organs. Following treatment with various *N*-nitroso derivatives, malignant tumors in animals have been found in the hepatobiliary system, urinary system, respiratory system, heart, alimentary tract, nasal cavities, nervous system, hematopoietic system, reproductive organs, pancreas, and mammary glands. The lowest daily dose resulting in tumors (liver) was 0.075 mg/kg *N*-nitrosodimethylamine administered for 600 days. Structure activity relationships have been found to correlate with the number of carbon atoms: as the molecule increases in size the carcinogenic activity decreases.

2 SELECTED EXAMPLES

The following examples are used to indicate the effects of selected *N*-nitrosamines. Included with each compound is a short discussion of the parent amine from which the *N*-nitroso compound may be formed.

2.1 *N*-Nitrosodimethylamine (*N*,*N*-Dimethylnitrosamine, DMN, DMNA)

2.1.1 Sources, Uses, and Industrial Exposure

N-Nitrosodimethylamine is a yellow liquid of low viscosity that has a boiling point of 149°C. It is soluble in water and is lipophilic. It is reasonably stable in neutral or alkaline solution, and the compound degrades under ultraviolet light. It is oxidized by strong oxidizing agents (to the nitramine) and can be reduced to the corresponding hydrazine or amine. The compound is used primarily in the production of 1,1-dimethylhydrazine, a rocket propellant. In addition, it has been either used or proposed for use as an industrial solvent and softener for plastics, as an antioxidant, and as an additive to lubricants.

The parent compound, dimethylamine, is used widely as an intermediate in the chemical and pharmaceutical industries, and in the treatment of hides in the manufacture of leather.

2.1.2 Determination

Several review articles have appeared in the literature on the determination of N-nitroso compounds (58–61). Thin layer chromatography (62–65), polarography (66–69), spectrometry (70–73), gas chromatography (GC) (74–77) gas chromatography/mas spectrometry (GC/MS) (78–82), high performance liquid chromatography (HPLC) (83, 84), and the Thermal Energy Analyzer, which is a detector specifically designed for the determination of N-nitroso compounds, have been used for the analysis of this class of compounds. As reflected in Table 43.1, N-nitrosodimethylamine has been the most frequently observed N-nitroso compound. The most applicable analytical methods have been GC/MS and the TEA analyzer when coupled with the appropriate separation mechanism (GC, HPLC). In all cases prior "chemical workup" has been required for determination of DMN.

2.1.3 Physiologic Response

N-Nitrosodimethylamine has been found to be carcinogenic in all species of animals tested. Inhalation studies of this compound with rats produced tumors of the kidney and nasal passages (85). Oral administration of DMN on several strains of rat produced renal tumors, liver tumors, and occasional lung tumors (86–88). Doses of DMN in drinking water as low as 89 mg/kg b.w. in mice produced tumors of the kidney (89). Guinea pigs dosed at 2 mg/kg b.w. for 40 to 55 weeks produced liver cell carcinomas (90). Subcutaneous injection of DMN on mice and hamsters induced haemangioendothelial sarcomas (91, 92).

A summary of toxicity and carcinogenicity data for DMN is presented in Table 43.2.

Exposure of man produces acute necrosis and cirrhosis. The compound is considered a carcinogen by the FDA and the EPA.

2.1.4 Metabolism

Studies indicate that DMN must undergo biochemical transformation to be carcinogenic (93, 94). The compound can be metabolized in vivo and in vitro

(95, 96), and strong evidence exists for its in vivo formation after intake of the parent amine, dimethylamine (97).

2.1.5   Hygienic Standard of Exposure

Standards are not available for DMN. The compound is considered a carcinogen.

2.1.6   Warning Properties

N-Nitrosodimethylamine has no distinct odor. No other warning properties have been reported; however, high levels of dimethylamine may indicate elevated exposure levels of DMN, particularly in the presence of nitrite or the nitrogen oxides.

2.2   *N*-Nitrosodiethylamine (*N,N*-Diethylnitrosamine, DEN, DENA)

2.2.1   Sources, Uses, and Industrial Exposure

N-Nitrosodiethylamine is a yellow liquid having a boiling point of 177°C. It is soluble in water and organic solvents and is considered lipophilic. It is stable in neutral and alkaline solutions, somewhat stable in strong acids, and it is ultraviolet light sensitive. It undergoes chemical reactions similar to those of DMN.

Nitrosodiethylamine is used as a gasoline and lubricant additive, as a solvent in the fiber industry, as a polymer softener, and in the synthesis of 1,1-diethylhydrazine. It is used also in condensers as a dielectric modifier.

The parent compound, diethylamine, is used in the rubber, petroleum, pharmaceutical, and plastics industries.

2.2.2   Determination

The methods discussed with respect to *N*-nitrosodimethylamine are applicable to the determination of DEN. Detection limits using the more sophisticated methods are reliable, generally, to approximately 10 ppb.

DEN has been observed in numerous matrices as shown in Table 43.1. Trace amounts of this compound have been observed in food products, tobacco smoke, and gastric juices.

2.2.3   Physiologic Response

Numerous animal studies have been published regarding the carcinogenicity of DEN in animals. Inhalation exposure of rats to this compound produced liver tumors whereas exposure of hamsters induced tracheal and lung tumors (99).

Table 43.1. Reported Occurrences of N-Nitroso Compounds (14)

Sample Source	N-Nitroso Compound (s) Identified[a]	Concentration[b]	Detection Methods[c]	Ref.
Meat Products				
Smoked sausages, salami, bacon (fried and unfried), ham, dried horsemeat and beef, souse, luncheon meats, frankfurters, and hamburger	DMN, DEN, DBN, NPYR, NPIP, NPRO, NSAR	0.4–440 ppb	TLC, HPLC, GC, GC-MS	15–29
Fish				
Salted, fresh, or smoked herring, haddock, mackerel, kipper, salmon, shad, sable, cod, hake, and fish meal (animal feed)	DMN, DEN	0.5–40 ppb	TLC, GC, GC-MS	30–36
Other foods				
Cheese, milk, flour (wheat)	DMN, DEN	1–10 ppb	GC, TLC, MS	37–39
Alcoholic beverages				
Wine, African spirits, cider distillates	DMN, DEN, DPN, MEN, EBN, NPIP, DBM	10 ppb–21 ppm	POL, TLC, GC	40–42
Miscellaneous				
Soybean oil	DMN	300 ppb	TLC, GC, MS	43, 44
Tobacco leaf and smoke condensate	DMN, DEN, DETN, MEN, MBN, DPN, DBN, DPIP, NPYR, NNN	85–180 ng per cigarette	TLC, GC, MS, HPLC	45–50
Water	Not identified	Est. 0.1 ppb	GC-TEA, HPLC-TEA	51, 52
Soil	DMN, DPN	1–6000 ppb	TEA-GC, TEA-HPLC	53, 54

Pesticide formulations	DMN, DPN	300 ppb–640 ppm	TEA-GC, TEA-HPLC	53
Synthetic cutting oils	DETN	1–3%	TEA-HPLC	55
Cosmetics	DETN	1 ppb–48 ppm	TEA-HPLC	56
Hydraulic fluids	Not identified	—	—	57

[a] Abbreviations:

DMN	=	N-nitrosodimethylamine
DEN	=	N-nitrosodiethylamine
DPN	=	N-nitrosodipropylamine
MEN	=	N-nitrosomethylethylamine
DBN	=	N-nitrosodibutylamine
EBN	=	N-nitrosoethylbutylamine
MBN	=	N-nitrosomethylbutylamine
NPIP	=	N-nitrosopiperidine
NPYR	=	N-nitrosopyrrolidine
NPRO	=	N-nitrosoproline
NSAR	=	N-nitrososarcosine
NNN	=	N-nitrosonornicotine
DETN	=	N-nitrosodiethanolamine

[b] Abbreviations:

ppb	=	parts per billion (µg/kg)
ppm	=	parts per million (mg/kg)
ng/m^3	=	nanograms per cubic meter
ND	=	not determined quantitatively

[c] Abbreviations:

TLC	=	thin layer chromatography
POL	=	polarography
GC	=	gas chromatography
MS	=	mass spectrometry
HPLC	=	high performance liquid chromatography
TEA	=	thermal energy analyzer

Table 43.2. Acute Toxicity and Carcinogenicity Data for Selected *N*-Nitroso Compounds (14)

N-Nitroso Compound	Acute Toxicity Data[a]		Animal Species	Carcinogenicity Data	
	Treatment Route[b]	LD$_{50}$[c]		Treatment Route[b]	Major Site of Cancer
N-Nitrosodimethylamine (dimethylnitrosamine)	Oral	26	Rat	Oral	Liver, kidney
	Inhln.	37		Inhln.	Nasal cavity
	SC	45			
	IV	40			
	Oral	21	Mouse	Oral	Liver, lung, kidney
	SC	28	Hamster	Oral	Liver, lung, kidney
				Inhln.	Lung, nasal cavity
			Rabbit	Oral	Liver
			Guinea pig	Oral	Liver
			Mink	Oral	Liver
			Monkey	Oral	Liver
			Newt	Oral	Liver
			Aquarium fish	Oral	Liver
			Trout	Oral	Liver
N-Nitrosodiethylamine (diethylnitrosamine)	Oral	280	Rat	Oral	Liver, esophagus
	IV	157		IV	Liver
	Oral	190–220	Mouse	Oral	Liver, esophagus
				IP	Liver, forestomach, lung, leukemia
				Topical	Nasal cavity
	SC	246	Hamster	Oral	Nasal cavity, lung, liver, trachea, larynx, esophagus, forestomach
		246–413			
			Hamster	SC	Olfactory neuroblastoma
					Nasal cavity, lungs

Compound	Route	Dose	Species	Route	Target Organ
	Oral	250	Guinea pig	Oral	Liver
			Rabbit	Oral	Liver
			Monkey	Oral	Liver
			Dog	Oral	Liver
			Pig	Oral	Liver
			Trout	Oral	Liver
			Parakeet	Oral	Liver
1-(Methoxymethyl)methylnitrosamine	Oral	540	Rat	Oral	Liver
1-(Methoxyethyl)methylnitrosamine	Oral	240	Rat	Oral	Esophagus, lung
1-(Methoxyethyl)ethylnitrosamine	Oral	1000	Rat	Oral	Liver, lung
N-Nitrosodi-n-propylamine (di-n-propylnitrosamine)	Oral	480	Rat	Oral	Liver, esophagus, tongue
	SC	600	Hamster	Oral	Nasal cavity, trachea
N-Nitrosodiisopropylamine	Oral	850	Rat	Oral	Liver
			Rat	Oral	Esophagus, liver, lung, nasal cavity
N-Nitroso-bis (2-acetoxypropyl)amine	SC	6500	Hamster	Oral	Pancreas
N-Nitrosomethyl-n-propylamine	SC	493	Hamster	Oral	Nasal cavity, lung, liver, trachea
			Rat	Oral	Liver, esophagus
N-Nitrosodi-n-butylamine	Oral	1200	Rat	Oral	Liver, esophagus, bladder
	SC	561	Hamster	SC	Liver, bladder
			Mouse	Oral	Forestomach, liver, bladder, lung
				Oral	Esophagus, liver, bladder, tongue
N-Nitrosodiisobutylamine	SC	5600	Guinea Pig	Oral	Liver, bladder
			Hamster	Oral	Trachea, lung
			Mouse	Oral	Nasal cavity, eyelid
N-Nitrosodi-n-pentylamine (di-n-amylnitrosamine)	Oral	1750	Rat	Oral	Liver
	SC	3000	Rat	SC	Lung

Table 43.2. (Continued)

N-Nitroso Compound	Acute Toxicity Data[a]			Carcinogenicity Data	
	Treatment Route[b]	LD$_{50}$[c]	Animal Species	Treatment Route[b]	Major Site of Cancer
N-Nitrosodicyclohexylamine	Oral	5000	Rat	Oral	None
N-Nitrosodiphenylamine	Oral	1650	Rat	Oral	none
	Oral	3850	Mouse		
N-Nitrosodibenzylamine	Oral	900	Rat	Oral	None
N-Nitrosomethylamine	Oral	90	Rat	Oral	Liver
N-Nitrosomethylvinylamine	Oral	22	Rat	Oral	Lung, esophagus, nasal cavity, pharynx
	Inhln.	24		Inhln.	
N-Nitrosomethylallylamine	Oral	340	Rat	Oral	Esophagus, kidney
	IV	320		IV	Kidney, lung, nasal cavity
N-Nitrosomethyl-n-butylamine	Oral	130	Rat	Inhln.	Esophagus, nasal cavity
	SC	90	Mouse	Oral	Eyelid, nasal cavity
			Hamster	SC	Trachea, lung
N-Nitrosomethyl-n-pentylamine	Oral	120	Rat	Oral	Esophagus
	SC	120			
N-Nitrosomethylcyclohexylamine	IP	30	Rat	Oral	Esophagus, lung
	Oral	28			
	Oral	57	Mouse		
	Oral	168	Hamster		

Compound	Route[b]	LD_{50}[c]	Species	Route	Carcinogenic target
N-Nitrosomethyl-n-heptylamine	SC	420	Rat	SC	Lung
N-Nitrosomethyl-n-dodecylamine	Oral	5400	Rat	Oral	Bladder
N-Nitrosomethylphenylamine (N-nitroso-N-methylamiline)	Oral	225	Rat	Oral	Esophagus, forestomach
	IP	200			
N-Nitrosomethylbenzylamine	Oral	150	Mouse	Oral	Lung
	Oral	18	Rat	Oral	Esophagus
			Mouse	Oral	Esophagus, forestomach
N-Nitrosoethylphenylamine (N-nitroso-N-ehtylaniline)	Oral	48(1)	Rat	Oral	Esophagus
N-Nitrosohexamethylenimine	Oral	336	Rat	Oral	Liver, esophagus, tongue, trachea
			Mouse	Oral	Lung
			Hamster	Oral	Liver, trachea
N-Nitrosooctamethylenimine	Oral	566	Rat	Oral	Lung, esophagus, trachea
N-Nitrosodecamethylenimine	Oral	NA	Mouse	Oral	Liver, stomach

[a] Toxicity data taken from *Registry of Toxic Effects of Chemical Substances*, H. E. Christensen and E. J. Fairchild, (Eds.), NIOSH. (1976), unless another reference is indicated in parentheses.

[b] Oral = oral administration;
SC = subcutaneous injection;
IV = intravenous injection;
Inhln. = inhalation;
Ip = intraperitoneal injection.

[c] LD_{50} = lethal dose, 50 percent kill, expressed in milligrams of substance per kilogram animal weight.

Topical application of DEN to mice did not produce local skin tumors, but induced carcinoma within the nasal cavity (100). In hamsters DEN produced tumors of the trachea (101); again no local cancers were observed. Oral administration of DEN induced liver (102) and esophagus (103) cancer in mice and rats (104), trachea and lung cancer in hamsters (105), hepatocellular and adenocarcinomas in guinea pigs (106), and liver cancer in rabbits (107), dogs (108), pigs (109), and monkeys (110).

2.2.4 Metabolism

Studies (93, 94) indicate that DEN has metabolic properties similar to those of DMN.

2.2.5 Hygienic Standard of Exposure

Standards are not available for DEN.

2.2.6 Warning Properties

N-Nitrosodiethylamine has no distinct odor. No other warning properties have been reported; however, high levels of diethylamine may indicate elevated exposure levels of DEM, particularly in the presence of nitrite or the nitrogen oxides.

2.3 N-Nitrosodiethanolamine [NDELA, N-Nitroso-bis(2-hydroxyethyl)amine]

2.3.1 Sources, Uses, and Industrial Exposure

N-Nitrosodiethanolamine is a yellow, highly viscous liquid that is miscible in water and soluble in polar organic solvents. It is stable in neutral and alkaline solutions and is very sensitive to ultraviolet light. NDELA has not been used in industry, but its existence as a product of cutting fluids is well documented (55). For cutting fluids containing high concentrations of triethanolamine and nitrite, concentrations up to 3 percent have been observed. Triethanolamine in the cutting fluids produce, or have as impurities, diethanolamine, which subsequently reacts with nitrite to form NDELA.

This nitrosamine has also been observed in products that contain triethanolamine as an emulsifying agent, such as selected pesticides, cosmetics, and body lotions.

2.3.2 Determination

Ethyl acetate is generally used as an extraction solvent for NDELA. The majority of observations have been made using HPLC coupled with a TEA analyzer. In addition, gas chromatographic and colorimetric methods have been applied in selected cases.

2.3.3 Physiologic Response

No information has been reported on the effect of NDELA on man in the occupational environment. However, animal data indicate that the nitrosamine is a carcinogen. Oral administration of NDELA to rats produced hepatocellular carcinomas (111) at 600 to 1000 mg/kg b.w. Subcutaneous injection in hamsters produced adenocarcinomas, tracheal tumors, and hepatocellular edemas (112).

2.3.4 Metabolism

No data have been published regarding the metabolism of NDELA.

2.3.5 Hygienic Standard of Exposure

Standards are not available for NDELA.

2.3.6 Warning Properties

N-Nitrosodiethanolamine has no distinct odor. No other warning properties are available; however, high levels of diethanol or triethanolamine may indicate elevated exposure levels of NDELA, particularly in the presence of nitrite, nitrate, or the nitrogen oxides.

3 OTHER COMPOUNDS

Numerous other *N*-nitroso compounds are known to be carcinogens. Of the more than 150 compounds tested, approximately 80 percent have been found to be carcinogenic in at least one species of animal. Few industrial exposures to man are known, except those previously mentioned.

The greatest potential exposure in the industrial setting is presumed to be through indirect assimilation of the precursors, nitrite and secondary amines. Information regarding numerous *N*-nitroso compounds has been tabulated by IARC (113).

REFERENCES

1. J. Graymore, *J. Chem. Soc.*, 1311 (1938).
2. E. Gowley and J. Partington, *J. Chem. Soc.*, 1252 (1933).
3. P. Brookes and K. Walker, *J. Chem. Soc.*, 4409 (1957).

4. H. Emeleus and B. Tatterschall, *Angew. Chem.,* **76,** 961 (1964).

5. F. Klages and H. Sitz, *Chem. Ber.,* **96,** 2394 (1963).

6. M. Weissler, *Angew. Chem. Int. Ed.,* **13,** 743 (1974).

7. M. Weissler, *Tetrahedron Lett.,* 2575 (1975).

8. R. E. Lyle, J. E. Saavedra, and G. G. Lyle, *Synthesis,* **7,** 462.

9. G. A. Olah, J. A. Olah, and N. A. Overchuk, *J. Org. Chem.,* **30,** 3373 (1965).

10. H. Druckery et al., *Z. Krebsforsch.,* **69,** 103 (1967).

11. D. Schmähl and R. Preussmann, *Naturwissenschaften,* **46,** 175 (1959).

12. J. Althoff, J. Hilfrich, F. W. Kruger, and B. Bertram, *Z. Krebsforsch.,* **81,** 23 (1974).

13. H. K. Herring, *Proc. Meat. Ind. Res. Conf.,* American Meat Institute Foundation, Chicago, 1973, p. 47.

14. G. S. Drescher, Ph.D. Dissertation, The University of Iowa, 1977.

15. P. J. Groenen et al., in "Environmental N-Nitroso Compounds Analysis and Formation," *IARC Sci. Publ. No. 14,* Lyon, 1976, p. 321.

16. N. P. Sen, J. R. Iyengar, W. F. Miles, and T. Panalaks, in "Environmental N-Nitroso Compounds Analysis and Formation," *IARC Sci. Publ. No. 14,* Lyon, 1976, p. 333.

17. R. W. Stephany, J. Freudenthal, and P. L. Schuller, in "Environmental N-Nitroso Compounds Analysis and Formation," *IARC Sci. Publ. No. 14,* Lyon, 1976, p. 343.

18. J. H. Dhont and C. van Ingen, in "Environmental N-Nitroso Compounds Analysis and Formation," *IARC Scientific Publ. No. 14,* Lyon, 1976, p. 355.

19. T. A. Gough and C. L. Walters, in "Environmental N-Nitroso Compounds Analysis and Formation," *IARC Sci. Publ. No. 14,* Lyon, 1976, p. 195.

20. J. Kann, O. Tauts, K. Raja, and R. Kalve, in "Environmental N-Nitroso Compounds Analysis and Formation," *IARC Sci. Publ. No. 14,* Lyon, 1976, p. 385.

21. N. P. Sen, *Food Cosmet. Toxicol.,* **10,** 219 (1972).

22. N. P. Sen, B. Donaldson, J. R. Iyengar, and T. Panalaks, *Nature,* **241,** 473 (1973).

23. N. P. Sen et al., *Nature,* **245,** 104 (1973).

24. D. C. Havery et al., *J. Assoc. Offic. Anal. Chem.,* **59,** 540 (1976).

25. N. P. Sen, S. Seaman, and W. F. Miles, *Food Cosmet. Toxicol.,* **14,** 167 (1976).

26. J. J. Wartheson, D. D. Bills, R. A. Scanlan, and L. M. Libbey, *J. Agric. Food Chem.,* **24,** 892 (1976).

27. A. Mirna et al., *Fleischwirtschaft,* **56,** 1014 (1976).

28. L. Kotter, A. Fischer, and H. Schmidt, *Fleischwirtschaft,* **56,** 997 (1976).

29. P. Cooper, *Food Cosmet. Toxicol.,* **14,** 205 (1976).

30. T. Juskiewicz and B. Kowalski, in "Environmental N-Nitroso Compounds Analysis and Formation," *IARC Sci. Publ. No. 14,* Lyon, 1976, p. 375.

31. Y. Y. Fong and W. C. Chan, in "Environmental N-Nitroso Compounds Analysis and Formation," *IARC Sci. Publ. No. 14,* Lyon, 1976, p. 465.

32. N. D. Gorelova and P. P. Dikun, *Vopr. Onkol.,* **21,** 77 (1975).

33. D. F. Gadpois, R. M. Ravesi, R. C. Lundstrom, and R. S. Maney, *J. Agric. Food Chem.,* **23,** 665 (1975).

34. M. Ishidate et al., in *Topics in Chemical Carcinogenesis,* W. Nakahara, et al., Eds., University of Tokyo Press, 1972, p. 313.

35. Y. Y. Fong and W. C. Chan, *Food Cosmet. Toxicol.,* **11,** 841 (1973).

36. T. Fazio et al., *J. Agric. Food Chem.,* **19,** 250 (1971).

37. B. H. Thewlis, *Food Cosmet. Toxicol.*, **6**, 822 (1968).

38. L. Hedler and P. Marquardt, *Food Cosmet. Toxicol.*, **6**, 341 (1968).

39. H. J. Petrowitz, *Arzneim.-Forsch.*, **18**, 1486 (1968).

40. P. Bogovski, E. A. Walker, M. Castegnaro, and B. Pignatelli, in "*N*-Nitroso Compounds in the Environment," *IARC Sci. Publ. No. 9*, Lyon, 1974, p. 192.

41. N. D. McGlasahan, C. L. Walters, and A. E. M. McLean, *Lancet*, II, 1017 (1968).

42. C. H. Collins, P. Cook, J. K. Foreman, and J. F. Palframan, *Gut*, **12**, 1015 (1971).

43. L. Hedler et al., in "*N*-Nitroso Compounds Analysis and Formation," *IARC Sci. Publ. No. 3*, Lyon, 1972, p. 71.

44. L. Hedler and P. Marquardt, in "*N*-Nitroso Compounds in the Environment," *IARC Sci. Publ. No. 9*, Lyon, 1974, p. 183.

45. D. Hoffmann, G. Rathkamp, and Y. Y. Liu, in "*N*-Nitroso Compounds in the Environment," *IARC Sci. Publ. No. 9*, Lyon, 1974, p. 159.

46. D. Hoffmann, S. S. Hecht, R. M. Ornaf, E. L. Wynder, and T. Tso, in "Environmental N-Nitroso Compounds Analysis and Formation," *IARC Sci. Publ. No. 14*, Lyon, 1976, p. 307.

47. I. Schmeltz, S. Abidi, and D. Hoffmann, *Cancer Lett.*, **2**, 125 (1976).

48. S. S. Hecht, R. M. Ornaf, and D. Hoffmann, *J. Natl. Cancer Inst.*, **54**, 1237 (1975).

49. G. Neurath, B. Pirmann, and H. Wichern, *Beitr. Tabakforsch.*, **2**, 311 (1964).

50. G. Neurath, B. Pirmann, H. Wichern, and W. Luttich, *Beitr. Tabakforsch.*, **3**, 251 (1965).

51. D. H. Fine, D. P. Roundehler, N. M. Belcher, and S. S. Epstein, in "Environmental N-Nitroso Compounds Analysis and Formation," *IARC Sci. Publ. No. 14*, Lyon, 1976, p. 401.

52. D. H. Fine et al., *Bull. Environ. Contam. Toxicol.*, **14**, 404 (1975).

53. D. H. Fine et al., paper presented at the 172nd ACS National Meeting, San Francisco, September 2, 1976.

54. Anonymous, *Pest. Chem. News*, 5 (September 1, 1976).

55. T. Y. Fan et al., *Science*, **196**, 70 (1977).

56. Anonymous, *Chem. Eng. News*, p. 7 (March 28, 1977).

57. Anonymous, *Chem. Mark. Rep.*, **210**, 4 (1976).

58. N. T. Crosby, *Residue Rev.*, **64**, 77 (1976).

59. N. T. Crosby and R. Sawyer, *Adv. Food Res.*, **22**, 1 (1976).

60. W. Fiddler, *Toxicol. Appl. Pharmacol.*, **31**, 352 (1975).

61. A. E. Wassermann, in "*N*-Nitroso Compounds Analysis and Formation," *IARC Sci. Publ. No. 3*, Lyon, 1972, p. 10.

62. R. Preussmann, D. Daiber, and H. Hengy, *Nature*, **201**, 502 (1964).

63. R. Preussmann et al., *Z. Anal. Chem.*, **202**, 187 (1964).

64. C. L. Walters, E. M. Johnson, N. Ray, and G. Woolford, in "*N*-Nitroso Compounds Analysis and Formation," *IARC Sci. Publ. No. 3*, Lyon, 1972, p. 79.

65. A. A. L. Gunatilaka, *J. Chromatogr.*, **120**, 229 (1976).

66. F. L. English, *Anal. Chem.*, **23**, 344 (1951).

67. C. L. Walters, E. M. Johnson, and N. Ray, *Analyst*, **95**, 485 (1970).

68. C. L. Walters, *Lab. Prac.*, **20**, 574 (1971).

69. W. F. Smyth et al., *Anal. Chim. Acta*, **78**, 81 (1975).

70. A. A. Forist, *Anal. Chem.*, **36**, 1338 (1964).

71. G. Eisenbrand and R. Preussmann, *Arzneim.-Forsch.*, **20**, 1513 (1970).

72. E. M. Johnson and C. L. Walters, *Analyt. Lett.*, **4**, 383 (1971).

73. M. J. Downes, M. W. Edwards, T. S. Elsey, and C. L. Walters, *Analyst*, **101**, 742 (1976).

74. T. F. Kelly and J. R. Nunn, in "*N*-Nitroso Compounds in the Environment," *IARC Sci. Publ. No. 9*, Lyon, 1974, p. 26.

75. J. F. Palframan, J. McNab, and N. T. Crosby, *J. Chromatogr.*, **76**, 307 (1973).

76. E. Von Rappardt, G. Eisenbrand, and R. Preussmann, *J. Chromatogr.*, **124**, 247 (1976).

77. N. P. Sen, *J. Chromatogr.*, **51**, 301 (1970).

78. T. A. Gough and K. S. Webb, *J. Chromatogr.*, **64**, 201 (1972).

79. T. A. Gough and K. S. Webb, *J. Chromatogr.*, **95**, 59 (1974).

80. T. A. Gough and K. Sugden, *J. Chromatogr.*, **109**, 265 (1975).

81. R. W. Stephany et al., *J. Agric. Food Chem.*, **24**, 536 (1976).

82. E. D. Pellizzari et al., *Analyt. Lett.*, **9**, 579 (1976).

83. G. B. Cox, *J. Chromatogr.*, **83**, 471 (1973).

84. H. J. Klimisch and D. Ambrosius, *J. Chromatogr.*, **121**, 93 (1976).

85. H. Druckrey, R. Preussmann, S. Ivankovic, and D. Schmähl, *Z. Krebsforsch.*, **69**, 103 (1967).

86. P. N. Magee and J. M. Barnes, *Brit. J. Cancer*, **10**, 114 (1956).

87. P. N. Magee and J. M. Barnes, *J. Path. Bact.*, **84**, 19 (1962).

88. R. Preussmann, G. Neurath, G. Wulf-Lorentzen, D. Daiber, and H. Hengy, *Z. anal. Chem.*, **202**, 187 (1964).

89. B. Terracini, G. Palestro, M. Ramella Gigliardi, and R. Montesano, *Brit. J. Cancer*, **20**, 871 (1966).

90. R. N. Le Page and G. S. Christie, *Pathology*, **1**, 49 (1969b).

91. H. Otsuka and A. Kuwahara, *Gann*, **62**, 147 (1971).

92. K. McD. Herrold, *J. Natl. Cancer Inst.*, **39**, 1099 (1967).

93. P. N. Magee and J. M. Barnes, *Advc. Cancer Res.*, **10**, 163 (1967).

94. P. N. Magee and P. F. Swann, *Brit. Med. Bull.*, **25**, 240 (1969).

95. I. J. Mizrahi and P. Emmelot, *Cancer Res.*, **22**, 339 (1962).

96. D. F. Heath, *Biochem. J.*, **85**, 72 (1962).

97. P. N. Magee, in "Occurrence of Alkylating Substances," *Alkylierend wirkende Verbindungen*, Hamburg, (1968) p. 79.

98. W. Dontenwill, U. Mohr, and M. Zagel, *Z. Krebsforsch.*, **64**, 499 (1962).

99. K. McD. Herrold and L. J. Dunham, *Cancer Res.*, **23**, 773 (1963).

100. F. Hoffmann and A. Graffi, *Acta Biol. Med. Germ.*, **12**, 623 (1964a).

101. K. McD. Herrold, *Arch. Pathol.*, **78**, 189 (1964b).

102. D. Schmähl, C. Thomas, and K. König, *Naturwissenschaften*, **50**, 407 (1963a).

103. I. N. Shvemberger, *Vopr. Onkol.*, **11**, 74 (1965).

104. D. Schmähl, R. Preussmann, and H. Hamperl, *Naturwissenschaften*, **47**, 89 (1960).

105. W. Dontenwill and U. Mohr, *Z. Krebsforsch.*, **64**, 305 (1961).

106. H. Druckrey and D. Steinhoff, *Naturwissenschaften*, **49**, 497 (1962).

107. D. Schmähl and C. Thomas, *Naturwissenschaften*, **52**, 165 (1965a).

108. D. Schmähl, C. Thomas, and G. Scheld, *Naturwissenschaften*, **51**, 466 (1964).

109. D. Schmähl, H. Osswald, and U. Mohr, *Naturwissenschaften*, **54**, 341 (1967).

110. R. W. O'Gara and M. G. Kelly, *Proc. Am. Assoc. Cancer Res.*, **6,** 50 (1965).

111. H. Druckrey, R. Preussmann, S. Ivankovic, and D. Schmähl, *Z. Krebsforsch.*, **69,** 103–201 (1967).

112. J. Hilfrich, I. Schmeltz, and D. Hoffmann, *Cancer Lett.*, **4,** 55 (1978).

113. *IARC Monograph*, Vol. 17, WHO, 1978.

CHAPTER FORTY-FOUR

Aliphatic and Alicyclic Amines

RODNEY R. BEARD, M.D., M.P.H., and
JOSEPH T. NOE, M.D.

This chapter presents information regarding the toxicity of some of this family of chemicals. In many instances the data are either limited or inconclusive. There is an apparent difficulty in applying pharmacological research data to industrial toxicology. It will be noted that there is only sparse information relating to humans. Recent growth of interest in carcinogenicity has led to important discoveries, but much confusion and controversy remain. Methods for screening for the detection of carcinogenic potentials are developing rapidly. Epidemiologic studies of human populations are also being done more frequently, but studies of both kinds need encouragement. At the same time it must be kept in mind that great care and skill must be applied to the design and execution of these studies and that inept research can be dangerously misleading.

1 PHYSICAL AND CHEMICAL PROPERTIES

The aliphatic amines are derivatives of ammonia in which one or more hydrogen atoms are replaced by an alkyl or alkanol radical. They have a characteristic fishlike odor and are strongly alkaline. They are prepared by the alkylation of ammonia or hydrogenation of the appropriate nitrite. The more common and widely used amines are gases or fairly volatile liquids. They are widely used as industrial intermediates. The lower amines are very soluble in water, the gaseous members of the series commonly being supplied as aqueous solutions.

RODNEY R. BEARD AND JOSEPH T. NOE

The higher molecular weight amines are less volatile, odorless, and only partly soluble in water. Their solutions can be highly irritating and cause damage on contact with eyes, skin, and the respiratory tract. Branching of the alkyl chain tends to enhance volatility, whereas hydroxy substitution as in the alkanolamines decreases volatility. Many of the lower aliphatic amines have low flash points and fall into the category of flammable liquids or gases. Skin absorption is a problem; many are capable of cutaneous hypersensitization. Some have physiological or pharmacological effects such as histamine liberation and vasodilatation. In industrial exposures local effects predominate.

The aliphatic amines are conveniently classified as primary, secondary, and tertiary amines according to the number of substitutions on the nitrogen atom. If only one radical is substituted, the amine is a primary amine, even though the alkyl substituent may have a secondary or tertiary structure. Further subdivision is as follows:

1. Monoamines
    a. Primary
    b. Secondary
    c. Tertiary
    d. Unsaturated
    e. Alicyclic
2. Polyamines
3. Alkanolamines

The alkylamines and alkanolamines behave as bases in organic solvents and aqueous solutions. One way for expressing the basicity of an amine focuses on its reaction with water. The equilibrium constant for the acid–base reaction of an amine with water is called K_b.

$$RNH_2 + H_2O \rightleftharpoons RNH_3^+ + OH^-$$

$$K_b = \frac{[RNH_3^+][OH^-]}{[RNH_2]}$$

and this is conveniently converted to pK_b, where

$$pK_b = -\log_{10} K_b$$

The more basic the compound, the lower the pK_b. Measured in this way, the alkylamines have pK_b values of 3 to 5; arylamines have pK_b values of 9 to 10, and pyrrole has a pK_b of 13.6. The pK_b of ammonia is 4.76 (1, 2). pK_b values for some of the more common amines are given in Table 44.1.

In general, primary amines are stronger bases than ammonia, and secondary amines are stronger bases than tertiary amines. As the length of the chain increases up to four or five carbon atoms, the base strength tends to decrease.

Table 44.1. Basicity of Ammonia and Some Amines[a]

Amine	Formula	pK_b	pK_a
Ammonia	NH_3	4.76	9.24
Methylamine	CH_3NH_2	3.35	10.65
Dimethylamine	$(CH_3)_2NH$	3.32	10.68
Trimethylamine	$(CH_3)_3N$	4.22	9.78
Ethylamine	$CH_3CH_2NH_2$	3.29	10.71
Diethylamine	$(CH_3CH_2)_2NH$	3	11
Triethylamine	$(CH_3CH_2)_3N$	3.25	10.75
n-Propylamine	$CH_3CH_2CH_2NH_2$	3.39	10.61
Di-n-propylamine	$(C_3H_7)_2NH$	3	11
Isopropylamine	$(CH_3)_2CHNH_2$	3.4	10.6
n-Butylamine	$C_3H_7NH_2$	3.32	10.68
Cyclohexylamine	$C_6H_{11}NH_2$	3.3	10.7
Hexamethylenediamine	$NH_2(CH_2)_6NH_2$	3.3	10.7

[a] Modified from H. K. Hall, Jr., *J. Phys. Chem.*, **60,** 63 (1956), and A. Streitwieser, Jr., and C. H. Heathcock, *Introduction to Organic Chemistry*, Macmillan, New York, 1976.

Diamines such as ethylenediamine also behave as strong bases. The alkanolamines are weaker bases than the corresponding substituted amines (1, 2).

Odor is a property possessed by many organic compounds, and it can be used in a variety of ways. Certain compounds and mixtures of compounds with odors pleasing to most human noses are used as perfumes. Other compounds with odors less pleasing, not only to human noses but to those of other species as well, are used by certain animals as defense mechanisms. Many amines, particularly the lower molecular weight ones, belong in the second category, possessing odors that range from somewhat unpleasant to absolutely obnoxious. Odor can sometimes be used for identification or detection by the organic chemist, for the nose is a remarkably sensitive detection device. Often the amine structure of a compound has been inferred from the characteristic fishlike odor of the compound itself or some of its degradation products (2).

It has been said that the unpleasant odor from ammonia or amines can be controlled by adding dihydroxyacetone to the source. Dihydroxyacetone is nontoxic to humans and domestic animals and reacts rapidly with ammonia or amines (3).

It is interesting to note the great variety of amines in the human environment. Forty primary and secondary amines with different gas chromatographic properties were detected in samples of fresh vegetables, preserves, mixed pickles, fish and fish products, bread, cheese, stimulants, animal feedstuffs, and surface waters, and 21 of these were identified by mass spectrometry.

Secondary amines, precursors for the carcinogenic *N*-nitrosamines, were generally found in concentrations below 10 ppm, although higher concentrations occurred in herring preparations, some cheeses, and samples of large radish and red radish. The highest content of secondary amines found so far was in red radishes (38 ppm pyrrolidine, 20 ppm pyrroline, 5.4 ppm *N*-methylphenethylamine, and 1.1 ppm dimethylamine (4).

The physical properties of some of the aliphatic and alicyclic amines are given in Tables 44.2 and 44.3 (5–8).

2 MANUFACTURE AND USES

2.1 Manufacture

In 1976, the commercial production of alkylamines in the United States was in excess of 232 thousand tons. Unspecified amounts of fatty amines and cyclic amines are produced in the United States.

The lower aliphatic amines are made from a variety of starting materials, mainly ammonia with alcohols, aldehydes, ketones, or alkyl halides, or hydrogen cyanide with an alkene (olefin). The most used methods are as follows. (1) Ammonia and an alcohol are passed continuously over a catalyst at a temperature of 300 to 500°C and a pressure of 790 to 3550 kPa (100 to 500 psig), producing a good yield of a mixture of primary, secondary, and tertiary amines with negligible by-products. (2) Ammonia, hydrogen, and alcohol are passed continuously over a different catalyst at 130 to 250°C and high pressure as before, giving high conversion, again with a mixed amine product that must be separated. (3) Ammonia and aldehyde or ketone and hydrogen are passed over a hydrogenation catalyst under conditions similar to those of method 2. (4) Hydrogen cyanide is reacted with an alkene $R_2C{=}CH_2$ to yield R_3CNH_2, at a temperature of 30 to 60°C, followed by further heating to hydrolyze the intermediate amide, $R_3CNHCNO$. Many other methods have been used (8).

2.2 Uses

The uses of the alkylamines are legion. They are useful solvents, alone or with other compounds. They are used very extensively as beginning materials for chemical syntheses. Methylamine and dimethylamine alone are the bases for more than 20 products, including antispasmodic, analgesic, anesthetic, and antihistaminic pharmaceuticals, insecticides, a soil sterilizer, surfactants, a photographic developer, several solvents, an explosive, a fungicide, a rubber accelerator, a rocket propellant, an ion exchange resin, a plastic monomer, and a catalyst. Other amines are used as catalysts for polymerization reactions, a poultry feed, a bactericide, corrosion inhibitors, drugs, and herbicides (8).

Table 44.2. Physical and Chemical Properties of Aliphatic and Alicyclic Monoamines.

Name	Formula	Mol. Wt.	M.P. (°C)	B.P. (°C)	Density (g/ml)	Solubility in Water (g/100 ml)	Vapor Pressure (torr) (°C)	Vapor Density (Air = 1)	Flash Point[a] (°F)	1 mg/liter (ppm)	1 ppm (mg/m³)
Methylamine	CH_3NH_2	31.06	−93.5	−6.3	0.7691 (−70/4)	Very sol.	2 atm (25)	1.07	34 (30% soln.)	783	1.27
Dimethylamine	$(CH_3)_2NH$	45.08	−93	7.4	0.6804 (0/4)	Very sol.	2 atm (10)	1.55	54 (25% soln.)	542	1.84
Trimethylamine	$(CH_3)_3N$	59.11	−117.2	2.87	0.6709 (0/4)	Very sol.	760 (2.9)	2.04	38 (25% soln.)	414	2.42
Ethylamine	$CH_2CH_2NH_2$	45.08	−81	16.6	0.6836 (20/20)	Complete	400 (2.9)	1.55	<0.0	542	1.84
Diethylamine	$(CH_3CH_2)_2NH$	73.14	−48	56.3	0.7400 (20/4)	Complete	195 (20)	2.52	−9	334	2.99
Triethylamine	$(CH_3CH_2)_3N$	101.19	−115.3	89.5	0.7275 (20/4)	Sol.	53.5 (20)	3.49	20	242	4.14
Propylamine	$CH_3CH_2CH_2NH_2$	59.11	−83	48.7	0.719 (20/20)	Sol.	400 (31/5)	2.04	−35	414	2.42
Di-n-propylamine	$(CH_3CH_2CH_2)_2NH$	101.19	63	110.7	0.7400 (20/4)	Sol.	30 (25)	3.49	63	242	4.14
Isopropylamine	$(CH_3)_2CHNH_2$	59.08	−101.2	34	0.694 (15/4)	Complete	460 (20)	2.04	−35	414	2.42
Diisopropylamine	$[(CH_3)_2CH]_2NH$	101.19	−61	84	0.720 (20/20)	Sl sol.	70 (20)	3.49	30	242	4.14
n-Butylamine	$CH_3(CH_2)_3NH_2$	73.14	−50.5	77.8	0.740 (20/4)	Complete	72 (20)	2.52	10	334	2.99
Di-n-butylamine	$[CH_3(CH_2)_3]_2NH$	129.24	−60	159.6	0.7613 (20/20)	Sol.	1.9 (20)	4.46	117	189	5.29
Tri-n-butylamine	$[CH_3(CH_2)_3]_3N$	185.34	<−70	214	0.775 (20/20)	Insol.	20 (100)	6.39	187	132	7.58
Isobutylamine	$(CH_3)_2CHCH_2NH_2$	73.14	−85.5	68	0.724 (25/4)	Complete	100 (18.8)	2.52	<15	334	2.99
n-Amylamine	$CH_3(CH_2)_4NH_2$	87.17	−55	104	0.7782 (20/20)	Sol.		3.01	30	281	3.56
Isoamylamine	$(CH_3)_2CHCH_2CH_2NH_2$	87.17		95	0.7505 (20/4)	Sol.		3.01		281	3.56
n-Hexylamine	$CH_3(CH_2)_5NH_2$	101.19	−19	132.7	0.767	1.2	6.5 (20)	3.49	85	242	4.14
2-Ethylbutylamine	$(CH_3CH_2)_2CHCH_2NH_2$	101.19		125	0.776 (20/20)			3.49	70	242	4.14
n-Heptylamine	$CH_3(CH_2)_6NH_2$	115.22	−18	156.9	0.7754 (20/4)	Sl sol.		3.97	130	212	4.71
Di-n-heptylamine	$[CH_3(CH_2)_6]_2NH$	213.4	30	271		Sl sol.		7.35		115	8.73
2-Ethylhexylamine	$CH_3(CH_2)_3(CH_3CH_2)CHCH_2NH_2$	130.23		142.3	0.7894 (20/20)	0.25		4.46	140	189	5.29
Di(2-ethylhexyl)-amine	$[CH_3(CH_2)_3(CH_3CH_2)CHCH_2]_2NH$	214.45		280.7	0.8062 (20/20)	Insol.	<0.01 (20)	8.33	270	101	9.88
Octadecylamine	$CH_3(CH_2)_{17}NH_2$	269.5		232		Insol.		9.29		91	11.02
Allylamine	$CH_2: CHCH_2NH_2$	57.09		58	0.7621 (20/4)	Complete		1.97	−20	428	2.23
Diallylamine	$(CH_2: CHCH_2)_2NH$	97.16	−88.4	111	0.7627 (10/4)	8.6		3.35	60	252	3.97
Triallylamine	$(CH_2: CHCH_2)_3N$	137.22	<−70	155.6	0.809 (20/4)	0.25		4.73	103	178	5.61
Cyclohexylamine	$C_6H_{11}NH_2$	99.17		134	0.8191 (20/4)	Sol.		3.42	88	247	4.06
Dicyclohexylamine	$(C_6H_{11})_2NH$	181.31	20	254	0.9104 (25/25)	Sl sol.		6.25	>210	135	7.42
N,N-Dimethylcyclo-hexylamine	$C_6H_{11}N(CH_3)_2$	127.22	<−77	159	0.849 (20/20)	1.1	3 (25)	4.39	110	192	5.20

[a] Open cup.

3139

Table 44.3. Physical and Chemical Properties of Aliphatic and Alicyclic Polyamines

Name	Formula	Mol. Wt.	M.P. (°C)	B.P. (°C)	Density (g/ml)	Solubility in Water (g/100 ml)	Vapor Pressure (torr) (°C)	Vapor Density (air = 1)	Flash Point[a] (°F)	Conversion Units	
										1 mg/liter (ppm)	1 ppm (mg/m³)
Ethylenediamine	$NH_2CH_2CH_2NH_2$	60.10	8.5	116.1	0.898 (25/4)	Sol.	10 (21.5)	2.07	93	407	2.46
N,N-Diethylethylenediamine	$(CH_3CH_2)_2NCH_2CH_2NH_2$	116.20		145.2	0.8211	Very sol.	4.1 (20)	4.01	115	210	4.75
Trimethylenediamine	$NH_2(CH_2)_3NH_2$	74.13	−23.5	135.5	0.884 (25/4)	Sol.		2.56	75	330	3.03
1,2 Propanediamine	$CH_3CH(NH_2)CH_2NH_2$	74.13	−37.2	120.9	0.864 (20/20)	Complete	8.0 (20)	2.56	92	330	3.03
Tetramethylenediamine	$NH_2(CH_2)_1NH_2$	88.15	27	158		Very sol.		3.04		277	3.60
1,3-Butanediamine	$CH_3CH(NH_2)CH_2CH_2NH_2$	88.15		142–150	0.85			3.04	125	277	3.60
Pentamethylenediamine	$NH_2(CH_2)_5NH$	102.18	9	178–180	0.9174 (0/4)	Sol.		3.52		239	4.18
Hexamethylenediamine	$NH_2(CH_2)_6NH_2$	116.21	41–42	204–205		Sol.		4.01		210	4.75
Diethylenetriamine	$(NH_2CH_2CH_2)_2NH$	103.17	−3	207.1	0.9586 (20/20)	Complete	0.2 (20)	3.56	215	237	4.22
Triethylenetetramine	$NH_2(CH_2CH_2NH)_2CH_2CH_2NH_2$	146.24	12	266–267	0.9818 (20/20)	Complete	<0.01 (20)	5.04	290	167	5.98
Tetraethylenepentamine	$NH_2(CH_2CH_2NH)_3CH_2CH_2NH_2$	189.31		340.3	0.9980 (20/20)	Complete	<0.01 (20)	6.53	325	129	7.74

[a] Open cup.

3 ABSORPTION, EXCRETION, AND METABOLISM

There has been relatively little study of the metabolism of the industrially important aliphatic amines; more interest has been directed toward the pharmacologically important substituted amines. A number of aliphatic amines have been identified as normal constituents of mammalian and human urine. These include methylamine, dimethylamine, trimethylamine, ethanolamine, ethylamine, and isoamylamine, as well as the catecholamines (hydroxytyramine and norepinephrine), histamine, and piperidine (9). Rechenberger states that man excretes approximately 10mg of volatile alkyl amine nitrogen per day (10). Davies found as many as eight aliphatic and ring-substituted primary amines in urine (11). These were excreted in amounts up to 100 µg/day.

The origin of these amines is not completely known, although it has been suggested that they may arise from the absorption of primary amines formed by decarboxylation of amino acids by intestinal bacteria, and Simenhoff (12) has shown that amine excretion is remarkably reduced in germfree animals or in those whose intestinal flora have been destroyed. Norephinephrine and hydroxytyramine, and the heterocyclic substituted ethylamine, histamine, are naturally occurring amines with a wide species distribution and considerable pharmacological importance (13).

The amines are well absorbed from the gut and the respiratory tract. The simple aliphatic amines can produce lethal effects by percutaneous absorption; the LD_{50} by this route is often the same as that for oral administration. Rechenberger (10) recovered little methylamine, propylamine, or n-butylamine in the urine after oral administration of the hydrochlorides of these compounds to humans, whereas a high proportion of dimethylamine or diethylamine was recovered, suggesting that the monoamines were metabolized and the diamines were not. Intermediate amounts of ethylamine and isobutylamine were recovered. Simenhoff (12) similarly reported the excretion of dimethylamine, but not of mono- or trimethylamine. He said that mono- and trimethylamines are converted to dimethylamine in the body. These findings correlate well with what is known about amine metabolism and enzymic deamination by amine oxidases (13, 14).

Monoamine oxidase and diamine oxidase (histaminase) occur widely in animal tissues, being most concentrated in the liver, kidney, and the intestinal mucosa. It is assumed that they play an important part in the "detoxication" of amines not normally present, as well as in the metabolism of pharmacologically important substituted amines. Monoamine oxidase catalyzes the deamination of primary, secondary, and tertiary amines according to the following overall reaction:

$$2 \text{ RCH}_2\text{NR}'\text{R}'' + O_2 + 2 \text{ H}_2\text{O} \rightarrow 2 \text{ RCHO} + 2 \text{ NH}_2\text{R}'\text{R}'' + \text{H}_2\text{O}_2$$

Diamine oxidase deaminates one end of the diamine molecule as follows:

$$R'CH_2NH_2 + O_2 + H_2O \rightarrow R'CHO + NH_3 + H_2O_2$$

(R′ contains the other basic group).

In these reactions the ammonia that eventually is formed is converted to urea. The hydrogen peroxide (H_2O_2) is acted on by catalase and the aldehyde formed is probably converted to the corresponding carboxylic acid by the action of aldehyde oxidase. The rate of oxidation of monoamines is faster with primary amines than with secondary. Tertiary amines and branched chains are more slowly oxidized than straight chains. The rate of oxidation varies with the number of carbon atoms in the carbon chain, methylamine not being attacked at all by monoamine oxidase, whereas ethylamine is slowly oxidized. The rate of oxidation increases to a maximum at five or six carbon atoms and falls off with further increase in chain length. Compounds of longer chain lengths, for example, octadecylamine, may inhibit the enzyme. Short-chain diamines are oxidized by diamine oxidase (histaminase). The four-carbon diamine, putrescine (tetramethylenediamine, 1,4-butanediamine), is most rapidly oxidized. The rate drops off with increasing chain lengths to a minimum around C_{10}. Monoamine oxidase, which shows a lack of affinity for the short-chain diamines, does oxidize the longer chains, with a maximum at 13 methylene groups (15).

Although the amine oxidases are enzymes of considerable biologic importance in the inactivation of amines that occur naturally in the human body, and probably also for oxidative deamination of the terminal amino groups of foreign amines, this is not the only means of metabolism. Methylamine, for example, is not oxidized by amine oxidases, yet it is rapidly absorbed and is not excreted in the urine. Simenhoff (12) suggests that it is methylated to dimethylamine. Trimethylamine is partly metabolized to ammonia and subsequently to urea, but is also oxidized to trimethylamine oxide by a specific enzyme, trimethylamine oxidase (9). Cadaverine (pentamethylenediamine), which may be formed naturally from the amino acid lysine, appears to be cyclized to piperidine, since oral administration of cadaverine to rabbits causes a severalfold increase in the normal piperidine excretion. Ring-substituted primary amines such as benzylamine and β-phenylamine are metabolized by deamination. Histamine is acetylated and methylated as well as deaminated. Many of the biologically and pharmacologically important secondary and tertiary substituted amines are metabolized by dealkylation, which may be carried out through the function of an enzyme system that is different from monoamine oxidase, and is located in the microsomes of liver cells (9).

Wirth and Thorgeirsson (16) reported that the activity of flavin-dependent amine oxidase is relatively high around the time of birth. They suggest that this may lead to increased susceptibility of newborn animals and humans to poisoning by the metabolic products of many drugs (and chemicals) that contain secondary or tertiary amine groups.

Ethanolamine has an important role in metabolism as one of the principal precursors of phosphoglycerides, which are important elements in the structure of biological membranes. There are several reports of its incorporation in a variety of cells. It appears to be less important for muscle cells than for nervous or liver tissues.

4 PHYSIOLOGICAL AND PATHOLOGICAL EFFECTS IN ANIMALS

There has not been a great deal of work on the toxicology of the aliphatic amines. A great part of our knowledge is derived from the studies done by Smyth, Carpenter, and their colleagues at the Mellon Institute. They initiated a series of "range finding tests" to quickly ascertain the principal toxic characteristics of previously untested compounds. Their reports are the main sources of the information summarized in Table 44.4. It should be understood and remembered that the figures cited and the indexes quoted are not the carefully refined products of detailed study, but are approximations.

In their fifth publication on the range finding test, Smyth et al. (22) described the procedures as they had become established over a span of several years:

Single dose oral toxicity for rats is estimated by intubation of dosages in a logarithmic series to groups of five male rats or at times to female rats. The animals are Carworth–Wistar rats, raised in our own colony, fed Rockland rat diet complete, weighing 90 to 120 gm., and not fasted before dosing.

Chemicals are diluted with water, corn oil, or a 1% solution of sodium 3,9-diethyl-6-tridecanol sulfate (Tergitol Penetrant 7) when necessary to bring the volume given one rat to between 1 and 10 ml. Fourteen days after dosing, morbidity is considered complete. The most probable LD_{50} value and its fiducial range are estimated by the method of Thompson using the tables of Weil. When fractional mortality is not observed, the LD_{50} value is recorded without a fiducial range.

Penetration of rabbit skin is estimated by a technique essentially the one-day cuff method of Draize and associates, using groups of four male New Zealand giant albino rabbits weighing 2.5 to 3.5 kg. The fur is closely clipped over the entire trunk, and the dose, retained beneath an impervious plastic film, contacts about $1/10$ of the body surface. Dosages greater than 20 ml. per kilogram of body weight cannot be retained in contact with the skin. After 24 hours' contact the film is removed, and mortality is considered complete after 14 additional days. . . .

Earlier papers . . . have spoken of saturated vapor inhalation. . . . we now speak of concentrated vapor inhalation. This consists in exposing six male albino rats to a flowing stream of air approaching saturation with vapors. The stream is prepared by passing dried air through a fritted disc gas washing bottle initially at room temperature. When carbon dioxide will react with a sample, the air stream is freed of that gas.

Inhalations are continued for periods of time in an essentially logarithmic series with a ratio of two, extending to eight hours, until the period killing half the rats within 14 days of inhalation is defined. The Table records the longest period which allowed all rats to survive 14 days or, in a few instances marked by a symbol, the shortest period tried when this killed all rats.

Inhalation of known vapor concentrations by rats is conducted with flowing streams of vapors prepared by various styles of proportioning pumps. Nominal concentrations are recorded, not confirmed by analytical methods. Exposures are four hours long, although rarely an eight-hour period is used.

Concentrations are in an essentially logarithmic series with a factor of two, and the Table records the concentration yielding fractional mortality among six rats within 14 days. Where no fractional mortality was observed, both the concentration yielding no mortality and that yielding complete mortality are indicated.

Primary skin irritation on rabbits records in a 10-grade ordinal series the severest reaction on the clipped skin of any of five albino rabbits within 24 hours of application of 0.01 ml. of undiluted sample or of dilutions in water, propylene glycol, or kerosene. Grade 1 in the Table indicates the least visible capillary injection from undiluted chemical. Grade 6 indicates necrosis when undiluted, and Grade 10 indicates necrosis from a 0.01% solution.

Eye injury in rabbits records the degree of corneal necrosis from various volumes and concentrations of chemical as detailed by Carpenter and Smyth. Grade 1 in the Table indicates at most a very small area of necrosis resulting from 0.5 ml. of undiluted chemical in the eye; Grade 5 indicates a so-called severe burn from 0.05 ml., and Grade 10 indicates a severe burn from 0.5 ml. of a 1% solution in water or propylene glycol.

Entries in Table 44.4 identified by reference numbers 18 through 25 are derived from reports that use the foregoing methods and codes, except that the fiducial ranges for oral toxicity have not been included.

The warning of the authors should not be overlooked or forgotten: "It should be emphasized that the range-finding test is relied upon only to allow predictions of the comparative hazards of handling new chemicals. Acute toxicity studies, no matter how carefully planned, yield no more than indications of the degree of care necessary to protect exposed workmen and indications that certain technically feasible applications of a chemical may or may not eventually be proved safe."

From an industrial hygiene point of view, the most important action of the amines is the strong local irritation they produce. Animals exposed to concentrated vapors exhibit signs and symptoms of mucous membrane and respiratory tract irritations. Single exposures to near-lethal concentrations and repeated exposures to sublethal concentrations result in tracheitis, bronchitis, pneumonitis, and pulmonary edema. For most of the amines listed in Table 44.5, a single skin application will cause deep necrosis, and a drop in a rabbit's eye results in severe corneal damage or complete eye destruction. These effects are undoubtedly the result of the alkalinity of these compounds, even though a perfect correlation between the K_b values and the degree of skin or eye irritation is not seen. The acute oral toxicities range from moderately high to slight. Some of

Table 44.4. Acute Toxicity of Selected Amines[a]

Name	CAS Number	Acute Oral Toxicity LD$_{50}$ Rat[a] (g/kg)	Acute Skin Toxicity LD$_{50}$ Rabbit[a] (ml/kg)	Inhalation Toxicity (Rats except as Noted)				Rabbit Skin Irritation	Rabbit Eye Irritation	Other Manifestations	Ref.
				ppm	Time	Mortality	"Saturated" Vapor Time for 0 Deaths				
Methylamine	74-89-5	0.1–0.20 (10% soln.)	0.1 ml survived 40% 1 died gp					40% soln. 0.1 ml, necrosis	40% soln. corneal damage		18
Dimethylamine	124-40-3	0.698 / 0.240 r / 0.240 gp							After 50 mg, 5 min		36
Ethylamine	75-04-7	0.4 / 0.4–0.8 (70% soln.)	0.39	8000	4	2/6	2 min; all died	Grade 1 Necrosis from 70% gp	Grade 9		22
									Severe at 50 ppm	Lung irritation severe at 50 ppm	26
Diethylamine	119-89-7	0.54 / 0.649 m	0.82	4000	4	3/6	5 min; all died	Grade 4	Grade 10		21 / 21 / 36
									Severe at 50 ppm	Lung irritation severe at 50 ppm	26
Triethylamine	121-44-8	0.46	0.57	1000 / 1000	4 / 4	1/6 / 2/6 gp		Grade 2	Grade 9 Severe at 50 ppm	Lung irritation severe at 50 ppm	21 / 26
Propylamine	107-10-8	0.57	0.56	8000	4	5/6	2 min	Grade 6	Grade 9		23
Dipropylamine	142-84-7	0.93	1.25 r	1000	4	2/6	5 min	Grade 6	Grade 9		23
Tripropylamine	102-69-2	0.096 r	0.57	250	4	3/6	8 hr	Grade 4	Grade 1		24
Isopropylamine	75-31-0	0.82 r	0.55	4000 / 8000	4 / 4	0/6 / 6/6	2 min	Grade 6	Grade 10		21
Diisopropylamine	108-18-9	0.77		8000	4	2/6	5 min	Grade 1	Grade 8		22
				261 / 597 / 261	b / c / d	4/5 r / 1/2 / 1/2					30

Table 44.4. (Continued)

Name	CAS Number	Acute Oral Toxicity LD50 Rat[a] (g/kg)	Acute Skin Toxicity LD50 Rabbit[a] (ml/kg)	Inhalation Toxicity (Rats except as Noted)				Rabbit Skin Irritation	Rabbit Eye Irritation	Other Manifestations	Ref.
				ppm	Time	Mortality	"Saturated" Vapor Time for 0 Deaths				
Butylamine	109-73-9	0.5	0.5 gp	4000	4	2–4/6	2 min				18 29
Dibutylamine	11-92-2	0.55	1.01	250	4	0/6		Grade 5	Grade 9		22
Tributylamine	102-89-9	0.54	0.25	75	4	4/6	50 min	Grade 4	Grade 1		25
Diisobutylamine	110-96-3	0.258 m 0.629 m 0.620 gp									36
Amylamine, mixed isomers	110-58-71	0.47	0.65	2000	4	4/6	30 min	Grade 6	Grade 9		24
Pentyl-1-pentamine, (dipentylamine)	2050-92-2	0.27	0.35	63	4	4/6	30 min	Grade 6	Grade 5		23
2,2'-Diethyldihexylamine	106-20-7	1.64					8 hr	Grade 5	Grade 8		20
Allylamine (propenylamine)	107-11-9	0.106	0.035	LC50		286 ppm	4 hr	Extreme irritation	Extreme irritation	Chronic inhalation effect on liver at 5 ppm	27
Diallylamine	24-02-7	0.578	0.356	LC50		2755	4 hr	Severe irritation	Severe irritation	Chronic inhalation effects on several organs at 50 ppm, deaths at 200 ppm	27
Triallylamine	1102-75-5	1.31	2.25	LC50		828 ppm	4 hr	Mild irritation	Mild irritation	Chronic inhalation effects on several organs at 50 ppm, deaths at 200 ppm	27
2-Ethylbutylamine	617-79-8	0.39	2.00	2200	4	0/6	422 min	Grade 6	Grade 9		22
Cyclohexylamine	108-91-8									Human skin severely irritated by 125 mg 48 hr.	36
Dicyclohexylamine	101-83-7	0.71 0.373	0.32	4000	4	0/6		Grade 7 Severe irritation	Grade 10		24 36

Compound	CAS No.				Exposure	Mild irritation	Severe irritation	Ref.
2-Aminoethanol (ethanolamine)	141-43-5	2.1						36
2,2'-Iminodiethanol (diethanolamine)	111-42-2	10.20; 12.76						32; 36
2,2',2''-Nitrilotriethanol (triethanolamine)	102-71-6	8.0						33; 34
Isopropanolamine (1-amino-2-propanol)	78-96-6	4.26; 5.24	1.61		8 hr	Mild irritation Grade 3	Severe irritation Grade 5	18; 20; 33
Triisopropanolamine	122-20-3	1.52 gp; 6.5					Grade 6	33; 19; 28; 33
2-Methylaminoethanol	109-83-1	6.50; 2.34	1.37		8 hr	Grade 3	Grade 9	33; 22
2-Dimethylaminoethanol	109-01-0	2.31			8 hr	Grade 1	Grade 8	21
2-Ethylaminoethanol	110-73-6	1.48; 1.0	0.36		8 hr	Grade 2	Grade 9	22
2-Diethylaminoethanol	100-37-8	1.3	1.08 gp		4 hr	Grade 6	Grade 5	35
2-Dibutylaminoethanol	102-81-8	1.07	1.68		8 hr	Grade 2	Grade 8	18
2,2'-(Methylimino) diethanol	105-59-9	4.57			8 hr	Grade 2	Grade 8	22
2,2'-(Ethylimino) diethanol	139-87-7	4.25			8 hr	Grade 5	Grade 9	22
3-Amino-1-propanol	156-87-6	2.83	1.25		8 hr	Grade 6	Grade 8	23
Ethylenediamine	107-15-3	1.46	0.73	2000/8/0-6; 4000/8/6-6; 255/7×30/11-15		200 ppm, 10 sec mild eye irritation in humans; 400 ppm intolerable		21
		0.47 gp						22; 22
1,3-Propanediamine	109-76-2	0.35	0.20		8 hr	Grade 7	Grade 8	28
1,2-Propanediamine (propylenediamine)	78-90-0	2.23	0.50		8 hr	Grade 6	Grade 9	23; 22
1,3-Butanediamine	590-88-5	1.35	0.43		8 hr	Grade 6	Grade 9	21
Diethylenetriamine	111-40-0	2.33; 1.08	1.09		8 hr	Grade 6	Grade 8	20; 31
Diethylenediamine piperazine	110-85-0						Severe irritation Grade 6	28
Triethylenetetramine	112-24-3	4.34	0.82		4 hr	Grade 6	Grade 5	20

a = Except as noted. r = rabbit; gp = guinea pig; m = mouse.
b = 7 hr/day, 2 to 20 days.
c = 7 hr/day, 5 days.
d = 7 hr/day, 19 days.

the effects observed result from the local corrosive action of the bases in the gastrointestinal tract. The salts are less irritating, and therefore, less toxic by mouth. They also show less skin and eye irritation when applied as solutions, as shown by Fassett (37).

Interest in the pharmacology of the simple aliphatic amines was initially stimulated by their structural relationship with epinephrine (adrenalin).

Barger and Dale (38) introduced the term sympathomimetic to describe effects similar to those of epinephrine, which serve to stimulate the sympathetic branch of the autonomic nervous system, leading to elevation of blood pressure, contraction of smooth muscle, salivation, and dilatation of the pupil of the eye. They found that aliphatic amine hydrochlorides, given intravenously, caused increasing blood pressure responses with increasing carbon chain length up to C_6. Branched chain members were less active. There was decreasing sympathomimetic activity above C_7, with increasing cardiac depression. Studies on the influence of structure on epinephrinelike activity indicate that primary amines have somewhat more pressor activity than secondary and tertiary; straight chains are more active than branched; a second amine group in the chain increases this activity; and an amine group on the second carbon gives maximum

Table 44.5. Relative Toxicity of Amine Bases and Their Hydrochloride Salts in Aqueous Solutions (37)

	Base[a]		Hydrochloride Salt[b]	
Amine	Approximate Oral LD_{50}, Rats (g/kg)	Eye irritation, Rabbit, 1 Drop	Approximate Oral LD_{50}, Rats (g/kg)	Eye irritation, Rabbit, 1 Drop
Methylamine	0.1–0.2 (40%)	Immediate, severe (40%)	1.6–3.2 (40%)	Mild, normal in 24 hr (40%)
Ethylamine	0.4–0.8 (70%)	Immediate, severe (70%)	>3.2 (10%)	Moderate, normal at 14 days (70%)
n-Propylamine	0.2–0.4 (10%)	Immediate, severe (undiluted)	3.2–6.4 (25%)	Mild, normal in 24 hr (crystals)
n-Butylamine	0.2–0.4 (10%)	Immediate, severe (undiluted)	1.6–3.2 (10%)	Moderate, normal at 14 days (crystals)
Ethylenediamine	0.7–1.4 (85%)	Severe (undiluted)	1.6–3.2 (10%)	Mild, normal at 14 days (crystals)

[a] Bases = vol/vol.
[b] Salts = wt/vol.

activity (40, 42). It has been generally observed that when amines are administered repeatedly, cardiac stimulation is replaced by vasodilatation and cardiac depression (42). Convulsions frequently occur after near-fatal doses.

Simple aliphatic amines can be both inhibitors and substrates of amine oxidases. Aliphatic amines can cause the release of histamine and can potentiate its action. The monoamines produce a typical histaminelike "triple response" (white vasoconstriction, red flare, wheal) in human skin at concentrations sufficient to cause release of histamine from guinea pig lung. Maximum histamine release from the series $C_nH_{2n+1}NH_2 \cdot HCl$ is at C_{10}. Straight chain diamines show increasing histamine release from C_6 to about C_{14}. Potent histamine releasors, such as Compound 48/80 and octylamine, cause a decrease in blood pressure, tachycardia, headache, itching, erythema, urticaria, and facial edema, when administered intravenously in man, as does histamine. It is possible that histamine-releasing agents will produce bronchoconstriction and wheezing by inhalation since histamine aerosol has this effect (43).

Changes in the lungs, liver, kidneys, and heart have been observed in pathological studies. Pulmonary edema has been produced, with hemorrhage and bronchopneumonia, nephritis, liver degeneration, and degeneration of heart muscle in rabbits repeatedly exposed to ethylamines (26). With the exception of myocardial degeneration, similar effects occur in animals exposed to ethylenediamine (44). Myocardial damage has been described after vapor exposures to allylamines (27). Tabor and Rosenthal (45) have shown that spermine (diaminopropyltetramethylenediamine) has a high degree of nephrotoxicity. Monoamines (C_1–C_{10}), diamines (C_4–C_{10}), and diethylenetriamine, triethylenetriamine, and tetraethylenepentamine were inactive. Ethylenediamine, ethyleneimine, 1,3-diaminopropane, and 1,2-diaminopropane produced proteinuria and tubular damage of a lesser degree than spermine.

5 PHYSIOLOGICAL AND PATHOLOGICAL EFFECTS IN MAN

The majority of studies of the effects of aliphatic amines have related to local action which is primarily irritative and sensitizing. Vapors of the volatile amines cause eye irritation with lacrimation, conjunctivitus, and corneal edema, which result in "halos" around lights (26, 46, 47). Inhalation causes irritation of the mucous membranes of the nose and throat, and lung irritation with respiratory distress and cough. The vapors may also produce primary skin irritation and dermatitis (28). Direct local contact with the liquids is known to produce severe and sometimes permanent eye damage, as well as skin burns. Cutaneous sensitization has been recorded, chiefly due to the ethyleneamines. Systemic symptoms from inhalation are headache, nausea, faintness, and anxiety. These systemic symptoms are usually transient and are probably related to the pharmacodynamic action of the amines.

Eckardt (48) reported that a "variety of dermatitises and acute and chronic pulmonary problems have been associated with the use of two plastics, namely epoxides and polyurethanes, especially polyurethane foams. In the case of epoxy resins, these effects have been associated with the curing agents namely ethylenediamine, diethylenetriamine and triethylenetriamine. These are highly alkaline compounds capable of causing extensive corrosive skin reactions. They are also known as skin sensitizers, and hence may cause allergic skin reactions. Because the reaction is exothermal, fumes may be produced which in sensitized individuals can lead to bronchial asthma. Since the curing agents may be used in' excess, in order to drive the polymerization to completion, subsequent grinding, sanding, or polishing of expoxy resins may produce dusts and fumes, with all of the above-described reactions occurring."

Brubaker et al. (49) described a dose-related reduction of several pulmonary function test results, and symptoms of cough, phlegm, wheezing, and chest tightness associated with exposure to 3-(dimethylamino)propylamine at 0.9 ppm or lower average concentration. Pulmonary functions declined during the course of a 10-hr work shift. Improved ventilation reduced the exposure level to 0.13 ppm. After that, the workers showed improvement of pulmonary functions during a work shift and lower respiratory symptoms disappeared.

6 THRESHOLD LIMIT VALUES AND MAXIMUM PERMISSIBLE EXPOSURES

Of the more than 50 compounds mentioned in this chapter, only 14 have had a Threshold Limit Value established. This is due primarily to the lack of industrial experience, too few chronic animal studies, and the scarcity of reliable information about actual exposures when physiological or toxic effects in humans have been fortuitously observed.

Those compounds for which Threshold Limit Values have been established by the American Conference of Industrial Hygienists by 1979 appear in Table 44.6. There are three categories of Threshold Limit Values (TLVs), as follows:

1. Threshold Limit Values—Time-Weighted Average (TLV-TWA): the Time-Weighted average concentration for a normal 8-hr workday or 40-hr work week to which nearly all workers may be repeatedly exposed, day after day, without adverse effect.

2. Threshold Limit Value—Short-Term Exposure Limit (TLV-STEL): the maximal concentration to which workers can be exposed for a period up to 15 min continuously without suffering from (1) irritation, (2) chronic or irreversible tissue change, or (3) narcosis of sufficient degree to increase accident proneness, impair self-rescue, or materially reduce work efficiency, provided that no more than four excursions per day are permitted, with at least 60 min between exposure periods, and provided that the daily TLV-TWA also is not exceeded.

3. Threshold Limit Values—Ceiling (TLV-C): the concentration that should not be exceeded even instantaneously.

For further information see *TLVs, Threshold Limit Values for Chemical Substances in Workroom Air Adopted by the ACGIH for 1981*, published by the American Conference of Governmental Industrial Hygienists, Cincinnati, Ohio. Also listed in Table 44.6 are the permissible exposure levels, established by the Occupational Safety and Health Administration (OSHA), and the levels "immediately dangerous to life or health," designated by The Standards Completion Program, a joint program of OSHA and the National Institute for Occupational Safety and Health (NIOSH). Some values attributed to the USSR are also listed.

7 SPECIFIC COMPOUNDS

7.1 Aliphatic and Alicyclic Monoamines

The physical properties of the monoamines are given in Table 44.2. Acute animal toxicity data are summarized in Table 44.4.

7.1.1 Methylamines

The methylamines are supplied commercially as aqueous solutions in concentrations from 25 to 40 percent. Methylamines are used in tanning and organic syntheses. Dimethylamines are used as accelerators in vulcanizing rubber, tanning, and the manufacture of soaps.

Secondary amines are prone to react with nitrite to form nitrosamines (see Chapter 43), some of which are potent animal carcinogens. Friedman et al. (50) found that oral administration of sodium nitrite within 1 hr of oral dimethylamine to mice yielded a marked dose-dependent inhibition of liver nuclear RNA synthesis. No inhibition of nuclear RNA synthesis was observed when sodium nitrite was given 30 min prior to the dimethylamine. This study suggests the possibility of in vivo biosynthesis of carcinogenic nitrosamines following ingestion of nitrites and secondary amines present in food.

Coon et al. (51) tested the toxicity of dimethylamine in rats, guinea pigs, rabbits, monkeys, and dogs, using the inhalation route. Animals continuously exposed to as little as 9 mg/m^3 of dimethylamine showed mild inflammatory changes, primarily in the lungs.

The odor of methylamine is faint but readily detectable at less than 10 ppm, becomes strong at from 20 to 100 ppm, and becomes intolerably ammoniacal at 100 to 500 ppm. Olfactory fatigue occurs readily. Brief exposures to 20 to 100 ppm produce transient eye, nose, and throat irritation. No symptoms of irritation are produced from longer exposures at less than 10 ppm. Additional

Table 44.6. Permissible Exposure Levels, Threshold Limit Values, and Maximal Allowable Concentrations of Some Aliphatic and Alicyclic Amines

Name	Permissible Exposure Levels[a] and Levels Immediately Dangerous to Life or Health,[b] U.S.		Threshold limit Values Adopted by ACGIH[c]				Maximal Allowable Concentrations Adopted by USSR[d] (mg/m)
			Adopted Values, TWA		Tentative Values, STEL		
	PEL (ppm)	IDLH (ppm)	(ppm)	(mg/m)	(ppm)	(mg/m)	
Methylamine	10	100	10	12	—	—	5
Dimethylamine	5	2000	10	18	—	—	1
Ethylamine	10	4000	10	18	—	—	—
Diethylamine	25	2000	25	75	—	—	—
Triethylamine	—	—	25	100	40	160	—
Isopropylamine	5	4000	5	12	10	24	—
Propylamine	—	—	—	—	—	—	5
Diisopropylamine	5	1000	5	20 (skin)	—	—	—

Butylamine	5	2000	5	15 (skin)	—	—	10
Cyclohexylamine	—	—	10	40 (skin)	—	—	1
Ethylenediamine	10	2000	10	25	—	—	2
Diethylenetriamine	—	—	1	1 (skin)	—	—	—
Ethanolamine	3	1000	3	8	6	15	—
Ethyleneimine	—	—	0.5	1 (skin)	—	—	—
Propyleneimine	2	500	2	5 (skin)	—	—	—
N-Nitrosodimethylamine	—	—	—	A2 (skin)	—	A2	—

Key: TWA = Time-weighted average
STEL = Short-term exposure limit
(Skin) = Hazardous by skin absorption
A2 = Industrial substances suspect of carcinogenic potential for man

[a] Regulations of Occupational Safety and Health Administration, CFR 1910.1000, January 1, 1977.
[b] Defined by Standards Completion Program, a joint program of OSHA and NIOSH, and described in *Respiratory Protection Reference Document for Chemical Hazards.* [a] and [b] from NIOSH/OSHA *Pocket Guide to Chemical Hazards*, 1978.
[c] *Threshold Limit Values for 1979*, American Conference of Governmental Industrial Hygienists.
[d] *Encyclopedia of Occupational Health and Safety*, International Labour Office, Geneva, 1971.

information concerning odor and the amines was reported by Amoore and Furrester (52). They stated that about 7 percent of humans are unable to smell (are anosmic to) trimethylamine. Odor threshold measurements on 16 aliphatic amines were made with panels of specific anosmics and normal observers. The anosmia was most pronounced with low molecular weight tertiary amines, but was also observed in lesser degree with primary and secondary amines. This specific anosmia apparently corresponds with the absence of a new olfactory primary sensation, the fishy odor.

7.1.2 Ethylamines

Ethylamines are used in resin chemistry, as stabilizers for rubber latex, as intermediates for dyestuffs and pharmaceuticals, and in oil refining. Ethylamines are primarily irritants to the eyes, mucous membranes, and the skin. Brieger and Hodes (26) exposed rabbits repeatedly to measured concentrations of ethylamine, diethylamine, and triethylamine in air. The three amines produced lung, liver, and kidney damage at 100 ppm. The triethylamine produced definite degenerative changes in the heart at 100 ppm, whereas this was an inconstant finding with the other two; 50 ppm of these amines was sufficient to produce lung irritation and corneal injury (delayed until 2 weeks with ethylamine).

It has been reported (53a) that, "Exposure of rats to 8000 ppm for four hours was fatal. Rabbits survived exposures to 50 ppm daily for six weeks but showed pulmonary irritation and some myocardial degneration; corneal damage was observed after two weeks of exposure. In the rabbit eye, one drop of 70% solution of ethylamine caused immediate, severe irritation. A 70% solution dropped on the skin of guinea pigs caused prompt skin burns leading to necrosis; when held in contact with guinea pig skin for two hours, there was severe skin irritation with extensive necrosis and deep scarring."

7.1.3 Propylamines

Isopropylamine and diisopropylamine are used in chemical syntheses of dyes and pharmaceuticals.

n-Propylamine (1-aminopropane) and di-*n*-propylamine show approximately the same degree of toxicity in animals. See Table 44.4 for a summary of acute toxicity data.

Humans exposed briefly to isopropylamine at 10 to 20 ppm experienced irritation of the nose and throat. Workers complained of transient visual disturbances (halos around lights) after exposure to the vapor for 8 hr, probably owing to mild corneal edema, which usually cleared within 3 to 4 hr. The liquid can cause severe eye burns and permanent visual impairment. Isopropylamine in either liquid or vapor form is irritating to the skin and may cause burns; repeated lesser exposures may result in dermatitis (53b).

Diisopropylamine is an eye irritant in humans; it is a pulmonary irritant in animals, and severe exposure is expected to produce the same result in humans. Workers exposed to concentrations between 25 and 50 ppm complained of disturbances of vision described as "haziness." In two instances, there were also complaints of nausea and headache. Prolonged skin contact is likely to cause dermatitis. Exposure of several species of animals to 2207 ppm for 3 hr was fatal; effects were lacrimation, corneal clouding, and severe irritation of the respiratory tract. At autopsy, findings were pulmonary edema and hemorrhage (53c).

7.1.4 Butylamines

Butylamines are used in pharmaceuticals, dyestuffs, rubber, chemicals, emulsifying agents, photography, desizing agents for textiles, and pesticides. Acute toxicity data for animals are shown in Table 44.4.

n-Butylamine (1-aminobutane) at measured concentrations of 3000 to 5000 ppm produces an immediate irritant response, labored breathing, and pulmonary edema, with death of all rats in minutes to hours. Ten and 50 percent v/v aqueous solutions and the undiluted base produce severe skin and eye burns in animals. The immediate skin and eye reactions are not appreciably altered by prolonged washing or attempts at neutralization when these are commenced within 15 sec after application. Direct skin contact with the liquid causes severe primary irritation and deep second-degree burns (blistering) in humans. The odor of butylamine is slight at less than 1 ppm, noticeable at 2 ppm, moderately strong at 2 to 5 ppm, strong at 5 to 10 ppm, and strong and irritating at concentrations exceeding 10 ppm. Workers with daily exposures of from 5 to 10 ppm complain of nose, throat, and eye irritation, and headaches. Concentrations of 10 to 25 ppm are unpleasant to intolerable for more than a few minutes. Daily exposures to less than 5 ppm (most often between 1 and 2 ppm) produce no complaints or symptoms (37).

7.1.5 Amylamines (Pentylamine, 1-Aminopentane)

Direct contact with the liquid causes first- and second-degree burns of the skin. Inhalation results in irritation of the respiratory tract and mucous membranes (37). Isoamylamine (1-amino-3-methylbutane) shows pressor activity in humans. It produces flushing and apprehension when injected intravenously. The reaction is mild at 83 mg/kg (11, 14). It stimulates salivary and lacrimal secretions and smooth muscle (38). Hartung reports that 250 mg/kg is not toxic to rabbits, 1.5 g of the sulfate kills rats, and 1.8 g of the hydrochloride kills rabbits (39).

7.1.6 Hexylamines

n-Hexylamine (1-aminohexane) is an irritant for the skin, eyes, and mucous membranes. It is an active pressor agent (38). The vapor toxicity is somewhat higher than that of the butylamines and skin irritation is at least as great.

2-Ethylbutylamine (1-amino-2-ethylbutane, isohexylamine) exhibits acute toxicity for animals that is roughly equivalent to that of n-hexylamine. See Table 44.4.

7.1.7 Heptylamines

Dunker and Hartung determined the acute intraperitoneal toxicity of four primary aminoheptanes for mice. The LD_{50} ranged from 110 to 60 mg/kg in the order 4-aminoheptane, 1-aminoheptane, 3-aminoheptane, and 2-aminoheptane (54). In rats, the sulfate has an intraperitoneal LD_{50} of 42 mg/kg. It produces a sustained elevation of blood pressure in dogs. Some depressant activity and vasodilatation appear after repeated doses. In humans 2 mg/kg by mouth results in palpitation, dry mouth, and headache, with slight rise in blood pressure (55). It has been used as a nasal vasoconstrictor (2-aminoheptane, Tuamine®).

Di-n-heptylamine has an approximate oral LD_{50} for rats (undiluted) of 0.2 to 0.4 g/kg. The approximate oral LD_{50} for mice (5 percent in corn oil) is 0.2 to 0.4 g/kg. Deaths occur within a few minutes with dyspnea and convulsions. One drop of the undiluted amine causes strong irritation of the eye and surrounding tissues and permanent corneal damage. It is a strong primary skin irritant (37).

7.1.8 Higher Alkylamines (C_8–C_{18})

Little toxicity information is available on amines with alkyl chains containing 8 to 18 carbons. Smyth, Carpenter, et al. published range-finding toxicity data on about three dozen amino compounds, including a number of large, complex molecules (24, 25). The evidence suggests that these higher alkylamines would be strong local irritants for eyes, skin, and mucous membranes. Their low vapor pressure should decrease the hazard from vapor exposures.

2-Aminooctane produces elevation of blood pressure in dogs at 1 mg/kg. The minimum lethal dose by injection in 0.135 g/kg in mice. Lethal doses result in dyspnea, excitation, convulsions, and death in respiratory paralysis (56).

2-Ethylhexylamine exhibits a high degree of acute vapor toxicity for rats and is a potent skin and eye irritant (Table 44.4). Di-(2-ethylhexyl)amine (dioctylamine) shows a lower degree of oral toxicity for rats and is slightly less irritating to the skin and eye. A single concentrated vapor exposure produces no deaths in 8 hr (Table 44.4).

Laurylamine (dodecylamine) is classified by Fleming et al. with compounds that produce severe burns and vesication of the skin (57).

Octadecylamine has been studied by Deichmann et al. (58) in connection with its possible use as an anticorrosive agent in live steam that could be used to cook food. Rats fed levels of 0 to 500 ppm in the diet for 2 years showed no detectable effects on growth, food consumption, hematology, or microscopic pathology. At 3000 ppm there was anorexia, weight loss, and some histological changes in the gastrointestinal tract, mesenteric nodes, and liver. The acute oral LD_{50} for mice and rats is approximately 1 g/kg. It is a primary skin irritant (59).

7.1.9 Allylamines

The allylamines (aminopentenes) are the only unsaturated alkylamines that have been studied. Unlike the lower saturated alkylamines, the secondary and tertiary allylamines are less toxic than monoallylamines. Hart performed exploratory inhalation studies in mice using very heavy dosages of allylamine and diallylamine; almost all the animals died within 10 min at concentrations of 1.27 and 0.88 mM, respectively (2.7 and 1.9 percent). The experimenters experienced transient irritation of the mucous membranes of the nose, eyes, and mouth, with lacrimation, coryza, and sneezing after accidental exposure to an unspecified concentration of the vapor of allylamine (60, 61). The acute toxicity of the allylamines in animals as reported by Hine et al. (27) is summarized in Table 44.7. He found decreasing acute and percutaneous

Table 44.7. Toxicity of Allylamines for Rats (27)

	Monoallylamine	Diallylamine	Triallylamine
Oral LD_{50}, mg/kg	106	578	1310
Percutaneous LD_{50} (rabbit, mg/ kg)	35	356	2250
Inhalation LC_{50}, ppm[a]			
4 hr	286	2755	828
8 hr	177	795	554
Repeated inhalation, 7 hr × 50, ppm[b]			
Change in liver or kidney weights	5	200	100
Reduced growth	10	200	200
Deaths	40	200	200[c]

[a] Calculated.
[b] Measured.
[c] $1/15$ occurred at 100 ppm.

toxicity from monoallylamine to triallylamine. However, triallylamine showed increased relative toxicity as compared to diallylamine on inhalation. Both mono- and diallylamines were severely irritating to skin and triallylamine was mildly irritating. Acute exposures to the vapors produced symptoms and findings of respiratory tract irritation. The pathological changes observed in rats following repeated inhalation included chemical pneumonias and some liver and kidney damage. The most prominent pathological effect described was myocarditis, which occurred after repeated inhalation of all three allylamines. Experimental human exposures gave the following results (27).

1. Monoallylamine: recognizable at 2.5 ppm, mucous membrane irritation and chest discomfort in some persons at 2.5 ppm, intolerable to most at 14 ppm.

2. Diallylamine: recognizable but not unpleasant at 2 to 9 ppm, mucous membrane irritation and chest discomfort in a few subjects at 22 ppm, not intolerable at 70 ppm.

3. Triallylamine: recognizable at 0.5 ppm, mucous membrane irritation or chest discomfort in some at 12.5 ppm, increasingly frequent symptoms to 50 ppm, irritant symptoms more severe at 75 to 100 ppm, with unpleasant systemic symptoms including nausea, vertigo, and headache.

Calandra (62) conducted a 1-year chronic vapor inhalation study with monoallylamine in which rats, rabbits, and dogs were exposed for 8 hr/day, five days/week, to 5 ppm or 20 ppm. No adverse effects on growth, behavioral reactions, or abnormal blood or urine changes were observed. Deaths from pneumonia occured in three of six rabbits exposed to 20 ppm. Lung changes consistent with chronic irritation were found at both exposure levels. However, no myocardial damage was found in rabbits or dogs and only a few rats showed slight changes, which were not considered different from those expected in unexposed rats. Periodic liver and kidney function tests, transaminase determinations, and electrocardiographic examinations of dogs did not reveal any abnormalities. Congestive changes in the liver and kidneys were noted in dogs at both exposure levels.

7.1.10 Cyclohexylamines

General. Cyclohexylamine (aminocyclohexane) is used as a corrosion inhibitor in boiler water, as a rubber accelerator, and as an intermediate in chemical syntheses. Its acute animal toxicity is summarized in Table 44.4.

Cyclohexylamine produced convulsant deaths in rabbits when injected in olive oil at doses of 0.5 g/kg. When it was given daily for 82 days in the drinking water at 100 mg/kg, pathological findings or weight loss appeared in rabbits, guinea pigs, and rats (63).

Watrous and Schulz (64) exposed rabbits, guinea pigs, and rats to cyclohexylamine vapors 7 hr/day, 5 days/week, at average concentrations of 1200, 800, and 150 ppm. At 1200 ppm, all animals except one rat showed extreme irritation and died after a single exposure. Fractional mortality occurred after repeated exposure at 800 ppm. At 150 ppm, four of five rats and two guinea pigs survived 70 hr of exposure, but one rabbit died after 7 hr. The chief effects were irritation of the respiratory tract and eye irritation with the development of corneal opacities. No convulsions were observed.

Watrous and Schultz (64) also reported three cases of transitory systemic toxic effects from acute accidental industrial exposures. The symptoms were light-headedness, drowsiness, anxiety and apprehension, and nausea. Slurred speech, vomiting, and pupillary dilatation occurred in one case. Operators exposed to 4 to 10 ppm had no symptoms (64). In human patch tests a 25 percent solution produced severe skin irritation and possible skin sensitization (65). Tests in guinea pigs did not give evidence of skin sensitization (37).

Khera and Stoltz (66) observed that male rats given 220 mg/kg/day of cyclohexylamine, a dose that had no apparent effect on growth or behavior, fathered significantly fewer litters than control rats, and the litters were smaller.

Kroes et al. (67) reported that cyclohexylamine led to growth retardation, especially in females, in the earlier of six generations of mice exposed to 0.5 percent of the compound in their food. Also, the pregnancy rate, the number of live-borne fetuses, the number of postnatal survivors, and the body weight of the offspring were all affected unfavorably, and the proportion of male offspring was diminished.

Oser et al. (68) concluded from an extensive study of rats exposed to cyclohexylamine, in doses up to 150 mg/kg/day, "Except for some nonprogressive growth retardation in the higher dosage groups, due to lower food consumption, the physical and clinical observations in the test groups fell substantially within normal limits and were not significantly different from the untreated controls." However, a significant incidence of testicular atrophy and reduction of litter size were seen at the highest dosage level.

Gaunt et al. (69) fed rats diets containing 600, 2000, or 6000 ppm of cyclohexylamine chloride for 13 weeks. At the highest dosages, there was a reduced rate of weight gain, not fully explained by diminished food intake. No specific organ changes were observed except that the testes showed reduced spermatogenesis. Reduction of testis weight was evident. However, the rats remained fertile and their offspring appeared normal.

Gaunt et al. (70) used the same dosages for 2-year studies in rats. They observed a slight anemia, failure to produce normally concentrated urine, and an increase in the number of animals with foamy macrophages in the pulmonary alveoli at the highest dosage level. Decreased food intake caused the lessened body weight gain and organ weights as compared to the controls. Animals that received 2000 or 6000 ppm showed testicular atrophy or tubules with few spermatids. The no-untoward-effect level in both of these studies was 600 ppm, equivalent to an intake of about 30 mg/kg per day.

Hardy et al. (71), working in the same laboratory, gave mice 300 to 3000 ppm of cyclohexylamine for 80 weeks. Except for some depression of weight gain in the males and minor hepatitis in the females at 3000 ppm, they saw no untoward effects; there was no testicular atrophy or degeneration.

There are several reports regarding the cardiac and blood pressure effects of cyclohexylamine. Eichelbaum et al. (72), in a study of humans, found a close correlation between plasma levels of cyclohexylamine and arterial blood pressure. However, it was some orders of magnitude less potent than related sympathomimetic substances.

Gondry (73) reported that cyclohexylamine has a diabetogenic effect when fed to rats at a level of 1 percent in the diet. It inhibits the growth of mice at a level of 0.1 percent, and shows marked growth inhibition at higher doses, up to 1 percent. Continued over six generations, growth inhibition persists, but a normal growth pattern is resumed when cyclohexylamine administration is discontinued.

Dicyclohexylamine appears to be somewhat more toxic than cyclohexylamine. Symptoms and death appear earlier in rabbits after injection of 0.5 g/kg. Doses of 0.25 g/kg are just sublethal, causing convulsions and temporary paralysis. It is a skin irritant (64).

N,N-Dimethylcyclohexylamine is finding use in polyurethane plastics, textiles, and as a chemical intermediate. Acute toxicity tests show that it is somewhat less irritating than cyclohexylamine and less toxic on oral and intraperitoneal administration in rats and mice. The symptoms produced by effective doses are similar: weakness, tremor, salivation, gasping, and convulsions. Inhalation of either compound causes respiratory irritation. Repeated skin applications of the diluted amine (1 percent) do not give evidence of sensitization in guinea pigs (37). Mason and Thompson (74) fed cyclohexylamine to rats in concentrations of 600, 2000, and 6000 ppm for 90 days. Only at the highest dosage did they observe testicular atrophy and reduction of spermatogenesis, one strain of rats being affected more than another. Mice exposed to cyclohexylamine for 80 weeks and dogs exposed for $8\frac{1}{2}$ years showed no testicular lesions. They suggest that ". . . the seminiferous epithelium of the rat is unusually susceptible to chemically-induced damage."

Carcinogenesis.   Suspicion that cyclohexylamine might be carcinogenic arose because it was known to appear in the urine of dogs and men after ingestion of the artificial sweetener cyclamate (75). Cyclamate was suspected of being carcinogenic and it appeared probable that its derivative, cyclohexylamine, was the active agent.

In a 2-year study of chronic toxicity of cyclohexylamine in rats, groups of 25 males and 25 females were given doses of 0, 0.15, 0.5, or 15 mg/kg per day.

During the first year, there was only a slight depression of weight gain in the males of the high dose group; no other signs of toxicity were observed. At the end of 2 years, eight males and nine females were alive in the high dose group; there were 13 to 16 survivors in each of the other three groups. No drug-related changes were found in any of the organs except the urinary bladder. An invasive transitional cell carcinoma was found in the bladder in one male. Spontaneous bladder tumors were very rare in the strain of rats used (76).

In another experiment, rats were fed a mixture of cyclamate and saccharin in their food. Many of the rats were found to convert cyclamate to cyclohexylamine. After 78 weeks, half of the animals were given supplementary feedings of cyclohexylamine in doses from 25 to 125 mg/kg per day. Among 240 rats receiving 2500 mg/kg per day of cyclamate–saccharin mixture, seven males and one female showed papillary tumors of the bladder; all but one of these rats had been shown to convert cyclamate to cyclohexylamine. Three of the animals had received additional cyclohexylamine (76).

Legator et al. (77) gave rats intraperitoneal injections of 1 to 50 mg of cyclohexylamine daily for five days. On the sixth day, colcemid was injected (to arrest cell division) and the animals were killed. Cell preparations from bone marrow cells and spermatogonia were examined for chromosome breaks, with the following results:

Percent of Cells with Chromosome Breaks	Cyclohexylamine (mg/kg per day)					
	0	5	50	100	200	250
Spermatogonia	1.8	4.4	7.6	11.2	16.2	19.2
Bone marrow	2.72	4.0	5.12	8.0	12.16	16.28

The authors concluded, "Our observations indicate potential mutagenesis, carcinogenic, or teratogenic effects that have yet to be determined."

Turner and Hutchinson (78) injected 50 to 250 mg/kg of cyclohexylamine into fetuses of unborn sheep for periods of 5 or 18 hr, and examined cultures of fetal lymphocytes harvested after 48 and after 68 hr. The percentage of fetal lymphocyte aberrations was increased by cyclohexylamine treatment, and a relatively high frequency of chromosomal abnormalities was observed. The incidence of aberrations was four or five times greater with the 18-hr than with the 5-hr infusion. A dose-related inhibition of cell growth was observed, but there was no evidence in bone marrow preparations of any overt damage to cellular organelles, membranes, or achromatic apparatus.

Several investigations with negative results for carcinogenicity, mutagenesis, or teratogenesis of cyclohexylamine have been reviewed by Cooper (79). Lorke and Machemer fed 0.11 percent cyclohexylamine to mice for 10 weeks or longer, and saw no effect on appearance, weight gain, or behavior, and neither fertility nor fetal loss was affected. The calculated effective dose was 136 mg/kg per day. Dick et al. found no effect on chromosome analysis of bone marrow specimens taken after five days of dosage with 50 mg/kg per day of cyclohexylamine. Dick et al. also looked for chromosome damage in humans exposed to cyclohexylamine and found none. Knapp et al. treated fruit flies with a heavy dosage of cyclohexylamine and found no mutagenic effects.

Oser et al. (68) gave cyclohexylamine in dosages of 15 to 150 mg/kg per day to five generations of mice. There was some mucosal thickening of bladder walls and renal calcifications at the highest dose, but no tumors were seen and there was no evidence of teratogenicity or mutation. Gaunt et al. (70) fed rats diets containing up to 6000 ppm of cyclohexylamine for 2 years and saw no evidence of carcinogenesis.

Hardy et al. (71) used diets containing up to 3000 ppm for 80 weeks and saw no evidence of carcinogenicity in mice. Kroes et al. (67), in their six-generation study with cyclamate, saccharin, and cyclohexylamine combined, found no evidence of carcinogenicity or teratogenicity. Becker and Gibson (80) injected pregnant mice with cyclohexylamine on the eleventh day of pregnancy, in doses of 61, 77, or 122 mg/kg. On day 19, 190 fetuses were examined; few anomalies were found, and their incidence was not dose-related nor significantly different from controls. Lorke and Machemer (81) looked for dominant lethal mutations in mice after treatment with 102 mg/kg per day of cyclohexylamine (as sulfate) for 5 days, and found no indication of mutagenic action.

Cattanach reported a very extensive study of cyclohexylamine, cyclamate, and saccharin (82) that concluded that none of the compounds could be considered mutagenic.

7.2 Aliphatic Polyamines

The physical properties of the aliphatic polyamines are given in Table 44.3. Acute animal toxicity data are in Table 44.4.

7.2.1   General

Introduction of a second amine group into the alkyl radical tends to decrease systemic toxicity. The shorter chain diamines have a sympatholytic effect upon the blood pressure rather than a sympathomimetic effect. Ethylenediamine, tetramethylenediamine (1,4-butanediamine, putrescine), and pentamethylenediamine (1,5-pentanediamine, cadaverine) cause depression of the blood pressure in animals. The longer chain diamines may exhibit sympathomimetic activity, for example, hexadecylmethylenediamine (13). The histamine-releasing activity of the diamines is slight at C_4 (tetramethylenediamine) and increases to a maximum at C_{10} (1,10-decanediamine). There is increasing toxicity to paramecia from C_6 to C_{15} (83). The diamines are strong bases and exhibit skin and eye irritant properties similar to the monoamines. In some cases (ethylenediamine, hexamethylenediamine) they exhibit skin sensitization properties not experienced with the corresponding monoamines. They are absorbed through the skin. The acute percutaneous toxicity is often approximately equivalent to that of the corresponding monoamine. Some renal tubular damage and proteinuria is produced by intraperitoneal injections in rats of 1,3-propanediamine and 1,2-propanediamine. Similar effects are not produced by tetramethylenediamine, pentamethylenediamine, hexamethylenediamine, or decamethylenediamine (45).

Hexamethylenediamine (1,6-hexanediamine) causes anemia, weight loss, and degenerative microscopic changes in the kidneys and liver and to a lesser degree in the myocardium of guinea pigs after repeated doses (84). Conjunctival and upper respiratory tract irritation have been observed in workers handling hexamethylenediamine. One worker, out of 20 studied, developed acute hepatitis followed by dermatitis which was attributed to hexamethylenediamine. No anemia was observed. Air concentrations varied from 2 and 5.5 mg/m^3 during normal operations to 32.7 and 131.5 mg/m^3 during autoclave operations in two plants (85).

7.2.2 Ethylenediamine

Ethylenediamine is a hygroscopic, fuming liquid that is used as a chemical intermediate in the preparation of dyes, inhibitors, resins, and pharmaceuticals. The results of acute animal toxicity studies are summarized in Table 44.4. Observations on animals show that the vapors are irritating to the eyes, mucous membranes, and respiratory tract and that the liquid causes severe skin corrosion and corneal injury (21, 29, 37, 86). Repeated exposures of rats to measured concentrations of ethylenediamine vapors produced hair loss and lung, kidney, and liver damage at 484 ppm, with lesser degrees of injury at 225 and 132 ppm. No injury was observed at 125 ppm continued for 37-hr exposures (Table 44.4). Renal tubular damage and proteinuria are produced in rats from intraperitoneal doses of 300 mg/kg (45). Dermatitis occurred in a high proportion of exposed operating personnel manufacturing mixed ethyleneamines. It is probable that both primary irritation and sensitization occur. Respiratory irritation and asthmatic symptoms may follow exposures to low vapor concentrations (87). Ethylenediamine is known to cause severe eye damage in man (28). Voluntary vapor inhalation for 5 to 10 sec produced tingling of the face and irritation of the nasal mucosa at 200 ppm and severe nasal irritation at 400 ppm. An instance of asthmatic sensitization has been described (88).

Skin sensitization was observed in a number of instances because of the use of ethylenediamine as a stabilizer in pharmaceutical skin creams. Sensitization is less likely in industrial exposures because the contact is less intimate and because damaged skin is not usually involved (53).

A substituted ethylenediamine, ethylenediaminetetraacetic acid (EDTA), has been used extensively in medicine as a chelating agent for the removal of toxic heavy metals, usually in the form of the calcium disodium salt (calcium disodium edetate). Large or repeated doses may cause kidney injury. Gastrointestinal upset, pain at the injection site, transient bone marrow depression, mucocutaneous lesions, fever, muscle cramps, and histamine-like reactions (sneezing, lacrimation, nasal congestion) have been reported (89).

N,N,N-Diethylenediamine is less volatile than ethylenediamine. Concentrated vapors (not measured) produced no deaths in rats following an 8-hr exposure. Skin and eye effects in animals approximate those of ethylenediamine (Table 44.4).

7.2.3 Diethylenetriamine

Diethylenetriamine is a skin, eye, and respiratory irritant. It is known to produce skin sensitization and probably pulmonary sensitization. The acute toxic effects for animals are listed in Table 44.4. Hine and associates found the oral LD_{50} for rats to be 1.08 g/kg. Application to rabbit skin produced maximum irritation (65). Rats exposed to concentrated vapors and to 300 ppm showed no effects (86). Solutions of 15 percent to undiluted caused severe corneal injury, but a 5 percent solution caused only minor injury. Human skin sensitization has been observed repeatedly, particularly during the use of diethylenetriamine as a catalyst for epoxy resins. It has been stated that in view of the relatively high frequency of cutaneous and pulmonary sensitization, great care must be used in handling diethylenetriamine. If a definite odor of this compound can be detected, process control may be inadequate (90).

The substituted diethylenetriamines, N-(hydroxyethyl)diethylenetriamine and N-(cyanoethyl)diethylenetriamine are less toxic orally and intraperitoneally than diethylenetriamine and are less irritating to the skin of rabbits on single or repeated applications (31).

Another substituted compound, diethylenetriaminepentaacetic acid (DPTA), has been recommended for use as a chelating agent for the removal of radionuclides deposited in the respiratory tract. Dudley et al. (91) found that this material can be administered effectively by inhalation in beagle dogs, and that the aerosol particles need not penetrate deeply into the lung to assure adequate absorption.

7.2.4 Triethylenetetramine

In common with the other ethyleneamines, triethylenetetramine causes skin sensitization as well as primary irritation. Exposure to the hot vapor results in respiratory tract irritation and itching of the face with erythema and edema (92). Grandjean was unable to detect triethylenetatramine in the air of a workroom where dermatitis was occurring. He concluded that the control problem was primarily one of preventing direct skin contact. Successful control requires good personnel training and scrupulous handling technique (93).

This compound is reported to cause increased enzyme activity in the kidneys of pregnant guinea pigs and in the livers of nonpregnant guinea pigs (94). Skin application daily for 10 days and every second day for 45 days caused cachexia, cutaneous alterations at the site of application, liver degeneration, and congestion of the kidneys and brain. It caused necrotic changes in the placenta and miscarriages or fetal death in pregnant animals (95).

This compound is used as a plasticizer in plastics production.

7.3   Alkanolamines and Alkylalkanolamines

7.3.1   General Considerations

Properties and Uses.   The alkanolamines or amino alcohols are substituted primary, secondary, and tertiary amines. Under appropriate conditions they enter into reactions characteristic of both alcohols and amines. Their solutions are alkaline. They form salts readily with inorganic and organic acids. The salts formed with fatty acids are technically important in emulsifying agents and special soaps. They find wide use in the chemical and pharmaceutical industries as intermediates for the production of emulsifiers, detergents, solubilizers, cosmetics, drugs, and textile-finishing agents. The physical properties of some of the more common amino alcohols appear in Table 44.8.

Metabolism.   The metabolism of the amino alcohols has received little attention. Ethanolamine is naturally formed in mammals from serine and is a normal constituent of mammalian urine (96). Forty percent of ^{15}N-labeled ethanolamine appears as urea within 24 hr when it is given to rabbits, suggesting that it is deaminized. It is also methylated to choline and converted to serine and glycine. Monomethylaminoethanol and dimethylaminoethanol are intermediates in the conversion to choline. Some 33 percent of diethylaminoethanol injected into man in 1-g doses is excreted unchanged. The transformation of the remaining portion is unknown. It could be de-ethylated to ethanolamine and thus enter the normal metabolic pathways (9).

Pharmacology.   The pharmacological properties of dimethylaminoethanol have been studied more extensively than those of the other amino alcohols because it has potential usefulness as a central nervous system stimulant. Pfeiffer et al. (97) found that large doses of the tartrate salt result in depression and pulmonary edema in rats. Intravenous injection in anesthetized dogs produces a transient fall in blood pressure with moderate doses whereas larger doses

Table 44.8. Physical Characteristics and Toxicologic Levels of Some Alkanolamines

Name	Formula	Mol. Wt.	B.P. (°C)	Specific Gravity (20/4)	Flash Point[a] (°F)	Solubility in Water	Approximate Oral LD$_{50}$, Rats (g/kg)	Approximate Percutaneous LD$_{50}$ (ml/kg)	Ref., Toxicity Data
Ethanolamine	HOCH$_2$CH$_2$NH$_2$	61.08	170.5	1.018	185	Complete	2.1		3
Diethanolamine	(HOCH$_2$CH$_2$)$_2$NH	105.14	268	1.09664	280	96.4% w/w	1.82		21
Triethanolamine	(HOCH$_2$CH$_2$)$_3$N	149.19	335	1.124	355	Complete	9.11		21
3-Amino-1-propanol	HOCH$_2$CH$_2$CH$_2$NH$_2$	75.11	187–188	0.9824	175	Miscible	2.83	1.25	23
Isopropanolamine (1-amino-2-propanol)	CH$_3$CH(OH)CH$_2$NH$_2$	75.11	160	0.9611	171	Complete	4.26	1.64	20
Triisopropanolamine	(CH$_3$CH(OH)CH$_2$)$_3$N	191.27	305	1.0	320	Very sol.	6.50		19
2-Methylaminoethanol (methylethanolamine)	CH$_3$NHCH$_2$CH$_2$OH	75.11	158	0.937	165	Complete	2.34	Absorbed	22
2-Dimethylaminoethanol (dimethylethanolamine)	(CH$_3$)$_2$NCH$_2$CH$_2$OH	89.14	134	0.8864	105	Complete	2.34	1.37	21
2-Ethylaminoethanol (ethylethanolamine)	CH$_3$CH$_2$NHCH$_2$CH$_2$OH	89.14	169–170	0.914	160	Complete	1.48 / 1.00	0.36	22 / 36
2-Diethylaminoethanol (diethylethanolamine)	(CH$_3$CH$_2$)$_2$NCH$_2$CH$_2$OH	117.19	163	0.8921	140	Complete	1.3	1.0	18
2-Dibutylaminoethanol (dibutylethanolamine)	(CH$_3$(CH$_2$)$_3$)$_2$NCH$_2$CH$_2$OH	173.29	229.7	0.859	200	0.4% w/w	1.07	1.68	22
N-Methyl-2,2'-iminodiethanol (methyldiethanolamine)	CH$_3$N(CH$_2$CH$_2$OH)$_2$	119.16	247.2	1.0418	260	Complete	4.78	5.99	22
N-Ethyl-2,2'-iminodiethanol (ethyldiethanolamine)	CH$_3$CH$_2$N(CH$_2$CH$_2$OH)$_2$	133.19	246–252	1.02	280	Complete	4.57		22
1-Dimethylamino-2-propanol (dimethylpropanolamine)	(CH$_3$)$_2$NCH$_2$CH(OH)CH$_3$	103.16	121–127	0.85		Complete	1.89		22

[a] Open cup.

(greater than 30 mg/kg) cause a pressor effect. It exhibits a low order of acute toxicity (LD_{50} 3.1 g/kg intraperitoneally in mice) and convulsions do not result from single doses. On chronic administration to rats in doses of 500 mg/kg per day central nervous system stimulation appears, with a lowered threshold for audiogenic seizures. Occasional deaths from maximal convulsions occur after 3 to 4 weeks. In man, oral doses of 10 to 20 mg of the base, as tartrate salt, produce mild mental stimulation. At 20 mg/day there is a gradual increase in muscle tone and an apparent increased frequency of convulsions in susceptible individuals. Large doses produce insomnia, muscle tenseness, and spontaneous muscle twitches (97). Triethanolamine is said to be a powerful vasodilator (98). Smyth's laboratory (99) has found that intravenous injections of mono-, di-, and triethanolamines in dogs resulted in increased blood pressure, diuresis, saliva-tion, and pupillary dilatation. These symptoms resemble those produced by the pharmacologically active aliphatic amines. Larger doses produced sedation, coma, and death following depression of blood pressure and cardiac collapse. Monoethanolamine was the most effective and triethanolamine the least effective.

Toxicity for Animals.   The toxicological properties of the alkanolamines gen-erally resemble those of the corresponding alkylamines. The most pronounced effects in animals are those related to the local irritant effects of the concentrated alkaline liquids or solutions. Their salts show reduced local irritant activity and the tertiary alkanolamines are less irritating than the primary compounds. The *n*-alkyldialkanol compounds are less irritating and less toxic orally than the dialkylalkanolamines. The liquids and alkaline solutions produce severe eye injury in animals. The eye irritation scores in rabbits as determined in the extensive series studied in Smyth's laboratory usually lie between 5 and 9 (18–25, 28, 29). Primary irritation of the skin varies from slight to moderately severe. This is enhanced by repeated application of the material under an occlusive dressing. The material is absorbed through the skin and when held in contact with the skin of small animals may cause death in doses that are less than those that produce death when given by mouth (Table 44.5). The acute oral toxicity levels for laboratory animals are generally low. Concentrated, unneutralized solutions of the more soluble alkanolamines cause intense gas-trointestinal irritation with hemorrhage and congestion of the intestine. Adhe-sions of visceral organs are frequently found in the survivors. Neutralization and increasing dilution reduce the oral toxicity. In the series reported by Smyth et al., single exposures of rats to "saturated" vapors seldom produced deaths in less than 8 hr. Except for monoethanolamine, repeated vapor exposures in animals have not been reported. The generally low vapor pressures of these compounds reduce the inhalation hazard in industry. Human injuries by inhalation have not been reported, despite the widespread use of some of the alkanolamines. (17, p. 2044).

7.3.2 Specific Compounds

Ethanolamine (2-Aminoethanol).   Although monoethanolamine has had wide use in industry, there have been no reports of human injury. Its physical and chemical properties are indicated in Table 44.8. It is a normal constituent of human urine. The excretion rate in men varies between 4.8 and 22.9 mg/day with a mean of 0.162 mg/kg. Eleven women were observed to excrete larger amounts, varying between 7.7 and 34.9 mg/day with a mean excretion rate of 0.492 mg/kg per day. The excretion rates in animals were, approximately, for cats, 0.47 mg/kg per day; rats, 1.46 mg/kg per day; rabbits, 1.0 mg/kg per day. From 6 to 47 percent of methanolamine administered to rats can be recovered in the urine (96).

Smyth obtained the following results in a 90-day subacute oral toxicity study in rats: maximum daily dose with no effect, 0.32 g/kg; dose at which altered liver or kidney weight was seen, 0.64 g/kg; dose at which microscopic pathological changes and deaths appeared, 1.28 g/kg. The acute LD_{50} was 2.74 g/kg (30). Treon et al. (100) studied the inhalation toxicity of monoethanolamine. The conditions were such that an unknown proportion of the ethanolamine was converted to the carbonate in the exposure chamber. Dogs and cats survived the exposures to concentrations of 2.47 mg/liter for 7 hr on each of four successive days. Four of six guinea pigs died following exposure to a concentration of 0.58 mg/liter for 1 hr. Rats, rabbits, and mice were less susceptible than guinea pigs, but more susceptible than cats or dogs. Sixty of 61 animals survived exposure to the inhalation of concentrations of 0.26 to 0.27 mg/liter for 7 hr on each of five consecutive days and 25 of 26 animals survived 25 7-hr exposures (over a period of 5 weeks) to concentrations of 0.26 mg/liter (104 ppm). The observed effects were primarily those of respiratory tract irritation. Eye irritation was negligible, presumably due to the formation of carbonate under the conditions of the experiment. Pathological changes in those animals exposed to higher concentrations were chiefly those of pulmonary irritation, with some nonspecific degenerative changes in the liver and kidneys. Survivors of the lower concentrations had normal autopsy findings. Weeks (101) reports that dogs, rats, and guinea pigs survived inhalation of 12 to 25 ppm for 90 days, whereas fractional mortality occurred in 24 to 30 days at 100 ppm (dogs) and 66 to 75 ppm (rodents). Skin irritation and lethargy occurred at 5 and 12 ppm. He found the median detectable (odor) concentration for humans to be 3 to 4 ppm. Ethanolamine instilled in the rabbit eye produces severe injury (28). Browning states that when undiluted monoethanolamine is applied to human skin on gauze for 1½ hr, only marked redness and infiltration of the skin result (98).

Diethanolamine (2,2′-Iminodiethanol).   The acute and subacute oral toxicity of diethanolamine for rats is somewhat greater than that for monoethanolamine.

The acute oral LD_{50} is 1.82 g/kg. The maximum daily dose having no effect over a 90-day period is 0.02 g/kg. A daily dose of 0.17 g/kg over the same period produces microscopic pathology and deaths, and 0.09 g/kg causes changes in liver and kidney weights. The undiluted liquid and 40 percent solutions produce severe eye burns, whereas 15 percent produces only minor damage. Ten percent solution applied to rabbit skin causes redness. Higher concentrations cause increasing injury (17, p. 2066; 28).

Triethanolamine (2,2',2"-Nitrilotriethanol).   Triethanolamine is generally considered to have a low acute and chronic toxicity. Lehman (102) states that if deleterious effects were to occur in man from triethanolamine, these would probably be acute in nature and due to its alkalinity rather than its inherent toxicity. Kindsvatter (34) found the acute oral LD_{50} in rats and guinea pigs to be 8 g/kg. The effects observed were confined to the gastrointestinal tract. He felt the toxic effects were probably from the alkaline irritation, since larger doses of the neutralized material produced no symptoms at levels where the free base would cause 100 percent mortality. Repeated feeding produced only slight, reversible pathology in the liver and kidneys. Applications on the skin gave evidence of skin absorption. Smyth and Carpenter (18) also found low acute oral toxicity, with the LD_{50} for rats being 9.11 g/kg. In a 90-day subacute feeding experiment with rats, the maximum dose producing no effect was 0.08 g/kg. Microscopic lesions and deaths occurred at 0.73 g/kg, and 0.17 g/kg produced alterations in liver and kidney weights. Applications of 5 or 10 percent solution to rabbit or rat skin did not produce irritation. No industrial injuries from triethanolamine have been reported. It appears to be free of skin sensitization effects in its extensive use in cosmetics.

Hoshino and Tanooka (103) reported that triethanolamine in the diet of mice at levels of 0.03 or 0.3 percent caused a significant increase in the occurrence of tumors, both benign and malignant. Females showed a 32 percent increase, mostly of thymic lymphomas. The increase of all other tumors, in both sexes, was 8.2 percent. They also found that triethanolamine reacted with sodium nitrite to produce N-nitrosodiethanolamine and that the product caused mutagenesis in bacteria.

In view of its low vapor pressure (less than 0.01 torr), significant exposure by inhalation appears unlikely and the chief risk in industry would be from direct local contact of the skin or eyes with the undiluted, unneutralized fluid.

REFERENCES

1. A. Streitweiser, Jr. and C. H. Heathcock, *Introduction to Organic Chemistry*. Macmillan, New York, 1976.
2. C. D. Gutsche and D. J. Pasto, *Fundamentals of Organic Chemistry*. Prentice-Hall, Englewood Cliffs, N.J., 1975.

3. K. Hasagawa, K. Sakamoto, N. Ohmura, and K. Sasaki, "Deodorization of Ammonia or Amines." Japanese patent notice cited in *Chem. Abstr.* **84**: 49858, 1976.

4. G. B. Neurath, M. Duenger, F. G. Pein, D. I. Ambrosius, and O. Schreiber, "Primary and Secondary Amines in the Human Environment," *Food Cosmetic Toxicol.* **15**(4), 275–282 (1977).

5. *Handbook of Chemistry and Physics*, 59th ed., Chemical Rubber Publishing Co., Cleveland, 1979–1980.

6. *The Condensed Chemical Dictionary*, 8th ed., Van Nostrand Reinhold, New York, 1971.

7. *Merck Index*, 9th ed., Merck, Rahway, N.J.

8. *Encyclopedia of Chemical Technology*, 3rd ed., Wiley, New York, 1978.

9. R. T. Williams, *Detoxication Mechanisms*, 2nd ed., Wiley, New York, 1959.

10. J. Rechenberger, *Z. physiol. Chem.*, **265**, 275 (1940).

11. D. F. Davies, *J. Lab. Clin. Med.*, **43**, 620 (1954).

12. M. L. Simenhoff, "Metabolism of Aliphatic Amines," *Kidney Int. Suppl.* **3**, 314–317 (1975).

13. H. Blaschko, *Pharmacol. Revs.*, **4**, 415 (1952).

14. D. Richter, *Biochem. J.*, **32**, 1763 (1938).

15. H. Blaschko and J. Hawkins, *Brit. J. Pharmacol.*, **5**, 625 (1950).

16. P. J. Wirth and S. S. Thorgeirsson, "Amine Oxidase in Mice. Sex Differences and Developmental Aspects," *Biochem. Pharmacol.*, **27**(4), 601–603 (1978).

17. W. L. Sutton, "Aliphatic and Alicyclic Amines," in *Patty's Industrial Hygiene and Toxicology*, 2nd ed., Interscience, New York, 1963.

18. H. F. Smyth and C. P. Carpenter, "The Place of the Range Finding Test in the Industrial Toxicology Laboratory," *J. Ind. Hyg. Toxicol.*, **26**, 269–273 (1944).

19. H. F. Smyth, Jr. and C. P. Carpenter, "Further Experience with the Range Finding Test in the Industrial Toxicology Laboratory," *J. Ind. Hyg. Toxicol.*, **30**, 63–68 (1948).

20. H. F. Smyth, Jr., C. P. Carpenter, and C. S. Weil, "Range-finding Toxicity Data, List III," *J. Ind. Hyg. Toxicol.*, **31**, 60–62 (1949).

21. H. F. Smyth, Jr., C. P. Carpenter, and C. S. Weil, "Range-finding Toxicity Data, List IV," *AMA Arch. Ind. Hyg. Occup. Med.*, **4**, 109–122 (1951).

22. H. F. Smyth, Jr., C. P. Carpenter, C. S. Weil, and U. C. Pozzani, "Range-finding Toxicity Data, List V," *AMA Arch. Ind. Hyg. Occup. Med.*, **10**, 61–68 (1954).

23. H. F. Smyth, Jr., C. P. Carpenter, C. S. Weil, U. C. Pozzani, and J. A. Striegel, "Range-finding Toxicity Data, List VI," *Am. Ind. Hyg. Assoc. J.*, **23**, 95–107 (1962).

24. H. F. Smyth, Jr., C. P. Carpenter, C. H. Weil, U. C. Pozzani, J. A. Striegel, and J. S. Nycum, "Range-finding Toxicity Data, List VII," *Am. Ind. Hyg. Assoc. J.*, **30**, 470–476 (1969).

25. C. P. Carpenter, C. S. Weil, and H. F. Smyth, Jr., "Range-finding Toxicity Data, List VIII," *Toxicol. Appl. Pharmacol.*, **28**, 313–319 (1974).

26. H. Brieger and W. A. Hodes, "Toxic Effects of Exposure to Vapors of Aliphatic Amines," *AMA Arch. Ind. Hyg. Occup. Med.*, **3**, 287–291 (1951).

27. C. H. Hine, J. K. Kodama, R. J. Guzman, and G. S. Loquvam, "The Toxicity of Allylamines," *Arch. Environ. Health*, **1**, 343–352 (1960).

28. C. P. Carpenter and H. F. Smyth Jr., "Chemical Burns of the Rabbit Cornea," *Am. J. Ophthalmol.*, **29**, 1363–1372 (1946).

29. C. P. Carpenter, H. F. Smyth, Jr., and U. C. Pozzani, "The Assay of Acute Vapor Toxicity, and the Grading and Interpretation of Results on 96 Chemical Compounds," *J. Ind. Hyg. Toxicol.*, **31**(6), 343–346 (1949).

30. J. F. Treon, H. Sigmon, K. V. Kitzmiller, and F. F. Heyroth, "The Physiological Response of Animals to Respiratory Exposure to the Vapors of Diisopropylamine," *J. Ind. Hyg. Toxicol.*, **31**(3), 142–145 (1949).

31. C. H. Hine, J. K. Kodama, H. H. Anderson, D. W. Simonson, and J. S. Wellington, "The Toxicology of Epoxy Resins," *AMA Arch. Ind. Health*, **17**, 129–144 (1958).

32. E. H. Vernot, J. D. MacEwen, C. C. Haun, and E. R. Kinkead, "Acute Toxicity and Skin Corrosion Data for Some Organic and Inorganic Compounds and Aqueous Solutions," *Toxicol. Appl. Pharmacol.*, **42**, 417–423 (1977).

33. H. F. Smyth Jr., J. Seaton, and E. L. Fischer, "The Single Dose Toxicity of Some Glycols and Derivatives," *J. Ind. Hyg. Toxicol.*, **23**, 259–268 (1941).

34. V. H. Kindsvatter, "Acute and Chronic Toxicity of Triethanolamine," *J. Ind. Hyg. Toxicol.*, **22**, 206–212 (1940).

35. H. H. Cornish and J. Adefuin, "Effects of 2-*N*-Mono- and 2-*N*-Diethylaminoethanol on Normal and Choline-deficient Rats," *Food Cosmet. Toxicol.*, **5**, 327–332 (1967).

36. *Registry of Toxic Effects of Chemical Substances,* National Institute of Occupational Safety and Health, January 1980; Eastman Kodak Company, Rochester, N.Y., unpublished observations cited in Reference 17.

37. D. W. Fassett, Laboratory of Industrial Medicine, Eastman Kodak Company, Rochester, N.Y., unpublished observations cited in Reference 17.

38. G. Barger and H. H. Dale, *J. Physiol.*, **41**, 19 (1910–1911).

39. W. H. Hartung, *Chem. Rev.*, **9**, 389 (1931).

40. E. E. Swanson and K. K. Chen, *J. Pharmacol. Exp. Therap.* **88**, 10 (1946).

41. M. F. W. Dunker and W. H. Hartung, *J. Am. Pharmacol. Assoc.*, Sci. Ed., **30**, 619 (1941).

42. R. P. Ahlquist, *J. Pharmacol. Exp. Therap.*, **85**, 283 (1945).

43. T. Sollman, *A Manual of Pharmacology*, 8th ed., Saunders, Philadelphia, 1957.

44. U. C. Pozzani and C. P. Carpenter, *Arch. Ind. Hyg. Occup. Med.*, **9**, 233 (1954).

45. C. W. Tabor and S. M. Rosenthal, *J. Pharmacol. Exp. Therap.*, **116**, 139 (1956).

46. L. B. Bourne, F. J. M. Miller, and K. B. Alberman, *Br. J. Ind. Med.*, **16**, 81 (1959).

47. A. J. Amor, *Mfg. Chem.*, **20**, 540 (1949).

48. R. E. Eckardt, "Occupational and Environmental Health Hazards in the Plastics Industry," *Environ. Health Perspect.*, **17**, 103–106 (1976).

49. R. E. Brubaker, H. J. Muranko, D. B. Smith, G. J. Beck, and G. Scovel, "Evaluation and Control of a Respiratory Exposure to 3-(Dimethylamino)propylamine," *J. Occup. Med.*, **21**(10), 688–690 (1979).

50. M. A. Friedman, G. N. Millar, and S. S. Epstein, "Acute Dose Dependent Inhibition of Liver Nuclear RNA Synthesis and Methylation of Guanine Following Oral Administration of Sodium Nitrite and Dimethylamine to Mice," *Int. J. Environ. Stud.*, **4**(3), 219–222 (1975).

51. R. A. Coon, R. A. Jones, L. J. Jenkins, Jr., and J. Siegel, "Animal Inhalation Studies on Ammonia, Ethylene Glycol, Formaldehyde, Dimethylamine and Ethanol," *Toxicol. Appl. Pharmacol.*, **16**(3), 646–655 (1970).

52. J. E. Amoore and L. J. Forrester, "Specific Anosmia to Trimethylamine. The Fishy Primary Odor," *J. Chem. Ecol.*, **2**(1), 49–56 (1976). Cited in *Chem. Abstr.*

53. N. H. Proctor and J. P. Hughes, *Chemical Hazards of the Workplace*, Lippincott, Philadelphia, 1978; (a) p. 249; (b) p. 303; (c) p. 221; (d) p. 134; (e) p. 255; (f) p. 388.

54. F. W. Dunker and W. Hartung, *J. Am. Pharmacol. Assoc. Sci. Ed.*, **30**, 623 (1941).

55. D. F. Marsh, *J. Pharmacol. Exp. Therap.*, **94**, 225 (1948).

56. H. Morin, *Therapie*, **7**, 57 (1952); cited in *Chem Abstr.*, **47**, 1850i (1953).

57. A. J. Fleming, C. A. D'Alonzo, and J. A. Zapp, *Modern Occupational Medicine*, Lea and Febiger, Philadelphia, 1954.

58. W. B. Deichmann, J. L. Radomski, W. E. MacDonald, R. L. Kascht, and R. L. Erdman, *AMA Arch. Ind. Health*, **18**, 483 (1958).

59. H. H. Anderson and G. H. Hurwitz, *Arch. Exp. Pathol. Pharmakol.*, **219**, 119 (1953).

60. E. R. Hart, *Univ. Calif. Publ. Pharmacol.*, **1**, 213 (1938–1941).

61. E. R. Hart and C. Leake, *J. Pharmacol. Exp. Therap.*, **66**, 18 (1939).

62. J. C. Calandra, Industrial Bio-Test Laboratory, report to Shell Chemical Co., July 1959.

63. T. C. Carswell and H. L. Morrill, *Ind. Eng. Chem.*, **11**, 1247 (1937).

64. R. M. Watrous and H. N. Schulz, *Ind. Med. Surg.*, **19**, 317 (1950).

65. F. S. Mallette and E. von Haam, *Arch. Ind. Hyg. Occup. Med.*, **5**, 311 (1952).

66. K. S. Khera and D. R. Stoltz, "Effects of Cyclohexylamine on Rat Fertility," *Experientia*, **26**(7), 761–762 (1970).

67. R. Kroes, P. W. J. Peters, J. M. Berkvens, H. G. Verschuuren, T. De Vries, and G. J. Van Esch, "Long Term Toxicity and Reproduction Study (Including a Teratogenicity Study) with Cyclamate, Saccharin and Cyclohexylamine," *Toxicology*, **8**, 285–300 (1977).

68. B. L. Oser, S. Carson, G. E. Cox, E. E. Vogin, and S. S. Sternberg, "Long-term and Multigeneration Toxicity Studies with Cyclohexylamine Hydrochloride," *Toxicology*, **6**(1), 47–65 (1976).

69. I. F. Gaunt, M. Sharratt, P. Grasso, A. B. Lansdown, and S. D. Gangolli, "Short-term Toxicity of Cyclohexylamine Hydrochloride in the Rat," *Food Cosmet. Toxicol.*, **12**(5–6), 609–624 (1974).

70. I. F. Gaunt, J. Hardy, P. Grasso, S. D. Gangolli, and K. R. Butterworth, "Long-term Toxicity of Cyclohexylamine Chloride in the Rat," *Food Cosmet. Toxicol.*, **14**(4), 255–267 (1976).

71. J. Hardy, I. F. Gaunt, J. Hooson, R. J. Hendy, and K. R. Butterworth, "Long-term Toxicity of Cyclohexylamine Hydrochloride in Mice," *Food Cosmet. Toxicol.*, **14**(4), 269–276 (1976).

72. M. Eichelbaum, J. H. Hengstmann, H. D. Rost, T. Brecht, and H. J. Dengler, "Pharmacokinetics, Cardiovascular and Metabolic Actions of Cyclohexylamine in Man," *Arch. Toxikol.*, **31**(3), 243–263 (1974). Cited in *Biol. Abstr.*

73. E. Gondry, "Research on the Toxicity of Cyclohexylamine, Cyclohexanone and Cyclohexanol, Cyclamate Metabolites," *J. Eur. Toxicol.*, **5**(4), 227–238 (1972). Cited in *Biol. Abstr.*

74. P. L. Mason and G. R. Thompson, "Testicular Effects of Cyclohexylamine Hydrochloride in the Rat," *Toxicol.*, **8**(2), 143–156 (1977).

75. S. Kojima and H. Ichibagase, "Studies on Synthetic Sweetening Agents: VIII, Cyclohexylamine, a Metabolite of Sodium Cyclamate," *Chem. Pharm. Bull. (Tokyo)*, **14**, 971–974 (1966).

76. J. M. Price, C. G. Biava, B. L. Oser, E. E. Vogin, J. Steinfeld, and II. L. Ley, "Bladder Tumors in Rats Fed Cyclohexylamine or High Doses of a Mixture of Cyclamate and Saccharin," *Science*, **167**, 1131–1132 (1970).

77. M. S. Legator, K. A. Palmer, S. Green, and K. W. Peterson, "Cytogenetic Studies in Rats of Cyclohexylamine, a Metabolite of Cyclamate," *Science*, **165**, 1139–1140 (1969).

78. J. H. Turner and D. L. H. Hutchinson, *Mutat. Res.*, **26**, 407 (1974), cited by P. Cooper, *Food Cosmet. Toxicol.*, **15**, 69 (1977).

79. P. Cooper, "Resolving the Cyclamate Question," *Food Cosmet. Toxicol.*, **15**, 69–70 (1977).

80. B. A. Becker and J. E. Gibson, "Teratogenicity of Cyclohexylamine in Mammals," *Toxicol. Appl. Pharmacol.*, **17**, 551–552 (1970).

81. D. Lorke and L. Machemer, "Investigation of Cyclohexylamine Sulfate for Dominant Lethal Effects in the Mouse," *Toxicol.*, **2**, 231–237 (1974).

82. B. M. Cattanach, "The Mutagenicity of Cyclamates and their Metabolites," *Mutat. Res.*, **39**(1), 1–28 (1976).

83. J. L. Mongar and H. Schild, *Br. J. Pharmacol.*, **8,** 103 (1953).

84. D. Ceresa and M. DeBlasis, *Med. Lav.*, **4,** 78 (1950).

85. G. Gallo and L. Ghiringhelli, *Med. Lav.*, **49,** 688 (1958).

86. L. E. Savitt, *AMA Arch. Dermatol.*, **71,** 212 (1955).

87. C. U. Dernehl, *Ind. Med Surg.*, **20,** 541 (1951).

88. S. Lam and M. Chan-Yeung, "Ethylenediamine-induced asthma," *Am. Rev. Resp. Dis.*, **121**(1), 151–155 (1980).

89. *AMA Drug Evaluations,* 1st ed., American Medical Association, Chicago, 1971.

90. American Industrial Hygiene Association, Hygienic Guide Series, "Diethylene Triamine and Diethylamine," *Am. Ind. Hyg. Assoc. J.* **21,** 266, 268 (1960).

91. R. E. Dudley, B. A. Muggenburg, R. G. Cuddihy, and R. O. McClellan, "Absorption of Diethylenetriaminepentaacetic Acid (DTPA) from the Respiratory Tracts of Beagle Dogs," *Am. Ind. Hyg. Assoc. J.*, **41**(1), 5–11 (1980).

92. L. B. Bourne, F. J. M. Milner, and K. B. Alberman, *Br. J. Ind. Med.*, **16,** 81 (1959).

93. E. Grandjean. *Z. Präventivmed.*, **2,** 77 (1957).

94. W. Dobryszycka, J. Kulpa, A. Woyton, J. Woyton, J. Szacki, and A. Dzioba, "Influence of Industrial Toxic Compounds on Pregnancy: VI. Some Tissue Enzymes in Pregnant Guinea Pigs Exposed to the Action of Triethylenetetramine," *Arch. Immunol. Ther. Exp.*, **23**(6), 867–870 (1975). Cited in *Biol. Abstr.* and *Chem. Abstr.*

95. J. Szacki, J. Woyton, A. Dzioba, J. Rabczynski, and A. Woyton, "Influence of Industrial Toxic Compounds on Pregnancy. II. Biochemical Changes in Organisms of Guinea Pigs Exposed to the Action of Triethylenetetramine (TETA) during Pregnancy," *Arch. Immunol. Ther. Exp.*, **22**(1), 123–128 (1974).

96. J. M. Luck and A. Wilcox, *J. Biol. Chem.*, **205,** 859 (1953).

97. C. C. Pfeiffer, E. H. Jenney, W. Gallagher, R. P. Smith, W. Bevan, K. F. Killam, E. K. Killam, and W. Blackmore, *Science*, **126,** 610 (1957).

98. E. Browning, *Toxicity of Industrial Organic Solvents*, Chemical Publishing Co., New York, 1953.

99. H. F. Smyth, Jr., Mellon Institute of Industrial Research, Report No. 19–5, 1956.

100. J. F. Treon, F. P. Cleveland, E. E. Larson, and J. Cappel, "The Response of Animals to Airborne Monoethanolamine," presented at the 1958 Industrial Health Conference of the American Industrial Hygiene Association, Atlantic City.

101. M. H. Weeks, *U. S. Chem. Warfare Lab. Spec. Publ. 2–10*, Army Chemical Center, Md., 1958.

102. A. J. Lehman, *Assoc. Food Drug Offic. U. S. Q. Bull.*, **14,** 82 (1950).

103. H. Hoshino and H. Tanooka, "Carcinogenicity of Triethanolamine in Mice and its Mutagenicity after Reaction with Sodium Nitrate," *Cancer Res.*, **38,** 3918–3921 (1978).

CHAPTER FORTY-FIVE

Aliphatic Hydrocarbons

ESTHER E. SANDMEYER, Ph.D.

1 ALIPHATIC, LINEAR, AND BRANCHED CHAIN SATURATED HYDROCARBONS (ALKANES, PARAFFINS)

1.1 General

1.1.1 Occurrence or Source

Many of the industrially important aliphatic saturated hydrocarbons are found naturally in earth gas or crude oil. Some family members are used and others are formed during combustion of fuels, by catalytic cracking, or in other specialized petrochemical processes (1). Many of the lower paraffins, such as C_1 to C_5 n-alkanes, the branched isobutane and isopentane, nonane, and decane, have been detected in large city atmospheres (2–4). The normal paraffins from C_1 to C_5 and the branched isomers C_4 to C_6 have been detected in Baldwin engine exhausts (5). Paraffin mixtures are used extensively as fuels, refrigerants, propellants, pesticides, lubricants, solvents for paints, protective coatings, and plastics, in degreasing operations, and in purified form as food additives. Some major hydrocarbon mixtures that contain paraffins are listed in Table 45.1.

1.1.2 Physical Characteristics

Some selected physical and chemical properties of the alkanes, as shown in Table 45.2 (6–15), indicate progressive changes with increasing molecular weight and volume. In addition, the boiling point, liquid and vapor density,

3175

Table 45.1.   Occurrence and Boiling Range of Major Alkanes

Principal Alkane Number of Carbons	Product	Boiling Range (°C)
C_1, C_2	Natural gas	Gas at 25°C
C_3, C_4	Liquid petroleum gas	Gas at 25°C
C_4–C_6	Petroleum ether	20–60
C_5–C_7	Petroleum benzin	40–90
C_6–C_8	Petroleum naphtha	65–120
C_5–C_{10}	Gasoline	36–210
C_7–C_9	Mineral spirits	150–210
C_9–C_{16}	Kerosine	170–300
C_5–C_{16}	Jet and turbo fuels	40–300
C_{17} and higher	Lubricating oils	300–700

critical temperature, and vapor pressure, individually or combined, largely determine the alkanes' toxicologic characteristics.

1.1.3   Toxicology

The toxicity characteristics of the alkanes are minimal for the gases and solids, but are moderate for the liquid materials. The paraffin gases C_1 to C_4 are practically nontoxic below the lower flammability limit, 18,000 to 50,000 ppm; above this, low to moderate incidental effects such as CNS depression and irritation occur, but are completely reversible upon cessation of the exposure. Therefore, prevention of gaseous accumulation beyond the threshold limit value (TLV) or lower flammability limit may automatically prevent physiological effects. At higher concentrations and when mixed with air the gases may become anesthetics and subsequently asphyxiants by diluting or decreasing the available oxygen (16). The C_3 to iso-C_5 hydrocarbons show increasing narcotic properties; branching of the chain also enhances this effect (17, 18). The C_4 hydrocarbons appear to be more highly neurotoxic than the C_3 and C_5 members. The fluid members, C_5 to C_9, have anesthetic and CNS depressant actions; the C_6, and to a lower degree, the C_5 and C_7 members, have neurotoxic properties. They are fat solvents and on repeated, prolonged direct skin contact may cause chemical dermatitis. Direct aspiration into the lungs of paraffins with carbon numbers C_6 to C_{16} may cause chemical pneumonitis, pulmonary edema, and hemorrhaging (19). According to Krantz et al. (20), several C_1 to C_5 paraffins have been shown to be cardiac sensitizers. Alkanes have been shown to be absorbed into human tissue. Boitnott and Margolis (21) report that hydrocarbon oil droplets have been detected in human liver at at 0.26 mg/kg tissue. Pennington and Fuerst (22) have shown that many of the alkanes and alkenes cause cellular changes in rabbit erythrocytes when the gases are suspended as 2 percent of a red cell preparation.

Based on solubility characteristics, Crisp et al. (23) extrapolate that the C_1 to C_5 paraffins would be comparatively nontoxic and C_6 and higher more so, as expressed in relative narcotic potencies. The relative solubility, although very low, increases with rising molecular weight from the C_1 to the C_5 alkanes (24). The branched chain derivatives appear less toxic than the parent linear chain members. The odorant properties (25) increase with the lengthening of the carbon chain. The analgesic activity is high for the C_1, decreases for C_2 and C_3, increases again for C_4 to C_7 members, with the peak at C_5, decreases for C_8, and is inactive with a C_9 and longer alkyl chain. In addition, the dermal and the pulmonary irritancy increase with increasing carbon chain length (26), particularly from the C_5 to the iso-C_8 alkane. When ingested, an aspiration hazard exists for the compounds hexane through tetradecane, to a lesser extent for the C_{16} derivative, and none for the higher members, as shown with rats (19). The lower members, C_6 to C_8, when aspirated into rat lungs, caused almost immediate death due to respiratory paralysis, asphyxia, and cardiac arrest. With higher members death occurred more slowly following the development of pulmonary edema. Signs were dyspnea, cyanosis, and nasal hemorrhaging (19).

Presently, studies indicate that none of the alkanes possess teratogenic, mutagenic, or carcinogenic properties. Some C_{10} to C_{12} paraffins may present an exception. Under specific conditions, and used as carriers or solvents for certain potent carcinogens, some carcinogenic effects may be promoted.

1.1.4 General Microbiological Characteristics

Some of the aliphatic alkanes and alkenes, C_1 to C_4, protect *Neurospora crassa* against gamma irradiation, but others enhance the irradiation damage (27). Most C_2 to C_{10} paraffins prevent germination of *Bacillus megaterium*, and the C_3 to C_9 members show total inhibition (28). Almost all hydrocarbons are susceptible to microbial oxidation. In the Atlantic Ocean, the hydrocarbon oxidizing bacteria are most abundant in the coastal areas (29). For example, *Pseudomonas aeruginosa* has been shown to oxidize *n*-paraffins and to epoxidize α-olefins (30). Petroleum fractions containing C_7 and C_{12} to C_{20} hydrocarbons are actively degraded by such oxidizing bacteria (31). A variety of hydrocarbons, C_6 to C_{19}, are readily oxidized by the enzyme ω-hydroxylase of *Pseudomonas oleovorans* (32). A decrease is recorded in a C_{10} to C_{18} crude oil fraction and a C_{20} to C_{25} range is documented to be actively degraded by selected paraffin-utilizing bacteria (33). Others up to C_{44} have been observed to be metabolized by a variety of microorganisms (34).

1.1.5 Industrial Hygiene

The lower molecular weight members of the series are directly collected as gas, from C_5 to C_{16} absorbed on charcoal, and the C_{17} members and above collected on filters as particulate matter (35). The individual compounds are quantitatively

Table 45.2. Physicochemical Properties of Alkanes (6–15)

Compound	B.P. (°C)	CAS Registry No.	Density (20/4°C)	Emp. Formula	Flamm. Limits (%)	Flash Pt. [°C (°F)]	Freezing Pt. (°C)	Mol. Wt.	Refr. Index, n^{20}_D	Solubility[a] w/al/et	Spc. Gr. (25°C)	Vapor Dens. (Air = 1)	Vapor Pres.[b] [mm Hg (°C)]	Visc. (SUS)	W/V Conv. (mg/m³ ⇔ 1 ppm)
Methane	−161.49	74-82-8		CH_4	5.3–14	−187.78 (−306)	−182.48	16.042	—	s/s/s	0.42	0.55	40ᶜ(−86.3)	N.A.	0.66
Ethane	−88.63	74-84-0		C_2H_6	3.2–12.45	−135 (−211)	−183.27	30.069	1.0377	i/d/—	0.374	1.05	40ᶜ(23.6)	N.A.	1.23
Propane	−42.07	74-98-6	0.5005	C_3H_8	2.37–9.5	−104.44 (−156)	−189.69	44.096	1.2898	s/s/v	0.5077	1.551	10ᶜ(26.9)	N.A.	1.80
Butane	−0.50	106-97-8	0.5788	C_4H_{10}	1.86–8.41	−60.0 (−76)	−138.35	58.123	1.3326	s/v/v	0.5844	2.07	2ᶜ(18.8)	N.A.	2.38
Isobutane	−11.73	75-28-5	0.5572	C_4H_{10}	1.8–8.44	−82.78 (−181)	−159.6	58.123	—	s/v/v	0.5631	2.07	2ᶜ(7.5)	N.A.	2.38
Pentane	36.07	109-66-0	0.6262	C_5H_{12}	1.42–7.80	−49.0 (−56.2)	−129.72	72.150	1.3575	d/v/v	0.6312	2.49	400 (18.5)	<32	2.95
Isopentane	27.85	78-78-4	0.6197	C_5H_{12}	1.32–8.3	−51.0 (−59.8)	−159.9	72.150	1.3537	i/v/v	0.2648	2.5	595 (21.1)	<32	2.95
Neopentane	9.50	463-82-1	0.5910	C_5H_{12}	1.4–8.3	−6.67 (20)	−16.55	72.150	1.342	i/s/s	0.5967	2.48	1100 (21.8)	<32	2.95
Hexane	68.74	110-54-3	0.6594	C_6H_{14}	1.18–7.80	−22 (−7.6)	−95.348	86.177	1.3749	i/v/s	0.6640	2.97	100 (15.8)	<32	3.52
Isohexane	60.27	107-83-5	0.6532	C_6H_{14}	1.2–7.7	−23.33 (−10)	−153.67	86.177	1.3715	i/s/s	0.6579	3.0		<32	3.52
3-Methylpentane	63.28	96-14-0	0.6643	C_6H_{14}	1.2–7.0	<−20 (<−29)	−118	86.177	1.3765	i/s/v	0.6690	2.97	400 (10.5)	<32	3.52
Neohexane	49.74	75-83-2	0.6492	C_6H_{14}	1.2–7.0	−47.8 (−54)	−99.87	86.177	1.3686	i/s/s	0.6540	3.0	400 (31.0)	<32	3.52
2,3-Dimethylbutane	57.99	79-29-8	0.6616	C_6H_{14}	1.2–7.7	−28.89 (−20)	−128.53	86.177	1.3750	i/s/s	0.6664	3.0	400 (39.0)	<32	3.52
Heptane	98.43	142-82-5	0.6838	C_7H_{16}	1.2–6.7	−4.4	−90.61	100.203	1.3876	i/v/v	0.6882	3.52	40 (22.3)	<32	4.10
Isoheptane	90.0	591-76-4	0.6786	C_7H_{16}	1.0–7.0	−27.8 (−18)	−118.286	100.203	1.38495	i/s/v	0.6830	—	40 (14.9)	<32	4.10
Neoheptane (2,2-dimethylpentane)	79.197	590-35-2	0.6739	C_7H_{16}			−123.82	100.203	1.3822	i/s/s				<32	4.10
2,3-Dimethylpentane	89.78	565-59-3	0.6951	C_7H_{16}	1.1–6.7	<−28.9 (<20)	−135	100.203	1.3920	i/s/s	0.6994	3.45	40 (13.9)	<32	4.10
2,4-Dimethylpentane	80.50	108-08-7	0.6727	C_7H_{16}		<−28.9	−119.14	100.203	1.3815	i/s/s	0.6772	3.48	8.2 (21.0)	<32	4.10

Compound															
Octane	125.665	111-65-9	0.7025	C_8H_{18}	0.96–4.66	13.0 (55.6)	−56.79	114.230	1.3974	i/v/s	0.7068	3.94	10 (19.2)	<32	4.67
Pentane 2,2,4-Trimethyl-	99.238	540-84-1	0.6919	C_8H_{18}	1.1–6.0	−12 (10)	−116	114.230	1.3915	i/v/s	0.6962	3.93	40.6 (21)	<32	4.67
2,3,4-Trimethyl-	113.47	565-75-3	0.7191	C_8H_{18}		5 (41)	−109.0	114.230	1.4042	i/v/v	0.7233		0.970	<32	4.67
Nonane	150.80	111-84-2	0.7176	C_9H_{20}	0.87–2.9	31.1 (88)	−53.52	128.257	1.4054	i/v/v	0.7217	4.41	10 (38)	<32	5.25
2,2,5-Trimethyl-hexane	124.08	3522-94-9	0.7072	C_9H_{20}		12.8 (55)	−105.79	128.257	1.3997	i/v/v	0.7174	4.7	12.9 (21)		5.25
Decane	174.1	124-18-5	0.7301	$C_{10}H_{22}$	0.78–2.6	46.1 (115)	−29.662	142.284	1.4119	i/v/s	0.7341	4.9	1 (16.5)	<32	5.82
2,7-Dimethyl-octane	159.87		0.7242	$C_{10}H_{22}$			−54.0	142.284	1.4086	—/—/s					5.82
Undecane	195.890	1120-21-4	0.7402	$C_{11}H_{24}$		65 (149)	−25.59	156.306	1.4172	i/v/v	0.7441	5.4		<32	6.39
Dodecane	216.278	112-40-3	0.7487	$C_{12}H_{26}$	0.6 lel	73.89 (165)	−9.55	170.337	1.4216	i/v/v	0.7526	5.96	1 (47.8)	<32	6.97
Tridecane	235.44	629-50-5	0.7564	$C_{13}H_{28}$		79.44 (175)	−5.39	184.362	1.4256	i/v/v	0.7603		1 (59.4)	33.8	7.54
Tetradecane	253.57	629-59-4	0.7628	$C_{14}H_{30}$	0.5 lel	100 (212)	5.86	198.392	1.4289	i/v/v	0.7667	6.83	1 (76.4)	35.4	8.11
Pentadecane	270.63	629-62-9	0.7685	$C_{15}H_{32}$			9.93	212.418	1.4319	i/v/v	0.7724		1 (91.6)	37.0	8.69
Hexadecane	286.793	544-76-3	0.7734	$C_{16}H_{34}$		135 (275)	18.17	226.444	1.4345	i/d/v	0.7773	7.8	1 (105.3)	39.1	9.26
Heptadecane	301.82	629-78-7	0.7780	$C_{17}H_{36}$		148.89 (300)	21.98	240.471	1.4369	i/d/v	0.7818		1 (115)	41.9	9.84
Octadecane	316.12	593-45-3	0.7819	$C_{18}H_{38}$		165.56 (330)	28.18	254.498	1.4390	i/d/v	0.7858		1 (119)		10.41
Nonadecane	329.7	629-92-5	0.7855	$C_{19}H_{40}$		168.33 (335)	32.1	268.525	1.4409	i/d/s	0.7893		1 (133.2)		10.98
Pristane	296.0	1921-70-6	0.7827	$C_{19}H_{40}$				268.525	1.4385	i/—/s	0.7925			39.1	10.98
Eicosane	342.7	112-95-8	0.7887	$C_{20}H_{42}$		182.22 (360)	36.8	282.552	1.4426	i/—/s				Solid	11.56

[a] Solubility in water/alcohol/ether: v = very soluble; s = soluble; d = slightly soluble; i = insoluble.

[b] At 760 mm Hg, 25°C.

[c] At atmospheric pressure.

determined by gas chromatography and other methods (36). In the field, portable devices such as combustible gas indicators or organic vapor analyzers can also be used. According to the classification by the Occupational Safety and Health Administration (OSHA) (37), the lower alkanes are simple asphyxiants, and air concentrations are recommended to be held below 1000 ppm. Commonly maintained threshold limit values (TLVs) for the C_1 to C_3 hydrocarbons are 1000 ppm, and for the C_4, 600 ppm, as recommended by the American Conference of Governmental Industrial Hygienists (ACGIH) (38).

1.2 Methane

Methane, CH_4, is designated by the Department of Transportation (DOT) (39) as a flammable gas. It occurs in natural gas at a concentration of 60 to 80 percent (40) and to some degree at coal mining or geologically similar earth deposit sites, evolves as marsh gas, and forms during certain fermentation and sludge degradation processes. It is usually accompanied by other low molecular weight hydrocarbons and sulfur compounds. Methane is odorless, but has practically no physiological effects below the flammability limits. Other selected physical data are presented in Table 45.2 (6–15).

Methane is practically inert, but at very high concentrations is a simple asphyxiant (16). A concentration of 87 percent has been demonstrated to cause asphyxiation, and 90 percent respiratory arrest in mice (41). However, methane would be a rather inefficient anesthetic agent, owing to its apparent lower protein-binding properties (42). Methane appears to be absorbed (43, 44) and readily metabolized by the mammalian system (45). When inhaled, the main portion is exhaled again in unchanged form. Uptake in man is less rapid than in the rat (43). The liquefied gas may cause frostbite on skin contact.

Methane inhibits germination of *Bacillus megaterium*; however, in combination with other gases, it appears more promoting to *Escherichia coli* and *Neurospora crassa* (47), but not for spore formation (28). Also, the utilization of methane by some cultures, such as *Methylococcus* and others, (27, 48) as a carbon source suggests that methane is biodegradable.

The 96-hr aquatic toxicity rating (TLm96) is more than 1000 ppm (49). OSHA (37) offers no suggestion as to the threshold limits. ACGIH (38) suggests that methane be treated as a simple asphyxiant. A threshold concentration or time-weighted average (TWA) of 1000 ppm is commonly assumed.

1.3 Ethane

Ethane, CH_3CH_3, is a flammable gas (DOT) (39), and some further selected properties are listed in Table 45.2 (6–15). Ethane occurs in natural gas at 5 to 9 percent (40) and is practically inert. Small quantities of ethane, along with other C_1 to C_4 alkanes and alkenes, have been detected on mined coal samples

(50). Additionally, Koester et al. (51) have demonstrated in rats that ethane is produced as a catabolic product of lipid peroxidation. Guinea pigs exposed to 2.2 to 5.5 percent for 2 hr show slight signs of irregular respiration, readily reversible on cessation of the exposure (52). At higher concentration ethane becomes a simple asphyxiant (16). At concentrations of 15 to 19 percent when mixed with oxygen ethane is a weak cardiac sensitizer (20). Some microorganisms, such as *Neurospora crassa*, appear to grow on ethane (27). It also actively promotes growth of *Mycobacterium vaccae* (53), and is documented to serve as a medium for petro-protein production (59). On the other hand, ethane prevents the germination of *Bacillus megaterium* spore cells (28). Industrially, ethane is handled similarly to methane, and a threshold limit of 1000 ppm is commonly assumed. A method to separate and identify air sample mixtures of C_2 to C_5 hydrocarbons chromatographically has been developed by Westberg et al. (55).

1.4 Propane

Propane, $CH_3CH_2CH_3$, occurs in natural gas at a concentration of about 3 to 18 percent (40). It is emitted into the atmosphere from furnaces, automobile exhausts, and natural gas sources. It occurs in traces of human expired air (56). For example, measurements in a medium-sized U.S. city in 1972 have shown community air concentrations of approximately 50 ppb (57).

Propane is odorless, highly flammable (DOT) (39), and explosive. Some selected physical data are presented in Table 45.2 (6–15). Propane is used as an aerosol component and as a fuel source. With sufficient oxygen, it burns to carbon dioxide and water, but to carbon monoxide when oxygen deficient (58), and at 650°C decomposes to ethylene and ethane. Propane is a simple anesthetic (16) and is nonirritant to the skin and eyes. Direct contact with the liquefied product causes burns and frostbite. At air concentration levels below 1000 ppm, propane exerts very little physiological action (17). It presently is on the FDA's GRAS list, *Substances Generally Accepted as Safe* (59). At very high levels, propane has narcotic and asphyxiating properties. Twenty cases of "sudden death" have been reported by Ikoma (60) in which propane and propylene were idenfified in blood, urine, and cerebrospinal fluid. Animal inhalation studies indicate a gas concentration of 89 percent to be below the anesthetic level but to depress the blood pressure of cats (61). Guinea pigs showed sniffing and chewing movement at 2.2 to 5.5 percent, with a rapidly reversible effect upon cessation of exposure (52). According to Aviado (62), 1 percent causes hemodynamic changes in dogs; 3.3 percent decreases inotropism of the heart, a decrease in mean aortic pressure, stroke volume and cardiac output, and increase in pulmonary vascular resistance. In the primate, 10 percent propane induces some myocardial effects, and at 20 percent aggravation of these parameters and respiratory depression (63, 64). Ten percent propane in the mouse (63) and 15 percent in the dog (20) appear to produce no arrhythmia but weak cardiac sensitization. According to Vestal and Perry (53), propane is utilized by *Microbacterium vaccae*, and is readily degraded by soil bacteria (65). According to O'Brien and Brown (66), *Mycobacterium phlei* is capable of growing on propane as the only carbon source. Propane is suggested to be metabolized by the various microorganisms via the malonyl succinate pathway (65). According to Landry and Fuerst (67), propane reduces the viable cell count of *Escherichia coli* and biochemical mutations are obtained. The aquatic TLm96 is greater than 100 ppm (49).

An 8-hr TWA of 1000 ppm (1800 mg/m^3) is suggested by OSHA (37) and ACGIH (38). For monitoring, dual range explosive gas meters are used, properly calibrated for propane, in the range of 481 to 2016 ppm at 20.5°C/760 mm Hg (NIOSH Sampling Sheet S87) (68). A method to determine air sample mixtures of C_2 to C_5 hydrocarbons chromatographically has been developed by Westberg et al. (55).

1.5 Butanes

1.5.1 *n*-Butane

Butane, $CH_3(CH_2)_2CH_3$, occurs in natural gas and in the ambient community air, in small concentrations, originating from various sources, such as combustion of gasoline or similar petroleum products. It has been measured as exhaust from diesel engines at 22 ppm (5). The pure or mixed product is used as an aerosol propellant, fuel source, or chemical feedstock for special chemicals in the solvent, rubber, and plastics industries. It is a flammable gas (DOT) (39), and other selected physical properties are listed in Table 45.2 (6–15).

Upon direct contact, the liquefied product may cause burns or frostbite to the eyes and skin. The inhalation of 10,000 ppm for 10 min may result in drowsiness, but no systemic effects (17). According to Shugaev (69), the inhalation LC_{50} for the rat is 658 mg/liter for a 4-hr exposure. The inhalation LC_{50} for the mouse is 680 g/m^3 for 2 hr (69). Stoughton (18) reports that *n*-butane is anesthetic to mice at 13 percent in 25 min, at 22 percent in 1 min. In the dog, 25 percent is required for anesthesia (18). Nuckolls (52), in experimenting with guinea pigs, shows that concentrations of 2.1 to 5.6 percent cause sniffing and chewing movements with rapid rate of breathing, but quick recovery after cessation of exposure. Mixing butane and isobutylene has an additive narcotic effect in rats (70). The mechanism concerning the anesthetic properties of butane is proposed to resemble that of ethane and propane as discussed in Section 1.2.

Shugaev (71) observed some parallelism between the toxicity and effective cerebral concentrations. Butane is a weak cardiac sensitizer in the dog (72). Aviado (62) reports that 5000 ppm in the anesthetized dog may cause hemodynamic changes, such as a decrease in cardiac output, left ventricular pressure and stroke volume, a decrease in myocardial contractility, and aortic pressure. Butane is partially absorbed by rat tissue and translocated to the brain, kidney, liver, spleen, and perinephric adipose tissue (69). Microsomal enzyme systems have been found that oxidize butane to its parent alcohol (73). Conversely, butane promotes lipid synthesis by the hydrocarbon *Mycobaterium vaccae* (53). Butane inhibits the growth of some bacteria, mold, fungi, and plant seeds (28). It also inhibits enzymatic lysis of bacterial spores (74). On the other hand, *Mycobacterium crassa* and *phlei* grow on butane (66). In combination with various concentrations of oxygen butane supports the growth of *Neurospora crassa* (47), as well as the germination of *N. ascrospores* and growth of *Escherichia coli* strains B and Sd4 (17, 67), thus rendering butane potentially biodegradable. The aquatic TLm96 has been determined to be above 1000 ppm (49). The ACGIH (38) recommends a TWA of 600 ppm or 1400 mg/m^3 and a short-term exposure level (STEL) of 750 ppm or 1610 mg/m^3 for butane. No odor is recognized in air below 50,000 ppm (17), but is in water at 6.16 ppm (25). For air monitoring, color detector tubes are available (75). A method to determine air sample mixtures of C_2 to C_5 hydrocarbons chromatographically has been developed (55, 76).

1.5.2 2-Methylpropane, Isobutane

Isobutane, 2-methylpropane, $(CH_3)_3CH$, a flammable gas (DOT) (39), occurs in small quantities in natural gas and crude oil. It has been detected in community atmospheres (2) at concentrations of 44 to 74 ppb (3); it also evolves from natural sources, has been measured from diesel exhausts at 1.4 to 11 ppm (4, 5), and occurs in cigarette smoke at 10 to 60 ppm (77). Isobutane is produced in refining processes and used as a raw material for petrochemicals (1), a component of aerosol propellants (78), an industrial carrier gas, and general fuel source. It represents one of the basic raw materials used in the chemical industry for the production of propylene glycols and oxides and polyurethane foams and resins. Its physical properties are listed in Table 45.2 (6–15).

Toxicologically, the vapor exerts no effect on skin and eyes, except as a liquid in direct contact, where it produces chemical burns. Human volunteers exposed to 250 to 1000 ppm for 1 min to 8 hr and 500 ppm for 1 to 8 hr/day for 10 days showed no deleterious effects (79). An inhalation LC$_{50}$ of 52 mg/kg for the mouse for a 1-hr exposure has been documented by Aviado (62). Near the LC$_{50}$, mice exhibit CNS depression, rapid and shallow respiration, and apnea. Isobutane may be a cardiac sensitizer (62, 72). At high concentrations it caused a decrease in pulmonary compliance and tidal volume in the rat (80). According to Aviado, the anesthetized dog showed no significant effects up to 2 percent, but decreased myocardial contractility at 2.5 percent, exaggerated effects at 5 percent, with a decrease in ventricular and aortic pressure, and at 10 percent decreased left ventricular pressure, mean arterial flow, and stroke volume, with increased pulmonary vascular resistance. In the mouse, isobutane is narcotic at 15 percent in 60 min, and at 23 percent in 26 min (18). At 22 to 27 percent it is anesthetic in the mouse in 8.7 min, but causes respiratory arrest in 15 min (80). In the dog, anesthesia occurs at 45 percent in 10 min (18). Isobutane is oxidatively metabolized by rat liver microsomes to its parent alcohol (73). The gas inhibits the growth of some bacteria (81) but supports others, such as *Mycobacterium phlei* (66).

The threshold limits suggested by ACGIH (38) are 600 ppm or 1400 mg/m^3 for a TWA and 750 ppm or 1610 mg/m^3 for a short-term exposure limit (STEL). Isobutane can be chemically separated and determined by gas chromatography.

1.6 Pentanes

1.6.1   *n*-Pentane

Pentane, $CH_3(CH_2)_3CH_3$, is the first liquid member of the alkanes. It has been detected in the urban atmosphere and is believed to have originated from natural sources (76). Selected physical properties are listed in Table 45.2 (6--15). It is used as a component of aerosol propellants and solvents and is an important component of engine fuel (10); it is also a raw material for the production of chlorinated pentanes and pentanols (10). Pentane has lower acute narcotic potency than the C_1 to C_4 gases. The intensity appears generally to decrease with increasing molecular weight, but increases for the highly symmetrical compounds (23). For humans, the lowest recorded lethal concentration (LC$_{LO}$) is 130,000 ppm, and the lowest concentration to produce toxic effects (TC$_{LO}$) is 90,000 ppm (82). It is a weak cardiac sensitizer of the dog heart to epinephrine (20, 72). However, only a small dose differential separates narcosis and lethal effect. Inhalation of 5000 ppm (0.5 percent) pentane for 10 min has not been irritant to mucous membranes or shown to have local or systemic effects. A concentration of 200 to 300 mg/liter (200 to 300,000 mg/m^3) caused incoordination and inhibition of the righting reflex of white mice (83). Stoughton and Lamson (18) report that pentane is anesthetic to mice at 7 percent concentration in 10 min and 9 percent concentration in 1.3 min. Swann et al. (26) report that 128,000 ppm caused deep anesthesia in mice. Also in the mouse, a 9 to 12 percent concentration for 5 to 60 min resulted in narcosis, but at 12.8 percent for 37 min in death. Similarly air concentrations of 10.4, 50.9, and 94.7 mg/m^3 showed neurohistological, structural changes in the developing cerebral cortex of the rat (84). Massive doses, such as about 40 percent, resulted in death of mice, which showed collapsed lungs on autopsy (85). Chronic exposure has resulted in anoximia (16). When applied to segments of the rat nerve, *n*-pentane had a slow inhibitory action on the myelin sheath of the peripheral nerve tissue (86). A nerve impulse blockage has been demonstrated in the squid axon and also in the frong sciatic nerve (87) that could be verified for pentane and hexane, but not for the higher alkanes (88). Similarly, a solvent mixture of 80 percent pentane, 5 percent hexane, and 14 percent heptane was responsible for several cases of polyneuritis (89), although normally pentane is metabolized by hydroxylation to pentanol (73), conjugated with glucuronate, and subsequently excreted (90). However, some isolated systemic effects indicate local affinity and destruction of the myelin sheath of peripheral nerve tissue (86). This, however, pertains more to hexane than pentane. Pentanes and higher alkanes can be used as a sole carbon source by a variety of microorganisms such as *Pseudomonas fluorescens* and *Corynebacterium* (91), *Mycobacterium vaccae* (53), *M. phlei* (66), and *Micrococcus lysodeikticus*. An aquatic TLm96 of 100 to 10 ppm has been established (49). Threshold limits recommended by OSHA (37) are 1000 ppm and by ACGIH (38) 600 ppm or 1800 mg/m^3. The recommended occupational exposure standard for pentane is 120 ppm or 350 mg/m^3 and a ceiling concentration of 1800 mg/m^3 for 15 min for a mixture of C_5 to C_8 n-alkanes (92). Saturated air contains 66 percent (v/v) pentane, at 25°C/760 mm Hg. The odor detection of pentane is around 300 to 500 ppm, and at 5000 ppm is assigned a moderate odor intensity on the 0 to 5 degree scale by Patty and Yant (17). An air monitoring procedure has been established by NIOSH (see Sampling Data Sheet S379.01) (68). Samples are collected on charcoal, desorbed with carbon disulfide, and gas chromatographically determined using a flame ionization detector. A method to determine air sample mixtures of C_2 to C_5 hydrocarbons chromatographically has been developed by Westberg et al. (55).

1.6.2 2-Methylbutane, Isopentane

Isopentane, 2-methylbutane, $(CH_3)_2CHCH_2CH_3$, exhibits very similar physical and physiological characteristics to those of pentane; however, less information is available. Isopentanes have been detected in community air (2–4). Isopentanes are used for the production of amylnaphthalenes and isoprene (10). Some physical data are listed in Table 45.2 (6–15). As an anesthetic, isopentane is less potent (18) than the C_1 to C_4 alkane members; however, metabolically it appears more active. Very high vapor concentrations are irritant to the skin and eyes. Inhalation of up to 500 ppm appears to have no effect on humans (17). The inhalation LC$_{50}$ in the mouse is estimated at about 1000 mg/liter. Methylbutane may be narcotic between 270 and 400 mg/liter, similar to pentane (41), and is a weak cardiac sensitizer (20). Isopentane is metabolized by hydroxylation to the parent alcohol (73). A TLm96 of 1000 to 100 ppm has been determined (49). According to O'Brien and Brown (66), *Mycobacterium phlei* is capable of growing on isopentane.

No TLVs are officially available; however, TWA and ceiling values established for pentane may be recommended. Similarly, air monitoring and chemical analyses can be carried out as for pentane. Isopentane should be labeled as a flammable liquid (DOT) (39).

1.6.3 2,2-Dimethylpropane, Neopentane

Neopentane, 2,2-dimethylpropane, $(CH_3)_4C$, is physically and physiologically similar to butane [see Table 45.2 (6–15)]. Neopentane, like pentane, is an important component of petroleum fuel mixtures. Limited toxicologic data are available (11). However, by structure–action relationship extrapolations from physical, chemical, and biologic data, it is estimated that its toxicity characteristics are lower than those of isopentane and pentane. This has been confirmed in experimental work by Stoughton and Lamson (18). Neopentane is hydroxylated by rat liver microsomes to the parent alcohol (73).

For industrial hygiene information, see Section 1.6.2.

1.7 Hexanes

1.7.1 *n*-Hexane

n-Hexane, $CH_3(CH_2)_4CH_3$, is isolated from natural gas and crude oil (10). It is widely used pure or as commercial grade solvent, which may be *n*-hexane mixed with isohexanes and cyclopentane. Some selected physical properties are listed in Table 45.2 (6–15).

Hexane may be the most highly toxic member of the alkanes. It is an anesthetic (87, 88, 93). When ingested, it causes nausea, vertigo, bronchial and general intestinal irritation, and CNS effects and presents an acute aspiration hazard (19). About 50 g may be fatal to humans. Acute inhalation effects are euphoria, dizziness, and numbness of limbs, as described by Nelson et al. (94). An exposure of 880 ppm for 15 min can cause eye and upper respiratory tract irritation in humans (95). Patty and Yant (17) report that exposure to 5000 ppm for 10 min causes marked vertigo. Concentration of 100 ppm had an effect on the righting reflex in the white mouse (83). Chronic industrial exposure is documented by Battistini et al. (96), Paulson and Waylonis (97), and Herskowitz et al. (98) to have caused motor polyneuropathy in workers. Exposure to 2000 ppm produced comparable neuropathy in workers (96, 97). Similarly, among 1264 workmen, 32 cases of neuropathy were observed in the shoe manufacturing industry, where hexane–cyclohexane solvent mixtures were handled (99). In 1662 Japanese workers, 93 cases of polyneuropathy were observed due to *n*-hexane (100). Hexane is irritant to the skin. When injected subcutaneously into mice, it produced pathological changes similar to those seen in humans (101). The lowest lethal concentration (LC_{LO}) for mice is 120 g/m^3 (83). The lowest lethal dose (LD_{LO}) of hexane for the rat is 9100 mg/kg (102). Taira (103) reports that the intramuscular injection of hexane in rabbits caused edema and hemorrhaging of the lungs and tissues, with polymorphon-uclear leukocytic reactions. The acute oral LD_{50} in the rat is 24 to 49 ml/kg (104). Dermal application of 2 to 5 ml/kg for 4 hr to rabbits has resulted in ataxia and restlessness (105). No deaths occurred at 2 ml/kg, but some did at 5 ml/kg. Inhalation of 1000 to 64,000 ppm for 5 min in the mouse showed irritation of the respiratory tract and anesthesia (26), 30,000 ppm produced narcosis (19), and 34,000 to 42,000 ppm was lethal (106). Subchronic exposure of rats for 14 weeks resulted in nervous system type disease, including cerebral peripheral distal axonopathy (106). Subchronic exposure of pigeons to 3000 ppm, 5 hr/day, for 82 days over 17 weeks has shown no pathological nerve tissue alterations (107). However, exposure of rats to 400 to 600 ppm, 5 days/week, has resulted in peripheral neuropathy in 45 days (108). Inhalation of 100 ppm hexane by mice gave no evidence of peripheral polyneuropathy in 7 months; however, 250 ppm resulted in peripheral nerve injury (109). Exposure of rats to 850 ppm for 143 days showed loss in weight and degeneration of myelin and axis cylinders in the sciatic nerve (110–112). Loss of axons and nerve terminals has also been observed as a result of "glue sniffing" (113). Chronic effects from glue sniffing over a period of 5 to 15 months have been described as distal symmetrical motor sensory polyneuropathy by Ashbury et al. (111), Gonzales and Downey (112), and Altenkirsch et al. (114). Hexane is actively absorbed by the mammalian system and is accumulated in cellular tissue proportional to the lipid content (115). It is metabolized at relatively high rates to hydroxy derivatives (32), as also determined by a cytochrome P-450 containing mixed function oxidase system (116, 117) before being converted to a possible keto form. Some metabolites, 5-hydroxy-2-hexanone and 2,5-hex-anedione, also common to the neurotoxic methyl butyl ketone (MBK), have been recognized (118). Scala (119) reports that exposure to 200 ppm of MBK comparable to about 175 ppm hexane for 4 to 6 months can result in histological and physiological nerve changes in rats. Hexane activates various enzyme systems, such as UDP-glucuronyl transferase (120). Some normal hydroxylated intermediates may be excreted as the glucuronides (90). Hexane was found to be inactive as a tumor-producing agent (121). Some toxic forms apparently have an affinity to nerve tissue, where it affects blockage of nerve impulses in the frog (87) and structural changes of the myelin sheath, as shown in the rat (86). A TLm96 was determined to be greater than 1000 ppm (49). A variety of microorganisms degrade hexane by oxidation mechanisms similar to the lower homologues (32). Hexane is also utilized by various microorganisms, such as *Mycobacterium vaccae* (53) and *M. phlei* (66). The recommended threshold limit values are, by OSHA (37), a TWA of 500 ppm or 1800 mg/m^3, and by ACGIH (38), a TWA of 100 ppm or 360 mg/m^3 and a STEL of 125 ppm or 450 mg/m^3. The recommended occupational exposure standard is 100 ppm for hexane as such, or for a mixture of C_5 to C_8 alkanes, 350 mg/m^3; the ceiling concentration is 1800 mg/m^3 for 15 min (92). The concentration in saturated air is 19.94

percent at 25°C/760 mm Hg. For detailed protection information, see the NIOSH criteria document (92).

For air monitoring and detection, charcoal tubes are used to collect the samples, the materials are desorbed with carbon disulfide, and the hexane or its isomers are determined by gas chromatography, using flame ionization detectors (see NIOSH Sampling Data Sheet 290-1) (68).

1.7.2   2-Methylpentane, Isohexane

Little information is available on isohexane. Some physical properties are listed in Table 45.2 (6–15). No physiological data are available. However, toxicologically, the isohexanes are expected to be mucous membrane irritants and to have a low oral toxicity, but to be aspirated into the lungs and absorbed through the skin. Isohexanes are predicted to have narcotic properties and are documented to be cardiac sensitizers (20), but are not expected to have neurotoxic properties. The TLm96 is over 1000 ppm (49). For industrial hygiene, the information for hexane applies.

1.7.3   Other Hexane Isomers

Information for 2-methylpentane also applies to 3-methylpentane, as listed in Table 45.2 (6–15). Very little information is available on 2,2-dimethylbutane, neohexane. For toxicology, see 2-methylpentane. It is a flammable liquid (DOT) (39); some physical data are listed in Table 45.2 (6–15) for neohexane.

Information for 2,3-dimethylbutane, as listed in Table 45.2 (6–15), also applies to 2-methylpentane.

1.8   Heptanes

1.8.1   *n*-Heptane

n-Heptane, $CH_2(CH_2)_5CH_3$, a flammable liquid (DOT) (39), is isolated from natural gas, crude oil, or pine extracts. It is used and is produced in petroleum refining processes (1). It is used as an industrial solvent, as a high octane starter fluid or blend, or as paraffinic naphtha in place of hexane, where high flash solvents with lower evaporation rates are required (10), and as a gasoline knock-testing standard. Its physical properties are listed in Table 45.2 (6–15). Physiologically, heptane appears more active than hexane (23). Humans exposed to 0.1 percent heptane showed slight vertigo in 6 min, at 0.2 percent in 4 min, and at 0.5 percent marked inability to maintain equilibrium and coordination in 7 min (17). Post-experimental lingering taste of gasoline for several hours has also been reported (17). Concentrations of 4.8 percent caused respiratory arrest within 3.0 min of exposure (26). Survivors show marked vertigo and incoordination requiring 30 min for recovery. They also show mucous membrane irritation, slight nausea, and lassitude, usually for a few hours (17). A 40 mg/liter concentration affected the righting reflexes of white mice, and 70 mg/liter was lethal (83). Direct skin exposure may cause pain, burning, and itching. The time for symptom reversal is longer than for pentane or hexane (92). According to Fuehner (122), narcosis in mice occurs at 1.0 to 1.5 percent within 30 to 60 min. However, tetanic convulsions occasionally occur even with low narcotic concentrations (122). Heptane is also classified as a cardiac sensitizer (72). A narrow margin appears to exist between the onset of narcosis or convulsions and cardiac sensitization and recovery or death. Heptane also appears to be metabolized via its alchohol (90), conjugated by glucuronates, and subsequently excreted. Heptane is metabolized at relatively high rates to hydroxy derivatives by a cytochrome P-450 containing mixed function oxidase system before being converted to the corresponding keto forms. Heptane is believed to have similar neurotoxic characteristics as hexane. Heptane is a lipid solvent, is readily absorbed through the skin, and thus may have systemic effects similar to those of hexane.

Petroleum-utilizing organisms have been shown to biodegrade heptane from 19 to 55 percent in 92 hr, depending on the cultures used (31). *Mycobacterium phlei* is capable of growing on heptane (66), and heptane promotes lipid synthesis by *M. vaccae* (55). The odor threshold is about 200 ppm (17). A 96-hr aquatic toxicity determination showed the TLm96 to be greater than 1000 ppm (49). OSHA (37) and ACGIH (38) recommend a TWA of 400 ppm or 1600 mg/m^3. For monitoring and chemical analysis a method, S-89, is suggested by NIOSH (68). A NIOSH criteria document (92) recommends an occupational exposure standard of 85 ppm or 350 mg/m^3 of a C_5 to C_8 mixture of *n*-alkanes with a ceiling concentration of 1800 mg/m^3 for 15 min.

1.8.2 2-Methylhexane, Isoheptane

Isoheptane, $(CH_3)_2CH(CH_2)_3CH_3$, is one of the isoparaffins lending certain desired properties to fuel mixtures. It is used as raw material for the production of plasticizers. Some physical characteristics are listed in Table 45.2 (6–15). Physiologically, isoheptane is somewhat less active than hexane. It is narcotic at about 0.8 to 1.1 percent (83). The loss of the righting reflex in mice for isoheptane occurred at a concentration of 50 mg/liter; that for the heptane was observed at 40 mg/liter (83), rendering the isomer somewhat less toxic. However, it is not believed to be a neurotoxin.

For industrial hygiene information, see heptane.

1.8.3 Other Heptane Isomers

The other branched isomers, such as 3-methylhexane, are expected to have toxicological properties similar to those of 2-methylhexane. For selected physical data on neoheptane, 2,2'-dimethylpentane, see Table 45.2 (6–15).

1.9   Octanes

1.9.1   *n*-Octane

Octane, $CH_3(CH_2)_6CH_3$, a flammable liquid (DOT) (39), occurs in natural gas
and crude oil. It is used as a solvent and widely as a chemical raw material, and
it serves as an important chemical agent in the petroleum industry. Octane,
along with other *n*- and isoparaffins, offers highly desirable blending values to
achieve certain preferred antiknock and combustion qualities for high compression engine fuels. Some physical properties are listed in Table 45.2 (6–15).

Orally, octane may be more toxic than its lower homologues. If the material
is aspirated into the lungs, like heptane, it may cause rapid death due to cardiac
arrest, respiratory paralysis, and asphyxia (19). The narcotic potency of octane
is approximately that of heptane, but does not appear to exhibit the CNS effect
as do the two lower homologues. A minimal concentration to cause loss of
righting reflexes in mice was 35 mg/liter, and total loss of reflexes at 50 mg/liter
(83). A concentration of 95 percent causes loss of reflexes in mice in 125 min;
however, up to 1.9 percent appears to be tolerated for 143 min, with the effects
being reversible (122). Octane may be metabolized via its hydroxy derivatives
at relatively high rates through a cytochrome P-450 oxidase system (116).
Microorganisms such as *Pseudomonas oleovarans* (32) and others degrade octane
(53) by oxidation or hydroxylation to octanol.

OSHA (37) recommends a TWA of 500 ppm or 2350 mg/m^3, and ACGIH
(38) a TWA of 300 ppm or 1450 mg/m^3 or STEL of 375 ppm or 1800 mg/m^3.
The recommended occupational threshold limit by NIOSH (92) is 75 ppm for
octane and 350 mg/m^3 for a C_5 to C_8 mixture. The odor is detectable at 400
ppm (14, 68). Samples are collected for monitoring by absorption on charcoal
and desorbed with carbon disulfide, and analyzed by flame ionization gas
chromatography. A NIOSH Sampling Data Sheet S378 (68) is available.

1.9.2   Octane Isomers

Isooctanes, blended with lower alkane homologues, play an important role as
preignition additives for high compression engine fuel (10). Physical data are
listed in Table 45.2 (6–15) for trimethyl isomers. Orally, the isooctanes are
moderately toxic, more so than the lower homologues, and appear to cause
pulmonary lesions if aspirated into the lungs of rats (19). The narcotic
concentration for 2,5-dimethylhexane is about 70 to 80 mg/liter for the mouse,
and the effects appear to be lower than those for octane (83). When injected
intramuscularly into rabbits, isooctane produces hemorrhage, edema, and
polymorphonuclear leukocytic reactions (103), such as angitis, interstitial pneumonitis, abscess formation, thrombosis, and fibrosis (103). About 16,000 ppm
of isooctane causes respiratory arrest in mice (26). The compounds are believed
to be metabolized by the mammalian system similarly as shown in certain

microorganisms where some enzyme systems have been recognized that readily hydroxylate isooctanes (32). Other homologues, 2,2,4-trimethylpentane, 2,3,4-trimethylpentane, and mixed dimethylhexanes, are used in the chemical industry. Some selected physical properties are listed in Table 45.2 (6–15); however, very little direct toxicological information is presently available. By structure–action extrapolation, these compounds can be physiologically compared to 2,5-dimethylhexane. None of the branched octanes are expected to have neurotoxic properties. Branched octanes, such as 2,5-dimethylhexane or 2,2,4-trimethylpentane, are used as the sole carbon sources by a variety of microorganisms, such as *Pseudomonas fluorescens, Corynebacterium* (91), and *P. oleovorans* (32). For industrial hygiene methods, see Section 1.8.

1.10 Nonanes

1.10.1 *n*-Nonane

Nonane, $CH_3(CH_2)_7CH_3$, is an important component of gasoline (10). Nonane has been detected in the Los Angeles atmosphere (4). Bertsch et al. (123) report that concentrations of 1.6 to 4.4 ppb of *n*-nonane have been determined analytically in polluted air. For physical data, see Table 45.2 (6–15).

Few toxicological data are available. Early investigators found it difficult to generate sufficient vapor for inhalation experiments, owing to the material's low vapor pressure. An acute LC_{50} of 17 mg/liter (3200 ppm) for a 4-hr exposure has been established for rats (124). Carpenter et al. (124), with inhalation experiments in rats, have established no-effect levels of 1.9 and 3.2 mg/liter, administered 6 hr/day for 5 days/week for 13 weeks. However, 8.1 mg/liter (1500 ppm) resulted in mild tremors, slight coordination loss, and low irritation of the eyes and extremities. According to Vinogradov et al. (125), chronic inhalation of nonane vapors may cause altered neutrophils, although histopathological examination revealed no pulmonary lesions. Nonane is used as a carbon source by some bacteria, but prevents spore formation in others, such as *Bacillus megaterium* (28), indirectly implying that it may be biodegradable. Nonane is metabolized in the rat at relatively high rates to hydroxyl derivatives prior to conversion into the corresponding keto form, as determined with a cytochrome P-450 containing mixed function oxidase system (116). The TWA recommended by ACGIH (38) is 200 ppm or 1050 mg/m^3, with a STEL of 250 ppm or 1300 mg/m^3. For monitoring and sampling, the same methods apply as recommended for octane; see Section 1.8.

1.10.2 2,2,5-Trimethylhexane

Trimethylhexane, neononane, $(CH_3)_3C(CH_2)_2CH(CH_3)_2$, is a component of engine fuel. Some physical data are presented in Table 45.2 (6-15). However, little toxicologic information is available. By extrapolation, this branched compound is expected to be less toxic than nonane. An estimated TC_{LO} would be above 4000 ppm or the LD_{LO} >4 g/kg. Industrial hygiene methods listed for octane (Section 1.8) may apply.

1.11 Decanes

1.11.1   n-Decane

Decane, $CH_3(CH_2)_8CH_3$, is a component of gasoline (10). Some physical and chemical properties are presented in Table 45.2 (6–15). Decane has been detected in the Los Angeles atmosphere (4). Concentrations of 1 to 2.7 ppb have been determined analytically in polluted air by Bertsch et al. (123). Decane is considered relatively nontoxic. By thermodynamic considerations and theo-retical extrapolation, Crisp et al. (23) predict that decane may exhibit similar but somewhat decreased narcotic activity from hexane or octane. Nau et al. (126) showed that rats exposed to 540 ppm of n-decane by inhalation for 18 hr/day, 7 days/week for 57 days demonstrated a significant positive effect on the expected weight gains and a significant fall in the total white blood count, but no bone marrow changes or organ changes of significance. The lowest dermal toxic dose (TD_{LO}) for mice is 25 g/kg for 52 weeks (121, 126). Decane possesses similar lipid solubility characteristics as hexane and octane, and thus causes pulmonary effects when aspirated. Gerarde (19) has shown that this holds for the C_{10} alkanes. Decane is readily hydroxylated to decanol by *Pseudomonas desmolytica* S_{11} (127) and *P. oleovorans*, (32) thus implying that it is potentially metabolizable and biodegradable. Lebeault et al. (128) have documented that decane is microbiologically degraded by *Candida tropicalis* via the 1,2-dehydro-genated and 1-hydroxylated intermediates. Minor alternate pathways proceed via the 1,2-diol and also show the possibility of terminal oxidation (128). Decane is metabolized at relatively high rates to hydroxy derivatives (32) before being converted to the respective keto form (116), as determined with a cytochrome P-450 containing mixed function and microsomal oxidase systems and aerobic spore-forming bacteria (129).

Industrial hygiene methods as discussed for octane (Section 1.8) may apply for decane. Analytical procedures for decane and isodecane are available (36, 130).

1.11.2   2,7-Dimethyloctane

In 1929, Lazarew (83) found 2,7-dimethyloctane, $(CH_3)_2CH(CH_2)_4CH(CH_3)_2$, diisoamyl, not to be narcotic. Some chemical and physical data are listed in Table 45.2 (6–15).

1.12   Undecane

1.12.1   *n*-Undecane

n-Undecane, hendecane, $CH_3(CH_2)_9CH_3$, is a component of gasoline (10). Physical data are presented in Table 45.2 (6–15). Rode and Foster (28) document that undecane, when tested with bacterial spores, causes inhibition of certain growth media.

1.13   Dodecane

Dodecane, $CH_3(CH_2)_{10}CH_3$, is a component of gasoline (10). Some physical and chemical properties are listed in Table 45.2 (6–15).

Dodecane was shown to be a potentiator of some carcinogens above an apparent 0.02 percent threshold dose when, for example, benzo(*a*)pyrene dissolved in decalin was applied to the mouse skin (131–133). According to another source, the lowest toxic dose (TD_{LO}) for mice is 11 g/kg for 22 weeks for the above mixture (127). Dodecane, which is not highly toxic, is a possible potentiator of skin tumorigenesis by benzo(*a*)pyrene, decreasing the effective threshold dose by a factor of 10 (132). A mixture of 50 percent dodecane and 50 percent decalin provides a suitable carrier for a strongly carcinogenic agent for the skin of mice, as reported by Horton and Christian (134). Dodecane is used as a carbon source by a variety of microorganisms, such as *Pseudomonas fluorescens*, *Corynebacterium* (91), and *P. oleovorans* (32). It may inhibit carbohydrate transport in *Candida* species (135).

1.14   Tridecane and Higher Homologues

The higher homologues, such as hexadecanes, are used as stock for hydrocracking processes (1). Some physical and chemical properties are listed in Table 45.2 (6–15) for the alkanes, *n*-tridecane C_{13}, *n*-tetradecane C_{14}, *n*-pentadecane C_{15}, *n*-hexadecane C_{16}, *n*-heptadecane C_{17}, *n*-octadecane C_{18}, *n*-nonadecane C_{19}, 2,6,10,14-tetramethylpentadecane (pristane) $C_{19}H_{40}$, and *n*-eicosane C_{20} (8). Pristane is used as a lubricant and transformer oil, and is also oxidatively degraded by *Pseudomonas oleovorans* (32). The lowest toxic dose (TD_{LO}) of tetradecane for mice is 9600 mg/kg for 20 weeks (121). A toxicologic study by Gerarde (19) indicates that C_{13} to C_{16} alkanes, when aspirated into the lungs, are asphyxiants similar to the C_6 to C_{10} members, but cause death more slowly.

Topical applications of hexadecane to the guinea pig skin caused marked hyperkeratosis, with an accompanying increase in acid phosphatase (136). Conversely, hexadecane in combination with 2-butanone or cyclohexane enhanced the activity of local anesthetics (137). With the inhalation of similarly high concentration of squalene, $C_{30}H_{62}$, lung changes may occur (85). Hexadecane is readily degraded through mineralization by marine bacteria (138). Oral administration of 1 g heptadecane to rats within 2 hr showed a distribution of 0.7, 1.4, 1.2, and 0.2 percent in the intestinal wall, liver, intestinal content, and feces, respectively (139). Rats fed a diet of 25 percent *Spirulina algea* accumulated heptadecane in adipose tissue at 80.2 and 272.0 µg/g in males and females and in lung and muscle at ~10 to 20 µg/kg (140). Pigs receiving up to 52 ppm for 12 months in the diet excreted some heptadecane in milk during lactation (140). The higher homologues of the alkanes have been detected as oil droplets in a variety of human tissues (21), and C_{20} and higher members occur naturally in a variety of waxes (40). Varying with the season, C_{14} to C_{35} alkanes have been isolated from milk fat (141).

A variety of microorganisms grow on higher alkanes. For example, yeasts such as *Candida lipolytica* (142) and *Nocardia corralina* (143) can use alkanes as the sole carbon source. For example, tetra- and hexadecanes can be utilized by *Pseudomonas* strains for the synthesis of phenazine carboxylic acids (144) and a *Candida* strain was found capable of using tetradecane as the sole carbon source (145). In *Candida tropicalis*, Lebeault (128) has detected some alkane, alcohol, and aldehyde dehydrogenase dependent synthetases, demonstrating the organism's independent alkane degrading and synthesizing capabilities. Horton and Christian (134) report that cocarcinogenic activity (but more likely a general solvent property) may be common to many C_{12} to C_{30} aliphatic hydrocarbons, the counterparts ranging from alkylbenzenes to dibenzanthracenes. Lu et al. (116) found that the tetradecane and hexadecane were metabolized by a cytochrome P-450 mixed function oxidase system and also hydroxylated by an enzyme from *Pseudomonas oleovorans*, (32) and the C_{12}, C_{14}, and C_{16} *n*-paraffins also by *P. aeruginosa*. Hexadecane, in addition, can be degraded by a variety of oceanic bacteria (147). Using radioactive tracers, branched chain alkenes, such as the 2-methylhexadecane, 2,2-dimethylhexadecane, or pristane, have been demonstrated to readily undergo microbial oxidation (148). Heptadecane was found by Lee et al. (149) to be taken up into mussel. The alkanes to C_{17} may be collected on charcoal, the higher members as particulate matter on filters, and determined by gas chromatography (36).

2 ALKENES (OLEFINS)

2.1 General Correlations

2.1.1 Occurrence and Use

Most industrially important olefins are produced by petroleum cracking processes (1). Small quantities of the lower alkenes have been detected in community air (2). In a comprehensive environmental sampling and identification study,

Bertsch et al. (123) tentatively identified only isobutene to occur in higher quantities among a variety of 98 alkyl and aromatic hydrocarbon derivatives. At a direct, experimental diesel engine emission source, considerable quantities of lower alkenes were detected (5); however, they had almost vanished at a 6-ft distance from the source. Linnell and Scott (150) also detected lower alkenes in diesel fume exhaust, but in small quantities only. Most industrially important olefins are produced by petroleum cracking processes and pyrolysis from paraffins. These alkenes, in turn, serve for the preparation of a wide selection of petrochemicals (1) which are extensively utilized in the pharmaceutical, cosmetic, chemical, and rubber processing industries.

2.1.2 Physical and Chemical Properties

Chemically, alkenes are more reactive than alkanes (40), and thus enter addition reactions rapidly. When heated or in the presence of catalysts, most olefins polymerize. They have, for example, higher boiling points, as shown in Table 45.3 (6–15). This type of characteristic may be the reason for their slightly higher toxicity than the alkanes, and their environmental interaction with ozone and other photochemical reactants to form chemiluminescent and other oxidation adducts (151).

2.1.3 General Toxicity

Toxicologically, the alkenes are not particularly active. The lower members of the series, ethylene, propylene, butylene, isobutylene, butadiene, and acetylene, are weak anesthetics and simple asphyxiants. Amylene has been used for surgical anesthesia. The characteristics of increasing mucous membrane irritancy and cardiac effects with increasing chain length render the hexylenes and higher members unsuitable for this purpose. The higher members may cause narcosis (152), although the propylene trimer and higher molecular weight derivatives are not sufficiently volatile to be considered vapor hazards at room temperature. Branching decreases the toxicity of the C_3 member, does not appreciably affect the C_4 and C_5 alkenes, and increases for the C_6 to C_{18} olefins. Unlike the alkanes, the olefins do not exhibit neurotoxic properties. Repeated exposure to high concentrations of the lower members of the alkenes have produced hepatic damage and hyperplasia of the bone marrow in animals. However, no corresponding effects have been recorded in humans. The α-olefins are more reactive and toxic than the β isomers. However, the alkadienes are more irritant than the corresponding alkanes (153); diunsaturation, in general, increases the toxicity.

Gerarde (19) has demonstrated that alkenes, similarly to alkanes, can be aspirated into rat lungs to cause death due to chemical pneumonitis. The rate of death appears practically parallel between alkanes and alkenes of equal

Table 45.3. Physicochemical Properties of Alkenes (6–15)

Compound	B.P. (°C)	CAS Registry No.	Density (20/4°C)	Emp. Formula	Flamm. Limits (%)	Flash pt. [°C (°F)]	Freezing Pt. [°C (°F)]	Mol. Wt.	Refr. Index, n^{20}_D	Solubility[a] w/al/et	Sp. Gr.	Vapor Dens.	Vapor Pres. [mm (°C)]	Visc. (SUS)	W/V Conv. (mg/m³ ⇔ 1 ppm)
Ethylene	−103.71	74-85-1	0.5674	C_2H_4	2.7–36	−136.11 (−213)	−169.15 (−272.47)	28.054	—	i/d/s	—	0.978	40ᶠ (1.5)	N.A.	1.15
Propene	−47.4	115-07-1	0.5139	C_3H_6	2.0–11.0	−108 (−162)	−185.25 (−301.45)	42.08	—	i/v/v	0.522	1.46	10ᶠ (19.8)	N.A.	1.72
Butene-1	−6.26	25167-67-3	0.5951	C_4H_8	1.6–10	−112 (−170)	−185.35 (−301.65)	56.11	1.3962	i/v/v	0.6013	1.93	3480 (21)	N.A.	2.30
cis-Butene-2	3.72	390-18-1	0.6213	C_4H_8	1.7–9.0	−73.3 (−100)	−138.91 (−218.0)	56.11	1.3931ᵇ	i/v/v	0.6272	1.9	760 (3.7)	N.A.	2.30
trans-Butene-2	0.88	624-64-6	0.6042	C_4H_8	1.8–9.7	−73.3 (−100)	−105.550	56.11	1.3848ᵇ	—	0.6100	1.9	760 (0.9)	N.A.	2.30
Isobutylene	−6.90	115-11-7	0.5942	C_4H_8	1.8–9.6	−76.11 (−112)	−140.3 (−220.54)	56.11	1.3926ᵇ	i/v/v	0.6002	2.01	400 (21.6)		2.30
Pentene-1	29.97	109-67-1	0.6405	C_5H_{10}	1.5–8.7	−51.11 (−60)	−165.22 (−263.39)	70.134	1.3715	i/v/v	0.6457	2.42	400 (12.8)	<32.6	2.87
cis-Pentene-2	36.94	627-20-3	0.6556	C_5H_{10}			−151.39 (−240.50)	70.134	1.3830	i/v/v	0.668	2.4			2.87
trans-Pentene-2	36.35	646-04-8	0.6482	C_5H_{10}			−140.244 (−220.43)	70.134	1.3793	i/v/v	0.6533	2.4			2.87
Hexene-1	63.485	592-41-6	0.6732	C_6H_{12}	1.2–6.9	−26.11 (−15)	−139.82 (−219.69)	84.161	1.3879	i/s/s	0.6780	3.0	310 (35)	<32.6	3.44
cis-Hexene-2	68.84	592-43-8	0.6869	C_6H_{12}		−20.6 (−5)	−141.13 (−222.04)	84.161	1.3977	i/s/s	0.6916	2.9			3.44
trans-Hexene-2	67.87	592-43-8	0.6784	C_6H_{12}		−17.8 (0)	−132.97 (−207.35)	84.161	1.3935	i/s/s	0.6832	3.0			3.44
Isohexene-2	67.29		0.6863	C_6H_{12}			−135.07 (−211.13)	84.161	1.4004	i/s/—					3.44
Heptene-1	93.64	592-76-7	0.6930	C_7H_{14}		−3.89 (25)	−119.03 (−182.15)	98.188	1.3988	i/s/s	0.7015	3.39	101.4 (21.1)	<32.6	4.02

cis-Heptene-2	97.95		0.7012	C_7H_{14}		−2.2 (28)	−109.48 (−165.5)	98.188	1.4045	i/s/s	0.7057	3.34			4.02
trans-Heptene-3	95.67		0.6981	C_7H_{14}		−6.1 (21)	−136.63 (−213.93)	98.188	1.4043	i/s/s	0.7075	3.38			4.02
Octene	121.28	111-66-0	0.7149	C_8H_{16}		21.11 (70)	−101.74	112.21	1.4087	i/v/s	0.719	3.9	36 (38)	<32.6	4.59
Nonene	146.67	124-11-8		C_9H_{18}		26 (78)	−81.11	126.24		i/v/v	0.733	4.35	12 (38)	<32.6	5.16
Decene	170.56	872-05-9	0.7408	$C_{10}H_{20}$		48.89 (118.69)	−66.3	140.268	1.4215	i/s/s	0.745	4.84	1 (95.7)	<32.6	5.74
Dodecene	213.4	25378-22-7	0.7584	$C_{12}H_{24}$		<100 (<212)	−35.23	168.312	1.4300	i/s/s	0.70	5.81	1 (47.2)	<32.6	6.88
Hexadecene	284.4	629-73-2	0.7811	$C_{16}H_{32}$		>100 (>212)	4.1	224.429	1.4412	i/s/s				35.5	9.18
Octadecene	179 (15)	112-88-9	0.7891	$C_{18}H_{36}$		>100 (>212)	17.5	252.482	1.4448	i/—/—	0.79	0.71		Solid	10.33
Propadiene	−34.5	463-49-0	1.787	C_3H_4	2.1 lel	−136.6 (−213.8)		40.0646	1.4168	i/d/v	0.657	1.40	10 (33)		1.66
1,2-Butadiene	10.85		0.652	C_4H_6	2–12	(−105)	−136.90	54.0914	1.4205	i/v/v	0.658	1.9	760 (18.5)		2.21
1,3-Butadiene	−4.413	106-99-0	0.6211	C_4H_6	2.0–11.5	(−105)	−108.92	54.0914	1.4292	i/s/s	0.6272	1.87	1840 (21)		2.21
Isoprene	34.07	78-79-5	0.6810	C_5H_8	1.5–8.9	−54 (−65)	−145.94	68.118	1.4219	i/v/v	0.6861	2.35	2 (15.3)		2.79
1,3-Hexadiene	73.0		0.7050	C_6H_{10}	~2.0–6.1	−21 (−6)		82.145	1.4380	i/s/s	0.710	~2.8			3.36
1,5-Hexadiene	59.5		0.6923	C_6H_{10}	~2.0–6.1	−21 (−6)	−140.68	82.145	1.4042	i/s/s	0.6970	~2.8			3.36
1,4-Heptadiene	93		0.7270	C_7H_{12}		(−6)		96.172	1.4370	i/—/s					3.93
1,7-Octadiene				C_8H_{14}				110.199							4.51
Squalene	280	111-02-4	0.8584	$C_{30}H_{50}$			<−20	410.725	1.4990	i/d/s				<32.6	16.80
Lycopene				$C_{40}H_{56}$			175	536.882		i/d/s					21.96
α-Carotene		432-70-2	1.00	$C_{40}H_{56}$			188	536.882		i/d/s					21.96

[a] Solubility in water/alcohol/ether: v = very soluble; s = soluble; d = slightly soluble; i = insoluble.
[b] 25°C.
[c] Atmospheric pressure.

carbon chain length. Experiments have shown that some unsaturated compounds can undergo addition or oxidation reactions to form adducts causing plant leaf damage, which is light for the ethylenes to butylenes, and decene to tetradecene, but more serious for isobutylene and the pentenes to nonenes (154).

A variety of selective microbacteria, but fewer in number than are substrate-specific to alkanes, have been found to biodegrade alkenes.

2.1.4 Industrial Hygiene

Industrial hygiene sampling of alkenes and chemical determinations are carried out similarly as for the alkanes. Some specific procedures and chromatographic methods are now available for the determination of selected alkenes.

2.2 Ethene (Ethylene)

Ethene, ethylene, $CH_2:CH_2$, is a colorless, flammable (DOT) (39), and explosive gas, with a faintly sweet odor, and is strikingly reactive. Some physical properties are listed in Table 45.3 (6–15). It occurs in illuminating gas up to 4 percent and in ripening fruit at very low concentrations (8). It has been detected in the average community air at very low levels (155), but is more prevalent in the atmosphere of large metropolitan areas (57). Patton and Touey (77) have determined ethylene in cigarette smoke to be between 0.093 and 0.097 percent, depending on the brand type. Ethene is prepared by cracking of ethane and propane and other petroleum gases, and by catalytic dehydration of ethanol (1). It is also used as a fruit and vegetable ripening agent. Ethene is one of the most important industrial raw materials for a variety of chemicals, petrochemicals, polymers, and resins. It is applicable as a surgical anesthetic (156).

The toxicity of ethylene is not remarkable. Concentrations of less than 2.5 percent, namely, below the lower explosive limit of 12.7 percent, are physiologically inert. However, very high concentrations may cause narcosis, unconsciousness, and asphyxia due to oxygen displacement (157).

It is one of the preferred anesthetic agents; its advantages over comparable human anesthetics are rapid onset and recovery time after exposure termination (156) and little or no effect on cardiac and pulmonary functions. That is, respiration, blood pressure, and pulse rates are rarely changed, even under anesthetic conditions. Cardiac arrhythmias occur infrequently and affect little the renal and hepatic functions (158). This has been substantiated by Reynolds (159), who reported that mice repeatedly exposed at minimal narcotic concentrations showed no histopathological changes in kidneys, adrenals, hearts, or lungs. However, the disadvantage as an anesthetic is its explosion and flammability properties, which may coincide with the most commonly applied concentration ranges (160).

Humans exposed to higher concentrations may experience subtle signs of intoxication, resulting in prolonged reaction time, as extrapolated from data by Riggs (161). Exposure at 37.5 percent for 15 min may result in marked memory disturbances (158). Humans exposed to as much as 50 percent ethylene in air, whereby the oxygen availability is decreased to 10 percent, experience loss of consciousness, and death may follow at 8 percent O_2 (162). Therefore, ethylene used as an anesthetic agent should be supplemented with the appropriate oxygen concentration. Cowles et al. (156) showed in dogs that at 1.4 percent ethene was a fast-acting anesthetic. It reached alveolar, arterial, brain, muscle, and CNS partial pressure in 2 to 8.2 min, even more rapidly than ethyl ether (156).

Male rats exposed to 10, 25 and 57×10^3 ppm for 4 hr showed increased serum pyruvate and liver weights (163). Olson and Spencer (164) observed that liver mitochondrial volume increased in rats treated with ethylene. Rats exposed to 90 percent ethylene and 10 percent oxygen are anesthetized in about 20 min and show only slight nervous symptoms (165). Deep anesthesia occurs in a few seconds with 95 percent ethylene and 5 percent O_2, accompanied by marked cyanosis and depression with a slow fall in blood pressure (165). According to Krantz et al. (20), ethylene is not a cardiac sensitizer in the dog. It is nonirritant to the skin and eyes (41).

In a chronic study, 1-day-old and adult rats exposed at 3 mg/m³ per day for 90 days exhibited hypotension, disruption of the subordination chronaxy, and inhibited cholinesterase activity, but no hematologic changes (166). Metabolically, ethylene may also cause disturbances of the carbohydrate metabolism, indicated by temporary hypoglycemia (167). Additionally, a reduction of inorganic phosphates has been recorded (41). Conversely, ethylene can potentiate the wound healing process in muscle injuries of mice (168). For aquatic tests, a TLm96 of 1000 to 100 ppm has been determined (49). Ethylene showed no mutagenic properties toward *E. coli* and several *Bacillus* species (67). It is a plant hormone, effective at concentrations as low as 0.06 mg/liter (169, 170). At higher concentrations, it may inhibit plant metabolism (171).

For industrial monitoring, no official TLV has been established. However, a value of 1000 ppm, as for other asphyxiants, has been suggested by ACGIH (38). An odor threshold of 20 mg/m³ has been recorded (172). A procedure for industrial hygiene monitoring and sampling (S286-1) has been published by NIOSH (68). Ethylene, for example, can be analytically determined by using chemiluminescent reactions with ozone (151). Owing to the highly flammable and explosive characteristics, extreme caution is advised when handling ethylene. Also, skin and eye contact with the compressed gas should be prevented.

2.3 Propene (Propylene)

Propene, propylene, methylethylene, $CH_2:CHCH_3$, is a colorless, flammable, practically odorless gas. Commercially, it is available in liquefied form with other lower alkanes and alkenes as trace impurities (173). Other physicochemical properties are listed in Table 45.3 (6–15). Propylene occurs in petroleum refining products (1) and is produced in the petroleum cracking process (10). Propene has been identified in cigarette smoke (77) and is found in diesel engine exhaust fumes (5). Thus it may occur in the ambient air of metropolitan areas, and has been detected in smog samples of various cities, such as Los Angeles (155), St. Louis (57), and Tokyo (174). It also occurs naturally in fruits, such as bananas and apples (175), but also in ocean sediments as a microbiological degradation product (176).

Owing to its high reactivity, propene is utilized as raw material for a variety of chemicals, including plastics (10, 173). It is a valuable feedstock for the production of gasoline (10) and synthetic rubber (40). It is also used as an aerosol propellant or component (177). Investigations in the early 1920s indicated its possible use as an anesthetic (178), and it was found twice as powerful as ethylene (179). Propylene is of general low toxicity. It is a simple asphyxiant and mild anesthetic, with physiological effect only at extremely high concentrations (8, 179). It is nonirritant to the skin, but may cause burns from direct contact with the liquefied product (158). Propene has been used in humans in dental surgery as a temporary anesthetic (180). At a concentration of 6.4 percent for 2.25 min, mild intoxication, paresthesias, and inability to concentrate were noted. However, the memory was not impaired (41). At 12.8 percent in 1 min, the same symptoms were markedly accentuated and at 24 and 33 percent, unconsciousness followed in 3 min (41, 181). Human exposure to 23 percent propylene for 3 to 4 min did not produce unconsciousness (182). Two subjects exposed to 35 and 40 percent propylene vomited during or after the experiment, and one complained of severe vertigo (179). Exposure to 40, 50, and 75 percent for a few minutes caused initial reddening of the eyelids, flushing of the face, lacrimation, coughing, and sometimes flexing of the legs. No variation in respiratory or pulse rates or electrocardiograms were noted (180). A concentration of 50 percent prompted anesthesia in 2 min, followed by complete recovery without any physiological indications (178).

Animal experiments with cats (178) have shown no toxic signs when anesthesia was induced with propylene concentrations of 20 to 31 percent, some subtle effects from 40 to 50 percent, blood pressure decrease and rapid pulse at 70 percent, and the unusual ventricular ectopic beat from 50 to 80 percent (178). A concentration of 40 percent produced light anesthesia in rats, with no toxic symptoms within 6 hr (165), and an exposure to 55 percent for 3 to 6 min, 65 percent for 2 to 5 min, and 70 percent for 1 to 3 min resulted in deep anesthesia with no CNS signs or symptoms. However, according to Krantz et al. (20), propylene was found to be a cardiac sensitizer in the dog. Chronic exposure of mice to minimal narcotic concentrations caused moderate to very slight fatty degeneration of the liver, but somewhat less than ethylene (183). Metabolically, propylene is expected to undergo various addition reactions, such as hydration to the alcohol, and to be excreted as the conjugated alcohol or propionic acid. These reactions may progress in a similar manner as recorded for an enzyme system of *Pseudomonas oleovorans*, which hydroxylates but does not epoxidize propylene (184). Propylene is not mutagenic when tested with *E. coli*, but, conversely, protects against mutation (67). Similarly to ethylene, propene also has a stimulating effect on plant growth at low concentrations, but is inhibitory at high levels (154). For industrial hygiene monitoring, no official TLV has been established. However, ACGIH (38) implies a level of 1000 ppm as for similar asphyxiants. An odor threshold of 17.3 mg/m^3 and a light sensitivity to the eyes of approximately 1 mg/m^3 have been reported (172). Methods for the analytic determination are available (36).

Owing to the highly flammable and explosive characteristics, the gas or the compressed liquid should be handled with all necessary precautions, especially since its explosive range is reached before any physiological effects occur.

2.4 Butenes

The mono-unsaturated butenes can occur in various forms, saturated in either position 1 or 2, that is, α or β, and as linear or branched derivatives. The β-unsaturated compound can occur in cis or trans configuration, the trans being normally the more stable form. All derivatives are chemically more reactive, but the α-olefins to a greater extent than the comparable alkane or butane. Some physicochemical data are listed in Table 45.3 (6–15). Toxicologically, these materials are similar to butane. They are simple asphyxiants that can be used as anesthetics (161).

2.4.1 Butene-1

Butene-1, butylene, α-butylene, butylene-1, ethylethylene, $CH_2:CHCH_2CH_3$, is a colorless, flammable gas (DOT) (39) with a slightly aromatic odor (11). For further physicochemical properties, see Table 45.3 (6–15). Butene occurs in refinery gases (10) and has been detected directly over diesel exhaust systems (5), but not in the general atmosphere, owing to its higher chemical reactivity. When thermally reacted, butene pyrolyzes to C_1 and C_2 alkanes and several C_5 and C_6 cyclanes, cyclenes, or toluene (1).

Butylene is produced by cracking of petroleum products. It is used for the production of a wide variety of chemicals in the gasoline and rubber processing areas (40). It is very reactive, may polymerize, and undergoes addition reactions readily. It is recovered from catalytically cracked petroleum oil (10) and is mainly used for the synthesis of polymers (185). Butene is liberated from tetra- and tributyllead or tin during oxidative dealkylation by hepatic microsomes (186). This explains, then, the occurrence of butene in human respiratory air (56). Butene-1 is a simple asphyxiant and classified as "DOL-nontoxic." At concentrations above the flammability range, it is an anesthetic (157). It has a low acute toxicity and is a low eye irritant. Liquid butene on direct eye and skin contact can cause burns and frostbite. As an anesthetic it appears about 4.5

times more potent than ethylene (41). Inhalation experiments with white mice at a concentration of 15 percent showed reversible signs of incoordination, confusion, and hyperexcitability; at 20 percent deep anesthesia in 8 to 15 min, with subsequent respiratory failure in 2 hr; and at 30 percent in 2 to 4 min and 40 min, respectively. A concentration of 40 percent resulted in profound anesthesia in 30 sec, no CNS symptoms, but death in 10 to 15 min (165). As shown by a model (187), butene is probably metabolized through its 1-hydroxy derivative, although converting at a slow rate. In plants, butene-1 has been recorded to cause slight leaf damage (154). Butene supports *Neurospora crassa*, spore germination (47), and growth of some *E. coli* strains (27), but inhibits germination of *Bacillus megaterium* (28).

Unofficial threshold concentrations of 0.4 percent (4000 ppm) have been suggested to adequately protect for the lower explosive level (188). Odor thresholds in air of 0.003, 5 (25), and 15.4 mg/m^3 (172) or 0.93 to 1.3 ppm (25) have been reported. Chemical identification by spectral methods are available (36). Butylene should be handled with care, owing to its flammability and reactivity with other chemicals or polymerization potential, especially when heated (40, 173).

2.4.2 Butene-2

Butene-2, β-butylene, butylene-2, dimethylethylene, $CH_3CH:CHCH_3$, can occur in trans or cis conformation, and is also known as pseudobutylene (11). It is a colorless gas. For physicochemical properties, see Table 45.3 (6–15). It is recovered from refining gases or made by petroleum cracking (10). It serves to produce gasolines, butadiene, or a variety of other chemicals (10). Butene-2, like butene-1, has been detected directly over diesel exhausts, but to a lesser extent (5). The more highly reactive *trans*-2-butene occurs at a much lower frequency in the atmosphere than other comparable hydrocarbons (189). Butylene-2 appears somewhat more highly narcotic than the butene-1. About 13 to 13.5 percent (300 to 400 mg/liter) causes deep narcosis in mice and about 19 percent (120 to 420 mg/liter) is fatal (41). It appears to be a low mucous membrane irritant (41). According to Krantz et al. (20), it is a cardiac sensitizer. Human respiratory gases have been observed to contain butene-2. In the majority of subjects, the cis form was preponderant over the trans isomer, except in one case (56).

As far as industrial hygiene measures are concerned, the discussion under butene-1 will hold also for butene-2 (36). Odor thresholds for butene-2 have been recorded for detection in the air as generally 0.05 mg/liter to 0.059 mg/liter, and 0.0048 mg/liter for the trans-2 isomer (25).

2.4.3 Isobutene

Isobutene, α-butylene, isobutylene, 2-methylpropene, $CH_2:C(CH_3)_2$, is a highly volatile liquid, or a flammable (DOT) (39), easily liquefied gas (188). Further chemical–physical properties, which resemble those of the linear butenes, but exhibit a somewhat lower chemical reactivity, are listed in Table 45.3 (6–15). It has been detected in the urban atmosphere at a low concentration (3). Isobutylene is mainly used as a monomer or copolymer for the production of synthetic rubber (10) and various plastics (8). Isobutylene is obtained from refinery streams by absorption on 65 percent H_2SO_4 at about 15°C, or by reacting with a lower aliphatic primary alcohol and decomposition of the resulting ether (8). About 95 percent of the available isobutylene is used to produce dimers, trimers, butyl rubber, and other polymers. The other 5 percent is used to produce antioxidants for food, food packaging, and supplements, and for plastics (8).

The toxicologic properties of isobutene are similar to those of the other lower alkenes. It is a simple asphyxiant and a narcotic at higher concentrations. At 30 percent, it produces no narcosis in white mice, and excitement and narcosis in 7 to 8 min at 40 percent, but no excitement and immediate narcosis in 2 to 2.25 min at 50 percent, or in 50 to 60 sec at 60 to 70 percent (41). An inhalation LC_{50} in the mouse was 415 mg/liter in a 2-hr trial and 620 g/m^3 (620 mg/liter) in a 4-hr study in the rat (70). These results indicate that isobutylene is practically nontoxic. This has been confirmed in a study where brain concentrations were determined (71).

Metabolic studies on rats and mice that had inhaled isobutylene in conjunction with other volatile hydrocarbons showed that the contents of the brain and parenchymatous organs were similar, but the level in the fatty tissue was significantly higher than in the brain, liver, kidneys, or spleen. There was a linear relationship between the degree of CNS depression, narcosis, and the cerebral concentrations (69, 71). Butane and butylene have a toxicologically additive effect (70).

Isobutene is nonhazardous below the lower explosive limit, and thus the industrial hygiene measures under butene apply here, too. A colorimetric method for the determination of isobutylene without interference by ethylene and propylene has been developed. No threshold limit value (TLV) has been established for isobutylene in workroom air (38). With the exception of asphyxia resulting from exposure to high concentrations for prolonged periods, to date isobutylene has not been documented to have untoward effects on the health of workers.

2.5 Pentenes (Amylenes)

2.5.1 Pentene-1

Pentene-1, α-n-amylene, 1-amylene, propylethylene, $CH_2:CH(CH_2)_2CH_3$, occurs in coal tar (8) and in petroleum cracking mixtures (1). It is a flammable (DOT) (39) liquid at ambient temperature with a highly disagreeable odor.

Some further physicochemical properties are listed in Table 45.3 (6–15). It polymerizes on extended storing.

The physiological properties of pentene resemble those of butylene (182), and its anesthetic action appeared about 15 times more potent than that of ethylene. However, near narcosis it causes more severe primary excitement (182). It produces anesthesia at 6 percent in 15 to 20 min, but is more active and cardiotoxic than the lower homologues (41). In the early 1920s, it was used unsuccessfully as a human anesthetic in general surgery, but it has brought somewhat better results in dentistry (161). Animal experiments have shown a variety of symptoms, including respiratory and cardiac depression and the primary excitation observed in humans (41). Absorbed amylene is believed to be metabolized, oxidized at the double bond, and excreted as the alcohol or its conjugate (41). This is verified by the finding (154) that 1-pentene, but not 2-pentene, causes specific smoglike damage to agricultural plants.

The odor of amylene is detectable in the air at concentrations of 0.54 to 6.6 $\times 10^{-3}$ mg/liter or 0.54 to 6.6 mg/m^3, 1-pentene specifically at 0.19 ppm (25).

2.5.2 Pentene-2

Pentene-2, β-n-amylene, n-amylene-2, ethylmethyl ethylene, CH_3CH: $CHCH_2CH_3$, has similar properties to those of its α isomer [see Table 45.3 (6–15)]. Pentene-2 is produced by petroleum cracking and is further reacted to produce a variety of high octane fuel components (1). Toxicologically, it acts similarly to pentene-1.

2.5.3 Isopentene

The isopentenes occur in two forms, as α- or β-isoamylene. Isopentene-1 is also α-isoamylene or 3-methyl-1-butene, CH_2:$CHCH(CH_3)_2$. Isopentene-2 is β-isoamylene or 2-methyl-2-butene, $(CH_3)_2C$:$CHCH_3$ [see Table 45.3 (6–15)]. Most toxicologic work appears to have been carried out with the β isomer.

Isoamylene-2 at 5.4 percent in white mice showed incoordination, marked hyperexcitability, no toxic symptoms, and light anesthesia in 20 min, with a tendency toward convulsions. At 6.12 percent, it produced deep anesthesia in 6 to 15 min, with respiratory failure 1 hr post-exposure (165). At 6.12 percent for 30 to 45 min, convulsions and death resulted (165).

By inhalation, a minimal narcotic concentration of 1.57×10^{-3} moles/liter (112 mg/liter) and minimal lethal concentrations of 2.96×10^{-3} moles/liter (212 mg/liter) for mice have been established (83). Searle (190) classifies β-amylenes as cilia toxins, which may play a role towards their respiratory paralytic action.

2.6 Hexenes

2.6.1 Hexene-1

Hexene-1, butylethylene, α-hexylene, hexylene-1, $CH_2:CH(CH_2)_3CH_3$, is a colorless liquid, which is highly volatile and flammable (DOT) (39). Further physicochemical properties are listed in Table 45.3 (6–15). Hexene is produced by olefin cracking (1) and is used primarily for organic syntheses and in fuels. Hexene is a low to moderate irritant to the skin and eyes. When ingested, owing to its high volatility, it presents a moderate aspiration hazard, shown as a 60 percent probability in a trial with mice (19). When inhaled, it may produce narcosis in humans at a concentration of about 0.1 percent, with accompanying CNS effects, mucous membrane irritation, vertigo, vomiting, and cyanosis. The minimal narcotic concentration for mice was observed as 1.19 mol/m^3 (2.9 percent 29,100 ppm) and the minimal fatal concentration as 1.67 mol/m^3 (4.08 percent 40,800 ppm) (83). In pollution simulation research, hexene-1 has been shown to cause smoglike damage to various agricultural plants (154).

2.6.2 Hexene-2

Hexene-2, β hexylene, hexylene-2, methylpropylethylene, $CH_3CH:CH(CH_2)_2CH_3$, is a highly flammable liquid (DOT) (39), somewhat less volatile than the α isomer; see Table 45.3 (6–15). Its toxicity also resembles that of the α isomer.

2.6.3 Isohexene

Isohexene, β-isohexylene, 2-methylpent-2-ene, $(CH_3)_2C:CHCH_2CH_3$, a component of cigarette smoke, is classified as a cilia toxin to the respiratory tract (190). Physicochemical properties are listed in Table 45.3 (6–15). Analytical methods for the determination of all hexene isomers are available (36).

2.7 Heptenes

n-Heptene-1, α-heptylene, ethylpentyl ethylene, $CH_2:CH(CH_2)_4CH_3$, is a colorless, flammable (DOT) (39) liquid; see Table 45.3 (6–15). The 2- and 3-heptenes, $CH_3CH:CH(CH_2)_3CH_3$ and $CH_3CH_2CH:CH(CH_2)_2CH_3$, occur in the cis and trans configurations. Heptene-2 is β-heptylene and heptene-3 is γ-heptylene; see Table 45.5 (6–15). The general toxicity of the heptenes is somewhat lower than that of the hexenes, except that its aspiration hazard is increased (19). Animal experiments have shown 60 mg/liter to cause loss of the righting reflex in mice (83). Minimal narcotic and a lethal concentration have been determined as, 60 and above 200 mg/liter, respectively (41). A plant physiological study showed that the 3-n-heptene exerted a stronger damaging effect on leaves than its 1-ene isomer, but lower than the hexenes in general (154).

2.8 Octenes and Higher Alkenes

n-Octene-1, α-octylene-1, $CH_2 : CH(CH_2)_5CH_3$; n-2-octene, β-octylene, $CH_3CH : CH(CH_2)_4CH_3$; isooctene, 2-methylheptene-2, $(CH_3)_2C : CH(CH_2)_3CH_3$; the trimethylpentenes; and ethylhexene-1, $CH_2 : C(C_2H_5)(CH_2)_3CH_3$, are flammable liquids (DOT) (39); see Table 45.3 (6–15). When inhaled at high concentrations, they may cause headache, inability to pay sustained attention, vertigo, nausea, and narcosis (191). Octenes may be more irritant to mucous membranes, skin, and eyes than the lower homologues. Octenes, when ingested, may rapidly be aspirated into the lungs (19). A range-finding 4-hr LC_{50} of 4000 ppm in the rat has been observed for 2-ethylhexene-1 (192). All octenes may be metabolically hydroxylated, conjugated, and readily excreted from the mammalian system (41). An aquatic TLm96 range of 1000 to 100 ppm (49) has been established. As an atmospheric pollutant, octene has less damaging effects to plant leaves than hexene and heptene (154). The odor of 1-octene may be detected at a concentration of 2.0 ppm (25). Various analytical determination procedures are available.

Nonene may present an aspiration hazard when ingested (19). An aquatic toxicity determination (TLm96) showed an effect level of over 1000 ppm (49). The isomer 4-n-nonene is less active toward plant leaves than the lower homologues, whereas the higher members do not cause any damage (154). 1-Decene, n-decylene, $CH_2 : CH(CH_2)_7CH_3$, is a colorless, flammable liquid (DOT) (39). It has general irritant and narcotic properties (11). Decene and dodecene also may present an aspiration hazard when ingested (19). The odor threshold of 1-decene in air is about 7 ppm (25).

Analytical methods for the determination of decene and dodecene are available; see also Table 45.3 (6–15). Hexa- and octadecene-1 have been found in land and ocean natural deposits and sediment cores (147); see also Table 45.3 (6–15). The long-chain olefins, 9-tricosene and related compounds, have been observed to possess insect attractant and psychedelic properties (193).

2.9 Propadiene

Propadiene, allene, dimethylenemethane, $CH_2 : C : CH_2$, is a colorless, flammable (DOT) (39), and unstable gas, with a sweetish odor. Further physicochemical properties are listed in Table 45.3 (6–15). Propadiene is produced in small quantities in petroleum cracking processes. It is of low general toxicity, but may be narcotic at very high concentrations (188). Experiments with mice have shown that 20 percent of propadiene caused restlessness and narcosis in

11 min, 30 percent caused narcosis in 3 min, and 40 percent in 1 to 2 min (41). For industrial use, adequate precautions in regard to inhalation and skin exposure are recommended (188).

2.10   Butadiene

Butadiene, biethylene, 1,3-butadiene, α-butadiene, divinyl, erythrene, 1-methylallene, vinylethylene, $CH_2:CHCH:CH_2$, is a colorless, flammable gas (DOT) (39). It is potentially explosive when mixed with air. It carries a characteristic, aromatic odor (173). For further physicochemical properties, see Table 45.3 (6–15). Butadiene is an intermediate formed in tobacco smoke (194). Chemically, it is produced from ethanol, or butane, but primarily from butylene or by catalytic cracking of light oil or naphtha (8). It does not occur naturally. However, several enzyme systems biosynthesize a precursor, acetoin, which then chemically can be converted to butadiene (195). Butadiene is used for the preparation of a variety of chemicals, but also is the major component for the production of synthetic rubber, often in combination with styrene (196). Butadiene is highly reactive, dimerizes to 4-vinylcyclohexene, and polymerizes easily (1). Owing to this reactivity, inhibitors have to be added for its storage and transport (173).

General experiments with humans and animals have shown the toxicity of butadiene to be of low order, and not cumulative. It has anesthetic action and at very high concentrations causes narcosis, respiratory paralysis, and death. Human signs and symptoms from overexposure owing to leaks or spills are initially the characteristic blurred vision, nausea, prickling and dryness of the mouth, throat, and nose, followed by fatigue, headache, vertigo, nausea, decreased blood pressure and pulse rate, unconsciousness, and respiratory paralysis. To date, no hematologic disorders have been attributable to butadiene, as pointed out by Benini et al. (197).

At lower concentrations it may cause slight irritation to the skin and eyes. Dermatitis, seen occasionally with butadiene, appears to represent a secondary effect of additives, accelerators, or inhibitors (198, 199). On direct dermal contact, the liquefied product causes burns and frostbite. Humans exposed to 1000 ppm showed no irritant effects (200); 2000 to 4000 ppm resulted in slight irritation of the eyes and difficulty in focusing on instrument scales (200). Volunteers exposed to 8000 ppm for 8 hr also showed no effects, other than some irritation of the eyes and upper respiratory tract, and 10,000 ppm for 5 min showed no effects on blood pressure or respiration (173).

In animal experiments an oral LD_{50} in the rat was 5480 mg/kg (201), in the mouse 3210 mg/kg (201). The lowest lethal concentration by inhalation in the rabbit was determined as 250,000 ppm, and to cause death in 23 min (200). Inhalation experiments with white mice showed that a concentration of 10 percent caused no symptoms, 15 percent light narcosis, 20 percent some excitement and narcosis in 6 to 12 min, 30 and 40 percent excitement with twitching in 1 to 1.2 min and 40 to 60 sec, respectively (41). As a narcotic agent it is thus more potent than propadiene, but only half as powerful as butylene. Rabbits exposed to 25 percent (250,000 ppm) butadiene exhibited light, relaxed anesthesia in 1.6 min, followed by loss of various reflexes and CNS effects and death in 23 min (200). Rats, orally administered 100 mg/kg butadiene for 2.5 months, showed some lymphohistocytic infiltration in the heart, liver, and kidney (202). Rats, guinea pigs, and a dog exposed to butadiene at 600, 2300, and 6700 ppm for 7.5 hr/day, 6 days/week, for 8 months showed slightly retarded growth at the highest dose, but no blood dyscrasias were found. Conversely, the fertility in the rat was increased (200). The toxicity of some olefins to the rat and mouse was found to be proportional to the translocation and occurrence in certain tissues, such as the brain. Butadiene was found at higher concentrations in the medulla oblongata than in the cerebellum or cerebral cortex (69, 71). Most literature sources reflect that toxicologic tests were conducted of butadiene as commonly used in mixtures or in connection with other copolymers, such as styrene. Some combinations have shown added irritancy. For example, children playing with butadiene-mineral or clay mixtures developed dermatitis or specific rashes a few days post-contact (157). Atmospheric trials have demonstrated butadiene to have plant-leaf damaging properties similar to C_5 to C_8 linear and branched monoolefins (154). These findings indicate that butadiene reacts with other hydrocarbons and dimerizes easily.

Currently OSHA (37) and ACGIH (38) recommend an 8-hr/day, 5-day TLV of 1000 ppm (2200 mg/m^3) for butadiene. NIOSH has published a monitoring procedure, S-91 (68). Various determination procedures are available, as, for example, the pentoxide method for air and blood sampling (200). Handling procedures should be strictly followed, since butadiene is stable below its boiling point of 23.54°F. The greatest danger exists when it is mixed with air to the explosive limit of 2 percent. The use of protective garments and equipment is advised.

2.11 Isoprene

Isoprene, 2-methyl-1,3-butadiene, $CH_2:C(CH_3)CH:CH_2$, is a colorless, volatile, flammable (DOT) (39) liquid. It is highly reactive, usually occurs as its dimer (203), and unless inhibited undergoes further explosive autopolymerization (13). For further physical and chemical data, see Table 45.3 (6–15). Industrially, isoprene is manufactured by dehydrogenation in the oxo-process from isopentene or by high temperature thermal cracking of petroleum oil (10), gas oils, and naphthas (40). It is used in the manufacturing of butyl and synthetic rubber, plastics, and a variety of chemicals (10).

Isoprene also has been detected in the vapor phase of tobacco smoke (204). However, it occurs more prevalently in a vast majority of biologic systems, mammalian tissue, and plant bacteria, as the "isoprene unit," probably as phosphate or pyrophosphate derivative (203) or as the radical. Isoprene possesses two reactive centers and thus can combine linearly or three-dimensionally to form polyenes, cyclics, aromatics, and diverse polymers. Much literature is available on its biologic occurrence, its di- and polymerization to farnesene, C_{10}, squalene, C_{30}, and a variety of C_{40} compounds, such as the lycopenes and carotenes. Therefore, the isoprene unit is the most important building block for lipids, steroids, terpenoids, and a wide variety of natural products, such as latex, the raw material for natural rubber (195, 205). Also, in tobacco smoke isoprene has been determined to be a precursor of a number of polycyclic aromatics, as demonstrated by thermal condensations in the range of 450 to 700°C (194).

Physiologically, isoprene exhibits properties similar to those of butadiene and butylene (198), although it is more irritant than the comparable alkene or alkane of similar volatility (153). At low concentrations, except as a result of cigarette smoking and in some occupational connections, few toxic effects have been documented, although at high concentrations it is a narcotic and asphyxiant (157). Isoprene has been detected in human expired air at 15 to 390 µg/hr in the smoker and from 40 to 250 µg/hr in some nonsmokers (56). Further human experiments have demonstrated that isoprene was absorbed at 20 percent in the upper respiratory tract, but 70 to 99 percent were retained in the lungs (206, 207). In 10 human volunteers, the average odor perception occurred at 10 mg/m³, and at 160 mg/m³ they experienced slight irritation of the upper respiratory mucosa, larynx, and pharynx (208). Isoprene was also observed to be effective in reducing the tracheal mucous flow (209).

Animal experiments using the rabbit showed that 0.19 to 0.75 mg/liter caused an increased respiratory rate (208). The concentration of 109 mg/liter inhaled by white rats for 2 hr caused loss of the righting reflex (208). LC_{50} values were 148 mg/liter for the female and 139 mg/liter for the male mouse; survivors of both groups 24 hr post-exposure exhibited improved swimming time. Enlarged lungs were observed in mice expiring during the inhalation (208). Acute inhalation studies have shown a no-effect level of 20,000 ppm in the mouse, deep narcosis at 35,000 to 45,000 ppm, and death at 50,000 ppm (210). Similar results showed loss of the righting and total reflexes in the white mouse at 120 mg/liter but no deaths at 200 mg/liter (83). These results indicate that isoprene might be more highly toxic than butadiene. Mice and rabbits repeatedly exposed to 2.2 to 4.9 mg/liter, 4 hr/day, for 4 months, and rats for 5 months, showed no weight differential from the control. However, after the third month the oxygen consumption by the rat decreased. Final analysis demonstrated an increased number of leukocytes in the rabbit, slightly decreased erythrocytes, and some increased organ weights (208).

Metabolic studies with isoprene (71) confirmed that concentrations in the rat brain tissue were indicative of the relative toxic potency. Conversely, isoprene, when applied to mouse skin, reduced the number of papillomas, but at a much lower rate than, for example, retinyl acetate (211). Occupational exposure to concentrations above the maximum permissible concentrations have allegedly resulted in CNS and cardiac alterations (212) and subtle immunologic changes (213). However, those signs and symptoms may also be due to a variety of other materials utilized and produced in the isoprene rubber industry. This was confirmed by the administration of synthetic rubber extracts to rats for 9 months, which exhibited hepatic and cardiac changes with leukocytes and hemoglobin increases (214). Fish experiments revealed an aquatic TLm96 of 100 to 10 ppm (49).

No official TLV has been established, although a maximum allowable concentration of 40 mg/m^3 has been suggested in the Soviet Union (208). Its odor can be detected somewhat below this limit at 5 mg/m^3 (25). Air concentrations may be determined by collecting the material on charcoal and specific determination by coulometric titration (36).

2.12 Hexadiene and Higher Alkadienes

The 1,3-hexadiene, ethylbutadiene, $CH_2:CHCH:CHCH_2CH_3$, which can be detected at an air concentration of 2.0 ppm (25); the 1,4-hexadiene, allyl allene, $CH_2:CHCH_2CH:CHCH_3$; and the 1,5-hexadiene, diallyl $(CH_2:CHCH_2)_2$, are all colorless, flammable liquids. Some analytical determination methods are available; see also Table 45.3 (6–15).

The odor of 1,4-heptadiene, $CH_2:CHCH_2CH:CHCH_2CH_3$, can be detected at 9.0 ppm (25) and that of the three octadienes, 1,3, 1,4, and 2,4, at 2.0, 1.5, and 1.2 ppm, respectively (25). For 1,7-octadiene, an oral LD_{50} of 19.7 ml/kg has been observed in the rat (215). A dermal LD_{50} of 14.1 ml/kg and the saturated vapor caused death in 15 min (215). It was moderately irritant to the rabbit skin and produced low corneal injury (215).

2.13 Alkatrienes and Polyenes (Tri- and Polyolefins)

The alkatrienes, or triolefins, are also named cumulenes, after cumulene, which is the simplest compound of the series. Higher members of this class are represented, for example, by squalene, $C_{30}H_{48}$, lycopene, and carotenes, $C_{40}H_{60}$. All of the latter occur naturally in animals, plants and lower organisms. Therefore, their toxic characteristics are negligible. Some physicochemical data are listed in Table 45.3 (6–15).

3 ALKYNES

3.1 General

The alkynes, similar to the alkanes and alkenes, do not exert any acute local toxicity. The lower members also are anesthetics, but indicate slightly higher narcotic effects. They are practically nonirritant to the skin, but cause pulmonary irritation and edema at very high concentrations. The higher molecular weight members can be aspirated into the lungs when ingested.

3.2 Acetylene

Acetylene, ethine, ethyne, narcylene, CH : CH, is a colorless, highly flammable (DOT) (39) and explosive gas (see Table 45.4) (6–15). Acetylene prepared from calcium carbide occasionally contains phosphine and some arsine as impurities, namely, up to 95 and 3 ppm, respectively (216, 217). These impurities lend the product both a subtle etheral to garlic-like odor and its secondary toxicity (152, 173). Acetylene is highly reactive and forms explosive mixtures with oxygen, chlorine, and fluorine (162). When heated, it may separate into its elements under explosive exothermic reactions (162). Air analyses have indicated acetylene to occur at an atmospheric concentration of 0.03 ppm in U.S. urban smog (2). In Japan, acetylene has been detected in the air near industrial and suburban areas, but particularly in sinter- and coke-oven emissions (218). Conversely, various plant and bacterial systems can reduce and inactivate atmospheric acetylene through their nitrogen-fixing mechanisms (219, 220). An important industrial raw material, acetylene serves as a base for a number of chemicals, such as solvents or some alkenes, which, in turn, serve as monomers for plastic production (10). Acetylene is also utilized in a number of occupational areas, such as in brazing, cutting, flame scarting, and metallurgical heating and hardening, and in the glass industry (152). In optometry, it serves as contact lens coating component (221). Similarly to ethylene, acetylene is used to ripen fruit (222–224) and mature rubber trees (225) or flowers (223).

Acetylene is nontoxic below its lower explosive limit of 2.5 percent. At higher concentrations, it has anesthetic action and at higher levels is a simple asphyxiant, owing to its oxygen displacement capability (16). In the 1920s, acetylene was used as an anesthetic, since it afforded immediate recovery without aftereffects (161). However, owing to its explosive characteristics, it now finds only limited application as an anesthetic. Its narcotic action is about 1.15 times that of ethylene (161). Humans can tolerate an exposure of 100 mg/liter for 30 to 60 min (173). Marked intoxication occurs at 20 percent, incoordination at 30 percent and unconsciousness at 35 percent in 5 min. There is no evidence that tolerable levels have any deleterious effects on health (210, 226), although two deaths (227, 228) and a near fatality at 40 percent (157) have been recorded, which occurred during acetylene manufacturing using calcium carbide. The intoxications were attributed to the possible phosphine and arsine content of the crude acetylene. Animal experiments with warm-blooded animals have shown tolerance at 10 percent, intoxication at 25 percent, and death at 50 percent in 5 to 10 min (229). Rodents exposed to 25, 50, and 80 percent acetylene in oxygen for 1 to 2 hr daily, up to 93 hr, showed no organ weight changes or cellular injuries (230). A rise in blood pressure was noted in the cat with an 80 percent acetylene–oxygen mixture.

Table 45.4. Physicoshemical Properties of Alkynes (6–15)

Compound	B.P. (°C)	CAS Registry No.	Density (20/4°C)	Emp. Formula	Flamm. Limits (%)	Flash Pt. [°C (°F)]	Freezing Pt. (°C)	Mol. Wt.	Refr. Index, n_D^{20}	Solubility[a] w/al/et	Sp. Gr.	Vapor Dens.	Vapor Pres. [mm (°C)]	Visc. (SUS)	W/V Conv. (mg/m³ ⇔ 1 ppm)
Acetylene	−84.00	74-86-2	0.6208	C_2H_2	2.5–100		−81.0	26.038	1.00051	d/d/—	0.65	0.91[c]	40[c] (16.8)	N.A.	1.07
Propyne	−23.22	74-99-7	0.7062	C_3H_1	2.4–11.7		−102.7	40.065	1.3863 (−40°C)	d/v/—		1.38	3876[c] (20)	N.A.	1.64
1-Butyne	8.07	107-00-6	0.650	C_4H_6		−28.9 (> −20)	−125.72	54.091	1.3962	i/s/s			760 (2.4)	N.A.	2.21
2-Butyne	27.0		0.6910	C_4H_6	1.4 lel	> −20 (> −4)	−32.26	54.091	1.3921	i/s/s	0.69	1.86	760 (8.7)	N.A.	2.21
1-Pentyne	40.18		0.6901	C_5H_8			−105.7	68.118	1.3852	i/v/v					2.79
Isopentyne	26.35		0.666	C_5H_8			−89.7	68.11	1.3723	i/v/v				N.A.	2.79
1-Decyne	174.10		0.7655	$C_{10}H_{18}$			−44.0	138.252	1.4265	i/s/s					5.65
Butadiyne	10.3	460-12-8	0.7364[b]	C_4H_2				50.057	1.4189 (5°C)	v/s/v					2.05
1,6-Heptadiyne	112.0		0.8164	C_7H_8			−85.0	92.140	1.451 (17°C)	i/—/—					3.77

[a] Solubility in water/alcohol/ether: v = very soluble; s = soluble; d = slightly soluble; i = insoluble.
[b] 0/4°C.
[c] Atmospheric pressure.

A NIOSH criteria document (231) recommends a TLV for acetylene of 2500 ppm for a 10 hr/day, 40-hr work week. Acetylene detection tubes are available, and can be analytically determined by gas chromatography to the ppb level. Various scrubber systems for acetylene wash removal have been designed. The handling of acetylene warrants special attention in view of its wide-range flammability and capability of forming explosive mixtures (198). Outside storage in a cool, well-ventilated area is advised.

3.3 Propyne

Propyne, methylacetylene, $CH_3C:CH$, is a colorless, highly flammable (DOT) (39) and explosive gas; see Table 45.4 (6–15). Similarly to acetylene, propene is used as a welding torch fuel (232). It is a highly reactive material, but its toxic properties are minimal. Methyl acetylene has been detected in the human respiratory air at 0.81 µg/hr in a nonsmoker, and at 1.1 and 2.3 µg/hr in two smokers (56). With regard to its narcotic properties, it is about 18 times as potent as acetylene (161).

A threshold limit value of 1000 ppm (1650 mg/m^3) has been recommended by OSHA (37) and ACGIH (38). The gas should be handled with caution, as described for acetylene.

3.4 Higher Alkynes

Practically no toxic data are available on butynes and above. Selected physical data are listed in Table 45.4 for 1-butyne, 2-butyne, 1-pentyne, isopentyne, 1-decene, butadiyne, and 1,6-heptadiyne. Some symmetrical butadiynes have fungistatic effect (233). Isopentyne, 3-methylbutyne, $(CH_3)_2CHC:CH$, has caused loss of righting reflexes in mice at 150 mg/liter and was fatal at 250 mg/liter (41, 83). Therefore, the introduction of a triple unsaturation appears to increase the narcotic potency of alkanes and alkenes. The C_6 to C_{10} alkynes, when ingested, may present aspiration hazards (19).

Oral LD_{LO} values of 350 and 639 mg/kg (234) have been recorded for 1-buten-3-yne. For 1,6-heptadiyne, $C:C(CH_2)_3C:CH$, an oral LD_{50} of 2300 mg/kg for the rat, 2620 mg/kg for the rabbit, and 3830 mg/kg for the dog have been recorded (235). The taste of 1-decyne can be recognized at 0.1 ppm, and the odor recognized at 4 ppm in air (25). Some analytic procedures are available for these higher alkynes.

REFERENCES

1. R. F. Gould, Ed., *Refining Petroleum for Chemicals*, American Chemical Society Publication, Washington, D.C., 1970.
2. A. P. Altshuller and T. A. Bellar, *J. Air Pollut. Control Assoc.*, **13**(2), 81 (1963).

3. R. J. Gordon, H. Mayrsohn, and R. M. Ingels, *Environ. Sci. Technol.*, **2**(12), 1117 (1968).

4. A. P. Altshuller, W. A. Lonneman, F. D. Sutterfield, and S. L. Kopczynski, *Environ. Sci. Technol.*, **5**(10), 1009 (1971).

5. M. C. Battigelli, *J. Occup. Med.*, **5**(1), 54 (1963).

6. R. C. Weast, Ed., *CRC Handbook of Chemistry and Physics*, CRC Press, Cleveland, Ohio, 1977–1978.

7. *Laboratory Waste Disposal Manual*, 2nd ed., revised, Manufacturing Chemists Association, Inc., Washington, D.C., 1974.

8. M. Windholy, Ed., S. Budavari, Assoc. Ed., *The Merck Index*, 9th ed., Merck and Company, Inc., Rahway, N.J., 1976.

9. *Registry of Toxic Effects of Chemical Substances*, U.S. Dept. Health, Education, and Welfare, Cincinnati, Ohio, September, 1977.

10. V. B. Guthrie, Ed., *Petroleum Products Handbook*, McGraw-Hill, New York, 1960.

11. N. Sax, *Dangerous Properties of Industrial Materials*, 4th ed., Van Nostrand Reinhold, New York, 1975.

12. *Toxicology and Hazardous Industrial Chemicals Safety Manual for Handling and Disposal with Toxicity Data*, International Technical Information Institute, publ. Japan, 1976.

13. *Hazardous Chemicals Data 1975*, National Fire Protection Association, NFPA No. 49, Boston, 1975.

14. *Handbook of Organic Industrial Solvents*, 4th ed., American Mutual Insurance Alliance, Chicago, 1972.

15. F. Rossini, K. Pitzer, R. Arnett, R. Braun, and G. Pimentel, *Selected Values of Physical Thermodynamic Properties of Hydrocarbons and Related Compounds*, Carnegie Press, Pittsburgh, Penn., 1953.

16. Y. Henderson and H. W. Haggard, *Noxious Gases*, 2nd ed., Reinhold, 1943.

17. F. A. Patty and W. P. Yant, *Odor Intensity and Symptoms Produced by Commercial Propane, Butane, Pentane, Hexane and Heptane Vapor*, reprinted from the U.S. Bureau of Mines Report of Investigation No. 2979 (1929).

18. R. W. Stoughton and P. D. Lamson, *J. Pharmacol. Exp. Ther.*, **58**, 74 (1936).

19. H. Gerarde, *Arch. Environ. Health*, **6**, 329 (1963).

20. J. C. Krantz, Jr., C. J. Carr, and J. F. Vitcha, *J. Pharmacol. Exp. Therap.*, **94**, 315 (1948).

21. J. K. Boitnott and S. Margolis, *Johns Hopkins Med. J.*, **127**, 65 (1970).

22. K. Pennington and R. Fuerst, *Arch. Environ. Health*, **22**, 476 (1971).

23. D. J. Crisp, A. O. Cristie, and A. F. A. Ghobashy, *Comp. Biochem. Physiol.*, **22**, 629 (1967).

24. C. Poyart, E. Bursaux, A. Freminet, and M. Bertin, *Biomed. Express*, **25**(6), 224 (1976).

25. W. H. Stahl, Ed., *Compilation of Odor and Taste Threshold Values Data*, American Society for Testing and Materials, Philadelphia, 1973.

26. H. E. Swann, Jr., B. K. Kwon, G. H. Hogan, and W. M. Snellings, *Am. Ind. Hygiene Assoc. J.*, **35**(9), 311 (1974).

27. R. Fuerst and S. Stephens, *Dev. Ind. Microbiol.*, **11**, 301 (1970).

28. L. J. Rode and J. W. Foster, *Microbiology*, **53**, 32 (1965).

29. C. E. ZoBell, *API Proc.—Joint Conf. Prev. Control Oil Spills*, 317 (1969).

30. A. C. Van Der Linden and R. Juybregtse, *Antonie van Leeuwenhoek J. Microbiol. Serol.*, **33**(4), 382 (1967).

31. H. I. Kator, C. H. Oppenheimer, and R. J. Miget, "Microbial Degradation," *API Proc.—Joint Conf. Prev. Control Oil Spills*, 287 (1971).

32. E. J. McKenna and M. J. Coon, *J. Soil. Chem.*, **245**(15), 3882 (1970).

33. R. J. Miget, C. H. Oppenheimer, H. I. Kator, and P. A. LaRock, *API Proc.—Joint Conf. Prev. Control Oil Spills*, 327 (1969).

34. J. R. Haines and M. Alexander, *Appl. Microbiol.*, **28**(6), 1084 (1974).

35. T. Nagata, T. Kojima and S. Makisumi, *Jap. J. Legal Med.*, **25**, 439 (1971).

36. L. Meites, *Handbook of Analytical Chemistry*, 1st ed., McGraw-Hill, New York, 1963.

37. "Occupational Health and Environmental Control," Subpart G, Sec. 1910.93, Air contaminants, *Fed. Reg.*, **39**(125), 23540 (1974).

38. *Threshold Limit Values for Chemical Substances and Physical Agents in the Workroom Environment*, American Conference of Governmental Industrial Hygienists, Cincinnati, Ohio, 1978.

39. "Hazardous Materials Regulations and Miscellaneous Amendments," Title 49, Chapter 1. Materials Transportation, Bureau of Transportation, *Fed. Reg.*, **41**(252), 57018 (1976).

40. C. R. Noller, *Chemistry of Organic Compounds*, 3rd ed., W. B. Saunders, Philadelphia, 1966.

41. W. R. Von Oettingen, *Toxicity and Potential Dangers of Aliphatic and Aromatic Hydrocarbons*, Publ. Health Bull. No. 255, 1940.

42. D. Balasubramanian and D. B. Wetlaufer, *Physiology*, **55**, 762 (1956).

43. E. A. Wahrenbrock, E. I. Eger, R. G. Laravuso, and G. Maruschak, *Anesthesiology*, **40**(1), 19 (1974).

44. A. C. Carles, T. Kawashire, and S. Pueper, *Arch. Ges. Physiol.*, **359**, 209 (1975).

45. R. W. Dougherty, J. J. O'Toole, and M. S. Allison, *Proc. Soc. Exp. Biol. Med.*, **124**, 1155 (1967).

46. S. S. Kappus, S. Qureshi, and R. Fuerst, *Dev. Ind. Microbiol.*, **15**, 397 (1974).

47. S. Stephens, C. DeSha, and R. Fuerst, *Dev. Ind. Microbiol.*, **12**, 346 (1971)

48. A. J. Lawrence, M. B. Kemp, and J. R. Quayle, *Biochem. J.*, **116**, 631 (1970).

49. K. W. Hann and P. A. Jensen, *Water Quality Characteristics of Hazardous Materials*, Vols. 1–4, Environmental Engineering Division, Civil Engineering Department, Texas A & M University, 1974.

50. A. G. Kim and L. J. Douglas, *U.S. Natl. Tech. Inf. Serv. PB Rep. No. 221575/4*, 1973, 13 pp.

51. U. Koester, D. Albrecht, and H. Kappus, *Toxicol. Appl. Pharmacol.*, **41**, 639 (1977).

52. A. H. Nuckolls, *Underwriters Laboratory Report No. 2375*, Nov. 13, 1933.

53. J. R. Vestal and J. J. Perry, *Can. J. Microbiol.*, **17**, 445 (1970).

54. B. Voleski and J. E. Zajic, *Appl. Microbiol.*, **21**(4), 614 (1971).

55. H. H. Westberg, R. A. Rasmussen, and M. Holdren, *Anal. Chem.*, **46**(12), 1852 (1974).

56. J. P. Conkle, B. J. Camp, and B. E. Welsh, *Arch. Environ. Health*, **30**(6), 290 (1975).

57. S. L. Kopczynski, W. A. Lonneman, T. Winfield, and R. Seila, *J. Air. Pollut. Control Assoc.*, **25**(3), 251 (1975).

58. H. Eyer, *Muench. Med. Wochenschr.*, **114**(13), 10 (1972).

59. FDA's GRAS List—*Substances Generally Accepted as Safe*—CFR Title 21, 121.

60. T. Ikoma, *Nichidai Igaku Zasshi*, **31**(2), 71 (1972).

61. W. E. Brown and V. E. Henderson, *J. Pharmacol. Exp. Ther.*, **27**, 1 (1925).

62. D. M. Aviado, S. Zakhari, and T. Watanabe, *Non-fluorinated Propellants and Solvents for Aerosols*, CRC Press, Cleveland, Ohio, 1977, pp. 49–81.

63. D. M. Aviado, *Toxicology*, **3**, 321 (1975).

64. D. M. Aviado and D. G. Smith, *Toxicology*, **3**, 241 (1975).

65. J. R. Vestal and J. J. Perry, *J. Bacteriol.*, **99**(1), 216 (1969).

66. W. E. O'Brien and L. R. Brown, *Dev. Ind. Microbiol.*, **9**, 389 (1967).

67. M. M. Landry and R. Fuerst, *Dev. Ind. Microbiol.*, **9**, 370 (1968).

68. *NIOSH Manual of Sampling Data Sheets*, 1977 ed., U.S. Dept. HEW, U.S. Printing Office, Washington, D.C.

69. B. B. Shugaev, *Arch. Environ. Health*, **18**, 878 (1969).

70. B. B. Shugaev, *Farmakol. Toksikol*, **30**, 102 (1967).

71. B. B. Shugaev, *Farmakol. Toksikol.*, **31**, 360 (1968).

72. C. F. Reinhardt, A. Azar, M. E. Maxfield, P. E. Smith, and L. S. Mullin, *Arch. Environ. Health*, **22**, 265 (1971).

73. U. Frommer, V. Ulrich, and H. J. Staudinger, *Z. Physiol. Chem.*, **351**, 913 (1970).

74. K. Watanabe and S. Takesue, *Enzymologia*, **41**, 99 (1971).

75. I. Sunshine, ed., *CRC Handbook of Analytical Toxicology*, The Chemical Rubber Company, Cleveland, Ohio, 1969.

76. G. Holzer, H. Shanfield, A. Zlatkis, W. Bertsch, P. Juarez, H. Mayfield, and H. M. Liebich, *J. Chromatogr.*, **142**, 755 (1977).

77. H. W. Patton and G. D. Touey, *Anal. Chem.*, **28**, 1865 (1956).

78. R. E. Gosselin, H. C. Hodge, R. P. Smith, and M. N. Gleason, *Clinical Toxicology of Commercial Products*, 4th ed., Williams and Wilkins, Baltimore, 1976.

79. R. D. Stewart, A. A. Herrmann, E. D. Baretta, H. V. Forster, J. J. Sikora, P. E. Newton, and R. J. Soto, *Scand. J. Work Environ. Health*, **3**(4), 234 (1977).

80. S. A. Friedmann, M. Cammarato, and D. M. Aviado, *Toxicology*, **1**, 345 (1973).

81. K. Watanabe and S. Takesue, *Agric. Biol. Chem.*, **36**, 825 (1972).

82. *Documentation of the Threshold Limit Values for Substances in Workroom Air*, 3rd ed., American Conference of Governmental Industrial Hygiene, Cincinnati, 1971.

83. N. W. Lazarew, *Arch. Exp. Pathol. Pharm.*, **143**, 223 (1929).

84. T. I. Bonashevskaya and D. P. Partsef, *Gig. Sanit.* **36**(9), 11 (1971).

85. R. E. Pattle, C. Schock, and J. Battensby, *Br. J. Anesth.*, **44**(11), 1119 (1972).

86. M. G. Rumsby and J. B. Finean, *J. Neurochem.*, **13**, 1513 (1966).

87. D. A. Haydon, B. M. Hendry, S. R. Levinson, and J. Requena, *Biochim. Biophys. Acta*, **470**, 17 (1977).

88. D. A. Haydon, B. M. Hendry, and S. R. Levinson, *Nature*, **268**, 356 (1977).

89. M. Gaultier, G. Rancural, C. Piva, and M. L. Efthymion, *J. Eur. Toxicol.*, **6**(6), 294 (1973).

90. W. R. F. Notten and P. W. Henderson, *Biochem. Pharm.* **24**, 1093 (1975).

91. K. M. Fredericks, *Nature*, **209**, 1047 (1966).

92. National Institute for Occupational Safety and Health, Criteria for a Recommended Standard, Occupational Exposure to Alkanes (C_5–C_8), U.S. Dept. HEW, Washington, D.C., 1977.

93. T. DiPaolo, *J. Pharm. Sci.* **67**(4), 566 (1978).

94. K. W. Nelson, J. F. Ege, M. Ross, L. E. Woodman, and L. Silverman, *J. Ind. Hyg. Toxicol.*, **25**, 282 (1943).

95. R. Korobkin, A. Ashbury, and S. Neilson, *Arch. Neurol.*, **32**, 158 (1975).

96. N. Battistini, G. L. Lenji, E. Zanette, C. Fieschi, F. Battista, A. Franzinelli, and E. Sartorelli, *Riv. Pat. Nerv. Ment.*, **95**, 871 (1974).

97. G. W. Paulson and G. W. Waylonis, *Arch. Intern. Med.*, **136**, 880 (1976).

98. A. Herskowitz, N. Ishii, and H. Schaumburg, *New Engl. J. Med.*, **28**, 82 (1971).

99. C. Carapella, *Ann. 1st Super Sanita*, **13**(1–2), 353 (1977).

100. M. Iida, Y. Yamamura, and I. Sobue, *Electromyography*, **9**, 247 (1969).

101. N. Ishii, A. Herskowitz, and H. Schaumburg, *J. Neuropathol. Exp. Neurol.,* **31,** 198 (1972).

102. M. L. Keplinger, G. E. Lanier, and W. B. Deichmann, *Toxicol. Appl. Pharmacol.,* **1,** 156 (1959).

103. M. Taira, *Tokyo Ika Daigaku Zasshi,* **33,** 4 (1975).

104. E. T. Kimura, D. M. Elery, and P. W. Dodge, *Toxicol. Appl. Pharmacol.,* **19,** 699 (1971).

105. C. H. Hine and H. H. Zuidema, *Ind. Med.,* **39**(5), 215 (1970).

106. P. S. Spencer, M. C. Bischoff, and H. H. Schaumburg, *Toxicol. Appl. Pharmacol.,* **44,** 17 (1978).

107. V. Foa, R. Gilioli, C. Bugheroni, M. Maroni, and G. Chiappino, *Med. Lavoro,* **67**(2), 136 (1976).

108. H. H. Schaumburg and P. S. Spencer, *Brain,* **99,** 182 (1976).

109. T. Inoue, S. Yamada, H. Miyagaki, and Y. Takeuchi, *Proc. XVI Int. Congr. Occup. Health (Tokyo),* 522 (1969).

110. K. Kurita, *Jap. J. Ind. Health Exp.,* **9,** 672 (1974).

111. A. K. Ashbury, S. L. Nielsen, and R. Telfer, *J. Neuropathol. Exp. Neurol.,* **33,** 191 (1974).

112. E. G. Gonzales and J. A. Downey, *Arch. Phys. Med. Rehab.,* **53,** 333 (1972).

113. J. Towfighi, N. K. Gonatas, D. Pleasure, H. S. Cooper, and L. McCree, *Neurology,* **26**(3), 238 (1977).

114. H. Altenkirsch, J. Mager, G. Stoltenburg, and J. Helmbrecht, *J. Neurol.,* **214**(2), 137 (1977).

115. P. Bohlen, U. P. Schlunegger, and E. Läuppi, *Toxicol. Appl. Pharmacol.,* **25,** 242 (1973).

116. A. Y. H. Lu, H. W. Strobel, and M. J. Coon, *Mol. Pharm.,* **6,** 213 (1970).

117. H. Kramer, H. Staudinger, and V. Ullrich, *Chem–Biol. Interactions,* **8,** 11 (1974).

118. G. D. Divicenzo, C. J. Kaplan, and J. Dedinas, *Toxicol. Appl. Pharm.,* **36,** 511 (1976).

119. R. A. Scala, *Ind. Occup. Hyg,* **19,** 293 (1976).

120. H. Vainio, *Acta Pharmacol. Toxicol.,* **34**(3), 152 (1974).

121. J. Sice, *Toxicol. Appl. Pharm.,* **9**(1), 70 (1966).

122. H. Fuehner, *Biochem. Z.,* **115,** 235 (1921).

123. W. Bertsch, R. C. Chang, and A. Zlatkis, *J. Chromatogr. Sci.,* **12,** 175 (1974).

124. C. P. Carpenter, D. L. Geary, Jr., R. C. Myers, D. J. Nachreiner, L. J. Sullivan, and J. M. King, *Toxicol. Appl. Pharm.,* **44,** 53 (1978).

125. G. I. Vinogradov, I. A. Chernichenko, and E. M. Makarenko, *Gig. Sanit.,* **8,** 10 (1974).

126. C. A. Nau, J. Neal, and M. Thornton, *Arch. Environ. Health,* **12,** 382 (1966).

127. D. L. Liu and B. J. Dutka, *J. Water Pollut. Control Fed.,* **45**(2), 232 (1973).

128. J. M. Lebeault, B. Roche, Z. Duvnjak, and E. Azoulay, *Arch. Microbiol.,* **72,** 140 (1970).

129. E. I. Kvasnikov, E. F. Solomko, A. M. Zhuravel, and V. M. Romanenko, *Mikrobiologiya,* **40**(5), 858 (1971).

130. Y. A. Guzhova and V. S. Fadeev, *Zh. Anal. Khim.,* **34**(1), 184 (1979).

131. K. Adachi and S. Yamasawa, *Nature,* **222,** 191 (1969).

132. E. Bingham and H. L. Falk, *Arch. Environ. Health,* **19,** 779 (1969).

133. A. W. Horton, D. N. Eshleman, A. R. Schuff, and W. H. Perman, *J. Natl. Cancer Inst.,* **56**(2), 387 (1976).

134. A. W. Horton and G. M. Christian, *J. Natl. Cancer Inst.,* **53**(4), 1017 (1974).

135. C. O. Gill and C. Ratledge, *J. Gen. Microbiol.,* **75**(1), 11 (1973).

136. T. Maruta, H. Koga, M. Kinoshita, K. Takei, and R. Ogura, *Kurume Med. J.,* **22**(3), 183 (1975).

137. J. Ziegenmeyer, N. Reuter, and F. Meyer, *Arch. Int. Pharm. Ther.,* **224**(2), 238 (1976).

138. J. D. Walker and R. R. Colwell, *Appl. Environ. Microbiol.,* **31**(2), 198 (1976).

139. M. Popovic, N. Aliaga, and N. Gerencevic, *Acta Pharm. Jugosl.,* **24**(1), 17 (1974).

140. J. Tulliez, G. Bories, C. Boudene, and C. Fevrier, *Ann. Nutr. Aliment.*, **29**(6), 563 (1976).

141. R. Ristow and H. Weiner, *Fette Seifen. Anstrichm.*, **70**, 273 (1968).

142. A. Tanaka and S. Fukui, *J. Ferment. Technol.*, **48**, 137 (1970).

143. V. W. Jamison, R. L. Raymond, and J. O. Hudson, *Appl. Microbiol.*, **17**(6), 853 (1969).

144. T. Higashihara and A. Sato, *Agric. Biol. Chem.*, **33**(12), 1802 (1969).

145. Z. Duvnjak, B. Roche, and E. Azoulay, *Arch. Mikrobiol.*, **72**(2), 135 (1970).

146. K. Morihara, *Appl. Microbiol.*, **13**(5), 793 (1965).

147. G. J. Mulkins-Phillips and J. E. Steward, *Canad. J. Microbiol.*, **20**, 955 (1974).

148. D. F. Jones and R. Howe, *J. Chem. Soc.*, **22**, 2809 (1968).

149. R. F. Lee, R. Sauerheber, and A. A. Benson, *Science*, **177**, 344 (1972).

150. R. H. Linnell and W. E. Scott, *Arch. Environ. Health*, **5**, 616 (1962).

151. J. N. Pitts, Jr., B. J. Finlayson, H. Akimoto, W. A. Kummer, and R. J. Steer, International Symposium on Identification and Measurement of Environmental Pollutants, Ottawa, Ontario, Canada, June 14–17, 1971.

152. M. M. Kay, A. F. Henschel, J. Butler, R. N. Ligo, and I. R. Tabershaw, *Occupational Diseases, A Guide to their Recognition*, U.S. Dept. HEW, Washington, D.C., 1977.

153. H. B. Elkins, *The Chemistry of Industrial Toxicology*, 2nd ed., Wiley, New York, 1959.

154. A. J. Haagen-Smit, E. F. Darley, M. Zaitlin, H. Hull, and W. Noble, *Plant Physiol.*, **27**, 18 (1952).

155. E. R. Stephens, *U.S. Natl. Tech. Inf. Serv. PB Rep. No. 230993/8GA*, 1973.

156. A. L. Cowles, H. H. Borgstedt, and A. J. Gillies, *Anesthesiology*, **36**(6), 558 (1972).

157. W. B. Deichmann and H. W. Gerarde, *Toxicology of Drugs and Chemicals*, Academic Press, New York, 1969.

158. Manufacturing Chemists Assoc., Inc., Chemical Safety Data Sheets, Washington, D.C.

159. C. Reynolds, *Anesth. Analg.*, **6**, 121 (1927).

160. L. S. Goodman and A. Gilman, *The Pharmacological Basis of Therapeutics*, 4th ed., Macmillan, New York, 1971.

161. L. R. Riggs, *Proc. Soc. Exp. Biol. Med.*, **22**, 269 (1924–1925).

162. *Encyclopaedia of Occupational Health and Safety*, Vols. I and II, McGraw-Hill, New York, 1972.

163. R. B. Conolly, R. J. Jaeger, and S. Szabo, *Exp. Mol. Pathol.*, **28**, 25 (1978).

164. A. O. Olson and M. Spencer, *Can. J. Biochem.*, **46**, 283 (1968).

165. L. K. Riggs, *J. Am. Pharm. Assoc.*, **14**, 380 (1925).

166. M. L. Krasovitskaya and L. Malyarova, *Gig. Sanit.*, **33**(5), 7 (1968).

167. P. Cazzamali, *Clin. Chirurg.*, **34**, 477 (1931).

168. P. Pietsch and M. Chenoweth, *Proc. Soc. Exp. Biol. Med.*, **130**, 714 (1968).

169. H. K. Pratt and J. D. Goeschl, *Ann. Rev. Plant Physiol.*, **20**, 542 (1969).

170. R. E. Holm and J. L. Key, *Plant Physiol.*, **44**, 1295 (1969).

171. G. D. Clayton and T. S. Platt, *Am. Ind. Hyg. Assoc. J.*, **28**, 151 (1967).

172. M. L. Krasovitskaya and L. K. Malyarova, *Biol. Deistvie Gig. Znachenie Atm. Zagryaz.*, 74 (1966).

173. W. Braker and A. Mossman, *Matheson Gas Data Book*, 5th ed., Matheson Gas Products, East Rutherford, N.J., 1971.

174. I. Watanabe, T. Okita, and S. Seino, *Koshu Eiseiin Kenkyu Hokoku*, **20**(2), 107 (1971).

175. D. F. Meigh, *Nature*, **184**(Suppl. 14), 1072 (1959).

176. J. W. Swinnerton and V. J. Linnenbom, *Science*, **156**, 1119 (1967).

177. Serta Ltd., Br. Patent 1,026,685 (C1.A OIN), April 20, 1966.

178. W. E. Brown, *J. Pharmacol. Exp. Therap.*, **23**, 485 (1924).

179. J. T. Halsey, C. Reynolds, and W. A. Prout, *J. Pharm. Exp. Therap.*, **26**, 479 (1926).

180. M. H. Kahn and L. K. Riggs, *Ann. Int. Med.*, **5**, 651 (1932).

181. B. M. Davidson, *J. Pharmacol. Exp. Therap.*, **26**, 33 (1926).

182. L. K. Riggs and H. D. Goulden, *Anesth. Analg.*, **4**, 299 (1925).

183. C. Reynolds, *J. Pharmacol. Exp. Therap.*, **27**, 93 (1926).

184. S. W. May, R. A. Schwartz, B. J. Abbott, and O. R. Zaborsky, *Biochim. Biophys. Acta*, **403**, 245 (1975).

185. M. W. Ranney, *Synthetic Lubricants*, Noyes Data Corp., Park Ridge, N.J., 1972.

186. J. E. Casida, E. C. Kimmel, B. Holm, and G. Widmark, *Acta Chem. Scand.*, **25**(4), 1497 (1971).

187. B. Testa and D. Mihailova, *J. Med. Chem.*, **21**(7), 683 (1978).

188. W. Braker, A. L. Mossman, and D. Siegel, *Effects of Exposure to Toxic Gases—First Aid and Medical Treatment*, 2nd ed., Matheson, Lyndhurst, N.J., 1977.

189. R. Gould, *Photochemical Smog and Ozone Reactions*, American Chemical Society, Washington, D.C., 1972.

190. C. E. Searle, *Chemical Carcinogens*, American Chemical Society, Washington, D.C., 1976.

191. U.S. Coast Guard, Dept. Transp., *Chemical Data Guide for Bulk Shipment by Water*, U.S. Government Printing Office, Washington, D.C., 1973.

192. C. P. Carpenter, H. F. Smyth, Jr., and U. C. Pozzani, *J. Ind. Hyg. Toxicol.*, **32**(6), 344 (1949)

193. I. Richter, H. Krain, and H. K. Mangold, *Experientia*, **32**(2), 186 (1976).

194. E. Gil-Av and J. Shabtai, *Nature*, **197**, 1065 (1963).

195. J. S. Fruton and S. Simmonds, *General Biochemistry*, 2nd ed., John Wiley, New York, 1958.

196. M. W. Ranney, *Lubricant Additives*, Noyes Data Corporation, Park Ridge, N.J., London, 1973.

197. F. Benini, V. Colamussi, and M. L. Zannoni, *Arcisp. S. Anna Ferrara*, **23**(6), 511 (1970).

198. A. Hamilton and H. L. Hardy, *Industrial Toxicology*, 3rd ed., PSG Publishing Company, Inc., Littleton, Mass., 1974.

199. I. M. Mirzoyan, *Gig. Tr. Prof. Zabol.*, **11**, 38 (1972).

200. C. P. Carpenter, C. B. Shaffer, C. S. Weil, and H. F. Smyth, Jr., *J. Ind. Hyg. Toxicol.*, **26**, 69 (1944).

201. *Gig. Tru. Prof. Zab.*, **13**, 42, (1969) (through Ref. 9).

202. E. I. Donetskaya and F. S. Shvartsapel, *Tr. Permsk. Med. Inst.*, **82**, 223 (1970).

203. P. deMayo, *Mono- and Sesquiterpenoids*, Vol. II, Interscience, New York, 1959.

204. T. Dalhamn and R. Rylander, *Arch. Environ. Health*, **20** (1970).

205. D. P. Gough and F. W. Hemming, *Biochem. J.*, **118**, 163 (1970).

206. T. Dalhamn, M. Edfors, and R. Rylander, *Arch. Environ. Health*, **17**, 252 (1968).

207. J. L. Egle, Jr., and B. J. Gochberg, *Am. Ind. Hyg. Assoc. J.*, **36**(5), 369 (1975).

208. V. D. Gostinskii, *Gig. Tr. Prof. Zab.*, **9**(1), 36 (1965).

209. L. Weissbecker, R. M. Creamer, and R. D. Carpenter, *Am. Rev. Respir. Dis.*, **104**(2), 182 (1971).

210. F. A. Patty, *Industrial Hygiene and Toxicology*, 2nd rev. ed., Vol. II, John Wiley, New York, 1963.

211. R. J. Shamberger, *J. Natl. Cancer Inst.*, **47**(3), 667 (1971).

212. S. A. Pigolev, *Gig. Tr. Prof. Zabol.*, **15**(2), 49 (1971).

213. A. A. Nikultseva, *Gig. Tr. Prof. Zabol.*, **11**(12) 41 (1976).

214. A. G. Pestova and O. G. Petrovskaya, *Vrach. Delo.*, **4,** 135 (1973).

215. H. F. Smyth, Jr., C. P. Carpenter, C. S. Weil, W. C. Pozzani, J. A. Streigel, and J. S. Nycum, *Am. Ind., Hyg. Assoc. J.*, **30,** 470 (1969).

216. R. N. Harger and L. W. Spolyar, *Arch. Ind. Health*, **18,** 497 (1958).

217. C. H. Theines and T. J. Haley, *Clinical Toxicology*, 5th ed., Lea & Febiger, Philadelphia, 1972.

218. M. Yoshioka, S. Tsujimoto, N. Oki, M. Tamaki, Y. Torihashi, and N. Takada, *Hyogo-ken Kogai Kenkyusho Kenkyu Hokoku*, **8,** 48 (1976).

219. M. J. Dilworth, *Biochim. Biophys. Acta*, **127,** 285 (1966).

220. M. Kelly, *Biochem. J.*, **107,** 1 (1968).

221. H. Yasuda, M. O. Bumgarner, H. C. Marsh, B. S. Yamanashi, D. P. DeVito, M. L. Wolbarsht, J. W. Reed, M. Bessler, M. B. Landers III, D. M. Hercules, and J. Carver, *J. Biomed. Mater. Res.*, **9**(6), 629 (1975).

222. E. M. J. Gifford, *Am. J. Bot.*, **56**(8), 892 (1969).

223. W. W. Aldrich and H. Y. Nakasone, *J. Am. Soc. Hortic. Sci.*, **100**(4), 410 (1975).

224. T. W. Speitel and S. M. Siegel, *Plant Cell Physiol.*, **16**(2), 383 (1975).

225. E. Yip, W. A. Southorn, and J. B. Gomez, *J. Rubber Res. Inst. Malays.*, **24**(2), 103 (1974).

226. B. M. Davidson, *J. Pharmacol.*, **25,** 119 (1925).

227. A. T. Jones, *Arch. Environ. Health*, **5,** 417 (1960).

228. D. S. Ross, *Ann. Occup. Hyg.*, **16,** 85 (1973).

229. F. Flury, *Arch. Exp. Pathol. Pharmakol.*, **138,** 65 (1925).

230. H. Franken and L. Miklos, *Zentralbl. Gynaekol.*, **42,** 2493 (1933).

231. National Institute for Occupational Safety and Health, *Criteria for a Recommended Standard, Occupational Exposure to Acetylene*, U.S. Dept. HEW, Washington, D.C., 1976.

232. T. R. Norton, "Metabolism of Toxic Substances," Chapter 4 in L. J. Casarett and J. Doull, Eds., *Toxicology The Basic Science of Poisons*, Macmillan, New York, 1975.

233. J. Reisch, W. Spitzner, and K. E. Schulte, *Arzneim.-Forsch.*, **17**(7), 816 (1967).

234. Shell Chemical Company, unpublished report, 1961 (through Ref. 9).

235. F. Sperling, *Fed. Proc.*, **19**(1), 389 (1960).

CHAPTER FORTY-SIX

Alicyclic Hydrocarbons

ESTHER E. SANDMEYER, Ph.D.

1 CYCLOPARAFFINS

To this chemical category belong the cycloparaffins, also named the cycloalkanes, the cyclanes, or naphthenes, the cycloolefins, also termed cycloalkenes or cyclenes, and a variety of their substituted single, sequential, or fused ring systems. Some naturally occurring alkanyl- or alkenyl-substituted mono-, di-, or polycyclenes are commonly known as terpenes, which occur in assorted plants. Other compounds are isolated from crude petroleum refinery distillates or catalytically cracked petroleum products. Cyclanes are extensively used to produce reformed aromatics (1). Some are utilized as inhalation anesthetics, and are synthesized in pure form by the reduction of dihalogenated propane precursors. Some physical properties of the cycloparaffins are listed in Table 46.1 (2–11) and physiological data in Tables 46.2 and 46.3 (12–23).

The lower cycloparaffins are gases and have been used as anesthetics, especially cyclopropane, the simplest cyclane. The C_5 and higher members are liquids with narcotic properties. However, from C_6 on, the margin between narcosis and death is very narrow and symptomatically barely recognizable. The alicyclics, in general, are CNS depressants with low acute and chronic toxicities, owing to their rapid excretion in unchanged form or prompt conversion into water-soluble metabolites. Inhalation by humans and laboratory animals at high concentrations may cause excitement, loss of equilibrium, stupor, and coma, but rarely death. Oral administration in animals has resulted in severe diarrhea and vascular collapse, in turn leading to heart, lung, liver, and brain degeneration. Cycloparaffins are dermal irritants; they defat the skin to cause morpho-

Table 46.1. Physical and Chemical Data for Cycloparaffins (2–11)

Compound	B.P. (°C) (mm Hg)	CAS Registry No.	Density (g/ml, 20°C)	Emp. Formula	Flamm. Limits (%)	Flash Pt. [°C(°F)]	Freezing Pt. (°C)	Mol. Wt.	Refr. Index	Solubility[a] w/al/et	Sp. Gr.	Vapor Dens.	Vapor Pres. [mm Hg (°C)]	Visc. (SUS)	W/V Conv. (mg/m³ ⇔ 1 ppm)
Cyclopropane	−32.7	75-19-4	0.6769[b]	C_3H_6	2.4–10.4	—	−127.6	42.08	1.3799[d]	s/v/v	0.720	1.45		N.A.	1.72
Cyclobutane	12.0	287-23-0	0.720[c]	C_4H_8		<10 (<50)	−50.0	56.11	1.426	i/v/v		1.93		N.A.	2.30
Cyclopentane	49.26	287-92-3	0.7454	C_5H_{10}		−7 (−35)	−93.88	70.14	1.4065	i/v/v	0.751	2.42	400 (31.0)	<32.6	2.87
Methylcyclopentane	71.81	96-37-7	0.7486	C_6H_{12}	1.2–8.4	−28.9 (<20)	−142.465	84.16	1.4097	i/v/v	0.754	2.90	100 (17.9)	<32.6	3.44
Ethylcyclopentane	103.47	1640-89-7	0.7665	C_7H_{14}	1.1–6.7		−138.44	98.19	1.4198	i/v/v	0.771	3.4		<32.6	4.02
Cyclohexane	80.74	110-82-7	0.7786	C_6H_{12}	1.33–8.4	−20 (−4)	6.55	84.16	1.4262	i/v/v	0.779	2.90	100 (60.8)	<32.6	3.44
Methylcyclohexane	100.93	108-87-2	0.7694	C_7H_{14}	1.2–6.7	−4 (25)	−126.59	98.19	1.4231	i/s/s	0.774	3.39	43 (25)	<32.6	4.02
Ethylcyclohexane	131.78	1678-91-7	0.7880	C_8H_{16}	0.9–6.6	35 (95)	−111.32	112.21	1.4330	i/s/s	0.792	3.9		<32.6	4.59
Dimethylcyclohexane	119.54	2207-64-7	0.7584	C_8H_{16}		11.1 (52)	−33.50	112.21	1.4290	i/s/s	0.767	3.86	10 (10.2)		4.59
Cycloheptane	118.48	291-64-5	0.8098	C_7H_{14}		<37.8 (<100)	−12.0	98.19	1.4436	i/s/s	0.783	3.3			4.02
Cyclooctane	149.0 (749)	292-64-8	0.8349	C_8H_{16}		14.3 (57.9)	14.3	112.21	1.4586	i/v/v					4.59
Cyclononane	353.1			C_9H_{18}		11.1 (52)	11.1	123.23		i/—/—	0.854				5.16

[a] Solubility in water/alcohol/ether: v = very soluble; s = soluble; i = insoluble.
[b] −30°C.
[c] 5°C.
[d] −42°C.

logical changes and hypothermia, as does, for example, methylcyclohexane. The liquid members up to C_8 and possibly to C_{12}, when ingested, represent lung aspiration hazards.

1.1 Cyclopropane

Cyclopropane, trimethylene, $\overline{CH_2CH_2CH_2}$, the simplest cycloparaffin, is a colorless, flammable (DOT) (24), and explosive gas. Its odor resembles that of petroleum ether. Some physicochemical data are presented in Table 46.1 (2–11). Reagent grade cyclopropane is prepared by the reduction of 1,2-dibromocyclopropane (25). It is stable at room temperature but undergoes ring opening to propene when heated (26). In human and animal anesthesiology (see Table 46.2), it is the preferred agent for a wide range of purposes (14, 27). Cyclopropane's popularity as an anesthetic stems from its potent action but low acute toxicity. It is readily absorbed through the pulmonary system and exerts potent CNS depressant action (28, 29). Although there is a wide margin between anesthetic effect and toxicity, it also exerts some subtle effect on the pulmonary and cardiac systems as have been documented, including mild myocardial sensitization to epinephrine (30, 31). Studies with mice showed anesthesia in 3 min at 5.8 mmol/liter without death, and recovery in 1.5 min, (12) and a concentration of 18 percent was lethal in 39 min (see Table 46.2). Experimentally, it was fast acting in the dog, which then served as a model for predicting anesthetic concentrations for human application (32). A study using the chicken showed that prolonged, high exposure did produce embryonic abnormalities at 10 to 20 percent with a 6-hr exposure and increased deaths at 20 to 30 percent with 12-hr exposures (33). In general, however, cyclopropane should not present an industrial hazard. At present, no official TLV is established, although in 1959 ACGIH (34) recommended a TLV of 400 ppm. Analytic determination procedures are available.

1.2 Cyclobutane

Cyclobutane, tetramethylene, $\overline{CH_2(CH_2)_2CH_2}$, is a colorless, flammable (DOT) (24), explosive gas; see Table 46.1 (2–11). Cyclobutane, produced from butene, is synthesized and reutilized in catalytic cracking processes (1). The gas is used as an anesthetic, but it is a weak cardiac sensitizer (35). Industrially, similarly to cyclopropane, it is not expected to be a health hazard.

1.3 Cyclopentane

Cyclopentane, pentamethylene, $\overline{CH_2(CH_2)_3CH_2}$, is the lowest liquid member of the cyclane series. It is flammable (DOT) (24) and its vapors are explosive. Further physicochemical data are presented in Table 46.1 (2–11). Cyclopentane

Table 46.2. Inhalation Effects of Some Cyclanes

Compound	Species	Concentration ppm	Concentration mg/liter	Time	Findings	Ref.
Cyclopropane	Mouse	142,000	5.8mM	3 min	Anesthesia in 3 min, recovery in 1.5 min	12
		18,000	309.6	39 min	Lethal	12
Cyclopentane	Human	10–15	0.029–0.043		Tolerable	13
	Mouse	38.3	0.110		Minimal narcotic concentration, loss of reflexes and lethal	14
	Rat	112–1139	0.039–0.397	6 hr/day × 3 weeks	No effects in female and male rats	13
		8110	2.328	6 hr/day × 12 weeks	Decreased body weight gains in females	13
Cyclohexane	Mouse	18,000	61.9	5 min	Trembling	15
				15 min	Disturbed equilibrium	15
				25 min	Recumbent	15
	Rabbit	427	1.47		No effect	16
		440	1.51		No effect	16
		780–3300	2.68–11.3		Some micropathological changes	16
		3300	11.46		No visiblle effects	15
		<7500	<25.8		Some deaths	17
		14,535	50		Loss of righting reflex	18
		18,000	61.9	6 min	Trembling	15
				15 min	Disturbed equilibrium	15
				30 min	Recumbent	15
		18,500	63.4	8 hr	Not lethal	15
		26,600	89.4	1 hr	Lethal	15
		17,400–20,350	60–70		Lethal	18
		26,050	89.6		Before death, changes occur in hemoglobin and erythrocyte quantities-leukocytes increase	16

Compound	Species				Effect	Ref.
	Guinea pig	18,000	61.9	11 min	Slight trembling	15
	Cat	18,000	61.9	18–25 min	Disturbed equilibrium	15
					Recumbent	15
Cyclohexane	Primate	1243	4.28	6 hr/day × 50	No effect	15
	Rabbit	443	1.49	8 hr/day × 26 wk	No effect, no pathological effects	17
		786	2.70	6 hr/day × 50	Minor microscopic changes in liver and kidney	15
		3330	11.46	6 hr/day × 50	No deaths, no signs of injury	15
		7,400–18,500	25.5–63.4	6 hr/day × 10	Some fatalities	15
Methylcyclohexane	Rabbit	1050	4.22		No effect	16
		1162	4.67		No effect	16
		2830	11.35		Very subtle cellular injury to kidney and liver	16
	Mouse	7463–9950	30–40		Loss of righting reflex	18
		7500–10,800	30.2–43.4		Lethal	16
		9950–12,440	40–50		Lethal	18
	Primate, rabbit	363	1.46	6 hr/day × 50	No effect	16
		9850	39.6	6 hr/day × 50	Tendency to light convulsions	16
		14,900	59.9	6 hr/day × 50	Convulsions	16
Dimethylcyclohexane	Mouse	4360–5450	20–25		Loss of righting reflex	18
		5450–6540	25–30		Death	18
Ethylcyclohexane	Mouse	3270	15		Loss of righting reflex	18
		7625	35		Death	18

Table 46.3.   Physiologic Effects of Some Liquid Cyclanes

Compound	Route	Species	Parameter	Result	Ref.
Cyclohexane	Oral	Rat	LD_{50}	8.0–39.0 ml/kg	19
			LD_{50}	29.82 g/kg	20
		Mouse	LD_{50}	1.30 g/kg	21
		Rabbit	LD_{LO}	5.5–6.0 g/kg	17
			TD_{LO}	1.5–5 g/kg	17
	Dermal	Rabbit	LD_{100}	>180.2 g/kg	17
	IV	Rabbit	LD_{LO}	77 mg/kg	17
	Toxicity	Aquatic	TLm96	100–1000 ppm	22
Methylcyclohexane	Oral	Rabbit	TD_{LO}	1.0–4.0 g/kg	17
			LD_{LO}	4.0–4.5 g/kg	17
			LD_{100}	4.5–10.0 g/kg	17
	Dermal	Rabbit	LD_{50}	>86.7 g/kg	17
Cyclododecane	SC	Mouse	LD_{50}	10 g/kg	23

is not sufficiently stable to occur naturally in large quantities. It is produced from petroleum products, and is found as an impurity in technical grade hexane (31). Industrially, it is used for cracking aromatics (1). Commercially, cyclopentane is used to produce a variety of chemicals used as analgesics, sedatives, hypnotics, antitumor agents, CNS depressants, prostaglandins, insecticides, and many more (36).

Cyclopentane is a CNS depressant and lipid solvent (29); see Table 46.2. Experiments with mice have demonstrated that no safety margin exists between minimal narcotic concentration, loss of reflexes, and lethal dose, all occurring at 110 mg/liter (14). When ingested, cyclopentane presented a low to moderate aspiration hazard in mice (37). Cyclopentane applied to guinea pig skin produced slight erythema and dry appearance (38). An aquatic TLm96 determination was above 1000 ppm (22). No official threshold limit values have been established. Proper precautions are recommended, however, when handling the material.

1.3.1   Methylcyclopentane

Methylcyclopentane, $CH_3\overline{CH(CH_2)_3CH_2}$, is a colorless, flammable (DOT) (24) liquid and has a sweetish odor. For further physicochemical data, see Table 46.1 (2–11). Methylcyclopentane has been detected in human respiratory air in smokers and traces in nonsmokers (39). It resembles cyclopentane in its toxicity; see Table 46.2.

Methylcyclopentane is an excellent solvent extractant for essential oils from plants (40). Chemically, the substituted cyclopentane ring plays an integral part in a number of natural products, such as prostaglandins (36, 41). It also possesses good selectivity for aromatization to benzene (1). Toxicologically, it exhibits no safety margin between onset of narcosis and death.

1.3.2   Ethylcyclopentane

In addition to causing narcosis and anesthesia, the methyl- and ethylcyclopentane, $CH_3\overline{CH_2CH(CH_2)_3}CH_2$, have convulsive properties (14). The toxicity to the mouse increases from the cyclopentane to the methyl to the ethyl derivative with a minimal narcotic concentration of 110 mg/liter (14) and a fatal range of 30 to 50 mg/liter having been recorded (18). The aquatic TLm96 value was above 1000 ppm (22). For physicochemical and physiological properties, see Tables 46.1 and 46.2.

1.4   Cyclohexane

Cyclohexane, hexahydrobenzene, hexamethylene, $\overline{CH_2(CH_2)_4}CH_2$, is a colorless, flammable (DOT) (24) liquid with a sweetish odor when pure. Saturated air at 760 mm and 26.3°C contains 13.66 percent cyclohexane with a density of 1.23 times that of air (15). It is fractionated from crude oil, where it occurs in fairly large proportions, but is also prepared synthetically from benzene (26) or by hydrocracking of cyclopentane (1). Some physicochemical properties are presented in Table 46.1 and physiological data in Tables 46.2 and 46.3. Cyclohexane is used as a solvent for resins, fats, and waxes, but particularly for rubber and adhesives (42), as a raw material for a number of chemicals, especially adipic acid, and as the precursor of nylon-66 (43). Cyclohexane can be converted to benzene, or, when acidified, to methylcyclopentane by petroleum reforming processes (1).

When used as a solvent or diluent, cyclohexane was observed to accelerate the penetration of local anesthetics through the intact guinea pig skin (44). A comparative study with humans and animals has demonstrated that cyclohexane may have therapeutic value for reducing granulocytes (45). Conversely, cyclohexane may not only promote drug action, but in some cases potentiate the effect of toxic agents, such as tri-o-cresyl phosphate (46).

Cyclohexane generally is a CNS depressant (29) and may cause dizziness, nausea, and unconsciousness (14). It is a lipid solvent and on repeated contact defats the skin. The vapor may be irritant to the skin, eyes, and respiratory tract, and acute oral administration of cyclohexane to rats demonstrated LD_{50} values from 8.0 to 39.0 ml/kg, varying with the age of the animal (19); see Tables 46.2 and 46.3. Extrapolating to the human, Kimura determined a no-effect level of 0.016 ml/kg or 1.5 ml per 60 kg weight (19). Animals administered lethal doses exhibited, within 1 to 1.5 hr, signs of severe diarrhea and widespread vascular damage and collapse, hepatocellular degeneration, and toxic glomerulonephritis (17). Cyclohexane is nominally absorbed by the skin, although massive applications to the rabbit skin have shown microscopic changes in the liver and kidneys (17). Inhalation experiments demonstrated low acute toxicity, but a narrow margin between narcosis, loss of reflexes, and death; see Table 46.2. Chronic animal studies demonstrated low inhalation toxicity in the rabbit, but somewhat higher in the mouse. Cyclohexane is absorbed by inhalation, but a small fraction again exhaled. Another portion is excreted in the urine in unchanged form; the balance is metabolized via the vascular, hepatic and nephric systems. Inhalation induces liver microsomal hydroxylases, which, in turn, oxidize the cyclane to cyclohexanol, as investigated by numerous authors (47–50). Cyclohexanol then appears to be excreted mainly as the sulfate or glucuronide conjugate (16). However, the action of cyclohexane differs from that of benzene, inasmuch as the cytochrome P-450 monooxygenase system is not affected (51). Various bacterial systems have also been observed to oxidize cyclohexane to cyclohexanol (49). In addition, adaption to alternative substrates, such as other cyclanes, and also photodecomposition and mineralization (52) indicate that cyclohexane is biodegradable.

Currently for cyclohexane, a TLV of 300 ppm (1050 mg/m^3) for an 8-hr workshift has been recommended by OSHA (57). A sampling method, S28.01, has been issued by NIOSH (53). It suggests air collection on charcoal and analysis by flame ionization chromatography. To date, few industrial morbidity reports have been published, except one where cyclohexane was suspected to have caused severe hematopoietic effects. However, this later was mainly attributed to high benzene contents of the solvents used. For preventive industrial programs, however, in addition to air analysis, urine should be analyzed for cyclohexane or its metabolites. The decrease in the ratio of inorganic to total sulfates is roughly proportional to concentration of inhaled cyclohexane. The ratio adjustment does not occur as rapidly as with benzene (15). Thus proper control groups should be available to thoroughly establish normal excretion values. In general, cyclohexane is a safe solvent when handled according to instructions (42).

1.4.1   Methylcyclohexane

Methylcyclohexane, hexahydrotoluene, $CH_3\overline{CH(CH_2)_4}CH_2$, is a colorless, flammable (DOT) (24) liquid; see Table 46.1 (2–11). Saturated air at 760 mm and 25°C contains 5.65 percent methylcyclohexane, with its density 1.14 times that of air (15). It occurs in certain crude petroleum oils, from which it is separated by distillation. It is also prepared by hydrogenation of toluene or by acidic hydrocracking of polycyclic aromatics (1). Methylcyclohexane is used as a solvent for cellulose ethers and as raw material in a variety of synthetic processes (54).

Similarly to cyclohexane, methylcyclohexane administered orally produced diarrhea in rabbits within 1 to 1.5 hr of the exposure (17). For further data, see Tables 46.2 and 46.3. Dermally applied, methylcyclohexane was not absorbed initially, but caused defatting of the skin, then hardening of the keratin layer, some cellular injury, and slight hypothermia (17). The inhalation no-effect level reaches about 1200 ppm for the rabbit and about 300 ppm for the primate (16). Lethal concentrations in the primate caused mucous secretion, lacrimation, salivation, labored breathing, and diarrhea (16). Should solvent droplets reach the lungs, methylcyclohexane represents a greater aspiration hazard than cyclohexane or cycloheptane (37).

When inhaled, methylcyclohexane is absorbed and translocated to the liver and kidneys via the vascular system, then hydroxylated to the *trans*-4-ol by hepatic microsomal enzymes (55), and finally excreted, mainly as the sulfate or the glucuronide (16). Some bacterial ω-hydroxylases are also capable of hydroxylating methylcyclohexane to methylcyclohexanol (49). In plant pathology, methylcyclohexane has been found active in tumor destruction (56).

OSHA (57) has established a TLV of 500 ppm (2000 mg/m^3) for an 8-hr time-weighted average, and ACGIH (58) recommends 400 ppm or 1600 mg/m^3. NIOSH has published a sampling procedure, S-94, and suggests analytic determination by flame ionization chromatography (53).

1.4.2 Other Alkylcyclohexanes

Dimethylcyclohexane, $CH_3\overline{CH(CH_2)_4CH}CH_3$, is used as a solvent. For further physicochemical data, see Table 46.1 (2–11). Dimethylcyclohexane appears more toxic than the monomethyl derivative. A concentration of 20 to 25 mg/liter inhaled by the mouse showed loss of righting reflex and 25 to 30 mg/liter caused death (18). Ethylcyclohexane [see Table 46.1 (2–11)] was initially even more toxic. The loss of righting reflex occurred at a concentration of 15 mg/liter and death at 35 mg/liter (18). Dimethyl- and ethylcyclohexane are used in catalytic reforming to produce C_8 aromatic compounds (1); see Tables 46.2 and 46.3.

1.5 Cycloheptane

Cycloheptane, heptamethylene, $\overline{CH_2(CH_2)_5CH_2}$, is a colorless, flammable (DOT) (24) liquid; see Table 46.1 (2–11). The toxicologic properties resemble those of methylcyclohexane. When ingested, it presents an aspiration hazard. Heptamethylene, when applied to the guinea pig skin, causes morphological changes and increases arginase activity (38).

The industrial hygiene sampling procedures recommended for cyclohexane may apply here, too. Mass spectral data are available for its analytic determination (59).

1.6   Cyclooctane and Higher Cycloparaffins

Cyclooctane, octamethylene, $\overline{CH_2(CH_2)_6CH_2}$, is a flammable liquid (DOT) (24); see Table 46.1 (2–11). When ingested, it presents an aspiration hazard (37). When applied to the guinea pig skin, cyclooctane causes morphological changes and increases arginase activity, as does cyclododecane (38). Cyclooctane and cyclononane can be analytically determined by gas chromatography (59). Cyclododecane is used as a mothproofing agent (60). An LD_{50} value of 10 mg/kg was determined subcutaneously in the mouse (23); see Table 46.3. Cyclododecane was not found to be a skin irritant (23).

2   CYCLOOLEFINS

The cyclic olefins are more highly reactive than their paraffin counterparts. They form photochemical smog by reaction with ozone and other small molecular or ionic moieties (61). Some physicochemical properties are listed in Table 46.4 (2–11) and physiological data in Tables 46.5–46.7 (62–84).

Cycloolefins or cyclenes, with increasing molecular weight in the C_4 to C_7 series, appear to increase in toxicity by inhalation. The di- and polycyclic members, the cycladienes and polyenes, appear to possess increasingly irritant toxic and sensitizing properties, peaking somewhat higher for the dicyclenes than for the monocyclic olefins. The aspiration hazard also appears higher for the cycloolefins than the cycloparaffins, including C_6 to C_{16} compounds, dependent on the specific series involved (37).

2.1   Cyclopentene

Cyclopentene, $\overline{CH:CH(CH_2)_2CH_2}$, is a highly flammable liquid. It has a low flash point and reacts readily with oxidizing agents (7). For further physical–chemical and physiological properties, see Tables 46.4 to 46.6. A range-finding study (62) has shown an oral LD_{50} in the rat of 2.14 ml/kg and a dermal LD_{50} in the rabbit of 1.59 ml/kg. These results indicate that cyclopentene is absorbed through the skin. Inhalation of the concentrated vapor was lethal to all six rats in 5 min, (62) and a 16,000 ppm, 4-hr exposure was lethal to four of six rats (62). Skin irritation and corneal injury in rabbits were moderate to severe (62). Short-term exposure of cyclopentene to humans revealed a tolerable level of 10 to 15 ppm only (63). Chronic inhalation of 112 to 1139 ppm for 12 weeks by female and male rats showed no effects, whereas 8110 ppm, 6 hr/day, 5 days/week for 3 weeks resulted in decreased body weight gains of female rats (13).

Table 46.4. Physical and Chemical Data for Cyclolefins (2–11)

Compound	B.P. (°C) (mm Hg)	CAS Registry No.	Density (g/ml, 20°C)	Emp. Formula	Flamm. Limits (%)	Flash Pt. [°C(°F)]	Freezing Pt. (°C)	M.P. (°C)	Mol. Wt.	Refr. Index, n_D^{20}	Solubility[a] w/al/et	Sp. Gr.	Vapor Density (Air = 1)	Vapor Pres. [mm Hg (°C)]	W/V Conv. (mg/m³ ⇔ 1 ppm)
Cyclopentene −37.22	44.24	142-29-0	0.7720	C₅H₈		−37.22 (−35.0)	−135.08	−93.3	68.12	1.4225	i/s/s	0.778			2.79
Cyclohexene	82.98	110-83-8	0.8110	C₆H₁₀		−6 (21)	−103.51	−103.51	82.15	1.4465	i/v/v	0.816	2.8	160 (38)	3.36
Cycloheptene	115.0		0.8228	C₇H₁₂			−56.0	−56.0	96.17	1.4552	i/s/s				3.93
1-Vinylcyclohexene	145.0	2622-21-1	0.8623	C₈H₁₂					108.18	1.4915	i/—/s				4.43
4-Vinylcyclohexene	128.9	100-40-3	0.8299	C₈H₁₂		21.11 (70)		−108.89	108.18	1.4639	i/—/s	0.834	3.76	25.8 (38)	4.43
Limonene	178.0	138-86-3	0.8411	C₁₀H₁₆				−74.35	136.24	1.4730	i/v/v				5.57
1,3-Cyclopentadiene	40.0	542-92-7	0.8021	C₅H₆				−97.2	66.10	1.440	i/v/v	0.80			2.70
5-Methylcyclohexadiene-1,3	101.5	1489-57-2	0.8354						94.16	1.4763	i/v/s				3.85
1,3,5-Cycloheptatriene	117.0	544-25-2	0.8875	C₇H₈				−79.49	92.154	1.5343	i/s/s				3.77
Cycloocta-1,5-diene	150.8	111-78-4	0.8818	C₈H₁₂				−69– −70	108.18	1.4905	i/—/—				4.43
Cyclooctatetraene	140.56		0.9206	C₈H₈			−4.68	−7	104.54	1.5381	—/—/s			7.9 (25)	4.26
cis-Decalin	195.65	91-17-8	0.8965	C₁₀H₁₈	0.7–4.9[c]	57 (136)	−43.01	−43.01	138.25	1.4810	i/v/v	0.874	4.76	1 (22.5)	5.65
trans-Decalin	186.7		0.8699			57 (136)	−30.40	−30.7	138.25	1.4695	i/v/v		4.76	10 (47.2)	5.65
Tetralin	207.57	119-64-2	0.9702	C₁₀H₁₂	0.8–5[c]	71 (160)	−35.79	−35.79	132.20	1.5414	i/v/v	0.975	4.55	10 (47.6)	5.41
Dicyclopentadiene	170.0[d]	77-73-6	0.9302[b]	C₁₀H₁₂			35 (±5)	32	132.20	1.5050	i/s/v	0.93	4.55	10 (47.6)	5.41
α-Pinene	156.2	80-56-8	0.8582	C₁₀H₁₆		33 (91)		−55.0	136.24	1.4658	i/v/v	0.86	4.7	10 (37.3)	5.57
Turpentine	153– −175	8006-64-2	0.854–0.868		0.8 lel	35–39 (95–102)	−50 to −60 (−45.6 to −51.1)		134.5		i/s/s	0.857	4.6– 4.84		5.50
Camphene	160.2	79-92-5	0.8450	C₁₀H₁₆				52.0	134.24	1.4750	i/d/s				5.49
α-Caryophyllene	123 (10)	6753-98-6	0.8905	C₁₅H₂₄					204.36	1.5038	i/—/—				8.36

[a] Solubility in water/alcohol/ether: v = very soluble; s = soluble; d = slightly soluble. i = insoluble.
[b] At 35°C.
[c] AT 212°F.
[d] Slight decomposition.

Table 46.5. Physiologic Response to Cycloolefins and Dicyclenes

Compound	Route	Species	Parameter	Result Time or Dose	Findings	Ref.
Cyclopentene	Oral	Rat	LD_{50}	2.14 ml/kg		62
	Dermal	Rabbit	LD_{50}	1.59 ml/kg	Skin irritation moderate to severe	62
	Eye	Rabbit	Draize eye test	0.1 ml	Corneal injury moderate to severe	62
Vinylcyclohexene-4	Oral	Rat	LD_{50}	3.08 g/kg		62
	Dermal	Rabbit	LD_{50}	20 ml/kg		62
		Mouse	Dermal	145 g/kg × 54wk	Light neoplastic signs	63
Limonene	Oral	Rat	LD_{50}	5.0 g/kg		64
Acute						
		Mouse	LD_{50}	5.6–6.6 g/kg		65
	IP	Mouse	LD_{50}	1.3		65
	SC	Mouse	LD_{50}	25.6 ml/kg		66
		Aquatic	TLm96	1000 ppm		22
Subchronic	Oral	Mouse		277–2770 mg/kg/ day, 1/day × 1 mo	Slight decrease in body weight and food consumption	65
		Dog		1.2–3.6 ml/kg/day, 1/day × 6 mo	Frequent vomiting, decrease in body weight, blood sugar, and cholesterol, some kidney effects	67
Teratogenic	Oral	Rat		2.87 g/kg/day, days 9–15 of gestation	Weak teratogen	68
		Mouse		2.36 g/kg/day, days 7–12 of gestation	Weak teratogen	69
				0.250 g/kg	Inhibits tumorigenesis	70
Cyclopentadiene						
Dimer	Oral	Rat	LD_{50}	0.82 g/kg		71
	Dermal	Rabbit	LD_{50}	6.72 ml/kg		71
Monomer	SC	Rabbit	LD_{00}	0.5–1.0 cm³	No effects	14
			LD_{100}	3.0 cm³	Narcosis with fatal convulsions	14
Cycloheptatriene	Oral	Rat	LD_{50}	57 mg/kg		72
		Mouse	LD_{50}	171 mg/kg		72
	SC	Rat	LD_{50}	442–884 mg/kg	Severe dermal irritant, not a sensitizer	72
Cyclooctadiene	SC	Rat			Skin sensitizer	73
Cyclododecatriene	SC	Rat			Skin sensitizer	73
Decalin	Oral	Rat	LD_{50}	4.17 g/kg		74
	Dermal	Rat	LD_{50}	5.90 ml/kg		74
Tetralin	Oral	Human	LT_{Lo}	1–1.5 ml/kg	Transient effects	75
		Rat	LD_{50}	2.86 g/kg		74
	Dermal	Rat	LD_{50}	17.3 g/kg		74
Dicyclopentadiene	Oral	Rat (M)	LD_{50}	520 mg/kg		76
		Rat (F)	LD_{50}	378 mg/kg		76
				0.353 ml/kg		77
		Mouse (M)	LD_{50}	190 mg/kg		76
		Mouse (F)		250 mg/kg		76
	IP	Rat	LD_{50}	0.31 ml/kg		77
	Dermal	Rabbit	LT_{Lo}	2.0 g/kg	Irritant, not a sensitizer	76
	Dermal	Rabbit	LD_{50}	5.08 ml/kg		77
	Eye	Rabbit		0.1 ml	Temporary irritation	76
	SC	Rat		1.0 ml/kg × 14 exp.	96 hr post-exposure leukocytosis	15
Turpentine	Oral	Man	LD_{Lo}	60–100 g	Fatal, but also recovery from 120 g	78
		Rat		1.8 mg/kg/day, × 3 days	Stimulation of liver microsomes	79, 80

Table 46.6. Inhalation of Cyclenes

Compound	Species	Concentration ppm	Concentration (mg/liter)	Time	Findings	Ref.
Cyclopentene	Human	10–15	28–42 mg/m^3		Tolerable	
	Rat	Conc. vapor		5 min	Lethal	62
	Rat	16,000	44.6	4 hr	Lethal to 4 of 6	62
	Rat	112–1139		6 hr/day × 12 wk	No effects	13
	Rat	8110		6 hr/day × 5 days/wk × 3 wk	Decreased body weight gain of females	13
Cyclohexene	Mouse	8930	30		Loss of righting reflexes	18
		13,400–14,900	45–50		Fatal	18
	Rat, guinea pig, rabbit	75	0.25	6 hr/day × 5 days/wk	Increased alkaline phosphatase	81
		150	0.540		Increased alkaline phosphatase	81
		300	1.008		Increased alkaline phosphatase	81
		600	2.016		Lower weight gain, increased alkaline phosphatase	81
Vinylcyclohexene-1	Mouse		7.5	LC$_{50}$	Narcosis	92
			13.7			92
Vinylcyclohexene-4	Mouse	6095	27.0	LC$_{50}$		82
		Satd. vapor		LC$_{100}$	Time to death, 15 min	62

2.2 Cyclohexene

Cyclohexene, 1,2,3-tetrahydrobenzene, $\overline{\text{CH:CH(CH}_2)_3\text{CH}_2}$, is a flammable (DOT) (24) liquid which occurs in coal tar (4). For physical data, see Table 46.4. Cyclohexene is prepared by dehydration of cyclohexanol (4), but primarily by thermal reaction of a mixture of ethylene–propylene–butadiene (1). It is used as an alkylation component and in the chemical synthesis of adipic, maleic, and hexahydrobenzoic acids and aldehydes, and as a stabilizer for high octane gasoline (4).

The general toxicity of cyclohexene, as for the comparable cycloparaffins, is low; see Table 46.5. The liquid is irritant, and defats the skin on direct contact. It is a general anesthetic and CNS depressant. When ingested, it represents a low to moderate pulmonary aspiration hazard (37). Lazarew (18) determined that 30 mg/liter produced loss of righting reflex in the white mouse, while 45 to 50 mg/liter proved fatal. Dogs showed tremors and staggering gait from inhalation of cyclohexene (85). A 6-month inhalation study was carried out with rats, guinea pigs, and rabbits, all exposed to 75, 150, 300, and 600 ppm cyclohexene for 6 hr/day, 5 days/week. The results indicated lower weight gain of the rats at 600 ppm and a significant increased alkaline phosphatase for all rats, but no other significant changes (81). Cyclohexene is believed to be of low toxicity to the mammal. This may be due to its hydrophobic characteristics,

Table 46.7. Physiologic Response to Inhalation of Dicyclenes

Compound	Species	No.	Concentration ppm	Concentration mg/liter	Time	Parameter	Result	Ref.
Decalin	Rat	4	500	2.83		LC_{65}	Lethal to 4 of 6 animals	74
		2	Satd. vapor			LC_{100}	Lethal	74
	Guinea pig	8	319	1.8	8 hr/day × 23 days		1 of 3 lethal on day 1, 1 on day 21, and 34 on day 23	15
Dicyclopentadiene	Rat	6M	359.4	1.90	4 hr	LC_{50}		77
		F	385.2	2.08		LC_{50}		77
	Rat	(2M, 2F)	1000	5.40	4 hr	LC_{100}	Eye, nose irritation	83
						LC_{100}	Dyspnea, muscular incoordination, tremors; autopsy: lung, liver congestion	
		(2M, 2F)	2500	13.53	1 hr	LC_{100}	Eye, nose irritation, dyspnea narcosis; autopsy: congestion of lung, liver, kidney	83
	Mouse	6	145.5	0.70		LC_{50}		77
	Guinea pig	6	770.5	3.721		LC_{50}		77
	Rabbit		771.0	3.723				77

Rat		19.7	0.093	7 hr × 89		No effects	77
		35.2	0.170	7hr × 89		Kidney lesions in male rats	77
		73.8	0.356	7 hr × 89			77
	4M	100	0.483	6 hr × 15	LC_{00}	No toxic signs; organs normal	77
	2M	250	1.207	6 hr × 10	LC_{25}	Weight loss, nose irritation, dyspnea tremors, hypersensitivity; hematology: organs normal	83
Turpentine	Human	720–1100	3.96–6.05		LT_{LO}	Complaint of eye irritation, headache, dizziness, nausea, chest pain, and visual disturbances	15
	Cat	746–782	4.1–4.3	3.5–4 hr	LT_{LO}	Lethargy, incoordination, nausea	84
		1090	6.0	6 hr		Collapse, later recovery	84
		1455	8.0	0.5–1 hr		Incoordination, tonic convulsions	84
		2909–4364	16–24	0.3–1.5 hr	LD_{80}	Lethal to 80% of test animals	84
	Dog	54.5–112	0.3–1	1 hr/day × 8 days	LD_{00}	No effects	84
		818	4.5	3.5–4.5 hr	LT_{LO}	Nausea, incoordination, weakness light paralysis	84

rapid hydroxylation, conjugation, and ease of elimination (86). Microsomal oxidases have been detected in the rabbit liver, which readily hydroxylate cyclohexene and a number of related aromatic compounds to the corresponding dihydroxy derivatives (87).

Experimental work with the rabbit has demonstrated that cyclohexene also forms sulfur-containing metabolites (88). Incubation of 1-methyl-1-cyclohexene with *Aspergillus niger* has resulted in the formation of stereospecific (+)-1-methyl-1-cyclohexene-6-ol (89).

For cyclohexene a TLV of 300 ppm or 1015 mg/m^3 for a time-weighted average has been recommended by OSHA (57) and ACGIH (58). NIOSH has published a sampling procedure, S82.01 (53), using charcoal as a collection medium. Numerous analytic methods are available, which include infrared and ultraviolet spectrography, coulometric titration, nuclear magnetic resonance techniques, and gas chromatography (59).

2.3 Cycloheptene

Cycloheptene, also named suberene or suberylene, $\overline{CH:CH(CH_2)_4}CH_2$, is a flammable liquid (DOT) (24). Little toxicology has been published, except it is known that cycloheptene forms sulfur-containing metabolites, such as N-acetyl-S-cycloheptyl-L-cysteine and N-acetyl-S-hydroxycycloheptyl-L-cysteine, as shown in the rabbit (88).

Analytically, cycloheptene can be determined by gas chromatography (59).

2.4 Cyclooctene

Cyclooctene, $\overline{CH:CH(CH_2)_5}CH_2$, is a flammable liquid (DOT) (24), utilized as raw material in the chemical industry. As shown with the rat and the rabbit, cyclooctene is metabolized to 2- and 3-hydroxycyclooctylmercapturic acid (90).

3 ALKENYL CYCLOOLEFINS

3.1 Vinylcyclohexenes

Vinylcyclohexene-1, cyclohexenylethylene, 1-ethenylcyclohexene, tetrahydrostyrene, $CH_2:CH\overline{C:CH(CH_2)_3}CH_2$, a flammable liquid (DOT) (24), is an important chemical intermediate. Vinylcyclohexene is a common component of tobacco smoke. It is theorized to be formed by dimerization from butadiene (91). For physicochemical and physiological data, see Tables 46.4 and 46.6.

1-Vinylcyclohexene, when administered to mice in an inhalation study, exerted a narcotic effect at a concentration of 7.5 mg/liter. The LC$_{50}$ was 13.7 mg/liter (92).

3.1.2   4-Vinylcyclohexene-1

4-Vinylcyclohexene-1, 4-ethenylcyclohexene, $CH_2:CH\overline{CH(CH_2)_2}CH:CHCH_2$, is a colorless liquid. Additional chemical and physical properties are listed in Table 46.4 and physiological data in Tables 46.5 and 46.6. This substance is probably an irritant and narcotic in high concentrations. It is possibly a CNS depressant. It has a low degree of toxicity via ingestion and skin penetration.

An oral LD_{50} of 4-vinylcyclohexene-1 in the rat was found to be 3080 mg/kg, the dermal LD_{50} in the rabbit 20 ml/kg (62). Time to death for rats was 15 min in the saturated vapor (62) and a concentration of 8000 ppm for 4 hr proved lethal to four of six rabbits (62). An inhalation LC_{50} for 4-vinylcyclohexene-1 for the mouse has been recorded as 27,000 mg/m^3 (82). Chronic application of 145 g/kg of 4-vinylcyclohexene-1 to the mouse skin for 54 weeks showed light neoplastic signs (63). Methylated cyclohexene is used in perfume formulations, indicating that the toxicity of the alkyl derivatives is rather minimal (93).

No threshold limit value has been established. However, it has been suggested that the vapor concentration of this substance be kept below 100 ppm in any workroom environment.

3.2   Terpenes

Further side chain substitution of cyclohexene, with a methyl, isopropenyl, ethenyl, or other groups, forms a class of chemicals named terpenes. One example is limonene, a C_{10} cyclic olefin. Chemically, it is classified as a monoterpene, or diisoprene, and occurs commonly in nature.

3.2.1   Limonene

Limonene, carvene, dipentene, *p*-mentha-1,8-diene, methyl-4-isopropenyl cyclohexene-1, $\overline{CH_2C(CH_3)}:CHCH_2CH[C(CH_3):CH_2]\overline{CH_2}$ or

is, with the possible exception of α-pinene, the most frequently occurring monoterpene (94). For physicochemical data, see Table 46.4. It represents the highly fragrant main constituent or up to 86 percent of the terpenoid fraction of fruit, flowers, and leaves, bark, or pulp, from shrubs or trees of many species, such as anise, mint, caraway, polystachia, and pine, but especially the lime and orange oil (64, 95). Limonene also occurs in the gas phase of tobacco smoke

(96) and has been detected in the atmosphere (97). Some of its functions are lending fragrance and taste to fruit and essence to flowers and leaves. Limonene has antimicrobial, antiviral (98), antifungal (99), antilarval (100), and insect attractant (101) or repellent (102) properties. In Japan it has been used as a gallstone dissolving agent (103) and in wound healing (104). It also has been used widely as an odorant (105), to a lesser extent as a solvent (106), as an aerosol stabilizer (107) and as a wetting and dispersing agent (4). Limonene is of low acute toxicity, orally and dermally; see Table 46.5. Chronically administered to dogs, some effects were noted at a concentration of 1.2 to 3.6 ml/kg per day for 6 months (67). Absorption from bath additives through the intact skin is possible, since limonene is absorbed at 100 times the rate of water and 10,000 times that of the electrolytes, the sodium and chloride ions (108). However, limonene may be a minor skin sensitizer (4). When orally administered to rats during days 9 to 15 of gestation, 2869 mg/kg decreased the body weight gain of the mother and prolonged ossification of the fetal metacarpal bone and proximal phalanx, with slightly decreased fetal spleen and ovarian weights (68). Mice administered 2363 mg/kg limonene orally for 6 days from days 7 to 12 of gestation produced decreased weight gain and increased incidence of abnormal bone formation in the fetus (69). Rabbits showed decreased body weight gain and six deaths of 21 animals when administered 1000 mg/kg during gestation; 250 mg/kg resulted in no teratogenic effects (109). Limonene and its hydroperoxide injected subcutaneously into C57BL/6 mice decreased the incidence of dibenzopyrene-induced tumors appreciably (70). Metabolic studies with radioactive d-limonene in man and animals demonstrated that 75 to 95 percent of the material was excreted in the urine and up to 10 percent in the feces (110). Limonene enhanced hepatic functions in rat liver (111); however, label distribution was highest in the blood serum. Limonene was excreted within 48 hr, 25 percent of it in the bile (112). A metabolite isolated from the rabbit urine was p-mentha-1,8-dien-10-ol (113). Limonene was readily degraded in soil (114, 115).

The odor of limonene was detectable in water at 10 ppb (116) and analytically determinable by gas chromatography. With proper handling precautions, this material presents no health hazard.

4 CYCLOOLEFINS AND POLYENES

4.1 Cyclopentadiene

Cyclopentadiene, p-pentine, pentole, pyropentylene, $\overline{CH:CHCH_2CH:CH}$, occurs in the C_6 to C_8 petroleum distillation fraction (1). It is also obtained from coke-oven light oil fractions (4). Some physicochemical and physiological data are listed in Tables 46.4 and 46.5. Cyclopentadiene polymerizes easily in the presence of peroxides and trichloroacetic acid (4). It is used as an intermediate in chemical syntheses, especially for Diels–Alder reactions (4).

Cyclopentadiene vapors produce narcosis in the frog in 10 min, but recovery is complete in 70 min (14). Signs and symptoms during narcosis include primary motor unrest and decreased, intermittent respiration rate prior to death. For the dimer, a range-finding LD_{50} value was determined as 0.82 g/kg in the rat and a dermal LD_{50} of 6.72 ml/kg for the rabbit (71). Cyclopentadiene injected subcutaneously into the rabbit at 0.5 to 1.0 cm^3 caused no effects, but 3.0 cm^3 caused narcosis with fatal convulsions (14).

OSHA (57) and ACGIH (58) suggest an 8-hr TLV of 75 ppm or 200 mg/m^3.

4.1.1 Methylcyclopentadiene

Methylcyclopentadiene, $CH_3\overline{CH:CHCH_2CH:}CH$, a flammable liquid, occurs in the C_6 to C_8 petroleum distillation fraction (1).

4.1.2 Methylcyclohexadiene

Methylcyclohexa-1,3-diene, $\overline{CH_2C:(CH_3)CHCH_2CH:}CH$, is steam distilled from natural lime oil and terpenes and used as a fragrance. A U.S. patent is available. Some physicochemical data are presented in Table 46.4.

4.1.3 Cyclooctadiene

Cycloocta-1,5-diene, $\overline{CH_2CH:CH(CH_2)CH:CH}CH_2$, is a flammable, highly reactive liquid; see Table 46.4. It is produced from petroleum distillation fractions and is used as an intermediate in the plastics industry as a synthetic lubricant and in numerous other applications. Toxicity studies in rabbits have shown severe skin effects, with necrosis of the epidermis and ulceration and marked inflammation of the dermis (73). In the uncovered application test, cyclooctadiene produced an immediate erythematous reaction in the rabbit, guinea pig, and hairless mouse, and was also a skin sensitizer (73).

Cycloocta-1,3-diene, $\overline{CH:CH(CH_2)_3CH:CH}CH_2$, exhibited the same properties as the 1,5 isomer, but was more highly reactive and slightly reduced the total rat liver glutathiones (90).

Both the 1,3 and 1,5 isomers were metabolized, in the rat and the rabbit, to dihydroxycyclooctylmercapturic acids, and, to a lower extent, to sulfate and glucuronide conjugates (90). Cyclooctadiene applied to the guinea pig skin on three alternate days resulted in skin irritancy with erythema, dry appearance, slight dermal weight increase, and increased arginase activity (38).

4.2 Cycloheptatriene and Cyclododecatriene

Cyclohepta-1,3,5-triene, tropilidene, $\overline{CH:CHCH:CHCH:CH}CH_2$, and cyclododecatriene, cyclododeca-1,5,9-triene, $\overline{CH_2[CH:CH(CH_2)_2]_2CH:CH}CH_2$, are flammable liquids; see Table 46.4. The acute oral LD_{50} of cyclohepta-1,3,5-

triene in the rat was determined as 57 mg/kg, and in the mouse as 171 mg/kg (72). The acute percutaneous LD_{50} for the rat lies between 442 and 884 mg/kg (72). The compound was found to be a severe dermal irritant but not a sensitizer. Both compounds, when applied to the guinea pig skin on three alternate days, caused erythema, thickening, and increased weight of the epidermal layer and increased dermal arginase activity. All dermal effects were less prominent than for the cyclooctadienes (38), although cyclododecatriene was a more potent skin sensitizer (73). Both the C_8 and C_{12} compounds were immediately irritant to the rabbit eye, producing mild conjunctivitis which healed within 48 hr. The eyelids became swollen and exuded a discharge. Blephanitis healed somewhat faster for the C_8, but was still apparent for the C_{12} homologue 1 week after the application.

Analytical procedures are available for the determination of a number of the cycloolefins. The higher cycladienes warrant caution when handling. Protective garments should be worn to prevent contact with the skin and eyes.

5   DICYCLIC PARAFFINS

5.1   Decalin

Of the dicyclic paraffins, decalin, bicyclo[4.4.0]decane, decahydronaphthalene, naphthalane, naphthane

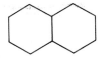

is the most important member industrially. It is a flammable material. For other physicochemical data, see Table 46.4. Decalin occurs naturally in crude oil. Commercially, it is produced by the catalytic hydrogenation of naphthalene. It is widely used as a solvent for naphthalene, fats, oils, resins, and waxes, as an alternate for turpentine in lacquers, shoe polish, and floor waxes, as a component in motor fuels and lubricants, and as a patent fuel for stoves (4). In research decalin has been utilized as a pharmacological, noncarcinogenic vehicle in long-term skin painting (117, 118) and somatic mutation evaluation studies (119).

Decalin is irritating to the eyes, skin, and mucous membranes; see Tables 46.5 and 46.7. Dermatitis, but without serious systemic poisoning, has been reported in painters working with decalin (15). By inhalation, the lowest dose to indicate an effect to man was 100 ppm (10).

An oral range-finding LD_{50} has been determined as 4.17 g/kg for the rat and a dermal LD_{50} of 5.90 ml/kg for the rabbit (74). Inhalation of the saturated vapor was lethal to rats in 2 hr (74). A 4-hr exposure to 500 ppm was lethal to four out of six rats (74). Of three guinea pigs exposed to a vapor concentration of 319 ppm (1.8 mg/liter) for 8 hr/day, one died on day 1, the second on day 21, and the third on day 23 (15). Gross and microscopic examination revealed lung congestion, kidney, and liver injury. Decalin applied to 6 cm² of guinea pig skin on two successive days resulted in deaths 10 days post-exposure. The systemic tissue injury was the same as from its inhalation (15).

An aquatic TLm96 was 1000 to 100 ppm (22). *cis-* and *trans-*Decalin gave rise to racemic decanols when metabolized in the rabbit (120), and were generally excreted in the urine, conjugated with glucuronic acid. Guinea pigs dosed orally, but not by inhalation or percutaneous administration, with decalin have shown a brownish green urine, an occurrence also reported in workers exposed to a mixture of decalin and tetralin (15). Conversely, marine organisms utilized decalin, but to a lesser extent than long chain alkanes and aromatics, as their carbon source (121). Fredericks (122) also observed that *Pseudomonas fluorescens* and *Corynbacterium* degraded decalin more slowly than alkanes and alkenes. Decalin may be collected on charcoal and determined by gas chromatography or mass spectroscopy.

6   DICYCLIC OLEFINS

6.1   Tetralin

Tetralin, 1,2,3,4-tetrahydronaphthalin, tetranap

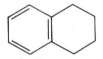

is a flammable liquid. Further physicochemical properties are listed in Table 46.4 and physiological data in Table 46.5. The odor of tetralin resembles that of benzene and menthol (4). Prolonged standing may produce peroxides, which cause explosions during the distillation of tetralin (4). It is prepared by catalytic hydrogenation of purified naphthalene and it is formed during acidic, catalytic hydrocracking of phenanthrene (1). Under further pyrolysis at 700°C, tetralin, along with, for example, butylbenzene, yields tars that contain appreciable quantities of 3,4-benzopyrene (123). Tetralin is used widely as a solvent for fats and oils and as an alternate for turpentine in polishes and paints (75).

Tetralin is irritant to the skin and eyes and mucous membranes, and may be narcotic in high concentrations. A human case involving the ingestion of ½ to ¾ of a pint of tetralin, that is, about 65 to 90 ml or 1 to 1.5 ml/kg, has been recorded (75). Effects consisted of nausea, vomiting, intragastric discomfort, transient liver and kidney damage, green-gray urine, and some clinical and enzymatic changes. All signs, symptoms, and effects returned to normal within 2 weeks (75). Acute range-finding studies revealed an oral LD_{50} in the rat of 2.68 g/kg and a dermal LD_{50} in the rabbit of 17.3 g/kg (74); see Table 46.5. Saturated vapor for 8 hr was not lethal to rats (74). In the guinea pig an inhalation LC_{LO} value of 275 ppm for an 8-hr exposure with 17 exposures has been reported (124). The compound is moderately irritant to the skin but mildly irritant to the eye (74). However, some animal experiments have shown the appearance of cataracts. For fish, an aquatic TLm96 of 100 to 10 ppm has been reported (22).

Tetralin has been found cytotoxic to ascites tumor cells when incubated in vitro for about 5 hr (125).

Radioactive tetralin, when administered to the rabbit, was excreted in the urine as glucuronide at 87 to 99 percent and in the feces at 0.6 to 1.8 percent of the original dose (126). Lin and Chen (127) observed that tetralin hydroperoxide, produced in rat liver, may be a metabolic intermediate. Conversely, tetralin was absorbed by mussel but not metabolized; indeed, up to 80 percent was released in unchanged form when the organism was transferred to fresh seawater (128). In plant pathology, tetralin has been successful in total destruction of crown gall and olive knot neoplasms (56).

Occupational exposure may occur in the paint, solvent, and varnish industries. No official threshold limits have been established. An odor threshold in water has been determined as 18.0 ppm (116). Analytically, tetralin can be determined by gas chromatography.

6.2 Dicyclopentadiene

Dicyclopentadiene, 3a,4,7,7a-tetrahydro-4,7-methanoindene, DCPD

is a colorless, crystalline combustible solid (7, 83). Further physicochemical properties are listed in Table 46.4. It is produced commercially by recovery from hydrocarbon streams from higher temperature cracked petroleum fractions, and is a by-product of the coke-oven industry. It is formed by spontaneous dimerization of cyclopentadiene. Both cyclopentadienes are used as bases for the synthesis of chlorinated hydrocarbon pesticides (76), to stabilize organophosphorus insecticides (129), and for related chemicals, such as water pond liners (130). Experimental work has demonstrated that DCPD is moderately to highly toxic; see Tables 46.5 and 46.7. Oral and intraperitoneal LD_{50} values in rodents are below 1 g/kg; the dermal are somewhat higher (77, 83). To the eyes, it represents a temporary hazard, since immediate washing with water does not shorten injury time (76). The principal pathological findings in orally dosed rats, typical for large doses of irritant hydrocarbons, were general congestion, hyperemia, and focal hemorrhage in many organ tissues. Affected were the kidney, intestine, stomach, bladder, and particularly the lungs. Gerarde (15) observed leukocytosis in the rat 96 hr post-subcutaneous administration of DCPD, in contrast to the leukopenias found in blood-forming tissues from benzene. Some of the inhaled DCPD is exhaled in unchanged form; the balance is absorbed, similar to other routes, translocated to the vascular system, and hydroxylated in the liver; the main portion is conjugated and excreted in the urine as glucuronide.

An aquatic TLM96 determination revealed a range of 10 to 1 ppm (22).

Temporary 1976 guidelines for purity of food and water for drinking, irrigation, recreation, and aquatic life recommend a 1.3 ppm limit for DCPD contamination (131). USSR guidelines of 1972 (132) suggest 0.001 mg/liter. No official TLV has been established in the United States. However, from their studies Kinkead et al. conclude that a TLV of 5 ppm may be reasonable (77). A variety of analytical determination methods using color detection indicators, thin layer, gas chromatographic, or ultraviolet spectral procedures are available.

The human sensory response detects DCPD at a concentration of 0.003 ppm (77). This may present some warning property for handling the material.

6.3 Pinenes

α-Pinene, 2-pinene, pinene, 2,6,6-trimethylbicyclo[3.1.1]hept-2-ene

is a fragrant, flammable liquid. Further physicochemical data are listed in Table 46.4. β- and γ-Pinene are position isomers of α-pinene and bear similar characteristics. They all occur naturally in a variety of trees and shrubs, and air concentrations near pine forests in the summer may reach 500 to 1200 μg/m³ (133). Pinene is obtained from turpentine or pine oil by distillation and is widely used in the manufacture of camphor as an insect attractant and repressant, insecticides, perfume bases, plasticizers, solvents, and synthetic pine oil (4). The pinenes' antibacterial action has been utilized medicinally.

Pinene is easily absorbed through the pulmonary system, the skin, and the intestine. It is irritant to the skin and mucous membranes and may cause dermal eruption and occasional benign tumors (4). Absorption of large doses may result in delirium, ataxia, and kidney damage. Inhalation may cause palpitation, dizziness, nervous disturbances, chest pain, bronchitis, and nephritis. About

150 ml may constitute a human oral fatal dose (29). Humans with cardiac diseases may experience increased olfactory sensitivity toward pinene (133). Pinene has been found lethal toward conifer needle-chewing insects (134), causing leukemic changes in fowl plasma (135) and deviations in avian plasma proteins accompanying erythroblastosis (136). Similar to other terpenes, such as limonene, pinene possesses choleretic action, as shown in male rats (137). It increased the microsomal protein content (138), cytochrome P-450, and the activities of some hydroxylases (139). Pinene is degraded by microbiological organisms in soil (140).

The odor of 2-pinene can be detected in water at 60 ppb (116). Occupationally, pinene vapors should not be inhaled or the liquid absorbed by the skin in excess. Analytically, it can be determined by gas chromatographic procedures.

6.4 Turpentines

Gum turpentine is the steam-volatile fraction of pine tree pitch. Chemically, it comprises 58 to 65 percent γ-pinene and the balance consists of β-pinene and other isomeric terpenes. Wood turpentine, obtained from waste wood chips or sawdust, contains 80 percent γ-pinene, 15 percent monocyclic terpenes, 1.5 percent terpene alcohols, and other terpenes. Sulfate turpentine is a by-product in paper manufacture. The turpentines are extensively used in the compounding of surface coatings, as solvents for oils, fats, waxes and resins, lacquers, and polishes. Turpentine is also used therapeutically as a human ointment and counterirritant (4) and in veterinary practice as an expectorant, rubifacient, and antiseptic (4), owing to its antimicrobial properties. Some physicochemical data are presented in Table 46.4.

Turpentine presents a moderate health hazard, since it is readily absorbed through inhalation and skin contact; see Tables 46.5 and 46.7. It is irritant to the skin, eyes, nose, and mucous membranes, and major systemic effects include kidney and bladder injury (31). The fatal oral dose lies between 60 and 100 g, but recoveries from 120 g have been recorded (78). Symptoms of intoxication consist of gastroenteric pain, with nausea and vomiting, toxic nephritis with hematuria, albuminuria, and possible oliguria. At very high doses, coma may be followed by death due to respiratory arrest. At lower doses, pronounced anemia occurs occasionally, probably due to bone marrow damage. Similar to other etheral oils, turpentine, or its volatile components, when eliminated through the respiratory system, can be recognized by the very specific odor of violets (78). Skin contact may cause eczema, possibly due to the formation of oxidation products (31). Occupational contact dermatoses are not only common among workers in the chemical, rubber, and welding industries (141–143) but also in the home, among housewives (144). Human cases have been recorded where inhalation of turpentine has caused extensive subacute and chronic glomerulonephritis (145). In the eye, it may cause corneal burns (31). Introduction into the skin produces chronic inflammation, with an effect on the collagen of the dermis (146). Fisher (147) has classified turpentine as a plant sensitizer. Many cases of dermatitis have been recorded due to turpentine, and approximately 14.8 percent of 1000 cases with eczema (148). Hypersensitivity to wood turpentine, more so than gum turpentine, may be due to impurities such as formic acid, formaldehyde, and phenols (149). Even dermal contact may cause allergic erythema with accompanying headaches, coughing, and sleeplessness (150). A case has been recorded where repeated injections with turpentine, along with insulin, caused necrotizing fasciitis (151). Inhalation experiments with rats showed LC_{50} values at concentrations of 12 to 20 mg/liter with 1 to 6 hr exposures, and with mice, 29 mg/liter with a 2-hr exposure (152). Signs and symptoms included a primary effect on the central nervous system, an increased respiration rate, and a decreased tidal volume, but no pulmonary lesions. Brain and spleen showed the highest tissue distribution (152). In workers in industry, acute concentrations of 0.010 to 5.2 mg/liter caused signs of irritation, and 5 to 10 mg/liter in 1 to 2 hr was lethal to several test animals (145). In an inhalation experiment, cats exposed to 4.1 to 4.3 mg/liter of turpentine for 3.5 to 4 hr produced lethargy, incoordination, and nausea; 6.0 mg/liter in 3 hr caused collapse, but recovery in 20 min; 8 mg/liter in 1.0 to 1.5 hr showed incoordination; and 16 to 24 mg/liter in 40 min to 1.5 hr was lethal to four of five cats (84). In dogs, 0.3 to 1 mg/liter daily for 8 days yielded no effects; 4.5 mg/liter in 3.5 to 4.5 hr caused nausea; and 6.0 mg/liter in 3 hr caused nausea, incoordination, weakness, and light paralysis, with fairly rapid recovery (84). There was no injury to the guinea pig after prolonged exposure to 715 ppm (153). Intradermal injection of turpentine in peanut oil 1:5 to rabbits produced erythematosis and granulocytes in connective tissue, which was still acutely inflamed 48 hr later. After 9 days it showed new vessel proliferation, round cell infiltration, and evidence of connective tissue remodeling (154). In the rat, intradermal abscesses are easily introduced (155), but have no immunologic effects in the mouse (156). In avian species, erythroblastosis has been observed (136).

The principal portal entry of turpentine in industrial use is inhalation. Part of this is eliminated unchanged in the expired air, and another portion is absorbed, metabolized, and excreted in the urine, conjugated with glucuronic acid (15).

Metabolic studies showed that certain protein synthesis inhibitors were capable of suppressing turpentine-induced inflammation (157). Turpentine at 1.8 mg/ kg per day orally for 3 days in the rat was observed to stimulate microsomal enzymes and reduce the toxicity toward parathion (79, 80). In the guinea pig, it protected the hypersensitivity to 6-mercaptopurine (158). When applied to the skin surface, turpene supported the growth of tumors in the rabbit, but not in the mouse (159).

A TLV of 100 ppm or 560 mg/m³ is recommended by OSHA (57) and

ACGIH (58). The sensory response to turpentine is satisfactory at 100 ppm, but causes throat irritation at 125 ppm and eye and nose irritation at 175 ppm (160). A procedure, S88, for air monitoring has been published by NIOSH, for charcoal absorption and analysis by gas chromatography (53). Procedures are available for the chemical removal of turpentine from air and water (161). Occupationally, turpentine constitutes no hazard, except on direct skin contact or when vapors are inhaled. Protection is advised, including the use of skin barriers and ointments, if feasible (162, 163).

7 OTHER CYCLIC OLEFINS

7.1 Ethylidenenorbornene

Ethylidenenorbornene, 5-ethylidenebicyclo[2.2.1]hept-2-ene, a compound closely related to pinene, has exhibited an oral LD_{50} of 3.2 g/kg in the rat (164). The LC_{50} by inhalation for a 4-hr exposure varied from 732 ppm for the female mouse to 3100 ppm for the male rabbit. Extrapolating from 89-day, 7-hr exposures of dogs, Kinkead suggests a 5 ppm threshold limit for an 8-hr workday, 5-day week (165).

7.2 Camphene

Camphene, 3,3-dimethyl-2-methylene norcamphone

is another related compound, also a terpene or diisoprene. It occurs naturally and is prepared from α-pinene and can be isolated as cubic crystals (4). It can form flammable vapors (7). It is used in tablet form for mothproofing (166), and in the cosmetic, perfume, and food flavoring industries (167). An acute oral LD_{50} value in the rat has been reported as >5 g/kg, and a dermal LD_{50} as >2.5 g/kg in the rabbit (167). In rabbits fed 1 g of cholesterol daily for 8 weeks, camphene potentiated the effects of cholesterol (168).

7.3 Caryophyllene

Caryophyllene, a dicyclic sesquiterpenoid, is a fragrant liquid and occurs naturally in many plants (4, 94). Chemically, it occurs in the α, β, and isoforms, and naturally as the iso and α forms (4). It is widely used in the perfume industry. In the boll weevil it suppresses intestinal bacteria (169). Analytically, it is determined by gas chromatography.

REFERENCES

1. R. F. Gould, Ed., *Refining Petroleum for Chemicals,* American Chemical Society, Washington, D.C., 1970.

2. R. C. Weast, Ed., *CRC Handbook of Chemistry and Physics,* CRC Press, Cleveland, Ohio, 1977.

3. *Laboratory Waste Disposal Manual,* 2nd ed., revised, Manufacturing Chemists Association, Inc., Washington, D.C., 1974.

4. M. Windholy, Ed., S. Budavari, assoc. Ed., *The Merck Index,* 9th ed., Merck and Company, Inc., Rahway, N.J., 1976.

5. *Registry of Toxic Effects of Chemical Substances,* U.S. Dept. of Health, Education and Welfare, Cincinnati, Ohio, September, 1977.

6. V. B. Guthrie, Ed., *Petroleum Products Handbook,* McGraw-Hill, Inc., New York, 1960.

7. N. I. Sax, *Dangerous Properties of Industrial Materials,* 4th ed., Van Nostrand Reinhold, New York, 1975.

8. *Toxicology and Hazardous Industrial Chemicals Safety Manual for Handling and Disposal with Toxicity Data,* International Technical Information Institute, Japan, 1976.

9. *Hazardous Chemicals Data 1975,* National Fire Protection Association, NFPA, No. 49, Boston, 1975.

10. *Handbook of Organic Industrial Solvents,* 4th ed., American Mutual Insurance Alliance, Chicago, 1972.

11. F. R. Rossini, K. S. Pitzer, R. L. Arnett, R. M. Braun, and G. C. Pimentel, *Selected Values of Physical Thermodynamic Properties of Hydrocarbons and Related Compounds,* Carnegie Press, Pittsburgh Pa., 1953.

12. R. W. Stoughton and P. D. Lamson, *J. Pharmacol. Exp. Ther.,* **58,** 74 (1936).

13. G. Kimmerle and J. Thyssen, *Int. Arch. Arbeitsmed.,* **34**(3), 177 (1975).

14. W. F. Von Oettingen, *Toxicity and Potential Dangers of Aliphatic and Aromatic Hydrocarbons,* Publ. Health Bull. No. 255 (1940).

15. F. A. Patty, *Industrial Hygiene and Toxicology,* Vol. II, John Wiley, New York, 1963.

16. J. F. Treon, W. E. Crutchfield, and K. V. Kitzmiller, *J. Ind. Hyg. Toxicol.,* **25**(8), 323 (1943).

17. J. F. Treon, W. E. Crutchfield, and K. V. Kitzmiller, *J. Ind. Hyg. Toxicol.,* **25**(6), 199 (1943).

18. N. W. Lazarew, *Arch. Exp. Path. Pharmakol.,* **143,** 223 (1929).

19. E. T. Kimura, D. M. Ebert, and P. W. Dodge, *Toxicol. Appl. Pharmacol.,* **19,** 699 (1971).

20. W. B. Deichmann and T. J. LeBlanc, *J. Ind. Health Toxicol.,* **25,** 415 (1943).

21. Shell Chemical Company, unpublished report, 1961 (through Ref. 5).

22. R. W. Hann and P. A. Jensen, *Water Quality Characteristics of Hazardous Materials,* Vols. 1–4, Environmental Engineering Division, Civil Engineering Department, Texas A & M University, 1974.

23. D. Irie, T. Sasaki, and R. Ito, *Toho Igakkai Zasshi,* **20**(5–6), 772 (1973).

24. "Hazardous Materials Regulations and Miscellaneous Amendments," Title 49, Chapter 1. Materials Transportation, Bureau of Transportation, *Fed. Reg.,* 4(252), 57018 (1976).

25. *Encyclopaedia of Occupational Health and Safety,* Vols. I and II, McGraw-Hill, New York, 1972.

26. J. C. Arcos, M. F. Argus, and G. Wolf, *Chemical Induction of Cancer,* Vol. I, Academic Press, New York, 1958.

27. E. A. Wahrenbrock, E. I. Eger, R. B. Laravuso, and G. Maruschak, Anesthesiology, **40**(1), 19 (1974).

28. L. S. Goodman and A. Gilman, *The Pharmacological Basis of Therapeutics,* 4th ed., MacMillan, London, 1971.

29. W. B. Deichmann and H. W. Gerarde, *Toxicology of Drugs and Chemicals*, Academic Press, New York, 1969.

30. G. W. Seuffert and K. F. Urbach, *Anesth. Analg. Curr. Res.*, **46**(2), 267 (1967).

31. M. M. Key, A. F. Henschel, J. Butler, R. N. Ligo, and I. R. Tabershaw, *Occupational Diseases, A Guide to their Recognition*, rev. ed. U.S. Dept. HEW, Washington, D.C., June, 1977.

32. A. L. Cowles, H. H. Borgstadt, and A. J. Gillies, *Anesthesiology*, **36**(6), 588 (1972).

33. N. B. Anderson, *Anesthesia*, **29**(1), 113 (1968).

34. *Threshold Limit Values for Chemical Substances and Physical Agents in the Workroom Environment*, American Conference of Governmental Industrial Hygienists, Cincinnati, Ohio, 1959.

35. J. C. Krantz, Jr., C. J. Carr, and J. F. Vitcha, *J. Pharmacol. Exp. Ther.*, **94**, 315 (1948).

36. J. K. Sugden and B. K. Razdan, *Pharm. Acta Helv.*, **47**(5), 257 (1972).

37. H. W. Gerarde, *Arch. Environ. Health*, **6**, 329 (1963).

38. V. K. H. Brown and V. L. Box, *Br. J. Dermatol.*, **85**, 432 (1971).

39. J. P. Conkle, B. J. Camp, and B. E. Welch, *Arch. Environ. Health*, **30**(6), 290 (1975).

40. A. M. Aliev, G. Y. Alikishi-Zade, and V. A. Shlyapnikov, *Otkrytiya, Izobret, Prom. Tovarnye Znaki*, **51**(44), 59 (1974).

41. P. Foss, G. Takeguchi, H. Tai, and C. Sik, *Ann. N.Y. Acad. Sci.*, **180**, 126 (1971).

42. H. B. Elkins, *The Chemistry of Industrial Toxicology*, 2nd ed., John Wiley, New York, 1959.

43. C. R. Noller, *Chemistry of Organic Compounds*, 3rd ed., W. B. Saunders, Philadelphia, 1966.

44. Z. Ziegenmeyer, M. Reuter, and F. Meyer, *Arch. Int. Pharmacodyn. Ther.*, **224**(2), 338 (1976).

45. L. Braier, *Haematologica*, **58**(7–8), 491 (1973).

46. I. Franchini, A. Cavatorta, M. D'Errico, M. DeSantis, G. Romito, R. Gatti, G. Juvarra, and G. Palla, *Experientia*, **34**(2), 250 (1978).

47. V. Ullrich, *Z. Physiol. Chem.*, **350**(3), 357 (1969).

48. W. Diehl, J. Schaedelin, and V. Ullrich, *Z. Physiol. Chem.*, **351**(11), 1359 (1970).

49. E. J. McKenna and M. Coon, *J. Biol. Chem.*, **245**(15), 3882 (1970).

50. G. Mohn, *Xenobiotica*, **7**(1–2), 96 (1977).

51. A. Kraemer, H. Staudinger, and V. Ullrich, *Chem-Biol. Interactions*, **8**, 11 (1974).

52. J. D. Walker and R. R. Colwell, *Appl. Environ. Microbiol.*, **31**(2), 198 (1976).

53. *NIOSH Manual of Sampling Data Sheets*, U.S. Dept. HEW, Cincinnati, Ohio, 1977.

54. T. R. Norton, "Metabolism of Toxic Substances," Chapter 4 in *Toxicology, the Basic Science of Poisons*, L. J. Casarett and J. Doull, Eds., Macmillan, New York, 1975.

55. U. Frommer, V. Ullrich, and H. Staudinger, *Z. Physiol. Chem.*, **351**, 913 (1970).

56. M. N. Schroth and D. C. Hildebrand, *Phytopathology*, **58**, 848 (1968).

57. "Occupational Health and Environmental Control," Subpart G, Sec. 1910.93 Air Contaminants, *Fed. Reg.*, **39**(125), 235–40 (1974).

58. *Threshold Limit Values for Chemical Substances and Physical Agents in the Workroom Environment*, American Conference of Governmental Industrial Hygienists, Cincinnati, 1978.

59. L. Meites, *Handbook of Analytical Chemistry*, 1st ed., McGraw-Hill, New York, San Francisco, 1963.

60. K. Masui and H. Kawauchi, Jap. Pat. 78:08777, March 31, 1978.

61. R. G. Gould, *Photochemical Smog and Ozone Reactions*, American Chemical Society, Washington, D.C., 1972.

62. H. F. Smyth, Jr., C. C. Carpenter, C. S. Weil, U. C. Pozzani, J. A. Striegel, and J. S. Nycum, *Am. Ind. Hyg. Assoc. J.*, **30**, 470 (1969).

63. J. Natl. Cancer Inst., **31,** 41 (1963) (through Ref. 5).

64. D. J. Opdyke, *Food Cosmet. Toxicol.,* **13,** 733 (1975).

65. M. Tsuji, Y. Fujisaki, Y. Arikawa, S. Masuda, S. Kinoshita, A. Okubo, K. Noda, H. Ide, and Y. Iwanaga, *Oyo Yakuri,* **9**(3), 387 (1975).

66. M. Tsuji, Y. Fujisaka, K. Yamachika, K. Nakagami, F. Fujisaki, M. Mito, T. Aoki, S. Kinoshita, A. Okubo, and I. Watanabe, *Oyo Yakuri,* **8**(10), 1439 (1974).

67. M. Tsuji, Y. Fujisaki, Y. Arikawa, S. Masuda, T. Tanaka, K. Sato, K. Noda, H. Ide, and M. Kikuchi, *Oyo Yakuri,* **9**(5), 775 (1975).

68. M. Tsuji, Y. Fujisaki, A. Okubo, Y. Arikawa, K. Noda, H. Ide, and T. Ikeda, *Oyo Yakuri,* **10**(2), 179 (1975).

69. R. Kodama, A. Okubo, E. Araki, K. Noda, H. Ide, and T. Ikeda, *Oyo Yakuri,* **13**(6), 863 (1977).

70. F. Homburger, A. Treger, and E. Boger, *Oncology,* **25**(1), 1 (1971).

71. H. F. Smyth, Jr., C. P. Carpenter, C. S. Weil, and U. C. Pozzani, *Arch. Ind. Hyg. Occup. Med.,* **10,** 61 (1954).

72. V. K. H. Brown, L. W. Ferrigan, and D. E. Stevenson, *Ann. Occup. Hyg.,* **10,** 123 (1967).

73. V. K. H. Brown and C. G. Hunter, *Br. J. Ind. Med.,* **25,** 75 (1968).

74. H. F. Smyth, Jr., C. P. Carpenter, and C. S. Weil, *Arch. Ind. Hyg. Occup. Med.,* **4,** 199 (1951).

75. D. E. Drayer and M. M. Reidenberg, *Drug Metab. Dispos.,* **1**(3), 577 (1973).

76. E. R. Hart and J. C. Dacre, *Proc. First Int. Congr. Toxicol., Toronto,* G. L. Plaà and W. A. M. Duncan, Eds., 1977.

77. E. R. Kinkead, U. C. Pozzani, D. L. Geary, and C. P. Carpenter, *Toxicol. Appl. Pharmacol.,* **20,** 552 (1971).

78. S. Moeschlin, *Poisoning, Diagnosis and Treatment,* 1st Am. ed., Grune & Stratton, New York, 1965.

79. F. Sperling, H. K. U. Ewenike, and T. Farber, *Environ. Res.,* **5**(2), 164 (1972).

80. R. Y. Omirov and A. Y. Aberkulov, *Med. Zh. Uzb.,* **1,** 50 (1972).

81. S. Laham, *Toxicol. Appl. Pharmacol.,* **37**(1), 155 (1976).

82. Monographs on the Evaluation of Carcinogenic Risks of Chemicals to Man, Vol. 11, 1976, p. 277 (through Ref. 5).

83. J. C. Gage, *Br. J. Ind. Med.,* **27**(1), 1 (1970).

84. K. B. Lehmann, *Arch. Hyg.,* **83,** 239 (1914).

85. J. Pohl, *Zentralbl. Gewerbehyg. Unfallverhuet.,* **12,** 91 (1925).

86. W. J. Canady, D. A. Robinson, and H. D. Colby, *Biochem. Pharmacol.,* **23**(21), 3075 (1974).

87. K. C. Leibman and E. Ortiz, *J. Pharmacol. Exp. Therp.,* **173**(2), 242 (1970).

88. S. P. James, D. J. Jeffery, R. H. Waring, and D. A. White, *Biochem. Pharmacol.,* **20,** 897 (1971).

89. K. Ganapathy, K. S. Khanchandani, and P. K. Bhattacharyya, *Indian J. Biochem.,* **3,** 66 (1966).

90. R. H. Waring, *Xenobiotica,* **1**(3), 303 (1971).

91. E. Gil-Av and J. Shabtai, *Nature,* **197,** 1065 (1963).

92. M. F. Savchenkov, *Gig. Sanit.,* **30**(7), 28 (1965).

93. D. Helmlinger and P. Naegeli, Ger. Pat. 2622611, January 13, 1977.

94. P. deMayo, *Mono- and Sesquiterpenoids,* Vol. II, Interscience, New York, 1959.

95. D. J. Opdyke, *Food Cosmet. Toxicol.,* **13,** 731 (1975).

96. H. Elmenhorst and H. P. Harke, *Z. Naturforsch.,* **23b,** 1271 (1968).

97. W. Bertsch, R. C. Chang, and A. Zlatkis, *J. Chromatogr. Sci.,* **12,** 175 (1974).

98. A. S. Bondarenko, B. E. Aizenman, L. A. Bakena, and I. S. Kozhina, *Rastit. Resur.*, **10**(4), 583 (1974).

99. H. T. Brodrick, *Phytophylactica*, **3**(2), 69 (1971).

100. K. L. Stevens and L. Jurd, U.S. Pat. 3954991, May 4, 1976.

101. G. O. Osborne and J. F. Boyd, *N. Z. J. Zool.*, **1**(3), 371 (1974).

102. R. P. Bordasch and A. A. Berryman, *Can. Entomol.*, **109**(1), 95 (1977).

103. H. Igimi, T. Hisatsugu, and M. Nishimura, *Am. J. Dig. Dis.*, **21**(11), 926 (1976).

104. M. O. Karryev, *Mater. Ybileinoi Resp. Nauchin. Konf. Farm.*, **62** (1972).

105. T. Yoshino, Jap. Pat. 7772825, June 17, 1977.

106. P. R. Perez, P. Carmona, B. Lafuente, J. Bellanato, and A. Hidalgo, *Rev. Agroquim. Tecnol. Aliment.*, **17**(1), 59 (1977).

107. S. Iscowitz, U.S. Pat. 3977826, August 31, 1976.

108. H. Roemmelt, A. Zuber, K. Dirnagl, and H. Drexel, *Muench. Med. Wochenschr.*, **116**(11), 537 (1974).

109. R. Kodama, A. Okubo, K. Sato, E. Araki, K. Noda, H. Ide, and T. Ikeda, *Oyo Yakuri*, **13**(6), 885 (1977).

110. R. Kodama, T. Yano, K. Furkawa, K. Noda, and H. Ide, *Xenobiotica*, **6**(6), 377 (1976).

111. T. Ariyoshi, M. Arakaki, K. Ideguchi, Y. Ishizuka, K. Node, and H. Ide, *Xenobiotica*, **5**(1), 33 (1975).

112. H. Igimi, M. Nishimura, R. Kodama, and H. Ide, *Xenobiotica*, **4**(2), 77 (1974).

113. R. Kodama, K. Noda, and H. Ide, *Xenobiotica*, **4**(2), 85 (1974).

114. N. R. Ballal, P. K. Bhattacharyya, and P. M. Rangachari, *Biochem. Biophys. Res. Commun.*, **23**(4), 473 (1966).

115. R. S. Dhavalikar, P. M. Rangachari, and P. K. Bhattacharyya, *Indian J. Biochem.*, **3**(3), 258 (1966).

116. W. H. Stahl, *Compilation of Odor and Taste Threshold Values Data*, American Society for Testing and Materials, Philadelphia, 1973.

117. A. W. Horton and G. M. Christian, *J. Natl. Cancer Inst.*, **53**(4), 1017 (1974).

118. E. Bingham, *Arch. Environ. Health*, **19**, 779 (1969).

119. K. Adachi, S. Yamasawa, R. R. Suskind, and G. Christian, *Nature*, **22**, 191 (1969).

120. T. H. Elliott, J. S. Robertson, and R. T. Williams, *Biochem. J.*, **100**, 403 (1966).

121. G. J. Mulkins-Phillips and J. E. Stewart, *Can. J. Microbiol.*, **20**, 955 (1974).

122. K. M. Fredericks, *Nature*, **209**, 1047 (1966).

123. G. M. Badger and J. Novotny, *Nature*, **198**, 1086 (1963).

124. *Med. Lavoro*, **33**, 145 (1942) (through Ref. 5).

125. H. Holmberg and T. Malmfors, *Environ. Res.*, **7**, 183 (1974).

126. T. H. Elliott and J. Hanam, *Biochem. J.*, **108**, 551 (1968).

127. C-C Lin and C. Chen, *Biochim. Biophys. Acta*, **192**(1), 133 (1969).

128. R. F. Lee, K. Sauerheber, and A. A. Benson, *Science*, **177**, 344 (1972).

129. Y. Tanimura and M. Yamada, Jap. Pat. 43860, December 12, 1973.

130. T. A. Sullivan and W. C. McBee, *Proc. Miner. Waste Util. Symp.*, [*Pap*], **4**, 245 (1974).

131. W. D. Burrows, *Joint Conf. Sens. Environ. Pollut.* [*Conf. Proc.*], **4**, 80 (1978).

132. Y. I. Taradin, G. V. Buravlev, G. S. Bokareva, N. Y. Kuchmina, and L. N. Shavrikova, *Toksikol. Gig. Prod. Neftekhim. Proizvod.*, **197** (1972).

133. L. Z. Geikhman and Z. A. Dubrovskii, *Vrach. Delo.*, **2**, 141 (1970).

134. A. K. Oshkaev, *Izhv. Vyssh. Uchebn. Zaved. Lesn. Zh.*, **20**(5), 28 (1977).

135. C. Q. Darcel, R. W. Bide, and M. Merriman, *Can. J. Biochem.*, **46**(5), 503 (1968).

136. M. Merriman and C. Q. Darcel, *Can. J. Biochem.*, **43**(10), 1667 (1965).

137. K. Moersdorf, *Chim. Ther.*, **7**, 442 (1966).

138. A. Pap and F. Szarvas, *Acta Med. Acad. Sci. Hung.*, **33**(4), 379 (1976).

139. J. Lesznyak, S. Benko, R. Szabo, and E. Muller, *Kiserl. Orvostud.*, **24**(6), 571 (1972).

140. O. P. Shukla, M. N. Moholay, and P. K. Bhattacharyya, *Indian J. Biochem.*, **5**(3), 79 (1968).

141. G. P. Elizarov, *Gig. Tr. Prof. Zabol.*, **1**(2), 32 (1967).

142. H. Duengeman, S. Borelli, and J. W. Wittmann, *Arbeitsmed. Sozialmed. Arbeitshyg.*, **7**(4), 85 (1972).

143. W. Schneider, *Berufsdermatosen*, **21**(2), 45 (1973).

144. C. Eberhartinger, *Wien. Med. Wochenschr.*, **123**(26–27), 449 (1973).

145. E. M. Chapman, *J. Ind. Hyg. Toxicol.*, **23**(7), 277 (1941).

146. A. J. Bailey, T. J. Sims, M. LeLous, and S. Bazin, *Biochem. Biophy. Res. Commun.*, **66**(4), 1160 (1975).

147. A. A. Fisher, *Dermatitis*, 2nd ed., Lea and Febiger, Philadelphia, 1973.

148. R. Brun, *Dermatologica*, **150**, 193 (1975).

149. C. P. McCord, *J. Am. Med. Assoc.*, **86**, 1978 (1926).

150. P. Mikhailov, N. Berova, A. Tsutsulova, *Allerg. Asthma*, **16**(4/5), 201 (1970).

151. C. Oh, F. Ginsberg-Fellner, and H. Dolger, *Diabetes*, **24**(9), 856 (1975).

152. F. Sperling, W. L. Marcus, and C. Collins, *Toxicol. Appl. Pharmacol.*, **10**(1), 8 (1967).

153. H. F. Smyth and H. F. Smyth, Jr., *J. Ind. Hyg.*, **10**(8), 261 (1928).

154. G. S. Lazarus, *J. Invest. Dermatol.*, **62**, 367 (1974).

155. R. C. Hays and G. L. Mandell, *Proc. Soc. Exp. Biol. Med.*, **147**, 29 (1974).

156. T. B. Wellington and J. V. Jones, *Immunology*, **27**, 125 (1974).

157. R. W. Schayer and M. Reilly, *Am. J. Physiol.*, **215**(2), 472 (1968).

158. S. M. Phillips and B. Zweiman, *J. Exp. Med.*, **137**(6), 1494 (1973).

159. F. Homburger and E. Boger, *Cancer Res.*, **28**, 2372 (1968).

160. K. W. Nelson, J. F. Ege, M. Ross, L. E. Woodman, and L. Silverman, *J. Ind. Hyg. Toxicol.*, **25**, 282 (1943).

161. M. Sittig, *How to Remove Pollutants and Toxic Materials from Air and Water*, Noyes Data Corporation, Park Ridge, N.J., 1977.

162. R. Schuppli, *Z. Haut-Geschlechtskr.*, **46**(20), 751 (1971).

163. C. Eberhartinger, *Wien Med. Wochenschr.*, **12**(25–26), 513 (1971).

164. V. V. Dobrynina and E. I. Lyublina, *Gig. Tr. Prof. Zabol.*, **10**, 52 (1974).

165. E. R. Kinkead, U. C. Pozzani, D. L. Geary, and C. P. Carpenter, *Toxicol. Appl. Pharmacol.*, **20**(2), 250 (1971).

166. V. N. Suchkov, S. N. Sakharova, E. A. Pogodina, and A. G. Shalatilova, *Otkrytiya Izobret. Prom. Obraztaz Tovarnye Znaki*, **50**(17), 11 (1973).

167. D. L. J. Opdyke, *Food Cosmet. Toxicol.*, **13**, 735 (1975).

168. J. Lesznyak and G. Lusztig, *Gefaesswand Blutplasma, Symp.*, **4**, 253 (1974).

169. P. A. Hedin, O. H. Lindig, P. P. Sikorowski, and M. Wyatt, *J. Econ. Entomol.*, **71**(3), 394 (1978).

CHAPTER FORTY-SEVEN

Aromatic Hydrocarbons

ESTHER E. SANDMEYER, Ph.D.

1 MONOCYCLIC AROMATIC COMPOUNDS (ARENES)

1.1 General Considerations

1.1.1 Introduction

The term "aromatic" can have several meanings, namely, (1) exhibiting an etheral, balsamy odor, or (2) a distillation grade by the refiner, or (3) the class of aromatic hydrocarbons that deals with benzene and its derivatives and homologues. Benzene should not be confused with benzin, which is a gasoline mixture and in several European languages translates to gasoline or fuel. The aromatic hydrocarbons are of considerable economic importance as industrial raw materials, solvents, and components of innumerable commercial and consumer products. However, the aromatics differ vastly in physical, chemical, and physiological characteristics from the aliphatic and alicyclic hydrocarbons discussed in the preceding two chapters. Additionally, the aromatics are more highly toxic to the mammalian system. The properties responsible are their higher volatility, accessibility, and absorbability through the respiratory system, and, to a lesser degree, through the skin, and particularly their unusual hematopoietic characteristics.

1.1.2 Chemical and Physical Properties

Chemically, the aromatic hydrocarbons can be divided into three groups: (*a*) the alkyl-, aryl-, and alicyclic-substituted benzene derivatives, (*b*) the di- and polyphenyls, and (c) the polynuclear compounds composed of two or more fused benzene ring systems. The basic chemical entity is the benzene nucleus, which occurs alone, substituted, joined, or fused.

The simplest compound is benzene, the nonsubstituted ring system. When a methyl group is attached, it constitutes toluene. Benzene with two methyl groups is xylene, which occurs in three isomeric forms. The hemimellitines and mesitylenes possess three methyl groups, durene four, and the penta- and hexamethylbenzenes, five and six methyl groups, respectively. Other industrially important compounds are the ethyl-substituted benzene; the isopropyl derivative, which is cumene; and styrene, which is vinylbenzene, all discussed in this chapter.

The aromatics are moderately reactive and in the atmosphere undergo photochemical changes. Synthetically, they are of prime value as chemical raw materials.

The aromatic compounds occur in liquid or solid form. The lower molecular weight derivatives possess higher vapor pressures, volatility, absorbability, and solubility in aqueous media than do the comparable aliphatic or alicyclic compounds; see Table 47.1 for physicochemical data of the benzenes and naphthalenes (1–10). These general properties appear to be the contributing factors for the lower aromatics' high biologic activities. They are characterized also by greater miscibility or conversion to compounds soluble in aqueous body fluids, high lipid solubility, and donor–acceptor and polar interactions (11). Owing to the low surface tension and viscosity, the aromatics may be aspirated into the lungs, where they can cause severe cellular injury.

1.1.3   Origin and Other Sources

Benzene and its alkyl derivatives, the polyphenyls and polynuclear aromatics (PNAs), are obtained as products or by-products in petroleum or coal refining, burning, or pyrolysis processes. From coke-oven operations, the aromatics are recovered from the gases and the coal tars. From crude oil distillation they are produced by fractionated distillation, solvent extraction, naphthenic dehydrogenation, alkylation of benzene or olefins, or from paraffins by catalytic cyclization or aromatization. Benzene has been detected in cigarette smoke (12) at 47 and 64 ppm, depending on the type used, as quoted by Newsome et al. (13). Benzene, toluene, and higher aromatics have been detected in rainwater, the former two at concentrations of 0.1 to 0.5 µg/liter, and about twice the quantity in the ambient air (14).

1.1.4   Utilization

The aromatic hydrocarbons are used widely as chemical raw materials, intermediates, or solvents, industrially in oil and rosin extractions, as components of multipurpose additives, and owing to their rapid drying characteristics, extensively in the glue and veneer industries. Aromatics serve in the dry-cleaning business, in the printing and metal processing industries, and for many more similar applications. They serve as important constituents of aviation and motor gasolines, and represent important raw materials in the preparation of pharmaceutical products.

1.1.5 Physiologic Response

General Considerations. The aromatics are primary skin irritants, and on repeated or prolonged skin contact may cause dermatitis, owing to their dehydrating and defatting properties. Eye contact with aromatic liquids may cause burns, lacrimation, and irritation. If contact is prolonged, tissue injury may ensue. Conjunctivitis and corneal burns have been reported from the C_6 to C_8 members. Naphthalene has been shown to cause cataracts in the eyes of experimental animals, and its vapors are respiratory and mucous membrane irritants and may cause severe systemic injury. Direct contact with most liquid hydrocarbons through aerosol incorporation or from ingestion and subsequent aspiration into the lungs can cause severe pulmonary edema, pneumonitis, and hemorrhaging (15). The alkylbenzenes, with side chains C_1 to C_4, when tested in rats for possible aspiration hazards, produced instant death, owing to cardiac arrest and respiratory paralysis. For example, death from hexylbenzene was delayed up to 18 min, which permitted time for extensive fluid infiltration into the alveolae (16), resulting in considerably increased lung weights. The higher alkylbenzenes showed few or no effects. In view of some biologic characteristics, namely in the effects on bone marrow and blood-forming mechanisms, benzene, with respect to other aromatics, stands virtually alone. Benzene also has only scant properties in common with the next higher member, toluene. For the alkylbenzenes in general, the acute toxicity is higher for toluene than for benzene, and decreases further with increasing chain length of the substituent, except for highly branched C_8 to C_{18} derivatives. The toxicity increases again for the vinyl derivatives. Pharmacologically, the alkylbenzenes are CNS depressants, owing to their particular affinity to nerve tissues.

Acute Toxicity. Overall, benzene is more highly toxic than any of the substituted benzene derivatives, except for toluene and vinylbenzene. Aromatic hydrocarbons cause local irritation and changes in endothelial cell permeability, and are absorbed rapidly. Secondary effects have been observed in the liver, kidney, spleen, bladder, thymus, brain, and spinal cord in animals after dosing with some alkylbenzenes (17). The aromatic hydrocarbons, even from a single dose, exhibit a special affinity to nerve tissue. Animals dosed with alkylbenzene may show signs of CNS depression, sluggishness, stupor, anesthesia, narcosis, and coma. This is in sharp contrast with benzene, which is a neuroconvulsant,

Table 47.1. Physicochemical Properties for Mono- and Dicyclic and Some Polyphenylic Hydrocarbons (1–10)

Compound	B.P. (°C)	CAS Registry No.	Density (20/4°C)	Emp. Formula	Flamm. Limits[a] (%)	Flash Pt. [°C (°F)]	Freezing Pt. (°C)	M.P. (°C)	Mol. Wt.	Refr. Index (n_D^{20})	Solubility[b] w/al/et	Sp. Gr. (25°C)	Vapor Dens. (Air = 1)	Vapor Pres. [mm Hg (°C)]	Visc. (SUS)	W/V Conv. (mg/m³ ⇔ 1 ppm)
Benzene	80.10	71-43-2	0.8787	C_6H_6	1.4–7.9	−11 (12)	5.53	5.5	78.11	1.5011	d/v/v	0.880	2.8	100 (26.1)	<32.6	3.19
Toluene	110.62	108-88-3	0.8869	C_7H_8	1.4–6.7	4.4 (40)	−94.99	−95	92.14	1.4961	i/v/v	0.87	3.1	36.7 (30)	<32.6	3.77
o-Xylene	144.41	95-47-6	0.8802	C_8H_{10}	1.0–6.0	32 (90)		−25.18	106.17	1.5055	i/v/v	0.90	1.1	6.8 (25)	<32.6	4.34
m-Xylene	139.10	108-38-3	0.8642	C_8H_{10}	1.1–7.0	29 (84)		−47.87	106.17	1.4972	i/v/v	0.87	1.03	8.3 (25)	<32.6	4.34
p-Xylene	138.35	106-42-3	0.8611	C_8H_{10}	1.1–7.0	27 (81)		13.26	106.17	1.4958	i/v/v	0.86	1.03	8.9 (25)	<32.6	4.34
Xylenes, mixed	138.3	1330-20-7	0.864	C_8H_{10}	1.0–7.0	37.6 (100)			106.17		i/v/v	0.9		6–16 (20)	<32.6	4.34
Trimethylbenzene																
1,2,3-	176.1	526-73-8	0.8944	C_9H_{12}	0.88 lel		−25.4 (−13.7)	−25.37	120.19	1.5139	i/v/v	0.899			<32.6	4.92
1,2,4-	169.35	96-63-6	0.8758	C_9H_{12}	0.88 lel	54.4 (130)	−42.2	−43.8	120.19	1.5067	i/s/s	0.880	4.1	341 (140.1)	<32.6	4.92
1,3,5-	164.7	108-67-8	0.8652	C_9H_{12}	0.88 lel		(−43.9)	−44.7	120.19	1.4994	i/v/v	0.870	4.1	1.82 (20)	<32.6	4.92
Tetramethylbenzene																
1,2,3,4-	205.0	488-23-3	0.9052	$C_{10}H_{14}$				−6.25	134.22	1.5203	i/v/v				<32.6	5.49
1,2,3,5-	198.0	527-53-7	0.8903	$C_{10}H_{14}$				−23.68	134.22	1.5130	i/v/v				<32.6	5.49
1,2,4,5-	196.8	95-93-2	0.8875	$C_{10}H_{14}$				79.24	134.22	1.5116	i/v/v				<32.6	5.49
Ethylbenzene	136.2	100-41-4	0.8670	C_8H_{10}	1.6–7	12.8 (55)		−94.97	106.17	1.4959	i/v/v	0.867	3.7	10 (25.9)	<32.6	4.34
Methylethylbenzene	161.3	620-24-4	0.8645	C_9H_{12}				−95.55	120.19	1.4966	i/v/v				<32.6	4.92
1,2-Diethylbenzene	183.4	135-01-3	0.8800	$C_{10}H_{14}$				−31.2	134.22	1.5035	i/v/v				<32.6	5.49
1,3-Diethylbenzene	181.0	141-93-5	0.8602	$C_{10}H_{14}$				−83.89	134.22	1.4955	i/v/v				<32.6	5.49
1,4-Diethylbenzene	183.8	105-05-5	0.8620	$C_{10}H_{14}$				−42.85	134.22	1.4967	i/v/v				<32.6	5.49

Name	Boiling point, °C	CAS No.	d	Formula	Flammable limits	Flash point, °C (°F)	Mol. wt.	Melting point, °C	n_D	Solubility	d_2				
Diethylbenzene, mixed	183.8	25340-17-4	0.868	$C_{10}H_{11}$		55.6 (132)	134.22			i/v/v	0.88	4.62	1 (20.7)	<32.6	5.49
n-Propylbenzene	159.2	103-65-1	0.8620	C_9H_{12}	0.8–6	30 (86)	120.19	−99.5	1.4920	i/v/v	0.86	4.14	10 (43.4)	<32.6	4.92
Cumene	152.4	98-82-8	0.8618	C_9H_{12}	0.9–6.5	36 (96)	120.19	−96	1.4915	i/v/v	0.86	4.1	10 (38.3)	<32.6	4.92
o-Cymene	178.15	527-84-4	0.8766	$C_{10}H_{14}$			134.22	−71.54	1.5006	i/v/v				<32.6	5.49
m-Cymene	175.14	535-77-3	0.8610	$C_{10}H_{14}$			134.22	−63.75	1.4930	i/v/v				<32.6	5.49
p-Cymene	177.1	99-87-6	0.8573	$C_{10}H_{11}$	0.7–5.6	47 (117)ᵈ	134.22	−67.94	1.4909	i/v/v	0.857	4.62	1 (17.3)	<32.6	5.49
n-Butylbenzene	183	104-51-8	0.8601	$C_{10}H_{14}$	0.8–5.8	71 (160)	134.22	−88.0	1.4898	i/v/v	0.8656	4.6	2.4 (37.8)	<32.6	5.49
sec-Butylbenzene	173.0	135-98-8	0.8621	$C_{10}H_{14}$	0.8–6.9	52.2 (126)	134.22	−75	1.4895	i/v/v	0.8664	4.62	4.02 (37.8)		5.49
Isobutylbenzene	172.8	538-93-2	0.8532	$C_{10}H_{14}$	0.8–6.0	52.2 (126)	134.22	−51.5	1.4366	i/v/v	0.8576	4.62	1.0 (14.1)		5.49
tert-Butylbenzene	169	98-06-6	0.8655	$C_{10}H_{14}$	0.7–5.7ᶜ	60 (140)	134.22	−57.85	1.4927	i/v/v	0.8710	4.62	5.7 (37.8)		5.49
tert-Butyltoluene	192.8	98-51-1	0.8575	$C_{11}H_{16}$		68.3 (155)	148.25	−52.4	1.4921	i/s/s		4.62	0.65 (25)		5.98
Dodecylbenzene	290–410	123-01-3	0.9	$C_{18}H_{30}$	1.1–6.1	140.6 (285)	246.44		—	i/s/s	0.905	8.47	—		10.08
Styrene	145.2	100-42-5	0.9060	C_8H_8		32 (90)	140.15	−30.63	1.5468	i/s/s		3.60	4.5 (20)		4.26
α-Methylstyrene	165.4	98-83-9	0.9062	C_9H_{10}	1.9–6.1	54 (129)	118.18	−23	1.5386	i/s/s	0.91	4.08			4.83
p-Methylstyrene	171	622-97-9		C_9H_{10}	0.8–11.0	52.8 (127)	118.18			i/s/s	0.90				4.83
o-Divinylbenzene	178.5 (11)	91-14-5	0.934	$C_{10}H_{10}$			130.19			i/d/d					5.33
m-Divinylbenzene	199.5	108-57-6	0.9289	$C_{10}H_{10}$	0.3	73.9 (165)	130.19	−66.9		i/d/d	0.93	4.48			5.33
p-Divinylbenzene	83.6 (16)	105-06-6	0.913 (40°C)	$C_{10}H_{10}$			130.19	31	1.5131	i/d/d			1 (32.7)		5.33
Allylbenzene	156		0.8920	C_9H_{10}			118.18	−40		i/s/s					4.83
1-Phenylbutene-2	175	1560-06-1	0.888	$C_{10}H_{12}$			132.21								5.41
Phenylacetylene	142.4	536-74-3	0.9300	C_8H_6			102.14	−44.8	1.5489						4.18
Diphenyl	255.9	92-52-4	0.8660	$C_{12}H_{10}$		71.1 (160)	154.21	71	1.558	i/s/s	0.933				6.31
Diphenylmethane	265.5	101-81-5	1.0060	$C_{13}H_{12}$		130 (266)	168.24	23.35	1.5723	i/s/s					6.88
cis-Stilbene	135 (10)	103-30-0		$C_{14}H_{12}$			180.25	−5	1.6188	i/s/s					7.37
trans-Stilbene	305 (720)	645-49-8	0.9707	$C_{11}H_{12}$			180.25	124.5–124.8	1.6264	i/d/s					7.37
o-Terphenyl	322	84-15-1	1.14	$C_{18}H_{11}$		163 (325)ᵞ	230.31	58		i/s/s	1.14	7.9			9.42
m-Terphenyl	365	92-06-8		$C_{18}H_{11}$		135 (675)ᵞ	230.31	89		i/s/s	1.16	7.9			9.42

Table 47.1. (*Continued*)

Compound	B.P. (°C)	CAS Registry No.	Density (20/4°C)	Emp. Formula	Flamm. Limits^a (%)	Flash Pt. [°C (°F)]	Freezing Pt. (°C)	M.P. (°C)	Mol. Wt.	Refr. Index (n^{20}_D)	Solubility^b w/al/et	Sp. Gr. (25°C)	Vapor Dens. (Air = 1)	Vapor Pres. [mm Hg (°C)]	Visc. (SUS)	W/V Conv. (mg/m³ ⟺ 1 ppm)
p-Terphenyl	405	92-94-1	1.236	$C_{18}H_{14}$		240 (465)^e		213	230.31		i/s/s	1.24				9.42
Naphthalene	217.9	91-20-3	1.0253	$C_{10}H_8$		87.8 (190)^e		80.55	128.17	1.4003	i/s/v	1.15	4.42			5.24
1-Methyl-naphthalene	244–64	90-12-0	1.0202	$C_{11}H_{10}$		79 (174)^d		−22	142.20	1.6170	i/v/v					5.82
2-Methyl-naphthalene	241.05	91-57-6	1.0058	$C_{11}H_{10}$				34.58	142.20	1.6019	i/v/v					5.82
1,4-Dimethyl-naphthalene	268		1.0166	$C_{12}H_{12}$				7.66	156.23	1.6127	i/v/v					6.39

^a Lower limit (lel) to upper.
^b Solubility in water/alcohol/ether: v = very soluble; s = soluble; d = slightly soluble; i = insoluble.
^c −100°C.
^d Closed cup.
^e Open cup.

producing stimulation characterized by tremors and convulsions. The narcotic potency of the alkylbenzenes depends on branching or side chain length. It diminishes with the number of substituents and increasing side chain carbon number up to dodecylbenzene, which practically has no narcotic activity (17). The concentration to produce narcosis in barnacle larvae after immersion for 15 min was 3.1 percent for benzene and 4.5 percent for toluene (18).

Chronic Toxicity.   Benzene in repeated doses is considered more toxic than any alkylbenzene, owing to its affinity to blood-forming tissue and myelotoxic activity. It appears that any type of substituted benzene derivative is devoid of this characteristic. This has been clearly demonstrated with toluene versus benzene, the latter of which does alter the leukocyte count or bone marrow nucleation in the rat (17).

Metabolism.   Benzene and its derivatives are readily hydroxylated and alkyl side chains further oxidized to carboxylic acids. Benzene may also be metabolized by a ring opening. Benzene immediately increases the urinary excretion of organic sulfate under diminution of the inorganic sulfates. Of the alkylbenzenes, only *m*-xylene and mesitylene follow this trend.

Marine Toxicology.   Aromatic hydrocarbons appear to be accumulated in marine animals to a greater extent and retained longer than the alkanes (19). In all species tested, the accumulation of aromatic hydrocarbons depended primarily on the water and tissue lipid partitioning coefficients. Higher molecular weight hydrocarbons were released more slowly (19).

Microbiological Toxicity.   To some bacteria as little as 0.01 percent of toluene, xylene, mesitylene, phenol, or cresol may be bacteristatic or bactericidal (20), whereas other microorganisms tolerate 0.5 percent of hydrocarbon concentrations, including 0.1 percent phenol. Benzene is least susceptible to bacterial oxidation; increasing substitution and chain length, especially to even numbers, promotes the ease of oxidation (20).

1.1.6   Industrial Hygiene

From an industrial hygiene standpoint, in view of monitoring and evaluation, the aromatics require close attention. This pertains particularly to benzene, toluene, xylene, ethylbenzene, cumene, *p-tert*-butyltoluene, and styrene. Within the past 10 years threshold limit values have been lowered incrementally for some of the aromatic compounds. This has been a consequence of the development of better sampling and analytic techniques. In addition, pathological findings and epidemiologic reports, particularly from southern European groups, have played a role in recognizing the importance of industrial hygiene surveys and evaluations.

1.1.7 Medical Aspect

Industrial monitoring programs should be continually evaluated. Where excursion values are found, biologic monitoring and follow-ups should be carried out in addition to the regular periodic health examination programs. According to a Polish survey, workers with long-term exposure to some aromatic hydrocarbons have shown changes in leukocyte alkaline phosphatase (21).

1.2 Benzene

1.2.1 General Aspects

Benzene is the simplest aromatic compound known. However, benzene should not be confused with *benzin*, which is a low-boiling petroleum fraction of predominantly aliphatic compounds, similar to gasoline. Benzin, in many non-English-speaking European countries, indicates gasoline. In industry and commerce, benzene has been one of the most important industrial chemicals. However, currently as an analytic agent and in household and many industrial products, benzene has been replaced largely by other solvents, such as toluene.

1.2.2 Chemical and Physical Data

Benzene, benzol, phene, $\overline{CH:CHCH:CHCH:CH}$, is a clear, colorless liquid with a characteristic pleasant odor at low concentrations, disagreeable at high levels. Some data are summarized in Table 47.1. Benzene forms a highly flammable (DOT) (22) and explosive mixture with air at 1.4 to 8.0 percent. It is an excellent solvent. Chemically, it is fairly stable, but readily undergoes substitution reactions to form halogen, nitrate, sulfonate, and alkyl derivatives. Commercial benzene, of which three grades have been standardized (ASTM D-16), usually contains varying concentrations of toluene, xylene, and phenol, and traces of carbon disulfide, thiophene, olefins, naphthalene, and related compounds.

The photochemical formation of nitrobenzenes and nitrophenols from benzene has been observed in the presence of nitrogen oxides. Benzene also combines photochemically with halogens to produce eye and mucous membrane irritants (23). Ozone reacts 10 to 20 times more slowly with benzene than with its methyl-substituted derivatives (24).

1.2.3 Occurrence, Industrial Sources, and Preparation

Benzene occurs in coal tar and petroleum naphtha from which it is commercially prepared (5). Benzene is a constituent of gasoline. A report from the U.S. Army in 1972 showed that their 37 unleaded and leaded gasolines averaged 0.8 percent benzene, with the highest value 2 percent. European gasolines even recently still contained up to 5 percent benzene, being produced from higher aromatic reformates (25). The benzene concentration in the gasoline vapor phase is lower than that in the liquid, but depends somewhat on the concentration of other hydrocarbon and metal additives (25). Some consumer products, such as paint removers, in the past have contained 50 percent benzene or more (26). Products with a benzene content above 0.1 percent now have to be adequately labeled with warnings to the consumer (27).

Benzene occurs in thermal degradation gases from high density polystyrene (28) and in solid waste gasification products (29). It has been identified as well in condensates of tobacco smoke (12).

Benzene has been detected in the expired air in three nonsmokers and two smokers of eight human volunteers (30). It occurs as a degradation product in the rat after administration of triphenyllead acetate (31) and *p*-toluic acid phenylhydrazide (32).

About 150 air samples analyzed from the areas around Houston, Texas over a period of 15 months revealed an average benzene concentration of 1.3 to 15 ppb (33). In the Toronto atmosphere, average benzene concentrations were detected at 2 to 98 ppb (34), and near Los Angeles 60 to 70 ppb in 1962 (35).

Benzene is produced in billion gallon quantities per year, mainly by fractionated distillation from crude oil, solvent extractions or crystallization, catalytic dehydrogenation, and re-forming of light naphthas, paraffins, and cycloolefins (36). It is produced as well in the coke-oven industry as a by-product.

1.2.4   Utilization

The benzene's ubiquitous use stems from its ready availability at relatively low cost. It is used mainly in chemical processes as a raw material and as a solvent in industry and commerce. Some important processes include the manufacture of ethylbenzene, styrene, cumene, phenolic resins, ketones, adipic acid, caprolactam, nylon (5), and various dyes (37). However, in most consumer products, benzene has been substituted, unless so labeled (27). Benzene comprises a high energy component of aviation and motor gasolines.

1.2.5   Physiologic Response

The physiological and toxicologic effects of benzene differ with the type of exposure. Some of the available data are summarized in Tables 47.2 to 47.5 (15–17, 38–100).

Response to Human Acute Exposure. *Oral Toxicity.*   Oral ingestion of 9 to 12 g benzene (38) has caused signs of staggering gait, vomiting, somnolence, shallow, rapid pulse, loss of consciousness, and later delirium, with subsequent chemical pneumonitis, serious collapse due to initial stimulation, then abrupt CNS depression. Generally, at moderate concentrations, symptoms are dizziness,

Table 47.2.   Human Acute Benzene Exposure

Route	Dose or Concentration	Results, Signs, or Symptoms	Ref.
Oral	9–12 g	Staggering gait, vomiting, somnolence, shallow, rapid pulse, loss of consciousness, delirium, death	38
	10 ml	May be approximate fatal dose	39
	30 g	May be approximate fatal dose	40
Inhalation	1.5 ppm (5 mg/m³)	Olfactory threshold	15
	25 ppm (0.08 mg/liter)/480 min	No effect, detectable in blood	17
	50–150 ppm (0.16–0.48 mg/liter)/300 min	Headache, lassitude, weariness	17
	500 ppm (1.6 mg/liter)/60 min	Headache	17
	1500 ppm (4.8 mg/liter)/60 min	Symptoms of illness	17
	3000 ppm (9.6 mg/liter)/30 min	May be tolerated for 0.5–1 hr	17, 38
	7500 ppm (24.0 mg/liter)/60 min	Signs of toxicity in 0.5–1 hr	38
	3100–5000 ppm (10–16 mg/liter)	Subtle signs of intoxication, absorbed 79.8–84.8%	38
	19,000–20,000 ppm (61–64 mg/liter)/5–10 min	May be fatal in 5 to 10 min	17

excitation, and pallor, followed by flushing, weakness, headache, breathlessness, constriction of the chest, and fear of impending death. Visual disturbances and convulsions are frequent. At higher concentrations, as shown above, the symptoms are excitement, euphoria, and hilarity, then quite suddenly change to weariness, fatigue, and sleepiness, followed by coma and death (101). Conversely, benzene was used therapeutically between 1910 and 1920 to reduce leukocytes in leukemia (38).

Dermal Effects.   Dermal contact presents a possible route of absorption, but generally at a much lower rate than through the mucous membranes of the respiratory system. Benzene is irritant to the skin (38), and by defatting the keratin layer may cause erythema, vesiculation, and dry and scaly dermatitis (102).

Inhalation Effects.   About 130 years ago, when benzene first was obtained in pure form, attempts were made to use this agent to produce anesthesia in man. However, owing to the unpleasant side and aftereffects, the practice was quickly abandoned (38).

Generally, as a consequence of benzene's high volatility, the inhalation route has been the most prevalent means of intoxication (103). Acute human exposure has occurred mainly during excessive promulgation of fumes in accidental spills, drying of soiled clothing in poorly ventilated areas in dry-cleaning businesses, in the home, or confined spaces where benzene was used as a solvent or product component.

Low benzene concentrations such as 25 ppm inhaled for 2 hr appeared quickly in the blood but cleared within 300 min (104). Absorption was determined as 79.8 to 84.8 percent by the human system (38) when inhaled at concentrations of 10 to 16 mg/liter. Some portion is normally exhaled in

Table 47.3. Acute Animal Experiments with Benzene

Acute	Species	Dose or Concentration	Results, Signs, or Symptoms	Ref.
Oral	Rat	3.4 g/kg	LD_{50}	41
	Rat	5.6 g/kg	LD_{50}	42
	Rat	<1.0–5 g/kg	LD_{50}, variation, age and strain dependent	41
	Rabbit	150–300 mg/kg (2.5–11.0 μCi)	Use of cold and radioactive benzene, no effect; 48.5% exhaled unchanged, 51.5% urinary excretion: 18.2–21.2% as phenol, 4.8% as quinol, 4.4% catechol	43
	Mouse	4.70 g/kg	LD_{50}	44
	Dog	2.00 g/kg	Lowest reported lethal dose	47
Intrapulmonary instillation	Rat	0.25 ml	LD_{100}, cardiac arrest, death	16
Eye	Rabbit	0.10 ml	Irritancy, moderate conjunctival irritant, causes transient corneal injury	42
SC	Mouse	0.088 g/kg	No effect	45
		0.44 g/kg	27% inhibition of circulatory erythrocytes	45
		2.20 g/kg	50% inhibition of circulatory erythrocytes	45
		2.70 g/kg	Found to be a teratogen	46
	Frog	1.40 g/kg	Lethal dose	47
	Rat	1.15 g/kg	Lethal dose	48
	Mouse	0.468 g/kg	LD_{50}	49
	Guinea pig	0.527 g/kg	Lethal dose	50
Inhalation	Rat	10,000 ppm (31.9 mg/liter)/7 hr	LC_{50}	17
		13,700 ppm (43.7 mg/liter)	LC_{50}, high levels of liver and lung congestion	51
		16,000 ppm (51.0 mg/liter)/4 hr	LC_{50}	52
	Mouse	2,195 ppm (7.0 mg/liter)	Narcosis	38
		9,980 ppm (31.8 mg/liter)	LC_{50}	53
	Rabbit	4,000 ppm (12.8 mg/liter)	Narcosis	54
		10,000 ppm (31.9 mg/liter)	Death	54
		35,000–45,000 ppm (111.6–144 mg/liter)		
		4–71 min	Slight anesthesia in 4 min and death in 22–71 min	55
		3.7 min	Light anesthesia, relaxed	55
		5.0 min	Excitation, tremors, running movements	55
		6.5 min	Loss of pupil reflex	55
		11.4 min	Loss of blinking reflex	55
		12.0 min	Pupillary contraction	55
		15.6 min	Involuntary blinking	55
		36.2 min	Death	55
	Guinea pig	6,270 ppm (20 mg/liter)/30 min peak	Benzene blood level 3.2 mg/100 ml, 1.7 mg/m³ phenol and clearing	56
		15,675 ppm (50 mg/liter)/30 min peak	Benzene blood level 8.0 mg/100 ml, 1.8 mg/m³ phenol and clearing	56
	Dog	45,800 ppm (146 mg/liter)	Lethal dose	50
	Cat	53,300 ppm (170 mg/liter)	Lethal dose	50
Aquatic	Striped bass	100–10 ppm	TLm96	57
		10.9 μl/liter	LC_{50}/96 hr toxicity similar to that of crude oil	57
	T. Californicus	0.1 ml/liter	Lethal and more toxic than other petroleum products	58

Table 47.4. Human Chronic Inhalation of Benzene

Years of Exposure	Year Published	No. of Cases	Type of Exposure	Exposure Conc.	Findings	Ref.
3–54 yr	1939	286/332	Rotogravure ink solvent application and personal dry-cleaning agent	11–1060 ppm/4–7 days/wk	Lowest concentration: fatigue and dizziness Medium concentration: fatigue, dryness of mucous membranes, hemorrhaging, nausea or vomiting, and lethargy Highest concentration: weakness, fatigue, epistaxis, dryness of mucous membranes, loss of appetite, nausea or vomiting, shortness of breath, dizziness, insomnia, and lethargy	59
1946–1956	1956	107/147	Shoe manufacturing	~400 ppm	Hematopathy and thrombocytopenia most common	60
Prior to 1964	1964	6/47	Paint thinners, printing inks, adhesives		Six myeloid, hemocytoblastic, or lymphatic effects, some reversible, of 47 cases with hemopathy	61
1945–1965	1965	20			Aplastic anemia	62
Prior to 1966	1966	3	Benzene vapor	—	Aplastic anemia with osmotic fragility	63
Prior to 1955	1966	125/147	Shoe adhesive solvents	~400 ppm 1955–1966	Nine-year follow-up of 125–147 cases with some decreased thrombocytosis from previous benzene exposure of 100/147 with abnormal hematology	64
1938–1968	1968	1	Occupational varnishing	—	Rare case of acute erythromyelosis	62

Table 47.4. (*Continued*)

Years of Exposure	Year Published	No. of Cases	Type of Exposure	Exposure Conc.	Findings	Ref.
Prior to 1971 3 mo–17 yr	1971	51/217	Shoe adhesive solvents	30–210 ppm	Leukopenia 9.7%, pancytopenia 2.8%, eosinophilia 2.3%, thrombocytopenia 1.8%, basophilia 0.5%, giant platelets 0.5%; anemia was reversible	65
Benzene prior to 1953, then toluene	1971	34	Rotogravure benzene and toluene	125–>525 ppm	Some higher chrosome aberration in peripheral blood lymphocytes	66
Prior to 1961 5–18 yr	1971	5/216	Watch industry	—	Ten-year follow-up: 3 thrombocytopenias, 1 thrombocytopenia and anemia, 1 death of aplastic anemia	67
4 mo–15 yr	1972	32	Shoe adhesive and solvents	150–650 ppm	Reversible pancytopenia, 28 thrombocytopenia, 14 macrocytic anemia, 3 megaloblastic erythropoiesis	68
Prior to 1973	1973	299/1000	—	—	Absolute monocytosis in an average of 14%, 19 reticuloendothelial hyperplasia	69
1 yr	1977	40	Shoe industry benzene and other solvents	—	Rate of leukemia-like effects of occupationally exposed workers 13.0/100,000 vs. 6.0/100,000 in general population	70
Prior to 1950 and later	1978	350/594	U.S. industrial benzene production and use	—	Various hematopathies and several cases with leukemia-like effects	71

Table 47.5. Chronic Animal Experiments with Benzene

Route	Species	Dose or Conc.	Time	Results or Findings	Ref.
Oral	Rat (F)	1 mg/kg/day	187 days	No effect	42
		10 mg/kg/day	187 days	Very slight leukopenia	42
		50 mg/kg/day	187 days	Leukopenia and erythrocytopenia	42
		100 mg/kg/day	187 days	Leukopenia and erythrocytopenia	42
		100 mg/liter[a]	4 mo	Development of leukopenia during 1st month, CNS disruption after 4 mo	72
		250 mg/liter[a]	40–50 days	Temporary changes in leukocyte counts and cholinesterase activity	73
		250 mg/liter[a] + 10 mg/m³	20 days	Temporary changes in leukocyte counts and cholinesterase activity	73
		1.6 ml/kg	3 days	Increased liver weight, decreased protein weight and enzyme activities	74
		1.5 mg/kg + 0.51 mg/m³	100 days	Sensitizing effect in 30–40 days	75
		9.0 mg/kg + 6.63 mg/m³	100 days	Very rapid sensitizing and intoxicating effect	75
	Rabbit	1 mg/kg	6 mo	Mild leukopenic, splenic, and testicular degeneration	42
Dermal	Rat	0.6 g/kg	4 hr/day × 4 mo	Plasmic cell increase in bone marrow, disruption of erythropoietic element maturation	76
	Mouse (hairless)	To cover dorsal skin	2 yr	Only slightly above spontaneous frequency of several tumor types, but some skin papillomas	77
		2 µl/appln.	3 ×/day × 3 days	Epidermal hyperplasia with neural invasion	78
	Rabbit	25% in vaseline	2 ×/wk	Tremors, excitement, cachexia	38
Dermal	Rabbit (ear)	Undiluted	10–20 appln.	Slight to moderate irritation, moderate necrosis	42
SC	Rat	1 mg/kg	12 days	Chromosomal damage in 50.94% of bone marrow cells	79
		2.5 mg/kg	30 days	Prolonged generation and transit time of granulocytes and reticulocytes, and decreased maturation time of neutrophilic granulocytes and erythroblasts	80
		0.2 g/kg	12 days	Lymphopenia, with toluene increased neutrophils and bacillonuclear neutrophils	81
		1 g/kg	12 days	Chromosomal analysis at metaphase; chromatid breaks 50.9%, gaps 44.7%, isochromatid breaks 4.34%, somewhat higher than from toluene	79
		2 ml/kg	21 days	Cellular injury, mitochondrial swelling, severe dilation of membrane system, rough and smooth endoplasmic reticulum; severely disintegrated hematopoietic cells	82
		1 ml/kg/day	14 days	Leukopenia, no significant change in hematocrit	83
		1 ml/kg/day	5 wk	Rapid decrease in femoral marrow nucleated cell count and DNA phosphorus % (of bone marrow dry wt.)	84
		2 ml/kg/day	3 wk		
	Mouse	0.025–0.1 ml	54 wk, 104 wk	Not a carcinogen	85
		0.5 ml/kg	2 appln./day × 1–10 days	Dose-dependent binding to liver, residue increasing with time; initial binding to bone marrow, disassociation after day 6	86
SC	Rabbit	0.1 ml/kg/day × 13 mo		Increased pseudoeosinopoiesis	86
		0.5 ml/kg/day × 10–15 days		Initial stimulation of pseudoeosinopoiesis, bone marrow promyelocyte, and myelocyte count	86

Table 47.5. (*Continued*)

Route	Species	Dose or Conc.	Time	Results or Findings	Ref.
		0.5 ml/kg/day × 3–4 wk		Decrease of blood leukocytes	86
		0.3 ml/kg/day × 1–9 wk		Pancytopenia, hypoplasia of bone marrow, with severe inhibition of the DNA synthesis	87
		0.5 ml/kg/day × 2–3 wk		Decrease in peripheral blood leukocyte count, decreases mitochondrial respiration	88
		1.0 ml/kg/day × 12 days		Increased erythrocyte permeability induction of porphyrin biosynthesis in brain gray matter	89
Inhaln.	Rat	4.8 ppm (15 mg/m³)	Continous	Hyperplasia and vacuolization of smooth reticulum, selective destruction of olfactory basal cells	90
	Rat, guinea pig, dog, primate	17 ppm (56 mg/m³) 30 ppm (98 mg/m³) 256 ppm (817 mg/m³)	8 hr/day × 5 day/wk continuously 9–127 days	Histopathology of all organs essentially negative, some slight weight reduction at the highest level	91
	Rat	94 ppm (350 mg/m³)	5.5 hr/day × 30 days	Increased α-aminobutyric acid level decreased in cerebellum but not mesencephalon, aspartic and glutamic acid decreased at both sites	92
		158 ppm (500 mg/m³)	4 hr/day interm.	Leukopenia and muscle antagonistic chromaxy	93
		400 ppm (1260 mg/m³)	13 wk	Increased thrombocytes, leukopenia, decreased segmented granulocytes	94
		158–1580 ppm (50–5000 mg/m³)	Continuous	Leukopenia, leukocytosis, reduced blood cholinesterase, alteration of reversible chromaxy ratio of muscle antagonists	95
Inhaln.	Rat	88 ppm (0.28 mg/liter)	7–8 hr/day × 5 day/wk × 6 mo for all concentrations	Slight splenomegaly	42
		2200 ppm (7.00 mg/liter)		Narcosis and growth reduction	
		4400 ppm (14.00 mg/liter)		Above symptoms amplified	
		6600 ppm (21.00 mg/liter)		Liver histopathology and bone marrow changes	
		9400 ppm (30.00 mg/liter)		Liver histopathology and bone marrow changes	
		1650 ppm (5.22 mg/liter)	6 hr/day × 5 day/wk × 12 wk	Leukopenia, partially reversed by phenobarbital	96
		8500 ppm (27.00 mg/liter)	6 hr/day × 10 days	Enzymatic changes in peripheral blood leukocytes, reduction of lymphocytes, granular lymphocytes	97
	Mouse	4680 ppm (14.80 mg/liter)	8 hr/day	Depletion of bone marrow colony-forming cells	98
	Rabbit	12,000 ppm (38.60 mg/liter)	1 hr/day × several days	Development of grayish white cornea, transient but persistent irritation	38
	Gpg	Vapor	6 hr/day × 5 wk	Increased blood serum aspartate and alanine aminotransferase activity, slightly decreased cholinesterase activity first week, decreased serum protein, increased prothrombin time	99
	Dog	500–800 ppm (1.6–2.55 mg/liter)	2–8 hr/day × 24–410 days	Base line benzene in blood increased from 5–6.5 times post-exposure	100

unchanged form (105). Inhaled benzene at a concentration of 0.340 mg/liter for 5 hr resulted in 33 to 65 percent body retention; 3.8 to 27.8 percent was exhaled unchanged through the lungs, and 0.1 to 0.2 percent excreted with body fluids. Retained benzene is converted 9.7 to 42 percent to urinary phenol, 0 to 5.4 percent to pyrocatechol, and 0.1 to 3.3 percent to hydroquinone (106), and via the hydroxy compound mainly to conjugated sulfates (106). Sato et al. (107) theorize that benzene distribution into tissues is high, owing to its cellular high water–fat partition coefficient.

Benzene intoxication, resembling that of gasoline (40) in the early acute stages, appears to be due primarily to CNS effects. The rate of recovery depends on the initial concentration and exposure time, but symptoms may persist for several weeks. The main signs of intoxication are manifested by drowsiness, dizziness, headache, vertigo, and delirium, and may proceed to loss of consciousness. Inhalation of a concentration of about 16 mg/liter may give a feeling of warmth (38). An acute first-time moderate exposure to benzene may produce typical prenarcosis syndrome, with headache, giddiness, and sometimes transient mild irritation of the respiratory and alimentary tract. A case is described by Moeschlin (40) where a workman experienced severe vomiting 6 hr after work; however, the symptoms subsided rapidly without aftereffects. Acute, high exposure may cause dyspnea inebriation with euphoria and tinnitus, which rapidly leads to typical deep anesthesia. If the victim is not treated at this stage, respiratory arrest rapidly ensues, often associated with muscular twitching and terminal convulsions (40). When inhaled by man, benzene has no effect at 25 ppm, but at 50 to 150 ppm produces headache, lassitude, and weariness, and at 500 ppm causes more exaggerated symptoms; 3000 ppm may be tolerated for 0.5 to 1.0 hr, 7500 ppm may result in toxic signs in 0.5 to 1.0 hr, and 20,000 ppm may be fatal if inhaled for 5 or 10 min (38). The latter facts have been substantiated by a number of accidental exposures with unknown air concentrations. However, in most cases mixtures of benzene and other hydrocarbons, mainly toluene, were involved. A case of sudden death has been described by Tauber (108), where a tank started to overflow in a light oil loading area. A combination of high air exposure concentrations of benzene and toluene, excitement due to the mishap, and running through the tank farm appeared to have contributed to the death of the worker. The benzene content was determined as 0.38 mg percent in the blood, 1.38 percent in the brain, and 0.26 percent in the liver. A sudden death occurred in a 16-year-old boy as a result of "sniffing" rubber cement containing benzene as a solvent (109, 110). A blood sample revealed 94 μg of benzene per 100 ml, and kidney tissue 0.55 mg/100 g. Similar "glue-sniffing" cases have been recorded from other U.S. sources and several European countries. In addition, a young laborer installing plastic tiles with liquid adhesives was found dead in the basement at the workplace and another worker unconscious nearby. The causes of the intoxication and death were attributed to fumes of benzene and toluene (111). Cause of deaths may have been a result of ventricular fibrillation, probably a result of myocardial sensitization to endogenous epinephrine (102). It is believed that the metabolism of benzene in man proceeds similarly to that observed in the rabbit, but at a more rapid rate than in the dog (38).

Cellular Effects. Acute erythromyelosis was diagnosed in a worker following his exposure to moderate benzene concentrations over a period of 30 years (62). Other effects of precursor cell hematopoiesis (98), peripheral blood and bone marrow changes (112), and autopsy reports with signs of hemorrhaging in the brain, pericardium, urinary tract, mucous membranes, and skin (102) have been described.

Acute Animal Experiments. Limited acute experiments with benzene have verified some of the above human findings.

Oral Administration. Oral LD_{50} values in the rat varied from 3.4 to 5.6 ml/kg, depending on the age or strain of the rat (41); see Table 47.3. Parke and Williams (43) investigated the fate of orally administered, radioactive benzene to rabbits. A dose of 150 to 350 mg/kg containing radioactive benzene at 2.5 to 11.0 μl was partially reexhaled and excreted, as shown in Table 47.3. Secondary to ingestion, aspiration into the lungs may occur. Gerarde (16) observed that direct instillation into the lungs of rats caused pulmonary edema and hemorrhaging at the site of contact. Cardiac arrest was observed when 0.25 ml of benzene was instilled into the rat lungs (16).

Eye Toxicology. Benzene in the rabbit eye is a moderate irritant, causes conjunctival irritation, and causes transient slight corneal injury (42).

Skin Effects and Permeability. When applied to the guinea pig skin, benzene elicits increased dermal permeability (113). Injected intracutaneously, it caused immediate local pallor and early ulceration (113). When benzene was injected subcutaneously, of 0.8 mg/kg in the adult rabbit, 25 percent was converted to urinary phenol, compared to only 12.5 percent in the young animal (114). The mouse epidermis, following 1 ml of benzene by subcutaneous injection, exhibited cellular injuries, such as nucleoplasmic vacuolation, condensation of nuclear chromatin, flattening of basal cells, and showed fewer organelles 24 hr postinjection. Recovery occurred within 2 weeks (115). Single subcutaneous doses of benzene decreased the iron (^{59}Fe) incorporation into circulating erythrocytes of the mouse; 440 mg/kg produced a 27 percent and 2200 mg/kg a 50 percent inhibition, but 88 mg/kg produced no effect (45).

Intraperitoneal Injection. The acute lethal doses for the rat, mouse, and guinea pig, when injected intraperitoneally (48, 50), are in the range of 0.5 to 1.0 g/kg;

see Table 47.3. Subsequently, 23 percent of 0.88 g/kg intraperitoneally administered benzene to the rat was observed to be excreted as phenol metabolites, 74 percent as sulfates, 9 percent as glucuronides, and 7 percent as free phenols (116). Intraperitoneally administered benzene induced reversible enzyme changes in the mouse liver within 6 hr (117). In the rat, benzene, gasoline, and related petroleum products elevated the liver but decreased the kidney alkaline phosphatase by 50 to 60 percent (118). Electronic measurements in the guinea pig revealed that 0.6 ml intraperitoneally injected benzene produced initial CNS excitation, then depression with muscle spasms and convulsions accompanying the hemolytic effect (119). In coho salmon, intraperitoneally injected benzene accumulated in the gallbladder, liver, brain, muscle, and in the carcass (120).

Inhalation of Benzene. In acute animal experimentation, adult rats and mice were more resistant to the effects of benzene than were young animals (121). The effects depend partly on the respiration rate and retention. The retention was 63 to 81 percent for the dog (122) and was independent of ventilation rates. The highest retention occurred at the lowest ventilatory rate. A LC_{50} value of 13,700 ppm has been recorded for the rat (123); see Table 47.3. Benzene inhaled by rabbits was absorbed at 37 and 54.5 percent (38). Rabbits exposed to concentrations of 35,000 to 45,000 ppm showed slight anesthesia in about 4 min and deaths in 22 to 71 min (55). Variations in time ranges until death were attributed to differences in tolerance and immunologic response of the laboratory animals (55). The inhalation of air saturated with benzene vapor resulted in ventricular extrasystole in the cat and the primate, with periods of ventricular tachycardia that occasionally terminated in ventricular fibrillation. In the mouse, 7 mg/liter produced narcosis, 15 mg/liter in 51 min, and 38 mg/liter in 8 min, with recovery. At 38 mg/liter, death occurred in 38 to 295 min, and at 77 mg/liter in about 50 min (38). In the rabbit, 4000 ppm produced narcosis, and more than 10,000 ppm proved fatal (54). At 35,000 to 45,000 ppm the effects were mainly on the central nervous system; anesthesia occurred in 3.7 min, with excitation and tremors in 5.0 min, and loss of pupil reflexes in 6.5 min (55). In the rabbit, sudden death from ventricular fibrillation has also been observed. Pharmacokinetic events include rapid release of adrenal hormones, epinephrine and norepinephrine, which then may sensitize the cardiac system, especially the myocardium, to the action of benzene (124). In acute inhalation by male rats, benzene-induced respiratory paralysis occurred, followed by ventricular fibrillation (125). In an elegant study, Peronnet (56) determined the benzene levels in blood of the guinea pig after an acute inhalation exposure. The author observed that different absorption and clearance kinetics prevailed for 20 and 50 mg/litter exposure levels. In 20 min, the blood level increased to 3.2 and 8.0 mg benzene/100 ml, respectively, and cleared to 1.9 and 5.1 mg/100 ml 5 min post-exposure. During the 20 min of the clearing period tested, the blood phenol level increased from 1.6 and 1.7 to 1.8 mg/100 ml, where it plateaued for both concentrations, practically independent of exposure concentration. However, these data also confirm that the CNS effects are due to benzene, rather than phenol. Histochemical studies revealed increased alkaline phosphatase in the spinal cord of the mouse, indicating disturbed neuronal transport characteristics (126). It appears that the central nervous system is more susceptible than kidney tissue to effects of benzene (127). The cellular action in acute exposures is overshadowed by CNS effects. However, Uyeki et al. (98) have demonstrated that precursor renewal systems are sensitive to inhaled benzene. An 8-hr inhalation experiment with mice at 4680 ppm and an in vitro cell culture showed significant depletion of bone marrow colony-forming cells.

Other Routes of Entry or Effects.   Benzene injected intravenously into rabbits at 0.1 ml/kg caused rapid CNS effects, intense hemolysis, and death (128). Das et al. (129), with subcutaneously injected benzene to guinea pigs, demonstrated the development of hematological changes, of leukocytosis, followed by leuko-penia and granulocytopenia in 6 to 8 days after exposure. Benzene, when inhaled or by other routes, can act as a potent sensitizer, skin or cardiac, even without epinephrine (130). Overshadowed by the intensive CNS effects, acute benzene intoxication does not appear to cause immediate cellular injury, except for early changes in liver endoplasmic reticulum (131).

Single Dose Metabolism.   In man and animals, 25 to 80 percent benzene from an acute exposure is partially reemitted in unchanged form through exhalation, depending on the species and the route of entry. Another portion is oxidized mainly to phenol, and in minor quantities to 1,2-di, 1,4-di, or 1,2,4-trihydroxy-benzenes, cresols, mercapturic, or muconic acids, and conjugated to the sulfate or glucuronide, depending on the nutritional state (116), route of entry (38), dose, or general pathway overload involved (132–134). Experimental young rabbits were shown to excrete mainly nonconjugated compounds (114). The conjugation appears to take place mainly in the liver and lung (51) but depends on the supply or induction of microsomal P-450 cytochrome and enzymes. Conversely, excessive phenol accumulation can inhibit microsomal activity by feedback inhibition (135). Since various hydrocarbons can induce microsomal enzymes (136), determinations such as serum or ornithine carbamyl transferase may represent an indicator for solvent exposure (137). Fasted animals show different pathways of metabolism, hydroxylation, and type of conjugations (114). High benzene concentrations increased the serotonin level in the rat brain (138). Another biologic indicator represents iron citrate, since its incor-poration into red blood cells is blocked by benzene (139). The combination of benzene and lead also inhibits both intact reticulocyte heme and protein synthesis (140). Some drugs, such as phenobarbital, stimulate microsomal enzyme production (141, 142), whereas toluene inhibits this activity and thus affects benzene oxidation (143). Subcutaneously injected into mice, 880 mg/kg toluene reversed the benzene-induced depression of iron uptake (144). Protection from intraperitoneal benzene toxicity to the rat occurred only when phenobarbital and benzene were injected simultaneously (123), but not with phenobarbital pretreatment for 3 days (123). This was confirmed by Timbrell and Mitchell (145), who found that phenobarbital pretreatment of rats increased the urinary metabolism but decreased respiratory benzene. Ascorbic acid appears to protect from intoxication by the facilitation of metabolic reactions (38).

Aquatic Studies.   Some TLm and LC_{50} studies listed in Table 47.3 reveal similar values for benzene and crude oil (57). Benzene was lethal the first and second day after exposure to a tidepool copepod *Tigriopus californicus* (58).

Human Chronic Benzene Exposure. *General.*   In the early 1930s, benzene was used extensively as a fast-drying solvent in a variety of trades. In the late 1930s, the hematopoietic potential effects of benzene were realized and benzene was controlled more strictly or subsequently replaced wherever feasible. Yet until recently its potential high inhalation exposure has been very common, also stemming from consumer product use where the exposure could not be controlled. Thus, in general, benzene intoxication stems from its use in improperly ventilated establishments. A summary in Table 47.4 illustrates some past high exposures with the chemical effects observed. However, diagnoses do not always agree, owing to variations in definitions and translations from assorted European languages, as, for example, the term leukemia. In addition, clinical determination procedures, such as chromosomal analyses, may represent 50 percent and more false positive results.

Inhalation.   Human chronic inhalation of benzene on a lower-order repeated basis appears to be related to a number of pathological conditions. The levels to produce effects vary widely with the individuals. Contributing factors are poor nutrition, certain immunologic tendencies, and consumption of alcohol or drugs. Some symptoms include headaches, dizziness, fatigue, anorexia, dyspnea, visual disturbances, and vague symptoms not connected usually with benzene poisoning. Signs also include fatigue, vertigo, pallor, visual disturbances, and loss of consciousness (106). Findings documented from occupational exposures have been recorded, such as membrane effects (59), hyperbilirubinemia (63), spleno- and adrenomegaly (146), blood dyscrasias with hemolytic effects (59), anemia (67), aplastic anemia (66, 68), hemocytoblastic and lymphatic involvement (61, 147) and reticulocytosis (63), leukopenia, pancytopenia, eosinophilia, basophilia and thrombocytopenia (65), monocytosis (69), and hyperplastic bone marrow effects (63). Also, an increase in chromosomal aberrations has been observed (66). There are three stages of involvement apparent: low, intermittent exposure resulting in very subtle hematopoietic changes; moderate to high exposures affecting enzyme synthesis, causing sensitizing effects and anemias; and high exposures producing irreversible blood dyscrasias. Further details have been presented in a summary and discussion by Goldstein (60).

Human Metabolism under Chronic Exposure. Benzene poisoning stems almost exclusively from its atmospheric occurrence, from where it is inhaled, absorbed, and translocated into the bloodstream if not emitted again with the exhaled air. It appears that with benzene in repeated administration metabolic reactions occur different from those observed in acute single exposure. It is apparent that repeated exposures to selected solvents such as benzene deplete a variety of protective enzyme systems. These include those active in iron incorporation or membrane transport, whereby cessation of action causes red blood cell hemolysis. This, in turn, leads to anemia. Another type concerns agglutination mechanisms and causes macrocell formation. Another influences cell mitosis and tubular aggregation, which also results in macromolecule formation or unlimited cell formation as, for example, with leukocytes. Benzene eventually also permeates bone marrow and depletes its red blood cell supply. Benzene per se is soluble in the body fluids to a very small degree only, but is rapidly deposited in any type of fatty tissue or absorbed through membranes, owing to its high lipid solubility. For example, the blood saturation equilibrium is reached at 2.1 mg/liter for each 100 ppm of benzene in the inhaled air (148). This appears to be a function of the fluid–air partition coefficient, which is 11.7 for blood and 5.5 for plasma, and the tissue–blood partition coefficient, which is 16.2 for bone marrow and 58.5 for fat (54). Ingested or inhaled benzene in part is reeliminated in unchanged form in the exhaled air or in the urine. However, the main portion in the human system is metabolized through a variety of major and minor pathways. The primary site of action is the liver, where benzene is oxidized to phenol (hydroxybenzene), catechol (1,2-dihydroxybenzene), or quinol (1,4-dihydroxybenzene). Phenol is subsequently conjugated with inorganic sulfate to phenylsulfate, the other hydroxybenzenes to a lesser extent, and all excreted in the urine (133). Minor pathways include further oxidation of catechol to hydroxyhydroquinol (1,2,4-trihydroxybenzene) or catabolism to *cis,cis-* or *trans,trans*-muconic acids, and phenol conjugation with glucuronic acid to form glucuronides, or with cysteine to produce 2-phenylmercapturic acid (133). Some Russian investigators determined that phenylsulfate peak excretion occurs about 4 to 8 hrs after exposure to benzene (149, 150). This has been substantiated with experiments using mice (151). If the metabolism is overwhelmed from excessive exposure concentrations, the excretion pathways via the glucuronides are used primarily (152). It appears that excessive benzene or phenol accelerates, but also inhibits, a variety of enzyme systems. Indeed, chronic exposure of various animal species to benzene has substantiated this point, as detailed in the next section. Pawelski et al. (153) observed that human granulocytic alkaline phosphatase was slightly reduced by benzene exposure. Benzene, owing to its affinity to lipids, may be stored and accumulated in human adipose tissue. According to Sato et al. (154), benzene from lipid storage is released much more slowly in the female than in the male.

Other Physiologic Effects.   The main effect, even in chronic benzene intoxication, appears to be the CNS involvement, which, however, may not be recognized owing to other visible effects. At low chronic exposure workers have exhibited signs of CNS lesions, abnormal caloric labyrinth irritability, and impairment of hearing (155). In combination with other solvents, workers revealed neurological syndromes of asthenoneurotic or astheno-vegetative polyneuritis with occasional neuronal progression, even after cessation of the exposure (156).

The possibility exists that benzene is a dermal sensitizer (157). Continued skin contact with benzene results in defatting of the skin which leads to erythema, dry scaling, and in some cases the formation of vesicular papules. Prolonged exposure may produce lesions resembling first- or second-degree burns. Systemic intoxication by cutaneous absorption is unlikely, owing to the overwhelming possibility of absorption by inhalation (106).

Intoxication at extremely high levels produces cardiac sensitization (158), a property not unique to benzene. Benzene has been noted to cause dyspnea and tachycardia (159). A later study revealed that benzene decreased arterial pressure and peripheral resistance, and caused diffuse-dystrophic myocardial alterations, apparently functional, which are reversible following termination of the exposure. The increased cardiac output, accelerated circulation rate, and greater potency of the precapillary bed are apparently of a compensatory–adaptive nature, to promote tissue oxygenation (160).

Benzene has appeared on "suspect carcinogen" lists; however, to date no confirming evidence has been found. Benzene is a suspected teratogen; however, only scant confirmative evidence is available. One study has revealed that benzene was determined at about the same concentration in maternal and cord blood (161), indicating that placental transfer does occur. For further details of metabolic events, see Reference 152.

Hematologic Effects.   Benzene's effect on the hematopoietic system and its unique myelotoxicity have been known for many years. In most cases the effects are not uniform; symptoms and laboratory findings are not parallel. However, generally, three stages can be diagnosed. Initially, there may occur blood-clotting defects, caused by functional, morphological, and quantitative platelet alteration (thrombocytosis), as well as generally reduced occurrence and production of all blood components (pancytopenia and aplastic anemia). At this stage, if diagnosed and treated, the effects are readily reversible. At a more advanced stage, the bone marrow becomes hyperplastic, then hypoplastic, the iron metabolism is disturbed, and internal hemorrhaging occurs. Sometimes the effects are associated with monocytosis or absolute lymphocytosis (69). At this stage, diagnosis and treatment should be prompt and intense, and all future exposure to benzene avoided. Indicative clinical findings are erythrocyte counts below 3.5 million, leukocytes below 4500, decreased platelet numbers, increased iron, and decreased transferrin. Unless treated, a third phase may be entered where bone marrow aplasia may become progressive (40). It is possible that bone marrow regeneration is excessively stressed through augmented destruction of peripheral cells, leading to final regenerative exhaustion, although individual variations in hematologic absolute values and continuous changes are extreme. For example, leukocyte counts may vary more during a life cycle than can be achieved through exposure to benzene (54). Over the years attempts have been made to correlate aplastic anemia and leukemia possibly caused by benzene or other solvent exposure. According to some Italian investigators, inhalation of benzene and other chemical agents may have caused observable hyporegenerative anemias or pancytopenias which, in some cases, may have evolved into myeloblastic (162, 163), granulocytic (164), or hemocytoblastic leukemia (68, 165). In France, Goguel et al. (166) in 1967 contemplated that benzene exposure, possibly in combination with other solvents, might have been the cause of 50 cases of leukemia, 13 of the myelocytic, and eight of the chronic lymphocytic diagnosed between 1950 and 1955.

Hematopoietic effects bordering on leukemia have been observed by Aksoy et al. in a number of cases in Turkey. Four shoemakers developed acute leukemia (167) from exposures to very high quantities of benzene and other adhesive components and solvents; one case of lymphoblastic and one of myeloblastic leukemia were observed (168). In the time frame 1967–1973, of 28,500 shoe workers, 31 were diagnosed with acute leukemia or preleukemia (169, 170). These incidents represent 13 cases per 100,000 workers, more than six cases of leukemia for 100,000 of the normal population (169, 170). According to Aksoy, however, the number of cases decreased in 1974–1975 and no new cases were reported in 1976 following the gradual substitution of benzene by gasoline (171). Gasoline in southeastern Europe, however, still contained a relatively high percentage of benzene. Aksoy believes that genetic and other extraneous factors may play an etiologic role, as observed in two families with the occurrence of leukemia in pairs of identical twins (172). This is substantiated by a case of familial chronic lymphocytic leukemia that occurred in a family that had worked in the dry-cleaning business for 20 to 30 years (173). Conversely, among 38,000 workers in the petroleum and petrochemical industries, the number of leukemia cases did not exceed the normal statistical occurrence (174).

Subcellular Effects. Chromosomal aberrations were found in peripheral blood lymphocytes of benzene workers (175) but not in personnel handling toluene (66). Khan and Khan (176) observed some chromosomal aberrations in bone marrow preparations. The aberrations were classified by chromatid and chromosome type, whereby both types were found to increase with years of exposure. Aberrations in some subjects were detected even several years after cessation of the benzene exposure (177). One female, following benzene exposure during pregnancy, showed an increase in chromosomal aberrations, particularly in group Gp-, but then delivered a normal, healthy boy and later a girl (177). In one case of granulocytic leukemia, after benzene and solvent exposure intermittently over 7 years, bone marrow showed 16 cells lacking one or more G-group chromosomes and 12 with one or two supernumerary C-group chromosomes (164). Another case with erythroleukemia showed inconsistent chromosomal changes (178).

Current analyses are complicated by preexisting familial chromosomal deviations and the natural changes that occur during normal life cycles (179). Therefore, at this time studies reveal some consistent but also inconsistent changes. Also, the direct mechanism involved in the formation of hematopathologically related chromosomal aberrations is unknown. However, according to Freedman (180), benzene interference on the molecular level appears to affect DNA replication, inhibition of hepatic polyribosomes, and inhibition of protein synthesis. More specifically, experiments with rabbits have demonstrated that benzene on the subcellular level interferes with the thymidine incorporation into bone marrow DNA (181). There exists the possibility that a very short-lived benzene metabolite could produce the subcellular effects, a benzene epoxide or so-called arene oxide (181). Such a compound may have the noted DNA and enzyme inhibitory properties, but is too unstable to be isolated and positively identified.

Immunologic Response.   It has long been suspected that benzene exposure may alter the immune response and general functions (54). Bone marrow functions appear to be closely related to immune mechanisms. Workers that had been exposed to benzene, isopropylbenzene, hydroperoxide, and phenol exhibited decreased phagocytic activity, diminished bactericidal activity of the skin, and lowered lysozyme and blood cholinesterase activity. Physiologically, they showed a tendency to hypotension, decreased pulse rate, and CNS changes (182). This was confirmed by the finding of decreased serum complement level in 62 of 79 workers exposed to benzene, toluene, and xylene (183).

Clinical Determinations.   Elevated urinary phenol levels normally indicate that benzene exposure occurred in the preceding 8 to 10 hr. Extensive salicylate, other drug intake, or certain foods may elevate the normal 30 mg/liter up to 75 mg/liter (184). A phenol excretion above 20 mg/liter following some benzene exposure is excessive, but not specific. An exposure of 25 to 30 ppm may raise the phenol excretion up to 100 or 200 mg/liter (39, 185). Conversely, with up to 5 ppm, no changes may occur. Ratio changes of urinary inorganic to organic sulfate excretion represents a further indicator. The normal value of inorganic total sulfate is about 86 percent (17), as determined by differential precipitation with barium salts (17). However, foods such as bananas, coffee, fish, and smoked meats can also alter the sulfate ratio (17). Exposures of 10 to 40 ppm benzene may lower the ratio to 72 percent, 40 to 75 ppm to 61 percent, 75 to 100 ppm to 43 percent, and 100 to 700 ppm to 38 percent (17). Early signs of benzene intoxication, but not diagnostic tools, are from phenol, sulfate, and direct blood changes. Better indicators are hematologic and cytological determinations on a continuous basis in order to detect progressive changes. Additionally, determination of Heinz bodies (186) and Feulgen's reaction in response to peripheral blood lymphocytes may be of value (187), as well as of serum replacement levels (183), exogenous colony formation (188), and bone marrow activity (80).

Unfortunately, clinical manifestations tend to be insidious at the time of onset. The earlier recorded cases were well advanced at the time of diagnosis and most complaints and visual signs did not correlate too well with the clinical picture, the characteristics, the exposure, or the prognosis. There is a possibility that benzene exposure can facilitate the onset of genetic leukemia. However, more epidemiologic, experimental, and statistical evidence is required to substantiate this point. Also, other chemical agents or drugs may induce similar signs and symptoms, as, for example, drug-induced thrombocytopenia (189) or agranulocytosis (190).

Prophylaxis and Follow-up. It appears that a variety of factors may predispose some individuals' sensitivity towards benzene effects. The young are more vulnerable; nutrition, genetics, and immunoresponse play a role. Vitamins aid in prophylaxis and treatment, such as vitamins of the B group, B_4, B_6, and B_{12}, and vitamin C (191). Recovery from moderate benzene exposure may take 1 to 4 weeks, whereby unsteady gait, nervous irritability, and breathlessness may persist for 2 weeks, and cardiac distress and the peculiar yellow color of the skin as long as a month, although evidence of chronic benzene intoxication still may be encountered later (106). Recovery from chronic high exposure may take from 2 months to several years (106). A Russian author comments on the hygienic effectiveness of the benzene and toluene levels in an industrial plant strictly controlled for 5 years. In 186 workers the nervous system changes returned to normal and urinary phenol levels were found at the base lines for 76 percent of the employees (192).

Benzene in Combination with Other Chemical Agents. Benzene, cyclohexane, and other chemical components used in the synthetic fiber manufacture exhibited early signs of nonspecific involuntary dysfunction in workers (193). Mixing benzene with toluene and gasoline appears to have additive effects as shown in young children exposed 2.5 to 10 years (194). The erythrocytes and thrombocytes were less affected than the leukocytes and circulating neutrophils. Similarly, glue sniffers were affected more severely with benzene, toluene, and gasoline mixtures, so that aside from asphyxiation, a number of deaths occurred (195). Definite immunologic effects were recognized in workers exposed to benzene, toluene, and xylene (196), but also periodontal problems (197). The effects of benzene and lead appear to exhibit additive inhibitory action on endogenous heme synthesis (180).

Chronic Experiments with Animals. *Oral Administration.* Animal experiments with oral dosing showed a no-effect level of 1 mg/kg in the rat and leukocyte changes with higher quantities, see Table 47.5. Simultaneous oral administration and inhalation exposure resulted in additive effects. Very high doses affected rat liver and protein fraction weights and decreased hepatic aminopyrine, *N*-demethylase, but increased acetanilide hydroxylase activity. Simultaneous administration of 9.0 mg/kg and an inhalation exposure of 6.63 mg/m^3 demonstrated sensitizing effects in the rat (75). This concentration might translate approximately into a 70-kg human exposure of 1500 ppm for 8 hr. In experiments with rabbits, Frash (86) has presented a model of "exhausted" bone marrow, with initial increases, then compensatory decreases of eosinophils, neutrophils, and myelocytes.

Dermal Application. Skin treatment of laboratory animals demonstrated that benzene can be absorbed to a moderate degree; see Table 47.5. Some papillomas and systemic hematopoietic effects were noted.

Subcutaneous Injections. Rats injected subcutaneously with benzene exhibited various hematopoietic and bone marrow effects, as shown in Table 47.5. The rat was more susceptible to benzene toxicity than the mouse and the rabbit, and the hamster least (198). Leukopenia has been recorded for the rat and the rabbit, increased DNA binding in the mouse, pancytopenia, and inhibition of DNA synthesis in the rabbit; also a significant decrease in granulocytes occurred, whereas toluene produced only transient granulocytopenia (199).

Experimental Inhalation. Young rats of 1.5 months when subjected to 1.6 ppm (5 mg/m^3) benzene exhibited a lower resistance than 4.5-month-old animals. At 9.5 ppm (0.03 mg/liter) the type of resistance was also independent of age (76). In general, continuous inhalation exposures exerted a more pronounced effect on the animals than did intermittent trials (95). The enzymatic activity of liver microsomes, and thus the metabolism of benzene, increased in rats exposed to benzene vapor for several weeks (200). Continuous inhalation of 4.8 ppm of benzene caused changes in the rat endoplasmic reticulum, associated with inflammation, as shown in Table 47.5. Histopathological, hematologic, cytological, and biochemical changes were induced. A distribution experiment in the dog showed that base line benzene increased, then plateaued from 2 to 8 hrs of exposure.

Metabolism.   Absorption, distribution, and metabolism of repeated benzene exposures proceed similarly in laboratory animals as observed for man, except the relative absorption appears higher in man as, for example, in the rabbit of 80 to 85 percent versus 37 to 55 percent at a concentration of 10 to 16 mg/liter (38). In a kinetic study (154) in the female and male rats with a large body fat content, benzene was eliminated more slowly and stored longer than in lean animals. Also, a decrease in white blood cells was observed only in rats with large body fat contents (154). Nutritional fats and paraffin oils stimulated the intestinal benzene absorption in the rat (201). Benzene distribution in the rabbit was highest in the adipose tissue, high for bone marrow, and lower for brain, heart, kidney, lung, and muscle (202), although direct binding was higher in the liver than in bone marrow (203). In the mouse, subcutaneously injected benzene formed mainly free phenol and phenylsulfate (203). Other metabolic changes from subcutaneous injections of benzene into the rabbit have been observed in the brain cortex and erythrocytes, in σ-aminolevulinic acid, porphobilinogen, and protoporphyrin (204). For hepatic benzene oxidation to phenol, involvement of a microsomal cytochrome P-450 iron complex is proposed (205). Hepatic microsomes are the main tools for benzene detoxification (136). A 4-hr benzene exposure at a concentration of 11.9 mg/liter enhanced hepatic microsomal cytochrome P-450 and NADPH oxidase activities in rats, while 35 mg/liter (the LC_{10}) inhibited the enzymes studied (206). This was accompanied by a decreased rate of hepatic hydroxylations, a drop of free SH groups, and reduced tissue oxygen tension (206). Pretreatment with phenobarbital in some cases protected the organism; in others it did not influence the rate of benzene detoxification (123, 207–209). So far, only benzene (80), but not its metabolites, has been demonstrated to interfere with myelogenic mechanisms (181). In mouse blood, benzene inhibits iron uptake into red cells (181).

Cytotoxic Effects.   Experimental work with cats has demonstrated that the primary action of benzene may be its capability to change vessel membrane permeability, and secondarily to affect the nutritional state of the cell and subcell, which also includes hydration and dehydration (210). Effects on the skin of the hairless mouse were signs of hyperplasia of the regenerative epithelium (78). Histochemical determinations indicated that benzene affected neuron enzymes of the mouse spinal cord (126) and pathomorphologically changed rat organ structures and the rabbit liver following injection of 0.05 g or 1 g/kg for 20 to 70 days (211). Degenerative muscle fibers were detected in the heart, bronchi, blood vessels, and bile duct in the rat, and degenerative and dystrophic changes in the central nervous system, cortex, thalamus, brain, and brainstem in the rabbit (211). Liver injury is restricted mainly to the endoplasmic reticulum (77, 131), as is true also in the renal cell, where it in addition decreased the number of mitochondria or caused their swelling (212).

Hematologic and Subcellular Effects.   Hematologic effects appear in female rats more rapidly than in the male (213). It induced the biosynthesis of porphyrins in rabbit erythrocytes (89). In the guinea pig, benzene initiated the destruction of erythrocytes in the spleen and lymph nodes, Kupffer cells in the liver, and enhanced phagocytic and hemolytic action (38). Chronic exposure of rats to benzene at 1 g/m^3, 4 hr daily for 6 months, caused an inhibition of the phagocytic activity of the leukocytes (214).

Subtle effects of decreased mitochondrial and lysosomal enzyme activities were observed in rabbits at 16 mg/liter, 6 hr/day for 14 days (215). No effect was detected at 10 mg/liter (215). The myelotoxic effect caused by chronic benzene exposure resembles that of ionizing radiation or exposure to radiomimetic substances (216). Both erythropoiesis and leukopoiesis of different cell lines are influenced (216). Various actions include hematopoiesis in the stem cells of the bone marrow and spleen and hypoplasia, which is a reduction of the hematopoietic stem cells (217). This is parallel to the reduction of erythrocytes, thrombocytes, and leukocytes in peripheral blood (217). Subcutaneous injection of 0.3 ml/kg to rabbits for 1 to 9 weeks was accompanied by pancytopenia and, in a portion of the animals, hypoplasia of the bone marrow (87). Rabbits subcutaneously injected with 0.2 ml/kg per day for 18 weeks exhibited induced peripheral pancytopenia and a high incidence of chromosomal aberrations, which persisted long past the hematologic recovery (218).

In vitro incubation of rabbit or human reticulocytes with iron transferrin at nonhemolytic concentrations of 56 to 113 mM resulted in cellular protein inhibition (219).

Sensitization.   Benzene appears to be a dermal sensitizer according to tests carried out with the female guinea pig (157).

Immunologic Response.   Experimental work with the rabbit indicates that benzene affects autoimmune processes leading to hemopathy (220).

Sex Differences, Teratology, and Reproduction.   In general, females or species with increased adipose tissue weight are more highly disposed to the effects of benzene intoxication. Benzene alone exerted no effect on fertility, conception, or embryonic development in rats exposed until days 17 to 19 of gestation. However, benzene, in combination with dichloroethane, for example, resulted in increased embryonic deaths and lower implantation rates (221). Mice, when injected subcutaneously with 3 ml/kg on days 11 to 15 of gestation, developed cleft palate, agnathia, or micrognathia (222).

Mutagenicity. Benzene injected subcutaneously into rats at 200 mg/kg once a day for 12 days produced some chromosomal changes in bone marrow. In the rat, some chromosomal aberrations, breaks, and gaps were noted (223). In the hairless mouse, reticulum cell neoplasms mounted to about 38 percent in the benzene-painted and the control animals (77). In the host-mediated assay, mutagenic reactions of benzene were negative (224).

Carcinogenicity. Benzene, when injected subcutaneously at 0.05 to 0.2 ml twice a week for 44 weeks, then once a week for 10 weeks, into weanling male C57Bl/6N mice, produced no carcinogenic effects during the 50-week observation period (85). Hairless mice painted twice weekly on the back skin up to 20 months yielded the same percentage of reticulum cell neoplasm as the control animals (77). Skin of white mice painted with benzene for 6 weeks showed ultrastructural changes in the dermis and epidermis, indicating nonspecific irritation of these tissues (225). However, these findings differed substantially from those produced by the carcinogen methylcholanthrene used as a positive control (225, 226). Conversely, when benzene is used as a solvent or carrier, the action of carcinogens such as epoxides is potentiated (227).

Metabolic Interactions. Some chemical or nutritional agents administered preceding, following, or in combination with benzene can have synergistic or antagonistic effects.

The primary antagonistic mechanism of the mammalian's system is the increase in hepatic microsomal production for the induction or promotion of hydroxylase activity in order to accelerate detoxification reactions. Phenobarbital does not affect the benzene LC_{50} in the female CD rat. It increases the hepatic phenol production from 0.84 to 4.43 nM/min per milligram of protein, but does not alter the lung microsomal phenol yield (123). Barbiturates partially reverse benzene-initiated decreased weight gains and lymphopenia in the rat; they also increase the rate of benzene metabolism up to tenfold. Intraperitoneally injected phenobarbital and, 4 days later, benzene, stimulated morphological, cytoenzymatic, and hepatic changes in the Wistar rat (207).

Ikeda and Ohtsuji found that sodium phenobarbital pretreatment increased hepatic hydroxylase activity three- to fourfold and increased tolerance to the leukopenic action of benzene (142). Toluene inhaled by rats several times a week increased the liver microsomal enzyme activity and the metabolism of benzene in vitro (200). Subcutaneous administration of 1720 mg/kg toluene reversed the benzene-induced depression of in vitro iron uptake into red cells and metabolic conversion in bone marrow (144). Toluene generally is a competitive inhibitor of the benzene metabolism in the rat (189), but also reduces the cytochrome C reductase activity (228). Human solvent intoxication usually concerns mixtures such as one case recorded by Moeschlin (40) where a black naphtha paint containing benzene, toluene, and xylene was used. After the paint was used for several weeks in nonventilated quarters, symptoms included headache, giddiness, loss of appetite, nausea, and vomiting. Clinical signs were hemoglobin, erythrocyte, leukocyte, lymphocyte, and thrombocyte changes, typical of benzene poisoning, which reversed after 2 years of treatment.

The action of benzene and lead for hematopoietic effects appears additive (140).

In addition, physical stress, such as vibration, can amplify the action of benzene or toxicants in general (229). Conversely, ascorbic acid, anhydrovitamin A, and retrovitamin A can counteract the toxic effects of benzene (230–232). Steroids, such as adrenocortical hormones, can favorably influence the action of benzene vapors (233). Iron sorbitol counteracted benzene toxicity by reversing DNA synthesis inhibitors in the rat and the rabbit (234), and protected against benzene when subcutaneously administered as sodium selenate by stomach tube to rats (235). A slow decrease of the blood benzene content was noted with the ingestion of 5 g cysteine HCl prior to benzene ingestion (236). When followed by ingestion of choline chloride and betaine aspartate, benzene levels in the blood were further reduced but toluene could be detected (236).

Toxicity to Marine and Plant Life.   When tested with a tide pool copepod, *Tigriopus californicus,* of a series of crude oil fractions, benzene was most lethal between days 1 and 2 (58).

Benzene is taken up by plants such as the avocado fruit (237) and grapes (238) and metabolizes it to carbon dioxide.

Biodegradation.   Various microorganisms, such as the *Pseudomonas* strains, can use benzene as their sole carbon source. The oxidative degradation is proposed to involve either the epoxide, with subsequent dehydrogenation, phenol, or peroxides as intermediates for the formation of catechol, which has been chromatographically identified (239). Benzene can also undergo ring opening with the aid of *Pseudomonas* and *Moraxella* species (240). In addition, benzene can be phytochemically degraded in reactions with ozone (24).

1.2.6   Industrial Hygiene

Threshold Air Concentrations.   In 1974, NIOSH (241) recommended lowering the previous threshold limit value (TLV) of 25 ppm to 10 ppm with an action level of 5 ppm. An emergency temporary new standard was published for a time-weighted average (TWA) of 1 ppm for an 8-hr period with a 5-ppm maximum peak for 15 min, proposed to be effective May 21, 1977 (241). NIOSH (242) has also published a monitoring method, S-311, for benzene, using charcoal tubes for absorption, carbon disulfide for desorption, sample separation, and quantitation by flame ionization gas chromatography. Earlier methods included benzene determination by a nitration procedure (100). The interferometer, the benzene indicator, and the aromatic hydrocarbon detector were used as field equipment for air analyses.

Benzene can be detected in the air at concentrations of 1.5 to 5 ppm (15, 39). Therefore, odor properties are inadequate for reliable benzene detection in air.

Biologic Monitoring. In biologic monitoring, quantitation is most important for benzene and other hydrocarbon solvent levels. Secondly, urinary phenol determinations are a valuable guide. Further, benzene in exhaled air presents an indication of blood benzene concentration and thus constitutes approximate exposure. Urinary phenol levels of 100 mg/liter represent an exposure of 200 ppm·hr or 8 hr of 25 ppm. Less than 20 to 25 mg/liter indicates normal phenol excretion values. Frequent determination of sulfate is advised; see the NIOSH criteria document (243). The current required regular physical examination should include blood pressure check, lung functions, blood chemistry, hematology, urinalysis, and skin examination (243). A variety of procedures are available for the determination of benzene and related compounds in the blood. The major method relies on extraction and gas chromatographic methods.

Customers or personnel occupationally exposed to benzene should be fully informed of the types and levels of hazards connected with the use of benzene-containing materials.

Precautionary Measures. A variety of procedures and devices are available for the removal of benzene from circulating air. Hands and face protection is required and respiratory equipment for levels above the threshold limits should be available. For a further review, see a discussion of criteria documents by Utidjian (244).

1.3 Toluene

1.3.1   Properties, Sources, and Use

Toluene is the lowest molecular weight member of the alkylbenzenes. Alkylbenzenes, in general, possess properties similar to those described for benzene, except they lack the benzene's hematopoietic characteristics. Alkyl substitution changes physical characteristics and, in turn, the absorption, partition, and metabolic properties.

The time required for hydroxylation, excretion, and sulfate ratio normalization increases with increasing doses and molecular weight of the side chain. The possibility of ring hydroxylation rises with increasing side chain size and probably utilization of alternate catabolic pathways (132).

Toluene, methylbenzene, methylbenzol, phenylmethane, $C_6H_5CH_3$, is a clear, colorless, noncorrosive, flammable (DOT) (22) liquid with a sweet, pungent, benzenelike odor. For further physicochemical properties, see Table 47.1.

Toluene has been measured in the urban air at 0.01 to 0.05 ppm (245), probably stemming from production facilities, automobile and coke oven emissions, gasoline evaporation (245), and cigarette smoke (12), and can occur in human respiratory air in smokers and nonsmokers (30). Toluene is found as a component of high-flash aromatic naphthas which are produced from crude oil by primary distillation. Toluene, in the past, has been recovered from by-products in the coal tar industry. Presently, toluene is produced from petroleum, and specifically from methylcyclohexane containing naphthas by the catalytic reforming process (5). Re-forming of n-heptane at 977°C yields about 62 percent toluene (36). Toluene is also a pyrolysis product of thermal cracking (36).

Toluene is used extensively as a solvent in the chemical, rubber, paint, and drug industries (102), as a thinner for inks (59), perfumes, dyes, and as a nonclinical thermometer liquid (246) and suspension solution for navigational instruments.

1.3.2 Physiologic Response

General Aspects. Toluene resembles benzene closely in its toxicological properties; however, it is devoid of benzene's chronic hematopoietic effects (see Tables 47.6 and 47.7). Findings of positive hematopoietic deviations in the past may have been caused by various degrees of benzene impurities in toluene. With toluene's higher volatility, their narcotic potencies are about equal, although by single application, toluene appears more highly toxic than benzene. Also, on skin and mucous membranes it exerts a stronger irritant action. Severe dermatitis may result from its drying and defatting action. Toluene is readily absorbed by inhalation, ingestion, and somewhat through skin contact. If ingested, toluene may present a lung aspiration hazard (16). By inhalation the no-effect range lies below 1250 ppm/4 hr for the guinea pig (257).

Human Acute Inhalation. The most rapid route of entry is through the pulmonary system. Respiratory elimination in volunteers directly reflected measured environmental solvent concentrations (105). Solvent retention, including toluene from cigarette smoke, was 86 to 96 percent (258). The blood desaturation curve was similar to that of benzene after inhalation of a concentration of 100 ppm. Volunteers exposed to low toluene concentrations exhibited transitory mild upper respiratory tract irritation at 200 ppm; mild eye irritation, lacrimation, and hilarity at 400 ppm; lassitude, hilarity, and slight nausea at 600 ppm; and rapid irritation, nasal mucous secretion, metallic taste, drowsiness, and impaired balance at 800 ppm (55).

Table 47.6. Physiologic Response to Toluene

Route	Species	Dose or Concentration	Results or Findings	Ref.
Inhalation	Human	100 ppm (0.38 mg/liter)	Psychological effects, transient irritation	247
		200 ppm (0.76 mg/liter)	Central nervous system effects	248
		400 ppm (1.52 mg/liter)	Mild eye irritation, lacrimation, hilarity	55
		600 ppm (2.3 mg/liter)	Lassitude, hilarity, slight nausea	55
		800 ppm (3.03 mg/liter)	Metallic taste, headache, lassitude, slight nausea	55
		100–1100 ppm (0.38–4.1 mg/liter)	Absence of illness, decreased erythrocytes (evaluated for 1940–1941)	249
Oral	Rat	2.5 g/kg	Lethal to ~30% of test animals	15
	Rat (14 day-old)	3.0 ml/kg	LD_{50}	41
	Rat (young adult)	6.4 ml/kg	LD_{50}	41
		7.0 g/kg	LD_{50}	42
		7.4 g/kg	LD_{50}	41
		7.53 ml/kg	LD_{50}	250
Inhalation	Rat	1700 ppm (6.4 mg/liter)/4 hr	Dose was tolerated	251
		4000 ppm (15.2 mg/liter)/4 hr	LC_{LO}	250
		8000 ppm (30.4 mg/liter)/4 hr	LC_{50}	52
		8800 ppm (35.0 mg/liter)/4 hr	LC_{50}	251
		15,000–25,000 ppm (40.0–66.5 mg/liter)/15–35 min	Lethal to 4 of 5 rats; blood, liver, and brain toluene concentration, 0.27, 0.64, and 0.87 mg/g, respectively	252
		15,000–25,000 ppm (40.0–66.5 mg/liter) + O_2/80–130 min	Lethal to 7 of 10 rats with O_2 supplied	252
		45,000–70,000 ppm (170–265 mg/liter)/2.9 min	Light anesthesia, relaxed	55
		45,000–70,000 ppm (170–265 mg/liter)/9.5 min	Pupillary contraction	55
		45,000–70,000 ppm (170–265 mg/liter)/14.8 min	Loss of blink reflex	55
		45,000–70,000 ppm (170–265 mg/liter)/16.1 min	Excitation, tremors, running movements	55
	Mouse	2650–3200 ppm (10–12 mg/liter)	Loss of reflexes	253
		8520 ppm (32.1 mg/liter)/8 hr	Lethal to 87.5%	53
		8000–9300 ppm (30–35 mg/liter)	Lethal to all test animals	253
	Cat	7800 ppm (31.0 mg/liter)/6 hr	CNS effect, mydriasis, mild tremors, prostration in 80 min, light anesthesia in 2 hr	251
	Dog	760 ppm (3.0 mg/liter)/6 hr	No signs of discomfort	251
Dermal	Rabbit	14 g/kg	LD_{50}	254
IP	Rat	800 mg/kg	LD_{50}	48
		1640 mg/kg	LD_{50}	7

High concentrations may result in paresthesia, disturbance of vision, dizziness, nausea, narcosis, and collapse, and one case of hematologic effects has been recorded (259). An employee was found unconscious after an exposure to high vapor concentrations for 18 hr. Tests indicated hepatic and renal involvement with myoglobinuria, with all effects reversible within 6 months (259). Some instant deaths have been recorded (109, 158) from "glue sniffing." It has been

Table 47.7. Chronic Animal Experiments—Toluene

Route	Species	Dose or Concentration	Results or Findings	Ref.
Oral	Rat (F)	118 mg/(kg)/(day) × 193 days	No effect	42
		354 mg/(kg)/(day) × 193 days	No effect	42
		590 mg/(kg)/(day) × 193 days	No effect	42
Dermal	Rabbit	Undiluted, 10–20 applications	Slight to moderate irritation, slight necrosis	42
Inhalation	Rat	7.7–255 ppm (0.03–1.0 mg/liter)	Reduced blood cholinesterase level	95
	Rat, guinea pig, dog, primate	107 ppm (0.39 mg/liter)/8 hr/day × 90–127 days	No histopathological or organ effects	91
		1085 ppm (4.10 mg/liter)/8 hr/day × 90–127 days	No histopathological or organ effects	91
	Rat, dog	245 ppm (0.95 mg/liter)/6 hr/day × 13 wk	No significantly different effects from the controls	251
		490 ppm (1.9 mg/liter)/6 hr/day × 13 wk	No significantly different effects from the controls	251
		1515 ppm (3.9 mg/liter)/6 hr/day × 13 wk	No significantly different effects from the controls	251
	Rat	390 ppm (1.0 mg/liter)/4 hr/day × 6 mo	Produced some inhibition of the phagocytic activity of leukocytes	214
	Rat (M)	1000 ppm (3.8 mg/liter)/8 hr/day × 4 wk	Increased adrenal weight and plasma hydrocorticoids; decreased eosinophiles	255
	Rat	6450 ppm (24.28 mg/liter)/5 hr/day × 4 mo	Decreased serum albumin, increased β- and γ-globulin and lipoprotein levels	256

determined that a glue-soaked cloth in a paper bag may create a toluene concentration from 200 ppm up to 50 times the TWA (260). Experiments with volunteers have shown that during exercise, the pulmonary toluene concentrations increase up to twofold (261, 262), whereas the mental productivity and reaction time decrease (263).

Human Chronic Exposure. *General Aspects.* Experience in the varnishing and painting industries has shown toluene to be a moderate toxicant. According to one case (264) it has proved to be a cardiac sensitizer and fatal cardiotoxin. It also causes hepatomegaly and bears hepatotoxic and nephrotoxic characteristics (39, 265). In several cases of habitual "glue sniffing," renal but also neural and especially cerebellar dystrophy occurred (263, 266, 267). Isolated cases with some cytochemical lymphocytic changes have been recorded (268, 269). Workers in a pharmaceutical plant in France exposed to toluene fumes developed leukopenia, and especially neutropenia. Within the following 6 months, those affected showed an increase in coagulation time and a decrease in the prothrombin level, with urinary hippuric acid excretion of 4′ g/liter (270). Normal values are around 0.6 g/liter (270). Other accompanying clinical changes were periodontal effects (197). Further cases have been described in the NIOSH Criteria Document of 1973 (271).

Skin and Inhalation Exposure. Prolonged contact of toluene with the skin may cause drying and defatting, leading to fissured dermatitis (102, 272). However, toluene is not a dermal sensitizer (272). Conversely, a 1940–1941 survey of 106 paint workers exposed to 100 to 1100 ppm toluene revealed the absence of illness. Subtle clinical findings were some cases of enlarged liver and decreased erythrocytes, with other hematologic tests normal (249).

Human Metabolism. Toluene, when inhaled, is retained 93 percent by the human system. This was demonstrated with toluene being a component of tobacco smoke (258). Absorption of toluene in the oral cavity was 29 percent (258). Both sources then contribute to the toluene's uptake into the vascular system (273), followed by distribution to various tissues and further metabolism (274). The uptake of toluene into the blood, apparently a function of the mutual solubility, was observed to proceed almost linearly, resulting in 0.55 mg/100 cm^3 from 200 ppm up to 2.23 mg/100 cm^3 stemming from an exposure to 800 ppm (15, 273).

Toluene in the human system is subsequently oxidized to benzoic acid and, in turn, conjugated with glycine to form hippuric acid, or with glucuronic acid to yield benzoylglucuronates. Both hippuric and benzoylglucuronic acid are excreted in the urine (133). Attempts have been made to correlate toluene blood levels with hippuric acid excretion rates. In a control group, the toluene base level was 15 μg/100 ml blood (275). However, in printing workers the hippuric acid excretion was not commensurate with the determined toluene air concentrations (275), but a somewhat better curve was described earlier by Von Oettingen (273). Data by Ogata et al. (276, 277), Elkins (278), and Ikeda and Ohtsuji (279) correlated well, whereby 3.5 g/liter represented an air concentration of 200 ppm. A biologic threshold concentration of 1100 mg/liter was proposed by DeRosa et al. (280) and >1 g/liter by Elkins (278). In general, urinary hippuric acid determination represents only an exposure approximation (281) and becomes increasingly reliable after exposures to 800 ppm and above (273). In painters, exposure to combinations of toluene, xylene, and glucuronates increased five times the normal hippuric acid concentration (149).

Another postulated but not isolated metabolite is toluene oxide (282), with the oxidation possibly potentiated by microsomal enzymes. Methylbenzenes have been shown to induce cytochrome P-450 in the Southern armyworm (283). Conversely, toluene inhibits mitochondrial oxidative phosphorylation (284).

Acute Animal Experiments. *Oral Toxicity.* With respect to acute oral experimental effects, toluene appears to be the least toxic of the alkylbenzenes (42), but is somewhat more toxic than benzene (38). For details, see Tables 47.6 and 47.7.

Inhalation. Inhaled toluene was retained 91 to 94 percent in the dog, with no direct respiratory rate dependence (122). Inhalation of 1500 to 2500 ppm toluene by the rat in 15 to 35 min resulted in recoveries of 0.27 mg in the blood, 0.64 in the liver, and 0.87 mg in the brain. When combined with oxygen, the stored quantities and also the toxicity tolerance were increased. Thus the above values increased to 0.43, 1.07, and 1.27 mg, respectively (252). These results were substantiated with a further study in 1975 (285).

Dermal and Eye.   Toluene has a marked irritant action on the skin. In the eye, it causes rapid and intense turbidity of the cornea and toxic effects on the conjunctiva (38). A dermal LD_{50} value of 14 g/kg was determined in the rabbit (254), indicating that toluene is practically not absorbed through the intact skin; see Table 47.6.

Intraperitoneal Injection.   The available LD_{50} values vary from 0.8 to 1.6 g/kg (7). Injection of 0.05 to 0.25 cm^3 in rats caused excitement, incoordination, clonic twitchings, and paralysis of hind legs (38). Higher doses in the guinea pig produced muscular twitching and narcosis, but little CNS irritation (38), and no clonic convulsions (119).

Acute Metabolism.   Toluene in the rabbit was readily taken up into body fluids. The partition coefficient for adipose tissue was higher than for any other organ tested (202). In the mouse, toluene also was distributed readily to adipose tissue, less to liver, kidney, and cerebrum (286). Metabolites from toluene in the rat included *o*- and *p*-, but not *m*-cresol (134), some benzyl alcohol, and hippuric acid (134).

Toxicity to Marine Life.   An aquatic threshold toxicity determination, using goldfish, was 23 ppm for toluene (287). Toluene was less toxic to the coho salmon (288) than benzene and xylene. Toluene was absorbed by the mussel, but to a lesser degree than the comparable C_7 paraffins (289).

Chronic Animal Experiments

Oral Tests.   Repeated oral feeding of toluene to female rats at 118, 354, and 590 mg/kg per day for 193 days produced no physiological changes (42); see Table 47.7.

Dermal Effects.   Repeated application of undiluted toluene to rabbit skin caused slight to moderate irritation and slight necrosis (42). Subcutaneous injection produced transient granulocytopenia in the rabbit (199).

Inhalation.   Long-term toluene inhalation in some experiments at 100 to 1000 ppm produced essentially no effects (91, 251); other authors found subtle cytological and clinical changes. Behavioral symptoms consisting of circling by rats have been reported by Ishikawa and Schmidt (290), and decreased learning capabilities have been observed by Ikeda and Miyake (291).

Other Effects and Sensitization.   Toluene, when administered subcutaneously to rats at 0.8 g/kg daily for 12 days, has been shown to stimulate neutropoiesis (81). Toluene caused transitory reduction in erythrocytes in the guinea pig and in the rabbit primary marked leukopenia, followed by leukocytosis (38). A dose of 300 mg/kg per day when administered subcutaneously produced aplastic anemia in 1 to 9 weeks in the rabbit (292).

According to Reinhardt et al. (158), toluene is a cardiac sensitizer, which caused "sudden death" from habitual sniffing of glue. The cardiotoxic effect of toluene appears somewhat lower than that of benzene (125).

Repeated Dose Metabolism.   The metabolic reactions appear to proceed from chronic exposures to toluene as those found in acute investigations. Although tissue distribution appears more extensive, as shown in the dog at 7.5 to 10 ml/14 hr per day for 6 months (15), the highest quantities were observed in liver, brain, bone marrow, cerebellum, and adrenals (15). Rats studied under adaptive inhalation at 186.6 mg/m^3 for 2 weeks to 6 months revealed increased peroxidase action and mobility of the thermoregulatory system but decreased catalase activity (293). Cellular and subcellular changes appear to be reversible contrary to those caused by benzene, except some chromosome gaps, chromatid and isochromatid breaks, which have been recorded for rat bone marrow cells (79).

Chemical–Biologic Interaction.   Certain combinations of toluene with other chemicals result in potentiated action over additive effects of the components when used alone. Other types of mixtures induce mutual inhibition of its individual biologic effect. Example pairs for potentiation are benzene and toluene (66, 294), in vivo and in vitro (295). Toluene with asphalt fumes (296) or chlorinated hydrocarbons, such as trichloroethylene and tetrachloroethene (297, 298), causes depression of hydroxylation and urinary conjugate excretions. However, a certain quantity of toluene administered to mice reversed specifically the iron uptake and benzene metabolite accumulation, without altering the bone marrow benzene level (144).

The combined exposure to toluene, xylene, formaldehyde, and aniline dyes decreased granulocytic alkaline phosphatase activity (153). Conversely, hydrogen peroxide administered to rats increased the blood peroxide activity and increased the tolerance to toluene (293), as did phenobarbital in specific quantities (142).

Effects on Plants.   Toluene is effectively metabolized by a variety of plants (299), such as grapes (239) and avocados (238).

Microbiologic Aspects.   Toluene exerts limited bacteriostatic, but pronounced fungistatic, effects toward some organisms and can reduce drastically *Actinomycetes* populations (300). It also possesses anthelmintic properties (301). Conversely, bacterial strains isolated from soil that closely resemble *Pseudomonas desmolytica* can utilize toluene as their sole carbon source (302), as can *Nocardia* cultures (303), *P. aeruginosa*, and *P. oleovorans* (304). The latter microorganisms require supplemental protein for maximal hydroxylation (304, 305). Toluene is biodegraded primarily through side chain hydroxylation (306).

1.3.3 Industrial Hygiene and Occupational Medical Aspects

General. Toluene is a popular solvent in the painting, varnishing, and adhesive industries, where in the past it has been measured in air at concentrations above the set threshold limits for many operations (278).

Threshold Limits. NIOSH (271) has recommended a TWA of 100 ppm (375 mg/m^3) for an 8-hr workday, with a ceiling of 200 ppm (750 mg/m^3) for a 10-min sampling period (307). For further details, see the discussion of the criteria document by Utidjian (308) and Weaver (309).

A TWA based on human experimental data suggests air concentration of up to 480 ppm (1.9 mg/liter) to represent no-effect levels (251). An odor threshold of 2.5 ppm (10 mg/m^3) has been determined by a human panel (251). An air sampling procedure, S-343, has been published by NIOSH (242); however, the listed TLVs are superseded.

Analytic Determinations. A method has been published by Elkins (278) using isooctane as a toluene desorbant and spectrophotometric sample determination at 268 nm. Newer methods suggest gas chromatographic analysis (310), either of air grab samples or of carbon disulfide desorbants from charcoal (311). A technique for the determination in water is available (312). Selected methods are available for the determination of toluene in blood (313, 314), other body fluids (314), and tissues (315). Methods also are available for the determination of urinary metabolites, such as hippuric acid, either as conjugate or the hydrolyzed components (276, 316–319).

Medical Aspects. Comprehensive preplacement and biennial medical examinations should be provided to all workers exposed to toluene. Laboratory tests should include hematology and urine analyses. For further details, see the criteria document on toluene (271). Exposed workers should be checked immediately. The average urinary biologic threshold of hippuric acid is 1100 mg/liter (280). A high protein diet is advised as a prophylactic measure (320); see Human Metabolism in Section 1.3.2 above.

Precautionary Measures. Eyes and skin should be protected when toluene is used in liquid form. Skin penetration of toluene may be prevented by using gloves or protective skin creams (321). Respiratory equipment should be provided and used above the TLVs. Safety and handling instructions should be followed, as provided; see also MCA SD-63 (322).

A variety of methods and procedures are available for the removal of toluene from gaseous or liquid waste media.

1.4 Xylenes

1.4.1 General Aspect and Physical Properties

The xylenes, xylols, or dimethylbenzenes, occur in three isomeric forms, namely as the o-, m-, or p-xylene, or the 1,2-, the 1,3-, and the 1,4-dimethylbenzene, respectively.

Xylene occurs in many petroleum products, in coal naphthas, and as an impurity in petrochemicals, such as benzene, toluene, and similar materials. o-, m-, and p-Xylene have been identified among the volatile products in tobacco smoke (12). m- and p-Xylene have been detected also in particulate matter samples of Houston community air (33).

The three isomers possess similar properties. They also are commercially available separated or mixed as colorless, flammable liquids (DOT) (22) (17, 323); see Table 47.1. Not listed are the evaporation rates of 9.2, 9.2, and 9.9 for the ortho, meta, and para forms, respectively, relative to ether (ether = 1) (9). Xylene readily dissolves fats, oils, and waxes. For further physicochemical and biological properties, see Tables 47.1, and 47.8, and 47.9.

1.4.2 Production and Use

The production and handling of xylene in refineries belong to the standard block units, BTX benzene–toluene–xylene (36). They are produced by catalytic reforming and, depending on the feedstock, xylene yields of above 85 percent can be achieved (36). Commercially, xylene is recovered also from coal tar, yielding a typical mixture of about 10 to 20 percent ortho, 40 to 70 percent meta, and 10 to 25 percent paraisomer, but also may contain 6 to 10 percent ethylbenzene, benzene, toluene, trimethylbenzene, phenol, thiophene, pyridine, and nonaromatic hydrocarbons (17, 323). Purification can be achieved by fractional distillation.

The xylenes are used widely as thinners (59), solvents for inks, rubber, gums, resins, adhesives, and lacquers, as paint removers, in the paper coating industry (278), as solvents and emulsifiers for agricultural products (283, 339), as fuel components (37, 340, 341), and commonly in the chemical industry as intermediates (342). Xylene is utilized widely to replace benzene, especially as a solvent. Specifically, the o-xylene serves as a raw material for the production of

Table 47.8. Effects of Xylene on Humans

Material	Route of Entry	Species (Number)	Dose or Concentration	Effects or Results	Ref.
Xylene	Eye	Human	460 ppm (1980 mg/m³)	Irritant to 4 of 6 subjects	325
	Dermal	Human	Undiluted	Burning effect, also drying and defatting of the skin	102, 324
	Inhalation (acute)	Human (6)	1 ppm (4.3 mg/m³)	Odor threshold	325
		(6)	40 ppm (17.2 mg/m³)	Identification threshold	325
		(10)	100 ppm (431.0 mg/m³)	Satisfactory for occupational 8-hr exposure	326
		(10)	200 ppm (860 mg/m³) 3–5 min	Irritant to eyes, nose, throat	326
			110–460 ppm (472–1980 mg/m³)	Irritant to eyes, nose, throat	325
		(8)	Unknown	Respiratory irritation to 6 of 8 workers, with clinical signs	327
		(3)	~10,000 ppm (~43.1 mg/liter) 18.5 hr	One death, lung congestion, brain hemorrhage, 2 workers unconscious 19–24 hr, retrograde amnesia and some renal effects, one hypothermia and lung congestion	328
	Inhalation (chronic)	Human	59 ppm (0.254 mg/liter)	Hippuric acid 1998 ± 1197 mg/liter	329
			93 ppm (0.40 mg/liter)	Hippuric acid 1812 ± 1275 mg/liter	329
			256 ppm (1.10 mg/liter)	Hippuric acid 3821 ± 1113 mg/liter	329
			398 ppm (1.71 mg/liter)	Hippuric acid 5500 ± 1690 mg/liter	329

Table 47.9. Acute Animal Experiments with Xylene

Material	Route of Entry	Species	Dose or Concentration	Results or Effects	Ref.
o-Xylene	Oral	Rat	2.5 ml/kg	Lethal to 7 of 10 animals	15
m-Xylene		Rat	2.5 ml/kg	Lethal to 3 of 10 animals	15
p-Xylene		Rat	2.5 ml/kg	Lethal to 6 of 10 animals	15
Xylene	Oral	Rat	4.3 g/kg	LD$_{50}$	42
			10.0 ml/kg	LD$_{50}$	330
	Eye	Rabbit	13.8 mg	Turbidity and irritation of the conjunctiva, lacrimation, edema	38
		Cat	Undiluted	Vacuoles in the cornea resembling "polishers' keratitis"	17
		Mouse	1.80 ml/kg	LD$_{50}$	331
o-Xylene	Inhalation	Rat	6350 ppm (27.4 mg/liter)/4 hr	LC$_{50}$	330
			6700 ppm (29 mg/liter)/4 hr	LC$_{50}$	325
		Cat	9500 ppm (41 mg/liter)/2 hr	LC$_{100}$, with typical CNS effects	325
		Mouse	3500–4600 ppm (15–20 mg/liter)	Produces narcosis	253
		Rat	6125 ppm (26.3 mg/liter)/12 hr	Lethal dose	332
		Mouse	6920 ppm (30 mg/liter)	Lethal dose	148
m-Xylene	Inhalation (acute)	Rat	8000 ppm (34.5 mg/liter)/4 hr	Lethal dose	333

Table 47.9. *(Continued)*

Material	Route of Entry	Species	Dose or Concentration	Results or Effects	Ref.
		Mouse	2010 ppm (8.7 mg/liter)/2 hr	Lethal dose	332
			2300–3500 ppm (10–25 mg/liter)	LT$_{LO}$, narcosis	148
			11,500 ppm (50 mg/liter)	LC$_{100}$	148, 253
p-Xylene		Rat	4912 ppm (21.1 mg/liter)/24 hr	Lethal	332
		Mouse	2300 ppm (10 mg/liter)	Produces narcosis	253
			3500–8100 ppm (15–35 mg/liter)	LC$_{100}$	148, 253
Xylene	Eye (chronic)	Cat	Undiluted	Corneal vacuoles	334
		Rabbit	Undiluted	No effect	325
Xylene	Dermal	Rabbit	Undiluted	Moderate to marked irritation moderate necrosis	42
	SC	Rabbit	300 mg/(kg)/(day) × 6 wk	No myelotoxic effects	335
			700 mg/(kg)/(day) × 9 wk	No myelotoxic effects	335
		Guinea pig	1–2 ml/(kg)/(day) × 10 days	No effects except slight reduction in red blood cells without affecting white blood cells	38
	Inhalation	Mouse	11.5 ppm (0.05 mg/liter) 4 hr/day × 12 mo	Hematologic and immunologic changes with eventual decomposition	336
			46.4 ppm (0.20 mg/liter) 2 hr/day × 12 mo	Hematologic and immunologic changes with eventual decomposition	336
		Rat, guinea pig, primate	78 ppm (0.337 mg/liter) 8 hr/day × 5 day/wk × 90 exp	Essentially no hematologic effects	91
Xylene	Inhalation	Rat, beagle	180 ppm (0.77 mg/liter) 6 hr/day × 5 day/wk × 13 wk	No statistically significant effects	323
		Guinea pig	300 ppm (1.3 mg/liter) 4 hr/day × 6 day/wk × 64 exp.	At necropsy some liver degeneration and inflammation of the lungs	257
		Rat, guinea pig, primate	780 ppm (3.36 mg/liter) 8 hr/day × 5 day/wk × 30 exp.	Essentially no hematologic effects	91
		Rat, beagle	460 ppm (2.0 mg/liter) 6 hr/day × 5 day/wk × 13 wk	No statistically significant effects	323
		Rat, rabbit	700 ppm (3.0 mg/liter) 8 hr/day × 6 day/wk × 130 day	No significant erythro- and thrombocyte changes, slight reduction of leukocytes	337
		Rat, beagle	810 ppm (3.5 mg/liter) 6 hr/day × 5 day/wk × 13 wk	No statistically significant effects	323
		Rabbit	1150 ppm (5.0 mg/liter) 8 hr/day × 6 day/wk × 55 days	Subtle reduction of erythro-, leuko-, and thrombocytes, some bone marrow hyperplasia without structural changes	337
			2300 ppm (10.0 mg/liter) 6 hr/day × 32 days (180 hr)	Subtle reduction of erythro- and lymphocytes, increase in leukocytes, death due to pneumonia	338
		Cat	2300 ppm (10.0 mg/liter) 6 hr/day × 9 days (46.5 hr)	Subtle reduction of erythro- and leukocytes also monocytes, death due to pneumonia	338

3293

plasticizers, alkyd resins, glass-enforced polyesters; the paraderivative for polyester fibers and films; and the metaisomer to produce isophthalic acid, polyester, and alkyd resins (36).

1.4.3 Physiologic Response

General Aspects. The majority of the experimental work indicates that xylene is more highly toxic than toluene; others indicate that toluene may be more toxic; see Tables 47.8 and 47.9. This discrepancy may be due to the existence of a dose–activity relationship with an inversion point. At low doses, toluene, and high doses, xylene, may be more toxic. Also, the fact that benzene occurs as an impurity has to be considered, and may render historic early toxicity determinations unreliable (148).

Human Acute Exposure. Ingestion of xylene will cause severe gastrointestinal distress. If aspiration into the lungs occurs, chemical pneumonitis, pulmonary edema, and hemorrhage ensue (17). One case has been reviewed by Ghislandi and Fabiani (343) where ingestion of small quantities of xylene produced urinary dextrose and urobilinogen excretion, with toxic hepatitis, which was reversible in 20 days.

Overall, xylene appears to be more highly toxic than benzene or toluene (17). A concentration of 460 ppm was classified by a human panel as an irritant (325); see Table 47.8. Conjunctivitis and corneal burns have been reported following direct eye contact with xylene (17).

Xylene, in direct contact with the intact skin, is an irritant and causes defatting, which may lead to dryness, cracking, blistering, or dermatitis (38, 102).

When inhaled at high concentrations, human acute signs may include a flushing and reddening of the face and a feeling of increased body heat, owing to the dilation of the superficial blood vessels (17). In addition, disturbed vision, dizziness, tremors, salivation, cardiac stress, CNS depression, confusion, and coma (344), as well as respiratory difficulties may be apparent (324). Xylene may occur as an air pollutant. As such, it has been classified as a cilia toxin and mucus coagulating agent (345). One death has been ascribed to the misuse of a shampoo–solvent mixture (195). Crooke classifies a variety of solvents, among them xylene, when inhaled at high concentrations, as causing instant death (195). This may be due to sensitization of the myocardium to epinephrine (158), so that even endogenous hormones precipitate sudden and fatal ventricular fibrillation (124), or respiratory arrest, and consequent asphyxia (53).

Inhalation of a lower concentration of waste vapors from epoxy resin–concrete disposal containing xylene is recorded to have caused upper respiratory irritation in six of eight workers (327). Clinical findings were temporary albuminuria, microhematuria, and pyuria (327). An acute exposure to paint, xylene thinner, and traces of toluene of three workers while painting in a tank caused one death with lung congestion, focal intra-alveolar hemorrhage, acute pulmonary edema, and hepatic, anoxic, and neuronal damage. The other two workers were found unconscious after 18.5 hr, but regained consciousness 1 and 5 hr post emergency treatment. Clinical findings included hepatic and renal impairment, with significant hypothermia in one case (328). In general, female workers appear more susceptible to solvent effects, such as produced by xylene (346).

Human Chronic Exposure. Signs and symptoms from chronic exposure resemble those from acute mishaps, but are in part systemically more severe.

Repeated, prolonged exposure to fumes may produce conjunctivitis of the eye and dryness of the nose, throat, and skin (102). Direct liquid contact may result in flaky or moderate dermatitis.

Inhalation of vapors may cause CNS excitation, then depression, characterized by paresthesia, tremors, apprehension, impaired memory, weakness, nervous irritation, vertigo, headache, anorexia, nausea, and flatulence (347, 348), and may lead to anemia and mucosal hemorrhage (339). Clinically, no bone marrow aplasia, but hyperplasia (38), moderate liver enlargement, necrosis, and nephrosis may occur (38, 348). Occupationally, a hazard exists where xylene and such solvents are used in nonventilated areas.

In a hematology laboratory where technicians were exposed to xylene, air concentrations measured 21 to 120 ppm (349). Hippuric acid, phenol, glucuronic acid, and uric acid determinations in groups of ship painters correlated with xylene exposure concentrations; see Table 47.8. Uric acid was decreased, the other parameters elevated (329). A variety of solvents in the intaglio printing industry, among them xylene, are thought to be related to the occurrence of anemia (350). Lymphocytosis and aplastic anemias possibly stem from the benzene contamination of the solvents. During several days of painting a water tank with an agent containing 65 percent xylene and 35 percent benzene, half of the work force experienced nausea and excreted red to coffee-brown urine. Almost all complained of headaches, loss of appetite, and extreme fatigue, and one death occurred (351). At present, when used industrially, commercially, or in the home, the most harmful effect on the health of the user occurs in spray painting. According to Kucera (352) through the NIOSH criteria document on xylene, in Czechoslovakia from 1959 to 1966, of 1.5×10^6 live and stillborn infants, more than 20,000 were malformed, nine with caudal regression syndrome. Five of the nine mothers had been exposed during pregnancy to xylene vapors (352). In 11 paired cord blood samples gas chromatographically analyzed, it was confirmed that xylene passed the placenta, since the quantitative determinations corresponded to that detected in the maternal blood (161).

Human Absorption and Metabolism. Xylene, when ingested, is readily absorbed by the human system, as has been shown in accidental ingestions. Absorption through the intact and broken skin occurs readily. Experimentally, m-xylene was absorbed by healthy human subjects with one or both hands immersed, at an approximate rate of 2 $\mu g/cm^2$ per minute (353), which was confirmed by Lauwerys et al. (354). Immersion of both hands in m-xylene for 15 min equals an estimated pulmonary retention at 100 ppm (353). Percutaneous exposure to 600 ppm xylene vapor for 3.5 hr corresponded to an equally long inhalation exposure of less than 10 ppm (355).

Xylene is absorbed mainly through the mucous membranes and pulmonary system (38). Pulmonary retention of vapors in human subjects amounted to about 64 percent for all participants and xylene isomers tested (356). Absorbed xylene is translocated through the vascular system, but to a lesser degree than benzene. Generally, the xylenes are metabolized to the corresponding o-, m-, or p-toluic acids ($CH_3C_6H_4COOH$) (133), and excreted in urine free or conjugated with glycine as methylhippuric acid ($CH_3C_6H_4COONHCH_2COOH$). Normally, urinary phenol is not elevated (348). Xylene absorbed through the skin was 80 to 90 percent eliminated as methylhippuric acid (353). This was confirmed by Lauwerys et al. (354), who showed that m-xylene absorbed through the skin was excreted as m-methylhippuric acid. Application of barrier creams did not significantly influence the rate of absorption (354). Inhaled m- and p-xylene by human volunteers were excreted as urinary m- and p-methylhippuric acid (277). A linear relationship was found between atmospheric xylene concentration and excreted toluic acid (357). Of an equal isomeric mixture, 63.6 percent was retained by humans, 95 percent metabolized, and 5 percent eliminated through the lungs (358). Volunteers aged 17 to 33 years exposed to 100, 300, and 600 ppm showed that the vapor retained in the lungs tended to decrease at the end of an exposure (359). Alanine aminotransferase increased and serum cholinesterase activity decreased in workers occupationally exposed to xylene and other organic solvent mixtures (360).

Acute Animal Experiments. According to Wolf et al. (42), xylene by single-dose administration at 4.3 g/kg in the rat is of higher toxicity than toluene, but only slightly more than benzene. An oral LD_{50} of about 10 ml/kg was determined for the three xylenes mixed with ethylbenzene (330); see Tables 47.9. Human visual disturbances can be compared with the exaggerated rotary and positional nystagmus observed in the rabbit. The animal responded to a lipid infusion which produced a xylene blood level of 30 ppm (361). In the rabbit eye, 13.8 mg xylene produced marked turbidity and severe irritation of the conjunctiva, with lacrimation and edema (38). Wolf et al. (42) document slight conjunctival irritation and very slight transient corneal injury in the rabbit eye.

Xylene has been found to increase the dermal permeability (113, 362) and thus the water content. Subcutaneously injected xylene was lethal to the rat and the mouse (363). For further information, refer to the criteria document on xylene; see Section 1.1.4. Xylene was less toxic than benzene when injected intraperitoneally (364). An intraperitoneal LD_{50} was 1.8 ml/kg in the mouse (331). In the guinea pig, 2.0 g/kg was lethal to three of four animals (137). Intraperitoneally injected xylene at 0.4 and 0.6 ml/kg initially increased, then decreased the irritancy threshold dose-related in the guinea pig (119).

When inhaled, 450 ppm xylene was very toxic to the guinea pig, but 300 ppm was tolerated for some time without harm (38). Acute experiments with the rat and the mouse demonstrated the variation between the three xylene isomers; see Table 47.9.

In fish, an aquatic TLm96 value was 100 to 10 ppm for mixed o- and p-xylene (365), but for mixed xylene a more precise LC_{50} was established as 17 ppm (287). Rainbow trout exposed to 7.1 ppm xylene for 2 hr survived, but suffered 100 percent mortality at 16.1 ppm. A 0.36 ppm exposure for 56 days yielded directly proportioned analytic residues (366).

Subacute and Chronic Animal Experiments. Corneal vacuoles observed in humans, upon repeated exposure to xylene vapor, were also shown in the cat (334), but not in the rabbit (325). Repeated application of undiluted xylene to the rabbit skin produced moderate to marked irritation and moderate necrosis (42). In rabbits injected subcutaneously, Speck and Moeschlin (335) found no myelotoxic effects; see Table 47.9. Erythrocytes, reticulocytes, leukocytes, and thrombocytes, tested twice per week, were all within normal limits and no DNA effects were noted (335). In earlier experiments, German investigators had demonstrated various hematologic and myelotoxic effects. These are now attributed to possible benzene and other alkylbenzene impurities. Rabbits injected intraperitoneally with 1 ml/day xylene for 3 days showed no effects (363). Early subacute and chronic inhalation experiments also demonstrated more drastic hematologic changes than some recent investigations; see Table 47.9. Fabre et al. (337) confirmed in 1960 subtle changes in the erythrocytes but not leukocytes in the exposed rat and the rabbit at 3 mg/liter. They also observed slight reductions in erythro-, leuko-, and thrombocytes, and some bone marrow hyperplasia at 5 mg/liter (337). Parts of the blood picture were also confirmed by Engelhardt (338). Several investigators mention that in contrast to benzene and other hydrocarbon exposures, xylene and toluene administration produces highly vicious laboratory animals. Kashin et al. (336) found in mice exposed to 0.05 mg/liter, 4 hr/day, a more highly significant three-phasic effect than at 0.2 mg/liter for 2 hr/day. Period 1, from 1 to 3 months, showed subcellular changes with decreased immunobiologic response, decompensation in phase 2, and normalization in phase 3 (336). Conversely, a study is reported where 158 to 1580 ppm (0.015 to 1.0 mg/liter) inhaled continuously by rats reduced the blood cholinesterase level, altered the reversible chromaxy ratio of muscle antagonists, and produced leukopenia or leukocytosis

(95). Another exposure of rats to 1 mg/liter of *m*-xylene, 4 hr/day for 6 months, caused an inhibition of the phagocytic activity of leukocytes (214). Several investigations confirm that xylene exerts little or no teratogenic action on the rat or chicken by implant (367). However, one report indicates a significant increase in chick malformations (352).

Metabolism. Xylene was absorbed rapidly by the rabbit and partitioned between the blood and mainly tissue with the highest lipid content (202). It has been known since 1867 that ingested xylene in the dog was excreted as the toluic acid–glycerine conjugate, methylhippuric acid (368). Further, Bray et al. (369) demonstrated that toluic acids, orally administered at 1 g/kg to the rabbit was excreted in the urine 37 to 76 percent as ether glucuronide, 8 to 17 percent as the sulfate, 0 to 15 percent as the ester glucuronide, and 0 to 5 percent in free form. It was further confirmed by Fabre et al. (370) that 0.4 ml/kg per day of *o*-xylene administered to the rabbit by stomach tube for 1 week was excreted as the *o*-toluic acid glucuronide ($C_{14}H_8O_7$). However, a further metabolite, 3,4-dimethylphenol ($C_8H_{10}O$) glucuronide, was identified by ether hydrolysis, elemental, and infrared analysis (370). The same metabolites were isolated and identified in the rat and guinea pig. In all three species, *m*-xylene was excreted as *m*-toluic acid derivatives. *p*-Xylene was excreted as *p*-toluic acid derivative, but also a 2,5-dimethylphenol glucuronide was isolated from the excretion product of all three species. These facts were confirmed by Bakke and Scheline (134), who also observed that *m*-xylene in the rat was converted to 2,4-dimethylphenol. According to one report (371), *o*-, *m*-, and *p*-xylene, in decreasing order, are demethylated to phenol. Inhaled *m*-xylene produced 57.3 to 63.9 mg of *m*-methylhippuric acid in the rabbit exposed to 1 mg/liter, 4 hr/day for 32 days (372). The metabolite was completely cleared 24 hr post-experimentally.

In vitro xylene hydroxylation by rat microsomal lung and liver preparations has demonstrated that the oxidation to toluic acid is mediated by hepatic but not the pulmonary microsomal enzyme systems (373). In vitro activation experiments demonstrated that also rabbit hepatic but not pulmonary microsomal enzyme systems were affected by phenobarbital pretreatment (374). Further, phenobarbital, 3-methylcholanthrene, and chlorpromazine raise the LC_{50} of inhaled *p*-xylene, but only 3-methylcholanthrene had any effect on the injected solvent (375). Intraperitoneal injection of 1 to 2 g/kg xylene increased serum OCT activity from 2 to 25.2 IU (137).

Microbiologic Xylene Utilization and Metabolism. Xylene can serve as substrate for the hydrocarbon-utilizing *Pseudomonas desmolytica* (302) and some *Pseudomonas* strains which produce large quantities of toluic acid (376, 377). *P. aeruginosa* converts *p*-xylene into *p*-methylbenzyl alcohol and possibly further to methylbenzoic acid (306). However, according to Skriabin et al. (378), *m*-xylene can be metabolized also to methylsalicylic acid and further to 3-methylcatechol. Another pathway is proposed by Omori and Yamada (379) for *p*-xylene via *p*-methylbenzyl alcohol to *p*-methylbenzoic acid, *p*-toluic acid, *p*-cresol, *p*-hydroxybenzoic alcohol, *p*-hydroxybenzaldehyde, *p*-hydroxybenzoic acid, and protocatechuic acid (3,4-dihydroxybenzoic acid). A similar sequence for *Pseudomonas* and *p*-xylene is proposed by Davis et al. (380), except for the last few steps, whereby *p*-toluic acid may convert to 4-methylcatechol and then to 2-hydroxy-5-methylmuconic semialdehyde. *P. putida* converts *p*-xylene to *cis*-3,6-methyl-3,5-cyclohexadiene-1,2-diol (381). A *Pseudomonas* strain, *P. pxy*, can grow on *m*- and *p*-xylene and utilize them as the sole source of energy. A mutant *P. pxy-82* can transform *m*-xylene to 3-methylcatechol and 3-methylsalicylic acid (382). From *Nocardia* cultures the xylene fermentation products 2,3-dihydroxy-*p*-toluic acid and 3,6-dimethylpyrocatechol have been identified (303, 383).

1.4.4 Industrial Hygiene and Biologic Monitoring

Threshold Limit Values. For occupational exposures OSHA (185) has established a TLV of 100 ppm (435 mg/m^3); ACGIH (384) specifies for skin, 100 ppm (435 mg/m^3), and NIOSII (368), in a criteria document, also recommends a TWA of 100 ppm for a 10-hr/day, 40-hr work week, with a ceiling of 200 ppm (868 mg/m^3) as determined by a 10 min sampling period. All standards apply to the mixed, the *m*-, *o*-, and the *p*-xylene.

The odor threshold in air is 0.47 ppm for *p*-xylene (385), and according to Carpenter et al. (325) as 1.0 ppm (4.5 mg/m^3) by a panel of six people. The odor threshold detection for xylene in water has been recorded as 2.12 ppm (385).

Air Sampling. An air sampling method, S-318, has been published by NIOSH (242), which includes charcoal tube collection, desorption with carbon disulfide, and analysis by flame ionization gas chromatography.

Analytic Methods. Xylene is preferentially determined by gas chromatography (242) based on the analytical work by White et al. (311). Another method involves simultaneous absorption of xylene and toluene from the air by impinger in ethanol (386) and analytical quantitation by UV absorbance (313, 386).

Biologic Monitoring and Medical Aspects. Xylene, without interference by benzene and toluene, can be determined in the blood by gas chromatography (386). Several methods have been published on the quantitation of methylhippuric acid in urine (276, 316, 317, 388–391) using paper, thin layer, and gas chromatography. Employees exposed to xylene should undergo comprehensive preplacement (102) and biennial medical checkups (368). Air and biologic monitoring programs should be established and evaluated regularly.

Occupational Exposure and Materials Handling. Occupational exposure to xylene vapors and direct liquid contact is possible in xylene manufacture and use. Incidental use may occur in histological technical work (102) or in paint spray operations. In all cases, proper precautionary measures are warranted. For labeling, warning precautions, and general protection, see the criteria document on xylene (368).

Methods for xylene waste gas treatment (392), waste purification (393), and solid waste xylene recycling (394) are available.

1.5 Tri- and Polymethylbenzenes

The properties of tri- and polymethylbenzenes resemble those of their lower homologues. Some physicochemical and physiologic data are presented in Tables 47.1 and 47.10.

1.5.1 Trimethylbenzenes

General. The trimethylbenzenes $(CH_3)_3C_6H_3$, are colorless, flammable liquids, which occur in three isomeric forms, the 1,2,3-trimethylbenzene, also known as hemimellitine, the 1,2,4-trimethylbenzene, pseudocumene, and the 1,3,5-derivative, also named mesitylene. Some physicochemical data are listed in Table 47.1. All isomers may occur in refined petroleum and coal tars (3). They are used in industry, mainly as chemical raw materials, as paint thinners, and solvents, and as motor fuel components (17). They have been detected in community air in trace quantities (33). Mixed trimethylbenzene and mesitylene have been identified in human exhaled air (30). Physiologically, they also present types of materials which pass through the placenta (161). In nature, the 1,2,4 isomer is an insect attractant.

Physiologic Response. Little experimental work has been carried out with trimethylbenzene since Lazarew (253) published his basic acute studies in 1929; see Table 47.10. He determined loss of righting response for 1,2,4-trimethylbenzene in the mouse at 40 mg/liter (8130 ppm) and loss of reflexes at 40 to 45 mg/liter (8130 to 9140 ppm); for the 1,3,5 isomer these losses were 25 to 35 mg/liter (7110 to 9140 ppm) and 35 to 45 mg/liter (7110 to 9140 ppm), respectively (253). The intraperitoneal fatal dose in the guinea pig is 1.788 g/kg (368).

Subacute experiments by Woronow in 1929, as quoted by Von Oettingen (38), demonstrated that daily subcutaneous injections of 2 to 3 g/kg 1,2,4-trimethylbenzene in olive oil in the rabbit caused local infiltration and necrosis. After three weeks a slight reduction of the erythrocytes occurred similarly as shown with xylene and toluene, with a temporary leukopenia and leukocytosis (38). Daily subcutaneous injections of 0.12 ml/kg of the 1,3,5 isomer caused a

Table 47.10.　Physiological Response of Tri- and Tetramethylbenzenes

Material	Route of Entry	Species	Dose or Concentration	Results or Effects	Ref.
Trimethylbenzene 1,2,3-	Oral (subacute)	Rat	1.2 g/kg	Urinary metabolites: 2,3-dimethylhippuric acid, 17.3% 2,3-dimethylbenzoic glucuronide, 19.4% 2,3-dimethylbenzoic sulfate, 19.9%	395
	Aquatic	Fsih	1000–100 ppm	TLm96	365
1,2,4-	IP (acute)	Guinea pig	1.788 g/kg	Minimum fatal dose (LD$_{LO}$)	368
	Inhalation	Mouse	8130 ppm (40 mg/liter) 8130–9140 ppm (40–45 mg/liter)	Loss of righting response (prostration) Loss of reflexes	252 252
	SC (subacute)	Rabbit	2–3 g/kg/day	Local infiltration and necrosis	368
	Oral	Rat	1.2 g/kg/day	Metabolites: 3,4-dimethylhippuric acid, 43.2% 3,4-dimethylbenzoic glucuronide, 6.6% 3,4-dimethylbenzoic sulfate, 12.9% 3,4-dimethylbenzoic acid, —	395
	Aquatic	Fish	1000–100 ppm	TLm96	365
1,3,5-	Oral (acute)	Rat	23 g/kg	Lethal to 7 of 10 test animals	15
	IP	Rat	1.5–2.0 g/kg	LD$_{100}$, minimal fatal dose	363
	Inhalation	Mouse	7110–9140 ppm (25–35 mg/liter)	Loss of righting response (prostration)	253
			8130 ppm (40 mg/liter)	Loss of reflexes	253
	Inhalation (subacute)	Rabbit	0.12 mg/kg/day	Moderate thrombocytosis	396
			0.2 ml/kg	Primary thrombopenia	396
		Rat	1 mg/liter/4 hr/day × 6 mo	Inhibition of phagocytic actions	214
	Oral (subacute)	Rat	1.2 g/kg/day	Metabolites: 3,5-dimethylhippuric acid, 78.0% 3,5-dimethylbenzoic glucuronide, 7.6% 3,5-dimethylbenzoic sulfate, 1.2%	395
	Aquatic	Fish	1000–100 ppm	TLm96	365
		Goldfish	13 ppm	96-hr LC$_{50}$	287
Durene	Oral (acute)	Rat	>5 g/kg	LD$_{50}$	15

moderate thrombocytosis in the rabbit, while 0.2 ml/kg produced a primary thrombopenia (396).

Inhalation of mixed trimethylbenzene by rats at 1 mg/liter for 4 hr/day during 6 months caused an inhibition of phagocytic activity of the leukocytes (214).

The aquatic toxicity rating for the three isomers fell between 1000 to 100 ppm (365). However, Brenniman et al. (287) determined a 96-hr LC$_{50}$ of 13 ppm for the mixed compound.

Metabolism.   From the gastrointestinal tract the trimethylbenzenes are readily absorbed into the vascular system and then further metabolized. In an elegant series of experiments, Mikulski and Wiglusz (395) determined the fate of the individual benzene trimethyl isomers. When 1.2 g/kg of 1,2,3-trimethylbenzene (hemimellitine) was orally administered to the rat, it was excreted as various urinary metabolites (see Table 47.10), with minor quantities as trimethylphenol. The 1,2,4 isomer was excreted as the corresponding 3,4-dimethylbenzoic acid conjugate, and the 1,3,5 isomers as the 3,5-dimethyl derivative; see Table 47.10. The excretion of the respective free trimethylphenols in minor quantities from the 1,2,4 and the 1,3,5 isomers was confirmed by Bakke and Scheline (134).

Mesitylene, when inhaled by the rat at 1.5, 3.0, and 6.0 mg/liter for 6 hr revealed dose-dependent increases in the percentage of neutrophilic granulocytes, but a decrease in lymphocytes (406) which persisted for 7 to 14 days. Long-term inhalation exposures at 3.0 mg/liter, 6 hr/day for 5 weeks did not significantly affect the hematologic picture (406). However, it elevated the blood serum alkaline phosphatase activity on short-term exposures, and the serum glutamic–oxalacetic transaminase activity in long-term experiments (407). Apparently, 3,5-dimethylbenzoic acid (mesitylenic acid) is formed also by the human metabolism, thus resembling that of the rat (408).

Industrial Hygiene.   A TLV of 25 ppm (125 mg/m^3) has been suggested by ACGIH (384).

Sampling procedures similar to those recommended for xylene may apply, and analytic methods are available (30).

A biologic microdetermination method for mesitylenic acid, the metabolite of 1,3,5-trimethylbenzene, has been published (408).

1.5.2   Tetramethylbenzene

Tetramethylbenzene, $(CH_3)_4C_6H_2$, can sterically form three isomeric compounds, the 1,2,3,4-tetramethylbenzene or prehnitine, the 1,2,3,5 isomer, also named isodurene, and the 1,2,4,5 derivative, also known as durene. Of these isomers, durene is commercially of highest importance.

Durene is a white, odorless solid, and further physicochemical properties are listed in Table 47.1. Durene is produced by methylation of low-boiling aromatics, principally pseudocumene (17). It is used as a solvent, as a constituent of motor fuels, and as chemical raw material. Physiologically, little information is available. However, durene appears to exhibit a toxicity close to that of the trimethylbenzenes; see Table 47.10. It is presumably metabolized similarly to the other alkylbenzenes. A saturated solution exhibited fungistatic properties (17, 409, 410).

No air threshold limit values have been established; however, the one for the trimethyl derivative may hold. Analytical methods are available; for example, gas chromatography and mass spectrometry have been suggested (411).

1.5.3 Pentamethylbenzene

Pentamethylbenzene, $(CH_3)_5C_6H_2$, is a white solid at ambient temperatures. Physiologically, it enhances microsomal drug-metabolizing enzymes (412) by induction of mixed function oxidases (413). Conversely, it also has fungistatic properties and decreases fruit mold (410). Some analytic methods are available.

1.5.4 Hexamethylbenzene

Hexamethylbenzene, $(CH_3)_6C_6$, is a solid material produced and used in refining of petroleum and in chemical syntheses. Hydrocracking by the paring reaction of polyalkylbenzenes, such as hexamethylbenzene, produces primarily light isoparaffins, cycloparaffin, and C_{10} and C_{11} methylbenzenes (36).

The hexamethylated ring appeared far more toxic than the lower alkylated benzenes. It proved lethal to nine of 10 rats, versus none of 10 for durene when 2.5 ml was administered orally (15). Long-term skin studies of mice rendered hexamethylbenzene noncarcinogenic (414).

Some analytic methods are available.

1.6 Ethylbenzenes

1.6.1 Ethylbenzene

General. Ethylbenzene, phenylethane, $CH_3CH_2C_6H_5$, is a colorless, flammable liquid (DOT) (22) with a pungent odor. Being heavier than air, its vapors may travel a considerable distance and ignite and backflash (8). It evaporates about 94 times more slowly than ether (9). Further physicochemical data are presented in Table 47.1.

Traces of ethylbenzene have been detected in the human expiratory air at somewhat higher concentrations in the smoker than in the nonsmoker (30). It also occurs in the gas phase of smoke condensate (12) and has been detected at 3.1 to 4.5 ppb in community air (33).

Ethylbenzene, a petrochemical, is prepared by dehydrogenation of naphthenes, or from catalytic cyclization and aromatization (148), but mainly by alkylation of benzene. It finds its greatest use in the production of styrene and synthetic polymers (36), but also as a solvent (148, 415) and component for automotive and aviation fuels (17).

Physiological Response. Ethylbenzene appears more highly irritant than its lower homologues; see Table 47.11. Human exposure to high vapor concentrations may present primarily an irritation hazard, but secondarily also may cause CNS effects. It is absorbed also through the skin at a low rate. Ethylbenzene has been detected in subcutaneous adipose tissue samples of workers 3 days after low to high exposure to styrene and related rubber manufacturing components (416).

Acute and chronic animal studies confirm the effects noted in humans; see Table 47.11. Ethylbenzene has been detected in cord blood samples, indicating its transport through the placenta (161).

Metabolism. Ethylbenzene is absorbed primarily through the respiratory system. When administered orally to the rabbit (17, 417), it was found to be converted metabolically to a number of oxidation products and subsequently excreted. The major urinary metabolite was hippuric acid. The oxidation products were benzoic acid, C_6H_5COOH, phenylacetic acid, $C_6H_5CH_2COOH$, and mandelic acid, $C_6H_5CH(OH)COOH$, excreted as the glycine conjugate, and also methylphenylcarbinol, 1-phenylethanol, $C_6H_5CH(OH)CH_3$, excreted as the glucuronide. From a dose of 100 mg/kg administered orally to the rat, Bakke and Scheline (134) identified the urinary metabolites p-ethylphenol, p-$C_6H_4(OH)(CH_2CH_3)$, about 0.3 percent, and smaller quantities of 1- and 2-phenylethanol, $C_6H_5CH(OH)CH_3$ and $C_6H_5CH_2CH_2OH$. When absorbed through the skin, mandelic acid was excreted at 4.6 percent, whereas after lung absorption the majority of the ethylbenzene was converted to mandelic acid and conjugated with glycine (397).

Urinary sulfate ratio decreases are normally a rough estimate of dose-related alkylbenzene hydroxylation due mainly to side chain oxidation activities (132). This, however, does not hold with a dose–action relationship for ethylbenzene. Namely, at high doses, ring hydroxylation increases, altering the sulfate ratio. Ethylbenzene stimulates microsomal enzyme synthesis and phenobarbital enhances its metabolic hydroxylation even further (418). Microorganisms, such as *Pseudomonas putida*, are capable of oxidizing ethylbenzene to (+)-*cis*-3-ethyl-3,5-cyclohexadiene-1,2-diol and related compounds (419).

Industrial Hygiene. OSHA (185) established a TLV for the occupational environment of 100 ppm (435 mg/m^3) and ACGIH in 1979 (384) concurs. For an occupational standard, OSHA in 1975 (420) retained these values, but added an action level of 50 ppm (217.5 mg/m^3) and more detailed occupational safety and medical instructions.

A sampling procedure, S-29, has been published by NIOSH (242), which includes collection in charcoal tubes, desorption with disulfides, and analytic determination of flame ionization gas chromatography. Other analytic methods have utilized ultraviolet spectrometry (313).

Occupationally, hands and face should be protected and respiratory devices should be available for air concentrations above the TLV. For biologic monitoring, see the section on xylene.

1.6.2   Methylethylbenzenes

Methylethylbenzene, $(CH_3)(CH_3CH_2)C_6H_4$, is a clear, flammable liquid. Limited physicochemical data are presented in Table 47.1 and physiologic observations in Table 47.11. It has been detected in the community atmosphere at 1.5 to 4.0 ppb (33). The toxicity of this material resembles that of ethylbenzene. Little information appears available on dimethylethylbenzene, $(CH_3)_2(CH_3CH_2)C_6H_3$. However, it has been documented to pass the placenta (161).

1.6.3   Diethylbenzene

Diethylbenzene, $(CH_3CH_2)_2C_6H_4$, (the mixed isomer), is a colorless, mobile, flammable liquid; see Table 47.1.

Physiologically, diethylbenzene appears slightly more toxic than the monoethyl derivative; see Table 47.11. The mixed compound was tested as 25 percent ortho, 40 percent meta, and 35 percent paraisomer (17). In 1971, Dyachkov and Severin (421) observed that diethylbenzene administered orally to the rat at $0.1 \times LD_{50}$ ($\sim$120 mg/kg) caused slight hemorrhages and dystrophic changes in the liver, gastric mucosa, duodenum, spleen, and kidneys, and also hepatic decreases in protein and glycogen. Several *Pseudomonas* strains have been found capable of using *m*- and *p*-diethylbenzene as the sole carbon source (302). Tanabe et al. (422) observed that the degradation proceeded by direct terminal side chain oxidation, rather than by dehydrogenation and subsequent hydration, yielding *p*-ethylphenylacetic acid.

No threshold limit values have been established for occupational exposure. However, the limits for ethylbenzene may apply here also.

The odor and taste threshold in water was 0.04 to 0.05 mg/liter (401), the base for which this range was used to establish a water threshold limit in the Soviet Union.

For handling diethylbenzene and worker protection, similar measures may apply as recommended for ethylbenzene.

1.7   Propylbenzenes

Some of the propylbenzene isomers, such as cumene and cymene, are of considerable industrial and commercial importance.

1.7.1   Propylbenzene

Propylbenzene, *n*-propylbenzene, $(CH_3CH_2CH_2)C_6H_5$, is a colorless, flammable liquid; see Table 47.1 for further physicochemical data. *n*-Propylbenzene in the mouse produces a loss of righting response at 10 to 15 mg/liter, loss of reflexes at 15 mg/liter, and death at 20 mg/liter; see Table 47.12 (17, 253). It is

Table 47.11.   Physiologic Response of Ethylbenzenes

Material	Route of Entry	Species	Dose or Concentration	Results or Effects	Ref.
Ethylbenzene	Dermal	Human	22–33 mg/m^3/hr	Absorption rate high on hands and forearms	397
	Inhalation	Human	1000–2000 ppm (4.92–9.84 mg/liter)/6 min	Severe eye irritation, lacrimation, gradual response, fatigue, but increasing vertigo, chest constriction, and dizziness when leaving the chamber	398
			5000 ppm (24.6 mg/liter)	Unacceptable concentration	398
	Inhalation (chronic)	Human	100 ppm (0.492 mg/liter)/8 hr	Irritant	399
	Oral (acute)	Rat	2.7 g/kg	Lethal to 7 of 10 test animals	15
			3.5 g/kg	LD$_{50}$	42
			5.46 ml/kg	LD$_{50}$	333
	Eye	Rabbit	Undiluted	Slight conjunctival irritation	42
			Undiluted	Corneal injury in some rabbits, classification 2/10	333
	Dermal	Rabbit	Undiluted	Irritant, classification 4/10	333
			5.0 g/kg	LD$_{50}$	400
			17.8 ml/kg	LD$_{50}$	333
	IP	Guinea pig	539 mg/kg	Slightly more toxic and irritant than benzene	38
	Inhalation (acute)	Rat	4000 ppm (19.7 mg/liter)/4 hr	~LC$_{50}$, 3 of 6 survived	333
		Mouse	3050 ppm (15 mg/liter)	Loss of righting response	253
			9150 ppm (45 mg/liter)	Death in 2 hr	253
		Guinea pig	1000 ppm (4.92 mg/liter)/3 min	Slight nasal irritation	398
			1000 ppm (4.92 mg/liter)/8 min	Eye irritation	398
			2000 ppm (9.84 mg/liter)/1 min	Moderate eye and nasal irritation	398
			2000 ppm (9.84 mg/liter)/345 min	One animal unconscious	398
			2000 ppm (9.84 mg/liter)/390 min	Apparent vertigo	398
			2000 ppm (9.84 mg/liter)/480 min	Static and motor ataxia	398

Compound	Route	Species	Dose	Effect	Ref.
			5000–10,000 ppm (24.6–49.2 mg/liter)	Immediate, intense irritation to conjunctiva and nasal mucous membranes, lacrimation, staggering gait. On pathological examination, intense cerebral congestion, lung edema and congestion, blood cyanotic	398
	Aquatic	Fish	100–10 ppm	TLm96	365
	Oral (subchronic)	Rat	13.6–136 mg/kg/day × 182 days	No effects	42
	Dermal	Rabbit	408–680 mg/kg/day × 182 days	Liver and kidney weight increases, slight pathological signs	42
	Dermal	Rabbit	Undiluted, repeated	Moderate irritation with slight necrosis	42
	Inhalation	Rat	400–2200 ppm (1.7–9.5 mg/liter)/7 hr/day × 144–214 days	Slight liver and kidney weight increases, slight pathological changes at two highest doses	42
	Inhalation	Rabbit, guinea pig, primate	400–600 ppm (1.7–2.6 mg/liter)/7 hr/day × 186–214 days	Little or no effect	42
Methylethyl-benzene	Inhalation	Mouse	3000 ppm (15 mg/liter)	Loss of righting reflex, no deaths	252
Diethylben-zene	Oral (acute)	Rat	1.2 g/kg	LD$_{50}$	42
o-, m-, p-	Oral (acute)	Rat	5.0 g/kg	Lethal to all 10 animals	17
	Oral (acute)	Rat	5.0 g/kg	Lethal to 8 of 10 animals	17
	Inhalation	Mouse	>5500 ppm (>30 mg/liter)	Loss of righting reflex, no deaths	252
	Aquatic	Fish	100–10 ppm	TLm96	365
	Oral	Rat, rabbit	0.0025 mg/kg/day	No effect	401
			0.25 mg/kg/day	No effect	401
			2.5 mg/kg/day	Some adrenal gland weight decrease	401
	Dermal	Rabbit	Undiluted	Moderate irritation, slight necrosis	42

Table 47.12. Physiologic Response to Propylbenzenes and Higher Alkylbenzenes

Material	Route of Entry	Species	Dose or Concentration	Results or Effects	Ref.
n-Propyl-benzene	Oral	Rat	4.830 g/kg	LD_{LO}	402
			5.0 g/kg	2 deaths of 10	17
	Inhalation	Mouse	2000–3000 ppm (10–15 mg/liter)	Loss of righting response	252
			3000 ppm (15 mg/liter)	Loss of reflexes	252
			4100 ppm (20 mg/liter)	Death	252
Cumene	Oral (acute)	Rat	1.4 g/kg	LD_{50}	42
			5.0 g/kg	6 deaths in 10 test animals	15
	Eye	Rabbit	Undiluted	Slight conjunctival irritation	42
	Inhalation	Human	200 ppm (1.0 mg/liter)	Irritant	9
		Rat	8000 ppm (39 mg/liter)/4 hr	50% mortality	17
		Mouse	2000 ppm (10 mg/liter)/7 hr	Lethal dose	148
			4100 ppm (20 mg/liter)	Loss of righting response	253
			5100 ppm (25 mg/liter)	Loss of reflexes, no deaths	253
	Aquatic	Fish	100–10 ppm	TLm96	365
	Oral (subchronic)	Rat (F)	154 mg/kg/day × 194 days	No effect	42
			462 mg/kg/day × 194 days	Slight kidney weight increase	42
			769 mg/kg/day × 194 days	Moderate kidney weight increase	42
	Skin	Rabbit	Undiluted	Moderate irritation, slight necrosis	42
	Inhalation	Rat, Rabbit	500 ppm (2.5 mg/liter) 8 hr/day × 6 days/wk × 150 days	Hyperemia and congestion of lungs, liver, and kidney	148
Cymene	Oral	Rat	4750 mg/kg	LD_{50}	402
	Inhalation	Rat	5000 ppm (27.5 mg/liter)/45 min	LC_{LO}	403
n-Butyl-benzene	Oral	Rat	5.00 g/kg	2 deaths of 10	17
sec-Butyl-benzene	Oral	Rat	2.240 g/kg	LD_{50}	404
			5.0 g/kg	Lethal to 8 of 10 animals	17
tert-Butyl-benzene	Oral	Rat	5.00 g/kg	Lethal to 7 of 10 animals	15
tert-Butyl-toluene	Inhalation (acute)	Human	10 ppm (0.06 mg/liter)/3 min	Irritant	405
			20 ppm (0.12 mg/liter)/5 min	CNS	405
	Oral	Rat	1.6 g/kg	Lethal to 50% of colony	17
			1.8 ml/kg	LD_{50}	405
		Mouse	0.9 ml/kg	LD_{50}	405
		Rabbit	2.0 ml/kg	LD_{50}	405
	Inhalation	Rat, mouse	248 ppm (1.5 mg/liter)/4 hr	LC_{50}	405
	Inhalation (subacute)	Rat (F)	25–50 ppm (0.15–0.3 mg/liter)/1–7 hr	Hemoglobin increase, erythrocyte, and leukocyte decreases	405
Dodecyl-benzene	Oral	Rat	5 g/kg	No deaths	17

metabolized by the mammalian system similarly to other alkylbenzenes, namely by several routes, as demonstrated by Gerarde and Ahlstrom (132) through urinary sulfate ratio determinations. An analytic determination procedure is available (313).

1.7.2 Isopropylbenzene (Cumene)

General. Isopropylbenzene, cumene, 1-methylethylbenzene, 2-phenylpropane, $(CH_3)_2CHC_6H_5$, is a colorless, flammable liquid with a sharp, aromatic odor. Further physicochemical properties are listed in Table 47.1, and occupational information is presented in a NIOSH/OSHA standard (420).

Cumene occurs in a variety of petroleum distillates and commercial solvents. It has been detected also in the expiratory air of human volunteers, but at a higher quantity in a smoker (30). Hemolytic effects may be produced whenever isopropylbenzene is permitted to oxidize to the peroxide (423).

Cumene is produced commercially by alkylation of benzene with propylene and recovered from petroleum by fractional distillation (148). Most of the commercially available material is used as a thinner for paints and enamels, as a constituent of some petroleum-based solvents, and in the past was extensively utilized as chemical raw material in the synthesis of phenol (17), and in the perfume industry, owing to its pronounced olfactory characteristics (424).

Physiologic Response. Cumene appears slightly less toxic than its n-propyl isomer, but more than benzene or toluene; see Table 47.12. Like its lower homologues, it may be irritant to the eyes and skin. It is a CNS depressant and narcotic, characterized by slow induction and long duration of effects (425). Thus inhalation of large vapor concentrations may cause dizziness, slight incoordination, and unconsciousness. Prolonged skin contact may result in skin rashes. In some short-term high dose experiments, animals exhibited damage to the spleen and fatty changes in the liver, but no renal nor pulmonary irritancy (148). Subacute inhalation experiments showed no significant changes in peripheral blood, but some liver, kidney, and lung effects; see Table 47.12.

Metabolism. Cumene is absorbed readily by the mammalian system and oxidized at the side chain, one of the metabolites being the dimethylphenylcarbinol glucuronide (426). Robinson et al. (427) also have observed increased urinary glucuronide excretions in orally dosed rabbits. In rats, following oral administration, no phenolic metabolites were detected (134), and less than 5 percent unchanged cumene was exhaled through the lungs in the rabbit (15).

A variety of bacterial strains, such as *Pseudomonas desmolytica* (302), *P. convexa*, and *P. ovalis* are capable of growing on cumene (302, 428). Oxidation products were identified as 3-isopropylcatechol and (+)-2-hydroxy-7-methyl-6-oxooctanoic acid (429). In soil, cumene inhibited ammonification and slightly affected the nitrification mechanisms (430).

Industrial Hygiene. OSHA (184) has published a proposed occupational safety and health standard with a TLV of 50 ppm (245 mg/m^3) for an average 8-hr work shift with an action of one-half of the TLV (420).

A sampling procedure, S-23.01, has been published by NIOSH (242), which recommends liquid air vapor collection with a charcoal tube and analysis by flame ionization gas chromatography. A gas chromatographic method to determine the urinary metabolite dimethylphenylcarbinol is available (426).

Exposure prevention includes proper eye, skin, and face protection and the availability of respirators for emergency situations. Appropriate medical surveillance programs should be available also.

1.7.3 Isopropyltoluene (Cymene)

Isopropyltoluene, isopropylmethylbenzene, cymene, $(CH_3)(CH_3)_2CHC_6H_4$, is a fragrant, flammable liquid. Cymene, which occurs in ortho, meta, or paraform is a constituent of a wide variety of essential oils. It is produced by catalytic cracking of petroleum and is one of the minor products of the wood pulp sulfite process (431). The paraisomer is used for the synthesis of *p*-cresol via the cymene hydroperoxide (431). The toxicity of cymene equals that of cumene and its lower homologues; see Table 47.1. It is readily absorbed into the mammalian system and distributed as similar lipid solvents (432).

p-Cymene is capable of serving as a carbon source for several strains of *Pseudomonas desmolytica*-like microbacteria (302) under the production of cumic acid. Isopropylbenzoic acid, *p*-isopropylbenzyl alcohol, and the aldehyde were also identified by Leavitt (433), indicating cymene's versatile biodegradability.

1.7.4 Diisopropylbenzenes

Diisopropylbenzene $[(CH_3)_2CH]_2C_6H_4$, occurs in three isomeric forms, namely as the ortho, the meta, and the para derivatives, all being flammable liquids. However, orally, they are of very low toxicity. Of 10 rats dosed with 5.0 ml/kg, no deaths for the meta, and one death for the ortho and the para isomers occurred (17). Although Russian reports of 1970 indicate that rats and rabbits exposed to 0.2 to 1.0 mg/liter (30 to 150 ppm) for 90 min daily for 5 weeks caused vascular hyperemia, hemorrhaging in most major organs, and fatty and protein dystrophy in the liver, kidney, and heart, and hyperplasia of the bone marrow (434). *meta*-Diisopropylbenzene caused decreased fertility in the rat and mouse at 1 to 3 mg/liter for 30 days (435).

1.8 Butylbenzenes

1.8.1 Monobutylbenzenes

Butylbenzene occurs in four isomeric forms, the n-butylbenzene, 1-phenylbutane, $CH_3(CH_2)_3C_6H_5$, the sec-butylbenzene, 2-phenylbutane, $CH_3CH_2CH(CH_3)C_6H_5$, the isobutylbenzene, 2-methyl-1-phenylpropane, $(CH_3)_2CHCH_2C_6H_5$, and $tert$-butylbenzene, (1,1-dimethylethyl)benzene, 2-methyl-2-phenylpropane, trimethylphenylmethane, $(CH_3)_3CC_6H_5$. They are odorous, flammable liquids, and further physicochemical properties are listed in Table 47.1 and some physiological data in Table 47.12. The neurotoxicity of p-$tert$-butylbenzene is believed to be a consequence of hemorrhage in the spinal cord due to vascular injury. A single oral dose of 0.075 ml produced an irreversible foreleg paralysis in the rat, but the animal was otherwise normal (17).

The data indicate a similar degree of toxic effects as noted for isopropylbenzene. All compounds are believed to be readily metabolized by side chain hydroxylation and conjugation for urinary excretion.

Isobutylbenzene is hydroxylated to isobutylcatechol by *Pseudomonas desmolytica* (429).

1.8.2 Polysubstituted Butylbenzenes

Tertiary butyltoluene, p-methyl-$tert$-butylbenzene, TBT, $(CH_3)(CH_3)_3CC_6H_5$, is the isomer of highest importance. It is a clear, colorless, combustible liquid, with a distinct aromatic odor (17). Saturated air contains about 800 ppm vapor. For further physicochemical data, see Table 47.1. Tertiary butyltoluene is produced by the alkylation of toluene with isobutylene (17, 420). It is used as a solvent in resin preparation and as a raw material in the chemical and pharmaceutical industries (17, 420). Some animal test data are listed in Table 47.12. Workers handling tertiary butyltoluene have experienced symptoms of nasal irritation, nausea, malaise, headache, and weakness (17). Tertiary butyltoluene may also decrease blood pressure, increase pulse rate, and cause CNS and hematopoietic effects (17).

A TLV of 10 ppm (60 mg/m^3) averaged over an 8-hr work shift has been recommended by OSHA (420). The OSHA standard also provides a sampling procedure, and industrial hygiene monitoring medical programs are suggested (420).

1.9 Other Alkylbenzenes

Very little information exists on amylbenzene but some does on hexylbenzene and dodecylbenzene, since the latter is used as a vehicle and solvent for polynuclear aromatics in carcinogenesis studies.

Dodecylbenzene is a flammable, odorous liquid. For further physicochemical properties, see Table 47.1. It is produced by the alkylation of benzene with propylene tetramer (5), and is used to produce arylalkyl sulfonates in the soap and detergent industries (5).

Biologically, it has been applied as a solvent and chemical vehicle. However, owing to its solvency characteristics, it may promote the tumorigenic and procarcinogenic potential of materials such as dimethylbenzanthracene (436). The oral toxicity of dodecylbenzene is very low; a 5 g/kg single oral dose caused no deaths in rats (17).

Industrially, dodecylbenzene may not present a problem, since it vaporizes slowly and can be contained rapidly. Also, it can be sulfonated in wastes and recycled (437).

1.10 Styrene

Styrene is the simplest member of the alkenylbenzenes. Of the alkenyl or olefinic benzenes, vinylbenzene or styrene is the most prevalent compound. Unfortunately, its physiological action has been predicted to equal that of vinyl chloride (102). However, no parallelism has been found to date.

Physicochemical properties and physiological characteristics are listed in Tables 47.1, 47.13, and 47.14.

1.10.1 General

Styrene, cinnamene, cinnamol, ethenylbenzene, phenylethylene, styrene monomer, styrol, styrolene, vinylbenzene (17), $CH_2:CHC_6H_5$, is a colorless to yellow, flammable liquid with a sweet, pleasant odor at low, but disagreeable at high concentrations (449). For further physicochemical and physiological properties, see Tables 47.1, 47.13, and 47.14. It occurs naturally in the sap of the styracaceous tree (148) and has been detected in the community atmosphere (33).

Styrene is produced by alkylation of ethylene and benzene, followed by catalytic dehydrogenation of the resulting ethylbenzene (5) or by demethylation of cumene (420). Chemically, styrene is highly reactive and polymerizes readily, with the possibility of being accompanied by violent explosions. The reactions occur rapidly at elevated, and more slowly at ambient, temperatures (450). Therefore, it is necessary to add polymerization inhibitors for transport and storage (450). Styrene, reacting with hypochloride followed by sodium hydroxide, may form styrene epoxide (431). With formaldehyde in sulfuric acid, it forms a diether (431). The styrene monomer is one of the world's major organic chemicals. It is used for the production of plastics, as a modifying additive for resins, as a dental filling component, as a chemical reaction intermediate, as a component in agricultural products (451), and as a stabilizing agent in a variety

Table 47.13. Physiological Action of Styrene

Route of Entry	Species	Dose or Concentration	Results or Effects	Ref.
Inhalation (acute)	Human	>10 ppm (0.04 mg/liter)	Odor not detectable	42
		60 ppm (0.26 mg/liter)	Detectable, but nonirritant	42
		100 ppm (0.43 mg/liter)	Strong odor, but without excessive discomfort	42
		200–400 ppm (0.85–1.7 mg/liter)	Objectionable strong odor	42
		376 ppm (1.6 mg/liter)/1 hr	Neurological impairment	438
		600 ppm (2.6 mg/liter)	Very strong odor, strong eye and nasal irritant	42
		800 ppm (3.4 mg/liter)/3 hr	Immediate eye and throat irritation, increased nasal mucous secretion, metallic taste, drowsiness, vertigo; after test termination, slight muscular weakness, accompanied by inertia and depression	55
Oral	Rat	5 g/kg	LD_{50}	42
		5 ml/kg	One mortality in 10 test animals	17
Eye	Rabbit	Undiluted	Moderate conjunctival irritation and slight, transient corneal injury	42
Inhalation	Rat	6000 ppm (26 mg/liter)/4 hr	Approximate LC_{50}	439
	Gpg	5200 ppm (22 mg/liter)/4 hr	Approximate LC_{50}	439
Aquatic	Fish	100–10 ppm	TLm96	365

of special products. The analysis of the oily fraction of gas phase cigarette smoke showed that styrene was one of the components (12). Various polymeric combinations contain styrene (452, 453), as was noted also in the rubber curing industry.

1.10.2 Physiologic Action

The major temporary effects of styrene is irritancy to eyes, skin, mucous membranes, and respiratory system. High dose levels may cause anesthesia and some systemic effects.

Acute Human Exposure. Human volunteers experienced no effect when inhaling styrene at less than 10 ppm, but strong irritation at 600 ppm; see

Table 47.14. Subacute and Chronic Exposure to Styrene

Route of Entry	Species	Dose or Concentration	Results or Effects	Ref.
Oral	Rat	66.7 mg/kg/day × 5 days/wk × 185 days	No effect	42
		100 mg/kg/day × 5 days/wk × 28 days	No effect	439
		133 mg/kg/day × 5 days/wk × 185 days	No effect	42
		400 mg/kg/day × 5 days/wk × 185 days	Growth, liver and kidney weight deviations	42
		500 mg/kg/day × 5 days/wk × 28 days	Poor weight gain, no significant pathology	439
		667 mg/kg/day × 5 days/wk × 185 days	Kidney weight, moderate growth and liver weight deviations	42
		1000 mg/kg/day × 5 days/wk × 28 days	Irritant to esophagus and GI tract; some deaths before 28 days	439
		2000 mg/kg/day × few days	Highly irritant to esophagus and GI tract, resulting in rapid death	439
	Rabbit	600 mg/kg/day × 3–10 days	Increased serum cholinesterase, carboxyl- and arylesterase activity	440
Dermal	Rabbit	Undiluted × 20 appln./4 wk	Moderate irritant, with blistering and hair loss	439
		Undiluted × 10–20 appln./2–4 wk	Moderate irritant, slight necrosis	42
SC	Rabbit	600 mg/kg/day × 3–10 days	Increased cholinesterase activity	441
	Rat	2.5 g/kg/day × 15–20 days	Decreased serotonin level in blood, lungs, intestine, and brain	442
Inhalation	Rat	6.5 ppm (35 mg/m³)/4 hr/day × 5 days/wk × 1–4 mo	Decreased neutrophil phagocytic activity, increased susceptibility towards staphylococcal infection	443
		300 ppm (7.9 mg/m³)/6 hr/day × 2–11 wk	Enzymatic adaptive changes	444
		1300 ppm (6.3 mg/liter)/7 hr/day × 139 exp./7 mo	Eye and nasal irritation	42
		2000 ppm (9.3 mg/liter)/day × 105 exp./5 mo	Eye and nasal irritation, moderate growth depression	42
	Mouse	6.5 ppm (35 mg/m³)/4 hr/day × 3 mo	Susceptibility towards staphylococcal infection initially decreased, then increased, then lowered again for 1 mo each	443
	Rabbit	6 ppm (29 mg/m³)/4 hr/day × 4 mo	Decreased phagocytic activity of leukocytes	445
		1300 ppm (6.3 mg/liter)/7 hr/day × 264 exp./12 mo	No effect	42
		2000 ppm (9.3 mg/liter)/7 hr/day × 126 exp./5 mo	No effect	42
	Guinea pig	650 ppm (3.0 mg/liter)/ 7 hr/day × 130 exp./6 mo	No effect	42
		1300 ppm (6.3 mg/liter)/7 hr/day × 139 exp./7 mo	Eye and nasal irritation, slight growth depression	42
		2000 ppm (9.3 mg/liter)/7 hr/day × 98 exp./5 mo	Eye and nasal irritation, moderate growth depression	42
	Rhesus monkey	1300 ppm (6.3 mg/liter)/7 hr/day × 264 exp./12 mo	No effect	42

Table 47.13. Healthy volunteers exposed to 50 and 150 ppm of styrene at rest and light physical exercise demonstrated that styrene in the alveolar air rose but varied with the exercise. The concentration of styrene in the arterial and venous blood increased sharply, and about 50 percent of the uptake was excreted as mandelic acid (454). When studied at 50 to 350 ppm, at the highest concentration a statistically significant impairment of the volunteers' reaction time was observed (455). In a series of exposures, volunteers at 376 ppm showed signs of transient neurological impairment (438). An accidental poisoning has been recorded (456). Styrene may be absorbed by the skin, as demonstrated with whole body vapor exposure for 3.5 hr (355).

Acute Animal Experiments. Oral administration of styrene to rats showed rather low toxicity; see Tables 47.13 and 47.14. Styrene in the rabbit eye caused moderate conjunctival irritation and slight, transient corneal injury. Nystagmus was demonstrated in rabbits, and during styrene exposure the directions of the rotatory nystagmus reversed. The blood levels indicated possible CNS involvement (457).

Intravenous injection of styrene caused the activation of several blood-clotting factors (458). Under ambient conditions, the vaporization of styrene is too low to be lethal to laboratory animals in a few minutes. The highest concentration to cause no serious systemic disturbances in 8 hr was 1300 ppm. A concentration of 2500 ppm was dangerous to life in 8 hr, and 10,000 ppm in 30 to 60 min (439). All animals showed eye and nasal irritation, those exposed to 2500 ppm showing varying degrees of weakness and stupor, followed by incoordination, tremors, and unconsciousness. Unconsciousness occurred at 2500 ppm in 10 hr, at 5000 ppm in 1 hr, and at 10,000 ppm in a few minutes (439). An inhalation LC_{50} was graphically determined from data by Spencer et al. (439); see Table 47.13. When administered as aerosol to mice, styrene caused low sensory irritation of the upper airways (459).

These animal experiments largely confirm the toxicologic signs and symptoms observed in humans.

Chronic Human Exposure. In 1956, Rogers and Hooper (460) described some signs and symptoms experienced by a group of employees handling styrene in a manufacturing plant as "styrene sickness." The duration of the symptoms was only a few hours, consisting of nausea, vomiting, loss of appetite, and general weakness.

An ocular examination of 345 workers in a styrene plant revealed conjunctival irritation in 22 percent of the work force, but no retrobulbar neuritis and central retinal vein occlusion, as previously had been reported (461).

Repeated or prolonged skin contact may lead to the development of dermatitis, marked by rough, dry, and fissured skin (17). In general, fair-skinned individuals appear to be less resistant than dark-skinned persons to the defatting and dehydrating action of styrene (460).

Inhalation exposures may occur during the synthesis and handling of styrene, in the production of block and emulsion polystyrene and its copolymers (462). An average exposure of 150 ppm or higher to humans resulted in prolonged simple reaction time (463). A clinical study of 494 production workers revealed initial findings of acute lower respiratory and prenarcotic symptoms, a very low percentage with FEV_1/FVC less than 75 percent and FVC <80 percent, but with normal liver function even at the high exposure level (464).

Some Russian reports indicate more severe problems than found elsewhere. One study of 50 petroleum workers producing synthetic rubber indicated one-half of the work force with dyspeptic signs of gastric acidity reduction, liver detoxification, and pancreatic changes (465, 466). A further consequence appeared as hematologic and clinical changes, indicated by moderate anemia, leukopenia, reticulocytosis, reduced coagulatility, and a rise in capillary permeability (467). Clinical chemical analyses indicated changes in blood protein composition (468, 469) and increased cholinesterase activities (440).

A group of 98 workers occupationally exposed to styrene showed psychological function changes parallel to air concentrations as determined by mandelic acid excretion (470, 471), abnormal brain waves (472), and some incidences of peripheral nerve lesions (472).

Styrene has been observed to pass the placenta (161), and some CNS defects were observed in children whose mother had been exposed to chemicals such as styrene during pregnancy (473).

Most of these signs, symptoms, and psychological and clinical changes are nonspecific; this could also relate to solvent effects. Exceptions are the neurological findings, the cholinesterase activity increase, and elevated incidence of sleeping activity. Most of the above quoted publications demonstrate such trends; however, little qualifying substantiation of air concentrations accompanies most observations.

Subacute and Chronic Animal Experiments.   Two oral experimental series with rats, using styrene in olive oil (42) and emulsified additionally in gum arabic solution (439), have shown a no-effect level of 133 to 667 mg/kg per day, low growth, and some organ weight deviation; 1 g/kg per day resulted eventually in some deaths. Additionally, 2 g/kg per day proved so highly irritant to the esophagus and stomach that death ensued quickly after only a few applications (439); see Table 47.14. Some enzymatic changes have been observed in the rabbit. An increased serum cholinesterase occurred similarly as observed in some styrene workers (440).

Styrene appears less irritant to the skin, even if applied undiluted (42, 439). Subcutaneous injection to the rat and the rabbit produced enzyme changes as seen in the human system (441, 442). Styrene increased cholinesterase, carboxylesterase, and arylesterase activities in the rabbit (440). Rats and guinea pigs exposed to 1300 ppm of styrene for 7 to 8 hr/day showed definite signs of eye and nasal irritation and appeared unkempt (42, 148), whereas the rabbit and the rhesus monkey exhibited no signs at this concentration.

Chronic inhalation experiments by Sidorov (474) indicate that styrene is accumulated in the rodent, increases the cholinesterase activity, and decreases spleen vitamin C content in young rats (475). Inhalation, even at low concentrations, may exhibit subtle immunologic changes.

Exposure of female rats to styrene vapor prolonged the estrus cycle (476) and affected offsprings embryotropically (477). Malformations were shown in chick embryos (478). Styrene, in some tests, has been found mutagenic in activated form; in others, negative (479) as tested by bacterial cultures (480). These results imply that styrene oxide possibly may be an active metabolite. Male rats exposed to 300 ppm, 6 hr/day, did show some increase in bone marrow chromosomal aberrations (481), from 8 to 12 percent versus 1 to 6 percent for the controls. In long-term feeding tests, so far styrene has been shown to be noncarcinogenic toward the Fischer 344 rat and the B6C3F1 mouse (482).

1.10.3   Absorption and Metabolism

The absorption of styrene in man and animals proceeds by all routes, but mainly through the respiratory tract. The partition between water and air and between air and blood has been determined (483), and the ratio between atmospheric concentration, inhalation exposure time, and initial blood levels for humans are proportional (484). The tissue uptake (485) is followed by metabolic transformations. In man, the two major metabolites of styrene are urinary mandelic acid $C_6H_5CH(OH)(COOH)$ and phenylglyoxylic acid $(C_6H_5COCOOH)$ (484) excreted as 85 and 10 percent, respectively, of the retained dosage (484), with about 2 percent exhaled in unchanged form (17). Below 200 to 250 ppm, the excretion can be related directly to exposure concentrations (486). An exposure elevation to 500 ppm, where the two acids plateau, increases the hippuric acid $(C_6H_5CONHCH_2COOH)$ excretion proportionally. Minor metabolites are 17-ketosteroids (487). Nonexcreted styrene appears to be accumulated in subcutaneous adipose tissue (416), forming the basis for the excretion lag (416). Usually, a weekend is not sufficient to excrete all the metabolites (488).

Various pathways have been proposed for the metabolism of styrene in man and animals. Experiments have shown that the excreted metabolites in the rat and the rabbit differ with the route and dose of administration, ranging from 9 to 32 percent for mandelic acid, 0 to 11 percent for phenylglyoxylic acid, 10 to 40 percent for hippuric acid, 6 to 8 percent for glucuronides, 5 to 9 percent for sulfur compounds, and traces for 1- and 2-phenylethanol (484) and 4-vinylphenol (134, 484).

In the rat, inhaled styrene is absorbed fairly rapidly and distributed into various tissues. Following the exposure to a 4-hr LC_{50} at 11.8 mg/liter, 25.0 mg/100 g was observed in the brain tissue, 20.0 in liver, 14.7 in the kidney, 19.1 in the spleen, and 133 mg/100 g in perirenal fat (484). After exposure to 21.0 mg/liter, a 2-hr LC_{50}, mouse brain contained 18.0 mg/100 g styrene (484).

When administered orally and subcutaneously to the rat, styrene is converted mainly to benzoic acid and excreted as hippuric acid (489); the β-carbon is exhaled primarily as carbon dioxide (490). The minor metabolites are mandelic acid $C_6H_5CH(OH)(COOH)$ and the glucuronide of phenylglycol $C_6H_5CH(OH)(CH_2OH)$ (489). The metabolism of intraperitoneally injected styrene into the rat appeared to proceed with the aid of microsomal enzymes (491) via styrene oxide (491, 492) to a more toxic metabolite, and to dihydrodiols (493). This sequence persisted especially when stimulated with phenobarbital (494, 495). Summaries of metabolic conversions are presented by several authors (17, 484, 492, 496).

1.10.4 Industrial Hygiene

Threshold Limit Values. OSHA published in the Federal Register a permissible styrene air concentration not in excess of 100 ppm (420 mg/m^3) averaged over an 8-hr work shift, and not in excess of 200 ppm at any time during the work shift, except for 5 min excursion of 600 ppm per 3 hr, provided the TWA is at or below 100 ppm (420), and an action level of one-half the 100 ppm level. ACGIH (384) also has proposed a TWA of 100 ppm (420 mg/m^3) for the styrene monomer. A concentration of 650 ppm has been determined as the human no-effect level by extrapolation of animal experiments (42).

Atmospheric Monitoring. A sampling method, S-30.01, has been published by NIOSH (242) for the exposure determination and measurement. The method includes charcoal absorption of the vapor, desorption with carbon disulfide, and analytic separation and quantitation by flame ionization gas chromatography. An air sampling method applicable to the rubber vulcanization industry also is available (497).

Analytic Procedures. A detailed determination method of styrene in air has been published by Rowe et al. (498). Methods for styrene determination in emission gases are also available. Most analytic methods involve gas chromatographic procedures; however, thin layer chromatography also has been utilized.

Biologic Monitoring. Medical monitoring and surveillance instructions are presented in detail in the Federal Register (420). A series of publications describe the correlation between styrene exposure and urinary monitoring and analytical determinations of mandelic acid and phenylglyoxylic acid (499–501). Since agents such as styrene or gasoline alter the metabolic amino acids, amino acid determination may present a direct indicator of the exposure (502).

Handling and Warning Properties. Precautionary measures when handling styrene should include the use of eye and skin protection, and respirators should be available for air concentrations above the recommended level. The odor of styrene renders the material detectable from 50 to 100 ppm, and very irritant above 600 ppm. Odor removal procedures for styrene in air and elimination from waste materials are available.

1.11 Vinyltoluene

1.11.1 General

Vinyltoluene, ethenylmethylbenzene, methylcthenylbenzene, methylstyrene, tolylethylene, $C_6H_4(CH_3)(CH:CH_2)$, is a colorless, combustible liquid with a strong, disagreeable odor (503). It usually occurs as a mixture of rarely the ortho, and mainly the meta and the para isomers at 50 to 70 and 30 to 45 percent, respectively (42). At elevated temperatures, vapors may form explosive mixtures with air, and polymerization may occur under explosive expansion. Further physicochemical data are listed in Table 47.1.

Vinyltoluene is produced by the dehydrogenation of *m*- and *p*-ethyltoluene and by catalytic re-forming. Its utilization is in the plastics industry, in resin production, and for special applications, such as its use as a block-packaging component for radioactive waste (504) and as an insecticide component (505).

1.11.2 Physiologic Response

Generally, vinyltoluene may cause eye, skin, and upper respiratory tract irritation. At high concentrations, it may exhibit anesthetic and systemic effects, the degree resembling that of styrene, as shown in Table 47.15. Exposures of 580 ppm were well tolerated by most laboratory animals; above 1130 ppm, some weight reduction was noted (42).

Results indicate that vinyltoluene may be metabolized by similar oxidative mechanisms, as shown earlier for styrene.

1.11.3 Industrial Hygiene

A technical standard published by OSHA (420) has established a threshold limit for vinyltoluene of 100 ppm (480 mg/m^3) averaged over an 8-hr work shift, and an "action level" of half the permissible exposure level. The 100 ppm (480 mg/m^3) TLV is also listed by ACGIH in 1979 (384).

A sampling procedure, S-25.0l, has been issued by NIOSH (242), which suggests charcoal absorption of the vapor, desorption with carbon disulfide, and analysis by flame ionization gas chromatography. Air and medical monitoring with detailed instructions are described in the federal standard for vinyltoluene (420).

Table 47.15. Physiological Response to Vinyltoluene and Other Alkenylbenzenes

Material	Route of Entry	Species	Dose or Concentration	Results or Effects	Ref.
Vinyltoluene	Inhalation (acute)	Human	<10 ppm (<0.05 mg/liter)	Odor not detectable	42
			50 ppm (0.24 mg/liter)	Detectable odor, but no irritation	42
			200 ppm (1.0 mg/liter)	Strong odor, but tolerated without discomfort	42
			300 ppm (1.5 mg/liter)	Objectionable, strong odor	42
			>400 ppm (>2.0 mg/liter)	Very potent odor, strong eye and nasal irritant	42
	Oral	Rat	2.5 ml[a]	Lethal to 4 of 10 animals	15
			4.0 g/kg	LD_{50}	42
			4.9 g/kg[b]	LD_{16}	446
			5.7 g/kg[b]	LD_{50}	446
		Mouse	3.16 g/kg[b]	LD_{50}	446
	Eye	Rabbit	Undiluted	Slight conjunctival irritation, no corneal injury	42
	Inhalation (acute)	Mouse	62 ppm (300 mg/liter)[c]	LC_{50}	446
	Inhalation (subchronic)	Guinea pig	6 ppm (29 mg/m³) × 4 mo	Teratogenic effects	446
		Rat	580 ppm (2.8 mg/liter)/7 hr/day[c]	No effect	42
			1130 ppm (5.5 mg/liter)	Moderate growth depression	42
			1350 ppm (6.5 mg/liter)	Moderate growth and liver weight depression; increased mortality	42
		Mouse	6200 ppm (30 mg/liter) × 1 mo	Slight weight reduction	446
		Rabbit, rhesus monkey	580–1350 ppm (2.8–6.5 mg/liter)/7 hr/day[c]	No significant effects	42

Compound	Species	Route	Concentration	Effect	Reference
	Guinea pig		580 ppm (2.8 mg/liter)/7 hr/day[c]	No effect	42
			1130 ppm (5.5 mg/liter)	Slight growth depression	42
			1350 ppm (6.5 mg/liter)	Slight growth depression, slight liver pathological effect	42
			6200 ppm (30 mg/m³) × 1 mo	Some teratogenic effects	446
Divinylbenzene	Rat	Oral	2.5 ml[a]	5 deaths of 10	15
			4.040 g/kg	LD$_{50}$	447
Allylbenzene	Rat	Oral	3.60 g/kg	LD$_{50}$	402
			4.620 g/kg	Lethal effects	402
	Mouse		2.90 g/kg	LD$_{50}$	448
α-Methylstyrene	Human	Inhalation	<10 ppm (<0.05 mg/liter)	Odor not detectable	42
			50 ppm (0.25 mg/liter)	Detectable odor, but no irritation	42
			100 ppm (0.5 mg/liter)	Strong odor, but tolerated without excessive discomfort	42
			200 ppm (1.0 mg/liter)	Objectionable, strong odor	42
			>600 ppm (>2.9 mg/liter)	Very potent odor, strong eye and nasal irritant	42
	Rat	Oral	4.9 g/kg	LD$_{50}$	42
	Rabbit	Eye	Undiluted	Slight conjunctival irritation	42
	Rat	Inhalation	3000 ppm (14.5 mg/liter)	Lethal effects	17
	Guinea pig		3000 ppm (14.5 mg/liter)	Lethal effects	17
	Fish	Aquatic	100–10 ppm	TLm96	365
	Rabbit	Dermal	Undiluted	Moderate to marked irritation, slight necrosis	42
	Rat	Inhalation (subchronic)	200 ppm (0.97 mg/liter)/7 hr/day × 139 exp.	No effect	42
			600–800 ppm (2.90–3.9 mg/liter)/7 hr/day × 28–149 exp.	Slight kidney and liver depression, also growth depression at 800 ppm	42
			3000 ppm (14.9 mg/liter)/7 hr/day × 3–4 exp	High mortality	42

Table 47.15. (*Continued*)

Material	Route of Entry	Species	Dose or Concentration	Results or Effects	Ref.
		Rabbit, rhesus monkey	200–600 ppm (0.97–2.9 mg/liter)/7 hr/day × 139–152 exp.	No effect except some growth depression and slightly increased mortality in the rabbit at 600 ppm	42
		Guinea pig	200–600 ppm (0.97–2.9 mg/liter)/7 hr/day × 139–144 exp.	No effect except some liver weight depression at 600 ppm	42
			800 ppm (3.9 mg/liter)/7 hr/day × 27 exp.	Slight growth, liver and kidney weight depression	42
			3000 ppm (14.5 mg/liter)/7 hr/day × 3–4 exp.	High mortality	42
1-Phenyl-butene-2	Oral	Rat	2.5 ml[a]	Lethal to 8 of 10 animals	15
4-Phenyl-butene-1	Oral	Rat	2.5 ml[a]	Lethal to all 10 rats	15
Phenyl-acetylene	Oral	Rat	5.00 g/kg	Lethal effects	4

[a] Per animal 1:1 in olive oil.
[b] Ortho and para isomers 28:72; other test for meta and para derivatives.
[c] 92–100 exposures.

Vinyltoluene should be handled with precautions similar to styrene, using eye and skin protection, and respiratory equipment above the TLV.

1.12 Divinylbenzene

1.12.1 General

Divinylbenzene, diethenylbenzene, vinylstyrene, $C_6H_4(CH:CH_2)_2$, is a water-white, combustible liquid, which occurs as the ortho, meta, and para isomers. For further physicochemical properties, see Table 47.1. Divinylbenzene is prepared by the dehydrogenation of diethylbenzenes. Usually, inhibitors are added to divinylbenzene to prevent the autopolymerization, which occurs readily at elevated temperatures for the meta and the para isomers (8). Its uses as a monomer or polymer are extensive. The monomer is utilized as an insecticide stabilizer, as an ion exchange resin, as a cross-linking agent in water purification and decolonization, as a sustained release agent, and as a dental filling component, applications for which patents are available. In biology, it has been applied as an experimental clotting agent for sustained life research (506).

1.12.2 Physiologic Response

Divinylbenzene is moderately irritant to the eyes and respiratory system, and to a lesser degree to the skin.

From the sparse data available (see Table 47.15), the toxicity of divinylbenzene appears to resemble that of styrene. An intravenous injection of 10 to 40 µm-diameter polystyrene–divinylbenzene copolymer particles was well tolerated by rats (507).

1.12.3 Industrial Hygiene

No air threshold concentration limits have been established; however, procedures for styrene may be adapted.

Eyes and skin should be protected, and respiratory equipment should be used above the irritant level.

1.13 Propenylbenzenes

Propenylbenzene occurs as the 1-propenyl or 2-propenyl derivative. The 1-propenylbenzene is also named β-methylstyrene, β-methylethenylbenzene, or phenyl-1-propene, $CH_3CH:CHC_6H_5$; the 2-propenylbenzene is also allylbenzene, or phenyl-2-propene, $CH_2:CHCH_2C_6H_5$. Allylbenzene occurs naturally in the essential oils of a variety of *Aniba* species (508).

The acute toxicity of 1-propenylbenzene is slightly below that of the vinyl-benzenes, indicating an increase in toxicity with the elongation of a side chain; see Table 47.15.

Acidic and neutral metabolites of allylbenzene have been identified as 1'-hydroxyallylbenzene, $CH_2:CHCH(OH)(C_6H_5)$, and cinnamyl alcohol, $C_6H_5CH:CHCH_2OH$; the former may be oxidized to phenyl vinyl ketone (509).

Industrial hygiene precautionary measures include eye, skin, and respiratory protection when handling liquids or vaporized materials at irritant levels.

1.14 α-Methylstyrene

1.14.1 General

α-Methylstyrene, isopropenylbenzene, *as*-methylphenylene, 1-methyl-1-phenylethene, β- or 2-phenylpropylene, 2-prop(-1-)enylbenzene, $(CH_3)(C_6H_5)C:CH_2$, is a colorless, combustible liquid with a sharp, aromatic odor. For further physicochemical data, see Table 47.1. Methylphenylene is synthesized by the catalytic alkylation of benzene with propylene in hydrofluoric and sulfuric acids (17) and by the dehydrogenation of cumene (5). Methylphenylene is used as a chemical raw material and as an intermediate in plastic and resin manufacture.

1.14.2 Physiologic Response

Generally, α-methylstyrene is irritant to the eyes, skin, and upper respiratory tract. Prolonged skin contact may cause dermatitis, and repeated inhalation may result in CNS depression.

The human tolerance has been investigated, and the odor was found tolerable at the threshold limit value (42); see Table 47.15. Russian reports mention hepatic dysfunction, enzyme and immunologic changes, and vitamin B_{12} deficiency in α-methylstyrene workers; however, no air concentrations were presented (510–512).

An oral LD_{50} value of 4.9 g/kg was found comparable to the results of the alkenylbenzenes discussed above; see Table 47.15. A mixture of butadiene and α-methylstyrene at 99.8 and 5.2 mg/m^3, respectively, inhaled by rats, resulted in a decreased level of leukocytes and phagocytic rate, respiration rate, and vitamin B and C levels of the blood and various organs (513). This was confirmed for α-methylstyrene in the rat and the rabbit by Makareva (514). Single-dose and repeated inhalation up to 6 months increased the activity of cholinesterases in several rat organs (515–517).

In dermal studies, α-methylstyrene was a moderate to marked irritant with slightly necrotic effects (42). It induced changes in connective tissue and neuron fibers in rat skin, although the effects were reversible 30 to 60 days post-treatment (518). α-Methylstyrene caused inflammation, hyperemia, edema, and hyperkeratosis after 20 applications to the rabbit skin (519).

In subchronic inhalation experiments, practically no effects were observed up to 600 ppm in several animal species.

1.14.3 Industrial Hygiene

According to the OSHA federal standard (420), the permissible exposure for α-methylstyrene has been established as 100 ppm (480 mg/m^3), averaged over an 8-hr work shift, with an action level of half that value.

A sampling procedure, S-26.01, for the determination of air concentrations, has been published by NIOSH (242). It includes absorption of the solute on charcoal, desorption with carbon disulfide, and analytical quantitation by flame ionization gas chromatography. Applied methods for the quantitation of phenylethylene in waste water and biomedia are also available.

Overall, α-methylstyrene appears to be somewhat less toxic than styrene and vinyltoluene. However, precautionary measures are advised. Detailed industrial hygiene and biological monitoring are presented in the OSHA federal standard of 1976 (420).

1.15 Other Alkenylbenzenes

Elongation of the side chain apparently increases the toxicity, as shown with phenylbutene orally in the rat (15); see Tables 47.1 and 47.15. Phenylbutene serves mainly as a raw material and intermediate in the chemical and pharmaceutical industries.

1.16 Alkynylbenzenes

1.16.1 Phenylacetylene

Phenylacetylene, ethynylbenzene, $C_6H_5C:CH$, is a highly odorous, flammable liquid; see Table 47.1. It is prepared from ω-bromostyrene and potassium hydroxide (431) and used as a chemical raw material. An oral LD$_{50}$ in the rat was determined as 5.00 g/kg (5) and an oral dose in the rabbit re-eliminated at 30 percent in three days (17).

2 DI-, TRI-, AND TETRAPHENYLIC COMPOUNDS

2.1 Diphenyls

2.1.1 Diphenyl

General. Diphenyl, bibenzene, biphenyl, phenylbenzene, $C_6H_5C_6H_5$ (Figure 47.1, **I**), is a colorless, solid crystallized in leaflets, with a pleasant, peculiar odor

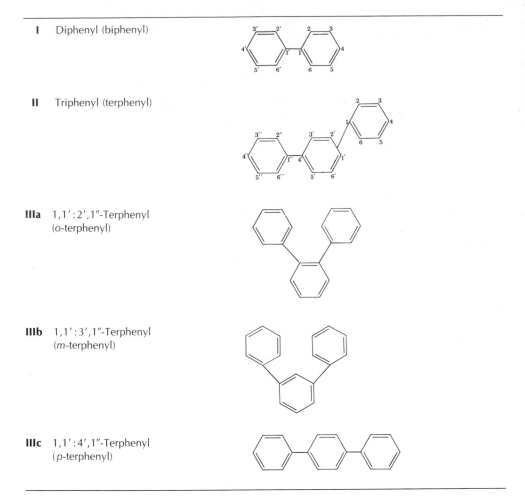

I   Diphenyl (biphenyl)

II   Triphenyl (terphenyl)

IIIa   1,1′:2′,1″-Terphenyl
       (*o*-terphenyl)

IIIb   1,1′:3′,1″-Terphenyl
       (*m*-terphenyl)

IIIc   1,1′:4′,1″-Terphenyl
       (*p*-terphenyl)

Figure 47.1   Di- and triphenylic aromatic compounds [according to rules of IUPAC, Paris, 1957 (1)].

(3). Further physicochemical properties are listed in Table 47.1 and physiological characteristics in Table 47.16.

On a small scale, diphenyl is prepared by heating phenyldiazonium chloride with copper (431); commercially it is manufactured by dehydrogenation of benzene (17), by passing benzene through an iron tube at 650 to 800°C (17), or by passing benzene over ferrous–ferric oxide and heating to 1000°C (37).

Diphenyl is thermally highly stable and thus used as a general heat transfer fluid (17, 520), and it has been utilized also as an organic coolant for nuclear reactors (520). It is used as a raw material in the chemical industry. In the citrus fruit and vegetable industries, it has been applied widely to fruit or vegetables or used in food packaging materials as a preservative, mold retardant (410), and fungistat (528, 529).

A residue tolerance of 110 ppm has been established in the United Sates for diphenyl on citrus fruit skin (17).

Physiologic Response.   In acute exposures, diphenyl may cause eye and skin irritation, and may exert a toxic action on the central and peripheral nervous systems and probably on the liver. Chronic human exposure is characterized by fatigue, headache, tremor, insomnia, sensory impairment, and mood changes, accompanied by clinical findings of cardiac or hepatic impairment, irregularities of the peripheral and central nervous systems, and possibly some brain lesions. One death has been recorded (522); see Table 47.16. In a follow-up study of earlier symptoms, some neurophysiological abnormalities were still observed 1 and 2 years after the initial investigation (523). Repeated skin contact may produce sensitization or dermatitis (102).

The major investigation carried out by Deichmann et al. (524) with the rat and the rabbit in acute tests (see Table 47.16) showed slight irritation of the gastrointestinal tract on oral administration, with the major cellular effects occurring in the hepatic and renal systems. A mild paralysis of the hind legs of some animals was observed (17).

In subchronic experiments, postmortem examinations demonstrated general hepatic, cardiac, and renal cell degeneration, with occasional spleen and other tissue involvement; see Table 47.16.

Metabolism.   Diphenyl is absorbed through the skin, the mucous membranes, and pulmonary system, then translocated to the liver through the blood vascular net, where it is converted to water-soluble hydroxy derivatives. Diphenyl is excreted unchanged by the biliary system of the rat when stimulated with phenobarbital intraperitoneally at 70 mg/kg for 4 days (530).

Diphenyl is primarily metabolized through various hydroxy derivatives. The 2-hydroxydiphenyl was identified in the urine of dogs after oral administration of diphenyl (531) in the rabbit (532), in vitro by the rat (533), hamster (534), young rat, mouse, hamster, cat, coypu, and the frog (535), and in various marine species (536). The 3-hydroxylation product has been identified in the urine of the rabbit after oral administration (531). Enzyme determinations have indicated that 4-hydroxylase differs from the 2-hydroxylation enzyme. The 4-hydroxylation was most prevalent and occurred in 11 laboratory animals tested (535), in rat microsomal preparations (533), hamster microsomes (534), the mouse (537), rabbit urine (531), and various marine animals (536).

In rabbit urine, additional 3,4- and 4,4'-dihydroxydiphenyl derivatives were identified (531, 532, 538). West et al. (539) had earlier determined 3,4- and 4,4'-dihydroxyphenyl, along with diphenylmercapturic acid and diphenylglucuronide, to be excreted by the orally fed rat. The related derivatives, 3-

Table 47.16. Physiologic Response of Diphenyl

Material	Route of Entry	Species	Dose or Concentration	Results or Effects	Ref.
Diphenyl (Dowtherm A[a])	Inhalation (acute)	Human	3–4 ppm (19–25 mg/m^3)	Irritation to eyes amd mucous membranes	17
Diphenyl	Inhalation (chronic)	47 workers	<1.6 ppm (<1 mg/m^3)	No symptoms and no deviations of cardiac or hepatic functions	520
		Workers	Unknown	Transient nausea, vomiting, bronchitis	521
		33 workers	4.4–128 ppm (28–800 mg/m^3) 5–15 years	Abdominal pain, headache, cardiac, hepatic, renal effects, peripheral and CNS abnormalities, 1 death	522, 523
	Oral (acute)	Rat	3.28 g/kg	LD$_{50}$	524
		Rabbit	2.41 g/kg	LD$_{50}$	524
	SC	Mouse	46 mg/kg	TD$_{LO}$, some neoplastic signs	525
	Inhalation	Rat	47.5 ppm (0.3 mg/liter)	No effects, no deaths	524
	Oral (chronic)	Rat	50–100 mg × 2 mo	Moderate degenerate changes in liver	526
		Rat	50–100 mg × 13 mo	Above changes enhanced, with affected thyroid and parathyroid functions, some papilloma and squamous cell carcinoma of the forestomach	17
			1% in the diet	Growth inhibition in weanlings	17
		Rabbit	1 g × 2–3 times/wk	Cumulative; total fatal dose 10–50 g	17

Diphenyl	Dermal (subacute)	Mouse	Dilution (with croton oil and acetone)	No significant effects	527
		Rabbit	0.5 g/kg × 2 hr/day × 5 days	Some deaths, growth depression, slight cardiac, hepatic, and renal tissue changes, follicular atrophy, necrosis and leukocytic infiltration of the spleen	524
Diphenyl-dust	Inhalation	Rat	6.5–47.5 ppm (0.04–0.3 mg/liter × 7 hr/day × 5 days/wk × 64 days	Irritation of nasal mucosa, respiration difficulties; some deaths, some bronchopulmonary lesions; hepatic and renal tissue effects	524
		Mouse	0.8 ppm (0.0005 mg/liter)/7 hr × 62 exp.	Same effect as rat	524
		Rabbit	6.5–47.5 ppm (0.04–0.3 mg/liter)/7 hr/day × 5 days/wk × 64 days	No effects	524

^{a} Diphenyl–ether mixture.

methoxy-4-hydroxy- and 3-hydroxy-4-methoxydiphenyl also have been identified as urinary metabolites in the orally fed rabbit (531). A study has shown that the 2-hydroxylase is prevalent in the young rat, but diminishes with aging, whereas the 4-hydroxylase is first of low, later in life of high activity (540). Differential hydroxylase stimulation has been observed also (541) with piperonyl butoxide, which suppresses the 4-hydroxy but stimulates the 2-hydroxy derivative formation (542).

Microbiology. Several organisms have been found that can grow on diphenyl and use it as the sole carbon source. These include *Pseudomonas desmolyticum, P. putida,* and *Acinetobacter* species (543) and 258 strains originating from Japanese natural resources (544). Several bacterial degradation products have been identified, such as 2,3-dihydroxydiphenyl (545), 2,3-dihydro-2,3-dihydroxydiphenyl, 3-hydroxydiphenyl and benzoic acid (546), 2-hydroxy-6-oxo-6-phenylhexa-2,4-dienoic acid (547), and γ-benzoylbutyric acid (544). These findings indicate that diphenyl is environmentally biodegradable.

Industrial Hygiene. An OSHA standard (548) for a diphenyl air concentration of 0.2 ppm (1 mg/m^3) has been set. ACGIH (348) recommends 0.2 ppm or 1.5 mg/m^3, although calculated, 0.2 ppm equals 1.08 mg/m^3. According to Deichmann et al. (524), prolonged exposure to a concentration of 0.75 ppm (0.005 mg/liter) should be considered a human health hazard.

Sampling techniques worked out by the Finnish team that investigated an accidental poisoning may be applied (522).

Spectrophotometric (17), thin layer (532), and gas and liquid chromatographic (538, 549) procedures have been developed for the separation and identification of the above diphenyl metabolites.

Occupational exposures should be kept below the threshold limits. Medical and industrial monitoring should be programmed. Eyes, skin, and lungs should be protected. Appropriate garments, gloves, and respiratory equipment should be available.

2.1.2 Alkyldiphenyls

Methyldiphenyl, *o*-benzyltoluene, $C_6H_5C_6H_4CH_3$, occurs naturally as a component of the essential oil of the flower from *Astragalus sinicus* (550).

Alkyldiphenyls, such as 4-*sec*-butyldiphenyl, the 4,4'-di-*sec* derivative, and related derivatives, are used as nonspreading lubricants (551).

2.1.3 Diphenylalkanes

Diphenylmethane, benzylbenzene, 1,1'-methylenebisbenzene, $(C_6H_5)_2CH_2$, occurs in orthorhombic needles, and possesses the odor of oranges (3). For further physicochemical data, see Table 47.1. It is prepared from methylene chloride and benzene, with aluminum chloride as the catalyst by the Friedel–Crafts reaction (3). It is used as a fragrance in the cosmetics industry (552). Amino-substituted diphenylmethane derivatives exhibit central depressive effects. Diphenylmethane may be one of the degradation products of DDT (553); however, it can serve as the sole carbon source to some *Hydrogenomas* (554) and *P. putida* strains, but also represents one of the most persistent compounds in nature by forming a tetraphenyl ether (553).

2.1.4 Diphenylalkenes

Diphenylethylene, α,α′-diphenylethene, stilbene, $C_6H_5CH:CHC_6H_5$, occurs in the trans or cis configuration, the trans form being prevalent owing to its higher stability. Stilbene is a solid at room temperature; see Table 47.1. It is used as a nutritional aid in agriculture and as a chemical intermediate, especially in the dye industry (431), where it serves as a coupler.

Stilbene is hydroxylated, primarily in the para position, secondarily in position 3, and readily metabolized in several species (555–558). No tissue damage has been reported so far; however, it is postulated that an arene oxide could be formed as an intermediate, which may possess tumorigenic properties. Conversely, Watabe and Akamatsu (559) have presented evidence that a variety of compounds can inhibit microsomal epoxidase activity.

Stilbene is covalently bound to rat liver microsomal protein and inhibited by some drug systems but accelerated by 3-methylcholanthrene pretreatment (560). According to Watabe and Akamatsu, rabbit liver microsomes are capable of cleaving the ethylenic linkage to produce benzoic acid (561).

The available data indicate that handling of diphenylethylene should be carried out using gloves. However, no human hazard is predicted.

2.2 Triphenyls

2.2.1 General

The open ring system of triphenyl is known also as terphenyl, $(C_6H_5)(C_6H_4)(C_6H_5)$ (Figure 47.1, **II**). It occurs naturally in petroleum oil. There are three chemical isomers (see Figure 47.1, **IIIa** to **IIIc**), the ortho, meta, and para derivatives, of which the ortho and the para forms appear industrially most prevalent. The latter is also known as *p*-diphenylbenzene or Santowax P. For specific product derivation and identification, consult the publication by Henderson and Weeks (527). All forms are solid at room temperature. Some physicochemical characteristics are listed in Table 47.1. The terphenyls are industrially important as chemical intermediates in the manufacture of nonspreading lubricants, as nuclear reactor coolants (520), and as heat storage and transfer agents (520). *p*-Terphenyl has been used as a sunscreen lotion component (562). Owing to their low vapor pressure, the terphenyls are considered a nonsignificant industrial hazard.

2.2.2   Physiologic Response

Human Toxicity.   The toxicity of the terphenyls to humans appears relatively low; see Table 47.17. No adverse effects have been detected in a work force except for some reversible skin rashes.

Animal Tests.   The rat and the mouse, when tested with terphenyls, show variable responses depending on the isomer used and the irradiation state; see Table 47.17. Although the nonirradiated mixture is nontoxic, the irradiated *o*-terphenyl appears most highly toxic.

Acute inhalation showed some effects in the guinea pig, depending on the particle size of the aerosol (566). Some sources indicate that the polyphenyls may be skin sensitizers. This, however, was definitely disproved for the terphenyls by Weeks et al. (520).

In chronic skin studies, one isolated papilloma was obtained. There is a possibility that terphenyl may have cocarcinogenic potential, as shown with tars (527). When terphenyl was inhaled by mice, soon after exposure cell debris was noted in the lungs, owing to nonspecific pulmonary membrane and cellular damage. This, however, cleared rapidly (568); see Table 47.17.

It appears that the long-term effects of the nonirradiated terphenyls are cumulative, thus increasing pulmonary and renal pathology, and irradiated terphenyls produce hepatic effects.

The hydroterphenyls are partially hydrogenated terphenyl mixtures. Their oral or intragastric LD_{50} values resemble closely those of the irradiated terphenyls; see Table 47.17. OSHA (569) has not established a TLV for these materials, but classifies them as causing cumulative liver, kidney, and lung damage.

Absorption, Distribution, and Metabolism.   An intragastric dose of ^{14}C-labeled *o*-terphenyl was rapidly absorbed, distributed, and almost completely excreted within 48 hr, in the rat mainly through the bile and in the rabbit mainly through urinary excretion (570). Temporary accumulation in the liver peaked at 4.5 hr in the mouse and was completely cleared in 1 week (571). Young rats when fed *o*- and *m*-terphenyl exhibited depressed body weight gain, and *m*-terphenyl induced liver hypertrophy (572). Following inhalation, hydrogenated terphenyls were rapidly eliminated through the lungs (571). Cholinesterase inhibition has been noted in rodents, which occurred at inhalation doses above 20 mg/m^3 of hydroterphenyl (565). *Pseudomonas desmolyticum* have been shown to grow on and degrade *m*-terphenyl (543).

2.2.3   Industrial Hygiene and Medical Surveillance

The OSHA standard shows an established ceiling value of 1 ppm (9 mg/m^3) for an 8-hr work shift (573). ACGIH (384) suggests a ceiling of 0.5 ppm (5 mg/m^3).

This limit has been recommended also in the Soviet Union based on data by Khromenko (567). A sampling procedure, S-27, for *o*-terphenyl has been published by NIOSH (242). Particulate matter collection on a cellulose membrane filter, extraction with carbon disulfide, and analytical quantitation by flame ionization chromatography is recommended.

Industrial hygiene monitoring and handling procedures are described and specified preemployment and periodic medical examinations are required; see the 1977 criteria document by NIOSH (574).

3 DINUCLEAR SYSTEMS

3.1 Naphthalene

3.1.1 General

Naphthalene is a white solid, also called naphthalin, naphthene, moth flake, tar camphor, or white tar, which exhibits a typical mothball odor. Chemically, it is composed of two fused benzene rings with the empirical formula $C_{10}H_8$, see (Figure 47.2, **I**).

Naphthalene occurs naturally in the essential oils of the roots of *Radix* and *Herba ononidis* (575). It is formed in cigarette smoke by pyrolysis (576, 577), and is a photodecomposition product of Sevin, a naphthyl carbamate, used in agriculture (578). Naphthalene also occurs in crude oil, from which it may directly be recovered as white flakes. It is also isolated from cracked petroleum (5), from coke-oven emission (148), or from high-temperature carbonization of bituminous coal (579). Naphthalene, both in solid and liquid form, is flammable. The vapors or dusts, with air, can produce explosive mixtures. Molten naphthalene (above 110°C), in contact with water, may result in violent foaming or the formation of explosive mixtures (580). Further physicochemical data are listed in Table 47.1 and physiologic characteristics in Table 47.18.

Naphthalene is used extensively as a raw material and intermediate in the chemical, plastics, and dye industries (148), as a moth repellent in the form of balls or disks, as an air freshener (246), and as a surface-active agent. It is utilized in the manufacture of insecticides, fungicides, lacquers, and varnishes (5), and to preserve wood and other materials (148). In medicine, it has been applied as an antiseptic, anthelminthic, and dusting powder in skin diseases.

3.1.2 Human Physiologic Response

Acute Oral Toxicity.   Ingestion of naphthalene in the form of mothballs has resulted in no ill effects in some of the cases described. The ingested material was excreted in the feces in unchanged form (40). Greater danger for toxic effects exists when naphthalene is ingested in combination with fats, which

Table 47.17 Physiologic Response to Terphenyls

Material	Route of Entry	Species	Dose or Concentration	Results or Effects	Ref.
Terphenyl	Skin (chronic)	Human	0.01–0.94 ppm (0.094–0.89 mg/m³)	Six of 200 workers developed nonspecific, readily reversible skin rashes; is not a skin sensitizer	520
	Inhalation (acute)	Human	Spills with short-term exposure	Headaches and sore throat, reversible in 24 hr	520
	Inhalation (chronic)	Human	0.01–0.94 ppm (0.094–0.89 g/m³)	No effect on blood pressure, pulmonary function even improved in exposed group; isocitric dehydrogenase borderline but not statistically elevated	520
Terphenyl (nonirradiated)	Oral (acute)	Rat	17.5 g/kg	LD₅₀	563
(irradiated)		Mouse	12.5 g/kg	LD₅₀	563
		Rat	6.0 g/kg	LD₅₀	563
		Mouse	6.0 g/kg	LD₅₀	563
o-Terphenyl		Rat	1.90 g/kg	LD₅₀	564
m-Terphenyl		Rat	2.40 g/kg	LD₅₀	564
p-Terphenyl		Rat	>10.0 g/kg	LD₅₀	564
Hydroterphenyl		Rat	6.6 g/kg	LD₅₀	565
		Mouse	4.2 g/kg	LD₅₀	565
Terphenyl	Inhalation (acute)	Rat	100 ppm (0.94 mg/liter) = (0.94 g/m³)	No mortality, but pulmonary pathology after a 1-hr exposure	566
			320 ppm (3.0 mg/liter)	4/8 deaths, with early asphyxial death due to crystalline plugs in the trachea	566

o-Terphenyl	Oral (chronic)	Rat	0.25–0.5 g/kg/day × 30 days	Increased liver and kidney to body weight ratios	520
m-Terphenyl			0.25–0.5 g/kg/day × 30 days	Increased liver to body weight ratios	520
p-Terphenyl			2.5–5.0 g/kg/day × 1 mo	Insignificant weight decreases, intensification of antitoxic functions of the liver	567
mixed Terphenyl (nonirradiated)		Mouse	0.25 g/kg/day × 8 wk	Changes in cytoplasm of hepatocytes, no lesions	566
			0.6 g/kg/day × 16 wk	Severe chemical nephrosis	563
			1.2 g/kg/day × 16 wk	Intensified nephritis, especially affecting the proximal tubules	563
(irradiated)			1.2 g/kg/day × 16 wk	Lethal	563
p-Terphenyl			0.3 ppm (3 mg/m^3) × 1 mo	No effect	567
	Inhalation (chronic)	Laboratory animals	3.7 ppm (35 mg/m^3) × 1 mo	Functional and morphological changes	567
			212 ppm (2000 mg/m^3) × 4 hr/day × 5 days/wk × 8 wk	Cell debris in lungs, but rapidly cleared	568

I Naphthalene

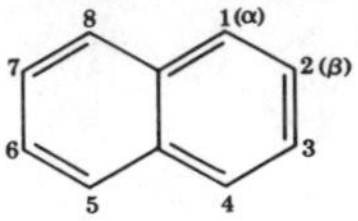

II Acenaphthene

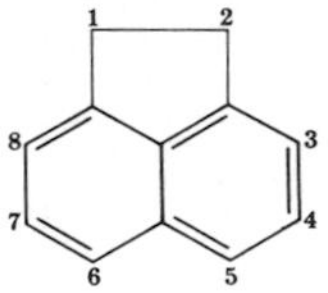

III Anthracene

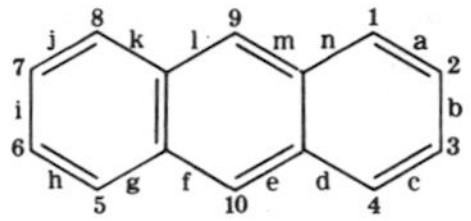

IV Phenanthrene

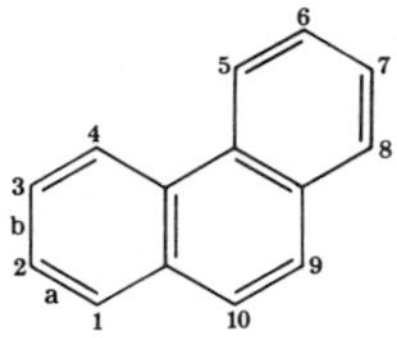

V Benz(a)anthracene (1,2-ben-
zanthracene) (then 7,12-di-
methylbenz(a)anthracene =
9,10-dimethylbenzanthracene)

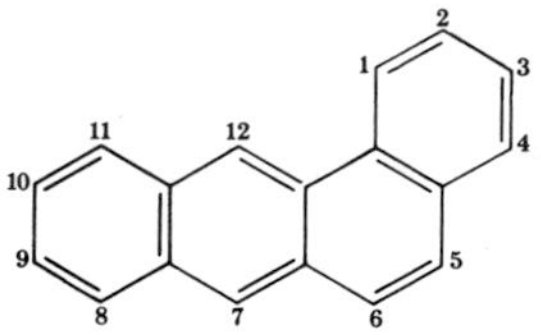

(Va)
Benz(a)anthracene (conventional numbering)

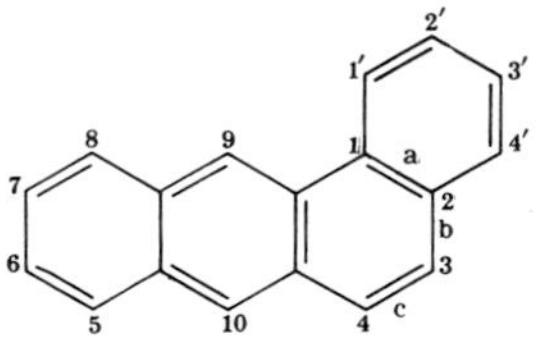

(Vb)
1,2-Benzanthracene (numbering by IUPAC rules)

VI Benzo(a)phenanthrene (1,2-
benzophenanthrene, chrysene)

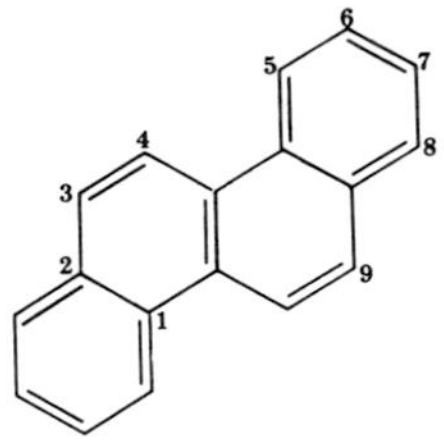

VII Benzo(c)phenanthrene (3,4-
benzophenanthrene)

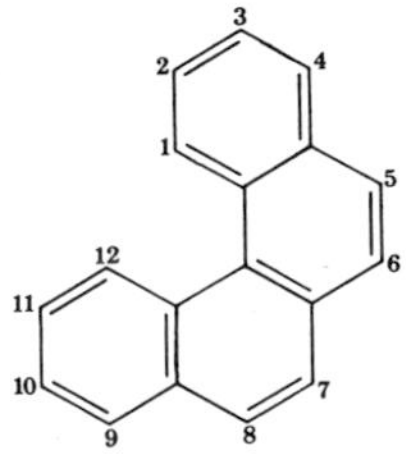

Figure 47.2 Di- and polynuclear fused aromatic hydrocarbons [according to rules of IUPAC, Paris, 1957 (1)].

VIII Pyrene

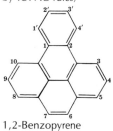

IX Benzo(a)pyrene (3,4-benzopyrene)

Benz(a)pyrene (conventional numbering)

3,4-Benzopyrene (numbering by IUPAC rules)

X Benzo(e)pyrene (1,2-benzopyrene)

Benzo(e)pyrene

1,2-Benzopyrene

XI 1,2 Cyclopentapyrene

XII 20-Methylcholanthrene (3-methylcholanthrene, 20-methylcyclopentabenzophenanthrene, 20-benz(j)aceanthrene)

20-MCH

3-MCH

XIII Dibenz(a,h)anthracene (1,2:5,6-dibenzanthracene)

Figure 47.2 *(Continued)*

3337

Table 47.18. Physiologic Response to Naphthalene

Material	Route of Entry	Species	Dose or Concentration	Results or Effects	Ref.
Naphthalene	Oral (acute)	Human (child)	2–3 g	Lethal dose	593–595
		(adult)	5–15 g	Lethal dose	17
	Oral	Rat	2.6 g/kg	All metabolized in 2 days	581
		albino	1 g/kg/day × 2 days	Slight eye effects	582
		pigmented	1 g/kg/day × 2 days	More severe eye effects	582
	Subcutaneous	Mouse	5.1 g/kg	LD_{50}	583
	IP	Rat	100 mg/kg	20–30% excreted in urinary metabolism	584
			1 g/kg	Lethal to 67.8% in 2–3 hr	585
		Mouse	150 mg/kg	LD_{LO}, lowest lethal dose published	586
	Aquatic	Fish	10–1 ppm	TLm96	365
		Crab	8–12 ppb	Lethal to all animals	587
	Oral (repeated)	Rabbit	1 g/kg/day	Browning of lens and eye humor, degeneration of retina and cataract formation	588
	Subcutaneous	Rat	820 mg/day	Not tumorigenic after >1000 days observation	589
Naphthalene					
Methyl-	Oral	Rat	4.360 g/kg	LD_{50}	447
1-Methyl-	Oral	Rat	5.00 ml/kg	Lethal to all animals	17
2-Methyl-	Oral	Rat	5.00 ml/kg	Lethal to all animals	17
1,6-Dimethyl-	Oral	Rat	5.00 ml/kg	Lethal to 7 of 10 animals	17

facilitate the absorption and subsequent systemic events (40). In severe cases, ingestion caused gastroenteric distress, tremors, and convulsions (324). Within 2 to 7 days moderate to severe anemia may develop (40). Heinz bodies appear, and the serum takes on a yellowish brown color. Peripheral blood may show eccentrically concentrated hemoglobin in the erythrocytes, hypochromia, and polynucleosis. The bone marrow may appear hyperplastic and show an increased proportion of nucleated cells of the erythrocyte series (17). In some cases, hemoglobinuria, possible occlusion of the renal tubules, and disturbed renal functions may occur (40). Death may ensue due to respiratory failure (324).

A 1-year-old child who had accidentally ingested naphthalene showed increasing lethargy and anorexia 2 weeks later (590), followed by hemolytic anemia (590). A case has been recorded by Pavlivoda (591) with early symptoms of toxic kidney attack, yellow skin, dark coloration of the urine, and sharp onset of pain. From 1961 to 1965, three cases of naphthalene poisoning were documented in India (592), and others have been recorded (593, 594). A fatal case has been recorded (324, 595) where 2 g administered over a 2-day period caused the death of a 6-year-old child. In most mortality cases, the lethal effects are acute hemolytic anemia, presence of Heinz bodies, and fragmented erythrocytes (593, 594). The probable lethal dose for an adult is 5 to 15 g (17); see Table 47.18.

Individuals with congenital erythrocyte glucose-6-phosphate dehydrogenase deficiency render themselves particularly susceptible to hemolytic agents, such as naphthalin, which may rapidly cause hemolytic anemia (324, 590, 596–599).

Acute Eye and Dermal Effects. Naphthalene may be acutely irritant to the eye (102), and retinotoxicity has been recorded (569, 600).

Upon direct skin contact, naphthalene is a primary irritant (102). Diapers or clothing stored with mothballs and used directly on infants have caused skin rashes and systemic poisoning (39, 324). The cutaneous absorption and effects are facilitated with the use of baby oil (39, 601, 602).

Acute Inhalation. Since naphthalene may volatilize and sublime at room temperature, it can be inhaled as particulate matter or it may deposit on the skin. Signs and symptoms due to inhalation of naphthalene vapors resemble those observed from ingestion or skin absorption.

Naphthalene vapors may cause eye and respiratory tract irritation, headache, nausea, and profuse perspiration, depending on the duration of the exposure concentration. Also, optic neuritis has been observed (17). Acute cases of inhalation, probably in addition to skin absorption, have been observed as the Chinese "full-moon disease" (603), stemming from dressing small infants in ceremonial clothing that had been stored in naphthalene. Such a case with twin brothers was observed in Australia (603).

Chronic Oral Effects.   Oral intoxication in industry is unlikely; however, in the general population, 50 cases of severe chronic effects from repeated ingestion of a naphthalene–isopropyl alcohol "cocktail" have been recorded (604). The symptoms resembled those of ethanol intoxication and consisted of tremors, restlessness, extreme apprehension, and hallucinations, but effects subsided in a few days (604).

Chronic Eye Effects.   Eye effects resembling those observed in humans over age 50 have been observed after repeated exposure to naphthalene vapor or dust. Corneal ulceration and cataracts have been observed (605), some lenticular opacities by Meyer (606), and general opacities in 8 of 21 employees exposed to naphthalene up to 5 years (607).

Dermal Effects.   On repeated contact, naphthalene may cause erythema and dermatitis, especially in hypersensitive individuals (102). Occasional allergic responses are rare.

Chronic Inhalation.   Repeated inhalation of vapors may produce malaise, headache, and vomiting, according to European sources. Toxic effects have been observed in industry where electric vaporizing devices were used (593, 608–610).

Other Human Effects.   One case of transplacental poisoning has been recorded (611).

Human Metabolism.   Naphthalene is rapidly absorbed by the human system when inhaled but more slowly through the intact skin and by ingestion. Apparently conversion to the hemolytic agents, α- and β-naphthol and α- and β-naphthoquinone, is rapid in the adult and very slow in the newborn (17, 612). The naphthols are partially excreted as the glucuronides (598). Naphthalene per se is nonhemolytic; the activity of the four metabolites decreases in the order as listed above (613). The most active metabolite is probably naphthalene-1,2-oxide (614), which may rapidly convert to hydroxy or oxy derivatives, unless the system is glucose-6-phosphate dehydrogenase or glutathione deficient.

3.1.3   Physiologic Response of Animals

Acute Oral Administration.   Experimental work with the rat has shown a LD_{50} of about 1.8 g/kg; see Table 47.18. A dose of 1.0 g/kg produced only light eye effects (582). Rabbits fed naphthalene exhibited browning of the lenses and eye humors (588), inhibition of the ciliary body and of the ascorbic acid transport across the blood–aqueous fluid barrier, and development of cataracts (615). Cataract formation was accompanied by glycolytic decreases and changes in the electrophoretic pattern of lens proteins (616) and the disappearance of oxidized glutathione (617). Dogs administered 3 g of naphthalene, simulating mothball portions, developed distemper-like attacks and moderate anemia (574).

Acute Subcutaneous and Intraperitoneal Effects.   Naphthalene, when injected subcutaneously into the mouse, caused little toxic effects (583); see Table 47.18. At 100 mg/kg intraperitoneally, 20 to 30 percent was excreted in the rat urine, 85 to 90 percent of these in the form of acid conjugates (584); 5 to 10 percent was excreted in the bile, of these also 70 to 80 percent as acid conjugates, with the major metabolite naphthalene-1,2-dihydrodiol.

Animal Experiments with Repeated Exposures.   Daily oral administration of 1 g/kg of naphthalene produced lenticular opacity and peripheral swelling of the lens, slightly visible after three doses, and causing definite changes after 20 doses (148, 588, 618). Modifications occur in the amino acid, ascorbic acid, protein, and carbohydrate metabolism of the eye, producing calcium oxalate crystals (588, 619).

At 1.5 g/kg per day by stomach tube, white spots appeared in the rabbit eye periphery, but were distributed over the whole retina in young animals (620).

Rats injected intraperitoneally with 40 mg/kg naphthalene for three days produced arylhydroxylase inhibition, whereas its 1- or 2-hydroxy derivatives had no influence (621).

In addition to its retinotoxic action, systemic absorption of naphthalene vapor may also lead to the formation of cataracts (600).

Toxicity to Insects.   Naphthalene and thiometon were the most effective of seven insecticides against fruit-piercing moths (622). Some resistant and susceptible strains of the housefly *Musca domestica* showed that single doses of naphthalene were excreted more rapidly by the male than the female (623). The resistance factor was found to depend on microsomal activity (624).

Toxicity to Marine Life.   Naphthalene was absorbed by the common marine mussel, *Mytilus edulis*, and released again in unchanged form (289); in several species, *G. mirabilis*, *O. maculosus*, and *C. stigmacus*, naphthalene was metabolized to 1,2-dihydro-1,2-dihydroxynaphthalene and excreted in the urine (625). A 96-hr median lethal dose was 10 to 1 ppm; see Table 47.18. In the rainbow trout, after an 8-hr exposure to 0.005 mg/liter, most tissues contained 20 to 100 times that of the water levels (626). Retention appeared to be concentrated in liver tissue (627). The half-life in fat was less than 24 hr (626). Naphthalene and its combination with serum albumin at 8 to 12 ppb in flowing seawater produced 100 percent mortality (587). Naphthalene was accumulated in marine animals and when transferred to oil-free seawater, the release to undetectable levels required 2 to 60 days and was species dependent (628). Naphthalene was extracted from No. 2 fuel oil by the polychaete *Neanthes arenaceodentata,* accumulated, and subsequently released within 400 hr by the male, but retained for 3 weeks by the female (629). Larvae of the exposed females contained up to 18 ppm naphthalene, which then decreased to zero during development (629).

Mutagenicity and Carcinogenicity.   Usually organic compounds that readily form epoxides are suspected mutagens or carcinogens. Naphthalene is not a mutagen and not a carcinogen (584), and is carcinogenically inactive (345). This has been substantiated by Skjaeggestad (630), who observed no carcinomas in hairless mice, owing to the naphthalene's low nucleic acid (DNA) (631, 632) and RNA binding capacities (633), to its specific fluorescence spectral characteristics (634), and also to the mammalian tissue's potent naphthalene-1,2-epoxide hydrase activity (614). On the contrary, naphthalene has been observed to inhibit the activity of Ehrlich carcinoma cells (635) and the tumorigenic potential of tobacco smoke condensate (636).

3.1.4   Absorption, Metabolism, and Excretion

The human system absorbs naphthalene most readily by the inhalation route (148). Experiments in the rat showed that naphthalene was readily converted to 1- or 2-naphthol and 1,2-dihydro-1,2-naphthalenediol, and excreted free or as the 1- or 2-hydroxyglucuronide or the 1-sulfate (637). It is now strongly suspected that the naphthalene-1,2-epoxide is a short-lived, first intermediate (638). The 1,2-oxide then may spontaneously be isomerized to the 1-naphthol (345, 639–641), also leading to naphthalene oxides and naphthoquinones (345, 642, 643), glutathione and mercapturic acid [*N*-acetyl-*S*-(1,2-dihydro-2-hydroxy-1-naphthyl)-L-cysteine] conjugates (345), and naphthalene dihydrodiol glucuronide as isolated from liver microsomes (644). Horning et al. (584) identified, besides hydroxydiols, the triol and tetrol metabolites, which indicated that multiepoxide systems may occur. A methylthio derivative has been identified also in rat urine (645). The 1,2-diol may be catalytically oxidized to 1,2-naphthoquinone, a reaction reversed by ascorbic acid (588) under elimination of calcium oxalate (588). A combination of the quinone with protein may be responsible for the browning of the eye preceding cataract formation (588).

Evidence has been presented by several investigators that the naphthalene 1,2-epoxidation progresses through a liver microsomal arene oxidase system (584, 646, 647) by forming a cytochrome P-450 complex (136, 648) and utilizing an ascorbic acid–iron–oxygen coenzyme system (641). The absorption of naphthalene, microsomal oxidation, and similar types of metabolism have been observed also in the housefly (649).

3.1.5　Microbiological Aspect

Several *Pseudomonas* strains are capable of growing on naphthalene (650, 651), decomposing (652), and utilizing it in soil or liquid media (653, 654). *P. desmolyticum* was tested for the degradation of industrial naphthalene in waste water (543) and petroleum effluents (655). A *Pseudomonas* strain degraded naphthalene to 1,2-dihydro-1,2-dihydroxynaphthalene (656), and *P. aeruginosa* produced salicylate (657). Conversely, naphthalene and Sevin, one of its derivatives, inhibited the microbial vitamin B_{12} formation in soil (658), but a derivative was selectively destructive to tomato tumors (659). Naphthalene reduced the photosynthetic processes in *Nitzschia palea* (660) and inhibited growth and the photosynthetic capabilities of the green algae, *Chlamydomonas angulosa* (661).

3.1.6　Industrial Hygiene and Biological Monitoring

OSHA (185) has suggested a threshold limit value of 10 ppm (50 mg/m^3); the same limit has been recommended by ACGIH (384) for an 8-hr working day. Saturated air at 25°C contains about 100 ppm naphthalene.

Naphthalene has a characteristic odor, recognizable at about 25 ppm (148).

NIOSII (242) has published a sampling procedure, S-292, recommending collection on charcoal, desorption with carbon disulfide, and quantitation by flame ionization gas chromatography. Ultraviolet spectrophotometric procedures have been described also (622). Biologically, urine can be monitored for naphthalene and 1-hydroxynaphthalene or related metabolites (17).

Workers handling naphthalene should wear protective clothing for hands and arms, and respiratory equipment should be available to prevent vapor inhalation.

3.2　Alkylnaphthalenes

Alkylnaphthalenes are formed as pyrolysis products in cigarette smoke (663). Some have been identified in commercial carbon paper (664). They also are the major components of the C_{10} to C_{13} alkylnaphthalene concentrate fraction, which distills at 400 to 550°F (36).

In a toxicologic investigation of catalytic reformates, Nau et al (665) observed that with a C_{11} to C_{12} petroleum mixture containing about 23 percent alkylnaphthalenes, skin and eye effects exceeded those observed with benzene. The alkylnaphthalenes appear more toxic to marine life than the alkylbenzenes (666). The toxicities and also the bioaccumulation rise with increasing molecular weight (628). Most hydrocarbons, as also the alkylnaphthalenes, are released eventually (283).

Nocardia cultures, isolated from soil, preferentially oxidized alkylnaphthalenes when methylated in the 2 position (303).

3.2.1 Methylnaphthalene

Methylnaphthalene can occur as the 1 or 2, the α or the β isomer, $C_{11}H_{10}$. Some physicochemical and physiological properties are listed in Tables 47.1 and 47.18. 1-Naphthalene, a flammable solid, also has been identified in the waste water of coking operations (667), of textile processing plants (688), and as a photodecomposition product of 1-naphthylacetic acid (669). Methylnaphthalene is used as a component in slow-release insecticides (670), in mole repellents (671), and often in combination with naphthalene.

In contrast to naphthalene, methylnaphthalene is not a human skin irritant or photosensitizer (17). Whole-body irradiation decreased the sensitivity of intragastrically administered material to the rat (672). Intraperitoneal injection of the 1-methyl derivative was about four times less lethal to rats than naphthalene, and the 2-methyl derivative showed no effect (673).

Methylnaphthalene accumulated in marine animals similar to naphthalene (629). However, biliary excretion of 2-methylnaphthalene in the trout was facilitated by pretreatment with 2,3-benzanthracene (674). Of the methyl derivatives, *Nocardia* cultures oxidized only the 2-methyl group to the carboxy function (303).

3.2.2 Di- and Polymethylnaphthalene

Dimethylnaphthalene, $C_{12}H_{12}$, can occur in various isomeric forms; see Table 47.1. The 1,2-dimethyl derivative has been used as a selective organic solvent (675). Similar to the lower homologues, it also is active in inhibiting crown gall and olive knot neoplasms (659). It accumulates in shrimp (283), clams, and other marine life to a greater extent than the lower homologues (676). Dimethyl derivatives with position 2 occupancy were oxidized to the acid by *Nocardia* species (303). Contrary to the lower homologues, the 2,3-dimethyl derivative, along with higher methylated compounds, appeared to act as a weak accelerator of skin tumor induction (636). The oral toxicity in the rat appeared lower for the 1,6-dimethyl derivative than for the monomethyls (17); see Table 47.18.

The trimethylnaphthalene was active in exterminating termites (677).

4 TRI- AND POLYNUCLEAR RING SYSTEMS

4.1 General Aspects

4.1.1 Polynuclear Aromatics in General

Fused ring systems occur in linear, staggered, or three-dimensional configurations. The simplest linear three-ring system is anthracene; when staggered, it is phenanthrene; see Figure 47.2, **III** and **IV**. Their biologic activities change with alkyl substitution, at which point the materials may inhibit or promote their own progressive oxidative metabolism. For physicochemical data see Table 47.19, and for physiological characteristics see Table 47.20.

Polynuclear aromatics, PNAs, or polyarylhydrocarbons, PAHs, are not synonymous with carcinogens. However, some of these compounds or their derivatives have carcinogenic or cocarcinogenic potential, depending on structure, cellular transport, storage potential, enzyme inducibility, oxidative metabolism, rate of excretion, and many more factors. These parameters are inherent in the type of biologic system, tissue, immunologic, and nutritional state involved.

4.1.2 Occurrence

In a study of commercially available meat, poultry, fish, and yeast, also fats and oils, the two polycyclic hydrocarbons, benzo(a)pyrene (BaP) and benz(a)anthracene were quantitated at 0.2 ppb in meat to 98 ppb in coconut oil (702). Even higher concentrations occur in cooked food, and, for example, 3.8 ppm of BaP in tarry filtrate from smoke rooms (702).

Polynuclear aromatics (PNAs) are formed in cigarette smoke (703). Smoking of a single test cigarette produced 9.7 to 11.1 ng/m³ BaP in a 20 m³ area (704). Environmental air concentrations of BaP varied with the season at 5 ng/m³ in September and 68 ng/m³ in March (704). Benzopyrene, dibenzopyrene, dibenzanthracene, and other PNAs have been identified in the exhaust gases of diesel engines (705). In peat deposit areas, anthracene, dibenzanthracene, benzopyrene, and methylcholanthrene have been observed (706). However, by far the greatest PNA quantities are emitted through domestic energy production, including gas, oil, wood, and refuse burning, mounting up to 6 g benzo(a)pyrene per person per year in the United States (707).

4.1.3 Physicochemical Relationships

The majorities of studies have concerned benzo(a)pyrene, possibly since it is more widely distributed in the environment and appears of relatively high biologic activity. The polynuclear aromatics are mainly solid materials, soluble in fats, oils, and organic solvents; see Table 47.19. The chemical structures of some of the compounds discussed in this section are shown in Figure 47.2. Various theories have been proposed linking mutagenic or carcinogenic properties to physicochemical properties, such as electronegativity (708) or K- and L-region reactivity indexes (709), electrophilic potency, dipole moment, intramolecular and subcellular binding (710), hydrophobicity, and others. However, it appears that these characteristics alone are not sufficiently specific for exclusive biologic prediction.

Table 47.19. Physicochemical Properties of Some Polynuclear Aromatics

Compound	B.P. [°C (mm Hg)]	CAS Registry No.	Density	Emp. Formula	Flamm. Limits (%)	Flash Pt. [°C(°F)]	M.P. (°C)	Mol. Wt.	Refr. Index	Solubility[a] w/al/et	Sp. Gr. (25°C)	Vapor Dens. (Air = 1)	Vapor Pres. [mm Hg (°C)]	W/V Conv. (mg/m ≏ 1 ppm)
Acenaphthene	279	83-32-9	1.189	$C_{12}H_{10}$	0.6–?		96.2	154.21	1.6048 (95°C)	i/s/s	1.02	5.32	10 (131.2)	6.31
Anthracene	340	120-12-7	1.283	$C_{14}H_{10}$	0.6–?	121 (250)	216.2–216.4	178.23		i/d/d	1.25	6.15	1 (145)	7.23
Phenanthrene	340	85-01-80	1.179	$C_{14}H_{10}$			101	178.23	1.59427	i/s/s	1.18	6.14	1 (118.3)	7.23
1,2-Benzanthracene	435 (subl.)	56-55-3		$C_{18}H_{12}$			162	228.29		i/d/s				9.34
6-Methyl-1,2-benzanthracene		316-14-3		$C_{19}H_{11}$			150.5–151.5	242.32		i/s/s				9.91
7-Methyl-1,2-benzanthracene		2541-69-7		$C_{19}H_{14}$			183–183.6	242.32		i/s/—				9.91
10-Methyl-1,2-benzanthracene		2541-69-7		$C_{19}H_{14}$			141	242.32		i/s/s				9.91
9,10-Dimethyl-1,2-benzanthracene		57-97-6		$C_{20}H_{16}$			122–123	256.35		i/d/—				10.49
1,2-Benzophenanthrene	448	218-01-9	1.274	$C_{18}H_{12}$			254	228.29		i/d/d				9.34
3,4-Benzophenanthrene		95-19-7		$C_{18}H_{12}$			68	228.29		i/d/—				9.34
3-Methylchrysene		—		$C_{19}H_{14}$			172.5–173.5	242.32		—/s/—				9.91
Pyrene	393	129-00-0	1.271	$C_{16}H_{10}$			156	202.26		i/s/s				8.27
3,4-Benzopyrene	312 (10)	50-32-8		$C_{20}H_{12}$			176.5–177.5	252.32		i/d/s				10.32
Benzo (e) pyrene		192-97-2		$C_{20}H_{12}$			178–179	252.32		i/v/s				10.32
Cyclopentapyrene		—		$C_{19}H_{12}$			170	240.31						9.76
Methylcholanthrene	280 (80)	56-49-5	1.28	$C_{21}H_{16}$			180	268.36		i/s/—				10.98

[a] Solubility in water/alcohol/ether: v = very soluble; s = soluble; d = slightly soluble; i = insoluble.

Table 47.20. Physiologic Response to Polynuclear Aromatics

Material	Route of Entry	Species	Dose or Concentration[a]	Results or Effects	Ref.
Phenanthrene	Oral (acute)	Mouse	700 mg/kg	LD_{50}	678
Benzo(a)pyrene	Subcutaneous	Rat	50 mg/kg	LD_{50}	679
	IP	Mouse	500 mg/kg	Lethal dose	680
Anthracene	Oral (subchronic)	Rat	4.5 g over >1000 days	No tumors	589
	Dermal	Mouse	10 μM/wk × 25 days	Few papillomas in a small number of animals	681
			0.5% sol. 3/wk × 25 days	Benign tumors not exceeding the control group	682
	Subcutaneous	Mouse	5 mg × 280 days	No tumors	683
			20 mg/wk × 33 wk	Sarcomas in 5 of 9 rats at injection site	589
Phenanthrene	Oral	Rat	1 ml/s.d./310 days obs.	No mammary tumors	684
	Dermal	Mouse	0.5% × 6 wk/tot. 20 appln.	Slight increase over control in benign tumors	682
Trimethylphenanthrene	Subcutaneous	Mouse	5 mg/appln. × 372 days	No tumorigenic effects	683
	Subcutaneous	Mouse	0.5 mg/(0.25 cm³ lard/s.d., obs. 17 mo	No tumors induced	685
Benz(a)anthracene	Oral	Rat	200 mg/s.d./310 days obs.	No mammary tumors	684
	SC	Mouse	0.05 mg/isc × 22–28 mo obs.	11% with tumors after 315 days	683
			0.2 mg/isc × 22–28 mo obs.	24% with tumors after 346 days	683
			1.0 mg/isc × 22–28 mo obs.	34% with tumors after 298 days	683
			5.0 mg/isc × 22–28 mo obs.	55% with tumors after 299 days, dose-action related	683
			2.2 μM/w × 35 wk	Multiple papillomas	681
			5 mg/s.d. × 15 mo obs.	No tumors	685
	IV	Rat	2.0 mg (13 mg/kg)/appln. at 50, 55, 56 days × 7 mo obs.	No tumors observed	686
		Mouse	0.25 ml/s.d. × 20 wk	Tumor incidence lower than for controls	687

Table 47.20. (Continued)

Material	Route of Entry	Species	Dose or Concentration[a]	Results or Effects	Ref.
4-Methyl-	SC	Mouse	2 mg/s.d. × 11 mo obs.	Some tumors in 4 months, 75 percent of animals affected in 11 mos	685
7-Methyl	SC	Mouse	230 µg × 3/wk × 12 mo 100 mg/s.d.	13 carcinomas/35 mice 31% with mammary tumors	688 684
8-Methyl-	SC	Mouse	3 mg at 1st mo/5 mg at 3 and 9 mo	Lung tumors, 5.5 per mouse at 7 mo observation time	685
10-Methyl-	Dermal	Mouse	300 µg/appln. × 2/wk × 20 wk	72% with papillomas, dose–action related	689
12 Methyl-	SC	Mouse	100 µg × 3/wk × 12 mo	17% with mammary tumors	684
1'-Methyl-	SC	Mouse	5 mg/s.d. × 15 mo obs.	Not a tumorigen	685
3,9-Dimethyl-	SC	Mouse	2 mg/s.d.× 13 mo obs.	Not a tumorigen	685
4,9-Dimethyl-	SC	Mouse	2 mg/s.d. × 13 mo obs.	Very weak tumorigenic response	685
7,12-Dimethyl-	Oral (gastric intubation)	Rat (F)	1 ml/dose with 0.1/0.5/1.0/5.0/10/50/100 mg/rat in sesame oil)	Lobular carcinoma	690
	Dermal	Mouse	0.2%, 2 × 1 drop/wk	Tumor induction in 10 wk on isc skin	691
Chrysene (Benzo(b)-phenanthrene)	SC Dermal	Mouse mouse	20 mg/s.d. × 12 mo 0.3–7.5% in benzene or mouse fat	100% with mammary tumors 2 of 5 samples tested weakly tumorigenic when dissolved in benzene	684 692
3,4-Phenanthrene	SC	Mouse	5 mg/s.d. × 271 days obs.	5 sarcomas in 23% of animals; weak carcinogen	683
2-Methyl-	IV	Mouse	0.25 mg/s.d. × 20 wk obs.	At 8 wk, 1.5 lung tumors in 2 of 10 mice; at 14 wk, 1.7 in 6 of 11 mice	687
2,9-Dimethyl	SC	Mouse	0.2 mg/s.d. × 18 mo obs.	No tumors observed	685
Pyrene	Dermal	Mouse	8.3% in croton oil × 25 days obs.	Tumor incidence slightly above controls	682
			10 µM/wk × 35 wk	Few papillomas in a low number of animals	681

Benzo(a)pyrene				
Oral	Rat	1 mg/g food during pregnancy	Teratogenic effects of stillbirths and reduced F_1 growth	693
	Mouse	100 mg/s.d. × >50 days obs.	Mammary tumors in 8 of 9 rats	684
		0.15 mg/g food × 80–140 days	Gastric papilloma, squamous cell carcinomas, pulmonary adenomas, and leukemia	694
Dermal	Mouse	1% soln. 2×/wk × 200 days	First tumor after 70 days; after 200 days all animals affected	695
	Rabbit	0.3% benzene solution 2×/wk × >400 days	At 400 days, 1 carcinoma, 10 of 12 animals exhibit skin tumors	695
SC	Mouse	0.09 mg/s.d. × 183 days obs.	Tumor yield within statistical range at 78%	683
		4 mg + 0.2 ml carbowax days 11, 13, 15	Pulmonary adenoma in progeny, general 2.36 adenomas/mouse	696
		4 mg as above + 2 drops 1% croton oil in acetone dermally 1/day × 28 wk	Skin papilloma in 23.6% of treated offspring	696
	Primate	10 mg in 0.2 ml olive oil s.d. × 7 mo obs.	One of 2 animals died within 24 hr, the second developed a palpable nodule in 6 mo, at 7 mo measured 30 × 40 × 21 mm at injection site, sarcoma on heart muscle	697
IP	Rat (M)	4 mg/s.d. × 9 mo obs.	Tumors of spleen and pancreas, 2/30 developed mammary, 2/30 uterine adenocarcinomata	698
	(F)	10 mg/s.d. × 9 mo obs.		
	Mouse	2 mg/s.d. × 33 w obs.	Intra-abdominal tumors at 15 wk adhering to internal organs	698
		2–4 mg/s.d. × 1 yr obs.	6/14 survivors showed lung adenomas	689
		750 mg/kg/s.d.	Moderate mutagenic index	700

Table 47.20. (Continued)

Material	Route of Entry	Species	Dose or Concentration[a]	Results or Effects	Ref.
	IV	Rat (F)	2 mg (13 mg/kg)/s.d. × 95 days obs.	Nine of 30 rats with mammary carcinomas at days 56–95	686
		Mouse	0.25 mg/s.d. × 20 wk obs.	At 8 wk, 8/10 mice with 2.3 lung tumors per mouse; at 20 wk, 10/10 with 3.7/mouse	687
Benzo(e)pyrene	Dermal	Mouse	10 μM/wk × 35 wk	Multiple papillomas	681
	SC	Mouse	1% in 0.20 cm^3 lard/s.d.	One liposarcoma at injection site in 8 days	701
Methylcholanthrene 3-Methyl-	SC	Mouse	0.02 mg/s.d. × 221 days obs.	14 sarcomas per 27 mice	683
	IV	Rat	2 mg at days 50, 53, 56	Carcinomas at days 44–98 in 7 of 30 rats	686
Dibenz (a, h)anthracene	Dermal	Mouse	2.5 μM/wk × 35 wk	Multiple papillomas	681
	SC	Mouse	0.04 mg/s.d.	6 sarcomas in 18 mice (33% with tumors), 195 days av. induction time	683
	IV	Mouse	0.25 mg/s.d.	After 8 and 20 wk, all mice showed lung tumors, av. 30.5/ms at 20 wk	687

[a] s.d. = single dose; isc = intrascapular.

4.1.4 Physiologic Response

Experimental work demonstrates that peroral, acute administration renders the polynuclear aromatics practically nontoxic; see Table 47.20. Oral LD_{50} values in the mouse range from about 1 g/kg for phenanthrene (Figure 47.2, **IV**) to above 18 g/kg for anthracene (Figure 47.2, **III**). Acute dermal application showed little effect, and the subcutaneous LD_{50} for benzo(a)pyrene was 50 mg/kg in the rat (679).

Six months following the oral administration of 50 different PNAs, cancer of the breast was observed in female Sprague–Dawley rats from 7,12-dimethyl-benz(a)anthracene and $4H$-cyclopenta(def)phenanthrene (690), and incidental carcinoma of the kidney in one of 20 rats receiving $4H$-cyclopenta(def)phenanthrene (690). Repeated and chronic administration of some PNAs has shown neoplastic, carcinogenic, or teratogenic effects. To date, compounds with linear ring structures have been neoplastically negative, whereas benzo(a)pyrene and some of its derivatives have been found active in animal experiments. Structure–action relationships have been described in detail by several authors (708, 711).

4.1.5 Metabolism

Similarly to naphthalene, polycyclic aromatic hydrocarbons are metabolized through epoxides, hydroxides, and their conjugates to be solubilized and excreted. Grover (712) has summarized some findings which showed that, for example, benz(a)anthracene was epoxidized to the K-region 5,6-epoxide by microsomal mixed function oxidases and NADPH when incubated with rat liver in the presence of DNA and protein. The epoxide then undergoes spontaneous rearrangement to the 5-hydroxide, is microsomally hydrated to the 5,6-diol by epoxide hydrase, is conjugated with reduced glutathione to the 5-S-glutathione 6-hydroxide with cytoplasmic glutathione S-epoxide transferase, or may react with cellular constituents to form complexes. A 7-hydroxy 5,6-epoxide was identified in rat lung preparations when 7-hydroxybenz(a)anthracene was added (712). Most related PNAs were shown to metabolize through an epoxide transition state (712). Other investigators have shown that direct monohydroxylation diol and triol formations occur with benzopyrene (713). In addition, as shown with arenes (282), various hydroxide–epoxide (713, 714) or hydroxide–oxide combinations have been identified. Rat liver microsomes can also produce 3- or 6-hydroxymethyl metabolites, as shown with benzo(a)pyrene (715, 716).

The enzyme systems, such as aryl hydrocarbon hydroxylase (AHH) to oxidize PNAs are present in almost all human (717) and animal cell tissues (718–721) and are hepatically inducible, if required for action (717, 722) by both noncarcinogenic and potentially carcinogenic hydrocarbons (715). The stability and cytochrome P-450 and others of the epoxides may depend on immunologic characteristics or state of the particular biologic system, as does the epoxide hydrase enzyme mechanism (723).

4.1.6 Teratogenicity and Mutagenicity

Teratogenic defects occur if an exogenous material passes the placenta to produce embryo- or fetotoxic effects (724). Mutagenic effects originate through a similar mechanism, but also may be transferred genetically from the carrier. Benzo(*a*)pyrene (Figure 47.2, **IX**) has been shown to be both teratogenic and mutagenic in rodents; see Table 47.20.

4.1.7 Carcinogenicity

A variety of animal and cellular carcinogenicity testing models have been described (725). In some cases, initiation of malignant tumors has been difficult, owing to the presence of natural inhibitors and repair mechanisms. Tumor initiation has been described by LaVoie et al. (726), Dipple (708), and many others, and its theoretical prerequisites by Herndon (727) and others (708). Cocarcinogenic or modifying action has been ascribed to concurrent excessive aryl hydroxylase production, increased DNA binding (728, 729), facilitated transport to target sites, promotion of lipid cooxidation (730), or chemical cytotoxicity (731). Conversely, a variety of inhibitors, acting directly on the chemical agent or indirectly on enzyme systems, have been described.

Negative events are metabolic detoxification and excretion, enzyme blocking and repair, and alternate DNA synthesis and repair. Detailed descriptions have been published (728, 732–734). Benzoflavones, particularly the 5,6 and 7,8 derivatives, have been found to possess inhibiting action by blocking the initiation of skin tumors by dimethylbenzanthracene and methylcholanthrene, and inhibition of the dermal NADPH-dependent covalent binding of 3-methylcholanthrene, benzopyrene, and dibenzanthracenes (735). Several direct acting inhibitors of PNAs have been observed, such as butylated hydroxyanisole (BHA), butylated hydroxytoluene (BHT) (736), and vitamin A (737). Some hydrocarbons themselves, biphenylanthracene, benzo(*a*)pyrene, and dimethyl-benzo(*a*)pyrene, are capable of deactivating the hydroxylation of zoxazolamine in the rat (738).

The mechanism of carcinogenic action is complex and completely overlaps with noncarcinogenic events. Many factors play a role and may become rate-limiting under extreme loading conditions. Therefore, extrapolation of bacterial and animal studies to human situations (739) and no-effect level determinations are difficult. *Bacillus megaterium* has been shown to degrade various PNAs (740) and other soil bacteria may convert PNAs to noncarcinogenic degradation products (741). Laboratory experiments have shown that immunologic factors also may play a role in susceptibility towards carcinogenic effects. However, well-nourished animals, preferably on a low fat diet, have shown a low disease incidence.

4.1.8 Industrial Hygiene

No threshold limit values have been established for polynuclear aromatics. ACGIH (384) suggests precautionary measures for workers exposed to benzo(*a*)pyrene.

Sampling techniques and analytical quantitations have been summarized by Strup et al. (742). They include collecting air particulates by sampling train, using an adsorbent glass sampler, recovery with pentane as a desorbant, and followed by spectral analysis (742). A method for air sampling with acrylonitrile–PVC filters is also available (743). Analytical quantitation also has been carried out by column (744) or thin layer chromatographic separation (745), gas chromatography–high resolution mass spectrometry (746), and chemiluminescence (747). The latter also can be used for single cell hydroxybenzo(*a*)pyrene analysis (748).

Methods for eliminating polynuclear aromatics from waste water are available (749).

4.2 Acenaphthene

Acenaphthene, 1,2-dihydroacenaphthylene, periethylenenaphthalene, 1,8-ethylenenaphthalene, $C_{12}H_{10}$, crystallizes in bipyramidal needles. The chemical structure and physicochemical properties can be found in Figure 47.2 (**II**) and Table 47.19, respectively.

Acenaphthene occurs in petroleum bottoms and is used as a dye intermediate, insecticide, fungicide, and in the manufacture of plastics (3).

Physiologic effects include irritation to the skin and mucous membranes (8). Treatment of *Allium cepa* root meristem cells with acenaphthalene vapor for 12 to 96 hr caused anomalies leading to random development of the cells (750). In *Allium cepa* and *Phloeum pratense*, it caused disorientation of microtubules, resulting in altered cellular expansion (751).

4.3 Anthracenes

4.3.1 Anthracene

General. Anthracene, anthracene oil, β-naphthalene, green oil, $C_{14}H_{10}$, is the simplest tricyclic aromatic compound; see Figure 47.2 (**III**) and Table 47.19. It crystallizes as monoclinic plates but sublimes. The crystals are clear white with violet fluorescence when pure, and are yellow and fluorescent green with tetracene and naphthacene as impurities (3). Anthracene occurs in coal tar naphtha (102), from which it is isolated by sublimation (3). Anthracene is used for the preparation of anthraquinone, alizarin dyes, (3) and fluorescent dyes for the evaluation of pesticides applied to cattle (752). Saturated air contains 0.13 percent anthracene (17).

Physiologic Response.   Anthracene asserts phototoxic and photoallergic action on the human skin (102, 324). This has been observed with workers occupied in coal tar and pitch processing. Coal tar fumes containing mainly anthracene and phenanthrene used in field roofing operations produce aerosol droplets of a mean diameter of 5.5 μm, a size that is respirable (753). Flue dust from coal tar pitch also contains respirable portions (754).

In the rat, 0.5 mg when injected subcutaneously decreased the antioxidative activity of the pancreas during the last 25 days after injection. The pancreatic insular cells showed increases in the cell, nucleus, and nucleolus size (755). When added to tocopherol solutions, the antioxidative and antiradical properties were low for anthracene, but were intensified for 7,12-dimethylbenzanthracene and even more for 20-methylcholanthrene (730). Anthracene is classified as carcinogenically inactive (708). This has been substantiated by Sung (756), who found low L- and K-region localization energies (756). This has been confirmed also by Purchase et al. (757) in five of six short-term mutagenicity tests, and by Styles (758) using human and rodent S9 fraction activated and nonactivated assays.

Daphnia pulex accumulated 760 times the anthracene quantity found in water, reaching an equilibrium at 4 hr (759).

Microbiologic Aspect.   Several *Pseudomonas* and *Nocardia* species were found to be capable of degrading anthracene in sediment cores and in shoreline waters in or near oil spills (760). In deep sediment cores, however, the PNA turnover was very slow (761).

Industrial Hygiene.   No official monitoring methods have been recommended. However, analytic procedures are available using colorimetric, polarographic, ultraviolet, spectral, and gas chromatographic techniques (310).

An antipollution procedure is also available (762).

4.3.2   Derivatives of Anthracene

The methyl, anthryl, dimethyl, dipropyl, dinaphthyl, trimethyl, and tetramethyl derivatives of anthracene were found carcinogenically inactive with the exception of 9,10-dimethyl- and 1,8,9-trihydroxyanthracene, which may have contained active impurities when tested (708).

In the rabbit, octahydro- and perhydroanthracenes were hydroxylated similarly to unsaturated hydrocarbon analogues (763).

4.4 Phenanthrenes

4.4.1 Phenanthrene

General. Phenanthrene, $C_{14}H_{10}$ (Figure 47.2, **IV**), an isomer of anthracene, is a crystalline solid. See Table 47.19 for other physicochemical properties.

It occurs in coal tar (3) and can be isolated from several types of crude petroleum (764).

Physiologic Response. Phenanthrene has been identified as a mild allergen (8) and human dermal photosensitizer (3). Limited acute and chronic animal experiments show that phenanthrene is of moderate to low toxicity and when dermally applied was weakly neoplastic (681). In an experimental sister chromatid exchange (SCE) using Chinese hamster cells, phenanthrene exhibited one dicentric chromosome and one gap in chromatid aberrations (765). However, SCE tests using Chinese hamsters in vivo bone marrow was not significant (766). K- and L-Region localization energy determinations predict that phenanthrene should be a noncarcinogen (756). Dipple (708) also concludes that phenanthrene is essentially inactive except for statistically insignificant skin papillomas (708); see also Table 47.20.

4.4.2 Derivatives of Phenanthrene

Little toxicologic information is available on substituted phenanthrenes. Carcinogenically, most derivatives are inactive, such as the 1-methyl, 3 isopropyl, 1-methyl-7-isopropyl, 1,9-dimethyl, and 1,2,3,4-tetramethyl derivatives (708). However, the first and the last compound of the above series and the 1,2,4-trimethyl-substituted phenanthrene have been found slightly active under some testing conditions (708).

Conversely, 1-8-octahydro- and 1-methylphenanthrene have been observed to reduce foliar wilt and vascular discoloration caused by *Fusarium oxysporum* in the tomato (767).

4.5 Benzanthracenes

4.5.1 Benzanthracene

General. Benzanthracene, BA, benz(*a*)anthracene, 1,2-benzanthracene, 2,3-benzophenanthrene, benzo(*b*)phenanthrene, $C_{18}H_{12}$ (Figure 47.2, **V**), is a crystalline solid at room temperature. Some physicochemical properties are listed in Table 47.19.

Benz(*a*)anthracene occurs in crude oil, coal tar, and flue dust (754), and as a pyrolysis product in tobacco smoke at 6.0 to 8.0 µg/100 cigarettes (768), and in coal-derived products (769). It is generated in the gasoline engine and emitted in exhaust gas at 17.3 μg/min from gasoline and up to 56.3 μg/min from other petroleum products (768). The atmosphere over U.S. cities in 1958 contained 0.1 to 21.6 μg/m^3 or an average of 4.0 μg benz(a)anthracene per m^3 (768). Common foods have analytically been determined to contain from 0.20 to 189 ppb (702).

Physiologic Response. In comparison to anthracene, benz(a)anthracene appears more highly toxic (see Table 47.20), but less so than phenanthrene by the dermal or subcutaneous route. Benzanthracene in the mammalian system appears to be metabolized through a 3,4-epoxide (770) to a 3,4-diol, a 3,4-diol-1,2-epoxide (771), or 8,9-dihydroxybenz(a)anthracene-10,11-oxide (772). Benzanthracene induces aryl hydrocarbon hydroxylase by increasing the normal value six to 12 times in the lung, four to nine times in the skin, and two to three times in the small intestine and the kidneys (283). Conversely, 7,8-benzoflavone inhibits this AHH activity (280).

On dermal long-term studies in the mouse, signs of carcinogenic effects have been recorded, but not when administered intravenously or by similar parenteral routes. When toluene was used as a solvent for repeated dermal administration, the tumor induction was insignificant below a concentration of 0.2 percent, whereas benzanthracene in dodecane was still slightly tumorigenic at 0.0002 percent (773). The K-region bond localization energy is relatively high (708), and no hydroxides are formed at positions 5 and 6 (708). Thus it is still a disputed carcinogen (708). Short-term mammalian cell transformation tests have demonstrated positive mutagenicity for 1,2-benzanthracene for the human and the rodent cell lines, but only when activated with S9-fraction homogenates (758).

Microbial transformations of benzanthracene in freshwater sediments have shown a higher turnover than for naphthalene and anthracene (761).

4.5.2 Methylbenzanthracenes

Methylbenzanthracene data are difficult to evaluate, since two types of molecular numbering systems have been used interchangeably in the past; see Figure 47.2, **V**. Therefore, 7-methylbenz(a)anthracene is assumed to be synonymous with 10-methyl-1,2-benzanthracene, whereas 7-methylbenzanthracene is assumed to be methylated β to the K-region.

Limited physicochemical data are listed in Table 47.19 for 6- and 10-methyl-1,2-benzanthracene (2).

The acute toxicity depends on the position of methyl groups, being somewhat lower than for the ring structure and lowest for the 10-methyl- derivatives; see Table 47.20.

Chronic testing has been shown differential properties owing to the variable stereochemical arrangements. *K*-Region bond localization energies predict the 5- and 6-methyl derivatives to be carcinogenically active and the 2- and 7-methyls moderately active (708). The sarcoma incidence was highest for the 6-, 7-, 8-, and 12-methyl compounds, low or negative for the 1-, 4-, 5-, 9-, and 10-methylbenz(*a*)anthracenes, see Table 47.20 (686, 774).

When 7-methylbenz(*a*)anthracene was metabolized in the rat liver and mouse skin preparations, all five possible *trans*-dihydrodiols were gas chromatographically identified (775). A carcinogenically active intermediate, 3,4-dihydro-3,4-dihydroxy-7-methylbenz(*a*)anthracene-1,2-oxide, is suggested by Tierney et al. (775) and confirmed by Chouroulinkov et al. (776). In single 25-µg mouse skin applications, the 3,4-dihydrodiol was the most active compound tested (776). In mouse embryo cells, the 7-methyl compound formed a 7-methylbenz(*a*)anthracene- 5,6-oxide, although it did not form the expected epoxide–ribonucleoside adduct (777), nor the DNA–epoxide complex (778). Slight capabilities in malignant transformations of hamster embryo cells were noted with 10 µg/ml substrate and a 7-month lag phase (779). Of the non-*K*-region diols, the 3,4-diol was most active in transforming mouse M2 fibroblast and V79 Chinese hamster lung cells (780).

4.5.3   Dimethylbenzanthracenes

General.   Many pairing combinations are possible for dimethylbenzanthracene, $C_{20}H_{16}$; however, most of the investigations have been carried out with the 7,12-dimethylbenz(*a*)anthracene, 7,12-DMBA, which is synonymous with 9,10-dimethyl-1,2-benzanthracene; see Figure 47.2, **Va** and **Vb**. Some physicochemical data are listed in Table 47.19. 7,12-DMBA crystallizes in platelets with a faint green-yellow tinge (3). It is produced by a number of synthetic routes (3).

Physiologic Response.   Repeated doses have been documented to have skin and sebaceous gland (226) and hepatic antioxidizing (781) effects, affecting enzymes of the carbohydrate metabolic scheme (782), and the hematopoietic (783) and the endocrinal systems (784). Cellular (785), extracellular (786), and teratogenic effects (787) have been reported also, including possible sterility, as shown in rodents (788). In oral administration to the rat, DMBA accumulated mainly in adipose tissue of the perirenal fat and the mammary gland (789). Orally administered 7,12-DMBA at 20 mg to rats was observed to be dissolved in the lipid fraction of the chylomicrons in the lymph (790). Following intraglandular injection, it was retained in the rat's mandibular gland (791), with liver homogenate 7,12-DMBA metabolized to various water-soluble mono-, di-, and hydroxymethyl- derivatives (792). The formation of glutathione conjugates has been proposed (793).

Tumorigenesis. Although benz(a)anthracene in male rats did not produce sarcomas, methyl substitution transforms it into compounds which, depending on the position of the methyl groups (774), may have high carcinogenic potential (686). For example, 7,12-dimethylbenz(a)anthracene, 7,12-DMBA, is one of the most powerful synthetic carcinogens (794). In humans, it may be an initiator when applied to the skin. Thereafter, a process may be promulgated by sunlight or ultraviolet rays, especially affecting the pale-skinned Caucasian type (795). It may be a direct acting agent; that is, it may produce tumors in situ (774), as shown with 100 mg 7,12-DMBA applied to the lip of rats, which provoked edema, muscle degeneration at the injection site, and fibrosarcoma of the lip 9 months after inoculation (796). DMBA may be transported also to various sites to cause the effect. For example, application to the skin may cause secondary metabolic changes, as shown in two prosimian species, which exhibited extensive enzyme changes and several benign tumors within the 2-year observation period (797). Similarly, oral administration of 9,10-dimethyl-1,2-benzanthracene at 15 mg/ml arachis oil to Chester Beatty strain and Wistar rats produced mammary tumors in 40 days and 4 to 8 weeks, respectively (641). Weekly intravenous injection of 7,12-DMBA to the Syrian hamster produced dermal melanocytomas and tumors of the forestomach, intestine, ovary, skin subcutis, and lymphoreticular tissue (798).

Single-dose applications of 7,12-DMBA may be effective if the material is subsequently permitted to remain in contact with the application site or is subsequently continually exposed to a promoting agent. Apparently rapid epoxidation and carbonylation is necessary for carcinogenic activity (799), which has a continued aftereffect. For example, with a single topical application, certain strains of mice developed ulcers in 5 to 15 days, whereas others showed little reaction (800). A definite dose–action relationship was noted in female mice, where a single topical application of 200 µg 7,12-DMBA had required 4 to 8 weeks, and 25 to 100 µg 36 weeks to induce papillomas (801), but some regressions were also observed. A single application of 7,12-DMBA, followed by long-wave length ionizating oxidation, induced carcinogenesis in hairless mice (802). Oral one-time administration caused a dose-related depigmentation in the golden hamster (803) and induced the formation of melanocyte tumors which appeared after 3 months (803). However, the overall inducibility of tumors by 7,12-DMBA appears to depend on the age, diet, and physiologic state of the rat (804).

A variety of factors have been observed to promote the effect of 7,12-DMBA. For example, ultraviolet irradiation preceding repeated dermal application of 7,12-DMBA increased the tumor incidence of mouse skin (805). Tobacco smoke condensate and phorbol esters extracted from condensates increased 7,12-DMBA-induced papilloma (806, 807). When followed by croton oil, 7,12-DMBA produced about twice as many papillomas in the mouse skin (808), and feeding of a 20 percent corn oil diet accelerated the mammary tumor growth (809).

However, a tumor regression also was observed (810). Indirect activation of 7,12-DMBA has been noted with increased metabolic potential through polyoma-virus cell transformation (811).

Concurrently, a variety of inhibitory factors may be available. For example, vitamin A has been shown to decrease papillomas in 7,12-DMBA fed (812) or dermally treated mice (813) or to decrease chromosomal breakage in blood leukocyte cultures, along with vitamin C (814). Vitamin A also decreased microsomal mixed function oxidases from mice and hamster liver and lung tissues (737). Vitamins C and E decreased the tumor incidence of mouse skin (814). Mouse skin tumors were inhibited by 7,8-benzoflavone (815, 816), as was also the mutation of Chinese hamster cells (817). Butylated hydroxyanisole (BHA) and hydroxytoluene (BHT) inhibited neoplastic effects in the mouse skin (736, 818). Actinomycin D inhibited the 7,12-DMBA uptake into Syrian hamster embryo cells (819) and inhibited mouse skin (820) and mammary tumors (821). A variety of other inhibitors are DDT [1,1,1-trichloro-2,2-bis(p-chlorophenyl)ethane] (822), cyproterone acetate (823), ergocornine (824), and chloramphenicol (825). The increase of the salt or water-loading of rats decreased the subcutaneously induced blastomas (826), but increased the liver microsomal activity with barbiturates (827) or the antioxidative properties of lipids (730). This decreased the general sarcoma incidence (828).

Immunologic responses may also play an important part in the resistance, susceptibility, and response to the carcinogenic effects of 7,12-DMBA. Thus neonatal injection of 7,12-DMBA into mice induced some tumors at 100 to 200 days and also depressed the immune response (829). This was observed also in the spleen of the rat (830). However, regional lymphadenectomy of induinal and axillary nodes did not significantly affect tumor induction (831). Subcutaneous injection of 7,12-DMBA into neonates, which has caused lymphocytic leukemia, may be correlated with thymus effects (832, 833) or B- and T-cell depression in the mouse (834).

Mutagenicity.   According to data by Styles (758), 7,12-dimethylbenzanthracene exhibits low mutagenic effects when tested with human cell cultures and S9-fraction activation, but practically no effects when tested without activation (758) as also found by Huberman and Sachs (817). In rodent cell tests, nonactivated it appears as a weak, and activated as a moderate, mutagen (758). The derivative 7,12-DMBA has also been shown to cause sister chromatid exchanges and chromosome aberrations (835). 9,10-Dimethylbenzanthracene short-term mammalian cell transformation tests showed a slightly positive reaction with the human cultures and moderate activity when activated with S9-fraction mixtures (758).

Cellular Mechanisms.   In both mutagenic and carcinogenic action, innumerable factors play a part, representing normal functional variables, such as 7,12-

DMBA binding to replicating and nonreplicating DNA (836), binding to main-band epidermal DNA (837), or reaction with K-region possible epoxides or diols, as shown with 7,12-DMBA in mouse fibroblasts (838). Cellular repair mechanisms are activated whenever the need arises, as has been described in detail for 7,12-DMBA by D'Ambrosio et al. (839). This also overlaps with regressions observed with actinomycin D (840) and smog extract-transformed cell lines (841).

Industrial Hygiene.   The benzanthracenes, especially the 4- to 10- or 12-methylated derivatives, warrant careful handling. The neoplastic no-effect level, for example, for 7,12-dimethylbenzanthracene, is very low for mice. However, the applicability to humans is presently unknown.

4.5.4   Trimethylbenzanthracenes

Syntheses of trimethylbenzanthracenes have been investigated and described by Newmann and Hung (842).

Cellular effects such as sister chromatid exchanges and chromosome aberrations of 7,8,12-trimethylbenz(*a*)anthracene appear to resemble those of 7,12-dimethylbenz(*a*)anthracene (835), as described above.

4.6   Benzophenanthrenes

4.6.1   1,2-Benzophenanthrene (Chrysene)

1,2-Benzophenanthrene, benzo(*a*)phenanthrene, chrysene, $C_{18}H_{12}$ (Figure 47.2, **VI**), is a solid material; see Table 47.19 for physicochemical data. Chrysene can be isolated from crude petroleum and coal tar. It occurs in cigarette smoke and has been detected at 1.5 to 13.3 ng/m^3 in community air (768).

Chrysene exerts low acute toxicity orally, dermally, to the eye, and by inhalation (843). It has been shown to exhibit low carcinogenic potential (768); however, this is presently disputed (708).

Industrially, chrysene should be handled with precaution, however, it presents a low risk unless contaminated. Antipollution methods are also available (762).

4.6.2   3,4-Benzophenanthrene

3,4-Benzophenanthrene, benzo(*c*)phenanthrene (Figure 47.2, **VII**), is also a solid material; see Table 47.19 for physicochemical data.

Theoretical NMR chemical shift determinations predict this to be a material of low carcinogenic potential (844). This also was suggested by bond localization energy evaluations (756).

4.6.3   Methylchrysenes

The 1-, 2-, 3-, and 6-methylchrysenes occur in cigarette smoke at 2 μg/100 cigarettes (768).

The 1-methyl, 2-isopropyl, and 2,3-dimethyl derivatives have shown no significant carcinogenic potential, the 6-methyl and the 4,5-dimethyl slight, and the 5-methyl and the 5,6-dimethyl moderate activity (730). These data and experiments with even further substitution (845–847) indicate that group introduction at position 5 is essential for basic carcinogenic potential. The 5-methylchrysene occurs in tobacco smoke at about 0.06 μg/100 cigarettes (768).

1,2-Dimethylchrysene is a solid material. For physicochemical properties, see Table 47.19. It is believed to possess low neoplastic potential. The 1,7-dimethyl chrysene has been isolated from coal tar pitch.

4.7   Pyrenes

4.7.1   Pyrene

General.   Pyrene, benzo(d,e,f)phenanthrene, $C_{16}H_{10}$ (Figure 47.2, **VIII**), is a colorless solid, soluble in organic solvents. Further physicochemical properties are listed in Table 47.19. It occurs in pyrolysis or cooking processes at the lower cooling temperatures (702), and has been detected in the U.S. community atmosphere in 12 cities from traces to 35 ng/m³, to 1.3 to 19.3 ng/m³ in Detroit in 1958 (768).

Physiologic Response.   According to Potapova et al. (848), rats at oral doses near the LD_{50} succumbed in 2 to 5 days, and at the inhalation LC_{50} in 1 to 2 days, although a dose of 10 g/kg on the mouse skin was of low toxicity; see Table 47.20. Inhalation also caused hepatic, pulmonary, and intragastric pathologic changes. The number of neutrophils, leukocytes, and erythrocytes decreased. Cutaneous applications for 10 days caused hyperemia, weight loss, and hematopoietic changes; applications for 30 days produced dermatitis; and chronic effects consisted of leukocytosis and lengthened chromaxia of the leg muscle flexors. Workers exposed to 3 to 5 mg/m³ noted disturbances that disappeared at levels below 0.1 mg/m³. Some teratogenic, but no blastomogenic or carcinogenic effects were noted (849, 850), except for an occasional papilloma (851). The low carcinogenic potential agreed with the calculated low K-region bond localization energy (756).

Rat liver microsomal systems were found capable of metabolizing pyrene to 1-hydroxy- and 4,5-dihydro-4,5-dihydroxypyrene, as well as 1,6- and 1,8-pyrenequinone (641).

4.7.2   Benzo(a)pyrene

General.   Benzo(a)pyrene, 3,4-benzopyrene, formerly named 1,2-benzpyrene, BaP, with an empirical formula $C_{20}H_{12}$ (Figure 47.2, **IX**), crystallizes as yellow needles. Further physicochemical properties are listed in Table 47.19.

Most experimental work on benzopyrenes has been carried out with benzo(a)pyrene, because of the findings that it is an animal carcinogen and suspected to be active in humans (283).

Benzo(a)pyrene occurs naturally in crude oils, shale oils (852), and coal tars (786), and is emitted with gases and fly ash from active volcanoes (853). In a variety of processes, BaP is pyrolytically formed and emitted into the atmosphere. Cigarette smoke and tar contain up to 0.1 percent 3,4-benzo(a)pyrene (584), pyrolyzed from isoprene and C_6 to C_{10} alkylbenzene precursors. In a 20-m^3 area where two persons smoked, a concentration of 9.7 to 11.1 ng/m^3 of benzo(a)pyrene was detected (704). The gasoline engine promulgates up to 0.170 ng benzo(a)pyrene/gal fuel but only 0.02 to 0.03 ng/gal in a 1968 emission controlled vehicle (768, 855). Equivalently, the greatest emissions occur from residential energy production in coal and wood furnaces (768), mounting to tons of BaP per year in the United States. Other sources represent industrial coke-oven emissions (856) and road abrasions, with the result that world atmospheric concentrations have been measured as 0.05 to 74 ng/m^3 in European and 100 U.S. cities in 1958 (768) and 18 ng/m^3 in Berlin (857). Subsequently, benzopyrene may enter the food chain. It was identified and found by one author at 7.0 ppb in 32 food products (858) and 0.4 to 99 ppb according to another review (702). Benzopyrene also is produced when edible fats are superheated (859). Conversely, baking and irradiation decrease the BaP content of foods (860). Soil contamination is generally proportional to prevailing air concentrations of BaP (861). From soil, migration into plants can occur (862), although BaP is degradable by some soil microorganisms (862).

Physiologic Response.   The acute toxicity of benzo(a)pyrene appears low when administered orally or dermally to laboratory animals; see Tables 47.20. A dose of 15 mg of BaP injected intraperitoneally to AKR and C57BL-6 mice was nontoxic (863).

Long-term effects in the human have been suspected to be bronchial carcinoma (864). Contributing factors may be smoking habits and environmental conditions (see preceding section). In an epidemiologic study by Sterling and Pollack (865), a relationship was observed between lung cancer, soot-borne benzo(a)pyrene, soot per se, and U.S. per capita cigarette consumption versus death rates. These factors all increased, whereas consumption of coal and lignite declined (865).

Experimental work with animals showed that the repeated oral BaP administration resulted in hypoplastic anemia in mice (866); intratrachial installation of 0.63 mg BaP once weekly for life into Syrian hamsters resulted in the development of bronchogenic adenomas, growth of epithelial cords of cells into lung tissues, and tumor formation (867), with changes to hyperplastic, then squamous metaplastic, epithelium, and papillomas (868). Subcutaneous injection of 0.5 mg BaP into rats increased the antioxidative activity of the pancreas with increases of the insular cell nucleus and nucleolus sizes (755). Experiments in mice have shown that transplacental passage of BaP to the embryonic tissue occurs 15 min after intravenous treatment (869). A dose of 12 mg administered to pregnant A and C57BL mice produced an incidence of lung tumors in 31.6 percent male and 9.1 percent female offspring versus 1.2 percent in control males, whereas pyrene produced no blastomogenic effects (849). Pregnant rats administered 4 mg/kg BaP during gestation produced no effects in the offspring, whereas 20 mg/kg resulted in a 20 percent tumor incidence (870). In a four-generation mouse study, the incidence of papillomas and carcinomas increased (871). In a detailed discussion, Juchau et al. (872) mention the increased sensitivity of fetal tissue, and in a study with human placental tissue show the increased benzo(a)pyrene hydroxylase activity in the early gestation period, and this especially in smokers (873).

Absorption, Distribution, and Metabolism.   Benzo(a)pyrene was readily absorbed by mussel gill in marine bodies (289). Distribution to striated muscle cells was rapid in the cells of *Xenopus laevis*.

Benzo(a)pyrene appears to be metabolized readily by the mammalian system or excreted in unchanged form. Leber et al. (874) have summarized experimental work with the rat, the primate, and the miniature swine. The rat hepatic microsomes convert BaP into a variety of metabolites, of which seven have been gas chromatographically identified as 9,10-, 4,5-, and 7,8-diol, 1,6- and 3,6-quinone, and 9- and 3-hydroxybenzo(a)pyrene by sequentially increasing retention time (283, 874–876). The synthesis and compound identification have been discussed by Jerina et al. (877), and a multiple pathway system, including possible epoxides, by others (707, 772, 875). The benzo(a)pyrene-4,5-oxide has been prepared and its physicochemical and biologic characteristics extensively studied by Glusker et al. (878). The 3- and the 6-hydroxy-, and 3- and the 5-hydroxymethylbenzo(a)pyrenes have been identified also as metabolic products (879–882). A major metabolite in marine fish has been revealed as 7,8-dihydro-7,8-dihydroxybenzene (290). Under the influence of Mn^{2+} at 10 μmol, the 9,10-, the 4,5-, and the 7,8-diol were produced; at 20 μmol, mainly the 7,8-diol appeared (883). Metabolites produced by human placental microsomes were 3-hydroxy-, 4,5-, 7,8-, and 9,10-dihydrodihydroxybenzo(a)pyrenes and some quinones, as well as other unidentified diols (884), with the majority of the 7,8-derivative.

Responsible for these metabolites are inducible enzyme systems. Rat liver microsomes have shown a portion of proteins belonging to the cytochrome P-

448, P-450, NADPH-cytochrome reductase C system (885). This system may be activated by hydrocarbons such as 3,4-benzopyrene (886, 887), which in turn induce BaP hydroxylases. Their activity decreases in the order, liver > intestine > lung > kidney (885). The mixed function oxidases responsible to convert lipid-soluble to water-soluble compounds appear responsible also for the promulgation of a variety of metabolites, such as the 7,8- and 9,10-diols (714, 888) and the 4,5- and 9,10-oxo derivatives (889, 890). One of the mixed function oxidase systems, aryl hydrocarbon hydroxylase (AHH), may be active in the conversion of BaP to the 3-hydroxy derivative (891). A variety of other hydroxylases (892) and monooxygenases are also activated by 3,4-BaP. Another specific enzyme system, the expoxide hydrase system, has been described by Lu et al. (893). This may be the rate-limiting, metabolizing enzyme. It can be inhibited by several chemical agents, such as 1,1,1-trichloropropene-2,3-oxide (894). Other chemical agents, such as styrene oxide and cyclohexene oxide, inhibit the general enzyme activities (895), whereas many materials such as lead are effective with BaP in inducing microsomal enzymes (896).

Mutagenicity. In cell transformation studies, 3,4-benzopyrene was slightly positive in the activated human WI-38 test and practically negative in the nonactivated, but considerably active in the rodent S9-activated system (758). Conversely, in the bacterial plating assay, human liver homogenate was inactive with 3,4-benzopyrene in *Salmonella typhimurium* mutagenicity testing (897). Wood et al. (898) determined the structural requirements of BaP derivatives for positive mutagenic characteristics. These included primarily benzo(*a*)pyrene with saturation at positions 7 and 8, or saturation and hydroxylation at 7 and 8, either cis or trans, and the 4,5-oxide.

Tumorigenic and Neoplastic Response. Bond localization energy determinations predict 3,4-benzopyrene to be of relatively high neoplastic potency (756). When it was injected subcutaneously into mice, a definite dose–response relationship was observed and also in tricaprylin a sigmoid dose–response curve was obtained (899). In the hamster, following intratracheal instillation, mainly respiratory tract and lung tumors were observed (900). However, 3,4-benzopyrene was optimally active primarily when administered with a carrier, such as the noncarcinogenic ferric oxide (901, 902), or in combination with α-radiation (903). For example, carbon black addition resulted in a benzo(*a*)pyrene dose-related response (904). The potentiation is physical, relying on the increased pulmonary retention rate. Hamster embryo cell-transformed foci have been reported (905) and some chromosomal and chromatid aberrations in Chinese hamster cells (766), but only slightly higher than observed for phenanthrene (765). Solvents such as croton oil (906, 907), *n*-dodecane (773), or C_{12} to C_{20} *n*-alkanes (908) promoted the effects of benzo(*a*)pyrene. Retinyl acetate weakly, but actively, enhanced the incidence of BaP-induced respiratory tract tumors (909); furfural (910) and pyran copolymer accelerated respiratory tract and skin tumors, respectively (911). Conversely, a variety of agents or conditions decelerate or inhibit BaP-induced tumor formation. For example, biphenylamine derivatives may delay the appearance of tumors (912), as does cetane. Olive oil practically inhibits mouse skin tumors. Aluminum oxide considerably decreased the tumor incidence (913), whereas the antioxidants BHA, BHT (914, 915), and α-tocopherol (916) decreased or completely abolished the neoplastic response to BaP-induced pulmonary adenoma formation (915), forestomach neoplasm, or mammary tumor development (914). A variety of other agents inhibit or greatly reduce the BaP-induced tumor incidence, as, for example, dimethylacetylene dicarboxylate (917), 7,8-benzoflavone (918), griseofulvin (919), or selenium (920).

No general mechanism of action has emerged yet, for the factors involved in normal benzopyrene metabolism are many. However, the reactions appear dose-related and nonselective (283), and depend on many normal mechanisms, such as enzymatic hydroxylations, oxidations, epoxidation, hydration at selected stereochemical positions (899), general induction of cytoxic response (921), or detoxification enzymes (922), lipid and carbohydrate metabolism adjustments (923), protein and serum albumin binding (924, 925), and benzopyrene–DNA (926–930) or RNA interaction (931).

The pancreas islet cells appear to possess a natural antioxidative capacity affecting PNAs, such as BaP (755). Immune systems may be involved also (932). Further aspects of mechanisms are discussed in detail by Ts'o (933). Single-dose treatment in the Syrian golden hamster appeared to be of reversible effect unless it was followed by a carrier or solubility-modifying agent.

Microbiologic Aspect.   Various soil types (934) and microorganisms are capable of degrading benzo(a)pyrene. Some are *Bacillus megaterium* (935, 936, 937) and *Pseudomonas aeruginosa* (938).

Industrial Hygiene.   For benzo(a)pyrene, no official air threshold limits have been established.

Sample collection by absorption on charcoal or silica may be used. A selection of analytic methods are available, the preferential techniques involved column chromatographic separation with gas chromatographic–mass spectral quantitation.

When benzo(a)pyrene-containing products are handled, protective garments should be worn, and adequate ventilation and respiratory equipment should be available.

4.7.3   Benzo(e)pyrene

Benzo(e)pyrene, 1,2-benzopyrene, BeP, $C_{20}H_{12}$ (Figure 47.2, **X**), is a position isomer of benzo(a)pyrene, BaP (**IX**). Its physicochemical properties resemble that of BaP; see Table 47.19.

Benzo(e)pyrene has been identified and can be isolated from coal tar (768).

It has been found in the atmosphere over U.S. cities at 1 to 25 ng/m³, at an average of 5 ng/m³ (912).

Whereas BaP has been found to act as a procarcinogen, BeP is inactive (875); see Table 47.20. This is explained by Lehr et al. (939), who summarized the findings that no chemically stable BeP epoxide has yet been isolated. A 9,10-dihydrobenzo(e)pyrene derivative has been observed to be very weakly active with an average of 0.5 papillomas per mouse (939). The inactivity also has been predicted with mathematical models, whereby Popp (926) determined that BeP possessed lower coupling affinity to DNA and RNA than BaP (926).

Some older studies carried out in 1936 and 1959 (701, 940) report neoplastic and carcinogenic effects in the mouse, but not in the guinea pig, possibly stemming from the presence of chemical impurities.

Analytic determination procedures are available (875).

4.7.4 Cyclopentapyrene

Cyclopenta(c,d)pyrene, acepyrene, 3,3a,4-cyclopentapyrene, $C_{17}H_{12}$ (Figure 47.2, **XI**), is a five-member fused ring system with a molecular weight of 226.28 and melting point of 170°C (941). Jacob and Grimmer (941) recently synthesized the compound from pyrene. Cyclopentapyrene has been isolated from carbon black by Neal and Trieff in 1972 (942) and later by Gold and identified by infrared, ultraviolet, and mass spectrometry (943).

Based on bacterial plating tests using *Salmonella* mutants, cyclopentapyrene was classified as a frame shift mutagen (944).

4.8 Cyclopenta- and Dibenzophenanthrenes

4.8.1 3-Methylcholanthrene

General. 3-Methylcholanthrene, 3-methylbenz(j)aceanthrene, 20-methylcholanthrene, 3-methyl-i,j-cyclopentabenz(a)anthracene, 3-methylcyclopentabenzophenanthrene, 3-MCH,$C_{21}H_{16}$ (Figure 47.2) **XII**), is a solid that crystallizes from benzene and ether in pale yellow prisms (3). For further physicochemical properties, see Table 47.19.

Physiologic Response. On direct contact, 3-methylcholanthrene is moderately, and on repeated exposure highly, irritant. When administered orally to inbred hamsters, 3-MCH produced neoplasms of the colon (732). When applied to mouse skin, the number of sebaceous glands dropped sharply, but regenerated after 7 to 9 days (945). Implantation in wax pellets provoked an initial inflammatory reaction which then subsided (946). About day 55, histocytes and lymphocytes started to proliferate, resulting in indignant growth (946). Generally, 3-MCH was more highly toxic than benzo(a)pyrene, and neoplasms appeared sooner in the mouse (947). 3-MCH administered to cystine-deficient mice resulted in sclerotic lesions of the aorta and other large arteries (948). On repeated or chronic exposure, 3-MCH proved neoplastic or carcinogenic by almost all routes tested in the rat, mouse, hamster, guinea pig, rabbit, and dog; see Table 47.20.

Metabolism. The metabolic products of 3-methylcholanthrene vary with the type of enzyme induction (949). In fetal rat liver, several compounds, such as the 1- or 2-hydroxy-, the *cis*- and *trans*-1,2-dihydroxy-, the 11,12-dihydroxy-11,12-dihydro-, and the 1- and 2-keto-3-cholanthrene were chromatographically identified (949). Most active as an enzyme inducer was the *cis*-11,12-dihydro-dihydroxy-3-methylcholanthrene. The 1, less than the 2, derivative was further metabolized to *trans*-9,10-dihydrodiols (713).

Methylcholanthrene is one of the prototypes of mixed function oxygenase inducers. It promotes especially the induction or activation of cytochrome P-448, also called P-450, of the aryl hydrocarbon hydroxylase (AHH) system. Therefore 3-MCH is often used as an experimental positive control. The AHH system activation has been studied in human lymphocyte cultures, where tissues from individuals with a history of lung and bladder neoplasms showed normal to very high AHH inducibility (950). Extensive studies in the mouse showed dose-related effects of 3-MCH and irradiation on AHH induction (951), gene-related effects with 3-MCH alone (952), induction in the rat with concurrent dimethylase increase (953), which also decreased the yield of toxic carbon tetrachloride metabolites (954), and induction in the chicken (955) and in the trout (674). Methylcholanthrene also was found to induce other enzyme systems, such as arene and alkene oxide monooxygenases (956).

Mutagenicity and Tumorigenesis. 3-Methylcholanthrene was practically inactive when tested for nonactivated mutagenic potential in the human cell system and was of low potency when S9-fraction induced, but proved highly active in rodent transformation systems (758). It caused slight chromosomal and sister chromatid aberrations in Chinese hamster cultures (765). The 1-hydroxy-9,10-diols appear to be mutagenically the most active of the metabolites tested (957).

3-Methylcholanthrene appears to be a rapid, all-around neoplastic agent (899) and a potent hepatotumorigen (912). It produced mammary carcinogenesis in hamsters when tested by gavage (958) and mammary tumors by 3-MCH embedded in paraffin pellets (959). Mineral oil (British definition) may cause a slight acceleration in tumor induction (960), whereas 1,1,1-trichloro-2-propene is a potent inhibitor of the 3-methylcholanthrene-11,12-oxide, stimulating tumor induction in the BALB/c mouse (961).

The exact mechanism of action is not quite clear; although it has been observed that 3-MCH tumor induction depends on aryl hydrocarbon hydroxylase, which, in turn, is related to dominant gene-regulating activities (962). It also has been recognized that 3-MCH binds to protein in the vascular system and changes liver lysosomal proteins (963), and in addition causes immunologic changes (964). In general, 3-MCH is a more potent tumorigen that dimethyl-benzanthracene (957).

4.9   Other Aryl Hydrocarbons

Dibenz(*a*,*h*)anthracene, 1,2 : 5,6-dibenzanthracene (Figure 47.2, **XIII**), has been isolated from coal tar pitch, and it occurs in coke-oven effluents (768). In Germany, it has been detected in the atmosphere at 3.2 to 32 ng/m^3 (768), and it occurs in cigarette smoke at 0.4 μg/100 g (768). On topical application to the mouse skin spontaneous tumors have occurred (899). The minimum effective dose was 0.0025 mg in olive oil (899). Most methyl-substituted dibenzanthracenes are tumorigenically active, the *db*(*a*,*c*)*a*-9,14-dimethyl derivative is inactive, the *db*(*a*,*j*)*a*-10-methyl compound slightly, the 2-, 3-, 7-, 14-methylated (*a*,*h*)anthracenes slightly, the 6-methyl moderately, and the 7-methyldibenz(*a*,*h*)anthracene highly active (708).

Picene and pentaphene are inactive. The triphenylenes are moderately active.

4.10   Polynuclear Aromatics in General

No general mechanism for tumorigenic action applies throughout the afore-mentioned compounds.

5   PETROLEUM AS A HYDROCARBON SOURCE

Petroleum or crude oil recovery and transport normally is carried out by mechanical means, using enclosed systems. Transfer between vessels may represent points of exposure which, however, constitute a relatively low human hazard. The health hazard is lower when handling crude oil than certain fractionated materials. A simplified list of crude oil fractions and their uses is presented in Table 47.21.

During refining exposures vary greatly, being low for the paraffin gases and moderate for kerosine or solvent type to light oil fractions; see Tables 47.22 and 47.23. A portion of still bottoms, residual, or pyrolysis oils, or asphalts, on direct dermal contact or on inhalation of fumes, may have tumorigenic potential. Permissive threshold concentrations can be calculated if the specific composition is known. Thus with proper precautions, the risks for toxic or tumorigenic effects are low; see Table 47.24.

A study of the deaths of refinery workers in Texas was carried out on the basis of OCAW membership from 1947 to 1977 (996). The data indicate that

Table 47.21. Petroleum Fractionation (5, 148)

Fraction	Organic Compound	Boiling Range [°C (°F)]	Use
Natural gas	C_1–C_2	-164 to -88 (-263 to -126.4)	Fuel, chemical
Liquefied, or bottled, gas	C_3–C_4	-44.4 to $+1.0$ (-48 to $+34$)	Fuel gas; for the synthesis of rubber components, petrochemicals
Petroleum ether	C_4–C_5	20–60 (68–140)	Solvents
Gasolines	C_5–C_{10}	32–149 (90–300)	Aviation fuel
		32–210 (90–410)	Motor gasoline
Naphthas	C_6–C_{10}	65–204.4 (149–400)	Cleaning fluids, solvents, refining stock
Kerosines	C_5–C_{16}	40–300 (104–572)	Jet and turbofuels
Kerosine		350–550 (176.7–287.8)	Stove oil, tractor and gas turbine fuel
Gas oil	C_9–C_{16}	204–371 (400–700)	Furnace oil Diesel oil
Lubricating stocks	C_{17}–higher	204–400 (400–750)	White oils Lubricating oils and greases
Waxes	C_{20}–higher	204–400 (400–750)	Sealing wax Food component
Bottoms	C_{20}–higher	400–higher (750–higher)	Heavy fuel oil Road oils Asphalts

deaths due to cancer and arterial diseases, according to some evaluations, are slightly higher in refinery and petrochemical production workers than for others in the plant and the control population. When the workers were divided into three work-year groups, the digestive cancer and respiratory system incidence appeared to increase with work-years, and brain and CNS malignancy declined at the 10 to 19 work-year range (996). However, the suspected causes and effects were not uniform whether between worker groups or plants (996).

For example, a petroleum fraction, Iomex, when administered orally to rats at 1 ml/kg for 4 weeks did not produce any toxic symptoms, but at higher doses

Table 47.22. Physicochemical Characteristics of Petroleum Naphthas and Thinners

Common Name	Alternate Name	B.P. [°C (°F)]	Flash Pt. [°C (°F)]	Mol. Wt.	Carbon Number	Class of Components	TLV [ppm (mg/m³)]	Ref.
Crude oil	Earth oil, petroleum	<0 to >1000 (<32 to >2000)	−7 to −32 (20−90)	N.A.	$C_1−>C_{50}$	About 300 organic substances identified, some heavy metals	5 mg/m³ for misting	965
Gasoline	Benzin, petrol, motor or aviation gasoline	32−210 (90−410)	40−70 (100−150)	~100	$C_4−C_{12}$	n- and isoparaffins, olefins, aromatics	250−500 (1000−2000)	5, 25
Petroleum ether	Ligroin, petroleum benzin	30−60 (86−140)	−57 to −46 (−70 to −50)	~77	$C_5−C_6$	Paraffins, (pentanes, hexanes, isohexanes)	200 (350) 120 (200)[c]	966
Rubber solvent		45−125 (113−257)	−13 (9)[a]	84 −97	$C_5−C_7$	Paraffins, monocycloparaffins, olefins (trace), benzene, alkylbenzenes	200 (350) 120 (200)[c]	966
VM&P naphtha	Range of 80 thinner	95−160 (203−320)	−7 to 13 (20−55)	87−114	$C_5−C_{11}$	Paraffins, mono- and dicycloparaffins, benzene (trace), alkylbenzenes	200 (350) 120 (200)[c]	966
Mineral spirits (petroleum spirits)	Refined petroleum solvent, White spirits	150−200 (302−392)	<0−35	~130	C_6	Paraffins, naphthenes, olefins, aromatics	200 (350) 200 (350)[c]	966

Name	Name	Boiling range, °C (°F)	Flash point, °C (°F)		Carbon number	Composition			Reference
Stoddard solvent	White spirits	160–210 (320–410)	38–43 (100–110)	~140	C$_7$–C$_{12}$	Paraffins, mono- and dicycloparaffins, benzene (trace), alkylbenzenes	200 (350)	200 (350)c	966
High flash naphtha		150–204 (302–400)		~140	C$_7$–C$_{12}$	Paraffins, naphthenes, aromatics	200 (350)		966
Aliphatic solvent naphtha	140° flash naphtha	185–207 (364–403)	59–60 (138–140)	154	C$_5$–C$_{13}$	Paraffins, mono- and dicycloparaffins, benzene (trace), alkylbenzene	200 (350)	200 (350)c	966, 967
Coal tar naphtha	Aromatic petroleum naphtha	93–315 (200–600)	2–38 (35–100)	~140	C$_8$–C$_{13}$	Paraffins, alkylbenzenes	200	(100)	968, 242
High aromatic naphtha	High aromatic solvent	184–206 (364–403)	62 (144)a	~140	C$_8$–C$_{13}$	Paraffins, mono-, di-, and tricyclic naphthenes, alkylbenzenes and naphthalenes, olefins			969
Thinner 40		186.7–230.6 (368–447)	49 (120)a	~148	C$_8$–C$_{13}$	Paraffins, mono-, and dicycloparaffins, mono- and diolefins			970
50		97.8–105 (208–221)	4.5 (40)	97	C$_6$–C$_8$	Paraffins, olefins, naphthenes, aromatics			971
60		128.3–159.4 (263–319)	30 (86)	120	C$_5$–C$_{10}$	Paraffins, monocycloparaffins, alkylbenzenes			972
70		157.2–210.6 (315–411)	—	132	C$_5$–C$_{12}$	Paraffins, monocycloparaffins, alkylbenzenes			973

Table 47.22. (*Continued*)

Common Name	Alternate Name	B.P. [°C (°F)]	Flash Pt. [°C (°F)]	Mol. Wt.	Carbon Number	Class of Components	TLV [ppm (mg/m^3)]	Ref.
80		96.7–142.2 (206–288)	30 (86)[b] 33 (38)[a]	106	C_6–C_9	Paraffins, dicycloparaffins, alkylbenzenes		974
Kerosine	Stove oil	163–288 (325–550)	49–52 (120–125)	~180	C_{10}–C_{16}	Aliphatics, mono- and dicycloparaffins, alkylbenzenes		2, 5
Deodorized kerosine		207.8–272.2 (406–522)	80 (176)[a] 88 (190)[b]	179	C_6–C_{14}	Paraffins, mono- and dicycloparaffins, aromatics		975

[a] Tagg closed cup.
[b] Tagg open cup.
[c] Action level.

Table 47.23. Physicochemical Data for Kerosine-type Distillates

Common Name	Alternate Name	B.P. [°C (°F)]	Flash Pt. [°C (°F)]	Freezing Pt.	Mol. Wt.	Carbon Number	Class of Components	Ref.
Jet fuel								
JP-1	SUS <32	(410–572)	35–63 (95–145)	−76		C_5–C_{16}	Aliphatics, mono- and dicycloparaffins, alkylbenzenes	2, 5
JP-3		(240–470)	−23 to −1 (−10 to 30)	−76		C_5–C_{16}	Aliphatics, mono- and dicycloparaffins, alkylbenzenes	2, 5
JP-4		(<290–470)		−76	~160	C_5–C_{16}	Aliphatics, olefins, mono- and dicyloparaffins, alkylbenzenes	5
JP-5		(400–550)	35–63 (95–145)	−40	~170	C_5–C_{16}	Aliphatics, olefins, mono- and dicycloparaffins, alkylbenzenes	2, 5
JP-6		(250–500)		−65		C_5–C_{16}	Aliphatics, olefins, mono- and dicycloparaffins, alkylbenzenes	5
Diesel fuel	SUS <32 to 45	177–400 (350–750)	38–54 (100–130)			C_5–>C_{16}	Aliphatics, olefins, mono- and dicycloparaffins, alkylbenzenes	5
Distillate heating oils								
No. 1	SUS ~35	216–330 (420–625)	38–74 (100–165)			C_{11}–>C_{16}		5
No. 2	SUS 35–55	184–334 (363–634)	38 (100) min			C_9–>C_{16}		5
Motor oil	SUS 60–530	366–588 (690–1090)				C_7–>C_{20}		5
White oil spray or medical	SUS 186.6 (37.8°C)	0.845–0.995[a]					Aliphatics	976
Petroleum wax	SUS 50–92.2 (210°C)	64–91[b] (425–580)			490–655	C_{20}–C_{32}	Aliphatics	5

[a] Specific gravity.
[b] Melting point.

the animals showed reduced water and food intake with loss of body weight, high mortality, and slight abnormalities in the lung, liver, and kidney, but significant decreases in erythrocytes, leukocytes, and hemoglobin (997).

A number of references point out the ability of marine animals to absorb, metabolize, and release crude petroleum or fractions; see discussion of individual alkanes. In general, aromatic hydrocarbons are retained longer than alkanes, and the highest molecular weight compounds released at the lowest rate (19).

Sampling procedures vary with the fractions under investigation.

Table 47.24. Physiologic Response to Petroleum Solvents

Material	Species	Route of Entry	Dose or Concentration	Result or Effect	Ref.
Crude oil	Fish	Aquatic	>1000 ppm	TLm96	365
Gasoline (b.p. <230°F)	Human	Oral	10–15 g	Lethal in children	977
	Human	Inhalation (acute)	20–50 g	Toxic effects in adults	40
			550 ppm (~2 mg/liter)/1 hr	No effects	148
			900 ppm (~3.5 mg/liter)/1 hr	Slight dizziness, irritation of eyes, nose, throat	148, 978
			2000 ppm (~7.6 mg/liter)/1 hr	Dizziness, mucous membrane irritation, and anesthesia	148
			10,000 ppm (~37 mg/liter)/1 hr	Nose and throat irritation in two minutes, dizziness in 4 min, signs of intoxication in 4–10 min	148
		Inhalation (chronic)	>500 ppm (~1.8 mg/liter)/day	May cause vomiting, diarrhea, insomnia, headache, dizziness, anemia, muscle and neurological symptoms	979
	Mouse	Inhalation	30,000 ppm (~110 mg/liter)/5 min	Lethal dose	980
Petroleum ether	Human	Dermal	Undiluted/30 min	Disruption of horny layer, peeling	966
			Undiluted/1 hr	Erythema, hyperemia, swelling pigmentation	966

	Species	Route	Concentration	Effect	
		Inhalation	Saturated 445–1250 ppm (1160–4400 mg/m³)	Cerebral edema	966
				Blurred vision, cold sensation in extremities fatiguability, headache; fatty degeneration of muscle fibers, demyelination and mild axonal degeneration	966
			500–2500 ppm (1.5–7.8 mg/liter)/day	Neurogenic atrophy	966
			1000–2500 ppm (3.1–7.8 mg/liter/day	Polyneuropathy in 6–9 months	966
Aromatic petroleum naphtha	Human	Inhalation	0.07 ppm (0.4 mg/m³)	Odor threshold	969
			0.5–2.5 ppm (2.2–11 mg/m³)	Identification threshold	665
	Rat	Inhalation	26 ppm (150 mg/m³)	Sensory threshold	969
			66 ppm (0.38 mg/liter)/8 hr	No observable physiological effects	969
			1500 ppm (8.7 mg/liter)/8 hr (2 exp.)	Lethal, erythrocyte fragility	969
b.p. 364–403°F	Mouse	Inhalation of aerosol	550 ppm (3.1 mg/liter)	Increased respiratory rate by 50%	969
	Cat		150 ppm (8.2 mg/liter)/6 hr	CNS depression	969
b.p. 311–392°F	Rat	Inhalation (subchronic)	50 ppm (300 mg/m³)/8 hr/day × 5 days/wk × 90 days	No detectable changes	665
			616 ppm (3.6 mg/liter)/18 hr/day × 7 days/wk × 150 days	Decreased weight gain, lung congestion, hemorrhaging	665
			1000 ppm (5.7 mg/liter)/18 hr/days × 7 days/wk × 78 days	Congestive changes in lung, liver, spleen, and kidney, decreased white blood count	665

Table 47.24. (*Continued*)

Material	Species	Route of Entry	Dose or Concentration	Result or Effect	Ref.
b.p. 392–480°F	Rat		50 ppm (300 mg/m³)/8 hr/day × 5 days/wk × 90 days	Slight bone marrow changes and DNA depression	665
			200 ppm (1.4 mg/liter)/8 hr/day × 5 days/wk × 90 days	Decreased white blood count and weight gain, one of 17 animals developed cataracts	665
			500 ppm (2.8 mg/liter)/18 hr/day × 7 days/wk	Lethal to 50% of the colony	665
b.p. 311–392°F	Primate	Inhalation	50 ppm (300 mg/m³)/7 hr/day × 5 days/wk × 90 days	No effect	665
			200 ppm (1.4 mg/liter)/7 hr/day × 5 days/wk × 90 days	Equilibrium disturbances, decreased white blood count, initially tremors, loss of hair, dry skin, some myelocytic depression, erythrocytic changes	665
b.p. 392–480°F	Primate	Inhalation	50 ppm (0.3 mg/liter)/7 hr/day × 5 days/wk × 90 days	Diarrhea, increased erythrocyte activity, irritation of face and eyes, stimulation of bone marrow erythrocyte activity, decreased myelocytes	665
			200 ppm (1.4 mg/liter)/7 hr/day × 5 days/wk × 90 days	Effects as for 50 ppm	665

Substance	Species	Route	Dose	Effect	Reference
Rubber solvent	Human	Inhalation	10 ppm (40 mg/m³)	Odor threshold concentration	981
	Rat	Inhalation	2800 ppm (10 mg/liter)	No effect	981
			15,000 ppm (6.1 mg/liter)	LC_{50}	981
	Dog	Inhalation	1500 ppm (0.6 mg/liter)	No effect	981
VM&P naphtha	Human	Inhalation	0.86 ppm (4 mg/m³)	Odor threshold	982
	Rat	Inhalation	3400 ppm (16 mg/liter)/4 hr	LC_{50}, eye irritant in 30 min	982
			1200 ppm (5800 mg/m³)/6 hr/day × 5 days/wk × 40 days	At day 40, increased neutrophils, decreased lymphocytes, normal at day 65	982
	Dog	Inhalation	1200 ppm (5800 mg/m³)/6 hr/day × 5 days/wk × 40 days	Increased reticulocyte count, increased alkaline phosphatase, decreased GOT	982
Mineral spirits	Human	Inhalation	Unknown/day × 4 mo	Aplastic anemia, lethal	961
			670–1670 ppm (1–2.5 mg/liter)/30 min	Nausea, vertigo during exercise or rest, one case of premature atrial beats, one case of T-wave inversion	983
	Dog	Inhalation	~140–360 ppm (238–619 mg/m³)/day × 23.5 hr/day × 90 days	Normal hematology	984
	Guinea pig	Inhalation	~200 ppm (363 mg/m³)/day × 7 days/wk × 60–90 days	Lowest lethal effect	984
			360 ppm (619 mg/m³)/day × 7 days/wk × 60–90 days	Increase in body weight	984
	Primate	Inhalation	~320 ppm (555 mg/m³)/day × 7 days/wk × 60–90 days	Decrease in body weight	984
	Human	Inhalation	(600 mg/m³)/8 hr	Irritant effect	985

Table 47.24. (Continued)

Material	Species	Route of Entry	Dose or Concentration	Result or Effect	Ref.
	Rat	IP	5680 mg/kg (50,000 mg/m³)/8 hr	Lethal	986
	Mouse				985
	Fish	Aquatic	>1000 ppm	TLm 96	365
Stoddard solvent	Human	Inhalation (acute)	9 ppm (50 mg/m³)	Odor threshold	987
			400 ppm (2300 mg/m³)	No eye, nose, or throat irritation	326
		Dermal	Undiluted 1–2/day × 6 mo	Fatal when used for washing hands	988
			Undiluted/2 yr	Fatal when used for removal of paint from hands	988
		Dermal + inhalation (chronic)	Undiluted/8 hr/day × 8 wk	Follicular dermatitis from daily contact in dry cleaning	989
	Human	Inhalation	Fumes × 3 mo	Hepatic involvement and sensitization, hepatic tests still elevated 1 year post-exposure	989
			Fumes 2–3 ×/mo × 2 yr	Lethal, moderate hypoplasia of bone marrow	988
			Fumes × 20 yr	Worker survived after spleenectomy	988
			Fumes × 17 yr	Aplastic anemia, intracerebral hemorrhage, death	990
	Rat, dog	Inhalation (acute)	410–1400 ppm (2.4–2.8 mg/ liter)	No significant difference between test and control	987
	Cat		1700 ppm (10 mg/liter)	CNS depression	987
	Rat	Inhalation (chronic)	84–330 ppm (0.48–1.9 mg/ liter)	Slight kidney pathological signs	987
	Dog		84–330 ppm (0.48–1.9 mg/ liter)	No significant effects	987

3378

	Species	Route	Dose	Effect	Ref.
140° Flash naphtha	Human	Inhalation	0.6 ppm (4 mg/m³)	Odor threshold level	967
	Human		17–49 ppm (110–310 mg/liter)/15 min/day × 2 days	Slight temporary dryness of eyes	967
	Rat, dog, cat	Inhalation (chronic)	37 ppm (230 mg/m³)	Suggested hygienic standard	967
		Inhalation	33–300 ppm (0.21–1.9 mg/liter)	Slight lacrimation of 1 dog, otherwise no significant clinical or observable effects	967
			37 ppm (230 mg/m³)/6 hr/day × 5 days/wk × 13 wk	No effects	967
Thinner 40	Rat	Inhalation (aerosol)	33 ppm (0.2 mg/liter)/7 hr	Dose tolerated	970
			~140 ppm (8.3 mg/liter)/7 hr (1 μ diameter)	Irritation to extremities, loss of coordination	970
	Dog		~41 ppm (0.25 mg/liter)/8 hr	Dose tolerated	970
	Cat	(Aerosol)	120 ppm (7.0 mg/liter)/6 hr (1 μ diameter)	No visible discomfort	970
Thinner 50	Human	Inhalation	2.5 ppm (10 mg/m³)	Odor threshold	971
			430 ppm (1.7 mg/liter)	4 of 5 people willing to work for 8 hr	971
			530 ppm (2.1 mg/liter)	Lightheadedness, headache in 30 min	971
	Rat	Inhalation	1300 ppm (5.2 mg/liter)	Dose tolerated	971
			8300 ppm (33.0 mg/liter)/4 hr	LC_{50}	971
	Dog	Inhalation	600 ppm (2.4 mg/liter)/6 hr	No discomfort	971
	Cat	Inhalation	7600 ppm (30.0 mg/liter)/6 hr	Signs of CNS effect, mydriasis, mild tremors, light anesthesia after 2–3 hr, but reversible	971
Thinner 60	Human	Inhalation	2 ppm (10 mg/m³)	Odor threshold	972
			170 ppm (0.85 mg/liter)	Tolerated	972
			350 ppm (1.7 mg/liter)	Tolerated	972

Table 47.24. (Continued)

Material	Species	Route of Entry	Dose or Concentration	Result or Effect	Ref.
	Rat		170 ppm (0.85 mg/liter)	No visible response	972
			690 ppm (3.4 mg/liter)	No visible response	972
			2500 ppm (12.0 mg/liter)	Slight effects, loss of coordination	972
			4900 ppm (24.0 mg/liter)/4 hr	LC_{50}	972
	Beagle dog		820 ppm (4.0 mg/liter)	No visible effects	972
			1900 ppm (9.5 mg/liter)	Loss of coordination	972
	Cat		4100 ppm (20 mg/liter)	Lethal in 4 hr	972
			7700 ppm (38 mg/liter)	Lethal in 150 min	972
Thinner 70	Human	Inhalation	0.7 ppm (4 mg/m³)	Odor threshold	973
			59 ppm (0.32 mg/liter)	Sensory response, minimal	973
			180 ppm (0.95 mg/liter)/15 min	Ocular and nasal irritation	973
	Rat		810 ppm (4.4 mg/liter)/8 hr	Lacrimation, loss of coordination, fine tremors	973
	Beagle		930 ppm (5.0 mg/liter)/4 hr	Convulsions in 2 hr	973
	Cat		370 ppm (2.0 mg/liter)/6 hr	CNS effects	973
	Rat, dog		200 ppm (1.0 mg/liter) × 13 wk	No ill effect level	973
			410 ppm (2.2 mg/liter) × 13 wk	No recognizable effects except for slight reduction in weight gain	973
Thinner 80	Human	Inhalation	0.9 ppm (4 mg/m³)	Odor threshold	974
			100 ppm (0.45 mg/liter)	Slight transitory eye irritation	974
			150 ppm (0.65 mg/liter)	Suggested hygienic standard	974
			230 ppm (1.0 mg/liter)	Response variable	974
	Rat		800 ppm (3.5 mg/liter)/4 hr	Dose tolerated	974
			6200 ppm (27 mg/liter)/4 hr	LC_{50}	974

Substance	Species	Route	Dose	Effect	Ref.
	Dog		480 ppm (2.1 mg/liter)/4 hr	Dose tolerated	974
	Cat		5500 ppm (24.0 mg/liter)/4 hr	CNS effects, animals prostrate in 41–65 min, recovery in 14 days	974
	Rat, dog		390 ppm (1.7 mg/liter)/6 hr/day × 5 days/wk × 14 wk	No visible effects	974
Kerosine (Kerosene)	Human	Oral	0.5 oz	Lowest lethal dose	991
			3–4 oz	Mean lethal dose	991
			8 oz	Highest nonlethal dose	991
		Oral (with aspiration into lungs)	<1 ml	May cause chemical pneumonitis systemic symptoms, including CNS effects	324
		Inhalation	14 ppm	No effect	975
			20 ppm (140 mg/m^3)	Odor threshold	975
Deodorized kerosine	Rabbit	Oral	28 g/kg	Lethal to some animals	992
Kerosine		Oral	64 ml/kg	Lethal to 4 of 10 animals	975
	Mouse	Oral	50 ml/kg	Intoxication in 12–15 min, with labored and rapid respiration, death 10 hr after second equal dose on 2nd day; congestion of renal tubules, lung surface hyperemia, liver yellow patches	993
Deodorized kerosine			64 ml/kg	Lethal to 1 of 10 animal	975
Kerosine	Rabbit	Intratracheal	28 g/kg	LD$_{50}$	992
	Guinea pig		20 g/kg	LD$_{50}$	992
	Rat		800 mg/kg	Lowest lethal dose	994
	Rabbit	Eye	Undiluted	Practically innocuous	992

Table 47.24. (Continued)

Material	Species	Route of Entry	Dose or Concentration	Result or Effect	Ref.
	Rat	Intraperitoneal	10.7 g/kg	Lowest lethal dose	986
	Rabbit		6.6 g/kg	LD_{50}	992
	Dog		50 ml/kg	Labored, rapid respiration, sedation	993
Deodorized kerosine	Rabbit	Intravenous	180 mg/kg	LD_{50}	992
	Rat	Inhalation	14 ppm (0.10 mg/liter)/8 hr	No signs of distress	975
Kerosine	Rabbit	Dermal (subacute)	3 ml/kg/day × 3 days	Hair loss, scaling, cracking of the epidermis, no systemic toxicity	992
	Guinea pig		0.5 ml/3 day × 2 wk	Carrier for dinitrochlorobenzene, increases the reactivity, swelling, infiltrations	995
Deodorized kerosine	Rat, dog, cat	Inhalation (subchronic)	14 ppm (0.1 ml/liter)/8 hr/day × 13 wk	No discomfort in saturated vapor	975
	Rat	Aerosol	7.4 mg/liter × 6 hr/day × 4 days	Skin irritation of extremities	975
	Cat	Aerosol	6.4 mg/liter × 6 hr/day × 4 days	No effect	975

5.1 Crude Oil

5.1.1 General

Crude oil or petroleum, a flammable liquid (DOT) (22), is a complex mixture of organic and some inorganic materials, varying with its geological origin. Some physicochemical properties are listed in Table 47.22. Depending on the composition, it is preferentially utilized to manufacture gasoline, such as the Venezuela crudes or the Near East oils, which are good sources of lubricating stock (998). Crude oil may vary from the distillate type, which is directly fractionable into gasoline and accompanying products, to highly viscous asphalts (5). The average composition includes paraffinic, naphthenic, aromatic, and sulfur-containing hydrocarbons, some nitrogen- and oxygen-containing compounds, and a variety of metals, such as cobalt, manganese (999), boron, chromium, nickel, sulfur, vanadium, and others (1000), as well as uranium in some samples from Utah (1001).

5.1.2 Physiologic Response

Similar to the action of paraffins, the major acute effect of crude oil on humans is narcosis, although reversible even at high concentrations. On inhalation of vapors, it may produce chemical pneumonitis.

On prolonged dermal contact or inhalation, it is an irritant and may cause systemic disorders. Two crudes, a low and a high sulfur-containing oil, were administered at 37.0 to 123.0 ml/kg once a day for 5 days to 11 cattle ranging in age from 6 months to 3.5 years. Toxic effects were vomiting, moderate to extreme bloating, aspiration pneumonia, more rapidly with the low than the high sulfur oil, anorexia, weight loss, mild mental depression, and a decreased plasma glucose level (1002). South Louisiana crude oil was more toxic to annelids than Kuwait crude (1003). Sublethal concentrations were reversibly toxic in the average marine animal (1004). Some research has concerned the fate of crude and fractionated oil in the ecosystem, as summarized by several authors (1005, 1006). Crude oil affects the growth and photosynthetic action of microalgae (661, 1007), but is degraded also by a variety of microorganisms.

5.1.3 Industrial Hygiene

When handling crude oil, vapors or mists should not be inhaled. Gloves and glasses should be worn for dermal and eye protection.

The European maximum allowable air concentration has been suggested as 500 ppm (40).

5.2 Natural and Liquefied Gases

Natural gas is a colorless, odorless, flammable gas (DOT) (22), which occurs naturally along with petroleum deposits in marshes or from waste decomposition. It consists mainly of methane, some ethane, propane, and butane, at 83 to 99, 1 to 13, 0.1 to 3, and 0.2 to 1.0 percent, respectively (8). The gas can be liquefied for transport and storage, and is primarily used as fuel. For physicochemical properties, see Methane to Butane in Chapter 45 on aliphatic hydrocarbons.

Petroleum gas, recovered during refining of crude oil, is a flammable gas (DOT) (22) that is easily compressed to LPG, liquefied petroleum or "bottled" gas. It consists of propane and butane, with minute quantities of mercaptans added for odorant warning properties (5). LPG is primarily used as fuel, as chemical raw material, and for refinery blending of a variety of materials. For further information, see Chapter 3 in the *Petroleum Products Handbook* (5).

Liquefied gases are practically nontoxic below the explosive limits, are narcotic at high concentrations, and may cause asphyxia by oxygen displacement (324).

For handling, proper precautions are recommended, see Methane to Butane in Chapter 45. Generally, flammability and explosive hazards outweigh the biologic effects.

5.3 Gasolines

5.3.1 General

Gasoline, petrol, is a flammable liquid (DOT) (22) produced from the light distillates during petroleum fractionation. Additional physicochemical data are listed in Table 47.22. The distillation ranges are specified for the particular application, mainly the reciprocating, spark ignition, and internal combustion engines. In order to serve specific purposes, various functional additives are blended into the gasolines. These consist of antiknock fluids, antioxidants, metal deactivators, corrosion inhibitors, anti-icing agents, preignition preventors, upper-cylinder lubricants, dyes, and decolorizers (5). More detailed properties and specifications are listed in the *Petroleum Products Handbook* (5) and ASTM Method D-86 (5). Probably the most critical property is the octane number, supplied with high octane hydrocarbons and lead compounds. The major components are primarily paraffins, olefins, naphthenes, and aromatics, and more recently 10 to 40 percent ethyl alcohol (5). The distillation from initial to final boiling point ranges from about 32 to 225°C (90 to 437°F) and the explosive limits, 1.3 to 6.0 percent (977).

The U.S. government is currently in the process of greatly reducing the allowable gasoline lead content. This, however, has placed the U.S. refineries in a dilemma, since no equivalent substitute is available to maintain the high efficiency of the gasoline engine (1008). Methylcyclopentadienyl manganese tricarbonyl (MMT) has been used as a partial replacement for lead; however, other adjustments, such as changes in organic composition, have been necessary (1009).

5.3.2 Physiologic Response

Although gasoline grades vary with octane number and engine requirement, the general toxic effects do not differ appreciably, except to some extent with volatility and the lead content. When gasoline is inhaled at high concentrations, the other additives, since in small quantities, exert only minor influence. The alcohol blends appear to be of even lower toxicity. Overall, few cases of intoxication have been reported in relation to gasoline quantities handled.

The toxicologic literature prior to 1966 has been summarized by Von Oettingen (38) and Machle (1010). Some selected data are presented in Table 47.24.

In children, death from accidental ingestion of as little as 10 to 15 g gasoline has been observed (40). In adults, ingestion of 20 to 50 g of gasoline may produce severe symptoms of poisoning. Accidental ingestion of gasoline from a pop bottle by an adult human caused immediate severe burning of the pharynx and gastric region. With immediate gastric lavage, no general symptomatic effects were noted, except for clinical findings of temporary galactose excretion of 10.6 g and slightly increased liver function results. The transient hepatic damage was probably due to the gasoline's lipid solubility (40). Symptoms in severe oral intoxication are mild excitation, loss of consciousness, occasionally convulsions, cyanosis, congestion, and capillary hemorrhaging of the lung and internal organs (40), followed by death due to circulatory failure (40); in milder cases, symptoms are inebriation, vomiting, vertigo, drowsiness, confusion, and fever (148). Unless prevented, aspiration into the lungs and secondary pneumonia may occur. Gasoline may cause hyperemia of the conjunctiva and other disturbances of the eye. It is a skin irritant and a possible allergen. On acute inhalation, humans experience intense burning of the throat and respiratory system, and possibly bronchopneumonia may develop. At extremely high concentrations where oxygen displacement is a factor, asphyxiation may occur. Severe intoxication is accompanied by CNS effects, coma, and convulsions with epileptiform seizures. The deaths of two occupants of a light aircraft wreckage appear to be attributable to systemic fat embolism following massive, acute gasoline inhalation and subsequent anesthesia (1011).

Repeated or chronic dermal contact may result in drying of the skin, lesions, and other dermatologic conditions (1012). Inhalation of gasoline during bulk handling operations produced no physiological effects (1013). Service station attendants in Finland tested for blood lead showed an average of 21 μg/100 ml versus 10 μg/100 ml of the general population (1014), and 29 μg/100 ml for garage workers compared to 37 μg/100 ml for traffic personnel in Lausanne (1015). In Lucknow, India, complaints from gasoline pump workers, possibly owing to the warmer climate, included headache, fatigue, disturbance of sleep, and loss of memory (1016). Urinary phenol levels of above 40 mg/liter could be directly related to quantities of gasoline handled per day (1016). Russian workers chronically exposed to gasoline vapors showed a decrease in the phagocytic activity of peripheral blood granulocytes, globulin, and total protein levels (1017). Occupational exposure to fumes of gasoline-powered equipment has been related to some nonlymphocytic leukemias in Sweden (1018). One case of acute hepatic and CNS effects due to high level gasoline inhalation has been reported by Moeschlin (40).

"Gasoline sniffing" has produced morbidity and mortality cases owing to acute and chronic inhalation (1019). One lethal case and one with signs of lead encephalopathy, elevated blood lead, and a marked decrease of erythrocytic S-aminolevulinic dehydratase (ALAD) levels have been described by Boeckx et al. (1020). It has been reported (1021) that gasoline, similar to C_4 and C_7 aliphatic hydrocarbons, can sensitize the myocardium and cause rapid central depression with respiratory failure, thus explaining sudden sniffing deaths (1022) and lethal effects to workers when cleaning storage tanks without proper respiratory protection (991). In acute animal experiments, oral administration of leaded gasoline resulted in the deposit of lead in rat femoral bone (1023). Intraperitoneal injection into rats showed mortality rates proportional to tetraethyllead content of Japanese gasolines (1024). Symptoms of intoxication included ataxia, drowsiness, nystagmus, convulsions, and hypothermia. Oral doses of leaded gasoline were nonteratogenic in the rat (1025). Intratracheal instillation of as little as 0.2 ml gasoline-type petroleum fractions have caused instant death in the rat (16). Exposure of rabbits to gasoline vapors (78 Octane) at 310 mg/liter for 2 hr resulted in circulatory effects, as observed in human intoxication (1026). Electrolytes were significantly decreased, as were also the heart muscle alkaline phosphatase, and blood α-1 and α-2 globulins. Rabbits chronically exposed to gasoline vapors showed circulatory effects with lipid metabolism and serum lipid changes (1027) and lymphoid cell decreases (1028). Rats exposed to 1 g/m³, 5 hr/day, 5 days a week, resulted in reversible damage of the eye blood vessels at 3 months, but atrophy and necrosis after 6 to 9 months (1029). Exposure of rats to 10 mg ethyl gasoline/liter for 6 hr/day caused disseminated degenerative changes in the neurons of the central nervous system (1030); at 49.7 mg/liter, 4 hr/day for 27 weeks, it caused alterations in the ovarian and pituitary functions (1031). When 0.25 ml/liter was added to 8.5 ml rat brain homogenate, inhibition of monoamine oxidase occurred at a higher rate by leaded than unleaded gasoline (1032).

Limited experiments demonstrated that gasolines exhibited no gonadotrophic nor mutagenic action (1033), nor did the organic phosphate additives (1034).

5.3.3   Industrial Hygiene

No official air threshold limits have been set for gasoline. However, calculated values range from 200 to 300 ppm (25, 1035), and 200 ppm is used most prevalently in Europe.

Gasoline can be collected on charcoal and for the determination, gas chromatographic procedures are preferred (1013). A combination gas chromatography–mass spectrometry can be used for the analytic quantitation of gasoline in the blood (1036).

Oil adsorbents can be used for the removal of gasoline from aqueous effluents (1037).

When working in an atmosphere where gasoline vapors may occur, precaution is advised. At no time should a facility be entered with levels at 2000 ppm or even above 500 ppm (148). Detailed precautionary measures are described in the *API Toxicological Review of Gasoline* (977).

5.4   Petroleum Naphthas

5.4.1   General

Petroleum naphthas or petroleum solvents are complex hydrocarbon mixtures which can be obtained from the petroleum light distillate or low boiling fraction (5). Regular organic solvents, such as acetates, alcohols, ketones, and chlorinated hydrocarbons, are not covered in this section, although in cases of intoxication, all possibilities should be considered. For details, see *API Toxicological Review of Petroleum Naphthas* (968), as well as Vol. 2C.

The light petroleum naphthas are composed mainly of paraffins, mono- and dicycloparaffins, some olefins, alkylbenzenes, naphthenes, and some benzenes, lending the mixture its specific physicochemical properties, such as boiling point range and flash point; see Table 47.22. The physicochemical characteristics, in turn, determine their specific use, such as rubber solvent, paint thinner, cleaning or degreasing agent, or petroleum refining stock. The hydrocarbon solvents are not used as food additives, as are other solvents (596). Unfortunately, there also has been wide abuse of glues and solvents, through the practice of inhalation or "sniffing" solvent fumes (260). Also, the belief has been prevalent in some groups of the population that naphthas, mineral spirits, or kerosine might possess medicinal value as a rubbing fluid. Several deaths have occurred from such misuses.

Physiologic Response.   Of the major component classes, the paraffins are least toxic; next are the naphthenes (cycloparaffins), such as cyclopentane and cyclohexane. The alkylbenzenes, as described in Section 1, may bear the most toxic characteristics, the degree depending on the volatility of the components. The lower-boiling naphthas, the petroleum benzin, mineral spirits, and naphthas, are more volatile and thus present a higher toxicity hazard than the higher boiling fractions. This has been observed with the chronic inhalation of petroleum distillate in a small manufacturing business where some employees experienced reversible, some irreversible, CNS damage (1038). Aside from CNS depression, myocardial (264) and hematopoietic effects have been recorded. Myelotoxic effects and hypoplasia have been ascribed to the benzene content of the solvents (1039). However, where possible, benzene now has been removed from most commercial materials.

Dermal single contact may cause erythema, blistering, and cellular damage, and naphthas may have allergenic potential (246). Repeated applications may cause dermatitis and lesions.

Experiments with rats have demonstrated that the oral LD_{50} values range from 4.5 to >25 ml/kg for a series of petroleum distillates from the rubber solvent type to the high flash naphthas (330). With accidental ingestion, aspiration into the lungs may occur, causing endothelial injury, edema, and hemorrhage (324). This has been demonstrated by Gerarde (16) in the rat, where aspiration of 0.2 ml of petroleum fractions of the solvent types, gasoline, fuels, and naphthas, with a viscosity of 39 SUS or less, resulted in eight of 10 deaths.

Range-finding inhalation LC_{50} values were approximately 2000 ppm to 73,680 ppm (330).

None of the petroleum solvents have carcinogenic potential, although some may aid in the translocation of carcinogenic agents by the solvent effect (899).

Industrial Hygiene. The threshold limits vary with the volatility of the solvent, from 100 to 1000 ppm. For sampling, activated charcoal collection has been recommended, and several analytic determination procedures are available (310, 311). Various patents have been granted for methods concerning the solvent removal from air and water (1040).

When petroleum hydrocarbon solvents are handled, gloves and respiratory protection are recommended. For special cases, barrier creams can be used. A method to test their efficiency has been described (321).

5.4.2 Petroleum Ether

Petroleum ether, ligroin, Skellysolve, petroleum benzin, also occasionally called petroleum distillate, is a flammable liquid (DOT) (22) and low-boiling cut. Further physicochemical properties are listed in Table 47.22. Petroleum ether is used as a universal solvent and extractant for chemicals, fats, waxes, paints, varnishes, and furniture polishes, and is used as a detergent, in photography, and as fuel (3).

Physiologic Response. Petroleum ether consists principally of *n*-pentane and *n*-hexane. Thus the general effects of intoxication are peripheral nerve disorders, CNS depression, and skin and respiratory irritation, discussed in detail in Chapter 45. Ingestion of furniture polish or lighter fluid which may contain ligroin has caused chemical pneumonia and pneumatoceles in children (1041). On human skin, it has caused erythema, edema, disruption of the horny layer, and peeling (966); see Table 47.24. Acute inhalation of petroleum ether, when mistakenly used as an anesthetic agent, caused reversible cerebral edema (966).

Numerous reports point to the neurotoxic effects on prolonged inhalation of petroleum ether in inadequately ventilated business establishments where employees experienced polyneuropathy (966). Signs and symptoms included loss of appetite, muscle weakness, impairment of motor action, and parethesia, similar to effects discussed for *n*-hexane.

Industrial Hygiene. NIOSH is recommending a TWA of 120 ppm (350 mg/m^3) for a 10-hr work shift (966) with an action level of 200 mg/m^3 and a ceiling of 590 ppm (1800 mg/m^3) (966).

5.4.3 Rubber Solvent

Rubber solvent is a clear, colorless, and flammable liquid, somewhat less volatile than petroleum ether. Some physicochemical and physiological data are presented in Table 47.22 and 47.24. It is used as a solvent in the manufacture of adhesives, brake linings, rubber cements, tires, intaglio inks, paints, and lacquers, and is used in degreasing operations (5).

When inhaled in large concentrations, rubber solvent causes disturbances similar to those observed with benzene. In six of eight recorded deaths, findings in rubber workers included myeloid leukemia in a nonsignificant number of cases (1042).

In an extensive study, Carpenter et al. (981) determined a 4-hr LC_{50} value in the rat of 15,000 ppm (61,000 mg/m^3) with a no-effect level of 2800 ppm and 1500 ppm in the rat and dog, respectively. Effective signs and symptoms included CNS depression and convulsions in the rat and the cat (981).

A human odor threshold of 10 ppm was determined (981). NIOSH (966) recommends a TWA of 350 mg/m^3 for a 10-hr workday, 40-hr work week, with an action level of 200 mg/m^3 and a ceiling of 454 ppm.

5.4.4 VM&P Naphtha

VM&P Naphtha, varnish makers and painters naphtha, also known as "light naphtha," "dry-cleaners' naphtha," and "spotting naphtha" (5), is a colorless to yellow, flammable and explosive liquid with an aromatic odor and of the boiling range 95 to 160°C (203 to 320°F); see Table 47.22 for physicochemical properties. VM&P naphtha is used extensively as a solvent for lacquers, varnishes, and quick-evaporating paint thinner. It is a direct distillation product containing C_5 to C_{11} hydrocarbons. This solvent is a mild eye and nose irritant. Exposure to VM&P naphtha, owing to overheating of a tank, caused labored breathing in 18 of 19 individuals; two were cyanotic with general excitation, tremors, and nausea, and hyperactivity (1043). The symptoms subsided in 30 min, except for one worker. In a human trial, 880 ppm (4.1 mg/liter) produced eye and throat irritation with temporary olfactory fatigue (982). In animal experiments (see Table 47.24) temporary hematologic effects were noted.

The threshold limit recommended by NIOSH is 200 ppm (350 mg/m³) for a 10-hr work shift, 40-hr work week (966), with an action level of 120 ppm (200 mg/m³) and a ceiling of 386 ppm (650 mg/m³) (966).

A NIOSH sampling procedure, S-380.1 (242) for solvents in the range of 120 to 147°C may be applied for solvent collection, using charcoal tubes and gas chromatography for quantitation.

The odor threshold was observed at 0.86 ppm (4 mg/m³) (981).

5.4.5 Petroleum Spirits

Petroleum spirits, refined petroleum solvent, white spirits, or mineral spirits, compose a fraction slightly lower in boiling point than Stoddard solvent; the names, however, are sometimes interchangeably used, including ligroin and alternate terms used for petroleum ether (3). It is a fraction containing paraffins, naphthenes, and aromatics (148). Pharmacologically and toxicologically, these mixtures compare with heptanes and octanes; see Tables 47.22 and 47.24.

Generally, mineral spirits are low irritants to the gastrointestinal tract on ingestion, to the skin on contact. Systemically, the central nervous (324) and cardiac (966) systems may be affected. Aspiration may occur, as has been reported from the ingestion of products containing mineral spirits (1041). Aplastic anemia and thrombocytopenia were diagnosed in a worker who had used white spirits for cleaning floors for 4 months. The case proved fatal 3 months later (966). In a clinical trial, 1000 to 2500 mg/m³ in adults caused slight effects (966); see Tables 47.24. Animal studies showed lesser effects.

The threshold limit TWA recommended by NIOSH (966) is 200 ppm (350 mg/m³) for a 10-hr day, 40-hr work week, with an action level of 200 ppm (350 mg/m³) as the TWA and a ceiling of 282 ppm (500 mg/m³) (966). Industrial hygiene sampling may be carried out as recommended for petroleum distillate; see VM&P naphtha, in Method S-380.01 (242) or S-382, for Stoddard solvent.

5.4.6 Stoddard Solvent

Stoddard solvent, also called white spirits, is a colorless, flammable fluid. Further physicochemical properties are listed in Table 47.22. There are three other solvent grades similar to Stoddard solvent, all belonging to the petroleum spirit category. These are 140 flash solvent with a flash point of 139 to 142°F, odorless Stoddard, flash point 121 to 130°F, and low end point Stoddard, flash point 100 to 108°F (966). They contain paraffins, naphthenes, and alkylbenzenes, with a trace of benzene. Stoddard solvent is used widely in dry-cleaning processes, and as a general cleaning and universal solvent.

Pharmacologically and toxicologically, Stoddard solvent resembles gasoline. It produced no effect on the human eye (600). Major manifestations are defatting, drying, scaling of the skin on direct contact, and possible development of dermatitis (770). On ingestion, aspiration into the lungs may occur, causing pneumonitis, pulmonary edema, and hemorrhage. Acute effects from inhalation of large concentrations are nausea, vomiting, cough, and pulmonary irritation (272). Chronic inhalation exposures in humans have also resulted in hepatic and hematopoietic changes similar to the effects of the lower naphthas; see Table 47.24.

NIOSH (966) recommends a threshold limit TWA of 200 ppm (350 mg/m^3) for a 10-hr day, 40-hr work week with the same action level (966), and a ceiling of 306 ppm (550 mg/m^3), superseding the earlier OSHA standard of 500 ppm (2900 m/m^3) (1044).

NIOSH (242) has published a sampling procedure, S-382, using activated charcoal as an adsorbent, carbon disulfide as a desorbant, and flame ionization gas chromatography for quantitation.

5.4.7   140° Flash Naphtha

140° Flash naphtha, or 140 flash aliphatic solvent, is a slightly higher boiling petroleum spirit fraction; see Table 47.22 for some physicochemical and composition data.

Physiological and pharmacological characteristics resemble those of the Stoddards and petroleum spirits; see Table 47.24.

No significant effects were observed in the rat, dog, and cat in chronic inhalation experiments (967).

A TLV at 37 ppm was concluded to be safe for an 8-hr day, 40-hr week exposure (967).

5.4.8   Aromatic Petroleum Naphthas

Aromatic petroleum naphthas, APN, coal tar or pyrolysis naphthas, are manufactured in three boiling ranges (968). Aromatic petroleum naphthas are processed from high-boiling distillate fractions, containing mainly alkylbenzenes, cumene, toluene, and xylene (968); see Table 47.22 for physicochemical data. These materials are used as chemical raw materials, as degreasing agents, in varnishes, lacquers, synthetic enamels, and lithography inks, and in textile printing (968). Aromatic petroleum naphthas are also used as solvents for herbicides, fungicides, and insecticides.

The physiological response of aromatic naphthas resembles that of benzene and the lower alkylbenzenes, as discussed in Section 1.

Acute signs of ingestion and inhalation, and to a lesser degree by dermal contact, are eye, nose, and throat irritations, vertigo, nausea, dyspnea, CNS depression, narcosis, and neurotoxicity if benzene is present. Chronic signs are CNS depression and slight to severe changes of the hematopoietic system, also depending on the benzene content. Upon ingestion, the naphthas present aspiration hazards.

A series of acute, subacute, and subchronic experiments with rodents, canines, and primates showed effects of decreased weight gain, white blood count, bone marrow effects, lung congestion, CNS depression, and isolated cataracts at high naphtha concentrations; see Table 47.24.

At present a threshold limit TWA of 100 ppm (400 mg/m^3) is suggested (242). A sampling procedure, S-86.01, has been recommended by NIOSH (242), using activated charcoal as an adsorbent, carbon disulfide as a desorbant, and flame ionization gas chromatography for analytic quantitation. Thresholds may be detected at 0.07 ppm for odor, 0.5 to 2.5 ppm for identification, and 26 ppm for sensory effects; see Table 47.24.

5.4.9 Thinners

A series of thinners or naphthas were tested for inhalation and odor recognition properties. The thinners, Nos. 40 to 80, are aliphatic-type naphthas of the petroleum spirits range. For physicochemical and composition data, see Table 47.22. They are clear to yellow, flammable liquids, used in paints, glues, varnishes, lacquers, and as general solvents or degreasing agents.

The physiological effects resemble these of the mineral spirits, hexane and benzene. A German team reports neurotoxicity in 18 juveniles who had sniffed glue thinner (1045), with motor defects still observed 8 months later. The general toxic effects are CNS and occasional myelotoxic effects; see Table 47.24.

Odor thresholds appear too low to serve as warning properties. An air threshold concentration limit of 150 ppm is suggested for thinner 80, a value also applicable to other thinner grades.

5.4.10 Naphthenic Aromatic Solvents

Similar to thinners, two naphthenic solvents with boiling ranges of 157 to 183 and 151 to 200°C have been tested for basic inhalation characteristics. Some physicochemical characteristics and toxicological data are listed in Tables 47.22 and 47.24.

5.5   Middle Distillate Products

5.5.1   Kerosine (Kerosene)

General.   Kerosine, astral oil, coal oil, No. 1 fuel oil, kerosene, also known as mineral seal, mineral colza, 300, and range oil (5), is a white to pale yellow, mobile, flammable, and combustible liquid. Kerosine is produced by direct fractionation from the "middle distillate fraction" (5). The individual kerosine composition varies widely, but consists mainly of linear and branched aliphatics, olefins, cycloparaffins, and aromatics in the C_{10} to C_{16} range; see Table 47.22.

For indoor heating fuels it is desirable to remove the olefins, aromatics, and sulfur compounds (5), since they provoke the evolution of soot and sulfur oxides. For some purposes, highly refined or "deodorized" kerosine is manufactured by treatment with activated charcoal or by clay filtration (17).

Kerosine is used widely as illuminating, heating, and cooking fuel, as a cleaning, degreasing, and mold release agent, as a solvent in asphalt coating, and for enamels, paints, polishes, thinners, and varnishes. The deodorized product is utilized mainly for household sprays, herbicides, insecticides, and pesticides. Of the heavier kerosines, the mineral seal oil also have been used as railway coach and caboose lamp fuel (5). It is used also medicinally for veterinary decontamination (1046); see Table 47.24.

Physiologic Response.   The physiological responses to kerosines vary vastly according to their origin and utilization. The deodorized and refined kerosines are least toxic. Others may contain benzene or alkylbenzenes, which result in hematopoietic or similar manifestations. Human ingestion of kerosine results in rapid absorption from the gastrointestinal tract, systemic effects, and possible aspiration into the lungs (224). A comparative ratio of oral to aspirated lethal doses may constitute 1 pt versus 5 ml (1047). Systemic effects are gastrointestinal irritation, vomiting, diarrhea, and in severe cases drowsiness and CNS depression, progressing to coma and death (991). Aspiration may cause hemorrhaging and pulmonary edema, progressing to pneumonitis and renal involvement.

Signs of lung involvement include increased rate of respiration, tachycardia, and cyanosis (324). Innumerable cases of accidental kerosine ingestion by children have been reported. In 1962, the Subcommittee on Accidental Poisoning listed 28,000 nonfatal poisoning cases in the United States attributable to petroleum distillates, but mainly kerosine (991). Complications were bacterial pneumonia (991) and pneumatoceles (1041). In 22 and 52 human cases, an increase of gastric fluid level has been observed (1048). In cases of acute ingestion, inducing vomiting and lavage are contraindicated, owing to the aspiration hazard (324). Preferred antidotes are charcoal (1049, 1050) and milk (1049). The administration of ipecac has aided in some cases (1051). For detailed discussion, see kerosine in *Clinical Toxicology of Commercial Products* (598) and the review by Goodman and Gilman (991). For mixtures containing 43 percent or more kerosine, the aspiration hazard is acute (16).

Kerosine even on single contact defats the skin, which may lead to irritation, infection, and dermatitis (324, 1052). Tagami and Ogino (1053) reported several cases of children developing blistering or diffuse redness with edema. Patch tests by Luplescu et al. (1054–1056) confirmed the acute blistering effect.

Kerosine is not irritant to the eye (600).

Kerosine, including most of the fuel oils, is not sufficiently volatile to constitute an acute inhalation hazard. Except when emitted as an aerosol or mist, kerosine may cause mucous membrane irritation and chemical pneumonitis (148). On occasions, kerosine has been misused by chronic dermal application or ingestion. Several such cases have been recorded, where kerosine was used to massage extremities, resulting in aplastic anemia and death (966, 1057). Another lethal case is reported where the effects of kerosine when used as a degreasing agent were amplified following earlier lead poisoning (966).

Animal experiments demonstrate the low oral toxicity to the rat, rabbit, and chicken, especially of deodorized kerosine. Aspiration into the lungs may increase the toxicity, LD_{50}, by a factor of $1:140$, as shown in the rat (1051). Experiments with primates have shown that aspiration into the lungs causes cellular damage (1058). Kerosine aerosols have varying effects, depending on droplet size and composition, causing mucous membrane irritation to polyemia (1059). Absorption through the intact skin is practically negligible (1060), but moderate through injured dermal surfaces. It does cause moderate to severe injury in prolonged, direct dermal contact. Sublethal doses injected intratracheally in rats resulted in two inflammatory processes. One, an acute exudative inflammation representing the reaction of the alveolar capillaries, reaches its maximum in 3 days and subsides in 7 days; the other, a chronic proliferative inflammation, which reaches its maximum in 10 days, subsides but is still present at the end of one month (1061).

To marine animals, kerosine is less toxic than diesel oil and lower fractions (58). Kerosine is readily absorbed through the gastrointestinal tract of the primate (1062) and by fish (1063), and readily distributed in the system.

Kerosine was utilized well as the sole carbon source by *Pseudomonas aeruginosa* (1064) and *P. pseudomallei* (1065) and mutants of *Candida lipolytica* (1066), but it inhibited the growth of *Blakeslea trispora* (1067).

Industrial Hygiene. The recommended TWA by NIOSH (966) for kerosine is 14 ppm (100 mg/m^3), which is also the action level. No ceiling value is believed to be necessary, since 14 ppm is the air saturation concentration. This concentration should be well tolerated for an 8-hr workday (975). The odor threshold has been estimated at 0.09 ppm (0.6 mg/m^3), the sensory threshold of 20 ppm (0.14 mg/liter) (975).

Vapors can be collected by charcoal absorption. Analytic methods are available, and ultraviolet spectral (313) and gas chromatographic analyses have been used.

When kerosine is handled, prolonged skin contact should be avoided, although the surface should be thoroughly washed in case of accidental contact. Kerosine should never be siphoned by mouth.

5.5.2 Jet Fuel

The term jet fuel encompasses the aircraft turbine engine and jet fuels. They are composed of hydrocarbons from the middle distillate fraction in the kerosine range, with some components from the light distillates. They are composed of C_5 to C_{16} aliphatics, monocycloparaffins, aromatics, and olefins; olefins are permitted for the turbine engines only, and the aromatics at a lower percentage for the jet fuels (5); see Table 47.23.

The physiologic effects resemble those of kerosine; see Table 47.24. However, in addition, neurologic effects have been recorded, indicating the presence of hexane-like constituents. One acute case has been published where a jet pilot became intoxicated owing to a fuel line leak (1068). The cockpit concentration retroactively was estimated at 3000 to 7000 ppm of JP4 (1068). Long-term worker exposure in the aircraft manufacture using jet fuels caused symptoms of dizziness, headache, nausea, palpitation, and pressure in the chest (966). Concentrations of the solvent in the air were estimated later at 500 to 3000 ppm, based on a molecular weight of 170. Clinical findings included neurasthenia, psychasthenia, and polyneuropathy (966).

Some animal studies have been conducted at Wright-Patterson Air Force Base. Subacute inhalation exposures with the Fischer 344 rat, the C57B1/6 female mouse, and the beagle dog at 0.15 and 0.75 mg/liter for 90 days produced some increased blood urea nitrogen in the rat and decreased serum albumin in the dog (1069). Jet fuel JP9 additives RJ-4 and RJ-5 in a 6-month inhalation study showed increased liver and kidney weights and some pulmonary irritation in the rat and the dog (1070); however, they produced no effects on chronic vapor exposure with the dog and the primate. The authors concluded that the JP9 additives were of low order toxicity (1070). Higher effects were noted with RJ-5 in fish. A concentration of >0.05 mg/liter (1071) decreased the hatchability of flagfish.

5.5.3   Diesel Fuel

Diesel fuel is a gas oil fraction obtained from the middle distillate in petroleum separation (5). It is available in various grades as required by different engine types, one of them being synonymous with fuel oil No. 2. Their compositions vary in ratios of predominantly aliphatic, olefinic, cycloparaffinic, and aromatic hydrocarbons, and also appropriate additives. The slightly viscous, brown fluids are flammable. For further physicochemical properties, see Table 47.23.

Diesel fuel is used for diesel or semidiesel, high-speed engines requiring a type of fuel with low viscosity and moderate volatility (5). The heavier grades are used for railroad and marine diesel engines.

Toxicologic effects are expected to resemble those of kerosine, but somewhat more pronounced, owing to additives, such as sulfurized esters. Sula and Krol (1072) were unable to detect any carcinogenic compounds in yeast cultivated on diesel fuel or in the unsaponifiable fraction of the muscle and liver of chickens and pigs that had consumed yeasts grown on the diesel media (1072). Conversely, mosquito larvae showed a general tissue response with chromatin clumping and loss of the granular matrix (1073). Triazine derivatives are used as effective biocides for petroleum fuel bacteria (1074).

5.5.4 Heating Oils

Heating oils are available in six grades, the selection depending on the type of use. No. 1 fuel oil, stove oil, which is mainly kerosine, is available for home heating using pot burners and stoves (5). It is a flammable liquid; see Table 47.23. Fuel oil No. 2 resembles diesel fuel, and is used in furnaces, burners, and semidiesel engines. Fuel oils 4, 5, and 6 are heavier grades, produced from the residual distillate fraction (5).

For physiological properties of No. 1 fuel oil, see kerosine. Fuel oil No. 2. resembles regular diesel oil and kerosine. Similarly, there is a low oral, moderate dermal, and high aspiration hazard. No. 2 oil was more highly toxic toward estuarine shrimp than crude oil (1075). The fuel was incorporated rapidly into mussel and still detectable after 35 days, with 14 days in clean seawater (1076). Similarly, it was taken up into clams and retained for more than 2 weeks (1077).

5.6 Products from Lubricating Stock Distillates

5.6.1 White Oils

The lubricating oils are vacuum distillation fractions. When further refined and treated, white oils or medicinal oils are obtained. In Great Britain, they are named paraffin oils. White oil, liquid paraffin, mineral oil, white mineral oil, is a mixture of middle aliphatic hydrocarbons. White oil is a flammable, oily, colorless liquid; for further physicochemical data, see Table 47.23.

White oils are physiologically inert, and therefore can be applied internally as a laxative, externally as a protectant and lubricant. Mineral oil does not possess the antidotal effects as believed earlier, since it appears to facilitate the absorption, for example, of benzene, toluene, and chlorinated hydrocarbons

(1078). In Great Britain, the term mineral oil is used interchangeably with petroleum oil or liquid petroleum products.

5.6.2 Paraffins (Waxes)

Upon dilution with naphtha or cooling the lubricating oil fraction, the paraffins or waxes solidify and thus can be separated from the oil. There are two types of waxes, the paraffin wax, found in the low-boiling petroleum fractions, and the microcrystalline wax, found in the high-boiling fractions (5). Paraffin wax, or hard wax, is a mixture of solid hydrocarbons, mainly alkanes. The wax is white, somewhat translucent, odorless, and flammable. For further physico-chemical data, see Table 47.23. Paraffin wax can be added to medicinal agents. Petroleum wax and petrolatum are the only hydrocarbons permitted for use in food products. Paraffin wax is used as a household wax and extensively as a coating for food containers and wrappers.

Physiologically, petroleum waxes are inert; wax fumes are mild eye, nose and throat irritants (569).

Paraffin wax is biodegradable, since a spectrum of microorganisms assimilate n-alkanes (1079).

An analytic procedure is available (310).

5.6.3 Lubricating Oils (Motor and Aviation Oils)

General. The lubricating oils are manufactured from the medium lube distillate and fall into several functional categories. These are the intermittently used motor, aviation, and tractor oils, the continuously servicing turbine oils, and specially prepared insulating and hydraulic fluids (5).

Physiologic Response. The lube oils are composed of aliphatic, olefinic, naphthenic (cycloparaffinic), and aromatic hydrocarbons, as well as additives, depending on their specific use. Lubricating oil additives include antioxidants, bearing protectors, wear resistors, dispersants, detergents, viscosity index im-provers, pourpoint depressors, and antifoaming and rust-resisting agents (5).

The overall viscosity ranges from 75 to 3000 SUS. Therefore, their oral and dermal toxicities are very low, owing to the low vapor pressure. Oral LD_{50} in rodents are normally above 10 g/kg and dermal LD_{50} values greater than 15 g/kg. Inhalation does not present a problem, except if misting occurs, although frequent and prolonged direct skin contact may produce skin irritation and dermatitis (148) in certain hypersensitive individuals, mainly owing to the presence of certain additives (599). Aspiration is also less likely to occur, except for the lower-boiling, light oils.

A metabolic study was undertaken of workers involved in machine-tool operations, including the handling of spindle oil and sulfate coolants. Results revealed a greater demand for oxygen as the worker's service year increased (1080).

Although the aromatic content of motor oil is relatively high, it does not include potentially active polynuclear aromatics; see Table 47.23. A study of workers in England showed scrotal tumors in men exposed to coal tar and polynuclear aromatics containing oil (1081). However, no analytic data were given.

A variety of microorganisms, such as *Pseudomonas, Aeromonas, Aerobacter, and Xanthomas,* grow on lubricating oils (1082). This fact requires the addition of bactericides to lubricants, but, conversely, exhibits the oils' universal biodegradability.

Industrial Hygiene. The threshold limit for oil mist is 5.0 mg/m^3 (185, 384). NIOSH (242) has issued a sampling procedure, S-272, recommending filter collection and chloroform extraction with fluorescence spectrophotometric quantitation. Emission photometric and gas chromatographic procedures also have been used (1083, 1084).

Several patents and reports are available on the use and reclaiming of lubricant oils (1085). Oil adsorbents are described in a Japanese patent (1086).

Handling of lubricant oils requires precautionary measures if misting occurs or fumes develop. The most toxic thermal degradation product was found to be carbon monoxide (1087).

5.6.4 Cutting Oils

General. Cutting oils are fluids that aid by lubrication and cooling the cutting of metal and similar machine operations. The cutting oils resemble the lubricating oils in physicochemical properties. They fall into three categories, (*a*) the water-insoluble petroleum blends (mineral oils), the straight cutting oils, and the concentrates; (*b*) the water-soluble petroleum sulfonates and the chemical emulsions; and (*c*) the synthetic fluids. They are composed of mineral and lard oil, sulfur compounds, and chlorine or chlorinated organic compounds (5). Additives such as formalin, mercurials, or phenolics serve as germicides and emulsifiers, are used as soaps and petroleum sulfonates, and serve as corrosion inhibitors, such as borates, dichromates, or amines (599).

Physiologic Response. Although the parent petroleum cut of the cutting oils is of low order toxicity, the final products, with the additives, have caused common dermal problems, such as contact dermatitis (1088, 1089). Possibly 1 percent of the work force may be affected (1090). This may be due to physical or physiologic blocking of hair follicles progressing to folliculitis (246, 599). The problem may start on dorsal surfaces of hands and arms, later include forearms, thighs, and the abdomen, and include formation of perifollicular papules and pustules. Melanosis may develop later (599). The nickel and chromium salt additives may be the causative elements (1091).

Separate from causing dermal effects, cutting oils have been documented to have been responsible for skin (1092) and scrotal cancer in Europe (1093). However, the occurrence in the United States has been very low (1090). This may have been due to continuous improvements of additives (1094) and decreasing the aromatic portions of the oils (1095). Also, the term mineral oil in Great Britain has been used interchangeably with heavy or aromatic oil.

Repeated application of commercial cutting oil to the mouse skin produced dysplasia or malignancy to 48 percent of the animal group, 8 percent to the controls (1095). The cutting oil bactericide, 1,3,5-tris(hydroxyethyl)-S-hexahydrotriazine, was not mutagenic when tested in Wistar rats (1096).

Industrial Hygiene. Reviews on cutting oil degradation are available (1097), including methods to reclaim the oil (1098). When cutting oils are handled, cutaneous reactions largely can be prevented by good personal hygiene. This consists of minimized contact, prompt removal of oil from the skin with soap and water, and wearing clean work clothing (148) and protective shields (1099). Protective skin creams may also be used (1089, 1100). If dermatitis should occur, *prompt* and expert medical advice should be sought before a chronic state is reached (1099). Prospective employees with a significant history of dermatitis or preexisting skin disorders should be excluded from employment where direct dermal exposure to cutting oils, lubricants, or coolants is likely to occur (1099).

5.6.5 Petrolatum

Petrolatum, mineral jelly, paraffin oil, petroleum jelly, vaseline, is the oldest marketed petroleum product. It is a viscous, yellow to amber mass, odorless and tasteless.

Petrolatum is used widely in the pharmaceutical, medicinal, and household areas. Industrially, it is used in skin protective coatings. Physiologically, it is inert, and is noncarcinogenic (1101). It is nonallergenic and nonirritating, and is thus utilized as a dermal test vehicle. Sample substances remain allergenic in petrolatum for at least 1 year (246).

5.7 Residual Oils

Residual or heavy oils comprise the basis for the heavier fuel oils Nos. 4, 5, and 6, the bunker C, and railroad oils.

Physiologically, they are of low order toxicity, owing to their low volatility. Although owing to their higher aromatic content, repeated prolonged dermal contact may have systemic effects.

REFERENCES

1. R. C. Weast, Ed., *CRC Handbook of Chemistry and Physics*, CRC Press, Cleveland, Ohio, 1977–78.

2. *Laboratory Waste Disposal Manual*, 2nd rev. ed., Manufacturing Chemists Association, Inc., Washington, D.C., 1974.

3. M. Windholz, Ed., S. Budavari, Assoc. Ed., *The Merck Index*, 9th ed., Merck and Company, Inc., Rahway, N.J., 1976.

4. *Registry of Toxic Effects of Clinical Substances*. U.S. Dept. Health, Education, and Welfare, Cincinnati, Ohio, September, 1977.

5. V. B. Guthrie, Ed., *Petroleum Products Handbook*, McGraw-Hill, 1960.

6. N. Sax, *Dangerous Properties of Industrial Materials*, 4th ed., Van Nostrand Reinhold, New York, 1975.

7. *Toxicology and Hazardous Industrial Chemicals Safety Manual for Handling and Disposal with Toxicity Data*, International Technical Information Institute, publ. Japan, 1976.

8. *Hazardous Chemicals Data 1975*, NFPA No. 49, National Fire Protection Association, Boston, 1975.

9. *Handbook of Organic Industrial Solvents*, 4th ed., American Mutual Insurance Alliance, Chicago, 1972.

10. F. Rossini, K. Pitzer, R. Arnett, R. Braun, and G. Pimental, *Selected Values of Physical Thermodynamic Properties of Hydrocarbons and Related Compounds*, Carnegie Press, Pittsburgh, Pa., 1953.

11. L. Mitterhauszerova, K. Fralova, A. Sulkova, and L. Krasnec, *Acta Fac. Pharm. Univ. Comenianae*, **25**, 9 (1974).

12. H. Elmenhorst and H. P. Harke, *Z. Naturforsch.*, **23b**, 1271 (1968).

13. J. R. Newsome, V. Norman, and C. H. Keith, *Tobacco Sci.*, **9**, 102 (1965).

14. E. Lahmann, B. Seifert, and D. Ulbrich, *Proc. Int. Clean Air Congr. 4th, Tokyo, Japan*, 595 (1977).

15. H. W. Gerarde, *Am. Med. Assoc. Arch. Ind. Health*, **19**, 403 (1959).

16. H. W. Gerarde, *Arch. Environ. Health*, **6**, 329 (1963).

17. H. W. Gerarde, *Toxicology and Biochemistry of Aromatic Hydrocarbons*, Elsevier Publishing Co., London, 1960.

18. D. J. Crisp, A. O. Christie, and A. F. A. Ghobasky, *Comp. Biochem. Physiol.*, **22**, 629 (1967).

19. J. M. Neff, B. A. Cox, D. Dixit, and J. W. Anderson, *Mar. Biol.*, **38**(3), 279 (1976).

20. C. E. ZoBell, *API Proce.—Joint Conf. Prev. Control Oil Spills*, 317 (1969).

21. R. Smolik, A. Lange, W. Zatonski, I. Juzwiak, and L. Andreasik, *Pol. Tyg. Lek.*, **28**(21), 769 (1973).

22. "Hazardous Materials Regulations," Department of Transportation, Materials Transportation Bureau, *Fed. Reg.*, **41**, 57018 (1976).

23. I. Kesy-Dabrowska, *Rocz. Panstw. Zakl. Hig.*, **24**(3), 337 (1973).

24. C. T. Pate, B. Atkinson, and J. N. Pitts, Jr., *J. Environ. Sci. Health, Part A*, **A11**(1), 1 (1976).

25. H. E. Runion, *Am. Ind. Hyg. Assoc. J.*, **36**, 338 (1975).

26. R. J. Young, R. A. Rinsky, P. F. Infante, and J. K. Wagoner, *Science*, **199**(4326), 248 (1978).

27. *Fed. Reg.*, **43**(98), 21838 (1978).

28. E. G. Ivanyuk and A. I. Kobzar, *Gig. Sanit.*, **2**, 63 (1974).

29. W. F. Feldman, U.S. Pat. 4005994, February 1, 1977.

30. J. P. Conkle, B. J. Camp, and B. E. Welch, *Arch. Environ. Health*, **30**(6), 290 (1975).

31. B. Williams, L. G. Dring, and R. T. Williams, *Biochem. J.*, **127**(2), 24 (1972).

32. P. S. Jaglan, J. L. Nappier, R. E. Hornish, and A. R. Friedmann, *J. Agric. Food Chem.*, **25**(4), 963 (1977).

33. W. Bertch, R. C. Chang, and A. Zlatkis, *J. Chromatogr. Sci.*, **12**, 175 (1974).

34. S. Pilar and W. F. Graydon, *Environ. Sci. Technol.*, **7**(7), 628 (1973).

35. I. R. Tabershaw, F. Ottoboni, and W. C. Cooper, Air Quality Monographs, #69-5, February, 1969.

36. R. F. Gould, Ed., *Refining Petroleum for Chemicals*, American Chemical Society, Washington, D.C., 1970.

37. *Encyclopaedia of Occupational Health and Safety*, Vols. I and II, McGraw-Hill, New York, 1972.

38. W. F. Von Oettingen, *Publ. Health Bull. No. 255*, U.S. Public Health Service, Washington, D.C., 1940.

39. H. Thienes and T. J. Haley, *Clinical Toxicology*, Lea & Febiger, Philadelphia, 1972.

40. S. Moeschlin, Ed., *Poisoning, Diagnosis and Treatment*, 1st Amer. ed., Grune and Stratton, New York, 1965.

41. E. T. Kimura, D. H. Ebert, and P. W. Dodge, *Toxicol. Appl. Pharmacol.*, **19**, 699 (1971).

42. M. A. Wolf, V. K. Rowe, D. D. McCollister, R. C. Hollingsworth, and F. Oyen, *Am. Med. Assoc. Arch. Ind. Health*, **14**, 387 (1956).

43. D. V. Parke and R. T. Williams, *Biochem. J.*, **55**, 337 (1953).

44. M. F. Savchenko, *Gig. Sanit.*, **32**, 349 (1967).

45. E. W. Lee, J. Kocsis, and R. Snyder, *Res. Commun. Chem. Pathol., Pharmacol.*, **5**(2), 547 (1973).

46. *Acta Med. Biol.*, **17**, 285 (1972) (through Ref. 4).

47. Abderhalden's *Handb. Biol. Arbeitsmethod.*, **4**, 1313 (1935) (through Ref. 4).

48. M. L. Keplinger, G. E. Lanier, and W. R. Deichmann, *Toxicol. Appl. Pharmacol.*, **1**, 156 (1959).

49. Shell Chemical Company, unpublished report, 1961 (through Ref. 4).

50. W. S. Spector, Ed., *Handbook of Toxicology*, Vol. 1, W. B. Saunders, Philadelphia and London, 1956.

51. C. Harper, R. T. Drew, and J. R. Fouts, *Drug Metab. Dispos.*, **3**(5), 381 (1973).

52. C. P. Carpenter, H. F. Smyth, and U. C. Pozzani, *J. Ind. Hyg. Toxicol.*, **31**, 343(1949).

53. J. L. Svirbely, R. C. Dunn, and W. F. Von Oettingen, *J. Ind. Hyg. Toxicol.*, **25**, 366 (1943).

54. B. K. Leong, *J. Toxicol. Environ. Health, Suppl.* **2**, 45 (1977).

55. C. P. Carpenter, C. B. Shaffer, C. S. Weil, and H. F. Smyth, Jr., *J. Ind. Hyg. Toxicol.*, **26**, 69 (1944).

56. M. Peronnet, *J. Pharm. Chim.*, **21**, 503 (1935).

57. R. D. Meyerhoff, *J. Fish. Res. Board Can.*, **32**(10), 1864 (1975).

58. C. J. Barnett and J. E. Kontogiannis, *Environ. Pollut.*, **8**(1), 45 (1975).

59. L. Greenburg, M. Mayers, L. Goldwater, and A. Smith, *J. Ind. Hyg. Toxicol.*, **21**(8), 395 (1939).

60. B. D. Goldstein, *J. Toxicol. Environ. Health*, **2**, 69 (1977).

61. E. C. Vigliani and G. Saita, *N. Engl. J. Med.*, **271**(17), 872 (1964).

62. C. Rozman, S. Woessner, and J. Saez-Serrania, *Acta Haematol.*, **40**(4), 234 (1968).

63. M. Aksoy, S. Erdem, T. Akgun, O. Okur, and K. Dincol., *Blut*, **13**, 85 (1966).

64. S. Hernberg, M. Savilahti, K. Ahlman, and S. Asp, *Br. J. Ind. Med.*, **23**, 204 (1966).

65. M. Aksoy, K. Dincol, T. Akgun, S. Erdem, and G. Dincol., *Br. J. Ind. Med.*, **28**, 296 (1971).

66. A. Forni, E. Pacifico, and A. Limonta, *Arch. Environ. Health,* **22,** 373 (1971).

67. E. Guberan and P. Kocher, *Schweiz. Med. Wochenschr.,* **101,** 1789 (1971).

68. M. Aksoy, K. Dincol, S. Erdem, T. Akgun, and G. Dincol, *Br. J. Ind. Med.,* **29,** 56 (1972).

69. L. Roth, P. Turcanu, I. Dinu, and G. Moise, *Folia Haematol.,* **100,** 213 (1973).

70. M. Aksoy, *New Istanbul Contrib. Clin. Sci.,* **12**(1), 3 (1977).

71. M. G. Ott, J. C. Townsend, W. A. Fishbeck, and R. A. Langnor, *Arch. Environ. Health,* **33**(1), 3 (1978).

72. V. V. Kosyakov, *Metod. Teor. Vopr. Gig. Atmos. Vozdukha,* 60 (1976).

73. V. V. Kosyakov, *Sb. Tr. Nauchno-Issled. Inst. Gig. Tr. Profzabol., Tiflis,* **15,** 186 (1976).

74. S. S. Pawar and A. M. Mungikar, *Indian J. Biochem. Biophys.,* **12**(2), 133 (1975).

75. L. A. Tepikina and M. S. Gordeeva, *Gig. Sanit.,* **3,** 23 (1978).

76. G. G. Avilova, I. P. Ulanova, E. E. Sarkisyants, and E. A. Karpukhina, *Gig. Sanit.,* **6,** 310774 (1974).

77. O. D. Laerum, *Acta Pathol. Microbiol. Scand.,* **81**(1), 57 (1973).

78. M. J. T. Fitzgerald, J. C. Folan, and T. M. O'Brien, *J. Invest. Dermatol.,* **64,** 169 (1975).

79. A. A. Lyapkalo, *Gig. Tr. Prof. Zabol.,* **3,** 14 (1973).

80. V. N. Frash, *Fiziol. Zh.,* **21**(5), 618 (1975).

81. V. B. Dobrokhotov, *Gig. Sanit.,* **10,** 36 (1972).

82. U. Ito, *Bull. Tokyo Med. Dent. Univ.,* **12,** 1 (1965).

83. H. W. Gerarde, *Arch. Ind. Health,* **13,** 468 (1956).

84. S. Koike, K. Kawai, and H. Sugimoto, *Bull. Natl. Inst. Ind. Health,* **2,** 1 (1959).

85. J. H. Ward, J. H. Weisburger, R. J. Yamamoto, T. Benjamin, C. A. Brown, and E. K. Weisburger, *Arch. Environ. Health,* **30,** 22 (1975).

86. V. N. Frash, *Patol. Fiziol. Eksp. Ter.,* **5,** 50 (1975).

87. B. Speck, T. Schnider, U. Gerber, and S. Moeschlin, *Schweiz. Med. Wochenschr.,* **96**(38), 1274 (1966).

88. A. P. Yastrebov and V. N. Frash, *Patol. Fiziol. Eksp. Ter.,* **2,** 76 (1975).

89. V. Muzyka, *Vopr. Med. Khim.,* **15**(5), 521 (1969).

90. T. I. Bonashevskaya, *Gig. Sanit.,* **9,** 14 (1975).

91. C. J. Jenkins, R. A. Jones, and J. Seigel, *Toxicol. Appl. Pharmacol.,* **16,** 818 (1970).

92. G. K. Kadyrov, M. I. Safarov, and I. A. Syntinsky, *Biochem. Pharmacol.,* **24,** 2083 (1975).

93. D. L. Coffin, D. E. Gardner, G. I. Sidorenko, and M. A. Pinigin, *J. Toxicol. Environ. Health,* **3**(5/6), 821 (1977).

94. H. Boje, W. Benkel, and H. J. Heiniger, *Blut,* **21**(4), 250 (1970).

95. Y. E. Yakushevich, *Gig. Sanit.,* **38**(4), 6 (1973).

96. C. Harper, R. J. Drew, and J. R. Fouts, *Biol. React. Intermed.* [*Proc. Int. Conf.*], 302 (1977).

97. P. Moszczynski and A. Starsk, *Pol. J. Pharmacol. Pharm.,* **30**(1), 21 (1978).

98. E. M. Uyeki, A. E. Ashkar, D. W. Shoeman, and T. U. Bisel, *Toxicol. Appl. Pharmacol.,* **40**(1), 49 (1977).

99. A. Rotter, *Arch. Immunol. Ther. Exp.,* **23**(6), 871 (1975).

100. H. H. Schrenk, W. P. Yant, S. J. Pearce, F. A. Patty, and R. R. Sayers, *J. Ind. Hyg. Toxicol.,* **23**(1), 20 (1941).

101. J. B. Lurie, *S. Afr. J. Clin. Sci.,* **3,** 212 (1952).

102. M. M. Key, A. F. Henschel, J. Butler, R. N. Ligo, and I. R. Tabershaw, Eds., *Occupational Diseases, a Guide to their Recognition,* NIOSH, Washington, D.C., 1977.

103. H. H. Cornish, Chap. 19 in *Toxicology, the Basic Science of Poisoning*, L. J. Casarett and J. Doull, Eds., Macmillan, New York, 1975.

104. A. Sato and Y. Fugiwara, *Sangyo Igaku*, **14**(3), 114 (1972).

105. K. Nomiyama and H. Nomiyama, *Int. Arch. Arbeitsmed.*, **32**(1–2), 85 (1974).

106. *API Toxicological Review of Benzene*, American Petroleum Institute, New York, 1960.

107. A. Sato, T. Nakajima, Y. Fugiwara, and K. Hirosawa, *Int. Arch. Arbeitsmed.*, **33**(3), 169 (1974).

108. J. Tauber, *J. Occup. Med.*, **12**(3), 91 (1970).

109. C. L. Winek and W. D. Collom, *J. Occup. Med.*, **13**, 5 (1971).

110. C. L. Winek, W. D. Collom, and C. H. Wecht, *Lancet*, **1**, 683 (1967).

111. B. Block and G. F. Tadjer, *Harefuah*, **89**(2), 74 (1975).

112. R. I. Volchkova, *Tr. Voronszh. Med. Inst.*, **87**, 29 (1972).

113. R. H. Steele and D. Wilhelm, *J. Exp. Pathol.*, **47**(6), 612 (1966).

114. I. D. Gadaskina, A. Z. Buzina, and O. N. Dorofeeva, *Gig. Sanit.*, **3**, 30 (1973).

115. R. A. Bhisey and S. M. Sirsat, *Indian J. Cancer*, **14**(1), 10 (1977).

116. H. H. Cornish and R. C. Ryan, *Toxicol. Appl. Pharmacol.*, **7**(6), 767 (1965).

117. J. Jonek, M. Kaminski, O. Kaminska, and H. Latkowska, *Rev. Roum. Embryol. Cytol. Ser. Cytol.*, **10**(2), 123 (1973).

118. A. Kala, G. S. Rao, and K. P. Pandye, *Bull. Environ. Contam. Toxicol.*, **19**(3), 287 (1978).

119. H. Mikiskova, *Arch. Gewerbepathol. Gewerbehyg.*, **18**, 300 (1960).

120. W. J. Roubal, T. K. Collier, and D. C. Malins, *Arch. Environ. Contam. Toxicol.*, **5**(4), 513 (1977).

121. Y. C. Manyashin, M. F. Savchenkov, and G. Sidnev, *Farmakol. Toksikol.*, **31**(2), 250 (1968).

122. J. L. Egle and B. J. Gochberg, *J. Toxicol. Environ. Health*, **1**(3), 531 (1976).

123. R. T. Drew and J. R. Fouts, *Toxicol. Appl. Pharmacol.*, **27**(1), 183 (1974).

124. L. H. Nahum and H. E. Hoff, *J. Pharmacol. Exp. Ther.*, **50**, 336 (1934).

125. V. Morvai, A. Hudak, G. Ungvary, and B. Varga, *Acta Med. Acad. Sci. Hung.*, **33**(3), 275 (1976).

126. J. Jonek, Z. Olkowski, and B. Zieleznik, *Acta Histochem.*, **20**, 286 (1965).

127. M. Kaminski, A. Karbowski, and J. Jonek, *Folia Histochem. Cytochem. (Krakow)*, **8**(1), 63 (1970).

128. L. Braier and M. Francone, *Arch. Mal. Prof.*, **11**, 367 (1950).

129. K. C. Das, N. N. Sen, and B. K. Aikat, *Indian J. Med. Res.*, **57**(4), 650 (1969).

130. C. F. Reinhardt, A. Azar, M. E. Maxfield, P. E. Smith, and L. S. Mullin, *Arch. Environ. Health*, **22**, 265 (1971).

131. E. S. Reynolds, *Biochem. Pharmacol.*, **21**(19), 2555 (1972).

132. H. W. Gerarde and D. B. Ahlstrom, *Toxicol. Appl. Pharmacol.*, **9**(1), 185 (1966).

133. S. Laham, *Ind. Med.*, **39**(5), 61 (1970).

134. O. H. Bakke and R. R. Scheline, *Toxicol. Appl. Pharmacol.*, **16**, 691 (1970).

135. F. U. Saito, J. J. Kocsis, and R. Snyder, *Toxicol. Appl. Pharmacol.*, **26**, 209 (1973).

136. W. J. Canady, D. A. Robinson, and H. D. Colby, *Biochem. Pharmacol.*, **23** 3075 (1974).

137. G. D. DiVincenzo and W. J. Krasavege, *Am. Ind. Hyg. Assoc. J.*, **35**(1), 21 (1974).

138. G. K. Kadyrow and M. I. Safarov, *Izv. Akad. Nauk. Azerb. SSR, Ser. Biol. Nauk.*, **3**, 109 (1972).

139. R. Snyder, in *Symposium on Toxicology of Benzene and Alkylbenzenes*, D. Braun, Ed., Industrial Health Foundation, Pittsburgh Pa., 1974, pp. 44–53.

140. J. M. Wildman, M. L. Freedman, J. Roseman, and B. Goldstein, *Res. Commun. Chem. Pathol. Pharmacol.*, **13**(3), 473 (1976).

141. A. M. Mungikar and S. S. Pawar, *Indian J. Exp. Biol.*, **13**(5), 439 (1975).

142. M. Ikeda and H. Ohtsuji, *Toxicol. Appl. Pharmacol.*, **20**, 30 (1971).

143. M. Ikeda, H. Ohtsuji, and T. Imamura, *Xenobiotica*, **2**(2), 101 (1972).

144. S. L. Andrews, E. W. Lee, C. H. Witmer, J. J. Kocsis, and R. Snyder, *Biochem. Pharmacol.*, **26**, 293 (1977).

145. J. A. Timbrell and J. R. Mitchell, *Xenobiotica*, **7**(7), 415 (1977).

146. G. P. Biscaldi, G. R. D. Cuna, and G. Pollini, *Haematol. Arch.*, **54**(8), 579 (1978).

147. M. Aksoy, S. Erdem, K. Dincol, T. Hepyuksel, and G. Dincol, *Blut*, **28**, 293 (1974).

148. F. A. Patty, Ed., *Industrial Hygiene and Toxicology*, Vol. II, Wiley-Interscience, New York, 1963.

149. P. Mikulski, R. Wiglusz, A. Bublewska, and J. Uselis, *Biul. Inst. Med. Morsk Gdansk*, **23**(1/2), 67 (1972).

150. N. L. Kanner, *Gig. Tr. Prof. Zabol.*, **15**(10), 60 (1971).

151. *Review of the Health Effects of Benzene*, National Research Council, Committee on Toxicology, National Academy of Sciences, Washington, D.C., 1975.

152. G. M. Rusch, B. K. J. Leong, and S. Laskin, *J. Toxicol. Environ. Health Suppl.*, **2**, 23 (1977).

153. S. Pawelski, E. Wechrzycka, B. Modzewski, and S. Roszkowski, *Pol. Tyg. Lek.*, **19**(38), 1433 (1964).

154. A. Sato, T. Nakajima, Y. Fujiwara, and N. Nuragama, *Br. J. Ind. Med.*, **32**, 321 (1975).

155. A. Brzecki, S. Misztel, and R. Kostolowski, *Ann. Acad. Med. Lodz*, **14**(1), 55 (1973).

156. E. A. Drogichina, L. A. Zorina, and I. A. Gribova, *Gig. Tr. Prof. Zabol.*, **15**(5), 18 (1971).

157. A. Björnberg and H. Mobackin, *Berufsdermatosen*, **21**(6), 245 (1973).

158. C. F. Reinhardt, L. S. Mullins, and M. E. Maxfield, *J. Occup. Med.*, **15**(12), 953 (1973).

159. A. M. Monaenkova and K. V. Glotova, *Gig. Tr. Prof. Zabol.*, **13**(11), 32 (1969).

160. A. M. Monaenkova and L. A. Zorina, *Gig. Tr. Prof. Zabol.*, **4**, 30 (1975).

161. B. J. Dowty and J. L. Laseter, *Pediatr. Res.*, **10**, 696 (1976).

162. A. Forni and L. Moreo, *Eur. J. Cancer*, 251 (1967).

163. E. C. Vigliani and A. Forni, *J. Occup. Med.*, **11**(3), 148 (1969).

164. M. Sellyei and E. Kelemen, *Eur. J. Cancer*, **7**(1), 83 (1971).

165. E. C. Vigliani, *Ann. N.Y. Acad. Sci.*, **271**, 143 (1976).

166. P. A. Goguel, A. Cavigneaux, and J. Bernard, *Nouv. Rev. Fr. Hematol.*, **7**(4), 465 (1967).

167. M. Aksoy, K. Dincol, S. Erdem, and G. Dincol, *Am. J. Med.*, 160 (1970).

168. M. Aksoy, S. Erdem, G. Erdogan, and G. Dincol, *Human Heredity*, **24**, 70 (1974).

169. M. Aksoy, S. Erdem, and G. Dincol, *Blood*, **44**(6), 837 (1974).

170. M. Aksoy, S. Erdem, and G. Dincol, *Acta Haematol.*, **55**, 65 (1976).

171. M. Aksoy, *Lancet*, **1**, 441 (1978).

172. M. Aksoy, S. Erdem, G. Erdogan, and G. Dincol, *Human Heredity*, **26**, 149 (1976).

173. F. A. Walker, *Ann. Intern. Med.*, **85**(3), 404 (1975).

174. J. J. Thorpe, *J. Occup. Med.*, **16**(6), 375 (1974).

175. I. M. Tough, P. G. Smith, W. M. C. Brown, and D. G. Harnden, *Eur. J. Cancer*, **6**(1), 49 (1970).

176. H. Khan and M. H. Khan, *Arch. Toxikol.*, **31**(1), 39 (1973).

177. A. M. Forni, A. Cappellini, E. Pacifico, and E. C. Vigliani, *Arch. Environ. Health*, **23**, 358 (1971).

178. A. Forni and L. Moreo, *Eur. J. Cancer*, **5**(5), 459 (1969).

179. S. R. Wolman, *J. Toxicol. Environ. Health Suppl.*, **2**, 63 (1977).

180. M. L. Freedman, *J. Toxicol. Environ. Health,* Suppl. 2, 37 (1977).

181. R. Snyder, E. W. Lee, J. J. Kocsis, and C. M. Witmer, *Life Sci.,* **21**(12), 1709 (1977).

182. R. V. Sapozhkov, *Tr. Sarat. Med. Inst.,* **71–88,** 18 (1970).

183. R. Smolik, K. Grzybek-Hryncewicz, A. Lange, and W. Zatonski, *Int. Arch. Arbeitsmed.,* **31,** 243 (1973).

184. W. A. Fishbeck, R. R. Langner, and R. J. Kociba, *Am. Ind. Hyg. Assoc. J.,* **36**(11), 820 (1975).

185. *Occupational Safety and Health Standards Subpart Z—Toxic and Hazardous Substances,* CFR, Title 29, sec. 1910.93, 1976.

186. J. N. George, D. R. Miller, and R. I. Weed, *N.Y. State J. Med.,* **70**(20), 2574 (1970).

187. G. Pollini, G. P. Biscaldi, and G. F. DeStefano, *Med. Lav.,* **67**(5), 506 (1976).

188. A. V. Karaulov, B. G. Yushkov, and V. N. Frash, *Radiobiologiya,* **16**(5), 791 (1976).

189. P. A. Miescher, *Semin. Haematol.,* **10**(4), 311 (1973).

190. E. J. Gralla, *Vet. Clin. North Am.,* **5**(4), 699 (1975).

191. E. Browning, *J. Occup. Med.,* **7,** 554 (1965).

192. V. I. Boiko, L. M. Makareva, Z. G. Podrez, M. Y. Burdygina, and V. A. Sukhanova, *Kazan. Med. Zh.,* **1,** 77 (1975).

193. Y. N. Pestrii and V. F. Budyak, *Vopr. Kurortol. Fizioter. Lech. Fiz. Kult.,* **36**(5), 455 (1972).

194. M. Barbacki, *Minerva Pediatr.,* **23**(12), 507 (1971).

195. S. T. Crooke, *Texas Med.,* **68,** 67 (1972).

196. A. Lange, R. Smolik, W. Zatonski, and J. Szymanska, *Int. Arch. Arbeitsmed.,* **31**(1), 37 (1973).

197. I. Juzwiak and W. Fedorowicz, *Czas Stomatol.,* **27**(8), 855 (1974).

198. M. Minai, Bull. *Tokyo Med. Dent. Univ.,* **14**(3), 327 (1967).

199. L. Braier, *Haematologica,* **58**(78), 491 (1973).

200. I. Gut, *Cesh. Hyg.,* **16**(6), 183 (1971).

201. W. Laase, *Pharmazie,* **28,** 10 (1973).

202. A. Sato, Y. Fujiwara, and T. Nakajima, *Sangyo Igaku,* **16**(1), 30 (1974).

203. J. J. Kocsis, E. W. Lee, M. D'Souza, and R. Snyder, *Fed. Proc. Abstr.,* **37**(3), 505 (1978).

204. H. Kahn and I. Muzyka, *Work Environ. Health,* **10**(3), 140 (1973).

205. R. Snyder and M. Ikeda, *Int. Workshop on Toxicity of Benzene, Paris, November 9–11, 1976.*

206. L. A. Tiunov, S. P. Nechiporenko, Z. I. Menshikova, N. M. Petushkov, V. A. Ivanova, T. S. Kolosova, and M. A. Akhmatova, *Farmakol. Toksikol.,* **40**(1), 97 (1977).

207. J. J. Jonek, M. Kaminski, H. Grzybek, P. Panz, and B. Gruszeczka, *Acta Histochem. (Jena),* **55**(1), 60 (1976).

208. I. Gut, *Arch. Toxicol.,* **35**(3), 195 (1975).

209. E. W. Lee, L. S. Andrews, C. M. Witmer, F. W. Deckert, J. J. Kocsis, and R. Snyder, *Fed. Proc.,* **34**(3), 227 (1975).

210. S. M. Gusman and V. A. Zhukov, *Strukt. Funkts. Gisto-Gematicheskikh Barerov, Mater. Soveshch. Probl. Gisto-Gematicheskikh Barerov,* 87 (1971).

211. V. N. Russkikh and A. V. Rodnikov, *Tr. Nauch. Konf. Nauch-Issled. Inst. Gig. Vod. Transp.,* **2,** 137 (1972).

212. J. Jonek, J. Sroczynski, M. Kaminski, and H. Grzybek, *Arch. Mal. Prof. Med.,* **32**(9), 517 (1971).

213. F. Ito, *Showa Igakkai Zasshi,* **22,** 278 (1962).

214. L. M. Bernshtein, *Vopr. Gig. Tr. Profzabol., Mater. Nauch. Konf.,* 53 (1972).

215. J. Sroczynski, K. Zapisz, G. Jonderko, and A. Wegiel, *Patol. Pol.*, **27**(1), 59 (1976).
216. M. Berlin, J. Gage, and E. Johnson, *Work Environ. Health*, **11**(1), 1 (1974).
217. V. N. Frash, B. K. Yushkov, and A. V. Karaulov, *Gig. Tr. Prof. Zabol.*, **12**, 44 (1976).
218. M. Kissling and B. Speck, *Helv. Med. Acta*, **36**(1), 59 (1971).
219. F. J. Fork, H. S. Cohen, J. Rosman, and M. C. Freedman, *Blood*, **47**(1), 145 (1976).
220. E. S. Tikhachek and V. N. Frash, *Gig. Tr. Prof. Zabol.*, **17**(8), 30 (1973).
221. M. A. Vozoyava, *Gig. Sanit.*, **6**, 100 (1976).
222. G. I. Watanabe and S. Yoshida, *Acta Med. Biol.*, **17**(4), 285 (1970).
223. V. B. Dobrokhotov and M. I. Enikeev, *Gig. Sanit.*, **1**, 32 (1977).
224. J. P. Lyon, *Diss. Abstr. Int. B.*, **36**(11), 5537 (1975).
225. D. Tarin, *Int. J. Cancer*, **3**(6), 734 (1968).
226. D. K. Garibyan and S. A. Papoyan, *Nekot. Itogi. Izuch. Zagryazneniya Vnesh. Sredy Kanstserogen. Veshchestvami*, 112 (1972).
227. B. L. Van Duuren, N. Nelson, L. Orris, E. D. Palmer, and F. L. Schmitt, *J. Natl. Cancer Inst.*, **31**(1), 41 (1963).
228. S. S. Pawar, A. M. Mungikar, and N. P. Galdhar, *Indian J. Biochem. Biophys.*, **12**, 252 (1975).
229. R. Y. Shlerengarts and N. I. Solomatina, *Gig. Tr. Prof. Zabol.*, **1**, 47 (1977).
230. T. R. Varma, P. Erdody, and T. K. Murray, *Biochim. Biophys. Acta*, **104**(1), 71 (1965).
231. T. K. Murray and P. Erdody, *Biochim. Biophys. Acta*, **124**(1), 190 (1966).
232. N. N. Pushkina, V. A. Gofmekler, and G. N. Klevtsova, *Byull. Eksp. Biol. Med.*, **66**(8), 51 (1968).
233. K. G. Kadyrov, E. A. Badullaeva, and M. I. Safarov, *Izv. Akad. Nauk. Azerb. SSR, Ser. Biol. Nauk.*, **2**, 102 (1972).
234. H. Cavusoglu, A. Kayseriligolu, M. Aksoy, and K. Dincol, *Istanbul Tip Fak. Mecm.*, **39**(3), 368 (1976).
235. J. Aleksandrowicz, A. Starek, and P. Moszczynski, *Med. Pr.*, **28**(6), 453 (1977).
236. L. Braier, *Rev. Bras. Pesqui. Med. Biol.*, **10**(5), 319 (1977).
237. F. Jansen and A. Olsen, *Plant Physiol.*, **44**(5), 786 (1969).
238. P. Tkhelidze, *Soobsch. Akad. Nauk. Gruz. SSR*, **56**(3), 697 (1969).
239. D. T. Gibson, J. R. Koch, and R. E. Kallio, *Biochem.*, **7**(7), 2653 (1968).
240. W. C. Evans, *Nature*, **270**(5632), 17 (1977).
241. *Fed. Reg.*, **42**(85), 22516 (1977).
242. *NIOSH Manual of Sampling Data Sheets*, U.S. Dept. Health, Education and Welfare, Cincinnati, Ohio, 1977.
243. *Criteria for Recommended Standard—Occupational Exposure to Benzene*, U.S. Dept. Health, Education and Welfare, Cincinnati, Ohio, 1974.
244. H. M. D. Utidjian, *J. Occup. Med.*, **18**, 7 (1976).
245. P. Walker, *U.S. Natl. Tech. Inf. Serv. Publ. Bull. Rep. Issue PL-256735* (1976).
246. A. A. Fisher, *Contact Dermatitis*, 2nd ed., Lea & Febiger, Philadelphia, 1973.
247. Work Environmental Health, **9**, 131 (1972) (through Ref. 4).
248. R. H. Wilson, *J. Am. Med. Assoc.*, **123**, 1106 (1943).
249. L. Greenberg, M. R. Mayers, H. Heimann, and S. Moskowitz, *J. Am. Med. Assoc.*, **118**, 573 (1942).
250. H. F. Smyth, Jr., C. P. Carpenter, C. S. Weil, U. C. Pozzani, J. A. Striegel, and J. S. Nycum, *Am. Ind. Hyg. Assoc. J.*, **30**, 470 (1969).

251. C. P. Carpenter, D. L. Geary, Jr., R. C. Myers, D. J. Nachreiner, L. J. Sullivan, and J. M. King, *Toxicol. Appl. Pharmacol.*, **26,** 473 (1976).

252. T. Kojima and H. Kobayashi, *Nippon Hoigaku Zasshi,* **27**(4), 258 (1973).

253. N. W. Lazarew, *Arch. Exp. Pathol. Pharmacol.*, **143,** 223 (1929).

254. *Union Carbide Data Sheet,* Industrial Medicine and Toxicology Dept., Union Carbide Corp., New York, 1976 (through Ref. 4).

255. Y. Takeuchi, T. Tanaka, T. Matsumoto, and T. Matsushita, *Sangyo Igaku,* **14**(6), 543 (1972).

256. E. I. Makshanova and M. S. Omelyanchik, *Zdravookh. Beloruss,* **4,** 81 (1977).

257. H. F. Smyth and H. F. Smyth, Jr., *J. Ind. Hyg.*, **10,** 261 (1928).

258. T. Dalhamn, M. C. Edfors, and R. Rylander, *Arch. Environ. Health,* **17,** 746 (1968).

259. E. Reisin, A. Teicher, R. Jaffe, and H. E. Eliahou, *Br. J. Ind. Med.*, **32,** 163 (1975).

260. M. G. Casarett, "Social Poisons," Chap. 25 in *Toxicology, the Basic Science of Poisons,* L. J. Casarett and J. Doull, Eds., Macmillan, New York, 1975.

261. J. Soderlund, *Int. J. Occup. Health Safety,* 42–55 (1975).

262. I. Astrand, H. Ehrner-Samuel, A. Kilbom, and P. Ovrum, *Work Environ. Health,* **9**(3), 119 (1972).

263. F. Gamberale and M. Hultengren, *Work Environ. Health,* **9**(3), 131 (1972).

264. I. Elster, *Dtsch. Med. Wochenschr.,* **97,** 1887 (1972).

265. S. M. Taher, R. J. Anderson, R. McCartney, M. M. Popovtzer, and R. W. Schrier, *N. Engl. J. Med.*, **290,** 765 (1974).

266. E. J. O'Brien, W. B. Yeoman, and J. A. E. Hobby, *Br. Med. J.*, **2,** 29 (1971).

267. T. W. Kelly, *Pediatrics,* **56,** 605 (1975).

268. A. Friborska, *Folia Haematol.,* **99**(2–3), 233 (1973).

269. A. Vlastiborova and A. Friborska, *Folia Haematol.,* **99**(2–3), 230 (1973).

270. D. Pey, *Arch. Mal. Prof. Med. Trav. Secur. Soc.,* **33**(10–11), 584 (1972).

271. *Criteria for a Recommended Standard—Occupational Exposure to Toluene,* U.S. Dept. Health, Education and Welfare, Cincinnati, Ohio, 1973.

272. S. R. Cohen and A. A. Maier, *J. Occup. Med.,* **16**(3), 201 (1974).

273. W. F. Von Oettingen, P. A. Neal, and D. D. Donahue, *J. Am. Med. Assoc.*, **118**(8), 579 (1942).

274. M. Sato, *Sangyo Igaku,* **15**(3), 261 (1973).

275. D. Szadkowski, R. Pett, J. Angerer, A. Manz, and G. Lehnert, *Int. Arch. Arbeitsmed.*, **31**(4), 265 (1973).

276. M. Ogata, K. Tomokuni, and Y. Takatsuka, *Br. J. Ind. Med.*, **27**(1), 43 (1970).

277. M. Ogata, Y. Takatsuka, and K. Tomokuni, *Br. J. Ind. Med.*, **28,** 382 (1971).

278. H. B. Elkins, *The Chemistry of Industrial Toxicology,* 2nd ed., John Wiley, New York, London, 1959.

279. M. Ikeda and H. Ohtsuji, *Br. J. Ind. Med.*, **26,** 244 (1969).

280. E. DeRosa, M. Mazzotta, F. Forin, and M. A. Corradina, *Lav. Um.,* **27**(1), 18 (1975).

281. A. Capellini and L. Alessio, *Med. Lav.,* **62**(4), 196 (1971).

282. D. M. Jerina, N. Kaubisch, and J. W. Daly, *Proc. Natl. Acad. Sci. U.S.,* **68**(10), 2545 (1971).

283. M. A. Q. Khan and J. P. Bederka, *Survival in Toxic Environments,* Academic Press, New York, 1974.

284. T. Hasegawa, S. Kira, and M. Ogata, *Igaku To Seibutsuzaku,* **89**(5), 291 (1974).

285. T. Kojima and H. Kobayashi, *Nippon Hoigaku Zasshi,* **29**(2), 82 (1975).

286. M. Ogata, T. Salki, S. Kira, T. Hasegewa, and S. Watanabe, *Sangyo Izoku,* **16**(1), 23 (1974).

287. G. Brenniman, R. Hartung, and W. J. Weber, Jr., *Water Res.,* **10**(2), 165 (1976).

288. J. E. Morrow, R. L. Gritz, and M. P. Kirton, *Copeia,* **2,** 326 (1975).

289. R. F. Lee, R. Sauerheber, and A. A. Benson, *Science,* **177**(4046), 344 (1972).

290. T. T. Ishikawa and H. Schmidt, Jr., *Pharmacol. Biochm. Behav.,* **1**(5), 593 (1973).

291. T. Ikeda and H. Miyake, *Toxicol. Letters,* **1**(4), 235 (1978).

292. S. Moeschlin and B. Speck, *Acta Haematol.,* **38,** 104 (1967).

293. G. P. Babanov, Y. A. Burov, N. A. Skolbei, A. L. Isakhanov, and I. A. Troitskaya, *Gig. Tr. Prof. Zabol.,* **16**(12), 57 (1972).

294. R. Girard and L. Revol, *Nouv. Rev. Fr. Hematol.,* **10**(4), 477 (1970).

295. A. Sato and T. Nakajima, *Toxicol. Appl. Pharmacol.,* **48,** 249 (1979).

296. M. H. Simmers, *Ind. Med. Surg.,* **34**(3), 255 (1965).

297. M. Ikeda, *Int. Arch. Arbeitsmed.,* **33**(2), 125 (1974).

298. R. J. Withey and J. W. Hall, *Toxicology,* **4**(1), 5 (1975).

299. S. V. Durmishidze, D. S. Ugrekhelidze, and A. N. Dzhikya, *Prik. Biokhim. Mikrobiol.,* **10**(5), 673 (1974).

300. N. R. Vishwanath, R. B. Patil, and G. Rangaswami, *Zentralbl. Bakteriol. Parasitenkd. Infektionskr. Hyg., Abt. 2,* **130**(4), 348 (1975).

301. T. Miller, *Am. J. Vet. Res.,* **27**(121), 1755 (1966).

302. K. Yamada, S. Horiguchi, and J. Takahashi, *Agric. Biol. Chem.,* **29**(10), 943 (1965).

303. R. L. Raymond, V. W. Jamison, and J. O. Hudson, *Appl. Microbiol.,* **15**(4), 857 (1967).

304. E. T. McKenna and M. J. Coon, *J. Biol. Chem.,* **245**(15), 3882 (1970).

305. J. Nozaka and M. Kusunose, *Agric. Biol. Chem.,* **32**(12), 1484 (1968).

306. J. Nozaka and M. Kusunose, *Agric. Biol. Chem.,* **32**(8), 1033 (1968).

307. *Fed. Reg.,* **40**(194), 46206 (1975).

308. H. Utidjian, *J. Occup. Med.,* **16**(2), 107 (1974).

309. N. K. Weaver, *J. Occup. Med.,* **16**(2), 109 (1974).

310. L. Meites, *Handbook of Analytical Chemistry,* 1st ed., McGraw-Hill, New York, 1963.

311. L. D. White, D. G. Taylor, P. A. Mauer, and R. E. Kupel, *Am. Ind. Hyg. Assoc. J.,* **31,** 225 (1970).

312. I. Viden, V. Kubelka, and J. Mostecky, *Z. Anal. Chem.,* **280**(5), 369 (1976).

313. D. L. Guertin and H. W. Gerarde, *Am. Med. Assoc. Arch. Ind. Health,* **20,** 262 (1959).

314. A. Sato, T. Nakajima, and Y. Fugiwara, *Br. J. Ind. Med.,* **32**(3), 210 (1975).

315. T. Kojima and H. Kobayashi, *Nippon Hoigaku Zasshi,* **27**(4), 255 (1973).

316. M. Ogata, K. Tomokuni, and Y. Takatsuka, *J. Ind. Med.,* **26**(4), 330 (1969).

317. J. P. Buchet and R. R. Lauwerys, *Br. J. Ind. Med.,* **30,** 125 (1973).

318. K. Engstrom, K. Husman, and J. Rantanen, *Int. Arch. Occup. Environ. Health,* **363,** 153 (1976).

319. P. B. Van Roosmalen and I. Drummond, *Br. J. Ind. Med.,* **35**(1), 56 (1978).

320. I. Gontea, E. Bistriceanu, M. Draghicesu, and M. Manea, *Arch. Sci. Physiol.,* **22**(3), 397 (1968).

321. M. Guillemin, J. C. Murset, M. Lob, and J. Riquez, *Br. J. Ind. Med.,* **31,** 310 (1974).

322. *Chemical Safety Data Sheet SD-63—Properties and Essential Information for Safe Handling and Use of Toluene,* Manufacturing Chemists' Assoc., Inc., Washington, D.C., 1956.

323. E. Browning, *Toxicity and Metabolism of Industrial Solvents,* Elsevier, New York, 1965.

324. W. B. Deichmann and H. W. Gerarde, *Toxicology of Drugs and Chemicals,* Academic Press, New York, 1969.

325. C. P. Carpenter, E. R. Kinkead, D. L. Geary, Jr., L. J. Sullivan, and J. M. King, *Toxicol. Appl. Pharmacol.*, **33**(3), 543 (1975).

326. K. W. Nelson, J. F. Ege, M. Ross, L. E. Woodman, and L. Silverman, *J. Ind. Hyg. Toxicol.*, **25**, 282 (1943).

327. R. E. Joyner and W. L. Pegues, *J. Occup. Med.*, **3**, 211 (1961).

328. R. Morley, D. W. Eccleston, C. P. Douglas, W. E. Greville, D. J. Scott, and J. Anderson, *Br. Med. J.*, **3**, 442 (1970).

329. P. I. Mikulski, R. Wiglusz, A. Bublewska, and J. Uselis, *Br. J. Ind. Med.*, **29**, 450 (1972).

330. C. H. Hine and H. H. Zuidema, *Ind. Med. Surg.*, **39**, 215 (1970).

331. H. Schumacher and E. Grandjean, *Arch. Gewerbepathol. Gewerbehyg.*, **18**, 109 (1960).

332. *J. Pathol. Bacteriol.*, **46**, 95 (1938) (through Ref. 4).

333. H. F. Smyth, Jr., C. P. Carpenter, C. S. Weil, U. C. Pozzani, and J. A. Striegel, *Am. Ind. Hyg. Assoc. J.*, **23**, 95 (1962).

334. W. Matthaus, *Klin. Monatsbl. Augenheilkd.*, **144**, 713 (1964).

335. B. Speck and S. Moeschlin, *Schweiz. Med. Wochenschr.*, **98** (42), 1684 (1968).

336. L. M. Kashin, I. L. Kulinskaya, and L. F. Mikhailovskaya, *Vrach. Delo.*, **8**, 109 (1968).

337. R. Fabre, R. Truhaut, and S. Laham, *Arch. Mal. Prof.*, **21**, 301 (1960).

338. W. E. Engelhardt, *Arch. Hyg. Bakteriol.*, **114**, 219 (1935).

339. W. J. Hayes, *Clinical Handbook on Economic Poisons*, U.S. Government Printing Office, Washington, D.C., 1971.

340. Hygienic Guide Series, *Am. Ind. Hyg. Assoc. J.*, **29**, 702 (1971).

341. L. T. Fairhall, *Industrial Toxicology*, 2nd ed., Hafner, New York, 1969.

342. D. Högger, *Schweiz. Med. Wochenschr.*, **97**, 368 (1967).

343. E. Ghislandi and A. Fabiani, *Med. Lav.*, **48**, 577 (1957).

344. S. Gitelson, L. Aladjemoff, S. Ben-Hador, and R. Katznelson, *J. Am. Med. Assoc.*, **197**(10), 165 (1966).

345. C. E. Searle, Ed., *Chemical Carcinogens*, American Chemical Society, Washington, D.C., 1976.

346. E. Lederer, *Muench. Med. Wochenschr.*, **114** (29/30), 1302 (1972).

347. V. Mathies, *Med. Klin.*, **63**, 463 (1970).

348. *API Toxicological Review of Xylene*, American Petroleum Institute, New York, 1960.

349. W. E. Clark, *Am. J. Clin. Path.*, **68**. 425 (1977).

350. L. Nelkin, *Zentralbl. Gewerbehyg.*, **18**, 182 (1931).

351. E. Rosenthal-Deussen, *Arch. Gewerbepathol. Gewerbehyg.*, **2**, 92 (1931).

352. J. Kucera, *J. Pediatr.*, **72**, 857 (1968).

353. K. Engstrom, K. Husman, and V. Riihimaki, *Int. Arch. Occup. Environ. Health*, **39**(3), 181 (1977).

354. R. R. Lauwerys, T. Dath, J. M. LaChapelle, J. P. Buchet, and H. Roels, *J. Occup. Med.*, **20**(1), 17 (1978).

355. V. Riihimaki and P. Pfaffli, *Scand. J. Work. Environ. Health*, **4**(1), 73 (1978).

356. V. Sedivec and J. Flek, *Int. Arch. Occup. Environ. Health*, **37**(3), 205 (1976).

357. V. Sedivec and J. Flek, *Prac. Lek.*, **27**(3), 68 (1975).

358. V. Sedivec and J. Flek, *Prac. Lek.*, **26**(7), 243 (1974).

359. W. Senczuk and J. Orlowski, *Br. J. Ind. Med.*, **35**(1), 50 (1978).

360. A. Winnicka, J. Chmielewski, and T. Mardkowicz, *Pol. Tyg. Lek.*, **32**(31), 1149 (1977).

361. G. Aschan, I. Bunnfors, D. Hyden, B. Larsby, L. M. Odkvist, and R. Tham, *Acta Otolaryngol.*, **84**(5–6), 370 (1977).

362. R. H. Rigdon, *Arch. Surg.*, **41**, 101 (1940).

363. G. R. Cameron, J. L. H. Paterson, G. S. W. deSaram, and J. C. Thomas, *J. Pathol. Bacteriol.*, **46**, 95 (1938).

364. J. J. Batchelor, *Am. J. Hyg.*, **7**, 276 (1927).

365. W. Hann and P. Jensen, *Water Quality Characteristics of Hazardous Materials*, Vols. 1–4, Environmental Engineering Division, Civil Engineering Dept., Texas A&M University, 1974.

366. D. F. Walsh, J. G. Armstrong, T. R. Bartley, H. A. Salman, and P. A. Frank, *U.S. Natl. Tech. Inf. Serv. PB Rep., Iss PB-267278* (1977).

367. Y. A. Krotov and N. A. Chebotar, *Gig. Tr. Prof. Zabol.*, **16**(6), 40 (1972).

368. *Criteria for a Recommended Standard—Occupational Exposure to Xylene*, U.S. Dept. Health, Education and Welfare, Natl. Inst. Occup. Safety and Health, Washington, D.C., 1975.

369. H. G. Bray, B. B. Humpris, and W. V. Thorpe, *Biochem. J.*, **47**, 395 (1950).

370. R. Fabre, R. Truhaut, and S. Laham, *C. R. Acad. Sci.*, **250**, 2655 (1960).

371. I. Fridlyand, *Farmakol. Toksikol.* (*Moscow*), **33**(4), 499 (1970).

372. W. Senczuk, S. Litewka, J. Orlowski, and H. Pogorzelska, *Ann. Pharm.* (*Poznan*), **9**, 3 (1971).

373. C. Harper, *Pharmacol. Fed. Proc.*, **34**(1), 785 (1975).

374. M. F. Carlone and J. R. Fouts, *Xenobiotica*, **4**(11), 705 (1974).

375. R. T. Drew and J. R. Fouts, *Toxicol. Appl. Pharmacol.*, **29**(1), 111 (1974).

376. T. Omori, S. Horiguchi, and K. Yamada, *Agric. Biol. Chem.*, **31**(11), 1337 (1967).

377. T. Omori and K. Yamada, *Agric. Biol. Chem.*, **33**(7), 979 (1969).

378. G. K. Skriabin, Kh. G. Ganbarov, L. A. Golovleva, I. I. Chervin, and V. M. Adanin, *Mikrobiologiia*, **45**(6), 951 (1976).

379. T. Omori and K. Yamada, *Agric. Biol. Chem.*, **34**(5), 659 (1970).

380. R. S. Davis, F. E. Hossler, and R. W. Stone, *Can. J. Microbiol.*, **14**(2), 1005 (1968).

381. D. T. Gibson, V. Mahadeven, and J. F. Davey, *J. Bacteriol.*, **119**(3), 930 (1974).

382. J. F. Davey and D. T. Gibson, *J. Bacteriol.*, **119**(3), 923 (1974).

383. V. W. Jamison, R. L. Raymond, and J. O. Hudson, *Appl. Microbiol.*, **17**(6), 853 (1969).

384. *Threshold Limit Values for Chemical Substances and Physical Agents in the Workroom Environment with Intended Changes for 1979*, American Conference of Governmental Industrial Hygienists, Cincinnati, Ohio, 1978.

385. W. H. Stahl, *Compilation of Odor and Taste Threshold Values*, Data, American Society for Testing and Materials, Philadelphia, 1973.

386. B. Sova, *Cesk. Hyg.*, **20**(4), 214 (1975).

387. A. Sato, *Sangyo Igaku*, **14**(2), 132 (1972).

388. J. Orlowski, *Bromatol. Chem. Toksykol.*, **7**, 87 (1974).

389. H. Matsui, M. Kasao, and S. Imamura, *J. Chromatogr.*, **145**(2), 231 (1978).

390. J. Angerer, *Int. Arch. Occup. Environ. Health*, **36**(4), 287 (1976).

391. S. Kira, *Sangyo Igaku*, **19**(3), 126 (1977).

392. T. Arima, T. Fukuda, and N. Tani, Jap. Pat. 75136282, October 29, 1975.

393. V. M. Bagnyuk, T. L. Oleinik, A. G. Brekhunets, and V. G. Koval, *Mater. Vses. Nauchn. Simp. Sovrem. Probl. Samoochishcheniya Regul. Kach. Vody, 5th*, **6**, 3 (1975).

394. R. J. Sperber and S. L. Rose, *Soc. Plast. Eng. Tech.*, **21**, 521 (1975).

395. P. I. Mikulski and R. Wiglusz, *Toxicol. Appl. Pharmacol.*, **31**(1), 21 (1975).

396. G. Hultgren, *C. R. Soc. Biol.*, **95**, 1060 (1926).

397. T. Dutkiewicz and H. Tyras, *Br. J. Ind. Med.*, **24**, 330 (1967).

398. W. L. Treadway, *Publ. Health Rep.*, **45**, 1239 (1930).

399. Z. Bardodej and E. Bardodejova, *Am. Ind. Hyg. Assoc. J.*, **31**, 206 (1970).

400. D. L. Opdyke, *Food Cosmet. Toxicol.*, **13**, 681 (1975).

401. K. F. Meleschenko, *Gig. Sanit.*, **6**, 90 (1975).

402. P. M. Jenner, E. C. Hagen, J. M. Taylor, E. S. Cook, and O. G. Fitzhugh, *Food Cosmet. Toxicol.*, **2**, 327 (1964).

403. D. W. Furnas and C. H. Hine, *Am. Med. Assoc. Arch. Ind. Health*, **18**, 9 (1958).

404. C. P. Carpenter, C. S. Weil, and H. F. Smyth, Jr., *Toxicol. Appl. Pharmacol.*, **28**, 313 (1974).

405. C. H. Hine, H. Unger, H. H. Anderson, J. K. Kodama, J. K. Critchlow, and N. W. Jacobsen, *Ind. Hyg. Occup. Med.*, **9**, 227 (1954).

406. R. Wiglusz, M. Kienitz, G. Delag, E. Galuszko, and P. Mikulski, *Bull. Inst. Marit. Trop. Med. (Gdynia)*, **26**(3–4), 315 (1975).

407. R. Wiglusz, G. Delag, and P. Mikulski, *Bull. Inst. Marit. Trop. Med. (Gdynia)*, **26**(3–4), 303 (1975).

408. S. Laham and E. O. Matutina, *Arch. Toxikol.*, **30**(3), 199 (1973).

409. E. Bateman and C. Henningsen, *Proc. Am. Wood Preservers' Assoc.*, 136 (1923).

410. R. E. McDonald and W. R. Buford, *Plant Dis. Rep.*, **58**(12), 1143 (1974).

411. S. H. Safe, H. Plugge, B. Chittim, and J. F. S. Crocker, *Publ. Environ. Secr. Natl. Res. Counc. Can., Iss. NRCC/CNRC, 16073,* Department of Chemistry, Guelph University, Guelph, Ontario, 1977.

412. L. B. Brattsten and C. F. Wilkinson, *Pestic. Biochem. Physiol.*, **3**(4), 393 (1973).

413. L. B. Brattsten, C. F. Wilkinson, and M. M. Root, *Insect. Biochem.*, **6**(6), 615 (1976).

414. H. Dannenberg, I. Brachmann, and C. Thomas, *Z. Krebsforsch.*, **74**(1), 100 (1970).

415. H. F. Uhlig and W. C. Pfefferle, Chapter 12 in *Refining Petroleum for Chemicals*, R. F. Gould, Ed., American Chemical Society, Washington, D.C., 1970.

416. M. S. Wolff, S. M. Daum, W. V. Lorimer, I. J. Selikoff, and B. B. Aubrey, *J. Toxicol. Environ. Health*, **2**(5), 997 (1977).

417. A. M. El Masri, J. N. Smith, and R. T. Williams, *Biochem. J.*, **64**, 50 (1956).

418. G. A. Maylin, M. J. Cooper, and M. H. Anders, *J. Med. Chem.*, **16**, 6 (1973).

419. D. T. Gibson, B. Gschwendt, W. K. Yeh, and M. Kobal, *Biochemistry*, **12**(8), 1520 (1973).

420. *Fed. Reg.*, **40**(196), 47262 (1976).

421. V. I. Dyachkov and M. V. Severin, *Nauch. Tr. Irkutsk Gos. Med. Inst.*, **113**, 118 (1971).

422. M. Tanabe, R. L. Dehn, and M. H. Kuo, *Biochemistry*, **10**(6), 1087 (1971).

423. K. A. Nikogosyan, *Zh. Eksp. Klin. Med.*, **12**(6), 76 (1972).

424. A. G. Ruhrchemie, Neth. Appl. Pat. 75 02553, August 17, 1976.

425. H. W. Werner, R. C. Dunn, and W. F. Von Oettingen, *J. Ind. Health Toxicol.*, **26**, 264 (1974).

426. W. Senczuk and B. Litewka, *Bromatol. Chem. Toksykol.*, **7**(1), 93 (1974).

427. D. Robinson, J. N. Smith, and R. F. Williams, *Biochem. J.*, **56**, XI (1954).

428. T. Omori, Y. Jigami, and Y. Minoda, *Agric. Biol. Chem.*, **39**(9), 1775 (1975).

429. Y. Jigami, T. Omori, and Y. Minoda, *Agric. Biol. Chem.*, **39**(9), 1781 (1975).

430. S. M. Fridman, S. M. Safonnikova, S. A. Sakaeva, and R. F. Daukaeva, *Gig. Sanit.*, **12**, 78 (1977).

431. C. R. Noller, *Chemistry of Organic Compounds*, 3rd ed., W. B. Saunders, Philadelphia, London, 1966.

432. J. Wepierre, Y. Cohen, and G. Valetti, *Eur. J. Pharmacol.*, **3**(1), 47 (1968).

433. R. I. Leavitt, *J. Gen. Microbiol.*, **49**(1), 411 (1967).

434. V. M. Abdullaev and L. P. Pavlova, *Gig. Tr. Prof. Zabol.*, **4**, 110 (1970).

435. R. V. Elisuiskaya, *Gig. Tr. Prof. Zabol.*, **4**, 145 (1970).

436. J. L. Palotay, K. Adachi, R. L. Dobson, and J. S. Pinto, *J. Natl. Cancer Inst.*, **57**(6), 1269 (1976).

437. L. Ahlstrom, *Kem. Tidskr.*, **89**(1–2), 28 (1977).

438. R. D. Stewart, H. C. Dodd, E. D. Baretta, and A. W. Shaffer, *Arch. Environ. Health*, **16**, 656–662 (1968).

439. H. C. Spencer, D. D. Irish, E. M. Adams, and V. K. Rowe, *J. Ind. Hyg. Toxicol.*, **24**, 295 (1942).

440. A. A. Askalonov, *Farmakol. Toksikol.*, **36**(5), 611 (1973).

441. A. A. Askalonov, *Mater. Povolzh. Konf. Fiziol. Uchastiem. Biokhim. Farmakol. Morfol.*, *6th*, **1**, 332 (1973).

442. A. A. Askalonov, *Sb. Mater. Nauch. Konf. Fiziol. Biokhim. Farmakol. Zapad-Sib. Obedin*, *5th*, 218 (1973).

443. V. V. Kazakova and M. B. Lis, *Farmakol. Toksikol.*, **34**, 615 (1971).

444. H. Savolainen and P. Pfäffli, *Acta Neuropathol.*, **40**(3), 237 (1977).

445. I. A. Bekeshev, *Azerb. Med. Zh.*, **51**(3), 59 (1974).

446. I. L. Krynskaya, L. I. Petrova, Z. G. Guricheva, E. G. Robachevskaya, and G. M. Bukevich, *Gig. Sanit.*, **34**, 334 (1969).

447. J. V. Marhold, Institut Pro Vychovu Vedoucicn Pracovniku Chemickeho Prumyclu Praha, 25 (1972) (through Ref. 4).

448. E. C. Hagan, P. M. Jenner, W. J. Jones, O. A. Fitzhugh, E. L. Long, J. G. Brouwer, and W. K. Webb, *Toxicol. Appl. Pharmacol.*, **7**, 18 (1965).

449. *API Toxicological Review of Styrene*, American Petroleum Institute, New York, 1962.

450. *Chemical Safety Data SD-37, Properties and Essential Information for Safe Handling and Use of Styrene Polymer*, Manufacturing Chemists' Association, Inc., Washington, D.C., 1971.

451. R. Aries, Fr. Pat. 2,108,829 November, 1972.

452. H. H. Cornish, K. J. Hahn, and M. L. Barth, *Environ. Health Perspect.*, **11**, 191 (1975).

453. M. Chaigneau and G. LeMona, *Ann. Pharm. Fr.*, **32**, 485 (1974).

454. I. Astrand, A. Kilbom, P. Ovrum, I. Wahlberg, and O. Vesterberg, *Work Environ. Health*, **11**(2), 69 (1974).

455. F. Gamberale and M. Hultengren, *Work Environ. Health*, **11**(2), 86 (1974).

456. J. M. Schwarzmann and N. P. Kutscha, *Res. Life Sci.*, **19**(3), 1 (1971).

457. B. Larsby, R. Than, L. M. Odkvist, D. Hyden, I. Bunnfors, and G. Ashan., *Scand. J. Work Environ. Health*, **4**(1), 60 (1978).

458. F. Nour-Eldin, *Nature*, **214**, 1362 (1967).

459. Y. Alarie, *Toxicol. Appl. Pharmacol.*, **24**(2), 279 (1973).

460. J. C. Rogers and C. C. Hooper, *Ind. Med. Surg.*, **26**, 32 (1957).

461. A. N. Kohn, *Am. J. Ophthalmol.*, **85**(4), 569 (1978).

462. N. I. Ponomareva and N. S. Zlobina, *Gig. Tr. Prof. Zabol.*, **15**(6), 22 (1971).

463. P. Gotell, O. Axelson, and B. Lindelof, *Work Environ. Health*, **9**(2), 76 (1973).

464. W. V. Lorimer, R. Lilis, W. J. Nicholson, H. Anderson, A. Fischbein, S. Daum, W. Rom, C. Rice, and I. J. Selikoff, *Environ. Health Perspect.*, **17**, 171 (1976).

465. A. A. Bashirov, *Ter. Arkh.*, **42**(12), 41 (1970).

466. Z. A. Volkova, L. E. Milkov, K. A. Lopukhova, L. M. Malyar, Y. L. Marenko, and T. K. Shakhova, *Gig. Tr. Prof. Zabol.*, **14**(1), 31 (1970).

467. I. I. Alekperov, *Gig. Tr. Prof. Zabol.*, **12**, 191 (1970).

468. A. A. Bashirov, *Gig. Tr. Prof. Zabol.*, **15**(5), 57 (1971).

469. L. P. Lukoshkina and I. I. Alekperov, *Gig. Tr. Prof. Zabol.*, **8**, 42 (1973).

470. K. Linström, H. Härkonen, and S. Hernberg, *Scand. J. Work Environ. Health*, **2**(3), 129 (1976).

471. H. Härkonen, K. Lindström, A. M. Seppälainen, A. Asp, and S. Hernberg, *Scand. J. Work Environ. Health*, **4**(1), 53 (1978).

472. A. M. Seppälainen and H. Härkonen, *Scand. J. Work Environ. Health*, **2**(3), 140 (1976).

473. P. C. Holmberg, *Scand. J. Work Environ. Health*, **3**(4), 212 (1977).

474. K. K. Sidorov, *Gig. Sanit.*, **37**(5), 93 (1972).

475. V. G. Lappo and I. I. Krasnikova, *Gig. Vop. Proizvod. Primen. Polim. Mater*, 222 (1969).

476. A. S. Izyumova, *Gig. Sanit.*, **37**(4), 29 (1972).

477. N. Ragule, *Gig. Sanit.*, **11**, 85 (1974).

478. H. Vainio, K. Hemminki, and E. Elovaara, *Toxicology*, **8**(3), 319 (1977).

479. P. Melvy and A. J. Garro, *Mutat. Res.*, **40**, 15 (1976).

480. H. Vainio, R. Pääkónen, K. Rönholm, V. Raunio, and O. Pelkonen, *Scand. J. Work Environ. Health*, **2**(3), 147 (1976).

481. T. Meretoja, H. Vainio, and II. Jarventaus, *Toxicol. Letters*, **1**(5–6), 315 (1976).

482. *Report NCI Bioassay of a Solution of Beta-Nitro-Styrene and Styrene for Possible Carcinogenicity-TR170*, Office of Cancer Communications, National Cancer Institute, Bethesda, Maryland (U.S. Government Printing Office, 1979).

483. H. Van Rees, *Int. Arch. Arbeitsmed.*, **33**(1), 39 (1974).

484. K. C. Leibman, *Environ. Health Perspect.*, **11**, 115 (1975).

485. I. Astrand, *Scand. J. Work Environ. Health*, **1**(4), 199 (1975).

486. M. Ikeda, T. Imamura, M. Bayashi, T. Tabuchi, and I. Hara, *Int. Arch. Arbeitsmed.*, **32**(1–2), 93 (1974).

487. A. Wink, *Ann. Occup. Hyg.*, **15**(2–4), 211 (1972).

488. C. Burkewicz, J. Rybkowska, and H. Zielinska, *Med. Prac.*, **25**(3), 305 (1974).

489. A. M. El Masri, J. N. Smith, and R. T. Williams, *Biochem. J.*, **68**, 199 (1958).

490. I. Danishefsky and M. Willhite, *J. Biol. Chem.*, **211**, 549 (1954).

491. K. C. Leibman and E. Ortiz, *J. Pharmacol. Exp. Ther.*, **173**(2), 242 (1970).

492. S. P. James and D. A. White, *Biochem. J.*, **104**, 914 (1967).

493. K. C. Leibman and E. Ortiz, *Biochem. Pharmacol.*, **18**, 552 (1969).

494. H. Ohtsuji and M. Ikeda, *Toxicol. Appl. Pharmacol.*, **18**(2), 321 (1971).

495. M. G. Parkii, J. Marmiemi, and H. Vainio, *Toxicol. Appl. Pharmacol.*, **38**(1), 59 (1976).

496. S. E. Ruvinskaya, *Gig. Tr. Prof. Zabol.*, **9**(11), 29 (1965).

497. S. M. Rappaport and D. A. Fraser, *J. Am. Ind. Hyg. Assoc.*, **5**, 205 (1977).

498. V. K. Rowe, G. J. Atchison, E. N. Luce, and E. M. Adams, *J. Ind. Hyg. Toxicol.*, **25**(8), 348 (1943).

499. A. Slob, *Br. J. Ind. Med.*, **30**, 390 (1973).

500. H. H. Härkonen, P. Kalliokoski, S. Hietala, and S. Hernberg, *Work, Environ. Health*, **11**(3), 162 (1974).

501. J. Sollenberg and A. Baldeston, *J. Chromatogr.*, **132**(3), 469 (1977).

502. F. M. Gadzhiev and T. V. Aliev, *Ser. Biol. Nauk.*, **4**, 112 (1974).

503. *Fire Protection Guide on Hazardous Materials*, 7th ed., National Fire Protection Association, Boston, Mass., 1979.

504. W. Boehr, S. Drobnik, W. Hild, R. Kroebel, A. Meyer, and G. Naumann, Ger. Pat. 2363474, June 26, 1975.

505. R. Aries, Fr. Pat. 2104717, May 26, 1972.

506. R. E. Fearon, *Am. Heart J.*, **75**(5), 634 (1968).

507. R. M. Gesler, P. J. Garvin, B. Klamer, R. U. Robinson, C. R. Thompson, W. R. Gibson, F. C. Wheeler, and R. G. Carlson, *Bull. Parenter. Drug Assoc.*, **27**(3), 101 (1973).

508. A. Alpande de Morais, C. M. Andrade da Mata Rezende, M. V. Von Buelow, J. C. Mourao, O. R. Gottlieb, M. C. Marx, A. I. Da Rocha, and M. T. Magalhaes, *An. Acad. Bras. Ciene*, **44**, 303 (1972).

509. J. D. Peele, Jr. and E. O. Oswald, *Biochim. Biophys. Acta*, **497**(2), 598 (1977).

510. L. M. Sergeta, M. K. Byalko, N. I. Alberton, and G. M. Balan, *Gig. Tr. Prof. Zabol.*, **12**, 51 (1975).

511. V. M. Sergeta, M. I. Alberton, and V. P. Fomenko, *Zdravookhr Kaz.*, **1**, 50 (1977).

512. V. N. Bravre, *Narusheniya Metab., Tr. Nauchn. Konf. Med. Inst. Zapadn. Sib., 1st*, 259 (1974).

513. Z. M. Fadeeva and Y. N. Eikhler, *Nauch. Tr. Omsk. Med. Inst.*, **107**, 166 (1971).

514. L. M. Makareva, *Farmakol. Toksikol.*, **35**(4), 491 (1972).

515. G. M. Klimina, *Narusheniya Metab., Tr. Nauchn. Konf. Med. Inst. Zapadn. Sib., 1st*, 46 (1974).

516. G. M. Klimina, *Narusheniya Metab., Tr. Nauchn. Konf. Med. Inst. Zapadn. Sib., 1st*, 291 (1974).

517. I. I. Solovev, *Narusheniya Metab., Tr. Nauchn. Konf. Med. Inst. Zapadn. Sib., 1st*, 255 (1974).

518. A. A. Nikiforova, *Mater. Nauch. Sess. Posvyashd. 50-Letuju Obrazov, USSR Omsk. Gos. Med. Inst.*, 871 (1972).

519. I. M. Mirzoyan and R. K. Zhakenova, *Vopr. Gig. Tr. Prof. Zabol.*, 247 (1972).

520. J. L. Weeks, M. B. Lentle, and B. C. Lentle, *J. Occup. Med.*, **12**(7), 246 (1970).

521. E. Weil, L. Kusterer, and M. H. Brogard, *Arch. Mal. Prof.*, **26**, 405 (1965).

522. I. Hakkinen, E. Siltanen, S. Hernberg, A. M. Seppalainen, P. Karli, and E. Vikkula, *Arch. Environ. Health*, **26**, 70 (1973).

523. A. M. Seppalainen and I. Hakkinen, *J. Neurol. Neurosurg. Psychiatry*, **38**, 248 (1975).

524. W. B. Deichmann, K. V. Kitzmiller, M. Dierker, and S. Witherup, *J. Ind. Hyg. Toxicol.*, **29**(1), 1 (1947).

525. *Natl. Tech. Inf. Serv. Publ. Bull. 223-159* (through Ref. 4).

526. L. Pecchiai and U. Saffiotti, *Med. Lav.*, **48**, 247 (1956).

527. J. S. Henderson and J. L. Weeks, *Ind. Med.*, **42**(2), 10 (1973).

528. J. Nordal, *Nord Vet. Med.*, **82**(9), 469 (1970).

529. R. Mestres, *Proc. Int. Citrus Cymp., 1st Riv.*, **2**, 1035 (1968).

530. W. G. Levine, P. Milburn, R. L. Smith, and R. Williams, *Biochem. Pharmacol.*, **19**(1), 235 (1970).

531. P. Von Raig and R. Ammon, *Arzneim. Forsch.*, **22**(8), 1399 (1972).

532. H. Von Berninger, P. Ammon, and I. Berninger, *Arzneim. Forsch.* **18**(2), 880 (1968).

533. F. McPherson, J. W. Bridges, and D. V. Parke, *Nature*, **252**(5483), 488 (1974).

534. M. D. Burke and J. W. Bridges, *Xenobiotica*, **5**(6), 357 (1975).

535. P. J. Creaven, D. W. Parke, and R. T. Williams, *Biochem. J.*, **96**, 879 (1965).

536. D. E. Willis and R. F. Addison, *Comp. Gen. Pharmacol.*, **5**(1), 77 (1974).

537. S. A. Atlas and D. W. Nebert, *Arch. Biochem. Biophys.*, **175**(2), 495 (1976).

538. P. Von Raig and R. Ammon, *Arzneim. Forsch.*, **9**, 1266 (1970).

539. H. D. West, J. R. Lawson, I. H. Miller, and G. R. Mathura, *Arch. Biochem. Biophys.*, **60**, 14 (1956).

540. F. J. McPherson, J. W. Bridges, and D. V. Parke, *Biochem. J.*, **154**, 773 (1976).

541. E. Bachman and L. Goldberg, *Exp. Mol. Pathol.*, **13**, 269 (1970).

542. H. Jaffe, K. Fujii, H. Guerin, M. Sengupta, and S. S. Epstein, *Biochem. Pharmacol.*, **18**, 1045 (1969).

543. D. Catelani, G. Mosselmans, J. Nienhaus, C. Sorlini, and V. Trecanni, *Experientia*, **26**(8), 922 (1970).

544. T. Ohmori, T. Ikai, Y. Minoda, and K. Yamada, *Agric. Biol. Chem.*, **37**(7), 1599 (1973).

545. D. Lunt and W. C. Evans, *Biochem. J.*, **118**, 54 (1976).

546. D. Catelani, C. Sorlini, and V. Treccani, *Experientia*, **27**(10), 1173 (1971).

547. D. Catelani, A. Colombi, C. Sorlini, and V. Treccani, *Biochem. J.*, **134**, 1063 (1973).

548. *Fed. Reg.*, **39**(125), 23540 (1974).

549. M. D. Burke, D. J. Benford, J. W. Bridges, and D. V. Parke, *Biochem. Soc. Trans.*, **5**(5), 1370 (1977).

550. H. Kameoka, K. Nishikawa, and H. Wada, *Nippon Nogei Kagaku Kaishi*, **49**(10), 557 (1975).

551. M. M. Ranney, *Synthetic Lubricants*, Noyes Data Corp., Park Ridge, N. J., 1972.

552. D. L. Opdyke, *Food Cosmet. Toxicol.*, **12**(5–6), 705 (1974).

553. R. V. Subba-Rao and M. Alexander, *Appl. Environ. Microbiol.*, **33**(1), 101 (1977).

554. D. D. Focht and M. Alexander, *Appl. Microbiol.*, **20**(4), 608 (1970).

555. S. W. Stroud, *J. Endocrinol.*, **2**, 55 (1940).

556. R. R. Scheline, *Experientia*, **30**(8), 880 (1974).

557. J. F. Sinsheimer and R. V. Smith, *Biochem. J.*, **111**(1), 35 (1969).

558. L. K. Tay and J. E. Sinsheimer, *Drug Metab. Dispos.*, **4**(2), 154 (1976).

559. T. Watabe and K. Akamatsu, *Biochem. Pharmacol.*, **23**(6), 1079 (1974).

560. E. L. Docks and G. Krishna, *Biochem. Pharmacol.*, **24**(21), 1965 (1975).

561. T. Watabe and K. Akamatsu, *Biochem. Pharmacol.*, **24**(3), 442 (1975).

562. H. Bradner, U. S. Pat. 3988437, October 26, 1976.

563. I. Y. R. Adamson and J. L. Weeks, *Arch. Environ. Health*, **27**(2), 69 (1973).

564. H. H. Cornish, E. B. Raymond, and R. C. Ryan, *Am. Ind. Hyg. Assoc. J.*, **23**, 372 (1962).

565. Z. F. Khromenko, V. D. Gostinskii, and N. G. Ivanov, *Nauch. Tr. Irkutsk. Med. Inst.*, **115**, 122 (1972).

566. M. O. Amdur and D. A. Creasia, *Am. Ind. Hyg. Assoc. J.*, **27**(4), 349 (1966).

567. Z. F. Khromenko, *Nauch. Tr. Irkutsk. Med. Inst.*, **115**, 121 (1972).

568. I. Y. R. Adamson, *Arch. Environ. Health*, **26**(4), 192 (1973).

569. *Industrial Hygiene Field Operational Manual*, U.S. Dept. Labor, OSHA Instruction CPL 2-2.20, Office of Field Coordination, Washington, D.C., April 2, 1979.

570. P. Scoppa and K. Gerbaulet, *Boll. Soc. Ital. Biol. Sper.*, **47**(7), 194 (1971).

571. I. Y. R. Adamson and J. M. Furlong, *Arch. Environ. Health*, **28**(3), 155 (1974).

572. S. Kiriyama, M. Banjo, and H. Matsushima, *Nutr. Rep. Int.*, **19**(2), 79 (1974).

573. *Fed. Reg.*, **37**, 23549 (1974).

574. *Criteria for a Recommended Standard—Occupational Exposure to Terphenyl*, U.S. Dept. Health, Education, and Welfare, Public Health Service, NIOSH, Washington, D.C., 1977.

575. C. Hesse, K. Hilp, H. Kating, and G. Schaden, *Arch. Pharm.*, **310**(10), 792 (1977).

576. R. A. Johnstone, *Nature*, **200**, 1184 (1963).

577. I. Schmeltz, J. Tosk, and D. Hoffmann, *Anal. Chem.*, **48**(4), 645 (1976).

578. J. B. Addison, P. J. Silk, and I. Unger, *Int. J. Environ. Anal. Chem.*, **4**(2), 135 (1975).

579. *API Toxicological Review of Naphthalene*, American Petroleum Institute, New York, 1959.

580. *Chemical Safety Data Sheet SD-58, Properties and Essential Information for Safe Handling and Use of Naphthalene*, Manufacturing Chemists' Association, Inc., Washington, D.C., 1956.

581. L. H. Chang, *J. Biol. Chem.*, **151**, 93 (1943).

582. H.-R. Koch, K. Doldi, and O. Hockwin, *Doc. Ophthalmol. Proc. Ser.*, **8**, 293 (1976).

583. D. Irie, T. Sasaki, and R. Ito, *Toho Igakkai Zasshi*, **20**(5/6), 772 (1973).

584. M. G. Horning, C. D. Kary, P. A. Gregory, and W. G. Stillwell, *Toxicol. Appl. Pharmacol.*, **37**(1), 18 (1976).

585. L. N. Rolonova, *Farmakol. Toksikol.*, **30**(4), 484 (1967).

586. *Natl. Tech. Inf. Serv. AD691-490* (through Ref. 4).

587. H. R. Sanborn and D. C. Malins, *Proc. Soc. Exp. Biol. Med.*, **154**, 151 (1977).

588. R. van Heyningen and A. Pirie, *Biochem. J.*, **102**, 842 (1967).

589. H. Druckrey and D. Schmähl, *Naturwissenschaften*, **42**, 159 (1955).

590. M. Sherer, *J. Am. Osteopath. Assoc.*, **65**(1), 60 (1965).

591. N. I. Pavlivoda, *Zdravookhr. Beloruss.*, **18**(7), 81 (1971).

592. N. L. Sharma, R. N. Singh, and N. K. Natu, *J. Indian Med. Assoc.*, **48**(1), 20 (1967).

593. T. Valaes, S. A. Doxiadis, and P. Fessas, *J. Pediatr.*, **63**, 904 (1963).

594. W. W. Zuelzer and L. Apt, *J. Am. Med. Assoc.*, **141**, 185 (1949).

595. T. Sollman, *A Manual of Pharmacology*, 8th ed., Saunders, Philadelphia, 1957.

596. J. Doull, Chapter 5 in *Toxicology, the Basic Science of Poisons*, L. J. Casarett and J. Doull, Eds., Macmillan, New York, 1975.

597. J. H. Sanderson, *Practitioner*, 216 (1976).

598. R. E. Gosselin, H. C. Hodge, R. P. Smith, and M. N. Gleason, *Clinical Toxicology of Commercial Products*, 4th ed., Williams and Wilkins, Baltimore, 1976.

599. A. Hamilton and H. L. Hardy, *Industrial Toxicology*, PSG Publishing Co., Littleton, Mass., 1974.

600. W. M. Grant, *Toxicology of the Eye*, 2nd ed., Charles C Thomas, Springfield, Ill., 1974.

601. J. P. Dawson, W. W. Thayer, and J. F. Desforges, *Blood*, **13**, 113 (1958).

602. W. B. Schafer, *Pediatrics*, **7**, 172 (1951).

603. W. G. Grigor, H. Robin, and J. D. Harley, *Med. J. Australia*, **2**(2), 1229 (1966).

604. R. H. Gadsden, R. R. Mellette, and W. C. Miller, Jr., *J. Am. Med. Assoc.*, **168**, 1220 (1958).

605. D. R. Adams, *Brit. J. Ophthalmol.*, **14**, 545 (1930).

606. R. J. Meyer, *New Engl. J. Med.*, **252**(15), 622 (1955).

607. G. Ghetti and L. Mariani, *Med. Lav.*, **57**, 533 (1956).

608. H. Hanssler, *Dtsch. Med. Wochenschr.*, **89**, 1794 (1964).

609. U. Irle, *Dtsch. Med. Wochenschr.*, **89**, 1798 (1964).

610. T. L. Naiman and M. H. Kosoy, *Can. Med. Assoc. J.*, **91**, 1243 (1964).

611. J. A. Anziulewicz, H. J. Dick, and E. E. Chiarulli, *Am. J. Obstet. Gynecol.*, **78**, 519 (1959).

612. A. K. Brown, *Am. J. Dis. Child*, **94**, 510 (1957).

613. J. V. Mackell, F. Rieders, H. Brieger, and E. L. Bauer, *Pediatrics*, **7**, 722 (1951).

614. M. R. Juchau and M. J. Namkung, *Drug Metab. Dispos.*, **2**(4), 380 (1974).

615. H. Ishizaka, *Showa Igakkai Zasshi*, **31**(9), 471 (1971).

616. K. Ikemoto, *Osaka City Med. J.*, **17**(1), 1 (1971).

617. S. K. Srivastava and E. Beutler, *Biochem. J.*, **112,** 421 (1969).

618. A. M. Potts and L. M. Gonasum, Chapter 13 in *Toxicology, the Basic Science of Poisons*, L. J. Casarett and J. Doull, Eds., Macmillan, New York, 1975.

619. A. Pirie, *Exp. Eye Res.*, **7**(3), 354 (1968).

620. M. Shimotori, *Acta Soc. Ophthalmol. Jap.*, **76**(11), 1545 (1972).

621. K. Alexandrov and C. Frayssinet, *J. Natl. Cancer Inst.*, **51**(3), 1067 (1973).

622. J. Yoon and K. Kim, *Hanguk Sikmul Poho Hakhoe Chi*, **16**(2), 127 (1977).

623. R. B. Boose and L. Terriere, *J. Econ. Entomol.*, **60**(2), 580 (1967).

624. J. A. Schafer and L. Terriere, *J. Econ. Entomol.*, **63**(3), 787 (1970).

625. R. F. Lee, R. Sauerheber, and G. W. Dobbs, *Mar. Biol.* (*Berl.*), **17**(3), 201 (1972).

626. M. J. Melancon, Jr. and J. J. Lech, *Arch. Environ. Contam. Toxicol.*, **7**(2), 207 (1978).

627. U. Varanasi, M. Uhler, and S. I. Stranahan, *Toxicol. Appl. Pharmacol.*, **44,** 277 (1978).

628. J. M. Neff, *Prepr. Div. Pet. Chem. Am. Chem. Soc.*, **20**(4), 839 (1975).

629. S. S. Rossi and J. W. Anderson, *Mar. Biol.*, **39**(1), 51 (1977).

630. O. Skjaeggestad, *Acta Pathol. Microbiol. Scand.*, **169,** 1 (1964).

631. P. Brookes and P. D. Lawley, *Br. Empire Cancer Campaign Res., 41st Ann. Rep., Part II*, 77 (1963).

632. P. Brookes and P. D. Lawley, *Nature*, **202,** 781 (1964).

633. P. O. Ts'o and P. Lu, *Proc. Natl. Acad. Sci. U.S.*, **51**(1), 17 (1964).

634. P. Daudel, M. Croisy-Delcey, P. Jacquignon, and P. Vigny, *C. R. Acad. Sci. |D|*, **277,** 2437 (1973).

635. M. Miko and L. Drobnica, *Folia Fac. Med. Univ. Comenianae Bratisl.*, **7**(1), 217 (1969).

636. I. Schmeltz, J. Tosk, J. Hilfrich, N. Hiroto, D. Hoffman, and E. L. Wynder, *Carcinog. Compr. Surv.*, **3,** 47 (1978).

637. F. D. S. Corner and L. Young, *Biochem. J.*, **61,** 132 (1955).

638. T. R. Norton, Chapter 4 in *Toxicology, the Basic Science of Poisons*, L. J. Casarett and J. Doull, Eds., Macmillan, New York, 1975.

639. J. W. Daly, D. M. Jerina, and B. Witkop, *Experientia*, **28**(10), 1129 (1972).

640. D. M. Jerina, J. W. Daly, B. Witkop, P. Zaltzman-Nirenberg, and S. Udenfriend, *Biochemistry*, **9**(1), 147 (1970).

641. E. Boyland, M. Kimura, and P. Sims, *Biochem. J.*, **92,** 631 (1964).

642. A. Pirie and R. van-Heyninger, *Biochem. J.*, **100**(3), 70 (1966).

643. R. van Heyningen, *Exp. Eye Res.*, **9,** 38 (1970).

644. K. W. Bock, G. V. Ackeren, F. Lorch, and F. W. Birke, *Biochem. Pharmacol.*, **25**(1), 2351 (1976).

645. W. G. Stillwell, G. W. Griffin, and M. G. Horning, *Biochem. Pharmacol.*, **4,** 1341 (1978).

646. D. M. Jerina, J. W. Daly, B. Witkop, P. Zaltzman-Nirenberg, and S. Udenfriend, *J. Am. Chem. Soc.*, **90**(23), 6525 (1968).

647. K. Netter, *Naunyn-Schmiedebergs Arch. Pharmakol. Exp. Pathol.*, **262**(3), 375 (1969).

648. H. D. Colby, R. E. Kramer, J. W. Greiner, D. A. Robinson, R. F. Krause, and W. J. Canady, *Biochem. Pharmacol.*, **24**(17), 1644 (1975).

649. R. D. Schonerod, M. A. Khan, L. C. Terriere, and F. W. Plapp, Jr., *Life Sci.*, **7**(13), 681 (1968).

650. B. Griffiths and W. C. Evans, *Biochem. J.*, **95**(3), 51 (1965).

651. E. I. Kvasnikov and N. Z. Tin'yanova, *Mikrobiol. Zh.* (*Kiev*), **32**(4), 416 (1970).

652. M. Malesset-Bras and E. Azoulay, *Ann. Inst. Pasteur,* **109**(6), 894 (1965).

653. M. S. Twefix and Y. Hamdi, *Acta Microbiol. Pol., Ser. B.,* **19**(2), 133 (1970).

654. A. M. Cundell and R. W. Traxler, *Mater. Org. (Berl.),* **11**(1), 1 (1976).

655. M. Martonova, B. Skarka, and Z. Radij, *Folia Microbiol.,* **17**(1), 63 (1972).

656. F. A. Catterall, K. Murray, and P. A. Williams, *Biochim. Biphys. Acta,* **237**(2), 361 (1971).

657. V. V. Modi and R. N. Patel, *Appl. Microbiol.,* **16**(1), 172 (1968).

658. O. Atlavinyte, A. Lugauskas, and J. Daciulyte, *Biosfera Chel., Mater. Vses. Simp., 1st,* 217 (1975).

659. M. N. Schroth and D. C. Hildebrand, *Phytopathology,* **58,** 848 (1968).

660. K. O. Kusk, *Physiol. Plant.,* **43**(1), 1 (1978).

661. C. Soto, J. A. Hellebust, and T. C. Hutchinson, *Verh.-Int. Ver. Theor. Angew. Limnol.,* **19**(3), 2145 (1975).

662. J. M. Neff and J. W. Anderson, *Bull. Environ. Contam. Toxicol.,* **14**(1), 122 (1975).

663. I. Schmeltz, D. Hoffman, and E. L. Wynder, *Trace Subst. Environ. Health,* **8,** 281 (1974).

664. K. Adachi, *Hyogo-ken Eisei Kenkyusho Kenkyu Hokoku,* **10,** 22 (1975).

665. C. A. Nau, J. Neal, and M. Thornton, *Arch. Environ. Health,* **12,** 382 (1966).

666. R. S. Caldwell, E. M. Calderone, and M. H. Mallon, *Fate Eff. Pet. Hydrocarbons Mar. Ecosyst. Org., Proc. Symp., 1976* (Mar. Sci. Cent., Oregon State Univ., Newport, Oreg.) 210 (publ. 1977).

667. L. G. Andreikova and L. A. Kogau, *Koks Khim.,* **8,** 47 (1977).

668. P. E. Gaffney, *J. Water Pollut. Control Fed.,* **48**(11), 2590 (1976).

669. D. G. Crosby and C-S Tang, *J. Agric. Food Chem.,* **17**(6), 1291 (1969).

670. B. Rabussier and J. P. Mandon, Ger. Pat. 2412368, September 19, 1974.

671. L. Ashino, Jap. Pat. 77 25021, February 14, 1977.

672. P. Scoppa, *Z. Naturforsch.,* **21**(11), 1054 (1966).

673. L. N. Bolonova, *Farmakol. Toksikol. (Moscow),* **30**(4), 484 (1967).

674. C. N. Statham, C. R. Elcombe, S. P. Szyjka, and J. J. Lech, *Xenobiotica,* **8**(2), 65 (1978).

675. K. Konya, T. Kitagaki, and Y. Konogai, Jap. Pat. 77 01022, January 6, 1977.

676. H. E. Tatem, *Fate Eff. Pet. Hydrocarbon. Mar. Ecosyst. Org. Proc. Symp., 1976* (Mar. Sci. Cen., Oregon State Univ., Newport, Oreg.) 210 (publ. 1977).

677. S. Itoh, K. Endo, Y. Nakajima, K. Shimizu, and H. Narita, Jap. Pat. No. 76 11020, September 29, 1976.

678. *Gig. Sanit.,* **27,** 19 (1964) (through Ref. 4).

679. *Z. Krebsforsch.,* **69,** 103 (1967) (through Ref. 4).

680. S. S. Epstein, E. Arnold, J. Andrea, W. Bass, and Y. Bishop, *Toxicol. Appl. Pharmacol.,* **23,** 288 (1972).

681. J. D. Scribner, *J. Natl. Cancer Inst.,* **50,** 1717 (1973).

682. M. H. Salaman and F. J. C. Roe, *Br. J. Cancer,* **10,** 363 (1956).

683. P. E. Steiner, *Cancer Res.,* **15,** 632 (1955).

684. C. Huggins and N. C. Yang, *Science,* **137,** 237 (1962).

685. M. J. Shear and J. Leiter, *J. Natl. Cancer Inst.,* **2,** 241 (1941).

686. J. Pataki and C. B. Huggins, *Cancer Res.,* **29**(3), 506 (1969).

687. H. B. Andervont and M. B. Shimkin, *J. Natl. Cancer Inst.,* **1,** 225 (1940).

688. M. Enomoto, E. C. Miller, and J. A. Miller, *Proc. Soc. Exp. Biol. Med.,* **136,** 1206 (1971).

689. J. A. Miller and E. C. Miller, *Cancer Res.,* **23,** 229 (1963).

690. D. P. Griswold, Jr., A. E. Casey, E. K. Weisburger, J. H. Weisburger, and F. M. Schabel, Jr., *Cancer Res., 26A*(4), 619 (1966).

691. W. T. Hill, D. W. Stanger, A. Pizzo, B. Riegel, P. Shubik, and W. W. Wartman, *Cancer Res.,* **11,** 892 (1951).

692. G. Barry, J. W. Cook, G. A. D. Haslewood, C. L. Hewett, I. Hieger, and E. L. Kennaway, *Proc. Royal Soc. London, Ser. B., Biol. Sci.,* **117,** 318 (1935).

693. R. H. Rigdon and E. G. Rennels, *Experientia,* **20,** 224 (1964).

694. R. H. Rigdon and J. Neal, *Proc. Soc. Exp. Biol. Med.,* **130,** 146 (1969).

695. O. Schürch and A. Winterstein, *Z. Physiol. Chem.,* **236,** 79 (1935).

696. O. M. Bulay and R. W. Wattenberg, *Proc. Soc. Exp. Biol. Med.,* **135,** 84 (1970).

697. W. F. Noyes, *Proc. Soc. Exp. Biol. Med.,* **127,** 594 (1968).

698. S. Payne, *Br. J. Cancer,* **12,** 65 (1958).

699. O. M. Bulay, *Acta Med. Turc. (Turkey),* **7,** 3 (1970).

700. S. S. Epstein and H. Shafner, *Nature,* **219,** 385 (1968).

701. C. D. Haagensen and O. F. Krehbiel, *Am. J. Cancer,* **27,** 474 (1936).

702. P. Grasso and C. O'Hare, Chapter 14 in *Chemical Carcinogens,* C. E. Searle, Ed., American Chemical Society, Washington, D.C., 1976.

703. I. Schmeltz and D. Hoffman, "Formation of Polynuclear Hydrocarbons," in *Carcinogenesis, A Comprehensive Survey, Volume 1, "Polynuclear Aromatic Hydrocarbons,"* R. Freudenthal and P. W. Jones, Eds., Raven Press, New York, 1976.

704. G. Grimmer, H. Boehnke, and H. P. Harke, *Int. Arch. Occup. Environ. Health,* **40**(2), 93 (1977).

705. M. Argirova, *Klig. Zdraeopaz.,* **18**(5), 18 (1975).

706. W. Ziechmann, *Therapiewoche,* **28**(7), 1199 (1978).

707. J. K. Selkirk, *J. Toxicol. Environ. Health,* **2,** 1245 (1977).

708. A. Dipple, Chapter 5 in *Chemical Carcinogens,* C. E. Searle, Ed., American Chemical Society, Washington, D.C., 1976.

709. S. Sung, *C. R. Acad. Sci., Ser. D.,* **274**(10), 1597 (1972).

710. D. L. Sanioto and S. Schreier, *Biochem. Biophys. Res. Commun.,* **67**(2), 530 (1975).

711. J. C. Arcos, *Am. Lab.,* July, 29 (1978).

712. P. L. Grover, "Polycyclic Hydrocarbon Epoxides: Formation and Further Metabolism by Animal and Human Tissues," in *Chemical Carcinogenesis Essays,* R. Montesano, L. Tomatis, and W. Davis, Eds., International Agency for Research on Cancer, Lyon, 1974.

713. D. R. Thakker, M. Nordqvist, H. Yagi, W. Levin, D. Ryan, P. Thomas, A. H. Conney, and D. M. Jerina, "Comparative Metabolism of PAH," in *Polynuclear Aromatic Hydrocarbons,* P. W. Jones and P. Leber, Eds., Third International Symposium on Chemistry and Biology— Carcinogenesis and Mutagenesis, Ann Arbor Science Publishers, Ann Arbor, Mich., 1979.

714. S. K. Yang, D. W. McCourt, J. C. Lentz, and H. V. Gelboin, *Science,* **196**(4295), 1199 (1977).

715. H. V. Gelboin, F. J. Wiebel, and N. Kinoshita, "Microsomal Aryl Hydrocarbon Hydroxylases," in *Biological Hydroxylation Mechanisms,* G. S. Boyd and R. M. S. Smellie, Eds., Academic Press, London and New York, 1972.

716. N. H. Sloane and T. K. Davis, *Arch. Biochem. Biophys.,* **163**(1), 46 (1974).

717. E. T. Cantrell, G. A. Warr, D. L. Busbee, and R. R. Martin, *J. Clin. Invest.,* **52**(8), 1181 (1973).

718. D. W. Nebert, *Clin. Pharmacol. Ther.,* **14**(4), 693 (1973).

719. P. H. Jellinck and G. Smith, *Biochem. Biophys. Acta,* **304**(2), 520 (1973).

720. A. Poland and A. Kende, *Cold Spring Harbor Conf. Cell Prolif.,* **4**(B), 847 (1977).

721. D. W. Nebert and H. V. Gelboin, *Arch. Biochem. Biophys.*, **134**(1), 76 (1969).

722. F. J. Wiebel and H. V. Gelboin, "Enzyme Induction and Metabolism," in *Chemical Carcinogenesis Essays, International Agency for Research on Cancer*, P. Montesano and L. Tomatis, Eds., Lyon, 1974.

723. P. E. Thomas, D. Ryan, and W. Levin, "Cytochrome P-450 and Epoxide Hydrase," in *Polynuclear Aromatic Hydrocarbons, Third International Symposium*, P. W. Jones and P. Leber, Eds., Ann Arbor Science Publishers, Ann Arbor, Mich., 1979.

724. R. Bass, G. Bochert, H. J. Merker, and D. Neubert, *J. Toxicol. Environ. Health*, **2**, 1353 (1977).

725. J. A. DiPaolo, "Mammalian Cell Models for Chemical Carcinogenesis," in *Chemical Carcinogenesis Essays, International Agency for Research on Cancer*, R. Montesano and L. Tomatis, Eds., Lyon, 1974.

726. E. LaVoie, V. Bedenko, N. Hirota, S. Hecht, and D. Hoffman, "Biological Effects of PAH," in *Polynuclear Aromatic Hydrocarbons, Third International Symposium*, P. W. Jones and P. Leber, Eds., Ann Arbor Science Publishers, Ann Arbor, Mich., 1979.

727. W. C. Herndon, *Trans. N.Y. Acad. Sci.*, **36**(2), 200 (1974).

728. E. Bresnick, T. A. Stoming, H. Muktar, and J. B. Vaught, "Polycyclic Hydrocarbon Oxides, Their Formation, Inactivation and Biological Effect," in *Carcinogenesis, A Comprehensive Survey, Vol. 1*, R. Freudenthal and P. W. Jones, Eds., Raven Press, New York, 1976.

729. R. A. Webster, H. L. Moses, and T. C. Spelsberg, *Cancer Res.*, **36**(8), 2896 (1976).

730. B. N. Tarusov, Yu. M. Petrusevich, N. G. Kozhanov, D. A. Makeev, and V. E. Novikov, *Tr. Mosk. O-va. Ispyt. Prir.*, **52**, 183 (1975).

731. G. D. Griffin, T. D. Jones, and P. J. Walsh, "Chemical Cytotoxicity-A Cancer Promoter," in *Polynuclear Aromatic Hydrocarbons, Third International Symposium*, P. W. Jones and P. Leber, Eds., Ann Arbor Science Publishers, Ann Arbor, Mich., 1979.

732. J. H. Weisburger, Chapter 1 in *Chemical Carcinogens*, C. E. Searle, Ed., American Chemical Society, Washington, D.C., 1976.

733. J. E. Cleaver, *J. Toxicol. Environ. Health*, **2**, 1387 (1977).

734. B. J. Strauss, P. Karran, and N. P. Higgins, *J. Toxicol. Environ. Health*, **2**, 1395 (1977).

735. T. J. Slaga, D. L. Berry, M. R. Juchau, S. Thompson, S. G. Buty, and A. Viaje, "PAH Metabolism in Skin Tumorigenesis," in *Carcinogenesis, A. Comprehensive Survey*, Vol. 1, R. Freudenthal and P. W. Jones, Eds., Raven Press, New York, 1976.

736. L. W. Wattenberg, *Digest Dis.*, **19**(10), 947 (1974).

737. D. L. Hill and T. W. Shih, *Cancer Res.*, **34**(3), 564 (1974).

738. N. P. Buu-Hoi and D. P. Hien, *C. R. Acad. Sci., Ser. D., Sin. Natur*, **268**(2), 423 (1969).

739. J. A. Zapp, Jr., *J. Toxicol. Environ. Health*, **2**, 1425 (1977).

740. G. E. Fedoseeva, A. Y. Khesina, M. N. Poglazova, L. M. Shabad, and M. N. Meisel, *Dokl. Akad. Nauk. SSSR*, **183**(1), 208 (1968).

741. E. J. McKenna and R. D. Heath, Ill. Univ. Urbana Champaign Water Resources Center, Champaign, Ill., Rep. Apr., 1976, 33 pp.

742. P. E. Strup, R. D. Giammar, T. B. Stanford, and P. W. Jones, "PAH in Combustion Effluents," in *Carcinogenesis, A Comprehensive Survey*, Vol. 1, R. Freudenthal and P. W. Jones, Eds., Raven Press, New York, 1976.

743. A. Bjorseth and G. Lunde, *Am. Ind. Hyg. Assoc. J.*, **38**, 224 (1977).

744. H. J. Klimisch and K. Fox, *J. Chromatogr.*, **120**(2), 482 (1976).

745. A. Candell, G. Morozzi, A. Paolacci, and L. Zoccolillo, *Atmos. Environ.*, **9**(9), 843 (1975).

746. A. Hase, P. H. Lin, and R. A. Hites, "Analysis of Complex Polycyclic Aromatic Hydrocarbon Mixtures by Computerized GC-MS," in *Carcinogenesis, A Comprehensive Survey*, Vol. 1, R. Freudenthal and P. W. Jones, Eds., Raven Press, New York, 1976.

747. A. E. Zakaryan, T. N. Akopyan, and G. A. Panosyan, *Biofizika*, **17**(5), 769 (1972).

748. H. W. Tyrer, E. T. Cantrell, and A. G. Swan, *Life Sci.*, **20**(10), 1723 (1977).

749. L. Weil, H. Berger, and K. E. Quentin, *Chem. Ing. Tech.*, **49**(5), 429 (1977).

750. J. F. Mesquita, *C. R. Acad. Sci. Paris, Ser. D.*, **265**(4), 322 (1967).

751. J. R. Brenna, *Phytomorphology*, **23**(3-4), 255 (1974).

752. J. E. Thornton and R. R. Bell, *Southwest. Vet.*, **26**(3), 227 (1973).

753. D. C. Hittle and J. J. Stukel, *J. Am. Ind. Hyg. Assoc.*, **37**(4), 199 (1976).

754. J. Neiser and V. Masek, *Zentralbl. Arbeitsmed. Arbeits. Prophyl.*, **26**(7), 127 (1976).

755. T. G. Samsonidze, M. A. Tsartsidze, and G. G. Samsonidze, *Soobsch. Akad. Nauk. Gruz. USSR*, **73**(1), 217 (1974).

756. S. Sung, *C. R. Acad. Sci., Ser. D.*, **273**(14), 1247 (1971).

757. I. Purchase, F. Longstaff, J. Ashby, J. Styles, D. Anderson, P. Lafevre, and F. R. Westwood, *Nature*, **264**, 624 (1976).

758. J. A. Styles, *Br. J. Cancer*, **37**(6), 931 (1978).

759. S. E. Herbes and G. F. Risi, *Bull. Environ. Contam. Toxicol.*, **19**(2), 147 (1978).

760. G. J. Mulkins-Phillips and J. E. Stewart, *Can. J. Microbiol.*, **20**, 955 (1974).

761. S. E. Herbes and L. R. Schwall, *Appl. Environ. Microbiol.*, **35**(2), 306 (1978).

762. A. S. Russell, N. Jarrett, M. J. Bruno, J. A. Remper, and L. K. King, U.S. Pat. 3977846, August 31, 1976.

763. J. S. Robertson and P. J. Dunstan, *Biochem. J.*, **124**(3), 543 (1971).

764. M. D. Kipling, Chapter 6 in *Chemical Carcinogens*, C. E. Searle, Ed., American Chemical Society, Washington, D.C., 1974.

765. N. C. Popescu, D. Turnbull, and J. A. DiPaolo, *J. Natl. Cancer Inst.*, **59**(1), 289 (1977).

766. U. Bayer and T. Bauknecht, *Experientia*, **33**(1), 25 (1977).

767. H. Buchenauer, *Phytopathol. Z.*, **72**(4), 291 (1971).

768. D. Hoffman and E. L. Wynder, Chapter 7 in *Chemical Carcinogens*, C. E. Searle, Ed., American Chemical Society, Washington, D.C., 1974.

769. H. Kubota, W. H. Griest, and M. R. Guerin, *Trace Subst. Environ. Health*, **9**, 281 (1975).

770. L. J. Casarett and J. Doull, Eds., *Toxicology, the Basic Science of Poisons*, Macmillan, New York, 1975.

771. D. M. Jerina, *Fed. Proc.*, **37**(6), 1383 (1978).

772. E. C. Miller and J. A. Miller, Chapter 16 in *Chemical Carcinogens*, C. E. Searle, Ed., American Chemical Society, Washington, D.C., 1974.

773. E. Bingham and H. L. Falk, *Arch. Environ. Health*, **19**, 779 (1969).

774. D. W. Jones and R. S. Matthews, Chapter 4 in *Progress in Medicinal Chemistry 10*, G. P. Ellis and G. B. West, Eds., Elsevier, New York, 1974.

775. B. Tierney, A. Hewer, C. Walsh, P. L. Grover, and P. Sims, *Chem. Biol. Interact.*, **18**(2), 179 (1977).

776. I. Chouroulinkov, A. Gentil, B. Tierney, P. Grover, and P. Sims, *Cancer Lett.*, **3**(5-6), 247 (1977).

777. W. Baird, P. L. Grover, P. Sims, and P. Brookes, *Cancer Res.*, **36**, 2306 (1976).

778. W. Baird, A. Dipple, P. L. Grover, P. Sims, and P. Brookes, *Cancer Res.*, **33**(10), 2386 (1973).

779. S. Levy, D. Papadopoulo, S. Nocentini, L. Chamaillard, O. Beesau, M. Hubert-Habart, and P. Makovits, *Eur. J. Cancer*, **12**(11), 871 (1976).

780. H. Marquardt, S. Baker, B. Tierney, P. L. Grover, and P. Sims, *Int. J. Cancer*, **19**(6), 828 (1977).

781. V. M. Bobr and Yu. P. Kozlov, *Tr. Mosk. Obshchest. Ispyt. Prir.*, **32**, 33 (1970).

782. G. Prodi, A. M. Ferreri, P. Rocchi, and S. Grilli, *Z. Krebsforsch. Klin. Onkol.*, **81**(2), 161 (1974).

783. B. Solymoss, A. Somogyi, and K. Kovacs, *Haematologia*, **5**(1–2), 87 (1971).

784. A. G. Schwartz and A. Perantoni, *Cancer Res.*, **35**(9), 2482 (1975).

785. F. Serri, G. Pisanu, and G. Cantu, *G. Ital. Dermatol./Minerva Dermatol.*, **108**(1), 73 (1973).

786. E. P. Shuba, *Ukr. Biokhim. Zh.*, **41**(3), 249 (1969).

787. C. W. Heizmann and H. J. Wyss, *Arch. Gynekol.*, **216**(1), 51 (1974).

788. R. W. Newburg, Chapter 12 in *Toxicology, the Basic Science of Poisons*, L. J. Casarett and J. Doull, Eds., Mcmillan, New York, 1975.

789. J. W. Flesher, *Biochem. Pharmacol.*, **16**(9), 1821 (1969).

790. C. J. Grubbs and R. C. Moon, *Cancer Res.*, **33**(7), 1785 (1973).

791. J. A. Schmutz, A. C. Brownie, and A. P. Chaudhry, *Cancer Res.*, **34**(3), 578 (1974).

792. F. B. Daniel, L. K. Wong, C. T. Oravec, F. D. Cazer, C'L. Wang, S. M. D'Ambrosio, R. W. Hart, and D. T. Witiak, "Metabolism and DNA Binding of PAH," in *Polynuclear Aromatic Hydrocarbons, Third International Symposium*, R. W. Jones and P. Leber, Eds., Ann Arbor Science Publishers, Ann Arbor, Mich., 1979.

793. J. Booth, G. R. Keysell, and P. Sims, *Biochem. Pharmacol.*, **22**(14), 1781 (1973).

794. J. H. Weisburger, Chapter 15 in *Toxicology the Basic Science of Poisons*, L. J. Casarett and J. Doull, Eds., Macmillan, New York, 1975.

795. R. L. Carter, *Br. J. Cancer*, **28**(1), 91 (1973).

796. V. DeCosta and R. F. Aguirre, *Rev. Cent. Cienc. Biomed., Univ. Fed. St. Maria*, **4**(3–4), 49 (1976).

797. K. Adachi, S. Yamasawa, and W. Montagna, *J. Natl. Cancer Inst.*, **42**(1), 61 (1969).

798. B. Toth, *Tumori*, **57**(3), 169 (1971).

799. R. Schoental, *Br. J. Cancer*, **29**(1), 92 (1974).

800. B. A. Taylor, *Life Sci.*, **10**(19), 1127 (1971).

801. V. S. Turusov and L. A. Andrianov, *Vopr. Onkol.*, **18**(1), 59 (1972).

802. J. H. Epstein, *Cancer Res.*, **32**(12), 2625 (1972).

803. C. Aubert and C. Bohuon, *C. R. Acad. Sci., Ser. D.*, **371**(2), 281 (1970).

804. D. To, L. Manning, and M. Carpenter, *Fed. Proc.*, **37**(3), 424 (1978).

805. F. Stenback, *J. Invest. Dermatol.*, **64**(4), 253 (1975).

806. B. L. Van Duuren, A. Sivak, and L. Langseth, *Br. J. Cancer*, **21**(2), 460 (1967).

807. B. L. Van Duuren, A. Sivak, C. Katz, and S. Melchionne, *J. Natl. Cancer Inst.*, **47**(1), 235 (1971).

808. A. C. Ritchie and H. Shinozuka, *J. Natl. Cancer Inst.*, **38**(4), 573 (1967).

809. S. Tominaga, *Kansai Ika Daigaku Zasshi*, **24**(1), 111 (1972).

810. R. J. Shamberger, *J. Natl. Cancer Inst.* **48**(5), 1491 (1972).

811. E. Huberman and M. Fogel, *Int. J. Cancer*, **15**(1), 91 (1975).

812. R. E. Davies, *Cancer Res.*, **27**(2), 237 (1967).

813. W. Bollag, *Experientia*, **27**(1), 90 (1971).

814. R. J. Shamberger, F. F. Baughman, S. L. Kalchert, C. E. Willis, and G. C. Hoffman, *Proc. Acad. Sci., U.S.*, **70**(5), 1461 (1973).

815. H. V. Gelboin, F. Wiebel, and L. Diamond, *Science*, **170**(3965), 169 (1970).

816. N. Kinoshita and H. V. Gelboin, *Proc. Natl. Acad. Sci., U.S.*, **69**(4), 824 (1972).

817. E. Huberman and L. Sachs, *Int. J. Cancer*, **13**(3), 326 (1974).

818. D. L. Berry, J. DiGiovanni, M. R. Juchau, W. M. Bracken, G. L. Gleason, and T. J. Slaga, *Res. Commun. Chem. Pathol. Pharmacol.*, **20**(1), 101 (1978).

819. D. I. Connell, L. A. Riechere, and J. A. DiPaolo, *J. Natl. Cancer Inst.*, **46**(1), 183 (1971).

820. A. Segal, T. Honohau, M. Schroeder, C. Katz, and B. L. Van Duuren, *Cancer Res.*, **32**(7), 1384 (1972).

821. H. A. Gardner, J. A. Kellen, and K. M. Anderson, *J. Natl. Cancer Inst.*, **50**(4), 915 (1973).

822. K. C. Silinskas and A. B. Olcey, *J. Natl. Cancer Inst.*, **55**(3), 653 (1975).

823. S. Szabo, G. Lazar, and K. Kovacs, *Experientia*, **29**(2), 185 (1973).

824. J. A. Clemens and C. J. Shaar, *Proc. Soc. Exp. Biol. Med.*, **139**(2), 659 (1972).

825. L. M. Shabad, G. A. Belitsky, and T. A. Bogush, *Z. Krebsforsch. Klin. Onkol.*, **82**(1), 13 (1974).

826. V. I. Arkhipenko and L. F. Kos, *Patol., Fiziol. Eksp. Ter.*, (2), 71 (1973).

827. P. Daudel, D. Papadopoulo, P. Markovits, M. Hubert-Habart, and L. Pichat, *C. R. Seances Soc. Boll. Ses. Fil.*, **169**(3), 507 (1975).

828. A. I. Aronskii and S. K. Nuryagdiev, *Izv. Akad. Nauk. Turkn. USSR, Ser. Biol. Nauk.*, (3), 85 (1975).

829. C. D. Baroni, R. Scelsi, P. C. Mingazzini, and S. Uccini, *Boll. Ist. Sieroter. Milan*, **50**(4), 303 (1971).

830. T. Sugiyama, T. O. Yoshida, and Y. Nishizuka, *Gann*, **64**(4), 397 (1973).

831. P. M. Bolton, *Oncologia*, **27**(5), 430 (1973).

832. H. Shisa and Y. Nishizuka, *Gann*, **62**(5), 407 (1971).

833. W. Pierpaoli and N. Haran-Ghera, *Nature*, **254**(5498), 334 (1976).

834. U. Yamashita, Y. Matsuoka and M. Kitagawa, *Gann*, **64**(3), 317 (1973).

835. N. Veda, H. Venaka, T. Akematsu, and T. Sugiyama, *Nature*, **262**(5569), 581 (1976).

836. G. T. Bowden, B. G. Shapas, and R. K. Boutwell, *Chem. Biol. Interact.*, **8**(6), 379 (1974).

837. R. S. Zeiger, R. Salomon, N. Kinoshita, and A. C. Peacock, *Cancer Res.*, **32**(3), 643 (1972).

838. H. Marquardt, P. L. Grover, and P. Sims, *Cancer Res.*, **36**(6), 2059 (1976).

839. S. M. D'Ambrosio, F. P. Daniel, and R. W. Hart, "DNA Repair of PAH," in *Polynuclear Aromatic Hydrocarbons, Carcinogenesis, A Comprehensive Survey, Vol. 1*, R. Freudenthal and P. W. Jones, Eds., Raven Press, New York, 1976.

840. H. S. Schwartz and J. E. Sodergren, *Cancer Res.*, **28**(3), 445 (1968).

841. J. S. Rihn, H. Y. Cho, L. Rabstein, R. J. Gordon, R. J. Bryan, M. B. Gardner, and R. J. Huebner, *Nature*, **239**(5367), 103 (1972).

842. M. S. Newmann and W. Hung, *J. Med. Chem.*, **20**(1), 179 (1977).

843. *CRC Handbook of Analytical Toxicology*, I. Sunshine, Ed., The Chemical Rubber Co., CRC Press, Cleveland, 1969.

844. K. D. Bartle, D. W. Jones, and R. S. Matthews, *J. Med. Chem.*, **12**, 1062 (1969).

845. S. S. Hecht, R. Mazzarese, S. Amin, E. LaVoie, and D. Hoffman, "Metabolic Activation of 5-Methylchrysene," in *Carcinogenesis, A Comprehensive Survey, Vol. 1*, R. Freudenthal and P. W. Jones, Eds., Raven Press, New York, 1976.

846. S. S. Hecht, M. Loy, R. Mazzarese, and D. Hoffman, *J. Med. Chem.*, **21**(1), 38 (1978).

847. S. S. Hecht, M. Loy, and D. Hoffman, "Carcinogenicity of Methylchrysenes," in *Carcinogens, Vol. 1, Polynuclear Aromatic Hydrocarbons*, R. Freudenthal and P. W. Jones, Eds., Raven Press, New York, 1976.

848. A. N. Potapova, V. B. Kapitul'skii, F. M. Kogan, and E. N. Pochashev, *Gig. Tr. Prof. Zabol.*, **15**(2), 59 (1971).

849. T. V. Nikonova, *Buyll. Eksp. Biol. Med.*, **84**(7), 88 (1977).

850. D. W. Lindasy, J. R. Jones, W. H. Higgins, and P. W. Brown, *Exp. Lung Cancer: Carcinog. Bioassays Int. Symp.* (Res. Div. Carreras Rothmans Ltd., Basildon/Essex, Engl.), 521 (1974).

851. B. L. Van Duuren and B. M. Goldshmidt, unpublished data in B. L. Van Duuren Chapter 2 in *Chemical Carcinogens*, C. E. Searle, Ed., American Chemical Society, Washington, D.C., 1974.

852. R. M. Coomes, *Colo. Sch. Mines*, **71**(4), 101 (1976).

853. A. P. Ilnitsky, V. S. Mischenko, and L. M. Shabad, *Cancer Lett.*, **3**(5-6), 227 (1977).

854. E. Gil-Av and J. Shabtol, *Nature*, **197,** 1065 (1963).

855. G. M. Badger and J. Novotny, *Nature*, **198,** 1086 (1963).

856. J. N. Neiser and V. Masek, *Sb. Pr. Pedagog. Fak. Ostrone Rada E*, **5,** 75 (1975).

857. W. Prietsch, K. Wettig, and H. Kahl, *Z. Ges. Hyg. Grenzgeb.*, **17**(8), 573 (1971).

858. J. Howard and T. Fazio, *Ind. Med. Surg.*, **39**(10), 46 (1970).

859. N. G. Turkiya, G. L. Chechelashivili, P. N. Krasnyanskaya, D. S. Beniashvili, and L. I. Dzagnidze, *Vopr. Pitan.*, **30**(1), 31 (1971).

860. M. Rohrlich and P. Suckow, *Getreide Mehl. Brot.*, **26**(4), 114 (1972).

861. A. Audere, Z. Lindbergs, and G. A. Smirnov, *Gig. Sanit.*, **4,** 98 (1975).

862. L. M. Shabad, Y. L. Cohan, A. P. Ilnitsky, A. Y. Khesina, N. P. Shcherbak, and G. A. Sonirnov, *J. Natl. Cancer, Inst.*, **47**(6), 1179 (1971).

863. F. A. Schmid, M. S. Demetriades, F. M. Schabel, and G. S. Tarnowski, *Cancer Res.*, **27**(3), 563 (1967).

864. L. M. Shabad, *Arch. Geschwulstforsch.*, **38**(3-4), 185, (1971).

865. T. D. Sterling and S. V. Pollack, *Am. J. Publ. Health*, **62**(2), 152 (1972).

866. R. C. Levitt, J. S. Felton, J. R. Robinson, and D. W. Nebert, *Pharmacologist*, **17**(2), 213 (1975).

867. H. Reznik-Schueller and U. Mohr, *Zentralbl. Bakteriol. Hyg.*, **159**(5-6), 493 (1974).

868. H. Reznik-Schueller and U. Mohr, *Zentralbl. Bakteriol. Hyg.*, **159**(5-6), 503 (1974).

869. I. A. Shendrikova, M. N. Ivanov-Golitsyn, and A. Y. Likhachev, *Vopr. Onkol.*, **20**(7), 53 (1974).

870. T. Tanaka, *Teratology*, **16**(86), (1977).

871. M. M. Andiranova, *Bull. Exp. Biol. Med.*, **71**(6), 677 (1971).

872. M. R. Juchau, D. L. Berry, P. K. Zachariah, M. J. Namkung, and T. J. Slaga, "Prenatal Biotransformation of Carcinogens," in *Carcinogenesis, A Comprehensive Survey, Vol. 1*, R. Freudenthal and P. W. Jones, Eds., Raven Press, New York, 1976.

873. M. R. Juchau, *Toxicol. Appl. Pharmacol.*, **18**(3), 655 (1971).

874. P. Leber, G. Kerchner, and R. I. Freudenthal, "Species Comparison of BP Metabolism," in *Carcinogenesis, A Comprehensive Survey, Vol. 1*, R. Freudenthal and P. W. Jones, Eds., Raven Press, New York, 1976.

875. J. K. Selkirk and M. C. McLeod, "Metabolism of B(a)P and B(e)P," in *Polynuclear Aromatic Hydrocarbons, Third International Symposium*, P. W. Jones and P. Leber, Eds., Ann Arbor Science Publishers, Ann Arbor, Mich., 1979.

876. C. A. Jones, B. P. Moore, G. M. Cohen, and J. W. Bridges, "Metabolism of PAH," in

Polynuclear Aromatic Hydrocarbons, Third International Symposium, P. W. Jones and P. Leber, Eds., Ann Arbor Science Publishers, Ann Arbor, Mich., 1979.

877. D. M. Jerina, H. Yagi, O. Hernandez, P. M. Dansette, A. W. Wood, W. Lefin, R. L. Chang, P. G. Wislocki, and A. H. Conney, "Biologic Activity of BP Metabolites," in *Carcinogenesis, A Comprehensive Survey*, Vol. 1, R. Freudenthal and P. W. Jones, Eds., Raven Press, New York, 1976.

878. J. P. Glusker, D. E. Zacharias, H. L. Carrell, P. P. Fu, and R. G. Harvey, *Cancer Res.*, **36** (11, Part 1), 3951 (1976).

879. G. A. Belitskii, T. P. Raybykh, and V. A. Kodlyakov, *Tsitologiya*, **19**(10), 1193 (1977).

880. J. W. Flesher and K. L. Sydnor, *Proc. Am. Assoc. Cancer Res.*, **13**, 55 (1972).

881. Y. Ioki, M. Kodama, Y. Tagashiar, and C. Nagata, *Gann*, **65**(4), 379 (1974).

882. N. H. Sloane, H. Chen, B. Divan, R. Bedigan, and H. Meier, "6-Hydroxymethylbenzo(a)pyrene Synthetase," in *Carcinogenesis, A Comprehensive Survey*, Vol. 1, R. Freudenthal and P. W. Jones, Eds., Raven Press, New York, 1976.

883. A. Currin and T. A. Kilroe-Smith, *Chem. Biol. Interact.*, **19**(3), 259 (1977).

884. I. Y. Wang, R. E. Rasmussen, R. Creasy, and T. T. Crocker, *Life Sci.*, **20**(7), 1265 (1977).

885. M. E. McManus, K. F. Ilett, *Drug Metab. Dispos.*, **5**(6), 503 (1977).

886. H. Vadi, Bengt. Jerstromm, and S. Orrenius, "BP Metabolism in Rat Liver and Lung," in *Carcinogenesis, A Comprehensive Survey*, Vol. 1, R. Freudenthal and P. W. Jones, Eds., Raven Press, New York, 1976.

887. D. I. Katz, R. J. Stenger, E. A. Johnson, R. K. Datta, and J. Rice, *Arch. Int. Pharmacodyn. Ther.*, **229**(2), 180 (1977).

888. N. Nemoto, S. Takayama, and V. H. Gelboin, *Biochem. Pharmakol.*, **26**(19), 1825 (1977).

889. W. A. Bornstein, B. Hassuck, H. A. Chuang, and E. Bresnick, *Fed. Proc.*, **37**(5), 1383 (1978).

890. S. K. Yang, D. W. McCourt, P. P. Noller, and V. H. Gelboin, *Proc. Natl. Acad. Sci., U.S.*, **73**(8), 2594 (1976).

891. R. E. Kouri, P. A. Lubet, and D. A. Brown, *J. Natl. Cancer Inst.*, **49**(4), 993 (1972).

892. R. E. Rasmussen and I. Y. Wang, *Cancer Res.*, **34**(9), 2290 (1974).

893. A. Y. Lu, W. Levin, M. Vore, A. H. Conney, D. R. Thakker, G. Holder, and D. M. Jerina, "BP Metabolism by P-448 and Epoxide Hydrase," in *Carcinogenesis, A Comprehensive Survey*, Vol. 1, R. Freudenthal and P. W. Jones, Eds., Raven Press, New York, 1976.

894. D. L. Berry, T. J. Slaga, A. Viaje, N. M. Wilson, J. DiGiovanni, M. R. Juchau, and J. K. Selkirk, *J. Natl. Cancer Inst.*, **58**(4), 1051 (1977).

895. W. E. Fahl, S. Nesnow, and C. R. Jefcoate, *Arch. Biochem. Biophys.*, **18**(2), 649 (1977).

896. C. P. Chow and H. H. Cornish, *Toxicol. Appl. Pharmacol.*, **43**(2), 219 (1978).

897. T. Tang and M. A. Friedman, *Mutat. Res.*, **46**(6), 387 (1977).

898. A. W. Wood, W. Levin, A. Y. H. Lu, D. Ryan, S. B. West, H. Yagi, H. D. Mah, D. M. Jerina, and A. H. Conney, *Mol. Pharmacol.*, **13**(6), 1116 (1977).

899. J. C. Arcos, M. F. Argus, and G. Wolf, *Chemical Induction of Cancer*, Vol. 1, Academic Press, New York and London, 1968.

900. A. R. Kennedy and J. B. Little, *Cancer Res.*, **35**(6), 1563 (1975).

901. U. Saffrotti, R. Montesano, A. R. Sellakumar, and D. G. Kaufman, *J. Natl. Cancer Inst.*, **49**(4), 1199 (1972).

902. M. C. Henry, C. D. Port, and D. G. Kaufman, *Cancer Res.*, **35**(1), 207 (1974).

903. J. B. Little and W. F. O'Toole, *Cancer Res.*, **34**(11), 3026 (1974).

904. B. R. Davis, J. K. Whitehead, M. E. Gill, P. N. Lee, A. D. Butterworth, and F. Roe, Jr., *Br. J. Cancer,* **31**(4), 443 (1975).

905. B. C. Castro, N. Janosko, and J. A. DiPaolo, *Cancer Res.,* **37**(10), 3508 (1977).

906. M. Kupfer, G. Kupfer, M. Waehmer, and W. Wuenscher, *Arch. Geschwulstforsch.,* **35**(2), 99 (1971).

907. O. S. Frankfurt and E. Raitcheva, *J. Natl. Cancer Inst.,* **51**(6), 1861 (1973).

908. A. W. Horton, D. N. Eshleman, A. R. Schaff, and W. H. Perman, *J. Natl. Cancer Inst.,* **56**(2), 387 (1976).

909. D. M. Smith, A. E. Rogers, B. J. Herndon, and P. M. Newberne, *Cancer Res.,* **35**(1), 11 (1975).

910. V. J. Feron, *Cancer Res.,* **32**(1), 28 (1972).

911. M. L. Kripke and T. Borsos, *J. Natl. Cancer Inst.,* **53**(5), 1409 (1974).

912. D. B. Clayson and R. C. Garner, Chapter 8 in *Chemical Carcinogens,* C. E. Searle, Ed., American Chemical Society, Washington, D.C., 1974.

913. E. H. Pfeiffer, *Zentralbl. Bakteriol.,* **160**(2), 99 (1975).

914. L. W. Wattenberg, *J. Natl. Cancer Inst.,* **48**(5), 1425 (1972).

915. J. L. Speier, L. K. T. Lam, and L. W. Wattenberg, *J. Natl. Cancer Inst.,* **60**(3), 605 (1978).

916. V. T. Vertushkov, *Biofiziko,* **23**(3), 427 (1978).

917. Y. Iwanami and S. Odashima, *Naturwissenschaften,* **61**, 509 (1974).

918. J. A. DiPaolo, P. J. Donovan, and R. L. Nelson, *Proc. Natl. Acad. Sci., U.S.,* **68**(12), 1953 (1971).

919. S. D. Vesselinovitch and N. Mihailovich, *Cancer Res.,* **28**(12), 2463 (1968).

920. M. V. Marshall, M. A. Rasco, and A. C. Griffin, *Fed. Proc.,* **37**(6), 1383 (1978).

921. R. A. Lubet, M. Turner-Lubet, and Y. L. Whitson, *Eur. J. Cancer,* **11**, 139 (1974).

922. N. Nemoto and S. Takayama, *Toxicol. Lett.,* **1**, 247 (1978).

923. M. Y. Akhalaya, V. A. Akhobadze, M. A. Tsartsidze, and B. A. Lomsadze, *Fiz-Khim. Osn-Funkts.,* **1**, 13 (1974).

924. L. J. Anghileri, *Biochim. Biophys. Acta,* **136**(2), 386 (1967).

925. P. Bothorel and J. P. Desmazes, *Biochim. Biophys. Acta,* **365**(1), 181 (1974).

926. F. A. Popps, *Z. Naturforsch.,* **27**(7), 850 (1972).

927. I. Radhakrishnan, L. L. Triplett, T. J. Slaga, and J. Papaconstantinou, "Interaction of Anti-BPDE with Eukaryotic DNA," in *Polynuclear Aromatic Hydrocarbons, Third International Symposium,* P. W. Jones and P. Leber, Eds., Ann Arbor Science Publishers, Ann Arbor, Mich., 1979.

928. H. W. S. King, M. R. Osborne, F. A. Reland, R. G. Harvey, and P. Brookes, *Proc. Natl. Acad. Sci., U.S.,* **73**(8), 1679 (1976).

929. W. Baird and L. Diamond, *Biochem. Biophys. Res. Commun.,* **77**(1), 162 (1977).

930. P. Cerutti, K. Shinohara, and J. Remsen, *J. Toxicol. Environ. Health,* **2**, 1375 (1977).

931. T. Kuroki and C. Heidelberger, *Cancer Res.,* **31**(12), 2168 (1971).

932. D. Schmaehl, *Z. Krebsforsch. Klin. Onkol.,* **81**(3), 211 (1974).

933. P. O. P. Ts'o, *J. Toxicol. Environ. Health,* **1**, 1306 (1977).

934. Yu L. Kogan, *Gig. Sanit.,* (7), 110 (1974).

935. M. N. Pozlazova, G. E. Fedoseeva, A. Y. Khesina, M. N. Meisel, and L. M. Shabad, *Life Sci.,* **6**(10), 1053 (1967).

936. M. N. Pozlazova, G. E. Fedoseeva, A. Y. Khesina, M. N. Meisel, and L. M. Shabad, *Dokl. Acad. Nauk. USSR Ser. Biol.,* **198**(5), 1211 (1971).

937. A. Y. Khesina, N. P. Scherbak, L. M. Shabad, and I. S. Vostrov, *Bull. Exp. Biol. Med.,* **68**(10), 70 (1969).

938. H. Lorbacher, H. D. Puels, and H. W. Schlipkoeter, *Zentralbl. Bakteriol. Parasitenk. Infektionskr.,* **155**(2), 168 (1971).

939. R. E. Lehr, C. W. Taylor, S. Kumar, W. Levin, R. Chang, A. W. Wood, A. H. Conney, D. R. Thakker, H. Yagi, H. D. Mah, and D. M. Jerina, "Biological Activity of Metabolites," in *Polynuclear Aromatic Hydrocarbons, Third International Symposium,* P. W. Jones and P. Leber, Eds., Ann Arbor Science Publishers, Ann Arbor, Mich., 1979.

940. E. L. Wynder and D. Hoffman, *Cancer,* **12,** 1079 (1959).

941. J. Jacob and G. Grimmer, *Zentralbl. Bakteriol. Parasitenkd. Infektionskr.,* **165**(3–4), 305 (1977).

942. J. Neal and N. M. Trieff, *Health Lab. Sci.,* **9**(1), 32 (1972).

943. A. Gold, *Environ. Aspects Chem. Use Rubber Process, Oper., Conf. Proc.,* 137 (1975).

944. A. Eisenstadt and A. Gold, *Proc. Natl. Acad. Sci., U.S.,* **75**(4), 1667 (1978).

945. N. N. Vasileeva, *Buyll. Eksp. Biol. Med.,* **71**(3), 116 (1971).

946. J. W. Orr, *J. Pathol. Bacteriol.,* **49,** 157 (1939).

947. W. Ho and A. Furst, *Toxicol. Appl. Pharmacol.,* **29**(1), 94 (1974).

948. J. White and G. B. Mider, *J. Natl. Cancer Inst.,* **2,** 95 (1941).

949. K. Buerki, R. A. Seibert, and E. Bresnick, *Biochim. Biophys. Acta,* **260**(1), 98 (1972).

950. G. Kellerman, E. Cantrell, and C. R. Shaw, *Cancer Res.,* **33**(7), 1654 (1973).

951. N. Prasad, R. Prasad, J. Thornby, S. C. Bushong, L. B. North, and J. E. Harrell, *Cancer Res.,* **37**(10), 3771 (1977).

952. P. E. Thomas, R. E. Kouri, and J. J. Hutton, *Biochem. Genet.,* **6**(2–3), 157 (1972).

953. N. E. Sladek and G. J. Mannering, *Mol. Pharmacol.,* **5**(2), 186 (1969)

954. B. Stripp, M. E. Hamrick, and J. R. Gillette, *Biochem. Pharmacol.,* **21**(5), 745 (1972).

955. S. J. Buynitzky, A. E. Wade, J. F. Munnell, and W. L. Ragland, *Drug Metab. Dispos.,* **6**(1), 1 (1978).

956. F. Oesch, *Xenobiotica,* **3**(5), 305 (1973).

957. A. W. Wood, W. Levin, R. L. Chang, H. Yagi, D. R. Thakker, R. E. Lehr, D. M. Jerina, and A. H. Conney, "Bay-Region Activation of Carcinogenic Polycyclic Hydrocarbons," in *Polynuclear Aromatic Hydrocarbons, Third International Symposium,* P. W. Jones and P. Leber, Eds., Ann Arbor Science Publishers, Ann Arbor, Mich., 1979.

958. M. A. Mehlman, R. E. Shapiro, and H. Blumenthal, *New Concepts in Safety Evaluation,* Hemisphere Publishing Corp., Washington, D.C., 1976.

959. W. F. Dunning, M. R. Curtis, and M. J. Eisin, *Am. J. Cancer,* **40,** 85 (1940).

960. H. P. Rusch, C. A. Baumann, and B. E. Kline, *Proc. Soc. Exp. Biol. Med.,* **42,** 508 (1939).

961. K. Buerki, T. A. Stoming, and E. Bresnick, *J. Natl. Cancer Inst.,* **52**(3), 785 (1974).

962. R. E. Kouri, *Am. Ind. Hyg. Assoc. J.,* **38,** 150 (1977).

963. M. G. Shengelia, M. A. Tsartidze, and B. A. Lomsadze, *Soobsch. Akad. Nauk. Gruz. SSSR,* **77**(1), 189 (1975).

964. Z. T. Halpin, J. Vaage, and P. B. Blair, *Cancer Res.,* **32**(10), 2197 (1972).

965. M. A. Rogers and C. B. Koons, Chapter 3 in *Origin and Refining of Petroleum,* R. F Gould, Ed., American Chemical Society, Washington, D.C., 1971.

966. *Criteria for a Recommended Standard—Occupational Exposure to Refined Petroleum Solvents,* U.S. Dept. Health, Education, and Welfare, Public Health Service, NIOSH, Washington, D.C., 1977.

967. C. P. Carpenter, E. R. Kinkead, D. L. Geary, Jr., L. J. Sullivan, and J. M. King, *Toxicol. Appl. Pharmacol.*, **34,** 413 (1975).

968. *API Toxicological Review of Petroleum Naphthas,* American Petroleum Institute, New York, 1969.

969. C. P. Carpenter, D. L. Geary, Jr., R. C. Myers, D. J. Nachreiner, L. J. Sullivan, and J. M. King, *Toxicol. Appl. Pharmacol.*, **41,** 235 (1977).

970. C. P. Carpenter, D. L. Geary, R. C. Myers, D. J. Nachreiner, L. J. Sullivan, and J. M. King, *Toxicol. Appl. Pharmacol.*, **36,** 457 (1976).

971. C. P. Carpenter, D. L. Geary, Jr., R. C. Myers, D. J. Nachreiner, L. J. Sullivan, and J. M. King, *Toxicol. Appl. Pharmacol.*, **36,** 427 (1976).

972. C. P. Carpenter, E. R. Kinkead, D. L. Geary, Jr., L. J. Sullivan, and J. M. King, *Toxicol. Appl. Pharmacol.*, **34,** 374 (1975).

973. C. P. Carpenter, E. R. Kinkead, D. L. Geary, Jr., R. C. Myers, D. J. Nachreiner, L. J. Sullivan, and J. M. King, *Toxicol. Appl. Pharmacol.*, **36,** 409 (1976).

974. C. P. Carpenter, E. R. Kinkead, D. L. Geary, Jr., L. J. Sullivan, and J. M. King, *Toxicol. Appl. Pharmacol.*, **34,** 395 (1975).

975. C. P. Carpenter, D. L. Geary, Jr., R. C. Myers, D. J. Nachreiner, L. J. Sullivan, and J. M. King, *Toxicol. Appl. Pharmacol.*, **36,** 443 (1976).

976. *The United States Pharmacopeia,* 19th ed., Mack, Easton, Pa., July, 1975.

977. *API Toxicological Review of Gasoline,* American Petroleum Institute, New York, 1967.

978. C. P. Yaglan and M. F. Warren, *J. Ind. Hyg. Toxicol.*, **25,** 225 (1943).

979. H. Peng, *Shish Yu T'ung Hsin*, **279,** 46 (1974).

980. F. Flury, *Arch. Exp. Pathol. Pharmakol.*, **138,** 65 (1928).

981. C. P. Carpenter, E. R. Kinkead, D. L. Geary, Jr., L. J. Sullivan, and J. M. King, *Toxicol. Appl. Pharmacol.*, **33,** 526 (1975).

982. C. P. Carpenter, E. R. Kinkead, D. L. Geary, L. J. Sullivan, and J. M. King, *Toxicol. Appl. Pharmacol.*, **32**(2), 263 (1975).

983. I. Astrand, A. Kilbom, and P. Ovrum, *Scand. J. Work Environ. Health*, **1,** 15 (1975).

984. D. E. Rector, B. L. Steadman, R. A. Jones, and J. Siegel, *Toxicol. Appl. Pharmacol.*, **9,** 257 (1966).

985. *Toksikol. Nov. Promys. Khim. Vesh. Akad. Medit. Nauk. USSR*, Moscow, **10,** 116 (1968), through NIOSH (4).

986. M. L. Keplinger, G. E. Lanier, and W. B. Deichmann, *Toxicol. Appl. Pharmacol.*, **1,** 156 (1959).

987. C. P. Carpenter, E. R. Kinkead, D. L. Geary, L. J. Sullivan, and J. M. King, *Toxicol. Appl. Pharmacol.*, **32**(2), 282 (1975).

988. J. L. Scott, G. E. Cartwright, and M. M. Wintrobe, *Medicine*, **38,** 119 (1959).

989. L. E. Brauenstein, *J. Am. Med. Assoc.*, **114,** 136 (1940).

990. D. Prager and C. Peters, *Blood*, **35,** 286 (1970).

991. L. S. Goodman and A. Gilman, *The Pharmacological Basis of Therapeutics*, 4th ed., Macmillan, New Haven, Conn., 1971.

992. W. B. Deichmann, K. V. Kitzmiller, S. Witherup, and R. Johansmann, *Ann. Intern. Med.*, **21,** 803 (1944).

993. M. J. Narasimhan, Jr. and V. G. Ganla, *Acta Pharmacol. Toxicol.*, **25,** 214 (1967).

994. H. W. Gerarde, *Toxicol. Appl. Pharmacol.*, **1,** 464 (1959).

995. J. A. Rebello and R. R. Suskind, *J. Invest. Dermatol.*, **41,** 67 (1963).

996. T. L. Thomas, P. Decoufle, and R. Moure-Eraso, *J. Occup. Med.*, **22**(2), 97 (1980).

997. P. K. Gupta, T. S. Dikshith, and K. K. Datta, *Toxicology*, **7**(1), 57 (1977).

998. H. M. Smith, *Bur. Mines Rep. Invest. 6542*, U.S. Dept. Interior, Washington, D.C., 1964.

999. D. I. Zul'Fugarly and N. S. Umakhanova, *Azerbaidzhan. Khim. Zhur.*, **1**, 65 (1960).

1000. S. M. Katchenov and E. I. Flegontova, *Vestsi Akad. Navuk. Belarus. USSR Ser. Khim. Navuk.*, (6), 95 (1970).

1001. R. L. Erickson, A. T. Myers, and C. A. Horr, *Bull. Am. Assoc. Petrol. Geol.*, **38**, 2200 (1954).

1002. L. D. Rowe, J. W. Dollahite, and B. J. Camp, *J. Am. Vet. Med. Assoc.*, **162**(1), 61 (1973).

1003. S. S. Rossi, J. W. Anderson, and G. S. Ward, *Environ. Pollut.*, **10**(1), 9 (1976).

1004. J. M. Neff, J. W. Anderson, B. A. Cox, R. B. Laughlin, Jr., S. S. Rossi, and H. E. Tatem, *Sources, Eff. Sinks. Hydrocarbons Aquat. Environ. Proc. Symp.*, 515 (1977).

1005. R. Knowles and C. Wishart, *Environ. Pollut.*, **13**(2), 133 (1977).

1006. R. F. Lee and M. Takashashi, *Cons. Int. Explor. Mer.*, **171**, 150 (1977).

1007. W. Pulich, Jr., K. Winters, and C. Van Baalen, *Mar. Biol.*, **28**(2), 87 (1974).

1008. B. S. Bailey, *J. Wash. Acad. Sci.*, **61**(2), 74 (1971).

1009. W. Moore, D. Hysell, R. Miller, M. Malanchuk, R. Hinners, Y. Yang, and J. F. Stara, *Environ. Res.*, **9**, 274 (1975).

1010. W. Machle, *J. Am. Med. Assoc.*, **117**, 1965 (1941).

1011. J. I. Tonge, R. N. Hurley, and J. Ferguson, *Lancet*, **1**, 1059 (1969).

1012. A. Rothe, *Z. Aerztl. Fortbild.* (Jena), **66**(15), 758 (1972).

1013. C. F. Phillips and R. K. Jones, *J. Am. Ind. Hyg. Assoc.*, **39**(2), 118 (1978).

1014. S. Tola, S. Hernberg, and J. Nikkanen, *Work Environ. Health*, **9**(3), 102 (1972).

1015. M. Lob, *Z. Praev. Med.*, **10**, 172 (1965).

1016. K. P. Pandya, G. S. Rao, A. Dhasmana, and S. H. Zaidi, *Ann. Occup. Hyg.*, **18**(4), 363 (1975).

1017. J. Przybylowski, J. Wysocki, Z. Szczepanski, A. Sychlowy, and A. Podolecki, *Bromatol. Chem. Toksykol.*, **9**(1), 33 (1976).

1018. L. Brandt, P. G. Nilsson, and F. Mitelman, *Br. Med. J.*, **1**, 553 (1978).

1019. A. Poklis and C. Burkett, *Clin. Toxicol.*, **11**(1), 35 (1977).

1020. R. L. Boeckx, B. Postl, and F. J. Coodin, *Pediatrics*, **60**(2), 140 (1977).

1021. M. B. Chenoweth, *J. Ind. Hyg. Toxicol.*, **28**, 151 (1946).

1022. M. Bass, *J. Am. Med. Assoc.*, **212**, 2075 (1970).

1023. F. Pott and A. Brockhaus, *Zentralbl. Bakteriol. Parasitenkd. Infektionskr. Hyg., Abt.* 1, **155**(1), 1 (1971).

1024. K. Saito, H. Inai, and E. Takakuwa, *Sangyo Igaku*, **14**(1), 9 (1973).

1025. R. M. McClain and B. A. Becker, *Toxicol. Appl. Pharmacol.*, **21**(2), 265 (1972).

1026. J. Przybylowski, *Arh. Hig. Rada*, **21**, 327 (1971).

1027. J. Przybylowski, W. Kowalski, and A. Podalecki, *Patol. Pol.*, **27**(2), 149 (1976).

1028. E. E. Gasanova and S. F. Fatalieva, *Azerb. Med. Zh.*, **48**(6), 29 (1971).

1029. R. S. Sunargulov, A. K. Giniyatullina, and T. S. Ivanova, *Oftalmol. Zh.*, **31**(1), 20 (1976).

1030. J. Karkos and J. Sikora, *Neuropatol. Pol.*, **11**(1), 99 (1973).

1031. N. A. Minkina, E. G. Berliner, and S. A. Chernova, *Probl. Adapt. Gig. Tr.*, 50 (1973).

1032. S. Urishibara, *Tokyo Jikeikai Ika Daigaku Zasshi*, **91**(2), 198 (1976).

1033. I. Feller, *Gig. Tr. Prof. Zabol.*, **16**(8), 25 (1972).

1034. G. Soderman, *Hereditas*, **71**(2), 335 (1972).

1035. H. J. McDermott and S. E. Killiany, Jr., *Am. Ind. Hyg. Assoc. J.*, **39,** 110 (1978).

1036. T. Nagata, M. Kagleura, K. Hara, and K. Totoki, *Nippon Hoegaku Zasshi*, **31**(3), 136 (1977).

1037. R. Takahashi, T. Sone, and T. Hirata, Jap. Pat. 77 19190, February 14, 1977.

1038. H. Hänninen, L. Eskelinen, K. Husman, and M. Nurminen, *Scand. J. Work Environ. Health*, **2**(4), 240 (1976).

1039. R. Rawson, F. Parker, and H. Jackson, *Science*, **93,** 2423 (1941).

1040. M. Sitting, *How to Remove Pollutants and Toxic Materials from Air and Water*, Noyes Data Corporation, Park Ridge, N.J., 1977.

1041. V. J. Harris and R. Brown, *Am. J. Roentgenol. Radium Ther. Nucl. Med.*, **125**(3), 531 (1975).

1042. A. J. McMichael, R. Spirtas, L. L. Kupper, and J. F. Gamble, *J. Occup. Med.*, **17**(4), 234 (1975).

1043. F. W. Wilson, *J. Occup. Med.*, **18,** 821 (1976).

1044. *Fed. Reg.*, **23,** 504 (1974).

1045. H. Altenkirch, J. Mager, G. Stoltenburg, and J. Helmbrecht, *J. Neurol.*, **214**(2), 137 (1977).

1046. G. L. Choules and W. C. Russell, *Vet. Hum. Toxicol.*, **19**(4), 253 (1977).

1047. J. A. Richardson and H. R. Pratt-Thomas, *Am. J. Med. Sci.*, **221**(5), 531 (1951).

1048. R. H. Daffner and J. P. Jiminez, *Radiology*, **106**(2), 383 (1973).

1049. L. Chin, A. Picchioni, and B. Duplisse, *J. Pharm. Sci.*, **58,** 1353 (1969).

1050. Anon., *Br. Med., J.*, **3,** 487 (1972).

1051. R. C. Ng, H. Darwish, and D. A. Stewart, *Can. Med. Assoc. J.*, **111,** 537 (1974).

1052. *API Toxicological Review of Kerosine*, American Petroleum Institute, New York, 1967.

1053. H. Tagami and A. Ogino, *Dermatologica*, **146,** 123 (1973).

1054. A. P. Luplescu, H. Pinkus, and D. J. Birmingham, *Proc. Electron Microsc. Soc. Am.*, **30,** 92 (1972).

1055. A. P. Luplescu, D. J. Birmingham, and H. Pinkus, *J. Invest. Dermatol.*, **60**(1), 32 (1973).

1056. A. P. Luplescu and D. J. Birmingham, *J. Invest. Dermatol.*, **65**(5), 419 (1975).

1057. D. E. Johnston, *J. Am. Med. Women's Assoc.*, **10,** 421 (1955).

1058. J. Wolfsdorf and H. Kundig, *S. Afr. Med. J.*, **46,** 619 (1972).

1059. A. Volkova, V. Tsetlin, E. Zhuk, and E. Izotova, *Gig. Sanit.*, **34,** 24 (1969).

1060. H. W. Gerarde, *Occup. Health Rev.*, **16**(3), 17 (1964).

1061. P. Gross, J. M. McNerney, and M. A. Baleyak, *Am. Rev. Respir. Dis.*, **88**(5), 656 (1963).

1062. M. D. Mann, D. J. Pirie, and J. Wolfsdorf, *J. Pediatr.*, **91**(3), 495 (1977).

1063. R. W. Lewis, *Int. J. Biochem.*, **2**(11), 609 (1971).

1064. K. Morihara, *Appl. Microbiol.*, **13**(5), 793 (1965).

1065. K. Katsuri and D. V. Tamhane, *Ind. J. Exp. Biol.*, **9**(2), 235 (1971).

1066. K. I. Markov and T. Kobarska, *Dokl. Akad. Sel-skokhoz. Nauk. Bolg.*, **4**(4), 413 (1971).

1067. N. Gerasimova and M. Bekhtereva, *Mikrobiologiya*, **39**(4), 616 (1970).

1068. N. E. Davies, *Aerosp. Med.*, **35,** 481 (1964).

1069. C. L. Gaworski and H. F. Leahy, *Proc. 9th Ann. Conf. Environ. Toxicol., Govt. Rept. AMRL-TR-79-68*, Aerospace Medical Research Laboratory, Wright-Patterson Air Force Base, Ohio, 1979.

1070. C. C. Haun, *Aerosp. Med. Res. Lab. Rep. AMRL-TR (U.S.)*, **125,** 287 (1975).

1071. S. A. Klein, D. Jenkins, and R. C. Cooper, *Aerosp. Med. Res. Lab. Rep., AMRL-TR (U.S.)*, **125,** 429 (1975).

1072. J. Sula and V. Krol, *Prot. Vitae*, **16**(6), 266 (1971).

1073. J. A. Berlin and D. W. Micks, *Ann. Entomol. Soc. Am.*, **66**(4), 775 (1973).

1074. D. G. Jones and S. H. Limaye, Fr. Pat. 2183533, January 25, 1974.

1075. H. E. Tatem, B. A. Cox, and J. W. Anderson, *Estuarine Coastal Mar. Sci.*, **6**(4), 365 (1978).

1076. R. C. Clark, Jr. and J. S. Finley, *Fish Bull.*, **73**(3), 508 (1975).

1077. D. M. Stainken, *J. Fish Res. Board Can.*, **35**(5), 637 (1978).

1078. J. Bothe, W. Braun, and A. Doenhardt, *Arch. Toxicol.*, **30**(3), 243 (1973).

1079. K. Yamada and M. Yogo, *Agric. Biol. Chem.*, **34**(2), 296 (1970).

1080. Z. Z. Bruskin and V. G. Demchenko, *Gig. Tr. Prof. Zabol.*, (4), 28 (1975).

1081. M. D. Kipling and H. A. Waldron, *Prev. Med.*, **5**, 262 (1976).

1082. H. E. Burmeister, *Berufsdermatosen*, **21**(2), 69 (1973).

1083. H. Luther and G. Bergmann, *Erdoel Kohle*, **8**, 298 (1955).

1084. E. G. Ivanyuk and V. V. Vasilenko, *Gig. Sanit.*, (7), 82 (1976).

1085. M. H. Whisman, J. W. Goetzinger, and F. O. Cotton, *U.S. Bur. Mines Rep. Invest. RI-7973*, U.S. Dept. Interior, Washington, D.C., 1974.

1086. K. Ohshima and T. Nakae, Jap. Pat. 77 42485, April 2, 1977.

1087. V. G. Litau, M. F. Obukhova, and V. I. Soloviev, *Gig. Tr. Prof. Zabol.*, **7**, 213 (1975).

1088. G. A. Gellin, *Ind. Med.*, **39**(2), 38 (1970).

1089. Anonymous, *Occup. Health Safety*, Sept./Oct., 16 (1976).

1090. G. A. Gellin, *J. Occup. Med.*, **11**(3), 128 (1969).

1091. M. H. Samitz and S. A. Katz, *Contact Dermatitis*, **1**, 158 (1975)

1092. W. Catchpole, E. MacMillan, and H. Powel, *Ann. Occup. Hyg.*, **14**(2), 171 (1971).

1093. J. A. Waterhouse, *Ann. Occup. Hyg.*, **14**(2), 161 (1971).

1094. T. H. F. Smith, *Ind. Med.*, **39**(2), 29 (1970).

1095. J. R. Jepsen, S. Stoyanov, M. Unger, J. Clausen, and H. Christensen, *Acta Pathol. Microbiol. Scand.*, **85**(5), 731 (1977).

1096. C. Urwin, J. Richardson, and A. Palmav, *Mutat. Res.*, **40**, 43 (1976).

1097. R. Cabridenc, *Microb. Mater.*, 123 (1974).

1098. J. Markind, J. Neri, and R. Stana, *AIChE Symp. Ser.*, **71**(151), 70 (1975).

1099. H. Ramos, *J. Occup. Med.*, **16**(4), 273 (1974).

1100. A. B. Balzan, *Fachh. Chemigr. Lithogr. Tiefdruck*, (3), 173 (1974).

1101. W. Lijinsky, U. Saffiotti, and P. Shubik, *Toxicol. Appl. Pharmacol.*, **8**(1), 113 (1966).

CHAPTER FORTY-EIGHT

Halogenated Aliphatic Hydrocarbons Containing Chlorine Bromine and Iodine

T. R. TORKELSON, Sc.D. and
V. K. ROWE, Sc.D. (hon.)

1 GENERAL CONSIDERATIONS

Many of the general considerations discussed in the chapter on aliphatic halogenated compounds in the second revised edition of this book are still applicable today (1). As Irish, the author of that chapter, indicated, the chlorinated, brominated, and iodinated aliphatic compounds have quite variable physical, chemical, and toxicologic properties that allow selection of the specific compound to fit the intended use. All are synthetically derived, although traces of naturally produced halomethanes and possible other halogenated compounds are found in air and water. Many have excellent and often specific solvent properties; some are used as chemical intermediates, monomers, aerosol solvents, blowing agents, and fumigants.

This edition utilizes data extensively from Irish's text, updating only those portions that we feel need significant alterations or additions. Sections on metabolism, teratogenesis, mutagenesis, and carcinogenesis have had to be altered or included because data in these areas have expanded rapidly. The reader is cautioned that data presented in these sections will be superseded rapidly by new data. Unfortunately, understanding has not kept abreast with generation of data and the significance of much of these data is not clear or universally acceptable.

Our understanding of carcinogenesis certainly has not kept pace with the data. Vinyl chloride is accepted as a carcinogen, having produced angiosarcoma of the liver in highly exposed workmen and these and other tumors in animals. In addition, carbon tetrachloride and chloroform have long been recognized as being carcinogenic in animals at hepatotoxic levels. However, we now find that many other hepatotoxic halogenated compounds also cause hepatocellular carcinomas in mice and some in rats. Human data, where available, do not support the carcinogenicity of these compounds in man, but the human data are in most cases from small studies with limited statistical confidence. The manner in which halogenated compounds are used makes it very difficult to find sizable populations for epidemiologic study.

The significance of mutagenic data, particularly those derived from in vitro systems, is equally obscure. Data have been cited for some compounds but extreme caution is needed in interpretation at this time. It is our opinion that mutagenicity is not of practical concern for any of the compounds discussed, at least at levels otherwise acceptable for occupational exposure.

It has been demonstrated that *in vitro* tests may not represent the total mammalian system, since the normal metabolic and protective mechanisms are not present. As discussed in more detail in the section on mutagenesis of 1,3-dichloropropene, rapid metabolism of that compound plus the presence of ample glutathione protect the rat and make it unlikely that mutagenesis will occur at reasonable doses.

The importance of pharmacokinetics and the influence of the size of the dose have become better recognized and accepted by toxicologists. Quantitative difference in response between laboratory rodents and humans may have provided safety factors for those old industrial hygiene guides that were based on animal data. There is little or no evidence of adverse effects in humans exposed to levels below those recommended by Irish in the second edition (1). It appears likely that pharmacokinetic studies will be the key to our understanding of species differences and to the carcinogenic studies in rodents.

It was not our intent to compile an exhaustive bibliography. References to halogenated aliphatic compounds are so numerous that it has been necessary to severely limit citations to those that appear most useful to understanding and controlling the toxicologic problems in the workplace. Obviously, this will not satisfy the needs of the specialists in medicine, biochemistry, pathology, oncology, or other disciplines. Certain bibliographies, review articles, and key references are cited if they appear to be useful to the specialists.

Whereas Irish felt several pages were necessary to discuss the differences in toxicity between members of this series of compounds, their proper use, the concepts of industrial hygiene standards, the duration of exposure, and the probability of air dispersion, awareness of these important concepts is much higher today than it was two decades ago; therefore, the reader is referred to Irish's text for an interesting discussion of these factors (1).

It was not considered practical to include the hygienic standards for each country. It is obvious that these values differ widely from one compound to another, even for the standards of different countries for the same compound. Several authors have commented on the differences in philosophies between the standards of various countries. The text generally includes the threshold limit value recommended by the American Conference of Governmental Industrial Hygienists (ACGIH) in 1980 (2), with an occasional opinion by us as to its appropriateness. Obviously legal standards, national, state, provincial, or other unit having jurisdiction, have precedence and must be adhered to in order to comply with local legal requirements.

1.1 Analytical Methods

So many factors are involved in sample collection and analysis for industrial hygiene purposes that no attempt is made to discuss specific methods. Therefore industrial hygiene methods are omitted from the discussion of each individual compound. Important basic considerations are treated in Chapter 17, Vol. I of the third revised edition of this book (3). Quality control and calibration, too often neglected by the industrial hygienist, are discussed in Chapters 24 and 25 of Vol. I.

For some halogenated hydrocarbons, specific methods for personal monitoring by charcoal tube collection and gas chromatographic analysis are discussed in the NIOSH *Manual of Analytical Methods*, Method 127 for organic solvents and Method 178 for vinyl chloride (4). The *American Industrial Hygiene Association Journal* publishes articles that describe other specific methods. Laboratory verification of published methods is recommended in all cases. Direct-reading "halide" meters are described in *Air Sampling Instruments*, 5th ed. (5).

Two recent developments in personal monitoring "badges" are 8-hr colorimetric tubes and passive dosimeters, which use charcoal absorption and diffusion or permeation rather than pumping air. Methods validation data for specific compounds have appeared in articles in the *American Industrial Hygiene Association Journal*, and other data are available from manufacturers. It is necessary for each user of any collection or analytical method to do adequate validation work to assure that the methods work satisfactorily under the specific conditions encountered in his surveys. Melcher et al. (6) have published criteria for methods validation.

1.2 Analysis of Biologic Materials

Analysis of biologic material has long been proposed as a technique for determining and quantitating exposure to many compounds and has had success particularly as a research tool. Blood, hair, nails, urine, and expired air have most commonly been analyzed because of the ease with which samples can be obtained. All these techniques have common benefits and disadvantages, but the appropriateness of using the worker as an indicator of his exposure instead of adequate environmental measurements is questionable. Properly applied, biologic monitoring may provide a verification of environmental monitoring, but must never replace it. If routes of exposure other than respiration, particularly skin contact, are important, biologic monitoring may be an essential supplement to environmental measurements and work practices.

Furthermore, monitoring of blood and urine provides at best only an estimate of integrated or averaged exposure for long-lasting materials and hence is most valuable when the effects of chronic exposure are most important. Expired air analysis for volatile compounds immediately after exposure provides primarily a measure of most recent exposures, with little information on exposure that may have taken place earlier during the day. Breath samples taken a long time after exposure ceases may provide an estimate of composite exposure but little information on duration of exposure or exposure concentration.

Biologic monitoring for individual compounds is discussed together with metabolism in that section for each compound.

1.3 Physical Properties

The physical properties cited represent a best estimate of those found in the literature. Although most of those compounds discussed are considered non-flammable, some compounds are quite flammable. It should be noted that all halogenated compounds can be broken down by heat or fire to produce halogenated acids and in some cases much smaller concentrations of carbonyl (phosgene-type) compounds. The halogen acids produced in fires or pyrolysis may be of toxicologic concern because of respiratory irritation and because of corrosion of metal and other materials of construction. Except for very unusual circumstances, "phosgene" concentrations have been shown to be much lower than the acids and hence acids provide warning to prevent exposure to the carbonyls. Care must be taken to prevent degradation of the halogenated aliphatics by heat or fire as exemplified by welding and gas heaters.

Similarly, certain compounds can be decomposed by caustic or strongly alkaline materials. An example has been shown in trichloroethylene, which can react to form dichloroacetylene, a highly toxic compound. This has occurred in recycling anesthesia machines used for surgery, in vapor degreasing machines, and in confined vessels such as submarines, where caustic or soda lime type absorbers were used.

2 SATURATED HALOGENATED HYDROCARBONS

2.1 Methyl Chloride, Monochloromethane, CAS 74-87-3

$$CH_3Cl$$

2.1.1 Uses and Industrial Exposure

About 4 percent of methyl chloride is used as a blowing agent for plastic foams; most is used as a chemical intermediate particularly in methylating reactions. Recent industrial exposure has been largely from operations in which foamed plastic is cut or shaped. Its current use as a refrigerant is rare, but occasional reports of exposure due to leaking refrigerators still occur. Since it is an odorless gas at room temperature, it must be used in closed or well ventilated systems. Inhalation is the only significant route of toxic exposure.

2.1.2 Physical and Chemical Properties

Physical state	Colorless gas
Molecular weight	50.49
Melting point	$-97.7°C$
Boiling point	$-24.22°C$
Solubility	0.9 g/100 ml water at 20°C; 7.8 g/100 ml ethanol at 20°C; soluble in ethyl ether, chloroform, acetone
Autoignition temperature	634°C
Flash point	Below 0°C
Flammability limits	8 to 19 percent in air

1 mg/liter ≈ 484 ppm and 1 ppm ≈ 2.06 mg/m³ at 25°C, 760 torr

2.1.3 Physiologic Response

Summary. Chronic and subacute exposure predominantly affect the nervous system. Symptoms observed are ataxia, staggering gait, weakness, tremors, vertigo, drowsiness, confusion, personality changes, loss of memory, difficulty in speech, and blurred vision. In severe acute poisoning, gastrointestinal disturbances such as nausea, vomiting, abdominal pain, and diarrhea may be observed. Acute animal experiments have indicated pulmonary congestion and edema. Histopathological changes of the internal organs are not common and laboratory values are unaffected. Literature reviews are available (7, 8).

In a practical sense, the major problem encountered in mild exposure is "drunkenness" or inebriation. The resulting incoordination and impaired judgment may lead to unsafe manual manipulation. The employee may injure himself or endanger others by mechanical misoperation. Because this condition may persist for some time, he is a hazard to himself and others if he drives his car after such exposure. The symptoms may be delayed in onset. They may also continue for some hours after the exposure has stopped. There are indications that with long, severe chronic exposure, symptoms may persist for several months.

Most experience would indicate complete recovery in a matter of hours following acute exposure; however, chronic effects following a massive almost lethal acute exposure have been reported to last as long as 5 to 13 years (9, 10).

The above practical experience on moderate exposure in humans is primarily from operations in which plastic foam cells are opened by sawing or cutting (1, 11). From this experience, The Dow Chemical Company concluded that exposure to fluctuating concentrations essentially below the 100 ppm time-weighted average was well tolerated. However, to provide a margin of safety, a maximum time-weighted average of 50 ppm is currently suggested as a hygienic guide by that company (11).

Single or Short-Term Exposure. *Oral, Eyes, and Skin.*   Since methyl chloride is a gas, ingestion is not of practical concern. Irritation of the skin and eyes has not been a significant problem but freezing due to evaporation could cause frostbite.

Inhalation.   Studies by Sayers et al (12) gave data from acute exposure of guinea pigs. Flury and Zernik (13) reported data from acute exposure of several animals. They also reported limited data from chronic exposure of mice and guinea pigs. They observed injury and deaths from exposure to approximately 3000 ppm for 15 min/day for a varying number of days (between 3 and 100). For the detailed data, the reader should consult the original publication. Table 48.1 summarizes the effects of acute exposure.

Evtushenko (14) reported an LC_{50} for 4-hr exposure of 2760 ppm (5800 mg/m^3) and stated the threshold of response was 230 mg/m^3 (about 110 ppm) based on conditioned reflex. The brains, lungs, kidneys, and livers of the dead animals were said to be markedly injured when examined histologically. This has not been confirmed in extensive, carefully conducted studies discussed below (11).

Repeated and Prolonged Exposure.   Smith and von Oettingen (15) reported data from chronic exposure of animals by inhalation. They exposed their animals 6 hr/day, 6 days/week. Guinea pigs, mice, dogs, rabbits, and rats showed

Table 48.1.   Acute Effects of Methyl Chloride

Single Exposure	Concentrations (ppm)
Kill most animals in a short time	150,000–300,000
Dangerous in 30–60 min	20,000–40,000
Maximum for 60 min without serious effect	7,000
Maximum for 8 hr	500–1,000

injury at 1000 ppm over varying periods up to 175 days. At 500 ppm, rats showed no effects but the other animals, including monkeys and dogs, showed significant response including marked neuromuscular damage and death. At 300 ppm, no effects were observed on any of the animals exposed for 64 weeks. They suggested that the "maximum allowable concentration" should be below 500 ppm.

Studies in which rats and rabbits were exposed 4 hr/day for up to 6 months to 116 ppm (240 mg/m^3) resulted in "derangement of the activity in a number of organs and systems (erythrocytes, liver, kidneys, nervous system)," according to the available translation (14). At 19.9 ppm (40 mg/m^3) no effects were noted until the sixth month of exposure, when change in the erythrocytes and nervous system were reported. Unpublished data (457) indicate that after 90 days of repeated exposure of rats and mice to 1500 ppm there were moderate effects in the liver, at 750 ppm minimal effects, and at 375 ppm no adverse effects. However, after 6 and 12 months of repeated exposure testicular degeneration was also noted in rats exposed to 1000 ppm. This lifetime study is still underway and histopathology has not been completed at lower concentrations.

Teratogenesis, Mutagenesis, and Carcinogenesis.   No reports of studies of teratogenesis or carcinogenesis were found but mutagenic studies in *Salmonella typhimurium* TA 1535 with and without microsomal enzyme activation were positive for reversions (16). Studies of teratogenesis, carcinogensis, and repro duction are reported to be underway (17).

Metabolism and Biologic Monitoring.   The metabolic fate of methyl chloride is not certain but considerable metabolism occurs. Smith (18) was unable to demonstrate methanol in the blood or significant increases of formic acid in the urine. He found no hematologic or biochemical changes.

Sperling et al (19) indicated that intravenously injected methyl chloride rapidly disappeared from the blood but only about 5 percent appeared in the expired air in 1 hr and only small amounts in the bile and urine. Similar results following subcutaneous injection were reported by Soucek (20). However, Bus (21) reported 63.9, 32.2, and 3.9 percent of the radioactivity of inhaled ^{14}C-methyl chloride was excreted by rats in exhaled air, urine, and feces during the first 24 hr. Very little radio-activity remained in the body 24 hr after exposure. The nature of the compound being excreted was not stated.

These results have been confirmed in rats inhaling 1500 or 375 ppm methyl chloride labeled with ^{14}C for 6 hr (personal communication, J. S. Bus). There was a rapid excretion of a small amount of unchanged methyl chloride (4.9 percent) via the lung for about 2 hr ($t_{1/2}$ 0.33 hr). Of the total excreted radioactivity, 60 percent was collected as carbon dioxide, with a longer half time. Urinary excretion was about 35 percent of the total excretion. Residual radioactivity at 48 hr was widely distributed and was not volatile, and hence was not likely to be in the form of methyl chloride, methanol, or formaldehyde.

The blood of rabbits exposed to 40 to 240 mg/m³ (19.4 to 116 ppm) was reported to contain 0.65 to 1.32 mg formaldehyde/100 ml (14), but several other investigators have not found formaldehyde or formate in human urine.

Pharmacokinetic and metabolic data have been reported by Landry et al. (474). According to their abstract,

Male Fischer 344 rats were exposed to methyl chloride (MeCl) for 6 hours. End exposure (apparent steady-state) blood MeCl concentrations were proportionate to exposure concentration in rats exposed to 50 and 1000 ppm. A 2-compartment-1st-order output model was used to describe the blood MeCl data: α and β phase $t\frac{1}{2}$'s corresponded to approximately 4 and 15 minutes respectively. Blood MeCl kinetics did not indicate altered relative rates of metabolism.

Rats exposed for 6 hr to 0, 50, 225, 600 or 1000 ppm ¹⁴C-MeCl were evaluated for tissue nonprotein sulfhydryl (NPSH), total ¹⁴C activity, non-extractable tissue ¹⁴C activity, and urinary metabolites. MeCl induced NPSH depletion was greatest in liver. Liver NPSH was 124, 86, 38 and 13% of control values in rats exposed to 50, 225, 600 and 1000 ppm, respectively. Metabolites in urine included two major ¹⁴C components. One has been identified as n-acetyl S-methyl cysteine. This metabolite is likely to be a product of a reaction between MeCl and glutathione. Kidneys, testis, and epididymis NPSH were less than control values in rats exposed to 225 ppm or higher, but were not diminished in rats exposed to 50 ppm. Total ¹⁴C in liver and kidney was approximately proportionate to exposure concentrations. Relative concentrations of non-extractable ¹⁴C decreased at higher exposure concentrations suggesting a possible dose-dependent metabolic pathway for MeCl in the rat.

It is unlikely that metabolites in the blood or urine will be useful in monitoring exposure to methyl chloride. Expired air falls below detectable levels within minutes after exposure to concentrations considered acceptable for industrial exposure (22).

Observations in Man. *Laboratory Studies.*   In an extensive study with human subjects, Stewart et al. (22) gave males single or repeated exposures to 0, 20, 100, or 150 ppm and females to 0 or 100 ppm. Exposures were generally held at a constant level but in one case was allowed to range from 50 to 150 ppm, averaging 100 ppm. Exposures were 1, 3, or 7½ hr/day, 5 days/week. Using a wide battery of behaviorial, neurological, electromyographic, and clinical chemical tests no significant decrements were found. No increase in methyl alcohol was found in the urine, and methyl chloride in expired air dropped so rapidly as to be of little or no value in quantitating exposure. There was a remarkable difference in individual response, with some subjects consistently showing several times the blood and expired air concentrations as the other subjects. This bimodal distribution has also been reported by Putz et al. (23). These authors reported a yet unconfirmed additive CNS impairment from methyl chloride and Valium® (Diazepam/Roche).

Industrial Experience.   Kegel et al. (24) and McNally (25) reported clinical cases of acute poisoning. The exposure resulted from leaking refrigerators.

Hansen et al. (26) observed the effects of excessive exposure after a spill. Fifteen workers manifested signs of dizziness, blurred vision, incoordination, and gastrointestinal complaints. Recovery was complete in 10 to 30 days. Although rare in the United States, some methyl chloride is still used as a refrigerant in other countries. A 1976 report describes poisoning of four members of a family due to a leaking refrigerator (389).

Klimkova-Dentschova (27) observed the neurological pictures in 100 workers. The report stated: "Involvement of the internal organs (kidney, optic disturbances) was absent even where nervous and mental changes indicated in a severe form of poisoning." Levels of exposure were not indicated.

Numerous other studies describe the effect of acute exposure but reports of chronic low level exposure are less common. One report described the nonspecific nature of six cases and reports many of the symptoms discussed earlier (468). Recovery seemed to occur in all subjects but often several months were required. Another report of eight cases is given by MacDonald (10), who described similar effects, ascribing them to exposures below 100 ppm in one subject and to have resulted in permanent injury in another. However, there is uncertainty in his exposure estimates.

A study of behavioral and neurological effects in 122 white males exposed to methyl chloride for several years has been reported (28). Numerous measurements of task performances, neurological function, electroencephalogram (EEG), and demographic data were collected and analyzed. Extensive industrial hygiene measurements during the study averaged 33.6 ppm; however, the subjects had had exposure to higher concentrations prior to the study period and therefore the study is not useful in establishing a dose–response relationship. Essentially none of the more than 80 parameters studied were related to exposure including the EEG, psychological tests, personality tests, and neurological examinations. Although the authors concluded that "comparison of the exposed to the non-exposed subjects indicated that exposure to methyl chloride adversely affects the performance of cognitive time-sharing tasks and increased the magnitude of finger tremors," the statistical methods used are questionable and the study ads little to knowledge of dose–response relationships.

2.1.4   Hygienic Standards

The threshold limit value for methyl chloride recommended by the ACGIH in 1980 is 50 ppm (103 mg/m^3).

2.1.5   Odor and Warning Properties

Methyl chloride has no odor or other warning property. This, and the fact that the material is a gas at normal temperature, increases the seriousness of the hazard. Charcoal used in most gas mask cartridges and canisters is not totally effective as a sorbant for methyl chloride and other small organic molecules.

2.2 Methyl Bromide, Monobromomethane, Bromomethane, CAS 74-83-9

$$CH_3Br$$

2.2.1 Uses and Industrial Exposures

The largest single use for methyl bromide is as a fumigant used to treat soil, a wide range of grains, and other commodities, mills, warehouses, and houses. The principal problems have been associated with the fumigating personnel and the control of other people who may enter the fumigated area.

Some methyl bromide is used as a chemical intermediate; one of the principal uses is as a methylating agent. A number of older publications indicate that it has been used as a refrigerant but such use is not significant in the United States. It has found use as a fire-extinguishing agent, particularly in automatic equipment for the control of engine fires on aircraft, but because of the toxicity of this material, its use as a fire estinguisher must be limited to such specialized applications. A number of reports of injury from use as a fire extinguisher can be found in the European literature.

Since methyl bromide is a gas at ordinary temperatures and has essentially no warning properties, dangerous concentrations may rapidly accumulate in a work area without warning to the operator.

In industrial operations regularly using methyl bromide, it is advisable to have some kind of warning or monitoring system for continuous analysis of the air. In fumigation operations, suitable analytical equipment is required and personnel must have proper protective equipment for that operation (see Section 2.2.5).

2.2.2 Physical and Chemical Properties

Physical state	Colorless gas
Molecular weight	94.95
Specific gravity	1.732 (0/0°C)
Melting point	−93.66°C
Boiling point	4.6°C
Solubility	0.09 g/100 ml water at 20°C; soluble in ethyl ether, ethanol, chloroform, carbon disulfide, benzene, carbon tetrachloride
Flammability	Practically nonflammable. Flame propagation is in the narrow range of 13.5 to 14.5 percent by volume in air. The ignition temperature is 537°C

1 mg/liter $\approx$ 257 ppm and 1 ppm $\approx$ 3.89 mg/m^3 at 25°C, 760 torr

2.2.3   Physiologic Response

Summary.   Inhalation is by far the most significant route of exposure, although serious skin burns may occur from confined contact, especially under clothing or in shoes and gloves. Unless the concentration is high enough to cause rapid narcosis and death from respiratory failure, the most striking response to exposure at high concentrations will be lung irritation with congestion and edema. These symptoms are observed in both animals and man and often develop into a typical confluent bronchial pneumonia. At lower levels of exposure, this lung condition may account for delayed deaths. If it leads to secondary infection, the delay may be a matter of days.

At threshold concentrations, this lung condition is not observed. The response is almost entirely referable to the nervous system and usually shows up only after prolonged and repeated exposures. Excitation and even convulsions have been observed in animals; but if they survive repeated exposures, the later signs are paralysis of the extremities. Paralysis of the extremities is most typical of threshold toxic response from repeated exposures over a long period of time. Animals that have been seriously paralyzed have recovered, although the recovery is somewhat slow. Human experience indicates that there is a high probability of complete recovery although the time necessary may be quite long, even months.

Owing to the high volatility one can readily attain high concentrations in a work atmosphere. Because methyl bromide essentially has no warning properties, such high concentrations can be attained without recognition. These factors and the fact that methyl bromide is quite toxic create a potentially high hazard. It must be used only by individuals who are well acquainted with proper methods of handling and fully cognizant of the consequences of exposure to excessive amounts.

Single or Short-Term Exposure. *Skin Absorption.*   A very limited study in which only the body of a single monkey was exposed to 5400 to 6000 ppm methyl bromide vapor for 3.5 hr resulted in no detectable increase in bromide ion in the blood immediately, or 24 or 48 hr after exposure (11).

Inhalation.   The 8-hr survival dose for rats is approximately 1 mg/liter (260 ppm). Rats survive 5200 ppm for 6 min and 2600 ppm for 24 min. The 6-hr survival dose for rabbits is approximately 2 mg/liter (520 ppm). They survive 5200 ppm for 6 min and 2600 ppm for 1 hr. These data are taken from Irish et al. (29). These authors studied rats, rabbits, guinea pigs, and monkeys. They described the response of most animals as typically one of lung irritation. If the exposure was severe enough, this resulted in lung edema and usually a typical confluent bronchial pneumonia.

Repeated or Prolonged Exposure. In chronic or repeated exposure to relatively low concentrations of methyl bromide, the picture differs from that of the single exposure to high concentrations. Unless the concentration is high enough to cause lung irritation, the response observed from repeated exposures will be essentially one of paralysis of the extremities as observed by Irish et al (29). Rats, rabbits, guinea pigs, and monkeys responded similarly with some quantitative differences. At 0.42 mg/liter (100 ppm), rats showed a varying pulmonary response from essentially normal lungs to quite severe pneumonia. Guinea pigs failed to show any significant pulmonary changes at this level. They survived up to 98 exposures and showed no histopathological changes. A monkey exposed at this level developed severe convulsions. At 0.25 mg/liter (66 ppm), rats showed essentially no response from a 6-month period of repeated exposures. Guinea pigs survived with similar lack of response. Rabbits, however, developed a characteristic paralysis and some pulmonary damage. Monkeys exposed at this level developed a paralysis comparable to that seen in rabbits. At 0.13 mg/liter (33 ppm), rabbits still showed pulmonary damage and paralysis. Monkeys appeared normal. At 0.065 mg/liter (17 ppm), all the animals survived without indications of response to the exposure.

Teratogenesis, Mutagenesis, and Carcinogenesis. No reports were found describing teratological or carcinogenic studies or bacterial or mammalian mutagenic test systems. Methyl bromide was apparently mutagenic to barley kernels (30).

Metabolism and Biologic Monitoring. Methyl bromide is readily absorbed through the lungs. There have been suggestions that it can be absorbed through the human skin, but experience so far has not shown absorption through the skin to be an important factor in methyl bromide intoxication. Certainly, the major problem is inhalation.

It is probable that excretion also is most predominantly by the lungs as unchanged methyl bromide. A significant amount of methyl bromide, however, is metabolized in the body and appears as inorganic bromide, which is excreted in the urine.

The mechanism of action of methyl bromide is by no means clear. A number of suggestions have been made but certainly nothing conclusively indicated. It would seem that methyl bromide, which is a very active methylating agent, could methylate any one of a number of substances.

Although determination of bromide ion in the blood is useful in establishing whether exposure to methyl bromide has occurred, bromide ion determination is not totally satisfactory for quantitating exposure. Numerous other sources of bromide, such as food, water, and medications, may interfere, particularly at the low blood levels that result from repeated exposure to concentrations considered acceptable for occupational exposure. "Normal" bromide ion concentration is below 1 mg/100 ml of blood serum, in the absence of the above dietary sources. Five milligrams/100 ml may be considered evidence that exposure to bromide from some source has occurred. If recent exposure to methyl bromide has occurred, 15 mg/100 ml of blood is consistent with toxic symptoms. Since bromide from ingested drugs may reach 150 mg/100 ml percent or more, it is obvious that the source of the bromide must be considered in interpreting the analytical results.

Observations in Man. *Skin and Eyes.* Watrous (31) described difficulties in handling methyl bromide in the drug industry. Repeated splashes on the skin resulted in severe skin lesions. Severe cases showed "vesicles or blebs." In a less rigorous exposure, severe itching dermatitis was observed. One report by Longley and Jones (32) describes serious systemic effects following gross skin contact with liquid methyl bromide, but inhalation may have been significant despite the absence of lung damage. Methyl bromide may cause difficulty when it is held in contact with skin by clothes. This is a special problem with gloves and shoes. In the case of shoes, it is suspected that a fairly high concentration of vapor near the floor may actually be absorbed into the leather and cause skin irritation. Spills onto or into a shoe may cause a severe burn if allowed to remain in contact with the skin.

When methyl bromide is handled, care should be taken that splashes do not get onto the clothing or shoes. When there is a fairly high vapor concentration, as in an accidental spill, care should be taken that it does not concentrate in the shoes or the protective clothing.

Liquid methyl bromide can cause severe corneal burns but the vapors do not appear to be irritating. Furthermore, a full-face respirator should always be used if vapor concentrations are significant; hence the eyes will be protected from exposure.

Inhalation. Numerous reports attest to the high toxicity of methyl bromide to humans. Many reports indicate failure to use reasonable recommended handling precautions. The symptoms observed in humans from acute exposure have been reported by a number of authors whose reports have been reviewed by von Oettingen (33). The early symptoms may be a feeling of illness, headache, nausea, and vomiting. Tremors and even convulsions may be observed, much as they are in animals, as well as lung edema and an associated cyanosis. It was indicated by von Oettingen that if the patient survived an acute exposure for the first 2 or 3 days, the probability of complete recovery was very high.

More recent observations of 10 cases acutely and chronically exposed have been summarized by Hine (34) and electroencephalographic studies on seven cases by Mellerio et al. (35).

Interesting experience from chronic exposure to methyl bromide was carefully studied during the early 1940s by several drug companies that were packaging methyl bromide to be used for fumigation of clothing of military personnel for vermin. During the investigations of people exposed in this operation, studies of blood bromide were made and were related to the probability of toxic response. It was indicated that concentrations of more than 10 mg/100 ml percent in the blood were indicative of probable difficulty. Levels of 15 or 20 mg/100 ml were associated with a quite severe response.

If blood bromide is to be used in determining the exposure to methyl bromide, it should be kept in mind that the figures can be very misleading if there are other sources of bromine. The most common are inorganic bromide medications, food, or water.

2.2.4 Hygienic Standards

The threshold limit of methyl bromide recommended by the ACGIH in 1980 is 5 ppm (20 mg/m^3).

2.2.5 Odor and Warning Properties

Methyl bromide has practically no odor or irritating effect and therefore no warning, even at physiologically hazardous concentrations. Some mixtures of methyl bromide used for fumigation contain chloropicrin as another active ingredient or as a warning agent. The chloropicrin may give warning of significant concentrations of methyl bromide from leaking containers; however, experience has shown that chloropicrin vapor may disappear before methyl bromide vapor and therefore the warning properties are lost. Charcoal gas mask canisters also preferentially remove chloropicrin. With prolonged use methyl bromide may penetrate the charcoal in harmful concentrations with no odor of chloropicrin.

2.3 **Methyl Iodide,** Monoiodomethane, Iodomethane, CAS 74-88-4

$$CH_3I$$

2.3.1 Uses and Industrial Exposures

Methyl iodide has had very limited use as a chemical intermediate. It has been proposed as a fire extinguisher and insecticidal fumigant.

2.3.2 Physical and Chemical Properties

Physical state	Colorless liquid
Molecular weight	141.95
Specific gravity	2.279 (20/4°C)

Melting point	$-66.1°C$
Boiling point	$42.50°C$
Vapor pressure	400 torr (25°C)
Refractive index	1.5293 (21.0°C)
Percent in "saturated" air	53 (25°C)
Solubility	1 vol/125 vol water at 15°C; soluble in ethanol, ethyl ether
Flammability	Methyl iodide is nonflammable by standard tests in air

1 mg/liter $\approx$ 172 ppm and 1 ppm $\approx$ 5.8 mg/m^3 at 25°C, 760 torr

2.3.3 Physiologic Response

Summary. Because methyl iodide has not been used extensively, toxicologic investigations and experience have been limited. It appears to be primarily a CNS depressant, but there are also indications of lung irritation and kidney involvement from acute exposures. Since radioactive iodine can be released as methyl iodide, the nuclear industry has investigated its absorption, distribution, and to some extent toxicity when inhaled, ingested, and injected.

Single and Short-Term Exposure. *Oral and Skin.* Buckell (36) determined the LD$_{50}$ for rats dosed either subcutaneously or orally to be 150 to 200 mg/kg of body weight. This author also reported that it would produce a "vestibular burn" if closely held to the human skin. It was indicated that clothing contaminated with methyl iodide should be removed immediately. Johnson reported that the acute oral LD$_{50}$ for mice was 76 mg/kg, but that repeated doses of 30 to 50 mg/kg were without effect (37). Single oral doses of 70 mg/kg were lethal to rabbits but several oral doses of 50 mg/kg were necessary before they caused death (11). Penetration of methyl iodide through the skin (species not indicated) was reported by Shugaev and Mazkova (38), who considered absorption to be of more concern than local irritative effects.
Inhalation. Bachem (39) studied the response of mice to methyl iodide. His work is summarized in Table 48.2 Chambers et al. (40) reported the lethal concentration for rats for a 15-min exposure to be 22 mg/liter in air (3790 ppm). The rats died within a period of 11 days. They showed lung irritation and pulmonary edema. Buckell (36) reported the LC$_{50}$ for mice to be 5 mg/liter in air from a 57-min exposure.

Teratogenesis, Mutagenesis, and Carcinogenesis. No data on teratogenesis were found.
Mutagenic tests on *Salmonella* have been reported by McCann et al. (41) to be weakly positive but details of the study are lacking.

Table 48.2. Acute Effects of Methyl Iodide

Concentration		Response
mg/liter	ppm	
454.4	78,700	Rapid narcosis; death after 10 min exposure
105.1	18,100	Death after 30 min exposure
42.6	7,340	After 15–50 min, side position, no complete narcosis, death 1 hr after beginning of exposure
21.3–31.6	3,670–5,370	Death after 2–2½ hr of exposure
0.43–4.26	73–730	Death of all animals within 24 hr
0.31	54	Death of all animals within 24 hr No marked toxic symptoms

Druckrey et al. (42) reported local sarcomas following weekly subcutaneous injection in BD-strain rats. Strain A mice (a susceptible strain) that were injected with methyl iodide were reported to have a slight but significant increase in the number of lung tumors per mouse (43).

Metabolism and Biologic Monitoring. The metabolism of methyl iodide is not yet certain but there appears to be rapid release of iodide ion and methylation of glutathione in rats dosed orally. The process is enzymatically catalyzed (37). Subcutaneous injection produced similar results (44).

Alveolar absorption of inhaled [132]I-methyl iodide by humans resulted in retention of 53 to 93 percent depending on the concentration inhaled and the rate of respiration (45).

Observations in Man. Garland and Camps (46) reported two cases of exposure of humans to the vapors of methyl iodide from industrial operations. The first case was quoted from Jaquet as reported in 1901. This case showed symptoms of vertigo, diplopia, and ataxia, there was evidence of urinary iodine, and the individual developed delirium and serious mental disturbances. The second case was one observed by Garland and Camps. The patient was found to be drowsy, unable to walk, and with slurred, incoherent speech. He had iodine in the urine. Death occurred 7 to 8 days after exposure. Serious involvement of the central nervous system was also reported in a chemical worker by Baselga-Monte et al. (45) but the route of exposure was not given. Depression and psychological disturbance persisted for several weeks with complete recovery requiring 122 days. A report by Appel et al. (47) of a 41-year-old chemist exposed to the vapor of methyl iodide presents similar symptoms. These authors summarize much of the available literature.

Skin irritation has been reported even while wearing protective gloves (48).

2.3.4 Hygienic Standards

The threshold limit value recommended by the ACGIH in 1980 was 5 ppm (28 mg/m^3) with a skin notation. There appear to be very limited data to support this or any other value.

2.3.5 Odor and Warning Properties

No data were found.

2.4 Methylene Chloride. Dichloromethane, Methylene Dichloride, CAS 75-09-2

$$CH_2Cl_2$$

2.4.1 Uses and Industrial Exposures

Methylene chloride is used as a blowing agent for foams and as a solvent for many applications, including coating photographic films, aerosol formulations, and to a large extent in paint stripping formulations. It is used as a solvent in a number of extraction processes, where its high volatility is desirable. It has high solvent power for cellulose esters, fats, oils, resins, and rubber. It is somewhat more water soluble than other chlorinated solvents.

Owing to its volatility, high concentrations may be rapidly attained in poorly ventilated areas. This property should be recognized in planning any operation using methylene chloride. It should also be remembered that formulations for paint stripping may contain other solvents as well as methylene chloride and that they are frequently found outside the workplace. These formulations often contain other ingredients that retard evaporation and in the process increase the likelihood of skin irritation.

2.4.2 Physical and Chemical Properties

Physical state	Colorless liquid
Molecular weight	84.94
Specific gravity	1.325 (20/4°C)
Melting point	−96.7°C
Boiling point	40.1°C
Vapor pressure	440 torr (25°C)
Refractive index	1.4237 (20°C)
Percent in "saturated" air	55 (25°C)
Solubility	2 g/100 ml water at 20°C; soluble in ethanol, ethyl ether, acetone
Flammability	No flash point or fire point by standard tests in air. The

flammability in oxygen is 15.5–66
volume percent

Autoignition temperature   624°C

1 mg/liter $\approx$ 288 ppm and 1 ppm $\approx$ 3.48 mg/m^3 at 25°C, 760 torr

2.4.3 Physiologic Response

Summary. Methylene chloride is the least toxic of the four chlorinated methanes. The toxic effect is predominantly narcosis, although apparently not effective enough for good anesthesia. Recently it has been recognized that methylene chloride and some other dichlormethanes are metabolized to carbon monoxide. The principal problem from use is the "drunkenness" that may cause inept operation, which may result in injury to the employee or others around him. The symptoms of excessive exposure may be dizziness, nausea, tingling or numbness of the extremities, sense of fullness in the head, sense of heat, stupor, or dullness, lethargy, and drunkenness. Exposure to very high concentrations may lead to rapid unconsciousness and death. Prompt removal from exposure prior to death usually results in complete recovery.

An extensive compilation of literature has been published (49). Industrial experience with methylene chloride has been remarkably free of serious adverse effects. Although dermatitis has been reported due to its common usage in paint remover formulations, only a few anesthetic deaths have occurred, all at very excessive concentrations. Reports of systemic injury are rare, and of questionable authenticity. Epidemiologic studies have been generally favorable. Methylene chloride is known to be metabolized to carbon monoxide but symptoms of carbon monoxide poisoning such as headaches have not been a common feature of methylene chloride exposure. This indicates that carboxy-hemoglobin levels alone are not a good measure of the toxic effect of methylene chloride.

Cardiac arrhythmia does not appear to be a practical problem with methylene chloride (58).

Single or Short-Term Exposure. *Oral.* Methylene chloride has a low to moderate acute oral toxicity in laboratory animals. The LD$_{50}$ for rats and rabbits fed undiluted methylene chloride is about 2000 mg/kg. Anesthetic deaths occurred rapidly after treatment, indicating rapid absorption from the gastrointestinal tract.

Skin and Eye Contact. Methylene chloride is mildly irritating to the skin of rabbits on repeated contact if allowed to evaporate. The problem in humans may be accentuated if the chemical is confined to the skin by shoes or tight clothing. The situation may be more severe with paint remover formulations that form a skin or film to prevent evaporation. Some irritation may be contributed by other constituents of the formulation. Limited absorption through skin occurs but probably is not of toxicologic significance in industrial situations (50). The pain produced by methylene chloride on the skin will limit dermal exposure to some extent.

Inhalation. Svirbely et al. (51) reported that the LD_{50} for mice was approximately 50 mg/liter or 15,000 ppm for an 8-hr exposure. Survival of all animals was observed at approximately 11,000 ppm. Lehmann and Schmidt-Kehl (52) reported levels of narcosis in cats; 32 mg/liter or 9000 ppm caused "displacement of equilibrium" in 20 min but no narcosis. At 37.5 mg/liter or 10,000 ppm, light narcosis occurred in 220 min and deep narcosis at 293 min. They reported that cats and rabbits tolerated 6 to 7 mg/liter for 8 to 9 hr/day for 4 weeks with no significant observable changes. Subsequent studies have confirmed these early data. Plaa and Larson (53) and Gehring (54) investigated the renal and hepatotoxicity of methylene chloride and concluded that the compound had a very low toxicity toward these organs.

Production of carbon monoxide and carboxyhemoglobin has been shown to occur in animals (55, 56) as well as man. The significance is discussed in the section below on observations in man. Cardiac sensitization has been demonstrated in animals only when adrenalin is injected (57); however, exposure to high concentrations was required and the practical significance in industry appears low (58).

Repeated or Prolonged Exposure. Heppel and associates (59) reported no pathology or growth depression in dogs, puppies, rats, guinea pigs, or rabbits exposed to 5000 ppm 7 hr/day, 5 days/week for 6 months. They did not detect depression of the central nervous system by ordinary observation but did not conduct specific tests. At 10,000 ppm they observed light to moderate narcosis. Several animals died, apparently from pulmonary congestion. Heppel and Neal (60) reported experiments with young male rats in an activity cage. At a concentration of 5000 ppm the animals showed definite reduction in activity,

Lifetime exposure of rats and hamsters to 3500, 1500, or 500 ppm 6 hr/day for 2 years resulted in "nonlife-shortening treatment related changes, similar to those seen in aging animals, in the livers of male and female rats at all exposure levels." The details of this study are discussed in the section on carcinogenesis (61). Hamsters were less affected than rats, with survival of female hamsters much higher than in the control group.

Haun et al. (62) summarized the data from exposure studies in which animals were exposed repeatedly and also described the results of continuous (24 hr daily) exposure of rats, mice, monkeys, and dogs for 100 days to either 25 or

100 ppm of methylene chloride vapors. Except for increased carboxyhemoglobin levels, dogs and monkeys were unaffected by this continuous exposure. Mice exposed to 25 ppm were likewise without effect, but the livers of the mice exposed to 100 ppm and rats exposed to both concentrations showed positive fat stains. Cytoplasmic vacuolization was noted in rats as well as degenerative and regenerative changes in the kidneys. Carboxyhemoglobin was more elevated in the monkeys than in the dogs, which showed no elevation at 25 ppm and about 1.5 percent elevation at 100 ppm. Levels in monkeys were increased about 1 percent at 25 ppm and 3 to 4 percent at 100 ppm.

Teratogenesis, Mutagenesis, and Carcinogenesis.   *Teratogenesis.*   Schwetz et al. (63) found no teratogenic effects in rats or mice exposed to 1225 ppm vapor, 7 hr/day on days 6 to 15 of gestation. Hardin and Manson (64) exposed rats to 4500 ppm by inhalation before and during gestation with no teratological effects. Carboxyhemoglobin was elevated to 7 to 10 percent, fetal body weight was decreased, and maternal liver weight was increased by exposure to methylene chloride. The authors concluded exposure to 4500 ppm methylene chloride vapors for 3 weeks before pregnancy plus the first 17 days of pregnancy caused a low degree of maternal and embryotoxicity but no teratological affect. However, the young of similarly exposed pregnant rats were found by Bornschein et al. (65) to show retarded behavioral habitation to novel environments. The authors did not attempt to determine if the effect was due to methylene chloride, to carboxyhemoglobin, or to the combination of the two. They were also hesitant to extrapolate to possible effects at concentrations of 200 to 500 ppm.

Wistar rats fed about 2.8 mg/kg of body weight per day in their drinking water (125 ppm) for 91 days were mated or sacrificed at the end of the exposure period; no adverse effects were noted on estrus or reproduction (66).

Mutagenesis.   Jongen et al. (67) have used methylene chloride in an Ames test with *Salmonella typhimurium* TA98 and TA100. Increased reversions occurred in both strains of bacteria. The activity was only slightly increased by the addition of rat liver homogenate.

Cytogenic aberrations in the bone marrow cells from rats were not increased after 6 months of repeated inhalation exposure to 500, 1500, or 3500 ppm of methylene chloride in the oncogenic study described in the next section (61).

Carcinogenesis.   Carcinogenic studies have been completed. Methylene chloride was not positive in a pulmonary tumor assay in Strain A mice (68). Rampy et al. (61) exposed Sprague-Dawley rats and Golden Syrian hamsters by inhalation to 3500, 1500, 500, or 0 ppm of methylene chloride, 6 hr/day, 5 days/week for 2 years. Interim sacrifices with clinical and pathological studies were carried out at 6, 12, 15 (rats only), and 18 months of the study, with a final sacrifice at

24 months. Animals dying during the study or sacrificed in a moribund condition were also examined. The authors' summary of this extensive study follows:

Female rats exposed to 3500 ppm had exposure-related increased mortality during the last 6 months of exposure; in contrast, female hamsters exposed to 1500 or 3500 ppm had decreased exposure-related mortality.

The liver appeared to be the primary target organ in the rat with slight exposure-related effects in both male and female rats exposed to 500, 1500 or 3500 ppm. Histopathologically, the liver effects in rats appeared to progress slightly between 12 and 18 months of exposure but did not progress further between 18 and 24 months. Also, liver weights were increased in male and female rats exposed to 3500 ppm at 18 months but not at 6, 12 or 24 months of exposure. Despite the presence of liver toxicity, there was no increase in malignant liver cell tumors in any exposure group.

A primary target organ was not identified in hamsters of either sex. The effects observed in male and female hamsters exposed to 500, 1500 or 3500 ppm of methylene chloride were primarily the result of decreased amounts of amylod (a naturally occurring geriatric disease).

During the course of the study, several parameters were evaluated including hematology, urinalysis, clinical chemistry and carboxyhemoglobin determinations. Slight increases in the mean corpuscular volumes (MCV) and mean corpuscular hemoglobin (MCH) were present in male and female rats, but not in the hamsters. Carboxyhemoglobin values were elevated in male and female rats and hamsters exposed to 500, 1500 or 3500 ppm with the percentage increase in hamsters greater than in rats.

While the number of female rats with a benign tumor was not increased, the total number of benign mammary tumors was increased in female rats exposed to 500, 1500 or 3500 ppm of methylene chloride. The total number of benign tumors increased in an exposure-related manner with 165 in the controls and 218, 245 and 287 in females exposed to 500, 1500 or 3500 ppm, respectively. This effect was evident in male rats, but to a lesser extent than in females and with apparent effects only in the 1500 and 3500 ppm exposure groups. This effect was not present in either male or female hamsters. Despite the observed increase in benign mammary tumors, there was no increase in malignant mammary tumors in any group exposed to methylene chloride.

Finally, male rats exposed to 1500 or 3500 ppm appeared to have an increased number of sarcomas (malignant mesenchymal tumors) in the ventral neck region that were in or around the salivary glands. This effect was statistically significantly increased in the 3500 ppm group with a trend (numerically increased but statistically not significantly increased) for an increase in the 1500 ppm group. The toxicological significance of this finding is unknown.

Therefore, in this two-year study, adverse effects were observed in male and female rats exposed to 500, 1500 or 3500 ppm of methylene chloride. In contrast, hamsters exposed to the same exposure concentrations did not have definite signs of toxicity. In fact, they had less extensive spontaneous geriatric changes, had exposure-related decreased mortality (females) and lacked evidence of definite target organ toxicity.

In our opinion methylene chloride does not present a practical risk of carcinogenesis in humans at currently acceptable levels of exposure. Several studies are currently underway in laboratory animals.

Metabolism and Biologic Monitoring.   Von Oettingen and associates (69) studied the absorption and excretion of methylene chloride. They reported that it was rapidly absorbed and largely excreted by the lungs. Although this appears to be partly true, a significant amount is metabolized to carbon dioxide and carbon monoxide; hence there has been considerable investigation of the absorption, excretion, and metabolism of methylene chloride in animals and man. It appears certain that the carboxyhemoglobin found in animals following exposure contains the carbon of the methylene chloride molecule (70, 71); that carboxyhemoglobin levels persist slightly longer after exposure ceases than carboxyhemoglobin levels from carbon monoxide itself (72); that absorption and metabolism are influenced by the concentration inhaled, following Michaelis–Menten kinetics in the rats (73, 74), and that concurrent exposure to toluene, ethanol, methanol, and isopropanol inhibited carboxyhemoglobin formation in animals (75).

Quantitation of the kinetic parameters for the formation and elimination of carboxyhemoglobin shows that in the rat saturation kinetics do influence its formation from methylene chloride exposures above 200 to 250 ppm. A steady-state carboxyhemoglobin concentration is not reached in men during an 8-hr exposure (11) and equilibrium requires at least 13 to 15 hr. The influence of methylene chloride on oxygen dissociation is not physiologically significant at concentrations of 500 ppm or less (11).

DiVincenzo et al. (76) and Stewart et al. (72) discussed biologic monitoring for methylene chloride in expired air. The latter report includes graphs of expired air concentrations following exposures to known concentrations of methylene chloride. Within certain limit these may be useful in determining the exposures of workers.

Observations in Man.   Experience in use of methylene chloride has generally been favorable, confirming the low toxicity observed in animals. Only six reports of deaths have been found, each the result of gross overexposure. Most commonly, recovery even from anesthetic concentrations has been rapid and without sequelae other than carboxyhemoglobin, which appears to persist somewhat longer than that resulting from carbon monoxide itself.

Several laboratory and epidemiologic studies in man have been reported. An extensive laboratory study of healthy adult test subjects exposed repeatedly 7½ hr/day to 250 ppm showed no untoward subjective or objective health responses (72). McKenna et al. (77) exposed six healthy male volunteers "to 100 and 350 ppm methylene chloride (CH_2Cl_2) during each of two 6-hr exposure periods. Measurement of blood CH_2Cl_2 and carboxyhemoglobin (COHb) levels and CH_2Cl_2 and carbon monoxide (CO) in expired air were performed during exposure and for the first 24 hr thereafter. Dose-dependent metabolism of CH_2Cl_2 was evident from both CH_2Cl_2 blood levels and the concentration of CH_2Cl_2 in the expired air when the data from the 100 and 350 ppm exposure experiments were compared. Blood COHb levels and exhaled CO were less than expected following exposure to 350 ppm CH_2Cl_2, indicative of dose-dependent or saturable metabolism of CH_2Cl_2 to CO. The rate of total CH_2Cl_2 metabolism for each subject was calculated from the inspired–expired air concentration and minute ventilation rate obtained at apparent steady-state. The relationship of the rate of CH_2Cl_2 metabolism during exposure to the CH_2Cl_2 exposure concentration followed apparent Michaelis-Menten kinetics." These authors derived a pharmacokinetic model to allow prediction of the extent of methylene chloride metabolism and production of carboxyhemoglobin following exposure to methylene chloride in laboratory animals and man.

An epidemiologic mortality analysis of a male population exposed to methylene chloride up to 30 years has been reported (78). Proportionate mortality ratio (PMR) analysis revealed no significant excess cause of death by either major diagnostic grouping or by malignancy subgrouping. No significant excessive standardized mortality ratios were found. However, the authors caution that the size and duration of the study limit its interpretation.

2.4.4 Hygienic Standards

The ACGIH recommended a reduction to a TLV of 100 ppm (360 mg/m^3) in 1980. This level appears unnecessarily low to prevent excessive concentrations of carboxyhemoglobin and certainly is lower than needed to prevent beginning anesthetic effects which occur after the exposure to about 1000 ppm for most people.

2.4.5 Odor and Warning Properties

Methylene chloride has a "not unpleasant," sweetish odor at concentrations above 300 ppm, but at about 1000 ppm the odor becomes unpleasant for most people. At 2300 ppm the odor is strong, intensely irritating; there may be dizziness after 5 min exposure.

2.5 Methylene Chlorobromide, Monobromomonochloromethane, CB, BCM, Halon 1011, CAS 74-97-5

$$CH_2ClBr$$

2.5.1 Uses and Industrial Exposures

Methylene chlorobromide has limited use as a fire extinguishing agent and to an even lesser extent as a chemical intermediate.

2.5.2   Physical and Chemical Properties

Physical state	Colorless liquid
Molecular weight	129.40
Specific gravity	1.991 (19/4°C)
Freezing point	$-88°C$
Boiling point	68 to 69°C
Vapor pressure	155 to 160 torr (25°C)
Refractive index	1.4850 (26°C)
Percent in "saturated" air	21 (25°C)
Solubility	Insoluble in water; soluble in organic solvents
Flammability	No flash or fire points by standard tests in air. It is an effective fire extinguisher

1 mg/liter $\approx$ 189 ppm and 1 ppm $\approx$ 5.3 mg/m^3 at 25°C, 760 torr

2.5.3   Physiologic Response

Summary.   Methylene chlorobromide is one of the least toxic halomethanes. It falls roughly in a class with methylene chloride. The primary response to this material is CNS depression. There appears to be very little organic injury following either acute or chronic exposure except possibly lung irritation from acute exposure at high levels. The compound is metabolized to carbon monoxide and produces carboxyhemoglobin as well as inorganic bromide. A review of the toxicity is available (469).

Single or Short-Term Exposure. *Oral.*   Highman et al. (79) reported that administration of methylene chlorobromide by stomach tube to mice caused no changes at doses of 500 mg/kg. Single doses of 3000 and 4500 mg/kg were followed by fatty degeneration of the liver and kidneys. These same authors observed no liver or kidney injury in animals exposed to the vapors. They commented that the difference in liver injury could be due to a different pathway of absorption from the gastrointestinal tract than from the lungs.

Single oral doses of 1 g/kg of body weight or less have no apparent effect on rats. An oral dose of 3 g/kg of body weight kills most rats within 24 hr (1).

Skin and Eyes.   Methylene chlorobromide, when applied repeatedly to the open skin of rabbits, resulted in some hyperemia and exfoliation. When bandaged on, it produced moderate irritation and hyperemia. This material would be expected to have a rather mild effect from ordinary contact, but gross and prolonged contact might cause dermatitis (80).

Inhalation.   Svirbely et al. (51) reported the LC_{50} by inhalation for mice exposed for 7 hr to be 12.03 mg/liter of air (2273 ppm). Comstock et al. (81) reported that concentrations as low as 3000 ppm produced light narcosis in rats. Transient pulmonary edema was observed at concentrations below 27,000 ppm. At higher concentrations, interstitial pneumonitis resulted in delayed deaths. Delayed deaths were also observed after exposure to 20,000 ppm. Deaths during exposure occurred only from exposures above 27,000 ppm. Matson and Dufour (82) reported limited acute studies on guinea pigs. The principal toxicologic observation from the exposure was lung injury. Van Stee (83) exposed dogs to 0.3 to 1.0 percent in oxygen in order to determine the effect on the cardiovascular system. Disturbances in myocardial energy metabolism occurred, including cardiac arrhythmias, but the studies were conducted on anesthetized animals and the industrial significance cannot be ascertained.

Repeated or Prolonged Exposure.   Repeated exposures by inhalation were reported by Svirbely et al. (51). They exposed animals 7 hr/day, 5 days/week for a period up to 14 weeks to a concentration of 1000 ppm. Rats, rabbits, and dogs survived these exposures and showed no evidence of toxic response. Growth was normal and there were no histopathological changes.

Torkelson et al. (84) indicate that female rats and dogs survived without significant effect, 370 ppm in air, 7 hr/day, 5 days/week for 6 months, but that some liver pathology was observed at 500 ppm. Male rats, male and female guinea pigs, and rabbits showed no effect except for elevated blood bromide at 500 ppm. However, at 1000 ppm several effects were noted including histopathological changes in the livers and testes in addition to increased blood bromide.

In a similar study, rats and dogs were exposed to methylene chlorobromide vapors 6 hr/day for a total of 124 exposures in a 6-month period (85). The exposure levels of 500 and 1000 ppm produced no adverse effect except for an apparent failure of rats to gain as much weight as their controls. The authors suggested that the lower weight might have been related to sedation due to the elevated levels of bromide ion in the blood. In about 20 days serum bromide levels in both rats and dogs reached an equilibrium of about 150 mg/100 ml in dogs and 140 mg/100 ml in rats exposed to 1000 ppm. Exposures to 500 ppm resulted in equilibrium concentrations of 125 and 100 mg/100 ml in the two species.

Teratogenesis, Mutagenesis, and Carcinogenesis.   No data on teratogenicity or carcinogenicity were found. When tested in *Salmonella typhimurium* and *Saccharomyces cerevisiae D3*, methylene chlorobromide did not cause a mutagenic response (11).

Metabolism and Biologic Monitoring.   Svirbely et al. (51) determined inorganic bromide in blood serum and urine in dogs exposed to 1000 ppm of methylene chlorobromide in air. These animals were exposed 7 hr/day, 5 days/week. During the third week, the blood serum inorganic bromide had increased from a normal of 5 to 10 mg/100 ml to more than 200 mg. By the thirteenth and fourteenth weeks, the concentration was greater than 300 mg of inorganic bromide per 100 ml of blood. The same authors determined the blood concentration of volatile bromide expressed as milligrams of methylene chlorobromide per 100 ml of blood. Taken immediately at the end of the exposure, concentrations between 5 and 9 mg of methylene chlorobromide per 100 ml were observed. At periods of 17 to 65 hr after the end of the last exposure, no volatile bromide was observed in one dog and concentrations less than 1 mg in the other. It would appear that methylene chlorobromide as such appears in the blood during exposure to vapors in air, but disappears rapidly on cessation of exposure. Apparently, a significant amount of material is hydrolyzed to yield inorganic bromide, which may be useful to determine if exposures to methylene chlorobromide have occurred.

Female rats exposed to 370 ppm 8 hr/day, 5 days/week, had a whole blood bromide ion concentration of about 70 mg/100 ml; dogs similarly exposed were found to have concentration of 25 to 30 mg Br^-/100 ml whole blood (84). Although not studied as extensively as methylene chloride, methylene chlorobromide also produces carboxyhemoglobin (70). Intraperitoneal doses of 3 mmol/kg (390 mg/kg) to rats resulted in a maximum carboxyhemoglobin of about 5 percent at 4 hr (methylene chlorobromide) and 8 percent at 2 hr (methylene chloride). Prior administration of phenobarbital, 3-methylcholanthrene, and SKF 525-A did not alter these levels of carboxyhemoglobin. These authors found elevations in carboxyhemoglobin with several dihalomethanes but not with other chloromethanes including carbon tetrachloride, chloroform, and monochloromethane, or with carbon disulfide, methanol, formaldehyde, dimethoxymethane, trichlorofluoromethane, and dichlorodifluoromethane.

Observations in Man.   Few reports of adverse effects in human from exposure to methylene chlorobromide are found in the literature. This is probably because of its low toxicity as well as its limited usage. One report describing exposure was to a mixture of materials including methyl chloride (86). The neurological effects described fit more closely those expected from methyl chloride than from methylene chlorobromide. A second report by Rutstein (80) describes an incident in which three men were grossly exposed during its use and misuse as a military fire extinguisher. All survived, but two required extensive therapy to prevent anesthetic deaths. The third victim, who was sprayed in the face with the liquid for over 40 sec did not lose consciousness. All subsequently recovered with no evidence of persistent sequelae. Liver biopsies of the first two victims revealed normal microanatomy, and liver function studies were normal a few days after exposure.

2.5.4   Hygienic Standards

The threshold limit value for methylene chlorobromide recommended by the ACGIH in 1980 is 200 ppm (1040 mg/m^3). The same figure, 200 ppm, was recommended by Torkelson et al. (84), who also recommended that a ceiling of 400 ppm be maintained to avoid beginning anesthetic effects.

2.5.5   Odor and Warning Properties

Methylene chlorobromide has a distinctive odor at 400 ppm, and hence has warning below any acutely hazardous concentration. Although the odor is distinctive at the acceptable concentration and so gives some warning, it is not disagreeable enough to drive anyone from the area. Employees may tolerate a level well above the acceptable level for chronic exposure.

2.6   Methylene Bromide, Dibromomethane, CAS 74-93-3

$$CH_2Br_2$$

2.6.1   Uses and Industrial Exposure

Methylene bromide has limited use as a solvent and chemical intermediate.

2.6.2   Physical and Chemical Properties

Physical state	Liquid
Molecular weight	173.82
Specific gravity	2.490 (25/25°C)
Melting point	$-52°C$
Boiling point	99°C (760 torr)
Vapor pressure	48 torr (25°C)
Refractive index	1.5381 (25°C)
Percent in "saturated" air	6 (25°C)
Solubility	1.18 g/100 ml water at 20°C; soluble in alcohol, ether, acetone
Flammability	Not flammable by standard test in air

1 mg/liter ≈ 136.7 ppm and 1 ppm ≈ 7.31 mg/m^3 at 25°C, 760 torr

2.6.3   Physiologic Response

Summary.   Methylene bromide is more toxic than either methylene chloride or methylene chlorobromide. It is metabolized to carbon monoxide and bromide, and has the capacity of producing significant liver and kidney injury in animals on repeated exposure. The vapors are anesthetic and may cause cardiac arrhythmias (87).

Single or Short-Term Exposure. Methylene bromide has a low acute oral toxicity, the LD_{50} for rats being greater than 1000 mg/kg. It is only slightly irritating to the eyes and skin of rabbits and does not appear to be absorbed significantly even when applied repeatedly (11). It is sufficiently volatile so that inhalation of the vapors can cause anesthesia and even death. A concentration of 17 to 20 mg/liter (2400 to 2800 ppm) caused "disorders in the central nervous system" (88). The duration of the exposure was not stated in the available abstract.

Repeated or Prolonged Exposure. *Oral.* In a very limited study methylene bromide was administered orally to a small group of rabbits at the rate of 300 mg/kg per day (60 doses in 92 days) with no alteration in weight gain, general appearance, or histopathology of the liver. Similar treatment with 400 mg/kg or more produced marked anesthesia (11).

Inhalation. In limited studies in which 10 rats and one rabbit of each sex were exposed to a nominal concentration of 1000 ppm (900 to 1000 ppm recovered analytically), there was no overt evidence of adverse effect in the rabbits, but liver and kidney degeneration were observed at autopsy following 54 exposures in 73 days (11). Blood bromide was elevated. Rats were much more affected. Incoordination and staggering were apparent during exposure. Failure to gain weight, possible increased mortality, and histopathological changes in the lungs, liver, and kidneys were observed in rats receiving 30 to 40 7-hr exposures.

In a second equally limited study by this same group, 79 exposures in 114 days to 200 ppm resulted in much less effect but evidence of stress was still present. The weight of the livers of male rat was elevated when compared to the controls, and histological changes were found in the livers and kidneys of rats and rabbits.

When inhaled 4 hr/day for 2 months, 0.25 mg/liter (35 ppm) methylene bromide was reported to cause "disorder in the protein-prothrombin and glycogenesis functions of the liver and the filtration capacity of the kidneys." It was considered less toxic than bromoform (89). The same investigator reported that 2.5 mg/liter of the vapor for 10 days (duration of exposure not stated) when inhaled by rabbits produced the same effects with dystrophic changes in these organs. A "threshold" concentration of 0.23 mg/liter (32 ppm) was claimed for chronic exposures. When doses of 100 to 200 mg/kg were injected subcutaneously for 10 days in guinea pigs, pronounced liver and kidney effects were reported.

Dykan's report is not consistent with a more extensive study on rats and dogs.

Sixty to 70 repeated 6-hr daily exposures were given to 25, 75, or 125 ppm in a 90-day period. Blood bromide was markedly elevated, as was carboxyhemoglobin. No gross or histological changes were seen in rats or dogs sacrificed at the end of the exposure period (11).

Teratogenesis, Mutagenesis, and Carcinogenesis. Methylene bromide has been tested in three strains of *Salmonella typhimurium* activated with liver homogenate and in *Saccharomyces cerevisiae* D3. It was at most weakly mutagenic in *Saccharomyces* and negative in *Salmonella* (11).

Metabolism and Biologic Monitoring. Methylene bromide was determined in the plasma of dogs following repeated 6-hr exposures to either 25, 75, or 150 ppm for 90 days. The plasma clearance was at least biphasic with an ill-defined alpha phase and a terminal phase ($t\frac{1}{2}$ of 103 ± 14 min) at all three concentrations. Blood bromide was elevated markedly in rabbits exposed to about 1000 ppm of the vapors, 7 hr/day for a month. After 54 exposures in 73 days the blood levels had reached 280 to 315 mg Br/100 ml of blood. After 79 exposures in 114 days to a vapor concentration of about 200 ppm, blood bromide levels of 85 to 100 mg Br/100 ml were found in rabbits (11).

Methylene bromide, like many of the dihalomethanes, is metabolized to produce carbon monoxide (70, 90, 55, 91). Intraperitoneal injection of 520 mg/kg of methylene bromide in corn oil elevated carboxyhemoglobin in rats to 14 percent 5 hr after injection (70). Repeated daily injection did not result in any significant difference in carboxyhemoglobin concentrations from single injection. In vitro studies (71) show that microsomal enzymes in the presence of NADPH and molecular oxygen convert methylene bromide to carbon monoxide and inorganic bromide.

Blood bromide and carboxyhemoglobin determinations may be of value in assessing whether exposure to methylene bromide has occurred. However, available data are inadequate to quantitate exposure.

Observations in Man. No data concerning human experience were found.

2.6.4 Hygienic Standard

No TLV has been recommended. A level of 25 ppm was recommended as a maximum by Turner (92). Dykan (88) recommended 0.01 mg/liter (1.4 ppm). No official standards were found but the recommendation of Dykan appears unnecessarily low.

2.6.5 Odor and Warning Properties

No data were found describing warning properties.

2.7   Chloroform, Trichloromethane, CAS 67-66-3

$$CHCl_3$$

2.7.1   Uses and Industrial Exposures

Chloroform was widely used for many years as an anesthetic. Owing to resultant, often delayed, liver injury and to a lesser extent to cardiac sensitization, this use is now considered obsolete. It has some use as a solvent; however, most is used as a chemical intermediate. Although the use of chloroform as a solvent in industry is not extensive, it may be found as a constituent in solvent mixtures. It is still commonly used as a laboratory solvent. Chloroform has some insecticidal value, although it is not widely used as a fumigant.

   Although the resultant levels are low, a few parts per billion, chloroform can be produced in the process of chlorinating water. Higher quantities can be produced during chlorination of sewage. Some chloroform appears to be produced by microorganisms in soil and water. It has been identified in fresh codfish (390).

2.7.2   Physical and Chemical Properties

Physical state	Colorless Liquid
Molecular weight	119.39
Specific gravity	1.49845 (15/4°C)
Melting point	$-63.5°C$
Boiling point	61.26°C
Vapor pressure	200 torr (25°C)
Solubility	1.0 g/100 ml water at 15°C; soluble in ethanol, ethyl ether, benzene, acetone, CS_2
Flammability	Not flammable by standard tests in air

   1 mg/liter $\approx$ 206 ppm and 1 ppm $\approx$ 4.89 mg/m^3 at 25°C, 760 torr

2.7.3   Physiologic Response

Summary.   Much of our toxicologic information has been developed because of the interest in chloroform as an anesthetic. The literature is replete with papers on anesthetic potency and liver injury. Most of these have limited value to the industrial toxicologist.

   High concentrations of chloroform result in narcosis and anesthesia. The most outstanding effect from acute exposure is depression of the central nervous system. Responses associated with exposure to concentrations below anesthetic or preanesthetic level are typically inebriation and excitation passing into narcosis. Vomiting and gastrointestinal upsets may be observed. Exposure to high concentrations may result in cardiac sensitization to adrenalin and similar compounds as well as liver and kidney injury. In cases of more chronic or repeated exposure to chloroform, liver injury is most typical. This is not unlike the effect of carbon tetrachloride. Although injury to the kidney is not as common as that to the liver, it may be observed from either acute or chronic exposure.

Numerous reviews are available, including, those by Zimmerman, 1968 (93); Challen et al., 1958 (94); Scholler, 1968 (95); Van Dyke et al., 1964 (96); NIOSH, 1974 (97); von Oettingen, 1955 (33); and Winslow and Gerstner, 1977 (98). The last-named review, with its 1387 references indicates the amount of research that has been done on chloroform.

Single or Short-Term Exposure. *Oral.* When chloroform was fed undiluted to male rats, an LD_{50} of 2000 mg/kg was determined (99). Deaths generally occurred in 2 to 4 hr but some were delayed as long as 2 weeks after treatment. Gross pathological examination showed liver and kidney changes at doses as low as 250 mg/kg. Gerbig and Warner (11) determined a LD_{50} of 1060 mg/kg in female rats. Clinical signs, including depression, profound sleep, dyspnea, anorexia, and hematuria, began 20 min after treatment and continued for several days. The last death occurred 8 days post-treatment. Thompson et al. (100) fed pregnant rats 20 mg/kg per day (10 doses) without effect on the dams. Fifty milligrams per kilogram appeared to cause fatty changes. These data are discussed in more detail in the section on teratology that follows.

Skin and Eyes. Oettel (101) indicated that chloroform was more irritating to the skin and eyes than many other chlorinated solvents. Oettel's conclusions have been confirmed by Torkelson et al. (99). One or two 24-hr applications on the skin of rabbits resulted in hyperemia and moderate necrosis. Healing of abraded skin appeared to be delayed by application of a cotton pad soaked in chloroform. Absorption through the intact skin of rabbits occurs as indicated by weight loss and degenerative changes in the kidney tubules, but doses as high as 3980 mg/kg were survived. When the liquid was instilled in the eyes of rabbits some corneal injury was evident in addition to conjunctivitis. The authors concluded that chloroform was more irritating to rabbit skin and eyes than most common organic solvents tested by the same technique in their laboratory.

Inhalation. Man and animals withstand very high concentrations of chloroform for a short period of time. Table 48.3 gives indications of the response to be expected in man (102). The response to acute exposure has been indicated in

Table 48.3. Physiologic Response to Various Concentrations of Chloroform in Man

Concentration		Response
mg/liter	ppm	
70–80	14,336–16,384	Narcotic limiting concentration
20	4096	Vomiting, sensation of fainting
7.2	1475	Dizziness and salivation after a few minutes
5	1024	Dizziness, intracranial pressure, and nausea after 7 min
5	1024	Definite after-effects; fatigue and head-ache still felt later
1.9	389	Endured for 30 min without complaint
1–1.5	205–307	Lowest amount that can be detected by smell

the summary as CNS depression, liver and kidney injury, and possible cardiac sensitization. At the levels of exposure currently considered acceptable for industrial workers, anesthesia is not of practical concern.

Repeated or Prolonged Exposure. *Ingestion.* Several studies describing long-term oral administration of chloroform are available, but they were designed to evaluate carcinogenic response and have very limited value in evaluating noncarcinogenic parameters. They are discussed in the section on teratogenesis, mutagenesis, and carcinogenesis.

Inhalation. Despite its long usage few reports of repeated exposure of laboratory animals are available. Repeated 7-hr exposures 5 days/week for 6 months to either 85, 50, or 25 ppm of the vapor of chloroform resulted in adverse effects in all or some of the species studied, rats, rabbits, guinea pigs, or dogs. The effects at 25 ppm were slight and reversible. Rats exposed 4, 2, or 1 hr/day were not adversely affected. Cloudy swelling of the kidneys and central lobular granular degeneration and necrosis of the liver are the principal adverse effects (99).

Teratogenesis, Mutagenesis, and Carcinogenesis. *Teratogenesis.* Chloroform appears to be unique among the smaller chlorinated aliphatics in that it is the only one that appears to be somewhat teratogenic and highly embryotoxic in animals. Schwetz et al. (103) exposed pregnant Sprague-Dawley rats 7 hr/day to either 0, 30, 100, or 300 ppm chloroform vapor. Exposures were given on days 6 to 15 of gestation and Caesarean sections made to evaluate embryonal and fetal development. According to the authors, "Exposure to chloroform caused an apparent decrease in the conception rate and a high incidence of fetal resorption (300 ppm), retarded fetal development (30, 100, 300 ppm), decreased fetal body measurements (30, 300 ppm) and a low incidence of acaudate fetuses with imperforate anus (100 ppm). Chloroform was not highly teratogenic but was highly embryotoxic. The results of this study disclosed no relationship between maternal toxicity and embryo or fetotoxicity as the result of exposure to chloroform by inhalation."

In another report on inhalation studies by Murray et al. (104), "The effect of inhaled chloroform on embryonal and fetal development was evaluated in CF-1 mice. Pregnant mice were exposed to 0 or 100 ppm of chloroform for 7 hr/day from days 6 through 15, 1 through 7, or 8 through 15 of gestation. Exposure to chloroform from days 6 through 15 or 1 through 7 produced a significant decrease in the incidence of pregnancy but did not cause significant teratogenicity. In comparison a significant increase in the incidence of cleft palate was observed among the offspring of mice inhaling chloroform from days 8 through 15 of gestation, but no effect on the incidence of pregnancy was discerned. A significant increase in SGPT activity was observed in mice exposed to chloroform from days 6 through 15 of gestation; among the mice exposed to chloroform, bred mice which were nonpregnant had a significantly higher SGPT activity than pregnant mice." Similar results were reported by Dilley et al. (105), who exposed pregnant rats to 20.1 mg/liter of chloroform vapor (duration of daily exposure not stated) and produced fetal mortality, decreased fetal weight gain but no teratological effects.

Thompson et al. (100) also failed to produce teratogenic effects in Sprague-Dawley rats by intubation of 0, 30, 50, or 126 mg/kg per day or in Dutch belted rabbits given doses of 0, 20, 35, or 50 mg/kg per day. These authors state: "The occurence of anorexia and weight gain suppression in dams of both species, as well as subclinical nephrosis in the rat and hepatotoxicity in the rabbit, indicated that maximum tolerated doses of chloroform were used. Fetotoxicity in the form of reduced birth weights was observed at the highest dose level in both species. There was no evidence of teratogenicity in either species at any dose tested." It appears that chloroform has more fetotoxic effect when inhaled than when given by gavage. Thompson et al. (100) speculate that doses given by gavage may result in different blood levels of chloroform which account for the apparent discrepancy with the effects seen on inhalation.

Mutagenesis. Chloroform failed to produce mutagenic changes in cultures of Chinese hamster lung fibroblast cells (106).

Carcinogenesis. The available data from carcinogenic studies in mice were summarized by IARC in 1972 (128). Chloroform has since been studied in the NCI bioassay program (107). The relationship of this study in which the maximum tolerated dose was given by repeated gavage to industrial exposure in which vapors are inhaled has been questioned (108, 109). Osborne-Mendel rats and B7C3F1 mice were used in this study. Male rats (but not female) developed kidney epithelial tumors at 180 and 90 mg/kg per day. Mice of both sexes developed hepatocellular carcinomas at dosages of 138 and 277 mg/kg per day (male mice) and 238 and 477 mg/kg per day (female mice). Limited data are available from long-term studies on rats, mice and dogs fed a toothpaste base containing chloroform (110–112). Rats, mice, and dogs were fed lower doses than those used in the NCI bioassay study. Table 48.4 summarizes the studies. The males of only one of four strains of mice developed excess tumors. No excess was found in the females of any strain, nor in the dogs and rats.

The tumors, adenomas, occurred in the renal cortex of male mice fed 60 mg/kg per day but not those fed 17/mg/kg per day. The significance of these tumors is not clear since they had not spread to other organs and have subsequently been observed in control mice of this same strain (ICI-Swiss).

The metabolic relationship of the data from high dose studies in animals to man as discussed by Reitz et al. (109) is presented in the section on metabolism.

Metabolism and Biologic Monitoring.   There are so many references to the absorption, excretion, and metabolism of chloroform that understandably the

Table 48.4.   Summary of Carcinogenicity Studies on Chloroform Carried out at Huntingdon Research Centre[a]

Species and Strain	Dose (mg CHCl₃/kg/day)[b]	No. of Animals		Duration (weeks)	Excess of Neoplasms
Rat (Sprague-Dawley)	60	50	50	95	None
Mouse					
ICI-Swiss	17	52	52	96	None
	60	52	52	96	Renal tumors none
C57BL	60	52	—	104	None
CBA	60	52	—	104	None
CF/1	60	52	—	93	None
Dog (beagle)	15	8	8	7½ yr + 20 weeks observation	None
	20	8	8		None

[a] Adapted from Roe *et al.* (110), Palmer *et al.* (111), and Haywood *et al.* (112).
[b] 6 days/week.

Table 48.5. Interspecies Comparisons: Chloroform[a]

Species	Oral Dose (mg/kg)	CHCl$_3$ Excreted Unchanged (% of dose)
Mouse	60	6[b]
Rat	60	20[b]
Monkey	60	78[b]
Man	7	17–66[c]

[a] From Reitz et al. (109).
[b] Brown et al., Xenobiotica, **4**, 151 (1974).
[c] Fry et al., Arch. int. Pharmacodyn. Ther., **196**, 98 (1972).

data are at times contradictory. There is no question that chloroform is rapidly absorbed through the lungs, gastrointestinal tract, and to some extent the skin; that some is metabolized; and that some is excreted unchanged in the expired air. However, as Irish commented in 1962, the mechanisms and paths of metabolism are still not certain. Van Dyke (113) considers both enzymatic and nonenzymatic metabolism to be important.

Based on similar compounds, it might be expected that the route of metabolism will be dose dependent, but this does not appear to have been investigated systematically. Differences in doses may account for the apparent discrepancies in routes of metabolism proposed by various investigators. Table 48.5, taken from Reitz et al. (109) illustrates wide differences between species in their ability to excrete unchanged chloroform.

Although the toxicity of chloroform and carbon tetrachloride has been ascribed to their solubility in cellular lipid, metabolism appears necessary to explain their toxicity. The comparative metabolism of carbon tetrachloride and chloroform has been reviewed by Hathway (114) who included the work of Recknagel (115) in his proposed metabolic path. According to Hathway's review, carbon tetrachloride must react to form the free radical ($\cdot$CCl$_3$), since chloroform and CCl$_3$–CCl$_3$ have both been identified as reaction products. The chloroform is oxidatively dechlorinated to carbon dioxide and chlorine. Since free radical formation is rapid compared to subsequent steps, carbon tetrachloride may cause hepatotoxicity due to epoxidation of lipid.

Recknagel (115) ascribes the accumulation of triglycerides by the liver following carbon tetrachloride intoxication to peroxidative decomposition of cytoplasmic membrane structural lipids of the endoplasmic reticulum, which are thereby rendered incapable of transferring triglycerides. Since chloroform does not form the free radical $\cdot$CCl$_3$, it would not be expected to be as toxic as carbon tetrachloride. Since most data suggest that chloroform is less hepatotoxic than carbon tetrachloride, there is support for this theory of toxic action.

No metabolite has been identified in the blood or urine that can be considered as a useful guide for evaluating occupational exposure to chloroform. Expired air has limited value because of analytic difficulties at the low concentrations in expired air at levels considered safe for occupational exposure.

Observations in Man. Considering the long history of chloroform, there is surprisingly little clinical literature on chronic exposure. There have been almost no quantitative toxicologic studies of the response from chronic exposure of humans to chloroform. Challen et al. (94) studied an industrial operation where chloroform was being used. Groups exposed to concentrations varying between 77 and 237 ppm exhibited definite symptoms. Apparently, there were also some high peak concentrations for very short periods of time. Symptoms were gastrointestinal distress and depression. Another group with shorter service was exposed to concentrations from 21 to 71 ppm. They also showed symptoms of comparable nature. Tests of both groups were made to determine possible liver injury but none were found. It should be remembered, however, that liver function tests are often insensitive to anything but severe liver injury. It is quite possible, as indicated by these authors, that there may have been mild liver injury in these cases. The recommendation of these authors that atmospheric exposure should be kept well below 50 ppm is entirely in order.

Bomski et al. (116) reported on an investigation of a pharmaceutical plant which used chloroform as a solvent. Smaller amounts of methanol and methylene chloride were also used. Estimates of the room atmosphere varied from 2 to 20 ppm (0.01 to 0.1 mg/liter). Actual time-weighted average exposures of the workers were not reported, and it is not clear if the room concentrations adequately described the workers' actual exposure.

Complaints of headache, nausea, eructation, and loss of appetite were reported as well as high incidence of enlargment of the liver and spleen. As in the previous study reported by Challen et al., liver function tests were not remarkable.

2.7.4   Hygienic Standards

The threshold limit of chloroform recommended by the ACGIH in 1980 is 10 ppm (48 mg/m^3).

2.7.5   Odor and Warning Properties

Chloroform has a sweetish odor. Lehmann and Flury (102) had indicated that the lowest concentration that could be detected was 200 to 300 ppm. This might be considered as some warning from exposure to acutely hazardous amounts, but it is by no means low enough to be considered a warning from chronic exposure.

2.8   Bromoform, Tribromomethane, CAS 75-25-2

$$CHBr_3$$

2.8.1   Uses and Industrial Exposures

Bromoform has limited use as a chemical intermediate and has been used as an antiseptic.

2.8.2   Physical and Chemical Properties

Physical state	Colorless liquid
Molecular weight	252.77
Specific gravity	2.890 (20/4°C)
Melting point	6 to 7°C
Boiling point	149.5°C
Vapor pressure	5.6 torr (25°C)
Refractive index	1.5980 (19°C)
Percent in "saturated" air	0.7 (25°C)
Solubility	0.3 g/100 ml water at 30°C; soluble in ethanol, ethyl ether, benzene
Flammability	Not flammable by standard tests in air

1 mg/liter $\approx$ 97 ppm and 1 ppm $\approx$ 10.34 mg/m^3 at 25°C, 760 torr

2.8.3   Physiologic Response

Summary.   Relatively little is known about bromoform, particularly from industrial experience. Animal studies are quite limited but indicate moderate acute and high chronic toxicity.

Single or Short-Term Exposure.   *Oral, Skin, and Eyes.*   Range-finding studies have been conducted on bromoform (11). It was found to have an LD$_{50}$ of 2500 mg/kg when fed to male rats as a 10 percent solution in corn oil. Depression, including anesthetic death, was observed at levels greater than the LD$_{50}$. Diarrhea and diuresis were noted. Examination for gross pathological changes in the livers indicated little effect; no histopathological examinations were made. Bowman et al. (458) reported oral LD$_{50}$s of 1400 and 1550 mg/kg for male and female mice.

The undiluted liquid was moderately irritating to rabbit eyes, but healing appeared complete in 1 to 2 days. It was only moderately irritating to rabbit skin even on repeated contact. Single doses of 2000 mg/kg under a cuff on the intact skin of rabbits were survived by two rabbits treated with undiluted bromoform. Lethargy and a slight weight loss were noted from these 24-hr applications (11).

Inhalation. Dykan (88) exposed rabbits to 11 to 13 mg/liter (1070 to 1270 ppm) of the vapor. The duration of the exposure was not stated. Bromoform was described as a narcotic and more toxic than methylene bromide, CH_2Br_2. Dykan states that exposure to 2.5 mg/liter for 10 days produced functional changes in the central nervous system, liver, and kidneys. Vascular and dystrophic changes were also reported. The threshold level (not defined) for chronic exposure was 0.05 mg/liter. Bromine containing metabolites were reported to be present.

Repeated and Prolonged Exposure. Only very limited inhalation data are available. Dykan (89) exposed rats 4 hr/day for 2 months to 0.25 mg/liter (25 ppm). Adverse effects were reported in the livers and in kidney function. The available abstract does not describe the nature of the injury.

Teratogenesis, Mutagenesis, and Carcinogenesis. Bromoform appears to be mutagenic when tested against three strains of *Salmonella typhimurium* (11).

Observations in Man. Much of the experience in poisoning cases in humans has been from the oral administration of the material. This was summarized by von Oettingen in 1955 (33). Anesthesia including respiratory failure has been reported.

2.8.4   Hygienic Standards

Dykan suggested 0.005 mg/liter. The ACGIH in 1980 recommended a TLV of 0.5 ppm (5 mg/liter). The basis for both these recommendations is obviously weak.

2.8.5   Odor and Warning Properties

Bromoform has a sweetish, chloroformlike odor. Because the material is probably quite toxic, the odor should not be considered an adequate warning property, even though the actual threshold of detection of the odor has not been determined.

2.9   Iodoform, Triiodomethane, CAS 75-47-8

$$CHI_3$$

2.9.1   Uses and Industrial Exposures

Idoform has limited use as a chemical intermediate and for medicinal purposes.

2.9.   Physical and Chemical Properties

Physical state	Yellow solid
Molecular weight	393.78
Specific gravity	4.008 (20/4°C)
Melting point	119°C
Flammability	Not inflammable by standard tests in air
Boiling point	210°C, sublimates, explodes
Refractive index	1.800 (20°C)
Solubility	0.01 g/100 ml water at 25°C

1 mg/liter ⋍ 62.1 ppm and 1 ppm ⋍ 16.1 mg/m^3 at 25°C, 760 torr

2.9.3   Physiologic Response

Summary.   Most of the problems associated with iodoform have been related to its topical application as an antiseptic material and to oral administration. Absorption of significant amounts of this material may result in CNS depression and injury to the heart, liver, and kidneys. In 1955 von Oettingen (33) reviewed its toxicity related to medicinal use. This compound is not of great significance in industry and is not discussed in detail.

Single or Short-Term Exposure.   Kutob and Plaa (117) administered iodoform subcutaneously to mice in an investigation of hepatotoxicity of a series of halogenated methanes. The LD_{50} by this route was 1.6 mmol/kg (630 mg/kg) compared to 27.5 mmol/kg for chloroform and 7.2 mmol/kg for bromoform. The doses causing changes in a phenobarbital sleeping time study were 0.59, 1.7, and 2.3 mmol/kg for iodoform, chloroform, and bromoform, respectively. Histological damage to the liver was noted at 1.28 mmol/kg (504 mg/kg) but not at 0.32 mmol/kg.

Repeated or Prolonged Exposure.   No adequate study to determine the chronic toxicity of iodoform has been reported. Although survival and body weight were affected in the NCI bioassay reported in the subsequent section (118), the doses fed were so high as to be improbable as vapor concentrations. They would indicate a low order of systemic toxicity but the studies give no indication of possible effect in the respiratory tract or of possible pain or discomfort that is likely to occur if high concentrations are inhaled.

Teratogenesis, Mutagenesis, and Carcinogenesis.   Iodoform has been included in the NCI bioassay program (118). According to their summary, "The high and low time-weighted average dosages of iodoform were, respectively, 142 and 71 mg/kg/day for male rats, 55 and 27 mg/kg/day for female rats, and 93 and 47 mg/kg/day for male and female mice. A significant positive association between dosage and mortality was observed in male rats but not in female rats or in mice of either sex. Adequate numbers of animals in all groups survived sufficiently long to be at risk from late-developing tumors. No statistical significance could be attributed to the incidences of any neoplasms in rats or mice of either sex when compared to their respective controls. Under the conditions of this bioassay, no convincing evidence was provided for the carcinogenicity of iodoform in Osborne-Mendel rats or B6C3F1 mice."

Observations in Man.   There appear to be no reports in the literature from industrial use of iodoform. Use as an antiseptic in medical applications has produced depression and other signs of toxicity.

2.9.4   Hygienic Standards

The ACGIH in 1980 recommended 0.6 ppm (10 mg/m^3) as a threshold limit value. The value is based primarily on analogy. The likelihood of systemic injury appears to be remote at this level but the irritant effects in the respiratory tract have not been adequately established.

2.9.5   Odor and Warning Properties

Although the odor is reported to be pungent, no quantitative data were found that would indicate the adequacy of the warning properties.

2.10   Carbon Tetrachloride, Tetrachloromethane, CAS 56-23-5

$$CCl_4$$

2.10.1   Uses and Industrial Exposures

Our understanding of the toxicity and industrial hygiene measures necessary to handle carbon tetrachloride has changed little since Irish prepared his chapter for the second revised edition. Awareness of its hepatotoxicity and the availability of less hazardous solvents have resulted in proportionately less being used as a solvent; most of current production is consumed in the production of fluorocarbons. However, occasional reports still appear indicating misuse and overexposure with resulting injury. Use of carbon tetrachloride in fire extinguishers has essentially disappeared, but a significant amount is used in fumigant mixtures, particularly in fumigating grain. It is an active insecticide and is also very effective in suppressing the flammability of more flammable fumigants.

2.10.2   Physical and Chemical Properties

Physical state	Colorless liquid
Molecular weight	153.84
Specific gravity	1.585 (25/4°C)
Freezing point	−22.8°C
Boiling point	76.75°C
Vapor pressure	113 torr (25°C)
Refractive index	1.46305 (15°C)
Percent in "saturated" air	15 (25°C)
Solubility	0.08 g/100 g water at 20°C; miscible with alcohol, diethyl ether, benzene
Flammable	Not flammable by standard tests in air; will not support combustion

1 mg/liter $\approx$ 159 ppm and 1 ppm $\approx$ 6.29 mg/m^3 at 25°C, 760 torr

2.10.3   Physiologic Response

Summary.   Exposure to high concentrations of carbon tetrachloride results in depression of the central nervous system. If the concentration is not high enough to lead to rapid loss of consciousness, other indications of CNS effects such as dizziness, vertigo, headache, depression, mental confusion, and incoordination are observed. Many individuals also show gastrointestinal responses such as nausea, vomiting, abdominal pain, and diarrhea. Some individuals may be nauseated by the "faintest smell of the stuff." Whether this is a conditioned reflex or a direct CNS effect is difficult to determine. Functional and destructive injury of the liver and kidney may occur from a single acute exposure, but it is much more likely to occur from repeated exposures. In a case of long-term chronic exposure to low concentrations, kidney and liver injury dominate the picture. The milder the exposure, the more tendency for the injury to be predominantly in the liver. At threshold concentrations, the injury of the liver appears mostly as malfunction and/or enlargement. Many enlarged livers are observed in animals at the threshold of response. The detection of an enlarged liver in humans should be considered of importance although enlargement may occur from a great many other causes. It has been recognized that the concurrent intake of significant amounts of alcohol with exposure to carbon tetrachloride may greatly increase the probability of injury.

Single or Short-Term Exposure.   *Oral.*   Availability of carbon tetrachloride in odd containers around the home and shop has made ingestion a serious problem. The LD$_{50}$ for rats is reported by McCollister et al. as 2920 mg/kg (119). The *NIOSH Registry of Toxic Effects of Chemical Substances*, 1979 (FG-49000), lists oral LD$_{50}$s of 2800 mg/kg for the rat; 12,800 mg/kg for the mouse; 6380 mg/kg for the rabbit; and 3680 mg/kg for the hamster. Numerous values for other routes of exposure are cited (120).

Skin and Eyes.   Because carbon tetrachloride is a good lipid solvent, it removes the fats from the skin and, in so doing, causes a dry disagreeable feeling and may facilitate secondary infection. Contact with the eyes may cause a transient disagreeable irritation but does not lead to serious injury.

Skin Absorption.   Absorption through the skin of monkeys from the vapor phase was studied by McCollister et al. (121). By using radioactive carbon tetrachloride, they were able to detect small amounts in the blood after exposure of the skin to vapor concentrations of 485 and 1150 ppm. Although traces were absorbed through the skin under these conditions, their conclusions were that "absorption through the intact skin would appear to be of no practical significance in considering the hazard to the health of industrial workers exposed to concentrations of at least as high as 1150 ppm in the air."

However, Stewart and Dodd (50) considered absorption of the liquid through the skin to present a potential problem based on a limited study on human subjects. In view of the high potential hepatotoxicity of carbon tetrachloride, contact of the skin with the liquid should be prevented, particularly if it could be repeated exposure.

Inhalation.   Adams et al. (122) reported the response of laboratory animals to single exposure to various concentrations of carbon tetrachloride. The maximum time–concentrations in air survived by rats were as follows: 12000 ppm for 15 min, 7300 ppm for 1.5 hr, 4600 ppm for 5 hr, and 3000 ppm for 8 hr. The maximum time–concentrations in air having no adverse effects upon male rats were as follows: 3000 ppm for 6 min, 800 ppm for 30 min, and 50 ppm for 7 hr. Similar data were reported for rabbits by Lehmann (123).

The data from these two sources indicated that rabbits and guinea pigs show a somewhat greater tolerance for carbon tetrachloride than rats do. In any case, the acute data given for the different animals are in the same range.

The response to a single exposure of carbon tetrachloride has been reviewed in the summary paragraph. The responses observed in animals and man are reasonably comparable. There seems to be a higher probability of significant kidney response in man than is observed in animals. Qualitatively, such injury is observed in both animals and man. Histopathological and biochemical studies of acutely injured animals show marked hepatic injury, as has been demonstrated by increased plasma prothrombin clotting time, an increase of serum phosphatase, an increase in liver weight, an increase of total lipid content of the liver, and central fatty degeneration of the liver. Renal injury was not apparent in the acute exposure of rats in the studies of Adams. Quite significant kidney injury has been reported, however, from what were thought to be single exposures in humans.

Repeated or Prolonged Exposure.   Although responses referable to the nervous system or gastrointestinal tract may still be observed in chronic exposure, they are much less important factors. These effects may not be noticed at all following a long period of chronic exposure to low concentrations; the organic and functional injury of the internal organs becomes predominant, particularly of the liver and the kidney. It is noticeable in the literature that carbon tetrachloride has become the classic agent for producing liver injury for laboratory investigations.

The most comprehensive toxicologic investigation in animals is that of Adams et al. (122). These investigators studied rats, guinea pigs, rabbits, and monkeys, which were given repeated 7-hr daily exposures 5 days/week. At a concentration of 400 ppm (2.52 mg/liter), rats and guinea pigs suffered severe intoxication. Less than half of them lived for 127 exposures during a period of 173 days. There was an increase in liver weight up to twice that of the controls and a moderate increase in kidney weight. Histological examination of the tissues showed central fatty degeneration with cirrhosis of the liver and slight parenchymatous degeneration of the tubular epithelium of the kidneys. Animals examined after 2 weeks of exposure demonstrated advanced liver and kidney changes by that time. At a concentration of 200 ppm (1.26 mg/liter), rats and guinea pigs still showed a definite response and high mortality. Biochemical and histological studies were comparable but less severe than in the 400 ppm exposure. At a concentration of 100 ppm (0.63 mg/liter), rats, rabbits, guinea pigs, and monkeys tolerated 146 to 163 exposures without evidence of adverse effect on gross appearance, behavior, growth, and other parameters. They all showed histopathological changes. The changes were equivocal in the monkey.

At a concentration of 50 ppm (0.32 mg/liter), rats, guinea pigs, and rabbits tolerated up to 134 exposures in 187 days, showing increased liver weight and moderate fatty degeneration of the liver. Monkeys tolerated 198 exposures in 277 days without evidence of gross, histopathological, or biochemical change. At a concentration of 25 ppm (0.16 mg/liter), rats, guinea pigs, and rabbits tolerated 137 exposures in 191 days. They showed a slight increase in liver weight and some fatty changes but no cirrhosis; otherwise, there was no difference from normal animals. At a concentration of 10 ppm (0.063 mg/liter), rats and guinea pigs showed a slight to moderate fatty degeneration and increased liver weight; rabbits were normal. At a concentration of 5 ppm (0.032 mg/liter), rats showed no effect. Guinea pigs also showed essentially no effect, although there was slight increase in liver weight in females. These authors proposed that a standard (ceiling) of 25 ppm would be acceptable if the average of the air analysis did not exceed 10 ppm.

Prendergast et al. (124) conducted two types of long-term inhalation studies on carbon tetrachloride which confirm Adams' findings. Thirty repeated 8-hr exposures to 80 ppm (515 mg/liter) were given during a 6-week period. Continuous (24 hr/day) exposures to 10 or 1 ppm (61 or 6.1 mg/liter) were given for 90 days. According to these investigators,

The repeated exposure study at 515 mg/m³ (80 ppm) reported in this paper resulted in the death of ³/₁₅ guinea pigs and ⅓ monkeys, and severe liver damage in all species. Surviving guinea pigs showed a marked increase in liver lipid content when compared with controls. This hepatic response confirms the observations reported by previous investigators.

A continuous exposure at 61 mg/m³ (10 ppm) resulted in the deaths of 3 guinea pigs, as well as growth depression and liver damage in the survivors of all species. A second continuous exposure at 6.1 mg/m³ (1 ppm) did not cause deaths or visible toxic signs in any species. All species except the rat exhibited slight growth depression but no hematologic or histopathologic evidence of toxicity, at 6.1 mg/m³.

Teratogenesis, Mutagenesis, and Carcinogenesis.   *Teratogenesis.*   Schwetz et al. (125) exposed pregnant rats 7 hr/day to either 330 or 1000 ppm carbon tetrachloride vapors from days 6 to 15 of pregnancy. These exposures had no effect on fetal resorptions but "at both concentrations fetal body weight and crown-rump length were significantly less than that of controls. No anomalies were seen upon gross examination of the fetuses. A significant incidence of subcutaneous edema was observed at 300 ppm but not at 1000 ppm. The incidence of sternal abnormalities was significantly increased in the fetuses of rats exposed to 1000 ppm CCl₄." Considerable hepatotoxicity was observed in the dams at both concentrations. They concluded that carbon tetrachloride is not teratogenic, but is embryotoxic and fetotoxic at these concentrations, and that these are not related to hepatotoxicity in the mothers. Smyth et al. (126) observed three generations of rats exposed to 50 to 400 ppm vapor. In these studies no evidence of reduced fertility or embryonic or fetal abnormalities were observed.

Mutagenesis.   Apparently carbon tetrachloride is not highly mutagenic. Simmon et al. (127) reported that carbon tetrachloride was not mutagenic in their Ames salmonella/microsome assay.

Carcinogenesis.   The International Agency for Research on Cancer (IARC) reviewed the available carcinogenicity data in 1972 and again in 1979 (128, 129). Carbon tetrachloride is grouped with 18 chemicals considered "probably carcinogenic for humans" and, according to IARC, is to be regarded "as if it presented a carcinogenic risk to humans." In a study by the National Cancer Institute Bioassay program, hepatocellular carcinomas developed in mice that received gavaged doses of 2500 or 1250 mg/kg per day for 78 weeks (107).

These massive doses also resulted in adrenal tumors. The amount of noncancerous pathology in the liver was not discussed, but may have been significant. Rats were fed doses of 100 (male) or 150 (female) mg/kg, 5 days/week for 78 weeks and then kept for 32 additional weeks. Half these doses were fed to second groups of rats. According to Weisburger, in rats "carbon tetrachloride caused neoplastic nodules and a few carcinomas of the liver. However, the incidence was lower than anticipated."

The relevance of these massive, hepatotoxic doses to industrial exposure to the vapors has been discussed in the section on chloroform carcinogenicity.

Metabolism and Biologic Monitoring. Early studies to measure absorption and excretion of carbon tetrachloride were handicapped by analytic difficulties. Studies were made only at very high concentrations. McCollister et al. (121) were among the first who studied the absorption, distribution, and elimination of carbon tetrachloride by using radioactive carbon. This allowed them to study these factors at concentrations which were physiologically significant from a chronic exposure point of view. They exposed monkeys to concentrations of 46 ppm (0.290 mg/liter) of carbon tetrachloride which was carbon-14 labeled. Approximately 30 percent was absorbed. The equivalent of at least 51 percent of the carbon tetrachloride absorbed during an inhalation period was estimated to have been eliminated in the expired air within 1800 hr. The remainder was excreted largely in the urine and feces. Approximately 4.4 percent was eliminated as carbon dioxide. Some 94.3 percent of the radioactivity in the urine was as a nonvolatile, unidentified intermediate. Small amounts occurred as urea and carbonate.

Numerous subsequent references to the metabolism of carbon tetrachloride have been published and are discussed with that of chloroform in an earlier section of this chapter.

Unchanged carbon tetrachloride can be measured in the expired air following exposures. Stewart et al. (130) have shown the relationship of exposure concentration and time to exhaled concentration, and it would appear that at the low levels of exposure considered safe for chronic exposure, breath samples will be of very limited value. Expired air may be of value in definitive diagnosis of acute exposure and possibly in semiquantitation of the magnitude of exposure.

At present no metabolites in blood or urine appear to be of value in monitoring exposure.

Observations in Man. Numerous reports of injury and death following acute and repeated exposure of humans to carbon tetrachloride can be found. These have been summarized and discussed in the preceding sections including the probable relationship of alcohol consumption toward predisposition of the individual toward the toxic effects of carbon tetrachloride.

However, it is apparent that few if any epidemiologic studies have been completed on an occupationally exposed population, nor have there been laboratory studies of long duration. Recognizing the care with which carbon tetrachloride is currently handled it seems doubtful that future mortality (carcinogenic) studies on an occupationally exposed population will be possible or fruitful.

Stewart and Dodd (50) and Stewart et al. (130) have conducted experimental human exposure studies with volunteer subjects. These reports indicate that absorption of the liquid through the skin may be significant, particularly in chronic exposure. They have also determined the concentration of carbon tetrachloride in expired air following exposure to the vapor.

2.10.4 Hygienic Standards

The ACGIH recommend a time-weighted average of 5 ppm as a threshold limit value in their 1980 list. Other sources recommend that a ceiling of 25 ppm also be maintained. These values appear to be low enough to avoid injury.

2.10.5 Odor and Warning Properties

Carbon tetrachloride has a sweetish odor. It is not considered particularly disagreeable by most people, although some people may be nauseated by small amounts. The odor is one to which the average individual becomes readily adapted. The odor would certainly not be considered a satisfactory warning of excessive exposure. According to one reference the threshold of detection of the odor of carbon tetrachloride is approximately 79 ppm and the odor is strong at 176 ppm (131). Another reference states that 100 percent of a small panel of test subjects recobnized 21 ppm of carbon tetrachloride produced from carbon disulfide and 100 ppm of carbon tetrachloride produced by chlorination of methane (132).

2.11 Carbon Tetrabromide, Tetrabromomethane, CAS 558-13-4

$$CBr_4$$

2.11.1 Uses and Industrial Exposures

Carbon tetrabromide is used to a limited extent as a chemical intermediate.

2.11.2 Physical and Chemical Properties

Physical state	Colorless solid
Molecular weight	331.67

Specific gravity	3.42 (20°C)
Melting point	α 48.4°C; β 90.1°C (slight decomposition on melting)
Boiling point	189.5°C (slight decomposition)
Refractive index	1.59998 (99.5°C)
Solubility	0.024 g/100 ml water at 30°C; soluble in ethanol, ethyl ether, chloroform
Flammability	Not flammable by standard tests in air

1 mg/liter ≈ 74 ppm and 1 ppm ≈ 13.58 mg/m³ at 25°C, 760 torr

2.11.3 Physiologic Response

Summary. Carbon tetrabromide is a highly toxic material, based on the limited amount of toxicologic data available. Acute exposure to high concentrations causes upper respiratory irritation and injury to the lungs, liver, and kidneys. The response to chronic exposure at very low concentrations is primarily liver injury. The material is a lacrimator, even at low levels.

Single and Long-Term Exposure.   *Oral.*   The LD_{50} by oral administration was found to be 1800 mg/kg body weight in the rat.

Eye and Skin.   In the eyes of rabbits, the undiluted material caused severe irritation and permanent corneal damage. When the material was promptly washed from the eyes, pain and irritation were noted but the corneal damage was temporary (11).

Skin contact causes relatively slight irritation in rabbits. If the material is confined tightly to the skin, it may cause hyperemia and a moderate edema. From observations made when the material was repeatedly bandaged onto the skin, there was no indication of toxic absorption, but no attempt was made to quantitate the dosage.

Inhalation.   There is little information published in the literature on the effect of exposure to carbon tetrabromide in air. Exposure of rats to carbon tetrabromide "fumes" (0.01 to 1 mg/liter) (0.07 to 74 ppm), 4 hr/day for 4 months was reported to cause metabolic changes in the livers. Even the lowest concentration caused irritation of the eyes and respiratory tract (133).

The above results are somewhat different from those of another limited study (11). Repeated exposures of rats 7 hr/day, 5 days/week for 6 months to the vapors of carbon tetrabromide were studied. When the concentration in air was determined by combustion of air samples and determination of halogen, the concentration without effect was found to be 0.1 ppm by volume. When the concentration in air was determined by a polarographic method, the concentration was 0.3 to 0.5 ppm by volume. Higher concentrations than this caused poor growth, and fatty and degenerative changes in the liver.

Metabolism and Biological Monitoring. The small amount of data that is available suggest that either hydrolysis or metabolism produces some bromide ion. Carbon tetrabromide would not be expected to produce physiologically significant quantities of bromide ion in the blood at levels of exposure considered acceptable by inhalation. Carbon tetrabromide is metabolized in vitro to produce carbon monoxide but the in vivo significance has not been established (459). Carbon tetrabromide has been isolated from red algae, *Asparagopsis toxiformis* found in the ocean near Hawaii (134).

2.11.4   Hygienic Standards

The ACGIH recommended a TLV of 0.1 ppm (1.4 mg/m³) in 1980. This value is obviously based on very limited data.

2.11.5   Odor and Warning Properties

Carbon tetrabromide has significant lacrimatory affect upon the eye at low concentrations. This may be reasonably good warning of acute exposure, but may not be adequate to prevent excessive repeated exposure.

2.12   Ethyl Chloride, Monochloroethane, CAS 75-00-3

$$CH_3CH_2Cl$$

2.12.1   Uses and Industrial Exposures

Ethyl chloride has been used as a chemical intermediate, as an anesthetic, and to a limited degree as a refrigerant. The most serious problems have been with its use as an anesthetic. Very little physiological difficulty has been encountered in industry. The major problems are fire and explosion.

2.12.2   Physical and Chemical Properties

Physical state	Colorless gas
Molecular weight	64.52
Specific gravity	0.8917 (25/25°C)
Melting point	−139°C
Boiling point	12.2°C
Vapor pressure	1200 torr (25°C)

Solubility	0.57 g/100 ml water at 20°C, 48 g/100 ml ethanol at 21°C; soluble in ethyl ether
Flammability limits	3.8 to 15.4 percent by volume in air
Autoignition temperature	519°C
Flash point	−50°C (closed cup); −43°C (open cup)

1 mg/liter ≈ 379 ppm and 1 ppm ≈ 2.64 mg/m^3 at 25°C, 760 torr

2.12.3 Physiologic Response

Summary. The principal problem in industrial use is that typical of an anesthetic material where "drunkenness" and incoordination may lead to inept operation and therefore, the possibility of an injury. Cardiac arrhythmia may be possible but high concentrations are required (143). The older data were summarized by von Oettingen (33). There appear to be few or no chronic data reported in the literature.

Single or Short-Term Exposure. *Skin.* Owing to the gaseous state of ethyl chloride at normal room temperatures, if large amounts of the liquid are spilled on the skin the evaporation may cause rapid cooling and possibly frostbite.

A single reference (391) reports allergic eczematous eruption in two subjects after ethyl chloride was sprayed on the skin in an allergy testing procedure. The wide use of ethyl chloride as a local anesthetic suggests that these may be rare subjects.

Inhalation. The acute toxicity for animals was reported by Sayers et al. (12) (see Table 48.6) and the narcotic concentrations for man were reported by Lehmann and Flury (102) (see Table 48.7).

A 2-hr LC$_{50}$ of 152 mg/liter (57600 ppm) has been reported (135). Deaths were anesthetic in nature but hyperemia, edema, and hemorrhages were reported in the internal organs, brain, and lungs. Repeated 2-hr exposures for 60 days to 14 mg/liter (5300 ppm) caused a decrease in the phagocytic activity of the leukocytes, lowered hippuric acid formation in the liver, and resulted in histological or pathological changes in the liver, brain, and lungs.

In a limited study, groups of six male and six female rats plus two male beagle dogs were exposed 6 hr/day, 5 days/week for 2 weeks to either 0, 1600, 4000, or 10,000 ppm ethyl chloride vapor (11). Except for possible CNS effects during exposure to 10,000 ppm, female rats and male dogs were unaffected as shown by histological and clinical chemical evaluation and organ and body weight measurements. Liver weight was slightly increased in male rats at 4000

Table 48.6. Response of Guinea Pigs to Ethyl Chloride Vapor in Air

Concentration (%)	Exposure time (min)	Response
23–24	5–10	Unconscious, some deaths
15.3	40	Some deaths in 30 min; some survived 40 min
9.1	30	Survived; histopathological changes in lungs, liver, etc.
5	40	Survived; lung congested
4	122	Survived; returned to normal
	270	Survived; histopathological changes in lungs, liver, and kidneys
	540	Some deaths
2	270	Survived; returned to normal
	540	Survived; histopathological changes in liver and kidneys
1	810	Survived; returned to normal

and 10,000 ppm. This was the only change in this sex of the rats except for CNS depression.

Perhaps the most serious problem from severe acute exposure, other than the anesthetic effect, is the possibility of potentiation of adrenalin. The resultant cardiac problems may be of serious consequence during accidental exposure to very high concentrations.

Table 48.7. Narcotic Ethyl Chloride Concentrations in Man

Concentration mg/liter	Concentration ppm	Response
105.6	40,000	After 2 inhalations, stupor, irritation of eyes, and stomach cramps
88.7	33,000	After 30 sec, quickly increased toxic effect
66.0	25,000	Lack of coordination
52.8	20,000	After 4 inhalations, dizziness and slight abdominal cramps
50.4	19,000	Weak analgesia after 12 min
34.3	13,000	Slight symptoms of poisoning

Metabolism and Biologic Monitoring. Ethyl chloride is readily absorbed through the lungs and excreted by the lungs. Thorough investigation of absorption, excretion, or metabolism has not been reported. The indications are that ethyl chloride is not metabolized to a significant degree.

2.12.4 Hygienic Standards

The threshold limit of ethyl chloride recommended by the ACGIH in 1980 is 1000 ppm (2600 mg/m³). There are very few data to support this or any other value.

2.12.6 Odor and Warning Properties

Ethyl chloride has an ethereal, somewhat pungent odor. Because it requires very high concentrations to have serious physiologic effects, this may be considered a possible warning property.

2.13 Ethyl Bromide, Monobromoethane, CAS 74-96-4

$$CH_3CH_2Br$$

2.13.1 Uses and Industrial Exposures

The principal use of ethyl bromide is as a chemical intermediate. Although it has been proposed occasionally as an anesthetic, it has not been used to any extent for that purpose in recent years.

2.13.2 Physical and Chemical Properties

Physical state	Colorless liquid
Molecular weight	108.98
Specific gravity	1.4505 (25/4°C)
Melting point	−119°C
Boiling point	38.4°C
Vapor pressure	475 torr (25°C)
Refractive index	1.42386 (25°C)
Percent in "saturated" air	62.5 (25°C)
Solubility	0.91 g/100 ml water at 20°C; soluble in ethanol, ethyl ether
Flammability limits	6.75 to 11.25 percent by volume in air

1 mg/liter ≈ 224.3 ppm and 1 ppm ≈ 4.46 mg/m³ at 25°C and 760 torr

2.13.3   Physiologic Response

Summary.   The primary response from exposure to ethyl bromide is CNS depression, as in the case of ethyl chloride. In contrast, however, ethyl bromide causes irritation of the lungs and injury to the liver and kidneys.

Many of the available acute data are from abstracts and give little detail. The material would appear to be rather low in acute toxicity. Quantitative chronic studies have been limited and there has been relatively little experience in industry with the use of ethyl bromide. It is therefore suggested that some caution be used even within the limits of the present proposed hygienic guide.

Single or Short-Term Exposure.   *Oral, Skin, and Eyes.*   The liquid is apparently irritating to the skin and eyes of rabbits and is reported to be absorbed through the skin (136). In a very limited study on rabbits, oral dosages as large as 200 mg/kg were administered daily (61 times) with no apparent injury. Paralysis resulted in the rabbits given 300 or 600 mg/kg per day 62 times, although they did survive (11).

Inhalation.   Sayers et al. (12) reported acute studies on the response of guinea pigs to ethyl bromide (see Table 48.8).

Vernot et al. (137) determined 1-hr LC_{50}s of 27,000 ppm and 16,200 ppm (120 and 72 mg/liter) for rats and mice, respectively.

Intraperitoneal.   The available abstract of a report by Kosenko (138) gives LD_{50} values of 2850 and 1750 mg/kg for mice and rats given single intraperitoneal injections. The abstract further states the material was rapidly detoxified and not accumulated significantly in mouse tissues.

Repeated or Prolonged Exposure.   According to the available abstract, rats were exposed 4 hr daily for 6 months to 2.4 mg/liter (about 540 ppm), with some evidence of liver injury and disrupted liver function (139).

Teratogenesis, Mutagenesis, and Carcinogenesis.   No data for teratogenicity were found. Because of its alkylating properties, ethyl bromide has been tested for mutagenicity in microbial systems. Simmon and Poirier (140) tested ethyl bromide against *S. typhimurium* TA 1535 and TA 100 and against *E. Coli* WP2 (her) and found it to be positive. When studied against yeast it was at most weakly mutagenic, but it was clearly mutagenic in *Salmonella* with and without rat liver homogenate (11). Vogel and Chandler (141) tested ethyl bromide against *Drosophila* and found it to be inactive.

Ethyl bromide did not increase lung adenomas in Strain A mice given 24 intraperitoneal injections totaling 55 mmol/kg (43). No other carcinogenic data were found.

Table 48.8. Physiologic Response of Guinea Pigs to Ethyl Bromide in Air

Concentration (%)	Exposure time (min)	Response
14	10	Unconscious; death in several days
5	98	Inconscious; died 1 hr later
6	10	Survived; lung injury
2.4	90	Died in 18 hr
	30	Some delayed deaths; pathological changes in lungs, liver, and spleen
	10	Dizzy; survived—slight congestion in lungs and liver
1.2	270	Some deaths; histopathological changes
	90	Survived; histopathological changes
	55	Survived; slight histopathological changes
0.65	270	Some deaths
	180	Survived; histopathological changes
0.07	810	One death and histopathological changes
	540	Survived; normal

Metabolism and Biologic Monitoring.   Relatively little study has been given to this compound. It has been indicated, however, that absorption and excretion through the lungs are rapid (142), but that it may be hydrolyzed to some degree in the body, resulting in the formation of inorganic bromide.

Because of the low systemic toxicity, determination of bromide ion in the blood may be of value in determining whether exposure to ethyl bromide has occurred. However, the available data do not permit a quantification of exposure based on blood bromide.

Observations in Man.   Limited human experience during surgical anesthesia has indicated, in addition to CNS depression, the possibility of lung congestion and degeneration of the liver and kidney tissues.

2.13.4 Hygienic Standards

The threshold limit of ethyl bromide recommended by the ACGIH in 1980 is 200 ppm (890 mg/m³). There are very few data to support this or any other value.

2.13.5 Odor and Warning Properties

Ethyl bromide has a definite though not particularly distinctive odor. A threshold of odor response has not been reported.

2.14 **Ethyl Iodide,** Monoiodoethane, CAS 75-03-6

$$CH_3CH_2I$$

2.14.1 Uses and Industrial Exposures

Ethyl iodide is used largely as a chemical intermediate. In the past, it had limited medicinal use.

2.14.2 Physical and Chemical Properties

Physical state	Colorless liquid which may become colored with iodine if exposed to the light
Molecular weight	155.98
Specific gravity	1.9245 (25/4°C)
Melting point	−108.5°C
Boiling point	72.2°C
Vapor pressure	137 torr (25°C)
Refractive index	1.5076 (25°C)
Percent in "saturated" air	18 (25°C)
Solubility	0.4 g/100 ml water at 20°C; soluble in ethanol, ethyl ether, benzene, chloroform
Flammability	No flash point by standard tests in air, but can be made to combust

1 mg/liter ≈ 156.7 ppm and 1 ppm ≈ 6.38 mg/m³ at 25°C, 760 torr

2.14.3   Physiologic Response

Summary.   Most of the information on the toxicity of ethyl iodide has been obtained because of the interest in this material for the treatment of fungus infections and for measurement of cardiac output. It has been administered to man as a vapor which, when inhaled, is absorbed by the lungs. It can cause CNS depression and may affect the kidneys, thyroid, lungs, and the liver.

Single or Short-Term Exposure.   *Oral.*   In a very limited study two rabbits survived an oral dose of 100 mg/kg in corn oil; three out of three died after a dose of 300 mg/kg (11).

Skin Absorption.   Von Oettingen (33) cites a 1936 report by Schwander which indicated ethyl iodide was absorbed through the skin with deaths occurring 7 days after treatment.

Inhalation.   Flury and Zernik (13) have reported the physiological effect of various concentrations on mice. These are shown in Table 48.9.

According to Reinhardt et al. (143), who cite data by Herman and Vial, it was not possible to classify ethyl iodide as to its potency as a cardiac sensitizer based on the available data.

Mutagenesis and Carcinogenesis.   Simmon and Poirier (140) tested ethyl iodide against *S. typhimurium* and *E. coli* with positive mutagenic results.

When injected intraperitonially two to four times in 24 weeks at a total dose of 38.4 mmol/kg, ethyl iodide did not result in an increase in lung adenomas in Strain A mice (43).

Observations in Man.   Discussions of the response of man to the therapeutic inhalation of the vapors of ethyl iodide are given in the publications of

Table 48.9.   Physiologic Response to Ethyl Iodide Vapors in Air (Mice)

Concentration		Duration of exposure (hr)	Response
mg/liter	ppm		
1.87	290	3	Fatal
0.94	150	24	Fatal
0.75	120	7	Tolerated

Blumgart et al. (144) and Schwartz (392). These reports describe skin lesions and peripheral neuritis in patients.

2.14.4 Hygienic Standards

In the second edition of this book, Irish stated: "While no standard has been established for ethyl iodide and there is very little basis to establish a standard, it should be recognized from the table of Flury and Zernik that mice tolerate a concentration of 120 ppm. A concentration of 150 ppm for 24 hours is fatal. It would therefore, be advisable to keep concentrations well below 100 ppm for relatively short periods of exposure. Such exposures should not be repeated frequently."

No standard has been established for ethyl iodide. Irish's (1) recommendation of keeping exposures well below 100 ppm even for short exposures still seems appropriate.

2.14.5 Odor and Warning Properties

Ethyl iodide has an ethereal odor but odor may not be an adequate warning of excessive exposure.

2.15 1,1-Dichloroethane, Ethylidene Chloride, Ethylidene Dichloride, CAS 75-34-3

$$CH_3CHCl_2$$

2.15.1 Uses and Industrial Exposures

1,1-Dichloroethane is flammable. It has limited use as a solvent and as a chemical intermediate. Formerly used as an anesthetic, it is of no importance in this field today.

2.15.2 Physical and Chemical Properties

Physical state	Colorless liquid
Molecular weight	98.97
Specific gravity	1.174 (20/4°C)
Melting point	$-96.7°C$
Boiling point	57.3°C
Vapor pressure	234 torr (25°C)
Refractive index	1.41655 (20°C)
Percent in "saturated" air	30.8 (25°C)

Solubility	0.5 g/100 ml water at 20°C; soluble in ethanol, ethyl ether
Flash point	14°C (open cup); −12°C (closed cup)
Flammability limits	5.6 to 11.4 percent by volume in air
Ignition temperature	493°C

1 mg/liter $\approx$ 247 ppm and 1 ppm $\approx$ 4.05 mg/m^3 at 25°C, 760 torr

2.15.3   Physiologic Response

Summary.   Much less has been published on the toxicity of 1,1-dichloroethane than on its more toxic isomer 1,2-dichloroethane (ethylene dichloride). Available data indicate that 1,1-dichloroethane is rather low in toxicity. It is capable of causing anesthesia, but has a relatively low capacity to cause liver or kidney injury even on repeated exposure.

Single and Short-Term Exposure.   *Oral.*   Although the 1979 *NIOSH Registry of Toxic Effects of Chemical Substances* (120) lists an oral LD$_{50}$ of 725 mg/kg for rats based on a 1967 article, this must be an error, since repeated daily doses higher than this were given by gavage for 78 weeks (145). No original reference is given for the report of an oral LD$_{50}$ of 14.1 g/kg for rats (146).

Skin.   When applied repeatedly to the skin of rabbits, 1,1-dichloroethane caused little effect unless confined to restrict evaporation. Even when confined, only a typical defatting action was noted with no evidence of absorption through the skin (11).

Inhalation.   Smyth (147) found that rats survived an 8-hr exposure to 4000 ppm but were killed by 16,000 ppm.

Intraperitoneal.   Little liver and kidney toxicity has been shown following intraperitoneal injection. Doses of 1000 mg/kg produced no renal necrosis in mice but some evidence of tubular swelling was reported (53). Urinary protein was increased after injection of 2000 mg/kg and urinary glucose after 4000 mg/kg.

Repeated or Prolonged Exposure.   Hofmann et al. (148) reported that rats, guinea pigs, rabbits, and cats tolerated 6-hr daily exposures to 500 ppm for 13 weeks (5 days/week) with no adverse effect. Rats, guinea pigs, and rabbits tolerated an additional 13 weeks at 1000 ppm but cats showed evidence of kidney injury histologically and by increased blood urea.

   Unpublished data from The Dow Chemical Company cited in the AIHA

Hygienic Guide (149) confirm that "1,1-dichloroethane has little capacity for causing liver damage, being similar to methylene chloride and 1,1,1-trichloroethane in this respect. Rats, guinea pigs, rabbits, and dogs were exposed to either 500 or 1000 ppm for seven hours per day, five days per week, for six months. Gross and microscopic pathological and hematological studies showed no evidence of changes attributable to the exposure."

Teratogenesis, Mutagenesis, and Carcinogenesis.   Pregnant female rats were exposed on days 6 to 15 of gestation to 3800 or 6000 ppm 1,1-dichloroethane vapors. Exposures were for 7 hr/day. Essentially no effect occurred in either the dams or fetuses except for slight but statistically significant decreases in food consumption and weight gain by the dams and delayed ossification in the fetus. No teratological effects were related to exposures. Liver weights of nonpregnant rats were increased by similar exposure but no histological changes were apparent grossly or microscopically (125).

When 1,1-dichloroethane was fed by gavage in the NCI bioassay program for a period of 78 weeks followed by an observation period of 33 weeks (rats) and 13 weeks (mice), survival was poor (145). The doses fed were very high, 764 and 382 mg/kg per day for male rats; 950 and 475 mg/kg per day for female rats; 2885 and 1442 mg/kg per day for male mice; and 3331 and 1665 mg/kg per day for female mice. It was reported that there was no conclusive evidence for the carcinogenicity of 1,1-dichloroethane in Osborne-Mendel rats or B6C3F1 mice, although marginal increases in mammary adenocarcinomas and hemangiosarcomas were noted in female rats and a statistically significant increase in the incidence of endometrial stromal polyps occurred in female mice.

Metabolism and Biologic Monitoring   No references to the metabolism of 1,1-dichloroethane were found. It is excreted in the expired air of dogs following inhalation exposure (11).

Observations in Man.   No reports of human experiments or experience were found.

2.15.4   Hygienic Standards

In 1980 the ACGIH recommended a TLV for 1,1-dichloroethane of 200 ppm (100 mg/m³). This value is probably unnecessarily low, based on the available data.

2.15.5   Odor and Warning Properties

1,1-Dichloroethane has been stated to have a "chloroformlike" odor. The level at which this odor is detectable has not been determined. It is possibly not a satisfactory warning of excess chronic exposure.

2.16   Ethylene Dichloride, 1,2-Dichloroethane, EDC, CAS 107-06-2

$$CH_2ClCH_2Cl$$

2.16.1   Uses and Industrial Exposures

In the United States about 80 percent of the current production of ethylene dichloride is used as the starting material for preparation of vinyl chloride monomer.

Other applications are much smaller, some being used in antiknock fluids for gasoline, in fumigant mixtures, and as a solvent. Because of its toxicity and flammability, usage as a solvent has decreased considerably as less hazardous replacements have become available.

2.16.2   Physical and Chemical Properties

Physical state	Colorless liquid
Molecular weight	98.97
Specific gravity	1.2529 (20/4°C)
Melting point	−35.3°C
Boiling point	83.5°C
Vapor pressure	87 torr (25°C)
Refractive index	1.44432 (20°C)
Percent in "saturated" air	11.5 (25°C)
Solubility	0.9 g/100 ml water at 20°C; soluble in ethanol, ethyl ether
Flash point	18.3°C (open cup); 13°C (closed cup)
Explosive limits	6.2 to 15.9 percent by volume in air
Ignition temperature	415°C

1 mg/liter $\approx$ 247 ppm and 1 ppm $\approx$ 4.05 mg/m^3 at 25°C, 760 torr

2.16.3   Physiologic Response

Summary.   The toxicity of ethylene dichloride has been extensively investigated both in animals and man. It has been reviewed by several authors (33, 150, 151, 165, 460). At very high concentrations, ethylene dichloride is irritating to the eyes, nose, and throat. The symptoms are largely related to CNS depression or gastrointestinal upset, that is, mental confusion, dizziness, nausea, and vomiting. Cardiac sensitization appears to be less important. At subacute levels, similar symptoms of CNS depression and gastrointestinal upset are observed. Definite liver, kidney, and adrenal injury may occur at these levels. From chronic exposure to lower concentrations, some indications of CNS depression are still observed. Nausea and vomiting are quite common in humans. The symptom of nausea and vomiting from ethylene dichloride is quite striking and similar to that often observed from carbon tetrachloride. The pathological picture from chronic exposure is injury of the liver, kidneys, and adrenals. This general pattern has been consistent in animal investigative work and in experience from human exposure.

Single or Short-Term Exposure. *Oral.* McCollister et al. (119) reported the LD_{50} of ethylene dichloride for rats to be 680 mg/kg of body weight. This would indicate that ethylene dichloride is several times more toxic than carbon tetrachloride when taken in a single oral dose, since the carbon tetrachloride LD_{50} indicated by the same authors was 2.98 g/kg. Rabbits and mice appear to be similarly affected as rats but dogs and humans may survive larger doses because of their ability to vomit (150).

Eyes. When liquid ethylene dichloride is splashed in the eyes, it may result in pain, irritation, and lacrimation. If it is promptly removed by washing, no significant injury should occur. If not removed, serious damage may be the result. Inhalation of ethylene dichloride vapors produces a clouding of the cornea of dogs and foxes but not of other species (151).

Skin. When ethylene dichloride is held close to the skin within a cuff, such as is used in skin absorption experiments, quite severe irritation and moderate edema and necrosis may be observed. When it is not held tight on the skin, ethylene dichloride does not cause serious irritation. Repeated or prolonged contact, however, may cause a rough, red, dry skin owing to extraction of fatty materials. It may result in cracking and chapping.

Skin Absorption. Ethylene dichloride is absorbed through the skin although it takes quite large doses to cause serious acute systemic poisoning. Rabbits survived a dose of 1.26 g/kg of body weight, and the LD_{50} was calculated to be 2.8 g/kg of body weight.

Inhalation. The response of guinea pigs to acute exposure to ethylene dichloride was reported by Sayers et al. (152), and Heppel et al. (153) reported on single exposures of rabbits, guinea pigs, dogs, cats, raccoons, mice, and rats. Although cats and raccoons seemed to be somewhat more resistant, there were otherwise no marked differences in sensitivity to acute exposure except for the eyes.

Spencer et al. (154) reported an investigation of the response of rats. The maximum time concentrations in air survived by rats from a single exposure were as follows: 12 min at 20,000 ppm, 1 hr at 3000 ppm, and 7 hr at 300 ppm. The maximum time–concentrations in air having no adverse effect on female rats were as follows: 6 min at 12,000 ppm, 1.5 hr at 1000 ppm, and 7 hr at 200 ppm. The responses of other species reported by other authors fall within the same general range. Some acute exposures produced considerable depression of the central nervous system. Concentrations below 12,000 ppm produced various degrees of CNS depression but not unconsciousness or death within the duration of the exposure. At concentrations of 3000 ppm or more, definite depression was observed in the form of inactivity or stupor. Organic changes were observed in the form of increased weight of liver and kidneys. Biochemical changes indicated liver malfunction. There were histopathological changes in the kidneys, liver, and adrenals.

Confirming the work of several investigators of the nineteenth century, Heppel et al. (151) demonstrated by inhalation studies the occurrence of a reversible clouding of the cornea in dogs. After several exposures, the dogs became resistant to this effect. These authors studied many different species but observed this effect only in the fox and the dog. This effect has not been observed in humans exposed to ethylene dichloride. A detailed study of this unique canine phenomenon has been published (155).

Repeated or Prolonged Exposure. Heppel et al. (156) studied the chronic toxicity of 1,2-dichloroethane by exposing animals 7 hr/day, 5 days/week to concentrations of 1000, 400, 200, and 100 ppm of the vapor in air. At a concentration of 1000 ppm rats, rabbits, and guinea pigs died after a few 7-hr exposures. Dogs and cats proved to be more resistant, but deaths eventually occurred among these animals. One monkey died after two exposures, a second after 43 exposures. Pathological examination of the various animals showed occasional pulmonary congestion, renal tubular degeneration, fatty degeneration of the liver, and less commonly, necrosis and hemorrhage of the adrenal cortex and fatty infiltration of the myocardium. Dogs appeared unaffected after 8 months of exposures to 400 ppm of dichloroethane. Functional tests of liver and kidney were negative. At autopsy there were slight fatty changes in the liver. Deaths occurred among guinea pigs, rabbits, and rats, although some of the animals survived many exposures. Pathological examination revealed lesions similar to those seen with 1000 ppm. A concentration of 200 ppm was well tolerated by two monkeys and five rabbits. Rats, guinea pigs, and mice died after a number of exposures and showed occasional lesions. When the concentration of solvent vapor was lowered to 100 ppm, even rats, guinea pigs, and mice survived exposures for 4 months and developed no demonstrable lesions.

A comparable chronic study was carried out by Spencer et al. (154), who exposed animals 7 hr/day, 5 days/week. They likewise showed high mortality at 400 ppm in rats and guinea pigs in periods of 14 to 56 days of exposure. The animals showed loss of weight and a slight increase in weights of the liver and kidneys, but relatively slight histopathological changes. Guinea pigs showed more definite histopathological changes in both the liver and kidneys. At 200 ppm, rats survived 151 exposures in 212 days without apparent adverse effect.

Growth was normal and there was no indication of injury. Guinea pigs survived comparable exposures, but about half of the animals showed some histological changes. At 100 ppm rats, guinea pigs, rabbits, and monkeys were given 120 exposures in 168 days without apparent effect.

Similarly Hofmann et al. (148) produced injury after 13 weeks of repeated 6 hr/day exposures of rats, guinea pigs, and rabbits to 500 ppm. Cats tolerated these exposures. All four species appeared unaffected by similar exposures to 100 ppm.

Teratogenesis and Reproduction, Mutagenesis, and Carcinogenesis. *Teratogenesis and Reproduction.* In the available abstract Vozovaya (157) states he produced decreased fertility and other adverse effects in pregnant female rats and the progeny of the first generation, but not of the second, by giving them repeated 4-hr/day exposures to 57 mg/m³ (14 ppm). However, an attempt to confirm this report failed to do so even though the rats were exposed to 100 ppm 7 hr/day for several months (461). Furthermore, pregnant Sprague-Dawley rats and New Zealand White rabbits were exposed to 0, 100, or 300 ppm of ethylene dichloride on days 6 through 15 (rats) and 6 through 18 (rabbits) of gestation. Severe maternal toxicity was observed among rats exposed to 300 ppm of ethylene dichloride; two-thirds of the dams died during the exposure period. No signs of toxicity were observed among rats at the 100 ppm dose level. Maternal toxicity was noted in rabbits as evidenced by maternal deaths at both dose levels. No adverse effects on embryonal or fetal development were observed among litters from the exposed rats at 100 ppm or among those from exposed rabbits. Owing to the severe maternal toxicity observed, no conclusions could be drawn concerning the teratogenic potential of inhaled ethylene dichloride in the rat at 300 ppm; ethylene dichloride was not embryotoxic or teratogenic in rats inhaling either 100 ppm or in rabbits inhaling 100 or 300 ppm of the compound during gestation (461).

The effect of ethylene dichloride in drinking water was determined by Riddle et al. (477). Male and female ICR Swiss mice received 0, 5.14, 15.4 or 49.7 mg/kg per day. There appeared to be no dose-dependent effects on fertility, gestation, viability, or lactation indices. Pup survival and weight gain were not adversely affected. Gross necropsy of male and female F/O generation mice treated with 1,2-DCE or 1,1,1-TCE failed to reveal compound or dose-related effects.

Mutagenesis. The predicted metabolites of ethylene dichloride, 2-chloroethanol, chloroacetaldehyde, and chloroacetic acid as well as the compound itself have been studied in mutagenic test systems. Drosophila (158–160) and barley (30) have been reported to be affected by ethylene dichloride but the details of the drosophila exposure conditions are not clear. Salmonella (TA 100), a susceptible strain, showed markedly increased reversions following exposure to chloroacetaldehyde, but other strains were not affected (41). Chloroethanol and vinyl chloride were rather ineffective in this system. Thus it appears that the ethylene dichloride is mutagenic in microbial test systems. Ethylene was shown to be negative in a dominant lethal test, which was part of the reproduction study described in the previous section (477).

Carcinogenesis. Ethylene dichloride has been included in the NCI bioassay program (161). In this study, in which corn oil solutions of ethylene dichloride were given by gavage to rats and mice, the doses were adjusted or omitted to permit survival at the top or maximum tolerated dose. Therefore, it was necessary to express the exposure concentrations as time-weighted averages. For male and female rats they were 95 and 47 mg/kg per day; for male mice they were 195 and 97 mg/kg per day; and for female mice the high and low doses were time-weighted as 299 and 149 mg/kg per day, respectively.

Under the severe conditions of this study,

A statistically significant positive association between dosage and the incidence of squamous cell carcinomas of the forestomach and hemangiosarcoma of the circulatory system occurred in the male rats, but not in the females. There was also a significantly increased incidence of adenocarcinomas of the mammary gland in female rats.

The incidence of mammary adenocarcinomas in female mice were statistically significant. There was a statistically significant positive association between chemical administration and the combined incidences of endometrial stromal polyps and endometrial stromal sarcomas in female mice. The incidence of alveolar/bronchiolar adenomas in both male and female mice was also statistically significant.

Another study in which animals were exposed to the vapors of ethylene dichloride has failed to produce an increase in tumors. Maltoni (162) reported that he had exposed rats and mice 7 hr/day for 18 months to 250 to 150, 10, or 5 ppm and then kept them for their lifetime. No specific type of tumor was increased in mice or rats. Benign mammary tumors were increased in all exposed groups of female rats, but these were ascribed to a general stress rather than a tumorigenic action.

Obviously the carcinogenic potential of ethylene dichloride has not been resolved by these two studies and a possible explanation is provided by Reitz et al. (475). This is discussed in the following section. In any case ethylene chloride does not appear to present a significant carcinogenic risk if exposures are controlled to levels acceptable for control of hepatotoxic effects.

Metabolism and Biologic Monitoring. In any case, ethylene dichloride is readily absorbed via the lungs and gastrointestinal tract. To a lesser extent, it is absorbed through the skin.

The extent or nature of the metabolism of ethylene dichloride has not been established, but some appears to go through 2-chloroethanol to monochloroacetic acid (163). Depending upon the dose, 10 to 42 percent of ^{14}C-ethylene dichloride which was injected intraperitoneally in mice was exhaled unchanged, and 12 to 15 percent exhaled as carbon dioxide. The urine contained 51 to 73 percent of the radioactivity, the feces very little, and 0.6 to 1.3 percent remained in the mouse.

Guengerich et al. reported on an extensive study on in vitro activation of ethylene dichloride by rat liver microsomal and cytosolic enzymes (462). They postulated "that several 1,2-dichloroethane activation pathways are operative which produce different adducts; the relative contribution of each pathway to total nonvolatile metabolites, mutagenic metabolites, and DNA and protein adducts under these *in vitro* assay conditions was estimated."

Reitz et al. (475) administered ^{14}C-EDC to male Osborne-Mendel rats by gavage (150 mg/kg in corn oil) or inhalation (150 ppm, 6 hr) to help understand the consequences of environmental exposure to EDC by either route. EDC was metabolized extensively following either exposure. The authors reported no significant differences in the route of excretion of nonvolatile metabolites. Approximately 85 percent of the total metabolites appear in the urine, with 7 to 8 percent, 4 percent, and 2 percent found in the CO_2, carcass, and feces, respectively, following each route of administration. The major urinary metabolites, thiodiacetic acid and thiodiacetic acid sulfoxide, were identified, suggesting a role for glutathione in biotransformation of EDC

Peak blood levels of EDC were approximately five times higher after gavage than after inhalation. It appears that elimination of EDC may become saturated when a critical blood level is exceeded. However, no marked differences in total macromolecular binding were noted between gavage/inhalation or between "target" and "nontarget" tissues after *in vivo* administration of ^{14}C-EDC.

Since the ability of EDC to induce mutations in *Salmonella typhimurium* (TA1535) is related directly to alkylation of DNA in these bacteria after incubation with ^{14}C-EDC and rat liver enzymes, alkylation of rat liver DNA was studied.

When DNA was purified from the organs of rats exposed *in vivo* to ^{14}C-EDC, very little DNA alkylation was observed after either gavage (150 mg/kg) or inhalation (150 ppm, 6 hr). The degree of alkylation ranged from 2 to 20 alkylations/10^6 nucleotides. Only slightly higher alkylation was seen after gavage versus inhalation (2 to 5 times), and no marked differences were noted between "target" and "nontarget" organs. Therefore the authors concluded *in vivo* genotoxicity (as measured by EDC/DNA binding) does not provide a satisfactory explanation for the differences noted in the two bioassays, suggesting that nongenetic factors may be important. One such factor may be the apparent saturation of EDC metabolism at the higher blood levels of EDC produced by gavage versus inhalation after nearly equivalent doses of EDC. (475)

No references were found that would indicate biologic monitoring to be satisfactory for industrial hygiene control.

Observations in Man. There have been clinical cases of acute exposure to ethylene dichloride reported in the literature. In general, they would confirm the picture observed in the animals. Menschick (164) reported four acute cases. The picture was dominated by an "hepato-renal syndrome." The author also lists 27 cases of poisoning by inhalation collected from the literature.

Ethylene dichloride does present a problem from oral ingestion. A significant number of the poisoning cases reported in the literature were by this route. The response is not unlike that observed from acute vapor exposure, that is, CNS depression, gastrointestinal upset, and injury to the liver, kidneys, and the lungs. Deaths may be delayed.

There have been some reports of chronic exposure of humans in industry that have been summarized (33, 165). Most of these reports suffer from lack of detail as to exposure concentrations, were of exposures to mixtures of solvents, or are otherwise difficult to interpret. McNally and Fosvedt (166) reported on two mild cases that had from 2 to 5 months of exposure. The symptoms were CNS depression and gastrointestinal upset with nausea and vomiting. The subjects recovered when removed from exposure. Only one report of an epidemiologic study was found and this involved a mixture of chemicals (394).

2.16.4 Hygienic Standards

The threshold limit of ethylene dichloride recommended by the ACGIH in 1980 was 10 ppm (40 mg/m^3). This is low enough to prevent systemic injury and well below a level causing subjective symptoms.

2.16.5 Odor and Warning Properties

Ethylene dichloride has a sweetish, not particularly disagreeable odor. The odor is barely detectable at 50 ppm in air and is definite but not unpleasant at 100 ppm. Although it is pronounced at 200 ppm, it still would not be considered unpleasant. Even though the odor may be definite enough to act as a warning of acutely hazardous concentrations, it is probably not sufficiently striking to be considered a significant warning of hazardous chronic exposure. This is particularly true since one can adapt to the odor at low concentrations.

2.17 Ethylene Dibromide, 1,2-Dibromoethane; EDB, CAS 106-93-4

$$CH_2BrCH_2Br$$

2.17.1 Uses and Industrial Exposure

Ethylene dibromide is used extensively in leaded antiknock gasoline and in fumigant mixtures. It is used to some extent as a chemical intermediate, special solvent, and gauge fluid. The most significant human exposure appears to have been the result of its use as a fumigant. The high boiling point, low vapor pressure, and high heat of vaporization greatly reduce the probability of high concentrations accumulating rapidly in the work atmosphere.

2.17.2   Physical and Chemical Properties

Physical state	Colorless liquid
Molecular weight	187.88
Specific gravity	2.1701 (25/4°C)
Melting point	9.97°C
Boiling point	131.6°C
Vapor pressure	12 torr (25°C)
Refractive index	1.53789 (20°C)
Percent in "saturated" air	1.5 (25°C)
Solubility	0.43 g/100 ml water at 30°C; soluble in ethanol, ethyl ether
Flammability	Essentially nonflammable by standard tests in air; it has been used as a fire extinguisher

1 mg/liter $\approx$ 130 ppm and 1 ppm $\approx$ 7.68 mg/m^3 at 25°C, 760 torr

2.17.3   Physiologic Response

Summary.   The toxicity of ethylene dibromide has long been recognized, and therefore human exposure, particularly during manufacture, has been carefully controlled. Exposure to high concentrations of the vapors of ethylene dibromide may result in some CNS depression, although the anesthetic action is weak. Deaths from acute exposure at high concentrations are usually due to pneumonia developed as a result of injury to the lungs. Following acute exposures, injury may be observed in the lungs, liver, and kidneys. Chronic exposure over a long period to levels significantly above the threshold also result in a response very similar to that seen from acute exposure. It should be particularly noted that in experimental animals the differences between the concentration causing severe injury and death and that which is tolerated for long-term exposure is not great.

Ethylene dibromide has been shown to be carcinogenic in animals and mutagenic in microbial systems and to cause reproductive effects in laboratory and domestic animals. These effects have not been observed in man, possibly because of careful control of exposure. Reviews are available (167, 168).

Single or Short-Term Exposure.   *Oral.*   Rowe et al. (169) reported the single oral dose LD$_{50}$ for several species of animals: female mice, 420 mg/kg; male rats, 148 mg/kg; female rats, 117 mg/kg; guinea pigs, 110 mg/kg; chicks, 79 mg/kg; and female rabbits, 55 mg/kg.

Eyes. Rowe et al. (169) also reported experiments on the effects of ethylene dibromide in the eyes of rabbits. Undiluted ethylene dibromide caused obvious pain and conjunctival irritation, clearing in 48 hr. A slight but superficial necrosis of the cornea was observed. Nevertheless, healing was prompt and complete. A 10 percent solution in propylene glycol was tested and produced a more severe reaction than did the undiluted material. However, the eye healed without scarring or apparent injury. It is obvious that, in handling ethylene dibromide, the eye should be protected. If the material gets into the eye, it should be washed out promptly.

Skin. Ethylene dibromide is definitely irritating and injurious to the skin. Rowe et al. (169) reported that a 1 percent solution of ethylene dibromide in butylcarbitol acetate applied 10 times in 14 days to a rabbit's ear caused slight irritation characterized by erythema and exfoliation. The same repeated applications, when bandaged on to the shaved abdomen, produced erythema and edema, progressing to necrosis and sloughing of the superficial layers of the skin.

Skin Absorption. Thomas and Yant (170) indicated that ethylene dibromide could be readily absorbed through the skin in toxic amounts. Rowe et al. reported quantitative measurements of the toxic dose absorbed through the skin in rabbits for a contact period of 24 hr. A dose of 0.21 g/kg body weight was survived by 14 out of 15 animals; 1.1 g/kg body weight killed five of five animals.

Inhalation. A quantitative expression of acute vapor toxicity of ethylene dibromide is found in the data of Rowe et al. (169). The maximum survival exposures of rats to ethylene dibromide vapor in air were as follows: 3000 ppm for 6 min, 400 ppm for 30 min, and 200 ppm for 2 hr. Guinea pigs survive 400 ppm for 2 hr and 200 ppm for 7 hr. The maximum exposures without adverse effect on female rats are 800 ppm for 6 min, 100 ppm for 2.5 hr, and 50 ppm for 7 hr. The pathological changes following acute exposures described by Rowe et al. were congestion, edema, hemorrhages, and inflammation of the lungs. The liver showed cloudy swelling and central lobular fatty degeneration and necrosis. The kidneys showed slight interstitial congestion and edema with slight cloudy swelling of the tubular epithelium in some cases.

Repeated or Prolonged Exposure. Chronic vapor exposure in animals has been reported by Rowe et al. (169). Animals were exposed 7 hr/day, 5 days/week for periods up to 6 months. At 100 ppm in air, rats and rabbits were in poor condition and some deaths occurred within the first week or two of exposure. The rats showed injury to the lungs, liver, and spleen. The rabbits showed definite liver injury. At 50 ppm, the rats showed a fairly high mortality (approximately 50 percent) due to pneumonia and infection of the upper respiratory tract, which may be related to the effect of ethylene dibromide on the lung. A number of them, however, lived through the full 6-month period. They had increases in lung, liver, and kidney weight and some histopathological changes in the lungs. Guinea pigs subjected to 57 7-hr exposures in 80 days showed some depression of growth but no increase in mortality. Lung, liver, and kidney weights were increased. There were slight histopathological changes in the liver and kidneys. An exposure of 25 ppm was tolerated by rats, guinea pigs, rabbits, and monkeys. The male rats, however, showed high mortality due to pneumonitis and infections of the upper respiratory tract, which the authors apparently considered the result of the chemical exposures. Testicular pathology was not exceptional in these animals and testicular weights were not consistently affected.

Plotnick et al. (171), however, saw adverse effects in Sprague-Dawley rats exposed to 20 ppm of the vapor 7 hr/day for their lifetime. Half the rats that also had 0.5 percent disulfiram (Antabuse) in their diet were very severely affected by 10 months of exposure. The section on carcinogenesis contains more details of this study. Changes in the testes of certain species are discussed in the next sections.

Teratogenesis and Reproductive Effects, Mutagenesis and Carcinogenesis. *Teratogenesis and Reproductive Effects.* Although it caused maternal toxicity at 20, 38, or 80 ppm when inhaled 23 hr/day, ethylene dibromide did not appear to be teratogenic in rats and mice (172). It has been shown to have rather marked effects on the reproductive system of some birds and mammals (173, 168). Ingestion of ethylene dibromide by chicken hens for a prolonged period reduced the number and size of the eggs. Male chickens were less affected (174) than hens, and bulls, but not cows, rams, or ewes, were markedly affected by repeated exposure to ethylene dibromide (175). The number of sperm, their morphology, and their motility were affected in bulls, as shown by examinations of testes and ejaculates. Rat spermatogenesis was reversibly impaired by five daily intraperitoneal injections of 10 mg/kg (176). When they studied a dominant lethal test system using mice, Epstein et al. (177) found ethylene dibromide to be negative. Doses of 50 or 100 mg/kg were given orally and intraperitoneally.

Mutagenesis. Ethylene dibromide has been shown to be mutagenic in several microbial test systems (168).

Carcinogenesis. When fed in the NCI bioassay at the "maximum tolerated dose" and one-half the "maximum tolerated dose," ethylene dibromide was shown to be carcinogenic to Osborne-Mendel rats and B6C3F1 mice (178). Initially, the mice were given 120 or 60 mg/kg per day 5 days/week. Subsequently, the doses were raised to 200 or 100 mg/kg and subsequently lowered to 60 mg/kg for both groups. Initially rats got 80 or 40 mg/kg per day but dosing was intermittent for the high dose because of poor survival. "The compound induced squamous-cell carcinomas of the forestomach in rats of both sexes, hepatocellular carcinomas in female rats, and hemangiosarcomas in male rats. In mice of both sexes the compound induced squamous-cell carcinomas of the forestomach and alveolar/bronchiolar adenomas." These doses fed were highly toxic and caused excess mortality in the rat and mice.

Hepatocellular tumors, hemangiosarcomas, and numerous other tumors have been produced in rats inhaling 20 ppm of ethylene dibromide 7 hr/day for up to 18 months (179). A marked enhancement of the toxic effects was produced by simultaneous ingestion of disulfiram (Antabuse®) at a dosage of 0.5 percent of the diet. Morbidity, mortality, and tumor incidence were markedly increased and testicular atrophy occurred in 90 percent of the rats receiving the combination. Interestingly disulfiram did not enhance the effect of inhaled vinyl chloride in a similar experiment by this group (395).

Metabolism and Biologic Monitoring.   Studies by many investigators showed that oral and intraperitoneal administration of ^{14}C-ethylene dibromide to rats and mice resulted mainly in enzymatic metabolism to S-(2 hydroxyethyl)cysteine and its N-acetate. Radioactivity was widely distributed throughout the animal, and glutathione levels were reduced. Plotnick and Conner (171) reported 12 percent of the dose was recovered unchanged in the expired air of guinea pigs given 30 mg/kg intraperitoneally.

Increased Bromide ion may be found in the blood after exposure but is of limited value in biologic monitoring because of the low concentrations present at acceptable levels of exposure.

Observations in Man.   Considering the high mammalian toxicity of ethylene dibromide there are few reports of serious human injury. No doubt, the small number of people exposed and careful handling as a result of its long recognized toxicity have helped to limit the number of injured persons. NIOSH (167) cited an early case, reported in 1910, of mistaken administration of ethylene dibromide as an anesthetic with subsequent death. Olmstead (180) reported on a death following ingestion of 4.5 ml or 140 mg/kg by a 43-year-old woman.

Serious skin injury can occur from clothing (particularly shoes) wet with ethylene dibromide. This is true not only when material is spilled inside the shoe but also when it wets the outside, for it penetrates leather. Pflesser (181) observed a case of prolonged contact with the skin when liquid ethylene dibromide was accidentally spilled into the shoes. There was reddening, blistering, and burning pain.

Calingaert and Shapiro (182) have observed that ethylene dibromide can penetrate through several types of protective clothing, particularly neoprene rubber and several types of plastic gloves. Nylon was found to be the most resistant material but lacking in good physical characteristics. These authors proposed a combination of neoprene and nylon.

Human experience during manufacture of ethylene dibromide has been summarized (183). The mortality of two populations of employees was compared to that expected in an industrial population. In neither study was there evidence of increased deaths due to cancer. However, the populations were small; hence the statistical confidence ranges are very broad. Nevertheless, it is obvious that a one-hit statistical model based on animal data severely overestimates the number of tumors predicted in human populations exposed during the production of ethylene dibromide (184).

The reproductive history of 297 male employees manufacturing ethylene dibromide was also summarized (185). In this limited study, reproduction generally appeared to be unaffected by exposure to ethylene dibromide, although one of four plants had fewer children than predicted.

2.17.4 Hygienic Standards

No threshold limit value for ethylene dibromide was recommended by the ACGIH in 1980 although it was included in their list of animal carcinogens. It would appear prudent to limit exposure to a maximum time-weighted average of 5 ppm or less with careful control to minimize dermal contact.

2.17.5 Odor and Warning Properties

The odor of ethylene dibromide is not unpleasant and has been termed "chloroformlike." The odor is not detectable at a low enough concentration to be considered a good warning of excessive exposure.

2.18 Methyl Chloroform, 1,1,1-Trichloroethane, CAS 71-55-6

$$CH_3CCl_3$$

2.18.1 Uses and Industrial Exposures

Methyl chloroform is used almost exclusively as a solvent. Worldwide consumption is about 1 billion lb/year. Increasing consumption is due to a favorable combination of chemical, flammability, physical, and toxicologic properties, compared to many other common solvents. Most commercial methyl chloroform, which is sold under several trade names, contains inhibitors to prevent reaction of the solvent with aluminum and aluminum alloys. This reaction produces hydrogen chloride and in confined vessels may produce high pressures. Extensive testing and human experience indicate 1,1,1-trichloroethane is probably the least toxic of the chlorinated solvents, but its high volatility and careless use and abuse have resulted in anesthetic deaths from gross exposure, usually in confined spaces.

2.18.2 Physical and Chemical Properties

Physical state	Colorless liquid
Molecular weight	133.42
Specific gravity	1.3249 (26/4°C)
Boiling point	74.1°C
Vapor pressure	127 torr (25°C)
Refractive index	1.43765 (21°C)
Percent in "saturated" air	16.7 (25°C)
Solubility	0.09 g/100 ml water at 20°C; soluble in ethanol, ethyl ether
Flammability	(see note below)

1 mg/liter $\approx$ 183 ppm and 1 ppm $\approx$ 5.46 mg/m^3 at 25°C, 760 torr

Note: The flammable characteristics of methyl chloroform are similar to those of trichloroethylene. It has no flash point or fire point by ASTM procedures for Tag closed cup and Cleveland open cup tests. Limits of flammability of vapors of inhibited 1,1,1-trichloroethane have been found to be 10 to 15.5 percent in air with hot wire ignition. A considerable amount of energy is required for ignition. It will not sustain combustion (186).

2.18.3 Physiologic Response

Summary. The principal and first response from acute or chronic exposure to excessive amounts of methyl chloroform is depression of the central nervous system. It has little capacity to produce organ injury from either single or repeated exposures but at high levels can sensitize the heart to epinephrine.

The 1,1,1-trichloroethane isomer is considerably less toxic than the 1,1,2-trichloroethane isomer. Owing to an error in the literature, several authors have indicated the reverse. Torkelson et al. (187) have explained the circumstances and documented the literature.

Concentrations in excess of 14,000 to 15,000 ppm are fatal to animals. Human fatalities due to anesthesia (and/or cardiac arrhythmia) have occurred in confined spaces when exposures to high concentrations have not been promptly terminated. In cases where the victim has been alive when removed from the high concentration, recovery has generally been rapid and complete. Abuse (sniffing) has also resulted in deaths. Several reviews have been published (188–190).

Single or Short-Term Exposure. *Oral.* Torkelson et al. (187) reported the oral LD_{50} for several species to be as follows: male rats, 12.3 g/kg; female rats, 10.3 g/kg; female mice, 11.24 g/kg; female rabbits, 5.66 g/kg; and male guinea pigs, 9.47 g/kg. Deaths were due primarily to anesthesia. Recovery was rapid and complete in surviving animals.

Eyes. The material caused only minor, transient irritation in the eyes of rabbits.

Skin. The skin shows only slight reddening and scaliness from contact. The reaction is somewhat increased on repeated exposures. Confinement of the liquid on the skin results in considerable pain and irritation (187).

Skin Absorption. Applied under a cuff for 24 hr, 3.9 g/kg was survived by all rabbits; 15.8 g/kg failed to kill all the rabbits treated with the undiluted liquid (187). Data on absorption of the liquid and vapor observations through human skin are presented in the section on observations in man.

Inhalation. Adams et al. (191) reported on the response of animals to acute exposure. Maximum time–concentrations in air survived by rats were as follows: 6 min at 30,000 ppm, 1½ hr at 15,000 ppm, and 7 hr at 8000 ppm. Maximum time–concentrations in air with no detectable injury in rats were as follows: 18 min at 18,000 ppm and 5 hr at 8000 ppm. It should be noted that the maximum level with no detectable injury is very close to the maximum level survived. This information has been confirmed by Torkelson et al. (187) using inhibited solvent. Others have reported similar low hepato- and renal toxicity in rats and other species (53, 192–194). Gehring (54) determined that at 13,500 ppm the LT_{50} was 595 min but that liver function as measured by SGOT was virtually unaffected unless exposures approached those causing anesthetic death.

Injection. Klaasen and Plaa reported an intraperitoneal LD_{50} of 3.9 g/kg in rats (194) and an LD_{50} of 120 mM/kg (16 g/kg) in mice (192).

Repeated or Prolonged Exposure. *Inhalation.* Adams et al. (191) exposed animals 7 hr/day, 5 days/week for 1 to 3 months. At 10,000 ppm rats showed staggering gait and weakness in 10 min. By 3 hr, they showed loss of color, irregular respiration, and semiconsciousness. Survivors had completely recovered by the following morning. For those that succumbed, death seemed to be due to either cardiac or respiratory failure. At 5000 ppm, there was definite but mild narcotic effect within 1 hr. There was reduced activity. Rats survived for 31 exposures over 41 days without apparent injury. Rabbits showed slight retardation of growth at 5000 ppm. At 3000 ppm, rabbits and monkeys showed no response over a 2-month period. Guinea pigs showed a barely significant retardation of growth at 650 ppm.

Torkelson et al. (187) confirmed and extended these results, using inhibited material containing 2.4 to 3 percent dioxane, 0.12 to 0.3 percent butanol, and small amounts of ethylene dichloride, water, etc. Laboratory animals were exposed repeatedly to 500, 1000, 2000, or 10,000 ppm in order to establish conditions safe for repeated exposure. Rats, guinea pigs, rabbits, and monkeys were unaffected after 6 months of repeated 7-hr exposures 5 days/week to 500 ppm. Female guinea pigs, which were found to be the most sensitive in previous experiments, were able to tolerate 1000 ppm for 0.6 hr/day with no detectable adverse effects. Male rats tolerated exposure of 0.5 hr/day to 10,000 ppm with no organ injury.

The section on tetrachloroethylene describes the result of repeated exposure to a mixture of 1,1,1-trichloroethane and tetrachloroethylene (404).

Even continuous (24 hr/day) exposure for 14 weeks produced remarkably little injury to rats, mice, dogs, and monkeys at 1000 ppm and essentially no effect at 250 ppm except minor, apparently reversible cytoplasmic alterations in the livers of mice (124). After exposure to 1000 ppm dogs and monkeys were considered unaffected by the exposure but the rats and mice exhibited significant increases in liver weight and liver triglycerides and fatty and necrotic changes visible microscopically in the livers.

Quast et al. (195) repeatedly exposed groups of 96 rats of each sex to 875 or 1750 ppm 1,1,1-trichloroethane vapor 6 hr/day, 5 day/week for 1 year with no adverse effects (see section on carcinogenicity).

Teratogenesis Reproduction, Mutagenesis, and Carcinogenesis. *Teratogenesis and Reproduction.* Pregnant female rats and mice were exposed 7 hr/day on days 6 to 15 of gestation and the mothers and fetuses examined for teratological effects (63). The inhaled vapor concentration was 875 ppm of an inhibited formulation containing 5.5 percent inhibitors. No effects related to exposure were observed on either the mothers or fetuses. York et al. (473) repeatedly exposed female Long-Evans rats to 2100 ± 200 ppm of methyl chloroform vapors 2 weeks before breeding and until day 20 of gestation without effect on teratogenicity or subsequent neurobehavioral studies of the offspring allowed to deliver normally.

Riddle et al. (477) modified a multigeneration reproduction study to include screening for dominant lethal and teratogenic effects. Mice were fed water containing methyl chloroform so that daily dosage levels were 0, 99.4, 2640, or 8520 mg/kg. According to the authors, "There appeared to be no dose-dependent effects on fertility, gestation, viability, or lactation indices. Pup survival and weight gain were not adversely affected. Gross necropsy of male and female F/O generation mice treated with 1,2-DCE or 1,1,1-TCE failed to reveal compound or dose-related effects."

Mutagenesis.   Experimental results of mutagenic testing in vitro indicate negative results (11), or, at most, a very weak action. Similar to other chlorinated solvents small amounts of additional chemicals are frequently added to formulated products of methyl chloroform to stabilize against decomposition during use. Some of these stabilizers are electrophilic in nature and they may be responsible for the weakly positive findings in the Ames test reported sporadically (127). For example, in the case of trichloroethylene stabilized with epichlorohydrin, the stabilizer has been implicated as being responsible for the occasional positive results in in vitro and longer term animal carcinogenic studies. This is likely to remain a controversial issue but data available to date indicate that methyl chloroform per se is not likely to be positive in mutagenic tests. Methyl chloroform has been shown to be negative in a dominant lethal assay at daily doses as high as 8,500 mg/kg (477).

Carcinogenesis.   Methyl chloroform has failed to demonstrate carcinogenic properties in two lifetime studies. When fed by gavage at daily doses of 1500 and 750 mg/kg/day for 2 years there was a significant increase in mortality in rats (197), possibly due to respiratory involvement related to the dosing technique. Mice were dosed in this NCI bioassay study with "time-weighted" average daily doses of 5615 and 2807 mg/kg. A moderate body weight depression occurred and survival was significantly decreased. No increase in tumors was observed but the massive doses fed caused so much mortality that the investigators were reluctant to assess the carcinogenicity from this study.

More reasonable yet high exposures were given by inhalation to male and female rats with no evidence of increased tumors (195). Male and female rats (96 of each sex) were exposed to either 1750 or 875 ppm of an inhibited 1,1,1-trichloroethane formulation 6 hr/day for 1 year and then kept for their lifetimes. No adverse effects on mortality, hematology, clinical chemistry, or gross and histopathology were observed, nor was there an increase in tumors of any type.

Extensive additional testing is currently underway on 1,1,1-trichloroethane. In our opinion the available data indicate it is very unlikely that carcinogenicity is of concern if exposure to 1,1,1-trichloroethane is kept below levels causing frank anesthesia.

Metabolism and Biologic Monitoring.   The low systemic toxicity of methyl chloroform appears related to the small amount of metabolism which occurs in animals and man. Because of its low toxicity it has been used as a model compound in absorption–excretion studies (187, 198). In one of the first metabolic studies on this compound, Hake et al. (199) reported 97.6 percent of radioactive methyl chloroform injected intraperitoneally was excreted unchanged by the lungs and only 0.85 percent of the radioactivity was found in the urine. This and subsequent studies on man and animals indicate that only small amounts of trichloroacetic acid and trichloroethanol may be found in the urine after injection or inhalation (200–203, 478), but that most of the dose is expired unchanged no matter what the route of administration. Enzymatic dechlorination by rat liver microsomes in vitro is very low and apparently not enhanced by enzyme inducers (204). The reaction is catalyzed by cytochrome P-450 in the presence of oxygen (205).

Stewart et al. (206), have produced a series of graphs making it possible to quantitate exposure based on expired air samples. Analysis of urine for metabolites probably has limited value. Monster (213) has also published extensive studies on biologic monitoring of 1,1,1-trichloroethane.

Observations in Man. Methyl chloroform has been evaluated as a surgical anesthetic but for several reasons, including lack of potency and effects on cardiac function, it was abandoned from serious consideration (207–209). It appears reasonable to assume that cardiac arrhythmias and fibrillations could be factors in some of the industrial deaths that have been reported from methyl chloroform. However, since industrial deaths have occurred in confined spaces as the result of exposures to obviously anesthetic levels, death could also have been due to anesthesia. If the subject has been alive when removed from exposure, recovery has generally been rapid and complete. In every fatality, misuse such as failure to prevent inhalation of high concentrations, failure to adequately ventilate confined spaces, or deliberate sniffing has apparently been involved.

Table 48.10 on the probable effects of exposure to humans to the vapor of methyl chloroform has been published for establishing emergency exposure limits (210).

Laboratory Studies. Numerous laboratory studies in man have permitted confirmation of much of the toxicologic data derived in animals. Stewart et al. in a series of papers have evaluated absorption and excretion following single and repeated inhalation exposures as well as dermal application. They have also evaluated the use of expired air for monitoring employee exposure and have produced a series of graphs based on expired air concentration following known exposures to various concentrations for different periods of time (206). These studies have also investigated metabolism and urinary excretion and evaluated clinical chemical parameters. Most important, however, are their studies to evaluate equilibrium, coordination, alertness, and other signs of anesthetic action. Based on these studies it has been concluded that, except for possible objections to odor, no untoward responses are observed even after repeated prolonged exposures of human subjects to 350 ppm. A few subjects

Table 48.10. Probable Result of Single Exposure to the Vapors of 1,1,1-Trichloroethane (210)

Exposure time (min)	Concentration in Air (ppm)	Expected Effect in Humans[a]
5	20,000	Complete incoordination and helplessness (R)
	10,000	Pronounced loss of coordination (R)
	5,000	Definite incoordination (R, M)
	2,000	Disturbance of equilibrium. Odor is unpleasant but tolerable (H)
15	10,000	Pronounced loss of coordination (R)
	2,000	Loss of equilibrium (H)
	1,000	Possible beginning loss of equilibrium (H)
30	10,000	Pronounced loss of coordination (R)
	5,000	Incoordination (R, M)
	2,000	Loss of equilibrium (H)
	1,000	Mild eye and nasal discomfort; possible slight loss of equilibrium (H)
60	20,000	Surgical anesthesia, possible death (R)
	10,000	Pronounced loss of coordination (R)
	5,000	Obvious loss of coordination (R, M)
	2,000	Loss of coordination (H)
	1,000	Very slight loss of equilibrium (H)
	500	No detectable effect, but odor is obvious (R, H)
	100	Apparent odor threshold (H)

[a] Expected effects are based on (H) human data, (M) monkey data, and (R) rat data.

may respond as concentrations approach 500 ppm, and if allowed to inhale 800 to 1000 ppm some subjects show minor CNS impairment. Other investigators (211) have found similar results but Gamberale and Hultengren (212) reported effects at lower vapor concentrations. Gamberale and Hultengren, however, used a mask to administer the vapors and this may have influenced their test results.

Monster (213) has published extensive studies on biologic monitoring of 1,1,1-trichloroethane, trichloroethylene, and perchloroethylene.

Liquid methyl chloroform has been shown by Stewart and Dodd (50) as well as Fukabori et al. (214) to be absorbed through the intact skin, but Riihimaki and Pfaffli (215) have shown that the vapors are not absorbed in toxicologically significant amounts.

Industrial Experience. Although chronic effects have been shown to be of little consequence in industrial exposure (216), failure to recognize the high vapor pressure of methyl chloroform and to prevent inhalation of high concentrations have resulted in reports of death. The number of deaths per year (two to three worldwide) appears to be decreasing relative to consumption, as proper precautions are being taken to avoid exposure. A few nonoccupational cases of deliberate sniffing have been reported, but generally a consistent pattern of use of the solvent in a confined space without regard for ventilation is the cause of industrial death. Stewart (188) discussed the consequences of overexposure and the treatment of overexposed subjects.

A rather extensive study of an industrial population exposed to methyl chloroform for up to 6 years confirms the data derived from animals (216). No effect of exposure was found in a matched-pair study of 151 subjects and 151 controls despite particular emphasis placed on hepato- and cardiotoxic effects.

2.18.4   Hygienic Standards of Permissible Exposure

The threshold limit value for methyl chloroform recommended by the ACGIH in 1980 was 350 ppm (1900 mg/m^3). This value provides a wide margin of safety from systemic injury. Peak concentrations should be limited to about 500 ppm to prevent beginning anesthesia in some subjects.

2.18.5   Odor and Warning Properties

Although methyl chloroform has a typical sweetish odor, it is not striking enough to be considered a good warning. The odor may be noticeable at concentrations near 100 ppm, well below those known to cause physiologic response. However, the odor at 500 ppm and even 1000 ppm is not so unpleasant as to discourage exposure. The odor has been described as strong and unpleasant at 1500 to 2000 ppm. Stewart et al. (206) reported female test subjects exposed to 350 ppm objected to the odor; however, this has not been an industrial problem.

2.19 1,1,2-Trichloroethane, Vinyl Trichloride, CAS 79-00-5

$$CHCl_2 - CH_2Cl$$

2.19.1 Uses and Industrial Exposure

The use of 1,1,2-trichloroethane is quite restrictive. It is used to a slight extent as a speciality solvent and chemical intermediate, but the availability of other less toxic solvents discourages its use.

2.19.2 Physical and Chemical Properties

Physical state	Colorless liquid
Molecular weight	133.42
Specific gravity	1.443 (20/4°C)
Melting point	−36.7°C
Boiling point	113.5°C
Vapor pressure	25 torr (25°C)
Refractive index	1.4711 (20°C)
Percent in "saturated" air	3.3 (25°C)
Solubility	0.44 g/100 g water at 20°C; soluble in ethanol and ethyl ether
Flammability	Not flammable by standard tests in air

1 mg/liter $\approx$ 183 ppm and 1 ppm $\approx$ 5.46 mg/m^3 at 25°C, 760 torr

2.19.3 Physiologic Response

Summary. The principal physiologic responses to 1,1,2-trichloroethane are depression of the central nervous system and liver injury.

Similarity in chemical names has resulted in some confusion in older literature which misquoted work by Lazarew (217) and stated that the 1,1,2 compound was less toxic than 1,1,1-trichloroethane. Torkelson et al. (187) have explained the source of the error, which could be serious since 1,1,2-trichloroethane is much more hepatotoxic when given to animals by single or repeated doses. Consistent with the small quantity of the material used in industry and the care with which it is now generally handled, no published reports of human injury were found.

Single or Short-Term Exposure. *Oral.* Wright and Shaffer (218) reported that the oral lethal dose for dogs was 0.5 ml/kg. Irish (1) indicated an LD$_{50}$ for rats of 0.1 to 0.2 g/kg. Liver and kidney pathology were seen at considerably lower doses. Gehring (54) included 1,1,2-trichloroethane in a series of chlorinated solvents evaluated for hepatotoxicity and considered it less hepatotoxic in the rat than carbon tetrachloride or chloroform, but markedly more toxic than 1,1,1-trichloroethane. Other investigators have confirmed this in mice as well as rats (194, 219). Kidney injury also occurs following oral and subcutaneous treatment of mice (196).

Skin and Eye.   1,1,2-trichloroethane is not highly irritating to the skin and eyes but, as with many other solvents, injures the skin by defatting.

Skin Absorption.   Absorption of the liquid through the skin has been shown to occur in rabbits (11) and guinea pigs (220). In rabbits, three of four survived single 24-hr applications of 1 or 2 g/kg, but these dosages and 0.5 g/kg caused illness, delayed recovery, and liver and kidney injury. Jakobson and other co-workers also investigated the kinetics of absorption through the guinea pig skin as well as the nature of injury to the skin with prolonged contact (220, 221). Wahlberg (222) reported that repeated applications of 0.5 ml or more to the skin killed all the guinea pigs in 3 days, but 0.25 ml killed only 5 of 20 animals treated for a longer period of time. Although absorption does not appear to be a serious problem from acute exposure, prolonged or repeated exposure of the skin may result in manifestations of chronic toxicity.

Inhalation.   Pozzani et al. (223) reported that a single 7-hr exposure to 500 ppm of the vapor was lethal to about half of the rats so exposed. Carpenter reported the acute lethal concentration for about half of the exposed rats to be 2000 ppm for a 4-hr exposure followed by a 14-day observation period.

Limited unpublished data (11) indicate a moderate to high toxicity in laboratory animals. Single 7-hr exposures to 500 and 250 ppm resulted in death of over half of groups of female rats. Autopsy of survivors showed marked injury to the liver and kidneys. Rats exposed to 250 ppm for 4 hr survived but also showed liver and kidney necrosis. Two- or 1-hr exposures apparently did not cause gross or microscopic injury. All rats survived a 7-hr exposure to 100 ppm but were not examined microscopically.

Repeated or Prolonged Exposure.   Six months of repeated 7-hr exposures, 5 days/week to 15 ppm was tolerated by male and female rats, guinea pigs, and rabbits. No histopathological changes were observed on necropsy and the usual parameters of growth, mortality, organ weights, hemotology, and clinical chemistry were not affected (11).

However, 16 7-hr exposures to 30 ppm resulted in minor fatty changes and cloudy swelling in the livers of female rats. Male rats appeared unaffected, although pneumonitis was slightly higher in the 10 exposed rats than in the controls.

Teratogenesis, Mutagenesis, and Carcinogenesis. *Teratogenesis.* No data were found on teratogenic studies.

Mutagenesis. Simmon et al. (127) tested 1,1,2-trichloroethane and found it not to be mutagenic in *S. typhimurium* and only weakly mutagenic in *S. cerevisiae.*

Rannug et al. (224) determined that 1,1,2-trichloroethane was not mutagenic in their tests on *S. typhimurium* TA 1535 even when rat mitochondrial liver fraction was added to the culture medium.

Carcinogenesis. 1,1,2-trichloroethane has been included in the NCI bioassay program, in which it was fed by gavage to rats and mice (225). As in many of these studies with hepatotoxic compounds, hepatocellular carcinomas occurred in mice but not rats fed for 78 weeks. Rats were kept an additional 35 weeks and mice 13 weeks following treatment. Pheochromocytomas were also observed in mice. The doses fed were 92 and 46 mg/kg per day for rats and 390 and 195 mg/kg per day for mice. Mortality was accelerated in female mice but not in the rats or male mice. The NCI report does not indicate the degree of noncarcinogenic histopathology produced by these doses.

Metabolism and Biologic Monitoring. Yllner (226) injected ^{14}C-labeled 1,1,2-trichloroethane intraperitoneally in mice. The site of labeling on the molecule was not stated. When collected over a 3-day period, 73 to 87 percent of the activity was recovered in the urine. Less than 2 percent was in the feces which may have been the result of contamination by urine. Expired air contained 16 to 22 percent of the radioactivity (60 percent ^{14}CO$_2$ and 40 percent unchanged 1,1,2-trichloroethane). One to three percent of the labeled compound remained in the animal. Metabolism appeared to proceed through chloroacetic acid, indicating oxidative dechlorination or dechlorination followed by oxidation of 1,1,2-trichloroethane at the carbon containing two chlorine atoms. This is followed by further oxidation of the chloroacetaldehyde to the corresponding acid. Cytochrome P 450 is probably involved in the dechlorination (205, 227).

1,1,2-trichloroethane is absorbed from the lungs and no doubt also is excreted in the breath. Although no data are available it seems probable that the low concentrations that could be present in expired air after exposure to levels considered safe for occupational exposure will not be useful for monitoring workers exposure. Qualitative analysis might verify that exposure to 1,1,2-trichloroethane has occurred.

Ikeda and Ohtsuji (200) have shown a positive Fujiwara reaction in the urine of rats and mice treated with 1,1,2-trichloroethane but the value of this determination in humans does not appear to have been investigated. At the low levels of exposure considered acceptable it may have limited value.

2.19.4   Hygienic Standards

The ACGIH in 1980 recommended a threshold limit value of 10 ppm with a notation warning against possible skin absorption.

2.19.5   Odor and Warning Properties

Although 1,1,2-trichloroethane is reported to have an odor similar to chloroform, no quantitative data were found concerning odor or warning properties. In view of its high systemic toxicity, odor cannot be used to protect against excessive acute or chronic exposure.

2.20   Acetylene Tetrachloride, 1,1,2,2-Tetrachloroethane, CAS 79-34-5

$$CHCl_2CHCl_2$$

2.20.1   Uses and Industrial Exposures

Acetylene tetrachloride was formerly used as a solvent for cleaning and extraction processes and is still used to some extent as a chemical intermediate. Present usage is quite limited since less toxic solvents are available.

2.20.2   Physical and Chemical Properties

Physical state	Colorless liquid
Molecular weight	167.86
Specific gravity	1.5869 (25°C)
Melting point	−42.5°C
Boiling point	146.3°C
Vapor pressure	6 torr (25°C)
Refractive index	1.4918 (25°C)
Percent in "saturated" air	0.79 (25°C)
Solubility	0.32 g/100 water at 25°C; soluble in ethanol and ethyl ether
Flammability	Not flammable by standard tests in air

1 mg/liter ≏ 145.8 ppm and 1 ppm ≏ 6.86 mg/m³ at 25°C, 760 torr

2.20.3   Physiologic Response

Summary.   Acetylene tetrachloride is considered to be among the more toxic of the smaller chlorinated hydrocarbons. The most significant injury from subacute or chronic exposure is to the liver. The first indication may be a greatly enlarged and palpable liver, which may progress to fatty degeneration and cirrhosis. Injury to the kidneys may also be observed.

This compound also causes CNS depression, dizziness, and incoordination, as do many chlorinated hydrocarbons. In very severe acute exposures, unconsciousness and death from respiratory failure may be seen. Central nervous system depression is not a striking part of the response to usual industrial exposure because of the low volatility and because other injurious effects predominate at lower levels. Respiratory irritation may be observed and may lead to pulmonary damage. A significant irritation in the gastrointestinal tract is also observed and may result in nausea, vomiting, and gastric pain.

Literature reviews are available (33, 396), but there is a remarkable lack of quantitative data in animals to support the apparent high toxicity in humans. It is possible, however, that inhaled concentrations have been underestimated or that dermal contact has been more significant than realized.

Limited data on the 1,1,1,2 isomer (228) indicate that this isomer is less toxic than the symmetrical isomer.

Single or Short-Term Exposure. *Oral.* Barsoum and Saad (229) reported that an oral dose of 0.7 g/kg body weight in dogs is a toxic dose but Wright and Schaffer (208) reported that a lethal dose for dogs was 0.3 ml/kg given orally.

Skin and Eyes. Schwander (230) indicated that acetylene tetrachloride may be absorbed through the intact skin. However, it does not have a high acute toxicity by this route; Smyth (147) reported an acute dermal LD_{50} of 6.4 g/kg in rabbits.

Inhalation. Although there are a number of animal experiments with this material, a clean-cut quantitative study of animal response from acute exposure has not been reported. Smyth (147) quoted unpublished work by his laboratory indicating that rats were found to survive a 4-hr exposure at 500 ppm but would not survive 4 hr at 1000 ppm. Carpenter et al. (231) reported an acute lethal concentration (approximate LC_{50}) for rats to be 2000 ppm from a 4-hr exposure followed by 14 days of observation.

Intraperitoneal Injection. One hundred to 800 μm/kg (17–130 mg/kg) per day was injected intraperitoneally for 7 days in male rats with no toxic symptoms; however, liver enzymes and possible hematologic changes were reported (232).

Repeated or Prolonged Exposure. *Inhalation.* Lehmann and Flury (102) reported data from limited vapor experiments on cats and rabbits. They were exposed to a concentration of from 100 to 160 ppm for 8 to 9 hr daily for 4 weeks. No typical organ changes were found. These observations are rather surprising when, as is discussed later, industrial experience indicates that injury to man has occurred at much lower concentrations.

Navrotskii et al. (233) exposed rabbits to tetrachloroethane (presumably the 1,1,2,2 isomer) vapor for 7 to 11 months. The 3 to 4 hr daily exposures were to 100 mg/m³ (14.6 ppm). According to the available abstract only slight effects on the liver were observed. Schmidt et al. (234) also reported only slight effects in rats exposed to 15 mg/m³ (2 ppm) for 265 days and considered their results somewhat inconclusive. Some toxicologic parameters deviated during the exposure but were not different from the controls at the end. Mortality was not affected, but histological examination appears to have been inadequate to evaluate chronic effects.

Teratogenesis, Mutagenesis, and Carcinogenesis. *Teratogenesis.* A very limited reproduction study by Schmidt et al. (234) indicated no gross reproductive or teratological effect in rats exposed daily for 9 months prior to mating. A second report by Schmidt (235) describes fetotoxic (lethal) effects in two strains of mice given intraperitoneal injections of tetrachloroethane, presumably the 1,1,2,2 isomer. The doses administered 300, 400, or 700 mg/kg per day were given singly or on several days of pregnancy. According to the author's summary the compound was fetotoxic and faintly teratogenic.

Mutagenesis. The mutagenicity of acetylene tetrachloride in S. *typhimurium* has been evaluated and found positive (248). Rosenkrantz (236) found it to be only weakly active in *Salmonella* strains TA 1535 and TA 1538.

Carcinogenesis. The National Cancer Institute (237) has included acetylene tetrachloride in their bioassay series using rats and mice. Their summary states, "The time weighted average doses (by gavage) were 108 and 62 mg/kg/day for male rats, 76 and 43 mg/kg/day for female rats and 282 and 142 mg/kg/day (78 wk) for all mice. There was a highly significant pos.(itive) dose related trend in the incidence of hepatocellular carcinoma in mice of both sexes. No statistically significant incidence of neoplastic lesions was observed in male or female rats. However, 2 hepatocellular carcinomas and 1 neoplastic nodule, which are rare tumors in the male Osborne-Mendel rat, were observed in the high dose males. Under the conditions of this bioassay, orally administered 1,1,2,2-tetrachloroethane was a liver carcinogen in B6C3F1 mice of both sexes. The results did not provide conclusive evidence for the carcinogenicity of 1,1,2,2-tetrachloroethane in Osborne-Mendel rats."

Metabolism and Biologic Monitoring. Studies have shown that acetylene tetrachloride is readily absorbed via the lungs or gastrointestinal tract. Some authors have indicated absorption by the skin. It is apparently readily excreted by the lungs.

Yllner (238) injected ¹⁴C-labeled 1,1,2,2-tetrachloroethane intraperitoneally in female albino mice at doses of 0.21 to 0.32 g/kg and studied the elimination for 3 days. About half the dose, 45 to 61 percent, was excreted as carbon dioxide with 28 percent excreted in the urine. About 16 percent remained in the animal and only 4 percent was expired unchanged. Several products were found in the urine with half the urinary activity unaccounted for. Both enzymatic and nonenzymatic activity was postulated. Enzymatic hydrolytic fission of chlorine–carbon bonds results in dichloroacetic acid and glycolic acid. Nonenzymatic dehydrochlorination results in trichloroethylene and subsequently trichloroethanol and trichloroacetic acid.

Ikeda and Ohtsuji (200) also found trichloro products in the urine of mice and rats exposed for 8 hr to 200 ppm of 1,1,2,2-trichloroethane vapor.

The high toxicity of acetylene tetrachloride would suggest that neither breath nor urine analysis would be suitable for monitoring exposures at the low level considered acceptable for occupational exposures.

Observations in Man.   The review by von Oettingen summarizes much of the old data on human experience and based on numerous cases describes the rather severe effect discussed earlier in this section.

Sherman (239) reported on eight humans who were given 3 ml of tetrachloroethane by mistake. It was given with 30 g of magnesium sulfate and water. Within 1.5 to 2.5 hr, all were comatose. Reflexes were absent and the pulse barely perceptible. Respiration was shallow and rapid. The patients all recovered and showed no aftereffects.

Gurney (240) reported a number of cases of chronic exposure to acetylene tetrachloride. He studied 277 individuals of whom 75 had symptoms and 55 had enlarged livers. There were nearly as many with enlarged livers among those who did not show symptoms as there were among those who did. Coyer (241) reported on six cases, one of which was fatal.

It is evident from clinical reports of exposure to acetylene tetrachloride that the principal effect involves the liver. Symptoms referable to gastrointestinal injury may also be observed.

Hygienic Standards.   The ACGIH in 1980 recommended 5 ppm (35 mg/m³) as a TLV for 1,1,2,2-tetrachloroethane, with a notation to avoid skin contact.

There appear to be few data to support this or any other value.

Odor and Warning Properties.   Acetylene tetrachloride had a mild, sweetish odor similar to several other chlorinated hydrocarbons. Lehmann and Schmidt-Kehl (52) reported 3 ppm to have a noticeable odor and 13 ppm to be tolerated for 10 min. Concentrations of 186 ppm inhaled for 30 min, or 335 ppm inhaled for 10 min, had a disagreeable and marked odor, causing upper respiratory irritation and CNS effects. Since the odor is not particularly striking, and since the acceptable level is 5 ppm, odor does not appear to be of value as a warning property.

2.21 Acetylene Tetrabromide, 1,1,2,2-Tetrabromoethane, CAS 79-27-6

$$CHBr_2CHBr_2$$

2.21.1 Uses and Industrial Exposures

Because of its high density acetylene tetrabromide is used as a gauge fluid, for balancing equipment, and for ore separation. It has had some use as a special solvent.

2.21.2 Physical and Chemical Properties

Physical state	Colorless to yellow liquid
Molecular weight	345.7
Specific gravity	2.9638 (20/4°C)
Melting point	0.13°C
Boiling point	Decomposition, 239 to 242°C
Vapor pressure	0.04 torr (24°C)
Refractive index	1.63795 (20°C)
ppm v/v in air "saturated" at 25°C	80 (theoretical)
Solubility	0.065 g/100 ml water at 30°C; soluble in ethanol, ethyl ether, chloroform
Flammable	Not flammable by standard tests in air

1 mg/liter $\approx$ 70.4 ppm and 1 ppm $\approx$ 14 mg/m^3 at 25°C, 760 torr

2.21.3 Physiologic Response

Summary. Acetylene tetrabromide is a CNS depressant. If given in sufficiently large doses, it may cause narcosis, coma, and eventually death from respiratory failure. There may be lung irritation and pathological changes may be observed in the liver and kidneys. Because the vapor pressure is exceedingly low at room temperatures, inhalation exposure may be controlled by reasonable precautions and ordinary ventilation. However, contact with the skin should be avoided as well. Very little new data has been reported since the second revised edition of this book.

Single or Short-Term Exposure. *Oral.* Gray (242) reported the LD$_{50}$ for rabbits and guinea pigs to be approximately 0.4 g/kg of body weight. Rats survived 0.6 g/kg and succumbed to 1.6 g/kg (243).

Skin.   Contact with the open skin does not result in any skin reaction if it is washed off in a reasonable period of time. If the material is bandaged onto the skin and allowed to remain there for a period of hours, a slight redness appears. In 24 hr, there are some edema and blistering. It can be concluded that ordinary contact does not represent a skin problem. However, wet clothing or shoes that may contain the material should be removed and cleaned before reuse.

Aerosols.   Arkhangel-Skaya and Yanushkevich (244) reported on the exposure of rats to aerosols of acetylene tetrabromide. A concentration varying from 3.7 to 4.2 mg/liter for a single exposure of 2 hr gave only slight and ill-defined symptoms of toxicity. Concentrations varying from 5.9 to 7.2 mg/liter caused excitation followed by sleepiness. Repeated daily exposure to aerosol concentrations of 3.7 to 4.2 mg/liter resulted in death of the animals.

Inhalation.   Merzbach (245) and Glaser and Frisch (246) exposed animals to vapors of acetylene tetrabromide. It is a bit difficult to determine the significance of their findings since Gray (242) indicated that the concentrations reported by these authors were, in most instances, well above a saturated atmosphere.

Gray reported exposure of animals to the vapors of acetylene tetrabromide at near saturation in a static chamber. Rabbits were exposed for up to $2\frac{1}{2}$ hr and rats for up to 3 hr with no deaths. Guinea pigs exposed $\frac{1}{2}$ hr survived. One out of two survived an exposure of 1 hr but all exposed succumbed to exposures of $1\frac{1}{2}$ hr or more. In the guinea pigs that died, injury was seen in the liver and kidneys. The guinea pigs recovered consciousness in a period of 5 hr but died in periods of from 1 to 5 days following the exposure.

Repeated or Prolonged Exposure.   Gray (242) exposed mice, rats, guinea pigs, and rabbits for 15 min daily for 47 to 92 days to a "saturated atmosphere." One mouse out of 48 died of an unknown cause; the rest survived without adverse effects. Four out of 36 albino rats were ill from lobar pneumonia, a complication that was considered unrelated to the exposure. The rest of the animals survived without injury. The rabbits were observed taking significant amounts of the condensed acetylene tetrabromide orally by licking it from their bodies. They were, therefore, not considered significant to the experiment. It should also be recognized that a certain amount of oral intake would have been possible in the case of the other animals as well, but even so the chronically exposed animals showed no significant pathological change.

Hollingsworth et al. (243) reported that the concentration in air could not be maintained above approximately 14 ppm by analysis. The theoretically saturated concentration should have been about 79 ppm. Rats, guinea pigs, rabbits, and a monkey were exposed 7 hr/day, 5 days/week for periods ranging from 100 to 106 days. The average concentration was 14 ppm by analysis. All animals survived and appeared normal. There was growth depression in guinea pigs.

All animals showed an increase in liver weight. Histopathological changes were observed in the liver and lungs. At 4 ppm, some animals showed slight histopathological changes in the liver and some in the lungs. At 1.1 ppm, all animals appeared normal.

Teratogenesis, Mutagenesis, and Carcinogenesis. No reports of completed teratogenic or carcinogenic studies were found. However, the material has been applied to mouse skin (247), although no report is yet available from their carcinogenicity study.

Brem et al. (248) reported that the compound was not mutagenic to *S. typhimurium* and that it inhibited the growth of *E. coli*.

Rosenkrantz et al. (249) subsequently reported that 1,1,2,2-tetrabromoethane was negative in the standard (Ames) assay but strongly mutagenic (against *S. typhimurium*) when tested in suspension.

Metabolism and Biologic Monitoring. Few data are available but some metabolism might occur; Hollingsworth et al. (243) reported a slight increase in blood bromide of chronically exposed animals.

Observations in Man. Van Haaften (250) describes the only available report of human injury alleged to be due to acetylene tetrabromide. Although there appears to be little doubt that the chemists involved had exposure to the vapors, skin exposure was very high and it is impossible to separate the effects. "Near-fatal liver injury" was reported with headaches, anorexia, stomach ache, and heartburn. Although there is no question of a serious response by the liver, the effects were similar to viral hepatitis. Unfortunately no determinations of blood bromide were reported to determine the magnitude of the exposures.

2.21.4 Hygienic Standards

The threshold limit value for acetylene tetrabromide recommended by the ACGIH in 1980 is 1 ppm (14 mg/m^3).

2.21.5 Odor and Warning Properties

The odor of acetylene tetrabromide is sweetish and has been compared with chloroform or camphor. The odor is not distinctive enough to be considered a good warning property, although it may give sufficient warning to avoid serious acute exposure.

2.22 Pentachloroethane, CAS 76-01-7

$$CHCl_2CCl_3$$

2.22.1   Uses and Industrial Exposures

Pentachloroethane has been used as a solvent and chemical intermediate but has had little commercial utilization.

2.22.2   Physical and Chemical Properties

Physical state	Liquid
Molecular weight	202.31
Specific gravity	1.6712 (25/4°C)
Melting point	−29°C
Boiling point	162°C
Vapor pressure	3.4 torr (25°C)
Refractive index	1.50250 (24°C)
Percent in "saturated" air	0.45 (25°C)
Solubility	0.05 g/100 ml at 20°C, soluble in ethanol, ethyl ether
Flammability	Not flammable by standard tests in air

1 mg/liter ≏ 120.8 ppm and 1 ppm ≏ 8.37 mg/m³ at 25°C, 760 torr

2.22.3   Physiologic Response

Summary.   Pentachloroethane has a strongly narcotic effect which has been indicated to be even greater than that of chloroform. Exposure to this material may also result in injury to the liver, lungs, and kidneys. It has a local irritating effect on the eyes and the upper respiratory tract.

Single or Short-Term Exposure.   *Oral.*   Barsoum and Saad (229) reported the lethal oral dose in dogs to be 1.75 g/kg of body weight.

Inhalation.   Lehmann and Flury (102) indicated that cats inhaled 1 mg/liter of air (120 ppm) 8 to 9 hr daily for 23 days without overt symptoms of poisoning. However, they showed significant pathological changes in the liver, lungs, and kidneys. Dogs exposed to the vapor for 3 weeks showed fatty degeneration of the liver and injury to the kidneys and lungs.

Repeated or Prolonged Exposure.   Navrotskii et al. (233) exposed rabbits to 100 or 10 mg pentachloroethane vapor per cubic meter of air 3 to 4 hr daily for 7 to 11 months. Only limited data are discussed in the available abstract. Antibody titers were reported to be elevated after 1 to 1.5 months exposure to

100 mg/m³ (12 ppm). The abstract also states that toxic effects occurred at 10 mg/m³ (1.2 ppm) but the nature of the toxic effects is not specified.

Metabolism and Biologic Monitoring. Yllner (251) injected ¹⁴C-pentachloroethane subcutaneously in mice at doses of 1.1 to 1.8 g/kg and determined the excretion over a 3-day period. About one-third of the dose (12 to 51 percent) was expired unchanged; 16 to 32 percent was excreted as 2,2,2-trichloroethanol and 9 to 18 percent as trichloroacetic acid in the urine. The expired air also contained trichloroethylene (3 to 9 percent of the dose), indicating both dechlorination and dehydrochlorination.

2.22.4 Hygienic Standards

No standards have been proposed for pentachloroethane. It is probably not possible to set a reliable standard because of the limited toxicologic information and experience reported on this material. It is obvious that a concentration safe for repeated exposure would be well below the 121 ppm that was shown to cause pathological change. The unconfirmed report of Navrotskii suggests that exposures be kept much lower.

2.22.5 Odor and Warning Properties

Pentachloroethane has a sweetish odor, not unlike chloroform. The threshold at which the odor is detected has not been determined and therefore its value as a warning is not known.

2.23 Hexachloroethane, CAS 67-72-1

$$CCl_3CCl_3$$

2.23.1 Uses and Industrial Exposures

Hexachloroethane has been used as a chemical intermediate, in pyrotechnics, as an insecticide, and as a parasiticide in animals. It is an undesired by-product of certain chlorination processes.

   The compound does not appear to have as high an acute toxicity as tetra- or pentachloroethane and the toxicity from repeated exposure is uncertain. The primary physiological response to hexachloroethane is variously reported to be depression and stimulation of the central nervous system. Because the material is a solid and has a rather low vapor pressure, the acute hazard of breathing the vapor in industrial handling is relatively low. There has been some experience in industry to indicate that an excessive amount of dust in the air can cause irritation. Irritation has been observed from fumes of the material when handled hot. In general, it would not be considered a significant acute industrial hazard if handled with reasonable cleanliness, but it may be a problem with repeated exposures.

2.23.2 Physical and Chemical Properties

Physical state	Solid—rhombic crystals
Molecular weight	236.76
Specific gravity	2.091 (20°C)
Melting point	Sublimes at 187°C
Solubility	0.005 g/100 ml water at 22°C; soluble in ethanol, ethyl ether
Flammability	Not flammable by standard tests in air

1 mg/liter $\approx$ 103.0 ppm and 1 ppm $\approx$ 9.68 mg/m^3 at 25°C, 760 torr

2.23.3 Physiologic Response

Summary. Most of the experience with this material has come about because of its use as a parasiticide in animals, although a few recent articles indicate renewed interest in this chemical. Von Oettingen (33) reviewed the literature available in 1955. In view of a recent report by Weeks et al. (252) it is possible that some of the effects ascribed to hexachloroethane in the older literature may have been due to impurities in the samples tested.

Single or Short-Term Exposure. *Oral.* Hexachoroethane has a low acute oral toxicity. Thorpe (253) reported the LD$_{50}$ for rats to be 5.9 g/kg, a value consistent with Reynolds and Yee (254), who reported rats survived 6.16 g/kg for up to 24 hr, and Barsoum and Saad (229), who reported dogs survived 6 g/kg. A dose of 6 g/kg over a 2-day period was lethal to cats (255). Similar figures are given in Table 48.11, taken from Weeks et al. (252).

Twelve daily oral doses of 1000 mg/kg to rabbits cause significant reduction in body weight and increased relative liver and kidney weights. Gross and histological injury occurred in the liver and kidneys. Daily doses of 320 mg/kg caused liver injury but 100 mg/kg caused no significant effect (252).

Eyes, Skin. Weeks et al. (252) reported reversible injury when crystalline hexachloroethane was applied to the cornea of rabbit eyes for a prolonged period. It produced little or no skin irritation and did not appear to be significantly absorbed percutaneously. The dermal LD$_{50}$ was greater than 32 g/kg by this route. The compound did not sensitize the skin of guinea pigs.

Table 48.11. Lethal Dosages for Animals Following Single Administration of Hexachloroethane[a]

Animal	Treatment[b]	Diluent	Dosage (mg/kg)
Rabbit, male	Oral ALD	Methylcellulose	>1000
Rat, male	IP ALD	Corn oil	2900
Rat, female	Oral LD_{50}	Corn oil	4460
		Methylcellulose	7080
Rat, male	Oral LD_{50}	Corn oil	5160
		Methylcellulose	7690
Guinea pig, male	Oral LD_{50}	Corn oil	4970
Rabbit, male	Dermal LD_{50}	Water paste	≥32000

[a] Adapted from Weeks et al. (252).
[b] ALD = Approximate lethal dose.

Inhalation. Rats exposed to 2.5 mg/liter (260 ppm) of the vapors for 8 hr showed no adverse effects but 57 mg/liter caused severe injury including death. The higher concentration was supersaturated and contained particles of hexachloroethane (252).

Injection. Barsoum and Saad (229) indicated that an intravenous dose of 325 mg/kg in dogs resulted in death. A corresponding dose for pentachloroethane was found to be 100 mg, and for chloroform, 90 mg. This would indicate that hexachloroethane is intrinsically less toxic than the other materials, or is absorbed much more slowly. Weeks et al. found the approximate lethal dose (ALD) when injected intraperitoneally in male rats to be 2900 mg/kg and Baganz et al. (256) determined an LD_{50} for white mice of 4500 mg/kg by this route.

Repeated or Prolonged Exposure. Weeks et al. (252) have reported considerable data from subacute and short chronic studies in several species. These authors conclude that hexachloroethane is less toxic than it had been thought to be (257), and that the older literature may be in error. This question must be resolved by additional testing.

Oral. Hexachloroethane was fed to rats for 110 days at dose levels of 0, 1.5, 20, or 80 mg/kg body weight per day (258). Female rats showed only slight histopathological changes in the livers at the highest dose. Male rats were much more affected. Males given 20 or 80 mg/kg per day had histopathological alterations of the liver and kidneys and increased urinary excretion of uroporphyrin, creatinine, and δ-aminolevulinic acid; the latter two parameters were also increased in males at the 1.5 mg/kg level.

Inhalation.   Rats, dogs, guinea pigs, and quail were exposed 6 hr/day, 5 days/ week for 6 weeks to hexachloroethane vapor (252). Extensive studies were conducted to determine possible injury including behavioral, reproductive, clinical chemical, hematologic, and histological examinations. Animals exposed to 260 ppm were seriously affected, including death (except for quail). Quail showed no adverse effect at any dosage. There was no evidence of injury of any type in any species at 15 ppm. Exposure to 48 ppm caused minimal injury. The most significant effects appeared to be irritation of the eyes and respiratory tract. Apparently liver and kidney injury were not significant at these levels of exposure.

Teratogenesis, Mutagenesis, and Carcinogenesis.   *Teratogenesis.*   Weeks et al. (252) carried out teratological studies in pregnant rats fed by gavage or exposed by inhalation. Oral doses of 50, 100, or 500 mg/kg were fed on days 6 through 16 of gestation. Separate groups were also exposed 6 hr/day to 15, 48, or 260 ppm of the vapors. Although body weight gain of the dams in the 500 mg/kg oral and 260 ppm inhalation groups was lower than the controls, there appeared to be no teratological effects on the fetuses. There were adverse effects on gestation indexes and on the number of fetuses alive at the highest oral dose. The fetal reabsorption rate was increased in this group.

Mutagenesis.   Hexachloroethane was tested against one yeast strain, *Saccharomyces cervisiae*, and five strains of *Salmonella typhimurium* with and without rat liver activation and with no evidence of mutagenic effect (252).

Carcinogenesis.   When hexachloroethane was fed by gavage in the NCI bioassay program at doses of 423 and 212 mg/kg per day to rats and 1179 and 590 mg/ kg per day to mice, increased mortality was observed in rats but not mice (259). Toxic tubular nephropathy was observed in all groups. Hepatocellular carcinoma occurred at the higher level in treated mice, but not in rats of either sex.

Metabolism and Biologic Monitoring.   Jondorf et al. (260) reported that a dose of 0.5 g/kg of body weight in a rabbit was slowly metabolized. Approximately 5 percent appeared in the urine in a period of 3 days and from 14 to 24 percent in the expired air. These authors used chromatographic and isotopic dilution techniques to determine the nature of the metabolites in the urine. These were reported as percent of the dose given: trichloroethanol, 1.3; dichloroethanol, 0.4; trichloroacetic acid, 1.3; dichloroacetic acid, 0.8; monochloroacetic acid, 0.7; and oxalic acid, 0.1.

Fowler reported different results for sheep and Leghorn cockerels (261, 262), but the differences in the methods used make it impossible to compare the data from these species with that from rats. Interestingly, hexachloroethane is reported to be a metabolite of carbon tetrachloride but not chloroform. For details see the section on metabolism of chloroform.

Gorzinski et al. (258) analyzed liver, kidneys, blood, and adipose tissue from rats fed hexachloroethane in a 110-day dietary feeding study described in a previous section. After 57 days the authors reported that the "concentration of HCE in the kidneys of male rats was significantly higher at all dose levels when compared to females (μg HCE/g of kidney with increasing dose—males: 1.4, 24.3, 95.1; females: 0.4, 0.7, 2.0); this is consistent with the more pronounced renal toxicity noted for male rats. However, the results of the tissue analysis indicated that HCE was cleared in an apparent first-order manner with a half-life estimated to be 2–3 days."

Observations in Man.   A single report of inhalation exposures of man describes the effects of a military smoke bomb in a fraternity house (263). However, it is probably not appropriate to ascribe the effects specifically to hexachloroethane since a variety of chemicals would be present in the smoke.

2.23.4   Hygienic Standards

The threshold limit value for hexachloroethane recommended by the ACGIH in 1980 is 1 ppm with a notation indicating possible skin absorption. Weeks et al. present considerable data indicating 1 ppm may be an unnecessarily low level.

2.23.5   Odor and Warning Properties

Hexachloroethane is reported to have a camphorous odor (264, 265)

2.24   1-Propyl Chloride, 1-Chloropropane; *n*-Chloropropane, CAS 540-54-5

$$CH_3CH_2CH_2Cl$$

2.24.1   Uses and Industrial Exposures

Propyl chloride has found little use in industry. It has been studied as an anesthetic and antiparasiticide.

2.24.2   Physical and Chemical Properties

Physical state	Colorless liquid
Molecular weight	78.54
Specific gravity	0.8910 (20/4°C)

Melting point	$-122.8°C$
Boiling point	$46.4°C$
Vapor pressure	350 torr (25°C)
Refractive index	1.38838 (20°C)
Percent in "saturated" air	44.5
Solubility	0.27 g/100 ml water at 20°C; soluble in ethanol, ethyl ether
Flammability	Flash point $< -18°C$; explosive limits 2.6 to 11.1 percent in air

1 mg/liter ≈ 311.5 ppm and 1 ppm ≈ 3.21 mg/m³ at 25°C, 760 torr

2.24.3 Physiologic Response

Summary. Very little information is available on n-propyl chloride since it has found so little use in industry. Von Oettingen (33) reviewed the data available prior to 1955.

Single or Short-Term Exposure. *Oral.* Limited data indicates rats survived a single oral dose of 1 g/kg but died when given 3 g/kg (11).

Skin. The liquid was only slightly irritating to the skin of a single rabbit, even when bandaged on 5 days/week for 2 weeks (11).

Inhalation. Abreu et al. (266) exposed rats to 1.7 mmol/liter (40,000 ppm) 1 hr/day for 4 days and then examined the livers and lungs. Slight alveolar hemorrhage and "significant" focal necrosis of the liver were observed. This concentration was reported to be anesthetic in mice.

Metabolism. Barnsley (267) found 2-hydroxypropylmercapturic acid in the urine of rats injected subcutaneously with a 40 percent w/v solution on 1-chloropropane in arachis oil. Van Dyke and Wineman (204) reported that enzymatic dechlorination occurs in vitro using rat liver microsomes.

2.24.4 Hygienic Standards of Permissible Exposure

None have been established for propyl chloride.

2.24.5 Odor and Warning Properties

No data were found concerning odor or warning properties.

2.25 Isopropyl Chloride, 2-Chloropropane, CAS 75-29-6

$$CH_3CHClCH_3$$

2.25.1 Uses and Industrial Exposures

Isopropyl chloride has been used as a solvent and chemical intermediate, and to some extent as an anesthetic. Its flammability has limited its use despite its apparent favorable toxicity.

2.25.2 Physical and Chemical Properties

Physical state	Colorless liquid
Molecular weight	78.54
Specific gravity	0.859 (20°C)
Melting point	−117°C
Boiling point	35.3°C
Vapor pressure	523 torr (25°C)
Percent in "saturated" air	68.7 (25°C)
Solubility	0.31 g/100 ml water at 20°C; soluble in alcohol, diethyl ether
Flammability	Flash point −45°C (Tag open cup)

1 mg/liter ⩰ 311 ppm and 1 ppm ⩰ 3.21 mg/m³ at 25°C, 760 torr

2.25.3 Physiologic Response

Summary. Isopropyl chloride has a potent anesthetic action and has been proposed for medical usage.

Von Oettingen in 1955 reviewed the literature on the use of this material as an anesthetic (33). Vomiting and cardiac arrhythmia have been observed. It may also cause histopathologic changes in the liver and kidneys but is rather low in potency.

Few references have been published and most of the data are from unpublished work by The Dow Chemical Company.

Oral. Guinea pigs survived 3 g/kg of body weight but succumbed to 10 g/kg of body weight.

Skin. On the open surface of the skin of rabbits, isopropyl chloride caused very little irritation. When it was bandaged on to the skin, some erythema and superficial irritation resulted.

Eye. The liquid material splashed into the eyes of rabbits was painful but caused only transient irritation.

Inhalation. Rats, rabbits, mice, guinea pigs, and monkeys exposed 7 hr/day, 5 days/week for a total of 127 exposures over a period of 181 days survived a level of 1000 ppm in air with normal growth and appearance (11). Histologic changes in the liver and kidneys and possibly the lungs were observed in some of the species tested. Similar 7-hr exposures of rats, guinea pigs, rabbits, and dogs for 6 months to 500 ppm produced no adverse effect by any of the criteria studied including demeanor, appearance, growth, final organ and body weights, hematologic and clinical chemical studies, and gross and histologic examination.

Similar results have been published from somewhat shorter studies in rats. Twenty 6-hr exposures to 1000 ppm caused no evidence of injury during exposure but extensive vacuolation and necrosis were seen in the liver of male and female rats. Similar exposures to 250 ppm caused no injury (268).

Mutagenesis. Simmon et al. (127) reported that isopropyl chloride was mutagenic in *S. typhimurium* when tested in desiccators.

Metabolism. Van Dyke and Wineman (204) report that 2-chloropropane was enzymatically dechlorinated in vitro by rat liver microsomes.

2.25.4 Hygienic Standards of Permissible Exposure

None has been established for isopropyl chloride. Based on available data from animals it is suggested that time-weighted average exposures should be controlled below 500 ppm.

2.25.5 Odor and Warning Properties

Warning properties may be inadequate to protect against excessive repeated exposure. Several subjects did not detect 500 ppm of the vapors.

2.26   1-Propyl Bromide, 1-Bromopropane, *n*-Propyl bromide, CAS 106-94-5

$$CH_3CH_2CH_2Br$$

2.26.1 Uses and Industrial Exposures

Propyl bromide has had but minor use in industry. It has been studied as an anesthetic but has not found significant use.

2.26.2 Physical and Chemical Properties

Physical state	Colorless liquid
Molecular weight	123
Specific gravity	1.3539 (20/4°C)
Melting point	−109.85°C
Boiling point	71.0°C
Vapor pressure	143 torr (25°C)
Refractive index	1.43411 (20°C)
Percent in "saturated" air	19.3 (25°C)
Solubility	0.25 g/100 ml water at 20°C; soluble in alcohol, ethyl ether
Flammability	No data were found

1 mg/liter $\approx$ 198.8 ppm and 1 ppm $\approx$ 5.03 mg/m^3 at 25°C, 760 torr

2.26.3 Physiologic Response

Summary. Propyl bromide has a depressant action on the central nervous system and, for that reason, has been considered as a possible anesthetic. Exposure of animals to anesthetic concentrations may result in injury to the lungs and liver.

The material is reported to be irritating to the skin and eyes of mice (136) and has an intraperitoneal LD$_{50}$ of 2.5 and 2.9 g/kg for mice and rats, respectively (269). This same author also reported that orally administered 1-bromopropane is excreted in expired air (142). Barnsley et al. (267) report that rats metabolize 1-bromopropane.

2.26.4 Hygienic Standards of Permissible Exposure

None has been established.

2.27 Propylene Dichloride, 1,2-Dichloropropane, CAS 78-87-5

$$CH_2ClCHClCH_3$$

2.27.1 Uses and Industrial Exposure

Propylene dichloride is used as a solvent, chemical intermediate, and fumigant.

2.27.2   Physical and Chemical Properties

Physical state	Colorless liquid
Molecular weight	112.99
Specific gravity	1.1593 (20/20°C)
Freezing point	− 100.44°C
Boiling point	96.6°C
Vapor pressure	50 torr (25°C)
Refractive index	1.437 (25°C)
Percent in "saturated" air	3.4 (25°C)
Solubility	0.27 g/100 ml water at 20°C; soluble in alcohol, ethyl ether
Exposure limits	3.4 to 14.5 percent in air
Flash point	60°F
Ignition temperature	557 to 570°C

1 mg/liter $\simeq$ 216.5 ppm and 1 ppm $\simeq$ 4.62 mg/m^3 at 25°C, 760
torr

2.27.3   Physiologic Response

Summary.   Propylene dichloride causes CNS depression and, following chronic exposure, injury to the liver and kidneys. Indication of adrenal injury in animal experiments has also been reported. At high concentrations the vapors are irritating to the eyes and respiratory tract. Von Oettingen in 1955 summarized the literature prior to that date (33).

Single or Short-Term Exposure.   *Oral.*   The lethal dose of propylene dichloride to guinea pigs is between 2 and 4 g/kg of body weight. Repeated oral feedings of doses as low as 0.2 g/kg of body weight are survived for some period of time, but the animal shows liver injury. An oral LD$_{50}$ for rats of 1.9 g/kg (1.7 to 2.1 g/kg) has been reported (270).

Skin.   Propylene dichloride on the open skin causes only mild irritation. Single short contact will probably be without any effects. The intensity of the reaction is greatly increased when it is held close to the skin by clothing. The LD$_{50}$ for percutaneous absorption has been reported as 8.15 g/kg (270).

Eye.   Propylene dichloride causes some pain and irritation when splashed into the eyes of rabbits but it would not be expected to cause serious or permanent injury. It should be washed out immediately with water.

Inhalation.   An acute inhalation LC$_{50}$ for 8-hr exposures of rats to propylene dichloride was reported by Pozzani et al. (223) as 14 mg/liter (3000 ppm).

In an unpublished study (11) mice were exposed to analytically determined concentrations of propylene dichloride vapor in order to study the relationship of lethality, to anesthesia and to liver injury (as measured by SGPT in the blood). Based on 10-hr exposures, an LC_{50} of 720 ppm was determined. The anesthetic ET_{50} (ET = effective time) at that concentration was 350 min and the ET_{50} for increased SGPT was 186 min. Based on these data propylene dichloride has much less potency toward causing liver injury than carbon tetrachloride, since its anesthetic, hepatotoxic, and lethal conditions are quite similar. Gehring reported the ratio of ET_{50}-anesthesia to ET_{50}-SGPT for carbon tetrachloride as 136, as compared to 1.9 for this ratio for propylene dichloride. The ratio of LT_{50} to ET_{50}-SGPT for carbon tetrachloride was 5480 and for propylene dichloride 2.7 (54).

It was concluded by the investigator that in mice, respiratory injury, rather than liver injury, was the primary cause of acute death. Anesthetic effects were observed but they were considered secondary to obstructive respiratory failure.

Repeated or Prolonged Exposure.   The principal toxicological laboratory studies on this material were published by Heppel and co-workers (271, 272). These data have been summarized by von Oettingen (33) for repeated exposure to concentrations of 2200 to 400 ppm. "Exposure for 7 hours daily on 5 days per week to 1,000 ppm (4.4 mg/l) caused in guinea pigs no signs other than drowsiness; rabbits were not affected; and rats showed some incoordination and it appeared that the animals developed some tolerance to the exposure. On the other hand many rats and mice died after a few hours' exposure. Liver function tests performed in dogs showed no definite abnormal values but nevertheless histological examination of their organs gave evidence of liver damage. None of the animals studied gave evidence of injury of the blood or blood-forming organs."

Heppel et al. (272) reported that "repeated exposure to 400 ppm for 7 hours daily on 5 days per week for a total of 128 to 140 (exposures), caused no other ill effects except a decrease in weight gain of rats, whereas mice showed a high incidence of mortality." A group of C_3H strain mice exposed to 400 ppm propylene dichloride showed rather high mortality. Hepatomas were observed in some of the animals that survived.

A single unconfirmed report (273) claims impaired spermatogenesis in rats exposed continuously to 9 mg/liter (3.6 ppm). The duration of the study was not given in the available abstract but is described as "long periods," possibly 7 days. This report is not consistent with data cited above, which indicate the testes were not affected.

Teratogenesis, Mutagenesis, and Carcinogenesis.   *Teratogenesis and Carcinogenesis.*   No reports were found on teratogenesis or carcinogenesis.

Mutagenesis.   When tested in *Salmonella typhimurium* strains TA 1535, TA 1978, and TA 100, it was found mutagenic at levels of 10 to 50 mg per plate (274).

Metabolism and Biologic Monitoring.   It has been shown that rat liver microsomes in vitro are capable of metabolizing 1,2-dichloropropane (204). More information on metabolism is discussed in that section of 1,3-dichloropropene (431).

Observations in Man.   No reports of injury to humans from industrial exposure were found.

2.27.4   Hygienic Standards of Permissible Exposure

The ACGIH in 1980 recommended a threshold limit value of 75 ppm (350 mg/m^3).

2.27.5   Odor and Warning Properties

The odor may be adequate to warn against acute injury, but it appears doubtful if it will prevent excessive repeated exposure.

2.28   1,2-Dibromo-3-chloropropane, DBCP, CAS 96-12-8

$$CH_2BrCHBrCH_2Cl$$

2.28.1   Uses and Industrial Exposure

1,2-Dibromo-3-chloropropane has been used as a soil fumigant but has been banned for agricultural use in the United States except for restricted use on pineapple in Hawaii. It is used in other parts of the world. It is a highly toxic material, with reports of injury to the testes (spermatogenesis) of workmen. It is also capable of causing severe liver and kidney injury and has been shown to be carcinogenic in animals.

2.28.2   Physical and Chemical Properties

Physical state	Amber to brown liquid
Molecular weight	236.4
Specific gravity	209 (20/4°C)
Melting point	5°C
Boiling point	196°C
Vapor pressure	0.8 torr (21°C)
Refractive index	1.5518 (25°C)

Percent in "saturated" air   0.13 (25°C)
Solubility   Less than 0.1 g/100 ml water at 20°C;
         soluble in hydrocarbons, alcohols
Flammability   Not flammable by standard tests in air

1 mg/liter $\simeq$ 103.4 ppm and 1 ppm $\simeq$ 9.71 mg/m^3 at 25°C, 760 torr

2.28.3 Physiologic Response

Summary. Laboratory studies and human experience have shown 1,2-di-bromo-3-chloropropane (DBCP) to be a highly toxic material, particularly from chronic exposure. The testes, liver, and kidneys appear to be particularly affected. A summary of the toxicity has been published (168). Much of the data on acute and repeated studies in animals has been taken from Torkelson et al. (275).

Single or Short-Term Exposure. *Oral.* Single-dose oral LD$_{50}$s for male rats have been reported as 170 and 300 mg/kg and female mice as 260 and 410 mg/kg by two laboratories. Oral LD$_{50}$s of 210, 180, and 60 mg/kg were reported for male guinea pigs, male rabbits, and unsexed chicks. Kidney degeneration and depressed body weight, which took a long time for recovery, were noted after single doses of 126 mg/kg to male rats (275). Slightly higher LD$_{50}$ values were reported by Rakhumatullaev (276), who noted CNS depression and incoordination following acute oral doses. Single oral doses of 100 mg/kg produced inhibition of the CNS system, prolonged weight loss, and decreased spermatogenesis.

Eyes. Slight pain and irritation of the conjunctiva and irises were observed in rabbits treated with a few drops of undiluted DBCP or a 1 percent solution in propylene glycol.

Skin. Single application did not appear to be more than slightly irritating to rabbit skin, but when applied repeatedly there appeared to be extensive necrosis in the dermis although the epidermis showed only slight crustiness.

Skin Absorption. When applied to rabbits' skin for 24 hr, LD$_{50}$s of 1.4 and 0.5 g/kg have been reported for undiluted DBCP and a 10 percent solution in propylene glycol.

Inhalation. LCT$_{50}$ values of 368 ppm, 1 hr; 323 ppm, 2 hr; 154 ppm, 4 hr; and 103 ppm, 8 hr were reported by one laboratory (275). In this same paper a second laboratory found DBCP to be somewhat more lethal to their strain of rats, with kidney injury and delayed deaths commonly observed. When rats were given 15 7-hr exposures to 40 ppm, almost all died. Extensive injury was noted in the kidneys and testes. Six exposures produced poor health, loss of weight, lung congestion, cloudy swelling of the liver, and nephritis in groups of five male or female rats. The testes of one male appeared atrophied.

Intramuscular Injections.   Rats were reported to have survived two series of intramuscular injections each given for 3 consecutive days (total six injections). The daily doses, 25 mg/kg in propylene glycol, did not produce significant alterations in body weight, lymphocyte counts, or in nucleated cells in femoral smears. These data appear somewhat inconsistant with the previously discussed effect of subacute inhalation exposure (275).

Repeated or Prolonged Exposure.   Several studies have been reported which show a high chronic toxicity for DBCP. The liver, kidneys, and testes appear to be the major target organs.

Oral.   The only study in which DBCP was fed as part of the diet would not indicate as high a chronic toxicity as other studies by gavage or inhalation. Rats fed diets containing 1350 ppm by weight of DBCP were injured but rats ingesting diets containing 450 mg DBCP per kilogram of diet showed only slight evidence of injury. A diet containing 150 mg/kg fed for 90 days produced retarded weight gain of female rats but otherwise was reported to have little effect on male and female rats. The daily doses in milligrams per kilogram of body weight and the possible loss of DBCP by evaporation from the food prior to eating were not discussed (277).

Inhalation.   Torkelson et al. (275) reported severe effects following 50 to 66 7-hr exposures of rats, guinea pigs, rabbits, and female monkeys to 12 ppm. Five ppm inhaled 7 hr/day for 70 days (50 exposures) produced severely decreased testicular weights in half the rats, with histological changes in the testes, kidneys, and bronchioles. Other chronic studies are discussed in the sections on reproduction and carcinogenesis.

Teratogenesis and Reproductive Effects, Mutagenesis, and Carcinogenesis.
Teratogenesis and Reproductive Effects.   Although DBCP has a pronounced effect upon the male reproductive system, it was not teratogenic when given to female rats at dosages of 0, 12.5, 25, or 50 mg/kg body weight on days 6 through 15 of pregnancy. The investigators reported it was toxic to dams and fetuses at 50 and 25 mg/kg but not at 12.5 mg/kg (278). Reznik and Sprinchan (279) reported that the estrus cycle of female rats was affected by doses of 10 mg/kg fed repeatedly for 4 to 5 months. Morphological changes have not been reported in the ovaries but these authors suggest that decreased hormonal secretion by the ovary produces the altered estrus cycle.

The report by Torkelson et al. (275) of severe testicular injury has been confirmed by numerous other investigators (276, 279–282, 479). The papers by Burek et al. (281, 479) and Rao et al. (282) indicate significant species differences, with rabbits more sensitive than rats. Male rabbits were exposed 6 hr/day, 5 days/week to 0, 0.1, or 1.0 ppm (14 weeks) or 10 ppm (2 and 8 weeks). Male and female rats were exposed for 14 weeks to 0, 0.1, 1 or 10 ppm DBCP vapor. Two weeks of exposure of rabbits to 10 ppm produced detectable testicular alterations principally on spermatogenesis. Sertoli cells were unaffected in these animals. Six to 8 weeks of exposures to 10 ppm produced nearly complete atrophy of the germinal tissue of rabbits. Rats, on the other hand, showed no detectable changes after similar exposure to 10 ppm for 5 weeks and only moderate alterations after 14 weeks. There was considerable variation between individual animals. At 1 ppm, rats showed no testicular atrophy after 14 weeks but rabbits were moderately affected. At 0.1 ppm, no effect was seen microscopically in either species, but an equivocal, apparently reversible increase in abnormal sperm was seen by electron microscopy of the testes of the rabbits but not of the rats (479). Reproduction was severely impaired at the higher concentrations, as would be expected from the marked effect observed on sperm counts. Furthermore, a dominant lethal effect was observed when male rats exposed to 10 ppm were mated with unexposed virgin females. This effect was reversible since subsequent matings did not show the effect.

Mutagenesis.   DBCP has been found mutagenic in bacterial test systems using *S. typhimurium* TA 1530 and *E. coli* but is quite toxic to the latter. The dominant lethal effects reported by Rao et al. (282) also suggest mutagenic changes in mammals.

Kapp et al. (283) have reported Y-chromosomal nondisjunction in DBCP-exposed employees, but these data must be confirmed. The Y-chromosome counts were made on ejaculates with extremely low sperm counts, the control sample was from a different location, and the statistical methodology is subject to question.

Carcinogenesis.   When fed by gavage to male and female rats and mice in the NCI bioasssay program, DBCP has been shown to produce tumors. The high exposure rats were given 24 to 30 mg/kg per day and the low dose rats 12 to 15 mg/kg per day. The dosage to mice was changed during the study so that daily doses ranged from 160 to 260 mg/kg per day as well as half these levels (low dose) in male mice, and 120 to 260 mg/kg per day and half these levels (low dose) in female mice. Treatments were given 5 days/week for 78 weeks. Tumors were present after 10 weeks of treatment. Ultimately 90 percent of the mice and 60 percent of the rats developed squamous cell carcinomas of the forestomach. A high incidence (54 percent) of mammary adenocarcinomas was observed in female rats but not mice (107). These results have been confirmed in subsequent studies in which DBCP was administered to rats and mice in the diet, by gavage or by inhalation.

Metabolism and Biologic Monitoring.   When a dose of 20 mg/kg 1,2-dibromo-3-chloropropane-3-^{14}C was given orally, it was metabolized rather rapidly by male rats (481). Only traces (0.04 percent) of the ^{14}C were excreted in the expired air as unchanged DBCP. Essentially all (98.8 percent) was absorbed from the gut and 90 percent of the activity was excreted in 3 days. During the first 24 hr, 49, 14, and 16.5 percent of the activity was excreted in the urine, feces, and expired air, respectively. The radioactivity in expired air was primarily carbon dioxide. The activity in the urine was in the form of an acidic metabolite (8 percent of the activity in the urine) or as a highly polar metabolite(s) (>90 percent of urinary activity). Some accumulation occurred in the body fat in the form of a metabolite.

Although an increase in bromide ion in the blood of individuals exposed to DBCP is likely, the small quantities expected at acceptable levels of exposure would not be meaningful for assessing worker exposures. Possibly hormonal analysis (FSH) may be useful in determining effects but at this time the applicability of the method has not been established or quantitated.

Observations in Man.   Investigations in man are recent and result from the observation of decreased spermatogenesis in a group of employees formulating DBCP-based soil fumigant in California (284). Subsequent studies on this and other populations have confirmed the finding (285, 286, 480). Lanham (286, 480) also reported decreased testicular size and increased follicle stimulating hormone (FSH) levels in the most highly exposed men.

At the time of writing it is premature to draw many conclusions but it is obvious that effects on the testes of man can be severe and are related to the degree of exposure. The route of exposure has not yet been established but may involve both dermal and respiratory absorption. Although recovery of the sperm count of some individuals has occurred and they have fathered normal children, it appears possible that destruction of the germinal tissue can be so severe as to preclude regeneration.

2.28.4   Hygienic Standards

The U.S. Department of Labor OSHA standard for exposure to DBCP is 1 ppb. It is obvious that dermal exposure must be prevented.

2.28.5   Odor and Warning Properties

Unpublished data (11) indicates the odor is perceptible at 0.01 to 0.03 ppm. It appears doubtful that odor will provide significant warning to discourage excessive exposure, since 0.2 ppm was tolerable although disagreeable to a small series of test subjects.

3 UNSATURATED HALOGENATED HYDROCARBONS

3.1 Vinyl Chloride, Monochloroethylene, chloroethene, CAS 75-01-4

$$CH_2{=}CHCl$$

3.1.1 Uses and Industrial Exposures

Vinyl chloride is used as a chemical intermediate primarily as a monomer in plastic manufacture. The fire and explosion hazards were generally considered to be the dominant problem in handling vinyl chloride until 1974, when it was reported that three cases of angiosarcoma of the liver had occurred in workmen grossly exposed to vinyl chloride. Subsequently many diseases have been alleged to have been caused by exposure to vinyl chloride. Because many of these alleged effects are poorly documented it is difficult to draw solid conclusions at this time and the reader is cautioned to evaluate the literature carefully. One thing is certain, vinyl chloride is significantly more toxic than was generally accepted prior to 1974. It is hepatotoxic, carcinogenic, and possibly mutagenic. Acroosteolysis has been associated with vinyl chloride and polyvinyl chloride production. It will be possible to discuss only a few of the rapidly growing number of references. Numerous bibliographies and literature surveys are available, including those by von Oettingen (33); Warren et al. (287); NCI (288); and IARC (289).

3.1.2 Physical and Chemical Properties

Physical state	Gas
Molecular weight	62.5
Specific gravity	0.9121 (20/4°C)
Melting point	−153.71°C
Boiling point	−13.8°C
Vapor pressure	2580 torr (20°C)
Solubility	0.25 g/100 ml at 25°C; soluble in ethanol, ethyl ether
Stability	May produce peroxides
Flash point	−78°C (open cup)
Autoignition temperature	472.22°C
Explosive limits	4 to 22 percent by volume in air

1 mg/liter ≈ 391 ppm and 1 ppm ≈ 2.56 mg/m^3 at 25°C 760 torr

3.1.3 Physiologic Response

Summary.   Because vinyl chloride is a gas the only significant route of toxic exposure is inhalation. If it is confined on the skin in a liquid, some might be expected to be absorbed, but the relative amount is small. The likelihood of acute effects is not nearly as significant as are liver injury, angiosarcoma, and probably acroosteolysis which may occur following excessive repeated exposures.

It appears that metabolism of vinyl chloride is necessary before many of its toxic effects occur.

Single or Short-Term Exposure.   The primary acute physiological effect of vinyl chloride is depression of the central nervous system, which begins to occur at concentrations of 8000 to 10,000 ppm (290). Early studies as a surgical anesthetic were discouraged by its high flammability and by cardiac and circulatory disturbances noted in animals and patients at the 10 to 20 percent concentration necessary to produce surgical anesthesia (291–293).

Acute liver toxicity even at these concentrations appeared to be low, but there have been suggestions of a delayed carcinogenic response following massive subacute exposures (294).

Repeated or Prolonged Exposure.   Essentially no investigations were published on the response to chronic vapor exposure prior to 1961. Schaumann exposed mice and rats and found that they tolerated a level sufficient for "light narcosis" for periods of 4 hr/day for 5 to 8 consecutive days or for 1 hr/day for 4 weeks without showing kidney or liver injury [as cited by Lehmann and Flury (102)]

Torkelson et al. (295) report on repeated exposures of animals for 7 hr/day, 5 days/week. At 500 ppm, rats showed increased liver weight and histopathology. At 200 and 100 ppm, rats showed increased liver weight, but no changes could be observed in dogs or guinea pigs. All species tolerated 50 ppm for 6 months with no adverse effect. Repeated exposures for 1 hr/day at 200 or 100 ppm were tolerated without observable effect. The effects on the liver were mild and apparently reversible, since they were not observed in rats kept 6 to 8 weeks after the 6 months exposures ceased. This was the longest period any animals were observed in this study.

Noncarcinogenic effects in the liver have been reported in the oncological studies reported later.

Teratogenesis, Mutagenesis, and Carcinogenesis.   *Teratogenesis.*   At least two studies have shown no teratological response in laboratory animals inhaling vinyl chloride during pregnancy; however, fetotoxic effects were observed particularly when ethanol was administered simultaneously in the drinking water.

John et al. (296) evaluated the effects on mice, rats, and rabbits. Groups of pregnant CF-1 mice, Sprague-Dawley rats, and New Zealand white rabbits were exposed to 500 ppm vinyl chloride 7 hr daily during organogenesis. Mice were also exposed to 50 ppm and rats and rabbits to 2500 ppm. Maternal toxicity was observed, most prominently in the mice, but the exposures did not cause significant embryonal or fetal toxicity or teratological effect. Simultaneous administration of 15 percent ethanol in the drinking water produced toxic effects greater than those of vinyl chloride alone. Maternal toxicity was increased more than embryotoxicity.

Ungvary et al. (297) have shown that vinyl chloride crosses the placenta and is found in the amniotic fluid and fetal blood as well as the maternal blood of rats. Pregnant rats were exposed continuously to 1500 ppm (4000 mg/m^3) during the first, second, or third trimester of pregnancy with no teratological effects. No embryotoxicity was noted when vinyl chloride was administered during the second or third trimester, but during the first trimester it increased fetal mortality and caused other fetotoxic effects. Fetal losses and induction of CNS malformation due to trypan blue administration were not potentiated by a combined exposure of pregnant rats to vinyl chloride and the dye.

Mutagenesis. Numerous references attest to the mutagenic effect of vinyl chloride or its metabolites in in vitro systems using bacteria, yeast, and hamster cells. Drosophila males treated with 1 to 20 percent vinyl chloride vapors had an increased frequency of complete and mosaic recessive lethals. Muratov and Guskova (298) reported that continuous exposures of rats to 0.15, 0.4, or 10 mg/m^3 (60, 160, or 3900 ppm) for 3.5 months increased the incidence of chromosomal fragmentation, agglutination, and bridging. The incidence was reportedly increased when the rats were housed at 35°C rather than at 22°C. No mutagenic effects were observed at 0.07 mg/m^3 (28 ppm).

Inhalation of 2500 or 5000 ppm 4 hr/day for 5 days or intraperitoneal injection of 300 or 600 mg vinyl chloride/kg/day resulted in chromosomal changes in the marrow cells of Chinese hamsters (299). However, Anderson et al. (300) reported the lack of dominant lethal effect in a study in which mice were exposed to 3000, 10,000, or 30,000 ppm for 5 days. Similarly Short et al. (301) found no evidence of pre- or post-implantation loss in female rats mated with males which had been exposed for 11 weeks to 0, 250, or 1000 ppm. Exposures were 6 hr/day, 5 days/week.

Carcinogenesis. Since Viola et al. in 1970 (302) reported the production of tumors in rats by exposing them to 30,000 ppm vinyl chloride, numerous other studies have confirmed the carcinogenicity to rats as well as mice and hamsters. Maltoni (303) in 1973 first reported angiosarcomas of the liver, which in 1974 were also observed in man (304).

Maltoni's group has exposed laboratory animals 4 hr/day for 1 year to vapor concentrations ranging from 30,000 to 1 ppm and has fed olive oil solutions by gavage (305). The animals were kept for their lifetimes following treatment. These studies and those of many other investigators have produced a wide range of tumors, particulary at high concentrations.

In a summary paper of his massive studies, Maltoni states that liver angiosarcoma, carcinoma of the Zymbal glands, nephroblastoma, neuroblastoma, mammary adenocarcinoma, forestomach papilloma and acanthoma, extrahepatic angiosarcoma, and hepatoma in the rat were correlated to exposure to vinyl chloride (305). Table 48.12 from Maltoni's summary paper describes those tumors that occurred at the lowest dosages.

None of the previously cited tumors were observed at 5 ppm by inhalation and 0.03 mg/kg by ingestion. There appear to be marked differences between strains of animals and between species, with mice more susceptible than rats, which are more susceptible than hamsters (306). Based on angiosarcoma of the liver, all species appear considerably more susceptible than man. The discussion by Gehring et al. (307) of the relationship between species as it relates to production of reactive metabolites is presented in the following section.

Metabolism and Biologic Monitoring.   Numerous studies indicate that probably a reactive metabolite, not vinyl chloride per se, is responsible for its toxicity. Although some inhaled vinyl chloride is excreted unchanged (308), depending on the dose, a varying amount is metabolized. The metabolism of vinyl chloride has been the subject of numerous recent studies which need confirmation and interpretation. Currently it is thought that vinyl chloride is metabolized by epoxidation with subsequent production of chloroacetaldehyde. Further oxi-

Table 48.12.   Onset of Tumors Considered as Vinyl Chloride Correlated, at the Lowest Dose (305)

Dose	No. of Animals	Results
25 ppm	120	5 liver angiosarcomas 4 Zymbal gland carcinomas 1 nephroblastoma
10 ppm	120	1 liver angiosarcoma 2 extrahepatic angiosarcomas 2 Zymbal gland carcinomas
1 mg/kg	150	3 liver angiosarcomas 1 extrahepatic angiosarcoma 1 hepatoma 5 Zymbal gland carcinomas
0.3 mg/kg	150	1 liver angiosarcoma 1 hepatoma

dation and conjugation with glutathione are responsible for the metabolites found in urine.

Gehring et al. (307, 309) analyzed the metabolic and carcinogenic data from man and laboratory animals. Using several models to predict the incidence in man from animal data, they found that all models overpredicted unless corrections were made for rates of metabolism and for surface area of the different species.

Observations in Man. Tamburro (310) has concluded that current clinical, biochemical, and hematologic tests are of limited value in predicting angiosarcoma but may be of value in other hepatotoxic effects.

Laboratory Studies. Limited human studies by Mastromatteo et al. (290) have shown that concentrations must approach 1 percent (10,000 ppm) before humans notice the anesthetic effects of vinyl chloride gas. Surgical anesthesia requires concentrations greater than 10 percent. Baretta et al. (308) studied expired air of subjects exposed to 50, 250, or 500 ppm for 7.5 hr with a half-hour lunch break. Although they concluded that expired air sampling might have utility in monitoring exposure at the occupational standards then in use, expired air can be expected to be of little value at currently accepted levels of exposure.

Industrial Experience. The early industrial history of vinyl chloride is remarkably free of adverse reports except for problems of fire and explosion. The earliest reports of adverse effect on industrial workers related to vinyl chloride were to anesthesia and anesthetic deaths. Danziger (311) reported two fatal cases, only one of which was definitely ascribed to a massive exposure to vinyl chloride. Other reports indicate exposures were poorly controlled, and probably were to thousands of parts per millions in many situations prior to 1974.

Cutaneous lesions (scleroderma), collagen disease, and Raynaud's syndrome were reported by Suciu et al. (312) to have been observed in a polyvinyl chloride (PVC) plant, but contrary to some reports these authors did not ascribe the effects to vinyl chloride per se; rather they considered them to be related to other materials used in the factory. Subsequently numerous authors reported acroosteolysis in PVC workers in Europe and the United States. This was accompanied by scleroderma and Raynaud's syndrome. Recently a vast number of alleged effects on the liver and other organs have been reported. Although it is possible that some of these effects associated with PVC production may be related to vinyl chloride, most reports have failed to eliminate other chemicals as possible causes of injury. Furthermore, there is rarely an indication of the concentrations inhaled.

There appears to be no question that angiosarcoma of the liver has been caused by exposures to vinyl chloride primarily while cleaning reactor vessels.

As of May 1980, approximately 85 cases have been found worldwide, with five to six others probable (313). Angiosarcomas are certainly consistent with the results of animal studies in which these tumors have been produced in several studies; however, the epidemiologic evidence is not consistent from one study to the next, in regard to other cancers or toxic effects. Some of the most careful studies on populations in which several angiosarcomas have occurred do not confirm liver injury, splenomegaly, blood changes, respiratory injury, or other effects alleged to have been seen in other studies. Acroosteolysis, scleroderma, Raynaud's syndrome, angiosarcoma of the liver, and mutagenic changes appear to have been most conclusively shown to occur as a result of exposure of workmen to vinyl chloride. It appears probable that other effects such as other tumors, hepatic injury, and splenomegaly may also be related, but other causative agents have not been eliminated. Other effects are not well established and must at this time be considered as associated with the PVC/VCM production process but not necessarily with the monomer itself. Continued insistence by some that vinyl chloride per se is the causative agent of all those effects is probably retarding our acquisition of knowledge of the true cause of injury.

3.1.4 Hygienic Standards

The ACGIH in 1980 recommended 5 ppm (10 mg/m^3) as threshold limit value for vinyl chloride. The U.S. Department of Labor (OSHA) standard is 1 ppm, and the standards in other countries vary, but generally require that exposures be limited to 5 to 10 ppm or less. It appears reasonable to assume that these levels established to protect against angiosarcoma of the liver will prevent other adverse effects as well.

3.1.5 Odor and Warning Properties

Although vinyl chloride has an odor at high concentrations, it is of no value in preventing excessive exposure. The actual vapor concentration that can be detected has never been adequately determined and varies from one individual to another, from impurities in the sample and probably from duration of exposure.

3.2 Vinyl Bromide, Monobromoethylene, Bromoethene, CAS 593-60-2

$$CH_2{=}CHBr$$

3.2.1 Uses and Industrial Exposure

Vinyl bromide is used as a copolymer in the production of flame resistant acrylic polymers and to a limited extent as a chemical intermediate. A toxicology review is available (289).

3.2.2   Physical and Chemical Properties

Physical state   Gas
Molecular weight   106.9
Density   1.493 g/cm^3 (20°C)
Melting point   − 139.5°C
Boiling point   15.8°C
Vapor pressure   1033 torr (20°C)
Refractive index   1.4410 (20°C)
Solubility   Insoluble in water at 20°C; soluble in ethanol,
                        ether, acetone, benzene, chloroform
Flammability   Vinyl bromide has no flash point by standard
                        tests in air but with a high energy ignition
                        source the explosive limits are 9 to 15
                        percent by volume in air

1 mg/liter ≋ 228.7 ppm and 1 ppm ≋ 4.4 mg/m^3 at 25°C 760 torr

3.2.3   Physiologic Response

Summary.   When inhaled at high concentrations vinyl bromide causes anesthesia and death. When inhaled by rats for an extended period of time, its carcinogenic action seems to be similar to that of vinyl chloride. Human experience has been limited, but it should also be assumed that vinyl bromide, like vinyl chloride, is capable of causing cancer in man.

Single or Short-Term Exposure.   *Oral, Skin, and Eyes.*   Vinyl bromide is a gas and the only significant route of toxic exposure is inhalation. Limited data have been published. According to Leong and Torkelson (314), "Preliminary evaluation of the toxicity of vinyl bromide showed that the compound had an acute oral LD$_{50}$ of approximately 500 mg/kg when a chilled 50% solution in corn oil was fed to male rats. It was found to be slightly to moderately irritating to the eyes, to be essentially nonirritating to the intact or abraded skin of rabbits, and to produce no frostbite from evaporation of the liquid."

Inhalation.   Abreu (266) reported that 7 mmol/liter was the highest tolerable concentration for mice exposed for 10 min and that half that concentration produced pronounced anesthesia. Leong and Torkelson (314) published very limited acute data. They reported "acute inhalation toxicity studies showed that exposure of rats to a nominal concentration of 100,000 ppm (437,636 mg/m^3) resulted in deep anesthesia and death within 15 minutes, but if the exposure was terminated before death, all animals recovered and survived. Exposure to a nominal concentration of 50,000 ppm (218,818 mg/m^3) resulted in unconsciousness within 25 minutes. All animals survived a 1½-hour exposure, but not a 7-hour exposure. At 25,000 ppm (109,409 mg/m³) rats were anesthetized, but they recovered rapidly when removed from the atmosphere even after 7 hours of exposure. Necropsy of survivors of the 50,000 ppm groups 2 weeks after exposure revealed slight to moderate liver and kidney damage. However, no abnormality was observed grossly in the survivors of the 25,000-ppm group."

Repeated or Prolonged Exposure.   Male and female rats, rabbits, and monkeys were repeatedly exposed 6 hr/day to either 250 or 500 ppm for 6 months with no apparent effect except for an increase in blood bromide ion concentration (314). There were no compound-related effects with respect to demeanor, body and organ weights, food consumption, a number of hematologic parameters, or mortality. Gross and microscopic examination of the major organs and tissues at the end of the exposure period revealed no remarkable abnormal changes. This study was not long enough to evaluate carcinogenicity.

Similarly results were seen in groups of six rats exposed to 50, 250, or 1250 ppm for 6 months as part of a carcinogenic study and killed as part of an interim sacrifice (289). After 1 year, however, tumors of the liver and Zymbal glands of the ears were observed. These results are discussed in the section on carcinogenicity.

Teratogenesis, Mutagenesis, and Carcinogenesis.   *Teratogenesis.*   No reports were found.

Mutagenesis.   Vinyl bromide vapors have been shown to be mutagenic in *Salmonella typhimurium* TA 1530 and TA 100. Addition of rats and human liver enzymes enhanced activity in the in vitro system (315, 316).

Carcinogenesis.   Vinyl bromide has been shown to be carcinogenic in male and female rats exposed to either 10, 50, 250, or 1250 ppm in a lifetime study (still in progress at time of writing). The effects were so severe at 1250 ppm that this group was terminated after 74 weeks; the remaining groups were kept for 24 months. Angiosarcomas, primarily of the liver, were found in both sexes at all four concentrations. Neoplasms of the ceruminal (Zymbal) glands of the ears occurred in both sexes at the highest concentration and in male rats at 250 ppm. The incidence of hepatocellular neoplasms was increased at 250 ppm in both sexes and in female rats at 10 ppm (317).

Metabolism and Biologic Monitoring.   Vinyl bromide is metabolized, since bromide ion is found in the blood of exposed animals (314). Although determination of blood bromide was considered useful for biologic monitoring at the then-current industrial hygienic levels, blood bromide is of very limited value at currently used standards of 1 to 5 ppm. Metabolism is expected to be similar to that of vinyl chloride, probably proceeding through epoxidation, with subsequent conjugation to macromolecules and other biologic compounds (318).

Observations in Man. Commercial development of vinyl bromide has been recent and no references to human exposures were found.

3.2.4 Hygienic Standards

The ACGIH in 1980 recommended 5 ppm (20 mg/m³) as a threshold limit for vinyl bromide based on analogy to vinyl chloride.

3.2.5 Odor and Warning Properties

No published data were found but, based on its high chronic toxicity, vinyl bromide can be assumed to have inadequate warning properties to prevent excessive chronic exposure.

3.3 Vinylidene Chloride, 1,1-Dichloroethylene, CAS 75-35-4

$$CH_2{=}CCl_2$$

3.3.1 Uses and Industrial Exposures

Vinylidene chloride is used as a chemical intermediate, particularly as a monomer in the production of plastics. Literature reviews have been published by von Oettingen (33); Haley (319); Huffman and Desai-Greenway (320); Warren and Ricci (470); and IARC (289).

3.3 Physical and Chemical Properties

Physical state	Clear colorless liquid
Molecular weight	96.95
Specific gravity	1.218 (20/4°C)
Freezing point	−122.5°C
Boiling point	31.7°C
Vapor pressure	591 torr (25°C)
Refractive index	1.427 (20°C)
Percent in "saturated" air	78
Solubility	0.25 percent by weight in water at 25°C; soluble in most organic solvents
Stability	May produce peroxide (see note below)

Flash point	$-16°C$ (open cup)
Autoignition temperature	570°C
Flammability limits	5.6 to 16 percent by volume in air

1 mg/liter $\approx$ 252 ppm and 1 ppm $\approx$ 3.97 mg/m^3 at 25°C, 760 torr

Note: Vinylidene chloride in the presence of air or oxygen, with the inhibitor removed, forms a complex peroxide compound at temperatures as low as $-40°C$. The peroxide is violently explosive. Reaction products formed with ozone are particularly dangerous.

3.3.3 Physiologic Response

Summary. Exposure to high concentrations results primarily in CNS depression and the associated symptoms of drunkenness which may progress to unconsciousness. Chronic exposure to low concentrations results primarily in injury to the liver and kidneys.

Single or Short-Term Exposure. *Oral.* When fed in a corn oil solution to male rats, an acute oral LD_{50} of 1500 mg/kg was determined (321). These workers and Jaeger et al. (322) have shown numerous changes in liver and plasma enzymes, consistent with histological changes observed in the livers of treated animals.

Forkert and Reynolds (456) reported unusual histological changes in the Clara cells and ciliated cells of the bronchiolar epithelium of the lungs of rats fed 1,1-dichloroethylene at doses of 100 or 200 mg/kg body weight as a solution in mineral oil.

Eyes. Vinylidene chloride is moderately irritating to the eyes of rabbits. It will cause pain, conjunctival irritation, and some transient corneal injury. Permanent damage is not likely. However, a high concentration of the phenolic (MEHQ) inhibitor used in vinylidene chloride may cause eye injury. A contaminated eye should be flushed immediately with large quantities of flowing water.

Skin. Liquid vinylidene chloride is irritating to the skin of rabbits after direct contact of only a few minutes. The inhibitor content of the vinylidene chloride may be partly responsible for this irritation. Where leaks occur, vinylidene chloride will evaporate and the phenolic (MEHQ) inhibitor may accumulate till it reaches a concentration capable of causing local burns. Particular caution should be used with regard to contaminated clothing, which should be removed immediately and thoroughly cleaned before reuse. The volatility of vinylidene chloride probably prevents absorption of significant quantities through the skin unless it is in solution.

Inhalation.   A 4-hr LC_{50} at 6350 ppm was reported for male rats by Siegel et al. (323), although Carpenter et al. (231) reported that about half of their rats survived a single 4-hr exposure to 32000 ppm. When rats received 20 repeated 6-hr exposures to 200 ppm only slight nasal irritation was observed; similar exposure of four male and four female rats to 500 ppm also produced decreased weight gain and liver cell degeneration (268). Rabbits, monkeys, rats, and guinea pigs were exposed 8 hr/day, 5 days/week, for 6 weeks to 395 mg/m^3 (100 ppm) with no deaths, overt signs of toxicity, or histopathological changes (124).

Seven days of essentially continuous exposure (22 to 23 hr/day) to 40 ppm was more toxic to male mice than female mice or male rats in terms of SGOT, SGPT, and lethality (323). Seven of 10 male mice died whereas no female mice or male rats died. This is consistent with the higher rate of metabolism of vinylidene chloride by mice, the higher carcinogenic effect in that species, as well as the higher susceptibility of that species to liver and kidney injury.

Although vinylidene chloride has been shown to sensitize the heart of rats to epinephrin, only limited data are available on its potency (324).

Repeated and Prolonged Exposure.   *Ingestion.*   The results of long-term feeding studies are described in the section on carcinogenicity. Norris reported no effect in dogs ingesting 25 mg/kg per day for 90 days (325).

Inhalation.   Animals exposed 5 days/week, 8 hr/day for 6 months were found to have injury of the liver and kidney at concentrations of 100 ppm and 50 ppm. There was minimal injury to the liver and kidney, even at a concentration as low as 25 ppm (11). In a continuous exposure for 90 days to 189 mg/m^3 (about 47 ppm) in a study simulating confined spaces such as in submarines, considerable mortality occurred in guinea pigs and monkeys (7/15 and 3/9, respectively) but surviving animals exhibited no overt signs of toxicity. Mottled livers were evident on gross examination and considerable injury was seen microscopically in that organ. Dog livers were most affected. Kidneys, adrenals, and lungs were also injured. Less effect was observed following continuous exposure to 101, 61, and 20 mg/m^3 (124).

Repeated 6-hr daily exposures 5 days/week for 90 days to 25 or 75 ppm resulted in minimal, apparently reversible changes in the livers of rats (325).

Teratogenesis, Mutagenesis, and Carcinogenesis.   *Teratogenesis.*   Rats and rabbits were exposed 7 hr/day to either 20 (rats only), 80, or 160 ppm of vinylidene chloride vapor during organogenesis (326). Pregnant rats were also fed drinking water containing 200 ppm (w/w) of vinylidene chloride. No teratogenic effects were observed in either species but toxic effects (decreased weight gain, decreased food consumption, increased water consumption, or increased liver weight) were observed in the dams of rats inhaling 80 or 160 ppm and rabbits inhaling 160 ppm. Although a teratogenic effect was not observed, embryotoxic and fetotoxic effects were noted in both rats and rabbits. These included delayed ossification, wavy ribs, and resorptions (rabbits). "At levels causing little or no maternal toxicity (20 ppm in rats and 80 ppm in rabbits), there was no effect on embryonal or fetal development. Among the rats given drinking water containing 200 ppm vinylidene chloride there was no evidence of toxicity to the dams or their offspring."

Short et al. (482) consider vinylidene chloride to be a weak teratogen in rats and mice on the basis of inhalation studies.

Data cited by Norris (325) indicate no effect on reproduction in a three-generation study in rats fed up to 200 ppm w/w in their drinking water. This concentration in water results in daily ingestion of about 40 mg/kg per day.

Mutagenesis.   Vinylidene chloride vapors were weakly positive in a *Salmonella typhimurium* TA 1535 test system, in which supernatant from normal mouse liver and kidney was added. Supernatant from normal rats was not effective but when rats and mice were pretreated with Arochlor 1254, adding the supernatants to the test system produced positive results (327). Bartsch et al. (316) showed similar results. When vinylidene chloride was studied in a dominant lethal test at 10, 30, and 50 ppm 5 hr/day, 5 days/week, no adverse effects were observed (328) nor were there cytogenetic changes observed in the lymphocytes of rats exposed 6 hr/day for 6 months to 0, 25, or 75 ppm of the vapors of vinylidene chloride (325).

Carcinogenesis.   Varying results have been obtained in carcinogenic studies in animals. Although carcinogenicity of this material has not been observed in several long-term studies in rats (329–331), tumors have been reported in mice (331). The tumors were observed primarily in the kidneys of male mice and were accompanied by significant injury to the kidney tissue. Reitz et al. (332), Hathway (397), and McKenna et al. (333) consider the effects to be the result of tissue injury and subsequent repair rather than due to a genetic mechanism involving DNA.

The National Toxicology Program has included vinylidene chloride, and a draft final report has recently become available (NTP 81–82. DHHS Publication #(NIH)81–1784).*

Rampy et al. (330) and unpublished data (11) indicate no increase in tumors in rats ingesting up to 19.3 to 25.6 mg/kg per day (male and female rats, respectively) or in rats inhaling 25 or 75 ppm (v/v) 6 hr/day for 18 months and then kept for their lifetimes.

Lee et al. (334, 335) reported that exposure of a group of 71 rats to 55 ppm

* F344 rats were fed 1 or 5 mg/kg bodyweight and $B_6C_3F_1$ mice 2 or 10 mg/kg bodyweight in a corn oil solution by gavage for 104 weeks. The report concludes that "vinylidene chloride was not carcinogenic for F344 rats or $B_6C_3F_1$ mice of either sex." High-dose male and female rats showed chronic renal inflammation and high-dose male mice showed necrosis of the liver.

vinylidene chloride 6 hr/day for 12 months resulted in two hemangiosarcomas (one of the mesenteric nodes, and one of the subcutaneous tissue). Although the authors consider the finding significant, the failure to produce this tumor in other long-term studies, together with the fact that vinyl chloride was being used simultaneously in a parallel study, has prompted questions about the significance of the observation.

Van Duuren et al. (336) applied low doses of vinylidene chloride to ICR/Ha Swiss mice skin by several methods. Repeated dermal application of vinylidene choride by itself produced no tumors but when a promoter, phorbol myristate acetate, was used, skin tumors resulted. Vinylidene chloride did not produce tumors when injected subcutaneously at weekly intervals at a dose of 2 mg/ mouse per injection.

Metabolism and Biologic Monitoring.　　At relatively high dosages most ingested or injected vinylidene chloride is excreted unchanged in the expired air of rats (337). Mice metabolize a higher proportion of orally administered vinylidene chloride than other species, consistent with the higher toxicity in mice (338). The metabolism of vinylidene chloride following inhalation and oral administration has been studied (339, 340). Metabolism is dose dependent. Following a 6-hr exposure to 10 ppm, 98 percent of the acquired body burden was metabolized by the rat to nonvolatile compounds. At 200 ppm only 92 to 96 percent was metabolized. Fasting prior to exposure significantly reduced the detoxifying pathways for vinylidene chloride and enhanced covalent binding of metabolites to liver and kidney tissue. Fasted rats exposed to 200 ppm sustained liver and kidney damage whereas fed rats did not; fasting had no effect on rats exposed to 10 ppm vapor. The two major urinary metabolites were N-acetyl-S-(2-hydroxyethyl)cysteine and thiodiglycolic acid, indicating that the major detoxification pathway in the liver is by conjugation with liver glutathione possibly through an epoxy intermediate. Jones and Hathway (337) found monochloroacetic acid in the urine of rats treated orally with vinylidene chloride. These results are consistent with reports by Jaeger et al. (322, 341, 342), which showed glutathione had a protective effect on the liver of rats treated with vinylidene chloride. Short et al. (343) also reported that disulfiram reduced the acute lethal effects and hepatotoxic effects of vinylidene chloride.

Observations in Man.　　Two studies on industrial populations handling vinylidene chloride have been published. The first mortality study (344) has limited value since the major interest was vinyl chloride, and only limited exposure to vinylidene chloride occurred. The second study, also by Ott et al. (345), looked at mortality and health examination findings in 138 employees exposed to measured levels of vinylidene chloride where vinyl chloride was not involved. Measured concentrations ranged from 5 to 20 ppm TWA. No effects on mortality or health parameters were found, in this rather small study population.

3.3.4 Hygienic Standards

The ACGIH recommended 10 ppm (40 mg/m^3) in their 1980 list of threshold limit values.

3.3.5 Odor and Warning Properties

Vinylidene chloride has a characteristic sweet smell that resembles carbon tetrachloride or chloroform. Most persons can detect a mild but definite odor at 1000 ppm in air. Some people can detect it at 500 ppm. Vapors containing decomposition products have a disagreeable odor and can be detected at concentrations considerably less than 500 ppm. Neither odor nor irritating properties of vinylidene chloride is adequate to warn of excessive exposure.

3.4 *cis-* and *trans-*1,2-Dichloroethylene, CAS 156-59-2 Cis Isomer, CAS 156-60-5 Trans Isomer, CAS 540-59-0 Mixed Isomer

$$CHCl=CHCl$$

3.4.1 Uses and Industrial Exposures

The cis and trans isomers of 1,2-dichloroethylene have had use as solvents and chemical intermediates. Neither of the isomers has developed wide industrial usage in the United States partly because of their flammability.

3.4.2 Physical and Chemical Properties

Cis Isomer

Physical state	Liquid
Molecular weight	96.95
Specific gravity	1.2743 (25/4°C)
Melting point	−80.5°C
Boiling point	60.25°C
Vapor pressure	208 torr (25°C)
Percent in "saturated" air	27.4 (20°C)
Solubility	0.35 g/100 ml water at 20°C; soluble in ethanol, ethyl ether
Explosive limits	9.7 to 12.8 percent in air (mixed isomers)
Autoignition temperature	460°C
Flash point	2.2 to 3.9°C (closed cup)

  1 mg/liter ≃ 252 ppm and 1 ppm ≃ 3.97 mg/m^3 at 25°C and 760 torr

trans Isomers

Physical state	Liquid
Molecular weight	96.95
Specific gravity	1.2489 (25/4°C
Melting point	−50°C
Boiling point	48.35°C
Vapor pressure	324 torr (25°C)
Refractive index	1.44234 (20°C)
Percent in "saturated" air	42.5 (25°C)
Solubility	0.63 g/100 ml water at 20°C; soluble in ethanol, ethyl ether

1 mg/liter ⪦ 252 ppm and 1 ppm ⪦ 3.97 mg/m³ at 25°C and 760 torr

3.4.3 Physiologic Response

Summary. Since neither isomer has found wide usage, toxicologic data are limited. They were studied as anesthetics, and von Oettingen (33) summarized the old data. However, some of these data must be considered suspect since the purity of the samples and the ratio of the isomers is not always indicated.

Single or Short-Term Exposure. *Oral.* Unpublished data indicate a low to moderate oral toxicity. The LD_{50} for rats fed a 60:40 cis–trans mixture was greater than 2000 mg/kg (11). Fruendt et al. (346) reported an oral LD_{50} of 1 ml/kg for rats fed an olive oil solution of *trans*-1,2-dichloroethylene.

Inhalation. The available data conflict on whether there is a significant difference between the toxicity of the two isomers. According to Smyth, the cis isomer did not kill or anesthetize rats in 4 hr at 8000 ppm. At 16,000 ppm, they were anesthetized in 8 min and were killed in 4 hr. He also states that the trans isomer was twice as toxic and anesthetic as the cis isomer (147), and that the important effect of inhalation is narcosis. Lehmann and Flury (102) reported that disturbance of equilibrium and prostration occur in approximately the same length of time from similar concentrations of the cis and trans isomers. Slight narcosis and deep narcosis both occurred with the cis isomer at concentrations of about half of the trans isomer.

More recently Gradiski et al. (347) reported a 6-hr LC_{50} of 22,000 ppm for mice exposed to the trans isomer but Fruendt et al. (346) claimed adverse effects in rats receiving a single 8-hr exposure to 200 ppm of the trans isomer. Fruendt et al. report five of six rats had livers that appeared normal when stained for fat accumulation but one showed fat deposition. Apparently all six rats showed effects in the lungs, whereas only one control rat was similarly affected. These workers reported fibrous swelling in the cardiac muscle of rats exposed for 8 hr to 3,000 ppm. These data must be confirmed since they are markedly inconsistent with other reports.

Fruendt and Macholz (348) reported that a single 8-hr exposure to 200 ppm of either *cis*- or *trans*-1,2-dichloroethylene inhibited mixed function oxidizes in rats, with the cis isomer more active than the trans at high concentrations.

The concentration of 1,2-dichloroethylene required to cause anesthesia is not clear. No gross effects were seen in rats at 200 ppm (346), in several species at 1000 ppm (11), or in rats at 3000 ppm. Old data suggest that concentrations approaching 1 to 2 percent are required to cause surgical anesthesia in mice or other species (33). Again the relative anesthetic potencies of the isomers have not been clearly established (33).

Reinhardt et al. (143) concluded there were inadequate data on which to rank the ability of 1,2-dichloroethylene to sensitize the heart to adrenaline.

Injection.   Intraperitoneal LD_{50}s of 6 ml/kg (rats) and 3.2 ml/kg (mice) for the trans isomer were reported by Fruendt et al. (346) who, surprisingly, also reported a lower oral LD_{50} (1.0 ml/kg) for rats.

Repeated or Prolonged Inhalation.   Lehmann and Flury (102) reported the results of repeated exposure of cats and rabbits to vapor concentrations of 0.16 to 0.19 percent in air. Animals exposed to the cis isomer, at this concentration, showed loss of appetite, decrease in body weight, and pathological changes in the lungs, liver, and kidneys. Animals exposed to the trans isomer at the same concentrations showed loss of appetite and some respiratory irritation but no histopathological changes in the organs.

Conflicting data exist on the chronic toxicity of 1,2-dichloroethylene. Torkelson, as cited in the Documentation for the ACGIH TLV (257), reported no adverse effect in rats, rabbits, guinea pigs, and dogs exposed to either 500 or 1000 ppm 7 hr/day, 5 days/week for 6 months. The sample consisted of 60 percent cis and 40 percent trans isomers. The parameters studied included growth, mortality, organ and body weights, hematology, clinical chemistry, and gross and microscopic examinations of the major organs. In contrast Fruendt et al. (346) reported marked effects in rats exposed 8 hr daily, 5 days/week for 16 weeks to 200 ppm of the vapors of the trans isomer. Liver and lungs were affected and the leukocyte count was decreased.

Teratogenesis, Mutagenesis, and Carcinogenesis.   When the cis and trans dichloroethylene isomers were incubated in the presence of metabolically active mouse liver enzymes, they were not mutagenic to *E. coli* K-12 or *S. typhimurium* (349, 350).

No data on teratogenesis or carcinogenesis were found. *cis*- and *trans*-1,2-Dichloroethylene have been scheduled in the NCI bioassay program but no data are available.

Observations in Man. 1,2-Dichloroethylene was studied as an anesthetic in man, apparently with some success. Only one old report of one fatality was found (351).

3.4.4 Hygienic Standards

The threshold limit value recommended in 1979 by the ACGIH was 200 ppm (790 mg/m^3).

3.4.5 Odor and Warning Properties

No definitive data appear available. The odor has been described as slightly acrid.

3.5 **Trichloroethylene,** Acetylene Trichloride, Tri, Trilene, CAS 79-01-6

$$CHCl = CCl_2$$

3.5.1 Uses and Industrial Exposures

Trichloroethylene is widely used as an industrial solvent, particularly in metal degreasing and extraction processes. It has some use as a chemical intermediate and also as an anesthetic. By far the most likely contact with trichloroethylene in industry is with the vapors of the material when it is used as a solvent. In some industrial processes, such as vapor degreasing or hot extraction, the material may be used at elevated temperatures which increase the problem of escape of vapors into the workroom atmosphere. Deliberate inhalation of the vapors has been reported. Under certain conditions, such as in recycling anesthesia machines and in submarines, trichloroethylene has been decomposed by caustic scrubbers to produce much more toxic compounds.

3.5.2 Physical and Chemical Properties

Physical state	Colorless liquid
Molecular weight	131.4
Specific gravity	1.45560 (25/4°C)
Freezing point	−86.8°C
Boiling point	87°C
Vapor pressure	77 torr (25°C)
Refractive index	1.4777 (20°C)
Percent in "saturated" air	10.2 (25°C)
Solubility	0.1 g/100 ml water at 20°C; soluble in ethanol, ethyl ether

Flash point   None by standard tests in air
Autoignition temperature   410°C

1 mg/liter ≃ 185.8 ppm and 1 ppm ≃ 5.38 mg/m^3 at 25°C, 760
torr

3.5.3 Physiologic Response

Summary.   There is a tremendous literature on the toxicologic and pharmacological effects of trichloroethylene. Numerous reviews are available, including those by von Oettingen (33), Defalque (352), NIOSH (353), and Waters et al. (354).

The predominant physiologic response to exposure to trichloroethylene is one of CNS depression. This is particularly true as a response to acute exposure. Visual disturbances, mental confusion, fatigue, and sometimes nausea and vomiting are observed. Nausea and other gastrointestinal disturbances are not as striking with trichloroethylene as they are with carbon tetrachloride or ethylene dichloride. Sensitization of the heart to adrenalin has been reported but apparently is not a significant problem unless anesthetic levels are attained.

Degreaser's flush, in which the skin of the face and arms become extremely red, occurs occasionally if alcohol is consumed shortly before or after exposure to trichloroethylene.

Deliberate inhalation of the vapors, often reported as addiction, has resulted in injury and death; however, physical dependence is probably not involved.

The dangers in industry may be accentuated by visual disturbances and incoordination, which may lead to poor manual manipulation and, therefore, unsafe mechanical operations. Although CNS depression is a dominating problem from exposure to trichloroethylene, some indications of injury to the liver and kidneys may also be observed. Although such organic injury has been observed both in experimental animals and in the clinic, it is qualitatively and quantitatively of much less significance than in the case of materials like carbon tetrachloride. Trichloroethylene produces hepatocellular carcinomas in mice but not rats; hence the significance of carcinogenesis due to trichloroethylene is obscure at this time.

Single or Short-Term Exposure.   Inhalation is by far the most significant route of acute exposure in industry.

Oral.   Trichloroethylene is moderate to low in acute oral toxicity when fed to laboratory animals. An LD$_{50}$ of 4920 mg/kg has been reported for rats (270). The lowest lethal dose for dogs, cats, and rabbits has been reported as 6000 to 7000 mg/kg (353). The effects from ingestion are similar to inhalation.

Skin and Eye Contact. Trichloroethylene is irritating to the eyes and should be washed out promptly. It would not be expected to cause permanent injury from ordinary contact. Trichloroethylene is only mildly irritating to the skin if allowed to evaporate. From continued use of the material in contact with the skin, defatting can take place, producing a rough, chapped skin, which may result in erythema and possibly secondary infection. A much more severe skin response would be expected when the material is held close to the skin under tight clothing, such as shoes.

Skin Absorption. Frant and Westendorp (355) reported their experiments in which individuals exposed themselves to severe contact with the liquid and checked for excretion of trichloroacetic acid in the urine. They concluded that there was no significant skin absorption. Stewart and Dodd (50) and Sato and Nakajima (356), however, found trichloroethylene in the expired air after immersion of the hands of human subjects, although they also concluded that dermal absorption was not likely to be of toxicologic significance under normal industrial use.

Inhalation. The response of laboratory animals to single exposures to various concentrations of trichloroethylene has been reported by Adams et al. (357).

The maximum time–concentrations in air for a single exposure survived by all rats were as follows: 18 minutes at 20,000 ppm; 1½ hr at 6400 ppm, and 8 hr at 3000 ppm. Deaths were primarily due to anesthesia. Full anesthesia was seen at concentrations of 4800 ppm or more, but not at 3000 ppm.

Several studies have shown trichloroethylene to be moderate to low in hepato- and renal toxicity (53, 54, 192, 219, 358–360). Generally, hepatotoxicity is observed only at doses that are marked by anesthesia, in contrast to chloroform and carbon tetrachloride, which caused histological and functional changes in the liver at subanesthetic levels (54).

Sensitization of the heart to adrenalin can be demonstrated but exposures to highly anesthetic levels are apparently required. One of 12 dogs exposed to 5000 ppm and simultaneously injected with large doses of adrenalin responded, whereas seven of 12 dogs similarly injected and exposed to 10,000 ppm responded (58). Obviously much higher concentrations would be required if adrenalin were not injected as part of the experiment.

Repeated or Prolonged Exposure. Animal investigative work was reported by Adams et al. (357). They exposed several species of animals 7 hr/day, 5 days/week for approximately 6 months. At 3000 ppm by volume in air, rats and rabbits both showed an increase in liver and kidney weight. At 400 ppm, rats showed an increase in liver and kidney weights and the male rats also showed significantly less growth. Guinea pigs had increased liver weights and the growth of the exposed males was less than the controls. Rabbits showed a slight increase in liver weight. An exposed monkey showed no response at 400 ppm. At 200 ppm, the only effect was depressed growth in guinea pigs. Rats, rabbits, and monkey showed no response. At a concentration of 100 ppm, none of the species showed any significant response. The maximum concentrations tolerated for the 6-month period were as follows: monkeys, 400 ppm; rats and rabbits, 200 ppm; and guinea pigs, 100 ppm.

Prendergast et al. (124) exposed rats, guinea pigs, dogs, rabbits, and monkeys 24 hr/day for 90 days to 35 ppm with no effect except slight growth depression. Repeated 8-hr daily exposure to 700 ppm for 90 days were also without effect.

Baker (361) exposed dogs to trichloroethylene vapor, "varying from 500 to 3000 ppm repeatedly for periods varying from 2 to 8 hours daily often for 5 days weekly." Some dogs were exposed to a total of up to 162 hr. He observed histological changes in a number of areas of the brain. The most striking and severe changes were observed within the cerebellum and involved primarily the Purkinje cell layer. It was implied that all of the chronically exposed dogs showed the same histological change and no relation to specific exposure was given. The ataxia reported clinically following trichloroethylene exposure correlates with injury to the cerebellum.

Teratogenesis, Mutagenesis, and Carcinogenesis. *Teratogenesis.* Schwetz et al. (63) studied the effect of trichloroethylene vapors on pregnant rats and mice and their offspring. The pregnant dams were exposed for 7 hr to 300 ppm on days 6 to 15 of pregnancy with no evidence of adverse effect on the dams, on reproduction, or on the offspring by any of the usual criteria of a teratogenic study. Similar results were obtained in a second study also with rats (362).

Mutagenesis. When activated with microsomal enzymes, trichloroethylene has been shown to be weakly positive in certain microbial mutagen test systems. Trichloroethylene itself may not be active until metabolized. Fishbein (363) cites data by Greim et al. (364) as indicating weak mutagenic actively in an *E. coli* K-12 test system. When studied in a *Saccharomyces cerevisiae* D7 suspension test system, trichloroethylene was toxic, but not genetically active without microsomal activation. Addition of mouse liver supernatant produced an increase in mutations and gene conversions. In a host-mediated assay, trichloroethylene produced point mutations and gene conversions in D7 and gene conversions in D4 strains (365).

Unpublished data (366) indicate no dominant lethal effect was observed when male rats that had been exposed to 300 ppm 6 hr/day for 9 months were mated to untreated females. Male mice exposed for 24 hours to 50, 202 or 450 ppm did not reveal mutagenic effects in a dominant lethal assay (483).

In our opinion trichloroethylene does not present a problem from mutagenesis at levels otherwise acceptable for exposure to humans.

Carcinogenesis. Trichloroethylene has been included in the NCI Bioassay test program for carcinogenesis (367). Rats were fed either 1000 or 500 mg/kg per day by gavage; male mice received either 2400 or 1200 mg/kg per day and female mice 1800 or 900 mg/kg per day. Little, if any, effect was seen in rats but hepatocellular carcinomas were quite common in both sexes of mice fed these high dosages. The IARC (129) placed trichloroethylene in a group of chemicals that could not be classified as to the carcinogenicity for humans.

Henschler et al. (368) exposed rats, mice, and Syrian hamsters 6 hr/day, 5 days/week for 18 months to either 0, 100, or 500 ppm of pure trichloroethylene stabilized only with an amine base. According to the authors' abstract, "No significant increase in tumor formation was observed in any species or dosing group, except in malignant lymphomas, which were increased in female mice in the following incidence rates: 9/29 (controls), 17/30 (100 ppm), and 18/28 (500 ppm). Whether or not this high occurrence of lymphomas, which is peculiar to this strain of mice (NMRI) has any relationship to tri-exposure, cannot be decided upon by the present experiment. It is concluded that from these findings no indication for a carcinogenic potential of pure trichloroethylene can be deduced."

When trichloroethylene was applied to mouse skin by several methods, Van Duuren et al. (336) found low doses to be inactive. One mg/kg of the material was applied three times weekly to the dorsal skin with and without phorbol myristate acetate as a promoter. It was also injected subcutaneously at weekly intervals at a dosage of 0.5 mg/kg with no significant effect and intragastrically with no increase in tumors of the forestomach.

In our opinion trichloroethylene does not present a practical problem of carcinogenicity at the currently accepted levels of exposure, 50 or 100 ppm.

Metabolism and Biologic Monitoring. It has long been recognized that metabolites of trichloroethylene can be found in the urine mostly in the form of trichloroacetate (trichloroacetic acid) and trichloroethanol. Metabolism may proceed through an epoxy intermediate and chloral hydrate to trichloroacetic acid or trichloroethanol. The latter may be excreted as the glucuronide. High blood ethanol concentrations inhibit metabolism (369). The actual percentage of trichloroethylene that is metabolized appears to depend on the route and amount administered. After oral dosing or exposure to vapor is terminated a significant amount is excreted unchanged in the expired air. Stewart et al. (370–374) developed a series of curves useful for estimating exposures based on expired air samples. Åstrand and Gamberale (375) have described a method for estimate uptake of trichloroethylene and other solvents. Monster (213) has reported on extensive studies of uptake and excretion of trichloroethylene as well as perchloroethylene and 1,1,1-trichloroethane.

The use of urinary trichloroacetic acid for monitoring exposure is discussed in the following section.

Observations in Man. *Laboratory Studies.* Several laboratory studies have investigated the anesthetic and psychophysiological effects of trichloroethylene at concentrations near the TLV. Generally these show no effect at 100 ppm, marginal effects at 200, and slight effects above 300 or 400 ppm (211, 213, 370–377). Perception of discomfort and complaints that follow this same pattern are described in Section 3.5.5 on odor and warning properties.

Numerous investigators including Ertle et al. (378) and Muller et al. (379, 380) have studied metabolism in humans. They exposed humans once or on five consecutive days to several concentrations of trichloroethylene vapor. Trichloroethanol in the blood had a half-life of 12 hr after exposure ceased but the concentration was not proportional to the concentration inhaled. They found more trichloroethanol than trichloroacetic acid in the urine. Roughly a 2 : 1 ratio was found after one exposure but the ratios declined during the week. These authors questioned the value of urinary trichloroethanol or trichloroacetic acid excretion in monitoring exposure of occupationally exposed subjects.

Others would disagree. Ahlmark and Forssman (381) have several publications on the excretion of trichloroacetate in relationship to exposure. Their general conclusion in 1951 was that excretion of less than 20 mg/liter was not likely to result in any complaints. Complaints from some of the people were noted when they excreted from 40 to 75 mg/liter. Almost all people excreting 100 mg/liter or more had complaints; and if it was 200 mg/liter or greater, the complaints were pronounced and the individuals often lost time from work. Frant and Westendorp (355) made similar studies. Their data indicate a somewhat variable relationship between the excreted trichloroacetate and inhaled trichloroethylene. Another clinical study was reported by Grandjean et al. (382). Forssman and Holmquist (383) also studied the relationship between absorption and excretion. They indicated that a significant amount of trichloroethylene was retained by the body for some period of time after exposure. Only a small quantity was exhaled from a short exposure (21 to 28 percent) and 32 to 69 percent from a long exposure. The remainder was excreted or converted in some other way. Only a small part (from 1.2 to 7.8 percent) was recovered as trichloroacetic acid in the urine.

Several other investigators including Boillet (384) have recommended monitoring urinary trichloroacetate; Boillet also suggests that 150 mg/liter be considered an "alert level" above which symptoms will be noted.

Irish's 1962 summary statement (1) still applies. "The total matter might be summed up as follows: The excretion of trichloroacetate is a reasonable index of exposure to trichloroethylene. The relation both in quantity and time of exposure between the inhaled trichloroethylene and the excreted trichloroacetate is quite variable. Good monitoring of air exposure seems still to be the best control of exposure. Trichloroacetate analysis in the urine, however, should be very valuable supplemental information, particularly in studying an individual in the clinic."

Bardodej et al. (385) have made some interesting suggestions with regard to the reason for the apparent potentiation of toxic effect observed when trichloroethylene and ethanol are absorbed concurrently. They indicated that the trichloroethylene or the metabolism of trichloroethylene to chlorohydrate may inhibit liver aldehyde dehydrogenase, thus blocking the metabolism of ethanol beyond the aldehyde stage. The relationship of metabolism to degreaser's flush is at a present still speculative, but it has been shown that blood levels of 0.06 percent ethanol interfere with metabolism of trichloroethylene to trichloroacetic acid and trichloroethanol, and that blood levels of trichloroethylene are increased (380).

Industrial Experience. Clinical experience from acute exposure has come as a great extent from the use of trichloroethylene as an anesthetic. Death may result from respiratory failure or cardiac arrest. The cardiac response may, in most instances, be due to the potentiation of endogenous epinephrine. Much of the literature on anesthetic use of trichloroethylene has come from Great Britain.

Kleinfeld and Tabershaw (386) reported several fatal cases of trichloroethylene poisoning in the United States. Four of the cases were exposed to significantly high concentrations of vapor in industrial operations. These patients had reported symptoms from a number of exposures over some period of time. Death occurred several hours after the last exposure, which may have been fairly severe. On autopsy, they showed no anatomical abnormalities. It was assumed that they died from cardiac arrhythmia, probably due to the potentiation of endogenous epinephrine. The fifth case accidentally drank trichloroethylene. Severe injury occurred in both the liver and kidneys.

Stahl et al. (387) also described deaths due to trichloroethylene, as have several other authors. Von Oettingen (33) summarizes much of the older data from human experience and fatalities and NIOSH (353) summarizes more recent experience.

Epidemiologic and case studies have been of limited size and value and generally do not present adequate descriptions of the exposure concentrations. Generally the results are consistent with the animal data previously discussed.

In one of the better studies Axelson et al. (388) investigated cancer mortality in a group of Swedish workers. They used an interesting approach to estimate exposure. Since past records of urinary trichloroacetic acid excretion were available, it was possible to estimate vapor exposure. These authors concluded, "The present, fairly small cohort study comprising 518 men with rather low levels of exposure as estimated through trichloroacetic acid in the urine, did not reveal any excess cancer mortality. Thus, requiring ten years of latency time, the subcohort with trichloroacetic acid in urine above 100 mg/l (i.e., exposure supposed to be above 30 ppm) showed a close agreement between expected and observed number of cancer cases as did the subcohort with lower exposure (548 and 3643 person-years of observation, respectively). The cancer risk to man from trichloroethylene can by no means be ruled out from this study, particularly with regard to uncommon malignancies such as liver cancer. Nevertheless there is probably no serious cancer hazard at low exposures."

Numerous case reports allege diverse effects due to exposure to high concentrations of trichloroethylene. It is difficult to evaluate these reports because of inadequately described exposure, mixed exposures, or failure to eliminate other causes. One poorly understood phenomena is degreaser's flush, which occasionally occurs in the skin of the face, arms, and trunk of someone exposed to trichloroethylene who drinks alcohol (372).

3.5.4 Hygienic Standards

The tentative threshold limit for trichloroethylene recommended by the ACGIH in 1980 is 50 ppm (260 mg/m^3). This level is probably lower than needed to prevent beginning anesthetic effects and, based on the data presented, will prevent liver injury or other adverse effects.

3.5.5 Odor and Warning Properties

Trichloroethylene has a typical odor that has been characterized as ethereal or chloroformlike. It could not, however, be considered an effective warning. The following response to various concentrations is based on industrial experience and human studies (11).

100 ppm	Odor threshold, barely perceptible to unacclimated individuals
200 ppm	Odor apparent, not unpleasant. Transient mild eye irritation disappearing upon cessation of exposure
400 ppm	Odor very definite, not unpleasant. Slight eye irritation. Minimal light-headedness after 3 hr
1000 to 1200 ppm	Odor very strong, unpleasant. Definite eye and nasal irritation. Definite light-headedness, dizziness after 6 min
2000 ppm	Odor very strong, not likely to be tolerated. Markedly irritating to the eyes and respiratory tract. Drowsiness, dizziness, nausea in 5 min

3.6 Tetrachloroethylene, Perchloroethylene, Perc, CAS 127-18-4

$$CCl_2=CCl_2$$

3.6.1   Uses and Industrial Exposures

Perchloroethylene is used as an industrial solvent for a number of purposes, particularly dry cleaning and degreasing. It has been used as an anthelmintic in humans and animals. It also finds limited use as a chemical intermediate.

3.6.2   Physical and Chemical Properties

Physical state	Colorless liquid
Molecular weight	165.85
Specific gravity	1.6226 (20/4°C)
Melting point	−23.35°C
Boiling point	121.2°C
Vapor pressure	19 torr (25°C)
Refractive index	1.50534 (20°C)
Percent in "saturated" air	2.5 (25°C)
Solubility	0.015 g/100 ml water at 20°C; soluble in ethanol, ethyl ether, chloroform, benzene
Flammability	Not flammable by standard tests in air

1 mg/liter $\approx$ 147.4 ppm and 1 ppm $\approx$ 6.78 mg/m^3 at 25°C, 760 torr

3.6.3   Physiologic Response

Summary.   The major response to perchloroethylene at high concentrations is CNS depression. It is not, however, sufficiently effective to be considered a useful anesthetic. Irritation of the eyes, nose, and throat may also be observed at high concentrations. There are some indications of nausea and gastrointestinal upset at high concentrations, and changes in the liver and kidneys may be seen following excessive exposure; however, the effects are not as severe or striking as they are from a material such as carbon tetrachloride. Relatively few incidences of industrial problems due to exposure to perchloroethylene have been reported. This is probably because of a number of circumstances including low vapor pressure and toxicity. Case reports relate a very inconsistent set of symptoms. Certainly anesthesia and, in cases of high exposure, liver injury appear to be related to perchloroethylene. Nausea, headache, anorexia, vertigo, dizziness, and other symptoms may be related to the CNS and hepatic effects. Sensitization of the heart to epinephrine does not appear to occur with this compound. As for many compounds, massive doses of perchloroethylene given by gavage have produced liver cancer in mice but not rats.

Numerous reviews are available, including those by von Oettingen (33), Defolgue (352), IARC (168, 289), and NIOSH (399).

Single or Short-Term Exposure. *Oral.* Perchloroethylene has moderate to low toxicity by single dose ingestion. When fed to laboratory mice, an LD_{50} of 8850 mg/kg was reported. Animals apparently survived at 4000 (dog and cat) and 5000 (rabbit) mg/kg. A dose of 500 mg/kg is reported as not causing death in humans (399). Pozzani et al. (223) report on acute oral LD_{50} of 2600 mg/kg for rats.

Eye. Liquid perchloroethylene may cause pain, lacrimation, and burning; however, permanent injury is unlikely. High concentrations of the vapors are uncomfortable to the eyes.

Skin. If perchloroethylene is allowed to evaporate, no significant effect on the skin is likely. However, if it is confined on the skin or if exposures are prolonged and frequently repeated, the solvent will cause dermatitis by defatting of the skin.

Skin Absorption. Absorption of the liquid through the skin is not likely to be a significant route of toxic exposure (50), although some absorption does take place. Riihimaki and Pfaffli (215) determined in human subjects that absorption of the vapors through the skin was not significant at 600 ppm.

Inhalation. The response of rats to single exposures was reported by Rowe et al. (400). They stated that 2000 ppm was tolerated for up to 14 hr, and 3000 ppm was tolerated for 4 hr with no deaths. Unconsciousness was observed in the rats within a few minutes at concentrations of 6000 ppm or more, and after several hours at 3000 ppm, but unconsciousness was not observed at 2000 ppm. At these high-level single exposures, the predominant response was one of depression of the nervous system. Deaths occurred during or immediately after exposure. There were slight changes in the liver, characterized by a slight increase in weight, slight increase in total lipid, and slight cloudy swelling.

Pozzani et al. (223) reported an 8-hr LC_{50} of 34.2 mg/liter (5040 ppm) for rats.

Several other investigations have studied the effect of perchloroethylene on the liver and confirmed the observation of Rowe et al (400). Kylin et al. (359) exposed mice for 4 hr to 200, 400, 800, or 1600 ppm. No liver necrosis was observed, but fatty infiltration consistent with dose was observed at all levels. Drew et al. (401) exposed rats to 500, 1000, or 2000 ppm perchloroethylene vapors for 4 hr and measured serum enzymes. Four liver enzymes (SGOT, SGPT, glucose-6-phosphatase, and ornithine carbamyl transferase) were markedly increased by exposure to 2000 ppm and moderately increased by exposure to 1000 ppm, but very little effected by exposure to 500 ppm.

Gehring (54) exposed mice to 3700 ppm and determined an anesthetic ED_{50} of about 24 min, an SGPT ED_{50} of 470 min, and an LT_{50} of 730 min. He concluded that liver toxicity was of relatively low importance compared to anesthesia.

Injection. The hepatotoxicity and renal toxicity of perchloroethylene have been studied following subcutaneous injection. Plaa et al. (219) determined an LD_{50} of 390 mg/kg when perchloroethylene was injected in male mice. The hepatatoxicity was low (rated 3 vs. 1 for 1,1,1-trichloroethane and 190 for carbon tetrachloride). Plaa and Larson (53) measured nephrotoxicity following intraperitoneal injection and concluded that perchloroethylene was only weakly nephrotoxic.

**Repeated or Prolonged Exposure.   *Oral.*   **The NCI has included perchloroethylene in their bioassay program but unfortunately their experimental design did not call for adequate histological examinations of noncancerous lesions. They did report toxic nephropathy as discussed below under carcinogenesis (402).

Inhalation.   The first chronic study of tetrachloroethylene vapors was apparently carried out by Carpenter (403). He exposed rats 8 hr/day, 5 days/week for periods up to 7 months to concentrations of 70, 230, and 470 ppm. All his animals survived with growth comparable to his controls. At 70 ppm, no pathological effects were observed. At 230 ppm, he observed some pathological changes in both the liver and kidneys. At 470 ppm, the pathological indications of injury to the liver and kidneys were more striking.

Rowe et al. (400) reported the results of chronic vapor exposures of several species of animals. At 1600 ppm, in air, rats showed a drowsy, depressed condition for the first week and later a definite "irritation." Enlargement of the liver and kidneys was noticeable. Guinea pigs showed a loss in body weight and an increase in liver weight with moderate histological changes. At 400 ppm, for 130 exposures of 7 hr/day over a period of 183 days, rats showed no evidence of adverse effect. Guinea pigs showed definite increase in liver and kidney weights and slight fatty degeneration of the liver. Rabbits and monkeys showed no evidence of injury. At 200 ppm, guinea pigs still showed an increase in liver weight, increase in total liver lipid, and slight to moderate histopathological changes in the liver. At 100 ppm, female guinea pigs showed an increase in liver weight. Histologically, the liver appeared normal.

Guinea pigs in these studies had displayed a particular susceptibility, showing changes even at 100 ppm. However, because human experience had been favorable, these authors were inclined to accept the previously used maximum allowable concentration of 200 ppm as satisfactory. They suggested, however, that this should be considered a ceiling and fluctuations should be around an average of 100 ppm.

In a similar study rats, mice, guinea pigs, rabbits, and dogs were exposed to a 4 : 1 mixture of 1,1,1-trichloroethane–perchloroethylene vapors (466). Animals exposed 7 hr/day for 6 months to 800 ppm, 1,1,1-trichloroethane plus 200 ppm perchloroethylene exhibited mild and reversible liver and kidney changes consistent with the effect expected from 200 ppm of perchloroethylene alone. Exposures to 400 ppm 1,1,1-trichloroethane plus 100 ppm perchloroethylene were without effect in all species (404).

Kylin et al. (404) exposed mice to 200 ppm 4 hr/day, 6 days/week for 1, 2, 4, or 8 weeks. Fatty degeneration was observed but no liver cell necrosis nor effects on the kidneys were observed.

Teratogenesis, Mutagenesis, and Carcinogenesis. *Teratogenesis.* Schwetz et al. (63) evaluated the teratogenic potential of perchloroethylene in rats and mice. Pregnant rats and mice were exposed to 300 ppm perchloroethylene, 7 hr/day on days 6 to 15 of gestation. They reported a significant decrease in the fetal body weight of mice and a slight, but statistically significant, increase in the incidence of resorptions among the rat fetal population. Soft tissue examination of fetuses from exposed mice revealed a significant increase in the incidence of subcutaneous edema. The incidence of skeletal anomalies was not different from that of controls, but among litters of mice the incidence of delayed ossification of skull bones and the incidence of split sternebrae was slightly increased compared to that of control. A teratogenic effect was not observed.

Mutagenesis. Perchloroethylene has been shown to be inactive when tested in *E. coli* and *S. typhimurium* even when mouse liver chromosomes were added (349).

NIOSH (405) and Bartsch et al. (316) also reported no mutagenic effects in tests with salmonella but Cerna and Kyperova (406) claimed it to be positive even without metabolic activation. The available abstract also indicated perchloroethylene was positive in a host-mediated assay in ICR mice but the response was at high dosages (LD_{50} and $\frac{1}{2}$ LD_{50}) and was not dose related.

Cerna and Kyperova (406) reported no cytogenetic changes in female ICR mice given half the LD_{50} once or one-sixth the LD_{50} five times by intraperitoneal injection.

Rampy et al. reported no mutagenic effects in lymphocytes of rats after 12 months of 6-hr daily exposures 5 days/week to either 300 or 600 ppm of the vapor (407).

Carcinogenesis. Several studies of different types have been conducted on perchloroethylene, indicating it has at most a low carcinogenic potential in mice but not rats.

When perchloroethylene was fed by gavage in the NCI bioassay program at doses of approximately 1000 and 500 mg/kg per day to Osborne Mendel rats and $B_6C_3F_1$ mice, hepatocellular carcinomas were increased in mice but not rats (402, 107). Toxic nephropathy was seen in both rats and mice.

Perchloroethylene was not active when tested for lung adenomas in Strain A cancer susceptible mice (68). The mice were injected intraperitoneally with 80 mg/kg (14 injections), 200 mg/kg (24 injections), and 400 mg/kg (24 injections).

Van Duuren et al. (336) treated the skin of mice in different manners and did not produce cancer of the skin or of distant sites. Doses of 54 and 18 mg/kg were applied three times per week to the dorsal skin of mice. A dose of 163 mg/kg was applied in conjunction with a promoter (phorbal myristate acetate).

In a toxicological and carcinogenic study in which rats were exposed 6 hr/day for 12 months to 600 or 300 ppm perchloroethylene vapor and then kept for their lifetimes, there was no evidence of a tumerigenic response in either sex (407). Other criteria examined included body weights, mortality, hematologic data, urinalyses, clinical chemistry determinations, lymphocyte cytogenetics, terminal organ weights, and gross and histopathologic changes, especially tumor incidence. An increase in mortality from the fifth to the 24th month of the study in male rats exposed to 600 ppm was the only deviation from controls that was considered to be related to exposure. The mortality increase appeared to be associated with an earlier onset of spontaneous advanced chronic renal disease. Studies have been conducted in an attempt to explain, mechanistically, the sensitivity of the $B_6C_3F_1$ mouse and the resistance of the rat to perchloroethylene-induced hepatocellular carcinoma (408). On the basis of body weight, mice were found to activate perchloroethylene to a greater extent than rats, resulting in a greater degree of hepatic injury in the mouse. The authors suggested that the predominant mechanism of tumorigenicity of perchloroethylene in the mouse was by recurrent hepatic cytotoxicity which enhanced the spontaneous incidence of hepatocellular carcinoma normally found in $B_6C_3F_1$ mice and not by direct interaction of perchloroethylene with hepatic DNA. They also suggested that levels of perchloroethylene which protect against organ toxicity should be effective in preventing any tumorigenic risk from perchloroethylene for man.

Metabolism and Biological Monitoring. Numerous papers have been published on uptake, distribution, metabolism, and excretion of perchloroethylene. Perchloroethylene appears to be poorly absorbed from the gastrointestinal tract but some is excreted in the expired air following gavage.

Urinary trichloroacetic acid and unchanged perchloroethylene in the expired air have both been suggested as biologic monitors of exposure. Perchloroethylene is much less water soluble than trichloroethylene and many of the other chlorinated solvents. It has a long half-life when measured in expired air probably because of deposition in fat and similar tissue. It appears to be less metabolized than trichloroethylene, yet its hepatotoxicity appears to be related to metabolism. There appears to be a rather marked difference between species in the amount of perchloroethylene that is metabolized (408, 409).

Hake and Stewart (410) summarized much of their own and others' data on the use of expired air for monitoring industrial exposures. This included the results of an extensive study (411) in which male and female subjects were exposed singly or repeatedly for various periods of time to several concentrations of perchloroethylene. Graphs are presented which are useful in estimating workers' exposure on the basis of expired air samples. Ikeda (412) reported that limited metabolism of perchloroethylene takes place in humans but that the process is apparently saturated well below 100 ppm of the vapor. At 50 ppm only one-fifth as much metabolism occurs as with trichloroethylene.

Many others have shown low levels of urinary trichloroacetic acid following exposure to perchloroethylene. Monster (213), in a summary of several studies in which human subjects were exposed to 35 ppm of trichloroethylene or perchloroethylene, showed a 26-fold greater urinary trichloroacetic acid level with trichloroethylene.

Metabolism of perchloroethylene is generally assumed to proceed through the epoxide, tetrachloroethylene oxide, which then is converted to trichloroacetaldehyde and trichloroacetic acid. Yllner (413) showed that ^{14}C-tetrachloroethylene given to mice resulted in ^{14}C activity in the urine. Of the radioactivity, 52 percent was present as trichloroacetic acid, 11 percent as oxalic acid, and a trace as dichloroacetic acid; 18 percent of the activity was not extractable with ether even after hydrolysis. Trichloroacetic acid is also found in the urine of rats treated with perchloroethylene but to a lesser degree than in mice.

It would appear that analysis of expired air is useful for monitoring occupational exposure to perchloroethylene but that analysis of urinary metabolites is of less value (213).

Observations in Man.   *Laboratory Studies.*   In addition to clinical evalution of perchloroethylene as an anesthetic, several other studies in human subjects are available. Hake and Stewart (410) have summarized their extensive studies on humans. The analysis of expired air and urine were discussed in the preceding section and the absorption of perchloroethylene in the section on skin absorption. Based on detailed studies of neurological, psychological, and behavioral effects and the volunteers' subjective response, they concluded that 100 ppm of perchloroethylene is probably without effect but that the margin of safety is small. Perchloroethylene plus alcohol or Valium® (Diazepam) had no more effect on performance than alcohol or Valium® alone.

At 200 ppm some subjects objected to the odor and eye irritation, and higher concentrations were unacceptable. Light-headedness began at 200 ppm or slightly above. Carpenter (403) exposed himself and three other subjects and determined that the odor was detectable at 50 ppm; that 500 ppm produced salivation, a metallic taste, eye irritation, and other objectionable reactions; and that 1000 and 2000 ppm produced beginning narcosis and discomfort. This is consistent with data by Rowe et al. (400), who found 280 ppm to be objectionable to test subjects and 216 ppm to be marginal.

Industrial Experience.   The older data on human exposure to perchloroethylene were summarized by von Oettingen (33) and more recent data by NIOSH (405). As expected, numerous case reports give conflicting data.

Deaths probably due to anesthesia have occurred but there is little documentation of the concentrations inhaled. In other cases, victims have been unconscious for hours and survived with no sequelae. In general, it appears that the following responses are likely; anesthesia ranging from slight inebriation to death; nausea, although less pronounced than carbon tetrachloride or ethylene dichloride; and headache, anorexia, and eye and nasal irritation. Liver injury following excessive subacute and chronic exposure has been reported in a few but not all subjects.

Blair et al. (414) conducted a preliminary analysis of death certificates of 330 former dry-cleaning and laundry workers. The authors were quite cautious in attaching significance to their proportionate mortality analysis of the data but suggest that lung and cervical cancer appeared excessive. Neither cancer cases nor noncancer cases were associated with years of employment. Other studies of large size and better design are apparently underway at this time.

3.6.4   Hygienic Standards

The threshold limit value tentatively recommended by the ACGIH in 1980 was 50 ppm (335 mg/m^3).

3.6.5   Odor and Warning Properties

Perchloroethylene has a not unpleasant ethereal or aromatic odor. Carpenter (403) indicated that this odor is detectable at 50 ppm. This is in agreement with data summarized by The Dow Chemical Company from human experiments and industrial experience (11).

50 ppm	Odor threshold (very faint) to unacclimated. No physiological effects (8 hr)
100 ppm	Odor (faint) definitely apparent to unacclimated; very faint to not perceptible during exposure. No physiological effects (8 hr)

200 ppm	Odor (definite) moderate to faint upon exposure.
	Faint to moderate eye irritation.
	Minimal light-headedness.
	(Eye irritation threshold 100–200 ppm)
400 ppm	Odor (strong) unpleasant.
	Definite eye irritation, slight nasal irritation.
	Definite incoordination (2 hr)
600 ppm	Odor (strong) very unpleasant but tolerable.
	Definite eye and nasal irritation.
	Dizziness, loss of inhibitions (10 min)
1000 ppm	Odor (very strong) intense, irritating.
	Markedly irritating to eyes and respiratory tract.
	Considerable dizziness (2 min)
1500 ppm	Odor (almost intolerable) "gagging."
	Irritation almost intolerable to eyes and nose.
	Complete incoordination within minutes to unconsciousness within 30 min

3.7 Allyl Chloride, 3-Chloropropene, 3-Chloropropene-1, CAS 107-05-1

$$CH_2{=}CH{-}CH_2Cl$$

3.7.1 Uses and Industrial Exposures

The major use of allyl chloride is as a chemical intermediate. Allyl chloride is flammable.

3.7.2 Physical and Chemical Properties

Physical state	Liquid, colorless
Molecular weight	76.53
Specific gravity	0.9376 at 20/4°C
Freezing point	-134.5°C
Boiling point	44.6°C
Vapor pressure	368 torr (25°C)
Refractive index	1.4155 (20°C)
Percent in "saturated" air	48 (25°C)
Solubility	Miscible with ethanol, ethyl ether, and chloroform; 0.36 percent in water by weight (20°C)
Flash point	-28.9°C open cup; -31.7 closed cup

Explosive limits   3.3 to 11 percent in air
Autoignition temperature   391°C

1 mg/liter $\approx$ 320 ppm and 1 ppm $\approx$ 3.13 mg/m^3 at 25°C, 760 torr

3.7.3 Physiologic Response

Summary.   Although older data indicate allyl chloride to be moderate in acute toxicity and high in chronic toxicity, data recently derived on rats and mice indicate considerably lower toxicity. The effects are primarily due to inhalation which can produce respiratory irritation as well as liver and kidney injury. Effects on the pancreas have been alleged but this has not been confirmed. Skin contact can produce irritation and a deep-seated pain referred to as "deep-bone ache." A literature review is available (415).

Single or Short-Term Exposure.   *Oral.*   Smyth and Carpenter (416) reported an oral LD$_{50}$ of 700 mg/kg of body weight as determined in rats. Karmazin (417) also reported a moderate oral toxicity; LD$_{50}$s of 450, 500, and 300 mg/kg were determined for rats, mice, and rabbits, respectively. Mild degenerative changes were observed in the myocardium, livers, and kidneys.

Skin.   Unless confined on the skin, allyl chloride is only mildly irritating. It may be absorbed through the skin in amounts sufficient to cause systemic intoxication in animals. Smyth and Carpenter reported an LD$_{50}$ of 2.2 g/kg determined for a 24-hr skin absorption on rabbits. The Shell Chemical Corporation's Industrial Hygiene Bulletin (418) states, "The absorption of the liquid through the skin of human beings is attended by deep-seated pain in the contact area."

Eyes.   Vapor concentrations of allyl chloride of 50 or 100 ppm are irritating to the eyes. Higher concentrations of vapor may cause very severe irritation and pain. Liquids splashed in the eye would be expected to cause severe irritation and should be washed out immediately with water.

Inhalation.   There are contradictory reports of the toxicity of allyl chloride vapors in animals. Acute vapor studies on animals were reported by Adams et al. (419). Maximum exposure time–concentrations in air for a single exposure survived by rats were as follows: 3 hr at 290 ppm, 1 hr at 2900 ppm, and 15 min at 29,300 ppm. Comparable figures for guinea pigs were 8 hr at 290 ppm, 3 hr at 2900 ppm, and ½ hr at 29,300 ppm.

The primary lesions observed were in the kidneys and the lungs. The kidney injury was severe, with congestion, hemorrhage, and marked parenchymatous degeneration. The pulmonary injury was also severe, particularly at higher concentrations. It consisted of marked congestion with frequent hemorrhages in the alveolar spaces.

A similar high acute and subacute toxicity has been shown by Shamilov and Abasov (463), who determined 4-hr and 2-hr LC_{50}s of 2100 and 2600 ppm (6.6 and 8.2 mg/liter) for rats and mice. Four-hour daily exposures to 0.4 mg/liter (130 ppm) for a month affected the central nervous system and kidney function of rats, as shown by increased excretion of chlorides and protein.

However, a recent study indicates significantly less acute effect on rats and mice (465). The reason for the difference in response is not clear but may be due to changes in sample purity, the method of handling the sample, the strains of animals, changes in diets, or combinations of these and other factors. The authors summarize their data as follows:

Groups of male and female Fischer 344 rats and $B_6C_3F_1$ mice were exposed to various concentrations (range 200–2000 ppm) for 6 hours. Parameters monitored included appearance and demeanor, body weights, clinical chemistry values, organ weights, and gross and histopathological examination.

Significant findings included mortality, body weight loss, increased blood urea nitrogen and serum glutamic pyruvic transaminase values, and liver and kidney pathology. Although both liver and kidneys were target organs, the primary effect of allyl chloride exposure in both rats and mice was renal toxicity. Rats were more susceptible to allyl chloride vapor than were mice based on mortality data, renal pathology, and associated clinical laboratory findings. Female rats were affected to a greater degree than were male rats, whereas male mice were affected to a greater degree than were female mice. Minimal effect levels for renal toxicity were 300 ppm for rats and 800 ppm for mice. No significant effects were seen in rats or mice exposed to 200 ppm allyl chloride.

Previous studies had reported mortality in rats at 300 ppm; hence these results are quite different since the first mortality did not occur until exposures to concentrations of 1000 ppm were given.

Repeated or Prolonged Exposure. *Oral.*   Almeev and Karmazin (420) stated that in the course of an 8-month study, repeated oral doses of 0.015 mg/kg per day did not lead to morphological changes or other effects in rats. Karmazin (421) reported changes in conditioned reflexes after 6 months at this same dosage which appears to have been the only dose level fed in these studies.

Inhalation.   Data on repeated exposure of animals are also contradictory. The effects of repeated inhalation of the vapors of allyl chloride by four species were reported by Torkelson et al. (422). They reported "animals were exposed repeatedly to either eight or three ppm of allyl chloride in air to establish conditions safe for repeated exposure. Rats, guinea pigs and rabbits had definite liver and kidney pathology after one month of repeated seven-hour exposures to eight ppm." Histopathological examination showed "significant evidence of ill effect of a severe degree in the liver and kidneys of essentially all the animals. These were characterized by dilation of the sinusoids, cloudy swelling and focal necrosis in the liver; and by changes in the glomeruli, necrosis of the epithelium of the convoluted tubules and proliferation of the interstitial tissues in the kidney."

These species and dogs tolerated similar exposures to 3 ppm for 6 months with only slight reversible liver pathology seen in female rats.

The report by Torkelson et al. (422) is at considerable variance with more recent data by the same laboratory.

In 90-day studies, rats and mice have shown no effect from repeated 7-hr exposures to 20 ppm but adverse effects have been observed as a result of exposures to 50, 100, or 250 ppm (11).

Teratogenesis, Mutagenesis, and Carcinogenesis. *Teratogenesis.* Allyl chloride was found to be fetotoxic in rats exposed to 300 ppm vapor but not 30 ppm (465). It was not fetotoxic in rabbits or embryotoxic or teratogenic in rats or rabbits following inhalation exposure to 30 or 300 ppm vapor 7 hr/day during the period of major organogenesis. Maternal toxicity in rats was slight at 30 ppm but considerable at 300 ppm as well as in rabbits at both 30 and 300 ppm. Liver and kidney injury were observed in rat dams and maternal deaths in rabbits.

Mutagenesis.   McCoy et al. (423) reported in vitro tests to determine the genetic activity of allyl chloride in bacteria and yeast. They concluded that allyl chloride is mutagenic for *Salmonella typhimurium* (Ames test), induces gene conversions in *Saccharomyces cerevisiae*, and displays DNA-modifying activity of *E. coli*. The data which were presented by the authors in their report, however, do not fully substantiate their conclusions, and the evidence, as published, suggests only a mildly positive response in the Ames test and inconclusive results for DNA-modifying capacity for *E. coli*.

Carcinogenesis.   Allyl chloride has been included in the NCI bioassay program and reported to show no evidence of carcinogenicity in Osborne-Mendel rats of either sex (393). There was a higher than background incidence of squamous cell carcinomas in the forestomachs of both sexes of mice and squamous cell papillomas in the forestomachs of female mice.

The doses fed by gavage were as follows: male rats, 77 and 57; female rats, 73 and 55; male mice, 199 and 172; and female mice, 258 and 129 mg/kg per day.

Carcinogenic studies on mouse skin are apparently underway but results are not yet available (423).

Metabolism and Biological Monitoring. Allyl mercapturic acid has been isolated from the urine of rats dosed with allyl chloride (424). These investigators also reported that 3-hydroxypropylmercapturic acid was identified as a metabolite.

Observations in Man. Because of its recognized toxicity, allyl chloride has generally been handled carefully during its manufacture. Eye irritation as a result of overexposure to the vapors has been the most frequent complaint (418, 415). The onset of orbital pain may be delayed several hours after the exposure. More intense vapor exposure produces conjunctivitis, reddening of the eyelids, and corneal burn.

One unique feature of dermal exposure is a so-called deep bone ache, which may occur following dermal contact.

A single report available only as an abstract indicates kidney disfunction in workers exposed above the 3 mg/m^3 standard. Although kidney injury is consistent with the response in animals, not enough data are given in the abstract to determine the significance of the report in terms of dosage (425).

Another report by Hausler and Lenich (426) also gives inadequate description of the employees' exposure levels but indicates many adverse effects on a population of 60 people. An odor of garlic on the breath and body was a common complaint. The skin and respiratory tracts appeared normal, but there appeared to be alterations in liver function tests. The above effects disappeared when better industrial hygiene control measures were effected.

3.7.4   Hygienic Standards

The threshold limit of allyl chloride recommended by the ACGIH in 1980 is 1 ppm (3 mg/m_3). Torkelson et al. (422) suggested that the peak concentration of the work atmosphere should be below 2 ppm. Based on newer data, these values may be unnecessarily low. Skin contact must be prevented.

3.7.5   Odor and Warning Properties

Shell Chemical Company published the following information in their bulletin (418). The odor threshold for half of the people was from 3 to 6 ppm. The odor threshold for essentially all people was 25 ppm. Eye irritation occurs between 50 and 100 ppm. Nose irritation and pulmonary discomfort may be observed at levels below 25 ppm.

Another study reports a 50 percent response to 0.21 ppm and a 100 percent response to 0.5 ppm. A third study reports that a definite odor was detected by 10 of 13 volunteers exposed to 3 ppm (422).

Odor may be considered a warning of levels hazardous from acute exposure but allyl chloride is not detected at a low enough concentration to be considered a warning of levels hazardous from chronic exposure.

3.8 1,3-Dichloropropene, *cis,trans*-1,3-Dichloropropene; DD® Soil Fumigant (Shell Chemical Company); Telone® Soil Fumigants (The Dow Chemical Company), Vorlex® Soil Fumigants (Schering, A. G. Berlin/Bergkamen) CAS 542-75-6

$$CHCl{=}CH{-}CH_2Cl$$

3.8.1 Uses and Industrial Exposure

The cis and trans isomers of 1,3-dichloropropene are used primarily in soil fumigants or as components of soil fumigant mixtures. A small amount is used as a chemical intermediate. Exposure occurs principally during manufacture or during bulk handling activities. Since it is generally injected into the soil at depths of 15 to 30 cm airborne concentrations are generally well below 0.5 ppm even when measured in the middle of a fumigated field. The material is flammable.

3.8.2 Physical and Chemical Properties

Because the 1,3-dichloropropene isomers are not commonly separated and because commercial products generally contain varying amounts of other C_3 compounds, the following data for a 92 percent material must be considered only approximate.

Physical state	Liquid
Molecular weight	110.98 (pure isomers)
Specific gravity	~1.2 (20/20°C)
Freezing point	−84°C
Boiling point	~104°C
	104.3°C (cis)
	112°C (trans)
Vapor pressure	~28 torr (25°C)
Percent in "saturated" air	3.7 (25°C)
Solubility	~0.1 g/100 ml water; miscible with most organic solvents
Flash point	~28°C
Explosive limits	~4.3 to 10.6 percent (80°C)

1 mg/liter ≎ 220 ppm, and 1 ppm ≎ 4.54 mg/m³ at 25°C and 760 torr

3.8.3   Physiologic Response

Summary.   The subchronic inhalation toxicity data on 1,3-dichloropropene are contradictory. Older data indicate the commercial product studied at that time (1958–1975) was quite irritating and hepatotoxic, but data developed on currently produced fumigants indicate considerably less hepatoxicity. Although impurities in the samples used in older toxicity studies may account for part of the discrepancy, different strains of animals, different animal diets, and different methods of handling the samples used in recent studies may also explain the difference. It would appear reasonable to put greater reliance on the more recent data, which has been obtained with improved methods and with samples of better characterized purity.

Single and Short-Term Exposure.   *Oral.*   A 10 percent solution of a cis and trans mixture of 92 percent purity (8 percent related compounds) was found to have acute oral LD_{50}s of 710 and 470 mg/kg in male and female rats, respectively. The livers and kidneys of treated animals were grossly affected and there was a suggestion of lung injury in surviving animals (427).

Eye.   Severe to moderate injury occurred in the eyes of rabbits in which a couple of drops of liquid 1,3-dichloropropene had been placed. Prompt washing with water greatly relieved the degree of irritation. The vapors were quite irritating to the eyes and caused lacrimation (427).

Skin.   Necrosis and edema occurred when liquid 1,3-cichloropropene was confined on the skin of rabbits but if allowed to evaporate the effect was greatly reduced.

Skin Absorption.   Absorption through the skin occurred particularly when the liquid was confined or when in a propylene glycol solution which retarded evaporation.

Inhalation.   According to Torkelson and Oyen (427), 2700 ppm was extremely irritating to the respiratory tract and caused liver and kidney injury in rats. Rats survived a 1-hr exposure but died from a 2-hr exposure to 1000 ppm of the vapor. These rats and others exposed to 700 ppm had a peculiar garlic (or skunk) odor following exposure, indicating probable absorption and possible reaction of the 1,3-dichloropropene with the hair or skin of the rats. Guinea pigs, but not rats, succumbed following a single 7-hr exposure to 400 ppm; the rats, however, were severely injured, lost weight, and required 8 days to recover lost weight. Lung injury was still present eight days after exposure.

    In an early subacute study to define the conditions for a subsequent chronic study, Torkelson and Oyen (427) found considerable liver and kidney injury evident grossly in small groups of rats exposed 19 times to 50 ppm, 7 hr/day in a 28-day period. Less injury was seen at 11 ppm with these early chemical samples.

Studies with a more recently produced product, Telone II® soil fumigants (47 percent cis, 45 percent trans, the balance related compounds), indicate a considerably lower toxicity. Fischer 344 rats and CD-1 mice were exposed 7-hr/day, 5 days/week for 13 weeks to 0, 93, 32, or 12 ppm. The exposures resulted only in a failure to gain weight (high dose rats and female mice) and focal histomorphological changes of the epithelium of the nasal septums and turbinates in high dose rats of both sexes as well as female rats exposed to 32 ppm. Female but not male mice exposed to 93 ppm showed similar effects (11).

Parker et al. (428) also found much less effect in CD-1 mice and Fischer 344 rats exposed to either 0, 5, 15, or 50 ppm 6 hr/day, 5 days/week, for 6 or 12 weeks. Their test material was Shell DD® soil fumigant which contained 55 percent 1,3-dichloropropene and 30 percent 1,2-dichloropropane. The balance was related compounds. According to the available abstract, "The following parameters were examined: pharmacotoxic signs, body weights, hematology (HGB, HCT, RBC, WBC, and Diff. Leukocyte Count), serum chemistry (BUN, GLU, ALB, GPT, and ALP), urinalysis, gross pathology, histopathology, organ weights and organ weight/body weight ratios of brain, heart, liver, kidneys, testes or ovaries, and adrenals. The only exposure-related effect observed was increased mean liver weight/body weight ratios of male mice and rats and mean kidney/body weight ratios of female rats at the 50 ppm exposure level. Other statistically significant differences were either inconsistent or not dose-related. Slight to moderate diffuse hepatocytic enlargement in 12/21 of the 50 ppm male mice after 12 weeks exposure was the only compound-related histopathologic change present."

Repeated or Prolonged Exposure. *Ingestion.* When fed by gavage to groups of rats at 0, 1, 3, 10, or 30 mg/kg body weight, 6 days/week for 13 weeks it was concluded the 3 to 10 mg/kg was a nontoxic effect level. No gross or microscopic pathological changes related to exposure were observed at any level nor were general condition, demeanor, survival hematologic indexes, serum exposure activity, and urinalysis affected by exposure. The relative weights of the kidneys of both sexes were increased at 30 mg/kg/day in both sexes and in males at 10 mg/kg per day (429).

1,3-Dichloropropene has been included in the NCI bioassay program discussed under carcinogenesis in the next section.

Inhalation. Only older samples of 1,3-dichloropropene have been used in long-term (6 months) inhalation studies (427). Four species of animals were exposed 7 hr/day, 5 days/week to either 3 or 1 ppm 1,3-dichloropropene in air for a period of 6 months. There was no effect in rats, rabbits, guinea pigs, or dogs exposed to 1 ppm (0.9 ppm recovered analytically) nor in any group except male rats exposed to 3 ppm (2.6 ppm recovered analytically) based on demeanor, general appearance, growth, mortality, hematologic examination, final body and organ weights, and gross and microscopic examination of all major organs. The kidneys of the males rats exposed to 3 ppm showed cloudy swelling, which was attributed to exposure. This effect was also seen in a small group of male rats exposed 4 hr/day, but not in rats exposed 2 or 1 hr/day. Based on the lower toxicity of recently produced 1,3-dichloropropene in subacute oral and inhalation studies, it seems reasonable to assume that the minor effects reported are maximum probable effects.

Teratogenesis, Mutagenesis, and Carcinogenesis. *Teratogenesis.* No published reports were found.

Mutagenesis. While mutagenic effects have been produced in several *in vitro* tests, including recombinant tests with *B. subtilis* and reversion tests with *S. typhimurium* TA-1535 and TA-100 with both isomers (274, 471), the practical significance to mammalian systems is not clear. It has been shown that *in vitro* microbiological systems may not be representative of intact mammals. Dean et al. (472) have shown that the mutagenic action of *cis* 1,3-dichloropropene is greatly reduced by adding glutathione to the bacterial system, and Climie et al. (473) have shown glutathione conjugation to be part of the detoxification mechanism in rats. Based on these data it appears probable that 1,3-dichloropropene is not a mammalian mutagen when given at doses that do not overwhelm the animal or cause severe toxicity.

Carcinogenesis. Telone II fumigant has been included in the NCI bioassay procedures but only incomplete data are available (430). Rats have been fed 50 or 100 mg/kg three times/week and mice 25 or 50 mg/kg (430).

Metabolism and Biologic Monitoring. Hutson et al. (431) fed ^{14}C-labeled 1,2-dichloropropane and both isomers of 1,3-dichloropropene to rats and found differences in their metabolism. With all compounds 80 to 90 percent of the radioactivity was eliminated in the first 24 hrs. The major route of excretion of radioactivity was in the urine, where 50.2, 80.7, and 56.5 percent of the 1,2-dichloropropane, *cis* 1,3-dichloropropene, and *trans*-1,3-dichloropropene activity were found, respectively. The amount of ^{14}C-carbon dioxide excreted was quite different for the two isomers. The cis isomer yielded only 3.9 percent of the dose and the trans isomer 23.6 percent, with correspondingly less in the radioactivity in the urine. As expected with volatile compounds, residual unreacted compounds were not present as significant residues, although metabolites entered the normal metabolic pool. Subsequently Climie, et al. (473) showed that 82–84 percent of the radioactivity of ^{14}C labeled on the second carbon was recovered in the urine of rats as N-acetyl-S-((cis)-3-chloroprop-2enyl) cysteine.

It seems unlikely that analysis of blood urine or expired air will be of significant value for industrial hygiene monitoring at currently accepted levels of exposure.

Observations in Man. Only one brief case report was found in the literature. This article is of dubious value since it contains a significant amount of subjective evidence and presents inadequate control data (432).

3.8.4 Hygienic Standards

In 1980 the ACGIH recommended 1 ppm (5 mg/m^3) as a threshold limit value for 1,3-dichloropropenes. This value is based on data from older samples; it may be lower than necessary based on data from new chemical samples.

3.8.5 Odor and Warning Properties

Concentrations capable of causing injury from a single exposure will probably have an odor detected by most people; however, the odor can not be relied upon to warn against concentrations capable of causing injury from prolonged repeated exposure (433).

3.9 Chloroprene, β-Chloroprene, 2-Chlorobutadiene, CAS 126-99-8

$$CH_2=CCl-CH=CH_2$$

3.9.1 Uses and Industrial Exposures

Chloroprene is used as a chemical intermediate, largely as a monomer for the manufacture of rubber. At room temperature it reacts with oxygen to form peroxides and polymerizes to produce cyclic dimers or open chain, high molecular weight products (398).

3.9.2 Physical and Chemical Properties

Physical state	Colorless liquid
Molecular weight	88.54
Specific gravity	0.9583 (20/20°C)
Boiling point	59.4°C
Vapor pressure	215.4 torr (25°C); 188 torr (20°C)
Refractive index	1.4583 (20°C)
Percent in "saturated" air	28 (25°C)

Solubility   Slightly soluble in water; soluble in
                                    alcohol, diethyl ether
Flash point $-22°C$
Explosive limits   4 to 20 percent in air

1 mg/liter $\approx$ 276.5 ppm and 1 ppm $\approx$ 3.62 mg/m^3 at 25°C, 760
torr

3.9.3 Physiologic Response

Summary. Reviews of the toxicity of β-chloroprene are available (289, 434). At high concentrations β-chloroprene has an anesthetic action, but this is not as important as eye and respiratory tract irritation and liver injury which result from excessive exposures. Hair loss has also been reported in humans and animals exposed to β-chloroprene.

The available literature is extremely contradictory as to the type of injury that has resulted from exposure to β-chloroprene and to the exposure levels causing them. Russian literature claims considerably different and/or greater effects than have been observed in studies from other places that have tried to confirm these claims. No clear explanation is available, but at least two possible explanations appear reasonable. β-Chloroprene is a very unstable compound, which, unless handled with extreme care, expoxidizes and polymerizes to toxic compounds (435, 436). This might explain the alleged effects in animals. Alleged effects in humans may be due to this same cause or to the use of different chemical processes which produce different types of impurities. Many other causes can be postulated, but in our opinion more credence must be given to animal studies in which the sample is known to have been handled with extreme care and to the results of experience in U.S. industry where the method of handling has been reported.

Single or Short-Term Exposure. *Oral.* β-Chloroprene is an irritant when intubated into the stomach of laboratory animals. It also produces CNS depression, generalized congestion, and edema. Von Oettingen et al. (437) reported the LD$_{100}$ for rats to be 0.67 g/kg of body weight. Death occurred in periods from 5 hr to 4 days after administration. LD$_{50}$s of 251 mg/kg for rats and 260 mg/kg for mice were reported by Asmangulian and Badalian (467).

Eye. Conjunctivitis and necrosis of the cornea have been reported. Von Oettingen (33) quoted Roubal to this effect. Apparently, nervousness and irritability are typical responses to exposure. A more thorough discussion of the clinical picture can be found in reports by Nystrom (435) and Ritter and Carter (438).

The vapors are reported to cause irritation and evidence of pain at 625 ppm but not at 160 or 40 ppm in rats exposed 6 hr/day for 4 weeks (438).

Skin. von Oettingen (437) indicated some systemic toxicity from repeated topical application of chloroprene to the skin of rats. Ritter and Carter (438) indicated systemic poisoning from the topical application of chloroprene to guinea pigs. It is uncertain how completely the animals were prevented from breathing the vapors when it was applied to the skin.

Perhaps the most surprising effect from topical application is the loss of hair. Because the hair grows readily when the individual is removed from contact, it would appear that the action may be directly on the hair. Apparently, this is observed both in humans in industrial operations and in animals exposed in the laboratory. Ritter and Carter question whether this action is of chloroprene itself or an intermediate in the polymerization of chloroprene, but it is agreed that the depilatory action does occur in workers that are handling chloroprene in the polymerization process.

Inhalation. von Oettingen (437) indicated that a 1-hr exposure of mice to chloroprene vapor at a concentration of 3 mg/liter of air (829.2 ppm) was fatal to all the animals exposed. A concentration of 1 mg/liter (277 ppm) killed none of the animals. They also reported that the LC_{100} for an 8-hr exposure for rabbits is 7.5 mg/liter of air and for cats is 2.5 mg/liter.

Clary et al. (439), using fresh, purified β-chloroprene, determined an approximate lethal concentration of 2280 ppm (8.42 mg/liter) for a 4-hr exposure of rats. All six rats exposed to 1690 ppm (6.24 mg/liter) survived. Injury to the respiratory tract was apparent in a few rats sacrificed 1 or 2 days after exposure. Some liver "changes" were reported.

They also exposed rats and hamsters to either 625, 160, or 40 ppm, 6 hr/day, 5 days/week for 4 weeks. Toxic effects were observed at all levels. The author concluded,

"The primary effects seen at the low level were skin and eye irritation and weight loss in rats. At higher levels tissue damage, especially lung and liver, and mortality were observed. Hair loss was observed, primarily at the two highest levels in female rats.

The range-finding studies indicate that repeated exposure at approximately 625 ppm in rats resulted in mortality as well as growth retardation. In hamsters, one exposure at 630 ppm was lethal. Midzonal liver degeneration and necrosis were also noted at both the high exposure and midexposure levels as well as increased liver and kidney weights in both species. The lower exposure level (40 ppm) was irritating in both species, and significant growth retardation was also noted in the rats. The alopecia observed in the present studies is consistent with earlier reports . . . (436)."

Teratogenesis, Mutagenesis, and Carcinogenesis. *Teratogenesis.* Numerous Soviet reports describe teratological and reproductive effects in animals. It would appear that these effects may have been due to improper handling of the samples with resulting toxic reaction products.

When the purity of the sample was carefully controlled, repeated exposures to 25 ppm or less of the vapor have caused no reproductive, teratological, or embryotoxic effects in rats (440). Despite frank clinical toxicity in exposed pregnant rats, fetuses showed no teratogenic effects at β-chloroprene levels as high as 175 ppm (441).

Mutagenesis. β-Chloroprene, like many other compounds, has been shown to be mutagenic in some but not all test systems. Several reports summarized by NIOSH (434) indicate mutagenic activity in drosophila, mice, and rats in addition to microorganisms. Whether this is due to impurities in the samples is not clear. Bartsch et al. (442) reported that liver enzymes were needed to produce mutations in *S. typhimurium*. They were able to trap an alkylating metabolite and considered the mutagenic activity to be the likely result of an oxirane metabolite. An unpublished study (441) found no dominant lethal effects in rats and mice exposed to 100 ppm.

Carcinogenesis. A large number of Soviet studies which claim to have produced cancer in animals (434) have not been reproducible in several other laboratories where the purity of the sample was assured even though the experimental dosages were much higher. After reviewing the human and animal data, IARC (129) placed chloroprene in a group of chemicals that "could not be classified as to their carcinogenicity for humans."

Metabolism and Biologic Monitoring. Bartsch et al. (316) isolated a metabolic capable of alkylation when they added human liver enzymes to *S. typhimurium* in their mutagenicity test system.

Observations in Man. Even more so than other chemicals, the extreme instability of β-chloroprene and the resulting products make it difficult to ascribe effects in humans to the chemical itself. Numerous reports of injury in production and use of β-chloroprene have been published and presented by NIOSH (434). Symptoms include hair loss, chest pains, fatigue, nausea, personality changes, conjunctivitis, dermatitis and chemical burns, unconsciousness, decreased spermatogenesis and other reproductive effects, sexual impotency, skin cancer, lung cancer, and liver and kidney injury. These and other symptoms, mostly from eastern European reports, are so diverse as to suggest they are the result of exposure to a variety of chemicals.

Careful study of two U.S. plants (443) did not confirm the lung cancer. No other cause of death appeared to be related to exposure to β-chloroprene, and overall mortality was equal to the experience throughout the rest of the company and typical of many industrial populations, being less than the average for U.S. white males.

3.9.4 Hygienic Standards

The threshold limit of chloroprene recommended by the ACGIH in 1980 was 10 ppm (45 mg/m^3). Skin contact should be prevented.

3.9.5   Odor and Warning Properties

β-Chloroprene has a pungent odor and is a lacrimator.

3.10   Hexachlorobutadiene, HCBD, Perchlorobutadiene, CAS 87-68-3

$$CCl_2 {=\!\!=} CCl {-\!\!-} CCl {=\!\!=} CCl_2$$

3.10.1   Uses and Industrial Exposures

Although hexachlorobutadiene (HCBD) has had use as a pesticide in other countries, exposure in the United States has mostly been as an unwanted by-product of certain processes associated with chlorination of hydrocarbons.

3.10.2   Physical and Chemical Properties

The physical properties listed for hexachlorobutadiene are somewhat uncertain.

Physical state	Colorless liquid
Molecular weight	261
Specific gravity	1.68 (25/4°C)
Melting point	-18.5 to -21°C
Boiling point	>200°C
Vapor pressure	About 0.3 torr (25°C)
Refractive index	1.5535 (25°C)
Percent in "saturated" air	0.037 (25°C)
Solubility	Insoluble water at 20°C
Flammability	Not flammable by standard tests in air

1 mg/liter ≎ 936 ppm and 1 ppm ≎ 10.7 mg/m^3 at 25°C, 760 torr

3.10.3   Physiologic Response

Summary.   Hexachlorobutadiene has a rather high chronic toxicity, and is capable of causing renal injury in rats including renal cancer. Most of the data for this section are taken from Kociba et al. (444), who reviewed the literature and published their own chronic toxicity and reproductive studies in rats.

Single or Short-Term Exposure.   *Oral.*   Kociba et al. cite literature values for acute oral LD$_{50}$s of 90 mg/kg for guinea pigs, 87 to 116 mg/kg for mice, and 200 to 350 mg/kg for rats. They also determined LD$_{50}$s of 580 and 200 to 400 mg/kg for adult male and female rats and 64 and 46 mg/kg for adult male and female rats and 64 and 46 mg/kg for 21-day weanlings. Most deaths occurred 2 to 3 days after treatment but some were delayed as long as 17 days. Short-term dietary studies in which rats were fed 0, 50, 150, and 450 ppm in their diet for 2 weeks resulted in growth depression and kidney injury similar to that seen in acute studies (445).

Kociba et al. also reported that no kidney injury was produced by daily doses of 3 mg/kg body weight but that higher levels in a 30-day study did. At 100 mg/kg body weight/day minimal liver injury was observed.

Skin. According to Kociba, "Dermal application of 126 mg/kg hexachlorobutadiene was lethal to half the rabbits in 7 hours and 4 of 4 after 24 hours. All rabbits survived the dermal application of 120 mg/kg for 4 hours or 63 mg/kg for 24 hr." According to an available abstract, a single subcutaneous injection of 20 mg/kg into female rats resulted in death of young born 3 months later to these dams (446). This must be confirmed.

Inhalation. Single 4- to 7-hr exposures to 133 to 500 ppm hexachlorobutadiene cause the death of some or all exposed rats. Exposure for 0.9 hr to 160 ppm or 3.5 hr to 35 ppm was survived by all rats. Guinea pigs and cats died from similar exposures to 160 ppm or from 7.5-hr exposures to 35 ppm (444).

Gage (268) exposed rats to 250 ppm for 4 hr and 100, 25, 10, or 5 ppm for 6 hr. Fifteen exposures to 5 or 10 ppm resulted in no observed toxic effect except for retarded weight gain at 10 ppm. Fifteen exposures to 25 ppm, two exposures to 100 ppm, and two exposures to 250 ppm resulted in respiratory irritation and injury as well as rather pronounced effects on the renal tubules. The adrenals showed injury at 100 and 250 ppm.

Limited unpublished data of The Dow Chemical Company (11) also indicate a rather high subacute toxicity for hexachlorobutadiene when inhaled. Small groups of rats and guinea pigs were exposed to about 8 or 30 ppm 7 hr/day, 5 days/week. The guinea pigs exposed to 30 ppm died after four exposures whereas nine of 10 rats survived. However, severe injury was grossly apparent in the lungs, livers, and kidneys of the survivors. Ten of 10 rats and four of five guinea pigs survived 19 7-hr exposures to 8 ppm, but liver and kidney injury were grossly apparent on necropsy.

Repeated or Prolonged Exposure. *Inhalation.* Small groups of rats, rabbits, and guinea pigs exposed 7 hr/day, 100 times to 3 ppm in a 143-day period were adversely affected, but those exposed 129 times in 184 days to 1 ppm were not. The livers and kidneys of the animals exposed to 3 ppm were the organs most affected (11).

An available abstract indicates mice and rats tolerated 0.024 mg/liter (22 ppm) of hexachlorobutadiene in air for 7 months. The details of number of hours per day and number of exposures are not available (447).

Ingestion. Kociba et al. (444) have presented the most complete study to date. Their summary follows:

In the chronic dietary study, as in shorter-term studies conducted previously with HCBD, the kidney appeared to be the primary target organ. Ingestion of 20 mg/kg-day of HCBD for up to 2 years caused multiple toxicological effects, including decreased body weight gain, increased mortality, increased urinary excretion of coproporhyrin, increased weights of kidneys, increased renal tubular epithelial hyperplasia, and renal tubular adenomas and adenocarcinomas, some of which metastasized to the lungs.

Ingestion of the intermediate dose level of 2 mg/kg-day of HCBD for up to 2 years caused lesser degrees of toxicity, including an increase in urinary coproporphyrin excretion and an increase in renal tubular epithelial hyperplasia. The fact that urinary excretion of coproporphyrin was increased at this dose level which produced no neoplasms may indicate the usefulness of this parameter as a biological monitor when included in a medical surveillance program for workers exposed to HCBD.

Ingestion of the lowest level doses of 0.2 mg/kg-day of HCBD for up to 2 years caused no effects that could be attributed to treatment. Whereas the intermediate dose level of 2.0 mg/kg-day caused a slight degree of renal toxicity, the highest dose level of 20 mg/kg-day for up to 2 years caused multiple and substantial toxicological effects, including renal tubular neoplasms. Thus, these data indicate a clear-cut dose-response relationship for HCBD-induced toxicity affecting primarily the kidney. HCBD-induced renal neoplasms occurred only at a dose level higher than that causing discernible renal injury.

Teratogenesis, Mutagenesis, and Carcinogenesis. *Teratogenesis.* Several studies of teratogenicity and reproduction of hexachlorobutadiene have been reported. Japanese quail kept on diets containing 30 ppm (about 5 mg/kg body weight per day) for 90 days had no deleterious effect on reproduction (448). Male and female rats were fed 20, 2, or 0.2 mg/kg per day for 90 days prior to mating and during gestation and lactation (449). Both sexes fed the two higher levels were adversely affected, primarily in the kidneys, but reproductive indexes were not. Only slight changes occurred at 2 mg/kg per day and no effects were observed at 0.2 mg/kg per day. The only reproductive finding was a slight nonsignificant decrease in neonatal weight at 21 days at the top dosage only. No toxicologic effects were observed among the adults at a dose level of 0.2 mg/kg per day or among the neonates at dose levels of 0.2 or 2.0 mg/kg per day.

Harleman and Seinen (445) also reported no effect on fertility or progeny except for decreased body weight at birth and weaning in rats born to mothers fed 150 ppm in their diet. Grossly observable malformations were not seen in their small study.

Carcinogenesis. The section on repeated or prolonged exposure summarizes the available carcinogenic data.

Metabolism and Biologic Monitoring. Kociba et al. (444) have suggested that urinary excretion of coproporphyrins be included as part of a medical surveillance program on employees.

Currently there appear to be no useful biologic monitors for hexachlorobutadiene.

Observations in Man. No published reports of effects in humans were found.

3.10.4 Hygienic Standards

The ACGIH in 1980 had no recommended TLV for hexachlorobutadiene but did include it in their appendix as an animal carcinogen. Based on the ingestion study by Kociba et al. (444), the equivalent of about 0.13 ppm in air would be without effect, and 1.3 ppm would produce only reversible changes.

Incomplete data from Gulko et al. (447) suggest no adverse effects in rats at 2.3 ppm.

Based on animal data, a level of 0.05 ppm is suggested by the authors as a hygienic guide with adequate precautions to prevent ingestion and skin contact.

3.10.5 Odor and Warning Properties

Old data on an inadequately characterized sample indicates that 1 ppm was detectable by only about half of those who attempted to smell it. Properties of hexachlorobutadiene should be considered inadequate to warn against excessive repeated exposures.

3.11 Dichloroacetylene, 1,2-Dichloroacetylene, Dichloroethyne, CAS 7572-29-4

$$CCl\!\!\equiv\!\!CCl$$

3.11.1 Uses and Industrial Exposure

Dichloroacetylene is a highly toxic, spontaneously combustible, undesired product of dehydrochlorination of trichloroethylene. It has resulted from exposure of trichloroethylene vapor to Hopcalite in a closed environmental system (submarine) and soda lime in closed circuit (rebreathing) anesthesia machines and from exposure to trichloroethylene liquid to caustic in degreaser tanks.

3.11.2 Physical and Chemical Properties

Physical state	Liquid
Molecular weight	94.92
Melting point	-66 to $-64.2°C$
Boiling point	Explodes

Solubility   Insoluble water; soluble in organic solvents
Flash point   Spontaneously combustible

1 mg/liter $\backsim$ 237 ppm and 1 ppm $\backsim$ 3.9 mg/m³ at 25°C, 760 torr

3.11.3 Physiologic Response

Summary. Dichloroacetylene is highly toxic causing headaches, nausea, neurological, and liver and kidney injury. Deaths possibly due to kidney injury have been reported. Defolgue (352) summarized the human experience during anesthesia with trichloroethylene prior to 1961.

Single and Short-Term Exposure. Reichert et al. (450) reported that the LC_{50}s for 1- and 4-hr exposures of mice were 124 and 19 ppm, indicating a highly toxic vapor. Deaths were due to kidney injury. Degenerative lesions were found in the brain, a finding the investigators thought consistent with rather pronounced sensory loss and effects in facial muscles of humans. Reichert et al. (451) reported similar injury in rabbits.

Rats were exposed 6 hr/day, 5 days/week for 6 weeks to dichloroacetylene at 2.9, 9.8, or 15.5 ppm in a dichloroacetylene–trichloroethylene mixture as well as a continuous 90-day exposure to 2.8 ppm dichloroacetylene (452). The trichloroethylene vapor concentrations were 3.2, 50, and 150 ppm during the repeated exposures and 5.3 ppm during the continuous. Approximately 4 ppm acetaldehyde vapor was also present. Although no deaths were reported, rats exposed repeatedly to 9.8 and 15.5 ppm were unkempt in appearance and showed some respiratory distress. No signs of toxicity were seen during repeated exposure to 2.8 ppm but rats exposed continuously were emaciated, showing weakness in the hind limbs. One rat appeared to be blind at the end of the exposure period. The most striking observation during microscopic examination was in the kidneys of rats exposed continuously to 2.8 ppm and repeatedly to 15.5 ppm. Effects in the lungs were less consistent and not clearly related to exposure. The course of development of the kidney changes with time has been studied in rats exposed continuously to 4.8 ppm for up to 28 days. In addition to rather marked injury to this organ, weakness in the hind legs, self-inflicted bite wounds, weight loss, and some deaths were observed. Reichert et al. (453) also reported severe kidney injury and some deaths in rabbits exposed for 1 hr to 126, 202, and 307 ppms or 6 hr to 17 to 23 ppm.

Teratogenesis, Mutagenesis, Carcinogenesis, and Metabolism. No reports were found.

Observations in Man. Deaths ascribed to dichloroacetylene have been reported following use of trichloroethylene with soda lime in closed-circuit anesthesia machines but the precise inhaled concentrations were not determined (452). Defolgue (352) has summarized much of these data. Saunders (454) described human experience in an incident in a closed environmental system in which alkaline materials were present in the life support system.

3.11.4   Hygienic Standards

The threshold limit value for dichloroacetylene recommended by the ACGIH in 1980 is a ceiling of 0.1 ppm (0.4 mg/m^3).

3.11.5   Odor and Warning Properties

Based on the experience in closed environmental systems, it is doubtful if dichloroacetylene has useful warning properties. No data were found in odor thresholds.

REFERENCES

1. D. D. Irish, in *Industrial Hygiene and Toxicology*, 2nd rev. ed., Vol. II, Frank A. Patty, Ed., Interscience, New York, 1962.

2. American Conference of Governmental Industrial Hygienists, *Threshold Limit Values for Chemical Substances in Workroom Air Adopted by ACGIH for 1980*, Cincinnati, Ohio, 1980.

3. G. D. Clayton and F. E. Clayton, Eds., *Patty's Industrial Hygiene and Toxicology*, Vol. I, General Principles, Wiley-Interscience, New York, 1978.

4. U.S. Department of Health Education & Welfare, National Institute for Occupational Safety and Health. D. G. Taylor, Coordinator, *Manual of Analytical Methods*, 2nd ed., DHEW (NIOSH) Publication No. 77-237A, 1977.

5. American Conference of Governmental Industrial Hygienists, *Air Sampling Instruments for Evaluation of Atmospheric Contaminants*, 5th ed., Cincinnati, Ohio, 1978.

6. R. G. Melcher, R. R. Langner, and R. O. Kegel, *Am. Ind. Hyg. Assoc. J.* **39,** 349 (1978).

7. J. D. Repko and S. M. Lasley, *Behavioral Neurological and Toxicological Effects of Methyl Chloride: A Review of the Literature*, HEW (NIOSH) ITR-76-33, March 1976.

8. J. D. Repko and S. M. Lasley, *Crit. Rev. Toxicol.*, **6,** 283 (1979).

9. G. Gudmundssen, *Arch. Environ. Health*, **32,** 236 (1977).

10. J. D. C. MacDonald, *Occup. Med.* **6,** 81–84 (1964).

11. The Dow Chemical Company, Midland, Mich., unpublished data.

12. R. R. Sayers, W. P. Yant, B. G. H. Thomas, and L. B. Berger, *U.S. Publ. Health Bull. No. 185*, 1929.

13. F. Flury and F. Zernik, *Schädliche Gase*, Springer, Berlin, 1931.

14. G. Y. Evtushenko, *Gig. Tr. Prof. Zabol.*, **10,** 20 (1966).

15. W. W. Smith and W. F. von Oettingen, *J. Ind. Hyg. Toxicol.*, **29,** 47 (1947).

16. A. W. Andrews, E. S. Zawistowski, and C. R. Valentine, *Mutat. Res.*, **40,** 273 (1976).

17. Anonymous, National Library of Medicine, Toxicology Information Program, *Tox Tips*, **30,** 7 (1978).

18. W. W. Smith, *J. Ind. Hyg. Toxicol.*, **29**, 185 (1947).

19. F. Sperling, F. J. Macri, and W. F. von Oettingen, *A.M.A. Arch. Ind. Hyg. Occup. Med.*, **1**, 215 (1950).

20. B. Soucek, *Arch. Gewerbepathol. Gewerbehyg.*, **18**, 370 (1961) (CA *69*, 1882h).

21. J. S. Bus, *The Pharmacologist*, **20**, 214 (1978).

22. R. D. Stewart, C. L. Hake, A. Wu, S. A. Graff, H. V. Forster, A. J. Lebrun, P. E. Newton, and R. J. Soto, *Methyl Chloride: Development of a Biological Standard for the Industrial Worker by Breath Analysis*, The Medical College of Wisconsin, Milwaukee, Wis., NIOSH-MCOW-ENVM-MCM-77-1, 1977.

23. R. Putz, J. Setzer, J. S. Croxton, and F. C. Phipps, *Accepted Scand. J. Work Env. Hlth.* **7**, (1981).

24. A. H. Kegel, W. D. McNally, and A. S. Pope, *J. Am. Med. Assoc.*, **93**, 353 (1929).

25. W. D. McNally, *J. Ind. Hyg. Toxicol.*, **28**, 94 (1946).

26. H. N. Hansen, K. Weaver, and F. S. Venable, *A.M.A. Arch. Ind. Hyg. Occup. Med.*, **8**, 328 (1953).

27. E. Klimkova-Dentschova, *Rev. Czech. Med.*, **3**, 1 (1957).

28. J. D. Repko, P. D. Jones, L. S. Garcia, E. J. Schneider, E. Roseman, and C. P. Corum, *Behavioral and Neurological Effects of Methyl Chloride*, U.S. HEW (NIOSH) Publ. No. 77-125, December 1976.

29. D. D. Irish, E. M. Adams, H. C. Spencer, and V. K. Rowe, *Ind. Hyg. Toxicol.*, **22**, 218 (1940).

30. L. Ehrenberg, S. Osterman-Golkar, D. Singh, and A. Lundquist, *Radiat. Bot.*, **14**, 185 (1974).

31. R. M. Watrous, *Ind. Med.*, **11**, 575 (1942).

32. E. O. Longley, and A. T. Jones, *Ind. Med. Surg.*, **34**, 499 (1965).

33. W. F. von Oettingen, *The Halogenated Aliphatic, Olefinic, Cyclic, Aromatic, and Aliphatic-Aromatic Hydrocarbons including the Halogenated Insecticides, Their Toxicity and Potential Dangers*, U.S. Publ. Health Serv. Publ. No. 414, 1955.

34. C. H. Hine, *Occup. Med.*, **11**, 1 (1969).

35. F. M. Mellerio, C. Gaultier, and C. Bismut, *Eur. Toxicol. Environ. Hyg.*, **7**, 119 (1974).

36. M. Buckell, *Brit. J. Ind. Med.*, **7**, 122 (1950).

37. M. K. Johnson, *Biochem. J.*, **98**, 38 (1966).

38. B. B. Shugaev and A. V. Mazkova, *Toksikol. Gig Prod. Neftekhim, Neftekhim. PROIZ VOD*, **1968**, 139, *Chem. Abstr.*, **71**, 89664.

39. C. Bachem, *Arch. Exp. Pathol. Pharmacol.*, **122**, 69 (1927).

40. W. H. Chambers, E. H. Krachow, F. P. McGrath, S. B. Goldberg, L. H. Lawson, and J. K. McNamee, *U.S. Army Chem. Corp. Med. Div. Rep. No. 23*, 1950.

41. J. McCann, V. Simmon and D. Streitwieser, *Proc. Natl. Acad. Sci. U.S.*, **72**, 3190 (1975).

42. H. Druckrey, H. Krauss, R. Pruessmann, S. Ivankovic, and C. Landschuetz, *Z. Krebforsch.*, **74**, 241 (1970) (CA 73, 85731).

43. L. A. Poirier, G. D. Stoner, and M. B. Shimkin, *Cancer Res.*, **35**, 1411 (1975).

44. E. A. Barnsley and L. Young, *Biochem. J.*, **95**, 77 (1965).

45. M. Baselga-Monte, *et al*, *Med. Lav.*, **56**, 592 (1965) (*Chem. Abstr.* **64**, 10295).

46. A. Garland and F. E. Camps, *Brit. J. Ind. Med*, **2**, 209 (1945).

47. G. B. Appel, R. Galen, J. O'Brien, and R. Schoenfeldt, *Ann. Int. Med.*, **82**, 534 (1975).

48. J. Skutilova, *Prac. Lek.*, **27**, 341 (1975) (*Chem. Abstr.* **87**, 089899S).

49. National Institute of Occupational Safety and Health. *Criterion for a Recommended Standard— Occupational Exposure to Methylene Chloride*, HEW Publ. No. (NIOSH) 76-138, 1976.

50. R. D. Stewart, and H. C. Dodd, *Am. Ind. Hyg. Assoc.*, **25**, 439, 1964.

51. J. L. Svirbely, B. Highman, W. C. Alford, and W. F. von Oettingen, *Ind. Hyg. Toxicol.*, **29**, 382 (1947).

52. K. B. Lehmann, and L. Schmidt-Kehl, *Arch. Hyg. Bakteriol.*, **116**, 131 (1936).

53. G. L. Plaa and R. E. Larson, *Toxicol. Appl. Pharmacol.*, **7**, 37 (1965).

54. P. J. Gehring, *Toxicol. Appl. Pharmacol.*, **13**, 287 (1968).

55. A. E. Ahmed and M. W. Anders, *Drug Metab. Dispos.*, **4**, 357 (1976).

56. G. Fodor, D. Prajsma, and H. W. Schlipkotter, *Staub-Reinhalt. Luft*, **33**, 259 (1973).

57. D. G. Clark and D. J. Tinston, *Br. J. Pharm.*, **49**, 355 (1973).

58. C. F. Reinhardt, L. S. Mullin, and M. E. Maxfield, *J. Occup. Med.*, **15**, 953 (1973).

59. L. A. Heppel, P. A. Neal, T. L. Perrin, M. L. Orr, and V. T. Porterfield, *Ind. Hyg. Toxicol.*, **26**, 8 (1944).

60. L. A. Heppel and P. A. Neal, *Ind. Hyg. Toxicol.* **26**, 17, (1944).

61. Dow Chemical U.S.A., *Methylene Chloride: A Two-Year Inhalation Toxicity and Oncogenicity Study in Rats and Hamsters.* Unpublished report of work done for Diamond Shamrock Corporation, Dow Chemical U.S.A., Imperial Chemical Industries, Ltd. (UK), Stauffer Chemical Company, and Vulcan Materials Company by the Toxicology Research Laboratory, Dow Chemical U.S.A., Midland, MI 48640, December 31, 1980.

62. C. C. Haun, E. H. Vernot, K. I. Daumer, and S. S. Diemondi, *Aerospace Med. Res. Lab. Rep. AMRL-TR-72-130* (1972).

63. B. A. Schwetz, B. K. J. Leong, and P. J. Gehring, *Toxicol. Appl. Pharmacol.*, **32**, 84 (1975).

64. R. D. Hardin and J. M. Manson, *Toxicol. Appl. Pharmacol.*, **52**, 22 (1980).

65. R. L. Bornschein, L. Hastings, and J. M. Manson, *Toxicol. Appl. Pharmacol*, **52**, 29 (1980).

66. G. Bornmann, and A. Loesser, *Z. Lebsnmittel-Untersuch Forsch.* **136**, 14 (1967); through FAO Nutrition Meetings Report Series No. 48A WHO/FAO Food Add/70.39 (1970).

67. W. M. F. Jongen, G. M. Alink, and J. M. Koeman, *Mutat. Res.*, **56**, 245 (1978).

68. J. C. Theiss, G. D. Stoner, M. B. Shimkin, and E. K. Weisburger, *Cancer Res.*, **37**, 2717 (1977).

69. W. F. von Oettingen, C. C. Powell, N. E. Sharpless, W. C. Alford, and L. J. Pecora, *Natl. Inst. Health Bull. No. 191* (1949).

70. V. L. Kubic, M. W. Anders, R. R. Engel, C. H. Barlow, and W. S. Caughey, *Drug Metab. Dispos.*, **2**, 53 (1974).

71. V. L. Kubic and M. W. Anders, *Drug Metab. Dispos.* **3**, 104 (1975).

72. R. D. Stewart, C. L. Hake, M. V. Forster, A. J. Lebrun, J. E. Peterson, and A. Wu, *Methylene Chloride, Development of a Biological Standard for the Industrial Worker by Breath Analysis*, The Medical College of Wisconsin, Milwaukee, Wis., NIOSH-MCOW-ENVM-MC-74-9, 1974.

73. M. J. McKenna, J. A. Zempel, and W. H. Braun, *Proc. Ninth Conf. Environ. Toxicol.*, Wright-Patterson Air Force Base, AMRL TR-79-68, Dayton, Ohio, 1979.

74. M. J. McKenna, J. A. Zempel, and W. H. Braun, Abstr. No. 19, *Toxicol. Appl. Pharmacol.*, **48**, Part 2, A-10 (1979).

75. H. P. Ciuchta, G. M. Savell, and R. C. Spiker, Abstr. No. 308, *Toxicol. Appl. Pharmacol.*, **48**, Part 2, A-154 (1979).

76. G. D. DiVincenzo, F. J. Yanna, and B. D. Astill, *Am. Ind. Hyg. Assoc. J.*, **33**, 125 (1972).

77. M. J. McKenna, J. H. Saunders, W. H. Boeckler, R. J. Karbowski, K. D. Nitschke, and M. B. Chenoweth, Abst. 176, *Soc. Toxicol. 19th Ann. Meet. March 9–13, 1980*.

78. B. R. Friedlander, T. Hearne, and S. Hall, *J. Occup. Med.*, **20**, 657 (1978).

79. B. Highman, J. L. Svirbely, W. F. von Oettingen, W. C. Alford, and L. J. Pecora, *A.M.A. Arch. Pathol.*, **45**, 299 (1948).

80. H. R. Rutstein, *Arch. Environ. Health*, **7**, 66 (1963).

81. C. Comstock, R. W. Fogleman, and F. W. Oberst, *A.M.A. Arch. Ind. Hyg. Occup. Med.*, **7**, 526 (1953).

82. A. F. Matson and R. E. Dufour, *Underwriters' Lab. Bull. Res.*, **42** (Aug. 1948).

83. E. W. Van Stee, A. M. Harris, M. L. Horton, and K. C. Back, *Proc. Ann. Conf. Environ. Toxicol.*, *5th*, ISS AMRL-TR-74-125, 125 (1974).

84. T. R. Torkelson, F. Oyen, and V. K. Rowe, *Am. Ind. Hyg. Assoc. J.*, **21**, 275 (1960).

85. J. D. MacEwen, J. M. McNerney, E. H. Vernot, and D. T. Harper, *J. Occup. Med.*, **8**, 251 (1966).

86. B. Hirokawa, I. Yokoyama, T. Ishikawa, and A. Kurisu, *Bull. Inst. Publ. Health (Tokyo)*, **2**, 1 (1952).

87. M. P. Slyusar and E. B. Volodchenko, *Gig. Tr. Prof. Zabol*, **12**, 27 (1968).

88. V. A. Dykan, *Gig. (Kiev: Zdorov'e) 5b*, **1964**, 100 (*Chem. Abstr.*, **63**, 15429g).

89. V. A. Dykan, *Nauchn. Tr. Ukr. Nauch.—Issled. Inst. Gig. Tr.; Protzabolevarii*, **29**, 82 (1962) (*Chem. Abstr.*, **60**, 8541d).

90. G. G. Fodor, and A. Roscovanu, *Zentralbl. Bakteriol. Parasitenkd. Infektionskr. Hyg. Abstr.* 1, **162**, 34 (1976) (*Chem. Abstr.*, **85**, 1175636).

91. F. L. Rodney and M. A. Collinson, *Toxicol. Appl. Pharmacol.*, **40**, 39 (1977).

92. F. M. Turner, *At. Energy Res. Establ. G.B. MED/M-21*, 1958 (*Chem. Abstr.* **52**, 8357f).

93. H. J. P. Zimmerman, *Perspect. Biol. Med.*, **12**, 135 (1968).

94. P. J. R. Challen, D. E. Hickish, and J. Bedford, *Br. J. Ind. Med.*, **15**, 243 (1958).

95. K. L. Scholler, *Acta Anesth Scand Suppl.*, xxxii (1968).

96. R. A. Van Dyke, M. B. Chenoweth, and A. Van Poznak, *Biochem. Pharmacol.*, **13**, 1239 (1964).

97. National Institute for Occupational Safety and Health, *Criteria for a Recommended Standard—Occupational Exposure to Chloroform*, HEW Publ. No. (NIOSH) 75-114, 1974.

98. S. G. Winslow and H. G. Gerstner, *Health Aspects of Chloroform—A Review*, Oak Ridge National Library ORNL/TIRC-77-4, 1977.

99. T. R. Torkelson, F. Oyen and V. K. Rowe, *Am. Ind. Hyg. Assoc. J.*, **37**, 697 (1976).

100. D. J. Thompson, S. D. Warner, and V. B. Robinson, *Toxicol. Appl. Pharmacol.*, **29**, 348 (1974).

101. H. Oettel, *Arch. Exp. Pathol. Pharmakol.*, **183**, 655 (1936).

102. H. B. Lehmann, and F. Flury, *Toxicology and Hygiene of Industrial Solvents*, Transl. by F. King and H. F. Smyth, Williams & Wilkins, Baltimore, 1943.

103. B. A. Schwetz, B. K. J. Leong, and P. J. Gehring, *Toxicol. Appl. Pharmacol.*, **28**, 442 (1974).

104. F. G. Murray, B. A. Schwetz, M. G. McBride, and R. E. Staples, *Toxicol. Appl. Pharmacol.*, **50**, 515 (1979).

105. J. V. Dilley, N. Chernoff, D. Kay, N. Winslow, and N. G. Newell, Abstr. No. 154, *Toxicol. Appl. Pharmacol.*, **41**, 196 (1977).

106. J. Sturrock, *Br. J. Anesth.*, **49**, 207 (1977).

107. E. K. Weisburger, *Environ. Health. Perspect.*, **21**, 7 (1977).

108. H. E. Stokinger, *J. Am. Water Works Assoc.*, **69**, 399 (1977).

109. R. H. Reitz, P. J. Gehring and C. N. Park, *Food Cosmet. Toxicol.*, **16**, 571 (1978).

110. F. J. C. Roe, A. K. Palmer, A. N. Worden, and N. J. VanAbeé, *J. Environ. Pathol. Toxicol.*, **2**, 799 (1979).

111. A. K. Palmer, A. B. Street, F. J. C. Roe, A. N. Worden, and N. H. VanAbeé, *J. Environ. Pathol. Toxicol.*, **2**, 821 (1975).

112. R. Heywood, R. J. Sortwell, P. R. B. Noel, A. E. Street, D. E. Prentice, F. J. C. Roe, A. N. Worden, and N. J. VanAbeé, *J. Environ. Pathol. Toxicol.*, **2,** 835 (1979).

113. R. A. Van Dyke, *Modern Inhalation Anesthetics*, Springer, Berlin, Heidelberg, New York (1972).

114. D. E. Hathway, *Arzneim-Forsch.*, **24,** 173 (1974).

115. R. O. Recknagel, *Pharmacol. Rev.*, **19,** 145 (1967).

116. Bomski, Sobelwska, and Strakowski, *Int. Arch. Gewerbpathol. Gewerbhyg*, **24,** 127, (1967) (through Ref. 97).

117. S. D. Kutob and G. L. Plaa, *Toxicol. Appl. Pharmacol.*, **4,** 354 (1962).

118. National Cancer Institute, *Bioassay of Iodoform for Possible Carcinogenicity*, NCI-CG-TR-110, 1978.

119. D. D. McCollister, R. L. Hollingsworth, F. Oyen, and V. K. Rowe, *A.M.A. Arch. Ind. Health*, **13,** 1 (1956).

120. National Institute for Occupational Safety and Health, *1979 Registry of Toxic Effects of Chemical Substances, Volume 1.* (NIOSH) Publication No. 80-111, September 1980.

121. D. D. McCollister, W. H. Beamer, G. J. Atchison, and H. C. Spencer, *J. Pharmacol. Exp. Therap.*, **102,** 112 (1951).

122. E. M. Adams, H. C. Spencer, V. K. Rowe, D. D. McCollister, and D. D. Irish, *A.M.A. Arch. Ind. Hyg. Occup. Med.*, **6,** 50 (1952).

123. K. B. Lehmann, *Arch. Hyg.*, **74,** 1 (1911).

124. J. A. Prendergast, R. A. Jones, L. J. Jenkins, and J. Siegel, *Toxicol. Appl. Pharmacol.* **10,** 270 (1967).

125. B. A. Schwetz, B. K. J. Leong, and P. J. Gehring, *Toxicol. Appl. Pharmacol.*, **28,** 452, (1974).

126. H. F. Smyth, H. F. Smyth, Jr., and C. P. Carpenter, *J. Ind. Hyg. Toxicol.*, **18,** 277 (1936).

127. V. F. Simmon, K. Kauhanen and R. G. Tardiff, *Progress in Genetic Toxicology*, D. Scott, B. A. Bridges, and F. H. Sobel, Eds., Elsevier/North Holland Biomedical Press, 1977, pp. 249–258.

128. International Agency for Research on Cancer, *IARC Monographs on the Evaluation of the Carcinogenic Risk of Chemicals to Man*, Vol. 1, Lyon, France (1978).

129. International Agency for Research on Cancer, *IARC Monographs on the Evaluation of the Carcinogenic Risk of Chemicals to Humans. IARC Monograph Supplement I.*, Lyon, France, 1979.

130. R. D. Stewart, H. H. Gay, D. S. Erley, C. L. Hake, and J. E. Peterson, *J. Occup. Med.*, **3,** 586 (1961).

131. Manufacturing Chemists Association, *MCA Chemical Safety Data Sheet, SD-3 (Carbon Tetrachloride)*, Washington, D.C., 1946.

132. Manufacturing Chemists Association, *Research on Chemical Odor*, Washington, D.C., 1968.

133. V. V. Paustovskaya and N. M. Petran, *Farmakol. Toksikol. (Moscow)*, **32,** 736 (1969) (*Chem. Abstr.*, **72,** 77920Q).

134. B. J. Burreson, R. E. Moore, and P. Roller. *Tetrahedron Lett.*, **7,** 473 (1975).

135. M. M. Troshina, *Toksikol. Novykk Prom Khim. Veshchestv.*, **6,** 45 (1964) (through *Chem. Abstr.*).

136. A. M. Kosenko, *Mater Povolzh Konf. Fiziol. Uchastiem Biokhim., Farmakol. Morfol., 6th*, **2,** 36 (1973) (*Chem. Abstr.*, **82,** 081353K).

137. E. H. Vernot, J. D. MacEwen, C. C. Haun, and E. R. Kinkaid, *Toxicol. Appl. Pharmacol.* **42,** 417 (1977).

138. A. M. Kosenko, *Toksikol Gig. Prod. Neftekhim. Neftekhim Proizvod. Vses. Konf., (DØKL.) 2nd, 1971*, 205–110 (1972) (*Chem. Abstr.* **80,** 91722c, 91722d).

139. N. K. Karimullina and A. A. Gizatullina, *Farmakol. Toksikol*, **32,** 165 (1969) (*Chem. Abstr.* 009, 007106h).

140. V. F. Simmon and L. A. Poirier, *Proc. Am. Assoc. Cancer Res.*, **17**, 85 (1976).

141. E. Vogel and J. Chandler, *Experientia*, **30**, 621 (1974).

142. A. M. Kosenko and V. N. Salyaev, *Farmakol. Toksikol. Nov. Prod. Khim. Sint. Mater. Resp. Konf. 3rd*, 174 (1975) (*Chem. Abstr.*, **87**, 079171J).

143. C. F. Reinhardt, A. Azar, M. E. Maxfield, P. E. Smith, and H. S. Mullins, *Arch. Environ. Health*, **22**, 265 (1971).

144. H. L. Blumgart, D. R. Gilligan, and J. H. Schwartz, *J. Clin. Invest.*, **9**, 635 (1931).

145. National Cancer Institute, *Bioassay of 1,1-Dichloroethane for Possible Carcinogenicity*, DHEW Publ. No. (NIH) 78-1316, 1978.

146. P. G. Stecher, ed., *The Merck Index*, 8th ed., Merck and Co. Inc., Rahway, NJ, 1968, p. 439.

147. H. F. Smyth, Jr., *Am. Ind. Hyg. Assoc. Q.*, **17**, 129 (1956).

148. H. T. Hofmann, H. Birnstiel, and P. Jobst, *Arch. Toxikol.*, **27**, 244 (1971).

149. American Industrial Hygiene Association, *Hygienic Guide Series, 1,1-Dichloroethane. Am. Ind. Hyg. Assoc. J.*, **32**, 67 (1971).

150. World Health Organization, *Toxicological Evaluation of Some Extraction Solvents and Certain Other Substances*, FAO Nutr. Meet. Rep. Ser. No. 48A WHO/FAO ADD/70.39 (1970).

151. L. A. Heppel, P. A. Neal, K. M. Endicott, and V. T. Porterfield, *A.M.A. Arch. Ophthalmol.*, **32**, 391 (1944).

152. R. R. Sayers, W. P. Yant, C. P. Waite, and F. A. Patty, *Publ. Health Rep. U.S. Repr. No. 1349* (January 31, 1930).

153. L. A. Heppel, P. A. Neal, T. L. Perrin, K. M. Endicott, and V. T. Porterfield, *J. Pharmacol Exp. Therap.*, **1**, 53 (1945).

154. H. C. Spencer, V. K. Rowe, E. M. Adams, D. D. McCollister, and D. D. Irish, *A.M.A. Arch. Ind. Hyg. Occup. Med.*, **4**, 482 (1951).

155. T. Kuwabara, R. Quevedo, and D. G. Cogan, *Arch. Ophthalmol.*, **79**, 321 (1968).

156. L. A. Heppel, P. A. Neal, T. L. Perrin, K. M. Endicott, and V. T. Porterfield, *J. Ind. Hyg. Toxicol.*, **28**, 113 (1946).

157. M. A. Vozovaya, *Gig. Sanit.*, **7**, 25 (1974) (through *Chem. Abstr.*).

158. V. Sakaris, *Genitika* **5**, 89 (1969) (*Chem. Abstr.*, **72**, 1296775).

159. V. Sakaris, *Vestn. Leningrad Univ. Biol. 1970*, 153 (*Chem. Abstr.*, **73**, 32801K).

160. I. A. Rapoport, *Dokl. Akad Nauh SSSR*, **134**, 1214–1217 (1960) (*Chem. Abstr.*, **55**, 8675h).

161. National Cancer Institute, *Bioassay of 1,2-Dichloroethane for Possible Carcinogenicity*, DHEW Publ. No. (NIH) 78-1361 (1978).

162. C. Maltoni, L. Valgimigli, and C. Scarnato, *Cold Spring Harbor Symposium, November 15–19, 1979*.

163. S. Yllner, *Acta Pharmacol. Toxicol.*, **30**, 257 (1971).

164. H. Menschick, *Arch. Gewerbepathol. Gewerbehyg.*, **15**, 241 (1957).

165. National Institute of Occupational Safety and Health, *Criteria for a Recommended Standard . . . Occupational Exposure to Ethylene Dichloride (1,2-Dichloroethane)*, DHEW (NIOSH) 76-139, 1976.

166. W. D. McNally and G. Fosvedt, *Ind. Med.*, **10**, 373 (1941).

167. National Institute of Occupational Safety and Health, *Criteria for a Recommended Standard . . . Occupational Exposure to Ethylene Dibromide*, DHEW (NIOSH) Publication No. 77-221, 1977.

168. International Agency for Research on Cancer, *IRAC Monographs on the Evaluation of the Carcinogenic Risk of Chemicals to Man*, Vol. 11, Lyon, France, 1976.

169. V. K. Rowe, H. C. Spencer, D. D. McCollister, R. L. Hollingsworth, and E. M. Adams, *A.M.A. Arch. Ind. Hyg. Occup. Med.*, **6**, 1958 (1952).

170. B. G. H. Thomas and W. P. Yant, *Publ. Health Rep. U.S. Repr. No. 1139,* 370 (1927).

171. H. B. Plotnick and W. L. Conner, *Res. Commun. Chem. Pathol. Pharmacol.,* **13,** 251 (1976).

172. R. D. Short, J. L. Minor, J. M. Winston, J. Seifter, and C. C. Lee, *Toxicol. Appl. Pharmacol.,* **46,** 173 (1978).

173. Anonymous, *U.S. Fed. Reg.,* **42,** 63134 (1977).

174. O. E. Alumot, E. Nachtomi, O. Klempenich-Pinto, E. Mandel, and H. Schindler, *Poultry Sci.,* **47,** 1979 (1968).

175. D. J. Amir, *Reprod. Fert.,* **35,** 519 (1973).

176. K. Edwards, H. Jackson, and A. O. Jones, *Biochem. Pharmacol.,* **19,** 1783 (1970).

177. S. S. Epstein, E. Arnold, J. Andrea, W. Bass, and Y. Bishop, *Toxicol. Appl. Pharmacol.,* **23,** 288 (1972).

178. National Cancer Institute, *Bioassay of 1,2-Dibromoethane for Possible Carcinogenicity,* Tech. Rep. Ser. No. 86, NCI-CG-TR-86, 1978.

179. H. B. Plotnick, W. W. Weigel, D. E. Richards, and K. L. Cheever, *Res. Commun. Chem. Pathol. Pharmacol.,* **26,** 535 (1979).

180. E. V. Olmstead, *A.M.A. Ind. Hyg. Occup. Med.* **21,** 45 (1960).

181. G. Pflesser, *Arch. Gewerbepathol. Gewerbehyg.,* **8,** 591 (1938).

182. G. Calingaert and H. Shapiro, *Ind. Eng. Chem.,* **40,** 332 (1948).

183. W. C. Hunt, *An Appraisal of Ethylene Dibromide,* Great Lakes Chemical Company, West Lafayette, Ind., 1977.

184. J. D. Ramsey, C. N. Park, M. G. Ott, and P. J. Gehring, *Toxicol. Appl. Pharmacol.,* **47,** 411 (1979).

185. O. Wong, H. M. D. Utijian, and V. S. Karten, *J. Occup. Med.,* **21,** 98 (1979).

186. W. B. Crummett and V. A. Stenger, *Ind. Eng. Chem.,* **48,** 434 (1956).

187. T. R. Torkelson, F. Oyen, D. D. McCollister, and V. K. Rowe, *Am. Ind. Hyg. Assoc. J.,* **19,** 353 (1958).

188. R. D. Stewart, *Ann. Occup. Hyg.,* **11,** 71 (1968).

189. National Institute for Occupational Safety and Health *Criteria For a Recommended Standard . . . Occupational Exposure to 1,1,1-Trichloroethane,* HEW Publ. No. (NIOSH) 76-184, 1976.

190. D. M. Aviado, *Prog. Drug Res.,* **18,** 365, (1974).

191. E. M. Adams, H. C. Spencer, V. K. Rowe, and D. D. Irish, *A.M.A. Arch. Ind. Hyg. Occup. Med.,* **1,** 225 (1950).

192. C. D. Klaasen and G. L. Plaa, *Toxicol. Appl. Pharmacol.,* **9,** 139 (1966).

193. C. D. Klaasen and G. L. Plaa, *Toxicol. Appl. Pharmacol.,* **10,** 119 (1967).

194. C. D. Klassen and G. L. Plaa, *Biochem. Pharmicol.,* **18,** 2019 (1969).

195. J. F. Quast, B. K. J. Leong, L. W. Rampy, and P. J. Gehring, interim report (1975) (as cited in Ref. 189).

196. W. M. Watrous and G. L. Plaa, *Toxicol. Appl. Pharmacol.,* **23,** 640 (1972).

197. National Cancer Institute, *Bioassay of 1,1,1-Trichloroethane for Possible Carcinogenicity,* NCI-CG-TR-3 (1977).

198. B. Holmberg, I. Jakobson, and K. Sigvardsson, *Scand. J. Work Environ. Health,* **3,** 43, (1977).

199. C. L. Hake, T. B. Waggoner, D. N. Robertson, and V. K. Rowe, *Arch. Environ. Health,* **1,** 101 (1960).

200. N. Ikeda and H. Ohtsuji, *Bri. J. Ind. Med.,* **29,** 99, (1972).

201. A. Eben and G. Kimmerle, *Arch. Toxicol.,* **31,** 233, (1974).

202. B. E. Humbert and J. G. Fernandez, *Arch. Mor. Prof. Med. Trav. Secur. Soc.*, **38**, 415 (1977) (*Chem. Abstr.*, **87**, 162391B).

203. R. D. Stewart, H. H. Gay, D. S. Erley, C. L. Hake, and A. W. Schaffer, *Am. Ind. Hyg. Assoc. J.*, **22**, 252, 1961.

204. R. A. Van Dyke and L. Wineman, *Biochem. Pharmacol.*, **20**, 463, (1971).

205. R. A. Van Dyke, *Environ. Health Perspect.*, **21**, 121 (1977).

206. R. D. Stewart, C. H. Hake, A. Wu, S. A. Graff, H. V. Forster, A. J. Lebrun, P. E. Newton, and R. J. Sojo, *1,1,1-Trichloroethane: Development of a Biological Standard for the Industrial Worker by Breath Analysis*, Medical College of Wisconsin, Milwaukee, Wis., NIOSH-MCOW-ENVM-1,1,1-T-75-4, 1975.

207. J. C. Kranz, C. S. Park, and J. S. L. Ling, *Anesthesiology*, **20**,635 (1959).

208. W. H. L. Dornette and J. P. Jones, *Anesth. Analg.*, **39**, 249 (1966).

209. K. L. Siebecker Jr., B. J. Bamford, J. E. Steinhaus, and O. S. Orth, *Anesth. Analg.*, **39**, 180 (1960).

210. American Industrial Hygiene Association, "1,1,1-Trichloroethane Emergency Exposure Limits," *Am. Ind. Hyg. Assoc. J.*, **25**, 585 (1964).

211. M. Salvini, S. Binaschi, and M. Riva. *Br. J. Ind. Med.*, **28**, 293 (1971).

212. F. Gamberale and M. Hultengren, *Work-Environ-Health*, **10**, 82 (1973).

213. A. C. Monster, *Kinetics of Chlorinated Hydrocarbon Solvents*, Organization for Health Research TNO, 1978.

214. S. Fukabori, K. Nakaaki, J. Yonemoto, and O. Tada, *Rodo Kagaku*, **53**, 89 (1977) (*Chem. Abstr.*, **87**, 112568U).

215. V. Riihimaki and P. Pfaffli, *Scand. J. Work Environ. Health*, **4**, 73, (1978) (*Biol. Abstr. HEBP*, **79**, 00239).

216. C. G. Kramer, M. G. Ott, J. E. Fulkerson, N. Hicks, and H. R. Imbus, *Arch. Environ. Health*, **33**, 331 (1978).

217. N. W. Lazarew, *Arch. Exp. Pathol. Pharmakol*, **141**, 19 (1929).

218. W. H. Wright and J. M. Shaffer, *Am. J. Hyg.*, **16**, 325 (1932).

219. G. L. Plaa, E. A. Evans, C. H. Hine, *J. Pharm. Exp. Ther.*, **123**, 224 (1958).

220. I. Jakobson, B. Holmberg, and J. E. Wahlberg, *Acta Pharmacol. Toxicol.*, **41**, 497 (1977).

221. T. Kronevi, J. Wahlberg, and B. Holmberg, *Acta Pharmacol. Toxicol.*, **41**, 298 (1977) (*Chem. Abstr.*, **89**, 174597c).

222. J. E. Wahlberg, *Ann. Occup. Hyg.*, **19**, 115 (1976) (*Chem. Abstr.*, **86**, 066415W).

223. U. C. Pozzani, C. S. Weil, and C. P. Carpenter, *Am. Ind. Hyg. Assoc. J.*, **20**, 364, (1959).

224. V. Rannug, A. Sundvall, and C. Ramel, *Chem. Biol. Interact*, **20**, 1 (1978).

225. National Cancer Institute, *Bioassay of 1,1,2-Trichloroethane for Possible Carcinogenicity*, NCI-CG-TR-74, 1978.

226. S. Yllner, *Acta Pharmacol. Toxicol.*, **30**, 248 (1971).

227. R. A. Van Dyke and A. J. Gandolphi, *Mol. Pharmacol.*, **11**, 809 (1975).

228. R. Truhaut, N. P. Lich, H. Dutertre-Catella, G. Molas, and V. N. Huygen, *Arch. Mal. Prof. Med. Trav. Secur. Soc.*, **35**, 593 (1974).

229. G. S. Barsoum and K. Saad, *Q. J. Pharm. Pharmacol.*, **7**, 205 (1934).

230. P. Schwander, *Arch. Gewerpathol. Gewenbehyg.* **7**, 109 (1936).

231. C. P. Carpenter, H. F. Smyth, and U. C. Pozzani, *J. Ind. Hyg. Toxicol.*, **31**, 343 (1949).

232. M. E. Chieruttini and C. S. Franklin, *Br. J. Pharmacol.* **57**, 421 (1976) (*Chem. Abstr.*, **85**, 154681Q).

233. V. K. Navrotskii, L. M. Kashin, I. L. Kulinskaya, L. F. Mikhailovskaya, L. M. Shmuter, Z. I. Burlaka-Vovk and B. V. Zadorozhnyi, *Tr. S'ezde Gig. Vkr. SSR*, **8**, 224 (1971) (*Chem. Abstr.*, **77**, 071060V).

234. P. Schmidt, S. Binnewics, R. Gohlke, and R. Rothe, *Int. Arch. Arbeits. Med.*, **30**, 283 (1972).

235. R. Schmidt, *Biol. Rundsch*; **14**, 220, (1976) (*Chem. Abstr.*, **85**, 138175x).

236. H. S. Rosenkrantz, *Environ. Health Perspect.*, **21**, 79 (1977).

237. National Cancer Institute, *Bioassay of 1,1,2,2-Tetrachloroethane for Possible Carcinogenicity*, U.S. NTIS, PB Rep 155PB-277453, NCI-CG-TR-27, 1978.

238. S. Yllner, *Acta Pharmacol. Toxicol.*, **29**, 499 (1971).

239. J. B. Sherman, *J. Trop. Med. Hyg.*, **56**, 139 (1953).

240. R. Gurney, *Gastroenterology*, **1**, 1112 (1943).

241. H. A. Coyer, *Ind. Med.*, **13**, 320 (1944).

242. M. G. Gray, *A.M.A. Arch. Ind. Hyg. Occup. Med.*, **2**, 407 (1950).

243. R. L. Hollingsworth, V. K. Rowe, and F. Oyen, *Am. Ind. Hyg. Assoc. J.*, **24**, 28, (1963).

244. I. N. Arkhangel-Skaya and R. I. Yanushkevich, *Mater. Vopr. Gig. Tr. Klin. Profess. Bolezni Sbornik*, **5**, 190 (1956); abstr. *Ind. Hyg. Digest*, **21**, 16, No. 1264 (1957).

245. L. Merzbach, *Z. Ges. Exp. Med.*, **63**, 383 (1928).

246. E. Glaser and S. Frisch, *Arch. Hyg.*, **101**, 48 (1929).

247. Anonymous, National Library of Medicine, Toxicology Programs, *Tox Tips*, **10**, 5 (1971).

248. H. Brem, A. B. Stein, and H. S. Rosenkrantz, *Cancer Res.*, **34**, 2576, (1974).

249. H. S. Rosenkrantz, B. Gutter, and W. T. Speck, *Mutat. Res.*, **41**, 61 (1976).

250. A. B. Van Haaften, *Am. Ind. Hyg. Assoc. J.*, **30**, 251 (1969).

251. S. Yllner, *Acta Pharmacol. Toxicol.*, **29**, 481, (1971).

252. M. H. Weeks, R. A. Angerhofer, R. Bishop, J. Thomasino, and C. R. Pope, *Am. Ind. Hyg. Assoc. J.*, **40**, 187 (1979).

253. E. J. Thorpe, *Comp. Pathol. Ther.*, **75**, 45, (1965).

254. E. S. Reynolds and A. G. Yee, *Lab Invest.*, **19**, 273, (1968).

255. N. N. Plotnikov and L. N. Sokolov, *Farmakol. Toksikol.*, **47**, (1947).

256. H. Baganz, W. Perkow, G. T. Lim, and F. Meger, *Arzneim. Forsch.*, **11**, 902 (1961).

257. American Conference of Governmental Industrial Hygienists, *Documentation of the Threshold Limit Values, Third Edition*, Cincinnati, Ohio, 1979.

258. S. J. Gorzinski, R. J. Nolan, R. J. Kociba, R. J. Karbowski, C. E. Wade, D. C. Morden, E. A. Herman, and B. A. Schwetz, Abstr. 215, *Toxicol. Appl. Pharmacol.*, **48**(1), Part 2, A108 (1979).

259. National Cancer Institute, *Bioassay of Hexachloroethane for Possible Carcinogenicity*, NCI-CG-TR-68, 1978.

260. W. R. Jondorf, D. V. Parke, and R. T. Williams, *Biochem. J.*, **65**, 14P (1957).

261. J. S. L. Fowler, *Br. J. Pharmacol.*, **35**, 530 (1969).

262. J. S. L. Fowler, *J. Comp. Pathol.*, **80**, 465 (1970).

263. W. J. Fitzpatrick and L. B. Yeager, *Q. Bull. Northwestern Univ. Med. Sch.*, **26**, 313 (1952).

264. J. E. Amoore, *Nature*, **198**, 271 (1963).

265. J. E. Amoore and D. Venstrom, *J. Food Sci.*, **31**, 118, (1966).

266. B. E. Abreu, S. H. Auerbach, J. M. Thuringer, and S. A. Peoples, *J. Pharmacol. Exp. Therap.*, **80**, 139 (1944).

267. E. A. Barnsley, *Biochem. J.*, **100**, 362 (1966).

268. J. C. Gage, *Brit. J. Ind. Med.*, **27**, 1 (1970).

269. A. M. Kosenko, *Toksikol. Gig. Prod. Neftekhim. Neftekhim Proizvod., Vses Knof., (Dokl) 2nd,* 108 (1972) (*Chem. Abstr.,* **80,** 091723D).

270. H. F. Smyth Jr., C. P. Carpenter, C. S. Weil, U. C. Pozzani, J. E. Striebel, and J. S. Nycum, *Am. Ind. Hyg. Assoc. J.,* **30,** 470 (1969).

271. L. A. Heppel, P. A. Neal, B. Highman, and V. T. Porterfield, *J. Ind. Hyg. Toxicol.,* **28,** 1 (1946).

272. L. A. Heppel, B. Highman and E. G. Peake, *J. Ind. Hyg. Toxicol.,* **30,** 189, (1948).

273. V. M. Shaipak, *Sb. Tr. Naucho-Issled. Inst Gig. Protzabol., Tiflis,* **15,** 194 (1976) (*Chem. Abstr.,* **89,** 018050P).

274. F. DeLorenzo, S. Degl'Inocenti, A. Ruocco, L. Silengo and R. Cortese, *Cancer Res.,* **37,** 1915 (1977).

275. T. R. Torkelson, S. E. Sadek, V. K. Rowe, J. K. Kodama, H. H. Anderson, G. S. Loguvam, and C. H. Hine, *Toxicol. Appl. Pharmacol.,* **3,** 545 (1961).

276. N. N. Rakhamutullaev, *Gig. Sanit.,* **36,** 19 (1971).

277. J. K. Kodama and M. K. Dunlap, *Fed. Proc.,* **15,** 1459 (1956).

278. J. A. Ruddick and W. H. Newsome, *Bull. Environ. Contem. Toxicol.,* **21,** 486, (1979).

279. Y. B. Reznick and G. K. Sprinchan. *Gig. Sanit.,* **40,** 101 (1975).

280. E. V. Faidysh and M. G. Avkhimenko, *Gig. Profzabol.,* **8,** 42 (1974).

281. J. D. Burek, F. J. Murray, K. S. Rao, A. A. Crawford, J. S. Beyer, R. R. Albee, and B. A. Schwetz, Abstr. 241, *Toxicol. Appl. Pharmacol.,* **48,** A121 (1979).

282. K. S. Rao, F. J. Murray, A. A. Crawford, J. A. John, W. J. Potts, B. A. Schwetz, J. D. Burek, and C. M. Parker, Abstr. 273, *Toxicol. Appl. Pharmacol.,* **48,** A121 (1979).

283. R. W. Kapp, P. J. Picciano, and C. B. Jacobson, *Mutat. Res.,* **64,** 47 (1979).

284. D. Whorton, R. M. Krauss, S. Marshall, and T. H. Milby, *Lancet,* **2,** 1259 (1977).

285. D. Whorton, T. H. Milby, R. M. Krauss, and H. A. Stubbs, *J. Occup. Med.,* **21,** 161 (1979).

286. J. M. Lanham, Abstract, 64th Annual Meeting American Occupational Medical Association, *J. Occup. Med.,* **21,** 294 (1979).

287. H. S. Warren, J. E. Huff and H. B. Gerstner, *Vinyl Chloride. A Review 1835–1975; An Annotated Literature Collection 1835–1975; and a Literature Compilation,* 1976–1977. Oak Ridge National Library ORNL/TIRC-78/3, November, 1978.

288. National Cancer Institute. *Vinyl Chloride. An information Resource,* T. M. Milby, Ed., DHEW Publ. No. 78-1599, 1978.

289. International Agency for Research on Cancer *IARC Monographs on the Evaluation of Carcinogenic Risk of Chemicals to Man,* Vol. 19, Lyon, France, 1979.

290. E. Mastromatteo, A. M. Fisher, H. Christie, and H. Danziger, *Am. Ind. Hyg. Assoc. J.,* **21,** 394 (1960).

291. F. A. Patty, W. P. Yant, and C. P. Waite, *Publ. Health Rep. U.S., Repr. No. 1405,* **45**(34) (Aug. 1930).

292. S. A. Peoples and C. D. Leake, *J. Pharmacol. Exp. Ther.,* **48,** 284 (1933).

293. R. H. Oster, C. J. Carr, J. C. Krantz, and M. J. Sauerwald, *Anesthesiology,* **8,** 358 (1947).

294. C. Maltoni, *Ambio,* **4,** 18 (1975).

295. T. R. Torkelson, F. Oyen, and V. K. Rowe, *Am. Ind. Hyg. Assoc. J.,* **22,** 354 (1961).

296. J. A. John, F. A. Smith, B. K. J. Leong, and B. A. Schwetz, *Toxicol. Appl. Pharmacol.,* **39,** 497 (1977).

297. G. Ungvary, A. Hudak, E. Tatrai, M. Lorincz, and G. Folly, *Toxicology,* **2,** 45 (1978) (*Chem. Abstr.,* **89,** 191946H).

298. M. M. Muratov and S. I. Guskova, *Gig. Sanit.,* **7,** 111 (1978) (*Chem. Abstr.,* **89,** 101251a).

299. I. Fleig and A. M. Theiss, *J. Occup. Med.*, **20,** 557 (1978).

300. D. Anderson, M. C. E. Hodge, and I. H. F. Purchase, *Mutat. Res.*, **40,** 359 (1976).

301. R. D. Short, J. L. Minor, J. M. Winston, and C. C. Lee, *Toxicol. Environ. Health*, **3,** 965 (1977).

302. P. L. Viola, A. Bigotti, and A. Caputo, *Cancer Res.*, **31,** 516 (1971).

303. C. Maltoni, Cancer Detection and Prevention, *Proc. 2nd. Int. Symp. April 9–12, 1973, Bologna,* Int. Congr. Ser. No. 322, Excerpta Medica, Amsterdam, 15BN90 219 0228 1, 1973.

304. J. L. Creech and M. N. Johnson. *J. Occup. Med.*, **16,** 150 (1974).

305. C. Maltoni, *Proc. Club de Cancerogenese Chim., Paris, November 10, 1979.*

306. M. L. Keplinger, J. W. Goode, D. E. Gordon, and J. C. Calandra, *Ann. NY Acad. Sci.*, **246,** 219 (1975).

307. P. J. Gehring, P. G. Watanabe, and C. N. Park, *Toxicol. Appl. Pharmacol.*, **49,** 15 (1979).

308. E. D. Baretta, R. D. Stewart, and J. E. Mutchler, *Am. Ind. Hyg. Assoc. J.*, **30,** 537 (1969).

309. P. J. Gehring, P. G. Watanabe, and C. N. Park, *Toxicol. Appl. Pharmacol.*, **44,** 581 (1978).

310. C. H. Tamburro, *Yale J. Biol. Med.*, **51,** 67 (1978).

311. H. Danziger, *Can. Med. Assoc. J.*, **82,** 828 (1966).

312. I. Suciu, I. Drejam, and M. Valaskai, *Med. Int.*, **15,** 967, (1963).

313. J. Stafford, Imperial Chemical Industries Limited, personal communication, 1980.

314. B. K. J. Leong and T. R. Torkelson, *Am. Ind. Hyg. Assoc. J.*, **31,** 1 (1970).

315. H. Bartsch, "Mutagenic Tests in Chemical Carcinogenesis," in C. Rosenfeld and W. Davis, Eds., *Environmental Pollution and Carcinogenic Risks* (*IARC Scientific Publications No. 13*) Lyon, 1976, pp. 229–240.

316. H. Bartsch, C. Malaville, A. Barbin, and G. Planche, *Arch. Toxicol.*, **41,** 249 (1979).

317. Huntington Research Center, New City, New York, unpublished data submitted to OSHA, Jan. 25, 1979.

318. H. M. Bolt, J. G. Filser, and R. K. Hinderer, *Toxicol. Appl. Pharmacol.*, **44,** 481 (1978).

319. T. J. Haley, *Clin. Toxicol.*, **8,** 633 (1975).

320. R. D. Huffman and P. Desai-Greenway, *Health and Environmental Impacts,* U.S. NTIS ISS 258855, 1976.

321. L. J. Jenkins Jr., M. J. Trabulus, and S. D. Murphy, *Toxicol. Appl. Pharmacol.*, **23,** 501 (1972).

322. R. J. Jaeger, R. B. Conally, and S. D. Murphy, *Toxicol. Appl. Pharmacol.*, **29,** 81 (1974).

323. J. Siegel, R. A. Jones, R. A. Coon, and J. P. Lyon, *Toxicol. Appl. Pharmacol.*, **18,** 168, 1971.

323. R. D. Short, J. M. Winston, J. L. Minor, J. Seifter, and C. C. Lee, *Environ. Health Perspect.*, **21,** 125 (1977).

324. L. M. Siletchnik and G. P. Carlson. *Arch. Int. Pharmacodyn Ther.*, **210,** 359 (1974) (*Biol. Abstr.* 75/06265).

325. J. M. Norris, *Paper Synthetic Conference, Technical Association of the Pulp and Paper Industry, Chicago, Ill., September 26–28, 1977,* p. 45.

326. F. J. Murray, K. D. Nitschke, L. W. Rampy, and B. A. Schwetz, *Toxicol. Appl. Pharmacol.*, **49,** 189 (1979).

327. B. K. Jones and D. E. Hathway, *Cancer Lett.*, **5,** 1 (1978).

328. D. Anderson, M. C. E. Hodge, and I. F. H. Purchase, *Environ. Health Perspect.*, **21,** 71 (1977).

329. P. L. Viola and A. Caputo, *Environ. Health Perspect.*, **21,** 45 (1977).

330. L. W. Rampy, J. F. Quast, C. G. Humiston, M. F. Balmor, and B. A. Schwetz, *Environ. Health Perspect.*, **21,** 33 (1977).

331. C. Maltoni, *Environ. Health Perspect.*, **21,** 1 (1977).

332. R. H. Reitz, P. G. Watanabe, M. J. McKenna, J. F. Quast, and P. J. Gehring, *Toxicol. Appl. Pharmacol.*, **52**, 357 (1980).

333. M. J. McKenna, P. G. Watanabe, and P. J. Gehring, *Environ. Health Perspect.*, **21**, 99 (1977).

334. C. C. Lee, J. C. Bhandari, J. M. Winston, W. B. House, P. J. Peters, R. L. Dixon, and J. S. Woods, *Environ. Health Perspect.*, **21**, 25 (1977).

335. C. C. Lee, J. C. Bhandari, W. B. House, P. J. Peters, J. S. Woods, and R. L. Dixon, *Pharmacologist*, **18**, 245 (1976).

336. B. L. Van Duuren, B. M. Goldschmidt, G. Loewengart, A. C. Smith, S. Melchionne, I. Seidman, and D. Roth, *J. Natl. Cancer Inst.*, **63**, 1433 (1979).

337. B. K. Jones and D. E. Hathway, *Chem. Biol. Interact.* **20**, 27 (1978).

338. B. K. Jones and D. E. Hathway, *Br. J. Cancer*, **37**, 411 (1978).

339. J. M. McKenna, J. A. Zempel, E. O. Madrid, and P. J. Gehring, *Toxicol. Appl. Pharmacol.*, **45**, 599 (1978).

340. M. J. McKenna, J. A. Zempel, E. O. Madrid, W. H. Braun, and P. J. Gehring, *Toxicol. Appl. Pharmacol.*, **45**, 821 (1978).

341. R. J. Jaeger, M. J. Trabulus, and S. D. Murphy, *Toxicol. Appl. Pharmacol.*, **24**, 457 (1973).

342. R. J. Jaeger, R. Connally, and S. D. Murphy, *Exp. Mol. Pathol.*, **20**, 187 (1974).

343. R. D. Short, J. M. Winston, J. L. Minor, C. B. Hong, J. Seifert, and C. C. Lee, *J. Toxicol. Environ. Health*, **3**, 913 (1977).

344. M. G. Ott, R. R. Langner, and B. B. Holder, *Arch. Environ. Health*, **30**, 333 (1975).

345. M. G. Ott, W. A. Fishbeck, J. C. Townsend, and E. J. Schneider, *J. Occup. Med.*, **18**, 735 (1976).

346. K. J. Fruendt, G. P. Liebaldt, and E. Lieberwirth, *Toxicology*, **1**, 141 (1977).

347. D. Gradiski, P. Bonnet, G. Raoult, J. L. Magadur, and J. M. Francin, *Arch. Mal. Prof. Med. Trav. Secur. Soc.*, **39**, 249 (1978).

348. K. J. Fruendt and J. Macholz, *Toxicology*, **10**, 131 (1978).

349. H. Greim, D. Bimboes, G. Egert, W. Goeggelmann, and M. Kraemer, *Arch. Toxicol.*, **39**, 159 (1977).

350. H. Greim, G. Bonse and D. Henschler, *Lebershaedin Vinylchloride: Vinylchlorid-Kr.* (Wiss. Tag. 1, 2nd), 36 (1977) (*Chem. Abstr.*, **89**, 054029x).

351. A. Hamilton, *Industrial Toxicology*, Harper, 1934.

352. R. I. Defalque, *Clin. Pharmacol. Ther.*, **2**, 665 (1961).

353. National Institute for Occupational Safety and Health, *Criteria for a Recommended Standard . . . Occupational Exposure to Trichloroethylene*, NIOSH-TR-043-73 (1973).

354. E. M. Waters, H. B. Gerstner, and J. E. Huff, *Trichloroethylene. I. An Impact Overview*. E. M. Waters and S. A. Black, *II. An Abstracted Literature Collection*, Oak Ridge National Laboratory, ORNL/TIRC-76/2, 1976.

355. R. Frant and J. Westendorp, *A.M.A. Arch. Ind. Hyg. Occup. Med.*, **1**, 308 (1950).

356. A. Sato and T. Nakajima, *Br. J. Ind. Med.*, **35.** 43 (1978).

357. E. M. Adams, H. C. Spencer, V. K. Rowe, D. D. McCollister, and D. D. Irish, *A.M.A. Arch. Ind. Hyg. Occup. Med.*, **4**, 469 (1951).

358. B. Kylin, H. Reichard, I. Sumegi, and S. Yllner, *Nature*, **193**, 395 (1962).

359. B. Kylin, H. Reichard, I. Sumegi, and S. Yllner, *Acta Pharmacol. Toxicol.*, **20**, 16 (1963).

360. T. Ikeda, C. Nagano, and A. Okada, *Igaku To Seibutsugaku*, **79**, 123 (1969) (*Chem. Abstr.*, **72**, 77043k).

361. A. B. Baker, *J. Neuropathol. Exp. Neurol.*, **17**, 649 (1958).

362. Anonymous, *Chemecology*, Manufacturing Chemists Association, Washington, D.C., October 1977.

363. L. Fishbein, *Mutat. Res.*, **32**, 267 (1976).

364. H. Greim, G. Bonze, Z. Radwan, D. Reichert, and D. Henschler, *Biochem. Pharmacol.*, **24**, 2013 (1975).

365. G. Bronzetti, E. Zeiger, and D. Frezza, *J. Environ. Pathol. Toxicol.*, **1**, 411 (1978).

366. Z. G. Bell, personal communication, 1980.

367. National Cancer Institute, *Bioassay of Trichloroethylene for Possible Carcinogenicity*, NCI-CG-TR-2, 1976.

368. D. W. Henschler, W. Roman, H. M. Elsässer, D. Reichert, E. Eder, and Z. Radwan, *Arch. Toxicol.*, **43**, 237 (1980).

369. L. L. Miller, *Occup. Med.*, **5**, 194 (1948).

370. R. D. Stewart, H. H. Gay, D. S. Erley, C. L. Hake, and J. E. Peterson, *Am. Ind. Hyg. Assoc. J.*, **23**, 167 (1962).

371. R. D. Stewart, H. C. Dodd, H. H. Gay, and D. S. Erley, *Arch. Environ. Health*, **20**, 64 (1970).

372. R. D. Stewart, C. L. Hake, and J. E. Peterson, *Arch. Environ. Health*, **29**, 1 (1974).

373. R. D. Stewart, C. L. Hake, and J. E. Peterson, *Arch. Environ. Health*, **29**, 6 (1974).

374. R. D. Stewart, C. L. Hake, H. V. Forster, A. J. Lebrun, and J. B. Peterson, *Trichloroethylene: Development of a Biological Standard for the Industrial Worker by Breath Analysis*. The Medical College of Wisconsin, Milwaukee, Wis., NIOSH-MCOW-ENVM-TCE-74-8, 1974.

375. I. Åstrand and F. Gamberale, *Environ. Res.*, **15**, 1 (1978).

376. R. J. Vernon and R. K. Ferguson, *Arch. Environ. Health*, **18**, 894 (1969).

377. G. J. Stopps and M. McLaughlin, *Am. Ind. Hyg. Assoc. J.*, **28**, 43 (1967).

378. T. Ertle, D. Kenschlor, G. Muller, and M. Spassowski, *Arch. Toxikol.*, **29**, 171 (1972).

379. G. Muller, M. Spassowski, and D. Henschler, *Arch. Toxikol.*, **32**, 283 (1974).

380. G. Muller, M. Spassowski, and D. Henschler, *Arch. Toxikol.*, **33**, 173 (1975).

381. A. Ahlmark and S. Forssman, *A.M.A. Arch. Ind. Hyg. Occup. Med.*, **3**, 386 (1951).

382. E. Grandjean, R. Munchinger, V. Turrian, P. Haas, H. K. Knoepfel, and H. Rosenmund, *Br. J. Ind. Med.*, **12**, 131 (1955).

383. S. Forssman and C. E. Holmquist, *Acta Pharmacol. Toxicol.*, **9**, 235 (1953).

384. M. A. Boillet, *Praeventivmedizin*, **15**, 447 (1970) (*Chem. Abstr.*, **74**, 138704s).

385. Z. Bardodej, M. Krivaucova, and F. Pokorny, *Pracovni Leukarstvi*, **7**, 263 (1955) (*Chem. Abstr.*, **49**, 16269).

386. M. Kleinfeld and I. R. Tabershaw, *A.M.A. Arch. Ind. Hyg. Occup. Med.*, **10**, 134 (1954).

387. C. J. Stahl, A. V. Fattesh, and A. M. Dominguez, *J. Forensic Sci.*, **14**, 393 (1969).

388. O. Alexson, K. Andersson, C. Hogstedt, B. Holmberg, G. Molinz, and A. DeVerdier, *J. Occup. Med.*, **20**, 194 (1978).

389. L. Spevak, V. Nadj, and D. Felle, *Br. J. Ind. Med.*, **33**, 272 (1976).

390. N. P. Wong, J. N. Daminco, and H. Salwin, *J. Ass. Off. Anal. Chem.*, **50**, 8 (1967).

391. W. G. Van Ketel, *Contact Dermatitis*, **2**, 115 (1976).

392. J. H. Schwartz, *Arch. Dermatol. Syphilol.*, **40**, 962 (1939).

393. National Cancer Institute, *Bioassay of Allylchloride for Possible Carcinogenicity*, DHEW Pub. No. (NIH) 78-1323, (1978).

394. C. Hogstedt, O. Rohlen, B. S. Berndtsson, O. Axelson, and L. Ehrenberg, *Br. J. Ind. Med.*, **36**, 276 (1979).

395. J. M. Winston, J. C. Bhandari, A. M. El-Hawari, R. D. Short Jr., and C. E. Lee, Abstr. 246, *Toxicol. Appl. Pharmacol.*, **45**, 324 (1978).

396. National Institute of Occupational Safety and Health, *Criteria for a Recommended Standard— Occupational Exposure to 1,1,2,2-Tetrachloroethane*, DHEW (NIOSH) Publication No. 77-121, (1976).

397. D. E. Hathway, *Environ. Health Perspect.*, **21**, 55 (1977).

398. C. A. Stewart Jr., *J. Amer. Chem. Soc.*, **93**, 9815 (1971).

399. National Institute of Occupational Safety and Health, *Criteria for a Recommended Standard . . . Occupational Exposure to Tetrachloroethylene (Perchloroethylene)*, HEW Publication No. (NIOSH) 76-155, 1976.

400. V. K. Rowe, D. D. McCollister, H. C. Spencer, E. M. Adams, and D. D. Irish, *A.M.A. Arch. Ind. Hyg. Occup. Med.*, **5**, 566 (1952).

401. R. T. Drew, J. M. Patel, and F. Lin, *Toxicol. Appl. Pharmacol.*, **45**, 809 (1978).

402. National Cancer Institute, *Bioassay of Tetrachloroethylene for Possible Carcinogenicity*, DHEW Publication No. (NIH)77-813, 1977.

403. C. P. Carpenter, *J. Ind. Hyg. Toxicol.*, **19**, 323 (1937).

404. B. Kylin, I. Sumegi, and S. Yllner, *Acta Pharmacol. Toxicol.*, **22**, 379 (1965).

405. NIOSH, Interagency Report, 1977.

406. M. Cerna and H. Kyperova, Abstr. 36, *Mutat. Res.*, **46**, 214 (1977).

407. L. W. Rampy, J. F. Quast, B. K. J. Leong, and P. J. Gehring, *Proceedings of the First International Congress on Toxicology*, G. L. Plaa and W. A. M. Duncan, Eds., Academic Press, New York, 1978, p. 562.

408. A. M. Schumann, J. F. Quast, and P. G. Watanabe, *Toxicol. Appl. Pharmacol.*, **55**, 207 (1980).

409. D. G. Pegg, J. A. Zempel, W. H. Baron, and P. G. Watanabe, *Toxicol. Appl. Pharmacol.*, **51**, 465 (1979).

410. C. L. Hake and R. D. Stewart, *Environ. Health Perspect.*, **21**, 231 (1977).

411. R. D. Stewart, C. L. Hake, H. V. Forster, A. J. Lebrun, J. E. Peterson, and A. Wu. *Tetrachloroethylene—Development of a Biologic Standard for the Industrial Worker by Breath Analysis*, The Medical College of Wisconsin, Milwaukee, Wis., Rep. No. NIOSH-MCOW-ENVM-PCE-74-6, 1974.

412. M. Ikeda, *Environ. Health Perspect.*, **21**, 239 (1977).

413. S. Yllner, *Nature*, **191**, 820 (1961).

414. A. Blair, P. Decoufle, and D. Grauman, *Am. J. Publ. Health*, **69**, 508 (1979).

415. National Institute of Occupational Safety and Health, *Criteria for a Recommended Standard . . . Occupational Exposure to Allyl Chloride*, HEW Publ. No. (NIOSH) 76-204, 1976.

416. H. F. Smyth Jr. and C. P. Carpenter, *J. Ind. Hyg. Toxicol.*, **30**, 63 (1948).

417. V. E. Karmazin, *Vopr. Kommunaz's Gig.*, **6**, 108 (1966) (*Chem. Abstr.*, **68**, 89771b).

418. Shell Chemical Company, *Allyl Chloride Toxicity and Safety Bulletin*, SC:195:76, 1976.

419. E. M. Adams, H. C. Spencer, and D. D. Irish, *J. Ind. Hyg. Toxicol.*, **22**, 79 (1940).

420. K. S. Almeev and V. E. Karmazin, *Faktory Vneshn Sredy Lkh Znach Zdorov/ya Naseleniya*, **1**, 31 (1969) (*Chem. Abstr.*, **77**, 64877K).

421. V. E. Karmazin, *Faktory Vneshn Sredy Lkh Znach Zdorov'ya Naseleniya*, **1**, 35 (1969).

422. T. R. Torkelson, M. A. Wolf, F. Oyen, and V. K. Rowe, *Am. Ind. Hyg. Assoc. J.*, **20**, 217 (1969).

423. E. C. McCoy, L. Burrows, and H. S. Rosenkrantz, *Mutat. Res.*, **57**, 11 (1978).

424. C. M. Kaye, J. J. Clapp, and L. Young, *Xenobiotica*, **2**, 129 (1972).

425. G. A. Alizade, F. G. Guseinov, L. P. Agamova, R. S. Guseinova, and F. A. Aleskerov, *Azerb. Med. Zm.*, **53,** 54 (1976) (*Chem. Abstr.*, **86,** 194353M).

426. M. Hausler and R. Lenich, *Arch. Toxikol.*, **23,** 209 (1968).

427. T. R. Torkelson and F. Oyen, *Am. Ind. Hyg. Assoc. J.*, **38,** 217 (1977).

428. C. M. Parker, W. B. Coate, and R. W. Voelker, Abstr. 128, *19th Ann. Meet. Soc. Toxicol., March 10, 1980.*

429. Unpublished data, Report No. R-402. Work done for The Dow Chemical Company by Central Institute for Nutrition and Food Research Zeist Holland, November, 1973.

430. Anonymous, National Library of Medicine, Toxicology Information Program, *Tox Tips*, **31,** 33 (1978).

431. D. H. Hutson, J. A. Moss and B. A. Pickering, *Food Cosmet. Toxicol.*, **9,** 677 (1971).

432. P. Flessel, J. R. Goldsmith, E. Kahn, J. J. Wosolowski, K. T. Maddy, and S. A. Peoples, *Morb. Mortal. Rep.*, 50 and 55 (Feb. 17, 1978).

433. Anonymous, *Telone® II Soil Fumigant Safe Handling Guide*, The Dow Chemical Company, Midland, Mich., 1978.

434. National Institute of Occupational Safety and Health, *Criteria for a Recommended Standard . . . Occupational Exposure to Chloroprene*, DHEW (NIOSH) Publ. No. 77-210, 1977.

435. A. R. Nystrom, *Acta Med. Scand.*, (*Suppl. No. 219*), **132,** 1 (1948).

436. J. J. Clary, *Environ. Health Perspect.*, **21,** 269 (1977).

437. W. F. von Oettingen, W. C. Hueper, W. Deichmann-Grubler, and F. H. Wiley, *J. Ind. Hyg. Toxicol.*, **18,** 240 (1936).

438. W. L. Ritter and A. S. Carter, *J. Ind. Hyg. Toxicol.*, **30,** 192 (1948).

439. J. J. Clary, V. J. Feron, and P. G. J. Reuzel, *Toxicol. Appl. Pharmacol.*, **46,** 375 (1978).

440. R. Culik, D. P. Kelly, and J. J. Clary, *Toxicol. Appl. Pharmacol.*, **44,** 81 (1978).

441. Unpublished data, E. I. du Pont de Nemours & Co., Inc., Wilmington, Del.

442. H. Bartsch, C. Malaveille, R. Montesano, and L. Tomatis, *Nature*, **255,** 641 (1975).

443. S. Pell, *J. Occup. Med.*, **20,** 21 (1978).

444. R. J. Kociba, B. A. Schwetz, D. G. Keyes, G. C. Jersey, J. J. Ballard, D. A. Dittenber, J. F. Quast, C. E. Wade, and C. G. Humiston, *Environ. Health Perspect.*, **21,** 49 (1977).

445. J. H. Harleman and W. Seinen, *Toxicol. Appl. Pharmacol.*, **47,** 1 (1979).

446. G. E. Poteryayeva, *Gig. Sanit.*, **31,** 33 (1966) (*Chem. Abstr.*, **65,** 1281f).

447. A. G. Gulko, N. I. Ziman and I. G. Shruit, *Vopr. Gig. Sanit. Ozdorovl. Vnesha Sredy. Kishinev,* **56,** 128 (1964) (*Chem. Abstr.*, **62,** 13757c).

448. B. A. Schwetz, J. M. Norris, R. J. Kociba, P. A. Keeler, R. F. Cornier, and P. J. Gehring, *Toxicol. Appl. Pharmacol.*, **30,** 255 (1974).

449. B. A. Schwetz, F. A. Smith, J. F. Quast, and R. J. Kociba, *Toxicol. Appl. Pharmacol.*, **42,** 387 (1977).

450. D. Reichert, D. Ewald, and D. Henschler, *Food Cosmet. Toxicol.*, **13,** 511 (1975).

451. D. Reichert, G. Liebaldt, and D. Henschler, *Arch. Toxicol.*, **37,** 23 (1976).

452. J. Siegel, R. A. Jones, R. A. Coon, and J. P. Lyon, *Toxicol. Appl. Pharmacol.*, **18,** 168 (1971).

453. D. Reichert, D. Henschler and P. Bannasch, *Food Cosmet. Toxicol.*, **16,** 227 (1978).

454. R. A. Saunders, *Arch. Environ. Health*, **14,** 380 (1967).

455. C. A. Stewart Jr., *J. Am. Chem. Soc.*, **93,** 9815 (1971).

456. P. Forkert and E. S. Reynolds, Abstr. 408, *19th Ann. Meet., Soc. Toxicol., March 9–13, 1980.*

457. R. I. Mitchell, K. L. Pavkov, R. M. Everett, and D. A. Holzworth, CIIT Docket No. 63059 Research Triangle Park, North Carolina (1979).

458. F. J. Bowman, J. F. Borzelleca, and A. E. Munson, *Toxicol. Appl. Pharmacol.*, **44**, 213 (1978).

459. C. R. Wolf, D. Mensury, W. Nastainczyk, G. Deutschmann, and V. Ulrich, *Mol. Pharmacol.*, **13**, 698 (1977).

460. Oak Ridge National Library, *Investigations of Selected Environmental Pollutants 1,2-Dichloroethane*, prepared for EPA, NTIS PB-295-865, April 1979.

461. K. S. Rao, J. S. Murray, M. M. Deacon, J. A. John, L. L. Calhoun and J. T. Young, Banbury Report 5, *Ethylene Dichloride: A Potential Health Risk?* Cold Spring Harbor Laboratory, 1980, p. 149.

462. F. P. Guengerich, W. M. Crawford, Y. D. Domoradski, T. L. MacDonald, and P. G. Watanabe, *Toxicol. Appl. Pharmacol.*, **55**, 303 (1980).

463. T. A. Shamilov and D. M. Abasov, *Tr. Azerb, Nauchno-Issled, Inst. Gig. Tr. Prof. Zabol.*, **8**, 12 (1973) (*Chem. Abstr.*, **82**, 093797W).

464. J. W. Henck, J. F. Quast, and M. J. McKenna, Abstr. 161, *Am. Ind. Hyg. Conf., Chicago, May 31, 1979*.

465. M. M. Deacon, J. A. Ayres, T. R. Henley, Jr., M. K. Pilny, J. F. Quast, and J. A. John, The Effects of Inhaled Allyl Chloride on Embryonal and Fetal Development in Rats and Rabbits, Work done for the Chemical Manufacturers Association by The Dow Chemical Co., August 13, 1980.

466. V. K. Rowe, T. Wujkowski, M. A. Wolf, S. E. Sadek, and R. D. Stewart, *Am. Ind. Hyg. Assoc. J.*, **24**, 541 (1963).

467. T. A. Asmangulian and S. O. Badalian, *Tr. Erevan Med. Inst.*, **15**, 461 (1971) (through Ref. 434).

468. H. C. Scharnweber, G. N. Spears, and S. R. Cowles, *J. Occup. Med.*, **16**, 112 (1974).

469. E. W. Van Stee, *Review of the Toxicology of the Halogenated Fire Extinguishing Agents*, U.S. NTIS AD-A Rep.; ISS. No. 011538, (1974).

470. H. S. Warren and B. E. Ricci, *Vinylidene Chloride: I. An Overview, II. A Literature Collection 1974–1977*, Oak Ridge National Library, ORNL/TIRC-77-/3, April 1978.

471. T. Neudecker, A. Stefani, and D. Henschler, *Experimentia* **33**, 1084 (1977).

472. B. J. Dean, D. H. Hutson, A. S. Wright, and D. E. Stevenson, paper presented to joint Society of Toxicology/ASPET Meeting, Houston, Texas, August 13–17, 1978.

473. I. J. G. Climie, D. H. Hutson, B. J. Morrison, and G. Stoydin, *Xenobiotica* **3**, 149 (1981).

474. T. D. Landry, T. S. Gushow, P. W. Langvardt, and M. J. McKenna, *The Toxicologist* **1**, 6 (1981).

475. R. H. Reitz, T. R. Fox, J. C. Ramsey, J. F. Quast, P. Langvardt, and P. G. Watanabe, submitted for publication *Toxicol. Appl. Pharmacol.*

476. R. York, B. Sowry, L. Hastings, and J. Manson, *The Toxicologist* **1**, 28 (1981).

477. B. L. Riddle, R. A. Carchman, and J. F. Borzelleca, *The Toxicologist* **1**, 26 (1981).

478. A. M. Schumann, T. R. Fox, and P. G. Watanabe, *The Toxicologist* **1**, 101 (1981).

479. J. D. Burek, F. J. Murray, K. S. Rao, A. A. Crawford, J. A. John, T. J. Bell, J. S. Beyer, R. R. Albee, W. J. Potts, C. M. Parker, and B. J. Schwetz, *The Toxicologist* **1**, 103 (1981).

480. J. M. Lanham, *The Toxicologist* **1**, 103 (1981).

481. Y. Kato, K. Sato, S. Maki, O. Matano, and S. Goto, *J. Pesticide Sci.* **4**, 195 (1979).

482. D. R. Short, Jr., J. L. Minor, J. M. Winston, B. Ferguson, and T. Unger, ISS/EPA/560/6-77-022, Order No. PB-281713, 1977.

483. R. Slacik-Erben, R. Roll, G. Franke, and H. Verhleke, *Arch. für Toxicol.* **45**, 37 (1980).

CHAPTER FORTY-NINE

Halogenated Cyclic Hydrocarbons

WILLIAM B. DEICHMANN, Ph.D., M.D. (hon.)

PART 1   Chlorinated Benzenes, Chlorodiphenyls, and Chlorinated Naphthalenes

Chlorinated benzenes are aromatic ring compounds with one or more chlorines substituted for a hydrogen. Compounds with only a few chlorines are usually colorless liquids at room temperature and have an aromatic odor. The more highly substituted compounds are crystals (typically monoclinic).

Some of the occupations in which exposure may be encountered are cellulose acetate workers, deodorant makers, disinfectant workers, dyers, dye makers, fumigant workers, insecticide makers and workers, lacquer workers, organic chemical synthesizers, paint workers, resin makers, and seed disinfectors.

Chlorodiphenyls are diphenyl ring compounds in which one or more hydrogen atoms are replaced by a chlorine atom. These compounds are light, straw-colored liquids with typical chlorinated aromatic odors.

Exposure to chlorodiphenyls may occur in occupations such as cable coaters, dye makers, electric equipment makers, herbicide workers, lacquer makers, paper treaters, plasticizer makers, resin makers, rubber workers, textile flame-proofers, transformer workers, and wood preservers.

Chlorinated naphthalenes are naphthalenes in which one or more hydrogen atoms have been replaced by chlorine to form waxlike substances, beginning with monochloronaphthalene and going on to the octochloro derivatives. Their physical states vary from mobile liquids to waxy solids, depending on the degree of chlorination.

Exposure in industry usually occurs from mixtures of two or more chlorinated naphthalenes. Some of the occupations in which exposure may occur include cable coaters, condenser impregnators, electric equipment makers, insecticide workers, petroleum refinery workers, plasticizer makers, rubber workers, solvent workers, transformer workers, wire coaters, and wood preservers.

1   CHLOROBENZENE, C_6H_5Cl, Monochlorobenzene

1.1   Uses and Industrial Exposures

Chlorobenzene, a solvent and chemical intermediate, has been used extensively in industry; there is potential for industrial exposure for personnel engaged in the manufacture and handling of this compound.

1.2   Physical and Chemical Properties

Physical state	Colorless liquid
Molecular weight	112.56
Specific gravity	1.1066 (20°C)
Melting point	−44.9°C
Boiling point	132.0°C
Vapor density	3.88 (air = 1)
Vapor pressure	11.8 mm Hg (25°C)
Refractive index	1.5216 (25°C)
Percent in "saturated" air	1.55 (25°C)
Density of "saturated" air	1.05 (air = 1)
Solubility	0.049 g/100 ml water at 20°C; soluble in alcohol, benzene, diethyl ether
Flash point	−90°F (closed cup)

1 mg/liter = 217 ppm and 1 ppm = 4.60 mg/m³ at 25°C; 760 mm Hg

1.3 Effect in Animals

Chlorobenzene is a CNS depressant. Degeneration of the liver and kidneys has been observed following absorption of toxic doses. The histopathological changes may progress as exposure becomes more severe or as the period of exposure is lengthened. Liver injury may progress to necrosis and parenchymatous degeneration.

1.3.1 Acute Toxicity

Hollingsworth et al. (1) reported that rats survived an oral dose of 1.0 g/kg, but 4.0 g/kg was lethal to all. Guinea pigs survived an oral dose of 1.6 g/kg, but administration of 2.8 g/kg was fatal. Varshavskaya (2) reported the following oral LD_{50}s: guinea pig, 5.06 g/kg; rat, 2.29 g/kg; rabbit, 2.25 g/kg; and mouse, 1.44 g/kg.

The intraperitoneal LD_{50} of chlorobenzene for rats was reported as 0.515 ml/kg by Kocsis et al. (3).

Subcutaneous injections of 4 to 5 g/kg rat caused death in a few days. Doses of 7 to 8 g/kg resulted in death within a few hours. Autopsy revealed necrosis of the liver and kidneys, as reported by von Oettingen (4).

Inhalation exposures of cats were conducted by Götzmann in 1904 and reported by Flury and Zernik (5). A dosage of 37 mg/liter (8000 ppm) of chlorobenzene resulted in severe narcosis in 30 min and death in 2 hr. Exposure to 17 mg/liter (3700 ppm) resulted in death in 7 hr, and 7 hr of exposure to 11 to 13 mg/liter (2400 and 2900 ppm) caused restlessness, tremor, and muscular spasms, but no serious injury or fatalities. Definite narcotic effects were produced by exposure to 5.5 mg/liter (1200 ppm) whereas concentrations of 1 to 3 mg/liter (200 to 660 ppm) were tolerated for hours without significant effects (8). Rozenbaum et al. (6) reported that mice died after 2 hr of exposure to 20 mg/liter.

According to the CMA Technical Report, *Worldwide Literature Search on Chlorobenzenes,* submitted to the Manufacturing Chemists Association, Washington, D.C., March 1976 (7), the consistently observed signs of acute intoxication reported by Varshavskaya (2) for chlorobenzene and *o*- and *p*-dichlorobenzene in mice, rats, and guinea pigs are as follows, depending on the route of administration, and varying in severity with the degree of exposure: hyperemia of the visible mucous membranes, increased salivation and lacrimation, initial excitation followed by drowsiness, adynamia, ataxia, paraparesis, paraplegia, and dyspnea. Acute mortality, which usually occurs within 3 days, results from respiratory paralysis. Changes revealed by gross postmortem examination are hypertrophy and necrosis of the liver and submucosal hemorrhages in the stomach. Histologically disclosed changes are edema of the brain and necrosis of the centrolobular region of the liver, the proximal convoluted tubules of the kidneys, the bronchial and bronchiolar epithelium of the lungs, and the stomach mucosa, the liver generally showed the most severe damage.

1.3.2   Skin

Irish (8) reported that local application of monochlorobenzene to the skin of rabbits caused slight reddening. Continuous contact for a week may result in moderate erythema and slight superficial necrosis. In these studies, there was no indication of absorption of a toxic dose. Prolonged skin contact is likely to be painful (9).

1.3.3   Eye

Eye contact with chlorobenzene may result in pain and transient conjunctival irritation, but usually clears up within 48 hr (9). No corneal injury has been observed (8).

1.3.4   Subacute Oral Toxicity

Chlorobenzene was administered repeatedly to rats 5 days/week for a total of 137 doses over a period of 192 days. A dose of 0.0144 g/kg body weight/day was survived without any observable effect. At 0.144 g/kg, there was a slight decrease in growth from which the rats recovered. At both 0.144 g/kg and 0.288 g/kg, there were significant increases in liver and kidney weights and slight liver pathology. Blood and bone marrow were normal as reported by Flury and Zernik (5, 8).

Similar observations were made by Hollingsworth et al. (1). No adverse effects were observed in rats following the repeated administration of 18.8 mg/kg per day, 5 days/week for 192 days. A dosage of 188 mg/kg per day caused a slight increase in liver and kidney weights, and a dosage of 376 mg/kg per day caused slight cirrhosis and focal necrosis of the liver and a slight decrease in average spleen weights. No other effects were observed.

Quoting from the CMA report (7), in a study conducted by Varshavskaya (2), groups of seven male albino rats were treated daily from 9 months with chlorobenzene or *o*-dichlorobenzene in doses of 0.1, 0.01, and 0.001 mg/kg by stomach tube. The lowest dosage level of both substances was completely ineffective. The middle level (0.01) produced effects similar to, but of less intensity than, the highest dose (0.1). In doses of 0.1 mg/kg, both substances inhibited higher nervous activity; both caused a statistically significant inhibition of erythropoiesis, thrombocytosis, and inhibition of mitotic activity in the marrow. Chlorobenzene also caused eosinophilia. The dichlorobenzene caused neutropenia, increased urinary 17-ketosteroids, increased adrenal weight, and reduced adrenal vitamin C. Both substances increased alkaline phosphatase and blood serum transaminase activity, as well as hepatic and renal acid phosphatase, but they reduced whole blood SH groups and hepatic and renal alkaline phosphatase, DPN, TPN, succinic dehydrogenase, glucose-6-phosphatase, and α-glycerophosphate. Specific macroscopic, histological, and histochemical examinations revealed no evidence of carcinogenic activity.

Knapp et al. (10) administered chlorobenzene (by capsule) to dogs in doses of 27.25, 54.5, and 272.5 mg/kg per day, 5 days/week for 93 days. At the low and intermediate levels, the treatment was without effect. The highest dosage produced a reduction of blood sugar, but an increase in immature leukocytes, elevated serum glutamic–pyruvic transaminase and alkaline phosphatase, and increases in total bilirubin and total cholesterol. Four of eight dogs treated with the highest dose died after 14 to 21 doses. Gross and/or microscopic pathology was evident in the liver, kidney, gastroenteric mucosa, and hematopoietic tissue (7).

1.3.5   Chronic Vapor Exposure

The following information was taken from data from the Biochemical Research Laboratory of The Dow Chemical Company, reported by Irish (8). Rats, rabbits, and guinea pigs were exposed 7 hr/day, 5 days/week for a total of 32 exposures over a period of 44 days. At a concentration of 1000 ppm in air, there were histopathological changes in the lungs, liver, and kidneys. There was a slight depression of growth. Although there was no mortality in rats or rabbits, the guinea pigs did show a higher-than-normal mortality. At 475 ppm, the same species of animals survived. There was a slight increase in liver weight and slight liver histopathology. The blood was essentially normal in all these animals. At a concentration of 200 ppm, all the animals appeared to be normal.

Khanin (11), after exposing rats to 24-hr inhalation of chlorobenzene in a concentration of 1.0 mg/m^3 for 70 to 82 days, noted encephalopathy and inflammation of the internal organs, protein dystrophy, and a scattering of regenerated cells in the liver lobes. Foci of giant-cell hyperplasia were present in the kidneys (7). Tarkhova (12) exposed groups of 15 male rats continuously for 60 days to chlorobenzene. The air concentration of 1.0 mg/m^3 produced a lowering and distortion of the ratio of the chronaxie of the antagonistic muscles, increased blood cholinesterase activity, and lowered the α-globulin blood serum content. Exposure to 0.1 mg/m^3 for 60 days did not produce these changes (7).

A steady reduction of blood catalase and leukocytic indophenol oxidase activities was noted in guinea pigs following exposure to concentrations of 0.1 to 1.5 mg/liter for 3 hr every second day for 62 weeks (13). Rozenbaum et al. (6, 7) reported that with 2 hr of exposure, the "absolutely lethal" air concentration of chlorobenzene for mice was 20 mg/liter. The results of an investigation were submitted by Skinner et al. to the Divison of Biomedical and Behavioral Sciences of the National Institute for Occupational Safety and Health (14); in this investigation male rats and rabbits were exposed to 75 or 250 ppm of chlorobenzene vapors for 7 hr/day, for up to 120 exposures over 24 weeks. Groups of animals from both species were sacrificed after 5 and 11 weeks of exposure and examined for hematologic and other changes. Statistical analyses of the data suggested some treatment-related effects on the red cell parameters including an increase in reticulocyte count, which was more apparent in the rats than in the rabbits. Clinical chemical changes in both species were nonspecific. Most of the tissues examined showed no remarkable changes; only the congestion of the liver and kidneys of the rabbits sacrificed after 5 weeks suggested a treatment relationship.

1.4 Absorption, Metabolism and Excretion

Monochlorobenzene is rapidly absorbed from the lungs. Absorption from the gastroenteric tract takes place; it is increased in the presence of fats or oils. Absorption through the skin appears minimal or negligible.

The metabolism and excretion of this compound have been studied and reported by Spencer and Williams (15) and by Smith et al. (16), who administered monochlorobenzene to rabbits in an oral dose of 0.5 g/kg of body weight. They found that 27 percent of the administered dose was excreted unchanged in the expired air; 25 percent appeared in the urine as the glucuronide; 27 percent as the ethereal sulfate; and 20 percent as the mercapturic acid. The total of 99 percent is an exceptionally good recovery for such a study.

Smith et al. (17) fed radioactive labeled monochlorobenzene to rabbits in doses of 0.5 g twice daily for 4 consecutive days. They analyzed the urine and feces of the animals during treatment and for 3 days thereafter. One rabbit was then killed, and its organs removed for ^{14}C analysis. In this experiment, loss of

Table 49.1. Excretion and Distribution of the Radioactive Metabolites from [^{14}C]-Chlorobenzene in Rabbits[a] (17)

Combined Results from Two Rabbits	Radioactivity (10^{-4} d.p.m.)	% of Dose
Urine	3164.0	19.6
Feces	170.0	1.05
Methanol extract		
Dry burn	250.0	1.55
Tissue (one animal)	8.2	5.29×10^{-2}

[a] Combined total urinary and fecal metabolites collected during 4 days dosing and for a further 3 days. Total dose 161.7×10^6 d.p.m. = 8.28 g.

Table 49.2. Distribution of Radioactive Metabolites in the Urine of Rabbits dosed with ^{14}C-Chlorobenzene (17)

Metabolite	Radioactivity (10^{-6} d.p.m.)	% of Total Urinary Radioactivity
3,4-Dihydro-3,4-dihydroxychlorobenzene	0.182	0.57
Monophenols	0.898	2.84
Diphenols	1.320	4.17
Mercapturic acids	7.530	23.80
Ethereal sulfates	10.720	33.88
Glucuronides	10.620	33.57
Total	31.270	98.83

the compound by respiration accounted for the overall low recovery (Tables 49.1 to 49.3).

The majority of the metabolites from ^{14}C-chlorobenzene-dosed rabbits is found in the urine and only small amounts of radioactivity are present in the tissues and feces, furthermore, the overall low recovery of radioactivity is almost

Table 49.3. Distribution of Chlorophenols Excreted as Conjugates and in the Free State in the Urine of Rabbits Dosed with [^{14}C]-Chlorobenzene (17)

Form of Chlorophenol	Ratio of Diphenols to Monophenols		Distribution of Isomeric Monochlorophenols (%)		
	Found	Previously Reported (17)	o	m	p
Free state	1.47	6.5	5.9	33.6	60.4
Ethereal sulfate	53.05	51.0	2.3	2.1	95.6
Glucuronide	42.48	44.0	3.2	nil	96.8
Combined total of free and conjugated phenols	15.72	—	4.9	22.9	72.2

certainly a consequence of loss of chlorobenzene by respiration. Similar observations were made by Parke and Williams (18), who studied the metabolism of ^{14}C-labeled benzene. The relative amounts of conjugates obtained in this study (mercapturic acids 25, ethereal sulfates 37, and glucuronides 37 percent) are in excellent agreement with those of Spencer and Williams (15) (mercapturic acids 25.9 to 30.0, ethereal sulfates 33.7 to 40.8, and glucuronides 30.2 to 38.1 percent). Further, the proportion of diphenolic conjugates to their monophenolic counterparts is virtually identical to the values reported by Smith et al. (16).

A small amount of the phenolic metabolites is excreted in the free state. In this study, Smith et al. (17) obtained proportionately more of these materials as monophenols than in the earlier study (16). It is possible that the analytical procedure of Smith et al. (16) which estimates the monophenols as toluene-*p*-sulfonate derivatives leaves a proportion of these isomers undetected.

For additional information on the metabolism of chlorobenzenes refer to the excellent volume, *Detoxication Mechanisms* by R. T. Williams (19), and to the CMA report (7). The possible mechanism of liver necrosis by aromatic organic compounds is presented in a report by Brodie et al. (20).

1.5   Human Experience

1.5.1   Intoxications

Reich (21) reported on the case of a 2-year-old boy who swallowed 5 to 10 ml of Puran, a cleaning agent containing monochlorobenzene, which demonstrated the effect of dietary fat in increasing the rate of absorption of chlorobenzene. He showed no ill effect for 2.5 hr, but after eating lunch he quickly lost consciousness and suffered vascular paralysis and heart failure, but recovered and survived. The odor of Puran in breath and urine persisted for 5 to 6 days.

Severe anemia and medullary aplasia in a 70-year-old woman was related to her employment in hat making, which required the use of glue containing 70 percent monochlorobenzene. Early complaints included headache and irritation of the upper respiratory tract and mucosa of the eyes (22).

1.5.2   Industrial Exposure

According to the CMA report for the Manufacturing Chemists Association (7) Rozenbaum et al. (6) reported in 1947 that they examined 52 people who were occupationally exposed to monochlorobenzene; 28 were employed in a factory where the only vapors they were exposed to were those of monochlorobenzene. Many of the individuals who had worked there for 1 to 2 years suffered from headache, dizziness, somnolence, and dyspeptic disorders. Examination revealed acroparesthesia in eight people, spastic contractions of finger muscles in nine, hypesthesia of the hands in four, spastic contraction of the gastrocnemius muscle in two, and a vasovegetative instability in eight. The remaining 24 individuals who were exposed to vapors of other chemicals in addition to monochlorobenzene displayed no characteristic abnormalities.

1.5.3 Experimental Subjects

To establish the threshold of action of monochlorobenzene on the electrical activity of the brain, Tarkhova (23) exposed four healthy persons, 17 to 24 years old, to the compound in concentrations of 0.1, 0.2, and 0.3 mg/m^3 for 13 min. The electrical potentials of the brain in response to light excitations were recorded on an EEG, and threshold of action was found to be 0.2 mg/m^3. The author suggested that the "maximum single-time permissible concentration" of monochlorobenzene in atmospheric air be set at 0.1 mg/m^3.

1.6 Odor and Taste

The odor of monochlorobenzene is generally described as like that of mothballs or benzene. The odor threshold is given as 0.21 ppm in air in the Arthur D. Little report (24). Varshavskaya (2) gave the threshold concentration for odor and taste in water as 0.02 mg/liter.

1.7 Teratology

Apparently, no reports have been published on the teratogenic potential of chlorobenzene.

1.8 Hygienic Standards of Permissible Exposure

The threshold limit of monochlorobenzene adopted by the American Conference of Governmental Industrial Hygienists (ACGIH) for 1979 is a TLV–TWA of 75 ppm (350 mg/m^3) (25).

The permissible exposure limit, according to the 1978 *NIOSH–OSHA Pocket Guide to Chemical Hazards,* is 75 ppm; the immediately dangerous to life or health (IDLH) concentration is 2400 ppm (26).

2   *o*-DICHLOROBENZENE, C$_6$H$_4$Cl$_2$, 1,2-Dichlorobenzene

2.1 Uses and Industrial Exposures

o-Dichlorobenzene is used as a solvent, fumigant, insecticide, and chemical intermediate. Industrial exposure has occurred in personnel engaged in the manufacturing and handling of this compound.

2.2 Physical and Chemical Properties

Physical state	Colorless liquid
Molecular weight	147.01
Specific gravity	1.2973 (25°C)
Melting point	− 17.6°C
Boiling point	180.48°C
Vapor density	5.07 (air = 1)
Vapor pressure	1.15 mm Hg (20°C)
	1.56 mm Hg (25°C)
Refractive index	1.5476 (25°C)
Percent in "saturated" air	0.2 (25°C)
Density of "saturated" air	1.01 (air = 1)
Solubility	Insoluble in water; soluble in ethanol, benzene, diethyl ether
Flash point	155°F (open cup); 151°F (closed cup)

1 mg/liter = 166.3 ppm and 1 ppm = 6.01 mg/m³ at 25°C and 760 mm Hg

2.3 Effect in Animals

The toxicologic effect of *o*-dichlorobenzene is injury primarily to the liver and secondarily to the kidneys. Although this material is weakly anesthetic, a brief exposure to a high concentration results in depression of the central nervous system.

2.3.1 Acute Toxicity

Varshavskaya (27) of the Sechenov First Moscow Medical Institute reported that the toxicity of chlorobenzene and dichlorobenzene (isomer not stated) is on the same level; increasing the number of chlorine atoms in a benzene molecule does not affect the toxic action, but only its degree of "expressivity." In large doses, the toxicity of dichlorobenzene depends more on the spatial distribution of chlorine atoms rather than their number; that is, the ortho isomer is more toxic than the para isomer. The effects of the compounds in rats were found to be essentially the same. Conditioned reflex activity was depressed, indicating a cerebral cortical effect; erythropoiesis was significantly decreased, chloroben-zene produced eosinophilia, and o-dichlorobenzene, neutropenia. o-Dichloro-benzene, more than chlorobenzene, led to a sharp rise in urinary steroids. Although both benzenes increased tissue acid phosphatase and sharply de-creased tissue alkali phosphatase, no sign of carcinogenic action was noted macroscopically, histologically, or histochemically. According to organoleptic effects, the maximum permissible concentration in water for chlorobenzene was determined at 0.01 mg/liter, and for o- and p-dichlorobenzene, 0.002 mg/liter for each.

Intraperitoneal injection of 0.03 ml of o-chlorobenzene in male 180 g Sprague-Dawley rats resulted in loss of liver glycogen and minimal necrosis. The same effects were noted after an intraperitoneal injection of 0.04 ml of monochlo-robenzene, but little or no effect was noted following the intraperitoneal injection of 0.1 g of p-dichlorobenzene (20).

Extensive renal necrosis (coagulation necrosis of the proximal convoluted tubules) was produced in 48 hr in C57 Black/6J mice after an intraperitoneal dose of 10 mmol of the ortho isomer/kg, but not by an equimolar dose of the para isomer. Similar histopathological changes were observed after intraperi-toneal dose of 6.75 mMol/kg of monochlorobenzene. Sprague-Dawley rats were less sensitive. The more pronounced toxicity to the liver of the ortho isomer has been associated with a more pronounced binding of the compound or its intermediate metabolites to liver proteins (28).

2.3.2 Subacute Oral Toxicity

Hollingsworth et al. (29) fed guinea pigs o-dichlorobenzene as a solution with olive oil. All guinea pigs survived 0.8 g/kg body weight but all succumbed to 2.0 g/kg body weight. These authors also administered o-dichlorobenzene to rats in acqueous gum arabic at a dose of 376 mg/kg body weight per day by stomach tube 5 days/week for a total of 138 doses in 192 days. Growth and mortality were not affected. There was a moderate increase in average liver weight and a slight increase in average kidney weight; slight histopathological changes were observed in the liver. At a dose of 188 mg/kg of body weight per day, there was a slight increase in the weights of the liver and kidneys but no apparent histopathology (8).

Oral administration of 2 or 4 mg/kg of o-dichlorobenzene per day for 12 weeks to rats resulted in the retention of 80 to 100 mg/kg of the compound in the fatty tissues (30).

According to the 1974 IARC report (31), the maximum tolerated dose for rats of o-dichlorobenzene administered by gavage 5 days/week for about 28 weeks ranged from 19 to 190 mg/kg of body weight per day. Minimal liver and kidney damage occurred at higher dosage levels. No adverse effect could be observed at 18.8 mg/kg of body weight per day.

2.3.3 Acute and Subacute Vapor Pressure

Hollingswoth et al. (29) reported that rats exposed to a concentration of 977 ppm in air survived for 2 hr, but succumbed to a 7-hr exposure period. Rats survived a single 7-hr exposure to 539 ppm. These animals showed drowsiness, unsteadiness, and eye irritation. Definite organic injury was observed. The weight of liver and kidneys showed an increase. Microscopic examination of the liver revealed marked central lobular necrosis; cloudy swelling of the tubular epithelium was noted in the kidneys (8).

The CMA technical report (7) referred to studies conducted by Czajkowska et al. (32), who exposed two groups of rats to o-chlorobenzene vapors in concentrations of 20 and 100 mg/m^3 for 4 hr daily, 5 days/week. This treatment decreased body weight gain, cholinesterase activity, and the number of thrombocytes, but increased the number of eosinophils. Catarrhal changes in bronchi and signs of subacute and chronic pulmonitis were observed. The effects were not marked in rats exposed to 100 mg/m^3. With the lower dosage, there was complete recovery within 1 month after cessation of exposure.

Dogs tolerated 1 hr exposure to o-dichlorobenzene in a concentration of 2 ml/m^3; exposure of one dog to this concentration for 2 hr/day for 14 days caused no obvious signs of intoxication; however, exposure of dogs to 4 ml/m^3 resulted in somnolence during exposure [Reidel (1941), reported by von Oettingen (4)].

2.3.4 Chronic Vapor Exposure

Apparently the only chronic vapor exposure studies reported are those of Hollingsworth et al. (29). They exposed animals for 7 hr/day, 5 days/week. At a concentration of 93 ppm of o-chlorobenzene in air, rats, guinea pigs, and rabbits survived for periods of 6 to 7 months. The animals showed no deleterious effects on growth, mortality, organ weights, hematology, and histopathology (8).

2.3.5 Eyes

Hollingsworth et al. (29) placed undiluted o-dichlorobenzene in the eye of the rabbit. The response was moderate pain and slight conjunctival irritation. There was no serious injury and the irritation cleared in a few days (8).

2.4 Metabolism

In 1923, Hele and Callow (33) reported that chlorobenzene and o-chlorobenzene are excreted in the dog as ethereal sulfates and as mercapturic metabolites.

They suggested that each of these metabolites contained an aromatic nucleus, to which the chlorine was still attached, and that in no case was the chlorine liberated before the metabolites were formed.

Following the oral administration of the ortho isomer (1.47 to 2.94 g) and monochlorobenzene (1.12 to 11.1 g) to dogs, Callow and Hele (34) observed a close correlation between the "organic" chlorine and the "extra" sulfur excreted in the urine. This observation was taken as evidence that mercapturic acids are formed from these compounds.

After its oral administration to rabbits, o-dichlorobenzene is metabolized mainly to 3,4-dichlorophenol; but 2,3-dichlorophenol, 3,4-dichlorophenylmercapturic acid, and 3,4- and 4,5-dichlorocatechol are also formed. p-Dichlorobenzene is converted to 2,5-dichlorophenol and 2,5-dichloroquinol conjugated with glucuronic and/or sulfuric acid (35).

δ-ALA synthetase activity was enhanced in rats treated with o-, m-, or p-dichlorobenzene, 63, 32, and 42 percent, respectively. Activities of aminopyrine demethylase and aniline hydroxylase were enhanced markedly in rats treated with monochlorobenzene, whereas the cytochrome content was not influenced significantly by treatment with any of the isomers. The effect of inducers and inhibitors of microsomal mixed function oxidases on the rate of metabolism and the extent of binding of o- and p-dichlorobenzene to cellular constituents suggest that arene oxides (epoxide) may be responsible for the differing biologic properties of the parent compounds (28).

2.5 Human Experience

A 15-year-old girl developed a fatal acute myeloblastic leukemia with a unilateral retroclavicular lymphadenopathy and peripheral leukoblastosis, which was traced to her habit of cleaning spots from her clothes, without removing them, with a product containing 37 percent of o-chlorobenzene. A 40-year-old workman developed chronic lymphatic leukemia with peripheral lymphadenopathies after 10 years of cleaning electrical contacts with a solvent containing 80 percent o-, 15 percent p-, and 2 percent m-dichlorobenzene. This same solvent was also implicated in the development of acute myeloblastic leukemia in a 55-year-old woman after an unspecified period of use in cleaning spots from her family's clothes (36). Tolot et al. in 1969 (37) presented the case of a 40-year-old worker who had been exposed to o-dichlorobenzene for 22 years in the preparation of dyestuffs. The man exhibited purpura, intense anemia, marked hepatomegalia, and discrete splenic enlargement. The disorder proved fatal in 4 months. The case was diagnosed as proliferating myelosis.

Hollingsworth et al. (29) recorded concentrations ranging from 1 to 44 ppm with an average of 15 ppm o-dichlorobenzene in a workroom atmosphere where large quantities of this compound were handled. The odor of o-dichlorobenzene was not detectable at these levels. Thorough physiological examination of all who worked in these areas failed to show any indication of injury (organic injury or hematological effects) from the exposure (8).

Six months of industrial inhalation exposure of a young man to Orthosol, a product containing 95 percent of the ortho and 5 percent of the para isomer, resulted in severe pallor, exhaustion, and vomiting, with intense gastric pain and headache. Blood tests revealed a rapidly developing hemolytic anemia. Rapid and complete recovery followed cessation of exposure. Thirteen co-workers similarly exposed suffered no injury (38).

2.5.1 Skin

Eczematoid dermatitis of the hands, arms, and face of a 47-year-old glazier who worked with *o*-dichlorobenzene was described by Downing (39). Intense erythema and edema appeared promptly when the compound was applied locally to the skin of one arm. A large bullous lesion developed somewhat later.

2.6 Flammability

The explosive limits of *o*-dichlorobenzene are 2 to 9 percent by volume in air (8).

2.7 Odor and Warning Properties

The odor of *o*-dichlorobenzene is detectable by the average person at 50 ppm in air. Eye and nose irritation are not noted at this level. The odor becomes strong and the irritation noticeable at concentrations around 100 ppm. The compound has fair warning properties at this level, but the possibility of adaptation must be recognized (8). According to Varshavskaya (27) the threshold concentrations for odor and taste in water for *o*-dichlorobenzene are 0.002 and 0.0001 mg/liter, respectively.

2.8 Cancer

According to the June 1974 IARC report (31) on *o*- and *p*-dichlorobenzene, "No adequate studies on which to have an evaluation of carcinogenicity were available to the working group. . . . One report has suggested an association between leukemia and exposure to dichlorobenzenes, but this is insufficient evidence from which to assess the carcinogenic risk of this compound . . . ."

2.9　Hygienic Standards of Permissible Exposure

The threshold limit of *o*-dichlorobenzene adopted by the ACGIH in 1979 is 50 ppm (300 mg/liter). This concentration should not be exceeded, even instantaneously (25).

According to the 1980 *NIOSH–OSHA Pocket Guide to the Chemical Hazards* the permissible exposure limit is 50 ppm (300 mg/m^3); the immediately dangerous to life or health concentration (IDLH) is 1700 ppm (26).

3　*p*-DICHLOROBENZENE, $C_6H_4Cl_2$, 1,4-Dichlorobenzene

3.1　Uses and Industrial Exposures

p-Dichlorobenzene has been widely used as an insecticide, disinfectant, deodorant, and chemical intermediate. The potential for exposure exists for those involved in the manufacture, handling, and use of this compound.

3.2　Physical and Chemical Properties

Physical state	Colorless or white crystals
Molecular weight	147.01
Specific gravity	1.248 (55°C)
Melting point	53°C
Boiling point	174°C
Vapor density	5.07 (air = 1)
Solubility	Insoluble in water; soluble in benzene, alcohol, diethyl ether
Flash point	53.8°C (open cup)

1 mg liter = 166.3 ppm and 1 ppm = 6.01 mg/m^3 at 25°C, 760 mm Hg

3.3 Effect in Animals

3.3.1 Acute Toxicity

Oral administration of 1 g of *p*-dichlorobenzene/kg body weight to dogs for the treatment of intestinal worms produced no deleterious effect (40); however, Dikmans (41) reported that single doses of 0.5 or 1.0 g/kg produced severe signs of intoxication, and in one dog, death.

Hollingsworth et al. (1) fed *p*-dichlorobenzene to rats as a 20 percent solution in olive oil. The rats survived single doses of 1 g/kg of body weight, but all animals succumbed to a dose of 4 g/kg of body weight. Guinea pigs were fed a 50 percent solution and survived a single dose of 1.6 g/kg body weight, but succumbed to a dose of 2.8 g/kg of body weight (8).

The subcutaneous LD_{50} in 22- to 26-g dd-y mice was reported as 5145 mg/kg. The animals developed tremors within 2 to 3 hr, which continued for 3 or more days, and died of respiratory failure (42).

According to Varshavskaya (43) the oral LD_{50}s of *p*-dichlorobenzene and related compounds for mice, rats, rabbits, and guinea pigs are as shown in Table 49.4.

3.3.2 Subacute Toxicity

The intramuscular injection of 20 daily doses of 125 mg of *p*-dichlorobenzene in guinea pigs produced a state of coagulative deficiency as a result of reduced activity of the prothrombin complex and thrombokinase. The concurrent administration of lipotropic agents resulted in a protective action (44).

Totaro and Licari (45) also injected 125 mg daily doses for 20 days in guinea pigs and found this treatment to cause loss of body weight and an increase in blood serum transaminase activity. The concurrent administration of betaine, chlorine, or vitamin B_{12} (lipotropic agents) produced protective action.

According to the CMA report (7), Rimington and Ziegler (46) reported that feeding 1140 mg monochlorobenzene/kg, 455 mg *o*-dichlorobenzene/kg, or 770

Table 49.4. Oral LD_{50}s for Chlorobenzenes (43)

| Species | LD$_{50}$ (mg/kg Body Weight) | | |
	Monochlorobenzene	*o*-Chlorobenzene	*p*-Chlorobenzene
White mice	1445	2000	3220
White rats	2390	2138	2512
Rabbits	2250	1875	2812
Guinea pigs	5060	3375	7595

mg *p*-dichlorobenzene/kg to male white rats by gastric intubation for 5, 15, and 5 days, respectively, induced experimental hepatic porphyria. The treatment caused increased urinary coproporphyrin excretion, followed by increases in porphobilinogen and γ-aminolevulinic acid excretion. The mono and ortho compounds caused severe liver damage with intense necrosis and fatty changes. Rats with livers thus damaged displayed significantly decreased activities of liver catalase.

3.3.3 Chronic Toxicity

An extensive vapor study in animals was reported by Hollingswoth et al. (1) in 1956. They exposed the animals 5 days/week, 7 hr/day for extended periods of time. Rats, guinea pigs, and rabbits exposed to 798 ppm in air showed definite reactions of toxicity. Exposures ran from a few to as many as 69; an occasional animal died. Signs of intoxication included tremors, weakness, weight loss, eye irritation, and unkempt appearance; some became comatose. Histopathologically, the liver showed cloudy swelling and central necrosis. There was also a slight cloudy swelling of the tubular epithelium of the kidneys in some animals. Rabbits showed some lungs changes.

At a vapor concentration of 341 ppm, rats and guinea pigs survived for a period of 6 months. There was slight growth depression in male guinea pigs, a slight increase in average liver weight in male rats, and slight pathologic changes in the liver of male guinea pigs. At a concentration of 158 ppm, animals survived exposures lasting from 137 to 219 days. No adverse effects on growth or mortality were observed in rats, mice, rabbits, or monkeys. There was slight growth depression of the guinea pigs. Liver weights were slightly increased in male and female rats and in female guinea pigs; there was some questionable histophatologic changes in the liver. At a concentration of 96 ppm, rats, guinea pigs, rabbits, mice, and a monkey were exposed for periods up to 6 or 7 months. No adverse effects on many species of animal were observed (8).

Zupko and Edwards (47) exposed groups of rabbits, rats, and guinea pigs 20 to 30 min/day for up to 34 days to a *p*-dichlorobenzene vapor concentration of approximately 100 mg/liter of air. Each exposure induced intense irritation of eyes and nose, tremors and twitches of the extremities, loss of righting reflex, definite nystagmus, and rapid but labored respiration. The animals, as a rule, recovered from these signs of intoxication in 30 min to 2 hr; the use of the hind legs was always the last function to return to normal. Some animals died before the end of the exposure period. The compound was found to have a selective action on the granulocytes of the blood, producing a granulocytopenia in a majority of the animals. The total leukocytic (showing some increase in lymphocytes) and erythrocytic counts were not generally affected. The blood picture returned to normal within 30 days after the last exposure. Histological studies of 12 rabbits, 13 rats, and eight guinea pigs revealed lung damage and a definite selective action on the kidneys. Every animal showed marked and extensive kidney damage. There was comparatively little liver damage or evidence of hepatitis.

Hollingsworth et al. (1) studied the effect of repeated oral administration of *p*-dichlorobenzene. Rats were fed 5 days/week, a total of 138 doses in 192 days. At a daily dose of 376 mg/kg, an increase in liver weight and a slight increase in kidney weight were observed. Microscopic examination of the liver revealed slight cirrhosis and focal necrosis. At 188 mg/kg, a slight increase in the average weight of the liver and kidneys occurred. No effects could be observed in rats at 18.8 mg/kg of body weight/dose. Similar liver and kidney changes were noted in rabbits following doses of 1000 mg/kg of body weight/dose; the effects were less intense, but there was definite injury at 500 mg/kg (8).

In 1958 Hollingsworth et al. (48) published their results on the effects of oral dosing (by stomach tube) of rats, five times a week for a total of 138 doses of 18.8, 188, and 376 mg of the para isomer per kilogram body weight. The lowest dose produced no deleterious effects. The 188 mg/kg dose caused a slight increase in liver and kidney weights. The repeated oral administration of the highest dose level resulted in a moderate increase in liver weight, together with slight cirrhosis and focal necrosis of the liver, a slight increase in kidney weight, and a slight decrease in the weight of the spleen.

Groups of rabbits were treated similarly with a dose of 500 mg/kg for as many as 263 doses over 367 days, and with 92 doses of 1000 mg/kg over a period of 219 days. Both dosage levels produced loss of body weight, tremors, weakness, cloudy swelling, and focal necrosis of the liver. The 1000 mg/kg dose caused some deaths (48).

3.4 Human Experience

3.4.1 Nonoccupational Exposure

The CMA report (7) summarized early *p*-dichlorobenzene intoxications, many of which were reported in the foreign literature.

A 62-year-old man, presumably because of home use of *p*-dichlorobenzene, began to suffer from dizziness, asthenia, anemia, and hypogranulocytosis. He recovered gradually after discontinuation of exposure (49).

A 19-year-old girl who worked for 18 months with an agent containing 90 percent of the para isomer and 10 percent of hexachloroethane suffered marked asthenia, dizziness, significant loss of body weight, a mild anemia, and hyperleukocytosis. She recovered rapidly after withdrawal from exposure. Two female workers who performed the same work suffered no abnormalities (50).

A 53-year-old woman who used *p*-dichlorobenzene crystals extensively in her home for 12 to 15 years suffered pulmonary granulomatosis. A lung biopsy revealed numerous small lesions which contained crystals physically similar to those of *p*-dichlorobenzene (51).

A number of similar intoxications were presented in some detail in the CMA report (7). Briefly, the signs and symptoms these individuals experienced included periorbital swelling, intense headache, and profuse rhinitis; or acute illness with nausea, headache, vomiting, weight loss, numbness, clumsiness, and a burning sensation in the legs. The latter subject lost 22.5 kg in 3 months, developed ascites, and died. One man, a 52-year-old trapper who used the compound on animal hides, suffered weakness, nausea, subacute yellow atrophy of the liver, jaundice with elevated serum bilirubin, and elevated alkaline phosphatase (52).

Campbell and Davidson (53) reported the unusual case history of a pregnant 21-year-old woman who developed a pica for the para isomer. She consumed, throughout pregnancy, one or two blocks of toilet air freshener per week, which were composed primarily of *p*-dichlorobenzene. She developed a severe hypochromic, microcytic anemia, with excessive polychromasia, and marginal nuclear hypersegmentation of the neutrophils. She recovered completely after withdrawal of the chemicals. Neonatal examination of the child revealed no abnormalities.

Statistics of poisonings in children disclosed *p*-dichlorobenzene to be the most frequent cause (4.9 percent) of all poisoning by household products in 1972 (54). For additional details of instances of intoxication refer to the CMA report (7).

According to Peterson and Liner (55) *p*-dichlorobenzene is the least toxic of the active ingredients in moth repellent products. Other ingredients are naphthalene and camphor; the latter was used particularly in the past. Naphthalene products look dry and are slowly soluble in cold ethanol whereas *p*-dichlorobenzene products appear wet and oily and are rapidly soluble in unheated ethanol. According to this report "ingestion of 20 g of *p*-dichlorobenzene has been well tolerated in man."

3.4.2 Occupational Exposure

The many incidents of intoxication that resulted from the household use of products containing *p*-dichlorobenzene contrast sharply with the scarcity of reports of industrial exposure to this compound.

According to the 1964 *Hygienic Guide Series* (56) on *p*-dichlorobenzene, voluntary industrial overexposure is unlikely. The most serious effects of overexposure are those of lung, liver, and kidney injury.

Hollingsworth et al. (1) reported on their extensive industrial experience in handling *p*-dichlorobenzene. They reported on 58 men who had worked continuously or intermittently on operations involving the handling of *p*-

dichlorobenzene for periods of 8 months to 25 years—an average of 4.75 years. Early investigations showed air concentrations ranging from 10 to 550 ppm, with an average of 85 ppm. A later study separated the job area into two ranges; one with concentrations from 100 to 725 ppm, with an average of 380 ppm; and another with concentrations from 5 to 275 ppm, with an average of 90 ppm.

A third survey was made after there had been some major changes in operations. The concentrations varied from 50 to 170 ppm, with an average of 105 ppm. Under these conditions, there were still some complaints of eye and nose irritation by the workmen. There were no complaints when air concentrations ranged from 15 to 85 ppm, with an average of 45 ppm. The men in these areas, throughout the different periods of study, were examined thoroughly. There was no evidence of organic injury, hematologic effects, or eye changes (8).

Von Oettingen in 1955 (4), and also Hollingsworth et al. in 1956 (1), reviewed the literature. According to Berliner (57) cataract formation followed exposure to a "mothproofing agent" containing p-dichlorobenzene. Considering also the publications of several other authors and the report by Pike (58) in particular, which stated that p-dichlorobenzene does not cause cataracts, and the report by Hollingsworth et al. (1) on industrial experience with this compound, it appears that Berliner's findings may have been due to chemicals other than p-dichlorobenzene, which were present in the "mothproofing agent" used (8).

Severe systemic effects were reported by Walgren (59) in eight workers employed in the production of moth deterrents with p-dichlorobenzene as the primary ingredient. The individuals developed irritation of the mucous membranes of eyes and throat, methemoglobinemia, and loss of appetite and body weight. Only one of the workers suffered an exposure that exceeded 1 year, and one worker only 1 month. Methemoglobin concentrations ranged between 0.12 and 0.29 g/100 ml blood. All cases showed increased muscle reflexes, ankle clonus, and fine finger tremors. Blood pressures were within the normal range. In the blood, slight diminution of erythrocytes and hemoglobin was observed, and relative lymphocytosis up to 63 percent. Four workers showed thrombocytopenia, and one workman had granulocytopenia. Sternal marrow was essentially normal. The interpretation of these data is difficult in the absence of the specific compounds involved other than p-dichlorobenzene in the working atmosphere.

3.5 Absorption, Metabolism, and Excretion

The compound is not absorbed through the intact skin in acutely hazardous amounts (4). Vapors of the compound are readily absorbed from the lungs. All indications are that absorption of the solid compound occurs readily from the gastroenteric tract. Fats and oils or organic solvents enhance absorption of the compound.

According to Azouz et al. (60), in the rabbit p-dichlorobenzene (0.5 g/kg) is oxidized mainly (approximately 30 percent) to 2,5-dichlorophenol, which is excreted conjugated. 2,5-Dichloroquinol is also formed (approximately 6 percent of the dose), but in contrast to the ortho derivative, no mercapturic acid or dichlorocatechol is formed. The excretion of the metabolites of o-dichlorobenzene is slow and appears to be complete in 5 to 6 days after feeding. With the para isomer the excretion of metabolites is incomplete, and is appreciable after 6 days.

Two metabolites appeared in the blood of male Wistar rats, fasted for more than 16 hr, then given orally 200 or 800 mg/kg of the para isomer (5.0 ml/kg dose, in a corn oil solution). The metabolites that remained in the blood for many hours after the para compound had practically disappeared were 2,5-dichlorophenyl methyl sulfoxide and 2,5-dichlorophenyl methyl sulfone. The authors (61) suggest that the storage and slow release of p-dichlorobenzene from the fatty body tissues is responsible for the prolonged presence of the metabolites in blood.

Carlson and Tardiff (62) studied the effect of chlorinated benzenes on the metabolism of foreign organic compounds. They found that 1,4-dichlorobenzene, 1,2,4-trichlorobenzene, 1-bromo-4-chlorobenzene, and hexachlorobenzene (but not chlorobenzene) decreased hexobarbital sleeping time immediately and/or to 14 days following treatment.

Cytochrome C reductase, cytochrome P-450, EPN detoxification, glucuronyl transferase, benzyopyrene hydroxylase, and azoreductase were increased to varying degrees by the administration for 14 days of these four chlorinated benzenes, at doses from 10 to 40 mg/kg per day. Administration of 1,4-dichlorobenzene or 1,2,4-trichlorobenzene for 90 days at these doses resulted in increases in EPN detoxification, benzopyrene hydroxylase, and azoreductase. The former two were still elevated 30 days after cessation of the administration of the compounds. 1,2,4-Trichlorobenzene was found to be a more potent inducer with cytochrome C reductase; cytochrome P-450 was also elevated and remained so through the 30-day recovery period. The authors concluded that even simple chlorinated benzenes can support the metabolism of foreign organic compounds (62).

Pagnotto and Walkley (63) investigated the value of measuring the urinary excretion of p-dichlorophenol as an index of industrial exposure to p-dichlorobenzene. Their study involved workers in a chemical manufacturing plant, in a household-products packaging factory, and in a plant where p-dichlorobenzene was employed in the manufacture of abrasive wheels. Exposure was almost exclusively due to vapor inhalation. The concentrations of p-dichlorobenzene in air ranged from 7.0 to 49.0 ppm (average 8 to 34 ppm). Spot air samples were collected during the work shift (usually 3 shifts/day).

Excretion of dichlorophenol in urine occurred as expected, starting very shortly after the exposure began and then rising to a maximum at the end of the exposure period. This was followed by an initial rapid decrease, then a more gradual reduction over a period of several days. In one instance, small but detectable amounts of dichlorophenol were found several weeks after cessation of exposure (63).

The presence of dichlorophenol in the urine is revealed also by its distinctive odor, which is noticeable at a level of 100 mg/liter and still detectable at about 20 mg/liter. The concentrations of dichlorophenol in the urine ranged from 10 to 233 mg/liter (average 14 to 103 mg/liter). No painful irritation of the eyes or nose was reported by the workers, except when there was direct contact with p-dichlorobenzene dust or crystals. These studies showed a fairly good correlation between the average air concentration of p-dichlorobenzene and the excretion of p-dichlorophenol at the end of an exposure period (63).

3.6 Eyes

Pike (58) exposed rabbits repeatedly for 8 hr to p-chlorobenzene vapor concentrations of 4.6 to 4.8 mg/liter (770 to 800 ppm). He also administered the compound orally to rabbits in repeated doses of 0.5 or 1.0 g/kg. His purpose for administering highly toxic doses was to observe possible harmful effects in the eyes of these animals. He noted "toxic eye-ground changes," but "no lens changes" in the rabbits. He concluded, based on the results in rabbits and his clinical experience, that these observations lead us to believe that cataracts are not produced by p-dichlorobenzene. He also administered naphthalene orally to rabbits and produced cataracts in these animals; hence it is very likely that earlier reports of cataract formation may have had their origin in other compounds used in the manufacture of "mothproofing agents."

Solid particles, vapors, or fumes of p-dichlorobenzene are very painful to the eyes and nose of man and animal. But in order for vapors to be painful, it is usually necessary for the material to be heated, or to be dispersed in such a way that there is a very large surface area for evaporation in a poorly ventilated area. p-Dichlorobenzene is painful to most people in concentrations between 50 and 80 ppm; the discomfort is quite severe at 160 ppm (8).

3.7 Skin

Solid p-dichlorobenzene causes very little irritation to the skin. It does produce a burning sensation when held in close contact with the skin for a prolonged period. Fumes from the surface of hot p-dichlorobenzene may irritate the skin slightly when the contact is repeated or prolonged. There is no evidence of significant absorption through the skin (8).

3.8 Odor, Taste, and Warning Properties

p-Dichlorobenzene has a very distinctive aromatic odor. The threshold of detection varies from 15 to 30 ppm in air. The odor becomes very strong at concentrations between 30 and 60 ppm. The vapors are painful to the eyes and nose at concentrations of 80 to 160 ppm. Above 160 ppm, they are intolerable to any person who has not been exposed to the compound long enough to have developed tolerance.

The odor and irritating effects are good warnings to prevent overexposure to *p*-dichlorobenzene. It should be recognized, however, that a person may become sufficiently accustomed to the odor to tolerate high concentrations (8).

Exposure of hens for 3 days to air containing 3.4 to 6.4 ppm of *p*-dichloro-benzene imparted an unpleasant, sweetish taste to the egg yolks. Ill effects in the hens were not noted and there was no reduction in egg production (64).

The threshold concentrations for odor and taste for *p*-dichlorobenzene in water were reported as 0.002 and 0.006 mg/liter, respectively (43).

3.9 Mutagenicity

Srivastava (65) studied the effect of *p*-dichlorobenzene on somatic chromasomes of *Vicia faba, V. narbonensis, V. hirsuta, Pisum arvense,* and *Lathyrus sativus.* Various mitotic anomalies were encountered. These included shortening and thickening of chromosomes, precocious separation of chromatids, tetraploid cells, binu-cleate cells, chromosome bridges, and chromosome breakage. Chromosome breaks generally took place at the heterochromatic regions.

3.10 Cancer

According to the International Agency for Research on Cancer (66), no adequate studies on which to base an evaluation of carcinogenicity were available to the "Working Group" at its meeting in Lyon, June 1974. ". . . One report has suggested an association between leukemia and exposure to dichlorobenzenes but this is insufficient evidence from which to assess the carcinogenic risk of this compound."

3.11 Hygienic Standards of Permissible Exposure

The threshold limits of *p*-dichlorobenzene adopted in 1979 by the ACGIH are TLV–TWA 75 ppm (450 mg/m^3) and TLV–STEL (tentative value) 110 ppm (675 mg/m^3). (25).

According to the 1978 *NIOSH–OSHA Pocket Guide to Chemical Hazards* the permissible exposure limit is 75 ppm, and the immediately dangerous to life and health concentration is 1700 ppm (26).

4 HEXACHLOROBENZENE, C_6Cl_6, Perchlorobenzene

4.1 Uses and Industrial Exposure

Hexachlorobenzene is used as a fungicide to control wheat bunt and smut fungi of other grains. The technical grade used in agriculture contains 98 percent hexachlorobenzene, 1.8 percent pentachlorobenzene, and 0.2 percent 1,2,4,5-tetrachlorobenzene. Commercial formulations applied as dusts contain 10 to 40 percent hexachlorobenzene. Other applications include use as an additive for pyrotechnic compositions for the military, porosity controller in the manufacture of electrodes, intermediate in dye manufacture and organic synthesis, and wood preservative. Individuals engaged in the manufacture or handling of this compound are prone to industrial exposure. Aerial dispersion appears to be the major pathway for this compound entering the marine environment (67).

4.2 Physical and Chemical Properties

Physical state	White needles (monoclinic prisms)
Molecular weight	284.80
Specific gravity	1.5691 at 23.6°C
Melting point	230°C
Boiling point	322 (sublimes)
Vapor pressure	1 mm at 114.4°C
Vapor density	9.83
Solubility	Insoluble in water; slightly soluble in cold alcohol; soluble in benzene, chloroform, ether
Flash point	242°C
Fire hazard	Slight when exposed to heat or flame
Stability	Very stable, unreactive compound. There is no evidence that hexachlorobenzene is broken down by physical or chemical processes in the environment. It is "dangerous when heated to decomposition,

it emits highly toxic fumes of chlorides"
(EPA) (67)

1 mg/liter = 85.8 ppm and 1 ppm = 11.65 mg/m^3 at 25°C, 760
mm Hg

4.3 Effect in Animals, Absorption, Metabolism, and Excretion

4.3.1 Acute Toxicity

Hexachlorobenzene is of a low degree of toxicity. It is absorbed slowly from the
gastroenteric tract, primarily via the lymphatic system. Oral LD$_{50}$s for laboratory
animals are for cats, 1700 mg/kg; rabbits, 2600 mg/kg; rats, 3500 mg/kg; and
mice, 4000 mg/kg. According to the EPA report, the average lethal oral dose
for guinea pigs is 3000 mg/kg. For bluegill fish, rainbow trout, and channel
catfish, the average lethal dose in water is given as more than 100 ppm (67).

4.3.2 Subacute Toxicity

Table 49.5 summarizes the subacute toxicity and signs of intoxication produced
in rats, rabbits, mice, guinea pigs, chickens, and Japanese quail. This table was
prepared by the EPA and is based on information from a 1975 report by the
National Academy of Sciences (68).

4.3.3 Effect in Rats

Grant et al. (1974) (69) fed dietary concentrations of 10, 20, 40, 60, 80, and 160
ppm for 9 or 10 months to weanling male and female Sprague-Dawley rats; the
20 ppm dosage was about equal to 1 mg/kg body weight.

The response in the sexes differed to some extent, with the exception of
hexachlorobenzene residues in the liver and the liver-to-body weight ratios in
rats fed 80 and 160 ppm; these were dose-related but not sex-related. Also, the
pharmacologic action of pentobarbital and zoxazolamine was shortened in both
males and females fed 20 ppm and higher levels.

In males fed 40 ppm and higher concentrations, hepatic aniline hydroxylase
and N-demethylase activities and cytochrome P-450 levels were increased; in
females, these activities remained unaltered at all dietary levels.

In females fed 80 or 160 ppm, weight gains were reduced, and females but
not males readily acquired chemical porphyria.

More recent reports include the investigation by Kuiper-Goodman et al. (70),
who studied the subacute toxicity of hexachlorobenzene in Charles River strain
rats fed a diet providing dosages of 0.5, 2, 8, and 32 mg of the compound in
corn oil/kg body weight per day. Subgroups of rats were killed at 3, 6, 9, 12,

Table 49.5. Subacute Toxicity of Hexachlorobenzene[a]

Species	Number of Animals	Dose	Test Duration	Effects Observed
Rats	5	2 mg/kg/day	13 days	No toxic effects
	5	6 mg/kg/day	13 days	Very light skin twitching and nervousness Significant incorporation into liver
	5	20 mg/kg/day	13 days	Neurotoxic symptoms. Increase in liver weight
	5	60 mg/kg/day	13 days	Neurotoxic symptoms. Increase in liver and kidney weight
	5	200 mg/kg/day	13 days	Neurotoxic symptoms. Increase in liver and kidney weight
Rats	4	10 mg/kg/day	30 days	No toxic effects
	4	30 mg/kg/day	30 days	Increase in food consumption and body weight gains; increase in coproporphyrin excretion in urine; increase in liver weight and liver; body weight ratio increased
	4	65 mg/kg/day	30 days	Same as at 30 mg/kg/day
	4	100 mg/kg/day	30 days	Same as 30 mg/kg/day plus elevation in excretion of uroporphyrin
Rats	33	100 mg/kg/day	51 days	13 deaths in 1 month; neurotoxic symptoms; increased liver weight; porphyria
Rats	10	300 mg/kg/day	10 days	30% mortality
	10	150 mg/kg/day	30 days	60% mortality
	10	50 mg/kg/day	30 days	30% mortality
Male rats	26	0.2%	12 weeks	Retardation in weight gain; porphyria degenerative changes in the liver
Rats	13	0.025 mg/kg/ day[b]	4–8 months	No toxic symptoms. Possible effect on conditioned reflexes
Guinea pigs		0.5%	8–10 days	Marked neurological symptoms
Mice		0.5%	8–10 days	Marked neurological symptoms
Rabbits		0.5%	6 weeks	Increase in urinary porphyrins
		0.5%	8–12 weeks	Death occurred
Japanese quail	15	1 ppm	90 days	No toxic effects
	15	5 ppm	90 days	Slight increase in liver weight; minimal porphyria
	15	20 ppm	90 days	Increased liver weight, decreased egg production; porphyria; liver and kidney pathological changes
	15	80 ppm	90 days	5 deaths (18- to 62-day period); neurotoxic symptoms; porphyria; increased liver weight; decreased egg production and hatchability; liver and kidney pathological changes
Japanese quail	12	2500 ppm	30 days	All died in 30 days (4 died in 7 days)
	12	500 ppm	30 days	All died within a month
	12	100 ppm	3 months	Mortality (1 on 20th day; 10 within 7 weeks; 1 in 10 weeks). Surviving cock showed marked loss of weight. Necrosis of liver cells; porphyria.
Chickens		120–480 ppm	3 months	No toxic effects

[a] This table was prepared by C. E. Mumma and E. W. Lawless for the EPA (67) and is based on information from a 1975 report by the National Academy of Sciences (68). The route is oral (in diet) unless otherwise indicated.
[b] Administered in water.

and 15 days of feeding. At 15 weeks the remaining rats were then fed a compound-free diet and sacrificed at 1, 2, 4, 16, and 33 weeks.

Signs of intoxication were dose-related and included excessive irritability, tremors, alopecia, and nonspecific dermal changes with slight scabbing due to itchiness accompanying alopecia. Ataxia with hind leg paralysis and loss of pain sensation in the legs was seen in a few females at the 32 mg/kg per day dose level after 6 and 9 weeks of treatment. In males and females tremors disappeared after 3 and 4 weeks of feeding, respectively; all other signs of intoxication had disappeared by 9 weeks.

Tissue hexachlorobenzene concentrations rose rapidly and, by 3 weeks of treatment, had more or less equilibrated. Concentrations in adipose tissue and liver were about 200- to 500-fold and ten- to twenty-fold higher, respectively, than the serum levels. At the 32 mg/kg per day level the serum level, by week 15, reached 19.4 ± 1.2 μg/g tissue in males, and 30.9 ± 4.4 μg/g in females. Concentrations, particularly in adipose tissue and liver, but also in brain, kidney, and spleen, were significantly higher in females than in males.

Histologic changes were confined to the liver and spleen. Hepatomegaly and an increase in the size of the centrolobular hepatocytes was noted. In males sorbitol dehydrogenase activity, an indicator of liver damage, was first increased (maximally at 6 weeks) at the highest dose level, then subsequently dropped to almost the control level. Histochemically, this was related to a depletion of this enzyme in the liver. Liver succinic dehydrogenase, sorbitol dehydrogenase, and glucose-6-phosphatase became equally depleted.

Females, particularly, developed porphyria (together with porphyrin excretion) with high porphyrin values persisting in some rats, but not in others, after the animals were placed on a hexachlorobenzene-free diet.

These studies support earlier investigations that female rats are more susceptible than males to the toxic effects of hexachlorobenzene. According to the investigators, a dose of 0.5 mg/kg per day appears to be the "no-effect" level in rats.

Boger et al. (71) administered similar dosages for a more prolonged period. They focused their investigation on liver changes in female Wistar rats after oral administration, by stomach tube, of 0.5, 2.8, or 32 mg of hexachlorobenzene/ kg body weight, twice a week, for 29 weeks. They reported a dose-related increased retention of the compound in the liver, which had increased in weight. There was also a dose-dependent enlargement of all hepatocytes, along with porphyrin deposition, an increase of glycogen, and alterations in the endoplasmic reticulum. At the 32 mg/kg dose level the liver contained 273 μg of hexachlorobenzene per gram of tissue.

Dose-related changes in the hepatic ultrastructure were reported by Mollenhauer et al. (72). They fed outbred Sprague-Dawley rats dietary concentrations of 5 to 25 ppm of hexachlorobenzene for 3 to 12 months. In some animals the smooth endoplasmic reticulum proliferated to become the predominant feature in the cell, whereas in others, the reticulum was apparently replaced by quantities of a storage product, presumably glycogen. The proliferation of the smooth endoplasmic reticulum occurred most often in animals fed the smaller doses, whereas storage product accumulations were most often observed in rats given the larger doses. The most significant changes were noted in the mitochondria, which became elongated and swollen. Storage bodies of 1 to 4 mm in diameter appeared in some cells. These bodies were surrounded by a double membrane derived from the endoplasmic reticulum and may have been the partly digested remains of generation mitochondria.

Lissner et al. (73) fed Wistar rats a diet containing 0.2 percent of hexachlorobenzene. During the first week, the levels of cytochrome P-450 and the activity of aniline hydroxylase in the liver were elevated. After this initial three-fold rise in cytochrome P-450, there was a second steep rise after the thirtieth day of treatment. Aniline hydroxylase activity showed a similar pattern. After 40 days of feeding the urinary excretion of 5-aminolevulinic acid and porphyrin were markedly elevated.

According to the authors, during an early period of hexachlorobenzene exposure, microsomal enzyme induction could lead to the oxidation of the compound, with the resulting metabolite possibly being the actual porphyrogenic agent, and a more potent inducer of the microsomal enzymes than unchanged hexachlorobenzene.

Koss et al. (74) studied the toxicity of hexachlorobenzene in rats over a 53-week period. From the first through the fifteenth week, the compound was given orally every other day at a dose of 178 μmol/kg (50 mg/kg). Nine weeks after the start of the treatment an equilibrium was reached between intake and elimination of the compound and its metabolites. At this time, 1 g of liver contained approximately 1 μmol hexachlorobenzene, 50 nmol pentachlorophenol, 5 nmol tetrachlorohydroquinone, and approximately 0.1 nmol pentachlorothiophenol. Thirty-eight weeks after cessation of administration, the biologic half-life was 4 to 5 months.

Although the body weights of experimental rats stayed within normal limits during the treatment period, the weights of the liver, spleen, kidney, and adrenal glands showed relative increases, followed by recovery during the following 38-week period.

Liver porphyrin, urinary porphyrin, δ-aminolevulinic acid, and porphobilinogen increased up to week 15. During the subsequent 38-week period the porphyrin content of the liver continued to increase, but the porphyrin content of the urine and porphyrin precursors had decreased to almost normal levels.

The metabolism of hexachlorobenzene also was studied by Engst et al. (75) in male Wistar rats, fed by stomach tube 8 mg of the compound per kilogram (in sunflower oil) for 19 days. Tissue concentrations at the termination of the treatment period were for body fat, 82 ppm; muscle, 17 ppm; total liver, 125 μg; total kidney, 21 μg; total spleen, 9 μg; total heart, 1.5 μg; and adrenal, 0.5 μg each.

The urine contained hexachlorobenzene and its primary metabolite, pentachlorophenol, which was present alone or sometimes associated with 2,3,4,6-tetrachlorophenol and/or 2,3,5,6-tetrachlorophenol, or 2,4,5-trichlorophenol. Trace amounts of related compounds were also detected. The feces contained high amounts of hexachlorobenzene and smaller amounts of pentachlorobenzene.

Schuster and Renner (76), who dosed rats with a total of 300 mg/kg of hexachlorobenzene for 10 months, found pentachlorophenol and 2,4,5-trichlorophenol to be the primary urinary metabolites. 1,2,4,5-Tetrachlorobenzenes and certain other chlorobenzenes were detected, but not completely identified.

Lui and Sweeney (77) also concurred that pentachlorophenol is the primary metabolite in rats. They also detected more polar compounds, presumably derivatives of hexachlorobenzene. Formation of pentachlorophenol from hexachlorobenzene may proceed either through a free radical mechanism or by initial formation of an arene oxide. In either event, reactive intermediates may form covalent bonds with cellular constituents leading to irreversible cell damage.

Koszo et al. (78) reported that when rats are given an oral dose (by stomach tube) of 0.2 g hexachlorobenzene, this compound can be found in urine and feces 1 day later in concentrations of 32 and 43,260 μg, respectively. The urinary and fecal excretion of pentachlorophenol was 30 μg and less than 1 μg, respectively.

In experiments by Koss et al. (79) with labeled hexachlorobenzene, they obtained evidence for an unexpectedly high extent of its conversion to metabolites. Four weeks after administration, 7 percent of the radioactivity was excreted in rats via the kidneys and 27 percent with the feces. Nearly the total in the urine was contained in metabolites of hexachlorobenzene, and 69 percent of the radioactivity in the feces was represented by the unchanged drug. It was calculated that the rat eliminates almost half the amount of hexachlorobenzene in the form of metabolites, including pentachlorophenol, which accounts for about one-fifth of the total excreted. The other major metabolites identified are tetrachlorohydroquinone and pentachlorothiophenol. These substances represent 3 and 16 percent, respectively, of the label excreted.

More recently, Koss et al. (80) described the identification of pentachlorothiophenol and pentachlorothioanisole in the livers of rats treated with hexachlorobenzene. In order to clarify the further fate of these metabolites, they were administered to rats, and the conversion products excreted in urine and feces were isolated. The metabolites of pentachlorothiophenol and pentachlorothioanisole were found to be excreted in both conjugated and free form. From extracts of the excreta, tetra- and trichlorobenzene with two or three sulfur-containing substituents on the ring were isolated; these were analogous compounds in which thiol groups were converted into sulfoxide and sulfone groups, as well as analogous compounds with a phenolic oxygen, in addition to sulfur and sulfur-containing compounds in which chlorine was replaced by hydrogen. Following administration of the sulfoxide and of the sulfone of pentachloro-thioanisole under analogous conditions, pentachlorothiophenol and pentach-lorothioanisole and their metabolites were detected in the excreta of the animals. No evidence was obtained that the parent compounds are excreted in the unchanged form.

Goerz et al. (81) studied the possible role of the two hexachlorobenzene metabolites, pentachlorophenol and pentachlorobenzene, in the production of porphyria in Wistar rats. They fed hexachlorobenzene and pentachlorobenzene, each in a dietary concentration of 0.05 percent, for 80 days. A third group of rats, prior to hexachlorobenzene feeding, ingested a diet containing 0.05 percent of pentachlorophenol for 40 days and subsequently hexachlorobenzene for 54 days.

Since the urinary porphyrin excretion remained the same throughout the entire experimental period in rats fed pentachlorobenzene and pentachloro-phenol, and since a distinct porphyria was established after 60 days by hexachlorobenzene, the authors considered it quite unlikely that the two hexachlorobenzene metabolites are the porphyrogenic agents of hexachloro-benzene. In spite of the pentachlorophenol pretreatment, hexachlorobenzene led to porphyria after only 50 days, the same as for the hexachlorobenzene controls. The investigators consider it unlikely that the hexachlorobenzene-induced porphyria has a distinct pathogenic connection to the induction of the cytochrome P-450 system.

To elucidate the relationship between chemical structure and their biologic activities, the contents of cytochromes and hepatic constituents in addition to the activities of drug-metabolizing enzymes and δ-aminolevulinic acid (δ-ALA) synthetase were examined in rats treated with various chlorinated benzenes, that is, monochlorobenzene (MCB), p-dichlorobenzene (p-DCB), 1,3,5-trichlo-robenzene (1,3,5-TRCB), 1,2,4,5-tetrachlorobenzene (1,2,4,5-TECB), pentach-lorobenzene (PECB), and hexachlorobenzene (HCB) (82).

The content of cytochrome P-450 and activities of aminopyrine demethylase and aniline hydroxylase were increased by oral administration of all chloroben-zenes except MCB as a daily dose of 250 mg/kg, once daily, for 3 days. The contents of microsomal protein and phospholipids also showed a similar tendency to those described above. The activity of δ-ALA synthetase was increased by treatment with all compounds used.

The content of cytochrome P-450 and activity of aminopyrine demethylase were decreased in 24 hr after a single administration of MCB in doses of 125, 250, 500, and 1000 mg/kg, whereas the activity of δ-ALA synthetase was increased markedly by all doses used. The activity of aniline hydroxylase was increased by a dosing of 1000 mg/kg MCB.

In the time course after a single administration of MCB in a dose of 250 mg/

kg, the activity of δ-ALA synthetase decreased in 6 hr after administration, was subsequently restored to normal levels in 12 hr, and then increased markedly in 24 hr. The opposite changes were noted in the content of cytochrome P-450.

Timme et al. (83) studied the effect of siderosis in rats (following the intraperitoneal injection of imferon, five doses of 10 mg each), after an interval of 2 months between imferon injections and hexachlorobenzene feeding of a dietary concentration of 0.3 percent. Tremors and urinary porphyrin were noted soon after the feeding was started, but porphyrins appeared earlier in the siderotic than in the control, nonsiderotic hexachlorobenzene-fed rats. In the nonsiderotic group, the liver contained only traces of stainable iron whereas in the siderotic rats, iron was initially diffuse in the lobule, but later the centrolobular areas contained only minimal quantities. In both groups, red porphyrin fluorescence appeared first in the centrolobular zone, but in the siderotic group it became widespread throughout the lobule. The authors concluded that the administration of iron prior to hexachlorobenzene feeding induced a more extensive development of liver lesions, and they suggest that this could account for the higher levels of excreted porphyrins in the iron-laden animals.

Food deprivation and its effect on hexachlorobenzene toxicity was studied by Villeneuve et al. (84) in female (30- to 90-g) rats. They fed dietary concentrations of 4, 20, 100, and 500 ppm *ad lib* to four groups of 10 rats each. A similar second group of rats was fed concentrations of 8, 40, 200, and 1000 ppm, but the food intake of these animals was reduced to 50 percent of that of the first group. The investigators found that the reduced food intake, combined with the higher dietary concentrations, increased the following: the degree of liver hypertrophy, microsomal enzyme activity and porphyrin accumulation in the liver, excretion of porphyrin in the urine, and accumulation of hexachlorobenzene in brain and liver, as well as its rate of excretion in the feces.

Zawirska and Dzik (85) reported that spectrophotometric and chromatographic investigations of porphyrins isolated from rat livers indicated that the most efficient effect of synporphyrinogenesis, which was manifested through increased levels of uroporphyrin, had occurred in the groups treated simultaneously with hexachlorobenzene and testosterone, and with hexachlorobenzene, testosterone, and stilbestrol.

4.3.4 Effect in Beagles

Studies in 6.3- to 10.3-kg male and female beagles were performed by Gralla et al. (86), who administered the compound daily for up to 12 months in gelatin capsules at dose levels of 1, 10, 100, and 1000 mg/dog. Signs of intoxication included anorexia, loss of body weight, and mortality at the 1000 mg/dog dose. These signs were observed to a lesser degree in the dogs treated with the 100-mg dose. Clinical laboratory changes in the animals on the two highest dose levels included anemia, hypoglycemia, and hypocalcemia. Testicular degeneration was considered by the investigators to be related to malnutrition.

After 5 to 12 months of feeding, the concentrations of hexachlorobenzene in adipose tissue of the dogs were 11 ppm at the 1 mg/kg dose, 67 ppm at 10 mg/kg, 714 ppm at 100 mg/kg, and 1216 ppm at the 1000 mg/kg dose. The concentrations in bile were 0.52, 4, 87, and 50 ppm, respectively.

Widespread pathological lesions were confined to the abdomen and included serositis, necrosis, fibrosis, and steatitis of the omentum. Nodular hyperplasia of gastric lymphoid tissue was found in all dogs including those at the 1 mg/dog per day dose. A dose-related neutrophilia appeared in the animals receiving the two highest dosages. Four severely affected dogs receiving the highest dose showed generalized vasculitis and one dog showed amyloidosis.

No hepatic fluorescence was found at necropsy, indicating that these dogs were free of porphyria and suggesting that the dog, unlike the rat, is insensitive to this well-known toxic effect of hexachlorobenzene.

Luthra et al. (87) studied the effect of hexachlorobenzene on the serum lipoprotein pattern in beagles. They administered the compound in gelatin capsules in doses of 1, 10, and 100 mg/dog per day for 13 weeks. Sera were collected at the termination of the treatment period following a 12-hr fast. They found that the percentage distribution in untreated control dogs was 37 ± 6 for β-lipoprotein and 47 ± 11 for α-lipoprotein. In the experimental dogs they found, utilizing statistical analysis by the Dunnett's multiple comparison method, a non-dose-related decrease of 13 to 40 percent in β-lipoprotein and an increase of 28 to 46 percent in the α-lipoprotein fractions. Their results support the suggestion that hexachlorobenzene interferes with the lipid metabolism and/or transport.

4.3.5 Effect in the Monkey

Mueller et al. (88) studied the kinetic, metabolic, and histopathological effects of hexachlorobenzene and related compounds in rhesus monkeys. One male monkey was fed, with the diet, a daily dose of 10 ppm of ^{14}C-labeled hexachlorobenzene for 540 days. The compound was absorbed more slowly from the gastroenteric tract than dieldrin or pentachloronitrobenzene (PCNB), which were studied concurrently. Absorption occurred with minor involvement of the portal venous system; major absorption was by the lymphatic system. Deposition of chlorobenzene took place in the fat, thymus, and bone marrow. The only affected organs in the dieldrin monkeys were the liver and kidneys, which showed in the liver the advanced parenchymatous degeneration associated with nodular hyperplasia. In the PCNB animals, the kidneys were the only organs involved. In the hexachlorobenzene monkey, degenerative changes were evident in the kidneys and cerebellum, and there was thymic cortical atrophy.

Yang et al. (89), who administered ^{14}C-hexachlorobenzene intravenously in three rhesus monkeys, found that this treatment lead to the highest level of radioactivity in body fat and bone marrow. One year after an intravenous injection, the cumulative fecal and urinary metabolites were estimated to be only 2.8 and 1.6 percent, respectively, of the dose administered. They explained the long-term storage of hexachlorobenzene in fat as the basic reason for the slow elimination of this compound from the body.

In the monkeys, pentachlorophenol was the major metabolite, but traces of pentachlorobenzene were also found in the feces. The entire urinary radioactivity was in the form of unidentified polar metabolites of hexachlorobenzene; they were neither pentachlorobenzene or pentachlorophenol. (In rats the polar metabolites formed were estimated by Yang et al. to be less than 0.2 percent of the injected dose.)

In another investigation with rhesus monkeys, Rozman et al. (90) concluded that none of the monitored parameters indicated harmful effects from a dose of 110 μg/day of hexachlorobenzene for 550 days.

4.3.6   Effect in Cattle

Studies conducted with 2-year-old beef steers fed 6, 36, or 216 mg of hexachlorobenzene daily for 10 weeks led to levels of the compound (at 10 weeks) of 3.1, 16.25, and 98.75 mg/kg subcutaneous fat, respectively. When the feeding of the compound was discontinued, its concentration in fat decreased exponentially, with an average half-life of 10.5 weeks.

The rates of retention of hexachlorobenzene in subcutaneous body fat and its excretion in milk were studied in three cows fed 5 or 25 mg of the compound/day for 60 days. Hexachlorobenzene concentrations in milk fat during the 40- to 60-day period were 2.1 and 9 ppm, respectively, while the concentrations in body fat reached 1.9 and 8.8 ppm at 60 days. During the 15 days following the ingestion of the experimental compound, milk fat concentrations declined 32 percent; thereafter the decline was slower. The milk fat to body fat concentration ratio was 0.87 to 1. The biologic half-life of the compound in the three cows ranged from 29 to 64 days.

4.3.7   Effect in Pigs

A 90-day pig toxicity study was conducted by Tonkelaar et al. (91) in which the animals were fed 0.05, 0.5, 5, and 50 mg/kg per day. Only the highest dose caused signs of intoxication, including porphyria, increased liver weights, and death during the experiment.

All pigs excreted coproporphyrin and showed histopathological liver changes. Retention of hexachlorobenzene was highest in body fat (approximately 500 times the concentration in blood), followed by the liver, kidneys, and brain, in the order given. Under the conditions of this experiment with pigs the no-effect level was judged to be 0.05 mg/kg per day.

Fassbender et al. (92) found that the ingestion of hexachlorobenzene (1 mg/

kg feed) by pigs did not adversely affect their growth or health. Residues of the compound were found in significantly higher concentrations in liver and kidneys than in fat or muscle.

Hexachlorobenzene distribution and retention in the tissues of swine were studied by Hansen et al. (93). They administered the compound to third-litter sows at dietary concentrations of 1 or 20 ppm throughout gestation and nursing. Swine receiving 1 ppm were not adversely affected and residue concentrations in tissues other than fat and bone marrow remained at or below the dietary concentration. Residues of hexachlorobenzene in the dissectable fat of these pigs accumulated to concentrations five- to seven-fold higher than the dietary concentration. Piglets accumulated fat residues that were higher than those of the sows, through both placental transfer and nursing. A similar proportional accumulation of hexachlorobenzene in fat occurred in sows receiving 20 ppm in the diet. Other signs of toxicity included neutrophilia, gastric irritation, fatty replacement of Brunner's gland, and pancreatic periductal fibrosis. Hepatotoxicity was not apparent.

There was a tendency toward neutrophilia in the sows fed 20 ppm, an effect that was also observed in beagle dogs, but at higher hexachlorobenzene doses (86). The propensity toward gastric irritation and ulceration was also previously observed in swine by Hansen et al. (94).

In the discussion, Hansen et al. pointed out that the dietary concentrations of 1 ppm, equivalent to about 0.025 mg/kg per day, at which toxicity was not observed is still well above the conditional upper human intake of 0.6 μg/kg per day suggested by FAO/WHO (95). They cautioned against the accumulation of hexachlorobenzene in fat above dietary concentrations by exposing food-producing animals to higher levels. At a dietary concentration of 20 ppm (approximately 0.5 mg/kg per day), fat residues accumulated to values considered hazardous. The effects noted in this investigation were very similar to those observed when 20 ppm Arochlor 1242 (PCB) was fed to swine (94), but hexachlorobenzene accumulated to concentrations that were severalfold higher than those of PCB. Kuiper-Goodman et al. (70) found that 0.5 mg/kg per day for 15 weeks is a "no-effect" level for hexachlorobenzene in rats, and Gralla et al. (86) found that this dose level was not toxic in dogs, but swine appear to be somewhat more susceptible, in spite of their large fat reserve for storage of this compound.

4.3.8 Effect in Lambs

Mull et al. (96) studied the distribution and excretion of hexachlorobenzene in castrated growing lambs fed dietary concentrations of 0.01, 0.1, or 1 ppm for 90-day periods. Biopsies were taken at days 7, 15, 30, 45, and 60, and every 30 days thereafter through day 300. The treatment did not affect the growth rate of the animals.

By the end of the 90-day exposure period adipose tissue concentrations of hexachlorobenzene had reached a level 10 times that in the diet. The omental fat was found to contain higher concentrations of hexachlorobenzene residues than did perenial fat at 90 days, but not at 300 days. The apparent decrease of the compound in the fat following cessation of treatment was primarily due to dilution brought about by increases in the quantities of fat in the carcass. The studies demonstrated a non-dose-related hexachlorobenzene half-life of approximately 90 days.

4.3.9   Effect in Chickens

In an investigation with laying hens, given seven consecutive daily doses ranging from 1 to 100 mg/kg, Hansen et al. (93) determined the half-time of hexachlorobenzene in fat to be 24 to 27 days. They found that more than half of the compound is excreted in the yolks of eggs. Tissue concentrations of the compound paralleled the concentrations in body fat. The skin provided a significant reservoir for hexachlorobenzene.

4.4   Reproduction and Teratogenicity

Placental transfer of hexachlorobenzene has been reported in mice and rats.

Courtney and Andrews (97) performed studies with CD-1 mice which were treated with hexachlorobenzene by gastric intubation at levels of 10, 50, or 100 mg/kg per day. Two groups were treated before implantation from day 0 through 5 of gestation and sacrificed at day 12 or 17. Two other groups were treated after implantation from day 6 through 11 or from day 6 through 16 and were sacrificed 24 hr after final treatment. The pesticide was mobilized from maternal depots and transferred across placenta and deposited in the fetus. Concentrations in the fetuses ranged from 0.76 ppm to as high as 19.09 ppm and appeared to be related to the level of dose received by the mother. Fetal hexachlorobenzene levels were higher in the mice treated before implantation than in those treated on days 6 to 11, indicating the effect of maternal body burden.

Villeneuve et al. (84) studied the placental transfer of hexachlorobenzene in pregnant nulliparous Wistar rats fed from 5 to 120 mg of hexachlorobenzene per kilogram per day in corn oil from day 6 through day 16 of pregnancy. They killed the rats on day 22 of gestation and found that this compound crosses the placenta and accumulates in the fetuses in a dose-related manner. At all dose levels, the highest concentrations were found in the maternal liver, followed by the fetal liver, whole fetus, and fetal brain. They noted no fetopathological effects, suggesting that hexachlorobenzene, in the concentrations fed to pregnant rats, caused no apparent adverse effect on fetal development.

In a four-generation test, groups of 10 male and 20 female Sprague-Dawley rats were treated with 10, 20, 40, 80, 160, 320, or 640 mg of hexachlorobenzene/kg from weaning. Suckling pups in the F1 generation were particularly sensitive, and many died prior to weaning when the mothers were fed concentrations of 320 or 640 mg/kg diet. No gross abnormalities were found (98).

Hexachlorobenzene decreased survival of chicks of Japanese quail treated with 20 mg/kg diet for 90 days (99). These results confirmed those of another study in which administration of 80 mg/kg diet for 90 days reduced egg production and hatchability (125, 129).

Iatropoulos et al. (101) studied the response of nursing rhesus monkeys to hexachlorobenzene given to their lactating mothers. The compound was given daily by gavage to three lactating monkeys in a dose of 64 mg/kg for 22, 38, or 60 days. Their infants were 140, 99, 81 days old, respectively, when necropsied.

The 99-day-old infant was listless for 24 hr before it died. Its mother was asymptomatic. The infant necropsy revealed lung edema and extensive engorgement of all CNS vessels with multiple extravasations. The 140-day-old infant showed lethargy, depression, and ataxia shortly before it died. Its mother was also asymptomatic. The major infant necropsy findings were lung edema with bronchopneumonia. The youngest of the infants was asymptomatic, although its body weight was below the normal range. The body weight of the 99-day-old infant was likewise below normal, whereas the body weight of the 140-day-old infant was within the normal range.

The detectable microscopic changes in all hexachlorobenzene infants were minor or negligible. They consisted mainly of a mild centrolobular hepatocellular hypertrophy, a diffuse fatty metamorphosis, a coarse vacuolation of slight degree within the proximal renal tubular cells, and a mild gliosis in the cerebrum.

The data indicate that infant rhesus monkeys are less susceptible to the toxic effects of hexachlorobenzene than their mothers.

4.5 Human Experience

4.5.1. Nonoccupational Exposure

An epidemic of hexachlorobenzene intoxication occurred during the years 1955 to 1959 in Turkey, where some of the wheat that had been treated with this fungicide, and that was intended for planting, was ground and used as flour for bread. Some 5000 people suffered a vesicular and bullous disease resembling porphyria cutana tarda which was frequently associated with hepatomegaly, porphyria, and death. The annual mortality ranged from 3 to 11 percent. According to Cabral et al. (102, 103), young children were predominantly affected.

In mild cases, lesions developed early, with the vesicles and bullae appearing particularly in skin areas exposed to sunlight. Peak instances occurred during the summer months. In children, the initial lesions resembled comedones and milia, while in adults, bullous lesions developed promptly. Frequently, the lesions became crusted and at times ulcerated.

There were many clinical variations which included hyperpigmentation, hypertrichosis, alopecia (in some instances permanent) or excessive growth of hair (occasionally the entire body showed excess hair growth), corneal opacities, deformation of the exposed parts, notably the digits, loss of body weight, wasting of skeletal muscles, and hepatomegaly. The urine frequently showed the characteristic port wine color associated with porphyria.

Eventually, and generally, following cessation of ingestion of contaminated bread, recovery occurred. The daily intake of hexachlorobenzene was estimated as having ranged from 0.05 to 0.2 g/person per day for months. [See Cam and Nigogosyan (104), Cetingil and Ozen (105), and DeMatteis et al. (106).]

Twenty years after the explosive incident in Turkey, some of the individuals were still suffering from the effects of hexachlorobenzene. Cripps et al. (107), who went on a site visit to eastern Turkey, examined 32 porphyric Turks, of an average age of 29.5 years. These people still suffered the following: hyperpigmentation (53 percent of the patients), hirsutism (41 percent), scarring of hands and face (50 percent), pinched facies and rhagades (22 percent), fragile skin on hands and face (12.5 percent), large liver (9 percent), ascites and jaundice (9.4 percent), recent red urine (9.4 percent), weakness and parathesias (43.7 percent), small hands, sclerodermoid thickening, shortening of distal phalanx and painless arthritis (44 percent), and enlarged thyroid (38 percent), compared to 5 percent in the general Turkish population of this area. In 10 patients the excretion of uroporphyrin in urine and feces was elevated. A porphyric's maternal milk was found to contain more than 0.7 ppm of hexachlorobenzene. The cutaneous effects were frequently precipitated by sunlight; differences in susceptibility could be expected and have been reported.

These studies and a report by Peters (108) demonstrate that symptoms of subacute or chronic hexachlorobenzene intoxication may involve cutaneous, hepatic, arthritic, visceral, urinary, and neurological effects that may persist for 20 and more years. Unfortunately, precise information is not available on the severity of the original intoxications (blood concentrations) of the individuals who showed symptoms after 20 years.

A therapeutic trial of EDTA administered orally (with initial intravenous administration in some patients) was performed in seven patients who, following treatment, remained symptomatic with porphyria. Four of the EDTA-treated patients became asymptomatic within 3 months, and all patients were completely asymptomatic within 1 year. After cessation of treatment one patient underwent relapse 10 months later, followed by abrupt recovery on resumption of EDTA therapy (108).

Pederson and Carlson (109), in seeking an explanation of the mode of action of hexachlorobenzene, suggest that this compound forms a complex with one or several porphyrin compounds in the heme biosynthetic pathway of the liver; the sizes of the porphyrin ring and the hexachlorobenzene molecule are very similar, and the flat nature of both, and the high polarizability of each, could lead to a considerable van der Waals interaction.

For concentrations of hexachlorobenzene in human tissue in various countries of the world, and in food and drink, refer to the 1979 IARC monograph (129).

4.5.2   Occupational Exposure

A 24-year-old farmworker who handled hexachlorobenzene and other pesticides was admitted to a Buenos Aires hospital with classic symptoms of porphyria and other complaints. His condition improved after removal from exposure and 1 year of symptomatic treatment combined with administration of EDTA, but even after 1 year, hexachlorobenzene was still detected in his blood, some enzyme activities were still at above normal levels, and there was still an excessive uroporphyrin excretion in the feces. Liver biopsy confirmed the persistence of porphyrin-type pigments in this organ.

The "toxic hazard" rating of hexachlorobenzene in the industrial environment is "1" as it applies to acute and chronic local and acute systemic effects, which means that the compound causes readily reversible changes that disappear after the end of exposure EPA (67). As a result of industrial exposure the chronic systemic effects are very limited. Ehrlicher (110) reported no serious illnesses or changes of liver function in the workers of a manufacturing plant who had been exposed to vapors of hexachlorobenzene over a 40-year period.

According to Sairtskii (111), a "hexachlorobenzene concentration of 0.1 mg/ liter could be assumed to represent the threshold toxicity value, and that 1/100 of that value may then represent the limit of permissible concentration of hexachlorobenzene in air for workers" (67).

Porphyria is one of the longest lasting effects of hexachlorobenzene intoxication in man. According to Hayes (112), porphyrins are probably the cause of more frequent and more serious photosensitization in man than all other materials combined. But photosensitization does not occur in all cases of increased porphyrin concentration in blood, urine, or feces. According to Hayes, in some people who suffer from latent acute intermittent porphyria, an acute attack may be brought on by alcohol or barbiturates. [Porphyria was apparently first produced in rabbits in 1895 by Stokvis (113) by sulfonethyl-methane; in 1941, Deichmann (114) produced it in rabbits, cats, rats, and mice by treatment with methyl, ethyl, and n-butyl methacrylate. Exposure of porphyric rats and mice to sunlight produced marked edema in areas of the body unprotected by hair.]

4.6 Environmental Aspects

Hexachlorobenzene has been detected in four river waters, eight drinking waters, and in effluent waters from seven chemical plants in various U.S. and European locations. In urban rainwater runoff in the United States, levels of 0 to 339 ng/liter have been detected (115). In most river water residues in an industrialized U.S. region, concentrations were less than 2 μ/liter, but in one sample, a concentration of 90 μg/liter was found (116, 129). Soil concentrations of 0.19 mg/kg have been found in the top 2 cm, and 0.11 mg/kg in the 2 to 4 cm layer of greenhouse soil 19 months after application. Levee and ditch soil in an industrialized U.S. region ranged from 0 to nearly 900 μg/kg (wet-weight basis) and 1677 μg/kg (dry-weight basis) (116, 129).

Aerial dispersion of hexachlorobenzene liberated at hexachlorobenzene manufacturing plants and as a by-product or waste from related chemical industries (carbon tetrachloride, perchloroethylene, trichloroethylene) is the major pathway for entry of this compound into the environment (67). Based on a model ecosystem, Metcalf et al (117) found that the labeled compound was readily transferred from water through several food chains, such as alga, snail, plankton, water flea, mosquito, and fish. The water phase, but not any of the organisms, contained an appreciable quantity of pentachlorophenol. Hydrolysis of polar products in water showed also a family of related chlorinated phenols.

Table 49.6 presents quantitative values for the ecological magnification and biodegradability for eight organochlorine pesticides in fish and snail. The

Table 49.6. Quantitative Values for Ecological Magnification (EM) and Biodegradability Index (BI) for Eight Organochlorine Pesticides in Fish and Snail[a]

Pesticide	H_2O Solubility (ppm)	Fish (Gambusia) EM	Fish (Gambusia) BI	Snail (Physa) EM	Snail (Physa) BI
Hexachlorobenzene	0.006	287	0.46	1,247	0.10
Aldrin					
as Aldrin	0.20	3,140	0.00014	44,600	0.0017
as Dieldrin		5,957	0.00013	11,149	0.00016
Dieldrin	0.25	2,700	0.0018	61,657	0.0009
Endrin	0.23	1,335	0.009	49,218	0.0124
Mirex	0.085	219	0.0145	1,165	0.006
Lindane	7.3	560	0.091	456	0.052
DDT	0.0012	84,545	0.015	34,545	0.044
DDE	0.0013	27,358	0.032	19,529	0.017
DDD or TDE	0.002	83,500	0.054	8,250	0.024

[a] Adapted by EPA (67) from information published by Metcalf (117).

information in this table was adapted by EPA (67) from the data published by Metcalf et al (117) and demonstrates that hexachlorobenzene was accumulated in the tissues of fish (gambusia) and snail (physa) to levels much greater than that in the water of the model systems.

4.7 Photodecomposition

Photodecomposition of hexachlorobenzene (as a crystalline material or on silica gel) exposed to sunlight was extremely slow; no photodecomposition product was identified in this photodecomposition experiment; diphenylamine, but not benzophenone, was used as a sensitizing agent. Photolysis in methanol or hexane with light of wavelengths greater than 260 or 220 nm, respectively, was rapid, and the anticipated products of reductive dechlorination (pentachloro-benzene and tetrachlorobenzene) were obtained in each case. In addition, an unexpected photochemical reaction between hexachlorobenzene and methanol resulted in the formation of small amounts of pentachlorobenzyl alcohol and traces of another photoproduct that was probably a terachlorodi-(hydroxymethyl)benzene (118).

4.8 Destruction and Disposal

Based on a pilot incineration study, hexachlorobenzene can be destroyed if the temperature is high enough and the transit time is long enough. For it to be destroyed, a temperature of 950°C was required. This resulted in a residue of less than 100 mg/kg. At 800°C, and a transit time of 2 sec, a residue of about 1200 mg/kg remained (119).

According to a report by Quinlivan and Ghassemi (120), the ultimate methods for the disposal of industrial wastes containing hexachlorobenzene include incineration (with or without by-product recovery), land disposal, resource recovery, discharge to municipal sewage treatment plants, and emission to the atmosphere. Currently (1977) most hexachlorobenzene waste handled by in-dustrial facilities is disposed of in two industrial landfills using a soil cover of 120 to 180 cm, with a polyethylene film placed at approximately mid-depth.

In France, hexachlorobenzene was successfully removed in four of five plants using the biologic waste-water treatment process (accumulation of the compound in activated sludge in aeration and decantation tanks). Hexachlorobenzene was removed from the effluent at the rate of 96 percent (121).

4.9 Cancer

In 1977 Cabral et al. (103) reported on the life-long feeding of this compound to Syrian golden hamsters. A dietary concentration of 50 ppm was fed to 36 males and 30 females, 100 ppm to 30 males and 30 females, and 200 ppm to

60 males and 60 females. The control group consisted of 40 males and 40 females.

The observations when the animals were at 80 weeks of age demonstrated a significant production of hepatomas and liver hemangioendotheliomas. The capacity of hexachlorobenzene to induce hepatomas, and their incidence, are as follows: low dose, males 27.2 and females 37.5 percent; medium dose, males 75 and females 42 percent; high dose, males 87 and females 81 percent. The latency period appeared to be dose-related. No metastases were found, and no hepatomas were noted in the controls.

The incidence of hemangioendotheliomas in the hamsters fed 200 ppm was 34 percent in males and 9 percent in females. Three of the hemangioendotheliomas occurring in the (200 ppm) males caused metastases. The investigators accepted these observations as evidence that hexachlorobenzene in the dosages fed is carcinogenic in the Syrian golden hamster.

In 1978 and 1979, Cabral et al. (103) reported on groups of Swiss mice fed for life a diet containing hexachlorobenzene, 50 ppm (30 males and 30 females), 100 ppm (30 males and 30 females), and 200 ppm (50 males and 50 females).

Hepatomas produced in the mice fed 200 ppm were 4.0 percent in the males and 29.7 percent in the females. In both males and females fed 100 ppm, the incidence of hepatomas was 10 percent. None of the hepatomas metastasized, and none was noted in the control mice (50 males and 50 females).

In the groups receiving 100, 200, and 300 mg/kg diet, the incidences of liver-cell tumors in the survivors (males and females) at the time the first liver-cell tumor was observed were 3/12 and 3/12, 7/29 and 14/26, and 1/3 and 1/10. The effective intake of hexachlorobenzene that induced liver-cell tumors was 12 to 24 mg/kg body weight per day (103, 100). The results of these studies support their observations in the Syrian golden hamster.

According to the IARC (129), there is sufficient evidence that hexachlorobenzene is carcinogenic in mice and hamsters. "In the absence of adequate data for humans, it is reasonable, for practical purposes, to regard this compound as if it presented carcinogenic risk to humans."

4.10 Mutagenicity

In a dominant lethal test conducted by Khera (122), male rats that received 20, 40, or 60 mg/kg body weight hexachlorobenzene orally for 10 days were mated sequentially with untreated females (one male × two females; 14 mating periods, each of 5 days' duration). There were no significant differences between the test and control groups with regard to the incidence of pregnancies, corpora lutea, live implants, or deciduomas, at any dose level or in any of the mating periods.

Also using the dominant lethal rat assay procedure, Simon et al. (123) treated groups of 10 male rats with oral doses of 70 or 221 mg of hexachlorobenzene/

kg body weight per day for 5 consecutive days. Ten additional rats received a single oral dose of 0.5 mg triethylenemelamine/kg. The test compounds were dissolved in olive oil. Each male was mated with two naive nulliparous females/ week for 14 weeks. The females were then sacrificed on the fourteenth day of gestation, and their uteri and ovaries were exposed and examined for the number of corpora lutea and total viable and nonviable implantations.

Hexachlorobenzene (and also Kepone, which was tested similarly) failed to induce any compound-related effects. The authors concluded that hexachlorobenzene (and Kepone) do not appear to be potent mutagens. Triethylenemelamine elicited dominant lethal mutations.

4.11   Immunotoxicity

The immunotoxic potential of hexachlorobenzene has been documented in experimental animals. Alteration of host resistance to infectious agents has been demonstrated in mice fed a diet containing 167 ppm hexachlorobenzene for 3 and 6 weeks and then infected with the malarial parasite *Plasmodium berghei*, resulting in manifested reductions in the mean survival time of 24 percent and 31 percent, respectively (124).

The pre- and postnatal exposure of rats to a diet containing 50 and 150 mg/ kg until 5 weeks of age resulted in a twofold and threefold increase, respectively, in the susceptibility to a lethal challenge with *Listeria monocytogenes* (125). Alteration of humoral immunity has also been reported. Dietary feeding of 167 ppm to mice for 6 weeks resulted in a twofold reduction in the number of spleen cells that produce specific antibodies during the primary antibody response to sheep erythrocyte antigen (126).

Alteration of cell-mediated immune responses has also been investigated. Vos et al. (125) reported that rats exposed pre- and postnatally to up to 150 mg/kg until 5 weeks of age did not show any alteration in skin transplant rejection times. However, in a chronic toxicity study, Silkworth (127) reported that dietary administration of 167 ppm hexachlorobenzene to mice for 37 weeks resulted in a 20 percent reduction in the graft-versus-host activity of isolated spleen cells. In the same study, an 80 percent reduction in the specific cell-mediated lymphotoxicity directed against cell surface alloantigen by sensitized spleen cells from mice exposed to hexachlorobenzene for 6 weeks was observed; this was associated with the point of time when the highest concentration of this compound was detected in the spleen. Lymphocyte blastogenesis induced by the nitrogens LPS, PHA, and ConA is not consistently influenced by hexachlorobenzene (125, 127); however, an increased rate of background DNA synthesis which increased with continued exposure was reported by Silkworth (127). Inhibition of the rosette-forming ability of alveolar macrophages has been reported in rats fed 250 ppm hexachlorobenzene for 10 weeks (128).

The immunotoxic potential of hexachlorobenzene may be related to the ability of the thymus and spleen to concentrate this compound to levels above the serum concentration and it has been suggested that the mechanism of action exists within the effector phase of the immune response (125, 127).

4.12 Hygienic Standards of Permissible Exposure

According to the 1978 Joint FAO/WHO Food Standards Programme Codex Committee on Pesticide Residues (95), the conditional acceptable daily intake of hexachlorobenzene is 0.0006 mg/kg body weight. The maximum residue limits of hexachlorobenzene in milled cereal products is 0.01 mg/kg, in raw cereals 0.05 mg/kg, in milk and milk products 0.5 mg/kg, and in eggs and in the carcass fat of cattle, goats, pigs, sheep, and poultry, 1 mg/kg (95). The ACGIH has not published a standard for this compound.

5 POLYCHLORINATED BIPHENYLS, Chlorinated Biphenyl, Chlorinated Diphenyl, Polychlorinated Polyphenyls, Arochlor, Kanechlor

5.1 Uses and Industrial Exposures

Commercial polychlorinated biphenyls (PCBs) are generally mixtures of many different chlorinated biphenyls, according to specific operational specifications. The approximate chlorine content of PCB mixtures is as follows: Arochlors, 1221 (21 percent), 1016 (42 percent), 1242 (42 percent), 1254 (54 percent), and 1260 (60 percent); Kanechlors, 300 (43 percent), 400 (48 percent), and 500 (53 percent) (130).

According to the *IARC Monographs on the Evaluation of the Carcinogenic Risk of Chemicals to Humans* (131), technical products vary in composition, in the degree of chlorination, and possibly according to batch. For example, Kanechlor 500 has an average content of 55.0 percent pentachlorobiphenyl, 26.5 percent tetrachlorobiphenyl, 12.8 percent hexachlorobiphenyl, and 5.0 percent trichlorobiphenyl (lot 360). Kanechlor 400 has an average content of 43.8 percent tetrachlorobiphenyl, 32.8 percent trichlorobiphenyl, 15.8 percent pentachlorobiphenyl, 4.6 percent hexachlorobiphenyl, and 3.0 percent dichlorobiphenyl (lot 471). Kanechlor 300 has an average content of 59.8 percent trichlorobiphenyl, 23.0 percent tetrachlorobiphenyl, 16.6 percent dichlorobiphenyl, and 0.6 percent pentachlorobiphenyl (lot 348).

The PCBs are used extensively in capacitors and transformers. They also found use in the manufacture of plasticizers, hydraulic fluids, lubricants, surface coatings, inks, sealants, adhesives, and pesticide extenders, and for the microencapsulation of dyes for carbonless duplicating paper.

Commercial manufacture of PCBs in the United States began approximately in 1929 by the Monsanto Company, the sole producer. According to an IARC report (131), it was estimated in 1970 that since the commercial introduction of PCBs, 454 million kg of these chemicals had been sold in North America. In 1971, the U.S. producer voluntarily agreed to limit the sale of PCBs to those applications that minimize their introduction into the environment. In 1975, approximately 12,000 people were occupationally exposed to PCBs, primarily in the production of capacitors (130).

5.2   Physical and Chemical Properties

The polychlorinated biphenyls are thermally stable, very resistant to degradation, and resistant to oxidation, acids, bases, and other chemical agents. PCBs are soluble in most of the common organic solvents and lipids, but only slightly soluble in water, glycerol, and glycols. See Tables 49.7 to 49.9 for physical and chemical properties.

Table 49.7. Properties of Pure Substances (131, 131a)

Compound	Melting Point (°C)	Boiling Point °C at 760 mm Hg	Boiling Point °C (mm Hg)
2-Chlorobiphenyl	54	267–268	154 (12)
3-Chlorobiphenyl	89	284–285	
2,2'-Dichlorobiphenyl	59, 61–62		
4,4'-Dichlorobiphenyl	148	315–319	
2,4,5'-Trichlorobiphenyl	78–79		
3,5,4'-Trichlorobiphenyl	88		
3,4,3',4'-Tetrachlorobiphenyl	172		230 (50)
2,4,2',4'-Tetrachlorobiphenyl	83		
2,4,5,3',4'-Pentachlorobiphenyl	179		195–220 (10)
3,4,5,3',4',5'-Hexachlorobiphenyl	198		
2,3,4,5,2',4',5'-Heptachlorobiphenyl			240–280 (20)
2,3,4,5,6,2',3',4',5',6'-Decachlorobiphenyl	310		

Table 49.8. Chemical and Physical Data on Arochlors (131)

Property	Arochlor 1221	Arochlor 1242	Arochlor 1248	Arochlor 1254
Appearance	Colorless, mobile oil[a]	Almost colorless, mobile oil[a]	Yellow-green-tinted, mobile oil[a]	Light yellow, viscous oil[a]
Density				
(At 20°C)	1.18–1.19[a]	1.38–1.39[b]	1.45–1.47[b]	—
(At 90°C)	—	—	—	1.47–1.49[b]
Refractive index (at 20°C)	1.617–1.618[a]	1.625–1.627[b]	1.6305–1.6325[b]	1.638–1.640[b]
Pour point (°C)	+1[a]	−18 max[b]	−6 max[b]	+12 max[b]

[a] From Hubbard (131a).
[b] From Monsanto Co. (131b).

Table 49.9. Chemical and Physical Data on Kanechlors (131)

	Kanechlor		
Property	300	400	500
Appearance	Colorless gummy oil	Colorless gummy oil	Colorless gummy oil
Specific gravity			
(At 15°C)	1.337–1.339	1.453–1.468	
(At 100°C)	1.310–1.322	1.376–1.389	1.460–1.475
Viscosity			
(At 75°C)	3.5–4.4	5.4–7.3	12–19
(At 98°C)	2.1–2.6	2.8–3.7	5.0–6.6
Refractive index (at 25°C)	1.6230–1.6260	1.6295–1.6325	1.6370–1.690
Pour point (°C)	(−19) to (−15)	(−8) to (−5)	8–12
Acid value KOH (mg/g)	0.005>	0.005>	0.005>
Distillation temperature range (°C at 760 mm Hg)	325–360	340–375	365–390

5.3   Effect in Animals

5.3.1   Acute Toxicity

LD_{50}s of orally administered Arochlor 1254 and 1260 in Sherman rats are 4 to 10 g/kg body weight (132), and of Arochlor 1242, 4.25 g/kg body weight in male Wistar rats (133).

Table 49.10 summarizes the oral and dermal toxicity of seven Arochlor mixtures in rats and rabbits as summarized by Fishbein (134). These materials are of a low order of acute toxicity. Although the oral toxicity in rats decreases with increased chlorination, there is no apparent trend of toxicity with chlorination in the data in rabbits (134). The data in the table are based on J. W. Cooke FDA *Status Report on the Chemistry and Toxicology of Polychlorinated Biphenyls or Arochlors* as of June 1, 1970 (135).

Vos (136) called attention to the fact that a proper evaluation of toxicity data can be hindered by the probability that PCB samples may differ owing to the presence of impurities. Three commercial PCB samples showed a resemblance in gas chromatographic and mass spectra determinations. However, in a comparative feeding test, it was found that 100 percent mortality with associated organ deterioration occurred only with two of the feeding samples; a third produced only occasional hydropericardial effects.

To more accurately distinguish between the effects of PCBs and their impurities, pure isomers with known positions of the chlorine atoms should be used for study. Because of the possible presence of polychlorinated dibenzo-furans or other more toxic impurities in crude PCB preparations, it is suggested that the induction of ALA synthetase could be used as a criterion in the evaluation of a no-effect level with PCBs (136).

5.3.2   Rat Feeding Studies

Oral Toxicity.   Drinker (137) reported on oral toxicity studies of chlorinated diphenyl containing 65 percent chlorine. When rats were fed at 0.5 g/day, the first animal died in 9 days; four more died by the end of the month. Liver lesions were found in these animals similar to those seen in vapor exposure (8).

Miller (138) fed a chlorinated diphenyl which contained 42 percent chlorine. Two doses of 0.05 ml (69 mg) each were administered 1 week apart to guinea pigs. Death occurred in 11 to 29 days. The livers showed metamorphosis and central atrophy. Rats received 25 daily doses of 0.1 ml (139 mg). These animals survived the administration and were killed at 10-day intervals from 30 to 90 days after the beginning of the feeding. They showed the typical liver injury seen in the other animals.

Fifty male Sherman rats were fed 500 ppm Arochlor 1254 in the diet for 6 months; five rats were killed after 0, 1, 2, 3, 4, 6, 8, and 10 months of

Table 49.10. Acute Toxicity of Arochlors (g/kg)[a]

Species Test	1221	1232	1242	1248	1260	1262	1268	4465	5442	5460	2565
Rats oral LD_{50}	3.98[b]	4.47[b]	8.65[b]	11.00[b]	10.00[c]	11.30[c]	10.90[c]	16.00[c]	10.60[c]	19.20[d]	6.31[d]
Rabbits, skin MLD	>2.00[b]	>1.26[b]	>0.79[b]	>7.94[b]	>1.26[c]	>1.26[c]	—	>2.00[c]	>1.26[c]	>7.94[d]	>2.00[d]
	<3.16[b]	<2.00[b]	<1.26[b]	<1.26[b]	<2.00[b]	<3.16[c]	<2.25[d]	<3.16[c]	<2.00[c]	—	<3.16

[a] Reproduced from Refs. 134 and 135 with permission of the author L. Fishbein, and the publisher, Annual Reviews, Inc., Palo Alto, Calif.

[b] Undiluted.

[c] Administered as 50 percent in corn oil.

[d] Administered as 33.3 percent corn oil.

termination of exposure to Arochlor 1254. In the rats killed 10 months after the end of exposure, adipose tissue contained about 1200 ppm and liver 22 ppm of PCBs. The compounds found were those containing mainly five to seven chlorine atoms (139).

The primary metabolites following oral administration of 2,2'-, 2,4-, 2,3-, 3,4-, 3,3'-, and 4,4'-dichlorobiphenyls to male rats were dichlorodihydroxybiphenyls. These were excreted via the urine and feces as such and not as conjugates.

In male Wistar rats 64 percent of a single oral dose of 25 mg 3,4,3',4'-tetrachlorobiphenyl (TCB) was excreted unchanged in the feces during the first 14 days after dosing, and a metabolite considered to be a 2- or 5-hydroxy-3,4,3'4'-TCB was also found in small quantities (3 percent) (140). Similarly, the metabolites of 2,4,3',4'-TCB were identified in the feces of rats as free 5- or as 3-hydroxy-2,4,3-,4'-TCB (141).

Low levels of PCBs in the diet or the intraperitoneal injection of PCBs increase the microsomal mixed function oxidase activity of the liver in rats (142, 143).

Oishi et al. (144) fed PCB (Kanechlor 500), 100 ppm, and polychlorinated dibenzofurans (PCDFs), 1 ppm and 10 ppm, to groups of 10 Sprague-Dawley rats for 4 weeks. Rats fed diets containing 1 or 10 ppm PCDF showed significant inhibition of food consumption and growth; less inhibition was noted in the rats fed PCB. In some of the rats fed PCDF 10 ppm, chloracne-like lesions became apparent on the ears after 3 weeks; at 4 weeks, six of the 10 rats were so affected. In rats fed 1 or 10 ppm PCDF hemoglobin, hematocrit and mean corpuscular volume were significantly decreased, and in rats fed 10 ppm the erythrocyte count was significantly lowered. Mean corpuscular hemoglobin concentrations were elevated in all treatment groups. Both PCB and PCDF increased serum cholesterol concentrations and cholinesterase activity and decreased triglyceride concentrations and leucine aminopeptidase activity. PCDF decreased serum glutamic pyruvic transaminase activity and testosterone concentrations in the testis, and increased serum glutamic–oxaloacetic transaminase activity.

5.3.3 Mouse Feeding Studies

In DDD mice, the various components of a single oral dose of Kanechlor 400 were equally absorbed and distributed, mainly in the skin, liver, and kidney. The tetrachlorobiphenyls, the major components of Kanechlor 400, were almost completely eliminated after 3 to 4 weeks; but the penta- and hexachlorobiphenyls, the minor components, were still retained after 9 to 10 weeks (145). In male Wistar rats, three days after a single oral dose of 25 mg ^{3}H-Kanechlor 400, radioactivity was distributed mainly in the liver, skin, and adipose tissue.

During the first 4 weeks, 70 percent of the radioactivity was excreted in the feces and 2 percent in the urine (146).

5.3.4 Cattle Feeding Studies

A cow was inadvertently fed silage containing PCBs. Within one month the PCB concentration in the milk fat rose to 10 ppm, then dropped to 4 ppm in the following 4 weeks, during which PCB-free silage was fed; during the following 2 months the concentration dropped to 2 ppm (147). In cows fed 200 mg/day Arochlor 1254 for 60 days, the average PCB concentration in milk fat was 60 mg/kg (measured between the fortieth and sixtieth days of feeding) (148).

5.3.5 Bird Feeding Studies

When Arochlor 1254 was fed to pheasants, either in a single dose of 50 mg or in 17 weekly doses of 12.5 or 50 mg, up to 82 percent was absorbed from the gastroenteric tract, and up to 50 mg/kg (wet weight) was found in their eggs (149).

Arochlor 1254 in a 1:9 dilution with corn oil was introduced via capsule into the esophagus of pheasants. The rate of egg production among hens receiving 50 mg/week dropped below the rate in the controls and the group receiving 12.5 mg/week. Hatchability was lowest in the 50-mg group. Behavioral differences were noted in the ability of the pheasants to avoid hand capture, the offspring of parents given PCB being less efficient. PCBs were rapidly absorbed, stored in the lipid fractions, and subsequently excreted slowly in feces and in eggs. Levels of 300 to 400 ppm in the brain were associated with death. Marked splenic atrophy was consistently found in birds that died following administration of lethal doses of PCBs; enlarged kidneys and livers were also observed in these birds (150).

Two-day old cockerels were given a diet containing 400 mg/kg Phenoclor DP 6, Clophen A60, or Arochlor 1260 for 60 days. Concentrations of 120 to 2900 mg PCBs/kg liver and 40 to 700 mg PCBs/kg brain were found between the 12th and 58th days (151). A maximum of 50 mg/kg Arochlor 1254 was found in eggs of white Leghorn hens fed for 12 weeks a diet containing 50 mg PCBs/kg (152).

Signs of PCB toxicity in 1-day-old chicks fed dietary levels of 30 and 40 ppm of PCBs (Arochlor 1248) for 2½ weeks were reversed or eliminated when a noncontaminated diet was fed for 3½ weeks. Other 1-day-old chicks fed diets containing 40 and 50 ppm PCBs for 5 weeks showed abnormal weight gain. The PCBs that had accumulated in adipose tissue and liver following PCB feeding were lost (excreted) when the chicks were fed a noncontaminated ration for 8 weeks. Calcium and magnesium concentrations in plasma remained unchanged when chicks were fed diets containing up to 150 ppm PCBs (153).

5.3.6 Vapor Exposure

Of 10 mice exposed for 6 hr/day on five consecutive days to air concentrations ranging from 0.003 to 0.005 mg/liter of Arochlor manufactured in 1930, one mouse became cyanotic, then dyspneic during the second exposure; the remaining nine mice showed no signs of intoxication (154).

Drinker et al. (137) exposed rats for 16 hr/day to a concentration which averaged 0.57 mg/m^3 of air. After 6 weeks, there was slight liver damage which advanced during the next 2 months. These authors concluded that it was difficult to determine the difference in toxicity of the chlorinated diphenyls and chlorinated naphthalenes, but that the chlorinated diphenyl was certainly capable of causing injury in very low concentrations and was probably more hazardous than the chlorinated naphthalenes. It should be noted that they were using a chlorinated diphenyl containing 65 percent chlorine. This is equivalent to a little more than seven chlorine atoms per molecule (8).

Treon et al. (155) exposed cats, guinea pigs, mice, rabbits, and rats to vapors of two different chlorinated diphenyl mixtures. One contained essentially three chlorine equivalents and had a chlorine concentration of 42 percent. The other contained approximately five chlorine equivalents and had a chlorine concentration of 54.3 percent (8).

Experiments with the 42 percent chlorine-containing mixture were conducted with a concentration of 8.6 μg/liter of air. The animals were exposed 7 hr/day over 24 days for a total of 17 exposures. This concentration was "approaching saturation." The guinea pigs showed poor growth, but there was no indication of a toxic effect in the other animals. Exposure to a concentration of 6.83 μg/liter, 7 hr/day for 84 exposures over 122 days had essentially no effect on the animals (8).

Exposing rats to the mixture containing 54.3 percent chlorine at a concentration of 5.4 μg/liter of air, 7 hr/day for 83 exposures over 121 days resulted in liver cell injuries and increased liver weights. At a concentration of 1.5 μg/liter for 150 exposures over 213 days, the rats showed distinct histological changes in the liver. These authors concluded that, because the saturation point was approached these exposures and heat was required to vaporize the material in order to obtain this concentration, these materials probably did not represent a significant vapor problem from cold operations (8, 155).

These observations are in general accord with the reports of Drinker (137). However, Drinker showed a slightly higher toxicity; he was using a chlorinated diphenyl with 65 percent chlorine (8).

Tombergs (156) exposed rats to the vapors of Pydroul A200 (a mixture of low chlorinated biphenyls) and found that the concentration of PCBs in the liver reached 35 µg/g in 15 min and 70 µg/g in 2 hr. After 30 min of exposure, fat contained 14 µg/g, and brain, 9 µg/g.

5.3.7   Skin Exposure

Systemic effects (general malaise, rough fur, loss of body weight, death with and without convulsions, coma) were noted in rats after five skin applications of 1 ml of a 20 percent mixture of an Arochlor with olive oil. To reduce ingestion by licking, the material was placed between and over the shoulder blades. Reduction of the daily dose to 0.2 ml of the mixture still caused malaise, but to a lesser degree. One of 10 rats died after 10 application and another after 18, at which time the experiment was discontinued. The material did not produce signs of local irritation (154).

Miller (157) applied the same chlorinated diphenyl which he used in the rat feeding study to the skin of guinea pigs, rats, and rabbits. Guinea pigs received 11 daily skin applications of 1/40 of 1 ml (34.5 mg) of undiluted chlorinated diphenyl. The animals died at various intervals up to 21 days after the first application. Histopathological examination of the liver showed fatty degeneration and central atrophy. The kidneys were essentially normal. There was a thickening of the epidermis at the site of application.

Rats were treated with 25 daily applications of the same dose, 1/40 ml (34.5 mg), of undiluted chlorinated diphenyl and survived the applications. The animals were killed at 10-day intervals, beginning with 30 days and ending at 90 days after the initial treatment. Histopathological examination of the liver showed only slight changes. The treated skin was much thickened and hair follicles were swollen.

Rabbits received skin applications at 2-day intervals. They were given 86 mg/day for the first seven applications and 172 mg for the last eight applications of undiluted chlorinated diphenyl. Death occurred between 17 and 98 days. Fatty degeneration and central atrophy of the liver were more striking than in the previously treated animals. The treated skin showed a thinning of the prickle cell layer and a relative thickening of the outer cornified layers (157) (8).

5.4   Absorption, Metabolism, Excretion

The polychlorinated biphenyls, depending on their chemical constitution, are more or less readily absorbed from the gastroenteric tract. The vapors are readily absorbed by way of the lungs. Depending on the composition of the material and period of skin contact, absorption may lead to systemic toxicity and mild local effects.

5.4.1 Beagle Dogs and Monkeys

Slocumb et al. (158) studied the tissue distribution, metabolism, and excretion of 4,4'-dichlorobiphenyl (4,4'-PCB) in beagles and cynomolgus monkeys (*M. Fascicularis*). Following a single intravenous dose of 4,4'-PCB-14C (0.6 mg/kg) excreta, blood, bile, and tissues were collected at time intervals ranging from 15 min to 28 days. Samples were oxidized to $^{14}CO_2$ and quantitated by scintillation counting. Selected tissues and fluids were extracted to separate the parent compound from metabolites.

Within 24 hr beagles excreted 50 percent of the dose as metabolites in the urine (7 percent) and feces (43 percent), with the remaining ^{14}C present largely in fat, muscle, and skin. Bile and blood collected at 24 hr consisted primarily of metabolites. By 5 days more than 90 percent was excreted.

Unlike the beagle, the monkey excreted less than 15 percent of the dose within 24 hr, with only 1 percent of this in feces. Blood and bile collected at 24 hr consisted of metabolites, 75 percent and 100 percent, respectively. The remaining 4,4'-PCB was localized as the parent compound primarily in fat (33 percent), and smaller amounts in skin and muscle. Metabolites were found in blood and bile. By 28 days, 59 percent of the dose had been excreted, primarily in the urine.

A comparison of the biologic half-life of PCBs in several species indicates that dog can eliminate highly chlorinated PCBs more rapidly than the mouse, rat, or monkey.

To further clarify the mechanism for this rapid elimination, Curley et al. (159) administered to beagles ($N = 3$) one intravenous dose (0.6 mg/kg) of ^{14}C-labeled 2,3,6,2',3',6'-hexa-(236-HCB), 2,4,5,2',4',5'-hexa-(245-HCB), or 2,4,-5,2',4',5'-hexabromobiphenyl (245-HBB). Daily excretion of ^{14}C in the urine and feces was quantitated by oxidation to $^{14}CO_2$ and/or direct scintillation counting.

The investigators found that 236-HCB was rapidly excreted with 70 percent recovered in 3 days. By day 15, 66 percent of 245-HCB was excreted. The 235-HCB was slowly excreted with only 14 percent recovered by day 20. Unlike 245-HCB, the hexabromobiphenyl (245-HBB) was excreted slowly with only 8 percent recovered by the 25th day. In all instances the major excretory route was via the feces. From these studies it is evident that the position of the chlorines governs the rate of elimination, probably by affecting the rate of metabolism. In addition, the specific halogen governs the degree of metabolism and subsequent excretion.

5.4.2 Rats

Curley et al. (160) studied the distribution and storage of PCBs in Sherman rats. One day after administration of a single oral dose of 1600 mg/kg of

Arochlor 1254, a mean PCB content of 1147 ppm was determined in the fat, 274 ppm in the kidney, 141 ppm in the liver, 138 ppm in the brain, 80 ppm in the muscle, 66 ppm in the lung, and 24 ppm in plasma. After feeding a dietary concentration of 1000 ppm of Arochlor 1254 for 98 days, the fat contained 11,278 ppm, and the remaining tissues contained less than 200 ppm. In the feeding studies, a steady buildup of Arochlors in tissues occurred.

Meyer et al. (161) studied the metabolic disposition of ^{14}C-biphenyl in adult male albino rats by liquid scintillation counting. The animals were given one oral dose of ^{14}C-biphenyl (100 mg/kg). Urine and feces were collected separately in 24-hr samples for 4 days.

The total excretion of radioactivity after 96 hr was 92.2 percent of the dose. Urinary excretion accounted for 84.8 percent and fecal excretion for 7.3 percent of the dose. Most of this radioactivity, 75.8 percent and 5.8 percent, respectively, was excreted within 24 hr. Only trace amounts of $^{14}CO_2$ were detected in the expired air and 0.6 percent of the dose was still present in the rats 96 hr after biphenyl administration.

Extraction and fractionation of the 24-hr urine samples showed that the largest fraction (nearly 30 percent of the dose) consisted of conjugated phenolic metabolites. Acidic metabolites accounted for a quarter of the dose.

In a second biphenyl study by Meyer and Scheline (162), adult male albino rats were given the test compound by stomach tube as a solution in soya oil and by intracecal injection in doses of 100 and 400 mg/kg. Urine and feces were collected as before.

Again the main route of excretion was via the urine, and again the major portion (22.3 percent) of the biphenyl metabolites was excreted in the first 24 hr. The total urinary recovery 96 hr after administration was 29.5 percent of the dose administered. The metabolites detected were conjugates of mono-, di-, and trihydroxy derivatives of biphenyl as well as the m- and p-methyl ethers of the catecholic compounds. The two main urinary metabolites were 4-hydroxy-biphenyl (7.7 percent) and 4,4'-dihydroxybiphenyl (11.4 percent).

The experiments also showed that biphenyl has to be hydroxylated and then conjugated before it appears in the bile of the rat. Thus 5.2 percent of the dose administered was found in the 24-hr bile as conjugates, mainly of 4-hydroxy-, 4,4'-dihydroxy-, and 3,4,4'-trihydroxybiphenyl. Fecal excretion of phenolic biphenyl derivatives was found to be of minor importance, but 4.7 percent of the dose was detected during the first 24 hr after dosing. The following previously undetected metabolites of biphenyl were found: 3,4'-dihydroxy-, 3,4,4'-trihydroxy-, 3,4'-dihydroxy-4-methoxy-, and 4,4'-dihydroxy-3-methoxy-biphenyl.

Human populations that have been exposed to polybrominated biphenyls (PBBs) are also likely to have been exposed to PCBs. Therefore, to assess potential hazard of simultaneous exposure to PCBs and PBBs McCormack et al. (163) decided to determine the distribution of PCBs (Arochlor 1254) and

PBBs (Firemaster BP6) in several tissues of lactating rats, to determine the concentrations of both materials in milk (see also Section on reproduction) and to study the effects of these agents on hepatic and extrahepatic arylhydrocarbon hydroxylase (AHH). They placed Sprague-Dawley rats on diets containing PBBs and/or PCBs from the eighth day of pregnancy to postpartum day 14.

Treatment with the highest dose of PBBs (200 ppm) retarded both dam and pup body weight gain. Stimulation of AHH was greater after treatment with PBBs alone than PCBs alone. Extrahepatic tissue concentrations of PCBs and PBBs were similar regardless whether these agents were administered together or alone. Liver and milk contained lower concentrations of PBBs after treatment with an equal mixture of PCBs than when PBBs were administered alone.

In the discussion, the authors point out that pretreatment of rats with PCBs or PBBs potentiates the toxicity of a variety of chemicals, including carbon tetrachloride, chloroform, and bromobenzene (164–166). Certain insecticides are more toxic when administered with PCBs (167), and since PCBs and PBBs may be additive in stimulating microsomal enzymes in liver and kidney, concomitant exposure to these compounds may increase the potential for subsequent interactions with drugs or environmental chemicals.

In addition, activity of AHH may indicate the capacity for metabolic transformation of some relatively inert polycyclic hydrocarbons to highly reactive electrophilic compounds (arene oxides). Certain arene oxides may be cytotoxic, mutagenic, or carcinogenic (168). Additive stimulation of AHH by PCBs and PBBs may result in increased formation of arene oxides from polycyclic hydrocarbons (169).

PCBs have been found to reduce the thiamine level in blood, liver, and the sciatic nerve of rats fed a diet containing 500 ppm for 48 days. DDT, at a dietary concentration of 583 ppm, had the same effect in blood, brain, and liver. Both compounds also showed decreased transketolase activity in erythrocytes and liver (170).

Consistent with the findings of other investigators, LaRocca and Carlson (171) found that Arochlors and other chlorinated hydrocarbons inhibit the activity of both Mg- and NaK-adenosine triphosphatases (ATPases) (172–176). Arochlor 1242 significantly inhibits in vivo ATPase activity in male rats. The studies by LaRocca and Carlson also demonstrated that different Arochlors at a concentration of 30 ppm significantly inhibit in vitro rat Mg-ATPase activity of brain and heart tissues. However, since Arochlors are mixtures of PCB isomers that are contaminated with polychlorinated terphenyls and naphthalenes, the active (inhibitory) compound(s) could not be identified.

LaRocca and Carlson demonstrated a strong correlation between PCB-induced inhibition of ATPase and the lipophilic properties of the PCB molecule. When the aqueous solubility of the PCB molecule is increased (as in the use of a solubilizing agent) the inhibitory activity is decreased. They consider the lipid portion of the ATPase enzyme to be a separate phase which strongly favors association with PCBs. Their data suggest that the mechanism of PCB-induced inhibition of ATPase may be the lipophilic partitioning of these compounds into the lipid fraction of the ATPase enzyme which produces an allosteric change resulting in decreased ATPase activity. The degree to which this occurs for each isomer would then determine its relative inhibitory activity (171).

5.4.3   Mice

The disposition of multiple oral doses of selected PCBs, 4-chloro-, 4,4'-dichloro-, 2,4,5,2',5'-pentachloro-, and 2,4,5,2',4',5'-hexachlorobiphenyl, were studied in mice by Morales et al. (177). The investigators found that the rate of metabolism and excretion decreased with increasing chlorination, but was most profoundly affected by the elimination of adjacent unsubstituted carbon atoms.

Accumulation occurred mainly in adipose tissue, skin, and muscle. The tissue distribution for a particular polychlorinated biphenyl was found to be similar regardless of the route of administration and the number of doses. The fecal excretion rates of pentachloro- or hexachlorobiphenyl following multiple oral doses could be predicted from the excretion rates obtained after a single intravenous dose.

Previous work from this laboratory indicated that the disposition and excretion of PCBs are influenced by their lipophilicity and rate of metabolism. Once in the bloodstream, PCBs are rapidly removed to liver and muscle from these depots and are redistributed to tissues of greater affinity and lower perfusion, primarily adipose tissue and skin. Since greater than 90 percent of the excreted PCBs are in the form of metabolites, the rate of excretion reflects the rate of metabolism. Metabolism of these compounds is influenced not only by the number of chlorine atoms on the biphenyl molecule, but by the position of the chlorine atoms as well.

5.4.4   Rabbits

Grant et al. (178) demonstrated transplacental transfer of PCBs in rabbits.

5.4.5   Birds

Bunyan and Page (179) studied the effect of the structure of PCBs on hepatic microsomal enzymes in quail. In a preliminary experiment 4,4'-dichlorophenyl was shown to be reapidly eliminated from the body even at 500 ppm in diet. Enzyme activity was affected to a variable extent depending on the duration and level of feeding. Birds were maintained on diets containing 50 ppm of chlorinated biphenyl for 20 days. The residues and effects observed generally increased with the degree of chlorination although individual variations appeared to depend on isomeric structure. The commercial mixtures, Arochlor

1016 and Arochlor 1242, were similar to the dichloro or less persistent tetrachlorobiphenyl isomers in causing minimal changes in enzyme activity. Arochlor 1254 and 2,4,6,4'-tetrachlorobiphenyl were both persistent and appeared to belong to a new class of microsomal enzyme inducer which produced a spectral shift in the observed cytochrome P-450 spectrum to 448 nm. Most of the other highly chlorinated isomers fed gave rise to large residues and generally to elevated enzyme activities. 3,4,5,3',4',5'-Hexachlorobiphenyl appeared to be particularly toxic.

According to these authors, the pattern that emerged from these studies suggests that in quail, absorption, metabolism, and excretion generally become more difficult as the degree of chlorination of the biphenyl nucleus increases, but there is considerable deviation from this rule. The effect of individual isomers on hepatic enzymes appears to be related to the degree of chlorination and to the ease of metabolism as estimated by the heart residue levels at least up to the hexachlorinated level, although apparently closely related isomers can give rise to remarkably different residues and enzyme levels. The hexachlorobiphenyls are generally the most biologically active of the range. The Arochlor mixtures can be fitted into the general pattern. Arochlor 1242 and particularly Arochlor 1016 act rather like the di- or the less persistent tetrachlorobiphenyls at this intake level, and since this is unlikely to be exceeded in practice they probably do not present a serious problem of persistence.

Arochlor 1254 behaves more like the persistent tetra- and hexachlorobiphenyls which give rise to significant residue effects. 3,4,5,3',4',5'-Hexachlorobiphenyl proved to be exceptionally persistent and undoubtedly toxic.

These observations underline the need to consider the properties of each chlorinated biphenyl isomer individually (179).

5.5 Reproduction

5.5.1 Rats

Arochlors 1242, 1254, and 1260 were fed by Burke and Fitzhugh (180) to rats at concentrations of 1, 10, and 100 ppm. No deleterious effects on reproduction were noted in rats fed concentrations of 1 and 10 ppm. At the feeding level of 100 ppm Arochlor 1242 reduced the mating indexes in the second generation; Arochlor 1254 reduced the number of pups born and those that survived of the second and third litters; Arochlor 1260 caused an increase in the number of stillborn.

Low mating indices were also reported by Keplinger et al. (181) in rats fed Arochlors 1242 and 1254 at 100 ppm. In addition, a reduced rate of survival was noted in rats fed Arochlor 1242. No deleterious effects on reproduction were noted in rats fed Arochlor 1242 or 1254 at 1 or 10 ppm, nor in rats fed

Arochlor 1260 at 1, 10, and 100 ppm. Gellert and Wilson (182) reported an increase in testicular weight in Sprague-Dawley rats fed Arochlor 1221.

According to Tombergs (183), since PCBs, depending on tissue concentrations, are capable of stimulating the microsomal enzyme system of the rat which metabolizes adrenal hormones, an effect on reproductive and adrenal hormones can be expected.

Sherman rats were given doses of 10 or 50 mg/kg per day of Arochlor 1254 on days 7 to 15 of pregnancy; the average PCB concentrations in fetuses taken by Caesarean section on day 20 of pregnancy were 0.63 and 1.38 mg/kg, respectively, compared with less than 0.12 mg/kg in control rats (160).

McCormack et al. (163) reported that in pregnant Sprague-Dawley rats fed concomitantly PCBs and polybrominated biphenyls (PBBs) until the pups were 14 days old, the concentrations of PCBs and PBBs in the milk were highest at parturition and decreased with time until 13 days. The concentrations of PCBs were higher (292 to 31 μg/ml) than the concentrations of PBBs (180 to 50 μg/ml) in milk from animals fed 50 ppm of each at parturition when the concentrations were highest and 13 days following birth.

5.5.2 Mice

A reproductive capacity study was conducted with NMRI mice which were nursed by mothers given 50 mg/kg of PCB or DDT once a week for 4 weeks beginning on the day of parturition. Kihlstrom et al. (184) determined the reproductive capacity of the offspring according to the frequency of implanted ova in the uterus.

When both the male and female of a mating pair had been nursed by PCB- or DDT-treated mothers, the frequency of implantation decreased from a control level of 94 percent to 75 percent for PCB and 79 percent for DDT. No significant decreases were found when only one of the animals in a mating pair had been nursed by a PCB- or DDT-treated mother.

The authors suggest that PCB and DDT, even in low doses, may be able to disturb the normal sexual development in mice by increasing the catabolism of steroids if the young ingest milk containing PCB or DDT.

5.5.3 Birds

Chickens fed 100 ppm of Arochlors 1242 and 1254 responded with reduced egg production; also the hatchability of eggs was lowered at feeding levels of 10 and 100 ppm of these materials. No such effects were noted with Arochlor 1260 at 100 ppm (180). Reduced hatchability of eggs at a feeding level of 10 and 20 ppm Arochlor 1254 was also reported by Scott et al. (185).

Female quails (39 days old) kept in the dark and fed Arochlor 1242 for 2 months did not lay eggs and consequently did not mobilize vitamin A for deposition in the egg yolk. Liver vitamin A content was found to be reduced 50 percent. Egg production was restored in the PCB-fed quails by addition of calcium to the diet. Bitman et al. (186), working with rats, also reported a reduction in the concentration and content of vitamin A in the liver following treatment with Arochlor 1242.

Peakall et al. (187) studied the breeding of six pairs of ring doves fed Arochlor 1254 at 10 ppm over two generations. A low hatching rate of the second generation was due to the marked embyonic toxicity of the material. Eggshell thinning was not noted.

Cytogenetic studies of dove embryos at 3 to 6 days of incubation were conducted on PCB-fed doves, on controls, and on one embryo irradiated with 155 X-rays which served as a positive control. One chromosome rearrangement occurred in a PCB-treated embryo. Thirteen of 17 PCB embryos had aberration rates exceeding the mean control rate and four PCB embryos exceeded the highest control rate (frequencies of 2.4, 5.6, 3.1, and 9.4 percent). The results indicate a possible chromosome breaking action of a toxic dose of PCBs.

5.6 Human Experience

5.6.1 Intoxications

In 1968, an outbreak of poisoning occurred in Yusho, Japan, involving some 1000 people of all ages (15,000 victims were referred to by Umeda) (188), who ingested for several months rice bran oil that had been contaminated with PCBs to the extent of 1500 to 2000 ppm. After a latent period of 5 to 6 months, signs and symptoms of intoxication in these people included nausea, lethargy, chloracne, brown pigmentation of skin areas and nails, subcutaneous edema of the face, distinctive hair follicles, a cheese-like discharge from eyes, swelling of eyelids and transient visual disturbances, gastroenteric distress, and jaundice. Infants born to poisoned mothers had decreased birth weight and showed skin discoloration due to placental passage of PCBs. Two stillbirths of PCB-exposed women were reported (130). For more extensive information refer to the report by Fishbein (134).

Since the contaminated rice oil was considerably more toxic than anticipated from the concentrations of PCBs found in the oil, it was considered that the commercial oil contained impurities in the form of highly toxic polychlorinated dibenzofurans. This was later substantiated.

Production of PCBs was started in Japan in 1954; it is now prohibited.

5.6.2 Occupational Exposure

In 1963, Irish (8) reported as follows in the second edition of *Industrial Hygiene and Toxicology*, Vol. 2:

It is difficult to determine the role played by the chlorinated diphenyls as, in most instances, they were mixed with chlorinated naphthalene. Jones and Alden [189] reported on human cases working with chlorinated diphenyls. They observed a typical acneform dermatitis among workers handling this material. As these workers were manufacturing the material from crude coal tar distillates, these authors were inclined to feel that other chlorinated aromatic materials present which were considered to be more unstable might be the more important cause of the dermatological response. The workers did not get a typical acneform dermatitis from patch tests. This is not surprising as a number of authors recognize that it is necessary to have a long-term exposure to this material in order to get this typical response. It is quite possible that high chlorinated aromatic materials from the crude coal tar distillates may also cause acneform dermatitis. It is quite probable, however, that the dermatitis can also be observed following exposure to the highly chlorinated diphenyls [8].

In industry, exposure to PCBs is primarily dermal, in contrast to the oral intake that occurred in Japan. Following both dermal and oral exposures to low doses, effects are delayed for months. This was borne out in Japan, and has been the experience with chloracne-like skin effects reported occasionally in industrial workers. However, from the scarcity of reports, it appears that industrial exposure in the United States to PCBs has not been a specific problem. The older literature referred to several fatal intoxications following overexposure to materials which apparently contained PCBs and/or naphthalenes. These cases revealed severe liver pathology.

Kanechlor 300 has been used in the manufacture of noncarbon copy paper, particularly in Japan, but this paper has also been found in Great Britain and the United States. Since PCBs adhere to the fingers when the paper is handled, or to foods when paper cartons or paper products containing PCBs are used for packing, the discontinuance of PCBs for this purpose has been urged.

5.7 Environmental Factors

Polychlorinated biphenyls have entered the environment in the same manner as some of the organochlorine pesticides: from manufacturing plants, during the destruction of manufactured articles containing PCBs in municipal and industrial waste-disposal burners, through the gradual wear and weathering of PCB-containing products, and through leaching from land-fill dumps.

Transfer of PCBs into the general environment has occurred through volatilization, aerial transport of particulates, fallout as dust or in rain and snow, and through the downward movement into groundwaters and rivers (leaching–diffusion) of PCBs and of PCBs adhering to particulates from soils.

From these sources, PCBs have entered the food chain. The recycling in lake bottoms starts by a variety of bottom-feeding organisms (benthic invertebrates) until eventually the PCBs enter fishes, man, birds, and other animals.

With the greatly reduced use of PCBs pollution will gradually decline. Concentrations of PCBs in the North Atlantic surface waters have dropped fortyfold since 1972. In 1974 Harvey et al. (190) reported a tenfold decrease of PCBs in the atmosphere over the Northwest Atlantic since 1973, and a fivefold decrease since 1972 in the PCB content of mixed plankton collected along 09°N latitude in September 1973. These data indicate that about 2×10^4 tons of PCBs were "lost" from the upper 200 m of water in less than 1 year. The investigators discussed the possible fate of the "lost" PCBs.

Air concentrations of 2.1 to 9.4 ng/m^3 were reported in 1974 in Rhode Island. The chromatograms identified peaks of Arochlor 1242 which contains mainly tri-, tetra-, and pentachlorobiphenyls (191). Considerably lower concentrations were recorded in Bermuda and Hawaii.

5.7.1 Soil, Sediment, and Water

No data on PCB residues in various soils appear to be available (131). Sediments collected in 1971 from Lake Erie, in 1970 and 1974 from Lake St. Clair, and in 1974 from the Detroit River showed that the concentrations of PCB residues were about three times higher than the total organochlorine insecticide residues. Pesticides residues decreased between 1970 and 1974, coinciding with the restrictions placed on the use of these compounds (192).

Sediment samples collected along the Severn Estuary in the United Kingdom in 1978 contained 38.8 to 1120 ppb of total biphenyl. Natural biphenyl ranged from 8.79 to 144 ppb. Total naphthalene ranged from 177 to 1143 ppb; natural naphthalene ranged from 163 to 1121 ppb. Arochlor 1260 ranged in concentration from 71 to 2322 ppb. In all instances, the highest concentrations were found in the coal dust layer of an Arlingham sample. No significant DDT levels were observed (193).

In 1972, U.S. river waters were reported to contain PCB concentrations ranging from less than 0.01 to 0.45 μg/liter, and North Atlantic surface seawaters from less than 1 to 35 ng/liter (194, 195). The peaks resembled chromatograms of Arochlors 1254 and 1260. Also in 1972, concentrations of 10 to 100 ng/liter were reported for drinking water in Japan (131).

5.7.2 Animals

Zitko (196) reported concentrations of 0.02 to 1 mg polychlorinated biphenyls/kg in eel, salmon, herring, mackerel, mussel, cod, hake, and plaice, and traces in ocean perch, taken from the St. John River system, New Brunswick, or from the Nova Scotia bank. Similar levels (0.36 to 1.54 mg/kg) were found in tuna from the Atlantic Coast of North America (197); much higher levels (up to 7 mg/kg Arochlor 1254) have been found in shrimp and crab caught in accidentally polluted bays, for example, Escambia Bay, Florida (198), and in fish from inland lakes: for example, 26 mg/kg were found in a 12-year-old trout caught in Cayuga Lake in Ithaca, New York (199), in which case the gas chromatogram was similar to that of Arochlor 1254. In 1972, Stalling and Mayer (200) also reported high levels of polychlorinated biphenyls (more than 20 mg/kg) in inland lake and river fish in the United States (131).

Residue levels of PCBs and DDT in fish, mollusk, crustacean, and echinoderm tissues collected in the vicinity of San Diego and Laguna Beach, California sewer outfalls contained from essentially zero to 1.4 ppm PCBs and from undetectable to 0.83 ppm DDT. Values for fish livers averaged 0.3 ppm, and for mollusk, and echinoderm tissues 0.6 ppm DDT (201). The author suggests that Kellet's whelk would make a suitable indicator organism for the determination of water pollution. The whelk from Orange County, California contained 0.077 ppm total DDT and 0.23 ppm PCBs.

Since 1971, when PCBs were eliminated from open use, the levels of PCBs in fish have declined only slightly, as reported by EPA in 1979 (202).

Gruger et al. (203), working with cohorts of juvenile coho salmon, reported that for the first time a pentachlorobiphenyl was found to cause enhanced activity of the hepatic aryl hydrocarbon hydroxylase system in treated fish.

Numerous reports have appeared on PCBs in birds of prey in Canada, Denmark, Finland, Sweden, the United Kingdom, and the United States. Concentrations ranged from traces to an upper limit of 190 mg/kg, and in eggs up to 44 mg/kg.

Seals living off the coast of Sweden showed PCB residues up to 310 mg/kg fat (131).

Concentrations ranging from 27 to 564 mg PCB/kg were found in the blubber of harbor seals in the North German Wattensee. Even young seals contained up to the maximum concentrations. Concentrations in the brain ranged from 0.25 to 2.96 mg/kg; in the liver, 0.06 to 2.7 mg/kg; and in the kidneys, 0.18 to 1.22 mg/kg. There was no clear evidence that any of the compounds in the concentrations found in the tissues of seals including heavy metals, PCBs, and organochlorine pesticides had a negative effect on the health of these animals (204).

Clausen and Berg (205) give the following concentrations of certain chemicals in the adipose tissue in many species of birds collected in the Arctic (west coast of Greenland): PCBs, from 2.00 (eider duck) to 37.1 ppm (raven); p,p'-DDE, from 0.8 (eider duck) to 13.9 ppm (raven); p,p'-DDT and p,p'-DDD were not found, and lindane was present in trace amounts in the body fat of one third of the birds.

The body fat of mammals shot in this area revealed the following chemicals: PCBs, from 0.9 (ringed seal) to 21.0 ppm (polar bear); lindane, 0.003 (hooded seal) to 0.053 ppm (bearded seal); heptachlor, 0.001 (ringed seal) to 0.039 ppm (bearded seal); aldrin, 0.028 (hooded seal) to 3.06 ppm (polar bear); heptachlor epoxide, 0.026 (ringed seal) to 0.49 ppm (polar bear); and p,p'-DDE, 0.14 (arctic fox) to 1.25 ppm (polar bear). In the adipose tissues of six female Greenlanders PCBs ranged from 0.25 to 5.58 ppm and p,p'-DDE 0.04 to 0.52 ppm. The remaining pesticides were found only in trace amounts.

5.7.3 Foods

Very little information is available on the quantities of PCBs ingested by the U.S. population.

On the basis of observations made in Japan at the 1968 PCB Kanechlor 400 intoxication, the minimum oral intake toxic to humans was estimated to be 200 µg/kg body weight per day (131, 206). Signs and symptoms of poisoning are expected to become evident after a minimum total oral intake of 0.5 g (207). In 1977, the average intake of PCBs by a Japanese adult was 72.4 µg/day.

According to the Panel on Hazardous Trace Substances (1972) (207), the daily U.S. intake of milk by breast-fed infants is 150 g/kg body weight, and the amount of PCBs ingested with this milk is on the order of 9 µg/kg body weight per day. This figure is slightly above the maximum daily intake that would occur in adults eating 150 g/week of fish containing 15 mg/kg PCBs (e.g., coho salmon from Lake Michigan) or 80 g of fish/day in Japan, where values of 5 mg/kg may be present (131).

In 1972 the U.S. Department of Agriculture (208) reported on selected foods. Fish contained PCBs from traces to 35.3 mg/kg (average 1.87 mg/kg); cheese, traces to 1 mg/kg (average 0.25 mg/kg); milk, traces to 27.8 mg/kg (average 2.27 mg/kg); and eggs, traces to 3.74 mg/kg (average 0.55 mg/kg). Of the total number of commodity samples examined, only 19 percent were positive with respect to PCB contamination, reflecting an average concentration of 1.14 mg/kg sample of food.

When grayboard cartons and boxes containing PCBs were used for goods, 22 of a total of 720 composite food samples of the FDA's market basket survey were found to be polluted. PCBs in these food samples ranged from traces to 0.36 mg/kg (131).

5.7.4 Human Tissues

In a 1972 Human Monitoring Survey, Yobs (209) found measurable amounts of PCBs (1 to 2 mg/kg) in 26 percent, and less than 2 mg/kg in 5 percent of 637 samples of human adipose tissue taken from the general population of 18 states in the United States and of Colombia, C.A. Price and Welch (210) estimated that 36 percent of the general population had levels in adipose tissues of 1 to 2 mg/kg (wet weight) PCBs, ranging from penta- to decachlorobiphenyls. Similar concentrations have been detected in populations dying in the Oslo area of

Norway (1.6 mg/kg fat) (211) and in Japan (0.5 and 0.8 mg/kg fat) (212). Seventy to 79-year-old Japanese men had maximum concentrations of 5.1 mg/kg fat, and women of similar age, 2.4 mg/kg fat (131, 213).

PCBs have been demonstrated in human milk in the United States (see also section on foods) and in the milk of women of other countries. In the Federal Republic of Germany concentrations were reported of 0.1 mg/kg (whole milk) and 3.5 mg/kg (fat) (183). In Japan the reported concentrations were lower, 0.03 mg/kg for whole milk and 1.1 mg/kg for milk fat (214). PCBs could not be detected in human milk in Texas, U.S., or New Guinea (131, 215). In Stockholm, Sweden, in spite of restrictions imposed on the use of PCBs since June 1972, the average concentration of PCB in milk has not decreased from 1972 to 1977 (0.024 to 0.03 mg/kg milk (216).

Blood, milk, and adipose tissue of mothers and fetuses and infants were collected at delivery, stillbirth, or lactatio by Masuda et al. (217) at hospitals in Fukuoka and Kurume, Japan and analyzed for PCBs. Their results published in 1978 included the following: maternal milk contained 13 (1 to 36) ppb PCBs, maternal milk fat 350 (30 to 870) ppb, and maternal blood 1.4 (0.5 to 3.4) ppb. Infant blood contained 2.5 (0.7 to 6.2) ppb. In fetuses, adipose tissue contained 470 (270 to 960) ppb in fat only, and in total adipose tissue 150 (1 to 380) ppb. Liver contained 7.3 (0.3 to 21) ppb and adrenal 26 (6 to 80) ppb.

Their study demonstrated a higher PCB level in the blood of infants but a lower level in the cord blood than that in maternal blood. This suggests that the transfer of PCBs via the milk is much more significant than placental transfer. Their previous study using mice supports this interpretation; furthermore, fetal tissues were demonstrated to be much lower in PCBs than the corresponding adult tissues. This also correlates with their results from animal experiments that showed that PCB concentrations were much higher in dams than in fetuses.

5.8 Cancer

Chronic feeding studies for detecting carcinogenicity of PCBs were conducted with mice and rats. Volume 20 (1979) of the *IARC Monographs on the Evaluation of the Carcinogenic Risk of Chemicals to Humans* (131) summarizes these studies as follows: groups of 12 6-week old male dd mice were administered Kanechlor 300, 400, or 500 in the diet at concentrations of 100, 250, or 500 ppm for 32 weeks. Six control mice received the basal diet. Gross examination after 32 weeks of treatment showed that 7/12 mice given 500 ppm Kanechlor 500 developed liver nodules, and histopathological examination showed the presence of hepatocellular carcinomas in 5/12 mice. No mestastases or tumors in other organs were seen. Amyloid degeneration of the liver was seen in the other groups, especially in those at the 100 ppm dosage level. No changes occurred in the six controls (218).

Two groups of 50 male BALB cj mice were fed 300 ppm Arochlor 1254 (1) in the diet continuously for 11 months or (2) for 6 months followed by a control diet for 5 months. Two additional groups of 50 mice were fed the basal diet. At the end of 11 months 1/24 surviving mice treated for 6 months with Arochlor 1254 had a hepatoma; of the mice surviving the continuous 11-month treatment, 9/22 developed hepatomas. In addition, adenofibrosis was observed in all 22 mice fed Arochlor 1254 continuously for 11 months (219).

In groups of 10 male and 10 female Sherman Rats fed 20, 100, 500, or 1000 ppm Arochlor 1260 or 20, 100, or 500 ppm Arochlor 1254 for 8 months, several rats died before 6 months at the two high-dose levels. Lesions described as adenofibrosis of the liver occurred in 2/10 males fed 1000 ppm Arochlor 1260 and in 1/10, 1/10, and 4/10 females fed 100, 500, and 1000 ppm Arochlor 1260. A higher incidence of this lesion occurred in rats fed Arochlor 1254, the incidences in males and females being 10/10 and 9/10 at 500 ppm, and 1/10 and 7/10 of 100 ppm, respectively (220).

A group of 10 male and 10 female Donryu rats was fed a diet containing 38.5 to 616 ppm Kanechlor 400 for 400 days. The initial concentration of 38.5 ppm was increased by factors of 2 in varied steps during the first 125 days until 616 ppm was reached. After 56 days at this level the concentration was reduced to 462 ppm for the remainder of the study except during two 28-day rest periods. A control group of five males and five females received a basal diet. In 6/10 treated females, multiple adenomatous nodules of the liver were observed; such lesions did not occur in male rats nor in the controls. No hepatocellular carcinomas were seen (221).

Groups of 30 male Wistar rats were fed for 52 weeks on diets containing 100, 500, or 1000 ppm Kanechlor 300, 400, or 500. One group received a control diet. Each Kanechlor, when fed at 1000 ppm in the diet, produced cholangiofibrosis in some of the rats, but lower dosage levels were ineffective. Hepatic nodular hyperplasia was found with all compounds, the incidence increasing with the degree of chlorination and with concentration in the diet (222).

In a more recent investigation Nagasaki et al. (223) concluded that technical PCB (Kanechlor 500) fed at 500 ppm for 32 weeks led to the production of typical nodular hyperplasia and well differentiated hepatocellular carcinoma; however, in female mice, only typical nodular hyperplasia was present. In mice fed 250 and 500 ppm of α-benzenehexachloride the incidence of neoplastic liver changes was also higher in male than in female mice.

The conclusions are as follows: according to Volume 18 of the IARC Monographs, "A slight increase in the incidence of cancer, particularly melanoma of the skin, has been reported in a small group of men exposed occupationally to Arochlor 1254, a mixture of polychlorinated biphenyls" (236). Volume 20 states that Kanechlor 500 and Arochlor 1254 are carcinogenic in mice, producing benign and malignant liver cell tumors following oral administration, the only route tested. In rats, Kanechlor 300, 400, and 500 induced multiple hyperplastic liver nodules following oral administration. Only a limited number of PCBs have been tested (131).

5.9 Teratogenic and Mutagenic Effects

In rabbits, Arochlor 1254 has caused fetotoxic effects, maternal death, abortion, and stillbirth, at oral doses of 12.5 mg/kg (224). McLaughlin et al. (225) reported that when Arochlor 1242 is injected into the yolk sac of chicken (10 mg/egg) deformities and retardation of growth of the chicks results. Injection of 25 mg into the chicken yolk sac resulted in "no hatch." The same material, when injected by Carlson and Duby (226) into chicken eggs, markedly inhibited hatchability and reduced the rate of development and growth in the chicks that did hatch. Arochlors 1254 and 1260 were less fetotoxic.

Arochlor 1242 in sufficiently high doses decreased the number of normally deriving spermatogonial cells in Osborne-Mendel rats. The material did not produce spermatogonia or chromosomal abnormalities of the bone marrow in rats treated with an oral dose of 5000 mg/kg (227).

Dougherty et al. (228), employing negative chemical ionization spectrometry, established the presence of DDE, tri-, tetra-, and pentachlorophenol, hexachlorobenzene, and PCBs in human seminal plasma. They compared and considered these data related to the apparent decrease in sperm density distribution of 132 college students. They presented data showing that there has been an apparent decline in male fertility in the general U.S. population since 1950.

Keplinger (229), in a dominant lethal assay, reported no evidence of mutagenic activity of Arochlors 1242, 1254, and 1260. Methyl methanesulfonate exerted a dominant lethal effect during the first and second weeks following treatment (see also section on reproduction).

5.10 Immunosuppression

In 1972, Vos and de Roij (230) reported immunosuppresive activity of PCB preparations (Arochlor 1260 at 10 and 50 ppm) on both the humoral and cellular response immune system in albino guinea pigs. Absence of effects on weight and histomorphology of the thymus and adrenal glands of the animals indicated that not stress, but PCB itself was responsible for the immunosuppressive activity.

Studies in 5731/6 mice revealed that spleen cells from animals fed PCB 1016 for 24 weeks demonstrated a 56 percent enhancement of the mixed lymphocyte culture response to H2 alloantigens (Balb/c, BDF_1), whereas spleen cells from hexachlorobenzene-treated mice demonstrated a 46 percent depression. Spleen cells from mice fed PCB for 13 to 41 weeks also demonstrated a time dependent enhancement of up to 64 percent (above control values) of mitogen responsiveness, whereas spleen cells from mice fed hexachlorobiphenyl demonstrated a transient suppression of up to 71 percent followed by recovery in 41 weeks (231).

Thomas and Hinsdill (232) studied the immune response in rhesus monkeys and antlered albino mice fed Arochlor 1248. Monkeys fed a chow containing

2.5 or 5 ppm for 6 months developed chloracne, alopecia, and facial edema. After 11 months when they were immunized with sheep red blood cells (SRBC) and tetanus toxoid (TT), those fed 5 ppm responded with a significantly lower anti-SRRC antibody titers than controls at only two intervals following primary immunization. Antibody response to TT was not measurably affected.

The investigators interpreted their results to demonstrate that sustained exposure to low levels of PCB could have a slight to modest immunosuppressive effect, which might be important depending on the general health of an individual. Their mouse studies, following feeding of doses that produced overt signs of toxicity, revealed that such animals appeared to have an impaired ability to withstand challenge by pathogens and have an increased sensitivity to endotoxins.

Individual Arochlors can consist of up to 100 different chlorobiphenyls. In view of the toxicity of various PCBs, Luster et al. (233) initiated an extensive program to develop radioimmunoassays for analyzing PCBs. Their 1979 report suggests that an immunoassay approach for analysis of PCB mixtures is highly feasible, although the assay depends on the generation of antisera specific for at least several PCB isomers, and requires multiple radioimmunoassays to be performed simultaneously, which is not always desirable. Antisera presently available can readily distinguish Arochlor 1242 from Arochlors 1248 and 1254. However, distinction between Arochlors 1248 and 1254 remains questionable.

5.11 Photochemical Degradation

Photolysis of chlorobiphenyls reveals a number of degradative reactions that occur on irradiation in sunlight and under a number of laboratory conditions (234). In hexane, the most prominent reaction was the progressive reductive dechlorination of these compounds, as well as formation of polymeric materials. Irradiation of Arochlor 1254 in hydroxylic solvents at pH 9 yielded compounds corresponding to the addition of water to the PCB molecules and a more polar carboxylic acid fraction. The same sample irradiated as a thin film did not yield the water-addition products, but mass spectrometry revealed the presence of new hydroxylated species.

Data for a number of different irradiation experiments with 2,2',5,5'-tetrachlorobiphenyl indicated that dechlorination, formation of polymers and carboxylic products, as well as hydroxylation do occur (234).

5.12 Incineration

Polychlorinated biphenyls can be destroyed by incineration if the transit time is long enough. During incineration hexachlorobenzene is formed; its rate of formation increases at high temperatures. In order for the hexachlorobenzene to be destroyed, a temperature of 950°C is needed. This results in the formation of a residue less than 100 mg of hexachlorobenzene per kilogram. At 800°C and a transit time of 2 sec, a residue of about 1200 mg/kg remains (235).

5.13 Hygienic Standards of Permissible Exposure

The threshold limit of chlorodiphenyl (42 percent chlorine) on skin, by the ACGIH for 1979 (25) is an adopted TLV—TWA value of 1 mg/m^3 and a tentative TLV–STEL value of 2 mg/m^3. The adopted TLV–TWA value for chlorodiphenyl (54 percent chlorine) on skin is 0.5 mg/m^3 and the tentative TLV–STEL value is 1 mg/m^3.

According to the 1978 *NIOSH–OSHA Pocket Guide to Chemical Hazards* (26), the permissible exposure limit of chlorodiphenyl (42 percent chlorine, $C_{12}H_7Cl_3$ approx.) is 1 mg/m^3 (NIOSH 1.0 μg/m^3, 10-hr work-shift time-weighted average (TWA). The immediately dangerous to life or health concentration (IDLH) is 10 mg/m^3.

For chlorodiphenyl (54 percent chlorine, $C_{12}H_5Cl_5$ approx.), the permissible exposure limit is 1 mg/m^3 (NIOSH 1.0 μg/m^3, 10-hr TWA). The IDLH is 5 mg/m^3 (26).

In 1973 the FDA (236) established temporary tolerances for PCBs in foods, animals feeds, and food-packaging materials as follows: 2.5 mg/kg (ppm) in milk (fat basis); 2.5 mg/kg (ppm) in manufacturing dairy products (fat basis); 5 mg/kg (ppm) in poultry (fat basis); 0.5 mg/kg (ppm) in eggs; 5 mg/kg (ppm) in fish and shellfish (edible portion); 0.2 mg/kg (ppm) in infant and junior foods; 10 mg/kg (ppm) in paper food-packaging material intended for or used with human food, finished animal feed, and any components intended for animal feeds; 0.2 mg/kg (ppm) in finished animal feed for food-producing animals (except in feed concentrates, feed supplements, and feed premixes); 2 mg/kg (ppm) in animal feed components of animal origin, including fish meal and other by-products of marine origin and in finished animal feed concentrates, supplements, and premixes intended for food-producing animals.

6 CHLORINATED NAPHTHALENES, $C_{10}H_{(8-n)}Cl_n$, Halowax

6.1 Uses and Industrial Exposures

Individual polychlorinated naphthalenes are compounds with properties and uses similar to those of polychlorinated diphenyls. The Halowaxes are composed of several chlorinated naphthalenes and impurities.

Relatively few investigations have been reported since publication of this chapter by Irish in the second edition (8). As Irish reported, any one product is a mixture but will have a specific range of chlorine substitution. Those with three or less chlorine equivalents are much less toxic than those with four or more.

These materials have been used in electric wire insulation. They have also been used as additives to special lubricants.

The first problems that were encountered were in the manufacture of insulation resulting in skin effects (chloracne) and systemic intoxications. Contact with the cold material and exposure to the vapors were primarily responsible for both local and systemic effects.

6.2 Physical and Chemical Properties

Physical state	Waxy solids
Properties vary with the degree of chlorine substitution	
Solubility	Insoluble in water; soluble in organic solvents

6.3 Effects in Animals

In toxic doses the higher chlorinated naphthalenes are likely to cause severe injury to the liver, characterized as acute yellow atrophy. The systemic injury appears to be exclusively one of liver injury. Such injury may occur from ingestion or from inhalation of vapors liberated particularly from heated Halowaxes. Some authors have indicated that chlorinated naphthalenes may also be absorbed through the skin (8).

Drinker et al. (237) fed a mixture of penta- and hexachloronaphthalene at a rate of 3 g/day to rats for 1 month. Nine out of 10 animals died. They all lost weight, were sick, and showed severe liver injury. When a mixture of tetra- and pentachloronaphthalene was fed at a dose of 0.5 mg/day for 2 months, some sickness, some mortality, and definite liver injury were observed. (8).

Drinker et al. (237) also exposed rats to vapors of chlorinated naphthalenes. Since the material has a high boiling point and had to be heated to vaporize, it is possible that a certain amount of material must have condensed either on the cage or on the animals so that there may have been some oral intake. Animals were exposed for 16 hr/day. In a case of lower chlorinated material represented largely by a mixture of tri- and tetrachloronaphthalene, an average concentration of 1.31 mg/m^3 of air produced essentially no effect other than a possible slight enlargement of the liver. The same material at a concentration of 10.97 mg/m^3 resulted in slight liver injury. Exposure was repeated daily for up to 4½ months (8).

Similar exposure to the vapors of a mixture of penta- and hexachlorona-phthalene at an average concentration of 1.16 mg/m^3 produced definite liver injury. At a concentration of 8.88 mg/m^3, there was some mortality, poor growth, and severe liver injury. The higher chlorinated material showed a greater toxicity than did the lower (8).

Adams et al. (238) studied the nature of the response of rabbit skin to this type of material. They concluded that the response takes the form of, first, epithelial hyperplasia, second, inflammatory and degenerative changes, and finally, regenerative processes.

Injury to farm animals has been noted when lubricants containing chloronaphthalenes were ingested. Lubricants containing chloronaphthalenes have been used in machinery for pelleting feeds. Externally, a thick keratinized layer developed on the skin of the neck. The gallbladder, liver, pancreas, and kidneys were affected. The condition was often fatal. Although the most common cause was from pellet feeds contaminated with grease from the pelleting machinery, the same condition can occur from ingestion of oils or greases used on farm machinery if they contain chloronaphthalenes (8).

The nature of the hyperkeratosis was described in some detail in Leaflet No. 355, U.S. Department of Agriculture, 1954. There was apparently a hyperplasia and a keratinization noticeable in the neck, shoulders, and withers. There were also growths on the gums of the animal (8).

This condition in the animals is supposedly due to ingestion of the material. The injury to the liver and obvious systemic effect certainly would indicate that this is an important factor. It is possible, however, considering the reaction in the tissues of the mouth, cheek, and neck, that some of the materials may act by direct contact with the mouth and skin (8).

Bell (239) elicited signs of intoxication in cattle by a dose as low as 1 mg of chlorinated naphthalene per kilogram body weight given over a 7-day period. At this dosage, signs of intoxication appeared on the twelfth day after administration began. These included excessive lacrimation and salivation, nasal discharge, emaciation, proliferative lesions in mouth, diarrhea, and in more chronic cases, the typical hyperkeratosis of the skin (240). Gross lesions found at necropsy included proliferative lesions in the esophageal mucosa, cirrhosis of liver, cystic swelling of mucosa of gallbladder and extra hepatic bile ducts, and striation in the renal cortex (241, 242). Histopathological findings have shown degeneration and cirrhosis of liver, proliferation of the intrahepatic bile ducts, cystic dilation of mucosal glands of gallbladder and bile ducts, dilatation of the straight tubules of the kidney, squamous cell metaplasia of cervix and Gartner's ducts, and hyperkeratosis of the skin (the mysterious "X disease").

Cattle poisoned with highly chlorinated naphthalene have consistently shown a decrease in vitamin A plasma levels to less than 10 μg/100 ml of blood. This change occurred a few days after administration of the naphthalene.

Brock et al. (243) studied the effect of chlorinated naphthalene (Halowax 1014) in sheep and found that an intoxication in sheep is distinctly different from that in cattle. The amount of the chemical required to produce death in sheep was much greater and the pathological changes differed from those in cattle.

The smallest amount of chlorinated naphthalene per kilogram of body weight required to cause death in a sheep was approximately 100 times greater than the smallest amount per kilogram body weight to cause death in a calf as reported by Bell (239). This difference between bovine and ovine lethal doses is much greater than that which might have been deduced from the results obtained with contaminated wheat concentrates by Olafson and McEntee (244).

The two most characteristic pathological changes in bovine chlorinated naphthalene intoxication are cutaneous hyperkeratosis and extreme depression of plasma vitamin A levels. Hyperkeratosis was not observed in the sheep in this experiment. Plasma vitamin A levels, although significantly lower in the sheep given 11 mg and 28 mg of the chemical per kilogram of body weight than in the control sheep, did not show the extreme and rapid depression to values less than 10.0 μg/100 ml of plasma characteristic of bovine intoxication. The sheep given 11 mg of chlorinated naphthalene per kilogram body weight did not show any significant depression in plasma vitamin A levels. Thus the clinical changes most characteristic of this condition in cattle are either absent or comparatively insignificant in sheep (243).

6.4   Human Experience

The most commonly observed problem from the use and handling of chlorinated naphthalenes is the chloracne. This has been observed on the skin of workers from a number of operations where it has been used. A detailed discussion of dermatological problems has been published by Good and Pensky (245). The clinical aspects of this skin problem were also discussed by Schwartz and Peck (246). Another extensive report was given by the special bulletin of the state of Pennsylvania (247). Chloracne usually occurs from long-term contact with the material or from a much shorter contact with hot vapors. The reaction is usally slow to appear and may take months to return to normal (8).

The term chloracne was coined in 1901 by Herxheimer (248), who believed that nascent chlorine was responsible for the skin effects; but according to Crow (249), Herxheimer soon agreed that chlorinated aromatic hydrocarbons were the most likely cause.

Until 1957 there was general agreement that as the degree of chlorination increased so did the acneigenic properties and the systemic toxicity of the chloronaphthalenes. According to Crow's Dowling Oration delivered April 17, 1970 (249) a result of Shelley and Klingman's (250) experiments on human volunteers, the unexpected fact emerged that whereas tri-, tetra-, hepta-, and octachloronaphthalenes were entirely nonacneigenic, the penta- and hexachloro derivatives produced very severe chloracne indeed. Since Shelley deliberately chose suspensions of fine particles rather than solutions, varying solvents and solubilities cannot have accounted for the surprising failure of the two most highly chlorinated derivatives to have any acneigenic effect. In 1957 Hambrick (251), employing the extremely sensitive experimental technique described in 1941 by Adams et al. (252) using the inner side of the rabbit ear, had already demonstrated that mono- and dichloronaphthalenes did not produce chloracne, thus clearly incriminating only the penta- and hexachloro derivatives. This work is strongly confirmed by the massive outbreaks of chloracne which occurred during the late 1930s and 40s largely in the manufacture and use of electric cables whose outer covering was a fabric impregnated with penta- and hexachloronaphthalene.

There is no reason to dispute the fact that mono-, di-, hepta-, and octochloronaphthalenes are nonacneigenic, but the position with regard to tri- and tetrachloronaphthalenes, which invariably exist as a mixture, is confusing. Several reports of chloracne from trichloronaphthalene used as a paper capacitor impregnant appeared in the literature but do not stand up well to critical analysis (249).

There are various reasons for believing that other factors were probably responsible for the acne in these cases. It is, for instance, well known that molten pitch was extensively used to seal capacitors after impregnation, and was also mixed with the trichloronaphthalene itself in certain cases. It therefore seems likely that pitch was at fault, a suggestion strongly supported by Duvoir (253). It is even possible that pitch may have conferred upon the mixture acneigenic properties not possessed by tri- and tetrachloronaphthalenes alone (249).

Crow's studies confirmed the observations of other investigators that there is no evidence that tri- and tetrachloronaphthalene mixtures are acneigenic or dermatitic under industrial conditions.

6.5 Absorption, Metabolism, and Excretion

Cleary et al. (254) published an article in 1939 on the absorption and metabolism of the chlorinated naphthalenes. They indicated that the material was readily absorbed from an olive oil solution but found no evidence of storage in the lung, liver, skin, or kidneys. They did not find any excretion via the urine and concluded that the chlorine might be removed from the ring and excreted as chloride. They observed a rise in urinary ethereal sulfate, but no change in neutral sulfur (8).

Ruzo et al. (1975) (255), who administered 1-chloronaphthalene (a major component of Halowax 1031)—30 mg/kg in corn oil, by retrocarotid injection of 10 kg in pigs—identified two metabolites in the urine 5 hr later, a monohydroxylated compound (4-chloro-1-naphthol) and a trace of a dihydroxy compound. 2-Chloronaphthalene (also a constituent of Halowax 1031), when administered similarly, led to the urinary excretion of 3-chloro-2-naphthol.

Chu and co-workers (256), studying the metabolism of representative isomers of polychloronaphthalenes which they administered orally to male Wistar rats, found that hydroxylation and/or, as an alternative pathway, hydroxylation–dechlorination were the common metabolic pathways of these compounds. The metabolites were isolated from urine and feces and identified.

1,2-Dichloronaphthalene (40 mg/kg orally) was biotransformed in the rat to the glucuronide conjugate of 5,6-dichloro-1,2-dihydroxy-1,2-dihydronaphthalene. 2,7-Dichloronaphthalene was metabolized to free and conjugate 7-chloro-2-naphthol, and 2,6-dichloronaphthalene gave rise to free and conjugated 6-chloro-2-naphthol and 2,6-dichloronaphthol.

Chu et al. (257), in a second investigation, administered a single oral dose of $(1,4,5,8-^{14}C)$-1,2-dichloronaphthalene (DCN) (20 μCi/kg; 400 mg/kg) in corn oil to male Wistar rats (250 to 350 g).

Based on analyses of blood, they found that DCN was rapidly absorbed with the highest level of radioactivity at 1 hr, gradually declining over the next 8 hr. After 48 hr, the levels of radioactivity in blood were approximately 30 and 15 percent, respectively, of the level observed after 1 hr.

Tissue distribution of DCN in 24 and 48 hr showed no significant differences in radioactivity levels. After 7 days, virtually no activity was detected in the tissues, except in the skin and adipose tissue.

Highest levels of radioactivity were found in liver, kidneys, intestine, bladder, and adipose tissue, but when expressed in percent of the total dose administered, only the liver and intestine were important in retention because of the comparative large size of these organs.

Most of the radioactivity was eliminated from the body by way of the urine and feces. Excretion was rapid: 64 percent of the dose administered was eliminated within 2 days.

Twenty-six percent of the total dose was excreted in the urine in the first day, 33 percent in two days, and a total of 35 percent in 7 days. The first day feces accounted for 19 percent of the total dose, and 31 percent was removed in 2 days. The amount of radioactivity excreted via the feces in 7 days amounted to 42 percent of the original dose.

To ascertain if there was biliary excretion of DCN or its derived materials, serial bile samples were analyzed for radioactivity. It was found that more than 62 percent of the radioactivity was removed in bile within 24 hr and excretion was insignificant thereafter. It was observed that only 42 percent of the total dose was excreted in the feces but 65 percent entered into the gastroenteric tract via bile. This suggests that DCN or its metabolites are reabsorbed from the intestine and excreted into the bile.

Analysis of the fecal extract by TLC showed that the radioactivity was due to unchanged DCN. No DCN or free chloronaphthol could be detected in the urine. The metabolite associated with urine is the glucuronide of a dihydrodiol

Table 49.11.

Compound	Adopted Values TLV–TWA (mg/m^3)	Tentative Values TLV–STEL (mg/m^3)
Trichloronaphthalene	5	10
Tetrachloronaphthalene	2	4
Hexachloronaphthalene (skin)	0.2	0.6
Pentachloronaphthalene	0.5	2.0
Octachloronaphthalene (skin)	0.1	0.3

of DCN since it was isolated after hydrolysis with β-glucuronidase. This finding (1972) is consistent with the reported metabolic pathway of aromatic compounds in which arene oxide is the proposed intermediate (258).

The results of chu et al. reveal, for the first time, a dihydrodiol formation in chlorinated napthalenes and provide an alternate metabolic pathway for these compounds (255). The position of hydroxylation is not known at the present.

In summary, DCN was found to be rapidly absorbed from gastroenteric tract and metabolized to dihydrodiol. Forty-two percent of the DCN was excreted unchanged via feces; 35 percent appeared in urine as the glucuronide of a dihydrodiol.

6.6 Hygienic Standards of Permissable Exposure

The threshold limits of certain chlorinated naphthalenes according to the ACGIH in 1979 (25) and the 1978 *NIOSH–OSHA Pocket Guide to Chemical Hazards* (26) are presented in Tables 49.11 and 49.12.

Table 49.12. (26)

Compound	Permissible Exposure Limit (mg/m^3)	IDLH Level (mg/m^3)
Trichloronaphthalene	5	50
Tetrachloronaphthalene	2	20
Hexachloronaphthalene (Halowax 1014)	0.2	2
Pentachloronaphthalene (Halowax 1013)	0.5	NA
Octachloronaphthalene (Halowax 1051)	0.1	200

REFERENCES

1. R. L. Hollingsworth, V. K. Rowe, F. Oyen, H. R. Hoyle, and H. C. Spencer, *A.M.A. Arch. Ind. Health,* **14,** 138 (1956), also through *Hygienic Guide Series,* American Industrial Hygiene Association, May–June, 1964.

2. S. P. Varshavskaya, *Hyg. Sanit.,* **33**(10), 17 (1967).

3. J. J. Kocsis, S. Harkaway, and R. Snyder, *Ann. N.Y. Acad. Sci.,* **243,** 104 (1975).

4. W. F. von Oettingen, *the Halogenated Hydrocarbons, Toxicity and Potential Dangers,* U.S. Publ. Health Serv. Publ. No. 414, 1955.

5. F. Flury and F. Zernik, *Schädliche Gase,* Springer, Berlin, 1931.

6. N. D. Rozenbaum, R. S. Block, S. N. Kremneva, S. L. Ginzburg, and I. V. Pozhariskii, *Gig. Sanit.,* **12**(1), 21, (1947) (through Ref. 7).

7. CMA, *Technical Report Worldwide Literature Search on Chlorobenzenes,* submitted to the Manufacturing Chemists Association, Washington, D.C., March 4, 1976.

8. D. D. Irish, "Halogenated Hydrocarbons: II. Cyclic," *Industrial Hygiene and Toxicology,* 2nd Rev. ed., Wiley-Interscience, New York, 1963, pp. 1333–1345.

9. *Hygienic Guide Series,* American Industrial Hygiene Association, May-June, 1964; American Conference of Governmental Industrial Hygienists, "*Threshold Limit Values for 1963,*" *A.M.A. Arch. Environ. Health,* **7,** 592 (1963).

10. W. K. Knapp, Jr., W. M. Busey, and W. Kundzins, *Toxicol. Appl. Pharmacol.,* **19**(2), 393 (1971).

11. A. G. Khanin, *Tr. Tsent. Inst. Usoversh. Vrachei.,* **135,** 97 (2969) (through Ref. 7).

12. L. P. Tarkhova, *Hyg. Sanit.,* **30**(3), 327 (1965) (through Ref. 7).

13. M. Lecca-Radu, *Igiena,* **8,** 231 (1959) (through Ref. 7).

14. W. A. Skinner, G. W. Newell, and J. V. Dilley, *Toxic Evaluation of Inhaled Chlorobenzene,* Final Report prepared for the Division of Biomedical & Behavioral Sciences, National Institute of Occupational Safety and Health, Cincinnati, Ohio, June 15, 1977.

15. B. Spencer and R. T. Williams, *Biochem J.,* **46**(4), XV (1950); **47,** 279 (1950).

16. J. R. Lindsay Smith, B. Spencer, and R. T. Williams, *Biochem J.,* **47**(3), 284 (1950).

17. J. R. Lindsay Smith, A. J. Shaw, and D. M. Foulkes, *Xenobiotica,* **2**(3), 215 (1972).

18. D. V. Parke and R. T. Williams, *Biochem. J.,* **54,** 231 (1953).

19. R. T. Williams, *Detoxication Mechanisms,* Wiley, 1959.

20. B. B. Brodie, W. D. Reid, A. K. Cho, G. Sipes, G. Krishna, and J. R. Gillette, *Proc. Natl. Acad. Sci.,* **68**(1), 160 (1971).

21. H. Reich, *Samml. Vergiftungsfällen,* **5,** 193 (1934).

22. R. Girard, T. Tolot, P. Martin, and P. Bourret, *J. Med. Lyon,* **50**(1164), 771 (1969) (through Reg. 7).

23. L. P. Tarkhova, *Hyg. Sanit.,* **30**(3), 327 (1965) (through Ref. 7).

24. Arthur D. Little, Inc., *Research on Chemical Odors,* Part 1, *Odor Threshold for 53 Commercial Chemicals,* research report for the Manufacturing Chemists Association, Washington, D.C., October, 1968.

25. *TLVs, Threshold Limit Values for Chemical Substances and Physical Agents in the Workroom Environment with Intended Changes for 1979,* American Conference of Governmental Industrial Hygienists, Cincinnati, Ohio.

26. *NIOSH/OSHA Pocket Guide to Chemical Hazards,* U.S. Department of Health, Education and Welfare, Public Health Service, Center for Disease Control, National Institute for Occupational Safety and Health, September, 1978.

27. S. P. Varshavskaya, *Gig. Sanit.*, **33**(10), 15 (1968) [through abstract in *Toxicol. Appl. Pharm.*, 69-0069 (1971)].

28. W. D. Reid, G. Krishna, J. R. Gillette, and B. Bernard, *Pharmacology* **10**(4), 193 (1973).

29. R. L. Hollingsworth, V. K. Rowe, F. Oyen, T. R. Torkelson, and E. M. Adams, *A.M.A. Arch. Ind. Health*, **17,** 180 (1958).

30. A. Jacobs, M. Blangetti, and E. Hellmund, *Vom Wasser*, **43,** 259 (1974) (through G. R. Pielmeier).

31. *IARC Monographs on the Evaluation of Carcinogenic Risk of Chemicals to Man*, Vol. 7, International Agency for Research on Cancer, World Health Association, Leon, 1974.

32. T. Czajkowska, U. Ruta, S. Szendzikowski, and Z. Swierchowski, *Med. Pracy,* **21**(5), 450 (1970) (through Ref. 7).

33. T. S. Hele and E. H. Callow, *Proc. Physiol. Soc. J. Physiol.*, **57,** xlii (1923).

34. E. H. Callow and T. S. Hele, *Biochem. J.*, **20,** 598 (1926).

35. W. M. Azouz, D. V. Parke, and R. T. Williams, *Biochem. J.*, **59**(3), 410 (1955).

36. R. Girard, F. Tolot, P. Martin, and P. Bourret, *J. Med. Lyon*, **50**(1164), 771 (1969) (through Ref. 7).

37. F. Tolot, B. Soubrier, J. R. Bresson, and P. Martin, *J. Med. Lyon*, **50**(1164), 761 (1969) (through Ref. 7).

38. J. Gadrat, J. Monnier, A. Ribet, and R. Bourse, *Arch. Mal. Prof. Med. Trav. Secur. Soc.*, **23**(10/11), 710 (1962) (through Ref. 7).

39. J. G. Downing, *J. Am. Med. Assoc.*, **112,** 1457 (1939).

40. T. Sollmann, *A Manual of Pharmacology and Its Applications to Therapeutics and Toxicology*, 8th ed., W. B. Saunders, Philadelphia, 1957.

41. G. Dikmans, *J. Agric. Res.*, **35,** 645 (1927).

42. D. Irie, T. Sasaki, and R. Ito, *Tohoku Igaku Zasshi*, **20**(5/6), 772 (1973) (through Ref. 7).

43. S. P. Varshavskaya, *Nauchn. Tr. Aspir. Ordinatorov. 1-i Mosk. Med. Inst.*, 175 (1967) (through Ref. 7 and *Chem. Abstr.*, **51,** 4816 (1970).

44. L. Salamone and A. Coppola, *Folia Med.*, **43,** 259 (1960) (through Ref. 7).

45. S. Totaro and G. Licari, *Folia Med.*, **67**(5), 507 (1964) (through Ref. 7).

46. C. Rimington and G. Ziegler, *Biochem. Pharmacol.*, **12**(12), 1387 (1963).

47. A. G. Zupko and L. D. Edwards, *J. Am. Pharmacol. Assoc. Sci. Ed.*, **38**(3), 124 (1949).

48. R. L. Hollingsworth, V. K. Rowe, F. Oyen, T. R. Torkelson, and E. A. Adams, *A.M.A. Arch. Ind. Health*, **17**(1), 180 (1958).

49. M. Perrin, *Bull. Acad. Med.*, **125**(302) (1941) (through Ref. 7).

50. G. Petit and J. Champaix, *Arch. Mal. Prof.*, **9,** 311 (1948) (through Ref. 7).

51. R. W. Weller and A. J. Crellin, *Arch. Intern. Med.*, **91,** 408 (1953).

52. L. H. Cotter, *N.Y. State J. Med.*, **53,** 1690 (1953).

53. D. M. Campbell and R. J. L. Davidson, *J. Obstet. Gynaecol. Br. Commonw.*, **77,** 657 (1970).

54. M. L. Efthymiou and P. Gervais, *Cah. Med.*, **13**(10), 831 (1972) (through Ref. 7).

55. W. H. Peterson, Jr., and M. H. Liner, *Bull. Natl. Clgh. Poison Control Cent.*, U.S. Dept. of HEW, Bureau of Drugs, Bethesda, Maryland, July–August, 1975.

56. American Industrial Hygiene Association, *Hygienic Guide Series*, *p*-Dichlorobenzene, May–June, 1964.

57. M. L. Berliner, *A.M.A. Arch. Ophthalmol.*, **22,** 1023 (1939).

58. M. H. Pike, *J. Mich. Med. Soc.*, **43,** 581 (1944).

59. K. Walgren, *Zent. Arbeitsmed. Arbeitsschutz,* **3,** 14 (1953).

60. W. M. Azouz, D. V. Parke, and R. T. Williams, *Biochem. J.,* **59,** 410 (1955).

61. R. Kimura, T. Kayashi, M. Sato, A. T. Aimoto, and T. Murata, *J. Pharmacobiodyn.,* **2**(4), 237 (1979).

62. G. P. Carlson and R. G. Tardiff, *Toxicol. Appl. Pharmacol.,* **36,** 383 (1976).

63. L. D. Pagnotto and J. E. Walkley, *Am. Ind. Hyg. Assoc. J.,* **26,** 137 (1965).

64. H. J. Langner and H. G. Hilliger, *Berl. Muench. Tieraerztl. Wochenschr.,* **84**(18), 351 (1971).

65. L. M. Srivastava, *Cytologia,* **31,** 166 (1966).

66. *IARC Monographs on the Evaluation of Carcinogenic Risk of Chemicals to Man,* Vol. 7, International Agency for Research on Cancer, Lyon, 1974.

67. *EPA Survey of Industrial Processing Data, Hexachlorobenzene and Hexachlorobutadiene Pollution from Chlorocarbon Processes,* 560/3-75-003, Environmental Protection Agency, Washington, D.C., prepared by C. E. Mumma and E. W. Lawless of the Midwest Research Institute, Kansas City, Mo., 1975.

68. National Academy of Sciences, *Assessing Potential Ocean Pollutants: A Report of the Study Panel on Assessing Potential Ocean Pollutants,* Ocean Affairs Board Commission on National Resources, National Research Council, Washington, D.C., 1975.

69. D. L. Grant, F. Iverson, G. V. Hatina, and D. C. Villeneuve, *Environ. Physiol. Biochem.,* **4**(4), 159 (1974).

70. T. Kuiper-Goodman, D. L. Grant, C. A. Moodie, G. O. Korsrud, and I. C. Munro, *Toxicol. Appl. Pharmacol.,* **40,** 529 (1977).

71. A. Boger, G. Koss, W. Koransky, R. Naumann, and H. Frenzel, *Arch. Pathol. Anat. Physiol.,* **382**(2), 127 (1979).

72. H. H. Mollenhauer, J. H. Johnson, R. L. Younger, and R. L. Clark, *Am. J. Vet. Res.,* **36**(12), 1777 (1975).

73. R. Lissner, G. Goerz, M. G. Eichenauer, and H. Ippen, *Biochem. Pharmacol.,* **24**(18), 1729 (1975).

74. G. Koss, S. Seubert, W. Koransky, and H. Ippen, *Arch. Toxicol.,* **40**(4), 285 (1978).

75. R. Engst, R. M. Macholz, and M. Kujawa, *Bull. Environ. Contam. Toxicol.,* **16**(2), 248 (1976).

76. K. P. Schuster and G. Renner, *N Arch. Exp. Pathol Pharmakol.,* **297**(2), R5 (1977).

77. H. Lui and G. D. Sweeney, *FEBS (Fed. Eur. Biochem. Soc.) Lett.,* **51**(1), 225 (1975).

78. F. Koszo, C. Ziklosi, and N. Simon, *Biochem. Biophys. Res. Commun.,* **80**(4), 781 (1978).

79. G. W. Koss, W. Koransky, and K. Steinbach, *Arch. Toxicol.,* **35,** 107 (1976).

80. G. Koss, W. Koransky, and K. Steinbach, *Arch. Toxicol.,* **42**(1), 19 (1979) [through *Pestic. Abstr. U.S. Environ. Prot. Agency,* 79-2108 (1979)].

81. G. Goerz, W. Vizethum, K. Bolsen, and Th Krieg, *Arch. Dermatol. Res.,* **263**(2), 189 (1978).

82. T. Ariyoshie, K. Idegichi, Y. Ishizika, K. Iwasaki, and M. Akakan, *Chem. Pharm. Bull.,* **23**(4), 817 (1975).

83. A. H. Timme, J. J. F. Taljaard, B. C. Shanley, and S. M. Joubert, *S. Afr. Med. J.,* **48**(43), 1833 (1974).

84. D. C. Villeneuve, G. J. A. Speijers, E. M. den Tonkelaar, M. J. von Logten, J. G. Vos, P. A. Greve, and J. G. van Esch, *Toxicol. Appl. Pharmacol.,* **45**(1), 341 (1978).

85. B. Zawirska and D. Dzik, *Patol. Pol.,* **28**(3), 349 (1977) [through *Pestic. Abstr. U.S. Environ. Prot. Agency,* 78-0386 (1978)].

86. E. J. Gralla, R. W. Fleischman, Y. K. Luthra, M. Hagopian, J. R. Baker, E. Esber, and W. Marcus, *Toxicol. Appl. Pharmacol.,* **40,** 227 (1977).

87. Y. K. Luthra, J. J. Esber, E. J. Gralla, M. Hagopian, and W. Marcus, *Fed. Proc. Fed. Am. Soc. Exp. Biol.*, **36**(3), 356 (1977).

88. W. F. Mueller, M. J. Iatropoulos, K. Rozman, F. Korte, and F. Coulston, *Toxicol. Appl. Pharmacol.*, **45**(1), 283 (1978).

89. R. S. Yang, K. A. Pittman, D. R. Rourke, and V. B. Stein, *J. Agric. Food Chem.*, **26**(5), 1076 (1978).

90. K. Rozman, W. F. Mueller, F. Coulston, and F. Korte, *Chemosphere*, **7**(2), 177 (1978).

91. E. M. den Tonkelaar, H. G. Verschuuren, J. Bankovska, T. DeVries, R. Kroes, and G. J. van Esch, *Toxicol. Appl. Pharmacol.*, **43**(1), 137 (1978).

92. C. P. Fassbender, A. R. Alvarez, and S. Wenzel, *Arch. Lebensmittelhyg.*, **28**(6), 201 (1977).

93. L. G. Hansen, S. B. Dorn, S. M. Sundlof, and R. S. Vogel, *J. Agric. Food Chem.*, **26**(6), 1369 (1978).

94. L. G. Hansen, C. S. Byerly, R. L. Metcalf, and R. F. Bevill, *Am. J. Vet. Res.*, **36**, 23 (1975).

95. Codex Alimentarius Commission, Joint FAO/WHO Food Standards Programme Codex Committee on Pesticides Residues, Tenth Session, The Hague, 29 May–5 June. 1978.

96. R. L. Mull, W. L. Winterlin, S. A. Peoples, and L. Ocampo, *J. Environ. Pathol. Toxicol.*, **1**(6), 865 (1978).

97. K. D. Courtney and J. E. Andrews, *Toxicol. Lett.*, **3**(6), 357 (1979).

98. D. L. Grant, W. E. J. Phillips, and G. V. Hatina, *Arch. Environ. Contam. Toxicol.*, **5**, 207 (1977).

99. B. A. Schwetz, J. M. Norris, R. J. Kociba, P. A. Keeler, R. F. Cornier, and P. J. Gehring, *Toxicol. Appl. Pharmacol.*, **30**, 255 (1974).

100. J. G. Vos, H. L. van der Maas, A. Musch, and E. Ram, *Toxicol. Appl. Pharmacol.*, **18**, 944 (1971).

101. M. J. Iatropoulos, J. Bailey, H. P. Adams, F. Coulston, and W. Hobson, *Environ. Res.*, **16**(1–3), 38 (1978).

102. J. R. P. Cabral, T. Mollner, F. Raitano, and P. Shubik, *Toxicol. Appl. Pharmacol., Abstr.*, **45**(1), 323 (1978).

103. J. R. P. Cabral, *Toxicol. Appl. Pharmacol., Abstr.*, **41**, 155 (1977); J. G. Vos, M. J. Van Logten, J. G. Kreeftenberg, and W. Kruizinga, *N.Y. Acad. Sci.* (June 21, 1978).

104. C. Cam and G. Nigogosyan, *J. Am. Med. Assoc.*, **183**, 88 (1963).

105. A. L. Cetingil and M. A. Ozen, *Blood*, **16**, 1002 (1960).

106. F. DeMatteis, B. E. Prior, and C. Rimington, *Nature*, **191**, 363 (1961).

107. D. J. Cripps, H. A. Peters, and A. Gocmen, *J. Invest. Dermatol.*, **71**(4), 277 (1978); *Clin. Res.*, **26**(3), 489 A (1978).

108. H. A. Peters, *Fed. Proc. Fed. Am. Soc. Exp. Biol.*, **35**(12), 2400 (1976).

109. L. G. Pederson and G. L. Carlson, *J. Chem. Phys.*, **62**(5), 2009 (1975).

110. H. Ehrlicher, *Zentrabl. Arbeitsmed. Arbeitsschutz*, **18**(7), 204 (1968).

111. I. V. Sairtskii, *Vopr. Prom. Sel'skokhoz. Toksikol. Kievsk. Med. Inst.*, **158**, 173 (1964) [through *Chem. Abstr.*, **63**, 8952d (1965) and Ref. 67].

112. W. J. Hayes, Jr., *Toxicology of Pesticides*, Williams & Wilkins, Baltimore, 1975.

113. B. J. Stokvis, *Z. Klin. Med.*, **28**, 1 (1895) (through Ref. 112).

114. W. B. Deichmann, *J. Ind. Hyg. Toxicol.*, **23**(7), 343 (1941), and unpublished studies conducted in 1935.

115. G. Dappen, Dept. of the Interior, Offices of Water Resources Research, Report No. PB-238 593, Springfield, Va., National Technical Information Service (through Ref. 129).

116. A. L. Laska, C. K. Bartell, and J. L. Laseter, *Bull. Environ. Contam. Toxicol.*, **5**, 253 (1976).

117. R. L. Metcalf, I. P. Kapoor, P. Y. Lu, C. K. Schuth, and P. Sherman, *Environ. Health Perspect.* (May 1963) (through Ref. 67).

118. J. R. Plimmer and V. I. Klingebiel, *J. Agric. Food Chem.*, **24**(4), 721 (1976); B. Ahling and A. Lindskog, *Sci. Total Environ.*, **10**(1), 51 (1978).

119. B. Ahling and A. Lindskog, *Sci. Total Environ.*, **10**(1), 51 (1978).

120. S. C. Quinlivan and M. Ghassemi, *J. Hazardous Mater.*, **1**(4), 343 (1977).

121. R. Mestres, M. Tolede, F. Sabon, and C. Francois, *Trav. Soc. Pharm. Montp.*, **37**(2), 103 (1977); L. G. Hansen, J. Simon, S. B. Dorn, and R. H. Teske, *Toxicol. Appl. Pharmacol.*, **51**, 1 (1979).

122. K. S. Khera, *Food Cosmet. Toxicol.*, **12**, 471 (1974).

123. G. S. Simon, B. R. Kipps, R. G. Tardiff, and J. F. Borzelleca, *Toxicol. Appl. Pharmacol.*, **45**(1), 330 (1978).

124. L. D. Loose, J. B. Silkworth, K. A. Pittman, K. F. Benitz, and W. Mueller, *Infect. Immunol.*, **20**, 30 (1978).

125. J. G. Vos, M. J. van Logten, J. G. Kreeftenberg, P. A. Steerenberg, and W. Kruizinga, *Drug. Chem.*, **2**, 61 (1979).

126. L. D. Loose, K. A. Pittman, K. F. Benitz, and J. B. Silkworth, *J. Reticuloendothel. Soc.*, **22**, 253 (1977).

127. J. B. Silkworth, "Modification of Cell-mediated Immunity by Polychlorinated Biphenyl (Aroclor 1016) and Hexachlorobenzene," Thesis Dissertation, Center for Experimental Pathology and Toxicology, Albany Medical College, Albany, New York. Condensed version to be published by the Environmental Protection Agency Office of Pesticides and Toxic Substances series 560/ Research and Development Series 600 (1979).

128. R. L. Ziprin and S. R. Fowler, *Toxicol. Appl. Pharmacol.*, **39**, 105 (1977).

129. *IARC Monograph on the Evaluation of the Carcinogenic Risk of Chemicals to Humans*, Vol. 20, *Hexachlorobenzene*, International Agency for Research and Cancer, World Health Organization, 1979.

130. NIOSH Current Intelligence Bulletin, *Polychlorinated Biphenyls (PCBs)*, J. W. Lloyd, R. M. Moore, B. S. Woolf, and H. P. Stein, November 3, 1975, Bulletin 1 through 18, March 1, 1978.

131. *IARC Monographs on the Evaluation of the Carcinogenic Risk of Chemicals to Humans*, considered by the Working Group in Lyon, June 1974, Vol. 20, *Polychlorinated Biphenyls*, 1979.

131a. H. L. Hubbard, "*Chlorinated Biphenyl and Related Compounds.*" in *Encyclopedia of Chemical Technology*, R. E. Kirk and D. F. Othmer, Eds., 2nd ed., Vol. 5, Wiley, New York, 1964, pp. 289–297.

131b. Monsanto Co., *Aroclor Plasticizers*, Tech. Bull. No. O/PL-306A, St. Louis, Mo., 1970.

132. R. D. Kimbrough, R. E. Linder, and T. B. Gaines, *Arch. Environ. Health*, **25**, 354, (1972).

133. J. V. Bruckner, K. L. Khanna, and H. H. Cornish, *Toxicol. Appl. Pharmacol.*, **24**, 434, (1973).

134. L. Fishbein, *Toxicology of Chlorinated Biphenyls*, Vol. 14, *Annual Review of Pharmacology*, Annual Reviews, Inc., Palo Alto, Calif., 1974, pp. 139–156.

135. J. W. Cooke, *Status Report on the Chemistry and Toxicology of PCBs*, FDA, Washington, D.C., June 1, 1970 (through Ref. 134).

136. J. G. Vos, *Environ. Health Perspect.*, **1**, 105 (1972).

137. C. K. Drinker, M. F. Warren, and G. A. Bennett, *J. Ind. Hyg. Toxicol.*, **19**, 283, (1937).

138. J. W. Miller, *Publ. Health Rep. U.S.*, **59**, 1085 (1944).

139. R. D. Kimbrough, R. E. Linder, V. W. Burse, and R. W. Jennings, *Arch. Environ. Health*, **27**, 390 (1973).

140. H. Yoshimura and H. Yamamoto, *Fukuoka Igaku Zasshi*, **62,** 5 (1971) (through Ref. 131).
141. F. Coulston, F. Korte, and M. Goto, Eds., *New Methods in Environmental Chemistry and Toxicology*, Academic Scientific Book, Tokyo, 1973, pp. 291–297.
142. J. V. Bruckner, K. L. Khanna, and H. H. Cornish, *Toxicol. Appl. Pharmacol.*, **24,** 434 (1973).
143. D. H. Norbach and J. P. Allen, *Environ. Health Perspect.*, **1,** 137, 1972.
144. S. Oishi, M. Morita, and H. Fukuda, *Toxicol. Appl. Pharmacol.*, **43,** 13 (1978).
145. H. Yoshimura and M. Oshima, *Fukuoka Igaku Zasshi*, **62,** 5 (1971) (through Ref. 131).
146. H. Yoshimura, H. Yamamoto, J. Nagai, Y. Yae, H. Uzawa, Y. Ito, A. Notomi, S. Minakami, A. Ito, K. Kato, and Tsuji, *Fukuoka Igaku Zasshi*, **62,** 12 (1971) (through Ref. 131).
147. G. F. Fries, G. S. Marrow, Jr., and C. H. Gordon, *Bull. Environ. Contam. Toxicol.*, **7,** 252 (1972).
148. G. F. Fries, G. S. Marrow, Jr., and C. H. Gordon, *J. Agric. Food Chem.*, **21,** 117 (1973).
149. R. B. Dahlgren, Y. A. Greichus, and R. L. Linder, *J. Wildlife Management*, **35,** 823, (1971).
150. R. B. Dahlgren, R. L. Linder, and C. W. Carlson, *Environ. Health Perspect.*, **1,** 89 (1972).
151. J. G. Vos and J. H. Koeman, *Toxicol. Appl. Pharmacol.*, **17,** 656 (1970) (through Ref. 131).
152. N. S. Platanow and B. S. Reinhart, *Can. J. Comp. Med.*, **37,** 341 (1973) (through Ref. 131).
153. B. M. Rehfeld, R. L. Bradley, Jr., and M. L. Sunde, *Poultry Sci.*, **51**(2), 435 (1972).
154. W. B. Deichman, unpublished observations, 1936.
155. J. P. Treon, F. P. Cleveland, J. W. Cappel, and R. W. Atchley, *Am. Ind. Hyg. Assoc. Q.*, **17,** 204 (1956).
156. II. P. Tombergs, *Environ. Health Perspect.*, **1,** 179, (1972).
157. J. W. Miller, *Publ. Health Rep. U.S.*, **59,** 1085 (1944).
158. M. Slocumb, D. Perry, and D. E. Carter, *Soc. Toxicol, Eighteenth Ann. Meet., New Orleans, La., March 11–15, 1979*, Abstr. 309.
159. A. Curley, V. W. Burse, M. E. Grim, R. W. Jennings, and R. E. Linder, *Environ. Res.*, **4**(6), 481 (1971).
160. A. Curley, V. W. Burse, and M. E. Grim, *Food Cosmet. Toxicol.*, **11,** 471 (1973).
161. T. Meyer, J. Aarbakke, and R. R. Scheline, *Acta Pharmacol. Toxicol.*, **39,** 412 (1976).
162. T. Meyer and R. R. Scheline, *Acta Pharmacol. Toxicol.*, **39,** 419 (1976).
163. K. M. McCormack, P. Melrose, D. E. Rickert, J. G. Dent, J. E. Gibson, and J. B. Hook, *Toxicol. Appl. Pharmacol.*, **47,** 95 (1979).
164. G. P. Carlson, *Toxicology*, **5,** 69 (1975).
165. U. Roes, J. G. Dent, K. J. Netter, and J. E. Gibson, *J. Toxicol. Environ. Health*, **3,** 663 (1977).
166. K. M. Kluwe, K. M. McCormack, and J. B. Hook, *Environ. Health Perspect.*, **23,** 241 (1978).
167. E. P. Lichtenstein, K. R. Schulz, T. W. Fuhrmann, and T. T. Liang, *J. Econ. Entomol.*, **62,** 761 (1969).
168. F. Oesch, *Xenobiotica*, **3,** 305 (1973).
169. J. W. Daly, D. M. Jerina, and B. Witkops, *Experientia*, **28,** 1129 (1972).
170. N. Yogi, K. Kamohara, and Y. Itokawa, *J. Environ. Pathol. Toxicol.*, **2**(4), 1119 (1979), (through *Pestic. Abstr. U.S. Environ. Prot. Agency*, 79-1093 (1979).
171. P. T. LaRocca and G. P. Carlson, *Toxicol. Appl. Pharmacol.*, **48,** 185 (1979).
172. R. D. Campbell, T. P. Leadem, and D. W. Johnson, *Bull. Environ. Contam. Toxicol.*, **11,** 425 (1974).
173. P. W. Davis, J. M. Friedhoff, and G. A. Wedemeyer, *Bull. Environ. Contam. Toxicol.*, **8,** 97 (1972).

174. R. B. Koch, D. Desaiah, H. H. Yap, and L. K. Cutkomp, *Bull. Environ. Contam. Toxicol.*, **7**, 87 (1972).

175. R. Schneider, *Biochem. Pharmacol.*, **24**, 939 (1975).

176. M. R. Wells, J. B. Phillips, and G. C. Murphy, *Bull. Environ. Contam. Toxicol.*, **11**, 572 (1974).

177. N. M. Morales, D. B. Tuey, W. A. Coburn, and H. B. Matthews, *Toxicol. Appl. Pharmacol.*, **48**, 397 (1979).

178. D. L. Grant, D. C. Villeneuve, K. A. McCully, and W. E. J. Phillips, *Environ. Physiol.*, **1**, 61 (1971).

179. P. J. Bunyan and J. M. J. Page, *Toxicol. Appl. Pharmacol.*, **43**, 507 (1978).

180. J. Burke and O. G. Fitzhugh, Suppl. No. 1, States Report of Chemistry and Toxicology of PCBs, Food and Drug Administration, Washington, D.C., December 1, 1970.

181. M. L. Keplinger, O. E. Fancher, and J. C. Calandra, *Toxicol. Appl. Pharmacol.*, **19**, 402, (1971).

182. R. J. Gellert and C. Wilson, *Environ. Res.*, **18**(2), 437, 1979 [through *Pestic. Abstr. U.S. Environ. Health Prot. Agency*, 79-2145 (1979)].

183. H. P. Tombergs, *Environ. Health Perspect.*, **1**, 179 (1972).

184. J. E. Kihlstrom, C. Lundberg, J. Orberg, P. O. Danielsson, and J. Sydhoff, *Environ. Physiol. Biochem.*, **5**(1) 54 (1975) [through *Pestic. Abstr., U.S. Environ. Prot. Agency*, 75-2431 (1975)].

185. M. L. Scott, Results of Experiments on the Effects of PCBs on Laying Hen Performance, Cornell Nutrition Conference of Feed Manufacturers, 1971 (through Ref. 134).

186. J. Bitman, H. C. Cecil, and S. J. Harris, *Environ. Health Perspect*, **1**, 145 (1972).

187. D. B. Peakall, J. L. Linder, and S. E. Bloom, *Environ. Health Perspect.*, **1**, 103, (1972).

188. U. Umeda, *Ambio*, **1**, 132 (1972) (through Ref. 134).

189. J. W. Jones and H. S. Alden, *Arch. Dermatol. Syphilol.*, **33**, 1022, (1936).

190. G. R. Harvey, W. G. Steinhauer, and H. P. Miklas, *Nature*, **252**, (5482), 387 (1974).

191. T. F. Bidleman and C. E. Olney, *Science*, **183**, 516 (1974).

192. R. Frank, M. Holdrinet, H. E. Brown, R. L. Thomas, A. L. W. Kemp, and J. M. Jaquet, *Sci. Total Environ.*, **8**(3), 205, (1977).

193. W. Cooke, G. Nickless, A. Povey, and D. J. Roberts, *Sci. Total Environ.*, **13**(1), 17 (1979) [through *Pestic. Abstr. U.S. Environ. Prot. Agency*, 79-2948 (1979)].

194. G. D. Veith, *Environ. Health Perspect.*, **1**, 51, (1972).

195. G. R. Harvey, W. G. Steinhauer, and R. J. Emerick, *Bull. Environ. Contam. Toxicol.*, **9**, 321 (1973).

196. V. Zitko, *Bull. Environ. Contam. Toxicol.*, **6**, 464 (1971).

197. V. Zitko and P. M. Choi, *Fish Res. Board Can. Tech. Rep.*, **272**, 1 (1971).

198. T. W. Duke, J. I. Lowe, and A. J. Wilson, Jr., *Bull. Environ. Contam. Toxicol.*, **5**, 171 (1970).

199. C. A. Bache, J. W. Serum, W. D. Youngs, and D. J. Lisk, *Science*, **177**, 1191, (1972).

200. D. L. Stalling and F. L. Mayer, Jr., *Environ. Health Perspect.*, **1**, 159 (1972).

201. T. O. Munson, *Bull. Environ. Contam. Toxicol.*, **7**(4), 223 [through *Pestic. Abstr. U.S. Environ. Prot. Agency*, 72-1419 (1972).

202. EPA, Environmental Health Letter, October 15, 1979; Panel on Hazardous Trace Substances, Polychlorinated Biphenyls–Environment Impact, *Environ. Res.*, **5**, 249 (1972).

203. E. H. Gruger, Jr., T. Hruby, and N. L. Karrick, *Environ. Sci. Technol.*, **10**(10), 1033 (1976).

204. H. E. Drescher, V. Harms, and E. Huschenbeth, *Mar. Biol.*, **41**, 99 (1977).

205. J. Clausen and O. Berg, *Pure Appl. Chem.*, **42**(1–2), 223 (1976).

206. M. Kuratsune, T. Yoshimura, J. Matsuzaka, and A. Yamaguchi, *Environ. Health Perspect.*, **1**, 119 (1972).

207. Panel on Hazardous Trace Substances, Polychlorinated Biphenyls—Environmental Impact, *Environ. Res.*, **5**, 249 (1972) (through Ref. 131).

208. U.S. Department of Agriculture, *Agriculture's Responsibility Concerning Polychlorinated Biphenyls (PCBs)*, U.S. Government Printing Office, Washington, D.C., 1972.

209. A. R. Yobs, *Environ. Health Perspect.*, **1**, 79 (1972).

210. H. A. Price and R. L. Welch, *Environ. Health Perspect.*, **1**, 73 (1972).

211. J. E. Bjerk, *Tidssk Norske Laegeforen.*, **92**, 15 (1972) (through Ref. 131).

212. A. Curley, V. W. Burse, R. W. Jennings, E. C. Villanueva, L. Tomatis, and K. Akazaki, *Nature*, **242**, 338 (1973).

213. M. Doguchi, "Chlorinated Hydrocarbons in the Environment in the Kanto Plain and Tokyo Bay, as Reflected in Fishes, Birds and Man," in *New Methods in Environmental Chemistry and Toxicology*, F. Coulston, F. Korte, and M. Goto, Eds., Academic Scientific Book, Tokyo, 1970, pp. 269–289.

214. H. Oura, H. Kobayashi, T. Oura, I. Senda, and K. Kubota, *Nippon Noson Igakui Zasshi*, **21**, 300 (1972) (through Ref. 131).

215. P. G. Dyment, L. M. Herbertson, E. D. Gomes, J. S. Wiseman, and R. W. Hornabrook, *Bull. Environ. Contam. Toxicol.*, **6**, 532 (1971).

216. G. Westoo and K. Noren, *Ambio*, **7**(2), 62 (1978).

217. Y. Masuda, R. Kagawa, H. Kuroki, M. Kuratsune, T. Yoshimura, I. Taki, M. Kusuda, F. Yamashita, and M. Hayashi, *Food Cosmet. Toxicol.*, **16**, 543 (1978).

218. H. Nagasaki, S. Tomii, T. Mega, M. Marugami, and N. Ito, *Gann*, **63**, 805 (1972) (through Ref. 131).

219. R. D. Kimbrough and R. E. Linder, in press, quoted from Ref. 131).

220. R. D. Kimbrough, R. E. Linder, and T. B. Gaines, *Arch. Environ. Health*, **25**, 354 (1972).

221. N. T. Kimura and T. Baba, *Gann*, **64**, 105 (1973) (through Ref. 131).

222. N. Ito, H. Nagasaki, S. Makiura, and M. Arai, in press, quoted from Ref. 131.

223. H. Nagasaki, S. Tomii, T. Mega, S. Sugihara, Y. Miyata, and N. Ito, *Nara Igaku Zasshi*, **25**(6), 635 (1974) [through *Pestic. Abstr. U.S. Environ. Prot. Agency*, 76-0476 (1976)].

224. D. C. Villanueve, D. L. Grant, and K. Khera, *Environ. Physiol.*, **1**, 67 (1971).

225. J. McLaughlin, J. P. Marliae, M. J. Verrett, M. J. Mutchler, and O. G. Fitzhugh, *Toxicol. Appl. Pharmacol.*, **5**, 760 (1963).

226. R. W. Carlson and R. T. Duby, *Bull. Environ. Contam. Toxicol.*, **9**, 261 (1973).

227. S. Green, K. A. Palmer, and E. J. Oswald, *Ann. Meet. Soc. Toxicol. New York, March 18–22, 1973*, Abstr. 12.

228. R. C. Dougherty, M. J. Whitaker, S. Y. Tang, R. Bottcher, and M. Keller, personal communication, April, 1980.

229. M. L. Keplinger, PCB Conference, Quail Roost, Rougemount, N.C., December 1971, personal communication, April 16, 1980.

230. J. G. Vos and T. de Roij, *Toxicol. Appl. Pharmacol.*, **21**, 549 (1972).

231. J. B. Silkworth and L. D. Loose, *Toxicol. Appl. Pharmacol.*, **48**(1), A86 (1979).

232. P. J. Thomas and R. D. Hindsdill, *Toxicol. Appl. Pharmacol.*, **44**(1), 41 (1978).

233. M. I. Luster, P. W. Albro, G. Clark, K. Chae, S. K. Chaudhary, L. D. Lawson, J. T. Corbett, and J. D. McKinney, *Toxicol. Appl. Pharmacol.*, **50**(1), 147 (1979).

234. O. Hutzinger, S. Safe, and V. Zitko, *Environ. Health Perspect., Environ. Issue*, **1,** 15 (1972).

235. B. Ahling and A. Lindskog, *Sci. Total Environ.*, **10**(1), (1978).

236. "Polychlorinated Biphenyls (PCBs)," *U.S. Fed. Reg.*, **38**(129), 18096–18104 (1973).

237. C. K. Drinker, M. F. Warren, and G. A. Bennett, *J. Ind. Hyg. Toxicol.*, **19,** 283 (1937).

238. E. M. Adams, D. D. Irish, H. C. Spencer, and V. K. Rowe, *Ind. Med. Ind. Hyg. Sect.*, **1,** 1 (1941).

239. W. B. Bell, *Va. J. Sci.*, **3,** 169 (1952); *Vet. Med.*, **48,** 135 (1953).

240. P. Olafson, *Cornell Vet.*, **37,** 279 (1947).

241. C. C. Morrill and R. F. Link, *Univ. Ill. Circ.*, 656 (1952).

242. C. Olson, Jr., and R. H. Cook, *Am. J. Vet. Res.*, **12,** 261 (1951).

243. W. E. Brock, E. W. Jones, R. MacVicar, and L. S. Pope, *Am. J. Vet. Res.*, **18,** 625 (1957).

244. P. Olafson and L. McEntee, *Cornell Vet.*, **41,** 107 (1951).

245. C. K. Good and N. Pensky, *Arch. Dermatol. Syphilol.*, **48,** 251, (1943).

246. L. Schwartz and S. M. Peck, *N.Y. State J. Med.*, **43,** 1711 (1943).

247. *Commonw. Pa. Dept. Labor Ind. Spec. Bull. No. 43* (1936).

248. K. Herxheimer, *Seventh Dermatatol. Kongr. Breslau, 1901* p. 152.

249. K. D. Crow, *Trans. St. Johns Hosp. Dermatol. Soc.*, **56,** 79 (1970).

250. W. B. Shelley and A. M. Klingman, *Arch. Dermatol.*, **75,** 689 (1957).

251. G. W. Hambrick, *J. Invest. Dermatol.*, **28,** 89 (1957).

252. E. M. Adams, D. D. Irish, H. C. Spencer, and V. K. Rowe, *Ind. Med. Surg.*, **2,** 1 (1941).

253. M. Duvoir, *Ann. Med. Leg.*, **14,** 539 (1934).

254. R. V. Cleary, J. Maier, and G. H. Hitchings, *J. Biol. Chem.*, **127,** 403 (1939).

255. L. O. Ruzo, S. Safe, O. Hutzinger, N. Platonov, and D. Jones, *Metabolism*, **4**(3), 121 (1975).

256. Ih. Chu, D. C. Villeneuve, V. Secours, and A. Viau, *Agric. Food. Chem.*, **25**(4), 881 (1977).

257. Ih. Chu, V. Secours, D. C. Villeneuve, and A. Viau, *Bull. Environ. Contam.*, **18**(2), 177 (1977).

258. J. M. Daly, D. M. Jerina, and B. Witkop, *Experientia*, **28,** 1129 (1972).

PART 2   Halogenated Hydrocarbon Insecticides

The organochlorine insecticides have been used worldwide for some 35 years. The first compound to be introduced was DDT in 1943, with related insecticides following shortly thereafter. The acute toxicity of these compounds varies considerably (Table 49.13).

Table 49.13. Acute Toxicity of Organochlorine Insecticides[a]

Acute Toxicity	Pesticide	Estimated Single Lethal Oral Dose For an Adult	Rat Oral LD_{50} (per kg Body Weight)		
			Toxic Substances List 1974	Hayes, 1963 (9) (Males and Females Combined)	
Highly	Endrin	$\frac{1}{2}$ tsp	10 mg	13 mg	
Toxic	Thiodan	1 tsp		31 mg	
	Dieldrin	1 tsp	46 mg	46 mg	
	Aldrin	1 tsp	39 mg	50 mg	
Moderately	Toxaphene	1 tsp	60 mg	85 mg	
Toxic	Lindane	2 tsp	76 mg	90 mg	
	Heptachlor	3 tsp	40 mg	131 mg	
	Kepone	3 tsp	95 mg	—	
	p,p'-DDT	1 oz	113 mg	116 mg	
	Chlordane	1 oz	285 mg	283 mg	
Slightly	Kelthane	2 oz		1.05 g	
Toxic	DDE	2 oz		1.16 g	
	Perthane	7 oz		4.0 g	
	Methoxychlor	12 oz	5.0 g	6.0 g	

[a] Reproduced with permission by American Medical Association, August 29, 1979.

These insecticides served two general purposes; they were used extensively, and in certain countries they are still used extensively, as agricultural insecticides and as agents in combating such vector-borne diseases as malaria, typhus, plague, Chagas' Disease, yellow fever, dengue, encephalitis, filariasis, and African trypanosomiasis (sleeping sickness) (1). Of these insecticides, DDT is credited as the primary compound which, *for the first time in the history of man, brought epidemics of malaria, typhus, and plague to a complete stop.*

The persistence of these insecticides in the environment and their prolonged potency against certain pests can be attributed to their insolubility in water, high solubility in fats, absorption and adsorption on particulate matter, and resistance to chemical, physical, and microbiologic degradation. From plants that were sprayed, and from soils and waters, these compounds have entered the food chain of mammals, birds, fishes, and other living matter. Some of the insecticides more than others (primarily because of their more extensive use) have moved through the food chain of lower forms of living matter, and of fishes and birds, with the result that some of the birds toward the summit of the chain, those preying on fish and other birds, have suffered acute insecticide intoxications and also fatalities. In 1970, at the height of environmental insecticide use in the United States, Canada, and European countries, significant decreases in eggshell thickness were found in 15 of 22 species of aquatic birds, particularly in those feeding in fresh and brackish waters near agricultural areas. According to King et al. (2), populations of the following species declined and deserve continued study: brown pelican, reddish egret, white-faced ibis, laughing gull, and Forster's tern.

Some reports have dealt with DDT as the cause for larval fish kills. Based on studies with the great barracuda (*Sphyraena barracuda*), this can not have been a general problem in fish, since it was found that at the beginning of the spawning season, body fat with its load of organochlorine insecticides began to disappear as the gonads began to develop. At the height of the spawning season and with maximal development of the gonads, body fat and pesticides had disappeared almost completely from the fish. The ripe roe contained a lower concentration of pesticides than other body tissues (3).

In the early 1970s, because of the general environmental effects of these pesticides, but primarily because they were declared to be ". . . potential human carcinogens," the use of most organochlorine pesticides was discontinued or markedly curtailed in the United States, Canada, and most European countries (4–6). The envisioned hazard, cancer of the liver, was based on observations in laboratory animals, almost exclusively in mice, in which these compounds produced hepatomas and hepatocellular carcinomas.

The industrial workers (mostly men) who have been engaged in the manufacture, handling, and spraying of DDT, aldrin, dieldrin, toxaphene, chlordane, and heptachlor have been exposed to very considerably higher concentrations and quantities of these insecticides than the general U.S. public. Among these population groups, aside from eye, skin, or respiratory irritation, particularly from dusty materials, or instances of acute overexposure occurring primarily during the early years of production of some of these compounds, or induction of microsomal enzymes and the ability of some highly exposed workers to increase their drug-metabolizing ability, frank and undisputed injury to the liver or other organs has not been reported in the U.S., Canadian, and western European literature. To the best of our knowledge the organochlorine insecticides (individually and in combination), that have been ingested with home- and restaurant-prepared food and drink by the general U.S. population for more than 25 years (DDT was introduced in 1943), followed by a period of greatly reduced intake and exposure of some 10 or more years, have caused no recognized or clearly defined harmful effects.

In 1975, Kraybill (7) wrote, "None of the pesticide chemicals thus far have been shown to be carcinogenic to man." To the best of information, no reports have appeared as of 1980 relating exposure to the pesticides (referred to here) to cancer in man.

From 1950 to 1972, the U.S. production of DDT in thousand kilogram units equaled 1,204,700. The production of the aldrin–toxaphene group (aldrin, chlordane, dieldrin, endrin, heptachlor, toxaphene) over the years 1952–1972 totaled 865,600 10^3/kg units; a total of 41,500 10^3/kg units of benzene hexachloride was manufactured in the United States from 1950 to 1963. Of these quantities, approximately 50 percent of the DDT and 80 percent of the aldrin–toxaphene group were used in the United States (8).

When it is remembered that the organochlorine insecticides were considered potential human carcinogens, it is of interest that there has been a significant, almost constant decrease in the total rate of liver cancer deaths (classified since 1949 as *primary, secondary,* and *not stated whether primary or secondary*) in the continental United States, namely, from 8.8 (per 100,000 population) in 1930, to 8.4 in 1944 (when DDT was introduced for use), and to 5.6 in 1972. Based on the *U.S. Vital Statistics* for the *general* U.S. population, this almost steady decline in total liver cancer deaths for a 42-year period is even more significant in light of the constantly increasing life-span of the people of the United States, which in turn has resulted in a constant increasing percentage of the population "at risk" to liver cancer (8).

1 DDT, Dichlorodiphenyltrichloroethane, 1,1,1-Trichloro-2,2-bis(p-chlorophenyl)ethane

1.1 Uses and Occupational Exposures

DDT was released for commercial use in the United States on August 31, 1945; 1946 was the first full year of use. At that time, DDT was introduced in the malaria-stricken regions of the world and remains today one of the primary insecticides for the control of malaria through the eradication of vectors. From 1946 to 1972, DDT was one of the most widely used agricultural insecticides in the United States and other countries.

Toxicologically, the manufacture, handling, and use of DDT have not presented unusual problems. The only confirmed serious incidents of injury were those resulting from massive accidental or suicidal ingestion of DDT. The many years of manufacture, agricultural use, and malaria spraying operations involving this insecticide have not proved hazardous to the health of the greatly diversified occupational population groups thus engaged.

1.2 Physical and Chemical Properties

Physical state	Colorless crystals or white to slight off-white powder with odorless or slight aromatic odor
Molecular weight	354.5

Specific gravity	0.98 to 0.99
Melting point	108.5 to 109°C
Boiling point	260°C
Vapor pressure	1.5×10^{-7} mm Hg (at 20°C)
Solubility	Practically insoluble in water, diluted acids, alkalies; 78 g/100 ml benzene; 58 g/100 ml acetone; 45 g/100 ml carbon tetrachloride; 116 g/100 ml cyclohexanone

$$1 \text{ mg/liter} = 69 \text{ ppm and } 1 \text{ ppm} = 14.5 \text{ mg/m}^3 \text{ at } 25 °C, 760 \text{ mm Hg}$$

1.3 Effect in Animals

DDT is slowly and incompletely absorbed from the gastroenteric tract. When present in the air in the form of a small-particle-sized aerosol or dust, it is likely to enter the alveoli of the lung, from which it is readily absorbed. Skin absorption of DDT depends on the solvent or vehicle employed: crystalline or powdered DDT is not easily absorbed through the skin, but emulsions of DDT are absorbed to some degree (9).

Animal studies have demonstrated that the primary effect of DDT is on the nervous system. Toxic doses produce vomiting, apprehension, excitement, muscle weakness, disturbance of equilibrium, and finally clonic or asphyxial convulsions, followed by death from respiratory failure or ventricular fibrillation. The appearance of tremors is the most noticeable response to chronic exposure (10).

1.3.1 Air Dispersion

Neal et al. (11) reported exposures of animals to aerosols of DDT. Dogs, rats, and guinea pigs were exposed to an initial concentration of 54.4 mg of DDT/liter of air for a period of 45 min. No indications of toxicity were observed. They also reported that the oil used as a solvent had some effect on the response. Mice tolerated 6.22 mg of DDT/liter of air without manifesting signs of toxicity when the solution contained 6 percent sesame oil; when the concentration of oil was 9.5 percent, toxic effects were observed.

Cameron and Burgess (12) exposed several species of animals to a concentration approximating 1000 ppm w/v in air (mg/m³) for a period of 2 hr daily. The animals showed signs of intoxication, and deaths occurred after four to 10 exposures.

A rhesus monkey showed no signs of intoxication during or after two 7-hr periods of exposure to 0.13 mg DDT (Neocid) per liter of air on the first day, and to 0.4 mg/liter on the second day. Six rats concurrently exposed on the first day showed mild tremors; six other rats exposed on the second day demonstrated tremors that lasted 3 days. All rats survived. A rabbit, a cat, and a guinea pig exposed 7 hr/day for 3 days to 0.2, 0.3, and 0.45 mg DDT/liter of air, respectively, demonstrated no ill effects (10).

1.3.2   Acute Oral Exposure

Woodard et al. (13) investigated the acute oral toxicity of DDT. They reported some deaths in rats at a dosage of 140 mg/kg. Rabbits survived 260 mg/kg, but some died at 400 mg/kg. Mice survived 399 mg/kg, but some died at 448 mg/kg. Guinea pigs survived 178 mg/kg, but some died at 224 mg/kg. Lehman (14) reported the approximate LD_{50} for rats as 250 mg/kg.

Deichmann et al. (10) reported the approximate LD_{50} of DDT (GNB-A) for rats (from Wistar Institute) to vary with the solvent, as follows (in mg/kg): olive oil, 240; cyclohexanone, 280; corn oil, 420; propylene glycol, mineral oil, or cream (18 percent fat), 940; and DDT in 40 percent ethanol, aqueous methylcellulose, or tri-o-cresyl phosphate, 1400 mg/kg. They further reported that a room temperature of 30°C retarded, whereas a temperature of 5°C shortened, the onset of tremors. The lethal dose was not altered by these environmental temperatures. The immediate toxicities of recrystallized DDT, the American product identified as GNB-A, GNB-S, the Swiss product, were very similar.

1.3.3   Chronic Oral Exposure

Fitzhugh and Nelson (15) fed rats a diet containing DDT for 2 years. At concentrations of 600 and 800 ppm in the diet, the animals showed moderately severe tremors, particularly during the early months. An occasional animal had tremors at 400, but rarely at 200 ppm. Concentrations of 400 ppm and above produced a higher than normal rate of mortality, and an increase in weight of the liver. At concentrations of 600 or 800 ppm, there was indication of increase in weight of the kidney. Histopathological studies showed moderate liver damage at concentrations of 200 ppm and above. There were slight indications of liver damage, even at 100 ppm.

Treon and Cleveland (16) indicated that rats maintained for 18 months to 2 years on a diet containing 25 ppm DDT showed an increase in liver weight. At 12.5 ppm in the diet for a similar period, they reported no effect. Dogs on a diet containing 30 ppm of DDT for a period of 15.7 months showed no effect. Laug et al. (17) indicated that hepatic cell alterations were seen at 5 ppm in the diet (and higher), but not at 1 ppm. These authors observed an accumulation of DDT in body fat, even at 1 ppm in the diet.

For additional information on early studies on the acute or chronic toxicity and the mechanism of action of DDT, refer to *The Toxicity of DDT* (10).

1.3.4   Skin Absorption

Experiments by Draize et al. (18) indicated that dusts and solutions of DDT may cause a slight to moderate erythema on the skin of rabbits. Using a dry dust of 5 percent DDT in talc, they were unable to find any indication of a systemic toxic effect due to absorption through the skin; neither was there indication of systemic effects following application of a 10 percent solution of DDT in corn oil in doses up to 940 mg of DDT/kg body weight. Skin application of solutions of 25 and 30 percent of DDT in dimethyl phthalate or in dibutyl phthalate in doses up to 9.4 ml/kg for a 24-hr period induced signs of toxicity but no fatalities.

1.3.5   Absorption, Metabolism, and Excretion

After absorption in mammals, including man, the degradation of DDT proceeds by dehydrochlorination to the unsaturated DDE and by substitution of hydrogen for one chlorine atom yielding DDD (TDE). DDD is further metabolized through a series of intermediates yielding DDA. DDA is relatively water soluble and is excreted primarily with the urine (19).

Roan et al. (20) studied the urinary excretions following ingestion of DDT and DDT metabolites in six volunteers given technical DDT (5, 10, 20 mg/day) or p,p'-DDE, p,p'-DDD, or p,p'-DDA (5 mg/day for 21 to 183 days), demonstrating that within 24 hr of ingestion of DDT, urinary DDA excretion increases detectably. According to these investigators, excretion of DDT as DDA appeared to be totally dependent on the preferential reductive dechlorination of DDT to DDD (rather than DDE) and thence to DDA.

DDT and DDE are fat soluble and may be retained in the body fat of man for years. It is generally recognized that of the DDT isomers and metabolites, DDE is the compound most widely distributed in nature.

Human hair has been found to contain chlorinated hydrocarbon insecticides and other halogenated compounds, probably biphenyls, in concentrations less than 1 ppm. In experiments with rats, chlorinated ^{14}C-hydrocarbon insecticides and PCBs were excreted in hair in all instances, suggesting that this route of excretion may be more important in eliminating certain chlorinated hydrocarbons than was formerly recognized (21).

1.4   Human Experience

1.4.1   Malaria

According to Khambata (22), up to 1974 more than 1000 million people had been "freed" from malaria. It is because of this enviable record that DDT will continue to be used as an antimalarial agent in countries where this and related diseases are endemic (despite the fact that some vectors have become resistant to DDT). Therefore, the millions of people for whose benefit DDT is being sprayed will continue to be the primary nonoccupational exposure group. In these individuals the compound is absorbed by inhalation of the spray materials in their homes (huts) and fields, as well as by ingestion with their food and water. The annual 1977 production of DDT worldwide was estimated to be 100,000 tons (23).

Because of the extensive use of DDT for the control of malaria, typhus, and related diseases, it will continue to be a universal pollutant; therefore, its role in the trophic chain will continue. Planktons, bivalves, fishes, birds, and mammals, as well as the populace in parts of the world distant from the areas sprayed will also suffer some degree of exposure. Table 49.14 presents a brief summary of concentrations of DDT in various biologic structures (man, fish, birds), soils, and sediments.

1.4.2 Occupational Exposure

Ortelee (66) reported on a well planned study of 40 men exposed to DDT during its manufacture and formulation. Their exposure was followed by the analysis of DDT concentrations in urine, which was compared with the excretion of DDT by men whose oral intake of DDT was known. Ortelee concluded that it is unlikely that any illness will occur from DDT at the current dietary level, since the men studied showed no effects from occupational exposure for up to 6.5 years, during which time they absorbed an average of 200 times the quantity absorbed by the 1950 general population in the daily diet.

A study conducted by Laws et al. (24) of 35 men exposed for 11 to 19 years in a plant that produced DDT continuously and exclusively (2.7 million kg/month) from 1947 to 1967 did not reveal any ill effects in the workers attributable to DDT. The overall range of storage of the sum of isomers and metabolites of DDT in the body fat of these men ranged from 38 to 647 ppm, as compared to an average of 8 ppm in the general population. Based on the storage of total DDT in fat and the excretion of DDA in the urine, it was estimated that the average daily intake of DDT by the 20 workers with "high" exposure was 17.5 to 18.0 mg/man per day, as compared to an average of 0.05 mg/man per day for the general population. The concentrations of DDT in body fat averaged 344 times the concentrations in serum. The workers stored less DDE than DDT, which is related primarily to the intensity rather than to the duration of the exposure. DDA is much more important as an excretory product in those occupationally exposed to DDT than it is in members of the general population who absorb DDT primarily with their diet.

1.4.3 Studies in Volunteers

Hayes (67, 68), who wrote an excellent and comprehensive review on DDT and other pesticides, indicates that acute poisoning in man has occurred, but that

Table 49.14. Some Trends in Environmental and Tissue Concentrations of DDT (DDT, DDE, or DDD)

Investigation conducted	Type of Exposure	Refs.
	Occupational Exposure to DDT	
Early years of DDT manufacture	35 male workers exposed for 11–19 years; daily exp. 90–450 times that of general population. No ill effects	24
	12 male workers exposed for 5 years. Body fat concn. of DDT 307 ppm (141–739 ppm). Intensive exposure stimulated hepatic drug and steroid metabolism	25
1945–1971	31 of 35 workers previously reported on in Ref. 24. Exp: 3.6–18.0 mg/day for period averaging 21 years. "No evidence of hepatic disease or liver function abnormalities"	26
	Volunteers Ingesting DDT	
1954–1955	Daily intake by 19 men of 3.5 or 35.0 mg of recrystallized and technical DDT above dietary intake for 12 or 18 months. Concn. in body fat reached an average of 340 ppm for the pure material and 234 ppm for the technical compound after an identical dosage of 35 mg/man per day. No signs or symptoms of illness	27
	Same daily doses as above; 24 men, but for 21.5 months. After treatment, men were observed up to 5 years. Highest intake equivalent to approx. 720 times the average daily 1965–1970 dietary dose by general population. Concn. in body fat after highest dose of technical or pure DDT reached a mean of 325 and 281 ppm. ". . . No definite chemical or laboratory evidence of injury by DDT." The studies performed ". . . indicate a high degree of safety of DDT for the general population"	28
	Nonoccupational Exposure to DDT	
1964–1965	Concn. in air, agricultural communities during periods of pesticide application 0.1–8.0 µg/m³	29
Early 1960s	Concn. in air, urban areas, Pittsburgh, 1.14 µg/m³	30
	Dietary Intake of DDT	
	General U.S. population, average daily intake by a 70-kg individual:	31
1965	0.063 mg	
1966	0.070 mg	
1967	0.056 mg	
1968	0.049 mg	
1969	0.035 mg	
1970	0.028 mg	

Table 49.14. (*Continued*)

Investigation conducted	Type of Exposure	Refs.
1971	Concentration in variety of fruit and vegetables: up to 0.51 ppm	32
1967	Concn. in dairy products: 0.8 ppm	33
1972	Concn. in dairy products: 0.3 ppm	
1967	Concn. in meat, fish, poultry, max. 3.2 ppm	32
1972	Concn. in meat, fish, poultry, max. 0.9 ppm	
1968	Concn. in fat of beef, Arizona, 0.96 ppm	34
1970	Conc. in fat of beef, Arizona, 0.49 ppm	
1962–1964	Mean daily intake per person based on restaurant plus household meals, 0.321 mg; restaurant meals, 0.082 mg	35
	Concentration of DDT in Human Body Fat	
1961–1964	Concn. in body fat of the general U.S. population. Total as: DDT 5.7–19.7 ppm (dieldrin 0.11–0.31 ppm) DDT 2.3–7.4 ppm plus DDE 54.3–12.5 ppm	36
1968	In Utah, autopsy body fat of general population—9.01 ppm	37
1969	Concn. reduced to 7.15 ppm	
1970	Concn. reduced to 5.33 ppm	
1972	Concn. in 4 Negroes averaged 12.7 ppm (dieldrin 0.8 ppm)	
1966–1968	Concn. in Florida and California General white population: 9.84–17.0 ppm Spanish origin: 20.4 ppm Nonwhite: 42.2 ppm	38–41
1966–1967	Concn. in population dying in Miami area of carcinoma, atherosclerosis and hypertension: 17.38–24.85 ppm	42
1967	Concn. in 944 specimens of autopsy body fat in Chicago: average 9.6 ppm (lindane 0.48 ppm, dieldrin 0.01–1.39 ppm).	43
	Concentrations of DDT in Human Milk	
1951	Women in Washington, D.C.: 0–770 ppb	44
1974	Women in Mississippi and Arkansas, low income black: 447 ppb (59–1900 ppb)	
1974	Women in Nashville, Tenn., middle class white: 75 ppb (15–133 ppb)	45
1975	Women in Mississippi, black: 323 ppb (185–721 ppb)	
	Concentrations of DDT in Whole Fish	
1971–1972	Caught off Atlantic Coast of Canada, pelagic finfish, 0.1 mg/g DDT and PCB; bluefin tuna, frequently less than 1.0 mg/g	46
1973–1974	17 species of mesopelagic fishes, Gulf of Mexico: Generally 1.0 ppm (PCBs were ubiquitous, concn. one- to twofold higher)	47

Table 49.14. (*Continued*)

Investigation conducted	Type of Exposure	Refs.
1964–1965	Various species, estuary near Pensacola, Fla.: 0.01–1.26 ppm	48
1965–1968	Various species,1310 fish, major watersheds in Massachusetts: 0.17–28.8 ppm	49
1965–1969	28 species, 3801 fish in Great Lakes: 0.21–13.28 ppm (aldrin and/or dieldrin 0–0.47 ppm)	50
1967–1968	62 species in Great Lakes: 0.03–45.0 ppm, and major river basins throughout United States (aldrin and/or dieldrin 0 to nearly 2 ppm)	51
1969	Various species, San Francisco Bay: 0.01–0.41 ppm	52
1970–1971	Great Barracuda, east Florida coast, before spawning: 0.03–75.2 ppm in body fat (0.1 mg/kg fish); during spawning: 0–10.5 ppm in body fat (0.025 mg/kg fish)	53
1973–1974	Black and silver mullet, east and west Florida coast: 0.02–0.24 ppm (Concn. were essentially the same in gravid and nongravid fish)	54
1975–1976	Freshwater mullet, upper Great Lakes: trace–0.30 ppm (dieldrin 0–0.23 ppm; PCBs 0.06–0.79 ppm)	55
1972–1976	Juvenile fishes, 154 species in 144 estuaries nationwide: 0–213 mg/kg (net wt). (0 concn. in Alaska, Hawaii, Virgin Isl.; highest concn. in Delaware, Maryland, Puerto Rico) (PCBs: 0–1674 mg/kg, dieldrin 10–60 mg/kg). Estuarine pollution levels continue to decline	56
1968–1975	17 species peaked at 1.72 in Lake Superior, and 7.6 ppm in Lake Huron (lake trout). By 1975 the highest concn.: 1.06 and 1.87 ppm (PCBs 2 ppm, dieldrin in 5% of fish, 0.3 ppm)	57
	Concentrations of DDT in Soil and Sediments	
1970–1971	Sediments in Monterey Bay, Calif. coast: mean 10.8 ppb (max. 48.5 ppb)	58
1973	Three years later: mean 23.2 ppb (max. 112.0 ppb)	
1972–1975	Organic soil, Holland Marsh, Ontario: average 28.6 ppm (degradation was limited)	59
1967 1973–1976	DDT levels with soil samples in 1973 and 1976 were not significantly different from levels (4.5 ppm) measured in 1967	60
1971	U.S. agricultural soils in 37 states—45% of samples contained (in decreasing order): dieldrin, DDT, aldrin, chlordane, hepachlor epoxide; most pesticide levels ranged from 0.01 to 0.25 ppm	61
	Concentrations of DDT in Birds	
1970	In eggs of white ibis in Texas: 0.4 ppm In eggs of great egret in Texas: 23.2 ppm	62

Table 49.14.   (*Continued*)

Investigation conducted	Type of Exposure	Refs.
1969–1972	In total carcass of old-squaws in Michigan: 2–42 ppm DDE (eggshell thickness had declined 4.5 percent when compared with eggs collected before 1947) (PCBs 2–42 ppm)	63
1975–1976	In carcasses of 45 species, South Dakota: 0–32 ppm (0.01–0.54 ppm)	64
1969–1976	In 71 carcasses of 14 species of birds of prey, northern Florida: extremes 0–72.93 ppm (from 1973 to 1976 no significant increase or decrease was detected) (dieldrin ranged from 0 to 6.2 ppm; one hawk 38.24 ppm)	65

chronic intoxication in man has not been confirmed. An oral dose of 10 mg/kg produced illness in some men but not all; 285 mg/kg has been ingested without fatal results, although with toxic response. The tolerated chronic dose in man is not known, but from animal experiments, 2.5 to 5 mg/kg body weight per day might produce mild illness.

In 1956 Hayes et al. (27) reported on the ingestion of repeated oral doses of DDT. Three men completed 1 year of a dosage of 3.5 mg/man per day, and seven men completed 1 year at 35 mg/man per day (5 mg/kg body weight). This larger dose was about 200 times the daily rate at which the average person in the United States ingested DDT in the diet at that time. In this limited study, no evidence of injury related to DDT was reported by the men or was found by careful medical examination.

Hayes et al. (28) repeated and extended this investigation. Twenty-four male volunteers, aged 24 to 29, were given again either technical DDT or p,p'-DDT at the rate of 0.05 or 0.5 mg/kg per day for 21.5 months. They were then observed for an additional 25.5 months; 16 were followed for 5 years. The men on the high dosage of DDT for 21.5 months ingested the equivalent of approximately 1680 times the 1965–1970 U.S. dietary intake of p,p'-DDT.

Biopsy body fat DDT concentrations of the volunteers ingesting 0.5 mg/kg per day of technical or p,p'-DDT reached a mean of 325 and 281 ppm, respectively, in 18.8 months. Loss of tissue DDT progressed slowly after the ingestion of the high dose of DDT was discontinued; 22.5 months after the intake of the last doses of DDT, the concentrations in body fat in the two respective groups were reduced to 32 and 35 percent of the original values, respectively. DDE fat concentrations increased throughout the dosing period and continued to rise slightly even after discontinuation of the 0.5 mg/kg per day dose. DDA excretion in the volunteers was marked. All men were subjected to extensive physical, neurological, biochemical, hematologic, and organ function tests, which showed ". . . no definite chemical or laboratory evidence of injury by DDT," indicating ". . . a high degree of safety of DDT for the general population."

In a similar investigation, Morgan and Roan (19) administered DDT and certain closely related compounds to nonoccupationally exposed men. One volunteer ingested 10 mg of DDT per day for 183 days. Another received a daily dose equivalent to approximately 400 times the average (1965–1970) daily dietary intake of total DDT, namely, 20 mg/day for 183 days. A third received 5 mg *p,p'*-DDE/day for 92 days, and a fourth ingested 5 mg *p,p'*-DDD for 81 days. Extensive tests were conducted before, during, and after the periods of pesticide intake, but abnormalities or harmful effects were not detected in any of the men.

Ensberg et al. (69) found ". . . no shortening of antipyrine half-life" in volunteers who ingested 100 or 200 μg DDT for 21 days. Enzyme stimulation to a degree that reduced the half-life of one dose of 400 mg phenylbutazone was reported by Poland et al. (25). This was noted in 18 men of the Montrose Chemical Corporation whose serum and fat total DDT concentrations reached levels 20 and 30 times those of matched controls. Serum half-life of phenyl-butazone dropped 19 percent, whereas the urinary excretion of 6-β-hydroxy-cortisol rose 57 percent above normal. There were considerable variations in the susceptibility of these men to enzyme stimulation.

Kraybill (70) estimated that 90 percent of the total persistent pesticides ingested by the general U.S. and European populations originate with the daily diet. Of all foods consumed, those of animal origin, meat (particularly fatty meat), seafood, poultry (eggs), and dairy products, provide the major portion of these pesticides, including DDT. This is significant, since these food sources have been subject to only minimal direct application of insecticides; therefore, the presence of DDT, DDE, and related organochlorine compounds in foods of animal origin are related almost entirely to environmental sources.

Concentrations of DDT and its metabolites in the fatty tissues of the general U.S. population have shown a steady decline, from approximately 5.3 to 20.0 ppm for total DDT at the height of the DDT use in the 1960s to approximately 8.0 ppm (total DDT) in the early 1970s (Table 49.14). Indications are that these concentrations are still declining. Body fat concentrations of the organochlorine insecticides and of DDT in particular have varied widely among the U.S. population. In general, they have been considerably lower in infants and the very young than in adults, higher in adult males than adult females, higher in Negroes than Caucasians, and higher in population groups who live in or near agricultural areas and in those who used the insecticide in home or garden. Generally, the more body fat, the greater the *body load* of DDT or other organochlorine pesticides, but the lower the *concentrations* of these compounds in body fat (71).

Human breast milk contains about 3 percent fat; thus infants are likely to ingest, over a short period of time, higher concentrations of DDT and related compounds than do their mothers. However, as Hayes (72) recently reported, "Infants are in no danger from DDT in their mother's milk unless the dosage of the mother is one that approaches or perhaps reaches a level toxic to her." A study is under way in North Carolina to determine whether undesirable health effects are likely to result from ingestion of insecticides by infants (72).

Smokers have absorbed organochlorine pesticides from cigarettes. In 1969, the total chlorinated insecticide residues in cigarettes ranged from 12.5 to 63.4 ppm. DDT and DDD residues accounted for more than 97 percent; the remainder consisted of dieldrin and endrin. In 1974 the residues ranged from 3.6 to 7.1 ppm. Smoke condensate from one brand contained an average of 8.3 nanogram of DDT–DDD residues per cigarette (74). Commercial filters and increased butt length reduce the mainstream smoke residues considerably (75).

An epidemiologic study was reported by Fowler (76) in which a sizable population group in Mississippi was studied before and after a number of years of widespread application of DDT, both on crops and for mosquito control. It was concluded that in general, the health of the population had improved over this period of time. This could be attributed to improved sanitation, but it should also be noted that a number of disease-carrying insects were effectively controlled by the use of DDT. There was no indication that DDT produced deleterious effects on the population. Misuse of the compound did cause a few cases of acute intoxication.

1.5 Cancer

At this time there is no documented evidence that the dietary absorption of DDT, alone or in combination with insecticides of the aldrin–toxaphene group, has caused cancer in the general population (71, 77, 78). No evidence has as yet been presented that DDT has caused cancer among millions of individuals (almost entirely men) who have been occupationally engaged for as long as 35 years in the manufacture and handling or spraying of this insecticide (as dust, solution, and suspension) in all parts of the world and under all possible climatic conditions. Therefore, at this time, the only conclusion that can be justified is that DDT is not carcinogenic in man.

It is recognized that general population studies are not likely to uncover a low potential carcinogen, but if we accept the data reported by *Vital Statistics of the United States,* which provides cancer incident statistics, then it will be of interest and some degree of comfort to know that the death rates from cancer of the liver and its biliary passages, classified as *primary, secondary,* and *not stated whether primary or secondary,* have shown an almost constant decline in the United States over the past 42 years, namely, from 8.8 per 100,000 population in 1930, to 8.4 in 1944 when DDT was introduced, to 5.6 in 1972 (79).

The extensive literature on the possible carcinogenicity of DDT in animal species reveals that DDT, when ingested in high concentrations with the diet (as compared to human intake), causes hepatocellular carcinomas in several strains of mice (80–86).

Non-metastasizing liver tumors, according to the 1979 *IARC Monographs on the Evaluation of the Carcinogenic Risk of Chemicals to Humans*, occur in rats fed DDT (87, 88).

A recent report entitled *Bioassays of DDT, TDE and p,p'-DDE for Possible Carcinogenicity* was published in 1978 by the Carcinogenesis Testing Program of the National Cancer Institute (89). The summary reads as follows:

... There was no evidence for the carcinogenicity of DDT in Osborne-Mendel rats of B6C3F1 mice, of TDE in female Osborne-Mendel rats, although *p,p'*-DDE was hepato-toxic in Osborne-Mendel rats. The finding suggests a possible carcinogenic effect of TDE in male Osborne-Mendel rats based on the induction of combined follicular-cell carcinomas and follicular-cell adenomas of the thyroid. Because of the variation of these tumors in control male rats in this study, the evidence does not permit a more conclusive interpretation of these lesions; *p,p'*-DDE was carcinogenic in B6C3F1 mice, causing hepatocellular carcinomas in both sexes.

According to the 1979 *WHO Environmental Health Criteria Report*,

The evidence for the carcinogenicity of DDT in rats is not convincing and is negative in hamsters, ... negative results in dogs and monkeys are inconclusive because of the small groups studied and short duration of treaments; ... the carcinogenicity of *p,p'*-DDE is similar to that of DDT, but TDE produces a significant incidence of lung tumors.

All available evidence indicates that man does not appear to be susceptible to the tumorigenic action of the organochlorine insecticides and phenobarbital. No increase in the occurrence of tumors has been found in heavily exposed populations. This includes groups of workers who manufacture and formulate DDT and dieldrin and who have been examined carefully for tumors ... (24, 90, 91).

Yet, based primarily on the carcinogenic effects in some strains of mice, NIOSH announced (in 1978) that DDT should be handled in the workplace as a suspected occupational carcinogen [*DHEW (NIOSH) Publ. No. 78-174*, p. 150].

1.6 Mutagenicity

As pointed out in the NCI carcinogenesis report (89) DDT and its metabolites have been tested for mutagenicity in a variety of systems. DDT and DDE failed to revert histidine-requiring strains of *S. typhimurium* to prototype and, along with DDA, proved nonmutagenic in host-mediated bioassays in mice (92). However, highly significant increases in back-mutation rates were observed in two of the strains of *S. marcesceus* (used in the above studies) in the host-mediated bioassay with TDE; DDA proved positive for mutagenicity in *D. melanogaster*. DDT itself may be a very weak mutagen in *Drosophila* (93).

Yoder et al. (94) examined lymphocyte cultures from agricultural workers handling various insecticides, including DDT, for chromosomal aberrations during the peak spraying season and again during the winter. In the nonexposed controls no apparent differences in the number of chromatid breaks/person/25 cells examined was noted, but in the pesticide workers a fivefold increase in these lesions was noted during the peak spraying season.

1.7 Teratology

"Teratogenic effects of DDT have not been seen in studies of reproduction, including those for 2 generations of rats, 6 generations of mice, and 3 generations of dogs" [WHO, (90)].

Neither have teratogenic effects of DDT been noted in a six-generation Swiss white mouse reproduction study by Keplinger et al. (95). In this investigation, feeding of the parent generation (four males and 14 females) was continued through weaning of the second litter. Individual records were kept of each mouse of the first and second litters of each generation, and from the data collected, indexes for fertility, gestation, viability, lactation, and survival were calculated. At a feeding level of 25 ppm, the effects on lactation were questionable in the second, fourth, and sixth generations. At 100 ppm, the lactation index was lowered significantly in the first (parent), third, and sixth generations. The survival index was lowered significantly at this feeding level in the first, third, fifth, and sixth generations, and the viability index was lowered in the third and fifth generations. The reproduction study with 250 ppm DDT was discontinued after the second generation because of a high rate of fatalities among the pups. In general, effects were more severe when DDT was fed in combination with related organochlorine pesticides.

1.8 Synergism–Antagonism

Street (96) demonstrated that, in the rat, the combined administration of DDT and aldrin leads to an increased excretion of dieldrin. In beagles, the repeated administration of DDT plus aldrin resulted in a markedly augmented retention of DDT in blood and body fat (97, 98). (See Section 3 on dieldrin.)

Pretreatment of rats and sheep with DDT has been shown to potentiate the hepatotoxicity of carbon tetrachloride (99, 100).

1.9 Volatility

When technical grade DDT was applied to moist soil at concentrations up to 20 μg/g, the atmosphere in and above the soil contained approximately equal quantities of *o,p'*-DDT and *p,p'*-DDT, but at higher concentrations the ratio of *o,p'*- to *p,p'*-DDT increased in the vapor phase. The primary breakdown product of DDT, *p,p'*-DDE, has a higher vapor pressure than the original compound, *p,p'*-DDT, which indicates that much of the DDT present in the soil may volatilize as DDE (101).

The levels of dry aerial fallout of DDT were determined by Young et al. (102) over a coastline area from Point Conception to the United States–Mexico border, and were found to range from approximately 50 to almost 700 ng/m^2 per day in 1973–1974. The quantities generally increased with increasing proximity of the sampling stations to Los Angeles rather than to the agricultural regions. Sampling stations established near the site of a plant that formerly produced DDT, and near a landfill that received the wastes from this plant, provided a fallout of at least an order of magnitude higher than elsewhere in Los Angeles. The authors estimated that about 1.3 metric tons/year of DDT compounds fell onto the coastal waters annually in 1973–1974.

1.10 Photodecomposition

Crosby and Moilanen (103) studied the vapor-phase photodecomposition of DDT and found that UV irradiation of this compound resulted in the formation of chiefly DDE, with much smaller amounts of DDD and traces of other volatile compounds. The half-life of DDT exposed to UV radiation was about six days. DDE vapor decomposed relatively slowly while DDD vapor appeared stable. Ambient air samples from Davis, California, provided gas chromatograms with peaks corresponding to those of DDT, DDE (or its isomer), DDD, and 4,4'-dichlorobenzophenone (DCB).

1.11 Decontamination

According to Deshmukh et al. of New Delhi (104), the removal of DDT residues from domestic animals involves decontamination, for instance, of poultry by the forced moult procedure, and of dairy cows through administration of thyroprotein or charcoal. For the removal of DDT residues from animal feeds, they recommended washing with water or solvents, solvent vapor treatment, warm air treatment, commercial dehydration, or drying in the sun or by UV light. According to Crookshank and Smally (105), active charcoal is selective in the reduction of insecticide residues in the tissues of sheep. They found that feeding of charcoal with fodder caused an increase of DDT levels in the omental fat of these animals, whereas dieldrin residues decreased, heptachlor residues decreased slightly, and BHC levels remained the same. For the effect of fasting, see Section 3 on dieldrin.

1.12 Persistence in Soil and Water

The persistence of organochlorine insecticides in soils varies widely. In a study conducted by Agnihotri et al. (106) at the Indian Agricultural Research Institute in New Delhi, it was found that in 180 days, of the quantities mixed as 5 percent dusts into the top 15-cm layer of soil, 94 percent of p,p'-DDT, 97 percent of lindane, 96 percent of aldrin or dieldrin, and 86 percent of heptachlor were lost. The initial residue levels of these compounds ranged from 2.4 to 3.35 ppm.

Based on studies conducted in Georgetown County in South Carolina, Achari et al. (107) consider it possible that the presence of organochlorine pesticides in groundwaters may reflect a general pesticidal pollution across the nation rather than pollution initiated by the local use of pesticides. In all instances, the pesticide levels in groundwater samples from various areas were well below the recommended federal limits for drinking water.

Kiigemagi and Terriere (108) analyzed soils in Oregon after a 5-year lapse of pesticide use, and found a substantial decrease (about 30 percent) in the quantities of total DDT in the top 15 cm, but a sharp increase in the concentration of pesticide residues at the 63 to 91 cm level.

DDT (50 mg dissolved in 1000 ml of acetone) diluted with seawater in concentrations of 0.5 and 5.0 µg/liter showed half-life periods of 90 to 95 and 150 days, respectively (109).

1.13 Adrenocortical Hyperfunction

Mitotane, o,p'-DDD (Lysodren) has found use as an adrenal cytotoxic agent in the treatment of adrenocortical hyperfunction. For more details see more specific texts (110–115).

1.14 Hygienic Standards of Permissible Exposure

The threshold limits for DDT, adopted in 1979 by the American Conference of Governmental Industrial Hygienists (ACGIH), are TLV–TWA 1.0 mg/m^3 and TLV–STEL (tentative) 3.0 mg/m^3 (116). No limit was set in 1977 for the concentration immediately dangerous to life and health (117).

According to the 1978 Joint FAO/WHO Food Standards Programme Codex Committee on Pesticide Residues, the conditional acceptable daily intake (ADI) of DDT is 0.005 mg/kg body weight. As of 1978, the maximum residue limits (MRL) of DDT, DDD, and DDE, singly or in any combination, are established by FAO/WHO as follows: for nuts (shelled), strawberries, and root vegetables, 1 mg/kg, and for milk and milk products, 1.25 mg/kg. For other agricultural products, the MRLs range from 3.5 to 7 mg/kg (118).

2 ALDRIN, $C_{12}H_8Cl_6$, Hexachlorohexahydro-*endo, exo*-Dimethanonaphthalene
(Principal Constituent, known as HHDN)

Aldrin contains not less than 95 percent of HHDN and not more than 5 percent
of insecticidally active related compounds.

2.1 Uses and Industrial Exposures

Aldrin is used as an insecticide and for the control of termites around buildings.
Industrial exposure has occurred among all groups that have been involved in
the manufacture of aldrin, and in the handling and spraying of suspensions
and emulsions of this compound (see also dieldrin).

2.2 Physical and Chemical Properties

Molecular weight HHDN (principal constituent)	364.93
Physical state at 25°C	Tan to dark brown, solid
Odor	Mild chemical
Setting range, °C	54.4 to 65.6
Vapor pressure, mm of Hg at 25°C	6×10^{-6}
Flammability	Nonflammable
Chlorine content	57 to 59 percent W
Solubility	Moderately soluble in paraffins, aromatics, halogenated solvents, esters, ketones; sparingly soluble in alcohols; 11 ppb in water at 20°C
Corrosive action	Noncorrosive to steel, brass, monel, copper, nickel, aluminum
Stability	Stable in presence of ordinary organic bases, inorganic bases, alkaline oxidizing agents; stable with dilute acids but reacts with concentrated mineral acids, acid catalysts, acid oxidizing agents, phenols, active metals

2.3 Effect in Animals

Aldrin is a CNS stimulant. Toxic or lethal doses produce nausea, vomiting,
hyperexcitability, convulsions, and/or coma, followed by death initiated by

respiratory failure. Responses from chronic administration may include anorexia, loss of body weight, and degenerative changes in the liver. The CNS stimulant effects are general, not specific for this compound.

2.3.1 Acute Oral Toxicity

Lehman (14) gave the approximate LD_{50} for aldrin in rats as 67 mg/kg body weight. Tremors and convulsions were the characteristic responses to a toxic dose. Death may be delayed for several days.

Treon and Cleveland (16) reported the LD_{50} of a solution of aldrin in peanut oil in female rats to be 45.9 mg/kg, and in male rats as 49 mg/kg body weight. They reported the LD_{50} in dogs to range from 65 to 95 mg/kg. The toxic dose varies with the solvent in which it is given. Ball et al. (119) indicated that the toxicity of aldrin varies with the quality of the product and the nature of the formulation. They gave LD_{50} figures from 10.6 to 59.6 mg/kg for various preparations.

2.3.2 Chronic Oral Toxicity

Treon and Cleveland (120) fed aldrin to rats for a period of 2 years at concentrations of 2.5, 12.5, and 25 ppm by weight in the diet. There was no increase in mortality or decrease in growth at any of the levels fed. At 12.5 ppm and above, some animals showed increased liver weights and some degenerative hepatic cell changes. Ball et al. (119) fed rats concentrations of 5, 10, and 20 ppm in the diet. They saw no response over a 6-week period in which they fed the pure compound. The diet was then changed to comparable doses of aldrin as a commercial wettable powder. At 20 ppm, the growth rate showed an initial drop, then a gain to a level higher than normal. They observed no signs of neuromuscular involvement during the first 6 weeks, but after adding the wettable powder they observed hyperexcitability and other nervous system responses at levels of 10 and 20 ppm. Ball et al. also reported some disturbance of the estrous cycle of rats at concentrations of 10 and 20 ppm. Kitselman (121) reported parenchymatous degeneration of the liver and kidneys of dogs fed 0.02 and 0.06 mg/kg body weight per day for periods up to 1 year.

2.3.3 Skin

Lehman (14) reported that a 4 percent aldrin solution in dimethyl phthalate had an approximate LD_{50} of less than 150 mg/kg body weight. It produced no skin irritation, but the animals developed severe convulsions and died.

In animals, the acute dermal toxicity of dieldrin in xylene is roughly 40 times that of DDT. Tests with certain other solvents indicate a factor of only about six. An important difference is that undissolved DDT is not absorbed from the skin but undissolved dieldrin is readily absorbed (9).

2.3.4 Metabolism

Bann et al. (122) reported that aldrin is readily converted to dieldrin in the body and that it is stored as such, primarily in the fatty tissues. This has been confirmed in many later investigations.

2.3.5 Reproduction in Mice

Deleterious effects on fertility, gestation, viability, lactation, and survival indexes were noted in the parent (first) and second generations and in their offspring (first and second litters) of mice fed aldrin and dieldrin at 25 ppm each, and aldrin 10 ppm plus chlordane 100 ppm. Less marked, but still significant, effects were found in the first and second generations and their offspring after feeding aldrin at 3, 5, and 10 ppm, dieldrin at 3 and 10 ppm, and DDT at 100 and 250 ppm (96).

2.3.6 Starvation

Weanling, adult, and old Osborne-Mendel strain rats were fed for 4 weeks diets supplemented with 50 ppm DDT, 7.5 ppm aldrin, or a combination of 50 ppm DDT plus 7.5 ppm aldrin. They were subsequently starved for 6 days, with free access to water. This period of starvation resulted in marked loss of body weight, marked loss of total body lipids, decreased liver to body weight ratio, and a decreased total body lipid to body weight ratio.

As a result of severe starvation, "total DDT" (simple summation of DDT, DDE, and DDD) and dieldrin concentrations decreased in the blood of male and female rats of all ages, regardless of the pesticide supplement fed before starvation, but only in weanling male and female rats were reductions marked and statistically significant.

There was no distinct pattern to the effects of starvation on the concentration and retention of pesticides in the brain and kidney of male and female rats of all ages. In general, in all rats, except for adult and old females, starvation induced a decrease in both the concentration and the total quantity of "total DDT" and dieldrin in the liver. In these two groups of females, the opposite occurred. In male rats particularly, there was a marked conversion of DDT to DDD in the liver as a result of starvation.

The total quantity of pesticides in the total body decreased during the period of severe starvation, regardless of sex, age, or the pesticide supplement fed before starvation. On the whole, the effects were most marked in weanlings and least marked in old rats.

In females of all ages, starvation induced a moderate to a marked increase in the concentrations of DDT, its metabolites, and dieldrin in the abdominal fat. In male rats, "total DDT" increased, but dieldrin decreased in the abdominal fat.

With the exception of weanling male rats, starvation increased hepatic microsomal enzyme activity for the substrates tested: EPN, p-nitroanisole, and methyl orange.

Feeding of the DDT and/or aldrin supplement for 4 weeks to male and female weanling rats resulted in a significant increase in growth rate above that of weanling rats fed the control diet (123).

2.3.7 Enzyme Activities of Aldrin plus Parathion

In female Osborne-Mendel rats, a single oral dose of 20 mg/kg of aldrin (approximately 25 percent of an LD_{50}) increased liver enzyme activities as follows: O-demethylase, 320 percent; O-dearylase, 410 percent; N-demethylase, 350 percent; azoreductase, 190 percent; and nitroreductase, 300 percent. Oral administration of 1.5 mg/kg of parathion 1 hr after administration of 20 mg/kg aldrin decreased the above enzyme activities significantly. The highest levels reached were 110, 100, 105, 95, and 81 percent, respectively. Similar studies with DDT, chlordane, methoxychlor, and toxaphene indicated that the net results of hepatic microsomal enzyme activities of a combination dose of an organochlorine compound plus parathion is not predictable (124). Pretreatment of male Osborne-Mendel rats with an oral dose of 30 mg/kg of aldrin or dieldrin provided a significant degree of protection against the toxic effects of an LD_{50} of parathion (125).

2.3.8 Effects of Aldrin and DDT

Deichmann et al. (98) fed (by capsule) aldrin and DDT alone and in combination (5 times a week) to pure-bred beagles for 10 months. The administration of aldrin (0.6 mg/kg per day) resulted in a constantly increasing concentration of dieldrin in blood and in body fat which, after 10 months, reached a body fat concentration of 75 ppm (Figure 49.1). Discontinuation of aldrin administration resulted in a gradual decline in dieldrin fat concentration to 25 ppm after 12 additional months. The administration of aldrin at half this dosage (0.3 mg/kg per day) but in combination with DDT (12 mg/kg per day) resulted in the retention of roughly the same concentration of dieldrin in body fat (70 ppm). Observations were equally unexpected when the feeding of DDT was compared with the feeding (by capsule) of aldrin plus DDT. At 10 months, DDT alone (24 mg/kg per day) had produced a retention of 550 ppm of p,p'-DDT in body fat. Feeding of only half this dose of DDT (12 mg/kg per day), but in combination with aldrin (0.3 mg/kg per day), resulted in a body fat concentration of 1290 ppm of DDT (Figure 49.2). These beagle studies demonstrated that in

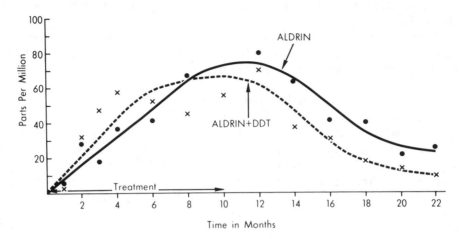

Figure 49.1   Retention of dieldrin in the body fat of dogs given 0.6 mg aldrin/kg per day (•) or 0.3 mg aldrin/kg per day plus 12 mg DDT/kg per day (×).

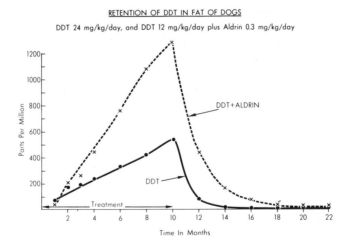

Figure 49.2   Retention of DDT in the body fat of dogs given 24 mg DDT/kg per day (•) or 12 mg DDT/kg per day plus 0.3 mg aldrin/kg per day (×).

3706

the dog, the fate of aldrin (dieldrin) is significantly influenced by the presence of DDT, and vice versa.

2.4 Human Experience

The effects of aldrin and dieldrin are similar, both quantitatively and qualitatively, in animals, and this, according to Hayes (9), appears to be true for man also:

Persons exposed to oral dosages which exceed 10 mg/kg frequently become acutely ill. A dosage of about 44 mg/kg led to convulsions in a child. Symptoms may appear within 20 minutes, and in no instance has a latent period of more than 12 hours been confirmed in connection with a single exposure.

The most thoroughly described related case involved an attempted suicide by ingesting aldrin at an estimated dosage of 25.6 mg/kg. There have been at least two deaths caused by the ingestion of undissolved dieldrin and several caused by drinking emulsions or solutions. The dosage in these cases is unknown.

No quantitative investigative work has been done with air dispersions of aldrin. Some clinical studies have been conducted on employees engaged in packaging and/or handling this material. Princi and Spurbeck (126) studied workers who were exposed through packaging the material, which undoubtedly led to some degree of absorption of aldrin through the skin and possibly by way of the respiratory tract. Actually, the exposure was to a mixture of chlordane, aldrin, and dieldrin. Analysis of the air showed concentrations of 5 to 57 mg/m^3, determined as chlorine and calculated as aldrin. By a special absorption technique, the actual aldrin concentration was determined to be between 1 and 2.6 mg/m^3. No evidence of a harmful response from the exposure was noted.

Nelson (127) studied a group of workers exposed to dusts of aldrin. He recorded complaints of headache, dizziness, nausea, and vomiting, but found no evidence of liver injury in these individuals.

According to Van Raalte (128), "It is certain that in industrial and other occupational (agricultural and public health use) situations, the principal route of intake is percutaneous. The amount deposited on the skin is much greater and much more important than the amount inhaled. This has been measured for endrin (Wolfe; Jegier), and, considering the physico-chemical properties, the same must be true for aldrin and dieldrin. The respiratory exposure is only a few percent of the total intake."

Only minor erythema is observed from skin contact with aldrin. It should be recognized that commercial preparations for agricultural use may contain other more irritating ingredients.

2.5 Carcinogenicity

See Section 3.7.

2.6 Hygienic Standards of Permissible Exposure

The threshold limits of aldrin (skin) adopted by the ACGIH in 1979 are TLV–TWA, 0.25 mg/m³, and TLV–STEL (tentative), 0.75 mg/m³ (116).

The IDLH, according to the 1978 *NIOSH–OSHA Pocket Guide to Chemical Hazards*, is 100 mg/m³ (117).

According to the 1978 FAO/WHO Standards Programme Codex Committee on Pesticide Residues, the ADI of aldrin and dieldrin (singly and in combination) is 0.0001 mg/kg body weight. The MRLs for agricultural commodities range from 0.02 to 0.15 mg/kg (118).

3 DIELDRIN, $C_{12}H_8Cl_6O$, Hexachloroepoxyoctahydro-*endo, exo*-Dimethanonaphthalene
(Principal Constituent, Known as HEOD)

Dieldrin contains not less than 85 percent by weight HEOD and not more than 15 percent by weight of insecticidally active related compounds.

3.1 Uses and Industrial Exposures

Dieldrin was first used by cotton growers in the 1950s; it has subsequently been used on other crops for the control of vector-borne diseases and for moth-proofing woolen goods. Dieldrin, as well as other cyclodiene insecticides, is uniquely suited for the control of termites. In 1974 the registration of products containing aldrin and dieldrin was canceled.

Occupational exposures have occurred among all groups that have been involved in the manufacture or handling of the compound, and in the spraying of dieldrin suspensions and emulsions. Overexposure, resulting in acute intox-ication, occurred primarily in the early days of dieldrin, aldrin, and endrin manufacture, and in spraying operations with these compounds in Kenya, India, Iran, and other malaria-ridden countries (9, 91).

3.2 Physical and Chemical Properties

Molecular weight HEOD (principal constituent)	380.93
Physical state at 25°C	Buff to light brown, solid dry flakes
Odor	Mild chemical
Melting point	95°C
Vapor pressure, mm Hg at 25°C	1.8×10^{-7}
Flammability	Nonflammable
Chlorine content	55 to 65 percent by weight
Solubility	Moderately soluble in aromatics, halogenated solvents, esters, ketones; sparingly soluble in aliphatic hydrocarbons and alcohols; 110 ppb in water at 20°C
Corrosive action	Noncorrosive to steel, brass, monel, copper, nickel, aluminum
Stability	Stable in presence of ordinary organic bases, inorganic bases, alkaline oxidizing agents; stable with dilute acids, but reacts with concentrated mineral acids, acid catalysts, acid oxidizing agents, phenols, active metals

3.3 Effect in Animals

Dieldrin is a neurotoxin. After absorption of a toxic dose in mammals, dieldrin acts as a nervous system stimulant. Dieldrin is stored unchanged, primarily in the fatty tissues. It is excreted as such, and in the form of several metabolites. Traces are secreted in the urine.

Histopathological organ and tissue changes have been studied primarily in rodents, following absorption of low to highly toxic doses. Many investigators have reported the production of characteristic hepatic "chlorinated insecticide" lesions (9, 10, 15, 129).

Virgo and Bellward (130), who fed various dietary concentrations of dieldrin to female Swiss-Vancouver mice, noted dose-related hepatomegaly, increases in cytochrome P-450 and microsomal protein, and a decrease in pentobarbital sleeping time. Hurkat (131) reported a decrease in liver glycogen and an increase in liver cholesterol in rabbits treated with dieldrin. The decrease in glycogen appeared to be due to the destruction of glucose-6-phosphatase in the membranes of the endoplasmic reticulum. These and related studies emphasize the specific toxicity of dieldrin for the liver in some species.

Wright and co-workers published two informative reports that dealt with the effects of dieldrin on the subcellular structure and function of mammalian liver cells, and on the effects of prolonged ingestion of dieldrin on the liver of male rhesus monkeys (132, 133). It is beyond the scope of this chapter to give the details of these publications. The first report discussed the significance of observations in rats, mice, beagles, and rhesus monkeys following administration of dieldrin and the carcinogen 4-amino-2,3-dimethylazobenzene (ADAB). Briefly, ingestion of dieldrin resulted in proliferation of the smooth endoplasmic reticulum of liver parenchymal cells, which was associated with an enhanced activity of the liver microsomal mixed function oxidative system. The dieldrin-induced alterations in liver subcellular structure and function were reversible in the rat, mouse, and dog. Regression was slow in the dog; it was not studied in the monkey. As expected, phenobarbitone elicited a similar response in the rat, mouse, and dog. The effects of ADAB on mouse liver contrasted with the effects of dieldrin and phenobarbitone, inducing a depression of liver glucose-6-phosphatase activity, with no increase in the activity of the liver microsomal mixed function oxidative system.

The second report deals with the biochemical aspects of an extensive study on the effects of an approximately 6-year exposure of rhesus monkeys to dieldrin at dietary concentrations of 0.01 to 5 ppm. Increases in the activity of liver microsomal monooxygenase system provided the most sensitive criteria effect. Significant increases were observed in the 1.75 and 5 ppm groups. The concentrations of dieldrin in the tissues of the 0.1 ppm group were very similar to those found in humans absorbing a similar daily dose of dieldrin per kilogram body weight.

It was unexpected that the concentrations of dieldrin in the monkey liver would be approximately 200 times higher than that found in the liver of the male mice which had ingested approximately 50 times more dieldrin per day than the monkeys. The dieldrin intake required for the induction of the rhesus monkey liver microsomal monooxygenase system was 25 to 30 μg/kg per day, which is a dose approximately 300 times greater than the daily intake of dieldrin by the general 1966–1967 population. The results obtained in these monkeys and the absence of detectable changes in human liver, particularly in the industrial population, point not only to a slow rate of metabolic clearance of dieldrin in these primate species, but also to a low degree of sensitivity of the liver to this compound (132, 133).

3.3.1 Reproduction

In a six-generation study, white Swiss mice were fed various concentrations of aldrin, dieldrin, DDT, chlordane, or toxaphene. Few or no adverse effects were noted through five or six generations fed toxaphene 25 ppm, chlordane 25 ppm, or DDT 25 ppm.

Marked effects in fertility, gestation, viability, lactation, or survival indexes were noted in the parent (first) and second generations and in their offspring (first and second litters) fed aldrin 25 ppm, dieldrin 25 ppm, and aldrin 10 ppm plus chlordane 100 ppm.

Less marked, but still significant, effects were found in the first and second generations and their offspring after feeding aldrin 3, 5, and 10 ppm, dieldrin 3 and 10 ppm, chlordane 50 and 100 ppm, and DDT 100 and 250 ppm.

Histological examination of the organs and tissues of mice revealed changes in the livers of all groups and in the kidneys, lungs, and brains of most groups. Typical liver changes included fatty metamorphosis with an increased amount of basophilic substances, hepatic cell necrosis throughout the parenchyma, but particularly near the central vein, and there was also moderate congestion. The kidneys demonstrated moderate vascular congestion, focal glomerulonephritis, and slight to moderate nodular lymphocytic infiltration. Also there were frequently slight dilatation of the convuluted tubules, cloudy swelling, desquamation, and pale basophilic masses in the tubular lumina. The lungs frequently showed moderate congestion and mild alveolar emphysema with minute hemorrhages. Some of the brain sections showed slight vascular congestion, edema in the parenchyma, and swollen upper motor neurons (96).

3.3.2 Forced Elimination from Tissues

Starvation proved to be the only practical method for augmenting dieldrin elimination in chickens (134). In another study, Sell et al. (135) found that when turkeys were subjected to three successive periods of fasting (7, 7, and 4 days), interrupted by periods of feeding (7, 12, and 24 days, respectively), there was acceleration in the decline of both the concentration of dieldrin in body fat and the total amount of dieldrin in the carcass. To be effective, starvation must be severe enough to reduce body lipids to approximately 10 percent or less of the carcass dry matter. Sodium barbital, charcoal, two anion exchange resins, Colestipol and cholestyramine, a high fiber diet, and a high energy protein diet were ineffective in augmenting dieldrin excretion in these birds.

3.4 Human Experience

3.4.1 Occupational Exposure

Dieldrin is readily absorbed through the skin and the gastroenteric tract, and by the respiratory tract following inhalation exposure. In man, overdoses have produced headache, vertigo, nausea, vomiting, and fatigue, followed somewhat later (depending on the dose) by muscle twitchings, myoclonic jerks, and convulsions (91). Hayes reported acute intoxications and some fatalities follow-ing excessive skin and inhalation exposures, plus some ingestion by spray personnel. Symptoms included ". . . sudden falls and convulsions with loss of consciousness." He also reported on men who had one or more "fits" 15 to 120 days following their last dieldrin exposure (136).

The potential hazards associated with the agricultural use of dieldrin and endrin in the Pacific Northwest were investigated by Wolfe et al. (137); in these studies these compounds were used as dust for the control of pests on potatoes, apples, and orchard cover sprays. Application was by hand gun attached to portable sprayers, by spray machines, and by air-blast machines. Few workers wore gloves during dusting operations. In all operations studied, calculations indicated that the potential dermal exposure was greater than the potential respiratory exposure. Respiratory exposure from dusting potatoes with 1 percent endrin dust was calculated to be 2.2 percent of dermal exposure. While spraying orchard cover crops with a liquid endrin formulation for mouse control, the spraymen were subjected to only 0.4 percent as much respiratory exposure as dermal exposure. When they sprayed pears with dieldrin, respi-ratory exposure was found to be 1.8 percent of the dermal contamination. The data presented indicate that the hazard from agricultural use of dieldrin and endrin, as practiced in the Pacific Northwest, is not particularly great when compared with the hazard associated with the use of the more toxic organic phosphorus compounds. The greatest hazard probably occurred when dieldrin and endrin, as emulsifiable concentrates, were measured and poured.

Jegier (138) arrived at similar conclusions, based on studies conducted with the agricultural use of endrin in the Canadian province of Quebec. He found the mean dermal exposure in six subjects to be 0.66 mg endrin/man per hour, whereas the respiratory exposure measured only 0.04 mg/man per hour.

Jager (91) summarized the data related to the medical supervision of workers exposed during the manufacture of dieldrin from 1954 to 1968: by 1968, 233 men had sustained exposure to aldrin or dieldrin for more than 4 years and 35 men for 10 to 13 years, totaling 1768 man-years of exposure. Geometric mean blood dieldrin concentrations (by gas–liquid chromatography) in these men for the six years 1964–1969 were 69, 59, 49, 31, 32, and 24 ppb, respectively. This, according to Hunter and Robinson (139), is equivalent to an approximate average daily dieldrin intake of 407 µg/man per day, or about 58 times the 1968 daily intake of dieldrin by the general U.K. population. Extensive clinical and laboratory tests conducted on these employees ". . . revealed no abnormal-ities other than those that would be expected in any group of 233 workers. Body weights, blood pressure, WBDs and SREs of the extreme exposure groups, compared before and after 10 years of exposure to those of a control group, did not show any effects from long-term intensive exposure with the insecticides aldrin, dieldrin, endrin, and telodrin" (91).

Jager established the half-time of dieldrin in blood at approximately 8.5

months, and a no-effect level for aldrin and dieldrin at a blood level of 0.105 mg/ml (105 ppb). This level was later modified by Van Raalte (140) to 200 ppb dieldrin in blood. This level corresponds to a total equivalent oral intake of 33 µg dieldrin/kg per day, or a total daily intake of more than 2000 µg/person per day (139).

3.4.2 Nonoccupational Exposure

General population groups have absorbed dieldrin primarily by way of food and drink. Of all foods consumed, those of animal origin, that is, meat (particularly fatty meat), seafood, poultry (eggs), and dairy products, provided the major portion of this compound. The home and garden use of this insecticide also contributed to the absorption and retention of dieldrin. For the years 1965–1970, the average daily dietary intake by a 70-kg individual in the United States was 0.7 µg aldrin, 4.9 µg dieldrin, and 0.3 µg endrin (31). There is no indisputable evidence that the dietary intake of dieldrin by these population groups has produced harmful effects; this includes nursing babies who have, at times, absorbed with the mothers' milk quantities or concentrations of dieldrin that considerably exceeded the ADI.

During the early 1970s the dieldrin blood level in the general U.S. population was approximately 0.3 ppb. The body fat concentration of the U.S. adult population (1966–1972) was approximately 0.18 ppm for dieldrin and less than 0.02 ppm for endrin (71).

3.5 Environmental Pollution

The local application and volatilization of dieldrin and related organochlorine pesticides contribute to their partial retention in soils and in surface and ground waters, thus contaminating the local environment as well as distant locations.

During its period of maximum use, dieldrin had a deleterious effect on susceptible wildlife, birds, marine life, and insects. Some species of insects developed resistance.

In countries where dieldrin is used for the control of malaria and related vector-borne diseases, the question will continue to be one of priority—disease control versus environmental pollution.

3.5.1 Photodieldrin

Dieldrin is partially converted to photodieldrin. Following the spraying of 5.6 kg/ha of dieldrin on pasture land, photodieldrin residues were detected in the grasses the day following the spraying. Five days after application, these residues had accumulated to a maximum concentration of 51 ppm, then declined to 9 ppm after 107 days. Photodieldrin accounted for one-third to one-half the total dieldrin residue after the first 23 days. About 26 g/ha of photodieldrin volatilized during the first 3 weeks after spraying, demonstrating that photo-dieldrin residues are less volatile than the parent compound dieldrin (141). Fifteen months after spraying the soil with 5 ppm of dieldrin (per soil dry weight), Weisberger et al. (142) identified three conversion products of pho-todieldrin in a sample of soil.

Photodieldrin appears to be metabolized by the plant after absorption into the leaf. One metabolic product was found in the leaves of kidney beans that had been exposed for 2 to 4 days to sunlight, but not to artificial light. At 20°C, only small losses occurred due to volatilization of photodieldrin; metabolites were not produced at this temperature. In water, with and without algae, ^{14}C-photodieldrin was persistent, but in the presence of algae, 40 percent of the added photodieldrin was adsorbed or absorbed by the algae (143).

3.6 Disposal by Incineration

One of the newest disposal methods of organochlorine wastes involves ocean incineration. The first officially sanctioned incident of ocean incineration in the United States occurred between October 1974 and January 1975 aboard the M/T Vulcanus in the Gulf of Mexico. A total of 16,800 metric tons of organoch-lorine waste was incinerated at a maximum rate of 25 tons/hr with 12,000°C minimum and 13,500°C average flame temperatures. More than 99.9 percent of the wastes was oxidized (144).

3.7 Carcinogenicity of Aldrin and Dieldrin

Aldrin and dieldrin are discussed together since further metabolic degradation of both compounds is similar.

In an investigation by Deichmann et al. (145), a total of 1100 Osborne-Mendel rats were fed a Purina diet supplemented with aldrin or dieldrin (20, 30, and 50 ppm per compound) or endrin (2, 6, and 12 ppm). The doses fed were intended to be excessive, short of producing signs of severe chronic intoxication. The overall tumor incidence (benign and malignant) in male and female rats fed aldrin or dieldrin was *significantly lower* than the tumor incidence of the control rats. However, there was no significant difference in the tumor incidence of the control rats and the experimental rats fed endrin. In the 956 rats examined histologically, *no* primary malignant hepatic tumor was found and only two benign hepatic tumors (hemangiomas), one in a male control rat and the other in a female rat fed endrin at 6 ppm.

The diet control male and female rats survived for a mean of 19.7 and 19.5 months, respectively. Because of chronic toxicity, six experimental groups showed a reduced life-span. The remaining 12 experimental groups of 50 animals each provided significant information on the noncarcinogenicity of these compounds to the rat. The mean survival (S) of these groups fed aldrin, dieldrin, or endrin is as follows: male and female rats fed aldrin: 20 ppm (S = 19.4 and 18.7 months, respectively), 30 ppm (S = 19.7 and 18.5 months), and male rats fed 50 ppm (S = 20.2 months); male and female rats fed dieldrin at 20 ppm (S = 19.5 and 20.5 months), male rats fed 30 ppm (S = 19.8 months); and finally, female rats fed endrin at 2 ppm (S = 20.8 months), and male and female rats fed this compound at 6 ppm (S = 19.6 and 19.3 months).

These negative carcinogenic effects support earlier reports on aldrin and dieldrin rat feeding studies by Cleveland (120), who fed 2.5 to 25 ppm, and by Song and Harville (146), who fed levels up to 285 ppm. Negative carcinogenic effects in rats were also reported by Walker et al. (147), who fed dieldrin for a period of 2 years in concentrations of 0.1, 1.0, and 10 ppm. They observed ". . . no tumorigenic activity which could be related to the feeding of dieldrin."

A second carcinogenic feeding study was undertaken by Deichmann et al. (148) with two strains of female rats (Osborne-Mendel and Sprague-Dawley), in the hope that these animals, at a dose level of 50 ppm, would survive for a longer period than those of the 1970 study, thus providing a better basis for assessing the effects of aldrin.

It was concluded that chronic feeding of female Osborne-Mendel and Sprague-Dawley rats with aldrin at 20 ppm and 50 ppm in the diet supported their previous rat studies; that is, aldrin does not produce benign or malignant liver tumors in these two strains of rats. Also, as noted before, aldrin at a feeding level of 50 ppm causes systemic toxicity as evidence by a reduced survival rate.

In the Osborne-Mendel rats, there was an elevated incidence of malignant lymphoreticular tumors in the 20 ppm group. The significance of this finding is questionable since a virus-related etiology cannot be excluded. These tumors were not noted in the female Osborne-Mendel rats fed 50 ppm, nor in the female Sprague-Dawley rats fed 20 ppm or 50 ppm of aldrin.

The most recent bioassay of aldrin and dieldrin for possible carcinogenicity was reported by the National Cancer Institute (149). Aldrin was fed to male and female Osborne-Mendel strain rats at dietary concentrations of 30 and 60 ppm, and to B6C3F1 hybrid male mice at 4 and 8 ppm concentrations, and to female B6C3F1 mice at 3 and 6 ppm. Dieldrin was fed to male and Osborne-Mendel rats at dietary concentrations of 29 and 65 ppm, and to male and female B6C3F1 mice at 2.5 and 5 ppm. All concentrations were time-weighted average doses.

3.7.1 Aldrin

In both male and female rats there was an increased combined incidence of follicular cell adenoma and carcinoma of the thyroid. The incidence was significant in low-dose, but not in high-dose groups, when compared with the pooled controls; however, when compared with matched controls, the incidence was not significant. In addition, cortical adenoma of the adrenal gland was observed in the aldrin-treated rats in significant proportions in the low-dose, but not in the high-dose females when compared with pooled controls. Because this increased incidence was not consistently significant, it is questionable whether the incidence of any of these adrenal tumors was related to the feeding of aldrin.

In male mice, there was a significant dose-related increase in the incidence of hepatocellular carcinomas. In the female mice, there was a trend in dose-related mortality, primarily with early deaths in the high-dose groups (149).

3.7.2 Dieldrin

In rats there was a significant difference in the combined incidence of adrenal cortical adenoma or carcinoma in the low-dose females and the pooled controls: "Although this tumor was also found in animals treated with aldrin, it is not clearly associated with treatment . . ." (149).

In male mice there was a significant increase in the incidence of hepatocellular carcinomas in the high-dose group, which may be associated with treatment. The incidence of neoplasm in the experimental female mice was much lower and ". . . probably not biologically significant." In the female mice there was a significant dose-related rate of mortality (149).

3.7.3 Conclusions

The NCI report (149) concluded:

Under the conditions of these bioassays none of the tumors occurring in Osborne-Mendel rats treated with aldrin or dieldrin could clearly be associated with treatment.

Aldrin was carcinogenic for the liver of male B6C3F1 mice, producing hepatocellular carcinomas. With dieldrin, there was a significant increase in the incidence of hepatocellular carcinomas in the high-dose males which may be associated with treatment.

Based on a review of the earlier aldrin–dieldrin rat or mouse cancer studies by Fitzhugh et al. (129), Deichmann et al. (145), Walker et al. (147), Stevenson et al. (150), Cleveland (120), Davis and Fitzhugh (151), and Thorpe and Walker (152), the NCI (149) report concluded that "there was no convincing evidence that aldrin or dieldrin was carcinogenic. However, several of the studies in mice showed an increase in liver lesions, usually termed 'hepatoma' in this species. Evaluation was not always possible because detailed data were lacking."

Aldrin and dieldrin have not been found to produce liver tumors in dogs (121) or Syrian hamsters (153).

3.8 Mutagenicity

The mutagenic potential of dieldrin was investigated through direct bacterial tests with and without microsomal activation, host-mediated assay, blood and urine analysis for active metabolites, micronuclei test, metaphase analysis, and dominant lethal test. The concentrations of dieldrin investigated were 0.08, 0.8, and 8 mg/kg per mouse. Overall evaluation of the data indicated that dieldrin was negative in all four animal tests. No increase in the number of mutants was found in any of the bacterial tests (154).

Dean et al. (155) conducted similar studies with HEOD, the major constituent of dieldrin, in mice and Chinese hamsters. Three test systems showed no evidence of induction of dominant lethality, chromosome breakage, or gene conversion in the animals. Additional studies were conducted by these investigators on short-term lymphocyte cultures from 21 workers currently or previously employed in the manufacture of dieldrin. The degree of chromosome damage in these workers did not differ significantly from that found in a control group of workers. These findings suggest that HEOD does not present a mutagenic hazard in mammals.

The results of these investigations were substantiated by Bidwell et al. (154) in comprehensive analyses designed to test the mutagenic potential of dieldrin.

3.9 Teratogenicity

Dix and Wilson (156) gave 32 pregnant rabbits 0.2 or 0.6 mg HEOD/kg daily from day 6 through day 18 of their gestation period. No indication of a teratogenic effect was noted. Ottolenghi et al. (157) administered aldrin 50 mg/kg or dieldrin 30 mg/kg each in a single dose to pregnant hamsters, and aldrin or dieldrin, 25 and 15 mg/kg, respectively, to pregnant CDI mice. They noted an increase in fetal deaths, congenital anomalies, and retardation of growth, but it must be noted that the vehicle used (corn oil) also produced fetotoxicity. In a later teratology study, dieldrin and photodieldrin were administered in doses of 1.5, 3.0, and 6.0 mg/kg per day on days 7 to 16 of gestation, to CDI mice and CD rats. In mice, the highest dose produced an increased percentage of supernumary ribs and a decrease in the number of caudal ossification centers. No such changes were observed in the rats. Both dieldrin and photodieldrin induced significant liver/body weight increases in the mothers.

In another teratology study by Dix et al. (quoted by Jager, 91), CFI mice were given oral doses of 0.25, 0.5, and 1.0 mg HEOD in corn oil. Control groups received the respective vehicles. Some maternal and fetal toxicity was noted in the mice dosed with DMSO and with HEOD in DMSO. No teratogenic effect occurred in either group.

3.10 Hygienic Standards of Permissible Exposure for Dieldrin

The threshold limits for dieldrin (skin) adopted by the ACGIH in 1979, are TLV–TWA, 0.25 mg/m^3, and TLV–STEL (tentative) 0.75/m^3 (116).

The IDLH concentration, according to the *NIOSH–OSHA Pocket Guide to Chemical Hazards* (1978) is 450 mg/m^3 (117).

According to the 1978 FAO/WHO Standards Programme Codex Committee on Pesticides Residues, the ADI of dieldrin and aldrin (singly and in combination) is 0.0001 mg/kg body weight. The MRLs for agricultural commodities, set by the Codex Committee in 1978, range from 0.02 to 0.15 mg/kg (118).

4 CHLORDANE (C$_{10}$H$_6$Cl$_8$), 1,2,4,5,6,7,8,8-Octachloro-3a,4,7,7a-tetrahydro-4,7-methanoindane

Technical chlordane contains chlordane isomers (approximately 60 percent) together with heptachlor (4 to 10 percent), and a variety of side-reaction products containing from six to nine chlorines (158, 159). The major chlordane isomers are designated as *cis*- and *trans*-chlordane (or α and γ), respectively; they occur in the ratio of approximately one-to-one (160).

4.1 Uses and Industrial Exposure

Chlordane has found use as an agricultural and home insecticide for the control of cutworms, ants, root weevils, rose beetles, grass hoppers, and grubbs, and is most effective in killing termites; a single application provides termite protection for more than 26 years. Related agents offer protection as follows: dieldrin, more than 25 years; heptachlor, more than 22 years; 8 percent DDT, maximum of 13 years; lindane, maximum of 12 years; and pentachlorophenol (in heavy oil), not more than 4 years (161).

4.2 Physical and Chemical Properties

Technical chlordane is a multicomponent organochlorine insecticide (162, 163).

    Physical state   Colorless to amber, viscous liquid; the commercial product is a mixture containing 60 to 75 percent of the pure compound and 25 to 40

	percent of related compounds. Chlorine content is 64 to 67 percent
Molecular weight	409.80
Specific gravity	1.59 to 1.63 (at 25°C)
Boiling point	175°C
Refractive index	1.56 to 1.57 (at 25°C)
Solubility	Insoluble in water; miscible with aliphatic and aromatic hydrocarbon solvents, including deodorized kerosine; decomposes in weak alkalies

$$1 \text{ mg/liter} = 59.7 \text{ ppm and } 1 \text{ ppm} = 16.76 \text{ mg/m}^3 \text{ at } 25°C; 760 \text{ mm Hg}$$

4.3   Effect in Animals

Response to the absorption of chlordane is not unlike that of other members of this group of chlorinated insecticides. The primary acute reponse is in the central nervous system. The signs of intoxication following absorption of a toxic or lethal oral dose include loss of appetite, irritability, hyperexcitability, vomiting, and tremors, leading to convulsions and death. Anorexia and loss of body weight may be marked if death is delayed. Poisoning from chronic exposure also produced effects on the central nervous system. According to Hyde and Falkenburg (164), electrocerebral disturbances serve as an early sensitive indicator of chlordane intoxication. Cellular changes in the liver may occur. Edema of the lungs and irritation of the gastroenteric tract have also been reported.

4.3.1   Acute Oral Toxicity

In 1952, Ingle (165) reported the LD_{50} for the rat to be 250 mg/kg body weight when dissolved in corn oil. Lehman (166) in 1952 reported the LD_{50} for the rat as 457 mg/kg body weight.

Chlordane manufactured before 1951 was more toxic than that manufactured during and after 1951. The greater toxicity of the early technical chlordane was partly due to the presence of hexachlorocyclopentadiene in the product. The rat oral LD_{50} value reported for purified chlordane in 1953 was 500 mg/kg (167), and for Velsicol technical chlordane (reported in 1954) was 570 mg/kg (168).

Chlordane, as well as heptachlor, is very toxic to many species of invertebrates. Both compounds are detrimental to populations of the pollinating species (CAST, 160).

4.3.2 Vapor Inhalation

According to Ingle's excellent monograph (169), Nickerson and Radeleff published reports on the effects of prolonged inhalation vapors when chlordane was used as a residual premise insecticide. Their conclusions were that ". . . no ill effect was observed in pigeons continuously exposed for 60 days to the vapors arising from surfaces treated with chlordane at the rate of 1,000 mg per square foot. . . . Histopathological studies revealed no lesions attributable to chlordane." Their conclusions were the same for chicks exposed to chlordane vapors for 30 days. Fog applications of 7 percent chlordane into a room housing rabbits, guinea pigs, white rats, mice, and poultry produced no obvious deleterious effects (170).

   The following toxicity tests were conducted at the Carworth Farms under the supervision of Dr. L. Ingle and Carworth Farms personnel. In these tests, there was not only inhalation of mists and vapors, but also dermal exposure and ingestion. The figures in parentheses in Table 49.15 show the percent of chlordane in deobase used in each test.

Table 49.15. Effects of Chlordane (L. Ingle) (169)

No. of Mice in cages	Single Exposures	Observations
10	Cages and mice sprayed (2%)	No signs of toxicity
10	Cages painted (5%); mice introduced subsequently	No signs of toxicity
10	Cages sprayed (2%); mice introduced subsequently	No signs of toxicity
10	Cages, shavings, and mice drenched by spraying 2%	1 death; 9 no toxic effects
10	Deobase (deodorized kerosine) only; cage, shavings, and mice drenched by spraying	1 death; 3 exhibited signs of toxicity; 6 no toxic effects

4.3.3   Skin Exposure

Eight cattle sprayed 12 times at 2-week intervals with 2 percent of the compound manufactured in 1953 showed no signs of local irritation or systemic intoxication. Spraying cattle with chlordane of earlier manufacture caused death in three of 10 cattle after three applications (171). Injury to the skin and mucous membranes by the early chlordane was significant. Later chlordane has not been shown to cause such irritation and is also more slowly and less completely absorbed (169).

4.3.4   Chronic Oral Toxicity

Ingle (169), in a 2-year chronic rat-feeding study, noted retardation of growth at concentrations of 150 and 300 ppm in the diet, but not at concentrations of 5, 10, and 30 ppm. Liver damage was marked at 150 and 300 ppm, but slight at 30, minimal at 10, and absent at 5 ppm. There was no injury to the kidneys at 5, 10, or 30 ppm, but there was marked injury at 150 and 300 ppm. The lung showed marked damage at 300 ppm, mild injury at 150, and no injury at lower concentrations.

In 1952, Lehman (166), after feeding the pre-1951 chlordane, reported that the minimal effective dose was 2.5 ppm and the maximum tolerated dose for rats over a 2-year period was 0.125 mg/kg per day. In 1953 Ambrose (167), in similar chronic rat-feeding studies, reported growth depression at 320 ppm, but normal growth at 160 ppm and less. He observed enlargement of the liver at 80 ppm, but no effect at 10 ppm.

A reinvestigation of technical chlordane was conducted by Ingle in 1955 (169)—when the material was fed to rats for 2 years at dosage levels of 2.5, 5, 10, 25, 50, 75, 150, and 300 ppm—and revealed the following:

1.   Hepatic cells (most sensitive indicator) showed cellular alterations at 50 ppm; there were no alterations and no variation from controls at 2.5, 5, 10, and 25 ppm.

2.   Growth retardation was apparent at 300 ppm, but not at 150 and lower concentrations.

3.   Mortality increased at 300 ppm, but not at 150 ppm or lower concentrations. The "no-effect level" was above 150 ppm.

4.   No tissue changes were noted that would suggest a carcinogenic effect.

For a more detailed review of chronic rat toxicity studies, refer to the monograph on chlordane by Ingle (169).

4.3.5   Absorption, Metabolism, and Excretion

Chlordane is absorbed through the skin, more readily via the lungs, and from the gastoenteric tract. It is retained primarily in body fat. Both isomers of chlordane are oxidatively degraded to a series of mono- and dihydroxy derivatives which are excreted in the feces and, to a lesser extent, in the urine (172). Nonachlor forms a small portion of the residue in the fat of rats (CAST, 160).

Balba and Saha (173) fed 1700 mg of ^{14}C-labeled alpha and gamma isomers of chlordane to four male rabbits, in four doses each. Of the alpha isomer, 77 percent was excreted in the feces and urine, and 84 percent of the gamma isomer. The alpha isomer was retained primarily in body fat, and successively less in kidney, muscle, liver, and brain; the highest concentrations of the gamma isomer were found in the kidney, followed by fat, liver, muscle, and brain. The concentrations of oxychlordane in the tissues (primarily in fat) were higher than those of the parent compound.

There has been little evidence that fish absorb chlordane from water, and there is equally little evidence of accumulation of chlordane residues in the tissues of aquatic invertebrates. Average concentrations of chlordane residues in freshwater fish examined ranged from 0.16 ppm to 1.01 ppm (174).

4.4 Human Experience

Chlordane is moderately toxic (Table 49.15). The very few reported incidents of acute chlordane intoxication in man or animal have resulted from gross negligence or misuse (9, 67, 71).

Use of early chlordane resulted, at times, in irritation of the eyes, mucous membranes, and/or skin of industrial and agricultural workers. This does not appear to have been a problem with the product manufactured since 1951.

Human exposure to vapors of 7 percent chlordane, for 15 min at 3-day intervals for periods of 12 weeks and repeated a year later, did not result in symptoms of toxicity (170).

According to CAST (160), the long-term effect on health due to chlordane in foods is "very small." Neither is there any evidence to indicate that the home and garden use of chlordane has constituted a significant hazard. Concentrations of chlordane in human body fat have been in the parts per billion range.

4.5 Volatility

Chlordane is translocated from plants and soils to which it is applied and enters the surrounding atmosphere and waters. Stauffer (159) found the vapor concentration over a 72 percent water emulsion of pooled liquid technical grade chlordane to be 213 ng/liter.

4.6 Persistence in Soil and Water

Chlordane is relatively persistent in the environment. It is adsorbed to soil solids; plants will absorb it from soils. Dorough and Pass (175) treated soils (at corn planting) with 455 and 910 g of "active ingredient" chlordane, or high purity chlordane. After 1 year, 50 to 70 percent of the residues had "dissipated" from the top 10 cm of soil. The whole corn plant harvested for silage 102 days after planting contained 0.03 to 0.04 ppm of chlordane (alpha and gamma isomers combined). The mature corn grain and cobs were free of detectable residues (less than 0.008 ppm). According to CAST (160), soil concentrations ranging from traces to mean values of 1.5 ppm have been reported. Residues rarely persist in detectable concentrations for more than 5 years.

A freshwater lake, initially free from detectable pesticide residues was treated with 10 ppb of technical chlordane. After 7 days, the lake water contained 4610 ppt total technical chlordane; by 421 days the level had dropped to 9.5 ppt or 0.095 percent of the initial concentration. Mean total chlordane residues in sediments of this lake were 35.3, 19.4, 33.9, 31.8, and 10.3 ppb (7, 24, 52, 279, and 421 days, respectively, after treatment). Neither heptachlor nor heptachlor epoxide was detected in sediments 279 days after treatment (176). Bioaccumulation of chlordane is similar to that of other organochlorine insecticides. In a terrestrial aquatic model system, algae, snail, mosquito, and fish accumulated ^{14}C-chlordane 98, 386, 132, 613, 6132, and 8258 times the ^{14}C-chlordane concentration in the water (177).

4.7 Cancer

In 1977, the National Cancer Institute (178) reported the results of a bioassay of chlordane which was examined for possible carcinogenicity. Groups of 50 male and female Osborne-Mendel rats were fed time-weighted average doses of chlordane: 203.5 and 407.0 ppm (males) and 120.8 and 241.5 ppm (females) for 80 weeks. They were then observed for 29 weeks. Groups of 50 B6C3F1 male and female mice were fed time-weighted average doses of 29.9 and 56.2 ppm (males) and 30.1 and 63.8 ppm (females).

The effect on survival rates indicated that mortality was dose-related for female rats and for male mice. Male control rats, for reasons unknown, showed an abnormally low survival rate.

In the experimental rats, there was significant statistical evidence for the induction of proliferative lesions of follicular cells of the thyroid, and of malignant fibrous histiocytoma, but these findings were discounted because the rates of incidence were comparatively low and/or are known to be variable in control rat populations.

Hepatocellular carcinoma failed to appear at a significant rate of incidence in the rats fed chlordane. Further, the number of lesions of the liver did not become significant with the addition of nodular hyperplasia, or with the application of life-table adjustment to the data.

In mice, hepatocellular carcinoma showed a highly significant dose-related trend. These high levels of significance were maintained when hepatocellular carcinoma was combined with nodular hyperplasia, or when the data were subjected to life-table adjustment. No other tumors were found in mice in sufficient numbers to justify analysis.

The report concluded that ". . . under conditions of this bioassay, chlordane is carcinogenic for the liver in mice."

On June 22, 1976, the Administrator of the United States Environmental Protection Agency initiated hearings on the continued use of chlordane and heptachlor, hearings that were terminated on February 21, 1978 (179). The testimony in the hearings dealt primarily with chronic laboratory studies involving chlordane and/or heptachlor with respect to whether one or both compounds cause cancer in laboratory animals, and what these results mean in terms of cancer risk in humans. Studies that were reviewed included those published by the National Cancer Institute (178, 180), U.S. Food and Drug Administration (181), International Research and Development Corporation (IRDC) (182), and Witherup et al. (183).

Since there was an obvious dispute among the pathologists in the interpretation of the experimental data and diagnosis of certain tissue slides, the Administrative Law Judge requested the National Academy of Sciences to review the data and render a report. This function was performed by the Pesticide Information Review and Evaluation Committee for the Advisory Center on Toxicology, Assembly of Life Sciences, National Research Council (184).

This committee concluded that ". . . chlordane and heptachlor epoxide, a metabolite of heptachlor, are carcinogens in the mouse." With respect to heptachlor, the evidence of hepatocellular carcinoma ". . . is not so clear." But studies of other possible neoplastic and preneoplastic changes in the liver ". . . suggest the probability that it too is carcinogenic in the mouse."

With respect to rats, the committee report stated that although a report of one previous bioassay "suggests that heptachlor epoxide is carcinogenic in the rat, examination of the slide made available to the committee did not confirm this" and that "There is no statistically significant evidence that any of the compounds are carcinogenic in rats" (184).

According to the Food and Agriculture Organization of the United Nations (185), in considering the production of hepatomas in certain strains of mice, "These liver tumors have not been found to develop in any species other than mouse as the result of exposure to dieldrin or chlordane, and these chemicals have been shown to be non-mutagenic in a variety of studies." In commenting on long-term studies, "Chlordane caused hepatocellular carcinoma in mice at a dose of 60 ppm in the diet, but not in rats at doses as high as 400 ppm for males and 240 ppm for females."

As to the question of risk to humans, the National Academy of Science Advisory Committee reported (184), in part: "There are no adequate data to show that these compounds are carcinogenic in humans, but because of their carcinogenicity in certain mouse strains and the extensive similarity of the carcinogenic action of chemicals in animals and in humans, the Committee concluded that chlordane, heptachlor and/or their metabolites may be carcinogenic in humans. Although the magnitude of risk is greater than if no carcinogenicity had been found in certain mouse strains, in the opinion of the Committee, the magnitude of risk cannot be reliably estimated because of the uncertainties in the available data and in the extrapolation of carcinogenicity data from laboratory animals to humans."

The National Academy of Science advisory committee (184), commenting on two ongoing epidemiological studies, stated that "The very limited data available do not indicate an increased risk of cancer in chlordane plant workers or in pest control operators," but emphasized that "the duration of these studies and the number of workers involved are not great enough to assess adequately the carcinogenic potential of chlordane and heptachlor."

Based on the scientific literature, there is no evidence that chlordane or heptachlor have caused cancer in man.

4.8 Hygienic Standards of Permissible Exposure

The threshold limits for chlordane (skin) adopted in 1979 by the American Conference of Governmental Industrial Hygienists were TLV–TWA, 0.5 mg/m^3, and TLV–STEL (tentative) 2.0 mg/m^3 (116).

The IDLH concentration of chlordane according to the 1978 *NIOSH-OSHA Pocket Guide to Chemical Hazards* is 500 mg/m^3 (117).

According to the 1978 Joint FAO/WHO Food Standards Programme Codex Committee on Pesticide Residues, the ADI of chlordane was recommended to be 0.001 mg/kg body weight. The MRLs of chlordane in various agricultural commodities range from 0.02 to 0.5 mg/kg (118).

On March 11, 1978, the Administrator of the U.S. Environmental Protection Agency signed a cancellation order for chlordane and heptachlor, but it provided for the continued use of chlordane and heptachlor for termite control, and the continued use of chlordane and/or heptachlor on a phase-out basis for a number of crops and insects (179).

5 HEPTACHLOR, 1,4,5,6,7,8,8-Heptachlor-3a,4,7,7a-tetrahydro-4,7,methanoindene (Principal Ingredient)

5.1   Uses and Industrial Exposures

Heptachlor is used as an insecticide. One of its most extensive uses (also of chlordane) has been in the control of certain soil-inhabiting insects that attack corn and other field crops. Heptachlor is also used for seed treatment. This insecticide and other cyclodiene insecticides are uniquely suited for the control of termites.

The potential for occupational exposure exists for personnel engaged in the manufacture, handling, and use of heptachlor.

5.2   Physical and Chemical Properties

Heptachlor is the chlorination product of chlordene. The technical grade product contains approximately 73 percent heptachlor, 22 percent *trans*-chlordane, 5 percent nonachlor.

Physical state	White crystalline solid
Molecular weight	373.32
Specific gravity	1.57 to 1.59
Melting point	95 to 96°C
Vapor density	3×10^{-4} mm at 25°C
Solubility	Insoluble in water; soluble in alcohol; slightly soluble in xylene, carbon tetrachloride, cyclohexane

1 mg/liter = 65.1 ppm and 1 ppm = 15.35 mg/m³ at 25°C, 760 mm Hg

5.3   Effect in Animals

Heptachlor is a chlorinated derivative of methanoindene, similar to chlordane. However, since the basic chemistry, biologic activity, and degradation products of heptachlor and chlordane differ, there are similarities as well as differences in the toxicity and effects of these two insecticides.

5.3.1   Acute Toxicity

Lehman (14) reported that when heptachlor was fed to rats in a single dose, the approximate LD_{50} was 90 mg/kg body weight. The principal responses were tremors and convulsions. Acutely toxic oral doses reported by Buck et al. (186) are for the rat 40 mg/kg (LD_{50}), mallard 2000 mg/kg (LD_{50}), rabbit 2000 mg/kg (LD_{50}), calf 20 mg/kg, and sheep 50 mg/kg.

The particle size and the solvents employed play a role in the toxicity of heptachlor. Lehman (2) reported (in 1952) that when the dried powder of heptachlor was applied to the skin of rabbits, the approximate LD_{50} was 2000 mg/kg body weight. When heptachlor was applied as a 20 percent solution in dimethyl phthalate, the approximate LD_{50} was less than 780 mg/kg, but when it was applied in repeated smaller doses, the approximate LD_{50} was less than 20 mg/kg body weight per day. There were no survivors after 14 doses of 28 mg/kg. No skin irritation from the materials was observed.

In earlier years, when heptachlor was being used on extensive areas in the southeast United States for fire ant control, ingestion of residues killed a number of birds, and resulted in the accumulation of sublethal doses in others. There seem to be no reports of heptachlor residues in fish-eating birds even when these birds have contained residues of other chlorinated hydrocarbon insecticides (160).

5.3.2 Chronic Oral Toxicity

The highest concentration that rats survived for 6 months was a dietary concentration of 30 ppm of heptachlor. At this level, a significant amount of heptachlor was stored in the fat of the animals (187).

Unpublished results from the Kettering Laboratory (188) and from the Food and Drug Administration (189) indicated that at least five chronic toxicity studies were conducted with rats, one with dogs, and one with mice. Liver damage similar to that produced by chlordane was produced in some of the animals. The maximum demonstrated no-effect level was a dietary concentration of 5 ppm heptachlor, which was equivalent ". . . to approximately 0.125 mg/kg per day for the dog, 0.25 mg/kg per day for the rat, and 0.75 mg/kg per day for the mouse" (160).

5.3.3 Absorption, Metabolism, and Excretion

Heptachlor can be absorbed through the skin and via the lungs and gastroenteric tract. In the rat, heptachlor is metabolized to heptachlor epoxide (187, 190, 191). Heptachlor epoxide is also the oxidation product of heptachlor in other animals, as well as in plants and microorganisms, but not all heptachlor is converted to the epoxide, which is more toxic and more stable than the parent compound (192–194). The epoxide is partially stored in body fat, where it may remain for prolonged periods. Dehydrochlorination of heptachlor epoxide, followed by hydroxylation and double-bond rearrangement, leads to the formation of a metabolite which is the principal form in which heptachlor is excreted in the feces (160, 195).

5.4 Human Experience

An acute intoxication by heptachlor is indicated by abnormal behavior, hyper-irritability, tremors, and convulsions. Acute heptachlor intoxications have been uncommon. Heptachlor has been handled and applied extensively in industry and agriculture (heptachlor since 1952, and chlordane since 1945) with an enviable record of safety. Although chlordane was used extensively in homes and gardens, heptachlor was not registered for this purpose in the United States.

Ingestion of heptachlor with the diet has not presented a problem. According to a report by the International Agency for Research on Cancer (196), the average daily intake of heptachlor and heptachlor epoxide in the United States decreased from 2.3 µg/day in 1965 to 1.4 µg in 1970; at present, the dietary concentrations are further decreasing. Only five of 79 foods of animal origin collected and analyzed in Louisiana from 1968 to 1972 contained heptachlor (including heptachlor epoxide) residues in detectable concentrations. The range was from 0.01 to 0.15 ppm, and the average was 0.06 ppm (197). There is little evidence of heptachlor accumulation in the food chain.

In the urine of the general population, the epoxide has been found in extremely low concentrations (198).

5.5 Persistence in Soil and Water

Heptachlor and chlordane are relatively persistent in the environment. Their half-life in the soil is approximately 0.8 and 1 year, respectively. When the insecticides are applied at agricultural rates, residues rarely persist in detectable quantities for more than 5 years after the last application (160).

Both heptachlor and chlordane are adsorbed on soil solids and hence tend to remain near the site of application; only small amounts are leached downward in the soil. Heptachlor can be absorbed by the roots of plants from the soil and transported to stems and leaves, but the amounts reported in plant tissues rarely exceeded 1 ppm (CAST). In surface waters, heptachlor and chlordane have been either undetectable or present in traces up to mean concentrations of 6 ppt (160).

The volatility of heptachlor exceeds that of chlordane; thus volatility is a major pathway of loss of these insecticides from soils (160).

5.6 Invertebrates

Hepachlor and chlordane are highly toxic to many species of invertebrates, and they are detrimental to populations of pollinating species. Both compounds are highly toxic to beneficial insects (160), and at very high rates of application highly detrimental to earthworms. Doane (199), Polinke (200), Schread (201), and Smith (202) found that field application at 1 kg/acre had no deleterious effect on earthworms.

For details on the hazard and toxicity of heptachlor to invertebrates and vertebrates, as well as on the behavior of heptachlor in the environment, refer to the excellent report prepared by the members of the Council for Agricultural Science and Technology (160).

5.7 Cancer

In 1977 the National Cancer Institute released a *Bioassay of Heptachlor for Possible Carcinogenicity* (180). This report, in its discussion, presents the following review of earlier feeding studies which were designed to investigate a possible carcinogenic action of this compound. In three of five unpublished long-term feeding studies, rats received diets containing 12.5 ppm of the test material for at least 2 years but showed no increase in tumor incidence attributable to treatment. In a fourth study, only an increase in liver weight was reported for rats receiving 10 and 20 ppm heptachlor. In a fifth study with rats, the incidences of tumors in the animals given 0.5 to 10 ppm [65/114 (57 percent) in males; 92/114 (81 percent in females)] were greater than in controls [8/23 (35 percent) in males; 13/24 (54 percent in females)], with most tumors located in endocrine organs and with liver tumors appearing in seven males and 12 females but not in the controls. In a study with C3Heb/Fe/J mice, the feeding of heptachlor or heptachlor epoxide at a level of 10 ppm in the diet has been reported to bring about an increase in liver tumors including carcinomas (203).

In the 1977 National Cancer Institute report on heptachlor (180), groups of 50 male Osborne-Mendel rats were fed time-weighted average doses of 38.9 and 77.9 ppm; similar groups of females were fed 25.7 and 51.3 ppm for 80 weeks. All animals were then observed for 30 weeks. Groups of 50 B6C3F1 mice of each sex were fed time-weighted average doses of 6.1 and 13.8 ppm (males) and 9 and 18 ppm (females).

No hepatic tumors were observed in rats administered heptachlor. There was significant statistical evidence for the induction of proliferative lesions of follicular cells of the thyroid in female experimental rats, but this finding was discounted because the rates of incidence were comparatively low and are known to be variable in the control rat population.

In mice fed heptachlor, hepatocellular carcinoma showed a highly significant dose-related trend in both males and females. No other tumors were found in these mice in sufficient numbers to justify analysis.

It was concluded that under the conditions of this bioassay, heptachlor is carcinogenic for the liver in mice. Based partially on this conclusion, EPA restricted its use (203).

5.8 Hygienic Standards of Permissible Concentrations

The threshold limits of heptachlor (skin) adopted in 1979 by the ACGIH were TLV–TWA 0.5 mg/m^3 and TLV–STEL (tentative) 2 mg/m^3 (116).

The IDLH concentration of heptachlor according to the 1978 *NIOSH–OSHA Pocket Guide to Chemical Hazards* is 100 mg/m^3 (117).

According to the 1978 FAO/WHO Standards Programme Codex Committee on Pesticide Residues, the ADI of heptachlor is 0.0005 mg/kg body weight. According to this 1978 report, the MRL of heptachlor and its epoxide, expressed as heptachlor for agricultural commodities, range from 0.01 to 0.2 mg/kg (118).

6   KELTHANE, 4,4'-Dichloro-a-(trichloromethyl)benzhydrol, Dicofol

6.1   Uses and Industrial Exposure

Kelthane is a synthetic nonsystemic organochlorine acaricide which has been used primarily for the control of mites on field crops, vegetables, citrus and noncitrus fruits, and in greenhouses.

6.2   Chemical and Physical Properties

Physical state	Crystals
Molecular weight	370.51
Melting point	77 to 78°C
Solubility	Very slightly soluble in water

1 mg/liter = 66.0 ppm, and 1 ppm = 15.15 mg/m^3 at 25°C; 760 mm Hg

6.3   Effects in Animals

The only information available on Kelthane is from animal experimentation. At high levels, animals usually showed a general weakness and coma prior to death. Tremors were not observed as they usually are with DDT, which is a closely related compound. Histopathological changes are limited to the liver and kidneys and are relatively mild in nature. It has been reported that Kelthane causes some suppression in adrenal cortical activity (204, 205).

6.3.1   Acute Oral Toxicity

The LD$_{50}$ for rats of a 20 percent solution of technical Kelthane in corn oil was 809 mg/kg for males and 684 mg/kg for females. The LD$_{50}$ for male rabbits was 1810 mg/kg and for dogs was more than 4000 mg/kg (204).

6.3.2 Chronic Oral Toxicity

Dogs fed for 1 year a diet containing 300 ppm Kelthane survived without effect. At a level of 900 ppm, two of four dogs died. There was no effect on body weight. Since this compound suppresses adrenal function, particular attention was given to the adrenal/body weight ratio, which was found to have remained within normal limits. At a dietary concentration of 300 ppm, there was a slight depression of plasma 17-hydroxycorticosteroids; at 900 ppm, this was marked. There was no evidence of micropathological organ changes. A dietary concentration of 300 ppm was found to be the no-effect level in the dog (204, 206).

Rats were fed dietary concentrations of 20, 100, 250, 500, and 1000 ppm of Kelthane for 2 years. Females fed levels of 250 ppm or higher, and males at 500 ppm and higher, showed depression of growth. Mortality increased at 1000 ppm. There was an increase in the liver to total body weight ratio in females fed 100 ppm; this was significant at 250 ppm and at higher concentrations. Kidney and heart to body weight ratios increased in females at 500 ppm; in males there was an increased ratio of testes to body weight at 1000 ppm. Hematologic changes remained within normal limits. Retention of Kelthane in adipose tissue was more marked in males than in females. The rate of disappearance of Kelthane was about 3/4 in male rats and in the females about 2/5 of the stored insecticide over a 13-week period. Hydropic changes in the liver were more marked in the experimental than in the control rats, but they were reversible. Carcinogenic effects were not noted. The no-effect level in rats lies between 20 and 100 ppm (204, 206).

6.3.3 Skin

Solutions of Kelthane in dimethyl phthalate caused erythema and superficial destruction of the skin of rabbits. Emulsions were more irritating, causing marked tissue destruction. When applied to the skin under a cuff, the compound was readily absorbed through the skin. The LD_{50} for the rabbit of a 30 percent solution of Kelthane in dimethyl phthalate was established at 2.1 g/kg body weight. Daily skin applications 5 days/week for 13 weeks of a 30 percent solution in dimethyl phthalate, at doses of 1 ml/kg and higher, caused death. Similarly, repeated applications of an emulsion of 18.5 percent active material caused some deaths in doses of 0.1 ml/kg and higher. A wettable powder containing 18.5 percent active ingredient with two parts water, applied repeatedly, caused deaths in doses of 0.5 g/kg and higher (204, 205).

6.4 Cancer

The results of a bioassay of Kelthane (technical grade dicofol) for possible carcinogenicity in Osborne-Mendel rats and B6C3F1 mice were reported by the National Cancer Institute (207). The high and low time-weighted average dietary concentrations of dicofol were, respectively, 942 and 471 ppm for male rats, 760 and 380 ppm for female rats, 528 and 264 ppm for male mice, and 243 and 122 ppm for female mice.

"Under the conditions of this bioassay, technical-grade dicofol was carcinogenic in male B6C3F1 mice, causing hepatocellular carcinomas. No evidence for carcinogenicity was obtained for the compound in Osborne-Mendel rats of either sex, or in female B6C3F1 mice" (207).

Review of the NCI *Bioassay of Dicofol* for carcinogenicity by the Data Evaluation/ Risk Assessment Subgroup of the Clearinghouse on Environmental Carcinogens (1978) (208) led to the following motion which passed unanimously: "Under the conditions of this bioassay, technical grade Dicofol produced no evidence of carcinogenicity in Osborne-Mendel rats of either sex or in female B6C3F1 mice; the failure to determine the stability of Dicofol throughout the study prohibits drawing any conclusion concerning its carcinogenicity."

6.5 Hygienic Standards for Permissible Exposure

According to the 1978 Joint FAO/WHO Standards Programme Codex Committee on Pesticide Residues, the ADI of dicofol is 0.025 mg/kg body weight. The MRLs of dicofol in vegetables and fruits (except strawberries) ranged from 1 to 5 mg/kg (118).

To date no TLVs have been adopted by the ACGIH.

7 CHLORDECONE, Decachlorooctahydro-1,3,4-metheno-2*H*-cyclobuta(c,d)pentalen-2-one,Kepone

7.1 Uses and Industrial Exposures

Chlordecone, a cyclodiene compound, was introduced under the trade name Kepone in 1958. In the United States, Kepone found some use as an ant and cockroach poison. Outside the United States, it was used primarily as a pesticide against the banana root borer. The compound is the ketone analogue of mirex; it is a contaminant of mirex and is also a product of mirex degradation. Production of Kepone in the United States was discontinued on July 24, 1975 because of severe industrial exposure and intoxications in men employed in the

Hopewell, Virginia plant, the only plant in the United States that manufactured Kepone.

7.2 Physical and Chemical Properties

Physical state	Crystals
Molecular weight	490.68
Melting point	Decomposes at 350°C
Vapor density	16.94
Vapor pressure	10^{-5} mm at 20 to 25°C
Solubility	Slightly soluble in water and hydrocarbon solvents; soluble in corn oil, alcohols, ketones, acetic acid

1 mg/liter = 49.8 ppm and 1 ppm = 20.07 mg/m^3 at 25°C, 760 mm Hg

7.3 Effect in Animals

7.3.1 Acute Oral Toxicity

Oral LD$_{50}$ values for Kepone are 132 mg/kg for male rats, 126 mg/kg for female rats, 71 mg/kg for male rabbits, and approximately 250 mg/kg for dogs. In these species, the outstanding sign of intoxication was the development of severe DDT-like tremors. These reached maximum intensity (in surviving animals) 2 to 3 days following treatment. The tremors gradually subsided over a period of a week or longer. Exacerbation of the tremors occurred whenever an animal became excited, for example, when handled (209).

Hyperexcitability and tremors were noted in mice the following day after one oral dose of 50 mg Kepone/kg; mortality occurred on the fifth day (210).

7.3.2 Skin Exposure

The percutaneous LD$_{50}$ for male rabbits was 410 mg/kg (209). Application of 0.25 and 0.50 g of the active ingredient (in ant bait) to the skin of rabbits (five 24-hr exposures over a period of 3 weeks) produced no signs of local irritation, no effect on body weight, and no significant micropathological organ changes (209).

Potentiation of hepatotoxicity of Kepone in rats was demonstrated by the subsequent administration of carbon tetrachloride (212). For an assessment of the neurotoxicity of Kepone, refer to the reports by End et al. (213) and Huang et al. (210).

Table 49.16. Effect of Dietary Concentrations of Kepone

Dietary Concentrations (ppm)	Species	Effects	Refs.
40	Rats	Feeding weanlings 40 ppm for 7 days caused a significantly exaggerated "startle response" which persisted throughout a 4-week feeding period	215
	Rats (M)	One oral dose of ^{14}C-labeled chlordecone. On day 1, distribution of Kepone ranged from 2.53 units in the adrenal gland to 8.38 units in blood. After 3 days, levels ranged from 6.6 units in the blood to 198 units in the liver; after 182 days, levels ranged from 0.03 units in the blood to 3.57 units in the liver. Half-life in the blood was 8.5 days during the first 4 weeks, approximately 24 days during the next 8 weeks, and 45 days during the following 14 weeks. A total of 12.7% of radioactive chlordecone was excreted in the feces during the first, 2.9% during the second, and 3.3% during the third 24 hr. During the first 7 days, a total of 0.7% was excreted in the urine. The total urinary excretion during 84 days was only 1.6%	216
	Mice	There were no fatalities during a 12-month feeding period. Feeding of 40 ppm for 5 days resulted in the retention of a concentration of 45 ppm of Kepone in the liver. Feeding of 40 ppm for 5 months resulted in liver Kepone concentrations up to 163 ppm; there were no further increases on continued feeding. Discontinuation of feeding resulted in a loss of 56% of Kepone from the liver in 24 days and 95% in 150 days	217
30	Mice	"Constant tremor syndrome," liver weights doubled in 60 to 90 days at this feeding level; micropathology included hypertrophy, hyperplasia, congestion, and local necrosis of the liver. Kepone was retained in the liver, brain, kidneys, and fat	217
25	Dogs	During 127 weeks of feeding, there was no proteinuria, but reduction in body weight and an increase in food consumption per unit body weight gain during the second year. There	209

Table 49.16. *(Continued)*

Dietary Concentrations (ppm)	Species	Effects	Refs.
		was no evidence of impaired liver function or micropathology of major organs after 25 months of feeding, but there was a significant increase in the ratio of the liver, kidney, and heart to total body weight	
	Rats	Near the end of the third month of feeding, tremors became evident in some rats. Depression of growth rate and proteinuria (particularly in males), and hepatotoxicity in a 2-year feeding study, also a significant increase in the ratio of liver to total body weight in both sexes	209
	Rats	Increase in ratio of testes to total body weight. Increase in oxygen consumption in females. Metabolic and other changes in rats not readily reversible after 12 months of feeding	217
	Mice	After daily oral dosing with 25 mg/kg, hyperexcitability and tremors were noted on the fourth day, and mortality on the seventh day	210
10	Rats	Increase in ratio of kidney to total body weight in females (217)	217
		After 3 months of feeding, minimal congestion in the liver; no significant micropathology after 12 months of feeding (209)	209
	Rats, Mice	Gastric intubation during the major period of organogenesis, days 7–16 of gestation. Fetal toxicity was noted in rats and mice at doses that caused significant reduction in maternal weight gains during gestation, also increased liver to body weight ratios. In rats, fetuses from dams receiving 10 mg/kg/day (which also caused 19% maternal mortality) exhibited a significant incidence of reduced fetal weight, reduced degree of ossification, edema, undescended testes, enlarged renal pelvis, and enlarged cerebral ventricles. Lower dose levels (6 or 2 mg/kg/day) also produced maternal weight loss, and liver to body weight ratio increase, but no mortality, and only a reduced degree of ossification and a reduced degree of	218

Table 49.16. *(Continued)*

Dietary Concentrations (ppm)	Species	Effects	Refs.
		fetal weight. Male rats born to dams treated with Kepone during gestation showed no impairment of reproduction. In the mouse, fetotoxicity occurred only in the highest dosage levels (12, 8, and 4 mg/kg/day) and was manifested by an increased fetal mortality and clubfoot	
	Mice	"pair days per litter" increased from 67 to 80 days. Litter size was reduced from 7.7 to 7.1 pups. Retention of Kepone was primarily in the liver, less in the brain, kidneys, and fat. After withdrawal of Kepone from the diet, tissue residues decreased rapidly	217
9	Rats	Increase in ratio of liver to total body weight in males at 12 months. (The levels of 9, 6, and 5.3 ppm were not actually fed, but were calculated from data at levels fed as minimally effective ones by analyses of regression with 95% confidence limits.) Hyperexcitability and tremors were noted on the ninth day, and mortality on the thirteenth day after a daily oral dose of 10 mg/kg	209, 210
6	Rats	Depression of growth rate in females at 26 weeks	209
5.3	Rats	Increase in ratio of liver to total body weight in females at 3 months of feeding	209
5	Dogs	No significant effect on body weight, hematocrit, hemoglobin, urinary sugar and protein excretion, liver function, and organ and body weights (124 weeks). No relatable histopathological findings after 127 weeks of feeding at 5 ppm	209
1	Dogs	No significant effect on food consumption, body weight, hematology, urinary sugar and protein excretion, and liver function (one experimental dog died during the 49th week, and one control dog during the 71st week); all dogs suffered from mange	209
	Rats	No significant effect on mortality, growth rate, food intake, hematology, urinary sugar and protein, and micropathology in males and females (2 years)	209

7.3.3   Chronic Oral Toxicity

Chronic Kepone effects may include some or all of the following: tremors, loss of body weight, increase in oxygen consumption, neurological impairment, disturbance of the estrous cycle, and fetotoxicity. Inhibition of muscle lactate dehydrogenase has been observed in in vitro studies in the rabbit (214). Anatomic organ changes may include liver injury (congestion, hyperplasia, hypertrophy, focal necrosis, hepatomegaly), nephropathy, atrophy of the testes, and proteinuria (Table 49.16).

7.4   Human Experience

7.4.1   Occupational Exposure

The first recorded incidents of Kepone poisoning were recognized between March 1974 and July 1975, when 76 of 113 workers (57 percent) at the Hopewell, Virginia plant showed signs and symptoms of intoxication that included skin rash, nervousness, brain damage (loss of memory and behavioral changes), weight loss, pleuritic and joint pains, oligospermia, tremors, ataxia, opsoclonus (rapid twitching of the eyes in horizontal, vertical, oblique, or rotary directions), and sterility (219). See also Cannon et al. (220), Cohn et al. (221), and Huff and Gerstner (222). The mean latency period between the start of employment and onset of symptoms was 6 weeks. When the effects were recognized, a detailed report was prepared in 1976 by a team of investigators of the U.S. Public Health Service (223). The symptoms persisted for as long as 6 months after termination of employment. The mean blood Kepone level for workers with illness was 2.53 ppm, and for those without disease, 0.60 ppm.

7.4.2   Nonoccupational Exposure

The U.S. Department of Agriculture (1977) reported 56 incidents of nonoccupational human exposure to Kepone. Children under 5 years of age were exposed in 52 of the incidents, two incidents involved adults, and two were of unspecified age. All but nine of the children suffered Kepone exposure at home, primarily by devices used for the control of ants and cockroaches (224).

Blood Kepone levels for workers in nearby businesses and for residents of a community located within a distance of 1.6 km of the Kepone manufacturing plant ranged from undetectable to 32.5 ppb. There was no apparent association between the frequency of symptoms and proximity to the plant. Illnesses attributable to Kepone were found in two wives of Kepone workers (220, 222).

It was estimated that consumption of seafood (finfish, crabs, oysters) caught in the James River and Chesapeake Bay led to an ". . . average daily intake on the order of 1 microgram of Kepone involving 5 to 10 million people" (219). According to calculations (of Lu and Deichmann), the estimated daily intake of

Kepone from finfish from the Chesapeake–James River waters during the years 1975–1976 were 0.3 µg for adults of the general population, and 2.1 µg/day for those whose diet consisted primarily of seafood (211).

Kepone residues in the tissues of the general U.S. population have apparently not been reported. Members of the EPA collected 298 samples of mothers' milk in nine southern states in 1976. Samples from three states (Alabama, Georgia, and North Carolina) showed Kepone levels ranging from less than 1 ppm to 5.8 ppb (226). Possible sources of exposure included the spraying of mirex for the control of fire ants and the subsequent degradation of mirex into Kepone (225) and the direct exposure to ant or cockroach traps.

7.5 Absorption, Metabolism, and Excretion

Kepone is absorbed from the gastroenteric tract and by way of the lungs and skin. It is retained primarily in the liver but also in the brain, kidneys, and fat. Blanke et al. (227) identified decachlorooctahydro-1,3,4-metheno-2H-cyclo-buta(c,d)pentalen-2-ol in stool specimens from patients diagnosed as suffering from Kepone poisoning.

Kepone crosses the placenta of mice and accumulates in the fetus. It is apparently not metabolized in the mouse (217). Kepone is rapidly excreted in the feces, with traces in the urine and milk of humans and cows (226).

According to Boylan et al. (228), the oral administration of Cholestyramine (an anion exchange resin) markedly increased the excretion of Kepone. According to Richter et al. (229) administration of 8 percent light liquid paraffin for 24 days to rats fed previously a diet containing 3.3 ppm of ^{14}C-Kepone for 3 days resulted in a significantly increased excretion of Kepone with the feces (61 percent) compared to controls (52 percent). The paraffin-treated rats had significantly lower concentrations of radioactivity in 14 of the 18 tissues analyzed. Excretion of radioactivity in the urine was of minor importance.

7.6 Biologic Magnification

Kepone is susceptible to transfer from particulate or food-web processes to higher trophic levels (226). Certain aquatic plants, animals, and fish readily accumulate the compound. The bioconcentration in finfish was found to be species-specific.

7.7 Decontamination

Analysis of soil samples indicated that in 1978 approximately 450 kg of residual Kepone existed around Hopewell, Virginia (230). Biodegradation of Kepone in the James River has been insignificant.

A process is being developed for the destruction of Kepone in which the compound is vaporized at a low temperature and then passed for 1 sec through a high temperature quartz tube at about 1000°C. Preliminary results are satisfactory (231). Other cleanup procedures under consideration range from dredging and destroying the compound by ozonation and ultraviolet light, adding activated carbon to the river to decrease the availability of Kepone to the food chain, and soil fixation with the sediment.

7.8 Cancer

In chronic feeding studies with Wistar strain rats, conducted at the Medical College of Virginia, Kepone was found to be hepatoxic and nephrotoxic. Liver tumors were noted in some of the animals. Representative liver sections submitted to four pathologists provided diagnoses for a particular tumor which ranged from "hyperplasia," "suspicious carcinoma," "no carcinoma," to "carcinoma." Based on these studies, the carcinogenicity of Kepone in the rat remained debatable (209, 211).

In 1976, the National Cancer Institute (232) published a report on the carcinogenesis of technical grade chlordecone (Kepone) with the following summary:

A carcinogenesis bioassay of technical grade chlordecone (Kepone) was conducted using Osborne-Mendel rats and B6C3F1 mice. Chlordecone was administered in the diet for 80 weeks at two dose levels, with the rats sacrificed at 112 weeks and the mice at 90 weeks. The starting dose levels were 15 and 30 ppm for male rats, 30 and 60 ppm for female rats, 40 ppm for male mice and 40 and 80 ppm for female mice. As these dose levels were not well tolerated, the dose levels were reduced during the course of the experiment such that the average dose levels were as follows: 8 and 24 ppm for male rats, 18 and 26 ppm for female rats, 20 and 23 ppm for male mice and 20 and 40 ppm for female mice.

Clinical signs of toxicity were observed in both species, including generalized tremors and dermatologic changes. A significant increase ($p < .05$) was found in the incidence of hepatocellular carcinomas of high dose level rats and of mice at both dose levels of chlordecone. The incidence in the high dose groups were 7 percent and 22 percent for male and female rats (compared with 0 in controls for both sexes) and 88 percent and 47 percent for male and female mice (compared with 16 percent for male room controls and 0 in females); for the low dose groups of mice the incidences were 81 percent for males and 52 percent for females. In addition, the time to detection of the first hepatocellular carcinoma observed at death was shorter for treated than control mice and, in both sexes and both species, it appeared inversely related to the dose. In chlordecone-treated mice and rats extensive hyperplasia of the liver was also found. The incidence of tumors other than in the liver for chlordecone-treated groups did not appear significantly different from that in controls.

The significance of the incidence of the hepatocellular carcinomas noted in the experimental male and female rats and mice is not clear. According to the report, the term "hepatocellular carcinoma" was used to diagnose proliferative lesions of the liver because, in the judgment of the pathologists, these lesions ". . . had the potential or the capacity for progressive growth, invasion, metastasis, and for causing the death of the host." However, neither in the experimental rats nor mice were vascular invasion and/or metastases observed "in the material examined." It appears, therefore, that the diagnosis of hepatocellular carcinoma for proliferative lesions was an overstatement.

Additional comments regarding this NCI report by Newberne (233) include the following:

It is unfortunate that dose levels were not determined more accurately before the chronic studies were initiated. One can, therefore, only speculate about the influence of early injury. . . . I do not, of course, agree with a diagnosis of "neoplastic nodule" since I don't know what this means . . . No vascular invasion nor metastases were observed in the material examined. . . . I am thus at a loss to know whether any of the tumors were cancer or simply nodular hyperplasia.

Based on the Medical College of Virginia, the NCI, and other reports of human and animal studies, it has been established that Kepone, in toxic doses, produces severe liver and kidney injury and damage to the central nervous system and reproductive organs. The risk of human cancer has not been established.

7.9 Hygienic Standards of Permissible Concentrations

No TLVs or tolerances have been established for Kepone. The production of Kepone was discontinued in the United States in 1975 after closing of the plant in Hopewell, Virginia.

8 HEXACHLOROCYCLOHEXANE, Gamma Isomer of 1,2,3,4,5,6-Hexachlorocyclohexane, Lindane; Mixed Isomers of 1,2,3,4,5,6-Hexachlorocyclohexane, Benzene Hexachloride, BHC

8.1 Uses and Industrial Exposure

Hexachlorocyclohexane is used as an insecticide for the control of insects on cotton and other foliar plants, for soil and seed treatment of fruit and vegetable crops, and for control of termites. Lindane has been found effective for the control of insects on livestock and pets, and is important to public welfare in the control of lice, mosquitoes, and flies that have become resistant to DDT.

The use of lindane and BHC do not present particularly difficult problems. These compounds are not considered greatly hazardous; they are in the same approximate category as DDT.

8.2 Physical and Chemical Properties

Hexachlorocyclohexane occurs as five isomers, with four of these found in sufficient quantity to be considered of insecticidal importance. A crude mixture of these four isomers is available under the common name, benzene hexachloride (BHC). The most effective isomer is the gamma, and this, highly purified, is available under the common name lindane. (Table 49.17).

8.3 Effects in Animals

The primary response to the alpha and gamma isomers is stimulation of the central nervous system, resulting in hyperexcitability and convulsions, whereas the beta and delta isomers cause central depression.

The toxicity of BHC varies and is related to the amount of lindane present, which is the most acutely toxic isomer. Chronically, lindane is the least cumulative isomer. The alpha, beta, and delta isomers have a low degree of acute toxicity, but are retained for a longer period than lindane; hence they have a higher degree of cumulative toxicity (particularly the beta isomer).

Micropathological changes are seen in the liver and kidneys and, to some degree, in the lungs following chronic exposure to toxic doses of these materials.

8.3.1 Acute Oral Toxicity

Lehman (14) reported the oral LD_{50} of lindane for rats as 125 mg/kg body weight, while Hayes (9) reported the oral LD_{50} for male rats to be 88 and for

Table 49.17 Physical and Chemical Properties of Hexachlorocyclohexane Isomers

Isomer	Melting Point (°C)	Vapor Pressure (20°C)	Solubility			
			Water (ppm)	Ethanol (g/100 g)	Ether (g/100 g)	Benzene (g/100 g)
Alpha	157.8–158	0.02	10	1.8	6.2	9.9
Beta	309	0.005	5	1.1	1.8	1.9
Gamma	112.5	0.03	10	6.4	20.8	28.9
Delta	138–139	0.02	10	24.4	35.4	41.1
Epsilon	217–218	—	—	—	—	—

female rats 91 mg/kg. The oral LD_{50} for the rabbit ranges from 50 to 60 mg/kg; the oral LD_{50} for the guinea pig has been given as 127 mg/kg.

8.3.2   Chronic Oral Toxicity

Lehman (234) indicated that the highest tolerated daily dose of lindane in rats over a 2-year period without effect was 5 mg/kg or a dietary feeding level of 50 ppm (235). According to Fitzhugh et al. (236), rats can tolerate long-term feeding of diets containing 800 ppm of lindane, but showed some liver enlargement at this level. A comparable response was observed from BHC at a level of 100 ppm in the diet. Lindane is stored at about the same level in fat as the level of intake, that is, a ratio of $1:1$ (237). Maximum storage is accomplished within 6 weeks and disappears from the fat depot within 3 weeks (238).

8.3.3   Air Dispersion

Queen (239) reported on the use of lindane by vaporization. Canaries were exposed to vapors of lindane (liberated as a cloud when a solid formulation was heated) in an air concentration of 0.34 mg/liter; the birds died after 6 to 16 days of exposure. Some lindane home generating devices produced air concentrations which exceeded those in manufacturing plants. The possibility of absorption of lindane or BHC by way of the lungs exists also when these materials are used as dusts or liquid sprays.

8.3.4   Skin

Lehman (166) reported that the approximate LD_{50} for rabbits, following absorption of lindane in dry form, was greater than 4000 mg/kg. Severe signs of intoxication and moderate skin irritation were observed at this level, but the animals survived. When lindane was applied as a 2 percent solution in dimethyl phthalate, the approximate LD_{50} was greater than 188 mg/kg. As a 1 percent solution in a vanishing cream base, the approximate LD_{50} was 50 mg/kg. Concentrations up to 4 ppm of BHC have been found in the milk of ewes after dipping with Entomoxan (0.5 percent BHC) (240).

8.3.5   Absorption, Metabolism, and Excretion

Van Asperen and Oppernoorth (241) reported that 1 mg of lindane injected subcutaneously in the mouse was eliminated in 4 days. After an intravenous injection of 200 gammas of lindane, the compound was not noted in the urine or feces, but traces were found in several organs.

Seidler et al. (242) studied the distribution, metabolism, and excretion of ^{14}C-

lindane in rats after they had become adapted to lindane administered orally. They found that body fat, kidneys, and the musculature were the primary sites of deposition. Marked differences in radioactivity were found between the cortex, brainstem, and cerebellum.

The metabolite δ-2,3,4,5,6-pentachlorocyclohexane and the sum of the metabolites penatachlorobenzene and hexachlorobenzene amounted to 1 to 3 percent of the hexane extract after 24 hr, and 5 to 8 percent after 72 hr, in certain organs. Feces, urine, and organs contained large amounts of conjugates and highly polar, hexane-soluble metabolites. The level of extractable (i.e., nonbound) ^{14}C-activity in the urine was 8 percent, and in the feces, 43 percent. The level of extractable activity was 35 percent in both urine and feces after 72 hr. An additional 3 and 8 percent after 24 hr, and 18 and 14 percent, respectively, after 72 hr, were present as glucuronides, and the remainder as unidentified water-soluble conjugates. In the urine, 1 to 3 percent of lindane was present in the free form; 3 to 6 percent of the total lindane was extracted unmetabolized. The γ-PCCH and pentachlorobenzene plus HCB accounted for 0 to 2.5 percent of the total excreted lindane. Half of the lindane administered was excreted within 3 to 4 days (242).

In 1975, Chadwick et al. (243) presented evidence for the hydrogenation of lindane by the hepatic mixed function oxidase system. A few years later, in 1978, Chadwick et al. (244) identified three lindane metabolites in the urine of Sprague-Dawley rats fed diets containing 400 ppm of lindane. Two of these, M-1 and M-2, had a shorter GLC retention time than a previously identified PCCOL metabolite. M-1 and M-2 were excreted primarily as glucuronide and sulfate conjugates. Mass spectra and dehydrogenation studies tentatively identified M-1 as 2,3,4,6-TCCOL, and M-2 as a configurational isomer of 2,4,5,6-tetrachloro-2-cyclohexan-1-ol (2,4,5,6-TCCOL). The studies demonstrated that this pesticide can be degraded in the environment by several pathways.

The release of enzyme by rat liver lysosomes was studied by Okazaki and Kimura (245) since the lysosomes play a role in the elimination of foreign compounds. When damaged by chemicals or radiation, the lysosomal membrane breaks, releasing the enzymes contained therein. It was found that BHC produced severe damage to lysosomes in vitro and in vivo, and that the ratio of "activity of released enzyme" to "total enzyme" of β-glucuronidase, after BHC application in vitro, amounted to almost 100 percent after 90 min at 37°C. In vitamin E deficient animals, the lysosomal membrane became more unstable, and the damage by chemicals increased.

Two opposite effects of lindane on the threshold for pentylenetetrazol-induced convulsions have been reported (246). The threshold is lowered a few hours after the administration of a single small oral dose of lindane to mice (247). In earlier experiments, Herken and Klempau (248) and Coper et al. (249) could show that rats given large, near-lethal doses of lindane are protected from the effects of convulsive doses of pentylenetetrazol when examined a couple of days after the lindane administration. An elevated threshold was also found after the administration of α-hexachlorocyclohexane, a structural isomer of lindane.

Compared to several other chlorinated hydrocarbon pesticides, lindane is rapidly metabolized in the mammalian body (250). A degradation scheme for lindane, based on experiments in the rat, has been proposed by Engst et al. (251).

In a more recent study Hulth et al. (246) studied the convulsive properties of lindane on the convulsive threshold of pentylenetetrazol and brain content of γ-aminobutyric acid (GABA) in the mouse. They found that following the administration of a single dose of lindane, the threshold first decreases and then, after 2 days, increases above the normal level. None of the tested metabolites of lindane altered the convulsive threshold. When lindane elevated the convulsive threshold it also elevated the content of GABA in the brain. The lindane isomer α-hexachlorocyclohexane elevated both convulsive threshold and the GABA content. The smallest single oral dose of lindane lowering the convulsive threshold for pentylenetetrazol in the mouse, 3 hr after administration, corresponded to a lindane concentration of 56 ppb in whole blood.

8.4 Human Experience

8.4.1 Tissue Concentrations

From 1965 to 1970, the average dietary intake of BHC (including lindane) averaged 5.6 μg for a 70-kg individual in the United States (31). Body fat concentrations in the U.S. adult population groups generally ranged from 0.2 to 0.6 ppm (71). Fiserova-Bergerova et al. (252), who analyzed (GLC) the body fat of stillborns and fetuses in Dade County, Florida (where pesticides are used more vigorously than in most parts of the United States), recorded concentrations of α-BHC of 0.14 ppm, β-BHC of 0.26 ppm, and γ-BHC of 0.11 ppm. Little or no lindane or BHC was detected in the blood of individuals not occupationally exposed to lindane (71).

In certain parts of Argentina, crude BHC and BHC containing a high ratio of the beta isomer were used extensively by the general population. As a result, people of all ages were found to have approximately 16 times the blood concentration of β-BHC found in the U.S. population. Similar high blood levels were found in Asiatics, principally in Japanese and Formosans (253).

In general, human milk in the United States contained only traces of BHC, even during the height of the lindane or BHC use. During the period 1972–1977, concentrations in the milk of lactating mothers in Okinawa increased because of the increased consumption of pork and American and Australian beef. In 1972 total BHC in 100-ml samples of human milk was 0.029 ppm; in 1977 it had risen to 0.075 ppm (254).

8.4.2 Intoxications

Danopoulos et al. (255) reported on clinical cases of exposure to technical BHC in Greece. A 40 percent dry powder of BHC, or the same powder mixed with either water or a petroleum solvent, was sprinkled on the ground, onto walls, and over clothing, bedding, and bodies of the people. Seventy-nine persons were affected, 18 seriously. Five were treated in a clinic; all survived. Symptoms were related primarily to the gastroenteric tract and the central nervous system. There were also indications of electrocardiographic and hematologic changes.

Heiberg and Wright (256) reported on an intoxication which occurred in a woman who "washed" two calves with a BHC solution. Her arms and hands were wet to the elbows with the material and her clothing was partially soaked. She showed a severe response with convulsions, but survived. On the day following her admission to the hospital, a concentration of 4.95 mg of BHC per 100 ml of urine was found.

The dermal application of 1 percent lotions of lindane (Kwellada), for the treatment of scabies resulted in severe intoxications in some children and infants. Marked mental and motor retardation was noted 2 days after treatment in a 4-month-old child (257). Local application of 1 percent lindane proved particularly dangerous following a hot soapy bath, since this increased the percutaneous absorption of the compound (258–260).

Accidental ingestion by a 1-year-old boy of one teaspoon of a 1 percent lindane solution, in addition to the local application of this solution for the treatment of scabies, resulted in irritability, hyperactivity, and vomiting, followed by recovery in the hospital (261).

A more severe intoxication was caused by the ingestion by an 8-year-old boy of about a dozen chocolate biscuits that had been sprayed with Cooper's Tip Dressing, containing 4 percent BHC, of which only 0.5 percent was in the form of the less toxic gamma isomer. His condition was critical for 48 hr, but he recovered fully after 72 hr (262). Eight cases of BHC poisoning followed ingestion of food containing 4 percent BHC. Three of the victims suffered convulsions, coma, pulmonary edema, and death. Post mortem examination revealed the presence of cerebral congestion and edema, and renal ischema with hemorrhages. Serious illness in the other victims was believed to have been prevented by vomiting (262).

Recently, eight similar incidents of nonfatal BHC poisoning (grand mal epilepsy) were reported by Nag et al. (263). The intoxications occurred in a family in a village in Uttar, India. These people had ingested wheat bread containing 0.005 percent BHC. It was found that the wheat flour had been contaminated with BHC during storage.

8.4.3   Occupational Exposure

According to Deichmann (264) in *Pesticides and the Environment* (reproduced with permission from Symposia Specialists, Inc., Miami, Florida), "Milby et al. (265) were apparently the first to establish that the presence of lindane in blood is a reflection of *recent* lindane absorption. Among groups occupationally exposed to this pesticide, blood concentrations did not appear to increase with increasing duration of exposure. However, blood lindane concentrations did increase as air and skin exposure became more intense. Air concentrations of lindane equal to 11 to 1170 micrograms/m³, in operations which also provided ample skin contact, to which men were exposed during working hours for years, were related to blood lindane concentrations of 30.6 ppm (6.0–93.0 ppb)."

Lindane has been implicated as an agent that may have been responsible for causing blood dyscrasias, primarily aplastic anemia (266–268). As Milby et al. (265) pointed out, in most cases the hematologic diagnosis is supported by convincing evidence, whereas the relationship to specific lindane exposure is less well documented.

Samuels and Milby (269) examined the effects a few weeks to several years of working in lindane processing plants might induce in 71 individuals, such as can be recognized by a battery of extended hematologic and biochemical blood and urine tests. Mean lindane blood concentrations of these workers were as follows: nonproduction workers, 0.93 ppb; production workers in two plants with little or no skin contact, 4.1 and 4.6 ppb, respectively; and working in a plant where ample skin contact was recognized, 30.6 ppb. (The study included seven children and one adult who lived in homes in which lindane vaporizing devices were operated intensively. The mean blood lindane of these was 2.2 ppb.) Considering all hematologic and biochemical abnormalities, but particularly diseases related to hematopoietic depression or renal or hepatic dysfunction, the authors could not detect any clinical symptomatology or physical evidence of disease that was attributable to lindane exposure. However, since certain isolated abnormalities were noted in occasional instances, they decided to conduct an additional study in which an exposed population group was matched by age, sex, and race, with individuals who suffered no exposure to lindane.

The Milby and Samuels' study (265, 270) dealt with 80 individuals, 40 of whom were employed in a plant in which lindane was processed. All were Caucasians; 12 were female. Each of these people was matched with a control having no occupational exposure.

The authors, in summarizing, pointed out that only blood lindane concentrations were found to differ in the control and occupational groups; they were 0.1 ppb and 11.9 ppb, respectively. Furthermore, "Significant differences were not observed in regard to blood uric acid, alkaline phosphatase, platelet count, hemoglobin and lymphocyte, eosinophil and monocyte counts. There was no evidence of pancytopenia with reticulopenia, the hallmark of hypoplastic or aplastic anemia." They did not rule out the possibility of an idiosyncratic or a hypersensitivity response in a few individuals. Although they questioned meaningful conclusions that are based on statistical analysis of relatively small groups of people, they concluded that ". . . within the limitations of this study, lindane does not appear to produce hematologic disorders on a basis of toxic suppression of hematopoiesis" (270).

8.5 Persistence in Waters and Soil

Benzene hexachloride is used extensively in certain countries. In Czechoslovakia, for instance, a 1975 survey showed that its increasing use has led to a gradual shift from DDT to lindane contamination of soils, waters, and animal foodstuffs (271).

In Japan, BHC accounted for about 90 percent of all organochlorine pesticides used from 1968 to 1969. This resulted in elevated concentrations of BHC in waters. In Tokyo, for instance, concentrations of α-BHC in rain water ranged from 45 to 930 ppt, and γ-BHC from 29 to 398 ppt. In snow, the concentrations were only 12 to 50 percent of the concentrations found in rainwater. This difference is undoubtedly related to the pesticide concentration in the atmosphere, not to rain or snow, as such. It was concluded that a large amount of BHC sprayed on farmlands or crops falls rather rapidly to the ground, particularly in rainfall, while the remaining portions remain in the atmosphere for prolonged periods. It was estimated that during the year 1968–1969, BHC dropping to the ground with rainfall amounted to 170 tons/year; this quantity was only 0.36 percent of the total quantity of BHC used per year in Japan (272). Like other organochlorine pesticides, BHC becomes partially absorbed and adsorbed to river bottom sediments.

Lindane has little effect on microorganisms in active sludge, at least not in the quantities which are soluble in water. Form a biologic treatment plant, considerable quantities of lindane were removed along with the sludge, after which the pesticide was concentrated, filtered, and incinerated; some quantities were decomposed by the sludge flora (273).

Blue-green algae (25 species) detoxified lindane by lowering its concentration in media containing 50, 60, or 80 μg of this compound per milliliter (274).

8.6 Cancer

Japanese scientists have conducted several chronic mouse-feeding studies to determine possible carcinogenic activities of BHC and lindane. Hanada et al. (275) fed groups of 7-week-old dd mice a diet supplemented with 600 ppm of α-BHC. Animals sacrificed after 1 month of feeding showed liver changes, including degeneration and necrosis; after 3 months, nodular proliferation of enlarged liver cells was noted, and after 5 months, liver nodules. Histologically, the arrangement of cells in the larger nodes became irregular with many nuclear divisions, suggesting malignancy.

In a study conducted by Hananouchi et al. (276), groups of male dd mice were fed a diet containing α-BHC at 500 ppm for periods of 16, 24, and 32 weeks, respectively. Tumor incidence increased with the feeding period of the experimental diet. Predominantly hepatic or nodular hypertrophies with large ovoid nuclei and enlarged nucleoli were noted in 24, 70, and 100 percent of the mice fed the diet for 16, 24, and 32 weeks, respectively.

When the feeding of the experimental diet containing 500 ppm α-BHC was discontinued at the twentieth week, and the animals killed 4 weeks later, the incidence of the liver changes had dropped from 70 to 40 percent and from 100 to 94.7 percent, respectively. When the animals were killed 8 weeks after the 20 weeks of feeding, the incidence was further reduced to 25 and 47 percent, respectively. The reduction in tumorigenic changes was associated with a marked reduction in liver weights. At 36 weeks (after a feeding period of 20 weeks of the 500 ppm diet, followed by 16 weeks of feeding a control diet), the liver changes were recognized as 100 percent hepatocellular hepatomas. "After the carcinostatic period, infiltration of interlobar cells between proliferated liver cells was seen. The findings suggest a major role for interlobar cells in the transition from precancerous changes to hepatocarcinogenesis" (216).

A bioassay of lindane for possible carcinogenicity was conducted for the Carcinogenesis Program of the National Cancer Institute, and published in 1977 (277). The compound was administered for 80 weeks to groups of Osborne-Mendel rats and B6C3F1 mice. Time-weighted average doses for male rats were 236 or 472 ppm, for female rats 135 or 270 ppm, and for male and female mice 80 or 160 ppm. Results were as follows:

In rats, no tumors occurred at a statistically significant incidence in the treated groups of either sex. In mice, the incidence of hepatocellular carcinoma in low-dose males was significant when compared with that in the pooled controls. (Controls 5/49, low dose 19/49, $P = 0.001$). This finding, by itself, is insufficient to establish the carcinogenicity of lindane. The incidence of hepatocellular carcinoma in high-dose male mice (9/46) was not significantly different from that in matched (2/10) or pooled controls.

It is concluded that under the conditions of this bioassay lindane was not carcinogenic for Osborne-Mendel rats or B6C3F1 mice.

No evidence has been presented that would link lindane or BHC to cancer in man.

8.7 Teratology

Tests for teratogenicity were conducted by Palmer et al. (278) using groups of rabbits and rats. Lindane was administered to these animals during pregnancy at dosages of 5, 10, or 20 mg/kg body weight daily for 12 days. Based on body weight changes, litter size, litter and mean fetal weights, and other observations, no evidence of teratogenicity or toxicity to the embryos by lindane was noted.

8.8 Hygienic Standards of Permissible Concentrations

The threshold limits of lindane (skin) adopted in 1979 by the ACGIH were TLV–TWA, 0.5 mg/m^3, and TLV–STEL (tentative), 1.5 mg/m^3 (116).

The IDLH concentration of lindane, according to the 1978 *NIOSH–OSHA Pocket Guide to Chemical Hazards*, is 1000 mg/m^3 (117).

According to the 1978 FAO/WHO Standards Programme Codex Committee on Pesticides Residues, the ADI of lindane is 0.01 mg/kg body weight. The MRLs of lindane for agricultural commodities range from 0.05 to 3 mg/kg (118).

9 HEXACHLOROCYCLOPENTADIENE, C_5Cl_6

9.1 Uses and Industrial Exposures

Hexachlorocyclopentadiene is used as a chemical intermediate in the manufacture of a variety of industrial compounds, including flame retardants and pesticides such as aldrin, chlordane, mirex, and others. The potential for skin or mucous membrane irritation or for inhalation exposure exists for those who are engaged in the manufacture or handling of this chemical.

9.2 Physical and Chemical Properties

Physical state	Liquid
Molecular weight	273
Freezing point	9 to 10°C
Boiling point	234°C
Vapor density	9.4 (air = 1)
Vapor pressure	0.080 mm Hg (25°C), 1 mm Hg (60°C)
Solubility in water	800 ppb

1 mg/liter = 89.6 ppm, and 1 ppm = 11.17 mg/m^3 at
25°C, 760 mm Hg

9.3   Effect in Animals

Hexachlorocyclopentadiene is highly irritating to the skin and mucous membranes, causing lacrimation, sneezing, and salivation. Toxic doses have produced
degenerative changes in the brain, heart, and adrenal glands, and degeneration
and necrosis in the liver and the kidney tubules. Pulmonary hyperemia and
edema have resulted from inhalation of hexachlorocyclopentadiene. Most
experimental data on this compound was taken from the report by Treon et al.
(8, 279).

9.3.1   Acute Vapor Exposure

Maximum time–concentrations by volume in air survived by guinea pigs were
0.25 hr at 20.2 ppm, 1 hr at 7.2 ppm, 3.5 hr at 3.1 ppm, and 7 hr at 1.5 ppm.
Levels survived by rats were 0.25 hr at 20.2 ppm, 0.5 hr at 7.2 ppm, 1 hr at 3.1
ppm, and 7 hr at 0.33 ppm (8).

9.3.2   Subacute and Chronic Vapor Exposure

Exposure of rats to concentrations of 0.0006 and 0.014 mg/liter of air 6 hr/day,
5 days/week for 4 weeks resulted in "toxic" liver changes (280). Guinea pigs,
rabbits, and rats were subjected to 150 7-hr exposures over 216 days. They
survived at 0.15 ppm by volume in air. Guinea pigs survived a concentration of
0.34 ppm, but rats and mice succumbed after 30 7-hr exposures. Micropathological tissue change revealed mild liver and kidney injury even at 0.15 ppm
(8).

9.3.3   Acute Oral Toxicity

The approximate lethal dose for rats and rabbits by single oral administration
of the 93.3 percent compound was between 420 and 620 mg/kg. The animals
showed diarrhea, lethargy, and retarded respiration. Rabbits showed diffuse
degenerative changes in the brain, heart, liver, and adrenals, necrosis of the
epithelium of the renal tubules, and severe hyperemia and edema of the lungs.
(8, 279). In a Velsicol Chemical Corporation report (281), the oral LD_{50} for rats
was given as 584 mg/kg. Lu et al. reported the oral LD_{50} of the compound in
mice as 430 mg/kg (282).

9.3.4   Skin

By skin absorption, the lethal dose in rabbits lies between 430 and 610 mg/kg. The material was extremely irritating (8).

9.4   Human Response and Environment

According to Ingle, the irritant effects of hexachlorocyclopentadiene in man are well known (283).

The potential for entry of this compound in biologic systems exists but there is apparently no information on residues in the environment. The compound is chemically volatile, but unfortunately there is, at present, no way to estimate free hexachlorocyclopentadiene in plastic products; there is no accepted analytical method available to detect residues; ". . . and finally, the forms of hex reaction products that are stable enough to leave terminal residues in the environment and that may be toxic to biologic systems are largely unknown, although they may result in critical impacts on human health and the environment" (232). According to Lu et al. (282), who worked with model terrestrial and aquatic ecosystems, hexachlorocyclopentadiene has considerable stability in the environment.

Fortunately the limited uses of this compound are not expected to result in significant levels of residues in humans or the environment, except in factory and other workers who may be directly exposed at work (232).

In March 1977, a large volume of industrial chemical hexachlorocyclopentadiene was dumped into a municipal sewage system in Kentucky. Morse et al. decided to evaluate the health effects of exposure to this compound in 145 sewage treatment plant workers. They found that 85 (59 percent) had noted eye irritation, 65 (45 percent) had headaches, and 39 (27 percent) had throat irritation. Symptoms occurred throughout the plant; however, highest attack rates occurred in primary sewage treatment areas. Medical examination of 41 employees 3 days after the plant was closed showed proteinuria and elevation of serum lactic dehydrogenase levels; these findings were not present 3 weeks later. This episode demonstrates the toxicity of hexachlorocyclopentadiene and emphasizes the vulnerability of sewage workers to chemical toxins in wastewater systems (284).

According to the National Academy of Sciences report, bioassays indicate that levels approximating 1 ppm are acutely toxic to fish (232).

9.5   Mutagenicity

Hexachlorocyclopentadiene was not mutagenic to *Salmonella typhimurium*, with or without a rat liver microsomal enzyme activating system (285).

9.6   Odor and Warning Properties

A faint odor is detected at 0.15 ppm by volume in air. A pronounced pungent odor is observed at 0.33 ppm (8).

9.7   Hygienic Standards of Permissible Exposure

The threshold limits for hexachlorocyclopentadiene adopted in 1979 by the ACGIH are TLV–TWA, 0.01 ppm (0.1 mg/m^3), and TLV–STEL (tentative value), 0.03 ppm (0.3 mg/m^3) (116).

10   TOXAPHENE, $C_{10}H_{10}Cl_{10}$, Chlorinated Camphene, Octachlorocamphene

10.1   Uses and Industrial Exposure

Toxaphene has been used extensively as an insecticide and for the control of livestock ectoparasites, mosquito larvae, leaf miners, bagworms, chinch bugs, yellow jackets, and caterpillars. The principal exposures have occurred during the manufacturing process, from dusts or mists during spraying, and from skin contact of solutions and emulsions.

10.2   Physical and Chemical Properties

Physical state	Thick amber-colored solid with a mild turpentine-like odor
Vapor pressure	0.00001 mm
Melting point	70 to 95°C
Solubility, in g/100 ml at 27°C	Acetone, benzene, carbon tetrachloride, toluene, xylene, hexane, more than 450 in each; deodorized kerosine, more than 280; mineral oil, 55 to 60; ethyl alcohol, 10 to 13; isopropyl alcohol, 15 to 18 (286)
Solubility in water	0.0003 percent at 20°C

Toxaphene, the final reaction product of camphene and chlorine, has a chlorine content of 67 to 69 percent. According to Casida et al. (287), toxaphene contains at least 175 different C_{10} polychloro compounds; Holmstead et al. (288) presented compositions of 177 compounds. This points out the complexity of the mixture referred to as toxaphene. Several components of toxaphene have been identified. See the excellent report and review on toxaphene by Pollock and Kilgore (289).

10.3 Effect in Animals

Toxaphene is absorbed through the skin and from the gastroenteric tract, or through the lungs. Signs and symptoms of toxicity following an acute exposure include stimulation of the cerebrospinal axis and, depending on the dose, may also include salivation, spasms, nausea, vomiting, hyperexcitability, tremors, clonic convulsions, and tetanic contractions of all skeletal muscles. These may continue until death, which is initiated by respiratory failure. Elimination of the material starts rather promptly and residues in fatty tissues disappear rapidly. Prior to excretion, toxaphene is dechlorinated; single components of toxaphene are dechlorinated and dehydrochlorinated.

Internal hemorrhages and temperatures in excess of 44°C have been reported following intoxications in farm animals (290). Animals that have been subjected to chronic high-dose oral intake of toxaphene have shown degenerative changes and necrosis in the liver and degenerative changes in the renal tubules.

10.3.1 Skin

Penumarthy et al. (290) reviewed the effects of toxaphene in livestock. When xylene was the vehicle, dips containing a 2.5 percent toxaphene emulsion caused toxicosis in cattle, and 1.5 percent dips resulted in comatose sheep within 3 min. Suckling calves showed responses of toxicity to toxaphene sprays ranging in concentration from 0.75 to 8.0 percent (289, 291).

10.3.2 Acute Oral Toxicity and Short-Term Studies

The oral LD_{50} of toxaphene has been reported to range from 60 to 120 mg/kg body weight in the rat, and from 290 to 365 mg/kg body weight in guinea pigs (292). Oral LD_{50}s and the solvents used in other mammals are as follows: mouse, 112 mg/kg (corn oil); dog, 49 (corn oil) and 250 mg/kg (kerosine); cat, 25 to 40 mg/kg (peanut oil); rabbit, 75 to 100 mg/kg (peanut oil) and 250 to 500 mg/kg (kerosine); cattle, 144 mg/kg (mixed in grain); and goat and sheep, 200 mg/kg (xylene) (293).

Following an oral administration as a solution or emulsion, toxaphene is more toxic in a digestive vegetable oil than in an oil such as kerosine (294). Some components of toxaphene are more toxic than the mixture itself, whereas other components are less toxic. The composition of toxaphene manufactured by the principal U.S. producer, Hercules ". . . was very similar over an extended period of production, but samples from other manufacturers varied significantly," according to Pollock and Kilgore (289).

Ortega et al. (295) fed groups of six male and six female rats 50 and 200 ppm toxaphene in the diet. One animal per group was sacrificed after 2, 4, and 6 months, and the remaining rats after 9 months. No signs of toxicity, change of food intake or body weight gain, or increase in liver weights were observed in these rats. Liver changes in six of the 12 rats fed 200 ppm consisted of centrolobular cell hypertrophy, peripheral migration of basophilic cytoplasmic granulation, and presence of liposphere inclusion bodies. There was no damage to the kidneys or spleen. Three of 12 rats fed 50 ppm showed slight changes such as those described above after being fed toxaphene for 6 to 9 months. It was concluded that this level produces possible borderline liver changes.

Toxaphene, administered by capsule in daily doses of 4 mg/kg to two dogs for 44 days, and to two (other) dogs for 106 days, induced CNS stimulation. This occurred occasionally and for a brief period after administration. Histological examination of many organs revealed some damage to the kidneys (degeneration of the tubular epithelium), and generalized hydropic degenerative liver changes, but no destruction of the cells (296).

Lehman (297) dissolved toxaphene in maize oil and administered it daily (5 days/week) to dogs in gelatin capsules. A dose of 25 mg/kg was fatal. Of two dogs given 10 mg/kg (equivalent to 400 ppm in the diet), one dog died after 33 days; the other lived and was sacrificed after 3.5 years. Four dogs received 5 mg/kg, and survived until sacrificed after almost 4 years.

10.3.3 Chronic Oral Toxicity

Groups of rats were fed 25, 100, and 400 ppm in the diet. The liver was the only organ that showed significant histologic changes at the 100 and 400 ppm levels (centrolobular hepatic cell enlargement with increased oxyphilia, peripheral margination of basophilic granules, and a tendency to hyalinization of the remainder of the cytoplasm). The effects at the various feeding levels were summarized by Lehman as follows: 400 ppm was the lowest level producing a gross effect of liver enlargement, 100 ppm the lowest level producing tissue damage, and 25 ppm the highest level not producing tissue damage (297, 298).

Treon et al. (299) fed four groups of rats (20 males and 20 females) 10, 100, 1000, and 1500 ppm of toxaphene in the diet. After 7.5 to 10 months of feeding, some of the rats fed 1500 ppm and a few of those fed 1000 ppm suffered occasional convulsions. There were no significant effects on mortality nor on the hematopoietic system. The liver weight and liver to body weight ratio were significantly increased only in the 1000 and 1500 ppm groups. Liver changes consisted of swelling and hemogeneity of the cytoplasm with a peripheral arrangement of the granules in the cytoplasm of the centrolobular hepatic cells. These changes occurred to a moderate degree in the 1500 ppm group and to a slight degree in the 1000 ppm group.

Lackey (296) reported that a dose of 5 mg of toxaphene/kg per day for 44 days to dogs caused convulsions after a few days of administration. Dogs fed 4 mg/kg per day for 106 days showed some reaction immediately after dosing, but seemed to recover between doses. At autopsy the kidneys of these animals showed degeneration of the tubular epithelium and generalized hydropic degenerative changes of the liver.

Toxaphene was also fed daily (6 days/week) for 2 years to eight 4-month-old dogs at levels of 10 and 50 ppm of the total wet diet, including liquids. The dogs received a daily dose of 0.60 to 1.47 mg/kg and 3.12 to 6.56 mg/kg (equivalent to approximately 40 and 200 ppm on a dry diet basis). There were no effects on behavior, body weight, mortality, or blood elements, but there were increases in liver weight and liver to body weight ratios, and there was a moderate liver degeneration at the 200 ppm level. At 40 ppm, one of three dogs showed slight liver enlargement and granularity, and vacuolization of the cytoplasm (299). However, reexamination of the liver sections of these animals failed to confirm any differences from control animals. All other tissues were normal at both feeding levels (300).

In an investigation conducted by Industrial Bio-Test Laboratories, Inc. (301), six male and six female dogs were fed toxaphene at dietary levels of 5, 10, and 20 ppm. Two male and two female dogs were sacrificed after 6, 12, and 24 months. None of the feeding levels produced alterations in organ weights, tissue changes or clinical or organ function tests.

10.3.4   Reproduction

A three-generation, six-litter reproduction rat study was conducted by Kennedy et al. (302). Before mating, weanling rats were fed 25 and 100 ppm of toxaphene for 79 days. All animals were then continued on these supplements, which were fed during mating, gestation, and weaning of two generations, or for a period of 36 to 39 weeks. Weanlings from the second litter were selected to serve as parents for the second generation and they were continued on their respective diets until after weaning of a second litter. Subsequent generations were selected in the same manner. The only pathological changes produced after 36 weeks of toxaphene administration were slight alterations in the livers of the 100 ppm group. Reproductive performance, fertility, and lactation were normal. The findings among all test animals, three parent generations and six litters of progeny, were comparable to those of the control rats.

More marked effects were reported in a more recent study. Quoting from Pollock and Kilgore's report (289), "Toxaphene orally administered to pregnant rats and mice during organogenesis caused definite fetal toxicity (303). At the highest dose in rats (35 mg/kg/day) etc., there was a maternal mortality of 31 percent, and at all doses (15, 25 and 35 mg/kg/day) there were significant reductions in maternal weight. Fetuses from dams (rats) fed 35 mg/kg/day weighed less than controls and there was a dose-response reduction in the average number of sternal and caudal ossification centers. Maternal mortality in mice dosed with 35 mg/kg/day was 8 percent and there was a dose-related reduction in maternal weight. All dosage groups showed an increase in the average maternal liver/body weight ratio. There was only one difference in the fetuses (mice) which occurred at the highest administered level. Five litters from this group had one or more fetuses with encephaloceles and none of the other groups showed this pathology."

In a five-generation reproduction study with Swiss white mice (14 females and 4 males per generation), in which toxaphene was fed in a dietary concentration of 25 ppm, few if any adverse effects were noted with regard to fertility, gestation, viability, lactation, and survival. The weights of fetuses and maternal weights were not determined in this study (304).

10.4 Human Experience

The fatal dose of toxaphene in man has been estimated to range from 2 to 7 g. Fatal human poisonings, however, have been rare. McGee and Reed (305) reported on six cases of poisoning from toxaphene, with recovery. They also reported four fatalities in children who had ingested toxaphene. In these instances, toxaphene produced congestion and edema in the lungs, dilatation of the heart, and petechial hemorrhages in the brain.

Despite the widespread application of toxaphene by formulators, sprayers, and agricultural workers, undisputed cases of occupational poisoning have apparently not been reported. According to Frawley (306), this record was achieved without special precautionary labeling, and he added, ". . . the absence of intoxications in men employed in the manufacture of toxaphene, as well as in applicators and agricultural workers, have convinced us of the low toxicity of toxaphene in man."

Members of the general population have absorbed insignificant quantities of toxaphene in the diet. Monitoring of food supplies in the United States has shown a very low incidence of toxaphene residues. In 1974, analyses of 2286 meat and 1921 poultry samples by the U.S. Department of Agriculture disclosed that one poultry and eight meat samples contained toxaphene residues (290).

Excretion of toxaphene in milk closely paralleled the level in body fat. Residues of toxaphene have, at various times, been found in the milk of cattle. Cows that were fed 20, 60, 100, and 140 ppm in the diet excreted the following concentrations in milk: 0.37, 0.74, 1.15, and 1.88 ppm, respectively (307). Toxaphene in human milk has been insignificant (308, 309).

10.5 Retention in Soil and Water

According to Nash and Woolson (310), the persistence of toxaphene in soil is similar to that of aldrin, dieldrin, dilan, or chlordane. In a related study, Nash and co-workers found the following percentages of pesticides in the soil 20 years after (soil) treatment: toxaphene, 45 percent; heptachlor, 35 percent; aldrin, dieldrin, and endrin, 12.8 percent; and chlordane, 7.7 percent.

Approximately 22 percent of toxaphene and 16 percent of DDT were recovered 10 years after application of these insecticides to the soil; 90 to 95 percent of the toxaphene and 60 to 75 percent of the DDT residues were in the top 30-cm layer (311).

According to Menzie (312) the half-life of pesticides in soil are for toxaphene, 10 years; DDT, 3 to 10 years; dieldrin, 1 to 17 years; and heptachlor, 7 to 12 years.

Paris and Lewis (313) published an extensive review on the persistence of toxaphene in various aquatic systems. Quoting from the review by Pollock and Kilgore (289), "Two lakes that differed in depth and relative amount of biological life were treated with toxaphene, and the persistencies of toxaphene residues in the two lakes were compared (314). A shallow lake 3,800 lb (1,721 kg) of toxaphene with a high biological activity was successfully restocked with trout within one year. The other deeper and less biologically active lake 5,800 lb (2,628 kg) of toxaphene remained toxic to trout for 6 years. Water samples from the shallow lake contained 0.6 ppb of toxaphene after 1 year."

According to studies by Hughes and Lee (315) and Lee et al. (316), four lakes treated with toxaphenes were found to be safe for restocking with fish within 7 to 10 months after the application of the chemical. Toxaphene levels in the water decreased rapidly for 4 months and then remained constant (1 to 4 ppb) for 10 months (end of sampling).

Toxaphene in anoxic salt marsh sediments was degraded within a few days to compounds having gas chromatographic retention times that were shorter than those of standard toxaphene components. This breakdown occurred in sterile as well as unsterile sediments, and also in sand–Fe(II)/Fe(III) systems (317).

According to Pollock and Kilgore (289), some components of toxaphene are extremely toxic to various freshwater fishes. Bioaccumulation does take place. Mayer et al. (318) reported a maximum toxaphene accumulation from water to fish of 69,000 times in fathead minnows and 50,000 times in channel catfish. The biomagnification of this pesticide in certain fishes was probably the reason for the deaths of some birds that fed on such fish.

10.6 Volatility

Air samples collected at various U.S. locations have shown residues of toxaphene. Concentrations from 16 to 2500 ng/m^3 were recorded by Stanley et al. (319) in the atmosphere of the southern United States. It was unexpected that air samples collected at locations in Bermuda and during cruises in the Atlantic Ocean were found to contain toxaphene residues in concentrations 10 times higher than those of other pesticides. The mean concentration for toxaphene was 0.63 ng/m³, while the mean p,p'-DDT level was 0.02 ng/m³ (289, 320). Toxaphene undergoes dehydrochlorination after prolonged exposure to sunlight, and at temperatures of about 155°C (286).

10.7 Cancer

The results of a bioassay of toxaphene for possible carcinogenicity were reported by the Carcinogenesis Testing Program of the National Cancer Institute (321). Male and female B6C3F1 mice were fed time-weighted average doses of 99 or 198 ppm of toxaphene for 80 weeks, then observed for 10 or 11 weeks. Male Osborne-Mendel rats ingested time-weighted average doses of 556 or 1112 ppm, and females 540 or 1080 ppm for 80 weeks followed by 28 or 30 weeks of observation. The report concluded that ". . . under the condition of this bioassay toxaphene was carcinogenic in male and female B6C3F1 mice, causing increased incidences of hepatocellular carcinomas. The test results also suggest carcinogenicity of toxaphene for the thyroid of male and female Osborne-Mendel rats."

Unpublished reports include oncogenic studies in mice and hamsters conducted for Hercules Incorporated by Litton Bionetics (322). In this investigation male and female B6C3F1 mice were fed 7, 20, or 50 ppm of toxaphene for 18 months; subsequently the mice were placed on the control diet for an additional 6-month period. Histopathological examination of tissues revealed a statistically significant increase in the incidence of hepatocellular adenomas and carcinomas in the male mice fed 50 ppm of toxaphene. The incidence was 35 percent (18/51) compared to 19 percent (10/53) for the untreated controls. The incidence for males fed 7 or 20 ppm toxaphene was 19 percent (10/54) and 23 percent (12/53), respectively. At all levels the incidence of hepatic tumors in treated female mice was similar to that in the untreated control female mice.

ARS Golden Syrian hamsters were fed 100, 300, or 1000 ppm of toxaphene. Females were treated for 18 months, males for 21.5 months. At sacrifice, a decrease in heart weight of males fed 300 and 1000 ppm and a decrease in thyroid weight of females fed 1000 ppm toxaphene were noted. Liver weight of males fed 1000 ppm was increased; this was related to the histopathological findings of megahepatocytes. No evidence of a tumorigenic effect related to toxaphene was seen in the hamsters at any level.

According to available information, there is no evidence that toxaphene has caused cancer in man.

10.8 Teratology

Camphechlor (toxaphene) was injected into the yolk of fertile eggs after 7 days incubation at doses ranging from 0 to 0.15 mg/egg. No effects on hatchability or teratogenic potential were observed, as quoted by WHO, 1974 (323).

10.9 Mutagenicity

Toxaphene is mutagenic in the *Salmonella* test without requiring liver homogenate for activity. According to Hooper et al. (324), some but not all of the mutagenic components of the more than 117 polychloroterpenes are easily separated from the insecticide ingredients.

10.10 Hygienic Standards of Permissible Exposure

The threshold limit value of chlorinated camphene for skin (toxaphene) adopted in 1979 by the ACGIH is a TLV–TWA of 0.5 mg/m^3. The tentative value for the TLV–STEL is 1 mg/m^3 (116).

According to the 1978 *NIOSH–OSHA Pocket Guide to Chemical Hazards* the permissible exposure limit is 0.5 mg/m^3. The IDLH of chlorinated camphene (toxaphene) is 200 mg/m^3 (117).

At the joint meeting of the FAO Working Party of Experts on Pesticide Residues and the WHO Expert Committee on Pesticides Residues, Geneva, Switzerland, November 26 to December 5, 1973, it was decided ". . . that it could not establish an ADI on the product available on a worldwide basis. The meeting expressed the hope that further research in progress will end in resolving the questions on the uniformity of the product conforming to FAO specifications . . ." (323).

REFERENCES

1. World Health Organization, Official Records No. 190, Geneva, April 1971.
2. K. A. King, E. L. Flickinger, and H. Hildebrand, *Pestic. Monit. J.* **12** (1), 1978, p. 16.
3. W. B. Deichmann, D. A. Cubit, W. E. MacDonald, and A. G. Beasley, *Arch. Toxikol.*, **29**, 287 (1972).
4. "Brief for Respondent in Environmental Defense Fund," *Fed. Reg.*, 439, 584 (1971).
5. Food and Drug Administration, *Fed. Reg.*, **37**, 13369 (July 7, 1972).
6. Environmental Protection Agency, Before the Administration, in re. Shell Chemical Company, Registrants FIFRA, Dockets No. 145, Recommended Decision, September 20 (1974).
7. H. F. Kraybill, U.S. Department of Health, Education and Welfare, National Institute of Health, Pest Control, Lecture: Houston, TE, October 22, 1975.
8. W. B. Deichmann and W. E. MacDonald, *Ecotoxicol. Environ. Safety*, **1**, 89 (1977); *Am. Ind. Hyg. Assoc. J.*, **37**, 495 (1976).
9. W. J. Hayes, Jr., *Clinical Handbook of Economic Poisons*, U.S. Dept. of Health, Education and Welfare, Communicable Disease Center—Toxicology Section, Atlanta, Ga., 1963.
10. W. B. Deichmann, S. W. Witherup, K. V. Kitzmiller, and F. F. Heyroth, *The Toxicity of DDT*, Kettering Laboratory, College of Medicine, University of Cincinnati, Ohio, 1950.
11. P. A. Neal, W. F. von Oettingen, W. W. Smith, R. B. Malmo, R. C. Dunn, H. E. Morann, T. R. Sweeney, D. W. Armstrong, and W. C. White, *Publ. Health Rep. Suppl.*, **177** (1944).

12. G. R. Cameron and F. Burgess, *Br. Med. J.*, **1**, 865 (1945).

13. G. Woodward, A. A. Nelson, and H. O. Calvary, *J. Pharmacol. Exp. Ther.*, **82**, 152 (1944).

14. A. J. Lehman, *Assoc. Food Drug Offic. U.S. Q. Bull.*, **15**, 122 (1951).

15. O. G. Fitzhugh and A. A. Nelson, *J. Pharmacol. Exp. Ther.*, **89**, 18 (1947).

16. J. R. Treon and F. P. Cleveland, *J. Agric. Food Chem.*, **3**, 402 (1955).

17. E. P. Laug, A. A. Nelson, O. G. Fitzhugh, and F. M. Kunze, *J. Pharmacol. Exp. Ther.*, **98**, 268 (1950).

18. J. H. Draize, A. A. Nelson, and H. O. Calvary, *J. Pharmacol. Exp. Ther.*, **82**, 159 (1944).

19. D. P. Morgan and C. C. Roan, *J. Occup. Med.*, **15** (1), 26 (1973).

20. C. Roan, D. Morgan, and E. H. Paschal, *Arch. Environ. Health*, **22**, 309 (1971).

21. H. B. Matthews, J. J. Domanski, and F. E. Guthrie, *Xenobiotica*, **6**(7), 425 (1976).

22. S. R. Khambata, *Pesticides*, **8**(12), 59 (1974).

23. M. J.-L. Monod, *Trav. Soc. Pharm. Montpel.* **37**(1), 9 (1977).

24. E. R. Laws, A. Curley, and F. J. Biros, *Arch. Environ. Health*, **15**, 766 (1967).

25. A. Poland, D. Smith, R. Kuntzman, M. Jacobson, and A. H. Conney, *Clin. Pharmacol. Ther.*, **11**(5), 724 (1970).

26. E. R. Laws, Jr., W. C. Maddrey, A. Curley, and V. W. Burse, *Arch. Environ. Health*, **27**, 318 (1973).

27. W. J. Hayes, Jr., W. F. Durham, and C. Cueto, *J Am. Med. Assoc.*, **162**(9), 890 (1956).

28. W. J. Hayes, Jr., W. E. Dale, and C. T. Pirkle, *Arch. Environ. Health*, **22**, 119 (1971).

29. E. C. Tabor, *N.Y. Acad. Sci.* **28**(5), 569 (1966).

30. P. M. Antommaria, M. Corn, and L. DeMaio, *Science*, **150**, 1476 (1965).

31. R. E. Duggan and P. E. Corneliussen, *Pestic. Monit. J.*, **5**(4), 313, 331 (1972).

32. P. E. Corneliussen, *Pestic. Monit. J.*, **2**(4), 140 (1969).

33. R. E. Duggan, H. C. Barry, and L. Y. Johnson, *Science*, **151**, 101 (1966); *Pestic. Monit. J.*, **11**(2), 2 (1977).

34. G. W. Ware, B. Estesen, N. A. Buck, and J. A. Marchello, *Bull. Environ. Cont. Toxicol.*, **20**(1), 28 (1978).

35. W. F. Durham, J. F. Armstrong, and G. E. Quinby, *Arch. Environ. Health*, **11**, 641 (1965).

36. W. T. Keane and M. R. Zavon, *Bull. Environ. Contam. Toxicol.*, **4**, 1 (1969).

37. S. Warnick, *Pestic. Monit. J.*, **6**, 9 (1972).

38. W. F. Edmundson, J. E. Davies, and W. Hull, *Pestic. Monit. J.*, **2**(2), 86 (1968).

39. J. E. Davies, W. F. Edmundson, A. Maceo, A. Barquet, and J. Cassady, *Am. J. Publ. Health*, **59**, 435 (1969).

40. V. Fiserova-Bergerova, J. Radomski, J. E. Davies, and J. Davis, *Ind. Med. Surg.*, **36**, 65 (1967).

41. R. T. Rappolt and W. E. Hale, *Clin. Toxicol.*, **1**(1), 57 (1968).

42. J. L. Radomski, W. B. Deichmann, and E. F. Clizer, *Food Cosmet. Toxicol.*, **6**, 209 (1968).

43. W. S. Hoffman, H. Adler, W. I. Fishbein, and F. C. Bauer, *Arch. Environ. Health*, **15**, 758 (1967).

44. E. P. Laug, F. M. Kunze, and C. S. Prickett, *A.M.A. Arch. Ind. Hyg.*, **3**, 245 (1951).

45. B. T. Woodard, B. B. Ferguson, and D. J. Wilson, *Natl. Tech. Info. Serv. PB-259*, 805 (1976).

46. G. G. Sims, J. R. Campbell, F. Zemlyak, and J. M. Graham, *Bull. Environ. Contam. Toxicol.*, **18**(6), 697 (1977).

47. R. C. Baird, N. P. Thompson, T. L. Hopkins, and W. R. Weiss, *Bull. Mar. Sci.*, **25**(4), 473 (1975).

48. D. J. Hanson, and A. J. Wilson, Jr., *Florida Pestic. Monit. J.*, **4**, 51 (1970).

49. L. D. Lyman, W. A. Tompkins, and J. A. McCann, *Pestic. Monit. J.*, **2**, 109 (1968).

50. R. E. Reinert, *Pestic. Monit. J.*, **3**, 233 (1970).

51. C. Henderson, W. L. Johnson, and A. Inglis, *Pestic. Monit. J.*, **3**, 145 (1969).

52. R. D. Earnest and P. E. Benville, Jr., *Pestic. Monit. J.*, **5**, 235 (1971).

53. W. B. Deichmann, D. A. Cubit, W. E. MacDonald, and A. G. Beasley, *Arch. Toxicol.*, **29**, 287 (1972).

54. D. A. Cubit, W. B. Deichmann, and W. E. MacDonald, *Am. Ind. Hyg. Assoc. J.*, **37**(1), 8 (1976).

55. M. E. Zabik, B. Olson, and T. M. Johnson, *Pestic. Monit. J.*, **12**(1), 36 (1978).

56. P. A. Butler and R. L. Schutzmann, *Pestic. Monit J.*, **12**(2), 51 (1978).

57. R. Frank, H. E. Braun, M. Holdrinet, D. P. Dodge, and G. E. Spangler, *Pestic. Monit. J.*, **12**(2), 60 (1978).

58. J. H. Phillips, E. E. Haderlie, and W. L. Lee, *Natl. Tech. Inf. Serv. PB-238*, 511 (1974).

59. J. R. W. Miles, C. R. Harris, and P. Moy, *J. Econ. Entomol.*, **71**(1), 91 (1978).

60. R. B. Owens, J. B. Dimond, and A. S. Getchell, *J. Environ. Qual.*, **6**(4), 359 (1977).

61. A. E. Carey, J. A. Gowen, H. Tai, W. G. Mitchell, and G. B. Wiersma, "National Soils Monitoring Program III," *Pestic. Monit. J.*, **12**(3), 117 (1978).

62. K. A. King, E. L. Flickinger, and H. H. Hildebrand, *Pestic. Monit. J.*, **12**(1), 16 (1978).

63. S. R. Peterson and R. S. Ellarson, *Pestic. Monit. J.*, **11**(4), 170 (1978).

64. Y. A. Greichus, B. D. Gueck, and B. D. Ammann, *Pestic. Monit. J.*, **12**(1), 4 (1978).

65. D. W. Johnston, *Pestic. Monit J.*, **12**(1), 8 (1978).

66. M. F. Ortelee, *A.M.A. Arch. Ind. Health*, **18**, 433 (1958).

67. W. J. Hayes, Jr., *Pharmacology and Toxicology of DDT: DDT Insecticides*, P. Muller, Ed., Vol. 2, Birkhauser, Basel, 1955, p. 11.

68. W. J. Hayes, *Toxicology of Pesticides*, Williams & Wilkins, Baltimore, 1975.

69. I. F. G. Ensberg, A. deBruin, and R. L. Zielhuis, Coronel Laboratory, University of Amsterdam, personal communication, 1971.

70. H. F. Kraybill, presentation at *Am. Meat Sci. Assoc. Conf., June 22–24, 1966, Cornell University, Ithaca, N.Y.*

71. W. B. Deichmann, *Environmental Problems in Medicine*, W. D. McKee, Ed., Charles C Thomas, St. Louis, Mo., 1974, pp. 347–420.

72. W. J. Hayes, Jr., *Toxicol. Appl. Pharmacol.*, **38**, 19 (1976).

73. B. Gladen, and W. Rogan, *Environ. Health Perspect.*, **20**, 248 (1977).

74. J. H. Thorstenson and H. W. Dorough, *Tob. Sci.*, **20**(5), 25 (1976).

75. H. W. Dorough and Y. H. Atallah, *Bull. Environ. Contam. Toxicol.*, **13**(1), 101 (1975).

76. F. E. L. Fowler, *J. Agric. Food Chem.*, **1**, 469 (1953).

77. T. H. Jukes, *Arch. Intern. Med.*, **138**(5), 772 (1978).

78. H. F. Kraybill, "Pesticide Toxicity and Potential for Cancer: A Proper Perspective," presented at the *Natl. Pest. Control. Assoc. Convention, Houston, Tex. October 1975.*

79. W. B. Deichmann and W. E. MacDonald, *Ecotoxical. Environ. Safety*, **I**, 89 (1977).

80. L. Tomatis and V. Turusov, *GNN*, **17**, 219 (1975).

81. S. K. Kashyap, S. K. Nigam, A. B. Karnik, R. C. Gupta, and S. K. Chatterjee, *Int. J. Cancer*, **19**(5), 725 (1977).

82. M. D. Reuber, *Sci. Total Environ.*, **10**(2), 105 (1978).

83. IARC, *Some Organochlorine Pesticides*, Vol. 5, International Agency for Research on Cancer, World Health Organization, Lyon, France, 1974.

84. E. Thorpe and A. I. T. Walker, *Food Cosmet. Toxicol.*, **11,** 433 (1973).

85. J. R. M. Innes, B. M. Ulland, M. G. Valerio, L. Petrucelli, L. Fishbein, E. R. Hart, A. J. Pallatta, R. R. Bates, H. L. Falk, J. J. Gart, M. Klein, I. Mitchell, and J. Peters, *J. Natl. Cancer Inst.*, **42,** 1101 (1969).

86. L. Rossi, M. Ravera, G. Repetti, and L. Santi, *Int. J. Cancer.*, **19,** 179 (1977).

87. *IARC Monographs*, Vol. 1–20, International Agency for Research Monographs Supplement, Lyon, September 1979.

88. J. R. P. Cabral, R. K. Hall, and P. Shubik, abstr. presented at *20th Congr. Eur. Soc. Toxicol., West Berlin, June 25–28, 1978*.

89. NCI, *Bioassays of DDT, TDE and p,p'-DDE for Possible Carcinogenicity*, CAS No. 50-29-3; 72-54-8, 72-55-9; NCI-CG-TR-131, U.S. Dept. of Health, Education and Welfare, Public Health Sevice, National Institute of Health, 1978.

90. WHO, *DDT and Its Derivatives*, publ. under joint sponsorship of United Nations Environmental Programme and World Health Organization, Geneva, 1979, pp. 106–107, 114–115.

91. K. W. Jager, *Aldrin, Dieldrin, Endrin, and Telodrin: An Epidemiological and Toxicological Study of Long Occupational Exposure*, Elsevier, New York, 1970.

92. W. Buselmaier, G. Roehrborn, and P. Propping, *Mutat. Res.*, **21,** 25 (1973).

93. E. Vogel, *Mutat. Res.*, **16,** 157 (1972).

94. J. Yoder, M. Watson, and V. V. Benson, *Mutat. Res.*, **21,** 335 (1973).

95. M. L. Keplinger, W. B. Deichmann, and F. Sala, *Effects of Combinations of Pesticides on Reproduction in Mice*, in *Pesticides Symposia*, Halos and Associates, Inc., Miami, Fla., 1970, pp. 125–138.

96. J. C. Street, *Science*, **146,** 1580 (1964).

97. W. B. Deichmann, M. Keplinger, I. Dressler, and F. Sala, *Toxicol. Appl. Pharmacol.*, **14,** 205 (1969).

98. W. B. Deichmann, W. E. MacDonald, D. A. Cubit, *Science*, **172,** 275 (1971).

99. M. T. Koeferl and R. E. Larson, *Toxicol. Appl. Pharmacol.*, **33**(1), 157 (1975).

100. M. A. Cawthorne, E. A. Murrell, J. Bunyan, J. W. G. Leiper, J. Green, and J. H. Watkins, *Res. Vet. Sci.*, **12**(6), 516 (1971).

101. W. F. Spencer and M. M. Cliath, *J. Agric. Food Chem.*, **20**(3), 645 (1972).

102. D. R. Young, D. J. McDermott, and T. C. Hessen, *Bull. Environ. Contam. Toxicol.*, **16**(5), 604 (1976).

103. D. G. Crosby and K. W. Moilanen, *Chemosphere*, **6**(4), 167 (1977).

104. S. N. Deshmukh, S. N. Singh, and R. L. Kalra, *Pesticides*, **11**(2), 51 (1977).

105. H. R. Crookshank and H. E. Smally, *Southwest Vet.*, **31**(1), 39 (1978).

106. N. P. Agnihotri, S. Y. Pandey, H. K. Jain, and K. P. Srivastava, *J. Entomol. Res.*, **1**(1), 89 (1977).

107. R. G. Achari, S. S. Sandhu, and W. J. Warren, *Bull. Environ. Contam. Toxicol.*, **13**(1), 94 (1975).

108. U. Kiigemagi and L. C. Terriere, *Bull. Environ. Contam. Toxicol.*, **7**(6), 348 (1972).

109. E. V. Grel and A. I. Ryabinin, *Tr. Gos. Okeanogr. Inst.*, **132,** 68 (1976).

110. V. S. Fang and O. Gomez, *Fed. Proc., Fed. Am. Soc. Exp. Biol.*, **37**(3), 243 (1978).

111. V. Lenner and F. Kuemmerle, *Therapiewoche*, **28**(30), 5569 (1978).

112. S. Leiba, H. Kaufman, G. Winkelsberg, and C. M. Bahary, *Acta Obstet. Gynecol. Scand.*, **57**(4), 373 (1978).

113. U. Bigler and J. Muller, *Schweiz. Med Wochenschr.*, **108**(2), 67 (1978).

114. T. F. Hogan, D. L. Citrin, B. M. Johnson, S. Nakamura, T. E. Davis, and E. C. Borden, *Cancer (Philadelphia)*, **42**(5), 2177 (1978).

115. Hospital Elm Park, Dublin, Ireland, *Ir. J. Med. Sci.*, **147**(12), 437 (1978).

116. *TLVs, Threshold Limit Values for Chemical Substances and Physical Agents in the Workroom Environment with Intended Changes for 1977*, American Conference of Governmental Industrial Hygienists, 1979.

117. *NIOSH/OSHA Pocket Guide to Chemical Hazards*, U.S. Dept. of Health, Education and Welfare, National Institute for Occupational Safety and Health and U.S. Dept. of Labor, Occupational Safety and Health Administration, September 1978.

118. *Codex Alimentarius Commission*, Joint Food and Agricultural Organization of the United Nations World Health Organization, Codex Committee on Pesticide Residues, Tenth Session, The Hague, May 29–June 5, 1978.

119. W. L. Ball, K. Kay, and J. W. Sinclair, *A.M.A. Arch. Ind. Hyg. Occup. Med.*, **7**, 292 (1953).

120. F. P. Cleveland, *Arch. Environ. Health*, **13**, 195 (1966).

121. C. H. Kitselman, *J. Am. Vet. Med. Assoc.*, **123**, 28 (1953).

122. J. M. Bann, T. J. DeCino, N. W. Earle, and Y. P. Sun, *J. Agric. Food Chem.*, **4**, 937 (1956).

123. W. B. Deichmann, W. E. MacDonald, E. Blum, M. Bevilacqua, J. Radomski, M. Keplinger, and M. Balkus, *Ind. Med.*, **39**, 37 (1970).

124. W. E. MacDonald, J. MacQueen, W. B. Deichmann, T. Hamill, and R. Copsey, *Int. Arch. Arbeitsmed.*, **26**, 31 (1970).

125. W. B. Deichmann and M. L. Keplinger, *Pesticides Symposia*, Halos and Associates, Inc., Miami, Fla., 1970, pp. 121–123.

126. F. Princi and G. H. Spurbeck, *Arch. Ind. Hyg. Occup. Med.*, **3**, 64 (1951).

127. E. Nelson, *Rocky Mt. Med. J.*, **50**, 483 (1953).

128. H. G. S. Van Raalte, *Ecotoxicol. Environ. Safety*, **1**, 203 (1977) and personal communication, 1979.

129. O. G. Fitzhugh, A. A. Nelson, and M. L. Quaife, *Food Cosmet. Toxicol.*, **2**, 551 (1964).

130. B. B. Virgo and G. D. Bellward, *Can. J. Physiol. Pharmacol.*, **53**(5), 903 (1975).

131. P. C. Hurkat, *Indian J. Anim. Sci.*, **47**(10), 671 (1977).

132. A. S. Wright, D. Potter, M. F. Wooder, and C. Donninger, *Food Cosmet. Toxicol.*, **10**, 311 (1972).

133. A. S. Wright, C. Donninger, R. D. Greenland, K. L. Stemmer, and M. R. Zavon, *Ecotox. Environ. Safety.*, **1**(4), 477 (1978).

134. K. L. Davison and J. L. Sell, *Arch. Environ. Contam. Toxicol.*, **7**(2), 245 (1978); **7**(3), 369 (1978).

135. J. L. Sell, K. L. Davison, and D. W. Bristol, *Poult. Sci.*, **56**(6), 2045 (1977).

136. W. J. Hayes, Jr., *Publ. Health Rep. U.S.*, **72**, 1087 (1957).

137. H. Wolfe, W. F. Durham, and J. F. Armstrong, *Arch. Environ. Health.*, **6**, 458 (1963).

138. Z. Jegier, *Arch. Environ. Health.*, **8**, 670 (1964).

139. C. G. Hunter and J. Robinson, *Food Cosmet. Toxicol.*, **6**, 253 (1968).

140. H. G. S. Van Raalte, *Ecotoxicol. Environ. Safety*, **1**(29), 203 (1977).

141. B. C. Turner, D. E. Glotfelty, and A. W. Turner, *Agric. Food Chem.*, **25**(3), 548 (1977).

142. I. Weisberger, D. Bieniek, J. Kohli, and W. Klein, *J. Agric. Food Chem.*, **23**(5), 873 (1975).

143. G. Reddy and M. A. Q. Khan, *Bull. Environ. Contam. Toxicol.*, **13**(1), 64 (1975).

144. T. A. Wastler, C. K. Offutt, C. K. Fitzsimmons, and P. E. Des Rosiers, *Natl. Tech. Inform. Serv. PB-253*, 979 (1975).

145. W. B. Deichmann, W. E. MacDonald, E. Blum, M. Bevilacqua, J. Radomski, M. Keplinger, and M. Balkus, *Ind. Med. Surg.*, **39**, 426 (1970).

146. J. Song and W. C. Harville, *Abstr., Fed. Proc.*, **23**, 336 (1964).

147. A. I. T. Walker, D. E. Stevenson, J. Robinson, E. Thorpe, and M. Roberts, *Toxicol. Appl. Pharmacol.*, **15**, 345 (1968).

148. W. B. Deichmann, W. E. MacDonald, and F. C. Lu, *Toxicology and Occupational Medicine*, W. B. Deichmann, Ed., Elsevier–North Holland, New York, 1977, pp. 407–413.

149. National Cancer Institute, *Bioassay of Aldrin and Dieldrin for Possible Carcinogenicity*, Carcinogenesis Tech. Rep. Ser. No. 21 (1978), CAS Nos. 309-00-2 and 60-57-1; NCI-CG-TR-21.

150. D. E. Stevenson, E. Thorpe, P. F. Hunt, and A. I. T. Walker, *Toxicol. Appl. Pharmacol.*, **38**(2), 247 (1976).

151. K. J. Davis and O. G. Fitzhugh, *Toxicol. Appl. Pharmacol.*, **4**, 187 (1962).

152. E. Thorpe and A. I. T. Walker, *Food Cosmet. Toxicol.*, **11**, 433 (1973).

153. J. R. P. Cabral, P. Shubik, S. A. Bronczyk, and R. K. Hall, *Assoc. Cancer Res. 68th Ann. Meet. Am. Soc. Clin. Oncology (13th Ann. Meet.), Denver, Colo., May 1977, Proc.*, Abstr. No. 111, p. 28.

154. K. Bidwell, E. Weber, I. Nienhold, T. Connor, and M. S. Legator, *Mutat. Res.*, **31**(5), 314 (1975).

155. B. J. Dean, S. M. A. Doak, and H. Somerville, *Food Cosmet. Toxicol.*, **13**(1), 317 (1975).

156. K. M. Dix and A. B. Wilson, Shell, *Tunstall Lab. Rep.* TLGR 0051.71 (1971).

157. A. D. Ottolenghi, J. K. Haseman and F. Suggs, *Teratol.*, **9**, 11 (1974).

158. J. G. Saha and Y. W. Lee, *Bull. Environ. Contam. Toxicol.*, **4**(5), 283 (1969).

159. T. B. Stauffer, *Natl. Tech. Inform. Serv. AD-A049*, 627 (1977).

160. Council for Agricultural Science and Technicology (CAST) (Iowa State University, Ames, Iowa), *Farm Chem.*, **138**(13), 36 (1975); *Rep. No. 59*, September 15, 1976.

161. V. K. Smith, personal communication to Dr. John V. Osmum, Purdue University (through Ref. 160).

162. W. P. Cochrane and R. Greenhalgh, *J. Assoc. Offic. Anal. Chem.*, **59**(3), 696 81976).

163. S. Gaeb, W. P. Cochrane, H. Parlar, and F. Korte, *Z. Naturforsch.*, **30b**(2), 238 (1975).

164. K. M. Hyde and R. L. Falkenburg, *Toxicol. Appl. Pharmacol.*, **37**(3), 499 (1976).

165. L. Ingle, *Arch. Ind. hyg. Occup. Med.*, **6**, 354 (1952).

166. A. Lehman, *Assoc. Food Drug Offic. U.S. Q. Bull.*, **16**, 3 (1952).

167. A. M. Ambrose, *Fed. Proc.*, **12**, 298 (1953).

168. L. Ingle, Chem. Specialties Mfrs. Assn. Proc. (December 1954).

169. L. Ingle, *A Monograph on Chlordane, Toxicological and Pharmacological Properties*, University of Illinois, Urbana, Illinois (1965).

170. D. W. DeLong and P. J. Ludwig, *J. Econ. Entomol.*, **47**, 1056 (1954).

171. R. D. Radeleff, G. T. Woodward, W. J. Nickerson, and R. C. Bushland, *USDA Tech. Bull.*, **1122**, 7 (1955).

172. N. H. Poonawalla and F. Korte, *J. Agric. Food Chem.*, **19**(3), 467 (1971).

173. H. M. Balba and J. G. Saha, *J. Environ. Sci. Health*, **B13**, 211 (1978).

174. C. Henderson, A. Inglis, and W. L. Johnson, *Pestic. Monit. J.*, **5**, 1 (1971).

175. H. W. Dorough and B. C. Pass, *J. Econ. Entomol.*, **65**(4), 976 (1972).

176. P. C. Oloffs, L. J. Albright, and S. Y. Szeto, Simon Fraser University, Burnaby, Canada [through *Pestic. Abstr. No. 5*, May 78-1020 (1978)].

177. J. L. Sanborn, R. L. Metcalf, W. N. Bruce, and P. Y. Lu, *Environ. Entomol.*, **5**(3), 533 (1976).

178. National Cancer Institute, Bioassay of Chlordane for Possible Carcinogenicity, Carcinogenesis Tech. Rep. Ser. No. 8, Cas No. 57-74-9, NCI-CG-TR-8 (1977).

179. Velsicol: Summary of the Toxicological Evidence of Heptachlor and Chlordane presented in

Administrative Hearings called by the United States Environmental Protection Agency, November 18, 1974–March 6, 1978.

180. National Cancer Institute, Bioassay of Heptachlor for Possible Carcinogenicity, Carcinogenesis Tech. Rep. Ser. No. 9, Cas No. 76-44-8, NCI-CG-TR-9 (1977).

181. U.S. Food and Drug Administration, internal memorandum Dr. Kent J. Davis to Dr. A. J. Lehman, "Pathology Report on Mice fed Aldrin, Dieldrin, Heptachlor and Heptachlor Epoxide for 2 Years" (G. Fitzhugh and K. J. Davis), July 19, 1965 (through Ref. 179).

182. International Research & Development Corporation, Mattowan, Michigan, Report No. 163-084, "75 Percent Heptachlor Epoxide–25 percent Heptachlor, 2-Acetoamidofluorine, Eighteen Month Oral Carcinogenic Study in Mice," September 26, 1973 (through Ref. 179).

183. S. Witherup, F. P. Cleveland, and K. Stemmer, "The Psychological Effects of the Introduction of Heptachlor Epoxide in Varying Levels of Concentration into the Diet of CFN Rats," unpublished report, The Kettering Laboratory, University of Cincinnati, November 10, 1959.

184. Pesticide Information Review and Evaluation Committee for the Advisory Center on Toxicology, Assembly of Life Sciences, National Research Council, National Academy of Science, Washington D.C., "An Evaluation of the Carcinogenicity of Chlordane and Heptachlor," October 1977 (through Ref. 179).

185. Food and Drug Organization of the United Nations, "FAO Plant Production and Protection Paper 10 Rev. Pesticides Residues in Food—1977," Report of the Joint Meeting of the FAO Panel of Experts on Pesticide Residues and Environment and the WHO "Expert Committee on Pesticide Residues," Rome, 1978.

186. W. B. Buck, G. D. Osweiler, and G. A. Van Gelder, *Clinical and Diagnostic Veterinary Toxicology,* Kendall Hunt, Dubuque, Iowa, 1973 (through Ref 160).

187. J. L. Radomski and B. Davidow, *J. Pharmacol. Exp. Ther.* **107**, 3, 259 (1953).

188. II. E. Fairchild, *Heptachlor—A Review of its Uses, Chemistry, Environmental Hazards and Toxicology,* Office of Pesticide Programs, Environmental Protection Agency, Washington, D.C., 1972.

189. K. J. Davis, unpublished internal memorandum, Food and Drug Administration, (through Ref. 160).

190. B. Davidow and J. L. Radomski, *J. Pharmacol. Exp. Ther.,* **107,** 259 (1953).

191. R. D. O'Brien, *Insecticides, Action and Metabolism,* Academic Press, New York, 1967.

192. G. T. Brooks, *Residue Rev.* **27,** 81 (1969).

193. K. P. Bovart, J. P. Fontenot, and B. M. Priode, *J. Anim. Sci.,* **33**(1), 127 (1971).

194. M. N. Melnikov, *Chemistry of Pesticides,* Springer, New York, 1971 (through Ref. 160).

195. F. Matsamura and J. O. Nelson, *Bull. Environ. Contam. Toxicol.,* **5**(6), 489 (1970).

196. IARC, International Agency for Research on Cancer, Report of Working Committees, Lyon, France, Vol. 5, 1974.

197. Louisiana Agric. Exp. Stn. Rep. No. 1, unpublished (1972) (through Ref. 160).

198. C. Cueto and F. J. Biros, *Toxicol. Appl. Pharmacol.,* **10,** 261 (1967).

199. C. C. Doane, *J. Econ. Entomol.,* **55,** 416 (1962).

200. J. B. Polinke, *Ohio J. Sci.,* **51,** 195 (1951).

201. J. C. Schread, *Coun. Agric. Exp. Stn. Bull,* 556 (1952).

202. R. D. Smith, M. S. Thesis, Louisiana State University (through Ref. 160).

203. Environmental Protection Agency, *Fed. Reg.* 39 (229); 41298 (1974).

204. R. B. Smith, Jr., P. S. Larson, J. K. Finnegan, H. B. Haag, G. R. Henningar, and F. Cobey, *Toxicol. Appl. Pharmacol.,* **1,** 119 (1959).

205. D. D. Irish, *Industrial Hygiene and Toxicology*, 2nd rev. ed. Vol. II, Wiley-Interscience, New York, 1963.

206. A. J. Lehman, *Summaries of Pesticide Toxicity*, The Association of Food and Drug Officials of the United States, Business Office, Evan Wright, Secretary-Treasurer, P.O. Box 1494, Topeka, Kansas, 1965.

207. National Cancer Institute, *Bioassay of Dicofol for Possible Carcinogenicity*, Carcinogenesis Tech. Rep. Ser. No. 90, CAS No. 115-32-2, NCI-CG-TR-90 (1978).

208. Data Evaluation/Risk Assessment Subgroup of the Clearinghouse on Environmental Carcinogens, p. 45.

209. P. S. Larson, J. L. Egle, Jr., G. R. Hennigar, R. W. Lane, and J. F. Borzelleca, *Toxicol. Appl. Pharmacol.*, **48**, 29 (1979), and personal communication from P. S. Larson, 1976.

210. D. T. P Huang, I. K. Ho, H. M. Mehendale, and A. S. Hume, *Fed. Proc. Fed. Am. Soc. Exp. Biol.*, **38**(3, Pt 1), 845 (1979).

211. W. B. Deichmann, *Kepone*, report prepared for Senator H. H. Bateman and the National Fisheries Institute, Inc., Washington, D.C., November 10, 1976.

212. L. R. Curtis and H. M. Mehendale, *Pharmacologist*, **20**(3), 187 (1978).

213. D. End. R. A. Carchman, and W. L. Dewey, *Fed. Proc. Fed. Am. Soc. Exp. Biol.*, **38**(3, Pt 1), 845 (1979).

214. B. M. Anderson, S. T. Kohler, and R. W. Young, *J. Agric. Food Chem.*, **26**(1), 130 (1978).

215. L. Reiter and K. Kidd, *Toxicol. Appl. Pharmacol.*, **45**(1), 357, (1978).

216. L. J. Egle, Jr., S. B. Fernandez, P. Guzelian, and J. F. Borzelleca, *Drug Metab. Dispos.*, **6**(1), 91, (1978).

217. J. J. Huber, *Toxicol. Appl. Pharmacol.*, **7**, 516 (1965).

218. N. Chernoff and E. H. Rogers, *Toxicol. Appl. Pharmacol.*, **38**(1), 189 (1976).

219. *Kepone/Mirex/Hexachlorocyclopentadiene: An Environmental Assessment. A Report prepared by the Panel on Kepone/Mirex/Hexachlorocyclopentadiene*, Coordinating Committee for Scientific and Technical Assessments of Environmental Pollutants, National Academy of Sciences, Washington, D.C., 1978.

220. S. B. Cannon, J. M. Veazey, Jr., R. S. Jackson, V. W. Burse, C. Hayes, W. E. Straub, P. J. Landrigan, and J. A. Liddle, *Am. J. Epidemiol.*, **107**, 529 (1978).

221. W. J. Cohn, R. V. Blanke, F. D. Griffith, Jr., and P. S. Guzelian, *Gastroenterology*, **71**, 901 (1976).

222. J. E. Huff and H. B. Gerstner, *J. Environ. Pathol. Toxicol.*, **1**(4), 377 (1978).

223. Public Health Service, Cancer and Birth Defects Division, Bureau of Epidemiology, CDC, Atlanta, EPI-76-7-2 (1976).

224. U.S. Department of Agriculture (1977), Comments of the Secretary of Agriculture in Response to the Notice of Intent to Cancel Pesticide Products Containing Chlordecone, Trade Name, Kepone, January 11, 1977, Washington, D.C. (through Ref. 219).

225. D. A. Carlson, K. D. Konyha, W. B. Wheeler, G. P. Marshall, and R. G. Zaylskie, *Science*, **194**(4268), 939 (1976) (through Ref. 219).

226. Environmental Protection Agency, *EPA News* (February 27, 1976) and *Review of the Environmental Effects of Mirex and Kepone* (1978). Prepared by the U.S. Environmental Protection Agency, Office of Research and Development, by Batelle Columbus Laboratories (M. A. Bell, R. A. Ewing, and G. A. Lutz), EPA 600/1-78-013, Washington, D.C. (through Ref. 219).

227. R. V. Blanke, M. W. Fariss, P. S. Guzelian, P. S. Patterson, and D. E. Smith, *Bull. Environ. Contam. Toxicol.*, **20**(6), 782 (1978).

228. J. J. Boylan, J. L. Egle, and P. S. Guzelian, *Science*, **199**, 893 (1978).

229. E. Richter, J. P. Lay, W. Klein, and F. Korte, *J. Agric. Food Chem.*, **27**(1), 187 (1979).

230. *Sci. News*, **113**(25), 405 (1978).

231. M. L. Motl., *Environ. Manage.*, **1**(6), 491 (1977).

232. National Cancer Institute, *Report on Carcinogenesis Bioassay of Technical Grade Chlordecone (Kepone)*, U.S. Department of Commerce, National Technical Information Service, January 1976.

233. P. M. Newberne, personal communication, November 6, 1979.

234. A. J. Lehman, *Assoc. Food Drug Offic. U.S. Q. Bull.*, **18**, 3 (1954).

235. A. J. Lehman, *Summary of Pesticide Toxicity*, Association of Food and Drug Officials of the United States, Topeka, Kansas, 1965.

236. O. J. Fitzhugh, A. A. Nelson, and J. P. Frawley, *J. Pharmacol. Exp. Ther.*, **100**, 59 (1950).

237. A. J. Lehman, *Assoc. Food Drug Offic. Q. Bull.*, **20**, 95 (1956).

238. B. Davidow and J. P. Frawley, *Proc. Soc. Exp. Biol. Med.*, **76**, 780 (1951).

239. W. A. Queen, *Assoc. Food Drug Offic. U.S. Q. Bull.*, **17**, 127 (1953).

240. S. Floru, A. Polizu, M. Ripeanu, and S. Cusa, *An. Inst. Cercet. Prot. Plant., Acad. Stiinte Agr. Silvice.*, **9**, 587 (1973) [through *Pestic. Abstr., U.S. Environ. Prot. Agency, No. 75-2974* (1975)].

241. K. van Asperen and F. J. Oppernoorth, *Nature*, **173**, 1000 (1954).

242. H. Seidler, R. M. Macholz, M. Haertig, M. Kujawa, and R. Engst, *Nahrung*, **19**(5/6), 473 (1975).

243. R. W. Chadwick, L. T. Chuang, K. Williams, *Pestic. Biochem. Physiol.*, **5**(6), 575 (1975).

244. R. W. Chadwick, C. C. Bryden, M. F. Copeland, and J. J. Freal, *Chemosphere*, **7**(8), 633 (1978).

245. H. Okazaki and S. Kimura, *Ecyo To Shokuryo (J. Jap. Soc. Food Nutr.)*, **25**(3), 201 (1972), [through *Pestic. Abstr., U.S. Environ. Prot. Agency No. 72-2174* (1972)].

246. L. Hulth, L. Hoglund, A. Bergman, and L. Moller, *Toxicol. Appl. Pharm.*, **46**, 101 (1978).

247. L. Hulth, M. Larson, R. Carlsson, and J. E. Kihlstrom, *Bull. Environ. Contam. Toxicol.*, **16**(2), 133 (1976).

248. H. Herken and I. Klempau, *Naturwissenchaften*, **37**, 493 (1950).

249. H. Coper, H. Herken, and I. Klempau, *Naturwissenschaften*, **38**, 69 (1951).

250. W. Koransky and J. Portig, *Arch. Exp. Pathol. Pharmakol.*, **244**, 564 (1963).

251. R. Engst, R. Macholz, M. Kujawa, H-J. Lewerenz, and R. Plass, *J. Environ. Sci. Health B.*, **11**(2), 95 (1976).

252. V. Fiserova-Bergerova, J. L. Radomski, J. E. Davies, and J. H. Davis, *Ind. Med. Surg.*, **36**, 65 (1967).

253. J. L. Radomski, E. Astolfi, W. B. Deichmann, and A. A. Rey, *Toxicol. Appl. Pharmacol.*, **20**, 186 (1971).

254. M. Tagami and N. Oshiro, *Okinawa-ken Kogai-Eisei Kenkyus hoho, Ann. Rept. Okinawa Pref. Inst. Publ. Health*, **11**, 76 (1977).

255. E. Danopoulos, K. Melissinos, and G. Katsas, *Arch. Ind. Hyg. Occup. Med.*, **8**, 582 (1953).

256. O. M. Heiberg and H. N. Wright, *A.M.A. Arch. Ind. Health*, **11**, 457 (1955).

257. J. J. Hutter, *Clin. Res.*, **27**(1), 97A (1979).

258. L. M. Solomon, L. Fahrner, and D. P. West, *Arch. Dermatol.*, **113**(3), 353 (1977).

259. W. E. Pace and J. Purres, *Can. Med. Assoc. J.*, **104**, 719 (1971).

260. C. M. Ginsberg, W. Lowry, and J. S. Reisch, *J. Pediatr.*, **91**(6), 998 (1977).

261. M. Wheeler, *West. J. Med.*, **127**(6), 518 (1977).

262. B. G. P. MacNamara, *Brit. Med. J.*, **3**(5722), 585 (1970).

263. D. Nag, G. C. Singh, and S. Senon, *Trop. Geogr. Med.*, **29**(3), 229 (1977).

264. W. B. Deichmann, *Pesticides and the Environment. A Continuing Controversy*, Symposia Specialists, Miami, Fla. 1973, p. 397–398.

265. T. H. Milby, A. J. Samuels, and F. Ottoboni, *J. Occup. Med.*, **10,** 584 (1968).

266. W. R. Best, *J. Am. Med. Assoc.*, **185,** 286 (1963).

267. L. Sanchez-Medal, J. P. Castenado, and F. Garcia-Rojas, *N. Engl. J. Med.*, **269,** 1365 (1963).

268. J. P. Loge, *J. Am. Med. Assoc.*, **193,** 110 (1965).

269. A. J. Samuels and T. H. Milby, *J. Occup. Med.*, **13**(3), 147 (1971).

270. T. H. Milby and A. J. Samuels, *J. Occup. Med.*, **13**(5), 256 (1971).

271. L. Rosival and A. Szokolay, *Pure Appl. Chem.*, **42**(1–2), 167 (1975).

272. O. Mashiro and H. Takahisa, *Environ. Pollut.*, **9**(4), 283 (1975).

273. R. C. Zeana and A. Zeana, *Rev. Chim. (Bucharest)*, **25**(11), 901 (1974).

274. B. Das and P. K. Singh, *Microbios Lett.*, **4**(14), 99 (1977).

275. M. Hanada, K. Kakudo, and T. Miyachi, *Nippon Gangakkai Kiji (Proc. Jap. Cancer Assoc. Ann. Meet.* **34,** 33, (1975) [through *Pestic. Abstr. U.S. Environ. Prot. Agency No. 76-0496* (1976)].

276. M. Hananouchi, S. Sugihara, T. Ogiso, H. Aoe, M. Hirose, Y. Nakamura, and N. Ito, *Nippon Gangakkai Kiji (Proc. Jap. Cancer Assoc. Ann. Meet.* **34,** (1975) [through *Pestic. Abstr. U.S. Environ. Prot. Agency No. 76-0495* (1976)].

277. National Cancer Institute, Carcinogenesis Tech. Rep. Ser. No 14, CAS No. 58-89-9; NCI-CG-TR-14 (1977).

278. A. K. Palmer, A. M. Bottomley, A. Warden, H. Frohberg, and A. Bauer, *Toxicology*, **9**(3), 239 (1978).

279. J. E. Treon, F. P. Cleveland, and J. Cappel, *A.M.A. Arch. Ind. Health*, **11,** 459 (1955).

280. Industrial Bio-Test Laboratories, Inc., Reports to Hooker Chemicals and Plastics Corp., April 18, June 13, and September 29 (1975) (through Ref. 232).

281. Velsicol Chemical Corporation, *Product Bull. No. 50101-2* (1976) (through Ref. 232).

282. P. Y. Lu, R. L. Metcalf, A. S. Hiriwe, and J. W. Williams, *J. Agric. Food Chem.*, **23**(5), 967 (1975).

283. L. Ingle, *Science*, **118,** 213 (1953) (through Ref. 232).

284. D. L. Morse, J. R. Kominsky, C. L. Wisseman, and P. J. Landrigan, *J. Am. Med. Assoc.*, **241,** 2177 (1979).

285. Industrial Bio-Test Laboratories, Inc., Northbrook Ill., Report to Velsicol Chemical Corporation, August 10, (1977), IBT No. 8536-10838 (through Ref. 232).

286. R. L. Metcalf, *Organic Insecticides, Their Chemistry and Mode of Action:* Interscience, New York, 1955.

287. J. E. Casida, R. L. Holmstead, S. Khalifa, J. R. Knox, T. Ohsawa, K. J. Palmer, and R. Y. Wong, *Science*, **183,** 520 (1974).

288. R. L. Holmstead, S. Khalifa, and J. E. Casida, *Agric. Food Chem.*, **22,** 939 (1974).

289. G. A. Pollock and W. W. Kilgore, *Toxaphene*, Springer, New York, 1978.

290. L. Penumarthy, F. W. Oehme, J. E. Spaulding, and W. A. Rader, *Vet. Toxicol.*, **18,** 60 (1976).

291. R. D. Radeleff and R. C. Bushland, *J. Econ. Entomol.*, **43,** 358 (1950).

292. FAO and WHO *1968 Evaluations of Some Pesticide Residues in Food*, FAO/PL: 1968/M/9/1, WHO/Food Add./69.35, issued jointly by FAO and WHO, Food and Agriculture Organization of the United Nations World Health Organization, Geneva, 1969, pp. 267–271.

293. *U.S. Environ. Prot. Agency Fed. Reg. 440/9-76-014* (1976).

294. *Hercules Toxicol. Data Bull. T-105D.*

295. P. Ortega, W. J. Hayes, and W. F. Durham, *A.M.A. Arch. Pathol.*, **64,** 614 (1957).

296. R. W. Lackey, *J. Ind. Hyg. Toxicol.*, **31,** 117 (1949).

297. A. J. Lehman, *Bull. Assoc. Food Drug Offic.*, **16,** 47 (1952).

298. G. Fitzhugh and A. A. Nelson, 1951 (through Ref. 292, p. 271).

299. J. F. Treon, F. Cleveland, B. Poynter, W. Wagner, and T. Gahegan, unpublished report of the Kettering Laboratory, University of Cincinnati, Cincinnati, Ohio, (1952) (through Ref. 304).

300. D. Brock and J. C. Calandra, 1964 (through Ref 292, p. 270).

301. Bio-Test Laboratories, Inc., unpublished report by J. C. Calandra (1965) (through Ref. 292).

302. G. Kennedy et al., 1968 (through Ref 292, p. 271).

303. N. Chernoff and B. D. Carver, *Bull. Environ. Contam. Toxicol.*, **15,** 660 (1976).

304. M. L. Keplinger, W. B. Deichmann, and F. Sala, presented at the 6th Inter-American Conference on Toxicology and Occupational Medicine, Miami, Florida, 1968, *Pesticides Symposia*, Halos and Associates, Inc., Miami, Fla. 1970.

305. J. C. McGee and H. L. Reed, *J. Am. Med. Assoc.*, **149,** 1124 (1952).

306. J. P. Frawley, personal Communication, 1972.

307. USDA, 1956 (through Ref. 292, p. 268).

308. H. V. Clayborn, H. D. Mann, M. C. Ivey, R. D. Radeleff, and G. T. Woodard, *J. Agric. Food Chem.*, **11,** 286 (1963).

309. G. E. Guyer, P. L. Adkisson, K. Dubois, C. Menzie, H. P. Nicholson, and G. Zweig, Special Report to Hazardous Materials Advisory Committee, U.S. Environmental Protection Agency, 1971

310. R. G. Nash and E. A. Woolson, *Science*, **157,** 924 (1967); *Soil Sci. Soc. Am. Proc.*, **32,** 525 (1973).

311. A. R. Swoboda, G. W. Thomas, F. B. Cady, R. W. Baird, and W. G. Knisel, *Environ. Sci. Technol.*, **5,** 141 (1971).

312. C. M. Menzie, *Sport Fish. Wildl. Spec. Sci. Rp.*, 1969.

313. D. F. Paris and D. L. Lewis, *Residue Rev.* **45,** 95 (1973).

314. L. C. Terriere and D. W. Ingalsbe, *J. Econ. Entomol.*, **46,** 751 (1953).

315. R. A. Hughes and G. F. Lee, *Environ. Sci. Technol.*, **7,** 934 (1973).

316. G. F. Lee, R. A. Hughes, and G. D. Veith, *Water Air Soil Pollut.*, **8**(4), 479 (1977).

317. R. R. Williams and T. F. Bidleman, *J. Agric. Food Chem.*, **26**(1), 280 (1978).

318. F. L. Mayer, P. M. Mehrle, and W. F. Dwyer, *Natl. Tech. Inform. Serv. PB-271,* 695 (1977); EPA-600/3-77-069 (June 1977).

319. C. W. Stanley, J. E. Barney, M. R. Helton, and A. R. Yobs, *Environ. Sci. Technol.*, **5,** 430 (1971).

320. F. F. Bidleman and C. E. Olney, *Nature*, **257,** 475 (1975).

321. NCI *Bioassay of Toxaphene for Possible Carcinogenicity*, U.S. Dept. of Health, Education and Welfare, National Institutes of Health, Tech. Rep. Ser. No. 37, CAS No. 8001-35-2; NCI-CG-TR-37, 1979.

322. Litton Bionetics, Kensington, Md., Unpublished report, through J. P. Frawley, Hercules, Inc., personal communication, 1979.

323. *WHO Pesticide Residues Series, No. 3, 1973 Evaluations of Some Pesticide Residues in Food, The Monographs*, World Health Organization, Geneva, 1974.

324. N. K. Hooper, B. N. Ames, M. A. Saleh, and F. E. Casida, *Science*, **205,** 591 (1979).

Subject Index

Refer to the Chemical Index for specific chemical compounds.

Chemical Index

Acenaphthene *[83-32-9]*, 3346, 3353

Acetanilide hydroxylase *[9059-06-7]*, 3278

Acetic acid *[64-19-7]*, 2960

N-Acetoxy-2-acetylaminofluorene *[6098-44-8]*, 2916-2917

2-Acetylaminofluorene *[53-96-3]*, 2881, 2893, 2901, 2921

N-Acetyl-S-cycloheptyl-L-cysteine *[13392-36-4]*, 3236

Acetylene *[74-86-2]*, 3195, 3211, 3212

Acetylene tetrabromide *[79-27-6]*, 3517-3519

Acetylene tetrachloride *[79-34-5]*, 3513-3516

N-Acetyl-S-(2-hydroxyethyl)cysteine *[15060-26-1]*, 2921, 3501

N-acetyl-S-hydroxycycloheptyl-L-cysteine, *[33525-15-4]*, 3236

Acid phosphatase *[9001-77-8]*, 3193

Acrylonitrile *[107-13-1]*, 2889

Actinomycin D *[50-76-6]*, 3359

Adenine *[73-24-5]*, 2926, 2927

Mg-Adenosine triphosphatase *[9000-83-8]*, 3656

NaK-Adenosine triphosphatase *[9000-83-8]*, 3656

Adipic acid, *[124-04-9]*, 3233, 3261

Alanine aminotransferase *[9000-86-6]*. 3296

Aldehyde oxidase *[9029-07-6]*, 3142

Aldrin *[309-00-2]*, 3664, 3686, 3702-3708, 3713

Allylamine *[107-11-9]*, 3139, 3146, 3158

Allylbenzene, *see* Propenylbenzene

Allyl chloride *[107-05-1]*, 3568-3572

Allyl mercapturic acid *[23127-41-5]*, 3572

Aluminum chloride *[7446-70-0]*, 2959, 3331

4-Aminobiphenyl *[92 67-1]*, 2889, 2901

γ-Aminobutyric acid *[56-12-2]*, 3744

1-Aminoheptane *[111-68-2]*, 3156

2-Aminoheptane *[123-82-0]*, 3156

3-Aminoheptane *[28292-42-4]*, 3156

4-Aminoheptane *[16751-59-0]*, 3156

σ-Aminolevulinic acid *[5451-09-2]*, 3279

γ-Aminolevulinic acid *[106-60-5]*, 3619

σ-Aminolevulinic acid synthetase *[9037-14-3]*, 3632

2-Aminooctane *[693-16-3]*, 3156

1-Amino-2-propanol *[78-96-6]*, 3147, 3166

3-Amino-1-propanol *[156-87-6]*, 3147, 3166

Aminopyrine demethylase *[9037-69-8]*, 3615, 3632

3-Amino-1, 2, 4-triazole *[61-82-5]*, 2892

Ammonia *[7664-41-7]*, 2967, 3045-3051 3073

Ammonium acid fluoride, *see* Ammonium bifluoride

Ammonium bifluoride *[1341-49-7]*, 2948-2952

Ammonium bromide *[12124-97-9]*, 2970

Ammonium chloride *[12125-02-9]*, 2939, 2959

Ammonium hydroxide *[1336-21-6]*, 3045-3051

Ammonium iodide *[12027-06-4]*, 2975, 2976

Ammonium nitrate *[6484-52-2]*, 3046

Ammonium phosphate *[7783-28-0]*, 3046